Normal curve areas

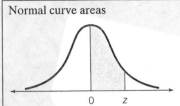

z	.00	.01	.02	.03	.04	.05	.06	.07	.08	.09
.0	.0000	.0040	.0080	.0120	.0160	.0199	.0239	.0279	.0319	.0359
.1	.0398	.0438	.0478	.0517	.0557	.0596	.0636	.0675	.0714	.0753
.2	.0793	.0832	.0871	.0910	.0948	.0987	.1026	.1064	.1103	.1141
.3	.1179	.1217	.1255	.1293	.1331	.1368	.1406	.1443	.1480	.1517
.4	.1554	.1591	.1628	.1664	.1700	.1736	.1772	.1808	.1844	.1879
.5	.1915	.1950	.1985	.2019	.2054	.2088	.2123	.2157	.2190	.2224
.6	.2257	.2291	.2324	.2357	.2389	.2422	.2454	.2486	.2517	.2549
.7	.2580	.2611	.2642	.2673	.2704	.2734	.2764	.2794	.2823	.2852
.8	.2881	.2910	.2939	.2967	.2995	.3023	.3051	.3078	.3106	.3133
.9	.3159	.3186	.3212	.3238	.3264	.3289	.3315	.3340	.3365	.3389
1.0	.3413	.3438	.3461	.3485	.3508	.3531	.3554	.3577	.3599	.3621
1.1	.3643	.3665	.3686	.3708	.3729	.3749	.3770	.3790	.3810	.3830
1.2	.3849	.3869	.3888	.3907	.3925	.3944	.3962	.3980	.3997	.4015
1.3	.4032	.4049	.4066	.4082	.4099	.4115	.4131	.4147	.4162	.4177
1.4	.4192	.4207	.4222	.4236	.4251	.4265	.4279	.4292	.4306	.4319
1.5	.4332	.4345	.4357	.4370	.4382	.4394	.4406	.4418	.4429	.4441
1.6	.4452	.4463	.4474	.4484	.4495	.4505	.4515	.4525	.4535	.4545
1.7	.4554	.4564	.4573	.4582	.4591	.4599	.4608	.4616	.4625	.4633
1.8	.4641	.4649	.4656	.4664	.4671	.4678	.4686	.4693	.4699	.4706
1.9	.4713	.4719	.4726	.4732	.4738	.4744	.4750	.4756	.4761	.4767
2.0	.4772	.4778	.4783	.4788	.4793	.4798	.4803	.4808	.4812	.4817
2.1	.4821	.4826	.4830	.4834	.4838	.4842	.4846	.4850	.4854	.4857
2.2	.4861	.4864	.4868	.4871	.4875	.4878	.4881	.4884	.4887	.4890
2.3	.4893	.4896	.4898	.4901	.4904	.4906	.4909	.4911	.4913	.4916
2.4	.4918	.4920	.4922	.4925	.4927	.4929	.4931	.4932	.4934	.4936
2.5	.4938	.4940	.4941	.4943	.4945	.4946	.4948	.4949	.4951	.4952
2.6	.4953	.4955	.4956	.4957	.4959	.4960	.4961	.4962	.4963	.4964
2.7	.4965	.4966	.4967	.4968	.4969	.4970	.4971	.4972	.4973	.4974
2.8	.4974	.4975	.4976	.4977	.4977	.4978	.4979	.4979	.4980	.4981
2.9	.4981	.4982	.4982	.4983	.4984	.4984	.4985	.4985	.4986	.4986
3.0	.4987	.4987	.4987	.4988	.4988	.4989	.4989	.4989	.4990	.4990

Source: Abridged from Table 1 of A. Hald, *Statistical Tables and Formulas* (New York: John Wiley & Sons, Inc.), 1952. Reproduced by permission of the publisher.

Critical values for Student's t

ν	$t_{.100}$	$t_{.050}$	$t_{.025}$	$t_{.010}$	$t_{.005}$	$t_{.001}$	$t_{.0005}$
1	3.078	6.314	12.706	31.821	63.657	318.31	636.62
2	1.886	2.920	4.303	6.965	9.925	22.326	31.598
3	1.638	2.353	3.182	4.541	5.841	10.213	12.924
4	1.533	2.132	2.776	3.747	4.604	7.173	8.610
5	1.476	2.015	2.571	3.365	4.032	5.893	6.869
6	1.440	1.943	2.447	3.143	3.707	5.208	5.959
7	1.415	1.895	2.365	2.998	3.499	4.785	5.408
8	1.397	1.860	2.306	2.896	3.355	4.501	5.041
9	1.383	1.833	2.262	2.821	3.250	4.297	4.781
10	1.372	1.812	2.228	2.764	3.169	4.144	4.587
11	1.363	1.796	2.201	2.718	3.106	4.025	4.437
12	1.356	1.782	2.179	2.681	3.055	3.930	4.318
13	1.350	1.771	2.160	2.650	3.012	3.852	4.221
14	1.345	1.761	2.145	2.624	2.977	3.787	4.140
15	1.341	1.753	2.131	2.602	2.947	3.733	4.073
16	1.337	1.746	2.120	2.583	2.921	3.686	4.015
17	1.333	1.740	2.110	2.567	2.898	3.646	3.965
18	1.330	1.734	2.101	2.552	2.878	3.610	3.922
19	1.328	1.729	2.093	2.539	2.861	3.579	3.883
20	1.325	1.725	2.086	2.528	2.845	3.552	3.850
21	1.323	1.721	2.080	2.518	2.831	3.527	3.819
22	1.321	1.717	2.074	2.508	2.819	3.505	3.792
23	1.319	1.714	2.069	2.500	2.807	3.485	3.767
24	1.318	1.711	2.064	2.492	2.797	3.467	3.745
25	1.316	1.708	2.060	2.485	2.787	3.450	3.725
26	1.315	1.706	2.056	2.479	2.779	3.435	3.707
27	1.314	1.703	2.052	2.473	2.771	3.421	3.690
28	1.313	1.701	2.048	2.467	2.763	3.408	3.674
29	1.311	1.699	2.045	2.462	2.756	3.396	3.659
30	1.310	1.697	2.042	2.457	2.750	3.385	3.646
40	1.303	1.684	2.021	2.423	2.704	3.307	3.551
60	1.296	1.671	2.000	2.390	2.660	3.232	3.460
120	1.289	1.658	1.980	2.358	2.617	3.160	3.373
∞	1.282	1.645	1.960	2.326	2.576	3.090	3.291

A SECOND COURSE IN STATISTICS
REGRESSION ANALYISIS

Seventh Edition

WILLIAM MENDENHALL
University of Florida

TERRY SINCICH
University of South Florida

Prentice Hall

Boston Columbus Indianapolis New York San Francisco
Upper Saddle River Amsterdam Cape Town Dubai
London Madrid Milan Munich Paris Montreal
Toronto Delhi Mexico City Sao Paulo Sydney
Hong Kong Seoul Singapore Taipei Tokyo

Editor in Chief: Deirdre Lynch
Acquisitions Editor: Marianne Stepanian
Associate Content Editor: Dana Jones Bettez
Senior Managing Editor: Karen Wernholm
Associate Managing Editor: Tamela Ambush
Senior Production Project Manager: Peggy McMahon
Senior Design Supervisor: Andrea Nix
Cover Design: Christina Gleason
Interior Design: Tamara Newnam
Marketing Manager: Alex Gay
Marketing Assistant: Kathleen DeChavez
Associate Media Producer: Jean Choe
Senior Author Support/Technology Specialist: Joe Vetere
Manufacturing Manager: Evelyn Beaton
Senior Manufacturing Buyer: Carol Melville
Production Coordination, Technical Illustrations, and Composition: Laserwords Maine

Many of the designations used by manufacturers and sellers to distinguish their products are claimed as trademarks. Where those designations appear in this book, and Pearson was aware of a trademark claim, the designations have been printed in initial caps or all caps.

1 2 3 4 5 6 7 8 9 10—**EB**—14 13 12 11 10

Prentice Hall
is an imprint of

ISBN-10: **0-321-74824-7**
ISBN-13: **978-0-321-74824-9**

CONTENTS

12   THE ANALYSIS OF VARIANCE FOR DESIGNED EXPERIMENTS   608

CASE STUDY 7 RELUCTANCE TO TRANSMIT BAD NEWS: THE MUM EFFECT   714

APPENDIX A DERIVATION OF THE LEAST SQUARES ESTIMATES OF β_0 AND β_1 IN SIMPLE LINEAR REGRESSION   720

APPENDIX B THE MECHANICS OF A MULTIPLE REGRESSION ANALYSIS   722

APPENDIX C A PROCEDURE FOR INVERTING A MATRIX   751

APPENDIX D USEFUL STATISTICAL TABLES   756

APPENDIX E FILE LAYOUTS FOR CASE STUDY DATA SETS   778

PREFACE

Overview

This text is designed for two types of statistics courses. The early chapters, combined with a selection of the case studies, are designed for use in the second half of a two-semester (two-quarter) introductory statistics sequence for undergraduates with statistics or nonstatistics majors. Or, the text can be used for a course in applied regression analysis for masters or PhD students in other fields.

At first glance, these two uses for the text may seem inconsistent. How could a text be appropriate for both undergraduate and graduate students? The answer lies in the content. In contrast to a course in statistical theory, the level of mathematical knowledge required for an applied regression analysis course is minimal. Consequently, the difficulty encountered in learning the mechanics is much the same for both undergraduate and graduate students. The challenge is in the application: diagnosing practical problems, deciding on the appropriate linear model for a given situation, and knowing which inferential technique will answer the researcher's practical question. This *takes experience*, and it explains why a student with a non-statistics major can take an undergraduate course in applied regression analysis and still benefit from covering the same ground in a graduate course.

Introductory Statistics Course

It is difficult to identify the amount of material that should be included in the second semester of a two-semester sequence in introductory statistics. Optionally, a few lectures should be devoted to Chapter 1 (A Review of Basic Concepts) to make certain that all students possess a common background knowledge of the basic concepts covered in a first-semester (first-quarter) course. Chapter 2 (Introduction to Regression Analysis), Chapter 3 (Simple Linear Regression), Chapter 4 (Multiple Regression Models), Chapter 5 (Principles of Model Building), Chapter 6 (Variable Screening Methods), Chapter 7 (Some Regression Pitfalls), and Chapter 8 (Residual Analysis) provide the core for an applied regression analysis course. These chapters could be supplemented by the addition of Chapter 10 (Time Series Modeling and Forecasting), Chapter 11 (Principles of Experimental Design), and Chapter 12 (The Analysis of Variance for Designed Experiments).

Applied Regression for Graduates

In our opinion, the quality of an applied graduate course is not measured by the number of topics covered or the amount of material memorized by the students. The measure is how well they can apply the techniques covered in the course to the solution of real problems encountered in their field of study. Consequently, we advocate moving on to new topics only after the students have demonstrated ability (through testing) to apply the techniques under discussion. In-class consulting sessions, where a case study is presented and the students have the opportunity to diagnose the problem and recommend an appropriate method of analysis, are very helpful in teaching applied regression analysis. This approach is particularly useful in helping students master the difficult topic of model selection and model building (Chapters 4–8) and relating questions about the model to real-world questions. The seven case studies (which follow relevant chapters) illustrate the type of material that might be useful for this purpose.

A course in applied regression analysis for graduate students would start in the same manner as the undergraduate course, but would move more rapidly over the review material and would more than likely be supplemented by Appendix A (Derivation of the Least Squares Estimates), Appendix B (The Mechanics of a Multiple Regression Analysis), and/or Appendix C (A Procedure for Inverting a Matrix), one of the statistical software Windows tutorials available on the Data CD (SAS®; SPSS®, an IBM® Company[1]; MINITAB®; or R®), Chapter 9 (Special Topics in Regression), and other chapters selected by the instructor. As in the undergraduate course, we recommend the use of case studies and in-class consulting sessions to help students develop an ability to formulate appropriate statistical models and to interpret the results of their analyses.

Features:

1. **Readability.** We have purposely tried to make this a teaching (rather than a reference) text. Concepts are explained in a logical intuitive manner using worked examples.

2. **Emphasis on model building.** The formulation of an appropriate statistical model is fundamental to any regression analysis. This topic is treated in Chapters 4–8 and is emphasized throughout the text.

3. **Emphasis on developing regression skills.** In addition to teaching the basic concepts and methodology of regression analysis, this text stresses its use, as a tool, in solving applied problems. Consequently, a major objective of the text is to develop a skill in applying regression analysis to appropriate real-life situations.

4. **Real data-based examples and exercises.** The text contains many worked examples that illustrate important aspects of model construction, data analysis, and the interpretation of results. Nearly every exercise is based on data and research extracted from a news article, magazine, or journal. Exercises are located at the ends of key sections and at the ends of chapters.

5. **Case studies.** The text contains seven case studies, each of which addresses a real-life research problem. The student can see how regression analysis was used to answer the practical questions posed by the problem, proceeding with the formulation of appropriate statistical models to the analysis and interpretation of sample data.

6. **Data sets.** The Data CD and the Pearson Datasets Web Site—www.pearsonhighered.com/datasets—contain complete data sets that are associated with the case studies, exercises, and examples. These can be used by instructors and students to practice model-building and data analyses.

7. **Extensive use of statistical software.** Tutorials on how to use four popular statistical software packages—SAS, SPSS, MINITAB, and R—are provided on

---

[1] SPSS was acquired by IBM in October 2009.

the Data CD. Printouts associated with the respective software packages are presented and discussed throughout the text.

New to the Seventh Edition

Although the scope and coverage remain the same, the seventh edition contains several substantial changes, additions, and enhancements. Most notable are the following:

1. **New and updated case studies.** Two new case studies (Case Study 1: Legal Advertising–Does it Pay? and Case Study 3: Deregulation of the Intrastate Trucking Industry) have been added, and another (Case Study 2: Modeling Sale Prices of Residential Properties in Four Neighborhoods) has been updated with current data. Also, all seven of the case studies now follow the relevant chapter material.

2. **Real data exercises.** Many new and updated exercises, based on contemporary studies and real data in a variety of fields, have been added. Most of these exercises foster and promote critical thinking skills.

3. **Technology Tutorials on CD.** The Data CD now includes basic instructions on how to use the Windows versions of SAS, SPSS, MINITAB, and R, which is new to the text. Step-by-step instructions and screen shots for each method presented in the text are shown.

4. **More emphasis on p-values.** Since regression analysts rely on statistical software to fit and assess models in practice, and such software produces p-values, we emphasize the p-value approach to testing statistical hypotheses throughout the text. Although formulas for hand calculations are shown, we encourage students to conduct the test using available technology.

5. **New examples in Chapter 9: Special Topics in Regression.** New worked examples on piecewise regression, weighted least squares, logistic regression, and ridge regression are now included in the corresponding sections of Chapter 9.

6. **Redesigned end-of-chapter summaries.** Summaries at the ends of each chapter have been redesigned for better visual appeal. Important points are reinforced through flow graphs (which aid in selecting the appropriate statistical method) and notes with key words, formulas, definitions, lists, and key concepts.

Supplements

The text is accompanied by the following supplementary material:

1. **Instructor's Solutions Manual** by Dawn White, California State University–Bakersfield, contains fully worked solutions to all exercises in the text. Available for download from the Instructor Resource Center at www.pearsonhighered.com/irc.

2. **Student Solutions Manual** by Dawn White, California State University–Bakersfield, contains fully worked solutions to all odd exercises in the text. Available for download from the Instructor Resource Center at www.pearsonhighered.com/irc.

3. **PowerPoint® lecture slides** include figures, tables, and formulas. Available for download from the Instructor Resource Center at www.pearsonhighered.com/irc.

4. **Data CD**, bound inside each edition of the text, contains files for all data sets marked with a CD icon. These include data sets for text examples, exercises, and case studies and are formatted for SAS, SPSS, MINITAB, R, and as text files. The CD also includes Technology Tutorials for SAS, SPSS, MINITAB, and R.

Technology Supplements and Packaging Options

1. **The Student Edition of Minitab** is a condensed edition of the professional release of Minitab statistical software. It offers the full range of statistical methods and graphical capabilities, along with worksheets that can include up to 10,000 data points. Individual copies of the software can be bundled with the text. (ISBN-13: 978-0-321-11313-9; ISBN-10: 0-321-11313-6)

2. **JMP® Student Edition** is an easy-to-use, streamlined version of JMP desktop statistical discovery software from SAS Institute, Inc., and is available for bundling with the text. (ISBN-13: 978-0-321-67212-4; ISBN-10: 0-321-67212-7)

3. **SPSS**, a statistical and data management software package, is also available for bundling with the text. (ISBN-13: 978-0-321-67537-8; ISBN-10: 0-321-67537-1)

4. Study Cards are also available for various technologies, including Minitab, SPSS, JMP, StatCrunch™, R, Excel® and the TI Graphing Calculator.

Acknowledgments

We want to thank the many people who contributed time, advice, and other assistance to this project. We owe particular thanks to the many reviewers who provided suggestions and recommendations at the onset of the project and for the succeeding editions (including the 7th):

Gokarna Aryal (Purdue University Calumet), Mohamed Askalani (Minnesota State University, Mankato), Ken Boehm (Pacific Telesis, California), William Bridges, Jr. (Clemson University), Andrew C. Brod (University of North Carolina at Greensboro), Pinyuen Chen (Syracuse University), James Daly (California State Polytechnic Institute, San Luis Obispo), Assane Djeto (University of Nevada, Las Vegas), Robert Elrod (Georgia State University), James Ford (University of Delaware), Carol Ghomi (University of Houston), David Holmes (College of New Jersey), James Holstein (University of Missouri–Columbia), Steve Hora (Texas Technological University), K. G. Janardan (Eastern Michigan University), Thomas Johnson (North Carolina State University), David Kidd (George Mason University), Ann Kittler (Ryerson Universtiy, Toronto), Lingyun Ma (University of Georgia), Paul Maiste (Johns Hopkins University), James T. McClave (University of Florida), Monnie McGee (Southern Methodist University), Patrick McKnight (George Mason University), John Monahan (North Carolina State University), Kris Moore (Baylor University), Farrokh Nasri (Hofstra University), Tom O'Gorman (Northern Illinois University), Robert Pavur (University of North Texas), P. V. Rao (University of Florida), Tom Rothrock (Info Tech, Inc.), W. Robert Stephenson (Iowa State University), Martin Tanner (Northwestern University), Ray Twery (University of North Carolina at Charlotte), Joseph Van Matre (University of Alabama at Birmingham),

William Weida (United States Air Force Academy), Dean Wichern (Texas A&M University), James Willis (Louisiana State University), Ruben Zamar (University of British Columbia)

We are particularly grateful to Charles Bond, Evan Anderson, Jim McClave, Herman Kelting, Rob Turner, P. J. Taylor, and Mike Jacob, who provided data sets and/or background information used in the case studies, Matthew Reimherr (University of Chicago), who wrote the R tutorial, and to Jackie Miller (The Ohio State University) and W. Robert Stephenson (Iowa State Unviersity), who checked the text for clarity and accuracy.

A Review of Basic Concepts (Optional)

Contents

Objectives

1. Review some basic concepts of sampling.
2. Review methods for describing both qualitative and quantitative data.
3. Review inferential statistical methods: confidence intervals and hypothesis tests.

Although we assume students have had a prerequisite introductory course in statistics, courses vary somewhat in content and in the manner in which they present statistical concepts. To be certain that we are starting with a common background, we use this chapter to review some basic definitions and concepts. Coverage is optional.

1.1 Statistics and Data

According to *The Random House College Dictionary* (2001 ed.), statistics is "the science that deals with the collection, classification, analysis, and interpretation of numerical facts or data." In short, statistics is the **science of data**—a science that will enable you to be proficient data producers and efficient data users.

> **Definition 1.1** **Statistics** is the science of data. This involves collecting, classifying, summarizing, organizing, analyzing, and interpreting data.

      Data are obtained by measuring some characteristic or property of the objects (usually people or things) of interest to us. These objects upon which the measurements (or observations) are made are called **experimental units**, and the properties being measured are called **variables** (since, in virtually all studies of interest, the property varies from one observation to another).

> **Definition 1.2** An **experimental unit** is an object (person or thing) upon which we collect data.

> **Definition 1.3** A **variable** is a characteristic (property) of the experimental unit with outcomes (data) that vary from one observation to the next.

All data (and consequently, the variables we measure) are either **quantitative** or **qualitative** in nature. Quantitative data are data that can be measured on a naturally occurring numerical scale. In general, qualitative data take values that are nonnumerical; they can only be classified into categories. The statistical tools that we use to analyze data depend on whether the data are quantitative or qualitative. Thus, it is important to be able to distinguish between the two types of data.

> **Definition 1.4 Quantitative data** are observations measured on a naturally occurring numerical scale.

> **Definition 1.5** Nonnumerical data that can only be classified into one of a group of categories are said to be **qualitative data**.

Example 1.1

Chemical and manufacturing plants often discharge toxic waste materials such as DDT into nearby rivers and streams. These toxins can adversely affect the plants and animals inhabiting the river and the riverbank. The U.S. Army Corps of Engineers conducted a study of fish in the Tennessee River (in Alabama) and its three tributary creeks: Flint Creek, Limestone Creek, and Spring Creek. A total of 144 fish were captured, and the following variables were measured for each:

1. River/creek where each fish was captured
2. Number of miles upstream where the fish was captured
3. Species (channel catfish, largemouth bass, or smallmouth buffalofish)
4. Length (centimeters)
5. Weight (grams)
6. DDT concentration (parts per million)

The data are saved in the FISHDDT file. Data for 10 of the 144 captured fish are shown in Table 1.1.

(a) Identify the experimental units.
(b) Classify each of the five variables measured as quantitative or qualitative.

Solution

(a) Because the measurements are made for each fish captured in the Tennessee River and its tributaries, the experimental units are the 144 captured fish.
(b) The variables upstream that capture location, length, weight, and DDT concentration are quantitative because each is measured on a natural numerical scale: upstream in miles from the mouth of the river, length in centimeters, weight in grams, and DDT in parts per million. In contrast, river/creek and species cannot be measured quantitatively; they can only be classified into categories (e.g., channel catfish, largemouth bass, and smallmouth buffalofish for species). Consequently, data on river/creek and species are qualitative.

🐟 FISHDDT

Table 1.1 Data collected by U.S. Army Corps of Engineers (selected observations)

River/Creek	Upstream	Species	Length	Weight	DDT
FLINT	5	CHANNELCATFISH	42.5	732	10.00
FLINT	5	CHANNELCATFISH	44.0	795	16.00
SPRING	1	CHANNELCATFISH	44.5	1133	2.60
TENNESSEE	275	CHANNELCATFISH	48.0	986	8.40
TENNESSEE	275	CHANNELCATFISH	45.0	1023	15.00
TENNESSEE	280	SMALLMOUTHBUFF	49.0	1763	4.50
TENNESSEE	280	SMALLMOUTHBUFF	46.0	1459	4.20
TENNESSEE	285	LARGEMOUTHBASS	25.0	544	0.11
TENNESSEE	285	LARGEMOUTHBASS	23.0	393	0.22
TENNESSEE	285	LARGEMOUTHBASS	28.0	733	0.80

1.1 Exercises

1.1 **College application data.** Colleges and universities are requiring an increasing amount of information about applicants before making acceptance and financial aid decisions. Classify each of the following types of data required on a college application as quantitative or qualitative.

(a) High school GPA
(b) Country of citizenship
(c) Applicant's score on the SAT or ACT
(d) Gender of applicant
(e) Parents' income
(f) Age of applicant

1.2 **Fuel Economy Guide.** The data in the accompanying table were obtained from the *Model Year 2009 Fuel Economy Guide* for new automobiles.

(a) Identify the experimental units.
(b) State whether each of the variables measured is quantitative or qualitative.

1.3 **Ground motion of earthquakes.** In the *Journal of Earthquake Engineering* (November 2004), a team of civil and environmental engineers studied the ground motion characteristics of 15 earthquakes that occurred around the world between 1940 and 1995. Three (of many) variables measured on each earthquake were the type of ground motion (short, long, or forward directive), earthquake magnitude (Richter scale), and peak ground acceleration (feet per second). One of the goals of the study was to estimate the inelastic spectra of any ground motion cycle.

(a) Identify the experimental units for this study.
(b) Identify the variables measured as quantitative or qualitative.

1.4 **Use of herbal medicines.** *The American Association of Nurse Anesthetists Journal* (February 2000) published the results of a study on the use of herbal medicines before surgery. Each of 500

MODEL NAME	MFG	TRANSMISSION TYPE	ENGINE SIZE (LITERS)	NUMBER OF CYLINDERS	EST. CITY MILEAGE (MPG)	EST. HIGHWAY MILEAGE (MPG)
TSX	Acura	Automatic	2.4	4	21	30
Jetta	VW	Automatic	2.0	4	29	40
528i	BMW	Manual	3.0	6	18	28
Fusion	Ford	Automatic	3.0	6	17	25
Camry	Toyota	Manual	2.4	4	21	31
Escalade	Cadillac	Automatic	6.2	8	12	19

Source: Model Year 2009 Fuel Economy Guide, U.S. Dept. of Energy, U.S. Environmental Protection Agency (www.fueleconomy.gov).

surgical patients was asked whether they used herbal or alternative medicines (e.g., garlic, ginkgo, kava, fish oil) against their doctor's advice before surgery. Surprisingly, 51% answered "yes."

(a) Identify the experimental unit for the study.
(b) Identify the variable measured for each experimental unit.
(c) Is the data collected quantitative or qualitative?

1.5 Drinking-water quality study. *Disasters* (Vol. 28, 2004) published a study of the effects of a tropical cyclone on the quality of drinking water on a remote Pacific island. Water samples (size 500 milliliters) were collected approximately 4 weeks after Cyclone Ami hit the island. The following variables were recorded for each water sample. Identify each variable as quantitative or qualitative.

(a) Town where sample was collected
(b) Type of water supply (river intake, stream, or borehole)

(c) Acidic level (pH scale, 1–14)
(d) Turbidity level (nephalometric turbidity units [NTUs])
(e) Temperature (degrees Centigrade)
(f) Number of fecal coliforms per 100 milliliters
(g) Free chlorine-residual (milligrams per liter)
(h) Presence of hydrogen sulphide (yes or no)

1.6 Accounting and Machiavellianism. *Behavioral Research in Accounting* (January 2008) published a study of Machiavellian traits in accountants. *Machiavellian* describes negative character traits that include manipulation, cunning, duplicity, deception, and bad faith. A questionnaire was administered to a random sample of 700 accounting alumni of a large southwestern university. Several variables were measured, including age, gender, level of education, income, job satisfaction score, and Machiavellian ("Mach") rating score. What type of data (quantitative or qualitative) is produced by each of the variables measured?

1.2 Populations, Samples, and Random Sampling

When you examine a data set in the course of your study, you will be doing so because the data characterize a group of experimental units of interest to you. In statistics, the data set that is collected for all experimental units of interest is called a **population**. This data set, which is typically large, either exists in fact or is part of an ongoing operation and hence is conceptual. Some examples of statistical populations are given in Table 1.2.

> **Definition 1.6** A **population data set** is a collection (or set) of data measured on all experimental units of interest to you.

Many populations are too large to measure (because of time and cost); others cannot be measured because they are partly conceptual, such as the set of quality

Table 1.2 Some typical populations

Variable	Experimental Units	Population Data Set	Type
a. Starting salary of a graduating Ph.D. biologist	All Ph.D. biologists graduating this year	Set of starting salaries of all Ph.D. biologists who graduated this year	Existing
b. Breaking strength of water pipe in Philadelphia	All water pipe sections in Philadelphia	Set of breakage rates for all water pipe sections in Philadelphia	Existing
c. Quality of an item produced on an assembly line	All manufactured items	Set of quality measurements for all items manufactured over the recent past and in the future	Part existing, part conceptual
d. Sanitation inspection level of a cruise ship	All cruise ships	Set of sanitation inspection levels for all cruise ships	Existing

measurements (population c in Table 1.2). Thus, we are often required to select a subset of values from a population and to make **inferences** about the population based on information contained in a **sample**. This is one of the major objectives of modern statistics.

Definition 1.7 A **sample** is a subset of data selected from a population.

Definition 1.8 A **statistical inference** is an estimate, prediction, or some other generalization about a population based on information contained in a sample.

Example 1.2

According to the research firm Magnum Global (2008), the average age of viewers of the major networks' television news programming is 50 years. Suppose a cable network executive hypothesizes that the average age of cable TV news viewers is less than 50. To test her hypothesis, she samples 500 cable TV news viewers and determines the age of each.

(a) Describe the population.
(b) Describe the variable of interest.
(c) Describe the sample.
(d) Describe the inference.

Solution

(a) The population is the set of units of interest to the cable executive, which is the set of all cable TV news viewers.

(b) The age (in years) of each viewer is the variable of interest.

(c) The sample must be a subset of the population. In this case, it is the 500 cable TV viewers selected by the executive.

(d) The inference of interest involves the *generalization* of the information contained in the sample of 500 viewers to the population of all cable news viewers. In particular, the executive wants to estimate the average age of the viewers in order to determine whether it is less than 50 years. She might accomplish this by calculating the average age in the sample and using the sample average to estimate the population average.   ■

Whenever we make an inference about a population using sample information, we introduce an element of uncertainty into our inference. Consequently, it is important to report the **reliability** of each inference we make. Typically, this is accomplished by using a probability statement that gives us a high level of confidence that the inference is true. In Example 1.2, we could support the inference about the average age of all cable TV news viewers by stating that the population average falls within 2 years of the calculated sample average with "95% confidence." (Throughout the text, we demonstrate how to obtain this measure of reliability—and its meaning—for each inference we make.)

Definition 1.9 A **measure of reliability** is a statement (usually quantified with a probability value) about the degree of uncertainty associated with a statistical inference.

The level of confidence we have in our inference, however, will depend on how **representative** our sample is of the population. Consequently, the sampling procedure plays an important role in statistical inference.

Definition 1.10 A **representative sample** exhibits characteristics typical of those possessed by the population.

The most common type of sampling procedure is one that gives every different sample of fixed size in the population an equal probability (chance) of selection. Such a sample—called a **random sample**—is likely to be representative of the population.

Definition 1.11 A **random sample** of n experimental units is one selected from the population in such a way that every different sample of size n has an equal probability (chance) of selection.

How can a random sample be generated? If the population is not too large, each observation may be recorded on a piece of paper and placed in a suitable container. After the collection of papers is thoroughly mixed, the researcher can remove n pieces of paper from the container; the elements named on these n pieces of paper are the ones to be included in the sample. Lottery officials utilize such a technique in generating the winning numbers for Florida's weekly 6/52 Lotto game. Fifty-two white ping-pong balls (the population), each identified from 1 to 52 in black numerals, are placed into a clear plastic drum and mixed by blowing air into the container. The ping-pong balls bounce at random until a total of six balls "pop" into a tube attached to the drum. The numbers on the six balls (the random sample) are the winning Lotto numbers.

This method of random sampling is fairly easy to implement if the population is relatively small. It is not feasible, however, when the population consists of a large number of observations. Since it is also very difficult to achieve a thorough mixing, the procedure only approximates random sampling. Most scientific studies, however, rely on computer software (with built-in random-number generators) to automatically generate the random sample. Almost all of the popular statistical software packages available (e.g., SAS, SPSS, MINITAB) have procedures for generating random samples.

1.2 Exercises

1.7 Guilt in decision making. The effect of guilt emotion on how a decision-maker focuses on the problem was investigated in the *Journal of Behavioral Decision Making* (January 2007). A total of 155 volunteer students participated in the experiment, where each was randomly assigned to one of three emotional states (guilt, anger, or neutral) through a reading/writing task. Immediately after the task, the students were presented with a decision problem (e.g., whether or not to spend money on repairing a very old car). The researchers found that a higher proportion of students in the guilty-state group chose not to repair the car than those in the neutral-state and anger-state groups.

(a) Identify the population, sample, and variables measured for this study.

(b) What inference was made by the researcher?

1.8 Use of herbal medicines. Refer to the *American Association of Nurse Anesthetists Journal* (February 2000) study on the use of herbal medicines before surgery, Exercise 1.4 (p. 3). The 500 surgical patients that participated in the study were randomly selected from surgical patients at several metropolitan hospitals across the country.

(a) Do the 500 surgical patients represent a population or a sample? Explain.

(b) If your answer was sample in part a, is the sample likely to be representative of the population? If you answered population in part a, explain how to obtain a representative sample from the population.

1.9 Massage therapy for athletes. Does a massage enable the muscles of tired athletes to recover from exertion faster than usual? To answer this question, researchers recruited eight amateur boxers to participate in an experiment (*British Journal of Sports Medicine*, April 2000). After a 10-minute workout in which each boxer threw 400 punches, half the boxers were given a 20-minute massage and half just rested for 20 minutes. Before returning to the ring for a second workout, the heart rate (beats per minute) and blood lactate level (micromoles) were recorded for each boxer. The researchers found no difference in the means of the two groups of boxers for either variable.

(a) Identify the experimental units of the study.

(b) Identify the variables measured and their type (quantitative or qualitative).

(c) What is the inference drawn from the analysis?

(d) Comment on whether this inference can be made about all athletes.

1.10 Gallup Youth Poll. A Gallup Youth Poll was conducted to determine the topics that teenagers most want to discuss with their parents. The findings show that 46% would like more discussion about the family's financial situation, 37% would like to talk about school, and 30% would like to talk about religion. The survey was based on a national sampling of 505 teenagers, selected at random from all U.S. teenagers.

(a) Describe the sample.

(b) Describe the population from which the sample was selected.

(c) Is the sample representative of the population?

(d) What is the variable of interest?

(e) How is the inference expressed?

(f) Newspaper accounts of most polls usually give a *margin of error* (e.g., plus or minus 3%) for the survey result. What is the purpose of the margin of error and what is its interpretation?

1.11 Insomnia and education. Is insomnia related to education status? Researchers at the Universities of Memphis, Alabama at Birmingham, and Tennessee investigated this question in the *Journal of Abnormal Psychology* (February 2005). Adults living in Tennessee were selected to participate in the study using a random-digit telephone dialing procedure. Two of the many variables measured for each of the 575 study participants were number of years of education and insomnia status (normal sleeper or chronic insomnia). The researchers discovered that the fewer the years of education, the more likely the person was to have chronic insomnia.

(a) Identify the population and sample of interest to the researchers.

(b) Describe the variables measured in the study as quantitative or qualitative.

(c) What inference did the researchers make?

1.12 Accounting and Machiavellianism. Refer to the *Behavioral Research in Accounting* (January 2008) study of Machiavellian traits in accountants, Exercise 1.6 (p. 6). Recall that a questionnaire was administered to a random sample of 700 accounting alumni of a large southwestern university; however, due to nonresponse and incomplete answers, only 198 questionnaires could be analyzed. Based on this information, the researchers concluded that Machiavellian behavior is not required to achieve success in the accounting profession.

(a) What is the population of interest to the researcher?

(b) Identify the sample.

(c) What inference was made by the researcher?

(d) How might the nonresponses impact the inference?

1.3 Describing Qualitative Data

Consider a study of aphasia published in the *Journal of Communication Disorders* (March 1995). Aphasia is the "impairment or loss of the faculty of using or understanding spoken or written language." Three types of aphasia have been identified by researchers: Broca's, conduction, and anomic. They wanted to determine whether one type of aphasia occurs more often than any other, and, if so, how often. Consequently, they measured aphasia type for a sample of 22 adult aphasiacs. Table 1.3 gives the type of aphasia diagnosed for each aphasiac in the sample.

APHASIA

Table 1.3 Data on 22 adult aphasiacs

Subject	Type of Aphasia
1	Broca's
2	Anomic
3	Anomic
4	Conduction
5	Broca's
6	Conduction
7	Conduction
8	Anomic
9	Conduction
10	Anomic
11	Conduction
12	Broca's
13	Anomic
14	Broca's
15	Anomic
16	Anomic
17	Anomic
18	Conduction
19	Broca's
20	Anomic
21	Conduction
22	Anomic

Source: Reprinted from *Journal of Communication Disorders,* Mar. 1995, Vol. 28, No. 1, E. C. Li, S. E. Williams, and R. D. Volpe, "The effects of topic and listener familiarity of discourse variables in procedural and narrative discourse tasks," p. 44 (Table 1) Copyright © 1995, with permission from Elsevier.

For this study, the variable of interest, aphasia type, is qualitative in nature. Qualitative data are nonnumerical in nature; thus, the value of a qualitative variable can only be classified into categories called *classes*. The possible aphasia types—Broca's, conduction, and anomic—represent the classes for this qualitative variable. We can summarize such data numerically in two ways: (1) by computing the *class frequency*—the number of observations in the data set that fall into each class; or (2) by computing the *class relative frequency*—the proportion of the total number of observations falling into each class.

Definition 1.12 A **class** is one of the categories into which qualitative data can be classified.

Definition 1.13 The **class frequency** is the number of observations in the data set falling in a particular class.

Definition 1.14 The **class relative frequency** is the class frequency divided by the total number of observations in the data set, i.e.,

$$\text{class relative frequency} = \frac{\text{class frequency}}{n}$$

Examining Table 1.3, we observe that 5 aphasiacs in the study were diagnosed as suffering from Broca's aphasia, 7 from conduction aphasia, and 10 from anomic aphasia. These numbers—5, 7, and 10—represent the class frequencies for the three classes and are shown in the summary table, Table 1.4.

Table 1.4 also gives the relative frequency of each of the three aphasia classes. From Definition 1.14, we know that we calculate the relative frequency by dividing the class frequency by the total number of observations in the data set. Thus, the relative frequencies for the three types of aphasia are

$$\text{Broca's:} \quad \frac{5}{22} = .227$$

$$\text{Conduction:} \quad \frac{7}{22} = .318$$

$$\text{Anomic:} \quad \frac{10}{22} = .455$$

From these relative frequencies we observe that nearly half (45.5%) of the 22 subjects in the study are suffering from anomic aphasia.

Although the summary table in Table 1.4 adequately describes the data in Table 1.3, we often want a graphical presentation as well. Figures 1.1 and 1.2 show two of the most widely used graphical methods for describing qualitative data—bar graphs and pie charts. Figure 1.1 shows the frequencies of aphasia types in a **bar graph** produced with SAS. Note that the height of the rectangle, or "bar," over each class is equal to the class frequency. (Optionally, the bar heights can be proportional to class relative frequencies.)

Table 1.4 Summary table for data on 22 adult aphasiacs

Class (Type of Aphasia)	Frequency (Number of Subjects)	Relative Frequency (Proportion)
Broca's	5	.227
Conduction	7	.318
Anomic	10	.455
Totals	22	1.000

Figure 1.1 SAS bar graph
for data on 22 aphasiacs

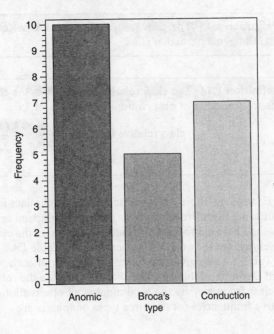

Figure 1.2 SPSS pie chart
for data on 22 aphasiacs

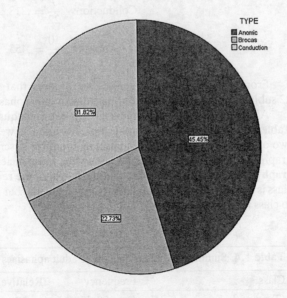

In contrast, Figure 1.2 shows the relative frequencies of the three types of aphasia in a **pie chart** generated with SPSS. Note that the pie is a circle (spanning 360°) and the size (angle) of the "pie slice" assigned to each class is proportional to the class relative frequency. For example, the slice assigned to anomic aphasia is 45.5% of 360°, or $(.455)(360°) = 163.8°$.

1.3   Exercises

1.13 **Estimating the rhino population.** The International Rhino Federation estimates that there are 17,800 rhinoceroses living in the wild in Africa and Asia. A breakdown of the number of rhinos of each species is reported in the accompanying table.

RHINO SPECIES	POPULATION ESTIMATE
African Black	3,610
African White	11,330
(Asian) Sumatran	300
(Asian) Javan	60
(Asian) Indian	2,500
Total	17,800

Source: International Rhino Federation, March 2007.

(a) Construct a relative frequency table for the data.
(b) Display the relative frequencies in a bar graph.
(c) What proportion of the 17,800 rhinos are African rhinos? Asian?

1.14 **Blogs for Fortune 500 firms.** Website communication through blogs and forums is becoming a key marketing tool for companies. The *Journal of Relationship Marketing* (Vol. 7, 2008) investigated the prevalence of blogs and forums at Fortune 500 firms with both English and Chinese websites. Of the firms that provided blogs/forums as a marketing tool, the accompanying table gives a breakdown on the entity responsible for creating the blogs/forums. Use a graphical method to describe the data summarized in the table. Interpret the graph.

BLOG/FORUM	PERCENTAGE OF FIRMS
Created by company	38.5
Created by employees	34.6
Created by third party	11.5
Creator not identified	15.4

Source: "Relationship Marketing in Fortune 500 U.S. and Chinese Web Sites," Karen E. Mishra and Li Cong, *Journal of Relationship Marketing,* Vol. 7, No. 1, 2008, reprinted by permission of the publisher (Taylor and Francis, Inc.)

1.15 **National Firearms Survey.** In the journal *Injury Prevention* (January 2007), researchers from the Harvard School of Public Health reported on the size and composition of privately held firearm stock in the United States. In a representative household telephone survey of 2,770 adults, 26% reported that they own at least one gun. The accompanying graphic summarizes the types of firearms owned.

(a) What type of graph is shown?
(b) Identify the qualitative variable described in the graph.
(c) From the graph, identify the most common type of firearms.

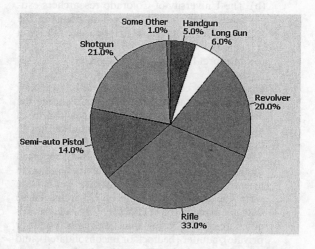

🌐 **PONDICE**

1.16 **Characteristics of ice melt ponds.** The National Snow and Ice Data Center (NSIDC) collects data on the albedo, depth, and physical characteristics of ice melt ponds in the Canadian arctic. Environmental engineers at the University of Colorado are using these data to study how climate impacts the sea ice. Data for 504 ice melt ponds located in the Barrow Strait in the Canadian arctic are saved in the PONDICE file. One variable of interest is the type of ice observed for each pond. Ice type is classified as first-year ice, multiyear ice, or landfast ice. A SAS summary table and horizontal bar graph that describe the ice types of the 504 melt ponds are shown at the top of the next page.

(a) Of the 504 melt ponds, what proportion had landfast ice?

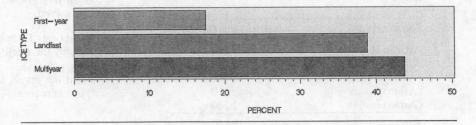

The FREQ Procedure

ICETYPE	Frequency	Percent	Cumulative Frequency	Cumulative Percent
First-year	88	17.46	88	17.46
Landfast	196	38.89	284	56.35
Multiyear	220	43.65	504	100.00

(b) The University of Colorado researchers estimated that about 17% of melt ponds in the Canadian arctic have first-year ice. Do you agree?

(c) Interpret the horizontal bar graph.

1.17 **Groundwater contamination in wells.** In New Hampshire, about half the counties mandate the use of reformulated gasoline. This has lead to an increase in the contamination of groundwater with methyl *tert*-butyl ether (MTBE). *Environmental Science and Technology* (January 2005) reported on the factors related to MTBE contamination in private and public New Hampshire wells. Data were collected for a sample of 223 wells. These data are saved in the MTBE file. Three of the variables are qualitative in nature: well class (public or private), aquifer (bedrock or unconsolidated), and detectible level of MTBE (below limit or detect). [Note: A detectible level of MTBE occurs if the MTBE value exceeds .2 micrograms per liter.] The data for 10 selected wells are shown in the accompanying table.

(a) Apply a graphical method to all 223 wells to describe the well class distribution.

(b) Apply a graphical method to all 223 wells to describe the aquifer distribution.

(c) Apply a graphical method to all 223 wells to describe the detectible level of MTBE distribution.

(d) Use two bar charts, placed side by side, to compare the proportions of contaminated wells for private and public well classes. What do you infer?

MTBE (selected observations)

WELL CLASS	AQUIFER	DETECT MTBE
Private	Bedrock	Below Limit
Private	Bedrock	Below Limit
Public	Unconsolidated	Detect
Public	Unconsolidated	Below Limit
Public	Unconsolidated	Below Limit
Public	Unconsolidated	Below Limit
Public	Unconsolidated	Detect
Public	Unconsolidated	Below Limit
Public	Unconsolidated	Below Limit
Public	Bedrock	Detect
Public	Bedrock	Detect

Source: Ayotte, J. D., Argue, D. M., and McGarry, F. J. "Methyl *tert*-butyl ether occurrence and related factors in public and private wells in southeast New Hampshire," *Environmental Science and Technology*, Vol. 39, No. 1, Jan. 2005. Reprinted with permission.

1.4 Describing Quantitative Data Graphically

A useful graphical method for describing quantitative data is provided by a relative frequency distribution. Like a bar graph for qualitative data, this type of graph shows the proportions of the total set of measurements that fall in various intervals on the scale of measurement. For example, Figure 1.3 shows the intelligence quotients (IQs) of identical twins. The area over a particular interval under a relative frequency distribution curve is proportional to the fraction of the total number of measurements that fall in that interval. In Figure 1.3, the fraction of the total number of identical twins with IQs that fall between 100 and 105 is proportional to the shaded area. **If we take the total area under the distribution curve as equal to 1, then the shaded area is equal to the fraction of IQs that fall between 100 and 105.**

Figure 1.3 Relative frequency distribution: IQs of identical twins

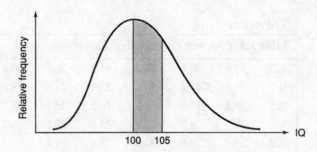

Throughout this text we denote the quantitative variable measured by the symbol y. Observing a single value of y is equivalent to selecting a single measurement from the population. The probability that it will assume a value in an interval, say, a to b, is given by its relative frequency or **probability distribution**. The total area under a probability distribution curve is always assumed to equal 1. Hence, the probability that a measurement on y will fall in the interval between a and b is equal to the shaded area shown in Figure 1.4.

Figure 1.4 Probability distribution for a quantitative variable

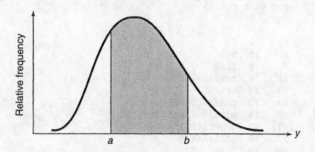

Since the theoretical probability distribution for a quantitative variable is usually unknown, we resort to obtaining a sample from the population: Our objective is to describe the sample and use this information to make inferences about the probability distribution of the population. **Stem-and-leaf plots** and **histograms** are two of the most popular graphical methods for describing quantitative data. Both display the frequency (or relative frequency) of observations that fall into specified intervals (or classes) of the variable's values.

For small data sets (say, 30 or fewer observations) with measurements with only a few digits, stem-and-leaf plots can be constructed easily by hand. Histograms, on the other hand, are better suited to the description of larger data sets, and they permit greater flexibility in the choice of classes. Both, however, can be generated using the computer, as illustrated in the following examples.

Example 1.3

The Environmental Protection Agency (EPA) performs extensive tests on all new car models to determine their highway mileage ratings. The 100 measurements in Table 1.5 represent the results of such tests on a certain new car model.

A visual inspection of the data indicates some obvious facts. For example, most of the mileages are in the 30s, with a smaller fraction in the 40s. But it is difficult to provide much additional information without resorting to a graphical method of summarizing the data. A stem-and-leaf plot for the 100 mileage ratings, produced using MINITAB, is shown in Figure 1.5. Interpret the figure.

📀 EPAGAS

Table 1.5 EPA mileage ratings on 100 cars

36.3	41.0	36.9	37.1	44.9	36.8	30.0	37.2	42.1	36.7
32.7	37.3	41.2	36.6	32.9	36.5	33.2	37.4	37.5	33.6
40.5	36.5	37.6	33.9	40.2	36.4	37.7	37.7	40.0	34.2
36.2	37.9	36.0	37.9	35.9	38.2	38.3	35.7	35.6	35.1
38.5	39.0	35.5	34.8	38.6	39.4	35.3	34.4	38.8	39.7
36.3	36.8	32.5	36.4	40.5	36.6	36.1	38.2	38.4	39.3
41.0	31.8	37.3	33.1	37.0	37.6	37.0	38.7	39.0	35.8
37.0	37.2	40.7	37.4	37.1	37.8	35.9	35.6	36.7	34.5
37.1	40.3	36.7	37.0	33.9	40.1	38.0	35.2	34.8	39.5
39.9	36.9	32.9	33.8	39.8	34.0	36.8	35.0	38.1	36.9

Figure 1.5 MINITAB stem-and-leaf plot for EPA gas mileages

```
Stem-and-leaf of MPG   N  = 100
Leaf Unit = 0.10

   1    30  0
   2    31  8
   6    32  5799
  12    33  126899
  18    34  024588
  29    35  01235667899
  49    36  01233445566777888999
 (21)   37  000011122334456677899
  30    38  0122345678
  20    39  00345789
  12    40  0123557
   5    41  002
   2    42  1
   1    43
   1    44  9
```

Solution

In a stem-and-leaf plot, each measurement (mpg) is partitioned into two portions, a *stem* and a *leaf*. MINITAB has selected the digit to the right of the decimal point to represent the leaf and the digits to the left of the decimal point to represent the stem. For example, the value 36.3 mpg is partitioned into a stem of 36 and a leaf of 3, as illustrated below:

$$\frac{\text{Stem} \mid \text{Leaf}}{36 \mid 3}$$

The stems are listed in order in the second column of the MINITAB plot, Figure 1.5, starting with the smallest stem of 30 and ending with the largest stem of 44.

The respective leaves are then placed to the right of the appropriate stem row in increasing order.* For example, the stem row of 32 in Figure 1.5 has four leaves—5, 7, 9, and 9—representing the mileage ratings of 32.5, 32.7, 32.9, and 32.9, respectively. Notice that the stem row of 37 (representing MPGs in the 37's) has the most leaves (21). Thus, 21 of the 100 mileage ratings (or 21%) have values in the 37's. If you examine stem rows 35, 36, 37, 38, and 39 in Figure 1.5 carefully, you will also find that 70 of the 100 mileage ratings (70%) fall between 35.0 and 39.9 mpg.   ▬

Example 1.4

Refer to Example 1.3. Figure 1.6 is a relative frequency histogram for the 100 EPA gas mileages (Table 1.5) produced using SPSS.

(a) Interpret the graph.

(b) Visually estimate the proportion of mileage ratings in the data set between 36 and 38 MPG.

Figure 1.6 SPSS histogram for 100 EPA gas mileages

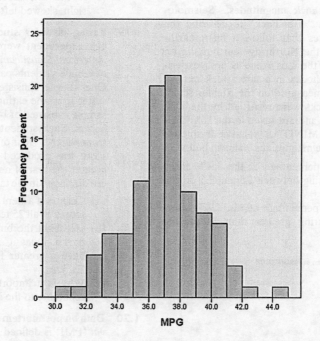

Solution

(a) In constructing a histogram, the values of the mileages are divided into the intervals of equal length (1 MPG), called **classes**. The endpoints of these classes are shown on the horizontal axis of Figure 1.6. The relative frequency (or percentage) of gas mileages falling in each class interval is represented by the vertical bars over the class. You can see from Figure 1.6 that the mileages tend to pile up near 37 MPG; in fact, the class interval from 37 to 38 MPG has the greatest relative frequency (represented by the highest bar).

Figure 1.6 also exhibits **symmetry** around the center of the data—that is, a tendency for a class interval to the right of center to have about the same relative frequency as the corresponding class interval to the left of center. This

---

* The first column in the MINITAB stem-and-leaf plot gives the cumulative number of measurements in the nearest "tail" of the distribution beginning with the stem row.

is in contrast to **positively skewed** distributions (which show a tendency for the data to tail out to the right due to a few extremely large measurements) or to **negatively skewed** distributions (which show a tendency for the data to tail out to the left due to a few extremely small measurements).

(b) The interval 36–38 MPG spans two mileage classes: 36–37 and 37–38. The proportion of mileages between 36 and 38 MPG is equal to the sum of the relative frequencies associated with these two classes. From Figure 1.6 you can see that these two class relative frequencies are .20 and .21, respectively. Consequently, the proportion of gas mileage ratings between 36 and 38 MPG is $(.20 + .21) = .41$, or 41%.

1.4 Exercises

EARTHQUAKE

1.18 Earthquake aftershock magnitudes. Seismologists use the term "aftershock" to describe the smaller earthquakes that follow a main earthquake. Following the Northridge earthquake on January 17, 1994, the Los Angeles area experienced 2,929 aftershocks in a three-week period. The magnitudes (measured on the Richter scale) for these aftershocks were recorded by the U.S. Geological Survey and are saved in the EARTHQUAKE file. A MINITAB relative frequency histogram for these magnitudes is shown below.

(a) Estimate the percentage of the 2,929 aftershocks measuring between 1.5 and 2.5 on the Richter scale.

(b) Estimate the percentage of the 2,929 aftershocks measuring greater than 3.0 on the Richter scale.

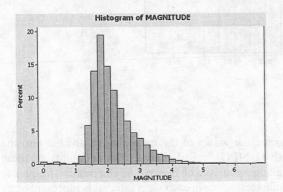

(c) Is the aftershock data distribution skewed right, skewed left, or symmetric?

1.19 Eating disorder study. Data from a psychology experiment were reported and analyzed in *American Statistician* (May 2001). Two samples of female students participated in the experiment. One sample consisted of 11 students known to suffer from the eating disorder bulimia; the other sample consisted of 14 students with normal eating habits. Each student completed a questionnaire from which a "fear of negative evaluation" (FNE) score was produced. (The higher the score, the greater the fear of negative evaluation.) The data are displayed in the table at the bottom of the page.

(a) Construct a stem-and-leaf display for the FNE scores of all 25 female students.

(b) Highlight the bulimic students on the graph, part a. Does it appear that bulimics tend to have a greater fear of negative evaluation? Explain.

(c) Why is it important to attach a measure of reliability to the inference made in part b?

1.20 Data on postmortem intervals. *Postmortem interval* (PMI) is defined as the elapsed time between death and an autopsy. Knowledge of PMI is considered essential when conducting medical research on human cadavers. The data in the table (p. 17) are the PMIs of 22 human brain specimens obtained at autopsy in a recent study (*Brain and Language*, June 1995). Graphically describe the PMI data with a stem-and-leaf plot. Based on the plot, make a summary statement about the PMI of the 22 human brain specimens.

BULIMIA

| Bulimic students: | 21 | 13 | 10 | 20 | 25 | 19 | 16 | 21 | 24 | 13 | 14 | | | |
| Normal students: | 13 | 6 | 16 | 13 | 8 | 19 | 23 | 18 | 11 | 19 | 7 | 10 | 15 | 20 |

Source: Randles, R. H. "On neutral responses (zeros) in the sign test and ties in the Wilcoxon-Mann-Whitney test," *American Statistician*, Vol. 55, No. 2, May 2001 (Figure 3).

BRAINPMI

Postmortem intervals for 22 human brain specimens

5.5	14.5	6.0	5.5	5.3	5.8	11.0	6.1
7.0	14.5	10.4	4.6	4.3	7.2	10.5	6.5
3.3	7.0	4.1	6.2	10.4	4.9		

Source: Reprinted from *Brain and Language,* Vol. 49, Issue 3, T. L. Hayes and D. A. Lewis, "Anatomical Specialization of the Anterior Motor Speech Area: Hemispheric Differences in Magnopyramidal Neurons," p. 292 (Table 1), Copyright © 1995, with permission of Elsevier.

1.21 **Is honey a cough remedy?** Coughing at night is a common symptom of an upper respiratory tract infection, yet there is no accepted therapeutic cure. Does a teaspoon of honey before bed really calm a child's cough? To test the folk remedy, pediatric researchers at Pennsylvania State University carried out a designed study conducted over two nights (*Archives of Pediatrics and Adolescent Medicine*, December 2007.) A sample of 105 children who were ill with an upper respiratory tract infection and their parents participated in the study. On the first night, the parents rated their children's cough symptoms on a scale from 0 (no problems at all) to 6 (extremely severe) in five different areas. The total symptoms score (ranging from 0 to 30 points) was the variable of interest for the 105 patients. On the second night, the parents were instructed to give their sick child a dosage of liquid "medicine" prior to bedtime. Unknown to the parents, some were given a dosage of dextromethorphan (DM)—an over-the-counter cough medicine—while others were given a similar dose of honey. Also, a third group of parents (the control group) gave their sick children no dosage at all. Again, the parents rated their children's cough symptoms, and the improvement in total cough symptoms score was determined for each child. The data (improvement scores) for the study are shown in the accompanying

table, followed by a MINITAB stem-and-leaf plot of the data. Shade the leaves for the honey dosage group on the stem-and-leaf plot. What conclusions can pediatric researchers draw from the graph? Do you agree with the statement (extracted from the article), "honey may be a preferable treatment for the cough and sleep difficulty associated with childhood upper respiratory tract infection"?

```
Stem-and-leaf of TotalScore   N   = 105
Leaf Unit = 0.10

   1     0   0
   4     1   000
   4     2
   7     3   000
  16     4   000000000
  20     5   0000
  28     6   00000000
  41     7   0000000000000
  52     8   00000000000
 (13)    9   0000000000000
  40    10   0000000000
  30    11   000000
  24    12   0000000000000
  11    13   0000
   7    14   0
   6    15   00000
   1    16   0
```

1.22 **Comparing voltage readings.** A Harris Corporation/University of Florida study was undertaken to determine whether a manufacturing process performed at a remote location could be established locally. Test devices (pilots) were setup at both the old and new locations, and voltage readings on the process were obtained. A "good" process was considered to be one with voltage readings of at least 9.2 volts (with larger readings better than smaller readings). The first table on p. 18 contains voltage readings for 30 production runs at each location.

HONEYCOUGH

Honey Dosage:	12 11 15 11 10 13 10 4 15 16 9 14 10 6 10 8 11 12 12 8
	12 9 11 15 10 15 9 13 8 12 10 8 9 5 12

DM Dosage:	4 6 9 4 7 7 7 9 12 10 11 6 3 4 9 12 7 6 8 12 12 4 12
	13 7 10 13 9 4 4 10 15 9

No Dosage (Control):	5 8 6 1 0 8 12 8 7 7 1 6 7 7 12 7 9 7 9 5 11 9 5
	6 8 8 6 7 10 9 4 8 7 3 1 4 3

Source: Paul, I. M., et al. "Effect of honey, dextromethorphan, and no treatment on nocturnal cough and sleep quality for coughing children and their parents," *Archives of Pediatrics and Adolescent Medicine*, Vol. 161, No. 12, Dec. 2007 (data simulated).

⊙ VOLTAGE

OLD LOCATION			NEW LOCATION		
9.98	10.12	9.84	9.19	10.01	8.82
10.26	10.05	10.15	9.63	8.82	8.65
10.05	9.80	10.02	10.10	9.43	8.51
10.29	10.15	9.80	9.70	10.03	9.14
10.03	10.00	9.73	10.09	9.85	9.75
8.05	9.87	10.01	9.60	9.27	8.78
10.55	9.55	9.98	10.05	8.83	9.35
10.26	9.95	8.72	10.12	9.39	9.54
9.97	9.70	8.80	9.49	9.48	9.36
9.87	8.72	9.84	9.37	9.64	8.68

Source: Harris Corporation, Melbourne, Fla.

(a) Construct a relative frequency histogram for the voltage readings of the old process.

(b) Construct a stem-and-leaf display for the voltage readings of the old process. Which of the two graphs in parts a and b is more informative?

(c) Construct a frequency histogram for the voltage readings of the new process.

(d) Compare the two graphs in parts a and c. (You may want to draw the two histograms on the same graph.) Does it appear that the manufacturing process can be established locally (i.e., is the new process as good as or better than the old)?

1.23 Sanitation inspection of cruise ships. To minimize the potential for gastrointestinal disease outbreaks, all passenger cruise ships arriving at U.S. ports are subject to unannounced sanitation inspections. Ships are rated on a 100-point scale by the Centers for Disease Control and Prevention. A score of 86 or higher indicates that the ship is providing an accepted standard of sanitation. The May 2006 sanitation scores for 169 cruise ships are saved in the SHIPSANIT file. The first five and last five observations in the data set are listed in the accompanying table.

(a) Generate a stem-and-leaf display of the data. Identify the stems and leaves of the graph.

(b) Use the stem-and-leaf display to estimate the proportion of ships that have an accepted sanitation standard.

(c) Locate the inspection score of 84 (*Sea Bird*) on the stem-and-leaf display.

(d) Generate a histogram for the data.

(e) Use the histogram to estimate the proportion of ships that have an accepted sanitation standard.

⊙ SHIPSANIT (selected observations)

SHIP NAME	SANITATION SCORE
Adventure of the Seas	95
Albatross	96
Amsterdam	98
Arabella	94
Arcadia	98
.	.
.	.
Wind Surf	95
Yorktown Clipper	91
Zaandam	98
Zenith	94
Zuiderdam	94

Source: National Center for Environmental Health, Centers for Disease Control and Prevention, May 24, 2006.

⊙ PHISHING

1.24 Phishing attacks to email accounts. *Phishing* is the term used to describe an attempt to extract personal/financial information (e.g., PIN numbers, credit card information, bank account numbers) from unsuspecting people through fraudulent email. An article in *Chance* (Summer 2007) demonstrates how statistics can help identify phishing attempts and make e-commerce safer. Data from an actual phishing attack against an organization were used to determine whether the attack may have been an "inside job" that originated within the company. The company setup a publicized email account—called a "fraud box"—that enabled employees to notify them if they suspected an email phishing attack.

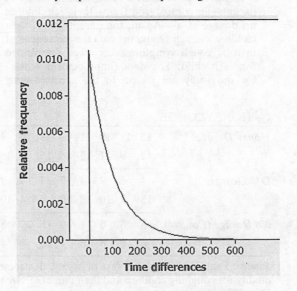

The interarrival times, that is, the time differences (in seconds), for 267 fraud box email notifications were recorded. *Chance* showed that if there is minimal or no collaboration or collusion from within the company, the interarrival times would have a frequency distribution similar to the one shown in the accompanying figure (p. 18). The 267 interarrival times are saved in the PHISHING file. Construct a frequency histogram for the interarrival times. Is the data skewed to the right? Give your opinion on whether the phishing attack against the organization was an "inside job."

1.5 Describing Quantitative Data Numerically

Numerical descriptive measures provide a second (and often more powerful) method for describing a set of quantitative data. These measures, which locate the center of the data set and its spread, actually enable you to construct an approximate mental image of the distribution of the data set.

Note: Most of the formulas used to compute numerical descriptive measures require the summation of numbers. For instance, we may want to sum the observations in a data set, or we may want to square each observation and then sum the squared values. The symbol Σ (sigma) is used to denote a summation operation.

For example, suppose we denote the n sample measurements on a random variable y by the symbols $y_1, y_2, y_3, \ldots, y_n$. Then the sum of all n measurements in the sample is represented by the symbol

$$\sum_{i=1}^{n} y_i$$

This is read "summation y, y_1 to y_n" and is equal to the value

$$y_1 + y_2 + y_3 + \cdots + y_n$$

One of the most common measures of central tendency is the **mean**, or arithmetic average, of a data set. Thus, if we denote the sample measurements by the symbols $y_1, y_2, y_3, \ldots$, the sample mean is defined as follows:

Definition 1.15 The **mean** of a sample of n measurements $y_1, y_2, \ldots, y_n$ is

$$\bar{y} = \frac{\sum\limits_{i=1}^{n} y_i}{n}$$

The mean of a population, or equivalently, the expected value of y, $E(y)$, is usually unknown in a practical situation (we will want to infer its value based on the sample data). Most texts use the symbol μ to denote the mean of a population. Thus, we use the following notation:

Notation

Sample mean: $\bar{y}$
Population mean: $E(y) = \mu$

The spread or variation of a data set is measured by its **range**, its **variance**, or its **standard deviation**.

> **Definition 1.16** The **range** of a sample of n measurements $y_1, y_2, \ldots, y_n$ is the difference between the largest and smallest measurements in the sample.

Example 1.5

If a sample consists of measurements 3, 1, 0, 4, 7, find the sample mean and the sample range.

Solution

The sample mean and range are

$$\bar{y} = \frac{\sum\limits_{i=1}^{n} y_i}{n} = \frac{15}{5} = 3$$

$$\text{Range} = 7 - 0 = 7$$

The variance of a set of measurements is defined to be the average of the *squares of the deviations* of the measurements about their mean. Thus, the population variance, which is usually unknown in a practical situation, would be the mean or expected value of $(y - \mu)^2$, or $E[(y - \mu)^2]$. We use the symbol σ^2 to represent the variance of a population:

$$E[(y - \mu)^2] = \sigma^2$$

The quantity usually termed the **sample variance** is defined in the box.

> **Definition 1.17** The **variance** of a sample of n measurements $y_1, y_2, \ldots, y_n$ is defined to be
>
> $$s^2 = \frac{\sum\limits_{i=1}^{n}(y_i - \bar{y})^2}{n - 1} = \frac{\sum\limits_{i=1}^{n} y_i^2 - n\bar{y}^2}{n - 1}$$

Note that the sum of squares of deviations in the sample variance is divided by $(n - 1)$, rather than n. Division by n produces estimates that tend to underestimate σ^2. Division by $(n - 1)$ corrects this problem.

Example 1.6

Refer to Example 1.5. Calculate the sample variance for the sample 3, 1, 0, 4, 7.

Solution

We first calculate

$$\sum_{i=1}^{n}(y_i - \bar{y})^2 = \sum_{i=1}^{n} y_i^2 - n\bar{y}^2 = 75 - 5(3)^2 = 30$$

where $\bar{y} = 3$ from Example 1.4. Then

$$s^2 = \frac{\sum\limits_{i=1}^{n}(y_i - \bar{y})^2}{n - 1} = \frac{30}{4} = 7.5$$

The concept of a variance is important in theoretical statistics, but its square root, called a **standard deviation**, is the quantity most often used to describe data variation.

Definition 1.18 The **standard deviation** of a set of measurements is equal to the square root of their variance. Thus, the standard deviations of a sample and a population are

Sample standard deviation: s
Population standard deviation: σ

The standard deviation of a set of data takes on meaning in light of a theorem (Tchebysheff's theorem) and a rule of thumb.* Basically, they give us the following guidelines:

Guidelines for Interpreting a Standard Deviation

1. For *any* data set (population or sample), at least three-fourths of the measurements will lie within 2 standard deviations of their mean.

2. For *most* data sets of moderate size (say, 25 or more measurements) with a mound-shaped distribution, approximately 95% of the measurements will lie within 2 standard deviations of their mean.

Example 1.7

Often, travelers who have no intention of showing up fail to cancel their hotel reservations in a timely manner. These travelers are known, in the parlance of the hospitality trade, as "no-shows." To protect against no-shows and late cancellations, hotels invariably overbook rooms. A study reported in the *Journal of Travel Research* examined the problems of overbooking rooms in the hotel industry. The data in Table 1.6, extracted from the study, represent daily numbers of late cancellations and no-shows for a random sample of 30 days at a large (500-room) hotel. Based on this sample, how many rooms, at minimum, should the hotel overbook each day?

NOSHOWS

Table 1.6 Hotel no-shows for a sample of 30 days

18	16	16	16	14	18	16	18	14	19
15	19	9	20	10	10	12	14	18	12
14	14	17	12	18	13	15	13	15	19

Source: Toh, R. S. "An inventory depletion overbooking model for the hotel industry," *Journal of Travel Research*, Vol. 23, No. 4, Spring 1985, p. 27. The *Journal of Travel Research* is published by the Travel and Tourism Research Association (TTRA) and the Business Research Division, University of Colorado at Boulder.

Solution

To answer this question, we need to know the range of values where most of the daily numbers of no-shows fall. We must compute $\bar{y}$ and s, and examine the shape of the relative frequency distribution for the data.

---

* For a more complete discussion and a statement of Tchebysheff's theorem, see the references listed at the end of this chapter.

Figure 1.7 is a MINITAB printout that shows a stem-and-leaf display and descriptive statistics of the sample data. Notice from the stem-and-leaf display that the distribution of daily no-shows is mound-shaped, and only slightly skewed on the low (top) side of Figure 1.7. Thus, guideline 2 in the previous box should give a good estimate of the percentage of days that fall within 2 standard deviations of the mean.

Figure 1.7 MINITAB printout: Describing the no-show data, Example 1.6

Stem-and-Leaf Display: Noshows

```
Stem-and-leaf of Noshows   N = 30
Leaf Unit = 0.10

   1     9 0
   3    10 00
   3    11
   6    12 000
   8    13 00
  13    14 00000
  (3)   15 000
  14    16 0000
  10    17 0
   9    18 00000
   4    19 000
   1    20 0
```

Descriptive Statistics: Noshows

Variable	N	Mean	Median	TrMean	StDev	SE Mean
Noshows	30	15.133	15.000	15.231	2.945	0.538

Variable	Minimum	Maximum	Q1	Q3
Noshows	9.000	20.000	13.000	18.000

The mean and standard deviation of the sample data, shaded on the MINITAB printout, are $\bar{y} = 15.133$ and $s = 2.945$. From guideline 2 in the box, we know that about 95% of the daily number of no-shows fall within 2 standard deviations of the mean, that is, within the interval

$$\bar{y} \pm 2s = 15.133 \pm 2(2.945)$$

$$= 15.133 \pm 5.890$$

or between 9.243 no-shows and 21.023 no-shows. (If we count the number of measurements in this data set, we find that actually 29 out of 30, or 96.7%, fall in this interval.)

From this result, the large hotel can infer that there will be at least 9.243 (or, rounding up, 10) no-shows per day. Consequently, the hotel can overbook at least 10 rooms per day and still be highly confident that all reservations can be honored.   ▪

Numerical descriptive measures calculated from sample data are called **statistics**. Numerical descriptive measures of the population are called **parameters**. In a practical situation, we will not know the population relative frequency distribution (or equivalently, the population distribution for y). We will usually assume that it has unknown numerical descriptive measures, such as its mean μ and standard deviation σ, and by inferring (using **sample statistics**) the values of these parameters, we infer the nature of the population relative frequency distribution. Sometimes we will assume that we know the shape of the population relative frequency distribution and use this information to help us make our inferences. When we do this, we are postulating a model for the population relative frequency distribution, and we must keep in mind that the validity of the inference may depend on how well our model fits reality.

Definition 1.19 Numerical descriptive measures of a population are called **parameters**.

Definition 1.20 A **sample statistic** is a quantity calculated from the observations in a sample.

1.5   Exercises

EARTHQUAKE

1.25   Earthquake aftershock magnitudes.   Refer to Exercise 1.18 (p. 16) and U.S. Geological Survey data on aftershocks from a major California earthquake. The EARTHQUAKE file contains the magnitudes (measured on the Richter scale) for 2,929 aftershocks. A MINITAB printout with descriptive statistics of magnitude is shown at the bottom of the page.

(a) Locate and interpret the mean of the magnitudes for the 2,929 aftershocks.

(b) Locate and interpret the range of the magnitudes for the 2,929 aftershocks.

(c) Locate and interpret the standard deviation of the magnitudes for the 2,929 aftershocks.

(d) If the target of your interest is these specific 2,929 aftershocks, what symbols should you use to describe the mean and standard deviation?

FTC

1.26   FTC cigarette rankings.   Periodically, the Federal Trade Commission (FTC) ranks domestic cigarette brands according to tar, nicotine, and carbon monoxide content. The test results are obtained by using a sequential smoking machine to "smoke" cigarettes to a 23-millimeter butt length. The tar, nicotine, and carbon monoxide concentrations (rounded to the nearest milligram) in the residual "dry" particulate matter of the smoke are then measured. The accompanying SAS

printouts describe the nicotine contents of 500 cigarette brands. (The data are saved in the FTC file.)

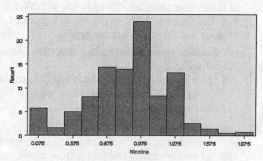

The MEANS Procedure
Analysis Variable : NICOTINE

N	Mean	Std Dev	Minimum	Maximum
500	0.8425000	0.3455250	0.0500000	1.9000000

(a) Examine the relative frequency histogram for nicotine content. Use the rule of thumb to describe the data set.

(b) Locate $\bar{y}$ and s on the printout, then compute the interval $\bar{y} \pm 2s$.

(c) Based on your answer to part a, estimate the percentage of cigarettes with nicotine contents in the interval formed in part b.

(d) Use the information on the SAS histogram to determine the actual percentage of nicotine contents that fall within the interval formed

---

Descriptive Statistics: MAGNITUDE

Variable	N	Mean	StDev	Variance	Minimum	Median	Maximum	Range
MAGNITUDE	2929	2.1197	0.6636	0.4403	0.0000	2.0000	6.7000	6.7000

Variable	Mode	N for Mode
MAGNITUDE	1.8	298

in part b. Does your answer agree with your estimate of part c?

SHIPSANIT

1.27 Sanitation inspection of cruise ships. Refer to the Centers for Disease Control and Prevention study of sanitation levels for 169 international cruise ships, Exercise 1.23 (p. 18). (Recall that sanitation scores range from 0 to 100.)

(a) Find $\bar{y}$ and s for the 169 sanitation scores.

(b) Calculate the interval $\bar{y} \pm 2s$.

(c) Find the percentage of scores in the data set that fall within the interval, part b. Does the result agree with the rule of thumb given in this section?

1.28 Most powerful business women in America. *Fortune* (October 16, 2008) published a list of the 50 most powerful women in business in the United States. The data on age (in years) and title of each of these 50 women are stored in the WPOWER50 file. The first five and last five observations of the data set are listed in the table below.

(a) Find the mean and standard deviation of these 50 ages.

(b) Give an interval that is highly likely to contain the age of a randomly selected woman from the *Fortune* list.

1.29 Ammonia from vehicle exhaust. Three-way catalytic converters have been installed in new vehicles in order to reduce pollutants from motor vehicle exhaust emissions. However, these converters unintentionally increase the level of ammonia in the air. *Environmental Science and Technology* (September 1, 2000) published a study on the ammonia levels near the exit ramp of a San Francisco highway tunnel. The data in the next table represent daily ammonia concentrations (parts per million) on eight randomly selected days during afternoon drive-time in the summer of 1999.

AMMONIA

1.53	1.50	1.37	1.51	1.55	1.42	1.41	1.48

(a) Find and interpret the mean daily ammonia level in air in the tunnel.

(b) Find the standard deviation of the daily ammonia levels. Interpret the result.

(c) Suppose the standard deviation of the ammonia levels during morning drive-time at the exit ramp is 1.45 ppm. Which time, morning or afternoon drive-time, has more variable ammonia levels?

1.30 Animal-assisted therapy for heart patients. Medical researchers at an *American Heart Association Conference* (November 2005) presented a study to gauge whether animal-assisted therapy can improve the physiological responses of heart patients. A team of nurses from the UCLA Medical Center randomly divided 76 heart patients into three groups. Each patient in group T was visited by a human volunteer accompanied by a trained dog; each patient in group V was visited by a volunteer only; and the patients in group C were not visited at all. The anxiety level of each patient was measured (in points) both before and after the visits. The next table (p. 25) gives summary statistics for the drop in anxiety level for patients in the three groups. Suppose the anxiety level of a patient selected from the study had a drop of 22.5 points. Which group is the patient more likely to have come from? Explain.

WPOWER50 (selected observations)

RANK	NAME	AGE	COMPANY	TITLE
1	Indra Nooyi	52	PepsiCo	CEO/Chairman
2	Irene Rosenfeld	55	Kraft Foods	CEO/Chairman
3	Pat Woertz	55	Archer Daniels Midland	CEO/Chairman
4	Anne Mulcahy	55	Xerox	CEO/Chairman
5	Angela Braley	47	Wellpoint	CEO/President
⋮	⋮	⋮		
46	Lorrie Norrington	48	eBay	CEO
47	Terri Dial	58	Citigroup	CEO
48	Lynn Elsenhans	52	Sunoco	CEO/President
49	Cathie Black	64	Hearst Magazines	President
50	Marissa Mayer	33	Google	VP

Source: Fortune, Oct. 16, 2008.

Summary table for Exercise 1.30

	SAMPLE SIZE	MEAN DROP	STD. DEV.
Group T: Volunteer + Trained Dog	26	10.5	7.6
Group V: Volunteer only	25	3.9	7.5
Group C: Control group (no visit)	25	1.4	7.5

Source: Cole, K., et al. "Animal assisted therapy decreases hemodynamics, plasma epinephrine and state anxiety in hospitalized heart failure patients," *American Heart Association Conference,* Dallas, Texas, Nov. 2005.

1.31 Improving SAT scores. The National Education Longitudinal Survey (NELS) tracks a nationally representative sample of U.S. students from eighth grade through high school and college. Research published in *Chance* (Winter 2001) examined the Standardized Admission Test (SAT) scores of 265 NELS students who paid a private tutor to help

them improve their scores. The table below summarizes the changes in both the SAT-Mathematics and SAT-Verbal scores for these students.

	SAT-MATH	SAT-VERBAL
Mean change in score	19	7
Standard deviation of score changes	65	49

(a) Suppose one of the 265 students who paid a private tutor is selected at random. Give an interval that is likely to contain this student's change in the SAT-Math score.
(b) Repeat part a for the SAT-Verbal score.
(c) Suppose the selected student's score increased on one of the SAT tests by 140 points. Which test, the SAT-Math or SAT-Verbal, is the one most likely to have the 140-point increase? Explain.

1.6 The Normal Probability Distribution

One of the most commonly used models for a theoretical population relative frequency distribution for a quantitative variable is the **normal probability distribution**, as shown in Figure 1.8. The normal distribution is symmetric about its mean μ, and its spread is determined by the value of its standard deviation σ. Three normal curves with different means and standard deviations are shown in Figure 1.9.

Figure 1.8 A normal probability distribution

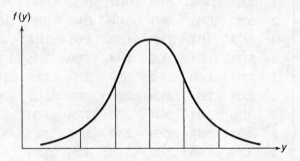

$f(y)$

Computing the area over an interval under the normal probability distribution can be a difficult task.* Consequently, we will use the computed areas listed in Table 1 of Appendix D. A partial reproduction of this table is shown in Table 1.7. As you can see from the normal curve above the table, the entries give areas under the normal curve between the mean of the distribution and a standardized distance

$$z = \frac{y - \mu}{\sigma}$$

* Students with knowledge of calculus should note that the probability that y assumes a value in the interval $a < y < b$ is $P(a < y < b) = \int_a^b f(y)dy$, assuming the integral exists. The value of this definite integral can be obtained to any desired degree of accuracy by approximation procedures. For this reason, it is tabulated for the user.

to the right of the mean. Note that z is the number of standard deviations σ between μ and y. The distribution of z, which has mean $\mu = 0$ and standard deviation $\sigma = 1$, is called a **standard normal distribution**.

Figure 1.9 Several normal distributions with different means and standard deviations

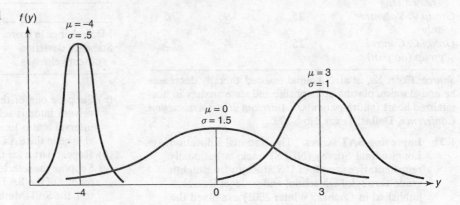

Table 1.7 Reproduction of part of Table 1 of Appendix D

z	.00	.01	.02	.03	.04	.05	.06	.07	.08	.09
0	.0000	.0040	.0080	.0120	.0160	.0199	.0239	.0279	.0319	.0359
.1	.0398	.0438	.0478	.0517	.0557	.0596	.0636	.0675	.0714	.0753
.2	.0793	.0832	.0871	.0910	.0948	.0987	.1026	.1064	.1103	.1141
.3	.1179	.1217	.1255	.1293	.1331	.1368	.1406	.1443	.1480	.1517
.4	.1554	.1591	.1628	.1664	.1700	.1736	.1772	.1808	.1844	.1879
.5	.1915	.1950	.1985	.2019	.2054	.2088	.2123	.2157	.2190	.2224
.6	.2257	.2291	.2324	.2357	.2389	.2422	.2454	.2486	.2517	.2549
.7	.2580	.2611	.2642	.2673	.2704	.2734	.2764	.2794	.2823	.2852
.8	.2881	.2910	.2939	.2967	.2995	.3023	.3051	.3078	.3106	.3133
.9	.3159	.3186	.3212	.3238	.3264	.3289	.3315	.3340	.3365	.3389
1.0	.3413	.3438	.3461	.3485	.3508	.3531	.3554	.3577	.3599	.3621
1.1	.3643	.3665	.3686	.3708	.3729	.3749	.3770	.3790	.3810	.3830
1.2	.3849	.3869	.3888	.3907	.3925	.3944	.3962	.3980	.3997	.4015
1.3	.4032	.4049	.4066	.4082	.4099	.4115	.4131	.4147	.4162	.4177
1.4	.4192	.4207	.4222	.4236	.4251	.4265	.4279	.4292	.4306	.4319
1.5	.4332	.4345	.4357	.4370	.4382	.4394	.4406	.4418	.4429	.4441

Example 1.8

Suppose y is a normal random variable with $\mu = 50$ and $\sigma = 15$. Find $P(30 < y < 70)$, the probability that y will fall within the interval $30 < y < 70$.

Solution

Refer to Figure 1.10. Note that $y = 30$ and $y = 70$ lie the same distance from the mean $\mu = 50$, with $y = 30$ below the mean and $y = 70$ above it. Then, because the normal curve is symmetric about the mean, the probability A_1 that y falls between $y = 30$ and $\mu = 50$ is equal to the probability A_2 that y falls between $\mu = 50$ and $y = 70$. The z score corresponding to $y = 70$ is

$$z = \frac{y - \mu}{\sigma} = \frac{70 - 50}{15} = 1.33$$

Therefore, the area between the mean $\mu = 50$ and the point $y = 70$ is given in Table 1 of Appendix D (and Table 1.7) at the intersection of the row corresponding to $z = 1.3$ and the column corresponding to .03. This area (probability) is $A_2 = .4082$. Since $A_1 = A_2$, A_1 also equals .4082; it follows that the probability that y falls in the interval $30 < y < 70$ is $P(30 < y < 70) = 2(.4082) = .8164$. The z scores corresponding to $y = 30$ ($z = -1.33$) and $y = 70$ ($z = 1.33$) are shown in Figure 1.11. ■

Figure 1.10 Normal probability distribution: $\mu = 50$, $\sigma = 15$

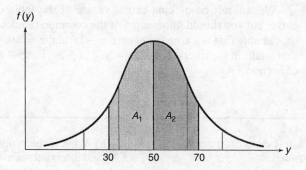

Figure 1.11 A distribution of z scores (a standard normal distribution)

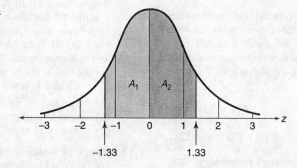

Example 1.9

Use Table 1 of Appendix D to determine the area to the right of the z score 1.64 for the standard normal distribution. That is, find $P(z \geq 1.64)$.

Solution

The probability that a normal random variable will fall more than 1.64 standard deviations to the right of its mean is indicated in Figure 1.12. Because the normal distribution is symmetric, half of the total probability (.5) lies to the right of the mean and half to the left. Therefore, the desired probability is

$$P(z \geq 1.64) = .5 - A$$

Figure 1.12 Standard normal distribution: $\mu = 0$, $\sigma = 1$

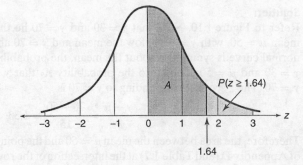

where A is the area between $\mu = 0$ and $z = 1.64$, as shown in the figure. Referring to Table 1, we find that the area A corresponding to $z = 1.64$ is .4495. So

$$P(z \geq 1.64) = .5 - A = .5 - .4495 = .0505$$

We will not be making extensive use of the table of areas under the normal curve, but you should know some of the common tabulated areas. In particular, you should note that the area between $z = -2.0$ and $z = 2.0$, which gives the probability that y falls in the interval $\mu - 2\sigma < y < \mu + 2\sigma$, is .9544 and agrees with guideline 2 of Section 1.4.

1.6 Exercises

1.32 Normal Probabilities. Use Table 1 of Appendix D to find each of the following:

(a) $P(-1 \leq z \leq 1)$ (b) $P(-1.96 \leq z \leq 1.96)$
(c) $P(-1.645 \leq z \leq 1.645)$ (d) $P(-3 \leq z \leq 3)$

1.33 Normal probabilities. Given that the random variable y has a normal probability distribution with mean 100 and variance 64, draw a sketch (i.e., graph) of the frequency function of y. Locate μ and the interval $\mu \pm 2\sigma$ on the graph. Find the following probabilities:

(a) $P(\mu - 2\sigma \leq y \leq \mu + 2\sigma)$ (b) $P(y \geq 108)$
(c) $P(y \leq 92)$ (d) $P(92 \leq y \leq 116)$
(e) $P(92 \leq y \leq 96)$ (f) $P(76 \leq y \leq 124)$

1.34 Transmission delays in wireless technology. Resource reservation protocol (RSVP) was originally designed to establish signaling links for stationary networks. In *Mobile Networks and Applications* (December 2003), RSVP was applied to mobile wireless technology (e.g., a PC notebook with wireless LAN card for Internet access). A simulation study revealed that the transmission delay (measured in milliseconds) of an RSVP linked wireless device has an approximate normal distribution with mean $\mu = 48.5$ milliseconds and $\sigma = 8.5$ milliseconds.

(a) What is the probability that the transmission delay is less than 57 milliseconds?

(b) What is the probability that the transmission delay is between 40 and 60 milliseconds?

1.35 Alkalinity of water. The alkalinity level of water specimens collected from the Han River in Seoul, Korea, has a mean of 50 milligrams per liter and a standard deviation of 3.2 milligrams per liter (*Environmental Science and Engineering*, September 1, 2000). Assume the distribution of alkalinity levels is approximately normal and find the probability that a water specimen collected from the river has an alkalinity level

(a) exceeding 45 milligrams per liter.
(b) below 55 milligrams per liter.
(c) between 51 and 52 milligrams per liter.

1.36 Range of women's heights. In *Chance* (Winter 2007), Yale Law School professor Ian Ayres published the results of a study he conducted with his son and daughter on whether college students could estimate a range for women's heights. The students were shown a graph of a normal distribution of heights and asked: "The average height of women over 20 years old in the United States is 64 inches. Using your intuition, please give your best estimate of the range of heights that would include 90% of women over 20 years old. Please make sure that the center of the range is the average height of 64 inches." The standard deviation of heights for women over 20 years old is known to be 2.6 inches. Find the range of interest.

1.37 **Psychological experiment on alcohol and threats.** A group of Florida State University psychologists examined the effects of alcohol on the reactions of people to a threat (*Journal of Abnormal Psychology*, Vol. 107, 1998). After obtaining a specified blood alcohol level, experimental subjects were placed in a room and threatened with electric shocks. Using sophisticated equipment to monitor the subjects' eye movements, the startle response (measured in milliseconds) was recorded for each subject. The mean and standard deviation of the startle responses were 37.9 and 12.4, respectively. Assume that the startle response y for a person with the specified blood alcohol level is approximately normally distributed.

(a) Find the probability that y is between 40 and 50 milliseconds.

(b) Find the probability that y is less than 30 milliseconds.

(c) Give an interval for y, centered around 37.9 milliseconds, so that the probability that y falls in the interval is .95.

1.38 **Modeling length of gestation.** Based on data from the National Center for Health Statistics, N. Wetzel used the normal distribution to model the length of gestation for pregnant U.S. women (*Chance*, Spring 2001). Gestation length has a mean of 280 days with a standard deviation of 20 days.

(a) Find the probability that gestation length is between 275.5 and 276.5 days. (This estimates the probability that a woman has her baby 4 days earlier than the "average" due date.)

(b) Find the probability that gestation length is between 258.5 and 259.5 days. (This estimates the probability that a woman has her baby 21 days earlier than the "average" due date.)

(c) Find the probability that gestation length is between 254.5 and 255.5 days. (This estimates the probability that a woman has her baby 25 days earlier than the "average" due date.)

(d) The *Chance* article referenced a newspaper story about three sisters who all gave birth on the same day (March 11, 1998). Karralee had her baby 4 days early; Marrianne had her baby 21 days early; and Jennifer had her baby 25 days early. Use the results, parts a–c, to estimate the probability that three women have their babies 4, 21, and 25 days early, respectively. Assume the births are independent events. [Hint: If events A, B, and C are independent, then $P(A \text{ and } B \text{ and } C) = P(A) \times P(B) \times P(C)$.]

1.39 **Mean shifts on a production line.** *Six Sigma* is a comprehensive approach to quality goal setting that involves statistics. An article in *Aircraft Engineering and Aerospace Technology* (Vol. 76, No. 6, 2004) demonstrated the use of the normal distribution in Six Sigma goal setting at Motorola Corporation. Motorola discovered that the average defect rate for parts produced on an assembly line varies from run to run, and is approximately normally distributed with a mean equal to 3 defects per million. Assume that the goal at Motorola is for the average defect rate to vary no more than 1.5 standard deviations above or below the mean of 3. How likely is it that the goal will be met?

1.7 Sampling Distributions and the Central Limit Theorem

Since we use sample statistics to make inferences about population parameters, it is natural that we would want to know something about the reliability of the resulting inferences. For example, if we use a statistic to estimate the value of a population mean μ, we will want to know how close to μ our estimate is likely to fall. To answer this question, we need to know the probability distribution of the statistic.

The probability distribution for a statistic based on a random sample of n measurements could be generated in the following way. For purposes of illustration, we suppose we are sampling from a population with $\mu = 10$ and $\sigma = 5$, the sample statistic is $\bar{y}$, and the sample size is $n = 25$. Draw a single random sample of 25 measurements from the population and suppose that $\bar{y} = 9.8$. Return the measurements to the population and try again. That is, draw another random sample of $n = 25$ measurements and see what you obtain for an outcome. Now, perhaps, $\bar{y} = 11.4$. Replace these measurements, draw another sample of $n = 25$ measurements, calculate $\bar{y}$, and so on. If this sampling process were repeated over and over again an infinitely large number of times, you would generate an infinitely large

Figure 1.13 Sampling distribution for $\bar{y}$ based on a sample of $n = 25$ measurements

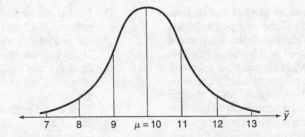

number of values of $\bar{y}$ that could be arranged in a relative frequency distribution. This distribution, which would appear as shown in Figure 1.13, is the probability distribution (or **sampling distribution**, as it is commonly called) of the statistic $\bar{y}$.

Definition 1.21 The **sampling distribution** of a sample statistic calculated from a sample of n measurements is the probability distribution of the statistic.

In actual practice, the sampling distribution of a statistic is obtained mathematically or by simulating the sampling on a computer using the procedure described previously.

If $\bar{y}$ has been calculated from a sample of $n = 25$ measurements selected from a population with mean $\mu = 10$ and standard deviation $\sigma = 5$, the sampling distribution shown in Figure 1.13 provides all the information you may wish to know about its behavior. For example, the probability that you will draw a sample of 25 measurements and obtain a value of $\bar{y}$ in the interval $9 \leq \bar{y} \leq 10$ will be the area under the sampling distribution over that interval.

Generally speaking, if we use a statistic to make an inference about a population parameter, we want its sampling distribution to center about the parameter (as is the case in Figure 1.13) and the standard deviation of the sampling distribution, called the **standard error of estimate**, to be as small as possible.

Two theorems provide information on the sampling distribution of a sample mean.

Theorem 1.1   If $y_1, y_2, \ldots, y_n$ represent a random sample of n measurements from a large (or infinite) population with mean μ and standard deviation σ, then, regardless of the form of the population relative frequency distribution, the mean and standard error of estimate of the sampling distribution of $\bar{y}$ will be

Mean: $E(\bar{y}) = \mu_{\bar{y}} = \mu$

Standard error of estimate: $\sigma_{\bar{y}} = \frac{\sigma}{\sqrt{n}}$

Theorem 1.2   **The Central Limit Theorem**   For large sample sizes, the mean $\bar{y}$ of a sample from a population with mean μ and standard deviation σ has a sampling distribution that is approximately normal, **regardless of the probability distribution of the sampled population**. The larger the sample size, the better will be the normal approximation to the sampling distribution of $\bar{y}$.

Theorems 1.1 and 1.2 together imply that for sufficiently large samples, the sampling distribution for the sample mean $\bar{y}$ will be approximately normal with mean μ and standard error $\sigma_{\bar{y}} = \sigma/\sqrt{n}$. The parameters μ and σ are the mean and standard deviation of the sampled population.

How large must the sample size n be so that the normal distribution provides a good approximation for the sampling distribution of $\bar{y}$? The answer depends on the shape of the distribution of the sampled population, as shown by Figure 1.14. Generally speaking, the greater the skewness of the sampled population distribution, the larger the sample size must be before the normal distribution is an adequate approximation for the sampling distribution of $\bar{y}$. For most sampled populations, sample sizes of $n \geq 30$ will suffice for the normal approximation to be reasonable. We will use the normal approximation for the sampling distribution of $\bar{y}$ when the sample size is at least 30.

Figure 1.14 Sampling distributions of $\bar{x}$ for different populations and different sample sizes

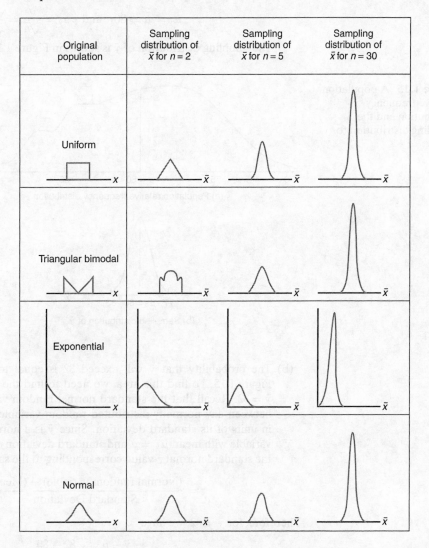

Example 1.10

Suppose we have selected a random sample of $n = 25$ observations from a population with mean equal to 80 and standard deviation equal to 5. It is known that the population is not extremely skewed.

(a) Sketch the relative frequency distributions for the population and for the sampling distribution of the sample mean, $\bar{y}$.

(b) Find the probability that $\bar{y}$ will be larger than 82.

Solution

(a) We do not know the exact shape of the population relative frequency distribution, but we do know that it should be centered about $\mu = 80$, its spread should be measured by $\sigma = 5$, and it is not highly skewed. One possibility is shown in Figure 1.15(a). From the central limit theorem, we know that the sampling distribution of $\bar{y}$ will be approximately normal since the sampled population distribution is not extremely skewed. We also know that the sampling distribution will have mean and standard deviation

$$\mu_{\bar{y}} = \mu = 80 \quad \text{and} \quad \sigma_{\bar{y}} = \frac{\sigma}{\sqrt{n}} = \frac{5}{\sqrt{25}} = 1$$

The sampling distribution of $\bar{y}$ is shown in Figure 1.15(b).

Figure 1.15 A population relative frequency distribution and the sampling distribution for $\bar{y}$

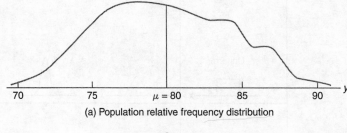

(a) Population relative frequency distribution

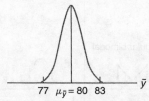

(b) Sampling distribution of $\bar{y}$

(b) The probability that $\bar{y}$ will exceed 82 is equal to the highlighted area in Figure 1.15. To find this area, we need to find the z-value corresponding to $\bar{y} = 82$. Recall that the standard normal random variable z is the difference between any normally distributed random variable and its mean, expressed in units of its standard deviation. Since $\bar{y}$ is a normally distributed random variable with mean $\mu_{\bar{y}} = \mu$ and standard deviation $\sigma_{\bar{y}} = \sigma/\sqrt{n}$, it follows that the standard normal z-value corresponding to the sample mean, $\bar{y}$, is

$$z = \frac{(\text{Normal random variable}) - (\text{Mean})}{\text{Standard Deviation}} = \frac{\bar{y} - \mu_{\bar{y}}}{\sigma_{\bar{y}}}$$

Therefore, for $\bar{y} = 82$, we have

$$z = \frac{\bar{y} - \mu_{\bar{y}}}{\sigma_{\bar{y}}} = \frac{82 - 80}{1} = 2$$

The area A in Figure 1.16 corresponding to $z = 2$ is given in the table of areas under the normal curve (see Table 1 of Appendix C) as .4772. Therefore, the tail area corresponding to the probability that $\bar{y}$ exceeds 82 is

$$P(\bar{y} > 82) = P(z > 2) = .5 - .4772 = .0228$$

Figure 1.16 The sampling distribution of $\bar{y}$

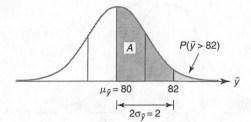

The central limit theorem can also be used to justify the fact that the *sum* of the sample measurements possesses a sampling distribution that is approximately normal for large sample sizes. In fact, since many statistics are obtained by summing or averaging random quantities, the central limit theorem helps to explain why many statistics have mound-shaped (or approximately normal) sampling distributions.

As we proceed, we encounter many different sample statistics, and we need to know their sampling distributions to evaluate the reliability of each one for making inferences. These sampling distributions are described as the need arises.

1.8 Estimating a Population Mean

We can make an inference about a population parameter in two ways:

1. Estimate its value.
2. Make a decision about its value (i.e., test a hypothesis about its value).

In this section, we illustrate the concepts involved in estimation, using the estimation of a population mean as an example. Tests of hypotheses will be discussed in Section 1.9.

To estimate a population parameter, we choose a sample statistic that has two desirable properties: (1) a sampling distribution that centers about the parameter and (2) a small standard error. If the mean of the sampling distribution of a statistic equals the parameter we are estimating, we say that the statistic is an **unbiased estimator** of the parameter. If not, we say that it is **biased**.

In Section 1.7, we noted that the sampling distribution of the sample mean is approximately normally distributed for moderate to large sample sizes and that it possesses a mean μ and standard error $\sigma/\sqrt{n}$. Therefore, as shown in Figure 1.17,

Figure 1.17 Sampling distribution of $\bar{y}$

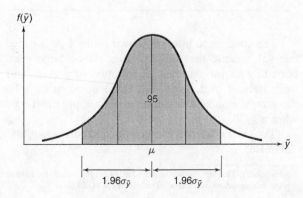

Figure 1.18 Locating $z_{\alpha/2}$ on the standard normal curve

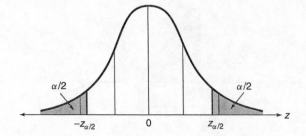

$\bar{y}$ is an unbiased estimator of the population mean μ, and the probability that $\bar{y}$ will fall within $1.96\,\sigma_{\bar{y}} = 1.96\,\sigma/\sqrt{n}$ of the true value of μ is approximately .95.[*]

Since $\bar{y}$ will fall within $1.96\sigma_{\bar{y}}$ of μ approximately 95% of the time, it follows that the interval

$$\bar{y} - 1.96\sigma_{\bar{y}} \quad \text{to} \quad \bar{y} + 1.96\sigma_{\bar{y}}$$

will enclose μ approximately 95% of the time in repeated sampling. This interval is called a 95% **confidence interval**, and .95 is called the **confidence coefficient**.

Notice that μ is fixed and that the confidence interval changes from sample to sample. The probability that a confidence interval calculated using the formula

$$\bar{y} \pm 1.96\sigma_{\bar{y}}$$

will enclose μ is approximately .95. Thus, the confidence coefficient measures the confidence that we can place in a particular confidence interval.

Confidence intervals can be constructed using any desired confidence coefficient. For example, if we define $z_{\alpha/2}$ to be the value of a standard normal variable that places the area $\alpha/2$ in the right tail of the z distribution (see Figure 1.18), then a $100(1 - \alpha)\%$ confidence interval for μ is given in the box.

Large-Sample $100(1 - \alpha)\%$ Confidence Interval for μ

$$\bar{y} \pm z_{\alpha/2}\sigma_{\bar{y}} \approx \bar{y} \pm z_{\alpha/2}\left(\frac{s}{\sqrt{n}}\right)$$

where $z_{\alpha/2}$ is the z-value with an area $\alpha/2$ to its right (see Figure 1.18) and $\sigma_{\bar{y}} = \sigma/\sqrt{n}$. The parameter σ is the standard deviation of the sampled population and n is the sample size. If σ is unknown, its value may be approximated by the sample standard deviation s. The approximation is valid for large samples (e.g., $n \geq 30$) only.

The confidence interval shown in the box is called a large-sample confidence interval because the sample size must be large enough to ensure approximate normality for the sampling distribution of $\bar{y}$. Also, and even more important, you will rarely, if ever, know the value of σ, so its value must be estimated using the sample standard deviation s. This approximation for σ will be adequate only when $n \geq 30$.

Typical confidence coefficients and corresponding values of $z_{\alpha/2}$ are shown in Table 1.8.

---

[*] Additionally, $\bar{y}$ has the smallest standard error among all unbiased estimators of μ. Consequently, we say that $\bar{y}$ is the **minimum variance unbiased estimator** (MVUE) for μ.

Table 1.8 Commonly used values of $z_{\alpha/2}$

Confidence Coefficient $(1 - \alpha)$	α	$\alpha/2$	$z_{\alpha/2}$
.90	.10	.05	1.645
.95	.05	.025	1.96
.99	.01	.005	2.576

Example 1.11

Psychologists have found that twins, in their early years, tend to have lower intelligence quotients and pick up language more slowly than nontwins (*Wisconsin Twin Research Newsletter*, Winter 2004). The slower intellectual growth of twins may be caused by benign parental neglect. Suppose we want to investigate this phenomenon. A random sample of $n = 50$ sets of $2\frac{1}{2}$-year-old twin boys is selected, and the total parental attention time given to each pair during 1 week is recorded. The data (in hours) are listed in Table 1.9. Estimate μ, the mean attention time given to all $2\frac{1}{2}$-year-old twin boys by their parents, using a 99% confidence interval. Interpret the interval in terms of the problem.

ATTENTIMES

Table 1.9 Attention time for a random sample of $n = 50$ sets of twins

20.7	14.0	16.7	20.7	22.5	48.2	12.1	7.7	2.9	22.2
23.5	20.3	6.4	34.0	1.3	44.5	39.6	23.8	35.6	20.0
10.9	43.1	7.1	14.3	46.0	21.9	23.4	17.5	29.4	9.6
44.1	36.4	13.8	0.8	24.3	1.1	9.3	19.3	3.4	14.6
15.7	32.5	46.6	19.1	10.6	36.9	6.7	27.9	5.4	14.0

Solution

The general form of the 99% confidence interval for a population mean is

$$\bar{y} \pm z_{\alpha/2}\sigma_{\bar{y}} = \bar{y} \pm z_{.005}\sigma_{\bar{y}}.$$

$$= \bar{y} \pm 2.576\left(\frac{\sigma}{\sqrt{n}}\right)$$

A SAS printout showing descriptive statistics for the sample of $n = 50$ attention times is displayed in Figure 1.19. The values of $\bar{y}$ and s, shaded on the printout, are $\bar{y} = 20.85$ and $s = 13.41$. Thus, for the 50 twins sampled, the 99% confidence interval is

$$20.85 \pm 2.576\left(\frac{\sigma}{\sqrt{50}}\right)$$

We do not know the value of σ (the standard deviation of the weekly attention time given to $2\frac{1}{2}$-year-old twin boys by their parents), so we use our best approximation, the sample standard deviation s. (Since the sample size, $n = 50$, is large, the approximation is valid.) Then the 99% confidence interval is

$$20.85 \pm 2.576\left(\frac{13.41}{\sqrt{50}}\right) = 20.85 \pm 4.89$$

or, from 15.96 to 25.74. That is, we can be 99% confident that the true mean weekly attention given to $2\frac{1}{2}$-year-old twin boys by their parents falls between 15.96 and

Figure 1.19 SAS descriptive statistics for $n = 50$ sample attention times

```
Sample Statistics for ATTIME

   N        Mean        Std. Dev.      Std. Error
--------------------------------------------------
   50       20.85         13.41           1.90

Hypothesis Test

   Null hypothesis:    Mean of ATTIME  =  0
   Alternative:        Mean of ATTIME ^= 0

With a specified known standard deviation of 13.41

            Z Statistic        Prob > Z
            -----------        --------
              10.993            <.0001

99% Confidence Interval for the Mean

            Lower Limit        Upper Limit
            -----------        -----------
              15.96              25.73
```

25.74 hours. [Note: This interval is also shown at the bottom of the SAS printout, Figure 1.19.] ∎

The large-sample method for making inferences about a population mean μ assumes that either σ is known or the sample size is large enough ($n \geq 30$) for the sample standard deviation s to be used as a good approximation to σ. The technique for finding a $100(1 - \alpha)\%$ confidence interval for a population mean μ for small sample sizes requires that the sampled population have a normal probability distribution. The formula, which is similar to the one for a large-sample confidence interval for μ, is

$$\bar{y} \pm t_{\alpha/2} s_{\bar{y}}$$

where $s_{\bar{y}} = s/\sqrt{n}$ is the estimated standard error of $\bar{y}$. The quantity $t_{\alpha/2}$ is directly analogous to the standard normal value $z_{\alpha/2}$ used in finding a large-sample confidence interval for μ except that it is an upper-tail t-value obtained from a student's t distribution. Thus, $t_{\alpha/2}$ is an upper-tail t-value such that an area $\alpha/2$ lies to its right.

Like the standardized normal (z) distribution, a Student's t distribution is symmetric about the value $t = 0$, but it is more variable than a z distribution. The variability depends on the number of **degrees of freedom**, **df**, which in turn depends on the number of measurements available for estimating σ^2. The smaller the number of degrees of freedom, the greater will be the spread of the t distribution. For this application of a Student's t distribution, df $= n - 1$.[†] As the sample size increases (and df increases), the Student's t distribution looks more and more like a z distribution, and for $n \geq 30$, the two distributions will be nearly identical. A Student's t distribution based on df $= 4$ and a standard normal distribution are shown in Figure 1.20. Note the corresponding values of $z_{.025}$ and $t_{.025}$.

The upper-tail values of the Student's t distribution are given in Table 2 of Appendix D. An abbreviated version of the t table is presented in Table 1.10. To find the upper-tail t-value based on 4 df that places .025 in the upper tail of the t distribution, we look in the row of the table corresponding to df $= 4$ and the column corresponding to $t_{.025}$. The t-value is 2.776 and is shown in Figure 1.17.

The process of finding a small-sample confidence interval for μ is given in the next box.

[†] Think of df as the amount of information in the sample size n for estimating μ. We lose 1 df for estimating μ; hence df $= n - 1$.

Figure 1.20 The $t_{.025}$ value in a t distribution with 4 df and the corresponding $z_{.025}$ value

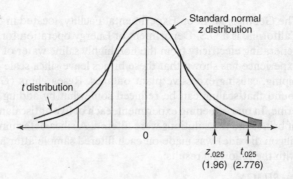

Standard normal
z distribution

t distribution

0

$z_{.025}$ $t_{.025}$
(1.96)   (2.776)

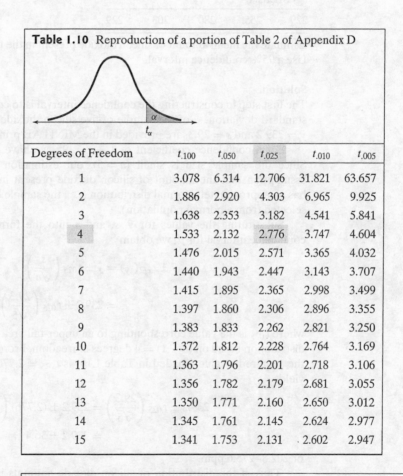

Table 1.10 Reproduction of a portion of Table 2 of Appendix D

α

t_α

Degrees of Freedom	$t_{.100}$	$t_{.050}$	$t_{.025}$	$t_{.010}$	$t_{.005}$
1	3.078	6.314	12.706	31.821	63.657
2	1.886	2.920	4.303	6.965	9.925
3	1.638	2.353	3.182	4.541	5.841
4	1.533	2.132	2.776	3.747	4.604
5	1.476	2.015	2.571	3.365	4.032
6	1.440	1.943	2.447	3.143	3.707
7	1.415	1.895	2.365	2.998	3.499
8	1.397	1.860	2.306	2.896	3.355
9	1.383	1.833	2.262	2.821	3.250
10	1.372	1.812	2.228	2.764	3.169
11	1.363	1.796	2.201	2.718	3.106
12	1.356	1.782	2.179	2.681	3.055
13	1.350	1.771	2.160	2.650	3.012
14	1.345	1.761	2.145	2.624	2.977
15	1.341	1.753	2.131	2.602	2.947

Small-Sample Confidence Interval for μ

$$\bar{y} \pm t_{\alpha/2}s_{\bar{y}} = \bar{y} \pm t_{\alpha/2}\left(\frac{s}{\sqrt{n}}\right)$$

where $s_{\bar{y}} = s/\sqrt{n}$ and $t_{\alpha/2}$ is a t-value based on $(n-1)$ degrees of freedom, such that the probability that $t > t_{\alpha/2}$ is $\alpha/2$.

Assumptions: The relative frequency distribution of the sampled population is approximately normal.

**Example
1.12**

The Geothermal Loop Experimental Facility, located in the Salton Sea in southern California, is a U.S. Department of Energy operation for studying the feasibility of generating electricity from the hot, highly saline water of the Salton Sea. Operating experience has shown that these brines leave silica scale deposits on metallic plant piping, causing excessive plant outages. Researchers (*Geothermics*, August 2002) found that scaling can be reduced somewhat by adding chemical solutions to the brine. In one screening experiment, each of five antiscalants was added to an aliquot of brine, and the solutions were filtered. A silica determination (parts per million of silicon dioxide) was made on each filtered sample after a holding time of 24 hours, with the following results:

💿 SILICA

229	255	280	203	229

Estimate the mean amount of silicon dioxide present in the five antiscalant solutions. Use a 95% confidence interval.

Solution

The first step in constructing the confidence interval is to compute the mean, $\bar{y}$, and standard deviation, s, of the sample of five silicon dioxide amounts. These values, $\bar{y} = 239.2$ and $s = 29.3$, are provided in the MINITAB printout, Figure 1.21.

For a confidence coefficient of $1 - \alpha = .95$, we have $\alpha = .05$ and $\alpha/2 = .025$. Since the sample size is small ($n = 5$), our estimation technique requires the assumption that the amount of silicon dioxide present in an antiscalant solution has an approximately normal distribution (i.e., the sample of five silicon amounts is selected from a normal population).

Substituting the values for $\bar{y}$, s, and n into the formula for a small-sample confidence interval for μ, we obtain

$$\bar{y} \pm t_{\alpha/2}(s_{\bar{y}}) = \bar{y} \pm t_{.025}\left(\frac{s}{\sqrt{n}}\right)$$

$$= 239.2 \pm t_{.025}\left(\frac{29.3}{\sqrt{5}}\right)$$

where $t_{.025}$ is the value corresponding to an upper-tail area of .025 in the Student's t distribution based on $(n - 1) = 4$ degrees of freedom. From Table 2 of Appendix D, the required t-value (shaded in Table 1.10) is $t_{.025} = 2.776$. Substituting this value yields

$$239.2 \pm t_{.025}\left(\frac{29.3}{\sqrt{5}}\right) = 239.2 \pm (2.776)\left(\frac{29.3}{\sqrt{5}}\right)$$

$$= 239.2 \pm 36.4$$

or 202.8–275.6 ppm.

Thus, if the distribution of silicon dioxide amounts is approximately normal, then we can be 95% confident that the interval (202.8–275.6) encloses μ, the true mean amount of silicon dioxide present in an antiscalant solution. Remember, the 95% confidence level implies that if we were to employ our interval estimator on repeated occasions, 95% of the intervals constructed would capture μ.

Figure 1.21 MINITAB descriptive statistics and confidence interval for Example 1.12

One-Sample T: PPM

Variable	N	Mean	StDev	SE Mean	95.0% CI
PPM	5	239.2	29.3	13.1	(202.8, 275.6)

The 95% confidence interval can also be obtained with statistical software. This interval is shaded on the MINITAB printout, Figure 1.21. You can see that the computer-generated interval is identical to our calculated one.   ■

Example 1.13

Suppose you want to reduce the width of the confidence interval obtained in Example 1.12. Specifically, you want to estimate the mean silicon dioxide content of an aliquot of brine correct to within 10 ppm with confidence coefficient approximately equal to .95. How many aliquots of brine would you have to include in your sample?

Solution

We will interpret the phrase, "correct to within 10 ppm ... equal to .95" to mean that we want half the width of a 95% confidence interval for μ to equal 10 ppm. That is, we want

$$t_{.025}\left(\frac{s}{\sqrt{n}}\right) = 10$$

To solve this equation for n, we need approximate values for $t_{.025}$ and s. Since we know from Example 1.12 that the confidence interval was wider than desired for $n = 5$, it is clear that our sample size must be larger than 5. Consequently, $t_{.025}$ will be very close to 2, and this value will provide a good approximation to $t_{.025}$. A good measure of the data variation is given by the standard deviation computed in Example 1.12. We substitute $t_{.025} \approx 2$ and $s \approx 29.3$ into the equation and solve for n:

$$t_{.025}\left(\frac{s}{\sqrt{n}}\right) = 10$$

$$2\left(\frac{29.3}{\sqrt{n}}\right) = 10$$

$$\sqrt{n} = 5.86$$

$$n = 34.3 \quad \text{or approximately } n = 34$$

Remember that this sample size is an approximate solution because we approximated the value of $t_{.025}$ and the value of s that might be computed from the prospective data. Nevertheless, $n = 34$ will be reasonably close to the sample size needed to estimate the mean silicon dioxide content correct to within 10 ppm.   ■

Important Note: Theoretically, the small-sample t procedure presented here requires that the sample data come from a population that is normally distributed. (See the assumption in the box, p. 37.) However, statisticians have found the one-sample t procedure to be **robust**, that is, to yield valid results even when the data are nonnormal, as long as the population is not highly skewed.

1.8 Exercises

1.40 Simulating a sampling distribution. The next table (p. 40) contains 50 random samples of random digits, $y = 0, 1, 2, 3, \ldots, 9$, where the probabilities corresponding to the values of y are given by the formula $p(y) = \frac{1}{10}$. Each sample contains $n = 6$ measurements.

(a) Use the 300 random digits to construct a relative frequency distribution for the data. This

relative frequency distribution should approximate $p(y)$.

(b) Calculate the mean of the 300 digits. This will give an accurate estimate of μ (the mean of the population) and should be very near to $E(y)$, which is 4.5.

(c) Calculate s^2 for the 300 digits. This should be close to the variance of y, $\sigma^2 = 8.25$.

(d) Calculate $\bar{y}$ for each of the 50 samples. Construct a relative frequency distribution for the sample means to see how close they lie to the mean of $\mu = 4.5$. Calculate the mean and standard deviation of the 50 means.

⊙ EX1_40

SAMPLE	SAMPLE	SAMPLE	SAMPLE
8, 1, 8, 0, 6, 6	7, 6, 7, 0, 4, 3	4, 4, 5, 2, 6, 6	0, 8, 4, 7, 6, 9
7, 2, 1, 7, 2, 9	1, 0, 5, 9, 9, 6	2, 9, 3, 7, 1, 3	5, 6, 9, 4, 4, 2
7, 4, 5, 7, 7, 1	2, 4, 4, 7, 5, 6	5, 1, 9, 6, 9, 2	4, 2, 3, 7, 6, 3
8, 3, 6, 1, 8, 1	4, 6, 6, 5, 5, 6	8, 5, 1, 2, 3, 4	1, 2, 0, 6, 3, 3
0, 9, 8, 6, 2, 9	1, 5, 0, 6, 6, 5	2, 4, 5, 3, 4, 8	1, 1, 9, 0, 3, 2
0, 6, 8, 8, 3, 5	3, 3, 0, 4, 9, 6	1, 5, 6, 7, 8, 2	7, 8, 9, 2, 7, 0
7, 9, 5, 7, 7, 9	9, 3, 0, 7, 4, 1	3, 3, 8, 6, 0, 1	1, 1, 5, 0, 5, 1
7, 7, 6, 4, 4, 7	5, 3, 6, 4, 2, 0	3, 1, 4, 4, 9, 0	7, 7, 8, 7, 7, 6
1, 6, 5, 6, 4, 2	7, 1, 5, 0, 5, 8	9, 7, 7, 9, 8, 1	4, 9, 3, 7, 3, 9
9, 8, 6, 8, 6, 0	4, 4, 6, 2, 6, 2	6, 9, 2, 9, 8, 7	5, 5, 1, 1, 4, 0
3, 1, 6, 0, 0, 9	3, 1, 8, 8, 2, 1	6, 6, 8, 9, 6, 0	4, 2, 5, 7, 7, 9
0, 6, 8, 5, 2, 8	8, 9, 0, 6, 1, 7	3, 3, 4, 6, 7, 0	8, 3, 0, 6, 9, 7
8, 2, 4, 9, 4, 6	1, 3, 7, 3, 4, 3		

1.41 Effect of n on the standard deviation. Refer to Exercise 1.40. To see the effect of sample size on the standard deviation of the sampling distribution of a statistic, combine pairs of samples (moving down the columns of the table) to obtain 25 samples of $n = 12$ measurements. Calculate the mean for each sample.

(a) Construct a relative frequency distribution for the 25 means. Compare this with the distribution prepared for Exercise 1.40 that is based on samples of $n = 6$ digits.

(b) Calculate the mean and standard deviation of the 25 means. Compare the standard deviation of this sampling distribution with the standard deviation of the sampling distribution in Exercise 1.40. What relationship would you expect to exist between the two standard deviations?

1.42 Using Table 2. Let t_0 be a particular value of t. Use Table 2 of Appendix D to find t_0 values such that the following statements are true:

(a) $P(t \geq t_0) = .025$ where $n = 10$
(b) $P(t \geq t_0) = .01$ where $n = 5$
(c) $P(t \leq t_0) = .005$ where $n = 20$
(d) $P(t \leq t_0) = .05$ where $n = 12$

1.43 Critical part failures in NASCAR vehicles. *The Sport Journal* (Winter 2007) published an analysis of critical part failures at NASCAR races. The researchers found that the time y (in hours) until the first critical part failure has a highly skewed distribution with $\mu = .10$ and $\sigma = .10$. Now, consider a random sample of $n = 50$ NASCAR races and let $\bar{y}$ represent the sample mean time until the first critical part failure.

(a) Find $E(\bar{y})$ and $Var(\bar{y})$.
(b) Although y has a highly skewed distribution, the sampling distribution of $\bar{y}$ is approximately normal. Why?
(c) Find the probability that the sample mean time until the first critical part failure exceeds .13 hour.

⊙ PERAGGR

1.44 Personality and aggressive behavior. How does personality impact aggressive behavior? A team of university psychologists conducted a review of studies that examined the relationship between personality and aggressive behavior (*Psychological Bulletin*, Vol. 132, 2006). One variable of interest to the researchers was the difference between the aggressive behavior level of individuals in the study who scored high on a personality test and those who scored low on the test. This variable, standardized to be between -7 and 7, was called "effect size." (A large positive effect size indicates that those who score high on the personality test are more aggressive than those who score low.) The researchers collected the effect sizes for a sample of $n = 109$ studies published in

Minitab Output for Exercise 1.44

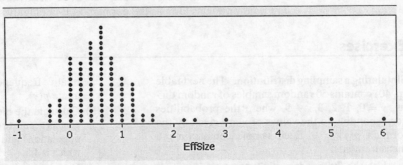

Variable	N	Mean	StDev	SE Mean	95% CI
EffSize	109	0.6477	0.8906	0.0853	(0.4786, 0.8167)

psychological journals. This data are saved in the PERAGGR file. A dot plot and summary statistics for effect size are shown in the accompanying MINITAB printouts (bottom of p. 40). Of interest to the researchers is the true mean effect size μ for all psychological studies of personality and aggressive behavior.

(a) Identify the parameter of interest to the researchers.

(b) Examine the dot plot. Does effect size have a normal distribution? Explain why your answer is irrelevant to the subsequent analysis.

(c) Locate and interpret a 95% confidence interval for μ on the accompanying printout.

(d) If the true mean effect size exceeds 0, then the researchers will conclude that in the population, those who score high on a personality test are more aggressive than those who score low. Can the researchers draw this conclusion? Explain.

1.45 Chicken pecking experiment. Animal behaviorists have discovered that the more domestic chickens peck at objects placed in their environment, the healthier the chickens seem to be. White string has been found to be a particularly attractive pecking stimulus. In one experiment, 72 chickens were exposed to a string stimulus. Instead of white string, blue-colored string was used. The number of pecks each chicken took at the blue string over a specified time interval was recorded. Summary statistics for the 72 chickens were $\bar{y} = 1.13$ pecks, $s = 2.21$ pecks (*Applied Animal Behaviour Science*, October 2000).

(a) Estimate the population mean number of pecks made by chickens pecking at blue string using a 99% confidence interval. Interpret the result.

(b) Previous research has shown that $\mu = 7.5$ pecks if chickens are exposed to white string. Based on the results, part a, is there evidence that chickens are more apt to peck at white string than blue string? Explain.

1.46 Impact of cooking on air particles. A group of Harvard University School of Public Health researchers studied the impact of cooking on the size of indoor air particles (*Environmental Science and Technology*, September 1, 2000). The decay rate (measured as μ m/hour) for fine particles produced from oven cooking or toasting was recorded on six randomly selected days. These six measurements are shown in the table.

(a) Find and interpret a 95% confidence interval for the true average decay rate of fine particles produced from oven cooking or toasting.

(b) Explain what the phrase "95% confident" implies in the interpretation of part a.

(c) What must be true about the distribution of the population of decay rates for the inference to be valid?

⊚ DECAY

.95	.83	1.20	.89	1.45	1.12

Source: Abt, E., et al. "Relative contribution of outdoor and indoor particle sources to indoor concentrations," *Environmental Science and Technology*, Vol. 34, No. 17, Sept. 1, 2000 (Table 3). Reprinted with permission from *Environmental Science.*

1.47 Accounting and Machiavellianism. Refer to the *Behavioral Research in Accounting* (January 2008) study of Machiavellian traits in accountants, Exercise 1.6 (p. 4). Recall that *Machiavellian* describes negative character traits that include manipulation, cunning, duplicity, deception, and bad faith. A Machiavellian ("Mach") rating score was determined for each in a sample of accounting alumni of a large southwestern university. Scores range from a low of 40 to a high of 160, with the theoretical neutral Mach rating score of 100. The 122 purchasing managers in the sample had a mean Mach rating score of 99.6, with a standard deviation of 12.6.

(a) From the sample, estimate the true mean Mach rating score of all purchasing managers.

(b) Form a 95% confidence interval for the estimate, part b.

(c) Give a practical interpretation of the interval, part c.

(d) A director of purchasing at a major firm claims that the true mean Mach rating score of all purchasing managers is 85. Is there evidence to dispute this claim?

1.48 Wearout of used display panels. Researchers presented a study of the wearout failure time of used colored display panels purchased by an outlet store (*Mathematical Sciences Colloquium*, December 2001). Prior to acquisition, the panels had been used for about one-third of their expected lifetimes. The failure times (in years) for a sample of 50 used panels are reproduced in the next table, followed by an SPSS printout of the analysis of the data.

(a) Locate a 95% confidence interval for the true mean failure time of used colored display panels on the printout.

(b) Give a practical interpretation of the interval, part a.

(c) In repeated sampling of the population of used colored display panels, where a 95% confidence interval for the mean failure time is computed for each sample, what proportion of all the confidence intervals generated will capture the true mean failure time?

PANELFAIL

0.01	1.21	1.71	2.30	2.96	0.19	1.22	1.75	2.30	2.98	0.51
1.24	1.77	2.41	3.19	0.57	1.48	1.79	2.44	3.25	0.70	1.54
1.88	2.57	3.31	0.73	1.59	1.90	2.61	1.19	0.75	1.61	1.93
2.62	3.50	0.75	1.61	2.01	2.72	3.50	1.11	1.62	2.16	2.76
3.50	1.16	1.62	2.18	2.84	3.50					

Source: Irony, T. Z., Lauretto, M., Pereira, C., and Stern, J. M. "A Weibull wearout test: Full Bayesian approach," paper presented at *Mathematical Sciences Colloquium*, Binghamton University, Binghamton, U.K., December 2001.

Descriptives

			Statistic	Std. Error
FAILTIME	Mean		1.9350	.13133
	95% Confidence Interval for Mean	Lower Bound	1.6711	
		Upper Bound	2.1989	
	5% Trimmed Mean		1.9454	
	Median		1.8350	
	Variance		.862	
	Std. Deviation		.92865	
	Minimum		.01	
	Maximum		3.50	
	Range		3.49	
	Interquartile Range		1.43	
	Skewness		-.008	.337
	Kurtosis		-.755	.662

1.49 Studies on treating Alzheimer's disease. Alzheimer's disease (AD) is a progressive disease of the brain. Much research has been conducted on how to treat AD. The journal *eCAM* (November 2006) published an article that critiqued the quality of the methodology used in studies on AD treatment. For each in a sample of 13 studies, the quality of the methodology was measured using the Wong Scale, with scores ranging from 9 (low quality) to 27 (high quality). The data are shown in the table below. Estimate the mean quality, μ, of all studies on the treatment of Alzheimer's disease with a 99% confidence interval. Interpret the result.

TREATAD

22	21	18	19	20	15	19	20	15	20	17	20	21

Source: Chiappelli, F., et al. "Evidence-based research in complementary and alternative medicine III: Treatment of patients with Alzheimer's disease," *eCAM*, Vol. 3, No. 4, Nov. 2006 (Table 1).

1.50 Reproduction of bacteria-infected spider mites. Zoologists in Japan investigated the reproductive traits of spider mites with a bacteria infection (*Heredity,* January 2007). Male and female pairs of infected spider mites were mated in a laboratory, and the number of eggs produced by each female recorded. Summary statistics for several samples are provided in the accompanying table. Note that in some samples, one or both infected spider mites were treated with an antibiotic prior to mating.

(a) For each female/male pair type, construct and interpret a 90% confidence interval for the population mean number of eggs produced by the female spider mite.

(b) Identify the female/male pair type that appears to produce the highest mean number of eggs.

FEMALE/ MALE PAIRS	SAMPLE SIZE	MEAN # OF EGGS	STANDARD DEVIATION
Both untreated	29	20.9	3.34
Male treated	23	20.3	3.50
Female treated	18	22.9	4.37
Both treated	21	18.6	2.11

Source: Reprinted by permission from Macmillan Publishers Ltd: *Heredity* (Gotoh, T., Noda, H., and Ito, S. "Cardinium symbionts cause cytoplasmic incompatibility in spider mites," Vol. 98, No. 1, Jan. 2007, Table 2). Copyright © 2007.

1.9 Testing a Hypothesis About a Population Mean

The procedure involved in testing a hypothesis about a population parameter can be illustrated with the procedure for a test concerning a population mean μ.

A statistical test of a hypothesis is composed of several elements, as listed in the box.

Elements of a Statistical Test of Hypothesis

1. **Null Hypothesis** (denoted H_0): This is the hypothesis that is postulated to be true.

2. **Alternative Hypothesis** (denoted H_a): This hypothesis is counter to the null hypothesis and is usually the hypothesis that the researcher wants to support.

3. **Test Statistic:** Calculated from the sample data, this statistic functions as a decision-maker.

4. **Level of significance** (denoted α): This is the probability of a *Type I error* (i.e., the probability of rejecting H_0 given that H_0 is true).

5. **Rejection Region:** Values of the test statistic that lead the researcher to reject H_0 and accept H_a.

6. **p-Value:** Also called the *observed significance level*, this is the probability of observing a value of the test statistic at least as contradictory to the null hypothesis as the observed test statistic value, assuming the null hypothesis is true.

7. **Conclusion:** The decision to reject or accept H_0 based on the value of the test statistic, α, the rejection region, and/or the p-value.

The test statistic for testing the null hypothesis that a population mean μ equals some specific value, say, μ_0, is the sample mean $\bar{y}$ or the standardized normal variable

$$z = \frac{\bar{y} - \mu_0}{\sigma_{\bar{y}}} \quad \text{where } \sigma_{\bar{y}} = \frac{\sigma}{\sqrt{n}}$$

The logic used to decide whether sample data *disagree* with this hypothesis can be seen in the sampling distribution of $\bar{y}$ shown in Figure 1.22. If the population mean μ is equal to μ_0 (i.e., if the null hypothesis is true), then the mean $\bar{y}$ calculated from a sample should fall, with high probability, within $2\sigma_{\bar{y}}$ of μ_0. If $\bar{y}$ falls too far away from μ_0, or if the standardized distance

$$z = \frac{\bar{y} - \mu_0}{\sigma_{\bar{y}}}$$

is too large, we conclude that the data disagree with our hypothesis, and we reject the null hypothesis.

The reliability of a statistical test is measured by the probability of making an incorrect decision. This probability, or *level of significance*, is (as stated in the box) $\alpha = $ P(Type I error) = P(Reject H_0 given H_0 is true). Prior to conducting the test, the researcher selects a value of α (e.g., $\alpha = .05$). This value is then used to find the appropriate rejection region.

Figure 1.22 The sampling distribution of $\bar{y}$ for $\mu = \mu_0$

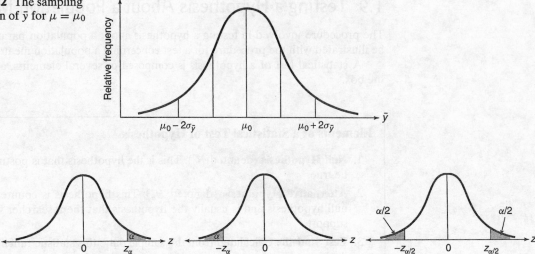

Figure 1.23 Location of the rejection region for various alternative hypotheses

(a) $\mu > \mu_0$ (b) $\mu < \mu_0$ (c) $\mu \neq \mu_0$

For example, if we want to test the null hypothesis H_0: $\mu = \mu_0$ against the alternative hypothesis H_a: $\mu > \mu_0$, we locate the boundary of the rejection region in the upper tail of the z distribution, as shown in Figure 1.23(a), at the point z_α. Note that α is the tail probability in Figure 1.23(a). We will reject H_0 if $z > z_\alpha$. Similarly, to test H_0: $\mu = \mu_0$ against H_a: $\mu < \mu_0$, we place the rejection region in the lower tail of the z distribution, as shown in Figure 1.23(b). These tests are called **one-tailed** (or **one-sided**) **statistical tests**. To detect either $\mu < \mu_0$ or $\mu > \mu_0$, that is, to test H_a: $\mu \neq \mu_0$, we split α equally between the two tails of the z distribution and reject the null hypothesis if $z < -z_{\alpha/2}$ or $z > z_{\alpha/2}$, as shown in Figure 1.23(c). This test is called a **two-tailed** (or **two-sided**) **statistical test**.

As an alternative to the rejection region approach, many researchers utilize the *p-value* to conduct the test. As stated above, the *p*-value is the probability of observing a value of the test statistic at least as contradictory to H_0 as the observed value of the test statistic, *assuming H_0 is true*. For example, if the test statistic for testing H_a: $\mu > \mu_0$ is $z = 2.12$, then the *p*-value is

$$p\text{-value} = P(z > 2.12) = .0170 \text{ (obtained from Table 1, Appendix D)}$$

Figure 1.24 Testing H_a: $\mu > \mu_0$ using a *p*-value

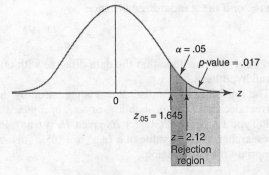

$\alpha = .05$

p-value = .017

$z_{.05} = 1.645$

$z = 2.12$
Rejection region

The researcher makes decisions about H_0 and H_a by comparing the p-value of the test to the selected value of α. If $\alpha > p$-value, we reject H_0 in favor of H_a. Figure 1.24 shows the p-value and rejection region at $\alpha = .05$ for testing $H_a: \mu > \mu_0$. Note that since $\alpha = .05$ exceeds p-value $= .0170$, we reject H_0. Also, the test statistic, $z = 2.12$, falls into the rejection region. Consequently, the two decision-making rules are equivalent. However, since statistical software automatically produces the p-value for a test of hypothesis, many researchers prefer the p-value approach.

The z test, summarized in the next box, is called a *large-sample test* because we will rarely know σ and hence will need a sample size that is large enough so that the sample standard deviation s will provide a good approximation to σ. Normally, we recommend that the sample size be $n \geq 30$.

Large-Sample ($n \geq 30$) Test of Hypothesis About μ

Test statistic: $z = (\bar{y} - \mu_0)/\sigma_{\bar{y}} \approx (\bar{y} - \mu_0)/(s/\sqrt{n})$

	ONE-TAILED TESTS		TWO-TAILED TEST
	$H_0: \mu = \mu_0$	$H_0: \mu = \mu_0$	$H_0: \mu = \mu_0$
	$H_a: \mu < \mu_0$	$H_a: \mu > \mu_0$	$H_a: \mu \neq \mu_0$
Rejection region:	$z < -z_\alpha$	$z > z_\alpha$	$|z| > z_{\alpha/2}$
p-*value:*	$P(z < z_c)$	$P(z > z_c)$	$2P(z > z_c)$ if z_c is positve
			$2P(z < z_c)$ if z_c is negative

Decision: Reject H_0 if $\alpha > p$-value, or if test statistic falls in rejection region

where $P(z > z_\alpha) = \alpha$, $P(z > z_{\alpha/2}) = \alpha/2$, $z_c =$ calculated value of the test statistic, and $\alpha = P(\text{Type I error}) = P(\text{Reject } H_0 | H_0 \text{ true})$.

We illustrate with an example.

Example 1.14

Humerus bones from the same species of animal tend to have approximately the same length-to-width ratios. When fossils of humerus bones are discovered, archeologists can often determine the species of animal by examining the length-to-width ratios of the bones. It is known that species A has a mean ratio of 8.5. Suppose 41 fossils of humerus bones were unearthed at an archeological site in East Africa, where species A is believed to have flourished. (Assume that the unearthed bones were all from the same unknown species.) The length-to-width ratios of the bones were measured and are listed in Table 1.11. Do these data present sufficient evidence to indicate that the mean ratio of all bones of this species differs from 8.5? Use $\alpha = .05$.

Solution

Since we wish to determine whether $\mu \neq 8.5$, the elements of the test are

$$H_0: \mu = 8.5$$
$$H_a: \mu \neq 8.5$$
$$\text{Test statistic: } z = \frac{\bar{y} - 8.5}{\sigma_{\bar{y}}} = \frac{\bar{y} - 8.5}{\sigma/\sqrt{n}} \approx \frac{\bar{y} - 8.5}{s/\sqrt{n}}$$
$$\text{Rejection region: } |z| > 1.96 \quad \text{for } \alpha = .05$$

The data in Table 1.11 were analyzed using SPSS. The SPSS printout is displayed in Figure 1.25.

⊘ BONES

Table 1.11 Length-to-width ratios of a sample of humerus bones

10.73	9.57	6.66	9.89
8.89	9.29	9.35	8.17
9.07	9.94	8.86	8.93
9.20	8.07	9.93	8.80
10.33	8.37	8.91	10.02
9.98	6.85	11.77	8.38
9.84	8.52	10.48	11.67
9.59	8.87	10.39	8.30
8.48	6.23	9.39	9.17
8.71	9.41	9.17	12.00
			9.38

Figure 1.25 SPSS printout for Example 1.14

One-Sample Statistics

	N	Mean	Std. Deviation	Std. Error Mean
LWRATIO	41	9.2576	1.20357	.18797

One-Sample Test

	Test Value = 8.5					
					95% Confidence Interval of the Difference	
	t	df	Sig. (2-tailed)	Mean Difference	Lower	Upper
LWRATIO	4.030	40	.000	.7576	.3777	1.1375

Substituting the sample statistics $\bar{y} = 9.26$ and $s = 1.20$ (shown at the top of the SPSS printout) into the test statistic, we have

$$z \approx \frac{\bar{y} - 8.5}{s/\sqrt{n}} = \frac{9.26 - 8.5}{1.20/\sqrt{41}} = 4.03$$

For this two-tailed test, we also find

$$p\text{-value} = 2P(z > 4.03) \approx 0$$

Both the test statistic and p-value are shown (highlighted) at the bottom of the SPSS printout. Since the test statistic exceeds the critical value of 1.96 (or, since $\alpha = .05$ exceeds the p-value), we can reject H_0 at $\alpha = .05$. The sample data provide sufficient evidence to conclude that the true mean length-to-width ratio of all humerus bones of this species differs from 8.5.

The *practical* implications of the result obtained in Example 1.14 remain to be seen. Perhaps the animal discovered at the archeological site is of some species other than A. Alternatively, the unearthed humeri may have larger than normal length-to-width ratios because they are the bones of specimens having unusual feeding habits for species A. **It is not always the case that a statistically significant result implies a**

practically significant result. The researcher must retain his or her objectivity and judge the practical significance using, among other criteria, knowledge of the subject matter and the phenomenon under investigation.   ■

Note: Earlier, we discussed the use of $\alpha = P(\text{Type I error})$ as a measure of reliability for the statistical test. A second type of error could be made if we accepted the null hypothesis when, in fact, the alternative hypothesis is true (a **Type II error**). As a general rule, you should never "accept" the null hypothesis unless you know the probability of making a Type II error. Since this probability (denoted by the symbol β) is often unknown, it is a common practice to defer judgment if a test statistic falls in the nonrejection region. That is, we say "fail to reject H_0" rather than "accept H_0."

A small-sample test of the null hypothesis $\mu = \mu_0$ using a Student's t statistic is based on the assumption that the sample was randomly selected from a population with a normal relative frequency distribution. The test is conducted in exactly the same manner as the large-sample z test except that we use

$$t = \frac{\bar{y} - \mu_0}{s_{\bar{y}}} = \frac{\bar{y} - \mu_0}{s/\sqrt{n}}$$

as the test statistic and we locate the rejection region in the tail(s) of a Student's t distribution with df $= n - 1$. We summarize the technique for conducting a small-sample test of hypothesis about a population mean in the box.

Small-Sample Test of Hypothesis About μ

Test statistic: $t = (\bar{y} - \mu_0)/(s/\sqrt{n})$

	ONE-TAILED TESTS		TWO-TAILED TEST
	$H_0: \mu = \mu_0$	$H_0: \mu = \mu_0$	$H_0: \mu = \mu_0$
	$H_a: \mu < \mu_0$	$H_a: \mu > \mu_0$	$H_a: \mu \neq \mu_0$
Rejection region:	$t < -t_\alpha$	$t > t_\alpha$	$\lvert t \rvert > t_{\alpha/2}$
p-value:	$P(t < t_c)$	$P(t > t_c)$	$2P(t > t_c)$ if t_c is positve
			$2P(t < t_c)$ if t_c is negative

Decision: Reject H_0 if $\alpha > p$-value, or if test statistic falls in rejection region

where $P(t > t_\alpha) = \alpha$, $P(t > t_{\alpha/2}) = \alpha/2$, $t_c = $ calculated value of the test statistic, and $\alpha = P(\text{Type I error}) = P(\text{Reject } H_0 | H_0 \text{ true})$.

Assumption: The population from which the random sample is drawn is approximately normal.

Example 1.15

Scientists have labeled benzene, a chemical solvent commonly used to synthesize plastics, as a possible cancer-causing agent. Studies have shown that people who work with benzene more than 5 years have 20 times the incidence of leukemia than the general population. As a result, the federal government lowered the maximum allowable level of benzene in the workplace from 10 parts per million (ppm) to 1 ppm. Suppose a steel manufacturing plant, which exposes its workers to benzene daily, is under investigation by the Occupational Safety and Health Administration (OSHA). Twenty air samples, collected over a period of 1 month and examined for benzene content, yielded the data in Table 1.12.

⊙ BENZENE

Table 1.12 Benzene content for 20 air samples

0.5	0.9	4.5	3.4	1.0
2.7	1.1	1.9	0.0	0.0
4.2	2.1	0.0	2.0	3.4
3.4	2.5	0.9	5.1	2.4

Is the steel manufacturing plant in violation of the changed government standards? Test the hypothesis that the mean level of benzene at the steel manufacturing plant is greater than 1 ppm, using $\alpha = .05$.

Solution

OSHA wants to establish the research hypothesis that the mean level of benzene, μ, at the steel manufacturing plant exceeds 1 ppm. The elements of this small-sample one-tailed test are

H_0: $\mu = 1$

H_a: $\mu > 1$

Test statistic: $t = \dfrac{\bar{y} - \mu_0}{s/\sqrt{n}}$

Assumptions: The relative frequency distribution of the population of benzene levels for all air samples at the steel manufacturing plant is approximately normal.

Rejection region: For $\alpha = .05$ and df $= n - 1 = 19$, reject H_0 if $t > t_{.05} = 1.729$ (see Figure 1.26)

Figure 1.26 Rejection region for Example 1.15

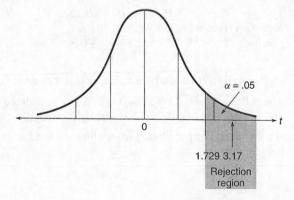

The SAS printout, Figure 1.27, gives summary statistics for the sample data. Substituting $\bar{y} = 2.1$ and $s = 1.55$ into the test statistic formula, we obtain:

$$t = \frac{\bar{y} - 1}{s/\sqrt{n}} = \frac{2.1 - 1}{1.55/\sqrt{20}} = 3.17$$

This test statistic value, as well as the p-value of the test, is highlighted at the bottom of the SAS printout. Since the calculated t falls in the rejection region (or, since $\alpha = .05$ exceeds p-value $= .0025$), OSHA concludes that $\mu > 1$ ppm and the plant

Figure 1.27 SAS output for testing benzene mean

```
                              One Sample t-test for a Mean

Sample Statistics for benzene

     N         Mean        Std. Dev.        Std. Error
------------------------------------------------------------
     20        2.10        1.55             0.35

Hypothesis Test

     Null hypothesis:    Mean of benzene <= 1
     Alternative:        Mean of benzene >  1

          t Statistic        Df        Prob > t
     ------------------------------------------------
             3.169           19          0.0025
```

is in violation of the revised government standards. The reliability associated with this inference is $\alpha = .05$. This implies that if the testing procedure were applied repeatedly to random samples of data collected at the plant, OSHA would falsely reject H_0 for only 5% of the tests. Consequently, OSHA is highly confident (95% confident) that the plant is violating the new standards. ■

1.9 Exercises

1.51 Key terms. Define each of the following:

(a) H_0 (b) H_a (c) Type I error
(d) Type II error   (e) α (f) β
(g) p-value

1.52 Key questions. In hypothesis testing,

(a) who or what determines the size of the rejection region?

(b) does rejecting H_0 prove that the research hypothesis is correct?

1.53 Rejection regions. For each of the following rejection regions, sketch the sampling distribution for z, indicate the location of the rejection region, and give the value of α:

(a) $z > 1.96$ (b) $z > 1.645$ (c) $z > 2.576$
(d) $z < -1.29$ (e) $|z| > 1.645$ (f) $|z| > 2.576$

1.54 Play Golf America program. The Professional Golf Association (PGA) and *Golf Digest* have developed the Play Golf America program, in which teaching professionals at participating golf clubs provide a free 10-minute lesson to new customers. According to *Golf Digest* (July 2008), golf facilities that participate in the program gain, on average, $2,400 in green fees, lessons, or equipment expenditures. A teaching professional at a golf club believes that the average gain in green fees, lessons, or equipment expenditures for participating golf facilities exceeds $2,400.

(a) In order to support the claim made by the teaching professional, what null and alternative hypothesis should you test?

(b) Suppose you select $\alpha = .05$. Interpret this value in the words of the problem.

(c) For $\alpha = .05$, specify the rejection region of a large-sample test.

1.55 Mercury poisoning of wading birds. According to a University of Florida wildlife ecology and conservation researcher, the average level of mercury uptake in wading birds in the Everglades is declining (*UF News*, December 15, 2000). Last year, the average level was 15 parts per million.

(a) Give the null and alternative hypotheses for testing whether the average level this year is less than 15 ppm.

(b) Describe a Type I error for this test.

(c) Describe a Type II error for this test.

1.56 Crab-spiders hiding on flowers. *Behavioral Ecology* (January 2005) published the results of an experiment on crab-spiders' use of camouflage to hide from predators (e.g., birds) on flowers. Researchers at the French Museum of Natural History collected a sample of 10 adult female crab-spiders, each sitting on the yellow central part of a daisy, and measured the chromatic contrast of each spider to the flower. The data (where higher values indicate a greater contrast, and, presumably, an easier detection by predators) are shown in the next table. The researchers discovered that a contrast of 70 or greater allows birds to see the spider. Of interest is whether or not true mean chromatic contrast of crab-spiders on daisies is less than 70.

(a) Define the parameter of interest, μ.

(b) Setup the null and alternative hypothesis of interest.

(c) Find $\bar{y}$ and s for the sample data, then use these values to compute the test statistic.

(d) Give the rejection region for $\alpha = .10$.

(e) Find the p-value for the test.

(f) State the appropriate conclusion in the words of the problem.

⊜ CRABSPIDER

57	75	116	37	96	61	56	2	43	32

Source: Data adapted from Thery, M., et al. "Specific color sensitivities of prey and predator explain camouflage in different visual systems," *Behavioral Ecology*, Vol. 16, No. 1, Jan. 2005 (Table 1).

1.57 Social interaction of mental patients. The *Community Mental Health Journal* (August 2000) presented the results of a survey of over 6,000 clients of the Department of Mental Health and Addiction Services (DMHAS) in Connecticut. One of the many variables measured for each mental health patient was frequency of social interaction (on a 5-point scale, where 1 = very infrequently, 3 = occasionally, and 5 = very frequently). The 6,681 clients who were evaluated had a mean social interaction score of 2.95 with a standard deviation of 1.10.

(a) Conduct a hypothesis test (at $\alpha = .01$) to determine if the true mean social interaction score of all Connecticut mental health patients differs from 3.

(b) Examine the results of the study from a practical view, then discuss why "statistical significance" does not always imply "practical significance."

(c) Because the variable of interest is measured on a 5-point scale, it is unlikely that the population of ratings will be normally distributed. Consequently, some analysts may perceive the test, part a, to be invalid and search for alternative methods of analysis. Defend or refute this position.

1.58 Heart rate during laughter. Laughter is often called "the best medicine," since studies have shown that laughter can reduce muscle tension and increase oxygenation of the blood. In the *International Journal of Obesity* (January 2007), researchers at Vanderbilt University investigated the physiological changes that accompany laughter. Ninety subjects (18–34 years old) watched film clips designed to evoke laughter. During the laughing period, the researchers measured the heart rate (beats per minute) of each subject with the following summary results: $\bar{y} = 73.5$, $s = 6$. It is well known that the mean resting heart rate of adults is 71 beats/minute. At $\alpha = .05$, is there sufficient evidence to indicate that the true mean heart rate during laughter exceeds 71 beats/minute?

1.59 Feminizing faces study. Research published in *Nature* (August 27, 1998) revealed that people are more attracted to "feminized" faces, regardless of gender. In one experiment, 50 human subjects viewed both a Japanese female and Caucasian male face on a computer. Using special computer graphics, each subject could morph the faces (by making them more feminine or more masculine) until they attained the "most attractive" face. The level of feminization y (measured as a percentage) was measured.

(a) For the Japanese female face, $\bar{y} = 10.2\%$ and $s = 31.3\%$. The researchers used this sample information to test the null hypothesis of a mean level of feminization equal to 0%. Verify that the test statistic is equal to 2.3.

(b) Refer to part a. The researchers reported the p-value of the test as $p \approx .02$. Verify and interpret this result.

(c) For the Caucasian male face, $\bar{y} = 15.0\%$ and $s = 25.1\%$. The researchers reported the test statistic (for the test of the null hypothesis stated in part a) as 4.23 with an associated p-value of approximately 0. Verify and interpret these results.

1.60 Analyzing remote-sensing data to identify type of land cover. Geographers use remote sensing data from satellite pictures to identify urban land-cover as either grassland, commercial, or residential. In *Geographical Analysis* (October 2006), researchers from Arizona State, Florida State, and Louisiana State Universities collaborated on a new method for analyzing remote sensing data. A satellite photograph of an urban area was divided into 4×4 meter areas (called pixels). Of interest is a numerical measure of the distribution of gaps or hole sizes in the pixel, called *lacunarity*. The mean and standard deviation of the lacunarity measurements for a sample of 100 pixels randomly selected from a specific urban area are 225 and 20, respectively. It is known that the mean lacunarity measurement for all grassland pixels is 220. Do the data suggest that the area sampled is grassland? Test at $\alpha = .01$.

1.61 Falsifying candy counts. "Hot Tamales" are chewy, cinnamon-flavored candies. A bulk vending machine is known to dispense, on average, 15 Hot Tamales per bag. *Chance* (Fall 2000) published an article on a classroom project in which students were required to purchase bags of Hot Tamales from the machine and count the number of candies per bag. One student group claimed they purchased five bags that had the following candy counts: 25, 23, 21, 21, and 20. There was some question as to whether the students had fabricated the data. Use a hypothesis test to gain insight into whether or not the data collected by the students are fabricated. Use a level of significance that gives the benefit of the doubt to the students.

1.62 **Cooling method for gas turbines.** During periods of high electricity demand, especially during the hot summer months, the power output from a gas turbine engine can drop dramatically. One way to counter this drop in power is by cooling the inlet air to the gas turbine. An increasingly popular cooling method uses high pressure inlet fogging.

The performance of a sample of 67 gas turbines augmented with high pressure inlet fogging was investigated in the *Journal of Engineering for Gas Turbines and Power* (January 2005). One measure of performance is heat rate (kilojoules per kilowatt per hour). Heat rates for the 67 gas turbines, saved in the GASTURBINE file, are listed in the table. Suppose that a standard gas turbine has, on average, a heat rate of 10,000 kJ/kWh.

(a) Conduct a test to determine if the mean heat rate of gas turbines augmented with high pressure inlet fogging exceeds 10,000 kJ/kWh. Use $\alpha = .05$.

(b) Identify a Type I error for this study, then a Type II error.

 GASTURBINE

14622	13196	11948	11289	11964	10526	10387	10592	10460	10086
14628	13396	11726	11252	12449	11030	10787	10603	10144	11674
11510	10946	10508	10604	10270	10529	10360	14796	12913	12270
11842	10656	11360	11136	10814	13523	11289	11183	10951	9722
10481	9812	9669	9643	9115	9115	11588	10888	9738	9295
9421	9105	10233	10186	9918	9209	9532	9933	9152	9295
16243	14628	12766	8714	9469	11948	12414			

1.10 Inferences About the Difference Between Two Population Means

The reasoning employed in constructing a confidence interval and performing a statistical test for comparing two population means is identical to that discussed in Sections 1.7 and 1.8. First, we present procedures that are based on the assumption that we have selected *independent* random samples from the two populations. The parameters and sample sizes for the two populations, the sample means, and the sample variances are shown in Table 1.13. The objective of the sampling is to make an inference about the difference $(\mu_1 - \mu_2)$ between the two population means.

Because the sampling distribution of the difference between the sample means $(\bar{y}_1 - \bar{y}_2)$ is approximately normal for large samples, the large-sample techniques are

Table 1.13 Two-sample notation

	Population	
	1	2
Sample size	n_1	n_2
Population mean	μ_1	μ_2
Population variance	σ_1^2	σ_2^2
Sample mean	$\bar{y}_1$	$\bar{y}_2$
Sample variance	s_1^2	s_2^2

based on the standardized normal z statistic. Since the variances of the populations, σ_1^2 and σ_2^2, will rarely be known, we will estimate their values using s_1^2 and s_2^2.

To employ these large-sample techniques, we recommend that both sample sizes be large (i.e., each at least 30). The large-sample confidence interval and test are summarized in the boxes.

Large-Sample Confidence Interval for $(\mu_1 - \mu_2)$: Independent Samples

$$(\bar{y}_1 - \bar{y}_2) \pm z_{\alpha/2}\sigma_{(\bar{y}_1 - \bar{y}_2)^*} = (\bar{y}_1 - \bar{y}_2) \pm z_{\alpha/2}\sqrt{\frac{\sigma_1^2}{n_1} + \frac{\sigma_2^2}{n_2}}$$

Assumptions: The two samples are randomly and independently selected from the two populations. The sample sizes, n_1 and n_2, are large enough so that $\bar{y}_1$ and $\bar{y}_2$ each have approximately normal sampling distributions and so that s_1^2 and s_2^2 provide good approximations to σ_1^2 and σ_2^2. This will be true if $n_1 \geq 30$ and $n_2 \geq 30$.

Large-Sample Test of Hypothesis About $(\mu_1 - \mu_2)$: Independent Samples

$$\text{Test statistic: } z = \frac{(\bar{y}_1 - \bar{y}_2) - D_0}{\sigma_{(\bar{y}_1 - \bar{y}_2)}} = \frac{(\bar{y}_1 - \bar{y}_2) - D_0}{\sqrt{\frac{\sigma_1^2}{n_1} + \frac{\sigma_2^2}{n_2}}}$$

	ONE-TAILED TESTS		TWO-TAILED TEST
	H_0: $\mu_1 - \mu_2 = D_0$	H_0: $\mu_1 - \mu_2 = D_0$	H_0: $\mu_1 - \mu_2 = D_0$
	H_a: $\mu_1 - \mu_2 < D_0$	H_a: $\mu_1 - \mu_2 > D_0$	H_a: $\mu_1 - \mu_2 \neq D_0$
Rejection region:	$z < -z_\alpha$	$z > z_\alpha$	$|z| > z_{\alpha/2}$
p-*value*:	$P(z < z_c)$	$P(z > z_c)$	$2P(z > z_c)$ if z_c is positve
			$2P(z < z_c)$ if z_c is negative

Decision: Reject H_0 if $\alpha > p$-value, or, if the test statistic falls in rejection region

where D_0 = hypothesized difference between means, $P(z > z_\alpha) = \alpha$, $P(z > z_{\alpha/2}) = \alpha/2$, z_c = calculated value of the test statistic, and $\alpha = P(\text{Type I error}) = P(\text{Reject } H_0 | H_0 \text{ true})$.

Assumptions: Same as for the previous large-sample confidence interval

---

Example 1.16

A dietitian has developed a diet that is low in fats, carbohydrates, and cholesterol. Although the diet was initially intended to be used by people with heart disease, the dietitian wishes to examine the effect this diet has on the weights of obese people. Two random samples of 100 obese people each are selected, and one group of 100 is placed on the low-fat diet. The other 100 are placed on a diet that contains approximately the same quantity of food but is not as low in fats, carbohydrates, and cholesterol. For each person, the amount of weight lost (or gained) in a 3-week period is recorded. The data, saved in the DIETSTUDY file, are listed in Table 1.14. Form a 95% confidence interval for the difference between the population mean weight losses for the two diets. Interpret the result.

---

*The symbol $\sigma_{(\bar{y}_1 - \bar{y}_2)}$ is used to denote the standard error of the distribution of $(\bar{y}_1 - \bar{y}_2)$.

⊙ DIETSTUDY

Table 1.14 Diet study data, example 1.16

Weight Losses for Low-fat Diet									
8	10	10	12	9	3	11	7	9	2
21	8	9	2	2	20	14	11	15	6
13	8	10	12	1	7	10	13	14	4
8	12	8	10	11	19	0	9	10	4
11	7	14	12	11	12	4	12	9	2
4	3	3	5	9	9	4	3	5	12
3	12	7	13	11	11	13	12	18	9
6	14	14	18	10	11	7	9	7	2
16	16	11	11	3	15	9	5	2	6
5	11	14	11	6	9	4	17	20	10

Weight Losses for Regular Diet									
6	6	5	5	2	6	10	3	9	11
14	4	10	13	3	8	8	13	9	3
4	12	6	11	12	9	8	5	8	7
6	2	6	8	5	7	16	18	6	8
13	1	9	8	12	10	6	1	0	13
11	2	8	16	14	4	6	5	12	9
11	6	3	9	9	14	2	10	4	13
8	1	1	4	9	4	1	1	5	6
14	0	7	12	9	5	9	12	7	9
8	9	8	10	5	8	0	3	4	8

Solution

Let μ_1 represent the mean of the conceptual population of weight losses for all obese people who could be placed on the low-fat diet. Let μ_2 be similarly defined for the regular diet. We wish to form a confidence interval for $(\mu_1 - \mu_2)$.

Summary statistics for the diet data are displayed in the SPSS printout, Figure 1.28. Note that $\bar{y}_1 = 9.31$, $\bar{y}_2 = 7.40$, $s_1 = 4.67$, and $s_2 = 4.04$. Using these values and noting that $\alpha = .05$ and $z_{.025} = 1.96$, we find that the 95% confidence interval is:

$$(\bar{y}_1 - \bar{y}_2) \pm z_{.025}\sqrt{\frac{\sigma_1^2}{n_1} + \frac{\sigma_2^2}{n_2}} \approx$$

$$(9.31 - 7.40) \pm 1.96\sqrt{\frac{(4.67)^2}{100} + \frac{(4.04)^2}{100}} = 1.91 \pm (1.96)(.62) = 1.91 \pm 1.22$$

or (.69, 3.13). This result is also given (highlighted) on the SPSS printout. Using this estimation procedure over and over again for different samples, we know that approximately 95% of the confidence intervals formed in this manner will enclose the difference in population means $(\mu_1 - \mu_2)$. Therefore, we are highly confident that

Group Statistics

	DIET	N	Mean	Std. Deviation	Std. Error Mean
WTLOSS	LOWFAT	100	9.31	4.668	.467
	REGULAR	100	7.40	4.035	.404

Independent Samples Test

		Levene's Test for Equality of Variances		t-test for Equality of Means					95% Confidence Interval of the Difference	
		F	Sig.	t	df	Sig. (2-tailed)	Mean Difference	Std. Error Difference	Lower	Upper
WTLOSS	Equal variances assumed	1.367	.244	3.095	198	.002	1.910	.617	.693	3.127
	Equal variances not assumed			3.095	193.940	.002	1.910	.617	.693	3.127

Figure 1.28 SPSS analysis for diet study, Example 1.16

the mean weight loss for the low-fat diet is between .69 and 3.13 pounds more than the mean weight loss for the other diet. With this information, the dietitian better understands the potential of the low-fat diet as a weight-reducing diet. ∎

The small-sample statistical techniques used to compare μ_1 and μ_2 with independent samples are based on the assumptions that both populations have normal probability distributions and that the variation within the two populations is of the same magnitude (i.e., $\sigma_1^2 = \sigma_2^2$). When these assumptions are approximately satisfied, we can employ a Student's t statistic to find a confidence interval and test a hypothesis concerning $(\mu_1 - \mu_2)$. The techniques are summarized in the following boxes.

Small-Sample Confidence Interval for $(\mu_1 - \mu_2)$: Independent Samples

$$(\bar{y}_1 - \bar{y}_2) \pm t_{\alpha/2}\sqrt{s_p^2 \left(\frac{1}{n_1} + \frac{1}{n_2} \right)}$$

where

$$s_p^2 = \frac{(n_1 - 1)s_1^2 + (n_2 - 1)s_2^2}{n_1 + n_2 - 2}$$

is a "pooled" estimate of the common population variance and $t_{\alpha/2}$ is based on $(n_1 + n_2 - 2)$ df.

Assumptions:

1. Both sampled populations have relative frequency distributions that are approximately normal.

2. The population variances are equal.

3. The samples are randomly and independently selected from the populations.

Example 1.17

Suppose you wish to compare a new method of teaching reading to "slow learners" to the current standard method. You decide to base this comparison on the results of a reading test given at the end of a learning period of 6 months. Of a random sample of 22 slow learners, 10 are taught by the new method and 12 are taught by the standard method. All 22 children are taught by qualified instructors under similar conditions for a 6-month period. The results of the reading test at the end of this period are given in Table 1.15.

Small-Sample Test of Hypothesis About $(\mu_1 - \mu_2)$: Independent Samples

Test statistic: $t = \dfrac{(\bar{y}_1 - \bar{y}_2) - D_0}{\sqrt{s_p^2\left(\frac{1}{n_1} + \frac{1}{n_2}\right)}}$ where $s_p^2 = \dfrac{(n_1 - 1)s_1^2 + (n_2 - 1)s_2^2}{n_1 + n_2 - 2}$

	ONE-TAILED TESTS		TWO-TAILED TEST

$H_0: \mu_1 - \mu_2 = D_0$ $H_0: \mu_1 - \mu_2 = D_0$ $H_0: \mu_1 - \mu_2 = D_0$

$H_a: \mu_1 - \mu_2 < D_0$ $H_a: \mu_1 - \mu_2 > D_0$ $H_a: \mu_1 - \mu_2 \neq D_0$

Rejection region: $t < -t_\alpha$ $z > t_\alpha$ $|t| > z_{\alpha/2}$

p-value: $P(t < t_c)$ $P(z > t_c)$ $2P(t > t_c)$ if t_c is positve

                                                              $2P(t < t_c)$ if t_c is negative

Decision: Reject H_0 if $\alpha > $ p-value, or, if test statistic falls in rejection region

where $D_0 = $ hypothesized difference between means, $P(t > t_\alpha) = \alpha$, $P(t > t_{\alpha/2}) = \alpha/2$, $t_c = $ calculated value of the test statistic, and $\alpha = P(\text{Type I error}) = P(\text{Reject } H_0 | H_0 \text{ true})$.

Assumptions: Same as for the previous small-sample confidence interval.

READING

Table 1.15 Reading test scores for slow learners

New Method				Standard Method			
80	80	79	81	79	62	70	68
76	66	71	76	73	76	86	73
70	85			72	68	75	66

(a) Use the data in the table to test whether the true mean test scores differ for the new method and the standard method. Use $\alpha = .05$.

(b) What assumptions must be made in order that the estimate be valid?

Solution

(a) For this experiment, let μ_1 and μ_2 represent the mean reading test scores of slow learners taught with the new and standard methods, respectively. Then, we want to test the following hypothesis:

$H_0: (\mu_1 - \mu_2) = 0$ (i.e., no difference in mean reading scores)

$H_a: (\mu_1 - \mu_2) \neq 0$ (i.e., $\mu_1 \neq \mu_2$)

To compute the test statistic, we need to obtain summary statistics (e.g., $\bar{y}$ and s) on reading test scores for each method. The data of Table 1.15 was entered into a computer, and SAS was used to obtain these descriptive statistics. The SAS printout appears in Figure 1.29. Note that $\bar{y}_1 = 76.4$, $s_1 = 5.8348$, $\bar{y}_2 = 72.333$, and $s_2 = 6.3437$

Figure 1.29 SAS output
for Example 1.17

```
                    Two Sample t-test for the Means of SCORE within METHOD

Sample Statistics

    Group        N      Mean      Std. Dev.    Std. Error
    ------------------------------------------------------------
    NEW         10      76.4       5.8348        1.8451
    STD         12    72.33333     6.3437        1.8313

Hypothesis Test

    Null hypothesis:     Mean 1 - Mean 2 =  0
    Alternative:         Mean 1 - Mean 2 ^= 0

    If Variances Are     t statistic       Df      Pr > t
    ------------------------------------------------------------
    Equal                   1.552          20       0.1364
    Not Equal               1.564        19.77      0.1336

95% Confidence Interval for the Difference between Two Means

    Lower Limit        Upper Limit
    ------------        ------------
      -1.40               9.53
```

Next, we calculate the pooled estimate of variance:

$$
\begin{aligned}
s_p^2 &= \frac{(n_1 - 1)s_1^2 + (n_2 - 1)s_2^2}{n_1 + n_2 - 2} \\[2mm]
&= \frac{(10 - 1)(5.8348)^2 + (12 - 1)(6.3437)^2}{10 + 12 - 2} = 37.45
\end{aligned}
$$

where s_p^2 is based on $(n_1 + n_2 - 2) = (10 + 12 - 2) = 20$ degrees of freedom. Now, we compute the test statistic:

$$
t = \frac{(\bar{y}_1 - \bar{y}_2) - D_0}{\sqrt{s_p^2 \left(\frac{1}{n_1} + \frac{1}{n_2} \right)}} = \frac{(76.4 - 72.33) - 0}{\sqrt{37.45 \left(\frac{1}{10} + \frac{1}{12} \right)}} = 1.55
$$

The rejection region for this two-tailed test at $\alpha = .05$, based on 20 degrees of freedom, is

$$
|t| > t_{.025} = 2.086 \qquad \text{(See Figure 1.30)}
$$

Figure 1.30 Rejection
region for Example 1.17

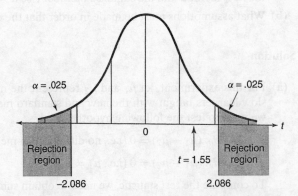

Since the computed value of t does not fall in the rejection region, we fail to reject H_0. There is insufficient evidence (at $\alpha = .05$) of a difference between the true mean test scores for the two reading methods.

This conclusion can also be obtained by using the p-value approach. Both the test statistic ($t = 1.552$) and p-value (.1364) are highlighted on the SAS printout, Figure 1.29. Since $\alpha = .05$ is less than the p-value, we fail to reject H_0.

(b) To properly use the small-sample confidence interval, the following assumptions must be satisfied:

(1) The samples are randomly and independently selected from the populations of slow learners taught by the new method and the standard method.

(2) The test scores are normally distributed for both teaching methods.

(3) The variance of the test scores are the same for the two populations, that is, $\sigma_1^2 = \sigma_2^2$.

The two-sample t statistic is a powerful tool for comparing population means when the necessary assumptions are satisfied. It has also been found that the two-sample t procedure is more robust against nonnormal data than the one-sample method. And, when the sample sizes are equal, the assumption of equal population variances can be relaxed. That is, when $n_1 = n_2$, σ_1^2 and σ_2^2 can be quite different and the test statistic will still have (approximately) a Student's t distribution. ■

In Example 1.17, suppose it is possible to measure the slow learners' "reading IQs" *before* they are subjected to a teaching method. Eight pairs of slow learners with similar reading IQs are found, and one member of each pair is randomly assigned to the standard teaching method while the other is assigned to the new method. The data are given in Table 1.16. Do the data support the hypothesis that the population mean reading test score for slow learners taught by the new method is greater than the mean reading test score for those taught by the standard method?

Now, we want to test

$$H_0: (\mu_1 - \mu_2) = 0$$

$$H_a: (\mu_1 - \mu_2) > 0$$

It appears that we could conduct this test using the t statistic for two independent samples, as in Example 1.17. However, *the independent samples* t-*test is not a valid procedure to use with this set of data*. Why?

The t-test is inappropriate because the assumption of independent samples is invalid. We have randomly chosen *pairs of test scores*, and thus, once we have chosen the sample for the new method, we have *not* independently chosen the sample for the standard method. The dependence between observations within pairs can be seen by examining the pairs of test scores, which tend to rise and fall together as we go from pair to pair. This pattern provides strong visual evidence of a violation of the assumption of independence required for the two-sample t-test used in Example 1.17.

We now consider a valid method of analyzing the data of Table 1.16. In Table 1.17 we add the column of differences between the test scores of the pairs of slow learners. We can regard these differences in test scores as a random sample of differences for all pairs (matched on reading IQ) of slow learners, past and present. Then we can use this sample to make inferences about the mean of the population of differences, μ_d, which is equal to the difference $(\mu_1 - \mu_2)$. That is, the mean of the population (and sample) of differences equals the difference between the population (and sample) means. Thus, our test becomes

$$H_0: \mu_d = 0 \quad (\mu_1 - \mu_2 = 0)$$

$$H_a: \mu_d > 0 \quad (\mu_1 - \mu_2 > 0)$$

PAIREDSCORES

Table 1.16 Reading test scores for eight pairs of slow learners

Pair	New Method (1)	Standard Method (2)
1	77	72
2	74	68
3	82	76
4	73	68
5	87	84
6	69	68
7	66	61
8	80	76

Table 1.17 Differences in reading test scores

Pair	New Method	Standard Method	Difference (New Method–Standard Method)
1	77	72	5
2	74	68	6
3	82	76	6
4	73	68	5
5	87	84	3
6	69	68	1
7	66	61	5
8	80	76	4

The test statistic is a one-sample t (Section 1.9), since we are now analyzing a single sample of differences for small n:

$$\text{Test statistic: } t = \frac{\bar{y}_d - 0}{s_d/\sqrt{n_d}}$$

where
$\bar{y}_d$ = Sample mean difference
s_d = Sample standard deviation of differences
n_d = Number of differences = Number of pairs

Assumptions: The population of differences in test scores is approximately normally distributed. The sample differences are randomly selected from the population differences. [*Note*: We do not need to make the assumption that $\sigma_1^2 = \sigma_2^2$.]

Rejection region: At significance level $\alpha = .05$, we will reject H_0: if $t > t_{.05}$, where $t_{.05}$ is based on $(n_d - 1)$ degrees of freedom.

Referring to Table 2 in Appendix D, we find the t-value corresponding to $\alpha = .05$ and $n_d - 1 = 8 - 1 = 7$ df to be $t_{.05} = 1.895$. Then we will reject the null hypothesis if $t > 1.895$ (see Figure 1.31). Note that the number of degrees of freedom decreases from $n_1 + n_2 - 2 = 14$ to 7 when we use the paired difference experiment rather than the two independent random samples design.

Figure 1.31 Rejection region for analysis of data in Table 1.17

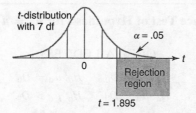

Summary statistics for the $n = 8$ differences are shown in the MINITAB printout, Figure 1.32. Note that $\bar{y}_d = 4.375$ and $s_d = 1.685$. Substituting these values into the formula for the test statistic, we have

$$t = \frac{\bar{y}_d - 0}{s_d/\sqrt{n_d}} = \frac{4.375}{1.685/\sqrt{8}} = 7.34$$

Because this value of t falls in the rejection region, we conclude (at $\alpha = .05$) that the population mean test score for slow learners taught by the new method exceeds the population mean score for those taught by the standard method. We can reach the same conclusion by noting that the p-value of the test, highlighted in Figure 1.32, is much smaller than $\alpha = .05$.

This kind of experiment, in which observations are paired and the differences are analyzed, is called a **paired difference experiment**. The hypothesis-testing procedures and the method of forming confidence intervals for the difference between two means using a paired difference experiment are summarized in the next two boxes for both large and small n.

Figure 1.32 MINITAB paired difference analysis of data in Table 1.17

```
Paired T for NEW - STANDARD

             N    Mean   StDev   SE Mean
NEW          8   76.00    6.93      2.45
STANDARD     8   71.63    7.01      2.48
Difference   8   4.375   1.685     0.596


95% lower bound for mean difference: 3.246
T-Test of mean difference = 0 (vs > 0): T-Value = 7.34   P-Value = 0.000
```

Paired Difference Confidence Interval for $\mu_d = \mu_1 - \mu_2$

Large Sample

$$\bar{y}_d \pm z_{\alpha/2}\frac{\sigma_d}{\sqrt{n_d}} \approx \bar{y}_d \pm z_{\alpha/2}\frac{s_d}{\sqrt{n_d}}$$

Assumption: Sample differences are randomly selected from the population.

Small Sample

$$\bar{y}_d \pm t_{\alpha/2}\frac{s_d}{\sqrt{n_d}}$$

where $t_{\alpha/2}$ is based on $(n_d - 1)$ degrees of freedom

Assumptions:

1. Population of differences has a normal distribution.
2. Sample differences are randomly selected from the population.

Paired Difference Test of Hypothesis for $\mu_d = \mu_1 - \mu_2$

ONE-TAILED TESTS   TWO-TAILED TEST

$H_0: \mu_d = D_0 \qquad H_0: \mu_d = D_0 \qquad H_0: \mu_d = D_0$

$H_a: \mu_d < D_0 \qquad H_a: \mu_d > D_0 \qquad H_a: \mu_d \neq D_0$

Large Sample

$$\text{Test statistic: } z = \frac{\bar{y}_d - D_0}{\sigma_d/\sqrt{n_d}} \approx \frac{\bar{y}_d - D_0}{s_d/\sqrt{n_d}}$$

Rejection Region:	$z < -z_\alpha$	$z > z_d$	$\lvert z \rvert > z_{\alpha/2}$
p-*value*:	$P(z < z_c)$	$P(z > z_c)$	$2P(z > z_c)$ if z_c positive
			$2P(z < z_c)$ if z_c negative

Assumption: The differences are randomly selected from the population of differences.

Small Sample

$$\text{Test statistic: } t = \frac{\bar{y}_d - D_0}{s_d/\sqrt{n_d}}$$

Rejection region:	$t < -t_\alpha$	$t > t_\alpha$	$\lvert t \rvert > t_{\alpha/2}$
p-*value*:	$P(t < t_c)$	$P(t > t_c)$	$2P(t > t_c)$ if t_c is positive
			$2P(t < t_c)$ if t_c is negative

Assumptions:

1. The relative frequency distribution of the population of differences is normal.
2. The differences are randomly selected from the population of differences.

1.10 Exercises

1.63 Describe the sampling distribution of $(\bar{y}_1 - \bar{y}_2)$.

1.64 To use the t statistic to test for differences between the means of two populations based on independent samples, what assumptions must be made about the two sampled populations? What assumptions must be made about the two samples?

1.65 **How do you choose to argue?** Educators frequently lament weaknesses in students' oral and written arguments. In *Thinking and Reasoning* (October 2006), researchers at Columbia University conducted a series of studies to assess the cognitive skills required for successful arguments. One study focused on whether students would choose to argue by weakening the opposing position or by strengthening the favored position. (Example: You are told you would do better at basketball than soccer, but you like soccer. An argument that weakens the opposing position is "You need to be tall to play basketball." An argument that strengthens the favored position is "With practice, I can become really good at soccer.") A sample of 52 graduate students in psychology was equally divided into two groups. Group 1 was presented with 10 items, where the argument always attempts to strengthen the favored position. Group 2 was presented with the same 10 items, but where the argument always attempts to weaken the nonfavored position. Each student then rated the 10 arguments on a 5-point scale from very weak (1) to very strong (5). The variable of interest was the sum of the 10 item scores, called total rating. Summary statistics for the data are shown in the table. Use the methodology of this chapter to compare the mean total ratings for the two groups, at $\alpha = .05$. Give a practical interpretation of the results in the words of the problem.

	GROUP 1 (SUPPORT FAVORED POSITION)	GROUP 2 (WEAKEN OPPOSING POSITION)
Sample size:	26	26
Mean:	28.6	24.9
Standard deviation:	12.5	12.2

Source: Kuhn, D., and Udell, W. "Coordinating own and other perspectives in argument," *Thinking and Reasoning*, October 2006.

1.66 Eating disorder study. The "fear of negative evaluation" (FNE) scores for 11 female students known to suffer from the eating disorder bulimia and 14 female students with normal eating habits, first presented in Exercise 1.19 (p. 16), are reproduced below. (Recall that the higher the score, the greater the fear of negative evaluation.)

(a) Find a 95% confidence interval for the difference between the population means of the FNE scores for bulimic and normal female students. Interpret the result.

(b) What assumptions are required for the interval of part a to be statistically valid? Are these assumptions reasonably satisfied? Explain.

1.67 Does rudeness really matter in the workplace? Studies have established that rudeness in the workplace can lead to retaliatory and counterproductive behavior. However, there has been little research on how rude behaviors influence a victim's task performance. Such a study was conducted and the results published in the *Academy of Management Journal* (October 2007). College students enrolled in a management course were randomly assigned to one of two experimental conditions: rudeness condition (45 students) and control group (53 students). Each student was asked to write down as many uses for a brick as possible in 5 minutes. For those students in the rudeness condition, the facilitator displayed rudeness by berating the students in general for being irresponsible and unprofessional (due to a late-arriving confederate). No comments were made about the late-arriving confederate for students in the control group. The number of different uses for a brick was recorded for each of the 98 students and the data saved in the RUDE file, shown below. Conduct a statistical analysis (at $\alpha = .01$) to determine if the true mean performance level for students in the rudeness condition is lower than the true mean performance level for students in the control group.

EVOS

1.68 Impact study of an oil spill. The *Journal of Agricultural, Biological, and Environmental Statistics* (September 2000) reported on an impact study of a tanker oil spill on the seabird population in Alaska. For each of 96 shoreline locations (called transects), the number of seabirds found, the length (in kilometers) of the transect, and whether or not the transect was in an oiled area were recorded. (The data are saved in the EVOS file.) Observed seabird density is defined as the observed count

BULIMIA

Bulimic students	21	13	10	20	25	19	16	21	24	13	14	
Normal students	13	6	16	13	8	19	23	18	11	19	7	10	15	20

RUDE

Control Group:
1 24 5 16 21 7 20 1 9 20 19 10 23 16 0 4 9 13 17 13 0 2 12 11 7
1 19 9 12 18 5 21 30 15 4 2 12 11 10 13 11 3 6 10 13 16 12 28 19 12 20 3 11

Rudeness Condition:
4 11 18 11 9 6 5 11 9 12 7 5 7 3 11 1 9 11 10 7 8 9 10 7
11 4 13 5 4 7 8 3 8 15 9 16 10 0 7 15 13 9 2 13 10

divided by the length of the transect. A comparison of the mean densities of oiled and unoiled transects is displayed in the following MINITAB printout. Use this information to make an inference about the difference in the population mean seabird densities of oiled and unoiled transects.

```
Two-Sample T-Test and CI: density, oil

Two-sample T for density

oil     N     Mean    StDev    SE Mean
no     36    3.27     6.70     1.1
yes    60    3.50     5.97     0.77

Difference = mu (no ) - mu (yes)
Estimate for difference: -0.22
95% CI for difference: (-2.93, 2.49)
T-Test of difference = 0 (vs not =): T-Value = -0.16  P-Value = 0.871  DF = 67
```

1.69 Family involvement in homework. Teachers Involve Parents in Schoolwork (TIPS) is an interactive homework process designed to improve the quality of homework assignments for elementary, middle, and high school students. TIPS homework assignments require students to conduct interactions with family partners (parents, guardians, etc.) while completing the homework. Frances Van Voorhis (Johns Hopkins University) conducted a study to investigate the effects of TIPS in science, mathematics, and language arts homework assignments (April 2001). Each in a sample of 128 middle school students was assigned to complete TIPS homework assignments, while 98 students in a second sample were assigned traditional, noninteractive homework assignments (called ATIPS). At the end of the study, all students reported on the level of family involvement in their homework on a 4-point scale ($0 =$ never, $1 =$ rarely, $2 =$ sometimes, $3 =$ frequently, $4 =$ always). Three scores were recorded for each student: one for science homework, one for math homework, and one for language arts homework. The data for the study are saved in the HWSTUDY file. (The first five and last five observations in the data set are listed in the next table.)

(a) Conduct an analysis to compare the mean level of family involvement in science homework assignments of TIPS and ATIPS students. Use $\alpha = .05$. Make a practical conclusion.

(b) Repeat part a for mathematics homework assignments.

(c) Repeat part a for language arts homework assignments.

(d) What assumptions are necessary for the inferences of parts a–c to be valid? Are they reasonably satisfied?

🔘 HWSTUDY (First and last five observations)

HOMEWORK CONDITION	SCIENCE	MATH	LANGUAGE
ATIPS	1	0	0
ATIPS	0	1	1
ATIPS	0	1	0
ATIPS	1	2	0
ATIPS	1	1	2
TIPS	2	3	2
TIPS	1	4	2
TIPS	2	4	2
TIPS	4	0	3
TIPS	2	0	1

Source: Van Voorhis, F. L. "Teachers' use of interactive homework and its effects on family involvement and science achievement of middle grade students." Paper presented at the annual meeting of the American Educational Research Association, Seattle, April 2001.

1.70 Comparing voltage readings. Refer to the Harris Corporation/University of Florida study to determine whether a manufacturing process performed at a remote location could be established locally, Exercise 1.22 (p. 17). Test devices (pilots) were setup at both the old and new locations, and voltage readings on 30 production runs at each location were obtained. The data are reproduced in the table below. Descriptive statistics are displayed in the SAS printout. [*Note*: Larger voltage readings are better than smaller voltage readings.]

🔘 VOLTAGE

OLD LOCATION			NEW LOCATION		
9.98	10.12	9.84	9.19	10.01	8.82
10.26	10.05	10.15	9.63	8.82	8.65
10.05	9.80	10.02	10.10	9.43	8.51
10.29	10.15	9.80	9.70	10.03	9.14
10.03	10.00	9.73	10.09	9.85	9.75
8.05	9.87	10.01	9.60	9.27	8.78
10.55	9.55	9.98	10.05	8.83	9.35
10.26	9.95	8.72	10.12	9.39	9.54
9.97	9.70	8.80	9.49	9.48	9.36
9.87	8.72	9.84	9.37	9.64	8.68

Source: Harris Corporation, Melbourne, Fla.

(a) Compare the mean voltage readings at the two locations using a 90% confidence interval.

(b) Based on the interval, part a, does it appear that the manufacturing process can be established locally?

1.71 Laughter among deaf signers. The *Journal of Deaf Studies and Deaf Education* (Fall 2006) published an article on vocalized laughter among deaf

SAS output for Exercise 1.70

```
-------------------------------- location=NEW --------------------------------

                         The MEANS Procedure

                    Analysis Variable : voltage

       N         Mean         Std Dev        Minimum         Maximum

       30      9.4223333      0.4788757      8.5100000      10.1200000


-------------------------------- location=OLD --------------------------------

                    Analysis Variable : voltage

       N         Mean         Std Dev        Minimum         Maximum

       30      9.8036667      0.5409155      8.0500000      10.5500000
```

users of American sign language (ASL). In video-taped ASL conversations among deaf participants, 28 laughed at least once. The researchers wanted to know if they laughed more as speakers (while signing) or as audience members (while listening). For each of the 28 deaf participants, the number of laugh episodes as a speaker and the number of laugh episodes as an audience member was determined. One goal of the research was to compare the mean number of laugh episodes of speakers and audience members.

(a) Explain why the data should be analyzed as a paired difference experiment.

(b) Identify the study's target parameter.

(c) The study yielded a sample mean of 3.4 laughter episodes for speakers and a sample mean of 1.3 laughter episodes as an audience. Is this sufficient evidence to conclude that the population means are different? Explain.

(d) A paired difference t-test resulted in $t = 3.14$ and p-value $< .01$. Interpret the results in the words of the problem.

1.72 **Impact of red light cameras on car crashes.** To combat red light–running crashes—the phenomenon of a motorist entering an intersection after the traffic signal turns red and causing a crash—many states are adopting photo-red enforcement programs. In these programs, red light cameras installed at dangerous intersections photograph the license plates of vehicles that run the red light. How effective are photo-red enforcement programs in reducing red light–running crash incidents at intersections? The Virginia Department of Transportation (VDOT) conducted a comprehensive study of its newly adopted photo-red enforcement program and published the results in a June 2007 report. In one portion of the study, the VDOT provided crash data both before and after installation of red light cameras at several intersections. The data (measured as the number of crashes caused by red light running per intersection per year) for 13 intersections in Fairfax County, Virginia, are given in the table. Analyze the data for the VDOT. What do you conclude?

REDLIGHT

INTERSECTION	BEFORE CAMERA	AFTER CAMERA
1	3.60	1.36
2	0.27	0
3	0.29	0
4	4.55	1.79
5	2.60	2.04
6	2.29	3.14
7	2.40	2.72
8	0.73	0.24
9	3.15	1.57
10	3.21	0.43
11	0.88	0.28
12	1.35	1.09
13	7.35	4.92

Source: Virginia Transportation Research Council, "Research Report: The Impact of Red Light Cameras (Photo-Red Enforcement) on Crashes in Virginia," June 2007.

1.73 **Light-to-dark transition of genes.** *Synechocystis*, a type of cyanobacterium that can grow and survive in a wide range of conditions, is used by scientists to model DNA behavior. In the *Journal of Bacteriology* (July 2002), scientists isolated genes of the bacterium responsible for photosynthesis and respiration and investigated the sensitivity of the genes to light. Each gene sample was grown to midexponential phase in a growth incubator in "full light." The lights were extinguished and growth measured after 24 hours in the dark ("full dark"). The lights were then turned back on for 90 minutes ("transient light") followed immediately by an additional 90 minutes in the dark ("transient dark"). Standardized growth measurements in each light/dark condition were obtained for 103 genes. The complete data set is saved in the GENEDARK file. Data for the first 10 genes are shown in the accompanying table.

(a) Treat the data for the first 10 genes as a random sample collected from the population of 103 genes and test the hypothesis of no difference between the mean standardized growth of genes in the full-dark condition and genes in the transient light condition. Use $\alpha = .01$.

(b) Use a statistical software package to compute the mean difference in standardized growth of the 103 genes in the full-dark condition and the transient-light condition. Did the test, part a, detect this difference?

(c) Repeat parts a and b for a comparison of the mean standardized growth of genes in the full-

dark condition and genes in the transient-dark condition.

🔘 GENEDARK (*first 10 observations shown*)

GENE ID	FULL-DARK	TR-LIGHT	TR-DARK
SLR2067	−0.00562	1.40989	−1.28569
SLR1986	−0.68372	1.83097	−0.68723
SSR3383	−0.25468	−0.79794	−0.39719
SLL0928	−0.18712	−1.20901	−1.18618
SLR0335	−0.20620	1.71404	−0.73029
SLR1459	−0.53477	2.14156	−0.33174
SLL1326	−0.06291	1.03623	0.30392
SLR1329	−0.85178	−0.21490	0.44545
SLL1327	0.63588	1.42608	−0.13664
SLL1325	−0.69866	1.93104	−0.24820

Source: Gill, R. T., et al. "Genome-wide dynamic transcriptional profiling of the light to dark transition in *Synechocystis Sp.* PCC6803," *Journal of Bacteriology,* Vol. 184, No. 13, July 2002.

(d) Repeat parts a and b for a comparison of the mean standardized growth of genes in the transient-light condition and genes in the transient-dark condition.

1.11 Comparing Two Population Variances

Suppose you want to use the two-sample t statistic to compare the mean productivity of two paper mills. However, you are concerned that the assumption of equal variances of the productivity for the two plants may be unrealistic. It would be helpful to have a statistical procedure to check the validity of this assumption.

The common statistical procedure for comparing population variances σ_1^2 and σ_2^2 is to make an inference about the ratio, σ_1^2/σ_2^2, using the ratio of the sample variances, s_1^2/s_2^2. Thus, we will attempt to support the research hypothesis that the ratio σ_1^2/σ_2^2 differs from 1 (i.e., the variances are unequal) by testing the null hypothesis that the ratio equals 1 (i.e., the variances are equal).

$$H_0: \frac{\sigma_1^2}{\sigma_2^2} = 1 \ (\sigma_1^2 = \sigma_2^2)$$

$$H_a: \frac{\sigma_1^2}{\sigma_2^2} \neq 1 \ (\sigma_1^2 \neq \sigma_2^2)$$

We will use the test statistic

$$F = \frac{s_1^2}{s_2^2}$$

To establish a rejection region for the test statistic, we need to know how s_1^2/s_2^2 is distributed in repeated sampling. That is, we need to know the sampling distribution of s_1^2/s_2^2. As you will subsequently see, the sampling distribution of s_1^2/s_2^2 depends on two of the assumptions already required for the t-test, as follows:

1. The two sampled populations are normally distributed.

2. The samples are randomly and independently selected from their respective populations.

When these assumptions are satisfied and when the null hypothesis is true (i.e., $\sigma_1^2 = \sigma_2^2$), the sampling distribution of s_1^2/s_2^2 is an **F distribution** with $(n_1 - 1)$ df and $(n_2 - 1)$ df, respectively. The shape of the F distribution depends on the degrees of freedom associated with s_1^2 and s_2^2, that is, $(n_1 - 1)$ and $(n_2 - 1)$. An F distribution with 7 and 9 df is shown in Figure 1.33. As you can see, the distribution is skewed to the right.

Figure 1.33 An F distribution with 7 and 9 df

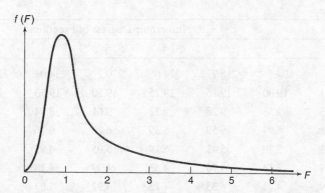

When the population variances are unequal, we expect the ratio of the sample variances, $F = s_1^2/s_2^2$, to be either very large or very small. Therefore, we will need to find F-values corresponding to the tail areas of the F distribution to establish the rejection region for our test of hypothesis. The upper-tail F-values can be found in Tables 3, 4, 5, and 6 of Appendix D. Table 4 is partially reproduced in Table 1.18. It gives F-values that correspond to $\alpha = .05$ upper-tail areas for different degrees of freedom. The columns of the tables correspond to various degrees of freedom for the numerator sample variance s_1^2, whereas the rows correspond to the degrees of freedom for the denominator sample variance s_2^2.

Thus, if the numerator degrees of freedom is 7 and the denominator degrees of freedom is 9, we look in the seventh column and ninth row to find $F_{.05} = 3.29$. As shown in Figure 1.34, $\alpha = .05$ is the tail area to the right of 3.29 in the F distribution with 7 and 9 df. That is, if $\sigma_1^2 = \sigma_2^2$, the probability that the F statistic will exceed 3.29 is $\alpha = .05$.

Suppose we want to compare the variability in production for two paper mills and we have obtained the following results:

Sample 1	Sample 2
$n_1 = 13$ days	$n_2 = 18$ days
$\bar{y}_1 = 26.3$ production units	$\bar{y}_2 = 19.7$ production units
$s_1 = 8.2$ production units	$s_2 = 4.7$ production units

To form the rejection region for a two-tailed F-test, we want to make certain that the upper tail is used, because only the upper-tail values of F are shown in Tables 3, 4, 5, and 6. To accomplish this, **we will always place the larger sample variance in the numerator of the F-test**. This doubles the tabulated value for α, since we double the probability that the F ratio will fall in the upper tail by always placing the larger sample variance in the numerator. In effect, we make the test two-tailed by putting the larger variance in the numerator rather than establishing rejection regions in both tails.

Table 1.18 Reproduction of part of Table 4 of Appendix D: $\alpha = .05$

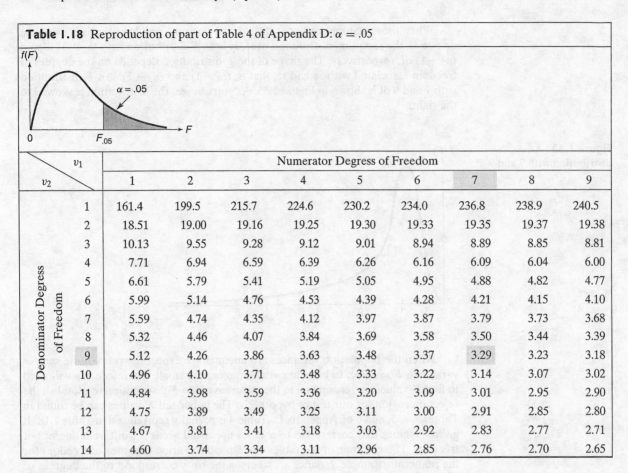

v_1		Numerator Degress of Freedom							
v_2	1	2	3	4	5	6	7	8	9
1	161.4	199.5	215.7	224.6	230.2	234.0	236.8	238.9	240.5
2	18.51	19.00	19.16	19.25	19.30	19.33	19.35	19.37	19.38
3	10.13	9.55	9.28	9.12	9.01	8.94	8.89	8.85	8.81
4	7.71	6.94	6.59	6.39	6.26	6.16	6.09	6.04	6.00
5	6.61	5.79	5.41	5.19	5.05	4.95	4.88	4.82	4.77
6	5.99	5.14	4.76	4.53	4.39	4.28	4.21	4.15	4.10
7	5.59	4.74	4.35	4.12	3.97	3.87	3.79	3.73	3.68
8	5.32	4.46	4.07	3.84	3.69	3.58	3.50	3.44	3.39
9	5.12	4.26	3.86	3.63	3.48	3.37	3.29	3.23	3.18
10	4.96	4.10	3.71	3.48	3.33	3.22	3.14	3.07	3.02
11	4.84	3.98	3.59	3.36	3.20	3.09	3.01	2.95	2.90
12	4.75	3.89	3.49	3.25	3.11	3.00	2.91	2.85	2.80
13	4.67	3.81	3.41	3.18	3.03	2.92	2.83	2.77	2.71
14	4.60	3.74	3.34	3.11	2.96	2.85	2.76	2.70	2.65

(Denominator Degress of Freedom — v_2 row labels)

Figure 1.34 An F distribution for 7 and 9 df: $\alpha = .05$

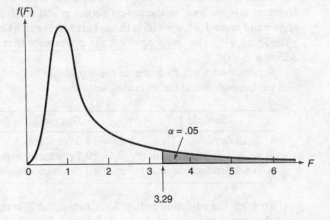

Thus, for our production example, we have a numerator s_1^2 with df $= n_1 - 1 = 12$ and a denominator s_2^2 with df $= n_2 - 1 = 17$. Therefore, the test statistic will be

$$F = \frac{\text{Larger sample variance}}{\text{Smaller sample variance}} = \frac{s_1^2}{s_2^2}$$

and we will reject $H_0: \sigma_1^2 = \sigma_2^2$ for $\alpha = .10$ if the calculated value of F exceeds the tabulated value:

$$F_{.05} = 2.38 \text{ (see Figure 1.34)}$$

Now, what do the data tell us? We calculate

$$F = \frac{s_1^2}{s_2^2} = \frac{(8.2)^2}{(4.7)^2} = 3.04$$

and compare it to the rejection region shown in Figure 1.35. Since the calculated F-value, 3.04, falls in the rejection region, the data provide sufficient evidence to indicate that the population variances differ. Consequently, we would be reluctant to use the two-sample t statistic to compare the population means, since the assumption of equal population variances is apparently untrue.

What would you have concluded if the value of F calculated from the samples had not fallen in the rejection region? Would you conclude that the null hypothesis of equal variances is true? No, because then you risk the possibility of a Type II error (accepting H_0 when H_a is true) without knowing the probability of this error (the probability of accepting $H_0: \sigma_1^2 = \sigma_2^2$ when it is false). Since we will not consider the calculation of β for specific alternatives in this text, when the F statistic does not fall in the rejection region, we simply conclude that **insufficient sample evidence exists to refute the null hypothesis that $\sigma_1^2 = \sigma_2^2$**.

The F-test for equal population variances is summarized in the box on p. 68.

Figure 1.35 Rejection region for production example F distribution

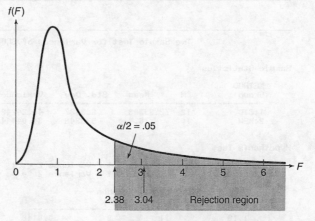

Example 1.18

In Example 1.17 we used the two-sample t statistic to compare the mean reading scores of two groups of slow learners who had been taught to read using two different methods. The data are repeated in Table 1.19 for convenience. The use of the t statistic was based on the assumption that the population variances of the test scores were equal for the two methods. Conduct a test of hypothesis to check this assumption at $\alpha = .10$.

READING

Table 1.19 Reading test scores for slow learners

New Method				Standard Method			
80	80	79	81	79	62	70	68
76	66	71	76	73	76	86	73
70	85			72	68	75	66

F-Test for Equal Population Variances: Independent Samples*

	ONE-TAILED TESTS		TWO-TAILED TEST
	$H_0: \sigma_1^2 = \sigma_2^2$	$H_0: \sigma_1^2 = \sigma_2^2$	$H_0: \sigma_1^2 = \sigma_2^2$
	$H_a: \sigma_1^2 < \sigma_2^2$	$H_a: \sigma_1^2 > \sigma_2^2$	$H_a: \sigma_1^2 \neq \sigma_2^2$

Test statistic: $\qquad F = s_2^2/s_1^2 \qquad\qquad F = s_1^2/s_2^2 \qquad F = \dfrac{\text{Larger sample variance}}{\text{Smaller sample variance}}$

Rejection region:		$F > F_\alpha$	$F > F_\alpha$	$F > F_{\alpha/2}$
Numerator df (ν_1):		$n_2 - 1$	$n_1 - 1$	$n - 1$ for larger variance
Denominator df (ν_2):		$n_1 - 1$	$n_2 - 1$	$n - 1$ for smaller variance
p-*value*:		$P(F > F_c)$	$P(F > F_c)$	$P(F^* < 1/F_c) + P(F > F_c)$

Decision: Reject H_0 if $\alpha > p$-value, or, if test statistic falls in rejection region

where F is based on ν_1 numerator df and ν_2 denominator df; F^* is based on ν_2 numerator df and ν_1 denominator df; F_c = calculated value of the test statistic; and $\alpha = P(\text{Type I error}) = P(\text{Reject } H_0 | H_0 \text{ true})$.

Assumptions:

1. Both sampled populations are normally distributed.
2. The samples are random and independent.

Figure 1.36 SAS F-test for the data in Table 1.19

```
                    Two Sample Test for Variances of SCORE within METHOD

Sample Statistics

    METHOD
    Group        N      Mean      Std. Dev.     Variance
    -------------------------------------------------------
    1:STD        12    72.33333     6.3437       40.24242
    2:NEW        10    76.4         5.8348       34.04444


Hypothesis Test

    Null hypothesis:      Variance 1 / Variance 2 =  1
    Alternative:          Variance 1 / Variance 2 ^= 1

              - Degrees of Freedom -
        F        Numer.    Denom.           Pr > F
    -------------------------------------------------------
       1.18        11         9             0.8148
```

Solution

We want to test

$$H_0: \frac{\sigma_1^2}{\sigma_2^2} = 1 \ (\text{i.e., } \sigma_1^2 = \sigma_2^2)$$

$$H_a: \frac{\sigma_1^2}{\sigma_2^2} \neq 1 \ (\text{i.e., } \sigma_1^2 \neq \sigma_2^2)$$

---

*Although a test of a hypothesis of equality of variances is the most common application of the F-test, it can also be used to test a hypothesis that the ratio between the population variances is equal to some specified value, $H_0: \sigma_1^2/\sigma_2^2 = k$. The test is conducted in exactly the same way as specified in the box, except that we use the test statistic

$$F = \left(\frac{s_1^2}{s_2^2}\right)\left(\frac{1}{k}\right)$$

The data were entered into SAS, and the SAS printout shown in Figure 1.36 was obtained. Both the test statistic, $F = 1.18$, and two-tailed p-value, .8148, are highlighted on the printout. Since $\alpha = .10$ is less than the p-value, we do not reject the null hypothesis that the population variances of the reading test scores are equal. ■

The previous examples demonstrate how to conduct a two-tailed F-test when the alternative hypothesis is $H_a: \sigma_1^2 \neq \sigma_2^2$. One-tailed tests for determining whether one population variance is larger than another population variance (i.e., $H_a: \sigma_1^2 > \sigma_2^2$) are conducted similarly. However, the α value no longer needs to be doubled since the area of rejection lies only in the upper (or lower) tail area of the F distribution. The procedure for conducting an upper-tailed F-test is outlined in the previous box. Whenever you conduct a one-tailed F-test, be sure to write H_a in the form of an upper-tailed test. This can be accomplished by numbering the populations so that the variance hypothesized to be larger in H_a is associated with population 1 and the hypothesized smaller variance is associated with population 2.

Important: As a final comment, we note that (unlike the small-sample t procedure for means) the F-test for comparing variances is not very robust against nonnormal data. Consequently, with nonnormal data it is difficult to determine whether a significant F-value implies that the population variances differ or is simply due to the fact that the populations are not normally distributed.

1.11   Exercises

1.74   Use Tables 3, 4, 5, and 6 of Appendix D to find F_α for α, numerator df, and denominator df equal to:

(a) .05, 8, 7   (b) .01, 15, 20
(c) .025, 12, 5   (d) .01, 5, 25
(e) .10, 5, 10   (f) .05, 20, 9

1.75   **Is honey a cough remedy?**   Refer to the *Archives of Pediatrics and Adolescent Medicine* (December 2007) study of honey as a children's cough remedy, Exercise 1.21 (p. 17). The data (cough improvement scores) for the 33 children in the DM dosage group and the 35 children in the honey dosage group are reproduced in the table below. The researchers want to know if the variability in coughing improvement scores differs for the two groups. Conduct the appropriate analysis, using $\alpha = .10$.

1.76   **How do you choose to argue?**   Refer to the *Thinking and Reasoning* (October 2006) study of the cognitive skills required for successful arguments, Exercise 1.65 (p. 60). Recall that 52 psychology graduate students were equally divided into two groups. Group 1 was presented with arguments that always attempted to strengthen the favored position. Group 2 was presented with arguments that always attempted to weaken the nonfavored position. Summary statistics for the student ratings of the arguments are reproduced in the table. In Exercise 1.65 you compared the mean ratings for the two groups with a small-sample t-test, assuming equal variances. Determine the validity of this assumption at $\alpha = .05$.

	GROUP 1 (SUPPORT FAVORED POSITION)	GROUP 2 (WEAKEN OPPOSING POSITION)
Sample size:	26	26
Mean:	28.6	24.9
Standard deviation:	12.5	12.2

Source: Kuhn, D., and Udell, W. "Coordinating won and other perspectives in argument," *Thinking and Reasoning*, October 2006.

💿 HONEYCOUGH

Honey Dosage:	12 11 15 11 10 13 10 4 15 16 9 14 10 6 10 8 11 12 12 8
	12 9 11 15 10 15 9 13 8 12 10 8 9 5 12
DM Dosage:	4 6 9 4 7 7 7 9 12 10 11 6 3 4 9 12 7 6 8 12 12 4 12
	13 7 10 13 9 4 4 10 15 9

Source: Paul, I. M., et al. "Effect of honey, dextromethorphan, and no treatment on nocturnal cough and sleep quality for coughing children and their parents," *Archives of Pediatrics and Adolescent Medicine*, Vol. 161, No. 12, Dec. 2007 (data simulated).

1.77 **Eating disorder study.** Refer to Exercise 1.66 (p. 61). The "fear of negative evaluation" (FNE) scores for the 11 bulimic females and 14 females with normal eating habits are reproduced in the table. The confidence interval you constructed in Exercise 1.66 requires that the variance of the FNE scores of bulimic females is equal to the variance of the FNE scores of normal females. Conduct a test (at $\alpha = .05$) to determine the validity of this assumption.

BULIMIA

Bulimic:	21 13 10 20 25 19 16 21 24 13 14
Normal:	13 6 16 13 8 19 23 18 11 19 7 10 15 20

Source: Randles, R. H. "On neutral responses (zeros) in the sign test and ties in the Wilcoxon-Mann-Whitney test," *American Statistician*, Vol. 55, No. 2, May 2001 (Figure 3).

1.78 **Human inspection errors.** Tests of product quality using human inspectors can lead to serious inspection error problems. To evaluate the performance of inspectors in a new company, a quality manager had a sample of 12 novice inspectors evaluate 200 finished products. The same 200 items were evaluated by 12 experienced inspectors. The quality of each item—whether defective or nondefective—was known to the manager. The next table lists the number of inspection errors (classifying a defective item as nondefective or vice versa) made by each inspector. A SAS printout comparing the two types of inspectors is shown in the next column.

(a) Prior to conducting this experiment, the manager believed the variance in inspection errors was lower for experienced inspectors than for novice inspectors. Do the sample data support her belief? Test using $\alpha = .05$.

(b) What is the appropriate p-value of the test you conducted in part a?

INSPECT

NOVICE INSPECTORS				EXPERIENCED INSPECTORS			
30	35	26	40	31	15	25	19
36	20	45	31	28	17	19	18
33	29	21	48	24	10	20	21

```
            Two Sample Test for Variances of ERRORS within INSPECT

Sample Statistics

    INSPECT
    Group       N     Mean    Std. Dev.    Variance
    -----------------------------------------------------
    1NOVICE     12   32.83333    8.6427     74.69697
    2EXPER      12   20.58333    5.7439     32.99242

Hypothesis Test

    Null hypothesis:      Variance 1 / Variance 2 <= 1
    Alternative:          Variance 1 / Variance 2 >  1

                - Degrees of Freedom -
        F         Numer.    Denom.            Pr > F
    -----------------------------------------------------
       2.26         11        11              0.0955
```

1.79 **Variation in wet sampler readings.** Wet samplers are standard devices used to measure the chemical composition of precipitation. The accuracy of the wet deposition readings, however, may depend on the number of samplers stationed in the field. Experimenters in The Netherlands collected wet deposition measurements using anywhere from one to eight identical wet samplers (*Atmospheric Environment*, Vol. 24A, 1990). For each sampler (or sampler combination), data were collected every 24 hours for an entire year; thus, 365 readings were collected per sampler (or sampler combination). When one wet sampler was used, the standard deviation of the hydrogen readings (measured as percentage relative to the average reading from all eight samplers) was 6.3%. When three wet samplers were used, the standard deviation of the hydrogen readings (measured as percentage relative to the average reading from all eight samplers) was 2.6%. Conduct a test to compare the variation in hydrogen readings for the two sampling schemes (i.e., one wet sampler vs. three wet samplers). Test using $\alpha = .05$.

Quick Summary/Guides

KEY IDEAS

Types of statistical applications

1. descriptive
2. inferential

Descriptive statistics

1. Identify **population** or **sample** (collection of **experimental units**)
2. Identify **variable(s)**

3. Collect **data**
4. **Describe** data

Inferential statistics

1. Identify **population** (collection of *all* **experimental units**)
2. Identify **variable(s)**
3. Collect **sample** data (*subset* of population)
4. **Inference** about population based on sample
5. **Measure of reliability** for inference

Types of data

1. qualitative (categorical in nature)
2. quantitative (numerical in nature)

Graphs for qualitative data

1. pie chart
2. bar graph

Graphs for quantitative data

1. stem-and-leaf display
2. histogram

Measure of central tendency

  mean (or **average**)

Measures of variation

1. range
2. variance
3. standard deviation

Percentage of measurements within 2 standard deviations of the mean

1. *any data set*: at least 3/4 (Tchebysheff's Theorem)
2. **normal (mound-shaped) distribution**: 95%

Key Formulas		
	Sample	**Population**
Mean:	$\bar{y} = (\Sigma y_i)/n$	μ
Variance:	$s^2 = \dfrac{\Sigma(y_i - \bar{y})^2}{n-1}$	σ^2
Std. Dev.	$s = \sqrt{s^2}$	σ

Properties of the sampling distribution of $\bar{y}$

1. $E(\bar{y}) = \mu$
2. $\text{Var}(\bar{y}) = \sigma^2/n$

Central Limit Theorem

For large n, the sampling distribution of $\bar{y}$ is approximately normal.

Formulation of Confidence Intervals for a Population Parameter θ and Test Statistics for $H_0: \theta = \theta_0$, where $\theta = \mu$ or $(\mu_1 - \mu_2)$

SAMPLE SIZE	CONFIDENCE INTERVAL	TEST STATISTIC
Large	$\hat{\theta} \pm z_{\alpha/2} s_{\hat{\theta}}$	$z = \dfrac{\hat{\theta} - \theta_0}{s_{\hat{\theta}}}$
Small	$\hat{\theta} \pm t_{\alpha/2} s_{\hat{\theta}}$	$t = \dfrac{\hat{\theta} - \theta_0}{s_{\hat{\theta}}}$

Note: The test statistic for testing $H_0: \sigma_1^2/\sigma_2^2 = 1$ is $F = s_1^2/s_2^2$ (see the box on page 68).

Population Parameters and Corresponding Estimators and Standard Errors

PARAMETER (θ)	ESTIMATOR ($\hat{\theta}$)	STANDARD ERROR ($\sigma_{\hat{\theta}}$)	ESTIMATE OF STANDARD ERROR ($s_{\hat{\theta}}$)
μ Mean (average)	$\bar{y}$	$\dfrac{\sigma}{\sqrt{n}}$	$\dfrac{s}{\sqrt{n}}$
$\mu_1 - \mu_2$ Difference between means (averages), independent samples	$\bar{y}_1 - \bar{y}_2$	$\sqrt{\dfrac{\sigma_1^2}{n_1} + \dfrac{\sigma_2^2}{n_2}}$	$\sqrt{\dfrac{s_1^2}{n_1} + \dfrac{s_2^2}{n_2}}, n_1 \geq 30, n_2 \geq 30$ $\sqrt{s_p^2\left(\dfrac{1}{n_1} + \dfrac{1}{n_2}\right)}$, either $n_1 < 30$ or $n_2 < 30$ where $s_p^2 = \dfrac{(n_1 - 1)s_1^2 + (n_2 - 1)s_2^2}{n_1 + n_2 - 2}$
$\mu_d = \mu_1 - \mu_2$, Difference between means, paired samples	$\bar{y}_d$	$\sigma_d/\sqrt{n}$	$s_d/\sqrt{n}$
$\dfrac{\sigma_1^2}{\sigma_2^2}$ Ratio of variances	$\dfrac{s_1^2}{s_2^2}$	(not necessary)	(not necessary)

Supplementary Exercises

1.80. Using Tchebysheff's theorem. Tchebysheff's theorem states that at least $1 - (1/K^2)$ of a set of measurements will lie within K standard deviations of the mean of the data set. Use Tchebysheff's theorem to find the fraction of a set of measurements that will lie within:

(a) 2 standard deviations of the mean ($K = 2$)
(b) 3 standard deviations of the mean
(c) 1.5 standard deviations of the mean

1.81. Computing descriptive statistics. For each of the following data sets, compute $\bar{y}$, s^2, and s.

(a) 11, 2, 2, 1, 9   (b) 22, 9, 21, 15
(c) 1, 0, 1, 10, 11, 11, 0   (d) 4, 4, 4, 4

1.82. Normal probabilities. Use Table 1 of Appendix D to find each of the following:

(a) $P(z \geq 2)$ (b) $P(z \leq -2)$ (c) $P(z \geq -1.96)$
(d) $P(z \geq 0)$ (e) $P(z \leq -.5)$ (f) $P(z \leq -1.96)$

1.83. Finding z-scores. Suppose the random variable y has mean $\mu = 30$ and standard deviation $\sigma = 5$. How many standard deviations away from the mean of y is each of the following y values?

(a) $y = 10$ (b) $y = 32.5$ (c) $y = 30$ (d) $y = 60$

1.84. Deep-hole drilling for oil. "Deep hole" drilling is a family of drilling processes used when the ratio of hole depth to hole diameter exceeds 10. Successful deep hole drilling depends on the satisfactory discharge of the drill chip. An experiment was conducted to investigate the performance of deep hole drilling when chip congestion exists (*Journal of Engineering for Industry*, May 1993). Some important variables in the drilling process are described here. Identify the data type for each variable.

(a) Chip discharge rate (number of chips discarded per minute)
(b) Drilling depth (millimeters)
(c) Oil velocity (millimeters per second)
(d) Type of drilling (single-edge, BTA, or ejector)
(e) Quality of hole surface

1.85. Finding misplaced items. Are men or women more adept at remembering where they leave misplaced items (like car keys)? According to University of Florida psychology professor Robin West, women show greater competence in actually finding these objects (*Explore*, Fall 1998). Approximately 300 men and women from Gainesville, Florida, participated in a study in which each person placed 20 common objects in a 12-room "virtual" house represented on a computer screen. Thirty minutes later, the subjects were asked to recall where they put each of the objects. For each object, a recall variable was measured as "yes" or "no."

(a) Identify the population of interest to the psychology professor.
(b) Identify the sample.
(c) Does the study involve descriptive or inferential statistics? Explain.
(d) Are the variables measured in the study quantitative or qualitative?

1.86. Salient roles of older adults. In *Psychology and Aging* (December 2000), University of Michigan School of Public Health researchers studied the roles that older adults feel are the most important to them in late life. The accompanying table summarizes the most salient roles identified by each in a national sample of 1,102 adults, 65 years or older.

MOST SALIENT ROLE	NUMBER
Spouse	424
Parent	269
Grandparent	148
Other relative	59
Friend	73
Homemaker	59
Provider	34
Volunteer, club, church member	36
Total	1,102

Source: Krause, N., and Shaw, B.A. "Role-specific feelings of control and mortality," *Psychology and Aging,* Vol. 15, No. 4, Table 2, Copyright © 2000, American Psychological Association, reprinted with permission.

(a) Describe the qualitative variable summarized in the table. Give the categories associated with the variable.
(b) Are the numbers in the table frequencies or relative frequencies?
(c) Display the information in the table in a bar graph.
(d) Which role is identified by the highest percentage of older adults? Interpret the relative frequency associated with this role.

1.87. Ancient Greek pottery. Archaeologists excavating the ancient Greek settlement at Phylakopi classified the pottery found in trenches (*Chance*, Fall 2000). The next table describes the collection of 837 pottery pieces uncovered in a particular layer at the excavation site. Construct and interpret a graph that will aid the archaeologists in understanding the distribution of the pottery types found at the site.

Table for Exercise 1.87

POT CATEGORY	NUMBER FOUND
Burnished	133
Monochrome	460
Slipped	55
Painted in curvilinear decoration	14
Painted in geometric decoration	165
Painted in naturalistic decoration	4
Cycladic white clay	4
Conical cup clay	2
Total	837

Source: Berg, I., and Bliedon, S. "The pots of Phylakopi: Applying statistical techniques to archaeology," *Chance,* Vol. 13, No. 4, Fall 2000.

🌑 OILSPILL

1.88. Tanker oil spills. Owing to several major ocean oil spills by tank vessels, Congress passed the 1990 Oil Pollution Act, which requires all tankers to be designed with thicker hulls. Further improvements in the structural design of a tank vessel have been proposed since then, each with the objective of reducing the likelihood of an oil spill and decreasing the amount of outflow in the event of a hull puncture. To aid in this development, *Marine Technology* (January 1995) reported on the spillage amount and cause of puncture for 50 recent major oil spills from tankers and carriers. The data are saved in the OILSPILL file.

(a) Use a graphical method to describe the cause of oil spillage for the 50 tankers.

(b) Does the graph, part a, suggest that any one cause is more likely to occur than any other? How is this information of value to the design engineers?

(c) Now consider the data on spillage amounts (in thousands of metric tons) for 50 major oil spills. An SPSS histogram for the 50 spillage

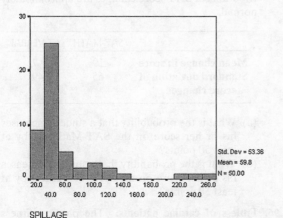

Std. Dev = 53.36
Mean = 59.8
N = 50.00

SPILLAGE

amounts is shown below. Interpret the histogram.

(d) Descriptive statistics for the 50 spillage amounts are also shown on the SPSS histogram. Use this information to form an interval that can be used to predict the spillage amount for the next major oil spill.

1.89. Eclipses of Saturnian satellites. Saturn has five satellites that rotate around the planet. *Astronomy* (August 1995) lists 19 different events involving eclipses or occults of Saturnian satellites during the month of August. For each event, the percent of light lost by the eclipsed or occulted satellite at midevent is recorded in the table.

🌑 SATURN

DATE	EVENT	LIGHT LOSS (%)
Aug. 2	Eclipse	65
4	Eclipse	61
5	Occult	1
6	Eclipse	56
8	Eclipse	46
8	Occult	2
9	Occult	9
11	Occult	5
12	Occult	39
14	Occult	1
14	Eclipse	100
15	Occult	5
15	Occult	4
16	Occult	13
20	Occult	11
23	Occult	3
23	Occult	20
25	Occult	20
28	Occult	12

Source: Astronomy magazine, Aug. 1995, p. 60.

(a) Construct a stem-and-leaf display for light loss percentage of the 19 events.

(b) Locate on the stem-and-leaf plot, part a, the light losses associated with eclipses of Saturnian satellites. (Circle the light losses on the plot.)

(c) Based on the marked stem-and-leaf display, part b, make an inference about which event type (eclipse or occult) is more likely to lead to a greater light loss.

1.90. Comparing voltage readings. Refer to the data on process voltage readings at two locations, Exercise 1.70. Use the SAS printout for Exercise 1.70 (p. 63) and the rule of thumb to compare the voltage reading distributions for the two locations.

1.91. Normal probabilities. The random variable y has a normal distribution with $\mu = 80$ and $\sigma = 10$. Find the following probabilities:

(a) $P(y \leq 75)$ (b) $P(y \geq 90)$ (c) $P(60 \leq y \leq 70)$
(d) $P(y \geq 75)$ (e) $P(y = 75)$ (f) $P(y \leq 105)$

1.92. Fluid loss in crab spiders. A group of University of Virginia biologists studied nectivory (nectar drinking) in crab spiders to determine if adult males were feeding on nectar to prevent fluid loss (*Animal Behavior*, June 1995). Nine male spiders were weighed and then placed on the flowers of Queen Anne's lace. One hour later, the spiders were removed and reweighed. The evaporative fluid loss (in milligrams) of each of the nine male spiders is given in the table.

⊚ SPIDERS

MALE SPIDER	FLUID LOSS
A	.018
B	.020
C	.017
D	.024
E	.020
F	.024
G	.003
H	.001
I	.009

Source: Reprinted from *Animal Behaviour*, Vol. 49, Issue 6, Simon D. Pollard, Mike W. Beck, and Gary N. Dodson, "Why do male crab spiders drink nectar?" p. 1445 (Table II), Copyright © 1995, with permission from Elsevier.

(a) Summarize the fluid losses of male crab spiders with a stem-and-leaf display.
(b) Of the nine spiders, only three drank any nectar from the flowers of Queen Anne's lace. These three spiders are identified as G, H, and I in the table. Locate and circle these three fluid losses on the stem-and-leaf display. Does the pattern depicted in the graph give you any insight into whether feeding on flower nectar reduces evaporative fluid loss for male crab spiders? Explain.

⊚ MTBE

1.93. Groundwater contamination in wells. Refer to the *Environmental Science and Technology* (January 2005) study of the factors related to MTBE contamination in 223 New Hampshire wells, Exercise 1.17 (p. 12). The data are saved in the MTBE file. Two of the many quantitative variables measured for each well are the pH level (standard units) and the MTBE level (micrograms per liter).

(a) Construct a histogram for the pH levels of the sampled wells. From the histogram, estimate the proportion of wells with pH values less than 7.0.

(b) For those wells with detectible levels of MTBE, construct a histogram for the MTBE values. From the histogram, estimate the proportion of contaminated wells with MTBE values that exceed 5 micrograms per liter.

(c) Find the mean and standard deviation for the pH levels of the sampled wells, and construct the interval $\bar{y} \pm 2s$. Estimate the percentage of wells with pH levels that fall within the interval. What rule did you apply to obtain the estimate? Explain.

(d) Find the mean and standard deviation for the MTBE levels of the sampled wells and construct the interval $\bar{y} \pm 2s$. Estimate the percentage of wells with MTBE levels that fall within the interval. What rule did you apply to obtain the estimate? Explain.

1.94. Dental Anxiety Scale. Psychology students at Wittenberg University completed the Dental Anxiety Scale questionnaire (*Psychological Reports*, August 1997). Scores on the scale range from 0 (no anxiety) to 20 (extreme anxiety). The mean score was 11 and the standard deviation was 3.5. Assume that the distribution of all scores on the Dental Anxiety Scale is normal with $\mu = 11$ and $\sigma = 3.5$.

(a) Suppose you score a 16 on the Dental Anxiety Scale. Find the z-value for this score.
(b) Find the probability that someone scores between a 10 and a 15 on the Dental Anxiety Scale.
(c) Find the probability that someone scores above a 17 on the Dental Anxiety Scale.

1.95. Improving SAT scores. Refer to the *Chance* (Winter 2001) study of students who paid a private tutor to help them improve their Standardized Admission Test (SAT) scores, Exercise 1.31 (p. 25). The table summarizing the changes in both the SAT-Mathematics and SAT-Verbal scores for these students is reproduced here. Assume that both distributions of SAT score changes are approximately normal.

	SAT-MATH	SAT-VERBAL
Mean change in score	19	7
Standard deviation of score changes	65	49

(a) What is the probability that a student increases his or her score on the SAT-Math test by at least 50 points?
(b) What is the probability that a student increases his or her score on the SAT-Verbal test by at least 50 points?

1.96. Fitness of cardiac patients. The physical fitness of a patient is often measured by the patient's maximum oxygen uptake (recorded in milliliters per kilogram, ml/kg). The mean maximum oxygen uptake for cardiac patients who regularly participate in sports or exercise programs was found to be 24.1 with a standard deviation of 6.30 (*Adapted Physical Activity Quarterly*, October 1997). Assume this distribution is approximately normal.

(a) What is the probability that a cardiac patient who regularly participates in sports has a maximum oxygen uptake of at least 20 ml/kg?

(b) What is the probability that a cardiac patient who regularly exercises has a maximum oxygen uptake of 10.5 ml/kg or lower?

(c) Consider a cardiac patient with a maximum oxygen uptake of 10.5. Is it likely that this patient participates regularly in sports or exercise programs? Explain.

1.97. Susceptibility to hypnosis. The Computer-Assisted Hypnosis Scale (CAHS) is designed to measure a person's susceptibility to hypnosis. In computer-assisted hypnosis, the computer serves as a facilitator of hypnosis by using digitized speech processing coupled with interactive involvement with the hypnotic subject. CAHS scores range from 0 (no susceptibility) to 12 (extremely high susceptibility). A study in *Psychological Assessment* (March 1995) reported a mean CAHS score of 4.59 and a standard deviation of 2.95 for University of Tennessee undergraduates. Assume that $\mu = 4.29$ and $\sigma = 2.95$ for this population. Suppose a psychologist uses CAHS to test a random sample of 50 subjects.

(a) Would you expect to observe a sample mean CAHS score of $\bar{y} = 6$ or higher? Explain.

(b) Suppose the psychologist actually observes $\bar{y} = 6.2$. Based on your answer to part a, make an inference about the population from which the sample was selected.

1.98. Data on postmortem intervals. In Exercise 1.20 (p. 17) you learned that postmortem interval (PMI) is the elapsed time between death and the performance of an autopsy on the cadaver. *Brain and Language* (June 1995) reported on the PMIs of 22 randomly selected human brain specimens obtained at autopsy. The data are reproduced in the table below.

💿 BRAINPMI

5.5	14.5	6.0	5.5	5.3	5.8	11.0	6.1
7.0	14.5	10.4	4.6	4.3	7.2	10.5	6.5
3.3	7.0	4.1	6.2	10.4	4.9		

Source: Reprinted from *Brain and Language,* Vol. 49, Issue 3, T. L. Hayes and D. A. Lewis, "Anatomical Specialization of the Anterior Motor Speech Area: Hemispheric Differences in Magnopyramidal Neurons," p. 292 (Table 1), Copyright © 1995, with permission of Elsevier.

(a) Construct a 95% confidence interval for the true mean PMI of human brain specimens obtained at autopsy.

(b) Interpret the interval, part a.

(c) What assumption is required for the interval, part a, to be valid? Is this assumption satisfied? Explain.

(d) What is meant by the phrase "95% confidence"?

1.99. Pigeon diet study. The *Australian Journal of Zoology* (Vol. 43, 1995) reported on a study of the diets and water requirements of spinifex pigeons. Sixteen pigeons were captured in the desert and the crop (i.e., stomach) contents of each examined. The accompanying table reports the weight (in grams) of dry seed in the crop of each pigeon. Use the SAS printout below to find a 99% confidence interval for the average weight of dry seeds in the crops of spinifex pigeons inhabiting the Western Australian desert. Interpret the result.

💿 PIGEONS

.457	3.751	.238	2.967	2.509	1.384	1.454	.818
.335	1.436	1.603	1.309	.201	.530	2.144	.834

Source: Table 2 from Williams, J. B., Bradshaw, D., and Schmidt, L. "Field metabolism and water requirements of spinifex pigeons *(Geophaps plumifera)* in Western Australia. "*Australian Journal of Zoology,* Vol. 43, no. 1, 1995, p. 7. Reprinted by permission of CSIRO Publishing. http://www.publish.csiro.au/nid/90/paper/ZO9950001.htm.

```
Sample Statistics for SEEDWT

     N        Mean      Std. Dev.      Std. Error
-----------------------------------------------------
     16       1.37        1.03            0.26

Hypothesis Test

   Null hypothesis:    Mean of SEEDWT  =   0
   Alternative:        Mean of SEEDWT  ^=  0

        t Statistic      Df        Prob > t
     -----------------------------------------
          5.312          15         <.0001


99 % Confidence Interval for the Mean

         Lower Limit:          0.61
         Upper Limit:          2.13
```

1.100. Psychological stress in victims of violence. Interpersonal violence (e.g., rape) generally leads to psychological stress for the victim. *Clinical Psychology Review* (Vol. 15, 1995) reported on the results of all recently published studies of the relationship between interpersonal violence and psychological stress. The distribution of the time elapsed between the violent incident and the initial sign of stress has a mean of 5.1 years and a standard deviation of 6.1 years. Consider a random sample of $n = 150$ victims of interpersonal violence. Let $\bar{y}$ represent the mean time elapsed between the violent act and the first sign of stress for the sampled victims.

(a) Give the mean and standard deviation of the sampling distribution of $\bar{y}$.
(b) Will the sampling distribution of $\bar{y}$ be approximately normal? Explain.
(c) Find $P(\bar{y} > 5.5)$.
(d) Find $P(4 < \bar{y} < 5)$.

1.101. Workers exposed to dioxin. The Occupational Safety and Health Administration (OSHA) conducted a study to evaluate the level of exposure of workers to the dioxin TCDD. The distribution of TCDD levels in parts per trillion (ppt) of production workers at a Newark, New Jersey, chemical plant had a mean of 293 ppt and a standard deviation of 847 ppt (*Chemosphere*, Vol. 20, 1990). A graph of the distribution is shown here. In a random sample of $n = 50$ workers selected at the New Jersey plant, let $\bar{y}$ represent the sample mean TCDD level.

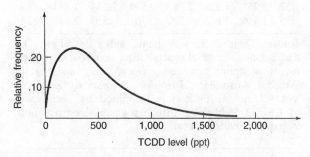

(a) Find the mean and standard deviation of the sampling distribution of $\bar{y}$.
(b) Draw a sketch of the sampling distribution of $\bar{y}$. Locate the mean on the graph.
(c) Find the probability that $\bar{y}$ exceeds 550 ppt.

1.102. Alkalinity levels in river water. The mean alkalinity level of water specimens collected from the Han River in Seoul, Korea, is 50 milligrams per liter (*Environmental Science and Engineering*, September 1, 2000). Consider a random sample of 100 water specimens collected from a tributary of the Han River. Suppose the mean and standard deviation of the alkalinity levels for the sample are $\bar{y} = 67.8$ mpl and $s = 14.4$ mpl. Is there sufficient evidence (at $\alpha = .01$) to indicate that the population mean alkalinity level of water in the tributary exceeds 50 mpl?

1.103. Temperature of molten iron. The Cleveland Casting Plant produces iron automotive castings for Ford Motor Company. When the process is stable, the target pouring temperature of the molten iron

is 2,550 degrees (*Quality Engineering*, Vol. 7, 1995). The pouring temperatures (in degrees Fahrenheit) for a random sample of 10 crankshafts produced at the plant are listed in the table below. Conduct a test to determine whether the true mean pouring temperature differs from the target setting. Test using $\alpha = .01$.

 IRONTEMP

2,543	2,541	2,544	2,620	2,560	2,559	2,562
2,553	2,552	2,553				

Source: Price, B., and Barth, B. "A structural model relating process inputs and final product characteristics," *Quality Engineering*, Vol. 7, No. 4, 1995, p. 696 (Table 2).

1.104. Mating habits of snails. *Genetical Research* (June 1995) published a study of the mating habits of hermaphroditic snails. The mating habits of the snails were identified as either self-fertilizing or cross-fertilizing. The effective population sizes of the two groups were compared. The data for the study are summarized in the table. Geneticists are interested in comparing the variation in population size of the two types of mating systems. Conduct this analysis for the researcher. Interpret the result.

SNAIL MATING SYSTEM	EFFECTIVE POPULATION SIZE		
	SAMPLE SIZE	MEAN	STANDARD DEVIATION
Cross-fertilizing	17	4,894	1,932
Self-fertilizing	5	4,133	1,890

Source: Jarne, P. "Mating system, bottlenecks, and genetic polymorphism in hermaphroditic animals." *Genetics Research*, Vol. 65, No. 3, June 1995, p. 197 (Table 4). 2009 © Cambridge Journals, reproduced with permission.

1.105. Heights of children who repeat grades. Are children who repeat a grade in elementary school shorter on average than their peers? To answer this question, researchers compared the heights of Australian schoolchildren who repeated a grade to those who did not (*Archives of Disease in Childhood*, April 2000). All height measurements were standardized using z-scores. A summary of the results, by gender, is shown in the table on p. 77.

(a) Conduct a test of hypothesis to determine whether the average height of Australian boys who repeated a grade is less than the average height of boys who never repeated. Use $\alpha = .05$.
(b) Repeat part a for Australian girls.
(c) Summarize the results of the hypothesis tests in the words of the problem.

Summary table for Exercise 1.105

	NEVER REPEATED	REPEATED A GRADE
Boys	$n = 1,349$	$n = 86$
	$\bar{x} = .30$	$\bar{x} = -.04$
	$s = .97$	$s = 1.17$
Girls	$n = 1,366$	$n = 43$
	$\bar{x} = .22$	$\bar{x} = .26$
	$s = 1.04$	$s = .94$

Source: Reproduced from *Archives of Disease in Childhood,* "Does height influence progression through primary school grades?" Melissa Wake, David Coghlan, and Kylie Hesketh, Vol. 82, Issue 4, April 2000 (Table 3), with permission from BMJ Publishing Group Ltd.

1.106. College students attitudes toward parents. Researchers at the University of South Alabama compared the attitudes of male college students toward their fathers with their attitudes toward their mothers (*Journal of Genetic Psychology*, March 1998). Each of a sample of 13 males was asked to complete the following statement about each of their parents: My relationship with my father (mother) can best be described as: (1) awful, (2) poor, (3) average, (4) good, or (5) great. The following data were obtained:

🔘 FMATTITUDES

STUDENT	ATTITUDE TOWARD FATHER	ATTITUDE TOWARD MOTHER
1	2	3
2	5	5
3	4	3
4	4	5
5	3	4
6	5	4
7	4	5
8	2	4
9	4	5
10	5	4
11	4	5
12	5	4
13	3	3

Source: Adapted from Vitulli, W. F., and Richardson, D. K. "College student's attitudes toward relationships with parents: A five-year comparative analysis," *Journal of Genetic Psychology*, Vol. 159, No. 1 (March 1998), pp. 45–52.

(a) Specify the appropriate hypotheses for testing whether male students' attitudes toward their fathers differ from their attitudes toward their mothers, on average.

(b) Conduct the test of part a at $\alpha = .05$. Interpret the results in the context of the problem.

1.107. Mathematics and gender. On average, do males outperform females in mathematics? To answer this question, psychologists at the University of Minnesota compared the scores of male and female eighth-grade students who took a basic skills mathematics achievement test (*American Educational Research Journal*, Fall 1998). One form of the test consisted of 68 multiple-choice questions. A summary of the test scores is displayed in the table.

	MALES	FEMALES
Sample size	1,764	1,739
Mean	48.9	48.4
Standard deviation	12.96	11.85

Source: Bielinski, J., and Davison, M. L. "Gender differences by item difficulty interactions in multiple-choice mathematics items," *American Educational Research Journal*, Vol. 35, No. 3, Fall 1998, p. 464 (Table 1). Reprinted by Permission of SAGE Publications.

(a) Is there evidence of a difference between the true mean mathematics test scores of male and female eighth-graders?

(b) Use a 90% confidence interval to estimate the true difference in mean test scores between males and females. Does the confidence interval support the result of the test you conducted in part a?

(c) What assumptions about the distributions of the populations of test scores are necessary to ensure the validity of the inferences you made in parts a and b?

(d) What is the observed significance level of the test you conducted in part a?

(e) The researchers hypothesized that the distribution of test scores for males is more variable than the distribution for females. Test this claim at $\alpha = .05$.

1.108. Visual search study. When searching for an item (e.g., a roadside traffic sign, a lost earring, or a tumor in a mammogram), common sense dictates that you will not reexamine items previously rejected. However, researchers at Harvard Medical School found that a visual search has no memory (*Nature*, August 6, 1998). In their experiment, nine subjects searched for the letter "T" mixed among several letters "L." Each subject conducted the search under two conditions: random and static. In the random condition, the location of the letters were changed every 111 milliseconds; in the static condition, the location of the letters remained unchanged. In each trial, the reaction time (i.e., the amount of time it took the subject to locate the target letter) was recorded in milliseconds.

(a) One goal of the research is to compare the mean reaction times of subjects in the two experimental conditions. Explain why the data should be analyzed as a paired-difference experiment.

(b) If a visual search has no memory, then the main reaction times in the two conditions will not differ. Specify H_0 and H_a for testing the "no memory" theory.

(c) The test statistic was calculated as $t = 1.52$ with p-value $= .15$. Make the appropriate conclusion.

🔵 MILK

1.109. Detection of rigged school milk prices. Each year, the state of Kentucky invites bids from dairies to supply half-pint containers of fluid milk products for its school districts. In several school districts in northern Kentucky (called the "tri-county" market), two suppliers—Meyer Dairy and Trauth Dairy—were accused of price-fixing, that is, conspiring to allocate the districts so that the winning bidder was predetermined and the price per pint was set above the competitive price. These two dairies were the only two bidders on the milk contracts in the "tri-county market" between 1983 and 1991. (In contrast, a large number of different dairies won the milk contracts for school districts in the remainder of the northern Kentucky market—called the "surrounding" market.) Did Meyer and Trauth conspire to rig their bids in the tri-county market? If so, economic theory states that the mean winning price in the rigged tri-county market will be higher than the mean winning price in the competitive surrounding market. Data on all bids received from the dairies competing for the milk contracts between 1983 and 1991 are saved in the MILK file.

(a) A MINITAB printout of the comparison of mean bid prices for whole white milk for the two Kentucky milk markets is shown below. Is there support for the claim that the dairies in the tri-county market participated in collusive practices? Explain in detail.

(b) In competitive sealed bid markets, vendors do not share information about their bids. Consequently, more dispersion or variability among the bids is typically observed than in collusive markets, where vendors communicate about their bids and have a tendency to submit bids in close proximity to one another in an attempt to make the bidding appear competitive. If collusion exists in the tri-county milk market, the variation in winning bid prices in the surrounding ("competitive") market will be significantly larger than the corresponding variation in the tri-county ("rigged") market. A MINITAB analysis of the whole white milk data in the MILK file yielded the

MINITAB Output for Exercise 1.109(a)

Two-Sample T-Test and CI: WWBID, Market

```
Two-sample T for WWBID

Market     N    Mean    StDev   SE Mean
SURROUND   254  0.1331  0.0158  0.00099
TRI-COUNTY 100  0.1431  0.0133  0.0013


Difference = mu (SURROUND) - mu (TRI-COUNTY)
Estimate for difference: -0.009970
95% upper bound for difference: -0.007232
T-Test of difference = 0 (vs <): T-Value = -6.02   P-Value = 0.000   DF = 213
```

MINITAB Output for Exercise 1.109(b)

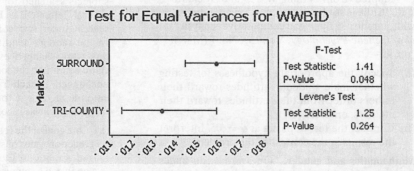

printout at the bottom of p. 78. Is there evidence that the bid price variance for the surrounding market exceeds the bid price variance for the tri-county market?

1.110. Alzheimer's and homophone spelling. A *homophone* is a word whose pronunciation is the same as that of another word having a different meaning and spelling (e.g., *nun* and *none*, *doe* and *dough*, etc.). *Brain and Language* (April 1995) reported on a study of homophone spelling in patients with Alzheimer's disease. Twenty Alzheimer's patients were asked to spell 24 homophone pairs given in random order, then the number of homophone confusions (e.g., spelling *doe* given the context, *bake bread dough*) was recorded for each patient. One year later, the same test was given to the same patients. The data for the study are provided in the table. The researchers posed the following question: "Do Alzheimer's patients show a significant increase in mean homophone confusion errors over time?" Perform an analysis of the data to answer the researchers' question. Use the relevant information in the SAS printout. What assumptions are necessary for the procedure used to be valid? Are they satisfied?

HOMOPHONE

PATIENT	TIME 1	TIME 2
1	5	5
2	1	3
3	0	0
4	1	1
5	0	1
6	2	1
7	5	6
8	1	2
9	0	9
10	5	8
11	7	10
12	0	3
13	3	9
14	5	8
15	7	12
16	10	16
17	5	5
18	6	3
19	9	6
20	11	8

Source: Neils, J., Roeltgen, D. P., and Constantinidou, F. "Decline in homophone spelling associated with loss of semantic influence on spelling in Alzheimer's disease," *Brain and Language*, Vol. 49, No. 1, Apr. 1995, p. 36 (Table 3). Copyright © 1995, with permission from Elsevier.

```
            Two Sample Paired t-test for the Means of TIME1 and TIME2

Sample Statistics

    Group      N    Mean    Std. Dev.   Std. Error
    ----------------------------------------------------
    TIME1      20   4.15    3.4985      0.7823
    TIME2      20   5.8     4.2128      0.942

Hypothesis Test

    Null hypothesis:      Mean of (TIME1 - TIME2) => 0
    Alternative:          Mean of (TIME1 - TIME2) <  0

    t Statistic     Df      Prob > t
    ----------------------------------------
    -2.306          19      0.0163

95% Confidence Interval for the Difference between Two Paired Means

        Lower Limit    Upper Limit
        -----------    -----------
        -3.15          -0.15
```

References

Freedman, D., Pisani, R., and Purves, R. *Statistics*. New York: W. W. Norton and Co., 1978.

Mcclave, J. T., and Sincich, T. *A First Course in Statistics*, 10th ed. Upper Saddle River, N.J.: Pearson Prentice Hall, 2009.

Mendenhall, W., Beaver, R. J., and Beaver, B. M. *Introduction to Probability and Statistics*, 12th ed. N. Scituate: Duxbury, 2006.

Tanur, J. M., Mosteller, F., Kruskal, W. H., Link, R. F., Pieters, R. S., and Rising, G. R. (eds.). *Statistics: A Guide to the Unknown*. San Francisco: Holden-Day, 1989.

Tukey, J. *Exploratory Data Analysis*. Reading, Mass.: Addison-Wesley, 1977.

INTRODUCTION TO
REGRESSION ANALYSIS

Contents

Objectives

1. To explain the concept of a statistical model

2. To describe applications of regression

Many applications of inferential statistics are much more complex than the methods presented in Chapter 1. Often, you will want to use sample data to investigate the relationships among a group of variables, ultimately to create a model for some variable (e.g., IQ, grade point average, etc.) that can be used to predict its value in the future. The process of finding a mathematical model (an equation) that best fits the data is part of a statistical technique known as **regression analysis**.

2.1 Modeling a Response

Suppose the dean of students at a university wants to predict the grade point average (GPA) of all students at the end of their freshman year. One way to do this is to select a random sample of freshmen during the past year, note the GPA y of each, and then use these GPAs to estimate the true mean GPA of all freshmen. The dean could then predict the GPA of each freshman using this estimated GPA.

Predicting the GPA of every freshman by the mean GPA is tantamount to using the mean GPA as a **model** for the true GPA of each freshman enrolled at the university.

In regression, the variable y to be modeled is called the **dependent** (or **response**) **variable** and its true mean (or **expected value**) is denoted $E(y)$. In this example,

$$y = \text{GPA of a student at the end of his or her freshman year}$$

$$E(y) = \text{Mean GPA of all freshmen}$$

> **Definition 2.1** The variable to be predicted (or modeled), y, is called the **dependent** (or **response**) **variable**.

The dean knows that the actual value of y for a particular student will depend on IQ, SAT score, major, and many other factors. Consequently, the real GPAs for all freshmen may have the distribution shown in Figure 2.1. Thus, the dean is modeling the first-year GPA y for a particular student by stating that y is equal to the mean GPA $E(y)$ of all freshmen plus or minus some random amount, which is unknown to the dean; that is,

$$y = E(y) + \text{Random error}$$

Since the dean does not know the value of the random error for a particular student, one strategy would be to predict the freshman's GPA with the estimate of the mean GPA $E(y)$.

This model is called a **probabilistic model** for y. The adjective *probabilistic* comes from the fact that, when certain assumptions about the model are satisfied, we can make a probability statement about the magnitude of the deviation between y and $E(y)$. For example, if y is normally distributed with mean 2.5 grade points and standard deviation .5 grade point (as shown in Figure 2.1), then the probability that y will fall within 2 standard deviations (i.e., 1 grade point) of its mean is .95. The probabilistic model shown in the box is the foundation of all models considered in this text.

In practice, we will need to use sample data to estimate the parameters of the probabilistic model—namely, the mean $E(y)$ and the random error ε. In Chapter 3, we learn a standard assumption in regression: the mean error is 0. Based on this assumption, our best estimate of ε is 0. Thus, we need only estimate $E(y)$.

General Form of Probabilistic Model in Regression

$$y = E(y) + \varepsilon$$

where
$y = $ Dependent variable
$E(y) = $ Mean (or expected) value of y
$\varepsilon = $ Unexplainable, or random, error

The simplest method of estimating $E(y)$ is to use the technique in Section 1.8. For example, the dean could select a random sample of freshmen students during the past year and record the GPA y of each. The sample mean $\bar{y}$ could be used as an estimate of the true mean GPA, $E(y)$. If we denote the predicted value of y as $\hat{y}$, the prediction equation for the simple model is

$$\hat{y} = \bar{y}$$

Therefore, with this simple model, the sample mean GPA $\bar{y}$ is used to predict the true GPA y of any student at the end of his or her freshman year.

Figure 2.1 Distribution of freshmen GPAs

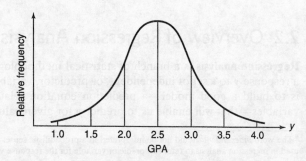

Figure 2.2 Relating GPA
of a freshman to SAT score

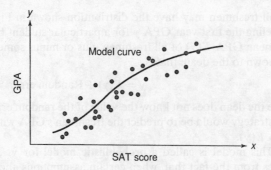

Unfortunately, this simple model does not take into consideration a number
of variables, called **independent variables**,* that are highly related to a freshman's
GPA. Logically, a more accurate model can be obtained by using the independent
variables (e.g., IQ, SAT score, major, etc.) to estimate $E(y)$. The process of finding
the mathematical model that relates y to a set of independent variables and best fits
the data is part of the process known as **regression analysis**.

> **Definition 2.2** The variables used to predict (or model) y are called **indepen-
> dent variables** and are denoted by the symbols x_1, x_2, x_3, etc.

For example, suppose the dean decided to relate freshman GPA y to a single
independent variable x, defined as the student's SAT score. The dean might select
a random sample of freshmen, record y and x for each, and then plot them on a
graph as shown in Figure 2.2. Finding the equation of the smooth curve that best
fits the data points is part of a regression analysis. Once obtained, this equation (a
graph of which is superimposed on the data points in Figure 2.2) provides a model
for estimating the mean GPA for freshmen with any specific SAT score. The dean
can use the model to predict the GPA of any freshman as long as the SAT score
for that freshman is known. As you can see from Figure 2.2, the model would also
predict with some error (most of the points do not lie exactly on the curve), but
the error of prediction will be much less than the error obtained using the model
represented in Figure 2.1. As shown in Figure 2.1, a good estimate of GPA for a
freshman student would be a value near the center of the distribution, say, the mean.
Since this prediction does not take SAT score into account, the error of prediction
will be larger than the error of prediction for the model in Figure 2.2. Consequently,
we would state that the model utilizing information provided by the independent
variable, SAT score, is superior to the model represented in Figure 2.1.

2.2 Overview of Regression Analysis

Regression analysis is a branch of statistical methodology concerned with relating
a response y to a set of independent, or predictor, variables $x_1, x_2, \ldots, x_k$. The goal
is to build a good model—a prediction equation relating y to the independent
variables—that will enable us to predict y for given values of $x_1, x_2, \ldots, x_k$, and to

---

* The word *independent* should not be interpreted in a probabilistic sense. The phrase *independent variable* is
used in regression analysis to refer to a predictor variable for the response y.

do so with a small error of prediction. When using the model to predict y for a particular set of values of $x_1, x_2, \ldots, x_k$, we will want a measure of the reliability of our prediction. That is, we will want to know how large the error of prediction might be. All these elements are parts of a regression analysis, and the resulting prediction equation is often called a **regression model**.

For example, a property appraiser might like to relate percentage price increase y of residential properties to the two quantitative independent variables x_1, square footage of heated space, and x_2, lot size. This model could be represented by a **response surface** (see Figure 2.3) that traces the mean percentage price increase $E(y)$ for various combinations of x_1 and x_2. To predict the percentage price increase y for a given residential property with $x_1 = 2,000$ square feet of heated space and lot size $x_2 = .7$ acre, you would locate the point $x_1 = 2,000$, $x_2 = .7$ on the x_1, x_2-plane (see Figure 2.3). The height of the surface above that point gives the mean percentage increase in price $E(y)$, and this is a reasonable value to use to predict the percentage price increase for a property with $x_1 = 2,000$ and $x_2 = .7$.

The response surface is a convenient method for modeling a response y that is a function of two quantitative independent variables, x_1 and x_2. The mathematical equivalent of the response surface shown in Figure 2.3 might be given by the deterministic model

$$E(y) = \beta_0 + \beta_1 x_1 + \beta_2 x_2 + \beta_3 x_1 x_2 + \beta_4 x_1^2 + \beta_5 x_2^2$$

where $E(y)$ is the mean percentage price increase for a set of values x_1 and x_2, and $\beta_0, \beta_1, \ldots, \beta_5$ are constants (or weights) with values that would have to be estimated from the sample data. Note that the model for $E(y)$ is deterministic because, if the constants $\beta_0, \beta_1, \ldots, \beta_5$ are known, the values of x_1 and x_2 determine exactly the value of $E(y)$.

Replacing $E(y)$ with $\beta_0 + \beta_1 x_1 + \beta_2 x_2 + \beta_3 x_1 x_2 + \beta_4 x_1^2 + \beta_5 x_2^2$ in the probabilistic model for y, we obtain the full equation for y:

$$y = \beta_0 + \beta_1 x_1 + \beta_2 x_2 + \beta_3 x_1 x_2 + \beta_4 x_1^2 + \beta_5 x_2^2 + \varepsilon$$

Now the property appraiser would obtain a sample of residential properties and record square footage, x_1, and lot size, x_2, in addition to percentage increase y in assessed value (see Section 2.4). Subjecting the sample data to a regression analysis will yield estimates of the model parameters and enable the appraiser to predict

Figure 2.3 Mean percentage price increase as a function of heated square footage, x_1, and lot size, x_2

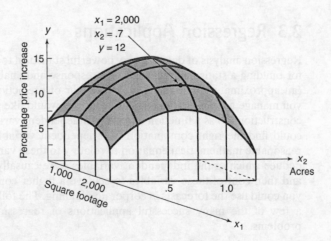

percentage increase y for a particular property. The prediction equation takes the form

$$\hat{y} = \hat{\beta}_0 + \hat{\beta}_1 x_1 + \hat{\beta}_2 x_2 + \hat{\beta}_3 x_1 x_2 + \hat{\beta}_4 x_1^2 + \hat{\beta}_5 x_2^2$$

where $\hat{y}$ is the predicted value of y, and $\hat{\beta}_0, \hat{\beta}_1, \ldots, \hat{\beta}_5$ are estimates of the model parameters.

In practice, the appraiser would construct a deterministic model for $E(y)$ that takes into account other quantitative variables, as well as qualitative independent variables, such as location and type of construction. In the following chapters, we show how to construct a model relating a response to both quantitative and qualitative independent variables, and we fit the model to a set of sample data using a regression analysis.

The preceding description of regression analysis is oversimplified, but it provides a preliminary view of the methodology that is the subject of this text. In addition to predicting y for specific values of $x_1, x_2, \ldots, x_k$, a regression model can be used to estimate the mean value of y for given values of $x_1, x_2, \ldots, x_k$ and to answer other questions concerning the relationship between y and one or more of the independent variables. The practical values attached to these inferences are illustrated by examples in the following chapters.

We conclude this section with a summary of the major steps involved in a regression analysis.

Regression Modeling: Six-Step Procedure

1. Hypothesize the form of the model for $E(y)$.
2. Collect the sample data.
3. Use the sample data to estimate unknown parameters in the model.
4. Specify the probability distribution of the random error term, and estimate any unknown parameters of this distribution.
5. Statistically check the usefulness of the model.
6. When satisfied that the model is useful, use it for prediction, estimation, and so on.

2.3 Regression Applications

Regression analysis of data is a very powerful statistical tool. It provides a technique for building a statistical predictor of a response and enables you to place a bound (an approximate upper limit) on your error of prediction. For example, suppose you manage a construction company and you would like to predict the profit y per construction job as a function of a set of independent variables $x_1, x_2, \ldots, x_k$. If you could find the right combination of independent variables and could postulate a reasonable mathematical equation to relate y to these variables, you could possibly deduce which of the independent variables were causally related to profit per job and then control these variables to achieve a higher company profit. In addition, you could use the forecasts in corporate planning. The following examples illustrate a few of the many successful applications of regression analysis to real-world problems.

Example 2.1	**Science (and a short history of regression)**

While studying natural inheritance in 1886, scientist Francis Galton collected data on heights of parents and adult children. He noticed the tendency for tall (or short) parents to have tall (or short) children, but not as tall (or short) on average as their parents. Galton called this phenomenon the "Law of Universal Regression" for the average heights of adult children tended to "regress" to the mean of the population. With the help of his friend and disciple Karl Pearson, Galton modeled a son's adult height (y) as a function of mid-parent height (x), and the term *regression model* was coined. ▪

Example 2.2	**Psychology**

"Moonlighters" are workers who hold two jobs at the same time. What are the factors that impact the likelihood of a moonlighting worker becoming aggressive toward his or her supervisor? This was the research question of interest in the *Journal of Applied Psychology* (July 2005). Based on data collected from a large sample of moonlighters, the researchers fit several multiple regression models for supervisor-directed aggression score (y). The most important predictors of y were determined to be age, gender, level of self-esteem, history of aggression, level of supervisor abuse, and perception of injustice at either job. ▪

Example 2.3	**Geography**

Can the population of an urban area be estimated without taking a census? In *Geographical Analysis* (January 2007) geography professors at the University of Wisconsin–Milwaukee and Ohio State University demonstrated the use of satellite image maps for estimating urban population. A portion of Columbus, Ohio, was partitioned into $n = 125$ census block groups and satellite imagery was obtained. Multiple regression was used to successfully model population density (y) as a function of proportion of block with low-density residential areas (x_1) and proportion of block with high-density residential areas (x_2). ▪

Example 2.4	**Music**

Writing in *Chance* (Fall 2004), University of Konstanz (Germany) statistics professor Jan Beran demonstrated that certain aspects of music can be described by a quantitative regression model. For famous compositions ranging from the 13th to the 20th century, the information content y of a musical composition (called *entropy*, and measured as how many times a certain pitch occurs) was found to be linearly related to the year of birth, x, of the composer. ▪

Example 2.5	**Accounting**

A study of Machiavellian traits (e.g., negative character traits such as manipulation, cunning, duplicity, deception, and bad faith) in accountants was published in *Behavioral Research in Accounting* (January 2008). Multiple regression was employed to model the Machiavellian ("Mach") rating score of an accountant as a function of the independent variables age, gender, education, and income. Of these, only income was found to have a significant impact on Mach rating score. ▪

Example 2.6

Engineering

During periods of high electricity demand, especially during the hot summer months, the power output from a gas turbine engine can drop dramatically. One way to counter this drop in power is by cooling the inlet air to the gas turbine. An increasingly popular cooling method uses high pressure inlet fogging. The performance of a gas turbine augmented with high pressure inlet fogging was investigated in the *Journal of Engineering for Gas Turbines and Power* (January 2005). One key performance variable, the heat rate (kilojoules per kilowatt per hour), was discovered to be related to several independent variables, including cycle speed (revolutions per minute), inlet temperature (°C), exhaust gas temperature (°C), cycle pressure ratio, and air mass flow rate (kilograms per second).

Example 2.7

Management

Do chief executive officers (CEOs) and their top managers always agree on the goals of the company? Goal importance congruence between CEOs and vice presidents (VPs) was studied in the *Academy of Management Journal* (February 2008). The researchers used regression to model a VP's attitude toward the goal of improving efficiency (y) as a function of the two independent variables, level of CEO leadership (x_1) and level of congruence between the CEO and the VP (x_2). They discovered that the impact of CEO leadership on a VP's attitude toward improving efficiency depended on level of congruence.

Example 2.8

Law

For over 20 years, courts have accepted evidence of "battered woman syndrome" as a defense in homicide cases. An article published In the *Duke Journal of Gender Law and Policy* (Summer 2003) examined the impact of expert testimony on the outcome of homicide trials that involve battered woman syndrome. Based on data collected on individual juror votes from past trials, the article reported that "when expert testimony was present, women jurors were more likely than men to change a verdict from not guilty to guilty after deliberations." This result was obtained from a multiple regression model for likelihood of changing a verdict from not guilty to guilty after deliberations, y, as a function of juror gender (male or female) and expert testimony (yes or no).

Example 2.9

Education

The Standardized Admission Test (SAT) scores of 3,492 high school and college students, some of whom paid a private tutor in an effort to obtain a higher score, were analyzed in *Chance* (Winter 2001). Multiple regression was used to successfully estimate the effect of coaching on the SAT-Mathematics score, y. The independent variables included in the model were scores on PSAT, whether the student was coached, student ethnicity, socioeconomic status, overall high school GPA, number of mathematics courses taken in high school, and overall GPA for the math courses.

Example 2.10

Mental Health

The degree to which clients of the Department of Mental Health and Addiction Services in Connecticut adjust to their surrounding community was investigated in the *Community Mental Health Journal* (August 2000). Multiple regression analysis was used to model the dependent variable, community adjustment y (measured quantitatively, where lower scores indicate better adjustment). The model contained a total of 21 independent variables categorized as follows: demographic (four variables), diagnostic (seven variables), treatment (four variables), and community (six variables). ▪

2.4 Collecting the Data for Regression

Recall from Section 2.2 that the initial step in regression analysis is to hypothesize a deterministic model for the mean response, $E(y)$, as a function of one or more independent variables. Once a model for $E(y)$ has been hypothesized, the next step is to collect the sample data that will be used to estimate the unknown model parameters (β's). This entails collecting observations on both the response y and the independent variables, $x_1, x_2, \ldots, x_k$, for each experimental unit in the sample. Thus, a sample to be analyzed by regression includes observations on several variables $(y, x_1, x_2, \ldots, x_k)$, not just a single variable.

The data for regression can be of two types: **observational** or **experimental**. Observational data are obtained if no attempt is made to control the values of the independent variables (x's). For example, suppose you want to model an executive's annual compensation y. One way to obtain the data for regression is to select a random sample of $n = 100$ executives and record the value of y and the values of each of the predictor variables. Data for the first five executives in the sample are displayed in Table 2.1. [Note that in this example, the x values, such as experience, college education, number of employees supervised, and so on, for each executive are not specified in advance of observing salary y; that is, the x values were uncontrolled. Therefore, the sample data are observational.]

> **Definition 2.3**   If the values of the independent variables (x's) in regression are uncontrolled (i.e., not set in advance before the value of y is observed) but are measured without error, the data are **observational**.

Table 2.1   Observational data for five executives

	Executive				
	1	2	3	4	5
Annual compensation, y ($\$$)	85,420	61,333	107,500	59,225	98,400
Experience, x_1 (years)	8	2	7	3	11
College education, x_2 (years)	4	8	6	7	2
No. of employees supervised, x_3	13	6	24	9	4
Corporate assets, x_4 (millions, $\$$)	1.60	0.25	3.14	0.10	2.22
Age, x_5 (years)	42	30	53	36	51
Board of directors, x_6 (1 if yes, 0 if no)	0	0	1	0	1
International responsibility, x_7 (1 if yes, 0 if no)	1	0	1	0	0

How large a sample should be selected when regression is applied to observational data? In Section 1.8, we learned that when estimating a population mean, the sample size n will depend on (1) the (estimated) population standard deviation, (2) the confidence level, and (3) the desired half-width of the confidence interval used to estimate the mean. Because regression involves estimation of the mean response, $E(y)$, the sample size will depend on these three factors. The problem, however, is not as straightforward as that in Section 1.8, since $E(y)$ is modeled as a function of a set of independent variables, and the additional parameters in the model (i.e., the β's) must also be estimated. In regression, the sample size should be large enough so that the β's are both estimable and testable. This will not occur unless n is at least as large as the number of β parameters included in the model for $E(y)$. To ensure a sufficiently large sample, a good rule of thumb is to select n greater than or equal to 10 times the number of β parameters in the model.

For example, suppose a consulting firm wants to use the following model for annual compensation, y, of a corporate executive:

$$E(y) = \beta_0 + \beta_1 x_1 + \beta_2 x_2 + \cdots + \beta_7 x_7$$

where $x_1, x_2, \ldots, x_7$ are defined in Table 2.1. Excluding β_0, there are seven β parameters in the model; thus, the firm should include at least $10 \times 7 = 70$ corporate executives in its sample.

The second type of data in regression, experimental data, are generated by designed experiments where the values of the independent variables are set in advance (i.e., controlled) before the value of y is observed. For example, if a production supervisor wants to investigate the effect of two quantitative independent variables, say, temperature x_1 and pressure x_2, on the purity of batches of a chemical, the supervisor might decide to employ three values of temperature ($100°C$, $125°C$, and $150°C$) and three values of pressure (50, 60, and 70 pounds per square inch) and to produce and measure the impurity y in one batch of chemical for each of the $3 \times 3 = 9$ temperature–pressure combinations (see Table 2.2). For this experiment, the settings of the independent variables are controlled, in contrast to the uncontrolled nature of observational data in the real estate sales example.

Definition 2.4 If the values of the independent variables (x's) in regression are controlled using a designed experiment (i.e., set in advance before the value of y is observed), the data are **experimental**.

Table 2.2 Experimental data

Temperature, x_1	Pressure, x_2	Impurity, y
100	50	2.7
	60	2.4
	70	2.9
125	50	2.6
	60	3.1
	70	3.0
150	50	1.5
	60	1.9
	70	2.2

In many studies, it is usually not possible to control the values of the x's; consequently, most data collected for regression applications are observational. (Consider the regression analysis in Example 2.2. Clearly, it is impossible or impractical to control the values of the independent variables.) Therefore, you may want to know why we distinguish between the two types of data. We will learn (Chapter 7) that inferences made from regression studies based on observational data have more limitations than those based on experimental data. In particular, we will learn that establishing a cause-and-effect relationship between variables is much more difficult with observational data than with experimental data.

The majority of the examples and exercises in Chapters 3–10 are based on observational data. In Chapters 11–12, we describe regression analyses based on data collected from a designed experiment.

Quick Summary

KEY IDEAS

Regression analysis

uses a mathematical model to predict a variable y from values of other variables, $x_1, x_2, \ldots, x_k$

Regression variables

1. **dependent (response), y**–variable to be modeled/predicted
2. **independent, $x_1, x_2, \ldots, x_k$**—variables used to predict y

Probabilistic model

$y = E(y) + \varepsilon$, where

1. $E(y) =$ mean (expected) y is a function of $x_1, x_2, \ldots, x_k$
2. ε is **random error**

Steps in regression

1. Hypothesize the form of the model for $E(y)$
2. Collect the sample data
3. Estimate the unknown parameters in the model
4. Specify the probability distribution of ε
5. Statistically check model adequacy
6. Use the model for prediction and estimation

Types of regression data

1. **observational** (values of x's are uncontrolled)
2. **experimental** (values of x's are controlled via a *designed experiment*)

SIMPLE LINEAR REGRESSION

Contents

Objectives

1. To introduce the straight-line (*simple linear regression*) model relating a quantitative response y to a single quantitative predictor x
2. To present the concept of *correlation* and its relationship to simple linear regression.

3. To assess how well the simple linear regression model fits the sample data.
4. To use the straight-line model for making predictions.

3.1 Introduction

As noted in Chapter 2, much research is devoted to the topic of **modeling**, (i.e., trying to describe how variables are related). For example, a physician might be interested in modeling the relationship between the level of carboxyhemoglobin and the oxygen pressure in the blood of smokers. An advertising agency might want to know the relationship between a firm's sales revenue and the amount spent on advertising. And a psychologist may be interested in relating a child's age to the child's performance on a vocabulary test.

The simplest graphical model for relating a response variable y to a single independent variable x is a straight line. In this chapter, we discuss **simple linear (straight-line) models**, and we show how to fit them to a set of data points using the **method of least squares**. We then show how to judge whether a relationship exists between y and x, and how to use the model either to estimate $E(y)$, the mean value of y, or to predict a future value of y for a given value of x. The totality of these methods is called a **simple linear regression analysis**.

Most models for response variables are much more complicated than implied by a straight-line relationship. Nevertheless, the methods of this chapter are very useful, and they set the stage for the formulation and fitting of more complex models in succeeding chapters. Thus, this chapter provides an intuitive justification for the techniques employed in a regression analysis, and it identifies most of the types of inferences that we want to make using a **multiple regression analysis** later in this book.

3.2 The Straight-Line Probabilistic Model

An important consideration in merchandising a product is the amount of money spent on advertising. Suppose you want to model the monthly sales revenue y of an appliance store as a function of the monthly advertising expenditure x. The first question to be answered is this: Do you think an exact (deterministic) relationship exists between these two variables? That is, can the exact value of sales revenue be predicted if the advertising expenditure is specified? We think you will agree that this is not possible for several reasons. Sales depend on many variables other than advertising expenditure—for example, time of year, state of the general economy, inventory, and price structure. However, even if many variables are included in the model (the topic of Chapter 4), it is still unlikely that we can predict the monthly sales *exactly*. There will almost certainly be some variation in sales due strictly to **random phenomena** that cannot be modeled or explained.

Consequently, we need to propose a probabilistic model for sales revenue that accounts for this random variation:

$$y = E(y) + \varepsilon$$

The random error component, ε, represents all unexplained variations in sales caused by important but omitted variables or by unexplainable random phenomena.

As you will subsequently see, the random error ε will play an important role in testing hypotheses or finding confidence intervals for the deterministic portion of the model; it will also enable us to estimate the magnitude of the error of prediction when the model is used to predict some value of y to be observed in the future.

We begin with the simplest of probabilistic models—a **first-order linear model*** that graphs as a straight line. The elements of the straight-line model are summarized in the box.

A First-Order (Straight-Line) Model

$$y = \beta_0 + \beta_1 x + \varepsilon$$

where

$y =$ **Dependent** variable (variable to be modeled—sometimes called the **response** variable)

$x =$ Independent variable (variable used as a **predictor** of y)

$E(y) = \beta_0 + \beta_1 x =$ Deterministic component

$\varepsilon =$ (epsilon) $=$ Random error component

$\beta_0 =$ (beta zero) $=$ **y-intercept** of the line, i.e., point at which the line intercepts or cuts through the y-axis (see Figure 3.1)

$\beta_1 =$ (beta one) $=$ **Slope** of the line, i.e., amount of increase (or decrease) in the mean of y for every 1-unit increase in x (see Figure 3.1)

In Section 3.4, we make the standard assumption that the average of the random errors is zero (i.e., $E(\varepsilon) = 0$). Then the deterministic component of the straight-line

---

* A general definition of the expression *first-order* is given in Section 5.3.

Figure 3.1 The straight-line model

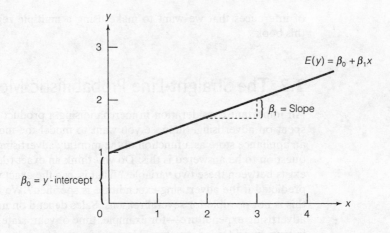

probabilistic model represents the line of means $E(y) = \beta_0 + \beta_1 x$. Note that we use Greek symbols β_0 and β_1 to represent the y-intercept and slope of the line. They are population parameters with numerical values that will be known only if we have access to the entire population of (x, y) measurements.

Recall from Section 2.2 that it is helpful to think of regression modeling as a six-step procedure:

Steps in Regression Analysis

Step 1. Hypothesize the form of the model for $E(y)$.

Step 2. Collect the sample data.

Step 3. Use the sample data to estimate unknown parameters in the model.

Step 4. Specify the probability distribution of the random error term, and estimate any unknown parameters of this distribution. Also, check the validity of each assumption made about the probability distribution.

Step 5. Statistically check the usefulness of the model.

Step 6. When satisfied that the model is useful, use it for prediction, estimation, and so on.

In this chapter, since we are dealing only with the straight-line model, we concentrate on steps 2–6. In Chapters 4 and 5, we discuss how to build more complex models.

3.2 Exercises

3.1 **Graphing lines.** In each case, graph the line that passes through the points.

(a) $(0, 2)$ and $(2, 6)$ (b) $(0, 4)$ and $(2, 6)$
(c) $(0, -2)$ and $(-1, -6)$ (d) $(0, -4)$ and $(3, -7)$

3.2 **Finding β_0 and β_1.** The equation for a straight line (deterministic) is

$$y = \beta_0 + \beta_1 x$$

If the line passes through the point $(0, 1)$, then $x = 0, y = 1$ must satisfy the equation. That is,

$$1 = \beta_0 + \beta_1(0)$$

Similarly, if the line passes through the point $(2, 3)$, then $x = 2, y = 3$ must satisfy the equation:

$$3 = \beta_0 + \beta_1(2)$$

Use these two equations to solve for β_0 and β_1, and find the equation of the line that passes through the points $(0, 1)$ and $(2, 3)$.

3.3   Finding the equation for a line.   Find the equations of the lines passing through the four sets of points given in Exercise 3.1.

3.4   Graphing lines.   Plot the following lines:

(a) $y = 3 + 2x$ (b) $y = 1 + x$ (c) $y = -2 + 3x$
(d) $y = 5x$ (e) $y = 4 - 2x$

3.5   Finding β_0 and β_1.   Give the slope and y-intercept for each of the lines defined in Exercise 3.4.

3.3   Fitting the Model: The Method of Least Squares

Suppose an appliance store conducts a 5-month experiment to determine the effect of advertising on sales revenue. The results are shown in Table 3.1. (The number of measurements is small, and the measurements themselves are unrealistically simple to avoid arithmetic confusion in this initial example.) The straight-line model is hypothesized to relate sales revenue y to advertising expenditure x. That is,

$$y = \beta_0 + \beta_1 x + \varepsilon$$

The question is this: How can we best use the information in the sample of five observations in Table 3.1 to estimate the unknown y-intercept β_0 and slope β_1?

ADSALES

Table 3.1 Appliance store data		
Month	Advertising Expenditure x, hundreds of dollars	Sales Revenue y, thousands of dollars
1	1	1
2	2	1
3	3	2
4	4	2
5	5	4

To gain some information on the approximate values of these parameters, it is helpful to plot the sample data. Such a graph, called a **scatterplot**, locates each of the five data points on a graph, as in Figure 3.2. Note that the scatterplot suggests a general tendency for y to increase as x increases. If you place a ruler on the scatterplot, you will see that a line may be drawn through three of the five points, as shown in Figure 3.3. To obtain the equation of this visually fitted line, notice that the line intersects the y-axis at $y = -1$, so the y-intercept is -1. Also, y increases exactly 1 unit for every 1-unit increase in x, indicating that the slope is $+1$. Therefore, the equation is

$$\tilde{y} = -1 + 1(x) = -1 + x$$

where $\tilde{y}$ is used to denote the predictor of y based on the visually fitted model.

One way to decide quantitatively how well a straight line fits a set of data is to determine the extent to which the data points deviate from the line. For example, to evaluate the visually fitted model in Figure 3.3, we calculate the magnitude of

Figure 3.2 Scatterplot for data in Table 3.1

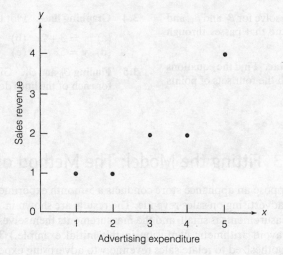

Figure 3.3 Visual straight-line fit to data in Table 3.1

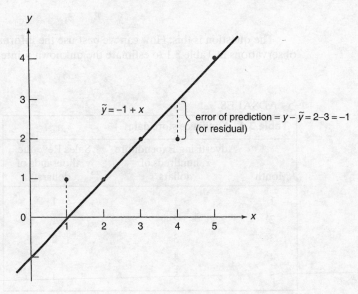

the **deviations** (i.e., the differences between the observed and the predicted values of y). These deviations or **errors of prediction**, are the vertical distances between observed and predicted values of y (see Figure 3.3). The observed and predicted values of y, their differences, and their squared differences are shown in Table 3.2. Note that the **sum of the errors (SE)** equals 0 and the **sum of squares of the errors (SSE)**, which gives greater emphasis to large deviations of the points from the line, is equal to 2.

By shifting the ruler around the graph, we can find many lines for which the sum of the errors is equal to 0, but it can be shown that there is one (and only one) line for which the SSE is a *minimum*. This line is called the **least squares line**, **regression line**, or **least squares prediction equation.**

To find the least squares line for a set of data, assume that we have a sample of n data points that can be identified by corresponding values of x and y, say, $(x_1, y_1), (x_2, y_2), \ldots, (x_n, y_n)$. For example, the $n = 5$ data points shown in Table 3.2

Table 3.2 Comparing observed and predicted values for the visual model

x	y	Prediction $\tilde{y} = -1 + x$	Error of prediction $(y - \tilde{y})$	Squared error $(y - \tilde{y})^2$
1	1	0	$(1 - 0) = 1$	1
2	1	1	$(1 - 1) = 0$	0
3	2	2	$(2 - 2) = 0$	0
4	2	3	$(2 - 3) = -1$	1
5	4	4	$(4 - 4) = 0$	0

Sum of errors (SE) = 0   Sum of squared errors (SSE) = 2

are $(1, 1)$, $(2, 1)$, $(3, 2)$, $(4, 2)$, and $(5, 4)$. The straight-line model for the response y in terms of x is

$$y = \beta_0 + \beta_1 x + \varepsilon$$

The line of means is

$$E(y) = \beta_0 + \beta_1 x$$

and the fitted line, which we hope to find, is represented as

$$\hat{y} = \hat{\beta}_0 + \hat{\beta}_1 x$$

The "hats" can be read as "estimator of." Thus, $\hat{y}$ is an estimator of the mean value of y, $E(y)$, and a predictor of some future value of y; and $\hat{\beta}_0$ and $\hat{\beta}_1$ are estimators of β_0 and β_1, respectively.

For a given data point, say, (x_i, y_i), the observed value of y is y_i and the predicted value of y is obtained by substituting x_i into the prediction equation:

$$\hat{y}_i = \hat{\beta}_0 + \hat{\beta}_1 x_i$$

The deviation of the ith value of y from its predicted value, called the ***i*th residual**, is

$$(y_i - \hat{y}_i) = [y_i - (\hat{\beta}_0 + \hat{\beta}_1 x_i)]$$

Then the sum of squares of the deviations of the y-values about their predicted values (i.e., the **sum of squares of residuals**) for all of the n data points is

$$\text{SSE} = \sum_{i=1}^{n} [y_i - (\hat{\beta}_0 + \hat{\beta}_1 x_i)]^2$$

The quantities $\hat{\beta}_0$ and $\hat{\beta}_1$ that make the SSE a minimum are called the **least squares estimates** of the population parameters β_0 and β_1, and the prediction equation $\hat{y} = \hat{\beta}_0 + \hat{\beta}_1 x$ is called the **least squares line**.

Definition 3.1 The **least squares line** is one that satisfies the following two properties:

1. $\text{SE} = \sum(y_i - \hat{y}_i) = 0$; i.e., the sum of the residuals is 0.
2. $\text{SSE} = \sum(y_i - \hat{y}_i)^2$; i.e., the sum of squared errors is smaller than for any other straight-line model with SE = 0.

The values of $\hat{\beta}_0$ and $\hat{\beta}_1$ that minimize the SSE are given by the formulas in the box.[*]

Formulas for the Least Squares Estimates

$$Slope: \hat{\beta}_1 = \frac{SS_{xy}}{SS_{xx}}$$

$$y\text{-}intercept: \hat{\beta}_0 = \bar{y} - \hat{\beta}_1\bar{x}$$

where

$$SS_{xy} = \sum_{i=1}^{n}(x_i - \bar{x})(y_i - \bar{y}) = \sum_{i=1}^{n}x_iy_i - n\bar{x}\bar{y}$$

$$SS_{xx} = \sum_{i=1}^{n}(x_i - \bar{x})^2 = \sum_{i=1}^{n}x_i^2 - n(\bar{x})^2$$

$$n = \text{Sample size}$$

Table 3.3 Preliminary computations for the advertising–sales example

x_i	y_i	x_i^2	x_iy_i
1	1	1	1
2	1	4	2
3	2	9	6
4	2	16	8
5	4	25	20
Totals: $\sum x_i = 15$	$\sum y_i = 10$	$\sum x_i^2 = 55$	$\sum x_iy_i = 37$
Means: $\bar{x} = 3$	$\bar{y} = 2$		

Preliminary computations for finding the least squares line for the advertising–sales example are given in Table 3.3. We can now calculate.[†]

$$SS_{xy} = \sum x_iy_i - n\bar{x}\bar{y} = 37 - 5(3)(2) = 37 - 30 = 7$$

$$SS_{xx} = \sum x_i^2 - n(\bar{x})^2 = 55 - 5(3)^2 = 55 - 45 = 10$$

Then, the slope of the least squares line is

$$\hat{\beta}_1 = \frac{SS_{xy}}{SS_{xx}} = \frac{7}{10} = .7$$

---

[*] Students who are familiar with calculus should note that the values of β_0 and β_1 that minimize SSE = $\sum(y_i - \hat{y}_i)^2$ are obtained by setting the two partial derivatives $\partial SSE/\partial\beta_0$ and $\partial SSE/\partial\beta_1$ equal to 0. The solutions to these two equations yield the formulas shown in the box. (The complete derivation is provided in Appendix A.) Furthermore, we denote the *sample* solutions to the equations by $\hat{\beta}_0$ and $\hat{\beta}_1$, whereas the "∧" (hat) denotes that these are sample estimates of the true population intercept β_0 and slope β_1.

[†] Since summations are used extensively from this point on, we omit the limits on $\sum$ when the summation includes all the measurements in the sample (i.e., when the summation is $\sum_{i=1}^{n}$, we write $\sum$.)

and the y-intercept is

$$\hat{\beta}_0 = \bar{y} - \hat{\beta}_1 \bar{x}$$
$$= 2 - (.7)(3) = 2 - 2.1 = -.1$$

The least squares line is then

$$\hat{y} = \hat{\beta}_0 + \hat{\beta}_1 x = -.1 + .7x$$

The graph of this line is shown in Figure 3.4.

Figure 3.4 Plot of the least squares line $\hat{y} = -.1 + .7x$

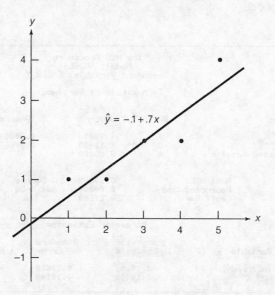

The observed and predicted values of y, the deviations of the y-values about their predicted values, and the squares of these deviations are shown in Table 3.4. Note that the sum of squares of the deviations, SSE, is 1.10, and (as we would expect) this is less than the SSE $= 2.0$ obtained in Table 3.2 for the visually fitted line.

Table 3.4 Comparing observed and predicted values for the least squares model

x	y	Predicted $\hat{y} = -.1 + .7x$	Residual (error) $(y - \hat{y})$	Squared error $(y - \hat{y})^2$
1	1	.6	$(1 - .6) = \quad .4$	.16
2	1	1.3	$(1 - 1.3) = -.3$	.09
3	2	2.0	$(2 - 2.0) = \quad 0$	.00
4	2	2.7	$(2 - 2.7) = -.7$	.49
5	4	3.4	$(4 - 3.4) = \quad .6$	.36
			Sum of errors (SE) $= 0$	SSE $= 1.10$

The calculations required to obtain $\hat{\beta}_0$, $\hat{\beta}_1$, and SSE in simple linear regression, although straightforward, can become rather tedious. Even with the use of a calculator, the process is laborious and susceptible to error, especially when the sample size is large. Fortunately, the use of statistical computer software can significantly reduce the labor involved in regression calculations. The SAS, SPSS, and MINITAB outputs for the simple linear regression of the data in Table 3.1 are displayed in Figure 3.5a–c. The values of $\hat{\beta}_0$ and $\hat{\beta}_1$ are highlighted on the printouts. These values, $\hat{\beta}_0 = -.1$ and $\hat{\beta}_1 = .7$, agree exactly with our hand-calculated values. The value of SSE = 1.10 is also highlighted on the printouts.

Whether you use a calculator or a computer, it is important that you be able to interpret the intercept and slope in terms of the data being utilized to fit the model.

Figure 3.5a SAS printout for advertising–sales regression

```
                              The REG Procedure
                                Model: MODEL1
                            Dependent Variable: SALES_Y

                              Analysis of Variance

                                        Sum of          Mean
Source                    DF           Squares         Square    F Value    Pr > F

Model                      1          4.90000        4.90000      13.36    0.0354
Error                      3          1.10000        0.36667
Corrected Total            4          6.00000


              Root MSE                0.60553     R-Square     0.8167
              Dependent Mean          2.00000     Adj R-Sq     0.7556
              Coeff Var              30.27650


                              Parameter Estimates

                            Parameter        Standard
         Variable     DF     Estimate          Error     t Value    Pr > |t|

         Intercept     1     -0.10000         0.63509     -0.16      0.8849
         ADV_X         1      0.70000         0.19149      3.66      0.0354
```

Figure 3.5b SPSS printout for advertising–sales regression

Model Summary

Model	R	R Square	Adjusted R Square	Std. Error of the Estimate
1	.904[a]	.817	.756	.606

a. Predictors: (Constant), ADV_X

ANOVA[b]

Model		Sum of Squares	df	Mean Square	F	Sig.
1	Regression	4.900	1	4.900	13.364	.035[a]
	Residual	1.100	3	.367		
	Total	6.000	4			

a. Predictors: (Constant), ADV_X

b. Dependent Variable: SALES_Y

Coefficients[a]

Model		Unstandardized Coefficients		Standardized Coefficients		
		B	Std. Error	Beta	t	Sig.
1	(Constant)	-1.00E-01	.635		-.157	.885
	ADV_X	.700	.191	.904	3.656	.035

a. Dependent Variable: SALES_Y

Figure 3.5c MINITAB printout for advertising–sales regression

```
The regression equation is
SALES_Y = - 0.100 + 0.700 ADV_X

Predictor      Coef      SE Coef        T       P
Constant     -0.1000      0.6351     -0.16   0.885
ADV_X         0.7000      0.1915      3.66   0.035

S = 0.6055      R-Sq = 81.7%      R-Sq(adj) = 75.6%

Analysis of Variance

Source          DF        SS        MS       F       P
Regression       1     4.9000    4.9000   13.36   0.035
Residual Error   3     1.1000    0.3667
Total            4     6.0000
```

In the advertising–sales example, our interpretation of the least squares slope, $\hat{\beta}_1 = .7$, is that the mean of sales revenue y will increase .7 unit for every 1-unit increase in advertising expenditure x. Since y is measured in units of $1,000 and x in units of $100, our interpretation is that mean monthly sales revenue increases $700 for every $100 increase in monthly advertising expenditure. (We will attach a measure of reliability to this inference in Section 3.6.)

The least squares intercept, $\hat{\beta}_0 = -.1$, is our estimate of mean sales revenue y when advertising expenditure is set at $x = \$0$. Since sales revenue can never be negative, why does such a nonsensical result occur? The reason is that we are attempting to use the least squares model to predict y for a value of x ($x = 0$) that is outside the range of the sample data and therefore impractical. (We have more to say about predicting outside the range of the sample data—called **extrapolation**—in Section 3.9.) Consequently, $\hat{\beta}_0$ will not always have a practical interpretation. Only when $x = 0$ is within the range of the x-values in the sample and is a practical value will $\hat{\beta}_0$ have a meaningful interpretation.

Even when the interpretations of the estimated parameters are meaningful, we need to remember that they are only estimates based on the sample. As such, their values will typically change in repeated sampling. How much confidence do we have that the estimated slope, $\hat{\beta}_1$, accurately approximates the true slope, β_1? This requires statistical inference, in the form of confidence intervals and tests of hypotheses, which we address in Section 3.6.

To summarize, we have defined the best-fitting straight line to be the one that satisfies the least squares criterion; that is, the sum of the squared errors will be smaller than for any other straight-line model. This line is called the **least squares line**, and its equation is called the **least squares prediction equation**. In subsequent sections, we show how to make statistical inferences about the model.

3.3 Exercises

3.6 Learning the mechanics. Use the method of least squares to fit a straight line to these six data points:

EX3_6

x	1	2	3	4	5	6
y	1	2	2	3	5	5

(a) What are the least squares estimates of β_0 and β_1?

(b) Plot the data points and graph the least squares line on the scatterplot.

3.7 Learning the mechanics. Use the method of least squares to fit a straight line to these five data points:

EX3_7

x	−2	−1	0	1	2
y	4	3	3	1	−1

(a) What are the least squares estimates of β_0 and β_1?

(b) Plot the data points and graph the least squares line on the scatterplot.

3.8 **Predicting home sales price.** Real estate investors, homebuyers, and homeowners often use the appraised (or market) value of a property as a basis for predicting sale price. Data on sale prices and total appraised values of 76 residential properties sold in 2008 in an upscale Tampa, Florida, neighborhood named Tampa Palms are saved in the TAMPALMS file. The first five and last five observations of the data set are listed in the accompanying table.

(a) Propose a straight-line model to relate the appraised property value x to the sale price y for residential properties in this neighborhood.

🔴 TAMPALMS

PROPERTY	MARKET VALUE (THOUS.)	SALE PRICE (THOUS.)
1	$184.44	$382.0
2	191.00	230.0
3	159.83	220.0
4	189.22	277.0
5	151.61	205.0
⋮	⋮	⋮
72	263.40	325.0
73	194.58	252.0
74	219.15	270.0
75	322.67	305.0
76	325.96	450.0

Source: Hillsborough County (Florida) Property Appraiser's Office.

MINITAB Output for Exercise 3.8

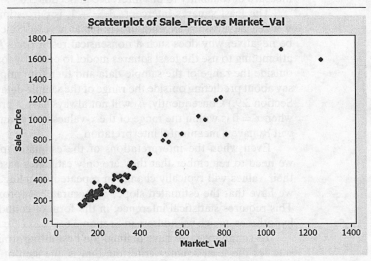

The regression equation is
Sale_Price = 1.4 + 1.41 Market_Val

Predictor	Coef	SE Coef	T	P
Constant	1.36	13.77	0.10	0.922
Market_Val	1.40827	0.03693	38.13	0.000

S = 68.7575   R-Sq = 95.2%   R-Sq(adj) = 95.1%

Analysis of Variance

Source	DF	SS	MS	F	P
Regression	1	6874024	6874024	1454.02	0.000
Residual Error	74	349842	4728		
Total	75	7223866			

(b) A MINITAB scatterplot of the data is shown on the previous page. [*Note*: Both sale price and total market value are shown in thousands of dollars.] Does it appear that a straight-line model will be an appropriate fit to the data?

(c) A MINITAB simple linear regression printout is also shown (p. 100). Find the equation of the best-fitting line through the data on the printout.

(d) Interpret the *y*-intercept of the least squares line. Does it have a practical meaning for this application? Explain.

(e) Interpret the slope of the least squares line. Over what range of *x* is the interpretation meaningful?

(f) Use the least squares model to estimate the mean sale price of a property appraised at $300,000.

3.9 Quantitative models of music. Writing in *Chance* (Fall 2004), University of Konstanz (Germany) statistics professor Jan Beran demonstrated that certain aspects of music can be described by quantitative models. For example, the information content of a musical composition (called *entropy*) can be quantified by determining how many times a certain pitch occurs. In a sample of 147 famous compositions ranging from the 13th to the 20th century, Beran computed the Z12-note entropy (*y*) and plotted it against the year of birth (*x*) of the composer. The graph is reproduced below.

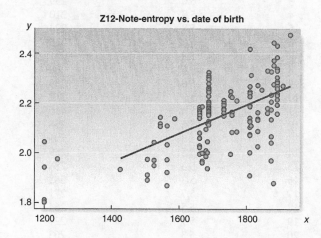

(a) Do you observe a trend, especially since the year 1400?

(b) The least squares line for the data since year 1400 is shown on the graph. Is the slope of the line positive or negative? What does this imply?

(c) Explain why the line shown is not the true line of means.

3.10 Wind turbine blade stress. Mechanical engineers at the University of Newcastle (Australia) investigated the use of timber in high-efficiency small wind turbine blades (*Wind Engineering*, January 2004). The strengths of two types of timber—radiata pine and hoop pine—were compared. Twenty specimens (called "coupons") of each timber blade were fatigue tested by measuring the stress (in MPa) on the blade after various numbers of blade cycles. A simple linear regression analysis of the data—one conducted for each type of timber—yielded the following results (where *y* = stress and *x* = natural logarithm of number of cycles):

Radiata Pine: $\hat{y} = 97.37 - 2.50x$
Hoop Pine: $\hat{y} = 122.03 - 2.36x$

(a) Interpret the estimated slope of each line

(b) Interpret the estimated *y*-intercept of each line

(c) Based on these results, which type of timber blade appears to be stronger and more fatigue resistant? Explain.

3.11 In business, do nice guys finish first or last? In baseball, there is an old saying that "nice guys finish last." Is this true in the business world? Researchers at Harvard University attempted to answer this question and reported their results in *Nature* (March 20, 2008). In the study, Boston-area college students repeatedly played a version of the game "prisoner's dilemma," where competitors choose cooperation, defection, or costly punishment. (Cooperation meant paying 1 unit for the opponent to receive 2 units; defection meant gaining 1 unit at a cost of 1 unit for the opponent; and punishment meant paying 1 unit for the opponent to lose 4 units.) At the conclusion of the games, the researchers recorded the average payoff and the number of times cooperation, defection, and punishment were used for each player. The scattergrams (p. 102) plot average payoff (*y*) against level of cooperation use, defection use, and punishment use, respectively.

(a) Consider cooperation use (*x*) as a predictor of average payoff (*y*). Based on the scattergram, is there evidence of a linear trend?

(b) Consider defection use (*x*) as a predictor of average payoff (*y*). Based on the scattergram, is there evidence of a linear trend?

(c) Consider punishment use (*x*) as a predictor of average payoff (*y*). Based on the scattergram, is there evidence of a linear trend?

(d) Refer to part c. Is the slope of the line relating punishment use (x) to average payoff (y) positive or negative?

(e) The researchers concluded that "winners don't punish." Do you agree? Explain.

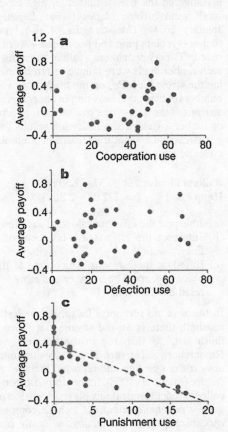

3.12 **Feeding behavior of blackbream fish.** In *Brain and Behavior Evolution* (April 2000), Zoologists conducted a study of the feeding behavior of blackbream fish. The zoologists recorded the number of aggressive strikes of two blackbream fish feeding at the bottom of an aquarium in the 10-minute period following the addition of food. The next table lists the weekly number of strikes and age of the fish (in days).

(a) Write the equation of a straight-line model relating number of strikes (y) to age of fish (x).

(b) Fit the model to the data using the method of least squares and give the least squares prediction equation.

(c) Give a practical interpretation of the value of $\hat{\beta}_0$, if possible.

(d) Give a practical interpretation of the value of $\hat{\beta}_1$, if possible.

BLACKBREAM

WEEK	NUMBER OF STRIKES	AGE OF FISH (days)
1	85	120
2	63	136
3	34	150
4	39	155
5	58	162
6	35	169
7	57	178
8	12	184
9	15	190

Source: Shand, J., et al. "Variability in the location of the retinal ganglion cell area centralis is correlated with onto-genetic changes in feeding behavior in the Blackbream, Acanthopagrus 'butcher',." *Brain and Behavior*, Vol. 55, No. 4, Apr. 2000 (Figure H).

3.13 **Sweetness of orange juice.** The quality of the orange juice produced by a manufacturer (e.g., Minute Maid, Tropicana) is constantly monitored. There are numerous sensory and chemical components that combine to make the best tasting orange juice. For example, one manufacturer has developed a quantitative index of the "sweetness" of orange juice. (The higher the index, the

OJUICE

RUN	SWEETNESS INDEX	PECTIN (ppm)
1	5.2	220
2	5.5	227
3	6.0	259
4	5.9	210
5	5.8	224
6	6.0	215
7	5.8	231
8	5.6	268
9	5.6	239
10	5.9	212
11	5.4	410
12	5.6	256
13	5.8	306
14	5.5	259
15	5.3	284
16	5.3	383
17	5.7	271
18	5.5	264
19	5.7	227
20	5.3	263
21	5.9	232
22	5.8	220
23	5.8	246
24	5.9	241

Note: The data in the table are authentic. For confidentiality reasons, the manufacturer cannot be disclosed.

sweeter the juice.) Is there a relationship between the sweetness index and a chemical measure such as the amount of water-soluble pectin (parts per million) in the orange juice? Data collected on these two variables for 24 production runs at a juice manufacturing plant are shown in the table on p. 102. Suppose a manufacturer wants to use simple linear regression to predict the sweetness (y) from the amount of pectin (x).

(a) Find the least squares line for the data.

(b) Interpret $\hat{\beta}_0$ and $\hat{\beta}_1$ in the words of the problem.

(c) Predict the sweetness index if amount of pectin in the orange juice is 300 ppm. [Note: A measure of reliability of such a prediction is discussed in Section 3.9.]

3.14 **Extending the life of an aluminum smelter pot.** An investigation of the properties of bricks used to line aluminum smelter pots was published in the *American Ceramic Society Bulletin* (February 2005). Six different commercial bricks were evaluated. The life length of a smelter pot depends on the porosity of the brick lining (the less porosity, the longer the life); consequently, the researchers measured the apparent porosity of each brick specimen, as well as the mean pore diameter of each brick. The data are given in the accompanying table

💿 SMELTPOT

BRICK	APPARENT POROSITY (%)	MEAN PORE DIAMETER (micrometers)
A	18.0	12.0
B	18.3	9.7
C	16.3	7.3
D	6.9	5.3
E	17.1	10.9
F	20.4	16.8

Source: Bonadia, P., et al. "Aluminosilicate refractories for aluminum cell linings," *American Ceramic Society Bulletin*, Vol. 84, No. 2, Feb. 2005 (Table II).

(a) Find the least squares line relating porosity (y) to mean pore diameter (x).

(b) Interpret the y-intercept of the line.

(c) Interpret the slope of the line.

(d) Predict the apparent porosity percentage for a brick with a mean pore diameter of 10 micrometers.

3.15 **Recalling names of students.** The *Journal of Experimental Psychology—Applied* (June 2000) published a study in which the "name game" was used to help groups of students learn the names of other students in the group. The "name game" requires the first student in the group to state his/her full name, the second student to say his/her name and the name of the first student, the third student to say his/her name and the names of the first two students, and so on. After making their introductions, the students listened to a seminar speaker for 30 minutes. At the end of the seminar, all students were asked to remember the full name of each of the other students in their group and the researchers measured the proportion of names recalled for each. One goal of the study was to investigate the linear trend between y = recall proportion and x = position (order) of the student during the game. The data (simulated based on summary statistics provided in the research article) for 144 students in the first eight positions are saved in the NAMEGAME2 file. The first five and last five observations in the data set are listed in the table. [Note: Since the student in position 1 actually must recall the names of all the other students, he or she is assigned position number 9 in the data set.] Use the method of least squares to estimate the line, $E(y) = \beta_0 + \beta_1 x$. Interpret the β estimates in the words of the problem.

💿 NAMEGAME2

POSITION	RECALL
2	0.04
2	0.37
2	1.00
2	0.99
2	0.79
⋮	⋮
9	0.72
9	0.88
9	0.46
9	0.54
9	0.99

Source: Morris, P.E., and Fritz, C.O. "The name game: Using retrieval practice to improve the learning of names," *Journal of Experimental Psychology—Applied*, Vol. 6, No. 2, June 2000 (data simulated from Figure 2). Copyright © 2000 American Psychological Association, reprinted with permission.

3.16 **Spreading rate of spilled liquid.** A contract engineer at DuPont Corp. studied the rate at which a spilled volatile liquid will spread across a surface (*Chemical Engineering Progress,* January 2005). Assume 50 gallons of methanol spills onto a level surface outdoors. The engineer used derived empirical formulas (assuming a state of turbulent free convection) to calculate the mass (in pounds) of the spill after a period of time ranging from 0 to 60 minutes. The calculated mass values are given in the next table. Do the data indicate that the mass of the spill tends to diminish as time increases? If so, how much will the mass diminish each minute?

🔵 LIQUIDSPILL

TIME (minutes)	MASS (pounds)	TIME (minutes)	MASS (pounds)
0	6.64	22	1.86
1	6.34	24	1.60
2	6.04	26	1.37
4	5.47	28	1.17
6	4.94	30	0.98
8	4.44	35	0.60
10	3.98	40	0.34
12	3.55	45	0.17
14	3.15	50	0.06
16	2.79	55	0.02
18	2.45	60	0.00
20	2.14		

Source: Barry, J. "Estimating rates of spreading and evaporation of volatile liquids," *Chemical Engineering Progress*, Vol. 101, No. 1, Jan. 2005.

3.4 Model Assumptions

In the advertising–sales example presented in Section 3.3, we assumed that the probabilistic model relating the firm's sales revenue y to advertising dollars x is

$$y = \beta_0 + \beta_1 x + \varepsilon$$

Recall that the least squares estimate of the deterministic component of the model $\beta_0 + \beta_1 x$ is

$$\hat{y} = \hat{\beta}_0 + \hat{\beta}_1 x = -.1 + .7x$$

Now we turn our attention to the random component ε of the probabilistic model and its relation to the errors of estimating β_0 and β_1. In particular, we will see how the probability distribution of ε determines how well the model describes the true relationship between the dependent variable y and the independent variable x.

We make four basic assumptions about the general form of the probability distribution of ε:

Assumption 1   The mean of the probability distribution of ε is 0. That is, the average of the errors over an infinitely long series of experiments is 0 for each setting of the independent variable x. This assumption implies that the mean value of y, $E(y)$, for a given value of x is $E(y) = \beta_0 + \beta_1 x$.

Assumption 2   The variance of the probability distribution of ε is constant for all settings of the independent variable x. For our straight-line model, this assumption means that the variance of ε is equal to a constant, say, σ^2, for all values of x.

Assumption 3   The probability distribution of ε is normal.

Assumption 4   The errors associated with any two different observations are independent. That is, the error associated with one value of y has no effect on the errors associated with other y values.

The implications of the first three assumptions can be seen in Figure 3.6, which shows distributions of errors for three particular values of x, namely, x_1, x_2, and x_3. Note that the relative frequency distributions of the errors are normal, with a mean of 0, and a constant variance σ^2 (all the distributions shown have the same amount

Figure 3.6 The probability distribution of ε

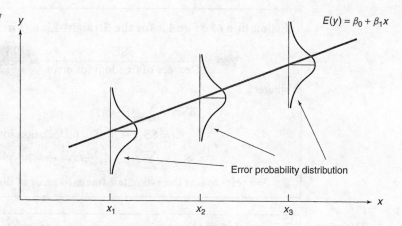

Error probability distribution

$E(y) = \beta_0 + \beta_1 x$

of spread or variability). A point that lies on the straight line shown in Figure 3.6 represents the mean value of y for a given value of x. We denote this mean value as $E(y)$. Then, the line of means is given by the equation

$$E(y) = \beta_0 + \beta_1 x$$

These assumptions make it possible for us to develop measures of reliability for the least squares estimators and to develop hypothesis tests for examining the utility of the least squares line. Various diagnostic techniques exist for checking the validity of these assumptions, and these diagnostics suggest remedies to be applied when the assumptions appear to be invalid. Consequently, *it is essential that we apply these diagnostic tools in every regression analysis.* We discuss these techniques in detail in Chapter 8. In actual practice, the assumptions need not hold exactly for least squares estimators and test statistics (to be described subsequently) to possess the measures of reliability that we would expect from a regression analysis. The assumptions will be satisfied adequately for many applications encountered in the real world.

3.5 An Estimator of σ^2

It seems reasonable to assume that the greater the variability of the random error ε (which is measured by its variance σ^2), the greater will be the errors in the estimation of the model parameters β_0 and β_1, and in the error of prediction when $\hat{y}$ is used to predict y for some value of x. Consequently, you should not be surprised, as we proceed through this chapter, to find that σ^2 appears in the formulas for all confidence intervals and test statistics that we use.

In most practical situations, σ^2 will be unknown, and we must use the data to estimate its value. The best (proof omitted) estimate of σ^2 is s^2, which is obtained by dividing the sum of squares of residuals

$$\text{SSE} = \sum (y_i - \hat{y}_i)^2$$

by the number of degrees of freedom (df) associated with this quantity. We use 2 df to estimate the y-intercept and slope in the straight-line model, leaving $(n - 2)$ df for the error variance estimation (see the formulas in the box).

Estimation of σ^2 and σ for the Straight-Line (First-Order) Model

$$s^2 = \frac{\text{SSE}}{\text{Degrees of freedom for error}} = \frac{\text{SSE}}{n-2}, \quad s = \sqrt{s^2}$$

where

$$\text{SSE} = \sum (y_i - \hat{y}_i)^2$$
$$= \text{SS}_{yy} - \hat{\beta}_1 \text{SS}_{xy} \text{ (calculation formula)}$$
$$\text{SS}_{yy} = \sum (y_i - \bar{y})^2 = \sum y_i^2 - n(\bar{y})^2$$

We refer to s as the **estimated standard error of the regression model**.

Warning

When performing these calculations, you may be tempted to round the calculated values of SS_{yy}, $\hat{\beta}_1$, and SS_{xy}. Be certain to carry at least six significant figures for each of these quantities to avoid substantial errors in the calculation of the SSE.

In the advertising–sales example, we previously calculated SSE = 1.10 for the least squares line $\hat{y} = -.1 + .7x$. Recalling that there were $n = 5$ data points, we have $n - 2 = 5 - 2 = 3$ df for estimating σ^2. Thus,

$$s^2 = \frac{\text{SSE}}{n-2} = \frac{1.10}{3} = .367$$

is the estimated variance, and

$$s = \sqrt{.367} = .61$$

is the estimated standard deviation of ε.

The values of s^2 and s can also be obtained from a simple linear regression printout. The SAS printout for the advertising–sales example is reproduced in Figure 3.7. The value of s^2 is highlighted on the printout (in the **Mean Square** column in the row labeled **Error**). The value, $s^2 = .36667$, rounded to three decimal places, agrees with the one calculated using the formulas. The value of s is also highlighted in Figure 3.7 (to the right of the heading **Root MSE**). This value, $s = .60553$, agrees (except for rounding) with the calculated value.

Figure 3.7 SAS printout for advertising–sales regression

The REG Procedure
Model: MODEL1
Dependent Variable: SALES_Y

Analysis of Variance

Source	DF	Sum of Squares	Mean Square	F Value	Pr > F
Model	1	4.90000	4.90000	13.36	0.0354
Error	3	1.10000	0.36667		
Corrected Total	4	6.00000			

Root MSE	0.60553	R-Square	0.8167	
Dependent Mean	2.00000	Adj R-Sq	0.7556	
Coeff Var	30.27650			

Parameter Estimates

Variable	DF	Parameter Estimate	Standard Error	t Value	Pr > |t|
Intercept	1	-0.10000	0.63509	-0.16	0.8849
ADV_X	1	0.70000	0.19149	3.66	0.0354

You may be able to obtain an intuitive feeling for s by recalling the interpretation given to a standard deviation in Chapter 1 and remembering that the least squares line estimates the mean value of y for a given value of x. Since s measures the spread of the distribution of y-values about the least squares line and these errors are assumed to be normally distributed, we should not be surprised to find that most (about 95%) of the observations lie within $2s$ or $2(.61) = 1.22$ of the least squares line. In the words of the problem, most of the monthly sales revenue values fall within $1,220 of their respective predicted values using the least squares line. For this simple example (only five data points), all five monthly sales revenues fall within $1,220 of the least squares line.

Interpretation of s, the Estimated Standard Deviation of ε

We expect most (approximately 95%) of the observed y-values to lie within $2s$ of their respective least squares predicted values, $\hat{y}$.

How can we use the magnitude of s to judge the utility of the least squares prediction equation? Or stated another way, when is the value of s too large for the least squares prediction equation to yield useful predicted y-values? A good approach is to utilize your substantive knowledge of the variables and associated data. In the advertising–sales example, an error of prediction of $1,220 is probably acceptable if monthly sales revenue values are relatively large (e.g., $100,000). On the other hand, an error of $1,220 is undesirable if monthly sales revenues are small (e.g., $1,000–5,000). A number that will aid in this decision is the **coefficient of variation (CV)**.

Definition 3.2 The **coefficient of variation** is the ratio of the estimated standard deviation of ε to the sample mean of the dependent variable, $\bar{y}$, measured as a percentage:

$$\text{C.V.} = 100(s/\bar{y})$$

The value of CV for the advertising–sales example, highlighted on Figure 3.7, is CV = 30.3. This implies that the value of s for the least squares line is 30% of the value of the sample mean sales revenue, $\bar{y}$. As a rule of thumb, most regression analysts desire regression models with CV values of 10% or smaller (i.e., models with a value of s that is only 10% of the mean of the dependent variable). Models with this characteristic usually lead to more precise predictions. The value of s for the advertising–sales regression is probably too large to consider using the least squares line in practice.

In the remaining sections of this chapter, the value of s will be utilized in tests of model adequacy, in evaluating model parameters, and in providing measures of reliability for future predictions.

3.5 Exercises

3.17 **Learning the mechanics.** Suppose you fit a least squares line to nine data points and calculate SSE = .219.

(a) Find s^2, the estimator of the variance σ^2 of the random error term ε.
(b) Calculate s.

3.18 **Learning the mechanics.** Find SSE, s^2, and s for the least squares lines in the following exercises. Interpret the value of s.
(a) Exercise 3.6   (b) Exercise 3.7

3.19 **Quantitative models of music.** Refer to the *Chance* (Fall 2004) study on modeling a certain pitch of a musical composition, Exercise 3.9 (p. 101). Recall that the number of times (y) a certain pitch occurs—called entropy—was modeled as a straight-line function of year of birth (x) of the composer. Based on the scatterplot of the data, the standard deviation σ of the model is estimated to be $s = .1$. For a given year (x), about 95% of the actual entropy values (y) will fall within d units of their predicted values. Find the value of d.

⊙ SMELTPOT
3.20 **Extending the life of an aluminum smelter pot.** Refer to the *American Ceramic Society Bulletin* (February 2005) study of bricks that line aluminum smelter pots, Exercise 3.14 (p. 103). You fit the simple linear regression model relating brick porosity (y) to mean pore diameter (x) to the data in the SMELTPOT file.

(a) Find an estimate of the model standard deviation, σ.
(b) In Exercise 3.14d, you predicted brick porosity percentage when $x = 10$ micrometers. Use the result, part a, to estimate the error of prediction.

3.21 **Structurally deficient highway bridges.** Data on structurally deficient highway bridges is compiled by the Federal Highway Administration (FHWA) and reported in the National Bridge Inventory (NBI). For each state, the NBI lists the number of structurally deficient bridges and the total area (thousands of square feet) of the deficient bridges. The data for the 50 states (plus the District of Columbia and Puerto Rico) are saved in the FHWABRIDGE file. (The first five and last five observations are listed in the table.) For future planning and budgeting, the FHWA wants to estimate the total area of structurally deficient bridges in a state based on the number of deficient bridges.

(a) Write the equation of a straight-line model relating total area (y) to number of structurally deficient bridges (x).
(b) The model, part a, was fit to the data using Minitab as shown in the accompanying printout. Find the least squares prediction equation on the printout.
(c) List the assumptions required for the regression analysis.

(d) Locate the estimated standard error of the regression model, s, on the printout.
(e) Use the value of s to find a range where most (about 95%) of the errors of prediction will fall.

⊙ FHWABRIDGE

STATE	NUMBER	AREA (thousands of sq. ft.)
Alabama	1899	432.7
Alaska	155	60.9
Arizona	181	110.5
Arkansas	997	347.3
California	3140	5177.9
⋮		
Washington	400	502.0
West Virginia	1058	331.5
Wisconsin	1302	399.8
Wyoming	389	143.4
Puerto Rico	241	195.4

Source: Federal Highway Administration, *National Bridge Inventory.*

Regression Analysis: SDArea versus NumberSD

```
The regression equation is
SDArea = 120 + 0.346 NumberSD


Predictor    Coef    SE Coef     T       P
Constant     119.9   123.0     0.97    0.335
NumberSD     0.34560 0.06158   5.61    0.000


S = 635.187   R-Sq = 38.7%   R-Sq(adj) = 37.4%


Analysis of Variance

Source           DF      SS         MS        F       P
Regression        1   12710141   12710141   31.50   0.000
Residual Error   50   20173111     403462
Total            51   32883252
```

3.22 **Thermal characteristics of fin-tubes.** A study was conducted to model the thermal performance of integral-fin tubes used in the refrigeration and process industries (*Journal of Heat Transfer*, August 1990). Twenty-four specially manufactured integral-fin tubes with rectangular fins made of copper were used in the experiment. Vapor was released downward into each tube and the vapor-side heat transfer coefficient (based on the outside surface area of the tube) was measured.

The dependent variable for the study is the heat transfer enhancement ratio, y, defined as the ratio of the vapor-side coefficient of the fin tube to the vapor-side coefficient of a smooth tube evaluated at the same temperature. Theoretically, heat transfer will be related to the area at the top of the tube that is "unflooded" by condensation of the vapor. The data in the table are the unflooded area ratio (x) and heat transfer enhancement (y) values recorded for the 24 integral-fin tubes.

(a) Fit a least squares line to the data.
(b) Plot the data and graph the least squares line as a check on your calculations.
(c) Calculate SSE and s^2.
(d) Calculate s and interpret its value.

⊚ HEAT

UNFLOODED AREA RATIO, x	HEAT TRANSFER ENHANCEMENT, y	UNFLOODED AREA RATIO, x	HEAT TRANSFER ENHANCEMENT, y
1.93	4.4	2.00	5.2
1.95	5.3	1.77	4.7
1.78	4.5	1.62	4.2
1.64	4.5	2.77	6.0
1.54	3.7	2.47	5.8
1.32	2.8	2.24	5.2
2.12	6.1	1.32	3.5
1.88	4.9	1.26	3.2
1.70	4.9	1.21	2.9
1.58	4.1	2.26	5.3
2.47	7.0	2.04	5.1
2.37	6.7	1.88	4.6

Source: Marto, P. J., et al. "An experimental study of R-113 film condensation on horizontal integral-fin tubes." *Journal of Heat Transfer*, Vol. 112, Aug. 1990, p. 763 (Table 2).

3.6   Assessing the Utility of the Model: Making Inferences About the Slope β_1

Refer to the advertising–sales data of Table 3.1 and suppose that the appliance store's sales revenue is *completely unrelated* to the advertising expenditure. What could be said about the values of β_0 and β_1 in the hypothesized probabilistic model

$$y = \beta_0 + \beta_1 x + \varepsilon$$

if x contributes no information for the prediction of y? The implication is that the mean of y (i.e., the deterministic part of the model $E(y) = \beta_0 + \beta_1 x$), does not change as x changes. Regardless of the value of x, you always predict the same value of y. In the straight-line model, this means that the true slope, β_1, is equal to 0 (see Figure 3.8.). Therefore, to test the null hypothesis that x contributes no information for the prediction of y against the alternative hypothesis that these variables are linearly related with a slope differing from 0, we test

$$H_0: \beta_1 = 0$$

$$H_a: \beta_1 \neq 0$$

If the data support the alternative hypothesis, we conclude that x does contribute information for the prediction of y using the straight-line model [although the true relationship between $E(y)$ and x could be more complex than a straight line]. Thus, to some extent, this is a test of the utility of the hypothesized model.

Figure 3.8 Graphing the model with $\beta_1 = 0$: $y = \beta_0 + \varepsilon$

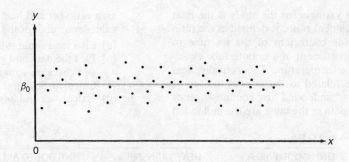

The appropriate test statistic is found by considering the sampling distribution of $\hat{\beta}_1$, the least squares estimator of the slope β_1.

Sampling Distribution of $\hat{\beta}_1$

If we make the four assumptions about ε (see Section 3.4), then the sampling distribution of $\hat{\beta}_1$, the least squares estimator of the slope, will be a normal distribution with mean β_1 (the true slope) and standard deviation

$$\sigma_{\hat{\beta}_1} = \frac{\sigma}{\sqrt{SS_{xx}}} \text{ (See Figure 3.9.)}$$

Figure 3.9 Sampling distribution of $\hat{\beta}_1$

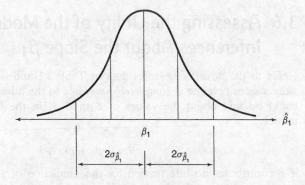

Since σ will usually be unknown, the appropriate test statistic will generally be a Student's t statistic formed as follows:

$$t = \frac{\hat{\beta}_1 - \text{Hypothesized value of } \beta_1}{s_{\hat{\beta}_1}}$$

$$= \frac{\hat{\beta}_1 - 0}{s/\sqrt{SS_{xx}}}$$

where $s_{\hat{\beta}_1} = \dfrac{s}{\sqrt{SS_{xx}}}$

Note that we have substituted the estimator s for σ, and then formed $s_{\hat{\beta}_1}$ by dividing s by $\sqrt{SS_{xx}}$. The number of degrees of freedom associated with this t statistic is the same as the number of degrees of freedom associated with s. Recall that this will be $(n-2)$ df when the hypothesized model is a straight line (see Section 3.5).

The test of the utility of the model is summarized in the next box.

Test of Model Utility: Simple Linear Regression

Test statistic: $t = \hat{\beta}_1/s_{\hat{\beta}_1} = \dfrac{\hat{\beta}_1}{s/\sqrt{SS_{xx}}}$

	ONE-TAILED TESTS		TWO-TAILED TEST
	$H_0: \beta_1 = 0$	$H_0: \beta_1 = 0$	$H_0: \beta_1 = 0$
	$H_a: \beta_1 < 0$	$H_a: \beta_1 > 0$	$H_a: \beta_1 \neq 0$
Rejection region:	$t < -t_\alpha$	$t > t_\alpha$	$|t| > t_{\alpha/2}$
p-*value:*	$P(t < t_c)$	$P(t > t_c)$	$2P(t > t_c)$ if t_c is positve
			$2P(t < t_c)$ if t_c is negative

Decision: Reject H_0 if $\alpha > p$-value, or, if test statistic falls in rejection region

where $P(t > t_\alpha) = \alpha$, $P(t > t_{\alpha/2}) = \alpha/2$, t_c = calculated value of the test statistic, the t-distribution is based on $(n - 2)$ df and $\alpha = P(\text{Type I error}) = P(\text{Reject } H_0 | H_0 \text{ true})$.

Assumptions: The four assumptions about ε listed in Section 3.4.

For the advertising–sales example, we will choose $\alpha = .05$ and, since $n = 5$, df $= (n - 2) = 5 - 2 = 3$. Then the rejection region for the two-tailed test is

$$|t| > t_{.025} = 3.182$$

We previously calculated $\hat{\beta}_1 = .7$, $s = .61$, and $SS_{xx} = 10$. Thus,

$$t = \frac{\hat{\beta}_1}{s/\sqrt{SS_{xx}}} = \frac{.7}{.61/\sqrt{10}} = \frac{.7}{.19} = 3.7$$

Since this calculated t-value falls in the upper-tail rejection region (see Figure 3.10), we reject the null hypothesis and conclude that the slope β_1 is not 0. The sample evidence indicates that advertising expenditure x contributes information for the prediction of sales revenue y using a linear model.

Figure 3.10 Rejection region and calculated t-value for testing whether the slope $\beta_1 = 0$

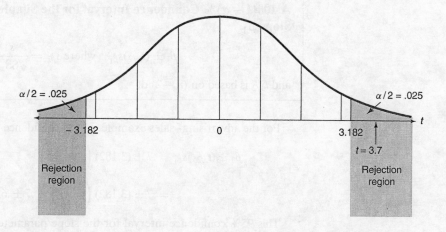

We can reach the same conclusion by using the observed significance level (*p*-value) of the test obtained from a computer printout. The SAS printout for the advertising–sales example is reproduced in Figure 3.11. The test statistic and two-tailed *p*-value are highlighted on the printout. Since *p*-value = .0354 is smaller than $\alpha = .05$, we will reject H_0.

```
                        The REG Procedure
                          Model: MODEL1
                     Dependent Variable: SALES_Y

                        Analysis of Variance

                                    Sum of          Mean
        Source             DF      Squares        Square      F Value     Pr > F

        Model               1      4.90000       4.90000       13.36      0.0354
        Error               3      1.10000       0.36667
        Corrected Total     4      6.00000


                    Root MSE            0.60553     R-Square      0.8167
                    Dependent Mean      2.00000     Adj R-Sq      0.7556
                    Coeff Var          30.27650


                           Parameter Estimates

                         Parameter      Standard
        Variable    DF    Estimate        Error     t Value   Pr > |t|     95% Confidence Limits

        Intercept    1    -0.10000       0.63509     -0.16     0.8849     -2.12112      1.92112
        ADV_X        1     0.70000       0.19149      3.66     0.0354      0.09061      1.30939
```

Figure 3.11 SAS printout for advertising–sales regression

What conclusion can be drawn if the calculated *t*-value does not fall in the rejection region? We know from previous discussions of the philosophy of hypothesis testing that such a *t*-value does *not* lead us to accept the null hypothesis. That is, we do not conclude that $\beta_1 = 0$. Additional data might indicate that β_1 differs from 0, or a more complex relationship may exist between *x* and *y*, requiring the fitting of a model other than the straight-line model. We discuss several such models in Chapter 4.

Another way to make inferences about the slope β_1 is to estimate it using a confidence interval. This interval is formed as shown in the next box.

A 100$(1 - \alpha)$% Confidence Interval for the Simple Linear Regression Slope β_1

$$\hat{\beta}_1 \pm (t_{\alpha/2})s_{\hat{\beta}_1} \quad \text{where } s_{\hat{\beta}_1} = \frac{s}{\sqrt{\text{SS}_{xx}}}$$

and $t_{\alpha/2}$ is based on $(n - 2)$ df

For the advertising–sales example, a 95% confidence interval for the slope β_1 is

$$\hat{\beta}_1 \pm (t_{.025})s_{\hat{\beta}_1} = .7 \pm (3.182)\left(\frac{s}{\sqrt{\text{SS}_{xx}}}\right)$$

$$= .7 \pm (3.182)\left(\frac{.61}{\sqrt{10}}\right) = .7 \pm .61 = (.09, 1.31)$$

This 95% confidence interval for the slope parameter β_1 is also shown (highlighted) at the bottom of the SAS printout, Figure 3.11.

Remembering that y is recorded in units of $1,000 and x in units of $100, we can say, with 95% confidence, that the mean monthly sales revenue will increase between $90 and $1,310 for every $100 increase in monthly advertising expenditure.

Since all the values in this interval are positive, it appears that β_1 is positive and that the mean of y, $E(y)$, increases as x increases. However, the rather large width of the confidence interval reflects the small number of data points (and, consequently, a lack of information) in the experiment. We would expect a narrower interval if the sample size were increased.

3.6 Exercises

3.23 Learning the mechanics. Do the data provide sufficient evidence to indicate that β_1 differs from 0 for the least squares analyses in the following exercises? Use $\alpha = .05$.
(a) Exercise 3.6   (b) Exercise 3.7

3.24 Predicting home sales price. Refer to the data on sale prices and total appraised values of 76 residential properties in an upscale Tampa, Florida, neighborhood, Exercise 3.8 (p. 100). An SPSS simple linear regression printout for the analysis is reproduced at the bottom of the page.

(a) Use the printout to determine whether there is a positive linear relationship between appraised property value x and sale price y for residential properties sold in this neighborhood. That is, determine if there is sufficient evidence (at $\alpha = .01$) to indicate that β_1, the slope of the straight-line model, is positive.

(b) Find a 95% confidence interval for the slope, β_1, on the printout. Interpret the result practically.

(c) What can be done to obtain a narrower confidence interval in part b?

3.25 Sweetness of orange juice. Refer to Exercise 3.13 (p. 102) and the simple linear regression relating the sweetness index (y) of an orange juice sample to the amount of water-soluble pectin (x) in the juice. Find a 90% confidence interval for the true slope of the line. Interpret the result.

3.26 English as a second language reading ability. What are the factors that allow a native Spanish-

speaking person to understand and read English? A study published in the *Bilingual Research Journal* (Summer 2006) investigated the relationship of Spanish (first language) grammatical knowledge to English (second language) reading. The study involved a sample of $n = 55$ native Spanish-speaking adults who were students in an English as a second language (ESL) college class. Each student took four standardized exams: Spanish grammar (SG), Spanish reading (SR), English grammar (EG), and English reading (ESLR). Simple linear regressions were used to model the ESLR score (y) as a function of each of the other exam scores (x). The results are summarized in the table.

INDEPENDENT VARIABLE(x)	p-VALUE FOR TESTING H_0: $\beta_1 = 0$
SG score	.739
SR score	.012
ER score	.022

(a) At $\alpha = .05$, is there sufficient evidence to indicate that ESLR score is linearly related to SG score?

(b) At $\alpha = .05$, is there sufficient evidence to indicate that ESLR score is linearly related to SR score?

(c) At $\alpha = .05$, is there sufficient evidence to indicate that ESLR score is linearly related to ER score?

$H_0 : \beta_1$

SPSS Output for Exercise 3.24

Two tails

		Unstandardized Coefficients		Standardized Coefficients			95% Confidence Interval for B	
Model		B	Std. Error	Beta	t	Sig.	Lower Bound	Upper Bound
1	(Constant)	1.359	13.768		.099	.922	-26.075	28.792
	Market_Val	1.408	.037	.975	38.132	.000	1.335	1.482

Coefficients[a]

a. Dependent Variable: Sale_Price

3.27 **Reaction to a visual stimulus.** How do eye and head movements relate to body movements when reacting to a visual stimulus? Scientists at the California Institute of Technology designed an experiment to answer this question and reported their results in *Nature* (August 1998). Adult male rhesus monkeys were exposed to a visual stimulus (i.e., a panel of light-emitting diodes) and their eye, head, and body movements were electronically recorded. In one variation of the experiment, two variables were measured: active head movement (x, percent per degree) and body plus head rotation (y, percent per degree). The data for $n = 39$ trials were subjected to a simple linear regression analysis, with the following results: $\hat{\beta}_1 = .88$, $s_{\hat{\beta}_1} = .14$

(a) Conduct a test to determine whether the two variables, active head movement x and body plus head rotation y, are positively linearly related. Use $\alpha = .05$.

(b) Construct and interpret a 90% confidence interval for β_1.

(c) The scientists want to know if the true slope of the line differs significantly from 1. Based on your answer to part b, make the appropriate inference.

3.28 **Massage therapy for boxers.** The *British Journal of Sports Medicine* (April 2000) published *a*

🔘 BOXING2

BLOOD LACTATE LEVEL	PERCEIVED RECOVERY
3.8	7
4.2	7
4.8	11
4.1	12
5.0	12
5.3	12
4.2	13
2.4	17
3.7	17
5.3	17
5.8	18
6.0	18
5.9	21
6.3	21
5.5	20
6.5	24

Source: Hemmings, B., Smith, M., Graydon, J., and Dyson, R. "Effects of massage on physiological restoration, perceived recovery, and repeated sports performance," *British Journal of Sports Medicine*, Vol. 34, No. 2, Apr. 2000 (data adapted from Figure 3).

study of the effect of massage on boxing performance. Two variables measured on the boxers were blood lactate concentration (mM) and the boxer's perceived recovery (28-point scale). Based on information provided in the article, the data in the table were obtained for 16 five-round boxing performances, where a massage was given to the boxer between rounds. Conduct a test to determine whether blood lactate level (y) is linearly related to perceived recovery (x). Use $\alpha = .10$.

🔘 NAMEGAME2

3.29 **Recalling names of students.** Refer to the *Journal of Experimental Psychology—Applied* (June 2000) name retrieval study, Exercise 3.15 (p. 103). Recall that the goal of the study was to investigate the linear trend between proportion of names recalled (y) and position (order) of the student (x) during the "name game." Is there sufficient evidence (at $\alpha = .01$) of a linear trend? Answer the question by analyzing the data for 144 students saved in the NAMEGAME2 file.

🔘 LIQUIDSPILL

3.30 **Spreading rate of spilled liquid.** Refer to the *Chemical Engineering Progress* (January 2005) study of the rate at which a spilled volatile liquid (methanol) will spread across a surface, Exercise 3.16 (p. 104). Consider a straight-line model relating mass of the spill (y) to elapsed time of the spill (x). Recall that the data are saved in the LIQUIDSPILL file.

(a) Is there sufficient evidence (at $\alpha = .05$) to indicate that the spill mass (y) tends to diminish linearly as time (x) increases?

(b) Give an interval estimate (with 95% confidence) of the decrease in spill mass for each minute of elapsed time.

3.31 **Pain empathy and brain activity.** Empathy refers to being able to understand and vicariously feel what others actually feel. Neuroscientists at University College London investigated the relationship between brain activity and pain-related empathy in persons who watch others in pain (*Science*, February 20, 2004). Sixteen couples participated in the experiment. The female partner watched while painful stimulation was applied to the finger of her male partner. Two variables were measured for each female: y = pain-related brain activity (measured on a scale ranging from -2 to 2) and x = score on the Empathic Concern Scale (0–25 points). The data are listed in the next table (p. 115). The research question of interest was: "Do people scoring higher in empathy show higher pain-related brain activity?" Use simple linear regression analysis to answer the research question.

🔘 BRAINPAIN

COUPLE	BRAIN ACTIVITY (y)	EMPATHIC CONCERN (x)
1	.05	12
2	−.03	13
3	.12	14
4	.20	16
5	.35	16
6	0	17
7	.26	17
8	.50	18
9	.20	18
10	.21	18
11	.45	19
12	.30	20
13	.20	21
14	.22	22
15	.76	23
16	.35	24

Source: Singer, T. et al. "Empathy for pain involves the affective but not sensory components of pain," *Science*, Vol. 303, Feb. 20, 2004 (data adapted from Figure 4).

🔘 HEAT

3.32 Thermal characteristics of fin-tubes. Refer to the *Journal of Heat Transfer* study of the straight-line relationship between heat transfer enhancement (y) and unflooded area ratio (x), Exercise 3.22 (p. 109). Construct a 95% confidence interval for β_1, the slope of the line. Interpret the result.

3.33 Does elevation impact hitting performance in baseball? The Colorado Rockies play their major league home baseball games in Coors Field, Denver. Each year, the Rockies are among the leaders in team batting statistics (e.g., home runs, batting average, and slugging percentage). Many baseball experts attribute this phenomenon to the "thin air" of Denver—called the "mile-high" city due to its elevation. *Chance* (Winter 2006) investigated the effects of elevation on slugging percentage in Major League Baseball. Data were compiled on players' composite slugging percentage at each of 29 cities for the 2003 season, as well as each city's elevation (feet above sea level). The data are saved in the MLBPARKS file. (Selected observations are shown in the table). Consider a straight-line model relating slugging percentage (y) to elevation (x).

(a) The model was fit to the data using MINITAB, with the results shown in the accompanying printout. Locate the estimates of the model parameters on the printout.

(b) Is there sufficient evidence (at $\alpha = .01$) of a positive linear relationship between elevation (x) and slugging percentage (y)? Use the p-value shown on the printout to make the inference.

(c) Construct a scatterplot for the data and draw the least squares line on the graph. Locate the data point for Denver on the graph. What do you observe?

(d) Remove the data point for Denver from the data set and refit the straight-line model to the remaining data. Repeat parts a and b. What conclusions can you draw about the "thin air" theory from this analysis?

🔘 MLBPARKS (Selected observations)

CITY	SLUG PCT.	ELEVATION
Anaheim	.480	160
Arlington	.605	616
Atlanta	.530	1050
Baltimore	.505	130
Boston	.505	20
Denver	.625	5277
Seattle	.550	350
San Francisco	.510	63
St. Louis	.570	465
Tampa	.500	10
Toronto	.535	566

Source: Schaffer, J. & Heiny, E.L. "The effects of elevation on slugging percentage in Major League Baseball," *Chance*, Vol. 19, No. 1, Winter 2006 (adapted from Figure 2).

Regression Analysis: SLUGPCT versus ELEVATION

```
The regression equation is
SLUGPCT = 0.515 + 0.000021 ELEVATION

Predictor      Coef      SE Coef      T       P
Constant     0.515140    0.007954   64.76   0.000
ELEVATION    0.00002074  0.00000719  2.89   0.008

S = 0.0369803   R-Sq = 23.6%   R-Sq(adj) = 20.7%

Analysis of Variance

Source          DF      SS        MS       F      P
Regression       1    0.011390  0.011390  8.33   0.008
Residual Error  27    0.036924  0.001368
Total           28    0.048314
```

3.7 The Coefficient of Correlation

The claim is often made that the crime rate and the unemployment rate are "highly correlated." Another popular belief is that IQ and academic performance are "correlated." Some people even believe that the Dow Jones Industrial Average and the lengths of fashionable skirts are "correlated." Thus, the term *correlation* implies a relationship or "association" between two variables.

The **Pearson product moment correlation coefficient** r, defined in the box, provides a quantitative measure of the strength of the linear relationship between x and y, just as does the least squares slope $\hat{\beta}_1$. However, unlike the slope, the correlation coefficient r is *scaleless*. The value of r is always between -1 and $+1$, regardless of the units of measurement used for the variables x and y.

Definition 3.3 The **Pearson product moment coefficient of correlation r** is a measure of the strength of the *linear* relationship between two variables x and y. It is computed (for a sample of n measurements on x and y) as follows:

$$r = \frac{SS_{xy}}{\sqrt{SS_{xx} \, SS_{yy}}}$$

Note that r is computed using the same quantities used in fitting the least squares line. Since both r and $\hat{\beta}_1$ provide information about the utility of the model, it is not surprising that there is a similarity in their computational formulas. In particular, note that SS_{xy} appears in the numerators of both expressions and, since both denominators are always positive, r and $\hat{\beta}_1$ will always be of the same sign (either both positive or both negative).

A value of r near or equal to 0 implies little or no linear relationship between y and x. In contrast, the closer r is to 1 or -1, the stronger the linear relationship between y and x. And, if $r = 1$ or $r = -1$, all the points fall exactly on the least squares line. Positive values of r imply that y increases as x increases; negative values imply that y decreases as x increases. Each of these situations is portrayed in Figure 3.12.

We demonstrate how to calculate the coefficient of correlation r using the data in Table 3.1 for the advertising–sales example. The quantities needed to calculate r are SS_{xy}, SS_{xx}, and SS_{yy}. The first two quantities have been calculated previously and are repeated here for convenience:

$$SS_{xy} = 7, SS_{xx} = 10, SS_{yy} = \sum y^2 - \frac{\left(\sum y\right)^2}{n}$$

$$= 26 - \frac{(10)^2}{5} = 26 - 20 = 6$$

We now find the coefficient of correlation:

$$r = \frac{SS_{xy}}{\sqrt{SS_{xx} SS_{yy}}} = \frac{7}{\sqrt{(10)(6)}} = \frac{7}{\sqrt{60}} = .904$$

The fact that r is positive and near 1 in value indicates that monthly sales revenue y tends to increase as advertising expenditures x increases—*for this sample*

Figure 3.12 Values of *r* and their implications

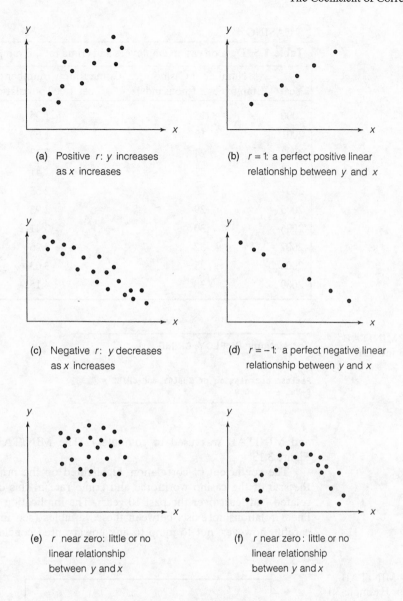

(a) Positive *r*: *y* increases as *x* increases

(b) *r* = 1: a perfect positive linear relationship between *y* and *x*

(c) Negative *r*: *y* decreases as *x* increases

(d) *r* = −1: a perfect negative linear relationship between *y* and *x*

(e) *r* near zero: little or no linear relationship between *y* and *x*

(f) *r* near zero: little or no linear relationship between *y* and *x*

of five months. This is the same conclusion we reached when we found the calculated value of the least squares slope to be positive.

Example 3.1

Legalized gambling is available on several riverboat casinos operated by a city in Mississippi. The mayor of the city wants to know the correlation between the number of casino employees and yearly crime rate. The records for the past 10 years are examined, and the results listed in Table 3.5 are obtained. Find and interpret the coefficient of correlation *r* for the data.

Solution

Rather than use the computing formula given in Definition 3.3, we resort to a statistical software package. The data of Table 3.5 were entered into a computer

⊙CASINO

Table 3.5 Data on casino employees and crime rate, Example 3.1

Year	Number of Casino Employees x (thousands)	Crime Rate y (number of crimes per 1,000 population)
2000	15	1.35
2001	18	1.63
2002	24	2.33
2003	22	2.41
2004	25	2.63
2005	29	2.93
2006	30	3.41
2007	32	3.26
2008	35	3.63
2009	38	4.15

Figure 3.13 MINITAB correlation printout for Example 3.1

Correlations: EMPLOY, CRIME

Pearson correlation of EMPLOY and CRIME = 0.987
P-Value = 0.000

---

and MINITAB was used to compute r. The MINITAB printout is shown in Figure 3.13.

The coefficient of correlation, highlighted on the printout, is $r = .987$. Thus, the size of the casino workforce and crime rate in this city are very highly correlated—at least over the past 10 years. The implication is that a strong positive linear relationship exists between these variables (see Figure 3.14). We must be careful, however, not to jump to any unwarranted conclusions. For instance, the

Figure 3.14 MINITAB scatterplot for Example 3.1

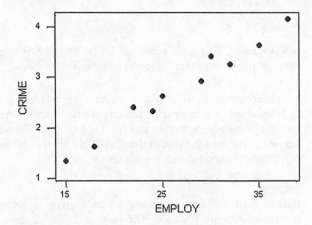

mayor may be tempted to conclude that hiring more casino workers next year will increase the crime rate—that is, that there is a *causal relationship* between the two variables. However, high correlation does not imply causality. The fact is, many things have probably contributed both to the increase in the casino workforce and to the increase in crime rate. The city's tourist trade has undoubtedly grown since legalizing riverboat casinos and it is likely that the casinos have expanded both in services offered and in number. *We cannot infer a causal relationship on the basis of high sample correlation. When a high correlation is observed in the sample data, the only safe conclusion is that a linear trend may exist between x and y.* Another variable, such as the increase in tourism, may be the underlying cause of the high correlation between x and y.   ▬

Warning

> High correlation does *not* imply causality. If a large positive or negative value of the sample correlation coefficient r is observed, it is incorrect to conclude that a change in x causes a change in y. The only valid conclusion is that *a linear trend may exist* between x and y.

Keep in mind that the correlation coefficient r measures the correlation between x-values and y-values in the sample, and that a similar linear coefficient of correlation exists for the population from which the data points were selected. The **population correlation coefficient** is denoted by the symbol ρ (rho). As you might expect, ρ is estimated by the corresponding sample statistic, r. Or, rather than estimating ρ, we might want to test

$$H_0: \rho = 0$$

against

$$H_a: \rho \neq 0$$

That is, we might want to test the hypothesis that x contributes no information for the prediction of y, using the straight-line model against the alternative that the two variables are at least linearly related. However, we have already performed this identical test in Section 3.6 when we tested $H_0: \beta_1 = 0$ against $H_a: \beta_1 \neq 0$.

It can be shown (proof omitted) that $r = \hat{\beta}_1 \sqrt{SS_{xx}/SS_{yy}}$. Thus, $\hat{\beta}_1 = 0$ implies $r = 0$, and vice versa. Consequently, the null hypothesis $H_0: \rho = 0$ is equivalent to the hypothesis $H_0: \beta_1 = 0$. When we tested the null hypothesis $H_0: \beta_1 = 0$ in connection with the previous example, the data led to a rejection of the null hypothesis for $\alpha = .05$. This implies that the null hypothesis of a zero linear correlation between the two variables, crime rate and number of employees, can also be rejected at $\alpha = .05$. The only real difference between the least squares slope $\hat{\beta}_1$ and the coefficient of correlation r is the measurement scale.* Therefore, the information they provide about the utility of the least squares model is to some extent redundant. Furthermore, the slope $\hat{\beta}_1$ gives us additional information on the amount of increase (or decrease) in y for every 1-unit increase in x. For this reason, we recommend using the slope to make inferences about the existence of a positive or negative linear relationship between two variables.

For those who prefer to test for a linear relationship between two variables using the coefficient of correlation r, we outline the procedure in the following box.

---

* The estimated slope, $\hat{\beta}_1$, is measured in the same units as y. However, the correlation coefficient r is independent of scale.

Test of Hypothesis for Linear Correlation

Test statistic: $t = r\sqrt{n-2}/\sqrt{1 - r^2}$

	ONE-TAILED TESTS		TWO-TAILED TEST
	$H_0: \rho = 0$	$H_0: \rho = 0$	$H_0: \rho = 0$
	$H_a: \rho < 0$	$H_a: \rho > 0$	$H_a: \rho \neq 0$
Rejection region:	$t < -t_\alpha$	$t > t_\alpha$	$|t| > t_{\alpha/2}$
p-*value*:	$P(t < t_c)$	$P(t > t_c)$	$2P(t > t_c)$ if t_c is positve
			$2P(t < t_c)$ if t_c is negative

Decision: Reject H_0 if $\alpha > p$-value or, if test statistic falls in rejection region

where $P(t > t_\alpha) = \alpha$, $P(t > t_{\alpha/2}) = \alpha/2$, $t_c =$ calculated value of the test statistic, the *t*-distribution is based on $(n - 2)$ df and $\alpha = P(\text{Type I error}) = P(\text{Reject } H_0| H_0 \text{ true})$.

Assumptions: The sample of (x, y) values is randomly selected from a normal population.

The next example illustrates how the correlation coefficient r may be a misleading measure of the strength of the association between x and y in situations where the true relationship is nonlinear.

Example 3.2

Underinflated or overinflated tires can increase tire wear and decrease gas mileage. A manufacturer of a new tire tested the tire for wear at different pressures, with the results shown in Table 3.6. Calculate the coefficient of correlation r for the data. Interpret the result.

 TIRES

Table 3.6 Data for Example 3.2

Pressure	Mileage
x, pounds per sq. inch	y, thousands
30	29.5
30	30.2
31	32.1
31	34.5
32	36.3
32	35.0
33	38.2
33	37.6
34	37.7
34	36.1
35	33.6
35	34.2
36	26.8
36	27.4

Figure 3.15 SPSS correlation analysis of tire data

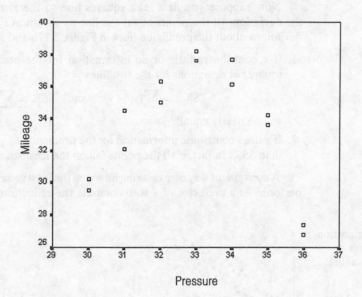

Correlations

		PRESSURE	MILEAGE
PRESSURE	Pearson Correlation	1	-.114
	Sig. (2-tailed)	.	.699
	N	14	14
MILEAGE	Pearson Correlation	-.114	1
	Sig. (2-tailed)	.699	.
	N	14	14

Figure 3.16 SPSS scatterplot of tire data

Solution

Again, we use a computer to find the value of r. An SPSS printout of the correlation analysis is shown in Figure 3.15. The value of r, shaded on the printout, is $r = -.114$. This relatively small value for r describes a weak linear relationship between pressure (x) and mileage (y).

The p-value for testing H_0: $\rho = 0$ against H_a: $\rho \neq 0$ is also shaded on the SPSS printout. This value (.699) indicates that there is no evidence of a linear correlation in the population at $\alpha = .05$. The manufacturer, however, would be remiss in concluding that tire pressure has little or no impact on wear of the tire. On the contrary, the relationship between pressure and wear is fairly strong, as the SPSS scatterplot in Figure 3.16 illustrates. Note that the relationship is not linear, but curvilinear; the underinflated tires (low pressure values) and overinflated tires (high pressure values) *both* lead to low mileage. ▬

A statistic related to the coefficient of correlation is defined and discussed in the next section.

3.8 The Coefficient of Determination

Another way to measure the utility of the regression model is to quantify the contribution of x in predicting y. To do this, we compute how much the errors of prediction of y were reduced by using the information provided by x.

To illustrate, suppose a sample of data produces the scatterplot shown in Figure 3.17a. If we assume that x contributes no information for the prediction of y, the best prediction for a value of y is the sample mean $\bar{y}$, which graphs as the horizontal line shown in Figure 3.17b. The vertical line segments in Figure 3.17b are the deviations of the points about the mean $\bar{y}$. Note that the sum of squares of deviations for the model $\hat{y} = \bar{y}$ is

$$SS_{yy} = \sum(y_i - \bar{y})^2$$

Now suppose you fit a least squares line to the same set of data and locate the deviations of the points about the line as shown in Figure 3.17c. Compare the deviations about the prediction lines in Figure 3.17b and 3.17c. You can see that:

1. If x contributes little or no information for the prediction of y, the sums of squares of deviations for the two lines

$$SS_{yy} = \sum(y_i - \bar{y})^2 \quad \text{and} \quad SSE = \sum(y_i - \hat{y}_i)^2$$

will be nearly equal.

2. If x does contribute information for the prediction of y, then SSE will be smaller than SS_{yy}. In fact, if all the points fall on the least squares line, then SSE = 0.

A convenient way of measuring how well the least squares equation $\hat{y} = \hat{\beta}_0 + \hat{\beta}_1 x$ performs as a predictor of y is to compute the reduction in the sum of squares of

Figure 3.17 A comparison of the sum of squares of deviations for two models

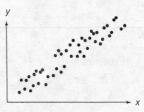

(a) Scatterplot of data

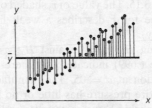

(b) Assumption: x contributes no information for predicting y; $\hat{y} = \bar{y}$

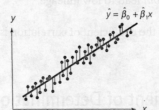

(c) Assumption: x contributes information for predicting y; $\hat{y} = \hat{\beta}_0 + \hat{\beta}_1 x$

deviations that can be attributed to x, expressed as a proportion of SS_{yy}. This quantity, called the **coefficient of determination** (and denoted), is

$$r^2 = \frac{SS_{yy} - SSE}{SS_{yy}}$$

In simple linear regression, it can be shown that this quantity is equal to the square of the simple linear coefficient of correlation r.

> **Definition 3.4**   The **coefficient of determination** is
>
> $$r^2 = \frac{SS_{yy} - SSE}{SS_{yy}} = 1 - \frac{SSE}{SS_{yy}}$$
>
> It represents the proportion of the sum of squares of deviations of the y-values about their mean that can be attributed to a linear relationship between y and x. (In simple linear regression, it may also be computed as the square of the coefficient of correlation r.)

Note that r^2 is always between 0 and 1, because $SSE \leq SS_{yy}$. Thus, an r^2 of .60 means that the sum of squares of deviations of the y-values about their predicted values has been reduced 60% by using the least squares equation $\hat{y}$, instead of $\bar{y}$, to predict y.

A more practical interpretation of r^2 is derived as follows. If we let SS_{yy} represent the "total sample variability" of the y-values around the mean, and let SSE represent the "unexplained sample variability" after fitting the least squares line, $\hat{y}$, then $(SS_{yy} - SSE)$ is the "explained sample variability" of the y-values attributable to the linear relationship with x. Therefore, a verbal description of r^2 is:

$$r^2 = \frac{(SS_{yy} - SSE)}{SS_{yy}} = \frac{\text{Explained sample variability}}{\text{Total sample variability}}$$

= Proportion of total sample variability of the y-values explained by the linear relationship between y and x.

> **Practical Interpretation of the Coefficient of Determination, r^2**
>
> About $100(r^2)\%$ of the sample variation in y (measured by the total sum of squares of deviations of the sample y-values about their mean $\bar{y}$) can be explained by (or attributed to) using x to predict y in the straight-line model.

Example 3.3

Calculate the coefficient of determination for the advertising–sales example. The data are repeated in Table 3.7.

Solution
We first calculate

$$SS_{yy} = \sum y_i^2 - n\bar{y}^2 = 26 - 5(2)^2$$
$$= 26 - 20 = 6$$

From previous calculations,

$$SSE = \sum (y_i - \hat{y}_i)^2 = 1.10$$

Table 3.7

Advertising Expenditure x, hundreds of dollars	Sales Revenue y, thousands of dollars
1	1
2	1
3	2
4	2
5	4

Then, the coefficient of determination is given by

$$r^2 = \frac{SS_{yy} - SSE}{SS_{yy}} = \frac{6.0 - 1.1}{6.0} = \frac{4.9}{6.0}$$

$$= .817$$

This value is also shown (highlighted) on the SPSS printout, Figure 3.18. Our interpretation is: About 82% of the sample variation in sales revenue values can be "explained" by using monthly advertising expenditure x to predict sales revenue y with the least squares line

$$\hat{y} = -.1 + .7x$$

Figure 3.18 Portion of SPSS printout for advertising–sales regression

Model Summary

Model	R	R Square	Adjusted R Square	Std. Error of the Estimate
1	.904[a]	.817	.756	.606

a. Predictors: (Constant), ADV_X

In situations where a straight-line regression model is found to be a statistically adequate predictor of y, the value of r^2 can help guide the regression analyst in the search for better, more useful models.

For example, a developer of an eight-story condominium complex in Florida used a simple linear model to relate the auction price (y) of a condo unit to the floor height (x) of the unit. Based on data collected for 106 units sold at auction, the least squares prediction equation given at the top of Figure 3.19 was derived. The analysis also concluded that auction price and floor height are linearly related, since the t statistic for testing H_0: $\beta_1 = 0$ was $t = -3.44$ and the associated p-value was .001. Thus, floor height should be useful for predicting the price of a condo unit sold at auction. However, the value of the coefficient of determination r^2 was found to be .102, indicating that only about 10% of the sample variation in prices is accounted for by the differences in floor heights of the units. This relatively small r^2 value led the developer to consider other independent variables (e.g., unit's distance from the elevator and view of the ocean) in the model to account for a significant portion of the remaining 90% of the variation in auction prices not explained by floor height. (See the "Condo Case" in Chapter 8.) In the next chapter, we discuss this important aspect of relating a response y to more than one independent variable.

Figure 3.19 MINITAB graph of simple linear model relating price (y) to floor height (x)

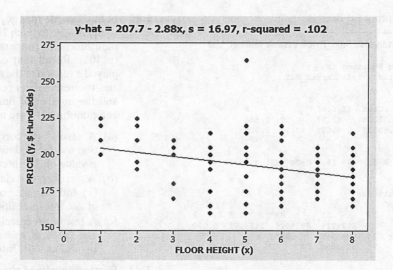

y-hat = 207.7 - 2.88x, s = 16.97, r-squared = .102

3.8 Exercises

3.34 Learning the mechanics. Describe the slope of the least squares line if

(a) $r = .7$ (b) $r = -.7$
(c) $r = 0$ (d) $r^2 = .64$

3.35 Learning the mechanics. Find the correlation coefficient and the coefficient of determination for the sample data of each of the following exercises. Interpret your results.

(a) Exercise 3.6   (b) Exercise 3.7

3.36 Examples of correlated variables. Give an example of two variables in your field of study that are

(a) positively correlated (b) negatively correlated

3.37 GPA and IQ. Do you believe that the grade point average of a college student is correlated with the student's intelligence quotient (IQ)? If so, will the correlation be positive or negative? Explain.

3.38 Crime rate and population. Research by law enforcement agencies has shown that the crime rate is correlated with the U.S. population. Would you expect the correlation to be positive or negative? Explain.

3.39 RateMyProfessors.com. A popular website among college students is RateMyProfessors.com (RMP). Established over 10 years ago, RMP allows students to post quantitative ratings of their instructors. In *Practical Assessment, Research and Evaluation* (May 2007), University of Maine researchers investigated whether instructor ratings posted on RMP are correlated with the formal

in-class student evaluations of teaching (SET) that all universities are required to administer at the end of the semester. Data collected for $n = 426$ University of Maine instructors yielded a correlation between RMP and SET ratings of .68.

(a) Give the equation of a linear model relating SET rating (y) to RMP rating (x).
(b) Give a practical interpretation of the value $r = .68$.
(c) Is the estimated slope of the line, part a, positive or negative? Explain.
(d) A test of the null hypothesis $H_0: \rho = 0$ yielded a p-value of .001. Interpret this result.
(e) Compute the coefficient of determination, r^2, for the regression analysis. Interpret the result.

TAMPALMS

3.40 Predicting home sales price. Refer to the data on sale prices and total appraised values of 76 residential properties recently sold in an upscale Tampa, Florida, neighborhood, Exercise 3.8 (p. 100). The MINITAB simple linear regression printout relating sale price (y) to appraised property (market) value (x) is reproduced on the next page, followed by a MINITAB correlation printout.

(a) Find the coefficient of correlation between appraised property value and sale price on the printout. Interpret this value.
(b) Find the coefficient of determination between appraised property value and sale price on the printout. Interpret this value.

MINITAB Output for Exercise 3.40

Regression Analysis: Sale_Price versus Market_Val

```
The regression equation is
Sale_Price = 1.4 + 1.41 Market_Val


Predictor      Coef  SE Coef      T      P
Constant       1.36    13.77   0.10  0.922
Market_Val  1.40827  0.03693  38.13  0.000


S = 68.7575    R-Sq = 95.2%    R-Sq(adj) = 95.1%


Analysis of Variance

Source          DF       SS       MS       F      P
Regression       1  6874024  6874024  1454.02  0.000
Residual Error  74   349842     4728
Total           75  7223866
```

Correlations: Sale_Price, Market_Val

```
Pearson correlation of Sale_Price and Market_Val = 0.975
P-Value = 0.000
```

3.41 Recognizing rotating objects. *Perception and Psychophysics* (July 1998) reported on a study of how people view the three-dimensional objects projected onto a rotating two-dimensional image. Each in a sample of 25 university students viewed various depth-rotated objects (e.g., hairbrush, duck, shoe) until they recognized the object. The recognition exposure time—that is, the minimum time (in milliseconds) required for the subject to recognize the object—was recorded for each. In addition, each subject rated the "goodness of view" of the object on a numerical scale, where lower scale values correspond to better views. The next table gives the correlation coefficient, r, between recognition exposure time and goodness of view for several different rotated objects,

OBJECT	r	t
Piano	.447	2.40
Bench	−.057	−.27
Motorbike	.619	3.78
Armchair	.294	1.47
Teapot	.949	14.50

(a) Interpret the value of r for each object.
(b) Calculate and interpret the value of r^2 for each object.
(c) The table also includes the t-value for testing the null hypothesis of no correlation (i.e., for testing $H_0: \beta_1 = 0$). Interpret these results.

3.42 In business, do nice guys finish first or last? Refer to the *Nature* (March 20, 2008) study of the use of punishment in cooperation games, Exercise 3.11 (p. 101). Recall that college students repeatedly played a version of the game "prisoner's dilemma" and the researchers recorded the average payoff and the number of times cooperation, defection, and punishment were used for each player.

(a) A test of no correlation between cooperation use (x) and average payoff (y) yielded a p-value of .33. Interpret this result.
(b) A test of no correlation between defection use (x) and average payoff (y) yielded a p-value of .66. Interpret this result.
(c) A test of no correlation between punishment use (x) and average payoff (y) yielded a p-value of .001. Interpret this result.

3.43 Physical activity of obese young adults. In a study published in the *International Journal of Obesity* (January 2007), pediatric researchers measured overall physical activity for two groups of young adults: 13 obese adults and 15 normal weight adults. The researchers recorded the total number of registered movements (counts) of each young adult over a period of time. *Baseline* physical activity was then computed as the number of counts per minute (cpm). Four years later, the physical activity measurements were taken again, called physical activity *at follow-up*.

(a) For the 13 obese young adults, the researchers reported a correlation of $r = .50$ between baseline and follow-up physical activity, with an associated p-value of .07. Give a practical interpretation of this correlation coefficient and p-value.
(b) Refer to part a. Construct a possible scatterplot for the 13 data points that would yield a value of $r = .50$.
(c) Refer to part a. Compute and interpret the coefficient of determination, r^2, relating baseline and follow-up physical activity for obese young adults in a simple linear regression.
(d) For the 15 normal weight young adults, the researchers reported a correlation of $r = -.12$ between baseline and follow-up physical activity, with an associated p-value of .66. Give a practical interpretation of this correlation coefficient and p-value.
(e) Refer to part d. Construct a possible scatterplot for the 15 data points that would yield a value of $r = -.12$.
(f) Refer to part d. Compute and interpret the coefficient of determination, r^2, relating

baseline and follow-up physical activity for normal weight young adults in a simple linear regression.

🐭 NAMEGAME2

3.44 Recalling student names. Refer to the *Journal of Experimental Psychology—Applied* (June 2000) name retrieval study, Exercise 3.15 (p. 103). Find and interpret the values of r and r^2 for the simple linear regression relating the proportion of names recalled (y) and position (order) of the student (x) during the "name game."

🐭 BOXING2

3.45 Massage therapy for boxers. Refer to the *British Journal of Sports Medicine* (April 2000) study of the effect of massage on boxing performance, Exercise 3.28 (p. 114). Find and interpret the values of r and r^2 for the simple linear regression relating the blood lactate concentration and the boxer's perceived recovery.

3.46 Snow geese feeding trial. Botanists at the University of Toronto conducted a series of experiments to investigate the feeding habits of baby snow geese (*Journal of Applied Ecology*, Vol. 32, 1995). Goslings were deprived of food until their guts were empty, then were allowed to feed for 6 hours on a diet of plants or Purina Duck Chow. For each of 42 feeding trials, the change in the weight of the gosling after 2.5 hours was recorded as a percentage of initial weight. Two other variables recorded were digestion efficiency (measured as a percentage) and amount of acid-detergent fiber in the digestive tract (also measured as a percentage). The data for selected feeding trials are listed in the table below.

(a) The botanists were interested in the correlation between weight change (y) and digestion efficiency (x). Plot the data for these two variables in a scatterplot. Do you observe a trend?

(b) Find the coefficient of correlation relating weight change y to digestion efficiency x. Interpret this value.

(c) Conduct a test to determine whether weight change y is correlated with a digestion efficiency x. Use $\alpha = .01$.

(d) Repeat parts b and c, but exclude the data for trials that used duck chow from the analysis. What do you conclude?

(e) The botanists were also interested in the correlation between digestion efficiency y and acid-detergent fiber x. Repeat parts a–d for these two variables.

3.47 Pain tolerance study. A study published in *Psychosomatic Medicine* (March/April 2001) explored the relationship between reported severity of pain and actual pain tolerance in 337 patients who suffer from chronic pain. Each patient reported his/her severity of chronic pain on a 7-point scale (1 = no pain, 7 = extreme pain). To obtain a pain tolerance level, a tourniquet was applied to the arm of each patient and twisted. The maximum pain level tolerated was measured on a quantitative scale.

(a) According to the researchers, "correlational analysis revealed a small but significant inverse relationship between [actual] pain tolerance and the reported severity of chronic pain." Based on this statement, is the value of r for the 337 patients positive or negative?

(b) Suppose that the result reported in part a is significant at $\alpha = .05$. Find the approximate value of r for the sample of 337 patients. [*Hint*: Use the formula $t = r\sqrt{(n-2)}/\sqrt{(1-r^2)}$.]

🐭 SNOWGEESE (First and last five trials)

FEEDING TRIAL	DIET	WEIGHT CHANGE (%)	DIGESTION EFFICIENCY (%)	ACID-DETERGENT FIBER (%)
1	Plants	−6	0	28.5
2	Plants	−5	2.5	27.5
3	Plants	−4.5	5	27.5
4	Plants	0	0	32.5
5	Plants	2	0	32
38	Duck Chow	9	59	8.5
39	Duck Chow	12	52.5	8
40	Duck Chow	8.5	75	6
41	Duck Chow	10.5	72.5	6.5
42	Duck Chow	14	69	7

Source: Gadallah, F. L., and Jefferies, R. L. "Forage quality in brood rearing areas of the lesser snow goose and the growth of captive goslings," *Journal of Applied Biology*, Vol. 32, No. 2, 1995, pp. 281–282 (adapted from Figures 2 and 3).

3.9 Using the Model for Estimation and Prediction

If we are satisfied that a useful model has been found to describe the relationship between sales revenue and advertising, we are ready to accomplish the original objectives for building the model: using it to estimate or to predict sales on the basis of advertising dollars spent.

The most common uses of a probabilistic model can be divided into two categories. The first is the use of the model for **estimating the mean value of y, $E(y)$, for a specific value of x.** For our example, we may want to estimate the mean sales revenue for *all* months during which $400 ($x = 4$) is spent on advertising. The second use of the model entails **predicting a particular y value for a given x.** That is, if we decide to spend $400 next month, we want to predict the firm's sales revenue for that month.

In the case of estimating a mean value of y, we are attempting to estimate the mean result of a very large number of experiments at the given x-value. In the second case, we are trying to predict the outcome of a single experiment at the given x-value. In which of these model uses do you expect to have more success, that is, which value—the mean or individual value of y—can we estimate (or predict) with more accuracy?

Before answering this question, we first consider the problem of choosing an estimator (or predictor) of the mean (or individual) y-value. We will use the least squares model

$$\hat{y} = \hat{\beta}_0 + \hat{\beta}_1 x$$

both to estimate the mean value of y and to predict a particular value of y for a given value of x. For our example, we found

$$\hat{y} = -.1 + .7x$$

so that the estimated mean value of sales revenue for all months when $x = 4$ (advertising = 400) is

$$\hat{y} = -.1 + .7(4) = 2.7$$

or $2,700 (the units of y are thousands of dollars). The identical value is used to predict the y-value when $x = 4$. That is, both the estimated mean value and the predicted value of y equal $\hat{y} = 2.7$ when $x = 4$, as shown in Figure 3.20.

The difference in these two model uses lies in the relative accuracy of the estimate and the prediction. These accuracies are best measured by the repeated sampling errors of the least squares line when it is used as an estimator and as a predictor, respectively. These errors are given in the box.

Figure 3.20 Estimated mean value and predicted individual value of sales revenue y for $x = 4$

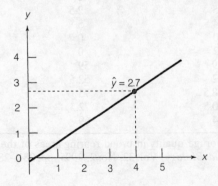

Sampling Errors for the Estimator of the Mean of y and the Predictor of an Individual y for $x = x_p$

1. The standard deviation of the sampling distribution of the estimator $\hat{y}$ of the mean value of y at a particular value of x, say, x_p, is

$$\sigma_{\hat{y}} = \sigma\sqrt{\frac{1}{n} + \frac{(x_p - \bar{x})^2}{SS_{xx}}}$$

where σ is the standard deviation of the random error ε. We refer to $\sigma_{\hat{y}}$ as the *standard error* of $\hat{y}$.

2. The standard deviation of the prediction error for the predictor $\hat{y}$ of an individual y-value for $x = x_p$ is

$$\sigma_{(y-\hat{y})} = \sigma\sqrt{1 + \frac{1}{n} + \frac{(x_p - \bar{x})^2}{SS_{xx}}}$$

where σ is the standard deviation of the random error ε. We refer to $\sigma_{(y-\hat{y})}$ as the *standard error of prediction*.

The true value of σ will rarely be known. Thus, we estimate σ by s and calculate the estimation and prediction intervals as shown in the next two boxes. The procedure is demonstrated in Example 3.4.

A $100(1 - \alpha)\%$ Confidence Interval for the Mean Value of y for $x = x_p$

$$\hat{y} \pm t_{\alpha/2} \text{ (Estimated standard deviation of } \hat{y})$$

or

$$\hat{y} \pm (t_{\alpha/2})s\sqrt{\frac{1}{n} + \frac{(x_p - \bar{x})^2}{SS_{xx}}}$$

where $t_{\alpha/2}$ is based on $(n - 2)$ df

A $100(1 - \alpha)\%$ Prediction Interval for an Individual y for $x = x_p$

$$\hat{y} \pm t_{\alpha/2} \text{ [Estimated standard deviation of } (y - \hat{y})]$$

or

$$\hat{y} \pm (t_{\alpha/2})s\sqrt{1 + \frac{1}{n} + \frac{(x_p - \bar{x})^2}{SS_{xx}}}$$

where $t_{\alpha/2}$ is based on $(n - 2)$ df

Example 3.4

Find a 95% confidence interval for mean monthly sales when the appliance store spends $400 on advertising.

Solution

For a \$400 advertising expenditure, $x_p = 4$ and, since $n = 5$, df $= n - 2 = 3$. Then the confidence interval for the mean value of y is

$$\hat{y} \pm (t_{\alpha/2})s\sqrt{\frac{1}{n} + \frac{(x_p - \bar{x})^2}{SS_{xx}}}$$

or

$$\hat{y} \pm (t_{.025})s\sqrt{\frac{1}{5} + \frac{(4 - \bar{x})^2}{SS_{xx}}}$$

Recall that $\hat{y} = 2.7$, $s = .61$, $\bar{x} = 3$, and $SS_{xx} = 10$. From Table 2 of Appendix C, $t_{.025} = 3.182$. Thus, we have

$$2.7 \pm (3.182)(.61)\sqrt{\frac{1}{5} + \frac{(4 - 3)^2}{10}} = 2.7 \pm (3.182)(.61)(.55)$$

$$= 2.7 \pm 1.1 = (1.6, 3.8)$$

We estimate, with 95% confidence, that the interval from \$1,600 to \$3,800 encloses the mean sales revenue for all months when the store spends \$400 on advertising. Note that we used a small amount of data for purposes of illustration in fitting the least squares line and that the width of the interval could be decreased by using a larger number of data points. ■

Example 3.5

Predict the monthly sales for next month if a \$400 expenditure is to be made on advertising. Use a 95% prediction interval.

Solution

To predict the sales for a particular month for which $x_p = 4$, we calculate the 95% prediction interval as

$$\hat{y} \pm (t_{\alpha/2})s\sqrt{1 + \frac{1}{n} + \frac{(x_p - \bar{x})^2}{SS_{xx}}} = 2.7 \pm (3.182)(.61)\sqrt{1 + \frac{1}{5} + \frac{(4 - 3)^2}{10}}$$

$$= 2.7 \pm (3.182)(.61)(1.14) = 2.7 \pm 2.2 = (.5, 4.9)$$

Therefore, with 95% confidence we predict that the sales next month (i.e., a month where we spend \$400 in advertising) will fall in the interval from \$500 to \$4,900. As in the case of the confidence interval for the mean value of y, the prediction interval for y is quite large. This is because we have chosen a simple example (only five data points) to fit the least squares line. The width of the prediction interval could be reduced by using a larger number of data points. ■

Both the confidence interval for $E(y)$ and the prediction interval for y can be obtained using statistical software. Figures 3.21 and 3.22 are SAS printouts showing confidence intervals and prediction intervals, respectively, for the advertising–sales example. These intervals (highlighted on the printouts) agree, except for rounding, with our calculated intervals.

Note that the confidence interval in Example 3.4 is narrower than the prediction interval in Example 3.5. Will this always be true? The answer is "yes." The error in estimating the mean value of y, $E(y)$, for a given value of x, say, x_p, is the distance between the least squares line and the true line of means, $E(y) = \beta_0 + \beta_1 x$. This

Figure 3.21 SAS printout showing 95% confidence intervals for $E(y)$

Dependent Variable: SALES_Y

Output Statistics

Obs	ADV_X	Dep Var SALES_Y	Predicted Value	Std Error Mean Predict	95% CL Mean	
1	1	1.0000	0.6000	0.4690	-0.8927	2.0927
2	2	1.0000	1.3000	0.3317	0.2445	2.3555
3	3	2.0000	2.0000	0.2708	1.1382	2.8618
4	4	2.0000	2.7000	0.3317	1.6445	3.7555
5	5	4.0000	3.4000	0.4690	1.9073	4.8927

Figure 3.22 SAS printout showing 95% prediction intervals for y

Dependent Variable: SALES_Y

Output Statistics

Obs	ADV_X	Dep Var SALES_Y	Predicted Value	Std Error Mean Predict	95% CL Predict	
1	1	1.0000	0.6000	0.4690	-1.8376	3.0376
2	2	1.0000	1.3000	0.3317	-0.8972	3.4972
3	3	2.0000	2.0000	0.2708	-0.1110	4.1110
4	4	2.0000	2.7000	0.3317	0.5028	4.8972
5	5	4.0000	3.4000	0.4690	0.9624	5.8376

error, $[\hat{y} - E(y)]$, is shown in Figure 3.23. In contrast, the error $(y_p - \hat{y})$ in predicting some future value of y is the sum of two errors—the error of estimating the mean of y, $E(y)$, shown in Figure 3.23, plus the random error that is a component of the value of y to be predicted (see Figure 3.24). Consequently, the error of predicting a particular value of y will always be larger than the error of estimating the mean value of y for a particular value of x. Note from their formulas that both the error of estimation and the error of prediction take their smallest values when $x_p = \bar{x}$. The farther x lies from $\bar{x}$, the larger will be the errors of estimation and prediction. You can see why this is true by noting the deviations for different values of x between the line of means $E(y) = \beta_0 + \beta_1 x$ and the predicted line $\hat{y} = \hat{\beta}_0 + \hat{\beta}_1 x$ shown in Figure 3.24. The deviation is larger at the extremities of the interval where the largest and smallest values of x in the data set occur.

A graph showing both the confidence limits for $E(y)$ and the prediction limits for y over the entire range of the advertising expenditure (x) values is displayed in Figure 3.25. You can see that the confidence interval is always narrower than the prediction interval, and that they are both narrowest at the mean $\bar{x}$, increasing steadily as the distance $|x - \bar{x}|$ increases. In fact, when x is selected far enough away from $\bar{x}$ so that it falls outside the range of the sample data, it is dangerous to make any inferences about $E(y)$ or y.

Figure 3.23 Error of estimating the mean value of y for a given value of x

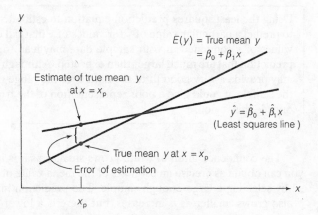

Figure 3.24 Error of predicting a future value of y for a given value of x

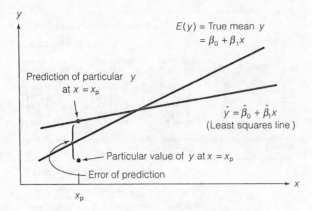

Figure 3.25 Comparison of widths of 95% confidence and prediction intervals

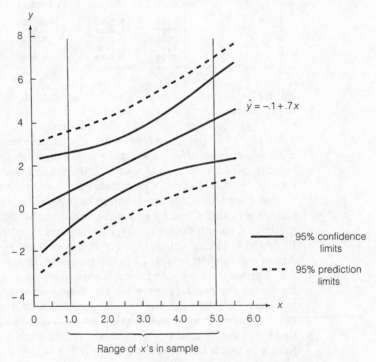

| | **Caution** | Using the least squares prediction equation to estimate the mean value of y or to predict a particular value of y for values of x that fall outside the range of the values of x contained in your sample data may lead to errors of estimation or prediction that are much larger than expected. Although the least squares model may provide a very good fit to the data over the range of x-values contained in the sample, it could give a poor representation of the true model for values of x outside this region. |

The confidence interval width grows smaller as n is increased; thus, in theory, you can obtain as precise an estimate of the mean value of y as desired (at any given x) by selecting a large enough sample. The prediction interval for a new value of y also grows smaller as n increases, but there is a lower limit on its width. If you examine the formula for the prediction interval, you will see that the interval can get no smaller than $\hat{y} \pm z_{\alpha/2}\sigma$.* Thus, the only way to obtain more accurate predictions for new values of y is to reduce the standard deviation of the regression model, σ. This can be accomplished only by improving the model, either by using a curvilinear (rather than linear) relationship with x or by adding new independent variables to the model, or both. Methods of improving the model are discussed in Chapter 4.

3.9 Exercises

3.48 Learning the mechanics.

(a) Explain why for a particular x-value, the prediction interval for an individual y-value will always be wider than the confidence interval for a mean value of y.

(b) Explain why the confidence interval for the mean value of y for a particular x-value, say, x_p, gets wider the farther x_p is from $\bar{x}$. What are the implications of this phenomenon for estimation and prediction?

3.49 Learning the mechanics. A simple linear regression analysis for $n = 20$ data points produced the following results:

$$\hat{y} = 2.1 + 3.4x \qquad SS_{xx} = 4.77$$
$$\bar{x} = 2.5 \qquad SS_{yy} = 59.21$$
$$\bar{y} = 10.6 \qquad SS_{xy} = 16.22$$

(a) Find SSE and s^2.

(b) Find a 95% confidence interval for $E(y)$ when $x = 2.5$. Interpret this interval.

(c) Find a 95% confidence interval for $E(y)$ when $x = 2.0$. Interpret this interval.

(d) Find a 95% confidence interval for $E(y)$ when $x = 3.0$. Interpret this interval.

(e) Examine the widths of the confidence intervals obtained in parts b, c, and d. What happens to the width of the confidence interval for $E(y)$ as the value of x moves away from the value of $\bar{x}$?

(f) Find a 95% prediction interval for a value of y to be observed in the future when $x = 3.0$. Interpret its value.

3.50 Predicting home sale price. Refer to the data on sale prices and total appraised values of 76 residential properties in an upscale Tampa, Florida, neighborhood, Exercise 3.8 (p. 100).

(a) In Exercise 3.8, you determined that appraised property (or market) value x and sale price y are positively related for homes sold in the Tampa Palms subdivision. Does this result guarantee that appraised value will yield accurate predictions of sale price? Explain.

(b) MINITAB was used to predict the sale price of a residential property in the subdivision with a market value of $300,000. Locate a 95% prediction interval for the sale price of this property on the printout shown below and interpret the result.

(c) Locate a 95% confidence interval for $E(y)$ on the printout and interpret the result.

```
Predicted Values for New Observations

New
Obs      Fit    SE Fit        95% CI             95% PI
  1   423.84     7.89   (408.12, 439.56)   (285.94, 561.74)


Values of Predictors for New Observations

New
Obs   Market_Val
  1         300
```

3.51 In business, do nice guys finish first or last? Refer to the *Nature* (March 20, 2008) study of the use of punishment in cooperation games, Exercises 3.11 and 3.42 (p. 126). Recall that simple linear regression was used to model a player's average payoff (y) as a straight-line function of the number of times punishment was used (x) by the player.

(a) If the researchers want to predict average payoff for a single player who used punishment 10 times, how should they proceed?

(b) If the researchers want to estimate the mean of the average payoffs for all players who used punishment 10 times, how should they proceed?

3.52 English as a second language reading ability. Refer to the *Bilingual Research Journal* (Summer 2006) study of the relationship of Spanish (first language) grammatical knowledge to English (second language) reading, Exercise 3.26 (p. 113). Recall

---

* The result follows from the facts that, for large n, $t_{\alpha/2} \approx z_{\alpha/2}$, $s \approx \sigma$, and the last two terms under the radical in the standard error of the predictor are approximately 0.

that three simple linear regressions were used to model the English reading (ESLR) score (y) as a function of Spanish grammar (SG), Spanish reading (SR), and English grammar (EG), respectively.

(a) If the researchers want to predict the ESLR score (y) of a native Spanish-speaking adult who scored $x = 50\%$ in Spanish grammar (x), how should they proceed?

(b) If the researchers want to estimate the mean ESLR score, $E(y)$, of all native Spanish-speaking adults who scored $x = 70\%$ in Spanish grammar (x), how should they proceed?

3.53 Sweetness of orange juice. Refer to the simple linear regression of sweetness index y and amount of pectin x for $n = 24$ orange juice samples, Exercise 3.13 (p. 102). The SPSS printout of the analysis is shown below (left). A 90% confidence interval for the mean sweetness index, $E(y)$, for each value of x is shown on the SPSS spreadsheet. Select an observation and interpret this interval.

NAMEGAME2

3.54 Recalling student names. Refer to the *Journal of Experimental Psychology—Applied* (June 2000) name retrieval study, Exercise 3.15 (p. 103).

(a) Find a 99% confidence interval for the mean recall proportion for students in the fifth position during the "name game." Interpret the result.

(b) Find a 99% prediction interval for the recall proportion of a particular student in the fifth position during the "name game." Interpret the result.

(c) Compare the two intervals, parts a and b. Which interval is wider? Will this always be the case? Explain.

LIQUIDSPILL

3.55 Spreading rate of spilled liquid. Refer to the *Chemical Engineering Progress* (January 2005) study of the rate at which a spilled volatile liquid will spread across a surface, Exercises 3.16 and 3.30 (p. 104). Recall that simple linear regression was used to model $y =$ mass of the spill as a function of $x =$ elapsed time of the spill

(a) Find a 90% confidence interval for the mean mass of all spills with an elapsed time of 15 minutes. Interpret the result

(b) Find a 90% prediction interval for the mass of a single spill with an elapsed time of 15 minutes. Interpret the result.

3.56 Predicting heights of spruce trees. In forestry, the diameter of a tree at breast height (which is fairly easy to measure) is used to predict the height of the tree (a difficult measurement to obtain). Silviculturists working in British Columbia's boreal forest conducted a series of spacing trials to predict the heights of several species of trees. The data in the next table are the breast height diameters (in centimeters) and heights (in meters) for a sample of 36 white spruce trees.

(a) Construct a scatterplot for the data.

(b) Assuming the relationship between the variables is best described by a straight line, use the method of least squares to estimate the y-intercept and slope of the line.

SPSS output for Exercise 3.53

Model Summary[b]

Model	R	R Square	Adjusted R Square	Std. Error of the Estimate
1	.478[a]	.229	.194	.2150

a. Predictors: (Constant), PECTIN
b. Dependent Variable: SWEET

ANOVA[b]

Model		Sum of Squares	df	Mean Square	F	Sig.
1	Regression	.301	1	.301	6.520	.018[a]
	Residual	1.017	22	.046		
	Total	1.318	23			

a. Predictors: (Constant), PECTIN
b. Dependent Variable: SWEET

Coefficients[a]

Model		Unstandardized Coefficients B	Std. Error	Standardized Coefficients Beta	t	Sig.
1	(Constant)	6.252	.237		26.422	.000
	PECTIN	-2.31E-03	.001	-.478	-2.554	.018

a. Dependent Variable: SWEET

run	sweet	pectin	lower90m	upper90m
1	5.2	220	5.64898	5.83848
2	5.5	227	5.63898	5.81613
3	6.0	259	5.57819	5.72904
4	5.9	210	5.66194	5.87173
5	5.8	224	5.64337	5.82560
6	6.0	215	5.65564	5.85493
7	5.8	231	5.63284	5.80379
8	5.6	268	5.55553	5.71011
9	5.6	239	5.61947	5.78019
10	5.9	212	5.66946	5.86497
11	5.4	410	5.05526	5.55416
12	5.6	256	5.58517	5.73592
13	5.8	306	5.43785	5.65219
14	5.5	259	5.57819	5.72904
15	5.3	284	5.50957	5.68213
16	5.3	383	5.15725	5.57694
17	5.7	271	5.54743	5.70434
18	5.5	264	5.56591	5.71821
19	5.7	227	5.63898	5.81613
20	5.3	263	5.56843	5.72031
21	5.9	232	5.63125	5.80075
22	5.8	220	5.64898	5.83848
23	5.8	246	5.60640	5.76091
24	5.9	241	5.61587	5.77454

WHITESPRUCE

BREAST HEIGHT DIAMETER x, cm	HEIGHT y, m	BREAST HEIGHT DIAMETER x, cm	HEIGHT y, m
18.9	20.0	16.6	18.8
15.5	16.8	15.5	16.9
19.4	20.2	13.7	16.3
20.0	20.0	27.5	21.4
29.8	20.2	20.3	19.2
19.8	18.0	22.9	19.8
20.3	17.8	14.1	18.5
20.0	19.2	10.1	12.1
22.0	22.3	5.8	8.0
23.6	18.9	20.7	17.4
14.8	13.3	17.8	18.4
22.7	20.6	11.4	17.3
18.5	19.0	14.4	16.6
21.5	19.2	13.4	12.9
14.8	16.1	17.8	17.5
17.7	19.9	20.7	19.4
21.0	20.4	13.3	15.5
15.9	17.6	22.9	19.2

Source: Scholz, H., Northern Lights College, British Columbia.

(c) Plot the least squares line on your scatterplot.

(d) Do the data provide sufficient evidence to indicate that the breast height diameter x contributes information for the prediction of tree height y? Test using $\alpha = .05$.

(e) Use your least squares line to find a 90% confidence interval for the average height of white spruce trees with a breast height diameter of 20 cm. Interpret the interval.

FHWABRIDGE

3.57 Structurally deficient highway bridges. Refer to the data on structurally deficient highway bridges compiled by the Federal Highway Administration (FHWA), Exercise 3.21 (p. 108). Recall that for future planning and budgeting, the FHWA wants to estimate the total area of structurally deficient bridges in a state based on the number of deficient bridges. Use the results of the simple linear regression relating total area (y) to number of structurally deficient bridges (x) to find a 95% prediction interval for y when $x = 350$ bridges. Interpret the result.

3.10 A Complete Example

In the previous sections, we have presented the basic elements necessary to fit and use a straight-line regression model. In this section, we assemble these elements by applying them in an example with the aid of computer software.

Suppose a fire safety inspector wants to relate the amount of fire damage in major residential fires to the distance between the residence and the nearest fire station. The study is to be conducted in a large suburb of a major city; a sample of 15 recent fires in this suburb is selected.

Step 1. First, we hypothesize a model to relate fire damage y to the distance x from the nearest fire station. We hypothesize a straight-line probabilistic model:

$$y = \beta_0 + \beta_1 x + \varepsilon$$

Step 2. Second, we collect the (x, y) values for each of the $n = 15$ experimental units (residential fires) in the sample. The amount of damage y and the distance x between the fire and the nearest fire station are recorded for each fire, as listed in Table 3.8.

Step 3. Next, we enter the data of Table 3.8 into a computer and use statistical software to estimate the unknown parameters in the deterministic component of the hypothesized model. The SAS printout for the simple linear regression analysis is shown in Figure 3.26.

The least squares estimates of β_0 and β_1, highlighted on the printout, are

$$\hat{\beta}_0 = 10.27793, \quad \hat{\beta}_1 = 4.91933$$

Thus, the least squares equation is (after rounding)

$$\hat{y} = 10.28 + 4.92x$$

⊙ FIREDAM

Table 3.8 Fire damage data

Distance from Fire Station x, miles	Fire Damage y, thousands of dollars
3.4	26.2
1.8	17.8
4.6	31.3
2.3	23.1
3.1	27.5
5.5	36.0
.7	14.1
3.0	22.3
2.6	19.6
4.3	31.3
2.1	24.0
1.1	17.3
6.1	43.2
4.8	36.4
3.8	26.1

This prediction equation is shown on the scatterplot, Figure 3.27.

The least squares estimate of the slope, $\hat{\beta}_1 = 4.92$, implies that the estimated mean damage increases by \$4,920 for each additional mile from the fire station. This interpretation is valid over the range of x, or from .7 to 6.1 miles from the station. The estimated y-intercept, $\hat{\beta}_0 = 10.28$, has the interpretation that a fire 0 miles from the fire station has an estimated mean damage of \$10,280. Although this would seem to apply to the fire station itself, remember that the y-intercept is meaningfully interpretable only if $x = 0$ is within the sampled range of the independent variable. Since $x = 0$ is outside the range, $\hat{\beta}_0$ has no practical interpretation.

Step 4. Now, we specify the probability distribution of the random error component ε. The assumptions about the distribution will be identical to those listed in Section 3.4:

(1) $E(\varepsilon) = 0$

(2) $\text{Var}(\varepsilon) = \sigma^2$ is constant for all x-values

(3) ε has a normal distribution

(4) ε's are independent

Although we know that these assumptions are not completely satisfied (they rarely are for any practical problem), we are willing to assume they are approximately satisfied for this example. The estimate of σ^2, shaded on the printout, is

$$s^2 = 5.36546$$

(This value is also called *mean square for error*, or *MSE*.)

Dependent Variable: DAMAGE

Analysis of Variance

Source	DF	Sum of Squares	Mean Square	F Value	Pr > F
Model	1	841.76636	841.76636	156.89	<.0001
Error	13	69.75098	5.36546		
Corrected Total	14	911.51733			

Root MSE	2.31635	R-Square	0.9235
Dependent Mean	26.41333	Adj R-Sq	0.9176
Coeff Var	8.76961		

Parameter Estimates

Variable	DF	Parameter Estimate	Standard Error	t Value	Pr > |t|	95% Confidence Limits	
Intercept	1	10.27793	1.42028	7.24	<.0001	7.20960	13.34625
DISTANCE	1	4.91933	0.39275	12.53	<.0001	4.07085	5.76781

Output Statistics

Obs	DISTANCE	Dep Var DAMAGE	Predicted Value	Std Error Mean Predict	95% CL Predict		Residual
1	3.4	26.2000	27.0037	0.5999	21.8344	32.1729	-0.8037
2	1.8	17.8000	19.1327	0.8340	13.8141	24.4514	-1.3327
3	4.6	31.3000	32.9068	0.7915	27.6186	38.1951	-1.6068
4	2.3	23.1000	21.5924	0.7112	16.3577	26.8271	1.5076
5	3.1	27.5000	25.5279	0.6022	20.3573	30.6984	1.9721
6	5.5	36.0000	37.3342	1.0573	31.8334	42.8351	-1.3342
7	0.7	14.1000	13.7215	1.1766	8.1087	19.3342	0.3785
8	3	22.3000	25.0359	0.6081	19.8622	30.2097	-2.7359
9	2.6	19.6000	23.0682	0.6550	17.8678	28.2686	-3.4682
10	4.3	31.3000	31.4311	0.7198	26.1908	36.6713	-0.1311
11	2.1	24.0000	20.6085	0.7566	15.3442	25.8729	3.3915
12	1.1	17.3000	15.6892	1.0444	10.1999	21.1785	1.6108
13	6.1	43.2000	40.2858	1.2587	34.5906	45.9811	2.9142
14	4.8	36.4000	33.8907	0.8450	28.5640	39.2175	2.5093
15	3.8	26.1000	28.9714	0.6320	23.7843	34.1585	-2.8714
16	3.5		27.4956	0.6043	22.3239	32.6672	.

Sum of Residuals		0
Sum of Squared Residuals		69.75098
Predicted Residual SS (PRESS)		93.21169

Figure 3.26 SAS printout for fire damage linear regression

Figure 3.27 Least squares model for the fire damage data

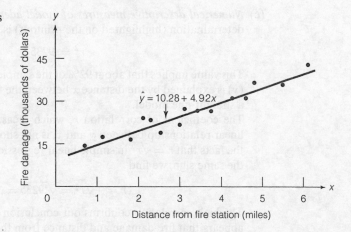

$\hat{y} = 10.28 + 4.92x$

The estimated standard deviation of ε, highlighted on the printout, is

$$s = 2.31635$$

The coefficient of variation C.V. $= 8.77\%$ (also highlighted on the printout) indicates that s is small relative to the mean fire damage $(\overline{y})$ in the sample. The value of s itself implies that most of the observed fire damage (y) values will fall within approximately $2s = 4.64$ thousand dollars of their respective predicted values when using the least squares line.

Step 5. We can now check the utility of the hypothesized model, that is, whether x really contributes information for the prediction of y using the straight-line model.

(a) *Test of model utility*: First, test the null hypothesis that the slope β_1 is 0, that is, that there is no linear relationship between fire damage and the distance from the nearest fire station, against the alternative that x and y are positively linearly related, at $\alpha = .05$. The null and alternative hypotheses are:

$$H_0: \beta_1 = 0$$

$$H_a: \beta_1 > 0$$

The value of the test statistic highlighted on the printout is $t = 12.53$, and the two-tailed p-value of the test also highlighted is less than .0001. Thus, the p-value for our one-tailed test is

$$p < \frac{.0001}{2} = .00005$$

Since $\alpha = .05$ exceeds this small p-value, there is sufficient evidence to reject H_0 and conclude that distance between the fire and the fire station contributes information for the prediction of fire damage and that mean fire damage increases as the distance increases.

(b) *Confidence interval for slope*: We gain additional information about the relationship by forming a confidence interval for the slope β_1. A 95% confidence interval for β (highlighted on the printout) is (4.07085, 5.76781). We are 95% confident that the interval from \$4,071 to \$5,768 encloses the mean increase (β_1) in fire damage per additional mile distance from the fire station.

(c) *Numerical descriptive measures of model adequacy*: The coefficient of determination (highlighted on the printout) is

$$r^2 = .9235$$

This value implies that about 92% of the sample variation in fire damage (y) is explained by the distance x between the fire and the fire station in a straight-line model.

The coefficient of correlation r, which measures the strength of the linear relationship between y and x, is not shown in Figure 3.26. Using the facts that $r = \sqrt{r^2}$ in simple linear regression and that r and $\hat{\beta}_1$ have the same sign, we find

$$r = +\sqrt{r^2} = \sqrt{.9235} = .96$$

The high correlation confirms our conclusion that β_1 differs from 0; it appears that fire damage and distance from the fire station are linearly correlated.

The results of the test for β_1, the high value of r^2, and the relatively small $2s$ value (STEP 4) all point to a strong linear relationship between x and y.

Step 6. We are now prepared to use the least squares model. Suppose the insurance company wants to predict the fire damage if a major residential fire were to occur 3.5 miles from the nearest fire station (i.e., $x_p = 3.5$). The predicted value, shaded at the bottom of the SAS printout, is $\hat{y} = 27.4956$, while the corresponding 95% prediction interval (also highlighted) is (22.3239, 32.6672). Therefore, we predict (with 95% confidence) that the fire damage for a major residential fire 3.5 miles from the nearest fire station will fall between \$22,324 and \$32,667.

Caution

We would not use this prediction model to make predictions for homes less than .7 mile or more than 6.1 miles from the nearest fire station. A look at the data in Table 3.8 reveals that all the x-values fall between .7 and 6.1. Recall from Section 3.9 that it is dangerous to use the model to make predictions outside the region in which the sample data fall. A straight line might not provide a good model for the relationship between the mean value of y and the value of x when stretched over a wider range of x values.

[*Note*: Details on how to perform a simple linear regression analysis on the computer using each of four statistical software packages, SAS, SPSS, R, and MINITAB are provided in tutorials that accompany this text.]

3.10 Exercises

3.58 Groundwater contamination in wells. In New Hampshire, a mandate to use reformulated gasoline has led to an increase in the contamination of groundwater. Refer to the *Environmental Science and Technology* (January 2005) study of the factors related to methyl *tert*-butyl ether (MTBE) contamination in 223 New Hampshire wells, Exercise 1.17 (p. 12). Two of the many quantitative variables measured for each well are the pH level (standard units) and the MTBE level (micrograms per liter). The data are saved in the MTBE file (with selected observations shown in the table). Consider a straight-line model relating MTBE level (y) to pH level (x). Conduct a complete simple linear regression analysis of the data. Interpret the results.

MTBE (Selected observations)

WELL	PH	MTBE
9	7.55	1.32
30	8.48	0.20
79	7.51	1.16
85	8.36	0.33
177	7.59	2.70
217	6.56	0.83

3.59 Cooling method for gas turbines. Refer to the *Journal of Engineering for Gas Turbines and Power* (January 2005) study of the performance of a gas turbine engine cooled by high pressure inlet fogging, Exercise 1.62 (p. 51). Recall that 67 gas turbines augmented with high pressure inlet fogging were tested and the heat rate (kilojoules per kilowatt per hour) of each measured. Among the other variables measured was speed (revolutions per minute) of the engine. The data for the 67 engines are saved in the GASTURBINE file

GASTURBINE (Selected observations)

ENGINE	HEAT RATE (kj/kwhr)	SPEED (rpm)
1	14622	27245
2	13196	14000
3	11948	17384
4	11289	11085
5	11964	14045
63	12766	18910
64	8714	3600
65	9469	3600
66	11948	16000
67	12414	14600

(with selected observations shown in the table on the previous page.). Consider a straight-line model relating heat rate (y) to speed (x). Conduct a complete simple linear regression analysis of the data. Interpret the results.

3.60 **Ranking driving performance of professional golfers.** A group of Northeastern University researchers developed a new method for ranking the total driving performance of golfers on the Professional Golf Association (PGA) tour (*Sport Journal,* Winter 2007). The method requires knowing a golfer's average driving distance (yards) and driving accuracy (percent of drives that land in the fairway). The values of these two variables are used to compute a driving performance index. Data for the top 40 PGA golfers (as ranked by the new method) are saved in the PGADRIVER file. (The first five and last five observations are listed in the table below.) A professional golfer is practicing a new swing to increase his average driving distance. However, he is concerned that his driving accuracy will be lower. Is his concern a valid one? Use simple linear regression, where y = driving accuracy and x = driving distance, to answer the question.

3.61 **An MBA's work–life balance.** The importance of having employees with a healthy work–life balance has been recognized by U.S. companies for decades. Many business schools offer courses that assist MBA students with developing good work–life balance habits and most large companies have developed work–life balance programs for their employees. In April 2005, the Graduate Management Admission Council (GMAC) conducted a survey of over 2,000 MBA alumni to explore the work–life balance issue. (For example, one question asked alumni to state their level of agreement with the statement "My personal and work demands are overwhelming.") Based on these responses, the GMAC determined a work–life balance scale score for each MBA alumni. Scores ranged from 0 to 100, with lower scores indicating a higher imbalance between work and life. Many other variables, including average number of hours worked per week, were also measured. The data for the work–life balance study is saved in the GMAC file. (The first 15 observations are listed in the following table.) Let x = average number of hours worked per week and y = work–life balance scale score for each MBA alumnus. Investigate the link between these two variables by conducting a complete simple linear regression analysis of the data. Summarize your findings in a professional report.

GMAC (First 15 observations)

WLB-SCORE	HOURS
75.22	50
64.98	45
49.62	50
44.51	55
70.10	50
54.74	60
55.98	55
21.24	60
59.86	50
70.10	50
29.00	70
64.98	45
36.75	40
35.45	40
45.75	50

Source: "Work–life balance: An MBA alumni report," *Graduate Management Admission Council (GMAC) Research Report* (Oct. 13, 2005).

PGADRIVER (Selected observations)

RANK	PLAYER	DRIVING DISTANCE (YARDS)	DRIVING ACCURACY (%)	DRIVING PERFORMANCE INDEX
1	Woods	316.1	54.6	3.58
2	Perry	304.7	63.4	3.48
3	Gutschewski	310.5	57.9	3.27
4	Wetterich	311.7	56.6	3.18
5	Hearn	295.2	68.5	2.82
36	Senden	291	66	1.31
37	Mickelson	300	58.7	1.30
38	Watney	298.9	59.4	1.26
39	Trahan	295.8	61.8	1.23
40	Pappas	309.4	50.6	1.17

Source: Wiseman, F. et al. "A new method for ranking total driving performance on the PGA tour," *Sport Journal*, Vol. 10, No. 1, Winter 2007 (Table 2).

3.11 Regression Through the Origin (Optional)

In practice, we occasionally know in advance that the true line of means $E(y)$ passes through the point $(x = 0, y = 0)$, called the **origin**. For example, a chain of convenience stores may be interested in modeling sales y of a new diet soft drink as a linear function of amount x of the new product in stock for a sample of stores. Or, a medical researcher may be interested in the linear relationship between dosage x of a drug for cancer patients and increase y in pulse rate of the patient 1 minute after taking the drug. In both cases, it is known that the regression line must pass through the origin. The convenience store chain knows that if one of its stores chooses not to stock the new diet soft drink, it will have zero sales of the new product. Likewise, if the cancer patient takes no dosage of the drug, the theoretical increase in pulse rate 1 minute later will be 0.

For situations in which we know that the regression line passes through the origin, the y-intercept is $\beta_0 = 0$ and the probabilistic straight-line model takes the form

$$y = \beta_1 x + \varepsilon$$

When the regression line passes through the origin, the formula for the least squares estimate of the slope β_1 differs from the formula given in Section 3.3. Several other formulas required to perform the regression analysis are also different. These new computing formulas are provided in the following box.

Formulas for Regression Through the Origin: $y = \beta_1 x + \varepsilon$

Least squares slope:*

$$\hat{\beta}_1 = \frac{\sum x_i y_i}{\sum x_i^2}$$

Estimate of σ^2:

$$s^2 = \frac{\text{SSE}}{n-1}, \quad \text{where SSE} = \sum y_i^2 - \hat{\beta}_1 \sum x_i y_i$$

Estimate of $\sigma_{\hat{\beta}_1}$:

$$s_{\hat{\beta}_1} = \frac{s}{\sqrt{\sum x_i^2}}$$

Estimate of $\sigma_{\hat{y}}$ for estimating $E(y)$ when $x = x_p$:

$$s_{\hat{y}} = s \left(\frac{x_p}{\sqrt{\sum x_i^2}} \right)$$

Estimate of $\sigma_{(y-\hat{y})}$ for predicting y when $x = x_p$:

$$s_{(y-\hat{y})} = s \sqrt{1 + \frac{x_p^2}{\sum x_i^2}}$$

---

*The derivation of this formula is provided in Appendix A.

Note that the denominator of s^2 is $n - 1$, not $n - 2$ as in the previous sections. This is because we need to estimate only a single parameter β_1 rather than both β_0 and β_1. Consequently, we have one additional degree of freedom for estimating σ^2, the variance of ε. Tests and confidence intervals for β_1 are carried out exactly as outlined in the previous sections, except that the t distribution is based on $(n - 1)$ df. The test statistic and confidence intervals are given in the next box.

Tests and Confidence Intervals for Regression Through the Origin

Test statistic for H_0: $\beta_1 = 0$:

$$t = \frac{\hat{\beta}_1 - 0}{s_{\hat{\beta}_1}} = \frac{\hat{\beta}_1}{s / \sqrt{\sum x_i^2}}$$

$100(1 - \alpha)\%$ *confidence interval for β_1:*

$$\hat{\beta}_1 \pm (t_{\alpha/2}) s_{\hat{\beta}_1} = \hat{\beta}_1 \pm (t_{\alpha/2}) \left(\frac{s}{\sqrt{\sum x_i^2}} \right)$$

$100(1 - \alpha)\%$ *confidence interval for $E(y)$:*

$$\hat{y} \pm (t_{\alpha/2}) s_{\hat{y}} = \hat{y} \pm (t_{\alpha/2}) s \left(\frac{x_p}{\sqrt{\sum x_i^2}} \right)$$

$100(1 - \alpha)\%$ *prediction interval for y:*

$$\hat{y} \pm (t_{\alpha/2}) s_{(y - \hat{y})} = \hat{y} \pm (t_{\alpha/2}) s \sqrt{1 + \frac{x_p^2}{\sum x_i^2}}$$

where the distribution of t is based on $(n - 1)$ df

Example 3.6

Graphite furnace atomic absorption spectrometry (GFAAS) is a method for measuring the amount of light absorbed by chemical elements in trace metals, blood, urine, steel, and petroleum products. In one experiment, samples of cadmium were deposited in a small graphite tube and then heated until vaporization. The amount of light absorbed was then measured using GFAAS. Researchers have discovered that the amount of light absorbed (y) can be linearly correlated to the concentration (x) of cadmium present in the graphite tube. Consider the data for $n = 6$ cadmium samples shown in Table 3.9, where the amount of light absorbed was measured as peak absorption spectroscopy (AS) value and concentration was measured in micrograms per milliliter. Now, since peak AS will be 0 when the concentration of cadmium is 0, use the data to fit a straight-line model through the origin and find SSE.

Solution

The model we want to fit is $y = \beta_1 x + \varepsilon$. Preliminary calculations for estimating β_1 and calculating SSE are given in Table 3.10. The estimate of the slope is

GFAAS

Table 3.9 GFAAS data for Example 3.6

Amount of Light Absorbed (peak AS), x	Concentration (ng/ml), y
0.000	0
0.125	6
0.240	12
0.350	18
0.600	24
0.585	30

Table 3.10 Preliminary calculations for Example 3.6

x_i	y_i	x_i^2	$x_i y_i$	y_i^2
0	0.000	0	0.00	0.000000
6	0.125	36	0.75	0.015625
12	0.240	144	2.88	0.057600
18	0.350	324	6.30	0.122500
24	0.600	576	14.40	0.360000
30	0.585	900	17.55	0.342225
		$\sum x_i^2 = 1980$	$\sum x_i y_i = 41.88$	$\sum y_i^2 = .89795$

$$\hat{\beta}_1 = \frac{\sum x_i y_i}{\sum x_i^2} = \frac{41.88}{1980}$$

$$= .02115$$

and the least squares line is

$$\hat{y} = .02115x$$

The value of SSE for the line is

$$SSE = \sum y_i^2 - \hat{\beta}_1 \sum x_i y_i$$

$$= .89795 - (.02115)(41.88)$$

$$= .01212$$

Both these values, $\hat{\beta}_1$ and SSE, are highlighted on the SPSS printout of the analysis, Figure 3.28. A MINITAB graph of the least squares line with the observations is shown in Figure 3.29.

Example 3.7

Refer to Example 3.6.

(a) Conduct the appropriate test for model adequacy at $\alpha = .05$.

(b) If the model is deemed adequate, predict the amount of light absorbed y for a cadmium sample with a concentration of $x = 18$ ng/ml with a 95% prediction interval.

Model Summary

Model	R	R Square[b]	Adjusted R Square	Std. Error of the Estimate
1	.993[a]	.986	.984	.049243

a. Predictors: CONC

b. For regression through the origin (the no-intercept model), R Square measures the proportion of the variability in the dependent variable about the origin explained by regression. This CANNOT be compared to R Square for models which include an intercept.

ANOVA[c,d]

Model		Sum of Squares	df	Mean Square	F	Sig.
1	Regression	.886	1	.886	365.303	.000[a]
	Residual	.012	5	.002		
	Total	.898[b]	6			

a. Predictors: CONC

b. This total sum of squares is not corrected for the constant because the constant is zero for regression through the origin.

c. Dependent Variable: PEAKAS

d. Linear Regression through the Origin

Coefficients[a,b]

Model		Unstandardized Coefficients		Standardized Coefficients	t	Sig.	95% Confidence Interval for B	
		B	Std. Error	Beta			Lower Bound	Upper Bound
1	CONC	.021	.001	.993	19.113	.000	.018	.024

a. Dependent Variable: PEAKAS

b. Linear Regression through the Origin

Figure 3.28 SPSS regression through origin printout for Example 3.6

Figure 3.29 MINITAB scatterplot for data in Example 3.6

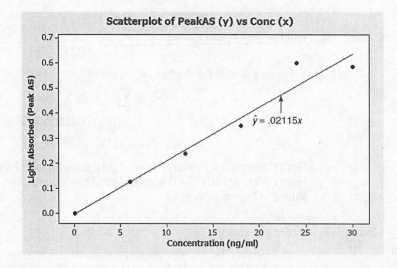

Scatterplot of PeakAS (y) vs Conc (x)

$\hat{y} = .02115x$

Solution

(a) To test for model adequacy, we conduct the test

$$H_0: \beta_1 = 0 \text{ versus } H_a: \beta_1 > 0$$

(We choose to perform an upper-tailed test since it is reasonable to assume that if a linear relationship exists between cadmium concentration x and amount of light absorbed y, it is a positive one.)

The test statistic, highlighted at the bottom of Figure 3.28, is $t = 19.1$. Of course, you could obtain this value using the formulas in the box, as shown:

$$s = \sqrt{\frac{SSE}{n-1}} = \sqrt{\frac{.01212}{5}} = .049$$

$$t = \frac{\hat{\beta}_1}{s/\sqrt{\sum x_i^2}} = \frac{.02115}{.049/\sqrt{1980}} = 19.1$$

To arrive at a conclusion, we compare the value of α selected ($\alpha = .05$) to the observed significance level (p-value) of the test. From the printout in Figure 3.28, the two-tailed p-value is .000; consequently, the upper-tailed p-value is $.000/2 = 0$. Since $\alpha = .05$ exceeds the p-value, there is sufficient evidence to conclude that the straight-line model is adequate for predicting amount of light absorbed y.

(b) To find a 95% prediction interval for y when $x = 18$, we could use the formulas in the box or obtain the result from statistical software. Figure 3.30 shows an SPSS spreadsheet with 95% prediction intervals for each observation in the data set. The interval for $x = 18$ (highlighted) is (.244, .517). Thus, we predict with 95% confidence that the amount of light absorbed for a cadmium specimen with a concentration of 18 ng/ml will fall between .244 and .517 peak AS.

Figure 3.30 SPSS spreadsheet with 95% prediction intervals

	PEAKAS	CONC	LOWER95PI	UPPER95PI
1	0.000	0	-0.12658	0.12658
2	0.125	8	-0.00082	0.25464
3	0.240	12	0.12271	0.38492
4	0.350	18	0.24418	0.51728
5	0.600	24	0.36381	0.65146
6	0.585	30	0.48188	0.78721

We conclude this section by pointing out two caveats of simple linear regression through the origin.

Caveat #1: The value of the coefficient of determination, r^2, for the regression-through-the-origin model $E(y) = \beta_1 x$ is computed as follows:

$$r^2 = 1 - \frac{\sum(y_i - \hat{y}_i)^2}{\sum y_i^2}$$

Recall, from Section 3.8, that the denominator of the r^2 statistic for the model $E(y) = \beta_0 + \beta_1 x$ is

$$\sum (y_i - \bar{y})^2$$

This value represents the total sample variation in the response y. Note that the denominators are different. Consequently, one should not attempt to compare directly the r^2 values from models with and without a y-intercept. (In fact, SPSS makes a note of this directly on the printout. See Figure 3.28.) Also, by using the regression-through-the origin formula for r^2, the intuitive interpretation discussed in Section 3.8 (and illustrated in Figure 3.17) is lost. No longer can we state that r^2 measures a percentage reduction in sum of squared errors, since the denominator does not represent the sum of squared errors for the model $E(y) = \beta_0$. Because of this, some statistical software packages (including MINITAB) will not report an r^2 value for regression through the origin.

Caveat #2: There are several situations where it is dangerous to fit the model $E(y) = \beta_1 x$. If you are not certain that the regression line passes through the origin, it is a safe practice to fit the more general model $E(y) = \beta_0 + \beta_1 x$. If the line of means does, in fact, pass through the origin, the estimate of β_0 will differ from the true value $\beta_0 = 0$ by only a small amount. For all practical purposes, the least squares prediction equations will be the same.

On the other hand, you may know that the regression passes through the origin (see Example 3.6), but are uncertain about whether the true relationship between y and x is linear or curvilinear. In fact, most theoretical relationships are *curvilinear*. Yet, we often fit a linear model to the data in such situations because we believe that a straight line will make a good approximation to the mean response $E(y)$ over the region of interest. The problem is that this straight line is not likely to pass through the origin (see Figure 3.31). By forcing the regression line through the origin, we may not obtain a very good approximation to $E(y)$. For these reasons, regression through the origin should be used with extreme caution.

Figure 3.31 Using a straight line to approximate a curvilinear relationship when the true relationship passes through the origin

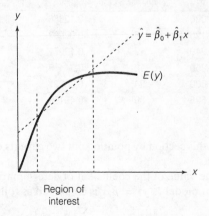

Region of interest

3.11 Exercises

3.62 Learning the mechanics. Consider the eight data points shown in the table.

⊚ EX3_62

x	−4	−2	0	2	4	6	8	10
y	−12	−7	0	6	14	21	24	31

(a) Fit a straight-line model through the origin; i.e., fit $E(y) = \beta_1 x$.

(b) Calculate SSE, s^2, and s.

(c) Do the data provide sufficient evidence to indicate that x and y are positively linearly related?

(d) Construct a 95% confidence interval for β_1.

(e) Construct a 95% confidence interval for $E(y)$ when $x = 7$.

(f) Construct a 95% prediction interval for y when $x = 7$.

3.63 Learning the mechanics. Consider the five data points shown in the table.

⊚ EX3_63

x	0	1	2	3	4
y	0	−8	−20	−30	−35

(a) Fit a straight-line model through the origin; i.e., fit $E(y) = \beta_1 x$.

(b) Calculate SSE, s^2, and s.

(c) Do the data provide sufficient evidence to indicate that x and y are negatively linearly related?

(d) Construct a 95% confidence interval for β_1.

(e) Construct a 95% confidence interval for $E(y)$ when $x = 1$.

(f) Construct a 95% prediction interval for y when $x = 1$.

3.64 Learning the mechanics. Consider the 10 data points shown in the table.

⊚ EX3_64

x	30	50	70	90	100	120	140	160	180	200
y	4	10	15	21	21	22	29	34	39	41

(a) Fit a straight-line model through the origin; i.e., fit $E(y) = \beta_1 x$.

(b) Calculate SSE, s^2, and s.

(c) Do the data provide sufficient evidence to indicate that x and y are positively linearly related?

(d) Construct a 95% confidence interval for β_1.

(e) Construct a 95% confidence interval for $E(y)$ when $x = 125$.

(f) Construct a 95% prediction interval for y when $x = 125$.

3.65 Drug designed to reduce smoking. A pharmaceutical company has developed a new drug designed to reduce a smoker's reliance on tobacco. Since certain dosages of the drug may reduce one's pulse rate to dangerously low levels, the product-testing division of the pharmaceutical company wants to model the relationship between decrease in pulse rate, y (beats/minute), and dosage, x (cubic centimeters). Different dosages of the drug were administered to eight randomly selected patients, and 30 minutes later the decrease in each patient's pulse rate was recorded. The results are given in the accompanying table. Initially, the company considered the model $y = \beta_1 x + \varepsilon$ since, in theory, a patient who receives a dosage of $x = 0$ should show no decrease in pulse rate ($y = 0$).

⊚ PULSE

PATIENT	DOSAGE x, cubic centimeters	DECREASE IN PULSE RATE y, beats/minute
1	2.0	12
2	4.0	20
3	1.5	6
4	1.0	3
5	3.0	16
6	3.5	20
7	2.5	13
8	3.0	18

(a) Fit a straight-line model that passes through the origin.

(b) Is there evidence of a linear relationship between drug dosage and decrease in pulse rate? Test at $\alpha = .10$.

(c) Find a 99% prediction interval for the decrease in pulse rate corresponding to a dosage of 3.5 cubic centimeters.

3.66 Shipments of flour bags. Consider the relationship between the total weight of a shipment of 50-pound bags of flour, y, and the number of bags in the shipment, x. Since a shipment containing $x = 0$ bags (i.e., no shipment at all) has a total weight of $y = 0$, a straight-line model of the relationship between x and y should pass through the point $x = 0$, $y = 0$. Hence, the appropriate model might be

$$y = \beta_1 x + \varepsilon$$

From the records of past flour shipments, 15 shipments were randomly chosen and the data in the following table recorded.

⊛ FLOUR

WEIGHT OF SHIPMENT	NUMBER OF 50-POUND BAGS IN SHIPMENT
5,050	100
10,249	205
20,000	450
7,420	150
24,685	500
10,206	200
7,325	150
4,958	100
7,162	150
24,000	500
4,900	100
14,501	300
28,000	600
17,002	400
16,100	400

(a) Find the least squares line for the given data under the assumption that $\beta_0 = 0$. Plot the least squares line on a scatterplot of the data.
(b) Find the least squares line for the given data using the model

$$y = \beta_0 + \beta_1 x + \varepsilon$$

(i.e., do not restrict β_0 to equal 0). Plot this line on the scatterplot you constructed in part a.
(c) Refer to part b. Why might $\hat{\beta}_0$ be different from 0 even though the true value of β_0 is known to be 0?

(d) The estimated standard error of $\hat{\beta}_0$ is equal to

$$s\sqrt{\frac{1}{n} + \frac{\bar{x}^2}{SS_{xx}}}$$

Use the t statistic,

$$t = \frac{\hat{\beta}_0 - 0}{s\sqrt{\dfrac{1}{n} + \dfrac{\bar{x}^2}{SS_{xx}}}}$$

to test the null hypothesis $H_0: \beta_0 = 0$ against the alternative $H_a: \beta_0 \neq 0$. Use $\alpha = .10$. Should you include β_0 in your model?

3.67 Projecting residential electricity customers. To satisfy the Public Service Commission's energy conservation requirements, an electric utility company must develop a reliable model for projecting the number of residential electricity customers in its service area. The first step is to study the effect of changing population on the number of electricity customers. The information shown in the next table was obtained for the service area from 2000 to 2009. Since a service area with 0 population obviously would have 0 residential electricity customers, one could argue that regression through the origin is appropriate.

(a) Fit the model $y = \beta_1 x + \varepsilon$ to the data.
(b) Is there evidence that x contributes information for the prediction of y? Test using $\alpha = .01$.
(c) Now fit the more general model $y = \beta_0 + \beta_1 x + \varepsilon$ to the data. Is there evidence (at $\alpha = .01$) that x contributes information for the prediction of y?
(d) Which model would you recommend?

⊛ PSC

YEAR	POPULATION IN SERVICE AREA x, hundreds	RESIDENTIAL ELECTRICITY CUSTOMERS IN SERVICE AREA y
2000	262	14,041
2001	319	16,953
2002	361	18,984
2003	381	19,870
2004	405	20,953
2005	439	22,538
2006	472	23,985
2007	508	25,641
2008	547	27,365
2009	592	29,967

Quick Summary/Guides

GUIDE TO SIMPLE LINEAR REGRESSION

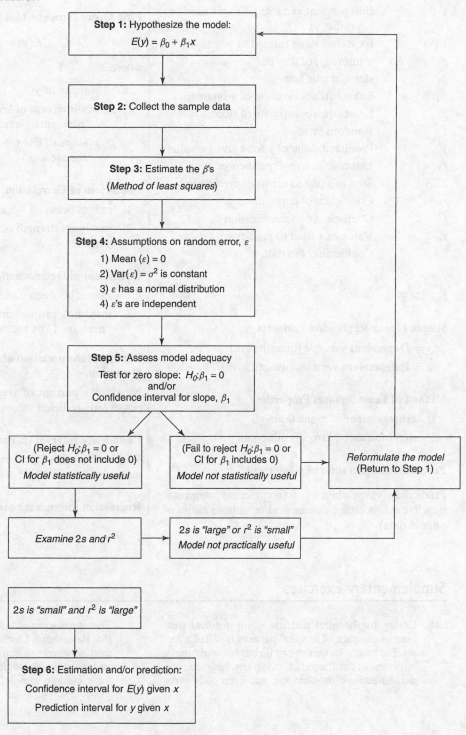

Step 1: Hypothesize the model:
$$E(y) = \beta_0 + \beta_1 x$$

Step 2: Collect the sample data

Step 3: Estimate the β's
(*Method of least squares*)

Step 4: Assumptions on random error, ε
1) Mean $(\varepsilon) = 0$
2) Var$(\varepsilon) = \sigma^2$ is constant
3) ε has a normal distribution
4) ε's are independent

Step 5: Assess model adequacy
Test for zero slope: $H_0: \beta_1 = 0$
and/or
Confidence interval for slope, β_1

(Reject $H_0: \beta_1 = 0$ or
CI for β_1 does not include 0)
Model statistically useful

(Fail to reject $H_0: \beta_1 = 0$ or
CI for β_1 includes 0)
Model not statistically useful

Reformulate the model
(Return to Step 1)

Examine 2s and r^2

2s is "large" or r^2 is "small"
Model not practically useful

2s is "small" and r^2 is "large"

Step 6: Estimation and/or prediction:
Confidence interval for $E(y)$ given x
Prediction interval for y given x

KEY SYMBOLS/NOTATION

y	Dependent variable (variable to be predicted)
x	Independent variable (variable used to predict y)
$E(y)$	Expected value (mean) of y
β_0	y-intercept of true line
β_1	slope of true line
$\hat{\beta}_0$	Least squares estimate of y-intercept
$\hat{\beta}_1$	Least squares estimate of slope
ε	Random error
$\hat{y}$	Predicted value of y for a given x-value
$(y - \hat{y})$	Estimated error of prediction
SSE	Sum of squared errors of prediction
r	Coefficient of correlation
r^2	Coefficient of determination
x_p	Value of x used to predict y
c.v.	Coefficient of variation

KEY IDEAS

Simple Linear Regression Variables

$y = $ **Dependent** variable (quantitative)

$x = $ **Independent** variable (quantitative)

Method of Least Squares Properties

1. average error of prediction $= 0$
2. sum of squared errors is minimum

Practical Interpretation of y-Intercept

Predicted y-value when $x = 0$ (no practical interpretation if $x = 0$ is either nonsensical or outside range of sample data)

Practical Interpretation of Slope

Increase (or decrease) in y for every 1-unit increase in x

First-Order (Straight-Line) Model

$$E(y) = \beta_0 + \beta_1 x$$

where

$E(y) = $ mean of y

$\beta_0 = $ **y-intercept** of line (point where line intercepts y-axis)

$\beta_1 = $ **slope** of line (change in y for every 1-unit change in x)

Coefficient of Correlation, r

1. ranges between -1 and $+1$
2. measures strength of *linear relationship* between y and x

Coefficient of Determination, r^2

1. ranges between 0 and 1
2. measures proportion of sample variation in y "explained" by the model

Practical Interpretation of Model Standard Deviation, s

Ninety-five percent of y-values fall within $2s$ of their respected predicted values

Confidence Interval vs. Prediction Interval

Width of *confidence interval for $E(y)$* will always be **narrower** than width of *prediction interval for y*

Regression through the origin model

$$E(y) = \beta_1 x$$

Supplementary Exercises

3.68. Caring for hospital patients. Any medical item used in the care of hospital patients is called a *factor*. For example, factors can be intravenous tubing, intravenous fluid, needles, shave kits, bedpans, diapers, dressings, medications, and even code carts. The coronary care unit at Bayonet Point Hospital (St. Petersburg, Florida) investigated the relationship between the number of factors per patient, x, and the patient's length of stay (in days), y. The data for a random sample of 50 coronary care patients are given in the following table, while a SAS print-out of the simple linear regression analysis is shown on page 152.

(a) Construct a scatterplot of the data.

(b) Find the least squares line for the data and plot it on your scatterplot.

(c) Define β_1 in the context of this problem.

(d) Test the hypothesis that the number of factors per patient (x) contributes no information for the prediction of the patient's length of stay (y) when a linear model is used (use $\alpha = .05$). Draw the appropriate conclusions.

(e) Find a 95% confidence interval for β_1. Interpret your results.

(f) Find the coefficient of correlation for the data. Interpret your results.

(g) Find the coefficient of determination for the linear model you constructed in part b. Interpret your result.

(h) Find a 95% prediction interval for the length of stay of a coronary care patient who is administered a total of $x = 231$ factors.

(i) Explain why the prediction interval obtained in part h is so wide. How could you reduce the width of the interval?

3.69. Arsenic linked to crabgrass killer. In Denver, Colorado, environmentalists have discovered a link between high arsenic levels in soil and a crabgrass killer used in the 1950s and 1960s (*Environmental Science and Technology*, September 1, 2000). The recent discovery was based, in part, on the scatterplot shown. The graph (p. 152) plots the level of arsenic, against the distance from a former smelter plant for samples of soil taken from Denver residential properties.

(a) Normally, the metal level in soil decreases as distance from the source (e.g., a smelter plant) increases. Propose a straight-line model relating metal level y to distance from the plant x. Based on the theory, would you expect the slope of the line to be positive or negative?

FACTORS

NUMBER OF FACTORS x	LENGTH OF STAY y, days	NUMBER OF FACTORS x	LENGTH OF STAY y, days
231	9	354	11
323	7	142	7
113	8	286	9
208	5	341	10
162	4	201	5
117	4	158	11
159	6	243	6
169	9	156	6
55	6	184	7
77	3	115	4
103	4	202	6
147	6	206	5
230	6	360	6
78	3	84	3
525	9	331	9
121	7	302	7
248	5	60	2
233	8	110	2
260	4	131	5
224	7	364	4
472	12	180	7
220	8	134	6
383	6	401	15
301	9	155	4
262	7	338	8

Source: Bayonet Point Hospital, Coronary Care Unit.

SAS Output for Exercise 3.68

Dependent Variable: LOS

Analysis of Variance

Source	DF	Sum of Squares	Mean Square	F Value	Pr > F
Model	1	126.58393	126.58393	28.68	<.0001
Error	48	211.83607	4.41325		
Corrected Total	49	338.42000			

Root MSE	2.10077	R-Square	0.3740
Dependent Mean	6.54000	Adj R-Sq	0.3610
Coeff Var	32.12193		

Parameter Estimates

Variable	DF	Parameter Estimate	Standard Error	t Value	Pr > |t|	95% Confidence Limits	
Intercept	1	3.30603	0.67297	4.91	<.0001	1.95293	4.65914
FACTORS	1	0.01475	0.00276	5.36	<.0001	0.00922	0.02029

Predictions

Obs	FACTORS	LOS	Predicted LOS	Lower prediction limit of LOS	Upper prediction limit of LOS
1	231	9	6.7144	6.11348	7.3153
2	354	11	8.5292	7.57292	9.4856
3	323	7	8.0718	7.24266	8.9010
4	142	7	5.4012	4.66664	6.1358
5	113	8	4.9733	4.13502	5.8116

(b) Examine the scatterplot for arsenic. Does the plot support the theory, part a? (*Note:* This finding led investigators to discover the link between high arsenic levels and the use of the crabgrass killer.)

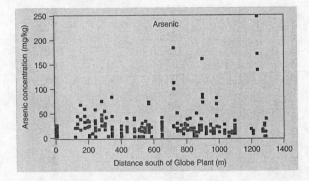

3.70. **Winning marathon times.** In *Chance* (Winter 2000), statistician Howard Wainer and two students compared men's and women's winning times in the Boston Marathon. One of the graphs used to illustrate gender differences is reproduced at right. The scatterplot graphs the winning times (in minutes) against the year in which the race was run. Men's times are represented by solid dots and women's times by open circles.

(a) Consider only the winning times for men. Is there evidence of a linear trend? If so, propose a straight-line model for predicting winning time (*y*) based on year (*x*). Would you expect the slope of this line to be positive or negative?

(b) Repeat part b for women's times.

(c) Which slope, men's or women's, will be greater in absolute value?

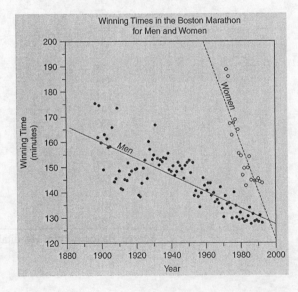

(d) Would you recommend using the straight-line models to predict the winning time in the 2020 Boston Marathon? Why or why not?

3.71. Automated materials handling. Modern warehouses use computerized and automated guided vehicles for materials handling. Consequently, the physical layout of the warehouse must be carefully designed to prevent vehicle congestion and optimize response time. Optimal design of an automated warehouse was studied in the *Journal of Engineering for Industry* (August 1993). The layout assumes that vehicles do not block each other when they travel within the warehouse (i.e., that there is no congestion). The validity of this assumption was checked by simulating (on a computer) warehouse operations. In each simulation, the number of vehicles was varied and the congestion time (total time one vehicle blocked another) was recorded. The data are shown in the accompanying table. Of interest to the researchers is the relationship between congestion time (y) and number of vehicles (x). Conduct a complete simple linear regression analysis.

⊙ WAREHOUSE

NUMBER OF VEHICLES	CONGESTION TIME, minutes	NUMBER OF VEHICLES	CONGESTION TIME, minutes
1	0	9	.02
2	0	10	.04
3	.02	11	.04
4	.01	12	.04
5	.01	13	.03
6	.01	14	.04
7	.03	15	.05
8	.03		

Source: Pandit, R., and U. S. Palekar. "Response time considerations for optimal warehouse layout design," *Journal of Engineering for Industry*, Transactions of the ASME, Vol. 115, Aug. 1993, p. 326 (Table 2).

3.72. Performance in online courses. Florida State University information scientists assessed the impact of online courses on student performance (*Educational Technology and Society*, January 2005). Each in a sample of 24 graduate students enrolled in an online advanced Web application course was asked, "How many courses per semester (on average) do you take online?" Each student's performance on weekly quizzes was also recorded. The information scientists found that the number of online courses and weekly quiz grade were negatively correlated at $r = -.726$.

(a) Give a practical interpretation of r.

(b) The researchers concluded that there is "significant negative correlation" between the number of online courses and weekly quiz grade. Do you agree?

3.73. English as a second language reading ability. Refer to the *Bilingual Research Journal* (Summer 2006) study of the relationship of Spanish (first language) grammatical knowledge to English (second language) reading, Exercise 3.26 (p. 113). Recall that each in a sample of $n = 55$ native Spanish-speaking adults took four standardized exams: Spanish grammar (SG), Spanish reading (SR), English grammar (EG), and English reading (ESLR). Simple linear regressions were used to model the ESLR score (y) as a function of each of the other exam scores (x). The coefficient of determination, r^2, for each simple linear model is listed in the table. Give a practical interpretation of each of these values.

INDEPENDENT VARIABLE (x)	r^2
SG score	.002
SR score	.099
ER score	.078

3.74. Agent Orange and Vietnam vets. *Chemosphere* (Vol. 20, 1990) published a study of Vietnam veterans exposed to Agent Orange (and the dioxin 2,3,7,8-TCDD). The next table (p. 154) gives the amounts of 2,3,7,8-TCDD (measured in parts per trillion) in both blood plasma and fat tissue drawn from each of the 20 veterans studied. One goal of the researchers is to determine the degree of linear association between the level of dioxin found in blood plasma and fat tissue. If a linear association between the two variables can be established, the researchers want to build models to predict (1) the blood plasma level of 2,3,7,8-TCDD from the observed level of 2,3,7,8-TCDD in fat tissue and (2) the fat tissue level from the observed blood plasma level.

(a) Find the prediction equations for the researchers. Interpret the results.
(b) Test the hypothesis that fat tissue level (x) is a useful linear predictor of blood plasma level (y). Use $\alpha = .05$.
(c) Test the hypothesis that blood plasma level (x) is a useful linear predictor of fat tissue level (y). Use $\alpha = .05$.
(d) Intuitively, why must the results of the tests, parts b and c, agree?

TCDD

VETERAN	TCDD LEVELS IN PLASMA	TCDD LEVELS IN FAT TISSUE
1	2.5	4.9
2	3.1	5.9
3	2.1	4.4
4	3.5	6.9
5	3.1	7.0
6	1.8	4.2
7	6.8	10.0
8	3.0	5.5
9	36.0	41.0
10	4.7	4.4
11	6.9	7.0
12	3.3	2.9
13	4.6	4.6
14	1.6	1.4
15	7.2	7.7
16	1.8	1.1
17	20.0	11.0
18	2.0	2.5
19	2.5	2.3
20	4.1	2.5

Source: Schecter, A., et al. "Partitioning of 2,3,7,8-chlorinated dibenzo-*p*-dioxins and dibenzofurans between adipose tissue and plasma-lipid of 20 Massachusetts Vietnam veterans," *Chemosphere*, Vol. 20, Nos. 7–9, 1990, pp. 954–955 (Tables I and II). Copyright © 1990, with permission from Elsevier.

3.75. Birds competing for nest holes. Two species of predatory birds, collard flycatchers and tits, compete for nest holes during breeding season on the island of Gotland, Sweden. Frequently, dead flycatchers are found in nest boxes occupied by tits. A field study examined whether the risk of mortality to flycatchers is related to the degree of competition between the two bird species for nest sites

CONDOR2

PLOT	NUMBER OF FLY-CATCHERS KILLED y	NEST BOX TIT OCCUPANCY x (%)
1	0	24
2	0	33
3	0	34
4	0	43
5	0	50
6	1	35
7	1	35
8	1	38
9	1	40
10	2	31
11	2	43
12	3	55
13	4	57
14	5	64

Source: Merila, J., and Wiggins, D. A. "Interspecific competition for nest holes causes adult mortality in the collard flycatcher," *The Condor*, Vol. 97, No. 2, May 1995, p. 449 (Figure 2), Cooper Ornithological Society.

(*The Condor*, May 1995). The above table gives data on the number y of flycatchers killed at each of 14 discrete locations (plots) on the island as well as the nest box tit occupancy x (i.e., the percentage of nest boxes occupied by tits) at each plot. SAS was used to conduct a simple linear regression analysis for the model, $E(y) = \beta_0 + \beta_1 x$. The printout is shown below. Locate and interpret the key numbers on the printout.

3.76. Hydraulic rock drilling. Two processes for hydraulic drilling of rock are dry drilling and wet drilling. In a dry hole, compressed air is forced down the drill rods to flush the cuttings and drive the hammer; in a wet hole, water is forced down. An

```
                    Dependent Variable: NOKILLED
                    Analysis of Variance

                                 Sum of        Mean
    Source         DF           Squares       Square    F Value   Pr > F
    Model           1          19.11669      19.11669    16.03    0.0018
    Error          12          14.31188       1.19266
    Corrected Total 13         33.42857


    Root MSE             1.09209     R-Square    0.5719
    Dependent Mean       1.42857     Adj R-Sq    0.5362
    Coeff Var           76.44618


                    Parameter Estimates

                    Parameter     Standard
    Variable   DF    Estimate      Error     t Value   Pr > |t|
    Intercept   1    -3.04686      1.15533    -2.64     0.0217
    TITPCT      1     0.10766      0.02689     4.00     0.0018
```

experiment was conducted to determine whether the time y it takes to dry drill a distance of 5 feet in rock increases with depth x. The results (extracted from *American Statistician*, February 1991) for one portion of the experiment are shown in the table below. Conduct a complete simple linear regression analysis of the data.

⊚ DRILLROCK

DEPTH AT WHICH DRILLING BEGINS x, feet	TIME TO DRILL 5 FEET y, minutes
0	4.90
25	7.41
50	6.19
75	5.57
100	5.17
125	6.89
150	7.05
175	7.11
200	6.19
225	8.28
250	4.84
275	8.29
300	8.91
325	8.54
350	11.79
375	12.12
395	11.02

Source: Penner, R., and Watts, D. G. "Mining information," *American Statistician*, Vol. 45, No. 1, Feb. 1991, p. 6 (Table 1).

3.77. Seed germination study. Researchers at the University of North Carolina–Greensboro investigated a model for the rate of seed germination (*Journal of Experimental Botany*, January 1993). In one experiment, alfalfa seeds were placed in a specially constructed germination chamber. Eleven hours later, the seeds were examined and the change

⊚ SEEDGERM

CHANGE IN FREE ENERGY, kj/mol	TEMPERATURE, °k
7	295
6.2	297.5
9	291
9.5	289
8.5	301
7.8	293
11.2	286.5

Source: Hageseth, G. T., and Cody, A. L. "Energy-level model for isothermal seed germination," *Journal of Experimental Botany*, Vol. 44, No. 258, Jan. 1993, p. 123 (Figure 9), by permission of Oxford University Press.

in free energy (a measure of germination rate) recorded. The results for seeds germinated at seven different temperatures are given in the table. The data were used to fit a simple linear regression model, with y = change in free energy and x = temperature.

(a) Graph the points in a scatterplot.
(b) Find the least squares prediction equation.
(c) Plot the least squares line, part b, on the scatterplot of part a.
(d) Conduct a test of model adequacy. Use $\alpha = .01$.
(e) Use the plot, part c, to locate any unusual data points (outliers).
(f) Eliminate the outlier, part e, from the data set, and repeat parts a–d.

3.78. Studying the art of walking. *American Scientist* (July–August 1998) reported on a study of the relationship between self-avoiding and unrooted walks. A self-avoiding walk is one where you never retrace or cross your own path; an unrooted walk is a path in which the starting and ending points are impossible to distinguish. The possible number of walks of each type of various lengths are reproduced in the table. Consider the straight-line model $y = \beta_0 + \beta_1 x + \varepsilon$, where x is walk length (number of steps).

⊚ WALKS

WALK LENGTH (number of steps)	UNROOTED WALKS	SELF-AVOIDING WALKS
1	1	4
2	2	12
3	4	36
4	9	100
5	22	284
6	56	780
7	147	2,172
8	388	5,916

Source: Hayes, B. "How to avoid yourself," *American Scientist*, Vol. 86, No. 4, July–Aug. 1998, p. 317 (Figure 5).

(a) Use the method of least squares to fit the model to the data if y is the possible number of unrooted walks.
(b) Interpret $\hat{\beta}_0$ and $\hat{\beta}_1$ in the estimated model, part a.
(c) Repeat parts a and b if y is the possible number of self-avoiding walks.
(d) Find a 99% confidence interval for the number of unrooted walks possible when walk length is four steps.
(e) Would you recommend using simple linear regression to predict the number of walks possible when walk length is 15 steps? Explain.

3.79. Dance/movement therapy. In cotherapy two or more therapists lead a group. An article in the *American Journal of Dance Therapy* (Spring/Summer 1995) examined the use of cotherapy in dance/movement therapy. Two of several variables measured on each of a sample of 136 professional dance/movement therapists were years of formal training x and reported success rate y (measured as a percentage) of coleading dance/movement therapy groups.

(a) Propose a linear model relating y to x.
(b) The researcher hypothesized that dance/movement therapists with more years in formal dance training will report higher perceived success rates in cotherapy relationships. State the hypothesis in terms of the parameter of the model, part a.
(c) The correlation coefficient for the sample data was reported as $r = -.26$. Interpret this result.
(d) Does the value of r in part c support the hypothesis in part b? Test using $\alpha = .05$.

3.80. Quantum tunneling. At temperatures approaching absolute zero ($-273°C$), helium exhibits traits that seem to defy many laws of Newtonian physics. An experiment has been conducted with helium in solid form at various temperatures near absolute zero. The solid helium is placed in a dilution refrigerator along with a solid impure substance, and the fraction (in weight) of the impurity passing through the solid helium is recorded. (This phenomenon of solids passing directly through solids is known as *quantum tunneling*.) The data are given in the table.

🔵 HELIUM

TEMPERATURE, x (°C)	PROPORTION OF IMPURITY, y
−262.0	.315
−265.0	.202
−256.0	.204
−267.0	.620
−270.0	.715
−272.0	.935
−272.4	.957
−272.7	.906
−272.8	.985
−272.9	.987

(a) Find the least squares estimates of the intercept and slope. Interpret them.
(b) Use a 95% confidence interval to estimate the slope β_1. Interpret the interval in terms of this application. Does the interval support the hypothesis that temperature contributes information about the proportion of impurity passing through helium?

(c) Interpret the coefficient of determination for this model.
(d) Find a 95% prediction interval for the percentage of impurity passing through solid helium at $-273°C$. Interpret the result.
(e) Note that the value of x in part d is outside the experimental region. Why might this lead to an unreliable prediction?

3.81. Loneliness of parents and daughters. Is there a link between the loneliness of parents and their offspring? Psychologists J. Lobdell and D. Perlman examined this question in an article published in the *Journal of Marriage and the Family* (August 1986). The participants in the study were 130 female college undergraduates and their parents. Each triad of daughter, mother, and father completed the UCLA Loneliness Scale, a 20-item questionnaire designed to assess loneliness and several variables theoretically related to loneliness, such as social accessibility to others, difficulty in making friends, and depression. Pearson product moment correlations relating a daughter's loneliness score to her parents' loneliness scores as well as other variables were calculated. The results are summarized below.

VARIABLE	CORRELATION (r) BETWEEN DAUGHTER'S LONELINESS AND PARENTAL VARIABLES	
	MOTHER	FATHER
Loneliness	.26	.19
Depression	.11	.06
Self-esteem	−.14	−.06
Assertiveness	−.05	.01
Number of friends	−.21	−.10
Quality of friendships	−.17	.01

Source: Lobdell, J., and Perlman, D. "The intergenerational transmission of loneliness: A study of college females and their parents," *Journal of Marriage and the Family*, Vol. 48, No. 8, Aug. 1986, p. 592. Copyright 1986 by the National Council on Family Relations. Reprinted with permission.

(a) Lobdell and Perlman conclude that "mother and daughter loneliness scores were (positively) significantly correlated at $\alpha = .01$." Do you agree?
(b) Determine which, if any, of the other sample correlations are large enough to indicate (at $\alpha = .01$) that linear correlation exists between the daughter's loneliness score and the variable measured.
(c) Explain why it would be dangerous to conclude that a causal relationship exists between a mother's loneliness and her daughter's loneliness.

(d) Explain why it would be dangerous to conclude that the variables with nonsignificant correlations in the table are unrelated.

3.82. **Can dowsers really detect water?** The act of searching for and finding underground supplies of water with the use of a divining rod is commonly known as "dowsing." Although widely regarded among scientists as a superstition, dowsing remains popular in folklore, and to this day there are individuals who claim to have this skill. A group of German physicists conducted a series of experiments to test the dowsing claim. A source of water was hidden in a random location along a straight line in a Munich barn, then each of 500 self-claimed dowsers was asked to indicate the exact location of the source (measured in decimeters from the beginning of the line). Based on the data collected for three (of the participating 500) dowsers who had particularly impressive results, the German physicists concluded that dowsing "can be regarded

as empirically proven." All three of these "best" dowsers (numbered 99, 18, and 108) performed the experiment multiple times and the best test series (sequence of trials) for each of these three dowsers was identified. These data are listed in the accompanying table. The conclusion of the German physicists was critically assessed and rebutted by Professor J.T. Enright of the University of California–San Diego (*Skeptical Inquirer*, January/February 1999). Enright applied simple linear regression to conclude the exact opposite of the German physicists.

(a) Let x = dowser's guess and y = pipe location for each trial. Graph the data. Do you detect a trend?

(b) Fit the straight-line model, $E(y) = \beta_0 + \beta_1 x$, to the data. Interpret the estimated y-intercept of the line.

(c) Is there evidence that the model is statistically useful for predicting actual pipe location? Explain.

(d) Keep in mind that the data in the DOWSING file represent the "best" performances of the three "best" dowsers, selected from among the 500 who participated in the Munich trials. Use this fact, and the results of part c, to critically assess the conclusion made by the German physicists.

3.83. **Do College administrators deserve their raises?** At major colleges and universities, administrators (e.g., deans, chairpersons, provosts, vice presidents, and presidents) are among the highest-paid state employees. Is there a relationship between the raises administrators receive and their performance on the job? This was the question of interest to a group of faculty union members at the University of South Florida called the United Faculty of Florida (UFF). The UFF compared the ratings of 15 University of South Florida administrators (as determined by faculty in a survey) to their subsequent raises in the year. The data for the analysis is listed in the accompanying table. [*Note*: Ratings are measured on a 5-point scale, where 1 = very poor and 5 = very good.] According to the UFF, the "relationship is inverse; i.e., the lower the rating by the faculty, the greater the raise. Apparently, bad administrators are more valuable than good administrators."[*] (With tongue in cheek, the UFF refers to this phenomenon as "the SOB effect.") The UFF based its conclusions on a simple linear regression analysis of the data in the next table, where y = administrator's raise and x = average rating of administrator.

🌀 DOWSING

TRIAL	DOWSER #	PIPE LOCATION	DOWSER'S GUESS
1	99	4	4
2	99	5	87
3	99	30	95
4	99	35	74
5	99	36	78
6	99	58	65
7	99	40	39
8	99	70	75
9	99	74	32
10	99	98	100
11	18	7	10
12	18	38	40
13	18	40	30
14	18	49	47
15	18	75	9
16	18	82	95
17	108	5	52
18	108	18	16
19	108	33	37
20	108	45	40
21	108	38	66
22	108	50	58
23	108	52	74
24	108	63	65
25	108	72	60
26	108	95	49

Source: Enright, J.T. "Testing dowsing: The failure of the Munich experiments," *Skeptical Inquirer*, Jan./Feb. 1999, p. 45 (Figure 6a). Used by permission of *Skeptical Inquirer*.

---

* *UFF Faculty Forum*, University of South Florida Chapter, Vol. 3, No. 5, May 1991.

⊙ UFFSAL

ADMINISTRATOR	RAISE[a]	AVERAGE RATING (5-pt scale)[b]
1	$18,000	2.76
2	16,700	1.52
3	15,787	4.40
4	10,608	3.10
5	10,268	3.83
6	9,795	2.84
7	9,513	2.10
8	8,459	2.38
9	6,099	3.59
10	4,557	4.11
11	3,751	3.14
12	3,718	3.64
13	3,652	3.36
14	3,227	2.92
15	2,808	3.00

Source: [a]Faculty and A&P Salary Report, University of South Florida, Resource Analysis and Planning, 1990. [b]Administrative Compensation Survey, *Chronicle of Higher Education*, Jan. 1991.

(a) Initially, the UFF conducted the analysis using all 15 data points in the table. Fit a straight-line model to the data. Is there evidence to support the UFF's claim of an inverse relationship between raise and rating?

(b) A second simple linear regression was performed using only 14 of the data points in the table. The data for administrator #3 was eliminated because he was promoted to dean in the middle of the academic year. (No other reason was given for removing this data point from the analysis.) Perform the simple linear regression analysis using the remaining 14 data points in the table. Is there evidence to support the UFF's claim of an inverse relationship between raise and rating?

(c) Based on the results of the regression, part b, the UFF computed estimated raises for selected faculty ratings of administrators. These are shown in the following table. What problems do you perceive with using this table to estimate administrators' raises at the University of South Florida?

(d) The ratings of administrators listed in this table were determined by surveying the faculty at the University of South Florida. All faculty are mailed the survey each year, but the response rate is typically low (approximately 10–20%). The danger with such a survey is that only disgruntled faculty, who are more apt to give a low rating to an administrator, will respond. Many of the faculty also believe that they are underpaid and that the administrators are overpaid. Comment on how such a survey could bias the results shown here.

RATINGS		RAISE
Very Poor	1.00	$15,939
	1.50	13,960
Poor	2.00	11,980
	2.50	10,001
Average	3.00	8,021
	3.50	6,042
Good	4.00	4,062
	4.50	2,083
Very Good	5.00	103

(e) Based on your answers to the previous questions, would you support the UFF's claim?

References

Chatterjee, S., Hadi, A., and Price, B. *Regression Analysis by Example*, 3rd ed. New York: Wiley, 1999.

Draper, N., and Smith, H. *Applied Regression Analysis*, 3rd ed. New York: Wiley, 1987.

Graybill, F. *Theory and Application of the Linear Model*. North Scituate, Mass.: Duxbury, 1976.

Kleinbaum, D., and Kupper, L. *Applied Regression Analysis and Other Multivariable Methods*, 2nd ed. North Scituate. Mass.: Duxbury, 1997.

Mendenhall, W. *Introduction to Linear Models and the Design and Analysis of Experiments*. Belmont, CA.: Wadsworth, 1968.

Montgomery, D., Peck, E., and Vining, G. *Introduction to Linear Regression Analysis*, 4th ed. New York: Wiley, 2006.

Mosteller, F., and Tukey, J. W. *Data Analysis and Regression: A Second Course in Statistics*. Reading, Mass.: Addison-Wesley, 1977.

Kutner, M., Nachtsheim, C., Neter, J., and Li, W. *Applied Linear Statistical Models*, 5th ed. New York: McGraw Hill, 2005.

Rousseeuw, P. J., and Leroy, A. M. *Robust Regression and Outlier Detection*. New York: Wiley, 1987.

Weisburg, S. *Applied Linear Regression*, 3rd ed. New York: Wiley, 2005.

LEGAL ADVERTISING — DOES IT PAY?

The Problem

According to the American Bar Association, there are over 1 million lawyers competing for your business. To gain a competitive edge, these lawyers are aggressively advertising their services. Advertising of legal services has long been a controversial subject, with many believing that it constitutes an unethical (and in some cases even illegal) practice. Nonetheless, legal advertisements appear in just about all media, ranging from the covers of telephone directories to infomercials on television, as well as a significant presence on the Internet. In fact, Erickson Marketing, Inc., reports that "attorneys are the #1 category of advertising in the Yellow Pages."

For this case study we present an actual recent court case involving two former law partners. One partner (A) sued the other (B) over who should pay what share of the expenses of their former partnership. Partner A handled personal injury (PI) cases, while partner B handled only workers' compensation (WC) cases. The firm's advertising was focused only on personal injury, but partner A claimed that the ads resulted in the firm getting more workers' compensation cases for partner B, and therefore that partner B should share the advertising expenses.

The Data

Table CS1.1 shows the firm's new personal injury and workers' compensation cases each month over a 42-month period for this partnership. Also shown is the total expenditure on advertising each month, and over the previous 6 months. These data are saved in the LEGALADV file.

Research Questions

Do these data provide support for the hypothesis that increased advertising expenditures are associated with more personal injury cases? With more workers' compensation cases? If advertising expenditures have a statistically significant association with the number of cases, does this necessarily mean that there is a causal relationship, that is, that spending more on advertising causes an increase in the number of cases? Based on these data, should Partner A or Partner B bear the brunt of the advertising expenditures?

LEGALADV

Table CS1.1 Legal advertising data

Month	Advertising Expenditure	New PI Cases	New WC Cases	6 Months Cumulative Adv. Exp.
1	$9,221.55	7	26	n/a
2	$6,684.00	9	33	n/a
3	$200.00	12	18	n/a
4	$14,546.75	27	15	n/a
5	$5,170.14	9	19	n/a
6	$5,810.30	13	26	n/a
7	$5,816.20	11	24	$41,632.74
8	$8,236.38	7	22	$38,227.39
9	-$2,089.55	13	12	$39,779.77
10	$29,282.24	7	15	$37,490.22
11	$9,193.58	9	21	$52,225.71
12	$9,499.18	8	24	$56,249.15
13	$11,128.76	18	25	$59,938.03
14	$9,057.64	9	19	$65,250.59
15	$13,604.54	25	12	$66,071.85
16	$14,411.76	26	33	$81,765.94
17	$13,724.28	27	32	$66,895.46
18	$13,419.42	12	21	$71,426.16
19	$17,372.33	14	18	$75,346.40
20	$6,296.35	5	25	$81,589.97
21	$13,191.59	22	12	$78,828.68
22	$26,798.80	15	7	$78,415.73
23	$18,610.95	12	22	$90,802.77
24	$829.53	18	27	$95,689.44
25	$16,976.53	20	25	$83,099.55
26	$14,076.98	38	26	$82,703.75
27	$24,791.75	13	28	$90,484.38
28	$9,691.25	18	31	$102,084.54
29	$28,948.25	21	40	$84,976.99
30	$21,373.52	7	39	$95,314.29
31	$9,675.25	16	41	$115,858.28
32	$33,213.55	12	48	$108,557.00
33	$19,859.85	15	28	$127,693.57
34	$10,475.25	18	29	$122,761.67
35	$24,790.84	30	20	$123,545.67
36	$36,660.94	12	27	$119,388.26

⊙ LEGALADV

Table CS1.1 *(continued)*

Month	Advertising Expenditure	New PI Cases	New WC Cases	6 Months Cumulative Adv. Exp.
37	$8,812.50	30	26	$134,675.68
38	$41,817.75	20	45	$133,812.93
39	$27,399.33	19	30	$142,417.13
40	$25,723.10	29	33	$149,956.61
41	$16,312.10	58	24	$165,204.46
42	$26,332.78	42	40	$156,725.72
43	$60,207.58	24	36	$146,397.56
44	$42,485.39	47	29	$197,792.64
45	$35,601.92	24	17	$198,460.28
46	$72,071.50	14	13	$206,662.87
47	$12,797.11	31	15	$253,011.27
48	$12,310.50	26	16	$249,496.28

Source: Info Tech, Inc., Gainesville, Florida.

The Models

One way to investigate the link between the increased advertising expenditures and number of new cases is to run a simple linear regression analysis on the data. We consider two straight-line models.

Model 1

$$E(y_1) = \beta_0 + \beta_1 x$$

where y_1 = number of new personal injury cases per month
    x = cumulative 6-month advertising expenditures (in thousands of dollars)

Model 2

$$E(y_2) = \beta_0 + \beta_1 x$$

where y_2 = number of new workers' compensation cases per month
    x = cumulative 6-month advertising expenditures (in thousands of dollars)

In Model 1, β_1 represents the change in the number of new personal injury cases per month for each additional $1,000 spent on advertising during the month. Similarly, in Model 2, β_1 represents the change in the number of workers' compensation cases per month for each additional $1,000 spent on advertising. Consequently, the values of these β's hold the key to understanding which partner benefits more (or at all) from increased advertising expenditures.

Descriptive Analysis

A MINITAB scatterplot of the data and a simple linear regression printout for Model 1 are shown in Figure CS1.1. Note that the least squares line is also displayed on the scatterplot. You can see that the line has a positive slope and, although there is some variation of the data points around the line, it appears that advertising expenditure (x) is fairly strongly related to the number of new personal injury cases (y_1). The estimated slope of the line (highlighted on Figure CS3.1.1) is $\hat{\beta}_1 = .113$. Thus, we estimate a .113 increase in the number of new personal injury cases for every \$1,000 increase in cumulative advertising expenditures.

Figure CS1.1 MINITAB analysis of new personal injury (PI) cases versus 6-month cumulative advertising expenditure

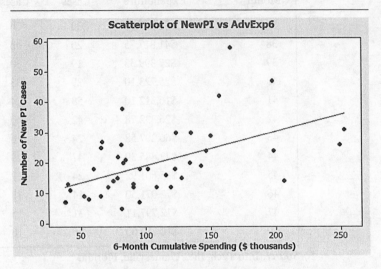

Regression Analysis: NewPI versus AdvExp6

```
The regression equation is
NewPI = 7.77 + 0.113 AdvExp6

Predictor       Coef   SE Coef      T      P
Constant       7.767     3.385   2.29  0.027
AdvExp6      0.11289   0.02793   4.04  0.000


S = 9.67521    R-Sq = 29.0%    R-Sq(adj) = 27.2%

Analysis of Variance

Source          DF      SS      MS      F      P
Regression       1  1529.5  1529.5  16.34  0.000
Residual Error  40  3744.4    93.6
Total           41  5273.9
```

Correlations: NewPI, AdvExp6

```
Pearson correlation of NewPI and AdvExp6 = 0.539
P-Value = 0.000
```

Does such a relationship also exist between number of new workers' compensation cases and advertising expenditure? A MINITAB scatterplot and simple linear regression analysis for Model 2 are shown in Figure CS1.2. Compared to the previous scatterplot, the slope of the least squares line shown in Figure CS1.2 is much flatter and the variation of the data points around the line is much larger. Consequently, It does not appear that number of new workers' compensation cases is very strongly related to advertising expenditure. In fact, the estimated slope (highlighted in Figure CS1.2) implies that a \$1,000 increase in cumulative advertising expenditures will lead to only a $\hat{\beta}_1 = .0098$ increase in the number of new workers' compensation cases per month.

Figure CS1.2 MINITAB analysis of new workers' compensation (WC) cases versus 6-month cumulative advertising expenditure

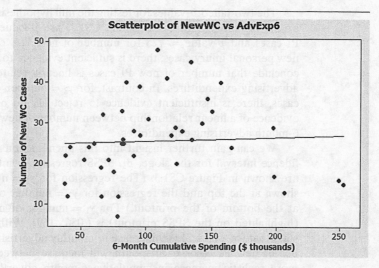

Regression Analysis: NewWC versus AdvExp6

```
The regression equation is
NewWC = 24.6 + 0.0098 AdvExp6


Predictor     Coef    SE Coef      T      P
Constant    24.574      3.367   7.30  0.000
AdvExp6    0.00982    0.02778   0.35  0.725

S = 9.62296   R-Sq = 0.3%   R-Sq(adj) = 0.0%

Analysis of Variance

Source           DF       SS      MS     F      P
Regression        1    11.58   11.58  0.13  0.725
Residual Error   40  3704.06   92.60
Total            41  3715.64
```

Correlations: NewWC, AdvExp6

```
Pearson correlation of NewWC and AdvExp6 = 0.056
P-Value = 0.725
```

Based on these descriptive statistics (scatterplots and least squares lines), it appears that partner A's argument that partner B should share the advertising expenses is weak, at best. Next, we employ confidence intervals and tests of hypothesis in order to provide a measure of reliability to this inference.

Testing the Models

To formally test the models, we conduct the one-tailed test of hypothesis for a positive slope:

$$H_0: \beta_1 = 0$$

$$H_a: \beta_1 > 0$$

The objective is to determine whether one or both of the dependent variables are statistically positively linearly related to cumulative 6-month advertising expenditures.

The two-tailed p-values for testing the null hypothesis, $H_0 : \beta_1 = 0$, (highlighted on the printouts, Figures CS1.1 and CS1.2) are p-value ≈ 0 for number of new PI cases and p-value $=. 725$ for number of new WC cases. For $y_1 = $ number of new personal injury cases, there is sufficient evidence to reject H_0 (at $\alpha = .01$) and conclude that number of new PI cases is linearly related to cumulative 6-month advertising expenditures. In contrast, for $y_2 = $ number of worker's compensation cases, there is insufficient evidence to reject H_0 (at $\alpha = .01$); thus, there is no evidence of a linear relationship between number of new WC cases and cumulative 6-month advertising expenditures.

We can gain further insight into this phenomenon by examining a 95% confidence interval for the slope, β_1. SPSS regression printouts for the two analyses are shown in Figure CS1.3. (The regression for $y_1 = $ number of new PI cases is shown at the top and the regression for $y_2 = $ number of new WC cases is shown at the bottom of the printout.) For $y_1 = $ number of new PI cases, the interval (highlighted on the SPSS printout) is (.056, .169). With 95% confidence, we can state that for every \$1,000 increase in monthly advertising expenditures, the number of new PI cases each month will increase between. 056 and. 169. Now, a more realistic increase in cumulative 6-month advertising expenditures is, say, \$20,000. Multiplying the endpoints of the interval by 20, we see that this increase in advertising spending leads to an increase of anywhere between 1 and 3 new PI cases.

Now, for $y_2 = $ number of new WC cases, the 95% confidence interval for the slope (also highlighted on the SPSS printout) is ($-.046$, .066). Since the interval spans the value 0, we draw the same conclusion as we did with the hypothesis test—there is no statistical evidence of a linear relationship between number of new WC cases and cumulative 6-month advertising expenditures. Consequently, these results do not support partner A's argument since there is no evidence (at the 95% confidence level) that partner B benefitted from advertising.

Coefficients[a]

Model		Unstandardized Coefficients B	Std. Error	Standardized Coefficients Beta	t	Sig.	95% Confidence Interval for B Lower Bound	Upper Bound
1	(Constant)	7.767	3.385		2.295	.027	.926	14.609
	CUM. ADV (thous)	.113	.028	.539	4.042	.000	.056	.169

a. Dependent Variable: New PI Cases

Coefficients[a]

Model		Unstandardized Coefficients B	Std. Error	Standardized Coefficients Beta	t	Sig.	95% Confidence Interval for B Lower Bound	Upper Bound
1	(Constant)	24.574	3.367		7.299	.000	17.770	31.379
	CUM. ADV (thous)	.010	.028	.056	.354	.725	-.046	.066

a. Dependent Variable: New WC Cases

Figure CS1.3 SPSS simple linear regressions for legal advertising data

More Supporting Evidence

Other statistics provided on the regression printouts support the conclusion that cumulative 6-month advertising expenditures is a statistically useful linear predictor of number of new personal injury cases, but not a useful linear predictor of number of new workers' compensation cases.

Consider the coefficients of correlation and determination (highlighted on the MINITAB printouts in Figures CS1.1 and CS1.2). For y_1 = number of new PI cases, the correlation coefficient value of $r = -.539$ is statistically significantly different from 0 and indicates a moderate positive linear relationship between the variables. The coefficient of determination, $r^2 = .29$, implies that almost 30% of the sample variation in number of new PI cases can be explained by using advertising expenditure (x) in the straight-line model. In contrast, for y_2 = number of new WC cases, $r = -.056$ is not statistically different from 0 and $r^2 = .3$ implies that only 0.3% of the sample variation in number of new WC cases can be explained by using advertising expenditure (x) in the straight-line model.

Conclusion

In court, a statistician presented the above results in support of the defendant (Partner B). Table CS1.2 is an exhibit that summarizes the regression results for the two models. Clearly the descriptive and inferential statistics provide support for the hypothesis that increased advertising expenditures are associated with more personal injury cases, but not with more workers' compensation cases. Ultimately, the court ruled that Partner A (not Partner B) should bear the brunt of the advertising expenditures.

Table CS1.2 Summary of linear regression results

	Model 1 y_1 = number of new PI cases	Model 2 y_2 = number of new WC cases
Estimated slope, $\hat{\beta}_1$	.113	.0098
p-value for testing $H_0 : \beta_1 = 0$	<.01	>.10
95% CI for β_1	(.056, .169)	(−.046, .066)
Coef. of Correlation, r	.539	.056
Coef. of Determination, r^2	.29	.003

Follow-up Questions

1. Access the data in the LEGALADV file and find the correlation between number of new personal injury cases (y_1) and number of new worker's compensation cases (y_2). Which partner (A or B) would benefit from reporting this correlation as evidence in the case? Explain.

2. Compare the standard deviations for the simple linear regression models of number of new personal injury cases (y_1) and number of new worker's compensation cases (y_2). Which partner (A or B) would benefit from reporting only these standard deviations as evidence in the case? Explain.

3. Access the data in the LEGALADV file and find the standard deviation for the number of new personal injury cases (y_1) and the standard deviation for the number of new worker's compensation cases (y_2). Compare these standard deviations to those you found in question 2. Which partner (A or B) would benefit from reporting this additional information as evidence in the case? Explain.

MULTIPLE REGRESSION MODELS

Chapter

4

Contents

Objectives

1. To develop a *multiple regression* procedure for predicting a response y based on two or more independent variables.
2. To assess how well the multiple regression model fits the sample data.

3. To introduce several different models involving both quantitative and qualitative independent variables.

4.1 General Form of a Multiple Regression Model

Most practical applications of regression analysis utilize models that are more complex than the first-order (straight-line) model. For example, a realistic probabilistic model for monthly sales revenue would include more than just the advertising expenditure discussed in Chapter 3 to provide a good predictive model for sales. Factors such as season, inventory on hand, sales force, and productivity are a few of the many variables that might influence sales. Thus, we would want to incorporate these and other potentially important independent variables into the model if we need to make accurate predictions.

Probabilistic models that include more than one independent variable are called **multiple regression models**. The general form of these models is shown in the box.

The dependent variable y is now written as a function of k independent variables, $x_1, x_2, \ldots, x_k$. The random error term is added to make the model probabilistic rather than deterministic. The value of the coefficient β_i determines the contribution of the independent variable x_i, given that the other $(k-1)$ independent variables are held constant, and β_0 is the y-intercept. The coefficients $\beta_0, \beta_1, \ldots, \beta_k$ will usually be unknown, since they represent population parameters.

General Form of the Multiple Regression Model

$$y = \beta_0 + \beta_1 x_1 + \beta_2 x_2 + \cdots + \beta_k x_k + \varepsilon$$

where y is the dependent variable

         $x_1, x_2, \ldots, x_k$ are the independent variables

         $E(y) = \beta_0 + \beta_1 x_1 + \beta_2 x_2 + \cdots + \beta_k x_k$ is the deterministic portion
of the model

         β_i determines the contribution of the independent variable x_i

Note: The symbols $x_1, x_2, \ldots, x_k$ may represent higher-order terms for quantitative predictors (e.g., $x_2 = x_1^2$) or terms for qualitative predictors.

At first glance it might appear that the regression model shown here would not allow for anything other than straight-line relationships between y and the independent variables, but this is not true. Actually, $x_1, x_2, \ldots, x_k$ can be functions of variables as long as the functions do not contain unknown parameters. For example, the carbon monoxide content y of smoke emitted from a cigarette could be a function of the independent variables

$$x_1 = \text{Tar content}$$

$$x_2 = (\text{Tar content})^2 = x_1^2$$

$$x_3 = 1 \text{ if a filter cigarette, 0 if a nonfiltered cigarette}$$

The x_2 term is called a **higher-order term** because it is the value of a quantitative variable (x_1) squared (i.e., raised to the second power). The x_3 term is a **coded variable** representing a qualitative variable (filter type). The multiple regression model is quite versatile and can be made to model many different types of response variables.

The steps we followed in developing a straight-line model are applicable to the multiple regression model.

Analyzing a Multiple Regression Model

Step 1. Collect the sample data (i.e., the values of $y, x_1, x_2, \ldots, x_k$) for each experimental unit in the sample.

Step 2. Hypothesize the form of the model (i.e., the deterministic component), $E(y)$. This involves choosing which independent variables to include in the model.

Step 3. Use the method of least squares to estimate the unknown parameters $\beta_0, \beta_1, \ldots, \beta_k$.

Step 4. Specify the probability distribution of the random error component ε and estimate its variance σ^2.

Step 5. Statistically evaluate the utility of the model.

Step 6. Check that the assumptions on σ are satisfied and make model modifications, if necessary.

Step 7. Finally, if the model is deemed adequate, use the fitted model to estimate the mean value of y or to predict a particular value of y for given values of the independent variables, and to make other inferences.

Hypothesizing the form of the model (step 2) is the subject of Chapter 5 and checking the assumptions (step 6) is the topic of chapter 8. In this chapter, we assume that the form of the model is known, and we discuss steps 3, 4, 5, and 7 for a given model.

4.2 Model Assumptions

We noted in Section 4.1 that the multiple regression model is of the form

$$y = \beta_0 + \beta_1 x_1 + \beta_2 x_2 + \cdots + \beta_k x_k + \varepsilon$$

where y is the response variable that you want to predict; $\beta_0, \beta_1, \ldots, \beta_k$ are parameters with unknown values; $x_1, x_2, \ldots, x_k$ are independent information-contributing variables that are measured without error; and ε is a random error component. Since $\beta_0, \beta_1, \ldots, \beta_k$ and $x_1, x_2, \ldots, x_k$ are nonrandom, the quantity

$$\beta_0 + \beta_1 x_1 + \beta_2 x_2 + \cdots + \beta_k x_k$$

represents the deterministic portion of the model. Therefore, y is made up of two components—one fixed and one random—and, consequently, y is a random variable.

$$y = \underbrace{\beta_0 + \beta_1 x_1 + \beta_2 x_2 + \cdots + \beta_k x_k}_{\substack{\text{Deterministic} \\ \text{portion of model}}} + \underbrace{\varepsilon}_{\substack{\text{Random} \\ \text{error}}}$$

We assume (as in Chapter 3) that the random error can be positive or negative and that for any setting of the x-values, $x_1, x_2, \ldots, x_k$, ε has a normal probability distribution with mean equal to 0 and variance equal to σ^2. Furthermore, we assume that the random errors associated with any (and every) pair of y-values are probabilistically independent. That is, the error ε associated with any one y-value is independent of the error associated with any other y-value. These assumptions are summarized in the accompanying box.

Assumptions About the Random Error ε

1. For any given set of values of $x_1, x_2, \ldots, x_k$, ε has a normal probability distribution with mean equal to 0 [i.e., $E(\varepsilon) = 0$] and variance equal to σ^2 [i.e., $\text{Var}(\varepsilon) = \sigma^2$].

2. The random errors are independent (in a probabilistic sense).

The assumptions that we have described for a multiple regression model imply that the mean value $E(y)$ for a given set of values of $x_1, x_2, \cdots, x_k$ is equal to

$$E(y) = \beta_0 + \beta_1 x_1 + \beta_2 x_2 + \cdots + \beta_k x_k$$

Models of this type are called **linear statistical models** because $E(y)$ is a *linear function* of the unknown parameters $\beta_0, \beta_1, \ldots, \beta_k$.

All the estimation and statistical test procedures described in this chapter depend on the data satisfying the assumptions described in this section. Since we will rarely, if ever, know for certain whether the assumptions are actually satisfied in practice, we will want to know how well a regression analysis works, and how much faith we can place in our inferences when certain assumptions are not satisfied. We have more to say on this topic in Chapters 7 and 8. First, we need to discuss the methods of a regression analysis more thoroughly and show how they are used in a practical situation.

4.3 A First-Order Model with Quantitative Predictors

A model that includes only terms for *quantitative* independent variables, called a **first-order model**, is described in the box. Note that the first-order model does not include any higher-order terms (such as x_1^2). The term *first-order* is derived from the fact that each x in the model is raised to the first power.

A First-Order Model in Five Quantitative Independent Variables

$$E(y) = \beta_0 + \beta_1 x_1 + \beta_2 x_2 + \beta_3 x_3 + \beta_4 x_4 + \beta_5 x_5$$

where $x_1, x_2, \ldots, x_5$ are all quantitative variables that *are not* functions of other independent variables.

Note: β_i represents the slope of the line relating y to x_i when all the other x's are held fixed.

Recall that in the straight-line model (Chapter 3)

$$y = \beta_0 + \beta_1 x + \varepsilon$$

β_0 represents the y-intercept of the line and β_1 represents the slope of the line. From our discussion in Chapter 3, β_1 has a practical interpretation—it represents the mean change in y for every 1-unit increase in x. When the independent variables are quantitative, the β parameters in the first-order model specified in the box have similar interpretations. The difference is that when we interpret the β that multiplies one of the variables (e.g., x_1), we must be certain to hold the values of the remaining independent variables (e.g., x_2, x_3) fixed.

To see this, suppose that the mean $E(y)$ of a response y is related to two quantitative independent variables, x_1 and x_2, by the first-order model

$$E(y) = 1 + 2x_1 + x_2$$

In other words, $\beta_0 = 1$, $\beta_1 = 2$, and $\beta_2 = 1$.

Now, when $x_2 = 0$, the relationship between $E(y)$ and x_1 is given by

$$E(y) = 1 + 2x_1 + (0) = 1 + 2x_1$$

A graph of this relationship (a straight line) is shown in Figure 4.1. Similar graphs of the relationship between $E(y)$ and x_1 for $x_2 = 1$,

$$E(y) = 1 + 2x_1 + (1) = 2 + 2x_1$$

and for $x_2 = 2$,

$$E(y) = 1 + 2x_1 + (2) = 3 + 2x_1$$

also are shown in Figure 4.1. Note that the slopes of the three lines are all equal to $\beta_1 = 2$, the coefficient that multiplies x_1.

Figure 4.1 Graphs of $E(y) = 1 + 2x_1 + x_2$ for $x_2 = 0, 1, 2$

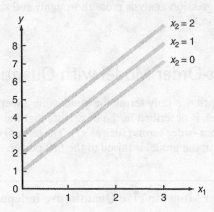

Figure 4.1 exhibits a characteristic of all first-order models: If you graph $E(y)$ versus any one variable—say, x_1—for fixed values of the other variables, the result will always be a *straight line* with slope equal to β_1. If you repeat the process for other values of the fixed independent variables, you will obtain a set of *parallel* straight lines. This indicates that the effect of the independent variable x_i on $E(y)$ is independent of all the other independent variables in the model, and this effect is measured by the slope β_i (as stated in the box).

The first-order model is the most basic multiple regression model encountered in practice. In the next several sections, we present an analysis of this model.

4.4 Fitting the Model: The Method of Least Squares

The method of fitting multiple regression models is identical to that of fitting the straight-line model in Chapter 3—namely, the method of least squares. That is, we choose the estimated model

$$\hat{y} = \hat{\beta}_0 + \hat{\beta}_1 x_1 + \cdots + \hat{\beta}_k x_k$$

that minimizes

$$\text{SSE} = \sum (y_i - \hat{y}_i)^2$$

As in the case of the straight-line model, the sample estimates $\hat{\beta}_0, \hat{\beta}_1, \ldots, \hat{\beta}_k$ will be obtained as solutions to a set of simultaneous linear equations.*

The primary difference between fitting the simple and multiple regression models is computational difficulty. The $(k + 1)$ simultaneous linear equations that

---

* Students who are familiar with calculus should note that $\hat{\beta}_0, \hat{\beta}_1, \ldots, \hat{\beta}_k$ are the solutions to the set of equations $\partial \text{SSE}/\partial \beta_0 = 0, \partial \text{SSE}/\partial \beta_1 = 0, \ldots, \partial \text{SSE}/\partial \beta_k = 0$. The solution, given in matrix notation, is presented in Appendix B.

must be solved to find the $(k+1)$ estimated coefficients $\hat{\beta}_0, \hat{\beta}_1, \ldots, \hat{\beta}_k$ are often difficult (tedious and time-consuming) to solve with a calculator. Consequently, we resort to the use of statistical computer software and present output from SAS, SPSS, and MINITAB in examples and exercises.

Example 4.1

A collector of antique grandfather clocks sold at auction believes that the price received for the clocks depends on both the age of the clocks and the number of bidders at the auction. Thus, he hypothesizes the first-order model

$$y = \beta_0 + \beta_1 x_1 + \beta_2 x_2 + \varepsilon$$

where

$$y = \text{Auction price (dollars)}$$

$$x_1 = \text{Age of clock (years)}$$

$$x_2 = \text{Number of bidders}$$

A sample of 32 auction prices of grandfather clocks, along with their age and the number of bidders, is given in Table 4.1.

(a) Use scattergrams to plot the sample data. Interpret the plots.

(b) Use the method of least squares to estimate the unknown parameters β_0, β_1, and β_2 of the model.

(c) Find the value of SSE that is minimized by the least squares method.

GFCLOCKS

Table 4.1 Auction price data

Age, x_1	Number of Bidders, x_2	Auction Price, y	Age, x_1	Number of Bidders, x_2	Auction Price, y
127	13	$1,235	170	14	$2,131
115	12	1,080	182	8	1,550
127	7	845	162	11	1,884
150	9	1,522	184	10	2,041
156	6	1,047	143	6	845
182	11	1,979	159	9	1,483
156	12	1,822	108	14	1,055
132	10	1,253	175	8	1,545
137	9	1,297	108	6	729
113	9	946	179	9	1,792
137	15	1,713	111	15	1,175
117	11	1,024	187	8	1,593
137	8	1,147	111	7	785
153	6	1,092	115	7	744
117	13	1,152	194	5	1,356
126	10	1,336	168	7	1,262

Solution

(a) MINITAB side-by-side scatterplots for examining the bivariate relationships between y and x_1, and between y and x_2, are shown in Figure 4.2. Of the two variables, age (x_1) appears to have the stronger linear relationship with auction price (y).

Figure 4.2 MINITAB side-by-side scatterplots for the data of Table 4.1

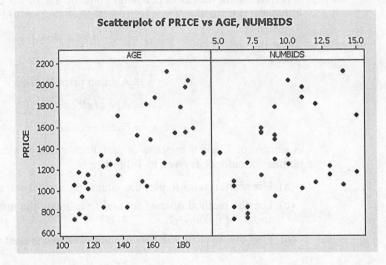

(b) The model hypothesized is fit to the data in Table 4.1 with SAS. A portion of the printout is reproduced in Figure 4.3. The least squares estimates of the β parameters (highlighted) are $\hat{\beta}_0 = -1,339$, $\hat{\beta}_1 = 12.74$, and $\hat{\beta}_2 = 85.95$. Therefore, the equation that minimizes SSE for this data set (i.e., the **least squares prediction equation**) is

$$\hat{y} = -1,339 + 12.74x_1 + 85.95x_2$$

Figure 4.3 SAS regression output for the auction price model, Example 4.1

```
                          The REG Procedure
                            Model: MODEL1
                       Dependent Variable: PRICE

              Number of Observations Read           32
              Number of Observations Used           32


                          Analysis of Variance

                               Sum of        Mean
  Source              DF      Squares       Square    F Value    Pr > F

  Model                2      4283063      2141531     120.19    <.0001
  Error               29       516727        17818
  Corrected Total     31      4799790


            Root MSE            133.48467    R-Square     0.8923
            Dependent Mean     1326.87500    Adj R-Sq     0.8849
            Coeff Var            10.06008


                          Parameter Estimates

                        Parameter      Standard
  Variable     DF        Estimate         Error    t Value    Pr > |t|

  Intercept     1     -1338.95134     173.80947      -7.70     <.0001
  AGE           1        12.74057       0.90474      14.08     <.0001
  NUMBIDS       1        85.95298       8.72852       9.85     <.0001
```

(c) The minimum value of the sum of the squared errors, also highlighted in Figure 4.3, is SSE $= 516,727$.

Example 4.2

Problem

Refer to the first-order model for auction price (y) considered in Example 4.1. Interpret the estimates of the β parameters in the model.

Solution

The least squares prediction equation, as given in Example 4.1, is $\hat{y} = -1,339 + 12.74x_1 + 85.95x_2$. We know that with first-order models, β_1 represents the slope of the line relating y to x_1 for fixed x_2. That is, β_1 measures the change in $E(y)$ for every one-unit increase in x_1 when the other independent variable in the model is held fixed. A similar statement can be made about β_2: β_2 measures the change in $E(y)$ for every one-unit increase in x_2 when the other x in the model is held fixed. Consequently, we obtain the following interpretations:

$\hat{\beta}_1 = 12.74$: We estimate the mean auction price $E(y)$ of an antique clock to increase \$12.74 for every 1-year increase in age (x_1) when the number of bidders (x_2) is held fixed.

$\hat{\beta}_2 = 85.95$: We estimate the mean auction price $E(y)$ of an antique clock to increase \$85.95 for every one-bidder increase in the number of bidders (x_2) when age (x_1) is held fixed.

The value $\hat{\beta}_0 = -1,339$ does not have a meaningful interpretation in this example. To see this, note that $\hat{y} = \hat{\beta}_0$ when $x_1 = x_2 = 0$. Thus, $\hat{\beta}_0 = -1,339$ represents the estimated mean auction price when the values of all the independent variables are set equal to 0. Because an antique clock with these characteristics—an age of 0 years and 0 bidders on the clock—is not practical, the value of $\hat{\beta}_0$ has no meaningful interpretation. In general, $\hat{\beta}_0$ will not have a practical interpretation unless it makes sense to set the values of the x's simultaneously equal to 0.

4.5 Estimation of σ^2, the Variance of ε

Recall that σ^2 is the variance of the random error ε. As such, σ^2 is an important measure of model utility. If $\sigma^2 = 0$, all the random errors will equal 0 and the prediction equation $\hat{y}$ will be identical to $E(y)$, that is, $E(y)$ will be estimated without error. In contrast, a large value of σ^2 implies large (absolute) values of ε and larger deviations between the prediction equation $\hat{y}$ and the mean value $E(y)$. Consequently, the larger the value of σ^2, the greater will be the error in estimating the model parameters $\beta_0, \beta_1, \ldots, \beta_k$ and the error in predicting a value of y for a specific set of values of $x_1, x_2, \ldots, x_k$. Thus, σ^2 plays a major role in making inferences about $\beta_0, \beta_1, \ldots, \beta_k$, in estimating $E(y)$, and in predicting y for specific values of $x_1, x_2, \ldots, x_k$.

Since the variance σ^2 of the random error ε will rarely be known, we must use the results of the regression analysis to estimate its value. Recall that σ^2 is the variance of the probability distribution of the random error ε for a given set of values for $x_1, x_2, \ldots, x_k$; hence, it is the mean value of the squares of the deviations of the y-values (for given values of $x_1, x_2, \ldots, x_k$) about the mean value $E(y)$.* Since

---

* Because $y = E(y) + \varepsilon$, then ε is equal to the deviation $y - E(y)$. Also, by definition, the variance of a random variable is the expected value of the square of the deviation of the random variable from its mean. According to our model, $E(\varepsilon) = 0$. Therefore, $\sigma^2 = E(\varepsilon^2)$.

the predicted value $\hat{y}$ estimates $E(y)$ for each of the data points, it seems natural to use

$$\text{SSE} = \sum (y_i - \hat{y}_i)^2$$

to construct an estimator of σ^2.

Estimator of σ^2 for Multiple Regression Model with k Independent Variables

$$s^2 = \text{MSE} = \frac{\text{SSE}}{n - \text{Number of estimated } \beta \text{ parameters}}$$

$$= \frac{\text{SSE}}{n - (k+1)}$$

For example, in the first-order model of Example 4.1, we found that SSE = 516,727. We now want to use this quantity to estimate the variance of ε. Recall that the estimator for the straight-line model is $s^2 = \text{SSE}/(n-2)$ and note that the denominator is (n − Number of estimated β parameters), which is $(n-2)$ in the straight-line model. Since we must estimate three parameters, β_0, β_1, first-order model in Example 4.1, the estimator of σ^2 is

$$s^2 = \frac{\text{SSE}}{n-3}$$

The numerical estimate for this example is

$$s^2 = \frac{\text{SSE}}{32-3} = \frac{516,727}{29} = 17,818$$

In many computer printouts and textbooks, s^2 is called the **mean square for error (MSE)**. This estimate of σ^2 is highlighted in the SAS printout in Figure 4.3.

The units of the estimated variance are squared units of the dependent variable y. Since the dependent variable y in this example is auction price in dollars, the units of s^2 are (dollars)^2. This makes meaningful interpretation of s^2 difficult, so we use the standard deviation s to provide a more meaningful measure of variability. In this example,

$$s = \sqrt{17,818} = 133.5$$

which is highlighted on the SAS printout in Figure 4.3 (next to **Root MSE**). One useful interpretation of the estimated standard deviation s is that the interval $\pm 2s$ will provide a rough approximation to the accuracy with which the model will predict future values of y for given values of x. Thus, in Example 4.1, we expect the model to provide predictions of auction price to within about $\pm 2s = \pm 2(133.5) = \pm 267$ dollars.[†]

For the general multiple regression model

$$y = \beta_0 + \beta_1 x_1 + \beta_2 x_2 + \cdots + \beta_k x_k + \varepsilon$$

we must estimate the $(k+1)$ parameters $\beta_0, \beta_1, \beta_2, \ldots, \beta_k$. Thus, the estimator of σ^2 is SSE divided by the quantity (n − Number of estimated β parameters).

[†] The $\pm 2s$ approximation will improve as the sample size is increased. We provide more precise methodology for the construction of prediction intervals in Section 4.9.

We use MSE, the estimator of σ^2, both to check the utility of the model (Sections 4.6 and 4.7) and to provide a measure of the reliability of predictions and estimates when the model is used for those purposes (Section 4.9). Thus, you can see that the estimation of σ^2 plays an important part in the development of a regression model.

4.6 Testing the Utility of a Model: The Analysis of Variance F-Test

The objective of step 5 in a multiple regression analysis is to conduct a test of the utility of the model—that is, a test to determine whether the model is adequate for predicting y. In Section 4.7, we demonstrate how to conduct t-tests on each β parameter in a model, where H_0: $\beta_i = 0, i = 1, 2, \ldots k$. However, this approach is generally **not** a good way to determine whether the overall model is contributing information for the prediction of y. If we were to conduct a series of t-tests to determine whether the independent variables are contributing to the predictive relationship, we would be very likely to make one or more errors in deciding which terms to retain in the model and which to exclude.

Suppose you fit a first-order model with 10 quantitative independent variables, $x_1, x_2, \ldots, x_{10}$, and decide to conduct t-tests on all 10 individual β's in the model, each at $\alpha = .05$. Even if all the β parameters (except β_0) in the model are equal to 0, approximately 40% of the time you will incorrectly reject the null hypothesis at least once and conclude that some β parameter is nonzero.* In other words, the overall Type I error is about .40, not .05!

Thus, in multiple regression models for which a large number of independent variables are being considered, conducting a series of t-tests may cause the experimenter to include a large number of insignificant variables and exclude some useful ones. If we want to test the utility of a multiple regression model, we will need a **global test** (one that encompasses all the β parameters).

For the general multiple regression model, $E(y) = \beta_0 + \beta_1 x_1 + \beta_2 x_2 + \cdots + \beta_k x_k$, we would test

$$H_0: \beta_1 = \beta_2 = \beta_3 = \cdots = \beta_k = 0$$

$$H_a: \text{At least one of the coefficients is nonzero}$$

The test statistic used to test this hypothesis is an F statistic, and several equivalent versions of the formula can be used (although we will usually rely on statistical software to calculate the F statistic):

$$\text{Test statistic: } F = \frac{(\text{SS}_{yy} - \text{SSE})/k}{\text{SSE}/[n - (k + 1)]} = \frac{\text{Mean Square(Model)}}{\text{MSE}}$$

---

* The proof of this result proceeds as follows:

$$P(\text{Reject } H_0 \text{ at least once} \mid \beta_1 = \beta_2 = \cdots = \beta_{10} = 0)$$

$$= 1 - P(\text{Reject } H_0 \text{ no times} \mid \beta_1 = \beta_2 = \cdots = \beta_{10} = 0)$$

$$\leq 1 - [P(\text{Accept } H_0: \beta_1 = 0 \mid \beta_1 = 0) \times P(\text{Accept } H_0: \beta_2 = 0 \mid \beta_2 = 0) \cdots$$

$$\times P(\text{Accept } H_0: \beta_{10} = 0 \mid \beta_{10} = 0)]$$

$$= 1 - [(1 - \alpha)^{10}] = 1 - (.95)^{10} = .401$$

Note that the denominator of the F statistic, MSE, represents the *unexplained* (or error) variability in the model. The numerator, MS(Model), represents the variability in y *explained* (or accounted for) by the model. (For this reason, the test is often called the "analysis-of-variance" F-test.) Since F is the ratio of the *explained* variability to the *unexplained* variability, the larger the proportion of the total variability accounted for by the model, the larger the F statistic.

To determine when the ratio becomes large enough that we can confidently reject the null hypothesis and conclude that the model is more useful than no model at all for predicting y, we compare the calculated F statistic to a tabulated F-value with k df in the numerator and $[n - (k + 1)]$ df in the denominator. Recall that tabulations of the F-distribution for various values of α are given in Tables 3, 4, 5, and 6 of Appendix D.

Rejection region: $F > F_\alpha$, where F is based on k numerator and $n - (k + 1)$ denominator degrees of freedom (see Figure 4.4).

However, since statistical software printouts report the observed significance level (p-value) of the test, most researchers simply compare the selected α value to the p-value to make the decision.

The analysis of variance F-test for testing the usefulness of the model is summarized in the next box.

Testing Global Usefulness of the Model: The Analysis of Variance F-Test

H_0: $\beta_1 = \beta_2 = \cdots = \beta_k = 0$ (All model terms are unimportant for predicting y)
H_a: At least one $\beta_i \neq 0$ (At least one model term is useful for predicting y)

Test statistic: $F = \dfrac{(SS_{yy} - SSE)/k}{SSE/[n - (k + 1)]} = \dfrac{R^2/k}{(1 - R^2)/[n - (k + 1)]}$

$\qquad\qquad = \dfrac{\text{Mean square (Model)}}{\text{Mean square (Error)}}$

where n is the sample size and k is the number of terms in the model.

Rejection region: $F > F_\alpha$, with k numerator degrees of freedom and $[n - (k + 1)]$ denominator degrees of freedom.

   or

$\alpha > p$-value, where p-value $= P(F > F_c)$, F_c is the computed value of the test statistic.

Assumptions: The standard regression assumptions about the random error component (Section 4.2).

Example 4.3

Refer to Example 4.2, in which an antique collector modeled the auction price y of grandfather clocks as a function of the age of the clock, x_1, and the number of bidders, x_2. The hypothesized first-order model is

$$y = \beta_0 + \beta_1 x_1 + \beta_2 x_2 + \varepsilon$$

Figure 4.4 Rejection region for the global *F*-test

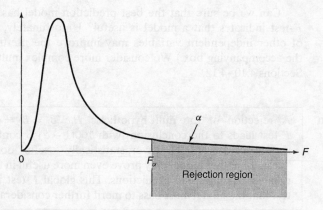

A sample of 32 observations is obtained, with the results summarized in the MINITAB printout in Figure 4.5. Conduct the global *F*-test of model usefulness at the $\alpha = .05$ level of significance.

Figure 4.5 MINITAB regression printout for grandfather clock model

---

Regression Analysis: PRICE versus AGE, NUMBIDS

```
The regression equation is
PRICE = - 1339 + 12.7 AGE + 86.0 NUMBIDS


Predictor      Coef   SE Coef       T      P
Constant    -1339.0     173.8   -7.70  0.000
AGE         12.7406    0.9047   14.08  0.000
NUMBIDS      85.953     8.729    9.85  0.000


S = 133.485    R-Sq = 89.2%    R-Sq(adj) = 88.5%


Analysis of Variance

Source          DF        SS       MS       F      P
Regression       2   4283063  2141531  120.19  0.000
Residual Error  29    516727    17818
Total           31   4799790
```

Solution

The elements of the global test of the model follow:

H_0: $\beta_1 = \beta_2 = 0$ [Note: $k = 2$]

H_a: At least one of the two model coefficients is nonzero

Test statistic: $F = 120.19$ (shaded in Figure 4.5)

p-value $= .000$ (shaded in Figure 4.5)

Conclusion: Since $\alpha = .05$ exceeds the observed significance level, $p = .000$, the data provide strong evidence that at least one of the model coefficients is nonzero. The overall model appears to be statistically useful for predicting auction prices.

Can we be sure that the best prediction model has been found if the global F-test indicates that a model is useful? Unfortunately, we cannot. The addition of other independent variables may improve the usefulness of the model. (See the accompanying box.) We consider more complex multiple regression models in Sections 4.10–4.12.

Caution

> A rejection of the null hypothesis $H_0: \beta_1 = \beta_2 = \cdots = \beta_k$ in the global F-test leads to the conclusion [with $100(1 - \alpha)\%$ confidence] that the model is statistically useful. However, statistically "useful" does not necessarily mean "best." Another model may prove even more useful in terms of providing more reliable estimates and predictions. This global F-test is usually regarded as a test that the model *must* pass to merit further consideration.

4.7 Inferences About the Individual β Parameters

Inferences about the individual β parameters in a model are obtained using either a confidence interval or a test of hypothesis, as outlined in the following two boxes.[*]

Test of an Individual Parameter Coefficient in the Multiple Regression Model

	ONE-TAILED TESTS		TWO-TAILED TEST
	$H_0: \beta_i = 0$ $H_0: \beta_i = 0$		$H_0: \beta_i = 0$
	$H_a: \beta_i < 0$ $H_a: \beta_i > 0$		$H_a: \beta_i \neq 0$

Test statistic: $t = \dfrac{\hat{\beta}_i}{s_{\hat{\beta}_i}}$

Rejection region: $t < -t_\alpha$ $t > t_\alpha$ $|t| > t_{\alpha/2}$

where t_α and $t_{\alpha/2}$ are based on $n - (k + 1)$ degrees of freedom and

n = Number of observations
$k + 1$ = Number of β parameters in the model

Note: Most statistical software programs report two-tailed p-values on their output. To find the appropriate p-value for a one-tailed test, make the following adjustment to P = two-tailed p-value:

$$\text{For } H_a: \beta_i > 0,\ p\text{-value} = \begin{cases} P/2 & \text{if } t > 0 \\ 1 - P/2 & \text{if } t < 0 \end{cases}$$

$$\text{For } H_a: \beta_i < 0,\ p\text{-value} = \begin{cases} 1 - P/2 & \text{if } t > 0 \\ P/2 & \text{if } t < 0 \end{cases}$$

Assumptions: See Section 4.2 for assumptions about the probability distribution for the random error component ε.

---

[*] The formulas for computing $\hat{\beta}_i$ and its standard error are so complex, the only reasonable way to present them is by using matrix algebra. We do not assume a prerequisite of matrix algebra for this text and, in any case, we think the formulas can be omitted in an introductory course without serious loss. They are programmed into all statistical software packages with multiple regression routines and are presented in some of the texts listed in the references.

A 100 $(1 - \alpha)$% Confidence Interval for a β Parameter

$$\hat{\beta}_i \pm (t_{\alpha/2})s_{\hat{\beta}_i}$$

where $t_{\alpha/2}$ is based on $n - (k + 1)$ degrees of freedom and

n = Number of observations

$k + 1$ = Number of β parameters in the model

We illustrate these methods with another example.

Example 4.4

Refer to Examples 4.1–4.3. A collector of antique grandfather clocks knows that the price (y) received for the clocks increases linearly with the age (x_1) of the clocks. Moreover, the collector hypothesizes that the auction price (y) of the clocks will increase linearly as the number of bidders (x_2) increases. Use the information on the SAS printout, Figure 4.6, to:

(a) Test the hypothesis that the mean auction price of a clock increases as the number of bidders increases when age is held constant, that is, $\beta_2 > 0$. Use $\alpha = .05$.

(b) Form a 95% confidence interval for β_1 and interpret the result.

Solution

(a) The hypotheses of interest concern the parameter β_2. Specifically,

$$H_0: \beta_2 = 0$$

$$H_a: \beta_2 > 0$$

```
                        The REG Procedure
                         Model: MODEL1
                    Dependent Variable: PRICE

              Number of Observations Read          32
              Number of Observations Used          32


                      Analysis of Variance

                             Sum of        Mean
  Source            DF      Squares      Square    F Value    Pr > F

  Model              2      4283063     2141531     120.19    <.0001
  Error             29       516727       17818
  Corrected Total   31      4799790


          Root MSE            133.48467    R-Square     0.8923
          Dependent Mean     1326.87500    Adj R-Sq     0.8849
          Coeff Var            10.06008


                      Parameter Estimates

               Parameter     Standard
  Variable  DF  Estimate        Error   t Value   Pr > |t|    95% Confidence Limits

  Intercept  1  -1338.95134   173.80947    -7.70    <.0001   -1694.43162    -983.47106
  AGE        1     12.74057     0.90474    14.08    <.0001      10.89017      14.59098
  NUMBIDS    1     85.95298     8.72852     9.85    <.0001      68.10115     103.80482
```

Figure 4.6 SAS regression output for the auction price model, Example 4.4

The test statistic is a t statistic formed by dividing the sample estimate $\hat{\beta}_2$ of the parameter β_2 by the estimated standard error of $\hat{\beta}_2$ (denoted $s_{\hat{\beta}_2}$). These estimates, $\hat{\beta}_2 = 85.953$ and $s_{\hat{\beta}_2} = 8.729$, as well as the calculated t-value, are highlighted on the SAS printout, Figure 4.6

$$\text{Test statistic: } t = \frac{\hat{\beta}_2}{s_{\hat{\beta}_2}} = \frac{85.953}{8.729} = 9.85$$

The p-value for the two-tailed test of hypothesis, $H_a: \beta_2 \pm 0$, is also shown on the printout under **Pr > |t|**. This value (highlighted) is less than .0001. To obtain the p-value for the one-tailed test, $H_a: \beta_2 > 0$, we divide this p-value in half. Consequently, the observed significance level for our upper-tailed test is p-value $= \dfrac{.0001}{2} = .00005$.

Since $\alpha = .05$ exceeds p-value $= .0005$, we have sufficient evidence to reject H_0. Thus, the collector can conclude that the mean auction price of a clock increases as the number of bidders increases, when age is held constant.

(b) A 95% confidence interval for β_1 is (from the box):

$$\hat{\beta}_1 \pm (t_{\alpha/2})s_{\hat{\beta}_1} = \hat{\beta}_1 \pm (t_{.05})s_{\hat{\beta}_1}$$

Substituting $\hat{\beta}_1 = 12.74$, $s_{\hat{\beta}_i} = .905$ (both obtained from the SAS printout, Figure 4.6) and $t_{.025} = 2.045$ (from Table C.2) into the equation, we obtain

$$12.74 \pm (2.045)(.905) = 12.74 \pm 1.85$$

or (10.89, 14.59). This interval is also shown (highlighted) on the SAS printout. Thus, we are 95% confident that β_1 falls between 10.89 and 14.59. Since β_1 is the slope of the line relating auction price (y) to age of the clock (x_1), we conclude that price increases between \$10.89 and \$14.59 for every 1-year increase in age, holding number of bidders (x_2) constant.

After we have determined that the overall model is useful for predicting y using the F-test (Section 4.6), we may elect to conduct one or more t-tests on the individual β parameters (as in Example 4.4). However, the test (or tests) to be conducted should be decided *a priori*, that is, prior to fitting the model. Also, we should limit the number of t-tests conducted to avoid the potential problem of making too many Type I errors. Generally, the regression analyst will conduct t-tests only on the "most important" β's. We provide insight in identifying the most important β's in a linear model in the next several sections.

Recommendation for Checking the Utility of a Multiple Regression Model

1. First, conduct a test of overall model adequacy using the F-test, that is, test
   $H_0: \beta_1 = \beta_2 = \cdots = \beta_k = 0$
   If the model is deemed adequate (i.e., if you reject H_0), then proceed to step 2. Otherwise, you should hypothesize and fit another model. The new model may include more independent variables or higher-order terms.

2. Conduct t-tests on those β parameters in which you are particularly interested (i.e., the "most important" β's). These usually involve only the β's associated with higher-order terms (x^2, x_1x_2, etc.). However, it is a safe practice to limit the number of β's that are tested. Conducting a series of t-tests leads to a high overall Type I error rate α.

We conclude this section with a final caution about conducting t-tests on individual β parameters in a model.

Caution

> Extreme care should be exercised when conducting t-tests on the individual β parameters in a *first-order linear model* for the purpose of determining which independent variables are useful for predicting y and which are not. If you fail to reject H_0: $\beta_i = 0$, several conclusions are possible:
>
> 1. There is no relationship between y and x_i.
> 2. A straight-line relationship between y and x exists (holding the other x's in the model fixed), but a Type II error occurred.
> 3. A relationship between y and x_i (holding the other x's in the model fixed) exists, but is more complex than a straight-line relationship (e.g., a curvilinear relationship may be appropriate). The most you can say about a β parameter test is that there is either sufficient (if you reject H_0: $\beta_i = 0$) or insufficient (if you do not reject H_0: $\beta_i = 0$) evidence of a *linear (straight-line)* relationship between y and x_i.

4.8 Multiple Coefficients of Determination: R^2 and R_a^2

Recall from Chapter 3 that the coefficient of determination, r^2, is a measure of how well a straight-line model fits a data set. To measure how well a multiple regression model fits a set of data, we compute the multiple regression equivalent of r^2, called the **multiple coefficient of determination** and denoted by the symbol R^2.

> **Definition 4.1** The **multiple coefficient of determination**, R^2, is defined as
>
> $$R^2 = 1 - \frac{\text{SSE}}{\text{SS}_{yy}} \quad 0 \le R^2 \le 1$$
>
> where $\text{SSE} = \sum(y_i - \hat{y}_i)^2$, $\text{SS}_{yy} = \sum(y_i - \bar{y})^2$, and $\hat{y}_i$ is the predicted value of y_i for the multiple regression model.

Just as for the simple linear model, R^2 represents the fraction of the sample variation of the y-values (measured by SS_{yy}) that is explained by the least squares regression model. Thus, $R^2 = 0$ implies a complete lack of fit of the model to the data, and $R^2 = 1$ implies a perfect fit, with the model passing through every data point. In general, the closer the value of R^2 is to 1, the better the model fits the data.

To illustrate, consider the first-order model for the grandfather clock auction price presented in Examples 4.1–4.4. A portion of the SPSS printout of the analysis is shown in Figure 4.7. The value $R^2 = .892$ is highlighted on the printout. This relatively high value of R^2 implies that using the independent variables age and number of bidders in a first-order model explains 89.2% of the total *sample variation* (measured by SS_{yy}) in auction price y. Thus, R^2 is a sample statistic that tells how well the model fits the data and thereby represents a measure of the usefulness of the entire model.

Figure 4.7 A portion of the SPSS regression output for the auction price model

Model Summary

Model	R	R Square	Adjusted R Square	Std. Error of the Estimate
1	.945[a]	.892	.885	133.485

a. Predictors: (Constant), NUMBIDS, AGE

A large value of R^2 computed from the *sample* data does not necessarily mean that the model provides a good fit to all of the data points in the *population*. For example, a first-order linear model that contains three parameters will provide a perfect fit to a sample of three data points and R^2 will equal 1. Likewise, you will always obtain a perfect fit ($R^2 = 1$) to a set of n data points if the model contains exactly n parameters. Consequently, if you want to use the value of R^2 as a measure of how useful the model will be for predicting y, it should be based on a sample that contains substantially more data points than the number of parameters in the model.

Caution

> In a multiple regression analysis, use the value of R^2 as a measure of how useful a linear model will be for predicting y only if the sample contains substantially more data points than the number of β parameters in the model.

As an alternative to using R^2 as a measure of model adequacy, the **adjusted multiple coefficient of determination**, denoted R_a^2, is often reported. The formula for R_a^2 is shown in the box.

> **Definition 4.2** The **adjusted multiple coefficient of determination** is given by
>
> $$R_a^2 = 1 - \left[\frac{(n-1)}{n-(k+1)}\right]\left(\frac{SSE}{SS_{yy}}\right)$$
>
> $$= 1 - \left[\frac{(n-1)}{n-(k+1)}\right](1-R^2)$$
>
> *Note*: $R_a^2 \leq R^2$ and, for poor-fitting models R_a^2 may be negative.

R^2 and R_a^2 have similar interpretations. However, unlike R^2, R_a^2 takes into account ("adjusts" for) both the sample size n and the number of β parameters in the model. R_a^2 will always be smaller than R^2, and more importantly, cannot be "forced" to 1 by simply adding more and more independent variables to the model. Consequently, analysts prefer the more conservative R_a^2 when choosing a measure of model adequacy. The value of R_a^2 is also highlighted in Figure 4.7. Note that $R_a^2 = .885$, a value only slightly smaller than R^2.

Despite their utility, R^2 and R_a^2 are only sample statistics. Consequently, it is dangerous to judge the usefulness of the model based solely on these values. A prudent analyst will use the analysis-of-variance F-test for testing the global utility of the multiple regression model. Once the model has been deemed "statistically" useful with the F-test, the more conservative value of R_a^2 is used to describe the proportion of variation in y explained by the model.

4.8 Exercises

4.1 Degrees of freedom. How is the number of degrees of freedom available for estimating σ^2, the variance of ε, related to the number of independent variables in a regression model?

4.2 Accounting and Machiavellianism. Refer to the *Behavioral Research in Accounting* (January 2008) study of Machiavellian traits (e.g., manipulation, cunning, duplicity, deception, and bad faith) in accountants, Exercise 1.47 (p. 41). Recall that a Machiavellian ("Mach") rating score was determined for each in a sample of accounting alumni of a large southwestern university. For one portion of the study, the researcher modeled an accountant's Mach score (y) as a function of age (x_1), gender (x_2), education (x_3), and income (x_4). Data on $n = 198$ accountants yielded the results shown in the table.

INDEPENDENT VARIABLE	t-VALUE FOR H_0: $\beta_i = 0$	p-VALUE
Age (x_1)	0.10	> .10
Gender (x_2)	−0.55	> .10
Education (x_3)	1.95	< .01
Income (x_4)	0.52	> .10

Overall model: $R^2 = .13$, $F = 4.74$ (p-value < .01)

(a) Write the equation of the hypothesized model relating y to x_1, x_2, x_3, and x_4.
(b) Conduct a test of overall model utility. Use $\alpha = .05$.
(c) Interpret the coefficient of determination, R^2.
(d) Is there sufficient evidence (at $\alpha = .05$) to say that income is a statistically useful predictor of Mach score?

4.3 Study of adolescents with ADHD. Children with attention-deficit/hyperactivity disorder (ADHD) were monitored to evaluate their risk for substance (e.g., alcohol, tobacco, illegal drug) use (*Journal of Abnormal Psychology*, August 2003). The following data were collected on 142 adolescents diagnosed with ADHD:

   y = frequency of marijuana use the past 6 months

   x_1 = severity of inattention (5-point scale)

   x_2 = severity of impulsivity–hyperactivity (5-point scale)

   x_3 = level of oppositional–defiant and conduct disorder (5-point scale)

(a) Write the equation of a first-order model for $E(y)$.

(b) The coefficient of determination for the model is $R^2 = .08$. Interpret this value.
(c) The global F-test for the model yielded a p-value less than .01. Interpret this result.
(d) The t-test for H_0: $\beta_1 = 0$ resulted in a p-value less than .01. Interpret this result.
(e) The t-test for H_0: $\beta_2 = 0$ resulted in a p-value greater than .05. Interpret this result.
(f) The t-test for H_0: $\beta_3 = 0$ resulted in a p-value greater than .05. Interpret this result.

4.4 Characteristics of lead users. During new product development, companies often involve "lead users" (i.e., creative individuals who are on the leading edge of an important market trend). *Creativity and Innovation Management* (February 2008) published an article on identifying the social network characteristics of lead users of children's computer games. Data were collected for $n = 326$ children and the following variables measured: lead-user rating (y, measured on a 5-point scale), gender ($x_1 = 1$ if female, 0 if male), age (x_2, years), degree of centrality (x_3, measured as the number of direct ties to other peers in the network), and betweenness centrality (x_4, measured as the number of shortest paths between peers). A first-order model for y was fit to the data, yielding the following least squares prediction equation:

$$\hat{y} = 3.58 + .01x_1 - .06x_2 - .01x_3 + .42x_4$$

(a) Give two properties of the errors of prediction that result from using the method of least squares to obtain the parameter estimates.
(b) Give a practical interpretation the estimate of β_4 in the model.
(c) A test of H_0: $\beta_4 = 0$ resulted in a two-tailed p-value of .002. Make the appropriate conclusion at $\alpha = .05$.

4.5 Runs scored in baseball. In *Chance* (Fall 2000), statistician Scott Berry built a multiple regression model for predicting total number of runs scored by a Major League Baseball team during a season. Using data on all teams over a 9-year period (a sample of $n = 234$), the results in the next table (p. 184) were obtained.

(a) Write the least squares prediction equation for y = total number of runs scored by a team in a season.
(b) Conduct a test of H_0: $\beta_7 = 0$ against H_a: $\beta_7 < 0$ at $\alpha = .05$. Interpret the results.
(c) Form a 95% confidence interval for β_5. Interpret the results.
(d) Predict the number of runs scored by your favorite Major League Baseball team last

year. How close is the predicted value to the actual number of runs scored by your team? (*Note*: You can find data on your favorite team on the Internet at www.mlb.com.)

INDEPENDENT VARIABLE	β ESTIMATE	STANDARD ERROR
Intercept	3.70	15.00
Walks (x_1)	.34	.02
Singles (x_2)	.49	.03
Doubles (x_3)	.72	.05
Triples (x_4)	1.14	.19
Home Runs (x_5)	1.51	.05
Stolen Bases (x_6)	.26	.05
Caught Stealing (x_7)	−.14	.14
Strikeouts (x_8)	−.10	.01
Outs (x_9)	−.10	.01

Source: Berry, S. M. "A statistician reads the sports pages: Modeling offensive ability in baseball," *Chance*, Vol. 13, No. 4, Fall 2000 (Table 2).

4.6 **Earnings of Mexican street vendors.** Detailed interviews were conducted with over 1,000 street vendors in the city of Puebla, Mexico, in order to study the factors influencing vendors' incomes (*World Development*, February 1998). Vendors were defined as individuals working in the street, and included vendors with carts and stands on wheels and excluded beggars, drug dealers, and prostitutes. The researchers collected data on gender, age, hours worked per day, annual earnings, and education level. A subset of these data appears in the accompanying table.

(a) Write a first-order model for mean annual earnings, $E(y)$, as a function of age (x_1) and hours worked (x_2).

STREETVEN

VENDOR NUMBER	ANNUAL EARNINGS y	AGE x_1	HOURS WORKED PER DAY x_2
21	$2841	29	12
53	1876	21	8
60	2934	62	10
184	1552	18	10
263	3065	40	11
281	3670	50	11
354	2005	65	5
401	3215	44	8
515	1930	17	8
633	2010	70	6
677	3111	20	9
710	2882	29	9
800	1683	15	5
914	1817	14	7
997	4066	33	12

Source: Adapted from Smith, P. A., and Metzger, M. R. "The return to education: Street vendors in Mexico," *World Development*, Vol. 26, No. 2, Feb. 1998, pp. 289–296.

(b) The model was fit to the data using SAS. Find the least squares prediction equation on the printout shown below.
(c) Interpret the estimated β coefficients in your model.
(d) Conduct a test of the global utility of the model (at $\alpha = .01$). Interpret the result.
(e) Find and interpret the value of R_a^2.
(f) Find and interpret s, the estimated standard deviation of the error term.
(g) Is age (x_1) a statistically useful predictor of annual earnings? Test using $\alpha = .01$.
(h) Find a 95% confidence interval for β_2. Interpret the interval in the words of the problem.

SAS output for Exercise 4.6

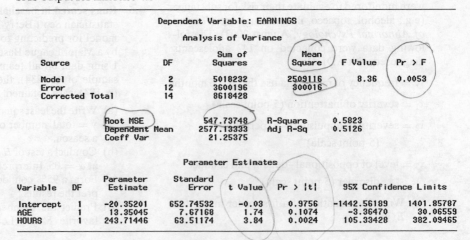

```
Dependent Variable: EARNINGS

                     Analysis of Variance

                              Sum of        Mean
  Source          DF         Squares      Square     F Value    Pr > F
  Model            2         5018232     2509116        8.36    0.0053
  Error           12         3600196      300016
  Corrected Total 14         8618428

            Root MSE              547.73748    R-Square     0.5823
            Dependent Mean       2577.13333    Adj R-Sq     0.5126
            Coeff Var              21.25375

                        Parameter Estimates

                    Parameter      Standard
  Variable    DF     Estimate         Error    t Value   Pr > |t|      95% Confidence Limits
  Intercept    1    -20.35201     652.74532      -0.03     0.9756    -1442.56189    1401.85787
  AGE          1     13.35045       7.67168       1.74     0.1074       -3.36470      30.06559
  HOURS        1    243.71446      63.51174       3.84     0.0024      105.33428     382.09465
```

4.7 Urban population estimation using satellite images. Can the population of an urban area be estimated without taking a census? In *Geographical Analysis* (January 2007) geography professors at the University of Wisconsin–Milwaukee and Ohio State University demonstrated the use of satellite image maps for estimating urban population. A portion of Columbus, Ohio, was partitioned into $n = 125$ census block groups and satellite imagery was obtained. For each census block, the following variables were measured: population density (y), proportion of block with low-density residential areas (x_1), and proportion of block with high-density residential areas (x_2). A first-order model for y was fit to the data with the following results:

$$\hat{y} = -.0304 + 2.006x_1 + 5.006x_2, \quad R^2 = .686$$

(a) Give a practical interpretation of each β-estimate in the model.
(b) Give a practical interpretation of the coefficient of determination, R^2.
(c) State H_0 and H_a for a test of overall model adequacy.
(d) Refer to part c. Compute the value of the test statistic.
(e) Refer to parts c and d. Make the appropriate conclusion at $\alpha = .01$.

4.8 Novelty of a vacation destination. Many tourists choose a vacation destination based on the newness or uniqueness (i.e., the novelty) of the itinerary. Texas A&M University professor J. Petrick investigated the relationship between novelty and vacationing golfers' demographics (*Annals of Tourism Research*, Vol. 29, 2002). Data were obtained from a mail survey of 393 golf vacationers to a large coastal resort in southeastern United States. Several measures of novelty level (on a numerical scale) were obtained for each vacationer, including "change from routine," "thrill," "boredom-alleviation," and "surprise." The researcher employed four independent variables in a regression model to predict each of the novelty measures. The independent variables were x_1 = number of rounds of golf per year, x_2 = total number of golf vacations taken, x_3 = number of years played golf, and x_4 = average golf score.

(a) Give the hypothesized equation of a first-order model for y = change from routine.
(b) A test of $H_0: \beta_3 = 0$ versus $H_a: \beta_3 < 0$ yielded a p-value of .005. Interpret this result if $\alpha = .01$.
(c) The estimate of β_3 was found to be negative. Based on this result (and the result of part b), the researcher concluded that "those who have played golf for more years are less apt to seek change from their normal routine in their golf vacations." Do you agree with this statement? Explain.
(d) The regression results for the three other dependent novelty measures are summarized in the table below. Give the null hypothesis for testing the overall adequacy of each first-order regression model.

DEPENDENT VARIABLE	F-VALUE	p-VALUE	R^2
Thrill	5.56	$< .001$	.055
Boredom-alleviation	3.02	.018	.030
Surprise	3.33	.011	.023

Source: Reprinted from *Annals of Tourism Research,* Vol. 29, Issue 2, J. F. Petrick, "An examination of golf vacationers' novelty," Copyright © 2002, with permission from Elsevier.

(e) Give the rejection region for the test, part d, using $\alpha = .01$.
(f) Use the test statistics reported in the table and the rejection region from part e to conduct the test for each of the dependent measures of novelty.
(g) Verify that the p-values in the table support your conclusions in part f.
(h) Interpret the values of R^2 reported in the table.

4.9 Highway crash data analysis. Researchers at Montana State University have written a tutorial on an empirical method for analyzing before and after highway crash data (Montana Department of Transportation, Research Report, May 2004). The initial step in the methodology is to develop a Safety Performance Function (SPF)—a mathematical model that estimates crash occurrence for a given roadway segment. Using data collected for over 100 roadway segments, the researchers fit the model, $E(y) = \beta_0 + \beta_1 x_1 + \beta_2 x_2$, where y = number of crashes per 3 years, x_1 = roadway length (miles), and x_2 = AADT (average annual daily traffic) (number of vehicles). The results are shown in the following tables.

Interstate Highways

VARIABLE	PARAMETER ESTIMATE	STANDARD ERROR	t-VALUE
Intercept	1.81231	.50568	3.58
Length (x_1)	.10875	.03166	3.44
AADT (x_2)	.00017	.00003	5.19

Non-Interstate Highways

VARIABLE	PARAMETER ESTIMATE	STANDARD ERROR	t-VALUE
Intercept	1.20785	.28075	4.30
Length (x_1)	.06343	.01809	3.51
AADT (x_2)	.00056	.00012	4.86

(a) Give the least squares prediction equation for the interstate highway model.

(b) Give practical interpretations of the β estimates, part a.

(c) Refer to part a. Find a 99% confidence interval for β_1 and interpret the result.

(d) Refer to part a. Find a 99% confidence interval for β_2 and interpret the result.

(e) Repeat parts a–d for the non-interstate highway model.

4.10 **Snow geese feeding trial.** Refer to the *Journal of Applied Ecology* (Vol. 32, 1995) study of the feeding habits of baby snow geese, Exercise 3.46 (p. 127). The data on gosling weight change, digestion efficiency, acid-detergent fiber (all measured as percentages) and diet (plants or duck chow) for 42 feeding trials are saved in the SNOWGEESE file. (The table shows selected observations.) The botanists were interested in predicting weight change (y) as a function of the other variables. The first-order model $E(y) = \beta_0 + \beta_1 x_1 + \beta_2 x_2$, where x_1 is digestion efficiency and x_2 is acid-detergent fiber, was fit to the data. The MINITAB printout is given below.

(a) Find the least squares prediction equation for weight change, y.

(b) Interpret the β-estimates in the equation, part a.

(c) Conduct the F-test for overall model adequacy using $\alpha = .01$.

(d) Find and interpret the values of R^2 and R_a^2. Which is the preferred measure of model fit?

(e) Conduct a test to determine if digestion efficiency, x_1, is a useful linear predictor of weight change. Use $\alpha = .01$.

(f) Form a 99% confidence interval for β_2. Interpret the result.

SNOWGEESE (First and last five trials)

FEEDING TRIAL	DIET	WEIGHT CHANGE (%)	DIGESTION EFFICIENCY (%)	ACID-DETERGENT FIBER (%)
1	Plants	−6	0	28.5
2	Plants	−5	2.5	27.5
3	Plants	−4.5	5	27.5
4	Plants	0	0	32.5
5	Plants	2	0	32
38	Duck Chow	9	59	8.5
39	Duck Chow	12	52.5	8
40	Duck Chow	8.5	75	6
41	Duck Chow	10.5	72.5	6.5
42	Duck Chow	14	69	7

Source: Gadallah, F. L., and Jefferies, R. L. "Forage quality in brood rearing areas of the lesser snow goose and the growth of captive goslings," *Journal of Applied Ecology*, Vol. 32, No. 2, 1995, pp. 281–282 (adapted from Figures 2 and 3).

4.11 **Deep space survey of quasars.** A quasar is a distant celestial object (at least 4 billion light-years away) that provides a powerful source of radio energy. The *Astronomical Journal* (July 1995) reported on a study of 90 quasars detected by a deep space survey. The survey enabled astronomers to measure several different quantitative characteristics of each quasar, including redshift range, line flux (erg/cm$^2 \cdot$ s), line luminosity (erg/s), AB$_{1450}$ magnitude, absolute magnitude, and rest frame equivalent width. The data for a sample of 25 large (redshift) quasars is listed in the table on p. 187.

(a) Hypothesize a first-order model for equivalent width, y, as a function of the first four variables in the table.

MINITAB output for Exercise 4.10

```
The regression equation is
WTCHANGE = 12.2 - 0.0265 DIGEST - 0.458 ADFIBRE


Predictor      Coef      SE Coef        T       P
Constant     12.180        4.402     2.77   0.009
DIGEST      -0.02654      0.05349    -0.50   0.623
ADFIBRE     -0.4578        0.1283    -3.57   0.001


S = 3.519     R-Sq = 52.9%    R-Sq(adj) = 50.5%

Analysis of Variance

Source           DF        SS        MS       F       P
Regression        2    542.03    271.02   21.88   0.000
Residual Error   39    483.08     12.39
Total            41   1025.12
```

QUASAR

QUASAR	REDSHIFT (x_1)	LINE FLUX (x_2)	LINE LUMINOSITY (x_3)	AB_{1450} (x_4)	ABSOLUTE MAGNITUDE (x_5)	REST FRAME EQUIVALENT WIDTH y
1	2.81	−13.48	45.29	19.50	−26.27	117
2	3.07	−13.73	45.13	19.65	−26.26	82
3	3.45	−13.87	45.11	18.93	−27.17	33
4	3.19	−13.27	45.63	18.59	−27.39	92
5	3.07	−13.56	45.30	19.59	−26.32	114
6	4.15	−13.95	45.20	19.42	−26.97	50
7	3.26	−13.83	45.08	19.18	−26.83	43
8	2.81	−13.50	45.27	20.41	−25.36	259
9	3.83	−13.66	45.41	18.93	−27.34	58
10	3.32	−13.71	45.23	20.00	−26.04	126
11	2.81	−13.50	45.27	18.45	−27.32	42
12	4.40	−13.96	45.25	20.55	−25.94	146
13	3.45	−13.91	45.07	20.45	−25.65	124
14	3.70	−13.85	45.19	19.70	−26.51	75
15	3.07	−13.67	45.19	19.54	−26.37	85
16	4.34	−13.93	45.27	20.17	−26.29	109
17	3.00	−13.75	45.08	19.30	−26.58	55
18	3.88	−14.17	44.92	20.68	−25.61	91
19	3.07	−13.92	44.94	20.51	−25.41	116
20	4.08	−14.28	44.86	20.70	−25.67	75
21	3.62	−13.82	45.20	19.45	−26.73	63
22	3.07	−14.08	44.78	19.90	−26.02	46
23	2.94	−13.82	44.99	19.49	−26.35	55
24	3.20	−14.15	44.75	20.89	−25.09	99
25	3.24	−13.74	45.17	19.17	−26.83	53

Source: Schmidt, M., Schneider, D. P., and Gunn, J. E. "Spectroscopic CCD surveys for quasars at large redshift," *Astronomical Journal*, Vol. 110, No. 1, July 1995, p. 70 (Table 1). Reproduced by permission of the American Astronomical Society.

(b) The first-order model is fit to the data using SPSS. The printout is provided on p. 188. Give the least squares prediction equation.

(c) Interpret the β estimates in the model.

(d) Test to determine whether redshift (x_1) is a useful linear predictor of equivalent width (y), using $\alpha = .05$.

(e) Locate R^2 and R_a^2 on the SPSS printout. Interpret these values. Which statistic is the preferred measure of model fit? Explain.

(f) Locate the global F-value for testing the overall model on the SPSS printout. Use the statistic to test the null hypothesis H_0: $\beta_1 = \beta_2 = \cdots = \beta_4 = 0$.

4.12 Arsenic in groundwater. *Environmental Science and Technology* (January 2005) reported on a study of the reliability of a commercial kit to test for arsenic in groundwater. The field kit was used to test a sample of 328 groundwater wells in Bangladesh. In addition to the arsenic level (micrograms per liter), the latitude (degrees), longitude (degrees), and depth (feet) of each well was measured. The data are saved in the ASWELLS file. (The first and last five observations are listed below.)

ASWELLS (Data for first and last five wells shown)

WELLID	LATITUDE	LONGITUDE	DEPTH	ARSENIC
10	23.7887	90.6522	60	331
14	23.7886	90.6523	45	302
30	23.7880	90.6517	45	193
59	23.7893	90.6525	125	232
85	23.7920	90.6140	150	19
.				
.				
.				
7353	23.7949	90.6515	40	48
7357	23.7955	90.6515	30	172
7890	23.7658	90.6312	60	175
7893	23.7656	90.6315	45	624
7970	23.7644	90.6303	30	254

SPSS output for Exercise 4.11

Coefficients[a]

Model		Unstandardized Coefficients		Standardized Coefficients	t	Sig.
		B	Std. Error	Beta		
1	(Constant)	21087.951	18553.161		1.137	.269
	REDSHIFT	108.451	88.740	1.102	1.222	.236
	LINEFLUX	557.910	315.990	2.786	1.766	.093
	LUMINOSITY	-340.166	320.763	-1.412	-1.060	.302
	AB1450	85.681	6.273	1.230	13.658	.000

a. Dependent Variable: RFEWIDTH

Model Summary

Model	R	R Square	Adjusted R Square	Std. Error of the Estimate
1	.955[a]	.912	.894	15.416

a. Predictors: (Constant), AB1450, REDSHIFT, LUMINOSITY, LINEFLUX

ANOVA[b]

Model		Sum of Squares	df	Mean Square	F	Sig.
1	Regression	49162.671	4	12290.668	51.720	.000[a]
	Residual	4752.769	20	237.638		
	Total	53915.440	24			

a. Predictors: (Constant), AB1450, REDSHIFT, LUMINOSITY, LINEFLUX

b. Dependent Variable: RFEWIDTH

(a) Write a first-order model for arsenic level (y) as a function of latitude, longitude, and depth.

(b) Fit the model to the data using the method of least squares.

(c) Give practical interpretations of the β estimates.

(d) Find the model standard deviation, s, and interpret its value.

(e) Interpret the values of R^2 and R^2_a.

(f) Conduct a test of overall model utility at $\alpha = .05$.

(g) Based on the results, parts d–f, would you recommend using the model to predict arsenic level (y)? Explain.

4.13 Cooling method for gas turbines. Refer to the *Journal of Engineering for Gas Turbines and Power* (January 2005) study of a high pressure inlet fogging method for a gas turbine engine, Exercise 3.59 (p. 139). Recall that the heat rate (kilojoules per kilowatt per hour) was measured for each in a sample of 67 gas turbines augmented with high pressure inlet fogging. In addition, several other variables were measured, including cycle speed (revolutions per minute), inlet temperature (°C), exhaust gas temperature (°C), cycle pressure ratio, and air mass flow rate (kilograms per

GASTURBINE (Data for first and last five gas turbines shown)

RPM	CPRATIO	INLET-TEMP	EXH-TEMP	AIRFLOW	HEATRATE
27245	9.2	1134	602	7	14622
14000	12.2	950	446	15	13196
17384	14.8	1149	537	20	11948
11085	11.8	1024	478	27	11289
14045	13.2	1149	553	29	11964
·					
·					
18910	14.0	1066	532	8	12766
3600	35.0	1288	448	152	8714
3600	20.0	1160	456	84	9469
16000	10.6	1232	560	14	11948
14600	13.4	1077	536	20	12414

Source: Bhargava, R., and Meher-Homji, C. B. "Parametric analysis of existing gas turbines with inlet evaporative and overspray fogging," *Journal of Engineering for Gas Turbines and Power*, Vol. 127, No. 1, Jan. 2005.

second). The data are saved in the GASTURBINE file. (The first and last five observations are listed in the table.)

(a) Write a first-order model for heat rate (y) as a function of speed, inlet temperature, exhaust temperature, cycle pressure ratio, and air flow rate.

(b) Fit the model to the data using the method of least squares.

(c) Give practical interpretations of the β estimates.

(d) Find the model standard deviation, s, and interpret its value.

(e) Find the adjusted-R^2 value and interpret it.

(f) Is the overall model statistically useful at predicting heat rate (y)? Test using $\alpha = .01$.

4.14 Removing oil from a water/oil mix. In the oil industry, water that mixes with crude oil during production and transportation must be removed. Chemists have found that the oil can be extracted from the water/oil mix electrically. Researchers at the University of Bergen (Norway) conducted a series of experiments to study the factors that influence the voltage (y) required to separate the water from the oil (*Journal of Colloid and Interface Science*, August 1995). The seven independent

variables investigated in the study are listed in the table below. (Each variable was measured at two levels—a "low" level and a "high" level.) Sixteen water/oil mixtures were prepared using different combinations of the independent variables; then each emulsion was exposed to a high electric field. In addition, three mixtures were tested when all independent variables were set to 0. The data for all 19 experiments are also given in the table.

(a) Propose a first-order model for y as a function of all seven independent variables.

(b) Use a statistical software package to fit the model to the data in the table.

(c) Fully interpret the β estimates.

(d) Find and interpret s.

(e) Find and interpret R^2.

(f) Conduct a test of overall model utility.

(g) Do you recommend using the model in practice? Explain.

4.15 Modeling IQ. Because the coefficient of determination R^2 always increases when a new independent variable is added to the model, it is tempting to include many variables in a model to force R^2 to be near 1. However, doing so reduces the degrees of freedom available for estimating σ^2, which adversely affects our ability to make reliable

WATEROIL

EXPERIMENT NUMBER	VOLTAGE y (kw/cm)	DISPERSE PHASE VOLUME x_1 (%)	SALINITY x_2 (%)	TEMPERATURE x_3 (°C)	TIME DELAY x_4 (hours)	SURFACTANT CONCENTRATION x_5 (%)	SPAN:TRITON x_6	SOLID PARTICLES x_7 (%)
1	.64	40	1	4	.25	2	.25	.5
2	.80	80	1	4	.25	4	.25	2
3	3.20	40	4	4	.25	4	.75	.5
4	.48	80	4	4	.25	2	.75	2
5	1.72	40	1	23	.25	4	.75	2
6	.32	80	1	23	.25	2	.75	.5
7	.64	40	4	23	.25	2	.25	2
8	.68	80	4	23	.25	4	.25	.5
9	.12	40	1	4	24	2	.75	2
10	.88	80	1	4	24	4	.75	.5
11	2.32	40	4	4	24	4	.25	2
12	.40	80	4	4	24	2	.25	.5
13	1.04	40	1	23	24	4	.25	.5
14	.12	80	1	23	24	2	.25	2
15	1.28	40	4	23	24	2	.75	.5
16	.72	80	4	23	24	4	.75	2
17	1.08	0	0	0	0	0	0	0
18	1.08	0	0	0	0	0	0	0
19	1.04	0	0	0	0	0	0	0

Source: Førdedal, H., et al. "A multivariate analysis of W/O emulsions in high external electric fields as studied by means of dielectric time domain spectroscopy," *Journal of Colloid and Interface Science*, Vol. 173, No. 2, Aug. 1995, p. 398 (Table 2). Copyright © 1995, with permission from Elsevier.

inferences. As an example, suppose you want to use the responses to a survey consisting of 18 demographic, social, and economic questions to model a college student's intelligence quotient (IQ). You fit the model

$$y = \beta_0 + \beta_1 x_1 + \beta_2 x_2 + \cdots + \beta_{17} x_{17} + \beta_{18} x_{18} + \varepsilon$$

where $y = $ IQ and $x_1, x_2, \ldots, x_{18}$ are the 18 independent variables. Data for only 20 students ($n = 20$) are used to fit the model, and you obtain $R^2 = .95$.

(a) Test to see whether this impressive-looking R^2 is large enough for you to infer that this model is useful (i.e., that at least one term in the model is important for predicting IQ). Use $\alpha = .05$.

(b) Calculate R_a^2 and interpret its value.

4.16 Urban/rural ratings of counties. *Professional Geographer* (February 2000) published a study of urban and rural counties in the western United States. University of Nevada (Reno) researchers asked a sample of 256 county commissioners to rate their "home" county on a scale of 1 (most rural) to 10 (most urban). The urban/rural rating (y) was used as the dependent variable in a first-order multiple regression model with six independent variables: total county population (x_1), population density (x_2), population concentration (x_3), population growth (x_4), proportion of county land in farms (x_5), and 5-year change in agricultural land base (x_6). Some of the regression results are shown in the next table.

(a) Write the least squares prediction equation for y.

(b) Give the null hypothesis for testing overall model adequacy.

(c) Conduct the test, part b, at $\alpha = .01$ and give the appropriate conclusion.

(d) Interpret the values of R^2 and R_a^2.

(e) Give the null hypothesis for testing the contribution of population growth (x_4) to the model.

(f) Conduct the test, part e, at $\alpha = .01$ and give the appropriate conclusion.

INDEPENDENT VARIABLE	β ESTIMATE	p-VALUE
x_1: Total population	0.110	0.045
x_2: Population density	0.065	0.230
x_3: Population concentration	0.540	0.000
x_4: Population growth	−0.009	0.860
x_5: Farm land	−0.150	0.003
x_6: Agricultural change	−0.027	0.580

Overall model: $R^2 = .44$ $R_a^2 = .43$
$F = 32.47$ p-value $< .001$

Source: Berry, K. A., et al. "Interpreting what is rural and urban for western U.S. counties," *Professional Geographer*, Vol. 52, No. 1, Feb. 2000 (Table 2).

4.17 Active caring on the job. An important goal in occupational safety is "active caring." Employees demonstrate active caring (AC) about the safety of their co-workers when they identify environmental hazards and unsafe work practices and then implement appropriate corrective actions for these unsafe conditions or behaviors. Three factors hypothesized to increase the propensity for an employee to actively care for safety are (1) high self-esteem, (2) optimism, and (3) group cohesiveness. *Applied and Preventive Psychology* (Winter 1995) attempted to establish empirical support for the AC hypothesis by fitting the model $E(y) = \beta_0 + \beta_1 x_1 + \beta_2 x_2 + \beta_3 x_3$, where

$y = $ AC score (measuring active caring on a 15-point scale)
$x_1 = $ Self-esteem score
$x_2 = $ Optimism score
$x_3 = $ Group cohesion score

The regression analysis, based on data collected for $n = 31$ hourly workers at a large fiber-manufacturing plant, yielded a multiple coefficient of determination of $R^2 = .362$.

(a) Interpret the value of R^2.

(b) Use the R^2 value to test the global utility of the model. Use $\alpha = .05$.

4.9 Using the Model for Estimation and Prediction

In Section 3.9, we discussed the use of the least squares line for estimating the mean value of y, $E(y)$, for some value of x, say, $x = x_p$. We also showed how to use the same fitted model to predict, when $x = x_p$, some value of y to be observed in the future. Recall that the least squares line yielded the same value for both the estimate of $E(y)$ and the prediction of some future value of y. That is, both are the result obtained by substituting x_p into the prediction equation $\hat{y} = \hat{\beta}_0 + \hat{\beta}_1 x$ and calculating $\hat{y}$. There the equivalence ends. The confidence interval for the mean $E(y)$

was narrower than the prediction interval for y, because of the additional uncertainty attributable to the random error ε when predicting some future value of y.

These same concepts carry over to the multiple regression model. Consider a first-order model relating sale price (y) of a residential property to land value (x_1), appraised improvements value (x_2), and home size (x_3). Suppose we want to estimate the mean sale price for a given property with $x_1 = \$15{,}000$, $x_2 = \$50{,}000$, and $x_3 = 1{,}800$ square feet. Assuming that the first-order model represents the true relationship between sale price and the three independent variables, we want to estimate

$$E(y) = \beta_0 + \beta_1 x_1 + \beta_2 x_2 + \beta_3 x_3 = \beta_0 + \beta_1(15{,}000) + \beta_2(50{,}000) + \beta_3(1{,}800)$$

After obtaining the least squares estimates $\hat{\beta}_0$, $\hat{\beta}_1$, $\hat{\beta}_2$ and $\hat{\beta}_3$, the estimate of $E(y)$ will be

$$\hat{y} = \hat{\beta}_0 + \hat{\beta}_1(15{,}000) + \hat{\beta}_2(50{,}000) + \hat{\beta}_3(1{,}800)$$

To form a confidence interval for the mean, we need to know the standard deviation of the sampling distribution for the estimator $\hat{y}$. For multiple regression models, the form of this standard deviation is rather complex. However, the regression routines of statistical computer software packages allow us to obtain the confidence intervals for mean values of y for any given combination of values of the independent variables. We illustrate with an example.

Example 4.5

Refer to Examples 4.1–4.4 and the first-order model, $E(y) = \beta_0 + \beta_1 x_1 + \beta_2 x_2$, where y = auction price of a grandfather clock, x_1 = age of the clock, and x_2 = number of bidders.

(a) Estimate the average auction price for all 150-year-old clocks sold at auctions with 10 bidders using a 95% confidence interval. Interpret the result.

(b) Predict the auction price for a single 150-year-old clock sold at an auction with 10 bidders using a 95% prediction interval. Interpret the result.

(c) Suppose you want to predict the auction price for one clock that is 50 years old and has two bidders. How should you proceed?

Solution

(a) Here, the key words *average* and *for all* imply we want to estimate the mean of y, $E(y)$. We want a 95% confidence interval for $E(y)$ when $x_1 = 150$ years and $x_2 = 10$ bidders. A MINITAB printout for this analysis is shown in Figure 4.8. The confidence interval (highlighted under "**95% CI**") is (1,381.4, 1,481.9). Thus, we are 95% confident that the mean auction price for all 150-year-old clocks sold at an auction with 10 bidders lies between \$1,381.40 and \$1,481.90.

(b) The key words *predict* and *for a single* imply that we want a 95% prediction interval for y when $x_1 = 150$ years and $x_2 = 10$ bidders. This interval (highlighted under "95% PI" on the MINITAB printout, Figure 4.8) is (1,154.1, 1,709.3). We say, with 95% confidence, that the auction price for a single 150-year-old clock sold at an auction with 10 bidders falls between \$1,154.10 and \$1,709.30.

(c) Now, we want to predict the auction price, y, for a single (*one*) grandfather clock when $x_1 = 50$ years and $x_2 = 2$ bidders. Consequently, we desire a 95% prediction interval for y. However, before we form this prediction interval,

Figure 4.8 MINITAB printout with 95% confidence intervals for grandfather clock model

---

Regression Analysis: PRICE versus AGE, NUMBIDS

```
The regression equation is
PRICE = - 1339 + 12.7 AGE + 86.0 NUMBIDS


Predictor      Coef   SE Coef      T      P
Constant    -1339.0     173.8  -7.70  0.000
AGE         12.7406    0.9047  14.08  0.000
NUMBIDS      85.953     8.729   9.85  0.000


S = 133.485   R-Sq = 89.2%   R-Sq(adj) = 88.5%

Analysis of Variance

Source           DF       SS       MS       F      P
Regression        2  4283063  2141531  120.19  0.000
Residual Error   29   516727    17818
Total            31  4799790


Predicted Values for New Observations

New
Obs    Fit  SE Fit        95% CI              95% PI
  1 1431.7    24.6  (1381.4, 1481.9)  (1154.1, 1709.3)

Values of Predictors for New Observations

New
Obs   AGE  NUMBIDS
  1   150     10.0
```

---

we should check to make sure that the selected values of the independent variables, $x_1 = 50$ and $x_2 = 2$, are both reasonable and within their respective sample ranges. If you examine the sample data shown in Table 4.1 (p. 171), you will see that the range for age is $108 \leq x_1 \leq 194$, and the range for number of bidders is $5 \leq x_2 \leq 15$. Thus, both selected values fall well *outside* their respective ranges. Recall the *Caution* box in Section 3.9 (p. 132) warning about the dangers of using the model to predict y for a value of an independent variable that is not within the range of the sample data. Doing so may lead to an unreliable prediction. If we want to make the prediction requested, we would need to collect additional data on clocks with the requested characteristics (i.e., $x_1 = 50$ years and $x_2 = 2$ bidders) and then refit the model.

As in simple linear regression, the confidence interval for $E(y)$ will always be narrower than the corresponding prediction interval for y. This fact, however, should not drive your choice of which interval to use in practice. Use the confidence interval for $E(y)$ if you are interested in estimating the average (mean) response; use the prediction interval for y if you want to predict a future value of the response.

4.9 Exercises

4.18 Characteristics of lead users. Refer to the *Creativity and Innovation Management* (February 2008) study of lead users of children's computer games, Exercise 4.4 (p. 183). Recall that the researchers modeled lead-user rating (y, measured on a 5-point scale) as a function of gender ($x_1 = 1$ if female, 0 if male), age (x_2, years), degree of centrality (x_3, measured as the number of direct ties to other peers in the network), and betweenness centrality (x_4, measured as the number of shortest paths between peers). The least squares prediction equation was

$$\hat{y} = 3.58 + .01x_1 - .06x_2 - .01x_3 + .42x_4$$

(a) Compute the predicted lead-user rating of a 10-year-old female child with five direct ties to other peers in her social network and with two shortest paths between peers.

(b) Compute an estimate for the mean lead-user rating of all 8-year-old male children with 10 direct ties to other peers and with four shortest paths between peers.

4.19 **Predicting runs scored in baseball.** Refer to the *Chance* (Fall 2000) study of runs scored in Major League Baseball games, Exercise 4.5 (p. 183). Multiple regression was used to model total number of runs scored (y) of a team during the season as a function of number of walks (x_1), number of singles (x_2), number of doubles (x_3), number of triples (x_4), number of home runs (x_5), number of stolen bases (x_6), number of times caught stealing (x_7), number of strikeouts (x_8), and total number of outs (x_9). Using the β-estimates given in Exercise 4.5, predict the number of runs scored by your favorite Major League Baseball team last year. How close is the predicted value to the actual number of runs scored by your team? [Note: You can find data on your favorite team on the Internet at www.mlb.com.]

4.20 **Earnings of Mexican street vendors.** Refer to the *World Development* (February 1998) study of street vendors' earnings (y), Exercise 4.6 (p. 184). The SAS printout below shows both a 95% prediction interval for y and a 95% confidence interval for $E(y)$ for a 45-year-old vendor who works 10 hours a day (i.e., for $x_1 = 45$ and $x_2 = 10$).

(a) Interpret the 95% prediction interval for y in the words of the problem.

(b) Interpret the 95% confidence interval for $E(y)$ in the words of the problem.

(c) Note that the interval of part a is wider than the interval of part b. Will this always be true? Explain.

4.21 **Snow geese feeding trial.** Refer to the *Journal of Applied Ecology* study of the feeding habits of baby snow geese, Exercise 4.10 (p. 186). Recall that a first-order model was used to relate gosling weight change (y) to digestion efficiency (x_1) and acid-detergent fiber (x_2). The MINITAB printout (top of p. 194) shows both a confidence interval for $E(y)$ and a prediction interval for y when $x_1 = 40\%$ and $x_2 = 15\%$.

(a) Interpret the confidence interval for $E(y)$.

(b) Interpret the prediction interval for y.

4.22 **Deep space survey of quasars.** Refer to the *Astronomical Journal* study of quasars, Exercise 4.11 (p. 186). Recall that a first-order model was used to relate a quasar's equivalent width (y) to redshift (x_1), line flux (x_2), line luminosity (x_3), and $AB_{1450}(x_4)$. A portion of the SPSS spreadsheet showing 95% prediction intervals for y for the first five observations in the data set is reproduced on p. 194. Interpret the interval corresponding to the fifth observation.

ASWELLS

4.23 **Arsenic in groundwater.** Refer to the *Environmental Science and Technology* (January 2005) study of the reliability of a commercial kit to test for arsenic in groundwater, Exercise 4.12 (p. 187). Using the data in the ASWELLS file, you fit a first-order model for arsenic level (y) as a function of latitude, longitude, and depth. Based on the model statistics, the researchers concluded that the arsenic level is highest at a low latitude, high longitude, and low depth. Do you agree? If so, find a 95% prediction interval for arsenic level for the lowest latitude, highest longitude, and lowest depth that are within the range of the sample data. Interpret the result.

SAS output for Exercise 4.20

Dependent Variable: EARNINGS

Output Statistics

Obs	AGE	HOURS	Dep Var EARNINGS	Predicted Value	Std Error Mean Predict	95% CL Predict	
16	45	10	.	3018	182.3519	1760	4275

Obs	AGE	HOURS	Dep Var EARNINGS	Predicted Value	Std Error Mean Predict	95% CL Mean	
16	45	10	.	3018	182.3519	2620	3415

MINITAB output for Exercise 4.21

```
The regression equation is
WTCHANGE = 12.2 - 0.0265 DIGEST - 0.458 ADFIBRE


Predictor       Coef     SE Coef        T       P
Constant       12.180       4.402     2.77   0.009
DIGEST       -0.02654     0.05349    -0.50   0.623
ADFIBRE       -0.4578      0.1283    -3.57   0.001


S = 3.519     R-Sq = 52.9%     R-Sq(adj) = 50.5%

Analysis of Variance

Source            DF         SS          MS       F       P
Regression         2     542.03      271.02   21.88   0.000
Residual Error    39     483.08       12.39
Total             41    1025.12


Predicted Values for New Observations

New Obs     Fit    SE Fit        95.0% CI              95.0% PI
1         4.251     0.776   (  2.683,   5.820)   ( -3.038,  11.541)

Values of Predictors for New Observations

New Obs   DIGEST   ADFIBRE
1           40.0      15.0
```

SPSS output for Exercise 4.22

	quasar	redshift	lineflux	linelum	ab1450	rfewidth	lcl_95	ucl_95
1	1	2.81	-13.48	45.29	19.50	117	101.29	172.22
2	2	3.07	-13.73	45.13	19.65	82	59.62	125.89
3	3	3.45	-13.87	45.11	18.93	33	-35.09	37.04
4	4	3.19	-13.27	45.63	18.59	92	63.76	139.25
5	5	3.07	-13.56	45.30	19.59	114	90.69	158.57

GASTURBINE

4.24 Cooling method for gas turbines. Refer to the *Journal of Engineering for Gas Turbines and Power* (January 2005) study of a high-pressure inlet fogging method for a gas turbine engine, Exercise 4.13 (p. 188). Recall that you fit a first-order model for heat rate (y) as a function of speed (x_1), inlet temperature (x_2), exhaust temperature (x_3), cycle pressure ratio (x_4), and air flow rate (x_5) to data saved in the GASTURBINE file. A MINITAB printout with both a 95% confidence interval for $E(y)$ and prediction interval for y for selected values of the x's is shown below.

MINITAB Output for Exercise 4.24

```
Predicted Values for New Observations

New
Obs     Fit  SE Fit        95% CI                95% PI
1   12632.5   237.3  (12157.9, 13107.1)   (11599.6, 13665.5)


Values of Predictors for New Observations

New
Obs   RPM   INLET-TEMP   EXH-TEMP   CPRATIO   AIRFLOW
1    7500        1000        525      13.5      10.0
```

Figure 4.11 MINITAB regression printout for grandfather clock model with interaction

Regression Analysis: PRICE versus AGE, NUMBIDS, AGEBID

```
The regression equation is
PRICE = 320 + 0.88 AGE - 93.3 NUMBIDS + 1.30 AGEBID


Predictor     Coef   SE Coef       T       P
Constant     320.5     295.1    1.09   0.287
AGE          0.878     2.032    0.43   0.669
NUMBIDS     -93.26     29.89   -3.12   0.004
AGEBID      1.2978    0.2123    6.11   0.000


S = 88.9145    R-Sq = 95.4%    R-Sq(adj) = 94.9%


Analysis of Variance

Source           DF        SS        MS       F       P
Regression        3   4578427   1526142  193.04   0.000
Residual Error   28    221362      7906
Total            31   4799790
```

(a) Test the overall utility of the model using the global F-test at $\alpha = .05$.

(b) Test the hypothesis (at $\alpha = .05$) that the price–age slope increases as the number of bidders increases—that is, that age and number of bidders, x_2, interact positively.

(c) Estimate the change in auction price of a 150-year-old grandfather clock, y, for each additional bidder.

Solution

(a) The global F-test is used to test the null hypothesis

$$H_0: \beta_1 = \beta_2 = \beta_3 = 0$$

The test statistic and p-value of the test (highlighted on the MINITAB printout) are $F = 193.04$ and $p = 0$, respectively. Since $\alpha = .05$ exceeds the p-value, there is sufficient evidence to conclude that the model fit is a statistically useful predictor of auction price, y.

(b) The hypotheses of interest to the collector concern the interaction parameter β_3. Specifically,

$$H_0: \beta_3 = 0$$

$$H_a: \beta_3 > 0$$

Since we are testing an individual β parameter, a t-test is required. The test statistic and two-tailed p-value (highlighted on the printout) are $t = 6.11$ and $p = 0$, respectively. The upper-tailed p-value, obtained by dividing the two-tailed p-value in half, is $0/2 = 0$. Since $\alpha = .05$ exceeds the p-value, the collector can reject H_0 and conclude that the rate of change of the mean price of the clocks with age increases as the number of bidders increases, that is, x_1 and x_2 interact positively. Thus, it appears that the interaction term should be included in the model.

(c) To estimate the change in auction price, y, for every 1-unit increase in number of bidders, x_2, we need to estimate the slope of the line relating y to x_2 when the age of the clock, x_1, is 150 years old. An analyst who is not careful may estimate this slope as $\hat{\beta}_2 = -93.26$. Although the coefficient of x_2 is negative, this does *not* imply that auction price decreases as the number of bidders increases. Since interaction is present, the rate of change (slope) of mean auction price with the number of bidders *depends* on x_1, the age of the clock. Thus, the estimated rate of change of y for a unit increase in x_2 (one new bidder) for a 150-year-old clock is

$$\text{Estimated } x_2 \text{ slope} = \hat{\beta}_2 + \hat{\beta}_3 x_1 = -93.26 + 1.30(150) = 101.74$$

In other words, we estimate that the auction price of a 150-year-old clock will *increase* by about \$101.74 for every additional bidder. Although the rate of increase will vary as x_1 is changed, it will remain positive for the range of values of x_1 included in the sample. Extreme care is needed in interpreting the signs and sizes of coefficients in a multiple regression model.

Example 4.6 illustrates an important point about conducting t-tests on the β parameters in the interaction model. The "most important" β parameter in this model is the interaction β, β_3. [Note that this β is also the one associated with the highest-order term in the model, $x_1 x_2$.*] Consequently, we will want to test $H_0: \beta_3 = 0$ after we have determined that the overall model is useful for predicting y. Once interaction is detected (as in Example 4.6), however, tests on the first-order terms x_1 and x_2 should *not* be conducted since they are meaningless tests; the presence of interaction implies that both x's are important.

Caution

> Once interaction has been deemed important in the model $E(y) = \beta_0 + \beta_1 x_1 + \beta_2 x_2 + \beta_3 x_1 x_2$, do not conduct t-tests on the β coefficients of the first-order terms x_1 and x_2. These terms should be kept in the model regardless of the magnitude of their associated p-values shown on the printout.

We close this section with a comment. The concept of interaction is not always intuitive and you will rarely know *a priori* whether two independent variables interact. Consequently, you will need to fit and test the interaction term to determine its importance.

4.10 Exercises

4.26 Role of retailer interest on shopping behavior. Retail interest is defined by marketers as the level of interest a consumer has in a given retail store. Marketing professors at the University of Tennessee at Chattanooga and the University of Alabama investigated the role of retailer interest in consumers' shopping behavior (*Journal of Retailing*, Summer 2006). Using survey data collected for $n = 375$ consumers, the professors developed an interaction model for y = willingness of the consumer to shop at a retailer's store in the future (called "repatronage intentions") as a function of x_1 = consumer satisfaction and x_2 = retailer interest. The regression results are shown below.

VARIABLE	ESTIMATED β	t-VALUE	p-VALUE
Satisfaction (x_1)	.426	7.33	< .01
Retailer interest (x_2)	.044	0.85	> .10
Satisfaction × Retailer interest ($x_1 x_2$)	−.157	−3.09	< .01

$R^2 = .65$, $F = 226.35$, p-value < .001

---

* The order of a term is equal to the sum of the exponents of the quantitative variables included in the term. Thus, when x_1 and x_2 are both quantitative variables, the cross product, $x_1 x_2$, is a second-order term.

(a) Is the overall model statistically useful for predicting y? Test using $\alpha = .05$.

(b) Conduct a test for interaction at $\alpha = .05$.

(c) Use the β-estimates to sketch the estimated relationship between repatronage intentions (y) and satisfaction (x_1) when retailer interest is $x_2 = 1$ (a low value).

(d) Repeat part c when retailer interest is $x_2 = 7$ (a high value).

(e) Sketch the two lines, parts c and d, on the same graph to illustrate the nature of the interaction.

4.27 Defects in nuclear missile housing parts. The technique of multivariable testing (MVT) was discussed in the *Journal of the Reliability Analysis Center* (First Quarter, 2004). MVT was shown to improve the quality of carbon-foam rings used in nuclear missile housings. The rings are produced via a casting process that involves mixing ingredients, oven curing, and carving the finished part. One type of defect analyzed was number y of black streaks in the manufactured ring. Two variables found to impact the number of defects were turntable speed (revolutions per minute), x_1, and cutting blade position (inches from center), x_2.

(a) The researchers discovered "an interaction between blade position and turntable speed." Hypothesize a regression model for $E(y)$ that incorporates this interaction.

(b) Practically interpret what it means to say that "blade position and turntable speed interact."

(c) The researchers reported a positive linear relationship between number of defects (y) and turntable speed (x_1), but found that the slope of the relationship was much steeper for lower values of cutting blade position (x_2). What does this imply about the interaction term in the model, part a? Explain.

4.28 Earnings of Mexican street vendors. Refer to the *World Development* (February 1998) study of street vendors in the city of Puebla, Mexico, Exercise 4.6 (p. 184). Recall that the vendors' mean annual earnings, $E(y)$, was modeled as a first-order function of age (x_1) and hours worked (x_2). Now, consider the interaction model $E(y) = \beta_0 + \beta_1 x_1 + \beta_2 x_2 + \beta_3 x_1 x_2$. The SAS printout for the model is displayed in the next column.

(a) Give the least squares prediction equation.

(b) What is the estimated slope relating annual earnings (y) to age (x_1) when number of hours worked (x_2) is 10? Interpret the result.

(c) What is the estimated slope relating annual earnings (y) to hours worked (x_2) when age (x_1) is 40? Interpret the result.

(d) Give the null hypothesis for testing whether age (x_1) and hours worked (x_2) interact.

(e) Find the p-value of the test, part d.

(f) Refer to part e. Give the appropriate conclusion in the words of the problem.

The REG Procedure
Dependent Variable: EARNINGS

Analysis of Variance

Source	DF	Sum of Squares	Mean Square	F Value	Pr > F
Model	3	5287427	1762476	5.82	0.0124
Error	11	3331000	302818		
Corrected Total	14	8618428			

Root MSE	550.28921	R-Square	0.6135
Dependent Mean	2577.13333	Adj R-Sq	0.5081
Coeff Var	21.35276		

Parameter Estimates

Variable	DF	Parameter Estimate	Standard Error	t Value	Pr > |t|
Intercept	1	1041.89440	1303.59326	0.80	0.4411
AGE	1	-13.23762	29.23395	-0.45	0.6595
HOURS	1	103.30564	162.01356	0.64	0.5368
AGEHRS	1	3.62096	3.84044	0.94	0.3660

4.29 Psychology of waiting in line. While waiting in a long line for service (e.g., to use an ATM or at the post office), at some point you may decide to leave the queue. The *Journal of Consumer Research* (November 2003) published a study of consumer behavior while waiting in a queue. A sample of $n = 148$ college students were asked to imagine that they were waiting in line at a post office to mail a package and that the estimated waiting time is 10 minutes or less. After a 10-minute wait, students were asked about their level of negative feelings (annoyed, anxious) on a scale of 1 (strongly disagree) to 9 (strongly agree). Before answering, however, the students were informed about how many people were ahead of them and behind them in the line. The researchers used regression to relate negative feelings score (y) to number ahead in line (x_1) and number behind in line (x_2).

(a) The researchers fit an interaction model to the data. Write the hypothesized equation of this model.

(b) In the words of the problem, explain what it means to say that "x_1 and x_2 interact to effect y."

(c) A t-test for the interaction β in the model resulted in a p-value greater than .25. Interpret this result.

(d) From their analysis, the researchers concluded that "the greater the number of people ahead, the higher the negative feeling score" and "the greater the number of people behind, the lower the negative feeling score." Use this information to determine the signs of β_1 and β_2 in the model.

4.30 Unconscious self-esteem study. Psychologists define *implicit* self-esteem as unconscious evaluations of one's worth or value. In contrast, *explicit* self-esteem refers to the extent to which a person consciously considers oneself as valuable and worthy. An article published in the *Journal of Articles in Support of the Null Hypothesis* (March 2006) investigated whether implicit self-esteem is really unconscious. A sample of 257 college undergraduate students completed a questionnaire designed to measure implicit self-esteem and explicit self-esteem. Thus, an implicit self-esteem score (x_1) and explicit self-esteem score (x_2) was obtained for each. (Note: Higher scores indicate higher levels of self-esteem.) Also, a second questionnaire was administered in order to obtain each subject's estimate of their level of implicit self-esteem. This was called estimated implicit self-esteem (x_3). Finally, the researchers computed two measures of accuracy in estimating implicit self-esteem: $y_1 = (x_3 - x_1)$ and $y_2 = |x_3 - x_1|$.

(a) The researchers fit the interaction model, $E(y_1) = \beta_0 + \beta_1 x_1 + \beta_2 x_2 + \beta_3 x_1 x_2$. The t-test on the interaction term, β_3, was "nonsignificant" with a p-value $> .10$. However, both t-tests on β_1 and β_2 were statistically significant (p-value $< .001$). Interpret these results practically.

(b) The researchers also fit the interaction model, $E(y_2) = \beta_0 + \beta_1 x_1 + \beta_2 x_2 + \beta_3 x_1 x_2$. The t-test on the interaction term, β_3, was "significant" with a p-value $< .001$. Interpret this result practically.

ASWELLS

4.31 **Arsenic in groundwater.** Refer to the *Environmental Science and Technology* (January 2005) study of the reliability of a commercial kit to test for arsenic in groundwater, Exercise 4.12 (p. 187). Recall that you fit a first-order model for arsenic level (y) as a function of latitude (x_1), longitude (x_2), and depth (x_3) to data saved in the ASWELLS file.

(a) Write a model for arsenic level (y) that includes first-order terms for latitude, longitude, and depth, as well as terms for interaction between latitude and depth and interaction between longitude and depth.

(b) Use statistical software to fit the interaction model, part a, to the data in the ASWELLS file. Give the least squares prediction equation.

(c) Conduct a test (at $\alpha = .05$) to determine whether latitude and depth interact to effect arsenic level.

(d) Conduct a test (at $\alpha = .05$) to determine whether longitude and depth interact to effect arsenic level.

(e) Practically interpret the results of the tests, parts c and d.

GASTURBINE

4.32 **Cooling method for gas turbines.** Refer to the *Journal of Engineering for Gas Turbines and Power* (January 2005) study of a high-pressure inlet fogging method for a gas turbine engine, Exercise 4.13 (p. 188). Recall that you fit a first-order model for heat rate (y) as a function of speed (x_1), inlet temperature (x_2), exhaust temperature (x_3), cycle pressure ratio (x_4), and air flow rate (x_5) to data saved in the GASTURBINE file.

(a) Researchers hypothesize that the linear relationship between heat rate (y) and temperature (both inlet and exhaust) depends on air flow rate. Write a model for heat rate that incorporates the researchers' theories.

(b) Use statistical software to fit the interaction model, part a, to the data in the GASTURBINE file. Give the least squares prediction equation.

(c) Conduct a test (at $\alpha = .05$) to determine whether inlet temperature and air flow rate interact to effect heat rate.

(d) Conduct a test (at $\alpha = .05$) to determine whether exhaust temperature and air flow rate interact to effect heat rate.

(e) Practically interpret the results of the tests, parts c and d.

WATEROIL

4.33 **Removing oil from a water/oil mix.** Refer to the *Journal of Colloid and Interface Science* study of water/oil mixtures, Exercise 4.14 (p. 189). Recall that three of the seven variables used to predict voltage (y) were volume (x_1), salinity (x_2), and surfactant concentration (x_5). The model the researchers fit is

$$E(y) = \beta_0 + \beta_1 x_1 + \beta_2 x_2 + \beta_3 x_5 + \beta_4 x_1 x_2 + \beta_5 x_1 x_5.$$

(a) Note that the model includes interaction between disperse phase volume (x_1) and salinity (x_2) as well as interaction between disperse phase volume (x_1) and surfactant concentration (x_5). Discuss how these interaction terms affect the hypothetical relationship between y and x_1. Draw a sketch to support your answer.

(b) Fit the interaction model to the WATEROIL data. Does this model appear to fit the data better than the first-order model in Exercise 4.14? Explain.

(c) Interpret the β estimates of the interaction model.

4.11 A Quadratic (Second-Order) Model with a Quantitative Predictor

All of the models discussed in the previous sections proposed straight-line relationships between $E(y)$ and each of the independent variables in the model. In this section, we consider a model that allows for curvature in the relationship. This model is a **second-order model** because it will include an x^2 term.

Here, we consider a model that includes only one independent variable x. The form of this model, called the **quadratic model**, is

$$y = \beta_0 + \beta_1 x + \beta_2 x^2 + \varepsilon$$

The term involving x^2, called a **quadratic term** (or **second-order term**), enables us to hypothesize curvature in the graph of the response model relating y to x. Graphs of the quadratic model for two different values of β_2 are shown in Figure 4.12. When the curve opens upward, the sign of β_2 is positive (see Figure 4.12a); when the curve opens downward, the sign of β_2 is negative (see Figure 4.12b).

Figure 4.12 Graphs for two quadratic models

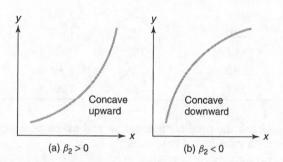

(a) $\beta_2 > 0$ Concave upward

(b) $\beta_2 < 0$ Concave downward

A Quadratic (Second-Order) Model in a Single Quantitative Independent Variable

$$E(y) = \beta_0 + \beta_1 x + \beta_2 x^2$$

where β_0 is the y-intercept of the curve
            β_1 is a shift parameter
            β_2 is the rate of curvature

Example 4.7

A physiologist wants to investigate the impact of exercise on the human immune system. The physiologist theorizes that the amount of immunoglobulin y in blood (called IgG, an indicator of long-term immunity) is related to the maximal oxygen uptake x (a measure of aerobic fitness level) of a person by the model

$$y = \beta_0 + \beta_1 x + \beta_2 x^2 + \varepsilon$$

To fit the model, values of y and x were measured for each of 30 human subjects. The data are shown in Table 4.2.

(a) Construct a scatterplot for the data. Is there evidence to support the use of a quadratic model?

(b) Use the method of least squares to estimate the unknown parameters β_0, β_1, and β_2 in the quadratic model.

AEROBIC

Table 4.2 Data on immunity and fitness level of 30 subjects

Subject	IgG y, milligrams	Maximal Oxygen Uptake x, milliliters per kilogram	Subject	IgG y, milligrams	Maximal Oxygen Uptake x, milliliters per kilogram
1	881	34.6	16	1,660	52.5
2	1,290	45.0	17	2,121	69.9
3	2,147	62.3	18	1,382	38.8
4	1,909	58.9	19	1,714	50.6
5	1,282	42.5	20	1,959	69.4
6	1,530	44.3	21	1,158	37.4
7	2,067	67.9	22	965	35.1
8	1,982	58.5	23	1,456	43.0
9	1,019	35.6	24	1,273	44.1
10	1,651	49.6	25	1,418	49.8
11	752	33.0	26	1,743	54.4
12	1,687	52.0	27	1,997	68.5
13	1,782	61.4	28	2,177	69.5
14	1,529	50.2	29	1,965	63.0
15	969	34.1	30	1,264	43.2

(c) Graph the prediction equation and assess how well the model fits the data, both visually and numerically.

(d) Interpret the β estimates.

(e) Is the overall model useful (at $\alpha = .01$) for predicting IgG?

(f) Is there sufficient evidence of concave downward curvature in the immunity–fitness level? Test using $\alpha = .01$.

Solution

(a) A scatterplot for the data in Table 4.2, produced using SPSS, is shown in Figure 4.13. The figure illustrates that immunity appears to increase in a curvilinear manner with fitness level. This provides some support for the inclusion of the quadratic term x^2 in the model.

(b) We also used SPSS to fit the model to the data in Table 4.2. Part of the SPSS regression output is displayed in Figure 4.14. The least squares estimates of the β parameters (highlighted at the bottom of the printout) are $\hat{\beta}_0 = -1,464.404$, $\hat{\beta}_1 = 88.307$, and $\hat{\beta}_2 = -.536$. Therefore, the equation that minimizes the SSE for the data is

$$\hat{y} = -1,464.4 + 88.307x - .536x^2$$

(c) Figure 4.15 is a MINITAB graph of the least squares prediction equation. Note that the graph provides a good fit to the data in Table 4.2. A numerical measure of fit is obtained with the adjusted coefficient of determination, R_a^2. From the SPSS printout, $R_a^2 = .933$. This implies that about 93% of the sample variation in IgG (y) can be explained by the quadratic model (after adjusting for sample size and degrees of freedom).

Figure 4.13 SPSS scatterplot for data of Example 4.7

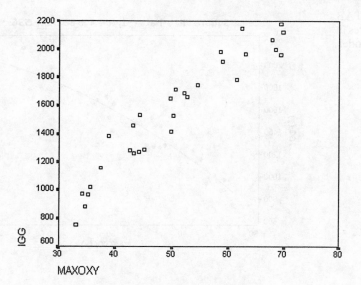

Figure 4.14 SPSS output for quadratic model of Example 4.7

Model Summary

Model	R	R Square	Adjusted R Square	Std. Error of the Estimate
1	.968[a]	.938	.933	106.427

a. Predictors: (Constant), MAXOXYSQ, MAXOXY

ANOVA[b]

Model		Sum of Squares	df	Mean Square	F	Sig.
1	Regression	4602211	2	2301105.316	203.159	.000[a]
	Residual	305818.3	27	11326.605		
	Total	4908029	29			

a. Predictors: (Constant), MAXOXYSQ, MAXOXY
b. Dependent Variable: IGG

Coefficients[a]

Model		Unstandardized Coefficients		Standardized Coefficients	t	Sig.
		B	Std. Error	Beta		
1	(Constant)	-1464.404	411.401		-3.560	.001
	MAXOXY	88.307	16.474	2.574	5.361	.000
	MAXOXYSQ	-.536	.158	-1.628	-3.390	.002

a. Dependent Variable: IGG

(d) The interpretation of the estimated coefficients in a quadratic model must be undertaken cautiously. First, the estimated y-intercept, $\hat{\beta}_0$, can be meaningfully interpreted only if the range of the independent variable includes zero—that is, if $x = 0$ is included in the sampled range of x. Although $\hat{\beta}_0 = -1,464.4$ seems to imply that the estimated immunity level is negative when $x = 0$, this zero point is not in the range of the sample (the lowest value of maximal oxygen uptake x is 33 milliliters per kilogram), and the value is nonsensical

Figure 4.15 MINITAB graph of least squares fit for the quadratic model

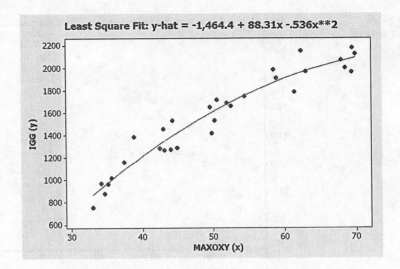

Least Square Fit: y-hat = -1,464.4 + 88.31x -.536x**2

(a person with 0 aerobic fitness level); thus, the interpretation of $\hat{\beta}_0$ is not meaningful.

The estimated coefficient of x is $\hat{\beta}_1 = 88.31$, but it no longer represents a slope in the presence of the quadratic term x^2.* The estimated coefficient of the first-order term x will not, in general, have a meaningful interpretation in the quadratic model.

The sign of the coefficient, $\hat{\beta}_2 = -.536$, of the quadratic term, x^2, is the indicator of whether the curve is concave downward (mound-shaped) or concave upward (bowl-shaped). A negative $\hat{\beta}_2$ implies downward concavity, as in this example (Figure 4.15), and a positive $\hat{\beta}_2$ implies upward concavity. Rather than interpreting the numerical value of $\hat{\beta}_2$ itself, we utilize a graphical representation of the model, as in Figure 4.15, to describe the model.

Note that Figure 4.15 implies that the estimated immunity level (IgG) is leveling off as the aerobic fitness levels increase beyond 70 milliliters per kilogram. In fact, the concavity of the model would lead to decreasing usage estimates if we were to display the model out to $x = 120$ and beyond (see Figure 4.16). However, model interpretations are not meaningful outside the range of the independent variable, which has a maximum value of 69.9 in this example. Thus, although the model appears to support the hypothesis that the *rate of increase* of IgG with maximal oxygen uptake *decreases* for subjects with aerobic fitness levels near the high end of the sampled values, the conclusion that IgG will actually begin to decrease for very large aerobic fitness levels would be a *misuse* of the model, since no subjects with x-values of 70 or more were included in the sample.

(e) To test whether the quadratic model is statistically useful, we conduct the global F-test:

$$H_0: \beta_1 = \beta_2 = 0$$

$$H_a: \text{At least one of the above coefficients is nonzero}$$

---

* For students with knowledge of calculus, note that the slope of the quadratic model is the first derivative $\partial y / \partial x = \beta_1 + 2\beta_2 x$. Thus, the slope varies as a function of x, rather than the constant slope associated with the straight-line model.

Figure 4.16 Potential misuse of quadratic model

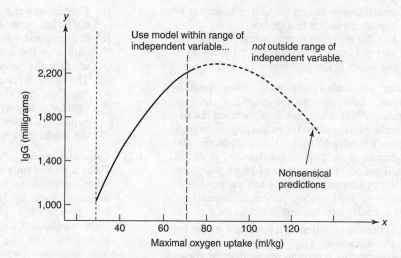

From the SPSS printout, Figure 4.14, the test statistic is $F = 203.159$ with an associated p-value of 0. For any reasonable α, we reject H_0 and conclude that the overall model is a useful predictor of immunity level, y.

(f) Figure 4.15 shows concave downward curvature in the relationship between immunity level and aerobic fitness level in the sample of 30 data points. To determine if this type of curvature exists in the population, we want to test

$$H_0: \beta_2 = 0 \text{ (no curvature in the response curve)}$$

$$H_a: \beta_2 < 0 \text{ (downward concavity exists in the response curve)}$$

The test statistic for testing β_2, highlighted on the SPSS printout (Figure 4.14), is $t = -3.39$ and the associated two-tailed p-value is .002. Since this is a one-tailed test, the appropriate p-value is $.002/2 = .001$. Now $\alpha = .01$ exceeds this p-value. Thus, there is very strong evidence of downward curvature in the population, that is, immunity level (IgG) increases more slowly per unit increase in maximal oxygen uptake for subjects with high aerobic fitness than for those with low fitness levels.

Note that the SPSS printout in Figure 4.14 also provides the t-test statistic and corresponding two-tailed p-values for the tests of $H_0: \beta_0 = 0$ and $H_0: \beta_1 = 0$. Since the interpretation of these parameters is not meaningful for this model, the tests are not of interest.

4.11 Exercises

4.34 **Assertiveness and leadership.** Management professors at Columbia University examined the relationship between assertiveness and leadership (*Journal of Personality and Social Psychology*, February 2007). The sample was comprised of 388 people enrolled in a full-time MBA program. Based on answers to a questionnaire, the researchers measured two variables for each subject: assertiveness score (x) and leadership ability score (y). A quadratic regression model was fit to the data with the following results:

INDEPENDENT VARIABLE	β ESTIMATE	t-VALUE	p-VALUE
x	.57	2.55	.01
x^2	−.88	−3.97	< .01
Model $R^2 = .12$			

(a) Conduct a test of overall model utility. Use $\alpha = .05$.

(b) The researchers hypothesized that leadership ability will increase at a decreasing rate with assertiveness. Set up the null and alternative hypothesis to test this theory.

(c) Use the reported results to conduct the test, part b. Give your conclusion (at $\alpha = .05$) in the words of the problem.

4.35 Urban population estimation using satellite images. Refer to the *Geographical Analysis* (January 2007) study that demonstrated the use of satellite image maps for estimating urban population, Exercise 4.7 (p. 185). A first-order model for census block population density (y) was fit as a function of proportion of block with low-density residential areas (x_1) and proportion of block with high-density residential areas (x_2). Now consider a second-order model for y.

(a) Write the equation of a quadratic model for y as a function of x_1.

(b) Identify the β term in the model that allows for a curvilinear relationship between y and x_1.

(c) Suppose that the rate of increase of population density (y) with proportion of block with low-density areas (x_1) is greater for lower proportions than for higher proportions. Will the term you identified in part b be positive or negative? Explain.

4.36 Cars with catalytic converters. A quadratic model was applied to motor vehicle toxic emissions data collected over 15 recent years in Mexico City (*Environmental Science and Engineering*, September 1, 2000). The following equation was used to predict the percentage (y) of motor vehicles without catalytic converters in the Mexico City fleet for a given year (x): $\hat{y} = 325{,}790 - 321.67x + 0.794x^2$.

(a) Explain why the value $\hat{\beta}_0 = 325{,}790$ has no practical interpretation.

(b) Explain why the value $\hat{\beta}_1 = -321.67$ should not be interpreted as a slope.

(c) Examine the value of $\hat{\beta}_2$ to determine the nature of the curvature (upward or downward) in the sample data.

(d) The researchers used the model to estimate "that just after the year 2021 the fleet of cars with catalytic converters will completely disappear." Comment on the danger of using the model to predict y in the year 2021.

4.37 Carp diet study. *Fisheries Science* (February 1995) reported on a study of the variables that affect endogenous nitrogen excretion (ENE) in carp raised in Japan. Carp were divided into groups of 2–15 fish, each according to body weight and each group placed in a separate tank. The carp were then fed a protein-free diet three times daily for a period of 20 days. One day after terminating the feeding

🐟 CARP

TANK	BODY WEIGHT x	ENE y
1	11.7	15.3
2	25.3	9.3
3	90.2	6.5
4	213.0	6.0
5	10.2	15.7
6	17.6	10.0
7	32.6	8.6
8	81.3	6.4
9	141.5	5.6
10	285.7	6.0

Source: Watanabe, T., and Ohta, M. "Endogenous nitrogen excretion and non-fecal energy losses in carp and rainbow trout." *Fisheries Science*, Vol. 61, No. 1, Feb. 1995, p. 56 (Table 5).

MINITAB output for Exercise 4.37

```
The regression equation is
ENE = 13.7 - 0.102 BODYWT +0.000273 BODYWTSQ

Predictor      Coef      SE Coef        T        P
Constant     13.713        1.306    10.50    0.000
BODYWT      -0.10184      0.02881    -3.53    0.010
BODYWTSQ    0.0002735    0.0001016     2.69    0.031

S = 2.194      R-Sq = 73.7%     R-Sq(adj) = 66.2%

Analysis of Variance

Source           DF        SS        MS        F        P
Regression        2    94.659    47.329     9.83    0.009
Residual Error    7    33.705     4.815
Total             9   128.364
```

experiment, the amount of ENE in each tank was measured. The table (p. 206) gives the mean body weight (in grams) and ENE amount (in milligrams per 100 grams of body weight per day) for each carp group.

(a) Graph the data in a scatterplot. Do you detect a pattern?

(b) The quadratic model $E(y) = \beta_0 + \beta_1 x + \beta_2 x^2$ was fit to the data using MINITAB. The MINITAB printout is displayed on p. 206. Conduct the test H_0: $\beta_2 = 0$ against H_a: $\beta_2 \neq 0$ using $\alpha = .10$. Give the conclusion in the words of the problem.

4.38 Estimating change-point dosage. A standard method for studying toxic substances and their effects on humans is to observe the responses of rodents exposed to various doses of the substance over time. In the *Journal of Agricultural, Biological, and Environmental Statistics* (June 2005), researchers used least squares regression to estimate the "change-point" dosage—defined as the largest dose level that has no adverse effects. Data were obtained from a dose–response study of rats exposed to the toxic substance aconiazide. A sample of 50 rats was evenly divided into five dosage groups: 0, 100, 200, 500, and 750 milligrams per kilograms of body weight. The dependent variable y measured was the weight change (in grams) after a 2-week exposure. The researchers fit the quadratic model $E(y) = \beta_0 + \beta_1 x + \beta_2 x^2$, where x = dosage level, with the following results: $\hat{y} = 10.25 + .0053x - .0000266x^2$.

(a) Construct a rough sketch of the least squares prediction equation. Describe the nature of the curvature in the estimated model.

(b) Estimate the weight change (y) for a rat given a dosage of 500 mg/kg of aconiazide.

(c) Estimate the weight change (y) for a rat given a dosage of 0 mg/kg of aconiazide. (This dosage is called the "control" dosage level.)

(d) Of the five groups in the study, find the largest dosage level x that yields an estimated weight change that is closest to but below the estimated weight change for the control group. This value is the "change-point" dosage.

4.39 Onset of a neurodegenerative disorder. Spinocerebellar ataxia type 1 (SCA1) is an inherited neurodegenerative disorder characterized by dysfunction of the brain. From a DNA analysis of SCA1 chromosomes, researchers discovered the presence of repeat gene sequences (*Cell Biology*, February 1995). In general, the more repeat sequences observed, the earlier the onset of the disease (in years of age). The following

scatterplot shows this relationship for data collected on 113 individuals diagnosed with SCA1.

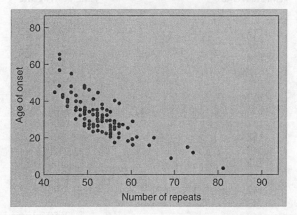

(a) Suppose you want to model the age y of onset of the disease as a function of number x of repeat gene sequences in SCA1 chromosomes. Propose a quadratic model for y.

(b) Will the sign of β_2 in the model, part a, be positive or negative? Base your decision on the results shown in the scatterplot.

(c) The researchers reported a correlation of $r = -.815$ between age and number of repeats. Since $r^2 = (-.815)^2 = .664$, they concluded that about "66% of the variability in the age of onset can be accounted for by the number of repeats." Does this statement apply to the quadratic model $E(y) = \beta_0 + \beta_1 x + \beta_2 x^2$? If not, give the equation of the model for which it does apply.

4.40 Failure times of silicon wafer microchips. Researchers at National Semiconductor experimented with tin-lead solder bumps used to manufacture silicon wafer integrated circuit chips (International Wafer Level Packaging Conference, November 3–4, 2005). The failure times of the microchips (in hours) were determined at different solder temperatures (degrees Centigrade). The data for one experiment are given in the next table (p. 208). The researchers want to predict failure time (y) based on solder temperature (x).

(a) Construct a scatterplot for the data. What type of relationship, linear or curvilinear, appears to exist between failure time and solder temperature?

(b) Fit the model, $E(y) = \beta_0 + \beta_1 x + \beta_2 x^2$, to the data. Give the least squares prediction equation.

(c) Conduct a test to determine if there is upward curvature in the relationship between failure time and solder temperature. (Use $\alpha = .05$.)

WAFER

TEMPERATURE (°C)	TIME TO FAILURE (hours)
165	200
162	200
164	1200
158	500
158	600
159	750
156	1200
157	1500
152	500
147	500
149	1100
149	1150
142	3500
142	3600
143	3650
133	4200
132	4800
132	5000
134	5200
134	5400
125	8300
123	9700

Source: Gee, S., & Nguyen, L. "Mean time to failure in wafer level–CSP packages with SnPb and SnAgCu solder bmps," International Wafer Level Packaging Conference, San Jose, CA, Nov. 3–4, 2005 (adapted from Figure 7).

4.41   Optimizing semiconductor material processing. Fluorocarbon plasmas are used in the production

RADICALS

RATE	TIME
1.00	0.1
0.80	0.3
0.40	0.5
0.20	0.7
0.05	0.9
0.00	1.1
−0.05	1.3
−0.02	1.5
0.00	1.7
−0.10	1.9
−0.15	2.1
−0.05	2.3
−0.13	2.5
−0.08	2.7
0.00	2.9

Source: Takizawa, K., et al. "Characteristics of C_3 radicals in high-density C_4F_8 plasmas studied by laser-induced fluorescence spectroscopy," *Journal of Applied Physics,* Vol. 88, No. 11, Dec. 1, 2000 (Figure 7). Reprinted with permission from Journal of Applied Physics. Copyright © 2000, American Institute of Physics.

of semiconductor materials. In the *Journal of Applied Physics* (December 1, 2000), electrical engineers at Nagoya University (Japan) studied the kinetics of fluorocarbon plasmas in order to optimize material processing. In one portion of the study, the surface production rate of fluorocarbon radicals emitted from the production process was measured at various points in time (in milliseconds) after the radio frequency power was turned off. The data are given in the accompanying table. Consider a model relating surface production rate (y) to time (x).

(a) Graph the data in a scattergram. What trend do you observe?

(b) Fit a quadratic model to the data. Give the least squares prediction equation.

(c) Is there sufficient evidence of upward curvature in the relationship between surface production rate and time after turnoff? Use $\alpha = .05$.

4.42   Public perceptions of health risks. In the *Journal of Experimental Psychology: Learning, Memory,*

INFECTION

INFECTION	INCIDENCE RATE	ESTIMATE
Polio	0.25	300
Diphtheria	1	1000
Trachoma	1.75	691
Rabbit Fever	2	200
Cholera	3	17.5
Leprosy	5	0.8
Tetanus	9	1000
Hemorrhagic Fever	10	150
Trichinosis	22	326.5
Undulant Fever	23	146.5
Well's Disease	39	370
Gas Gangrene	98	400
Parrot Fever	119	225
Typhoid	152	200
Q Fever	179	200
Malaria	936	400
Syphilis	1514	1500
Dysentery	1627	1000
Gonorrhea	2926	6000
Meningitis	4019	5000
Tuberculosis	12619	1500
Hepatitis	14889	10000
Gastroenteritis	203864	37000
Botulism	15	37500

Source: Hertwig, R., Pachur, T., & Kurzenhauser, S. "Judgments of risk frequencies: Tests of possible cognitive mechanisms," *Journal of Experimental Psychology: Learning, Memory, and Cognition,* Vol. 31, No. 4, July 2005 (Table 1). Copyright © 2005 American Psychological Association, reprinted with permission.

Example 4.8

Although a regional express delivery service bases the charge for shipping a package on the package weight and distance shipped, its profit per package depends on the package size (volume of space that it occupies) and the size and nature of the load on the delivery truck. The company recently conducted a study to investigate the relationship between the cost, y, of shipment (in dollars) and the variables that control the shipping charge—package weight, x_1 (in pounds), and distance shipped, x_2 (in miles). Twenty packages were randomly selected from among the large number received for shipment and a detailed analysis of the cost of shipment was made for each package, with the results shown in Table 4.3.

💿 EXPRESS

Table 4.3 Cost of shipment data for Example 4.8

Package	Weight x_1 (lbs)	Distance x_2 (miles)	Cost y (dollars)	Package	Weight x_1 (lbs)	Distance x_2 (miles)	Cost y (dollars)
1	5.9	47	2.60	11	5.1	240	11.00
2	3.2	145	3.90	12	2.4	209	5.00
3	4.4	202	8.00	13	.3	160	2.00
4	6.6	160	9.20	14	6.2	115	6.00
5	.75	280	4.40	15	2.7	45	1.10
6	.7	80	1.50	16	3.5	250	8.00
7	6.5	240	14.50	17	4.1	95	3.30
8	4.5	53	1.90	18	8.1	160	12.10
9	.60	100	1.00	19	7.0	260	15.50
10	7.5	190	14.00	20	1.1	90	1.70

(a) Give an appropriate linear model for the data.

(b) Fit the model to the data and give the prediction equation.

(c) Find the value of s and interpret it.

(d) Find the value of R_a^2 and interpret it.

(e) Is the model statistically useful for the prediction of shipping cost y? Find the value of the F statistic on the printout and give the observed significance level (p-value) for the test.

(f) Find a 95% prediction interval for the cost of shipping a 5-pound package a distance of 100 miles.

Solution

(a) Since we have no reason to expect that the relationship between y and x_1 and x_2 would be first-order, we will allow for curvature in the response surface and fit the complete second-order model

$$y = \beta_0 + \beta_1 x_1 + \beta_2 x_2 + \beta_3 x_1 x_2 + \beta_4 x_1^2 + \beta_5 x_2^2 + \varepsilon$$

The mean value of the random error term ε is assumed to equal 0. Therefore, the mean value of y is

$$E(y) = \beta_0 + \beta_1 x_1 + \beta_2 x_2 + \beta_3 x_1 x_2 + \beta_4 x_1^2 + \beta_5 x_2^2$$

(b) The SAS printout for fitting the model to the $n = 20$ data points is shown in Figure 4.18. The parameter estimates (highlighted on the printout) are:

$$\hat{\beta}_0 = .82702 \quad \hat{\beta}_1 = -.60914 \quad \hat{\beta}_2 = .00402$$

$$\hat{\beta}_3 = .00733 \quad \hat{\beta}_4 = .08975 \quad \hat{\beta}_5 = .00001507$$

Therefore, the prediction equation that relates the predicted shipping cost, $\hat{y}$, to weight of package, x_1, and distance shipped, x_2, is

$$\hat{y} = .82702 - .60914x_1 + .00402x_2 + .00733x_1x_2 + .08975x_1^2$$
$$+.00001507x_2^2$$

Figure 4.18 SAS multiple regression output for Example 4.8

Dependent Variable: Y

Analysis of Variance

Source	DF	Sum of Squares	Mean Square	F Value	Pr > F
Model	5	449.34076	89.86815	458.39	<.0001
Error	14	2.74474	0.19605		
Corrected Total	19	452.08550			

Root MSE	0.44278	R-Square	0.9939
Dependent Mean	6.33500	Adj R-Sq	0.9918
Coeff Var	6.98940		

Parameter Estimates

Variable	DF	Parameter Estimate	Standard Error	t Value	Pr > |t|
Intercept	1	0.82702	0.70229	1.18	0.2586
X1	1	-0.60914	0.17990	-3.39	0.0044
X2	1	0.00402	0.00800	0.50	0.6230
X1X2	1	0.00733	0.00063743	11.49	<.0001
X1SQ	1	0.08975	0.02021	4.44	0.0006
X2SQ	1	0.00001507	0.00002243	0.67	0.5127

Output Statistics

Obs	X1	X2	Dep Var Y	Predicted Value	Std Error Mean Predict	95% CL Predict	
1	5.9	47	2.6000	2.6114	0.2990	1.4655	3.7573
2	3.2	145	3.9000	4.0964	0.2111	3.0443	5.1486
3	4.4	202	8.0000	7.8238	0.1994	6.7822	8.8654
4	6.6	160	9.2000	9.4828	0.1818	8.4562	10.5093
5	0.75	280	4.4000	4.2666	0.3992	2.9880	5.5452
6	0.7	80	1.5000	1.2730	0.2453	0.1874	2.3587
7	6.5	240	14.5000	13.9229	0.2127	12.8693	14.9764
8	4.5	53	1.9000	1.9063	0.2284	0.8378	2.9748
9	0.6	100	1.0000	1.4862	0.2259	0.4201	2.5523
10	7.5	190	14.0000	13.0560	0.2155	11.9998	14.1122
11	5.1	240	11.0000	10.8562	0.1962	9.8175	11.8949
12	2.4	209	5.0000	5.0559	0.1999	4.0140	6.0979
13	0.3	160	2.0000	2.0332	0.2602	0.9316	3.1347
14	6.2	115	6.0000	6.3863	0.1954	5.3482	7.4243
15	2.7	45	1.1000	0.9383	0.2572	-0.1600	2.0366
16	3.5	250	8.0000	8.1527	0.2117	7.1001	9.2054
17	4.1	95	3.3000	3.2101	0.1803	2.1847	4.2356
18	8.1	160	12.1000	12.3066	0.3015	11.1577	13.4555
19	7	260	15.5000	16.3603	0.3109	15.1999	17.5207
20	1.1	90	1.7000	1.4749	0.1993	0.4335	2.5163
21	5	100	.	4.2414	0.1837	3.2133	5.2695

(c) The value of s (shaded on the printout) is .44278. Since s estimates the standard deviation σ of the random error term, our interpretation is that approximately 95% of the sampled shipping cost values fall within $2s = .886$, or about 89¢, of their respective predicted values.

(d) The value of R_a^2 (highlighted) is .9918. This means that after adjusting for sample size and the number of model parameters, about 99% of the total sample variation in shipping cost (y) is explained by the model; the remainder is explained by random error.

(e) The test statistic for testing whether the model is useful for predicting shipping cost is

$$F = \frac{\text{Mean square for model}}{\text{Mean square for error}} = \frac{\text{SS (Model)}/k}{\text{SSE}/[n - (k + 1)]}$$

where $n = 20$ is the number of data points and $k = 5$ is the number of parameters (excluding β_0) contained in the model. This value of F, highlighted on the printout, is $F = 458.39$. The observed significance level (p-value) for the test also highlighted is less than .0001. This means that if the model contributed no information for the prediction of y, the probability of observing a value of the F statistic as large as 458.39 would be only .0001. Thus, we would reject the null hypothesis for all values of α larger than .0001 and conclude that the model is useful for predicting shipping cost y.

(f) The predicted value of y for $x_1 = 5.0$ pounds and $x_2 = 100$ miles is obtained by computing

$$\hat{y} = .82702 - .60914(5.0) + .00402(100) + .00733(5.0)(100)$$
$$+ .08975(5.0)^2 + .00001507(100)^2$$

This quantity is shown (shaded) on the printout as $\hat{y} = 4.2414$. The corresponding 95% prediction interval (shaded) is given as 3.2133 to 5.2695. Therefore, if we were to select a 5-pound package and ship it 100 miles, we are 95% confident that the actual cost will fall between \$3.21 and \$5.27.   ■

Models with Qualitative x's   Multiple regression models can also include **qualitative** (or **categorical**) independent variables. (You have encountered some of these models in the exercises.) Qualitative variables, unlike quantitative variables, cannot be measured on a numerical scale. Therefore, we need to code the values of the qualitative variable (called **levels**) as numbers before we can fit the model. These coded qualitative variables are called **dummy variables** since the numbers assigned to the various levels are arbitrarily selected.

For example, consider a salary discrimination case where there exists a claim of gender discrimination—specifically, the claim that male executives at a large company receive higher average salaries than female executives with the same credentials and qualifications.

To test this claim, we might propose a multiple regression model for executive salaries using the gender of an executive as one of the independent variables. The dummy variable used to describe gender may be coded as follows:

$$x_3 = \begin{cases} 1 & \text{if male} \\ 0 & \text{if female} \end{cases}$$

The advantage of using a 0–1 coding scheme is that the β coefficients associated with the dummy variables are easily interpreted. To illustrate, consider the following model for executive salary y:

$$E(y) = \beta_0 + \beta_1 x$$

where

$$x = \begin{cases} 1 & \text{if male} \\ 0 & \text{if female} \end{cases}$$

This model allows us to compare the mean executive salary $E(y)$ for males with the corresponding mean for females:

Males $(x = 1)$: $E(y) = \beta_0 + \beta_1(1) = \beta_0 + \beta_1$

Females $(x = 0)$: $E(y) = \beta_0 + \beta_1(0) = \beta_0$

First note that β_0 represents the mean salary for females (say, μ_F). When using a 0–1 coding convention, β_0 will always represent the mean response associated with the level of the qualitative variable assigned the value 0 (called the **base level**). The difference between the mean salary for males and the mean salary for females, $\mu_M - \mu_F$, is represented by β_1—that is,

$$\mu_M - \mu_F = (\beta_0 + \beta_1) - (\beta_0) = \beta_1$$

Therefore, with the 0–1 coding convention, β_1 will always represent the difference between the mean response for the level assigned the value 1 and the mean for the base level. Thus, for the executive salary model we have

$$\beta_0 = \mu_F$$

$$\beta_1 = \mu_M - \mu_F$$

If β_1 exceeds 0, then $\mu_M > \mu_F$ and evidence of sex discrimination at the company exists.

The model relating a mean response $E(y)$ to a qualitative independent variable at two levels is shown in the next box.

A Model Relating $E(y)$ to a Qualitative Independent Variable with Two Levels

$$E(y) = \beta_0 + \beta_1 x$$

where

$$x = \begin{cases} 1 & \text{if level A} \\ 0 & \text{if level B} \end{cases}$$

Interpretation of β's:

$$\beta_0 = \mu_B \text{ (Mean for base level)}$$

$$\beta_1 = \mu_A - \mu_B$$

A Model Relating $E(y)$ to a Qualitative Independent Variable with Three Levels

$$E(y) = \beta_0 + \beta_1 x_1 + \beta_2 x_2$$

where

$$x_1 = \begin{cases} 1 & \text{if level A} \\ 0 & \text{if not} \end{cases} \qquad x_2 = \begin{cases} 1 & \text{if level B} \\ 0 & \text{if not} \end{cases} \qquad \text{Base level} = \text{Level C}$$

Interpretation of β's:

$$\beta_0 = \mu_C \text{ (Mean for base level)}$$

$$\beta_1 = \mu_A - \mu_C$$

$$\beta_2 = \mu_B - \mu_C$$

For models that involve qualitative independent variables at more than two levels, additional dummy variables must be created. In general, the number of dummy variables used to describe a qualitative variable will be one less than the number of levels of the qualitative variable. The box (bottom, p. 214) presents a model that includes a qualitative independent variable at three levels.

Example 4.9

Refer to the problem of modeling the shipment cost, y, of a regional express delivery service, described in Example 4.8. Suppose we want to model $E(y)$ as a function of cargo type, where cargo type has three levels—fragile, semifragile, and durable. Costs for 15 packages of approximately the same weight and same distance shipped, but of different cargo types, are listed in Table 4.4.

(a) Write a model relating $E(y)$ to cargo type.

(b) Interpret the estimated β coefficients in the model.

(c) A MINITAB printout for the model, part a, is shown in Figure 4.19. Conduct the F-Test for overall model utility using $\alpha = .05$. Explain the practical significance of the result.

⊗ CARGO

Table 4.4 Data for Example 4.9

Package	Cost, y	Cargo Type	x_1	x_2
1	$17.20	Fragile	1	0
2	11.10	Fragile	1	0
3	12.00	Fragile	1	0
4	10.90	Fragile	1	0
5	13.80	Fragile	1	0
6	6.50	Semifragile	0	1
7	10.00	Semifragile	0	1
8	11.50	Semifragile	0	1
9	7.00	Semifragile	0	1
10	8.50	Semifragile	0	1
11	2.10	Durable	0	0
12	1.30	Durable	0	0
13	3.40	Durable	0	0
14	7.50	Durable	0	0
15	2.00	Durable	0	0

Solution

(a) Since the qualitative variable of interest, cargo type, has three levels, we must create $(3 - 1) = 2$ dummy variables. First, select (arbitrarily) one of the levels

Figure 4.19 MINITAB multiple regression output for Example 4.9

```
The regression equation is
Y = 3.26 + 9.74 X1 + 5.44 X2

Predictor      Coef      SE Coef        T         P
Constant      3.260        1.075      3.03     0.010
X1            9.740        1.521      6.41     0.000
X2            5.440        1.521      3.58     0.004

S = 2.404       R-Sq = 77.4%      R-Sq(adj) = 73.7%

Analysis of Variance

Source            DF         SS        MS        F         P
Regression         2     238.25    119.13    20.61     0.000
Residual Error    12      69.37      5.78
Total             14     307.62
```

to be the base level—say, durable cargo. Then each of the remaining levels is assigned the value 1 in one of the two dummy variables as follows:

$$x_1 = \begin{cases} 1 & \text{if fragile} \\ 0 & \text{if not} \end{cases} \qquad x_2 = \begin{cases} 1 & \text{if semifragile} \\ 0 & \text{if not} \end{cases}$$

(Note that for the base level, durable cargo, $x_1 = x_2 = 0$.) The values of x_1 and x_2 for each package are given in Table 4.4. Then, the appropriate model is

$$E(y) = \beta_0 + \beta_1 x_1 + \beta_2 x_2$$

(b) To interpret the β's, first write the mean shipment cost $E(y)$ for each of the three cargo types as a function of the β's:

Fragile $(x_1 = 1, x_2 = 0)$:

$$E(y) = \beta_0 + \beta_1(1) + \beta_2(0) = \beta_0 + \beta_1 = \mu_F$$

Semifragile $(x_1 = 0, x_2 = 1)$:

$$E(y) = \beta_0 + \beta_1(0) + \beta_2(1) = \beta_0 + \beta_2 = \mu_S$$

Durable $(x_1 = 0, x_2 = 0)$:

$$E(y) = \beta_0 + \beta_1(0) + \beta_2(0) = \beta_0 = \mu_D$$

Then we have

$$\beta_0 = \mu_D \text{ (Mean of the base level)}$$

$$\beta_1 = \mu_F - \mu_D$$

$$\beta_2 = \mu_S - \mu_D$$

Note that the β's associated with the non–base levels of cargo type (fragile and semifragile) represent differences between a pair of means. As always, β_0 represents a single mean—the mean response for the base level (durable). Now, the estimated β's (highlighted on the MINITAB printout, Figure 4.19) are:

$$\hat{\beta}_0 = 3.26, \quad \hat{\beta}_1 = 9.74, \quad \hat{\beta}_2 = 5.44$$

Consequently, the estimated mean shipping cost for durable cargo ($\hat{\beta}_0$) is \$3.26; the difference between the estimated mean costs for fragile and durable cargo ($\hat{\beta}_1$) is \$9.74; and the difference between the estimated mean costs for semifragile and durable cargo ($\hat{\beta}_2$) is \$5.44.

(c) The F-Test for overall model utility tests the null hypothesis

$$H_0: \beta_1 = \beta_2 = 0$$

Note that $\beta_1 = 0$ implies that $\mu_F = \mu_D$ and $\beta_2 = 0$ implies that $\mu_S = \mu_D$. Therefore, $\beta_1 = \beta_2 = 0$ implies that $\mu_F = \mu_S = \mu_D$. Thus, a test for model utility is equivalent to a test for equality of means, that is,

$$H_0: \mu_F = \mu_S = \mu_D$$

From the MINITAB printout, Figure 4.19, $F = 20.61$. Since the p-value of the test (.000) is less than $\alpha = .05$, the null hypothesis is rejected. Thus, there is evidence of a difference between any two of the three mean shipment costs, that is, cargo type is a useful predictor of shipment cost y. ■

Multiplicative Models   In all the models presented so far, the random error component has been assumed to be *additive*. An additive error is one for which the response is equal to the mean $E(y)$ plus random error,

$$y = E(y) + \varepsilon$$

Another useful type of model for business, economic, and scientific data is the **multiplicative model**. In this model, the response is written as a *product* of its mean and the random error component, that is,

$$y = [E(y)] \cdot \varepsilon$$

Researchers have found multiplicative models to be useful when the change in the response y for every 1-unit change in an independent variable x is better represented by a percentage increase (or decrease) rather than a constant amount increase (or decrease).* For example, economists often want to predict a percentage change in the price of a commodity or a percentage increase in the salary of a worker. Consequently, a multiplicative model is used rather than an additive model.

A multiplicative model in two independent variables can be specified as

$$y = (e^{\beta_0})(e^{\beta_1 x_1})(e^{\beta_2 x_2})(e^{\varepsilon})$$

where β_0, β_1, and β_2 are population parameters that must be estimated from the sample data and e^x is a notation for the antilogarithm of x. Note, however, that the multiplicative model is not a linear statistical model as defined in Section 4.1. To use the method of least squares to fit the model to the data, we must transform the model into the form of a linear model. Taking the natural logarithm (denoted ln) of both sides of the equation, we obtain

$$\ln(y) = \beta_0 + \beta_1 x_1 + \beta_2 x_2 + \varepsilon$$

which is now in the form of a linear (additive) model.

When the dependent variable is $\ln(y)$, rather than y, the β parameters and other key regression quantities have slightly different interpretations, as the next example illustrates.

Example 4.10

Towers, Perrin, Forster & Crosby (TPF&C), an international management consulting firm, has developed a unique and interesting application of multiple regression analysis. Many firms are interested in evaluating their management salary structure,

---

* Multiplicative models are also found to be useful when the standard regression assumption of equal variances is violated. We discuss this application of multiplicative models in Chapter 8.

and TPF&C uses multiple regression models to accomplish this salary evaluation. The Compensation Management Service, as TPF&C calls it, measures both the internal and external consistency of a company's pay policies to determine whether they reflect the management's intent.

The dependent variable y used to measure executive compensation is annual salary. The independent variables used to explain salary structure include the variables listed in Table 4.5. The management at TPF&C has found that executive compensation models that use the natural logarithm of salary as the dependent variable are better predictors than models that use salary as the dependent variable. This is probably because salaries tend to be incremented in *percentages* rather than dollar values. Thus, the multiplicative model we propose (in its linear form) is

$$\ln(y) = \beta_0 + \beta_1 x_1 + \beta_2 x_2 + \beta_3 x_3 + \beta_4 x_4 + \beta_5 x_5 + \beta_6 x_1^2 + \beta_7 x_3 x_4 + \varepsilon$$

We have included a second-order term, x_1^2, to account for a possible curvilinear relationship between ln(salary) and years of experience, x_1. Also, the interaction term $x_3 x_4$ is included to account for the fact that the relationship between the number of employees supervised, x_4, and corporate salary may depend on gender, x_3. For example, as the number of supervised employees increases, a male's salary (with all other factors being equal) might rise more rapidly than a female's. (If this is found to be true, the firm will take steps to remove the apparent discrimination against female executives.)

⊜ EXECSAL

Table 4.5 List of independent variables for executive compensation example

Independent Variable	Description
x_1	Years of experience
x_2	Years of education
x_3	1 if male; 0 if female
x_4	Number of employees supervised
x_5	Corporate assets (millions of dollars)
x_6	x_1^2
x_7	$x_3 x_4$

A sample of 100 executives is selected and the variables y and $x_1, x_2, \ldots, x_5$ are recorded. (The data are saved in the CD file named EXECSAL.) The multiplicative model is fit to the data using MINITAB, with the results shown in the MINITAB printout, Figure 4.20.

(a) Find the least squares prediction equation, and interpret the estimate of β_2.

(b) Locate the estimate of s and interpret its value.

(c) Locate R_a^2 and interpret its value.

(d) Conduct a test of overall model utility using $\alpha = .05$.

(e) Is there evidence of gender discrimination at the firm? Test using $\alpha = .05$.

(f) Use the model to predict the salary of an executive with the characteristics shown in Table 4.6.

Figure 4.20 MINITAB multiple regression output for Example 4.10

```
The regression equation is
LNSAL = 9.86 + 0.0436 EXP + 0.0309 EDUC + 0.117 GENDER + 0.000326 NUMSUP
        + 0.00239 ASSETS - 0.000635 EXPSQ + 0.000302 GEN_SUP


Predictor        Coef      SE Coef        T       P
Constant      9.86182      0.09703   101.64   0.000
EXP           0.043643     0.003761    11.60   0.000
EDUC          0.030936     0.002950    10.49   0.000
GENDER        0.11661      0.03696      3.16   0.002
NUMSUP        0.00032594   0.00007850   4.15   0.000
ASSETS        0.0023911    0.0004439    5.39   0.000
EXPSQ        -0.0006348    0.0001383   -4.59   0.000
GEN_SUP       0.00030196   0.00009238   3.27   0.002


S = 0.0659583    R-Sq = 94.0%    R-Sq(adj) = 93.6%

Analysis of Variance
Source           DF       SS        MS        F       P
Regression        7  6.28215   0.89745   206.29   0.000
Residual Error   92  0.40025   0.00435
Total            99  6.68240


Predicted Values for New Observations

New
Obs      Fit  SE Fit          95% CI                  95% PI
  1  11.3021  0.0140  (11.2742, 11.3300)   (11.1681, 11.4360)

Values of Predictors for New Observations

New
Obs   EXP  EDUC    GENDER  NUMSUP  ASSETS  EXPSQ   GEN_SUP
  1  12.0  16.0  0.000000     400     160    144  0.000000
```

Table 4.6 Values of independent variables for a particular executive

$x_1 = 12$ years of experience

$x_2 = 16$ years of education

$x_3 = 0$ (female)

$x_4 = 400$ employees supervised

$x_5 = \$160 \times$ million (the firm's asset value)

$x_1^2 = 144$

$x_3 x_4 = 0$

Solution

(a) The least squares model (highlighted on the MINITAB printout) is

$$\widehat{\ln(y)} = 9.86 + .0436x_1 + .0309x_2 + .117x_3 + .000326x_4 + .00239x_5$$
$$- .000635x_6 + .000302x_7$$

Because we are using the logarithm of salary as the dependent variable, the β estimates have different interpretations than previously discussed. In general, a parameter β in a multiplicative (log) model represents the percentage increase (or decrease) in the dependent variable for a 1-unit increase in the corresponding independent variable. The percentage change is calculated by taking the antilogarithm of the β estimate and subtracting 1 (i.e., $e^{\beta} - 1$).[†] For example, the percentage change in executive compensation associated with a 1-unit (i.e., 1-year) increase in years of education x_2 is $(e^{\beta_2} - 1) = (e^{.0309} - 1) = .031$. Thus, when all other independent variables are held constant, we estimate executive salary to increase 3.1% for each additional year of education.

A Multiplicative (Log) Model Relating y to Several Independent Variables

$$\ln(y) = \beta_0 + \beta_1 x_1 + \beta_2 x_2 + \cdots + \beta_k x_k + \varepsilon$$

where $\ln(y)$ = natural logarithm of y

Interpretation of β's

$(e^{\beta_i} - 1) \times 100\%$ = Percentage change in y for every

1 unit increase in x_i, holding all other x's fixed

(b) The estimate of the standard deviation σ (shaded on the printout) is $s = .066$. Our interpretation is that most of the observed $\ln(y)$ values (logarithms of salaries) lie within $2s = 2(.066) = .132$ of their least squares predicted values. A more practical interpretation (in terms of salaries) is obtained, however, if we take the antilog of this value and subtract 1, similar to the manipulation in part a. That is, we expect most of the observed executive salaries to lie within $e^{2s} - 1 = e^{.132} - 1 = .141$, or 14.1% of their respective least squares predicted values.

(c) The adjusted R^2 value (highlighted on the printout) is $R_a^2 = .936$. This implies that, after taking into account sample size and the number of independent variables, almost 94% of the variation in the logarithm of salaries for these 100 sampled executives is accounted for by the model.

(d) The test for overall model utility is conducted as follows:

H_0: $\beta_1 = \beta_2 = \cdots = \beta_7 = 0$

H_a: At least one of the model coefficients is nonzero.

Test statistic: $F = \dfrac{\text{Mean square for model}}{\text{MSE}} = 206.29$ (shaded in Figure 4.20)

p-value $= .000$ (shaded in Figure 4.20)

Since $\alpha = .05$ exceeds the p-value of the test, we conclude that the model does contribute information for predicting executive salaries. It appears that at least one of the β parameters in the model differs from 0.

(e) If the firm is (knowingly or unknowingly) discriminating against female executives, then the mean salary for females (denoted μ_F) will be less than the

---

[†] The result is derived by expressing the percentage change in salary y, as $(y_1 - y_0)/y_0$, where $y_1 =$ the value of y when, say, $x = 1$, and $y_0 =$ the value of y when $x = 0$. Now let $y^* = \ln(y)$ and assume the log model is $y^* = \beta_0 + \beta_1 x$. Then

$$y = e^{y^*} = e^{\beta_0} e^{\beta_1 x} = \begin{cases} e^{\beta_0} & \text{when } x = 0 \\ e^{\beta_0} e^{\beta_1} & \text{when } x = 1 \end{cases}$$

Substituting, we have

$$\frac{y_1 - y_0}{y_0} = \frac{e^{\beta_0} e^{\beta_1} - e^{\beta_0}}{e^{\beta_0}} = e^{\beta_1} - 1$$

mean salary for males (denoted μ_M) with the same qualifications (e.g., years of experience, years of education, etc.) From our previous discussion of dummy variables, this difference will be represented by β_3, the β coefficient multiplied by x_3 if we set number of employees supervised, x_4, equal to 0. Since $x_3 = 1$ if male, 0 if female, then $\beta_3 = (\mu_M - \mu_F)$ for fixed values of x_1, x_2 and x_5, and $x_4 = 0$. Consequently, a test of

$$H_0: \beta_3 = 0 \quad \text{versus} \quad H_a: \beta_3 > 0$$

is one way to test the discrimination hypothesis. The p-value for this one-tailed test is one-half the p-value shown on the MINITAB printout (i.e., $.002/2 = .001$). With such a small p-value, there is strong evidence to reject H_0 and claim that some form of gender discrimination exists at the firm.

A test for discrimination could also include testing for the interaction term, $\beta_7 x_3 x_4$. If, as the number of employees supervised (x_4) increases, the rate of increase in ln(salary) for males exceeds the rate for females, then $\beta_7 > 0$. To see this, hold x_1, x_2, and x_5 constant in the model (e.g., $x_1 = 10$ years of experience, $x_2 = 15$ years of education, and $x_5 = \$120$ million in corporate assets), then substitute $x_3 = 0$ into the equation to obtain:

$$x_3 = 0 \text{ (Females): } E\{\ln(y)\} = \beta_0 + \beta_1(10) + \beta_2(15) + \beta_5(120) + \beta_6(15^2) + \beta_3(0)$$
$$+ \beta_4(x_4) + \beta_7(0)(x_4)$$
$$= \underbrace{\beta_0 + \beta_1(10) + \beta_2(15) + \beta_5(120) + \beta_6(15^2)}_{y\text{-intercept (constant)}} + \underbrace{\beta_4}_{\text{slope}} (x_4)$$

Similarly, substituting $x_3 = 1$ into the equation we obtain:

$$x_3 = 1 \text{ (Males): } E\{\ln(y)\} = \beta_0 + \beta_1(10) + \beta_2(15) + \beta_5(120) + \beta_6(15^2) + \beta_3(1)$$
$$+ \beta_4(x_4) + \beta_7(1)(x_4)$$
$$= \underbrace{\beta_0 + \beta_1(10) + \beta_2(15) + \beta_5(120) + \beta_6(15^2) + \beta_3}_{y\text{-intercept (constant)}}$$
$$+ \underbrace{(\beta_4 + \beta_7)}_{\text{slope}} x_4$$

Thus, the slope for males $(\beta_4 + \beta_7)$ exceeds the slope for females (β_4) only if $\beta_7 > 0$.

The one-tailed p-value for testing $H_0: \beta_7 = 0$ against $H_a: \beta_7 > 0$ (highlighted on the printout) is $.002/2 = .001$. Consequently, we reject H_0 and find further evidence of gender discrimination at the firm.

(f) The least squares model can be used to obtain a predicted value for the logarithm of salary. Substituting the values of the x's shown in Table 4.6, we obtain

$$\widehat{\ln(y)} = \hat{\beta}_0 + \hat{\beta}_1(12) + \hat{\beta}_2(16) + \hat{\beta}_3(0) + \hat{\beta}_4(400) + \hat{\beta}_5(160x)$$
$$+ \hat{\beta}_6(144) + \hat{\beta}_7(0)$$

This predicted value is given at the bottom of the MINITAB printout, Figure 4.20, $\widehat{\ln(y)} = 11.3021$. The 95% prediction interval, from 11.1681 to 11.4360, is also highlighted on the printout. To predict the salary of an executive with these characteristics, we take the antilog of these values. That is, the predicted salary is $e^{11.3021} = \$80,992$ (rounded to the nearest dollar) and the 95% prediction interval is from $e^{11.1681}$ to $e^{11.4360}$ (or from \$70,834 to

$92,596). Thus, an executive with the characteristics in Table 4.6 should be paid between $70,834 and $92,596 to be consistent with the sample data. ∎

Warning

To decide whether a log transformation on the dependent variable is necessary, naive researchers sometimes compare the R^2 values for the two models

$$y = \beta_0 + \beta_1 x_1 + \cdots + \beta_k x_k + \varepsilon$$

and

$$\ln(y) = \beta_0 + \beta_1 x_1 + \cdots + \beta_k x_k + \varepsilon$$

and choose the model with the larger R^2. But these R^2 values *are not comparable* since the dependent variables are not the same! One way to generate comparable R^2 values is to calculate the predicted values, $\widehat{\ln(y)}$, for the log model and then compute the corresponding $\hat{y}$ values using the inverse transformation $\hat{y} = e^{\widehat{\ln(y)}}$. A pseudo-$R^2$ for the log model can then be calculated in the usual way:

$$R^2_{\ln(y)} = 1 - \frac{\Sigma(y_i - \hat{y}_i)^2}{\Sigma(y_i - \bar{y}_i)^2}$$

$R^2_{\ln(y)}$ is now comparable to the R^2 for the untransformed model. See Maddala (1988) for a discussion of more formal methods for comparing the two models.

4.12 Exercises

4.43 **First-order model.** Write a first-order linear model relating the mean value of y, $E(y)$, to

(a) two quantitative independent variables
(b) four quantitative independent variables

4.44 **Second-order model.** Write a complete second-order linear model relating the mean value of y, $E(y)$, to

(a) two quantitative independent variables
(b) three quantitative independent variables

4.45 **Qualitative predictors.** Write a model relating $E(y)$ to a qualitative independent variable with

(a) two levels, A and B
(b) four levels, A, B, C, and D
    Interpret the β parameters in each case.

4.46 **Graphing a first-order model.** Consider the first-order equation

$$y = 1 + 2x_1 + x_2$$

(a) Graph the relationship between y and x_1 for $x_2 = 0, 1$, and 2.
(b) Are the graphed curves in part a first-order or second-order?
(c) How do the graphed curves in part a relate to each other?
(d) If a linear model is first-order in two independent variables, what type of geometric relationship will you obtain when $E(y)$ is

graphed as a function of one of the independent variables for various values of the other independent variable?

4.47 **Graphing a first-order model.** Consider the first-order equation

$$y = 1 + 2x_1 + x_2 - 3x_3$$

(a) Graph the relationship between y and x_1 for $x_2 = 1$ and $x_3 = 3$.
(b) Repeat part a for $x_2 = -1$ and $x_3 = 1$.
(c) If a linear model is first-order in three independent variables, what type of geometric relationship will you obtain when $E(y)$ is graphed as a function of one of the independent variables for various values of the other independent variables?

4.48 **Graphing a second-order model.** Consider the second-order model

$$y = 1 + x_1 - x_2 + 2x_1^2 + x_2^2$$

(a) Graph the relationship between y and x_1 for $x_2 = 0, 1$, and 2.
(b) Are the graphed curves in part a first-order or second-order?
(c) How do the graphed curves in part a relate to each other?
(d) Do the independent variables x_1 and x_2 interact? Explain.

4.49 Graphing a second-order model. Consider the second-order model

$$y = 1 + x_1 - x_2 + x_1x_2 + 2x_1^2 + x_2^2$$

(a) Graph the relationship between y and x_1 for $x_2 = 0, 1,$ and 2.

(b) Are the graphed curves in part a first-order or second-order?

(c) How do the graphed curves in part a relate to each other?

(d) Do the independent variables x_1 and x_2 interact? Explain.

(e) Note that the model used in this exercise is identical to the noninteraction model in Exercise 4.48, except that it contains the term involving x_1x_2. What does the term x_1x_2 introduce into the model?

4.50 Improving SAT scores. *Chance* (Winter 2001) published a study of students who paid a private tutor (or coach) to help them improve their Scholastic Assessment Test (SAT) scores. Multiple regression was used to estimate the effect of coaching on SAT–Mathematics scores. Data on 3,492 students (573 of whom were coached) were used to fit the model $E(y) = \beta_0 + \beta_1 x_1 + \beta_2 x_2$, where $y =$ SAT–Math score, $x_1 =$ score on PSAT, and $x_2 = \{1$ if student was coached, 0 if not$\}$.

(a) The fitted model had an adjusted R^2 value of .76. Interpret this result.

(b) The estimate of β_2 in the model was 19, with a standard error of 3. Use this information to form a 95% confidence interval for β_2. Interpret the interval.

(c) Based on the interval, part b, what can you say about the effect of coaching on SAT–Math scores?

4.51 Production technologies, terroir, and quality of Bordeaux wine. In addition to state-of-the-art technologies, the production of quality wine is strongly influenced by the natural endowments of the grape-growing region—called the "terroir." *The Economic Journal* (May 2008) published an empirical study of the factors that yield a quality Bordeaux wine. A quantitative measure of wine quality (y) was modeled as a function of several qualitative independent variables, including grape-picking method (manual or automated), soil type (clay, gravel, or sand), and slope orientation (east, south, west, southeast, or southwest).

(a) Create the appropriate dummy variables for each of the qualitative independent variables.

(b) Write a model for wine quality (y) as a function of grape-picking method. Interpret the β's in the model.

(c) Write a model for wine quality (y) as a function of soil type. Interpret the β's in the model.

(d) Write a model for wine quality (y) as a function of slope orientation. Interpret the β's in the model.

4.52 Detecting quantitative traits in genes. In gene therapy, it is important to know the location of a gene for a disease on the genome (genetic map). Although many genes yield a specific trait (e.g., disease or not), others cannot be categorized since they are quantitative in nature (e.g., extent of disease). Researchers at the University of North Carolina at Wilmington developed statistical models that link quantitative genetic traits to locations on the genome (*Chance*, Summer 2006). The extent of a certain disease is determined by the absence (A) or presence (B) of a gene marker at each of two locations, L1 and L2 on the genome. For example, AA represents absence of the marker at both locations whereas AB represents absence at location L1 but presence at location L2.

(a) How many different gene marker combinations are possible at the two locations? List them.

(b) Using dummy variables, write a model for extent of the disease, y, as a function of gene marker combination.

(c) Interpret the β-values in the model, part b.

(d) Give the null hypothesis for testing whether the overall model, part b, is statistically useful for predicting extent of the disease, y.

ACCHW

4.53 Homework assistance for accounting students. The *Journal of Accounting Education* (Vol. 25, 2007) published the results of a study designed to gauge the best method of assisting accounting students with their homework. A total of 75 accounting students took a pretest on a topic not covered in class, then each was given a homework problem to solve on the same topic. The students were assigned to one of three homework assistance groups. Some students received the completed solution, some were given check figures at various steps of the solution, and some received no help at all. After finishing the homework, the students were all given a posttest on the subject. The dependent variable of interest was the knowledge gain (or test score improvement). These data are saved in the ACCHW file.

(a) Propose a model for the knowledge gain (y) as a function of the qualitative variable, homework assistance group.

(b) In terms of the β's in the model, give an expression for the difference between the mean knowledge gains of students in the "completed solution" and "no help groups."

(c) Fit the model to the data and give the least squares prediction equation.

(d) Conduct the global F-Test for model utility using $\alpha = .05$. Interpret the results, practically.

4.54   Predicting oil spill evaporation. The *Journal of Hazardous Materials* (July 1995) presented a literature review of models designed to predict oil spill evaporation. One model discussed in the article used boiling point (x_1) and API specific gravity (x_2) to predict the molecular weight (y) of the oil that is spilled. A complete second-order model for y was proposed.

(a) Write the equation of the model.

(b) Identify the terms in the model that allow for curvilinear relationships.

4.55   Impact of race on football card values. University of Colorado sociologists investigated the impact of race on the value of professional football players' "rookie" cards (*Electronic Journal of Sociology*, 2007). The sample consisted of 148 rookie cards of National Football League (NFL) players who were inducted into the Football Hall of Fame (HOF). The researchers modeled the natural logarithm of card price (y, dollars) as a function of the following independent variables:

Race: $x_1 = 1$ if black, 0 if white
Card availability: $x_2 = 1$ if high, 0 if low
Card vintage: $x_3 =$ year card printed
Finalist: $x_4 =$ natural logarithm of number of times player on final HOF ballot
Position-QB: $x_5 = 1$ if quarterback, 0 if not
Position-KR: $x_6 = 1$ if kicker, 0 if not
Position-RB: $x_7 = 1$ if running back, 0 if not
Position-WR: $x_8 = 1$ if wide receiver, 0 if not
Position-TE: $x_9 = 1$ if tight end, 0 if not
Position-DL: $x_{10} = 1$ if defensive lineman, 0 if not
Position-LB: $x_{11} = 1$ if linebacker, 0 if not
Position-DB: $x_{12} = 1$ if defensive back, 0 if not
[Note: For Position, offensive lineman is the base level]

(a) The model $E\{\ln(y)\} = \beta_0 + \beta_1 x_1 + \beta_2 x_2 + \beta_3 x_3 + \beta_4 x_4 + \beta_5 x_5 + \beta_6 x_6 + \beta_7 x_7 + \beta_8 x_8 + \beta_9 x_9 + \beta_{10} x_{10} + \beta_{11} x_{11} + \beta_{12} x_{12}$ was fit to the data with the following results: $R^2 = .705$, adj-$R^2 = .681$, $F = 26.9$. Interpret the results practically. Make an inference about the overall adequacy of the model.

(b) Refer to part a. Statistics for the race variable were reported as follows: $\hat{\beta}_1 = -.147$, $s_{\hat{\beta}_1} = .145$, $t = -1.014$, p-value $= .312$. Use this information to make an inference about the impact of race on the value of professional football players' rookie cards.

(c) Refer to part a. Statistics for the card vintage variable were reported as follows: $\hat{\beta}_3 = -.074$, $s_{\hat{\beta}_3} = .007$, $t = -10.92$, p-value $= .000$. Use this information to make an inference about the impact of card vintage on the value of professional football players' rookie cards.

(d) Write a first-order model for $E\{\ln(y)\}$ as a function of card vintage (x_4) and position ($x_5 - x_{12}$) that allows for the relationship between price and vintage to vary depending on position.

4.56   Buy-side versus sell-side analysts' earnings forecasts. Harvard Business School professors carried out a comparison of earnings forecasts of buy-side and sell-side analysts and published the results in the *Financial Analysts Journal* (July/August 2008). The professors used regression to model the relative optimism (y) of the analysts' 3-month horizon forecasts. One of the independent variables used to model forecast optimism was the dummy variable $x = \{1$ if the analyst worked for a buy-side firm, 0 if the analyst worked for a sell-side firm$\}$.

(a) Write the equation of the model for $E(y)$ as a function of type of firm.

(b) Interpret the value of β_0 in the model, part a.

(c) The professors write that the value of β_1 in the model, part a, "represents the mean difference in relative forecast optimism between buy-side and sell-side analysts." Do you agree?

(d) The professors also argue that "if buy-side analysts make less optimistic forecasts than their sell-side counterparts, the [estimated value of β_1] will be negative." Do you agree?

4.57   Predicting red wine prices. The vineyards in the Bordeaux region of France are known for producing excellent red wines. However, the uncertainty of the weather during the growing season, the phenomenon that wine tastes better with age, and the fact that some Bordeaux vineyards produce better wines than others encourages speculation concerning the value of a case of wine produced by a certain vineyard during a certain year (or vintage). As a result, many wine experts attempt to predict the auction price of a case of Bordeaux wine. The publishers of a newsletter titled *Liquid Assets: The International Guide to Fine Wine* discussed a multiple-regression approach to predicting the London auction price of red Bordeaux wine in *Chance* (Fall 1995). The natural logarithm of the price y (in dollars) of a case containing a dozen bottles of red wine was modeled as a function of weather during growing season and age of vintage using data collected for the vintages

Results for Exercise 4.57

	BETA ESTIMATES (STANDARD ERRORS)		
INDEPENDENT VARIABLES	MODEL 1	MODEL 2	MODEL 3
x_1 = Vintage year	.0354 (.0137)	.0238 (.00717)	.0240 (.00747)
x_2 = Average growing season temperature (°C)	(not included)	.616 (.0952)	.608 (.116)
x_3 = Sept./Aug. rainfall (cm)	(not included)	−.00386 (.00081)	−.00380 (.00095)
x_4 = Rainfall in months preceding vintage (cm)	(not included)	.0001173 (.000482)	.00115 (.000505)
x_5 = Average Sept. temperature (°C)	(not included)	(not included)	.00765 (.565)
	$R^2 = .212$	$R^2 = .828$	$R^2 = .828$
	$s = .575$	$s = .287$	$s = .293$

Source: Ashenfelter, O., Ashmore, D., and LaLonde, R. "Bordeaux wine vintage quality and weather," *Chance*, Vol. 8, No. 4, Fall 1995, p. 116 (Table 2).

of 1952–1980. Three models were fit to the data. The results of the regressions are summarized in the table above.

(a) For each model, conduct a *t*-test for each of the β parameters in the model. Interpret the results.

(b) Interpret the β estimates of each model.

(c) The three models for auction price (y) have R^2 and s values, as shown in the table. Based on this information, which of the three models would you use to predict red Bordeaux wine prices? Explain.

4.58 **Effectiveness of insect repellents.** Which insect repellents protect best against mosquitoes? *Consumer Reports* (June 2000) tested 14 products that all claim to be an effective mosquito repellent. Each product was classified as either lotion/cream or aerosol/spray. The cost of the product (in dollars) was divided by the amount of the repellent needed to cover exposed areas of the skin (about 1/3 ounce) to obtain a cost-per-use value.

Effectiveness was measured as the maximum number of hours of protection (in half-hour increments) provided when human testers exposed their arms to 200 mosquitoes. The data from the report are listed in the table below.

(a) Suppose you want to use repellent type to model the cost per use (y). Create the appropriate number of dummy variables for repellent type and write the model.

(b) Fit the model, part a, to the data.

(c) Give the null hypothesis for testing whether repellent type is a useful predictor of cost per use (y).

(d) Conduct the test, part c, and give the appropriate conclusion. Use $\alpha = .10$.

(e) Repeat parts a–d if the dependent variable is maximum number of hours of protection (y).

4.59 **RNA analysis of wheat genes.** Engineers from the Department of Crop and Soil Sciences at Washington State University used regression to

REPELLENT

INSECT REPELLENT	TYPE	COST/USE	MAXIMUM PROTECTION
Amway Hourguard 12	Lotion/Cream	$2.08	13.5 hours
Avon Skin-So-Soft	Aerosol/Spray	0.67	0.5
Avon Bug Guard Plus	Lotion/Cream	1.00	2.0
Ben's Backyard Formula	Lotion/Cream	0.75	7.0
Bite Blocker	Lotion/Cream	0.46	3.0
BugOut	Aerosol/Spray	0.11	6.0
Cutter Skinsations	Aerosol/Spray	0.22	3.0
Cutter Unscented	Aerosol/Spray	0.19	5.5
Muskol Ultra 6 Hours	Aerosol/Spray	0.24	6.5
Natrapel	Aerosol/Spray	0.27	1.0
Off! Deep Woods	Aerosol/Spray	1.77	14.0
Off! Skintastic	Lotion/Cream	0.67	3.0
Sawyer Deet Formula	Lotion/Cream	0.36	7.0
Repel Permanone	Aerosol/Spray	2.75	24.0

Source: "Buzz off," *Consumer Reports*, June 2000. Copyright 2000 by Consumers Union of U.S., Inc. Yonkers, NY 10703-1057, a nonprofit organization. Reprinted with permission from the June 2000 issue of Consumer Reports® for educational purposes only. No commercial use or reproduction permitted. www.ConsumerReports.org.

⊚WHEATRNA

RNA PROPORTION (x_1)	NUMBER OF COPIES (Y, THOUSANDS)	
	MNSOD	PLD
0.00	401	80
0.00	336	83
0.00	337	75
0.33	711	132
0.33	637	148
0.33	602	115
0.50	985	147
0.50	650	142
0.50	747	146
0.67	904	146
0.67	1007	150
0.67	1047	184
0.80	1151	173
0.80	1098	201
0.80	1061	181
1.00	1261	193
1.00	1272	187
1.00	1256	199

Source: Baek, K. H., and Skinner, D. Z. "Quantitative real-time PCR method to detect changes in specific transcript and total RNA amounts," *Electronic Journal of Biotechnology*, Vol. 7, No. 1, April 15, 2004 (adapted from Figure 2).

estimate the number of copies of a gene transcript in an aliquot of RNA extracted from a wheat plant (*Electronic Journal of Biotechnology*, April 15, 2004). The proportion (x_1) of RNA extracted from a cold-exposed wheat plant was varied, and the transcript copy number (y, in thousands) was measured for each of two cloned genes: Mn Superoxide Dismutose (MnSOD) and Phospholipose D (PLD). The data are listed in the accompanying table. Letting x_1 = RNA proportion and x_2 = {1 if MnSOD, 0 if PLD}, consider the second-order model for number of copies (y):

$$E(y) = \beta_0 + \beta_1 x_1 + \beta_2 x_1^2 + \beta_3 x_2 + \beta_4 x_1 x_2 + \beta_5 x_1^2 x_2$$

(a) MINITAB was used to fit the model to the data. Locate the least squares prediction equation for y on the printout shown below.

(b) Is the overall model statistically useful for predicting transcript copy number (y)? Test using $\alpha = .01$.

(c) Based on the MINITAB results, is there evidence to indicate that transcript copy number (y) is curvilinearly related to proportion of RNA (x_1)? Explain.

4.60 Stop, look, and listen. Where do you look when you are listening to someone speak? Researchers have discovered that listeners tend to gaze at the eyes or mouth of the speaker. In a study published in *Perception and Psychophysics* (August 1998), subjects watched a videotape of a speaker giving a series of short monologues at a social gathering (e.g., a party). The level of background noise (multilingual voices and music) was varied during the listening sessions. Each subject wore a pair of clear plastic goggles on which an infrared corneal detection system was mounted, enabling the researchers to monitor the subject's

MINITAB Output for Exercise 4.59

```
The regression equation is
Y = 80.2 + 156 X1 - 42 X1SQ + 273 X2 + 760 X1X2 + 47 X1SQX2


Predictor    Coef    SE Coef      T      P
Constant    80.22     30.39    2.64   0.013
X1         156.5     128.6     1.22   0.233
X1SQ       -42.3     123.4    -0.34   0.734
X2         272.84     42.98    6.35   0.000
X1X2       760.1     181.8     4.18   0.000
X1SQX2      47.0     174.5     0.27   0.790


S = 54.4116   R-Sq = 98.6%   R-Sq(adj) = 98.3%


Analysis of Variance

Source           DF       SS        MS       F      P
Regression        5  6173670   1234734  417.05  0.000
Residual Error   30    88819      2961
Total            35  6262489
```

eye movements. One response variable of interest was the proportion y of times the subject's eyes fixated on the speaker's mouth.

(a) The researchers wanted to estimate $E(y)$ for four different noise levels: none, low, medium, and high. Hypothesize a model that will allow the researchers to obtain these estimates.

(b) Interpret the β's in the model, part a.

(c) Explain how to test the hypothesis of no differences in the mean proportions of mouth fixations for the four background noise levels.

4.13 A Test for Comparing Nested Models

In regression analysis, we often want to determine (with a high degree of confidence) which one among a set of candidate models best fits the data. In this section, we present such a method for **nested models**.

> **Definition 4.3** Two models are **nested** if one model contains all the terms of the second model and at least one additional term. The more complex of the two models is called the **complete** (or **full**) model. The simpler of the two models is called the **reduced** (or **restricted**) model.

To illustrate, suppose you have collected data on a response, y, and two quantitative independent variables, x_1 and x_2, and you are considering the use of either a straight-line interaction model or a curvilinear model to relate $E(y)$ to x_1 and x_2. Will the curvilinear model provide better predictions of y than the straight-line model? To answer this question, examine the two models, and note that the curvilinear model contains all the terms in the straight-line interaction model plus two additional terms—those involving β_4 and β_5:

$$\text{Straight-line interaction model:} \quad E(y) = \beta_0 + \beta_1 x_1 + \beta_2 x_2 + \beta_3 x_1 x_2$$

$$\text{Curvilinear model:} \quad E(y) = \overbrace{\beta_0 + \beta_1 x_1 + \beta_2 x_2 + \beta_3 x_1 x_2}^{\text{Terms in interaction model}} + \overbrace{\beta_4 x_1^2 + \beta_5 x_2^2}^{\text{Quadratic terms}}$$

Consequently, these are nested models. Since the straight-line model is the simpler of the two, we say that the *straight-line model is nested within the more complex curvilinear model*. Also, the straight-line model is called the **reduced** model while the curvilinear model is called the **complete** (or **full**) **model**.

Asking whether the curvilinear (or *complete*) model contributes more information for the prediction of y than the straight-line (or *reduced*) model is equivalent to asking whether at least one of the parameters, β_4 or β_5, differs from 0 (i.e., whether the terms involving β_4 and β_5 should be retained in the model). Therefore, to test whether the quadratic terms should be included in the model, we test the null hypothesis

$$H_0: \beta_4 = \beta_5 = 0$$

(i.e., the quadratic terms do not contribute information for the prediction of y) against the alternative hypothesis

$$H_a: \text{At least one of the parameters, } \beta_4 \text{ or } \beta_5, \text{ differs from 0.}$$

(i.e., at least one of the quadratic terms contributes information for the prediction of y).

The procedure for conducting this test is intuitive. First, we use the method of least squares to fit the reduced model and calculate the corresponding sum of squares for error, SSE_R (the sum of squares of the deviations between observed and predicted y-values). Next, we fit the complete model and calculate its sum of squares for error, SSE_C. Then, we compare SSE_R to SSE_C by calculating the difference, $SSE_R - SSE_C$. If the quadratic terms contribute to the model, then SSE_C should be much smaller than SSE_R, and the difference $SSE_R - SSE_C$ will be large. The larger the difference, the greater the weight of evidence that the complete model provides better predictions of y than does the reduced model.

The sum of squares for error will always decrease when new terms are added to the model since the total sum of squares, $SS_{yy} = \Sigma(y - \bar{y})^2$, remains the same. The question is whether this decrease is large enough to conclude that it is due to more than just an increase in the number of model terms and to chance. To test the null hypothesis that the curvature coefficients β_4 and β_5 simultaneously equal 0, we use an F statistic. For our example, this F statistic is:

$$F = \frac{\text{Drop in SSE/Number of } \beta \text{ parameters being tested}}{s^2 \text{ for larger model}}$$

$$= \frac{(SSE_R - SSE_C)/2}{SSE_C/[n - (5 + 1)]}$$

When the assumptions listed in Sections 3.4 and 4.2 about the error term ε are satisfied and the β parameters for curvature are all 0 (H_0 is true), this F statistic has an F distribution with $v_1 = 2$ and $v_2 = n - 6$ df. Note that v_1 is the number of β parameters being tested and v_2 is the number of degrees of freedom associated with s^2 in the larger, second-order model.

If the quadratic terms *do* contribute to the model (H_a is true), we expect the F statistic to be large. Thus, we use a one-tailed test and reject H_0 if F exceeds some critical value, F_α, as shown in Figure 4.21.

F-Test for Comparing Nested Models

Reduced model: $E(y) = \beta_0 + \beta_1 x_1 + \cdots + \beta_g x_g$

Complete model: $E(y) = \beta_0 + \beta_1 x_1 + \cdots + \beta_g x_g$
$$+\beta_{g+1} x_{g+1} + \cdots + \beta_k x_k$$

H_0: $\beta_{g+1} = \beta_{g+2} = \cdots = \beta_k = 0$

H_a: At least one of the β parameters being tested in nonzero.

Test statistic: $F = \dfrac{(SSE_R - SSE_C)/(k-g)}{SSE_C/[n-(k+1)]}$

$$= \frac{(SSE_R - SSE_C)/\text{Number of } \beta\text{'s tested}}{MSE_C}$$

where SSE_R = Sum of squared errors for the reduced model
SSE_C = Sum of squared errors for the complete model
MSE_C = Mean square error for the complete model
$k - g$ = Number of β parameters specified in H_0
(i.e., number of β's tested)
$k + 1$ = Number of β parameters in the complete model
(including β_0)

n = Total sample size

Rejection region: $F > F_\alpha$ where $v_1 = k - g =$ degrees of freedom for the numerator and $v_2 = n - (k + 1) =$ degrees of freedom for the denominator
or
$\alpha > p$-value, where p-value $= P(F > F_c)$, F_c is the calculated value of the test statistic.

Figure 4.21 Rejection region for the F-Test H_0: $\beta_4 = \beta_5 = 0$

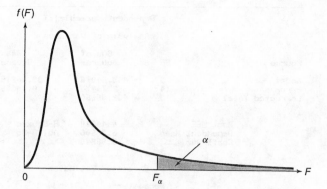

Example 4.11

In Example 4.8, we fit the complete second-order model for a set of $n = 20$ data points relating shipment cost to package weight and distance shipped. The SAS printout for this model, called the *complete* model, is reproduced in Figure 4.22. Figure 4.23 shows the SAS printout for the straight-line interaction model (the *reduced* model) fit to the same $n = 20$ data points. Referring to the printouts, we find the following:

Straight-line interaction (reduced) model:

$$\text{SSE}_R = 6.63331 \text{ (shaded on Figure 4.23)}$$

Second-order (complete) model:

$$\text{SSE}_C = 2.74474 \text{ (shaded on Figure 4.22)}$$

Test the hypothesis that the quadratic terms, $\beta_4 x_1^2$ and $\beta_5 x_2^2$, do not contribute information for the prediction of y.

Solution

The null and alternative hypotheses for our test are:

$$H_0: \beta_4 = \beta_5 = 0$$

$$H_a: \text{Either } \beta_4 \text{ or } \beta_5 \text{ (or both) are nonzero.}$$

The test statistic is

$$F = \frac{(\text{SSE}_R - \text{SSE}_C)/2}{\text{SSE}_C/(20 - 6)}$$

$$= \frac{(6.63331 - 2.74474)/2}{2.74474/14} = \frac{1.94428}{.19605} = 9.92$$

The critical value of F for $\alpha = .05$, $\nu_1 = 2$, and $\nu_2 = 14$ is found in Table 4 (Appendix D) to be

$$F_{.05} = 3.74$$

Since the calculated $F = 9.92$ exceeds 3.74, we reject H_0 and conclude that the quadratic terms contribute to the prediction of y, shipment cost per package. The curvature terms should be retained in the model. [*Note*: The test statistic and p-value for this nested model F-Test can be obtained in SAS with the proper SAS commands. Both the F-value and p-value are highlighted at the bottom of the

Figure 4.22 SAS output for complete model, Example 4.11

Dependent Variable: Y

Analysis of Variance

Source	DF	Sum of Squares	Mean Square	F Value	Pr > F
Model	5	449.34076	89.86815	458.39	<.0001
Error	14	2.74474	0.19605		
Corrected Total	19	452.08550			

Root MSE	0.44278	R-Square	0.9939
Dependent Mean	6.33500	Adj R-Sq	0.9918
Coeff Var	6.98940		

Parameter Estimates

Variable	DF	Parameter Estimate	Standard Error	t Value	Pr > |t|
Intercept	1	0.82702	0.70229	1.18	0.2586
X1	1	-0.60914	0.17990	-3.39	0.0044
X2	1	0.00402	0.00800	0.50	0.6230
X1X2	1	0.00733	0.00063743	11.49	<.0001
X1SQ	1	0.08975	0.02021	4.44	0.0006
X2SQ	1	0.00001507	0.00002243	0.67	0.5127

Test CURVE Results for Dependent Variable Y

Source	DF	Mean Square	F Value	Pr > F
Numerator	2	1.94428	9.92	0.0021
Denominator	14	0.19605		

Figure 4.23 SAS output for reduced model, Example 4.11

Dependent Variable: Y

Analysis of Variance

Source	DF	Sum of Squares	Mean Square	F Value	Pr > F
Model	3	445.45219	148.48406	358.15	<.0001
Error	16	6.63331	0.41458		
Corrected Total	19	452.08550			

Root MSE	0.64388	R-Square	0.9853
Dependent Mean	6.33500	Adj R-Sq	0.9826
Coeff Var	10.16385		

Parameter Estimates

Variable	DF	Parameter Estimate	Standard Error	t Value	Pr > |t|
Intercept	1	-0.14050	0.64810	-0.22	0.8311
X1	1	0.01909	0.15821	0.12	0.9055
X2	1	0.00772	0.00391	1.98	0.0656
X1X2	1	0.00780	0.00089766	8.68	<.0001

SAS printout, Figure 4.22. Since p-value $= .0021$ is less than $\sigma = .05$, we arrive at the same conclusion: reject H_0.]

Suppose the F-Test in Example 4.11 yielded a test statistic that did not fall in the rejection region. That is, suppose there was insufficient evidence (at $\alpha = .05$) to say that the curvature terms contribute information for the prediction of product quality. As with any statistical test of hypothesis, we must be cautious about accepting H_0 since the probability of a Type II error is unknown. Nevertheless, most practitioners of regression analysis adopt the principle of **parsimony**. That is, in situations where two competing models are found to have essentially the same predictive power, the model with the lower number of β's (i.e., the more **parsimonious model**) is selected. The principle of parsimony would lead us to choose the simpler (reduced) model over the more complex complete model when we fail to reject H_0 in the F-Test for nested models.

> **Definition 4.4** A **parsimonious model** is a model with a small number of β parameters. In situations where two competing models have essentially the same predictive power (as determined by an F-Test), choose the more parsimonious of the two.

When the candidate models in model building are nested models, the F-Test developed in this section is the appropriate procedure to apply to compare the models. However, if the models are not nested, this F-Test is not applicable. In this situation, the analyst must base the choice of the best model on statistics such as R_a^2 and s. It is important to remember that decisions based on these and other numerical descriptive measures of model adequacy cannot be supported with a measure of reliability and are often very subjective in nature.

4.13   Exercises

4.61   **Nested models.**   Determine which pairs of the following models are "nested" models. For each pair of nested models, identify the complete and reduced model.

(a) $E(y) = \beta_0 + \beta_1 x_1 + \beta_2 x_2$
(b) $E(y) = \beta_0 + \beta_1 x_1$
(c) $E(y) = \beta_0 + \beta_1 x_1 + \beta_2 x_1^2$
(d) $E(y) = \beta_0 + \beta_1 x_1 + \beta_2 x_2 + \beta_3 x_1 x_2$
(e) $E(y) = \beta_0 + \beta_1 x_1 + \beta_2 x_2 + \beta_3 x_1 x_2 + \beta_4 x_1^2 + \beta_5 x_2^2$

4.62   **Testing for curvature.**   Consider the second-order model relating $E(y)$ to three quantitative independent variables, $x_1, x_2,$ and x_3:

$$E(y) = \beta_0 + \beta_1 x_1 + \beta_2 x_2 + \beta_3 x_3 + \beta_4 x_1 x_2 + \beta_5 x_1 x_3$$
$$+ \beta_6 x_2 x_3 + \beta_7 x_1^2 + \beta_8 x_2^2 + \beta_9 x_3^2$$

(a) Specify the parameters involved in a test of the hypothesis that no curvature exists in the response surface.
(b) State the hypothesis of part a in terms of the model parameters.
(c) What hypothesis would you test to determine whether x_3 is useful for the prediction of $E(y)$?

4.63   **Study of supervisor-targeted aggression.** "Moonlighters" are workers who hold two jobs at the same time. What are the factors that impact the likelihood of a moonlighting worker becoming aggressive toward his/her supervisor? This was the research question of interest in the *Journal of Applied Psychology* (July 2005). Completed questionnaires were obtained from $n = 105$ moonlighters and the data were used to fit several multiple regression models for supervisor-directed aggression score (y). Two of the models (with R^2 values in parentheses) are given below:

Model 1:

$$E(y) = \beta_0 + \beta_1(\text{Age})$$
$$+ \beta_2(\text{Gender})$$
$$+ \beta_3(\text{Interaction injustice at second job})$$
$$+ \beta_4(\text{Abusive supervisor at second job})$$
$$(R^2 = .101)$$

Model 2:

$$E(y) = \beta_0 + \beta_1(\text{Age}) + \beta_2(\text{Gender})$$
$$+ \beta_3(\text{Interactional injustice at second job})$$
$$+ \beta_4(\text{Abusive supervisor at second job})$$
$$+ \beta_5(\text{Self-esteem})$$
$$+ \beta_6(\text{History of aggression})$$
$$+ \beta_7(\text{Interactional injustice at primary job})$$
$$+ \beta_8(\text{Abusive supervisor at primary job})$$
$$(R^2 = .555)$$

(a) Interpret the R^2 values for the models.

(b) Give the null and alternative hypotheses for comparing the fits of Models 1 and 2.

(c) Are the two models nested? Explain.

(d) The nested F-test for comparing the two models resulted in $F = 42.13$ and p-value $< .001$. What can you conclude from these results?

(e) A third model was fit, one that hypothesizes all possible pairs of interactions between Self-esteem, History of aggression, Interactional injustice at primary job, and Abusive supervisor at primary job. Give the equation of this model (Model 3).

(f) A nested F-test to compare Models 2 and 3 resulted in a p-value $> .10$. What can you conclude from this result?

GASTURBINE

4.64 Cooling method for gas turbines. Refer to the *Journal of Engineering for Gas Turbines and Power* (January 2005) study of a high-pressure inlet fogging method for a gas turbine engine, Exercise 4.13 (p. 188). Consider a model for heat rate (kilojoules per kilowatt per hour) of a gas turbine as a function of cycle speed (revolutions per minute) and cycle pressure ratio. The data are saved in the GASTURBINE file.

(a) Write a complete second-order model for heat rate (y).

(b) Give the null and alternative hypotheses for determining whether the curvature terms in the complete second-order model are statistically useful for predicting heat rate (y).

(c) For the test in part b, identify the "complete" and "reduced" model.

(d) Portions of the MINITAB printouts for the two models are shown below. Find the values of SSE_R, SSE_C, and MSE_C on the printouts.

(e) Compute the value of the test statistics for the test of part b.

(f) Find the rejection region for the test of part b using $\alpha = .10$.

(g) State the conclusion of the test in the words of the problem.

4.65 Students' perceptions of science ability. The *American Educational Research Journal* (Fall 1998) published a study of students' perceptions of their science ability in hands-on classrooms. A first-order, main effects model that was used to predict ability perception (y) included the following independent variables:

Control Variables
x_1 = Prior science attitude score
x_2 = Science ability test score

MINITAB Output for Exercise 4.64

___

Complete Model

```
The regression equation is
HEATRATE = 15583 + 0.078 RPM - 523 CPRATIO + 0.00445 RPM_CPR - 0.000000 RPMSQ
           + 8.84 CPRSQ


S = 563.513   R-Sq = 88.5%   R-Sq(adj) = 87.5%

Analysis of Variance

Source           DF         SS          MS        F       P
Regression        5   148526859    29705372    93.55   0.000
Residual Error    61    19370350      317547
Total             66   167897208
```

Reduced Model

```
The regression equation is
HEATRATE = 12065 + 0.170 RPM - 146 CPRATIO - 0.00242 RPM_CPR


S = 633.842   R-Sq = 84.9%   R-Sq(adj) = 84.2%

Analysis of Variance

Source           DF         SS          MS        F       P
Regression        3   142586570    47528857   118.30   0.000
Residual Error    63    25310639      401756
Total             66   167897208
```

___

4.14 A Complete Example

The basic elements of multiple regression analysis have been presented in Sections 4.1–4.13. Now we assemble these elements by applying them to a practical problem.

In the United States, commercial contractors bid for the right to construct state highways and roads. A state government agency, usually the Department of Transportation (DOT), notifies various contractors of the state's intent to build a highway. Sealed bids are submitted by the contractors, and the contractor with the lowest bid (building cost) is awarded the road construction contract. The bidding process works extremely well in competitive markets, but has the potential to increase construction costs if the markets are noncompetitive or if collusive practices are present. The latter occurred in the 1970s and 1980s in Florida. Numerous contractors either admitted to or were found guilty of price-fixing (i.e., setting the cost of construction above the fair, or competitive, cost through bid-rigging or other means).

In this section, we apply multiple regression to a data set obtained from the office of the Florida Attorney General. Our objective is to build and test the adequacy of a model designed to predict the cost y of a road construction contract awarded using the sealed-bid system in Florida.

Step 1. Based on the opinions of several experts in road construction and bid-rigging, two important predictors of contract cost (y) are the DOT engineer's estimate of the cost (x_1) and the fixed-or-competitive status of the bid contract (x_2). Since x_2 is a qualitative variable, we create the dummy variable

$$x_2 = \begin{cases} 1 & \text{if fixed} \\ 0 & \text{if competitive} \end{cases}$$

⊗ FLAG

Data collected on these two predictors and contract cost for a sample of $n = 235$ contracts are saved in the file named FLAG. Contract cost y and DOT estimate x_1 are both recorded in thousands of dollars.

Step 2. In Chapter 5, we learn that a good initial choice is the complete second-order model. For one quantitative variable (x_1) and one qualitative variable (x_2), the model has the following form:

$$E(y) = \beta_0 + \beta_1 x_1 + \beta_2 x_1^2 + \beta_3 x_2 + \beta_4 x_1 x_2 + \beta_5 x_1^2 x_2$$

The SAS printout for the complete second-order model is shown in Figure 4.24. The β estimates, shaded on the printout, yield the following least squares prediction equation:

$$\hat{y} = -2.975 + .9155 x_1$$
$$+ .00000072 x_1^2 - 36.724 x_2$$
$$+ .324 x_1 x_2 - .0000358 x_1^2 x_2$$

Step 3. Before we can make inferences about model adequacy, we should be sure that the standard regression assumptions about the random error ε are satisfied. For given values of x_1 and x_2, the random errors ε have a normal distribution with mean 0, constant variance σ^2, and are independent. We learn how to check these assumptions in Chapter 8. For now, we are satisfied with estimating σ and interpreting its value.

The value of s, shaded on Figure 4.24, is $s = 296.65$. Our interpretation is that the complete second-order model can predict contract costs to within $2s = 593.3$ thousand dollars of its true value.

Figure 4.24 SAS output for complete second-order model for road cost

Dependent Variable: COST

Analysis of Variance

Source	DF	Sum of Squares	Mean Square	F Value	Pr > F
Model	5	866723202	173344640	1969.85	<.0001
Error	229	20151771	87999		
Corrected Total	234	886874973			

Root MSE	296.64625	R-Square	0.9773
Dependent Mean	1268.70221	Adj R-Sq	0.9768
Coeff Var	23.38187		

Parameter Estimates

Variable	DF	Parameter Estimate	Standard Error	t Value	Pr > |t|
Intercept	1	-2.97527	30.89144	-0.10	0.9234
DOTEST	1	0.91553	0.02917	31.39	<.0001
DOTEST2	1	7.186888E-7	0.00000340	0.21	0.8330
STATUS	1	-36.72420	74.77310	-0.49	0.6238
STA_DOT	1	0.32421	0.11917	2.72	0.0070
STA_DOT2	1	-0.00003576	0.00002478	-1.44	0.1504

Test CURV Results for Dependent Variable COST

Source	DF	Mean Square	F Value	Pr > F
Numerator	2	9.15987	1.04	0.3548
Denominator	229	8.79990		

Is this a reasonable potential error of prediction? The DOT can gauge the magnitude of s by examining the coefficient of variation. The value of C.V. (also shaded on Figure 4.24) implies that the ratio of s to the sample mean contract cost is 23%. This relatively high proportion is a forewarning that prediction intervals for contract cost generated by the model may be deemed too wide to be of practical use.

Step 4. To check the statistical adequacy of the complete second-order model, we conduct the analysis of variance F-Test. The elements of the test are as follows:

H_0: $\beta_1 = \beta_2 = \beta_3 = \beta_4 = \beta_5 = 0$

H_a: At least one $\beta \neq 0$

Test statistic: $F = 1{,}969.85$ (shaded in Figure 4.24)

p-value: $p = .0001$ (shaded in Figure 4.24)

Conclusion: The extremely small p-value indicates that the model is statistically adequate (at $\alpha = .01$) for predicting contract cost, y.

Are all the terms in the model statistically significant predictors? For example, is it necessary to include the curvilinear terms, $\beta_2 x_1^2$ and $\beta_5 x_1^2 x_2$, in the model? If not, the model can be simplified by dropping these curvature terms. The hypothesis we want to test is

H_0: $\beta_2 = \beta_5 = 0$

H_a: At least one of the curvature β's is nonzero.

To test this subset of β's, we compare the complete second-order model to a model without the curvilinear terms. The reduced model takes the form

$$E(y) = \beta_0 + \beta_1 x_1 + \beta_3 x_2 + \beta_4 x_1 x_2$$

The results of this nested model (or partial) F-test are shown at the bottom of the SAS printout, Figure 4.24. The test statistic and p-value (shaded) are $F = 1.04$ and p-value $= .3548$. Since the p-value exceeds, say, $\alpha = .01$, we fail to reject H_0. That is, there is insufficient evidence (at $\alpha = .01$) to indicate that the curvature terms are useful predictors of construction cost, y.

The results of the partial F-Test lead us to select the reduced model as the better predictor of cost. The SAS printout for the reduced model is shown in Figure 4.25. The least squares prediction equation, highlighted on the printout, is

$$\hat{y} = -6.429 + .921x_1 + 28.671x_2 + .163x_1x_2$$

Note that we cannot simplify the model any further. The t-test for the interaction term $\beta_3x_1x_2$ is highly significant (p-value $< .0001$, shaded on Figure 4.25). Thus, our best model for construction cost proposes interaction between the DOT estimate (x_1) and status (x_5) of the contract, but only a linear relationship between cost and DOT estimate.

To demonstrate the impact of the interaction term, we find the least squares line for both fixed and competitive contracts.

Competitive ($x_2 = 0$): $\hat{y} = -6.429 + .921x_1 + 28.671(0) + .163x_1(0)$
$= -6.429 + .921x_1$
Fixed ($x_2 = 1$): $\hat{y} = -6.429 + .921x_1 + 28.671(1) + .163x_1(1)$
$= 22.242 + 1.084x_1$

Figure 4.25 SAS output for reduced model for road cost

Dependent Variable: COST

Analysis of Variance

Source	DF	Sum of Squares	Mean Square	F Value	Pr > F
Model	3	866540004	288846668	3281.22	<.0001
Error	231	20334968	88030		
Corrected Total	234	886874973			

Root MSE	296.69878	R-Square	0.9771	
Dependent Mean	1268.70221	Adj R-Sq	0.9768	
Coeff Var	23.38601			

Parameter Estimates

Variable	DF	Parameter Estimate	Standard Error	t Value	Pr > |t|
Intercept	1	-6.42905	26.20856	-0.25	0.8064
DOTEST	1	0.92134	0.00972	94.75	<.0001
STATUS	1	28.67148	58.66234	0.49	0.6255
STA_DOT	1	0.16328	0.04043	4.04	<.0001

A plot of the least squares lines for the reduced model is shown in Figure 4.26. You can see that the model proposes two straight lines (one for fixed contracts and one for competitive contracts) with different slopes. The estimated slopes of the $y - x_1$ lines are computed and interpreted as follows:

Competitive contracts ($x_2 = 0$):   Estimated slope $= \hat{\beta}_1 = .921$
For every \$1,000 increase in DOT estimate, we estimate contract cost to increase \$921.

Fixed contracts ($x_2 = 1$):   Estimated slope $= \hat{\beta}_1 + \hat{\beta}_4 = .921 + .163 = 1.084$
For every \$1,000 increase in DOT estimate, we estimate contract cost to increase \$1,084.

Before deciding to use the interaction model for estimation and/or prediction (step 5), we should check R_a^2 and s for the model. $R_a^2 = .9768$ (shaded on Figure 4.25) indicates that nearly 98% of the variation in the sample

Figure 4.26 Plot of the least squares lines for the reduced model

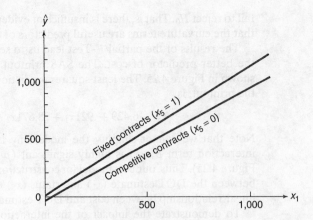

of construction costs can be "explained" by the model. The value of s (also shaded) implies that we can predict construction cost to within about $2s = 2(296.7) = 593.40$ thousand dollars of its true value using the model. Although the adjusted R^2 value is high, the large $2s$ value suggests that the predictive ability of the model might be improved by additional independent variables.

Step 5. A portion of the SAS printout for the interaction (reduced) model not shown earlier is presented in Figure 4.27. The printout gives predicted values and 95% prediction intervals for the first 10 contracts in the sample. The shaded portion gives the 95% prediction interval for contract cost when the DOT estimate is \$1,386,290 ($x_1 = 1,386.29$) and the contract is fixed ($x_2 = 1$). For a contract with these characteristics, we predict the cost to fall between 933.72 thousand dollars and 2,118.0 thousand dollars, with 95% confidence.

Figure 4.27 SAS output showing prediction intervals for reduced model

Dependent Variable: COST

Output Statistics

Obs	DOTEST	STATUS	Dep Var COST	Predicted Value	Std Error Mean Predict	95% CL Predict	
1	1386.29	1	1379	1526	47.7923	933.7200	2118
2	85.71	1	134.0300	115.2050	50.5334	-477.7949	708.2050
3	248.89	0	202.3300	222.8822	24.9485	-363.7625	809.5269
4	467.49	0	397.1200	424.2862	23.9891	-162.2031	1011
5	117.72	1	158.5400	149.9236	49.8442	-442.8499	742.6971
6	1008.91	1	1128	1117	42.7285	525.9115	1707
7	472.98	1	400.3300	535.2449	43.9161	-55.7058	1126
8	785.39	0	581.6400	717.1788	22.8762	130.8621	1303
9	370.02	0	353.9600	334.4836	24.3987	-252.0713	921.0385
10	174.25	0	138.7100	154.1137	25.3087	-432.5909	740.8183

We caution that a more complete analysis of the model is required before we recommend its use by the Florida DOT. Subsequent chapters extend the methods of this chapter to special applications and problems encountered during a regression analysis. One of the most common problems faced by regression analysts is the problem of multicollinearity (i.e., intercorrelations among the independent variables). Many of the potential independent variables considered by the DOT are highly correlated. Methods for detecting and overcoming multicollinearity, as well as other problems, are discussed in Chapter 7. Another aspect of regression analysis is the analysis of the residuals (i.e., the deviations between the observed and the predicted values of y). An analysis of residuals (Chapter 8) may indicate that the DOT data do not comply with the assumptions of Section 4.2 and may suggest appropriate procedures for modifying the model for contract cost.

Quick Summary/Guides

KEY FORMULAS

Estimator of σ^2 for a model with k independent variables

$$s^2 = \text{MSE} = \frac{\text{SSE}}{n - (k + 1)}$$

Test statistic for testing H_0: β_i

$$t = \frac{\hat{\beta}_i}{s_{\hat{\beta}_i}}$$

$100(1 - \alpha)\%$ confidence interval for β_i

$\hat{\beta}_i \pm (t_{\alpha/2}) s_{\hat{\beta}_i}$ –where $t_{\alpha/2}$ depends on $n - (k + 1)$df

Multiple coefficient of determination

$$R^2 = \frac{\text{SS}_{yy} - \text{SSE}}{\text{SS}_{yy}}$$

Adjusted multiple coefficient of determination

$$R_a^2 = 1 - \left[\frac{(n - 1)}{n - (k + 1)} \right] (1 - R^2)$$

Test statistic for testing H_0: $\beta_1 = \beta_2 = \cdots = \beta_k = 0$

$$F = \frac{\text{MS(Model)}}{\text{MSE}} = \frac{R^2/k}{(1 - R^2)/[n - (k + 1)]}$$

Test statistic for comparing reduced and complete models

$$F = \frac{(\text{SSE}_R - \text{SSE}_C)/\text{number of } \beta' \text{s tested}}{\text{MSE}_C}$$

KEY SYMBOLS

x_1^2	Quadratic form for a quantitative x
$x_1 x_2$	Interaction term
MSE	Mean square for error (estimates σ^2)
$\hat{\varepsilon}$	Estimated random error (residual)
SSE_R	Sum of squared errors, reduced model
SSE_C	Sum of squared errors, complete model
MSE_C	Mean squared error, complete model
$\ln(y)$	Natural logarithm of dependent variable

KEY IDEAS

Multiple Regression Variables

y = **Dependent** variable (quantitative)

$x_1, x_2, \ldots, x_k$ are **independent** variables (quantitative or qualitative)

First-Order Model in k Quantitative x's

$$E(y) = \beta_0 + \beta_1 x_1 + \beta_2 x_2 + \cdots + \beta_k x_k$$

Each β_i represents the change in y for every 1-unit increase in x_i, holding all other x's fixed.

Interaction Model in 2 Quantitative x's

$$E(y) = \beta_0 + \beta_1 x_1 + \beta_2 x_2 + \beta_3 x_1 x_2$$

$(\beta_1 + \beta_3 x_2)$ represents the change in y for every 1-unit increase in x_1, for fixed value of x_2

$(\beta_2 + \beta_3 x_1)$ represents the change in y for every 1-unit increase in x_2, for fixed value of x_1

Quadratic Model in 1 Quantitative x

$$E(y) = \beta_0 + \beta_1 x + \beta_2 x^2$$

β_2 represents the rate of curvature in for x

($\beta_2 > 0$ implies *upward* curvature)

($\beta_2 < 0$ implies *downward* curvature)

Complete Second-Order Model in 2 Quantitative x's

$$E(y) = \beta_0 + \beta_1 x_1 + \beta_2 x_2 + \beta_3 x_1 x_2 + \beta_4 x_1^2 + \beta_5 x_2^2$$

β_4 represents the rate of curvature in for x_1, holding x_2 fixed

β_5 represents the rate of curvature in for x_2, holding x_1 fixed

Dummy Variable Model for 1 Qualitative x

$$E(y) = \beta_0 + \beta_1 x_1 + \beta_2 x_2 + \cdots + \beta_{k-1} x_{k-1}$$

$x_1 = \{1$ if level $1, 0$ if not$\}$

$x_2 = \{1$ if level $2, 0$ if not$\}$

$x_{k-1} = \{1$ if level $k - 1, 0$ if not$\}$

$\beta_0 = E(y)$ for level k (base level) $= \mu_k$

$\beta_1 = \mu_1 - \mu_k$

$\beta_2 = \mu_2 - \mu_k$

Multiplicative Model in Quantitative x's

$$E\{\ln(y)\} = \beta_0 + \beta_1 x_1 + \beta_2 x_2 + \cdots + \beta_k x_k$$

$(e^{\beta_i} - 1)$ represents the % change in y for every 1-unit increase in x_i

Adjusted Coefficient of Determination, R_a^2

Cannot be "forced" to 1 by adding independent variables to the model.

Interaction between x_1 and x_2

Implies that the relationship between y and one x depends on the other x.

Parsimonious Model

A model with a small number of β parameters.

Recommendation for Assessing Model Adequacy

1. Conduct global F-test; if significant then:
2. Conduct t-tests on only the most important β's (*interaction* or *squared terms*)
3. Interpret value of $2s$
4. Interpret value of R_a^2

Recommendation for Testing Individual β's

1. If *curvature* (x^2) deemed important, do not conduct test for first-order (x) term in the model.
2. If *interaction* (x_1x_2) deemed important, do not conduct tests for first-order terms (x_1 and x_2) in the model.

Extrapolation

Occurs when you predict y for values of x's that are outside of range of sample data.

Nested Models

Models where one model (the *complete model*) contains all the terms of another model (the *reduced model*) plus at least one additional term.

GUIDE TO MULTIPLE REGRESSION

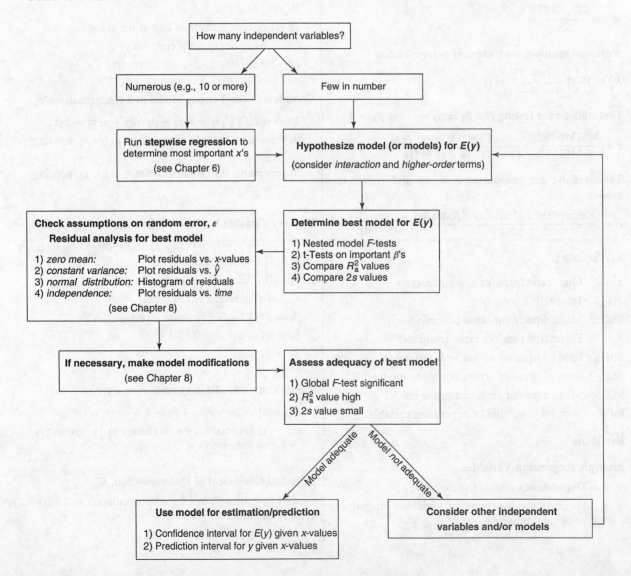

A Complete Example 241

Supplementary Exercises

4.70. Width of confidence intervals in regression. After a regression model is fit to a set of data, a confidence interval for the mean value of y at a given setting of the independent variables will *always* be narrower than the corresponding prediction interval for a particular value of y at the same setting of the independent variables. Why?

4.71. Assertiveness in disabled adults. A disabled person's acceptance of a disability is critical to the rehabilitation process. The *Journal of Rehabilitation* (September 1989) published a study that investigated the relationship between assertive behavior level and acceptance of disability in 160 disabled adults. The dependent variable, assertiveness (y), was measured using the Adult Self Expression Scale (ASES). Scores on the ASES range from 0 (no assertiveness) to 192 (extreme assertiveness). The model analyzed was $E(y) = \beta_0 + \beta_1 x_1 + \beta_2 x_2 + \beta_3 x_3$, where

$x_1 =$ Acceptance of disability (AD) score

$x_2 =$ Age (years)

$x_3 =$ Length of disability (years)

The regression results are shown in the table.

INDEPENDENT VARIABLE	t	TWO-TAILED p-VALUE
AD score (x_1)	5.96	.0001
Age (x_2)	0.01	.9620
Length (x_3)	1.91	.0576

(a) Is there sufficient evidence to indicate that AD score is positively linearly related to assertiveness level, once age and length of disability are accounted for? Test using $\alpha = .05$.

(b) Test the hypothesis $H_0: \beta_2 = 0$ against $H_a: \beta_2 \neq 0$. Use $\alpha = .05$. Give the conclusion in the words of the problem.

(c) Test the hypothesis $H_0: \beta_3 = 0$ against $H_a: \beta_3 > 0$. Use $\alpha = .05$. Give the conclusion in the words of the problem.

4.72. Chemical composition of rain water. Researchers at the University of Aberdeen (Scotland) developed a statistical model for estimating the chemical composition of water (*Journal of Agricultural, Biological, and Environmental Statistics*, March 2005). For one application, the nitrate concentration y (milligrams per liter) in a water sample collected after a heavy rainfall was modeled as a function of water source (groundwater, subsurface flow, or overground flow).

(a) Write a model for $E(y)$ as a function of the qualitative independent variable.

(b) Give an interpretation of each of the β parameters in the model, part c.

4.73. Scenic beauty of pines. *Artificial Intelligence (AI) Applications* (January 1993) discussed the use of computer-based technologies in building explanation systems for regression models. As an example, the authors presented a model for predicting the scenic beauty (y) of southeastern pine stands (measured on a numeric scale) as a function of age (x_1) of the dominant stand, stems per acre (x_2) in trees, and basal area (x_3) per acre in hardwoods. A user of the AI system simply inputs the values of x_1, x_2, and x_3, and the system uses the least squares equation to predict the scenic beauty (y) value.

(a) The AI system generates information on how each independent variable can be manipulated to effect changes in the dependent variable. For example, "If all else were held constant in the stand, allowing the age (x_1) of the dominant trees in the stand to mature by 1 year will *increase* scenic beauty (y)." From what portion of the regression analysis would the AI system extract this type of information?

(b) The AI system is designed to check the values of the input variables (x_1, x_2, and x_3) with the sample data ranges. If the input data value is "out-of-range," a warning is issued about the potential inaccuracy of the predicted y-value. Explain the reasoning behind this warning.

4.74. Erecting boiler drums. In a production facility, an accurate estimate of hours needed to complete a task is crucial to management in making such decisions as the proper number of workers to hire, an accurate deadline to quote a client, or cost-analysis decisions regarding budgets. A manufacturer of boiler drums wants to use regression to predict the number of hours needed to erect the drums in future projects. To accomplish this, data for 35 boilers were collected. In addition to hours (y), the variables measured were boiler capacity ($x_1 =$ lb/hr), boiler design pressure ($x_2 =$ pounds per square inch, or psi), boiler type ($x_3 = 1$ if industry field erected, 0 if utility field erected), and drum type ($x_4 = 1$ if steam, 0 if mud). The data (p. 242) are saved in the BOILERS file. A MINITAB printout for the model $E(y) = \beta_0 + \beta_1 x_1 + \beta_2 x_2 + \beta_3 x_3 + \beta_4 x_4$ is given on p. 242.

(a) Conduct a test for the global utility of the model. Use $\alpha = .01$.

(b) Both a 95% confidence interval for $E(y)$ and a 95% prediction interval for y when $x_1 = 150{,}000$, $x_2 = 500$, $x_3 = 1$, and $x_4 = 0$ are shown at the bottom of the MINITAB printout. Interpret both of these intervals.

BOILERS (First and last five observations)

HOURS y	BOILER CAPACITY x_1	DESIGN PRESSURE x_2	BOILER TYPE x_3	DRUM TYPE x_4
3,137	120,000	375	1	1
3,590	65,000	750	1	1
4,526	150,000	500	1	1
10,825	1,073,877	2,170	0	1
4,023	150,000	325	1	1
.				
.				
.				
4,206	441,000	410	1	0
4,006	441,000	410	1	0
3,728	627,000	1,525	0	0
3,211	610,000	1,500	0	0
1,200	30,000	325	1	0

Source: Dr. Kelly Uscategui, University of Connecticut.

MINITAB Output for Exercise 4.74

```
The regression equation is
MANHRS = - 3783 + 0.00875 CAPACITY + 1.93 PRESSURE + 3444 BOILER + 2093 DRUM


Predictor       Coef     SE Coef         T       P
Constant        -3783        1205     -3.14   0.004
CAPACITY    0.0087490   0.0009035      9.68   0.000
PRESSURE       1.9265      0.6489      2.97   0.006
BOILER         3444.3       911.7      3.78   0.001
DRUM           2093.4       305.6      6.85   0.000

S = 894.6      R-Sq = 90.3%     R-Sq(adj) = 89.0%

Analysis of Variance

Source           DF          SS          MS       F       P
Regression        4   230854854    57713714   72.11   0.000
Residual Error    31   24809761      800315
Total            35   255664615

Predicted Values for New Observations

New Obs    Fit    SE Fit         95.0% CI           95.0% PI
1         1936       239   (  1449,    2424) (    48,    3825)

Values of Predictors for New Observations

New Obs  CAPACITY  PRESSURE    BOILER      DRUM
1          150000       500      1.00  0.000000
```

(c) Which of the intervals, part b, would you use if you want to estimate the average hours required to erect all industrial mud boilers with a capacity of 150,000 lb/hr and a design pressure of 500 psi?

4.75. Distress in EMS workers. The *Journal of Consulting and Clinical Psychology* (June 1995) reported on a study of emergency service (EMS) rescue workers who responded to the I-880 freeway collapse during a San Francisco earthquake. The goal of the study was to identify the predictors of symptomatic distress in the EMS workers. One of the distress variables studied was the Global Symptom Index (GSI). Several models for GSI, y, were considered based on the following independent variables:

x_1 = Critical Incident Exposure scale (CIE)

x_2 = Hogan Personality Inventory— Adjustment scale (HPI-A)

x_3 = Years of experience (EXP)

x_4 = Locus of Control scale (LOC)

x_5 = Social Support scale (SS)

x_6 = Dissociative Experiences scale (DES)

x_7 = Peritraumatic Dissociation Experiences Questionnaire, self-report (PDEQ-SR)

(a) Write a first-order model for $E(y)$ as a function of the first five independent variables, $x_1 - x_5$.

(b) The model of part a, fitted to data collected for $n = 147$ EMS workers, yielded the following results: $R^2 = .469$, $F = 34.47$, p-value $< .001$. Interpret these results.

(c) Write a first-order model for $E(y)$ as a function of all seven independent variables, $x_1 - x_7$.

(d) The model, part c, yielded $R^2 = .603$. Interpret this result.

(e) The t-tests for testing the DES and PDEQ-SR variables both yielded a p-value of .001. Interpret these results.

4.76. Growth rate of algae. The *Journal of Applied Phycology* (December 1994) published research on the seasonal growth activity of algae in an indoor environment. The daily growth rate y was regressed against temperature x using the quadratic model $E(y) = \beta_0 + \beta_1 x + \beta_2 x^2$. A particular algal strain was grown in a glass dish indoors at temperatures ranging from $10°$ to $32°$ Celsius. The data for $n = 33$ such experiments were used to fit the quadratic model, with the following results:

$$\hat{y} = -2.51 + .55x - .01x^2, \quad R^2 = .67$$

(a) Sketch the least squares prediction equation. Interpret the graph.

(b) Interpret the value of R^2.

(c) Is the model useful for predicting algae growth rate, y? Test using $\alpha = .05$.

4.77. Parasites in school children. *Trichuristrichiura*, a parasitic worm, affects millions of school-age children each year, especially children from developing countries. A study was conducted to determine the effects of treatment of the parasite on school achievement in 407 school-age Jamaican children infected with the disease (*Journal of Nutrition*, July 1995). About half the children in the sample received the treatment, while the others received a placebo. Multiple regression was used to model spelling test score y, measured as number correct, as a function of the following independent variables:

$$\text{Treatment (T): } x_1 = \begin{cases} 1 & \text{if treatment} \\ 0 & \text{if placebo} \end{cases}$$

$$\text{Disease intensity (I): } x_2 = \begin{cases} 1 & \text{if more than 7,000} \\ & \text{eggs per gram of stool} \\ 0 & \text{if not} \end{cases}$$

(a) Propose a model for $E(y)$ that includes interaction between treatment and disease intensity.

(b) The estimates of the β's in the model, part a, and the respective p-values for t-tests on the β's are given in the next table. Is there sufficient evidence to indicate that the effect of the treatment on spelling score depends on disease intensity? Test using $\alpha = .05$.

(c) Based on the result, part b, explain why the analyst should avoid conducting t-tests for the treatment (x_1) and intensity (x_2) β's or interpreting these β's individually.

VARIABLE	β ESTIMATE	p-VALUE
Treatment (x_1)	$-.1$	.62
Intensity (x_2)	$-.3$	.57
T $\times$ I $(x_1 x_2)$	1.6	.02

4.78. Child-abuse report study. Licensed therapists are mandated, by law, to report child abuse by their clients. This requires the therapist to breach confidentiality and possibly lose the client's trust. A national survey of licensed psychotherapists was conducted to investigate clients' reactions to legally mandated child-abuse reports (*American Journal of Orthopsychiatry*, January 1997). The sample consisted of 303 therapists who had filed a child-abuse report against one of their clients. The researchers were interested in finding the best predictors of a client's reaction (y) to the report, where y is measured on a 30-point scale. (The higher the value, the more favorable the client's response to the report.) The independent variables found to have the most predictive power are listed here.

x_1: Therapist's age (years)

x_2: Therapist's gender (1 if male, 0 if female)

x_3: Degree of therapist's role strain (25-point scale)

x_4: Strength of client–therapist relationship (40-point scale)

x_5: Type of case (1 if family, 0 if not)

$x_1 x_2$: Age $\times$ gender interaction

(a) Hypothesize a first-order model relating y to each of the five independent variables.

(b) Give the null hypothesis for testing the contribution of x_4, strength of client–therapist relationship, to the model.

(c) The test statistic for the test, part b, was $t = 4.408$ with an associated p-value of .001. Interpret this result.

(d) The estimated β coefficient for the $x_1 x_2$ interaction term was positive and highly significant ($p < .001$). According to the researchers, "this interaction suggests that ... as the age of the therapist increased, ... male therapists were less likely to get negative client reactions than were female therapists." Do you agree?

(e) For this model, $R^2 = .2946$. Interpret this value.

4.79. Violence in Cartoons. Newspaper cartoons, although designed to be funny, often evoke hostility, pain, and/or aggression in readers, especially those cartoons that are violent. A study was undertaken to determine how violence in cartoons is related to aggression or pain (*Motivation and*

Emotion, Vol. 10, 1986). A group of volunteers (psychology students) rated each of 32 violent newspaper cartoons (16 "Herman" and 16 "Far Side" cartoons) on three dimensions:

y = Funniness (0 = not funny, ... , 9 = very funny)

x_1 = Pain (0 = none, ... , 9 = a very great deal)

x_2 = Aggression/hostility (0 = none, ... , 9 = a very great deal)

The ratings of the students on each dimension were averaged and the resulting $n = 32$ observations were subjected to a multiple regression analysis. Based on the underlying theory (called the *inverted-U theory*) that the funniness of a joke will increase at low levels of aggression or pain, level off, and then decrease at high levels of aggressiveness or pain, the following quadratic models were proposed:

Model 1: $E(y) = \beta_0 + \beta_1 x_1 + \beta_2 x_1^2$,
$R^2 = .099, F = 1.60$

Model 2: $E(y) = \beta_0 + \beta_1 x_2 + \beta_2 x_2^2$,
$R^2 = .100, F = 1.61$

(a) According to the theory, what is the expected sign of β_2 in either model?

(b) Is there sufficient evidence to indicate that the quadratic model relating pain to funniness rating is useful? Test at $\alpha = .05$.

(c) Is there sufficient evidence to indicate that the quadratic model relating aggression/hostility to funniness rating is useful? Test at $\alpha = .05$.

4.80. Using lasers for solar lighting. Engineers at the University of Massachusetts studied the feasibility of using semiconductor lasers for solar lighting in spaceborne applications (*Journal of Applied Physics*, September 1993). A series of $n = 8$ experiments with quantum-well lasers yielded the data on solar pumping threshold current (y) and waveguide A1 mole fraction (x) shown below.

⊚ LASERS

THRESHOLD CURRENT y, A/cm^2	WAVEGUIDE A1 MOLE FRACTION x
273	.15
175	.20
146	.25
166	.30
162	.35
165	.40
245	.50
314	.60

Source: Unnikrishnan, S., and Anderson, N. G. "Quantum-well lasers for direct solar photopumping," *Journal of Applied Physics*, Vol. 74, No. 6, Sept. 15, 1993, p. 4226 (adapted from Figure 2).

(a) The researchers theorize that the relationship between threshold current (y) and waveguide A1 composition (x) will be represented by a U-shaped curve. Hypothesize a model that corresponds to this theory.

(b) Graph the data points in a scatterplot. Comment on the researchers' theory, part a.

(c) Test the theory, part a, using the results of a regression analysis. What do you conclude?

4.81. Studying the art of walking. Refer to the *American Scientist* (July–August 1998) study of the relationship between self-avoiding and unrooted walks, Exercise 3.78 (p. 155). Recall that in a self-avoiding walk you never retrace or cross your own path, while an unrooted walk is a path in which the starting and ending points are impossible to distinguish. The possible number of walks of each type of various lengths are reproduced in the table below. In Exercise 3.78 you analyzed the straight-line model relating number of unrooted walks (y) to walk length (x). Now consider the quadratic model $E(y) = \beta_0 + \beta_1 x + \beta_2 x^2$. Is there sufficient evidence of an upward concave curvilinear relationship between y and x? Test at $\alpha = .10$.

⊚ WALKS

WALK LENGTH (number of steps)	UNROOTED WALKS	SELF-AVOIDING WALKS
1	1	4
2	2	12
3	4	36
4	9	100
5	22	284
6	56	780
7	147	2,172
8	388	5,916

Source: Hayes, B. "How to avoid yourself," *American Scientist*, Vol. 86, No. 4, July–Aug. 1998, p. 317 (Figure 5).

4.82. Extracting crude oil using carbon dioxide. One of the most promising methods for extracting crude oil employs a carbon dioxide (CO_2) flooding technique. When flooded into oil pockets, CO_2 enhances oil recovery by displacing the crude oil. In a microscopic investigation of the CO_2 flooding process, flow tubes were dipped into sample oil pockets containing a known amount of oil. The oil pockets were flooded with CO_2 and the percentage of oil displaced was recorded. The experiment was conducted at three different flow pressures and three different dipping angles. The displacement test data are recorded in the table (p. 245).

(a) Write the complete second-order model relating percentage oil recovery y to pressure x_1 and dipping angle x_2.

⊚ CRUDEOIL

PRESSURE x_1, pounds per square inch	DIPPING ANGLE x_2, degrees	OIL RECOVERY y, percentage
1,000	0	60.58
1,000	15	72.72
1,000	30	79.99
1,500	0	66.83
1,500	15	80.78
1,500	30	89.78
2,000	0	69.18
2,000	15	80.31
2,000	30	91.99

Source: Wang, G. C. "Microscopic investigation of CO_2 flooding process," *Journal of Petroleum Technology*, Vol. 34, No. 8, Aug. 1982, pp. 1789–1797. Copyright © 1982, Society of Petroleum Engineers, American Institute of Mining. First published in *JPT*, Aug. 1982.

(b) Graph the sample data on a scatterplot, with percentage oil recovery y on the vertical axis and pressure x_1 on the horizontal axis. Connect the points corresponding to the same value of dipping angle x_2. Based on the scatterplot, do you believe a complete second-order model is appropriate?

(c) Fit the interaction model

$$y = \beta_0 + \beta_1 x_1 + \beta_2 x_2 + \beta_3 x_1 x_2 + \varepsilon$$

to the data and give the prediction equation for this model.

(d) Construct a plot similar to the scatterplot of part b, but use the predicted values from the interaction model on the vertical axis. Compare the two plots. Do you believe the interaction model will provide an adequate fit?

(e) Check model adequacy using a statistical test with $\alpha = .05$.

(f) Is there evidence of interaction between pressure x_1 and dipping angle x_2? Test using $\alpha = .05$.

4.83. Predicting assassination risk. *New Scientist* (April 3, 1993) published an article on strategies for foiling assassination attempts on politicians. The strategies are based on the findings of researchers at Middlesex University (United Kingdom), who used a multiple regression model for predicting the level y of assassination risk. One of the variables used in the model was political status of a country (communist, democratic, or dictatorship).

(a) Propose a model for $E(y)$ as a function of political status.

(b) Interpret the β's in the model, part a.

4.84. Study of age and height in elementary school children. *Archives of Disease in Childhood* (April 2000) published a study of whether height influences a child's progression through elementary school. Australian schoolchildren were divided into equal thirds (tertiles) based on age (youngest third, middle third, and oldest third). The average heights of the three groups (where all height measurements were standardized using z-scores), by gender, are shown in the table below.

	SAMPLE SIZE	YOUNGEST TERTILE MEAN HEIGHT	MIDDLE TERTILE MEAN HEIGHT	OLDEST TERTILE MEAN HEIGHT
Boys	1,439	0.33	0.33	0.16
Girls	1,409	0.27	0.18	0.21

Source: Wake, M., Coghlan, D., and Hesketh, K. "Does height influence progression through primary school grades?" *Archives of Disease in Childhood*, Vol. 82, Apr. 2000 (Table 3), with permission from BMJ Publishing Group Ltd.

(a) Propose a regression model that will enable you to compare the average heights of the three age groups for boys.

(b) Find the estimates of the β's in the model, part a.

(c) Repeat parts a and b for girls.

4.85. Deregulation of airlines. Since 1978, when the U.S. airline industry was deregulated, researchers have questioned whether the deregulation has ensured a truly competitive environment. If so, the profitability of any major airline would be related only to overall industry conditions (e.g., disposable income and market share) but not to any unchanging feature of that airline. This profitability hypothesis was tested in *Transportation Journal* (Winter 1990) using multiple regression. Data for $n = 234$ carrier-years were used to fit the model

$$E(y) = \beta_0 + \beta_1 x_1 + \beta_2 x_2 + \beta_3 x_3 + \ldots + \beta_{30} x_{30}$$

VARIABLE	β ESTIMATE	t-VALUE	p-VALUE
Intercept	1.2642	.09	.9266
x_1	−.0022	−.99	.8392
x_2	4.8405	3.57	.0003
$x_3 - x_{30}$	(not given)	—	—

$R^2 = .3402$
F(Model) $= 3.49$, p-value $= .0001$
F(Carrier dummies) $= 3.59$, p-value $= .0001$

Source: Leigh, L. E. "Contestability in deregulated airline markets: Some empirical tests," *Transportation Journal*, Winter 1990, p. 55 (Table 4).

where

$$y = \text{Profit rate}$$
$$x_1 = \text{Real personal disposable income}$$
$$x_2 = \text{Industry market share}$$
$$x_3 \text{ through } x_{30} = \text{Dummy variables (coded 0–1)}$$
$$\text{for the 29 air carriers}$$
$$\text{investigated in the study}$$

The results of the regression are summarized in the table (p. 245). Interpret the results. Is the profitability hypothesis supported?

4.86. Emotional distress of firefighters. The *Journal of Human Stress* (Summer 1987) reported on a study of "psychological response of firefighters to chemical fire." It is thought that the following complete second-order model will be adequate to describe the relationship between emotional distress and years of experience for two groups of firefighters—those exposed to a chemical fire and those unexposed.

$$E(y) = \beta_0 + \beta_1 x_1 + \beta_2 x_1^2 + \beta_3 x_2 + \beta_4 x_1 x_2 + \beta_5 x_1^2 x_2$$

where
$$y = \text{Emotional distress}$$
$$x_1 = \text{Experience (years)}$$
$$x_2 = 1 \text{ if exposed to chemical fire,}$$
$$\quad\quad 0 \text{ if not}$$

(a) What hypothesis would you test to determine whether the *rate* of increase of emotional distress with experience is different for the two groups of firefighters?

(b) What hypothesis would you test to determine whether there are differences in mean emotional distress levels that are attributable to exposure group?

(c) Data for a sample of 200 firefighters were used to fit the complete model as well as the reduced model, $E(y) = \beta_0 + \beta_1 x_1 + \beta_2 x_1^2$. The results are $\text{SSE}_R = 795.23$ and $\text{SSE}_C = 783.90$. Is there sufficient evidence to support the claim that the mean emotional distress levels differ for the two groups of firefighters? Use $\alpha = .05$.

4.87. Sorption rate of organic compounds. *Environmental Science and Technology* (October 1993) published an article that investigated the variables that affect the sorption of organic vapors on clay minerals. The independent variables and levels considered in the study are listed here. Identify the type (quantitative or qualitative) of each.

(a) Temperature (50°, 60°, 75°, 90°)
(b) Relative humidity (30%, 50%, 70%)
(c) Organic compound (benzene, toluene, chloroform, methanol, anisole)
(d) Refer to part c. Consider the dependent variable sorption rate (y). Write a model for $E(y)$ as a function of organic compound at five levels.

(e) Interpret the β parameters in the model, part d.
(f) Explain how to test for differences among the mean sorption rates of the five organic compounds.

4.88. Cost of modifying a naval air base. A naval base is considering modifying or adding to its fleet of 48 standard aircraft. The final decision regarding the type and number of aircraft to be added depends on a comparison of cost versus effectiveness of the modified fleet. Consequently, the naval base would like to model the projected percentage increase y in fleet effectiveness by the end of the decade as a function of the cost x of modifying the fleet. A first proposal is the quadratic model

$$E(y) = \beta_0 + \beta_1 x + \beta_2 x^2$$

The data provided in the table were collected on 10 naval bases of similar size that recently expanded their fleets.

💿 NAVALBASE

PERCENTAGE IMPROVEMENT AT END OF DECADE y	COST OF MODIFYING FLEET x, millions of dollars	BASE LOCATION
18	125	U.S.
32	160	U.S.
9	80	U.S.
37	162	U.S.
6	110	U.S.
3	90	Foreign
30	140	Foreign
10	85	Foreign
25	150	Foreign
2	50	Foreign

(a) Fit the quadratic model to the data.
(b) Interpret the value of R_a^2 on the printout.
(c) Find the value of s and interpret it.
(d) Perform a test of overall model adequacy. Use $\alpha = .05$.
(e) Is there sufficient evidence to conclude that the percentage improvement y increases more quickly for more costly fleet modifications than for less costly fleet modifications? Test with $\alpha = .05$.
(f) Now consider the complete second-order model

$$E(y) = \beta_0 + \beta_1 x_1 + \beta_2 x_1^2 + \beta_3 x_2 + \beta_4 x_1 x_2$$
$$+ \beta_5 x_1^2 x_2$$

where
$$x_1 = \text{Cost of modifying the fleet}$$
$$x_2 = \begin{cases} 1 & \text{if U.S. base} \\ 0 & \text{if foreign base} \end{cases}$$

Fit the complete model to the data. Is there sufficient evidence to indicate that type of base (U.S. or foreign) is a useful predictor of percentage improvement y? Test using $\alpha = .05$.

4.89. Gender differences in industrial sales. The *Journal of Personal Selling and Sales Management* (Summer 1990) published a study of gender differences in the industrial sales force. A sample of 244 male sales managers and a sample of 153 female sales managers participated in the survey. One objective of the research was to assess how supervisory behavior affects intrinsic job satisfaction. Initially, the researchers fitted the following reduced model to the data on each gender group:

$$E(y) = \beta_0 + \beta_1 x_1 + \beta_2 x_2 + \beta_3 x_3 + \beta_4 x_4$$

where

y = Intrinsic job satisfaction (measured on a scale of 0 to 40)
x_1 = Age (years)
x_2 = Education level (years)
x_3 = Firm experience (months)
x_4 = Sales experience (months)

To determine the effects of supervisory behavior, four variables (all measured on a scale of 0 to 50) were added to the model: x_5 = contingent reward behavior, x_6 = noncontingent reward behavior, x_7 = contingent punishment behavior, and x_8 = noncontingent punishment behavior. Thus, the complete model is

$$E(y) = \beta_0 + \beta_1 x_1 + \beta_2 x_2 + \beta_3 x_3 + \beta_4 x_4 + \beta_5 x_5$$
$$+ \beta_6 x_6 + \beta_7 x_7 + \beta_8 x_8$$

(a) For each gender, specify the null hypothesis and rejection region ($\alpha = .05$) for testing whether any of the four supervisory behavior variables affect intrinsic job satisfaction.

(b) The R^2 values for the four models (reduced and complete model for both samples) are given in the accompanying table. Interpret the results. For each gender, does it appear that the supervisory behavior variables have an impact on intrinsic job satisfaction? Explain.

| | R^2 | |
MODEL	MALES	FEMALES
Reduced	.218	.268
Complete	.408	.496

Source: P. L. Schul, et al. "Assessing gender differences in relationships between supervisory behaviors and job related outcomes in industrial sales force," *Journal of Personal Selling and Sales Management*, Vol. X, Summer 1990, p. 9 (Table 4).

(c) The F statistics for comparing the two models are $F_{males} = 13.00$ and $F_{females} = 9.05$. Conduct the tests, part a, and interpret the results.

4.90. Gender differences in industrial sales (cont'd). Refer to Exercise 4.89. One way to test for gender differences in the industrial sales force is to incorporate a dummy variable for gender into the model for intrinsic job satisfaction, y, and then fit the model to the data for the combined sample of males and females.

(a) Write a model for y as a function of the independent variables, x_1 through x_8, and the gender dummy variable. Include interactions between gender and each of the other independent variables in the model.

(b) Based on the model, part a, what is the null hypothesis for testing whether gender has an effect on job satisfaction?

(c) Explain how to conduct the test, part b.

References

Chatterjee, S., Hadi, A., and Price, B. *Regression Analysis by Example*, 3rd ed. New York: Wiley, 1999.

Draper, N., and Smith, H. *Applied Regression Analysis*, 3rd ed. New York: Wiley, 1998.

Graybill, F. *Theory and Application of the Linear Model*. North Scituate, Mass.: Duxbury, 1976.

Kutner, M., Nachtsheim, C., Neter, J., and Li, W. *Applied Linear Statistical Models*, 5th ed. New York: McGraw-Hill/Irwin, 2005.

Maddala, G. S., *Introduction to Econometrics*, 3rd ed. New York: Wiley, 2001.

Mendenhall, W. *Introduction to Linear Models and the Design and Analysis of Experiments*. Belmont, Ca.: Wadsworth, 1968.

Montgomery, D., Peck, E., and Vining, G. *Introduction to Linear Regression Analysis*, 4th ed. New York: Wiley, 2006.

Mosteller, F., and Tukey, J. W. *Data Analysis and Regression: A Second Course in Statistics*. Reading, Mass.: Addison-Wesley, 1977.

Weisberg, S. *Applied Linear Regression*, 3rd ed. New York Wiley, 2005.

2

MODELING THE SALE PRICES OF RESIDENTIAL PROPERTIES IN FOUR NEIGHBORHOODS

The Problem

This case study concerns a problem of interest to real estate appraisers, tax assessors, real estate investors, and homebuyers—namely, the relationship between the appraised value of a property and its sale price. The sale price for any given property will vary depending on the price set by the seller, the strength of appeal of the property to a specific buyer, and the state of the money and real estate markets. Therefore, we can think of the sale price of a specific property as possessing a relative frequency distribution. The mean of this distribution might be regarded as a measure of the fair value of the property. Presumably, this is the value that a property appraiser or a tax assessor would like to attach to a given property.

The purpose of this case study is to examine the relationship between the mean sale price $E(y)$ of a property and the following independent variables:

1. Appraised land value of the property
2. Appraised value of the improvements on the property
3. Neighborhood in which the property is listed

The objectives of the study are twofold:

1. To determine whether the data indicate that appraised values of land and improvements are related to sale prices. That is, do the data supply sufficient evidence to indicate that these variables contribute information for the prediction of sale price?

2. To acquire the prediction equation relating appraised value of land and improvements to sale price and to determine whether this relationship is the same for a variety of neighborhoods. In other words, do the appraisers use the same appraisal criteria for various types of neighborhoods?

The Data

🔘 TAMSALES4

The data for the study were supplied by the property appraiser's office of Hillsborough County, Florida, and consist of the appraised land and improvement values and sale prices for residential properties sold in the city of Tampa, Florida, during the period May 2008 to June 2009. Four neighborhoods (Hyde Park, Cheval, Hunter's Green, and Davis Isles), each relatively homogeneous but differing sociologically and in property types and values, were identified within the city and surrounding area. The subset of sales and appraisal data pertinent to these four neighborhoods—a total of 176 observations—was used to develop a prediction equation relating sale prices to appraised land and improvement values. The data (recorded in thousands of dollars) are saved in the TAMSALES4 file and are fully described in Appendix E.

The Theoretical Model

If the mean sale price $E(y)$ of a property were, in fact, equal to its appraised value, x, the relationship between $E(y)$ and x would be a straight line with slope equal to 1, as shown in Figure CS2.1. But does this situation exist in reality? The property appraiser's data could be several years old and consequently may represent (because of inflation) only a percentage of the actual mean sale price. Also, experience has shown that the sale price–appraisal relationship is sometimes curvilinear in nature. One reason is that appraisers have a tendency to overappraise or underappraise properties in specific price ranges, say, very low-priced or very high-priced properties. In fact, it is common for realtors and real estate appraisers to model the natural logarithm of sales price, $\ln(y)$, as a function of appraised value, x. We learn (Section 7.7) that modeling $\ln(y)$ as a linear function of x introduces a curvilinear relationship between y and x.

Figure CS2.1 The theoretical relationship between mean sale price and appraised value x

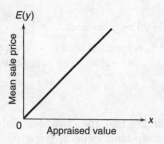

To gain insight into the sales–appraisal relationship, we used MINITAB to construct a scatterplot of sales price versus total appraised value for all 176 observations in the data set. The plotted points are shown in Figure CS2.2, as well as a graph of a straight-line fit to the data. Despite the concerns mentioned above, it appears that a linear model will fit the data well. Consequently, we use y = sale price (in thousands of dollars) as the dependent variable and consider only straight-line models. Later in this case study, we compare the linear model to a model with $\ln(y)$ as the dependent variable.

The Hypothesized Regression Models

We want to relate sale price y to three independent variables: the qualitative factor, neighborhood (four levels), and the two quantitative factors, appraised land value and appraised improvement value. We consider the following four models as candidates for this relationship.

Model 1 is a first-order model that will trace a response plane for mean of sale price, $E(y)$, as a function of x_1 = appraised land value (in thousands of dollars)

Figure CS2.2 MINITAB
scatterplot of
sales-appraisal data

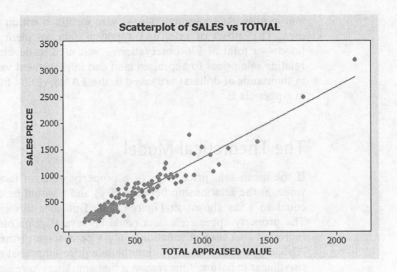

and x_2 = appraised improvement value (in thousands of dollars). This model will
assume that the response planes are identical for all four neighborhoods, that is, that
a first-order model is appropriate for relating y to x_1 and x_2 and that the relationship
between the sale price and the appraised value of a property is the same for all
neighborhoods. This model is:

**Model
1**

First-order model, identical for all neighborhoods

$$E(y) = \beta_0 + \overbrace{\beta_1 x_1}^{\substack{\text{Appraised land} \\ \text{value}}} + \overbrace{\beta_2 x_2}^{\substack{\text{Appraised} \\ \text{improvement value}}}$$

In Model 1, we are assuming that the change in sale price y for every \$1,000
(1-unit) increase in appraised land value x_1 (represented by β_1) is constant for a
fixed appraised improvements value, x_2. Likewise, the change in y for every \$1,000
increase in x_2 (represented by β_2) is constant for fixed x_1.

Model 2 will assume that the relationship between $E(y)$ and x_1 and x_2 is first-
order (a planar response surface), but that the planes' y-intercepts differ depending
on the neighborhood. This model would be appropriate if the appraiser's procedure
for establishing appraised values produced a relationship between mean sale price
and x_1 and x_2 that differed in at least two neighborhoods, but the differences
remained constant for different values of x_1 and x_2. Model 2 is

**Model
2**

First-order model, constant differences between neighborhoods

$$E(y) = \beta_0 + \overbrace{\beta_1 x_1}^{\substack{\text{Appraised land} \\ \text{value}}} + \overbrace{\beta_2 x_2}^{\substack{\text{Appraised} \\ \text{improvement value}}} + \overbrace{\beta_3 x_3 + \beta_4 x_4 + \beta_5 x_5}^{\substack{\text{Main effect terms} \\ \text{for neighborhoods}}}$$

where

$$x_1 = \text{Appraised land value}$$

$$x_2 = \text{Appraised improvement value}$$

$$x_3 = \begin{cases} 1 & \text{if Cheval neighborhood} \\ 0 & \text{if not} \end{cases}$$

$$x_4 = \begin{cases} 1 & \text{if Davis Isles neighborhood} \\ 0 & \text{if not} \end{cases}$$

$$x_5 = \begin{cases} 1 & \text{if Hyde Park neighborhood} \\ 0 & \text{if not} \end{cases}$$

The fourth neighborhood, Hunter's Green, was chosen as the base level. Consequently, the model will predict $E(y)$ for this neighborhood when $x_3 = x_4 = x_5 = 0$. Although it allows for neighborhood differences, Model 2 assumes that change in sale price y for every \$1,000 increase in either x_1 or x_2 does not depend on neighborhood.

Model 3 is similar to Model 2 except that we will add interaction terms between the neighborhood dummy variables and x_1 and between the neighborhood dummy variables and x_2. These interaction terms allow the change in y for increases in x_1 or x_2 to vary depending on the neighborhood. The equation of Model 3 is

Model 3

First-order model, no restrictions on neighborhood differences

$$E(y) = \beta_0 + \overbrace{\beta_1 x_1}^{\substack{\text{Appraised land} \\ \text{value}}} + \overbrace{\beta_2 x_2}^{\substack{\text{Appraised improvement} \\ \text{value}}}$$

$$+ \overbrace{\beta_3 x_3 + \beta_4 x_4 + \beta_5 x_5}^{\text{Main effect terms for neighborhoods}}$$

$$+ \overbrace{\beta_6 x_1 x_3 + \beta_7 x_1 x_4 + \beta_8 x_1 x_5}^{\text{Interaction, appraised land by neighborhood}}$$

$$+ \overbrace{\beta_9 x_2 x_3 + \beta_{10} x_2 x_4 + \beta_{11} x_2 x_5}^{\text{Interaction, appraised improvement by neighborhood}}$$

Note that for Model 3, the change in sale price y for every \$1,000 increase in appraised land value x_1 (holding x_2 fixed) is (β_1) in Hunter's Green and $(\beta_1 + \beta_6)$ in Cheval.

Model 4 differs from the previous three models by the addition of terms for x_1, x_2-interaction. Thus, Model 4 is a second-order (interaction) model that will trace (geometrically) a second-order response surface, one for each neighborhood. The interaction model follows:

Model 4

Interaction model in x_1 and x_2 that differs from one neighborhood to another

$$E(y) = \overbrace{\beta_0 + \beta_1 x_1 + \beta_2 x_2 + \beta_3 x_1 x_2}^{\text{Interaction model in } x_1 \text{ and } x_2} + \overbrace{\beta_4 x_3 + \beta_5 x_4 + \beta_6 x_5}^{\text{Main effect terms for neighborhoods}}$$

$$\left. \begin{array}{l} + \beta_7 x_1 x_3 + \beta_8 x_1 x_4 + \beta_9 x_1 x_5 + \beta_{10} x_2 x_3 \\ + \beta_{11} x_2 x_4 + \beta_{12} x_2 x_5 + \beta_{13} x_1 x_2 x_3 \\ + \beta_{14} x_1 x_2 x_4 + \beta_{15} x_1 x_2 x_5 \end{array} \right\} \begin{array}{l} \text{Interaction terms: } x_1, \\ x_2, \text{ and } x_1 x_2 \text{ terms by} \\ \text{neighborhood} \end{array}$$

Unlike Models 1–3, Model 4 allows the change in y for increases in x_1 to depend on x_2, and vice versa. For example, the change in sale price for a \$1,000 increase in appraised land value in the base-level neighborhood (Hunter's Green) is $(\beta_1 + \beta_3 x_2)$. Model 4 also allows for these sale price changes to vary from neighborhood to neighborhood (due to the neighborhood interaction terms).

We fit Models 1–4 to the data. Then, we compare the models using the nested model F test outlined in Section 4.13. Conservatively, we conduct each test at $\alpha = .01$.

Model Comparisons

The SAS printouts for Models 1–4 are shown in Figures CS2.3–CS2.6, respectively. These printouts yield the values of MSE, R_a^2, and s listed in Table CS2.1.

Table CS2.1 Summary of regressions of the models

Model	MSE	R_a^2	s
1	12,748	.923	112.9
2	12,380	.926	111.3
3	11,810	.929	108.7
4	10,633	.936	103.1

Figure CS2.3 SAS regression output for Model 1

Dependent Variable: SALES

Number of Observations Read   176
Number of Observations Used   176

Analysis of Variance

Source	DF	Sum of Squares	Mean Square	F Value	Pr > F
Model	2	26894548	13447274	1054.82	<.0001
Error	173	2205464	12748		
Corrected Total	175	29100013			

Root MSE	112.90859	R-Square	0.9242
Dependent Mean	549.89318	Adj R-Sq	0.9233
Coeff Var	20.53282		

Parameter Estimates

Variable	DF	Parameter Estimate	Standard Error	t Value	Pr > |t|
Intercept	1	-16.17597	14.99022	-1.08	0.2820
LAND	1	1.39329	0.05868	23.75	<.0001
IMP	1	1.33023	0.04640	28.67	<.0001

Test #1   *Model 1 versus Model 2*

To test the hypothesis that a single first-order model is appropriate for all neighborhoods, we wish to test the null hypothesis that the neighborhood parameters in Model 2 are all equal to 0, that is,

$$H_0: \beta_3 = \beta_4 = \beta_5 = 0$$

Figure CS2.4 SAS regression output for Model 2

```
                          Dependent Variable: SALES

                      Number of Observations Read        176
                      Number of Observations Used        176

                               Analysis of Variance

                                    Sum of           Mean
        Source              DF     Squares          Square    F Value    Pr > F

        Model                5    26995406         5399081     436.11    <.0001
        Error              170     2104606           12380
        Corrected Total    175    29100013


                Root MSE            111.26562    R-Square     0.9277
                Dependent Mean      549.89318    Adj R-Sq     0.9255
                Coeff Var            20.23404

                               Parameter Estimates

                             Parameter      Standard
        Variable      DF      Estimate         Error    t Value    Pr > |t|

        Intercept      1      -6.74999      17.55057      -0.38      0.7010
        LAND           1       1.59419       0.09116      17.49     <.0001
        IMP            1       1.30108       0.04688      27.75     <.0001
        CHEV           1     -10.18687      22.73174      -0.45      0.6546
        DAVIS          1     -93.16413      34.22825      -2.72      0.0072
        HYDE           1     -57.46750      29.18426      -1.97      0.0506
```

Test NBHD Results for Dependent Variable SALES

```
                                     Mean
        Source              DF     Square    F Value    Pr > F

        Numerator            3      33619       2.72     0.0464
        Denominator        170      12380
```

That is, we want to compare the complete model, Model 2, to the reduced model, Model 1. The test statistic is

$$F = \frac{(\text{SSE}_R - \text{SSE}_C)/\text{Number of } \beta \text{ parameters in } H_0}{\text{MSE}_C}$$

$$= \frac{(\text{SSE}_1 - \text{SSE}_2)/3}{\text{MSE}_2}$$

Although the information is available to compute this value by hand, we utilize the SAS option to conduct the nested model F test. The test statistic value, shaded at the bottom of Figure CS2.4, is $F = 2.72$. The p-value of the test, also shaded, is .0464. Since $\alpha = .05$ exceeds this p-value, we have evidence to indicate that the addition of the neighborhood dummy variables in Model 2 contributes significantly to the prediction of y. The practical implication of this result is that the appraiser is not assigning appraised values to properties in such a way that the first-order relationship between (sales), y, and appraised values x_1 and x_2 is the same for all neighborhoods.

Test # 2

Model 2 versus Model 3

Can the prediction equation be improved by the addition of interactions between neighborhood and x_1 and neighborhood and x_2? That is, do the data provide sufficient evidence to indicate that Model 3 is a better predictor of sale price than Model 2? To answer this question, we test the null hypothesis that the parameters associated

Figure CS2.5 SAS regression output for Model 3

Dependent Variable: SALES

Number of Observations Read   176
Number of Observations Used   176

Analysis of Variance

Source	DF	Sum of Squares	Mean Square	F Value	Pr > F
Model	11	27163118	2469374	209.09	<.0001
Error	164	1936895	11810		
Corrected Total	175	29100013			

Root MSE	108.67536	R-Square	0.9334	
Dependent Mean	549.89318	Adj R-Sq	0.9290	
Coeff Var	19.76300			

Parameter Estimates

Variable	DF	Parameter Estimate	Standard Error	t Value	Pr > |t|
Intercept	1	-65.58007	33.35042	-1.97	0.0509
LAND	1	1.59236	0.82050	1.94	0.0540
IMP	1	1.57589	0.16048	9.82	<.0001
CHEV	1	75.73165	53.95609	1.40	0.1623
DAVIS	1	4.64307	49.64671	0.09	0.9256
HYDE	1	-42.72927	81.61473	-0.52	0.6013
LAN_CHEV	1	-1.00467	1.12349	-0.89	0.3725
LAN_DAV	1	0.00398	0.82577	0.00	0.9962
LAN_HYDE	1	0.10687	0.91388	0.12	0.9070
IMP_CHEV	1	0.00524	0.30161	0.02	0.9862
IMP_DAV	1	-0.41413	0.17253	-2.40	0.0175
IMP_HYDE	1	-0.20389	0.20476	-1.00	0.3208

Test QN_NBHD Results for Dependent Variable SALES

Source	DF	Mean Square	F Value	Pr > F
Numerator	6	27952	2.37	0.0322
Denominator	164	11810		

with all neighborhood interaction terms in Model 3 equal 0. Thus, Model 2 is now the reduced model and Model 3 is the complete model.

Checking the equation of Model 3, you will see that there are six neighborhood interaction terms and that the parameters included in H_0 will be

$$H_0: \beta_6 = \beta_7 = \beta_8 = \beta_9 = \beta_{10} = \beta_{11} = 0$$

To test H_0, we require the test statistic

$$F = \frac{(SSE_R - SSE_C)/\text{Number of }\beta\text{ parameters in }H_0}{MSE_C} = \frac{(SSE_2 - SSE_3)/6}{MSE_3}$$

This value, $F = 2.37$, is shaded at the bottom of the SAS printout for Model 3, Figure CS2.5. The p-value of the test (also shaded) is .0322. Thus, there is sufficient evidence (at $\alpha = .05$) to indicate that the neighborhood interaction terms of Model 3 contribute information for the prediction of y. Practically, this test implies that the rate of change of sale price y with either appraised value, x_1 or x_2, differs for each of the four neighborhoods.

Test #3

Model 3 versus Model 4

We have already shown that the first-order prediction equations vary among neighborhoods. To determine whether the (second-order) interaction terms involving the appraised values, x_1 and x_2, contribute significantly to the prediction of y, we test

Figure CS2.6 SAS regression output for Model 4

```
                          Dependent Variable: SALES

              Number of Observations Read        176
              Number of Observations Used        176

                        Analysis of Variance

                              Sum of         Mean
Source              DF       Squares       Square    F Value   Pr > F

Model               15      27398705      1826580    171.78   <.0001
Error              160       1701307        10633
Corrected Total    175      29100013


              Root MSE           103.11726    R-Square     0.9415
              Dependent Mean     549.89318    Adj R-Sq     0.9361
              Coeff Var           18.75223


                        Parameter Estimates

                    Parameter      Standard
Variable     DF      Estimate         Error    t Value   Pr > |t|

Intercept     1      22.78669      74.40962       0.31     0.7598
LAND          1       0.10877       1.37278       0.08     0.9369
IMP           1       1.21342       0.31543       3.85     0.0002
LAN_IMP       1       0.00493       0.00376       1.31     0.1914
CHEV          1     132.46780     131.91123       1.00     0.3168
DAVIS         1      72.29217      89.88332       0.80     0.4224
HYDE          1    -133.43136     122.25288      -1.09     0.2767
LAN_CHEV      1      -0.93599       1.84088      -0.51     0.6118
LAN_DAV       1       1.07632       1.37896       0.78     0.4362
LAN_HYDE      1       1.59815       1.44160       1.11     0.2693
IMP_CHEV      1      -0.25255       0.58771      -0.43     0.6680
IMP_DAV       1      -0.45018       0.33429      -1.35     0.1800
IMP_HYDE      1       0.16672       0.40934       0.41     0.6843
L_I_CHEV      1       0.00024652     0.00521      0.05     0.9623
L_I_DAV       1      -0.00403       0.00376      -1.07     0.2856
L_I_HYDE      1      -0.00495       0.00381      -1.30     0.1952


       Test LAN_IMP_NBHD Results for Dependent Variable SALES

                                Mean
Source              DF        Square    F Value   Pr > F

Numerator            4         58897       5.54    0.0003
Denominator        160         10633
```

the hypothesis that the four parameters involving $x_1 x_2$ in Model 4 all equal 0. The null hypothesis is

$$H_0: \beta_3 = \beta_{13} = \beta_{14} = \beta_{15} = 0$$

and the alternative hypothesis is that at least one of these parameters does not equal 0. Using Model 4 as the complete model and Model 3 as the reduced model, the test statistic required is:

$$F = \frac{(SSE_R - SSE_C)/\text{Number of } \beta \text{ parameters in } H_0}{MSE_C} = \frac{(SSE_3 - SSE_4)/4}{MSE_4}$$

This value (shaded at the bottom of the SAS printout for Model 4, Figure CS2.6) is $F = 5.54$. Again, the p-value of the test (.0003) is less than $\alpha = .05$. This small p-value supports the alternative hypothesis that the $x_1 x_2$ interaction terms of Model 4 contribute significantly to the prediction of y.

The results of the preceding tests suggest that Model 4 is the best of the four models for modeling sale price y. The global F value for testing

$$H_0: \beta_1 = \beta_2 = \cdots = \beta_{15} = 0$$

is highly significant ($F = 171.78$, p-value $< .0001$); the R^2 value indicates that the model explains approximately 94% of the variability in sale price. You may notice that several of the t tests involving the individual β parameters in Model 4 are nonsignificant. Be careful not to conclude that these terms should be dropped from the model.

Whenever a model includes a large number of interactions (as in Model 4) and/or squared terms, several t tests will often be nonsignificant even if the global F test is highly significant. This result is due partly to the unavoidable intercorrelations among the main effects for a variable, its interactions, and its squared terms (see the discussion on multicollinearity in Section 7.5). We warned in Chapter 4 of the dangers of conducting a series of t tests to determine model adequacy. For a model with a large number of β's, such as Model 4, you should avoid conducting any t tests at all and rely on the global F test and partial F tests to determine the important terms for predicting y.

Before we proceed to estimation and prediction with Model 4, we conduct one final test. Recall our earlier discussion of the theoretical relationship between sale price and appraised value. Will a model with the natural log of y as the dependent variable outperform Model 4? To check this, we fit a model identical to Model 4—call it Model 5—but with $y^* = \ln(y)$ as the dependent variable. A portion of the SAS printout for Model 5 is displayed in Figure CS2.7. From the printout, we obtain $R_a^2 = .90, s = .18$, and global $F = 110.73$ (p-value $< .0001$). Model 5 is clearly "statistically" useful for predicting sale price based on the global F test. However, we must be careful not to judge which of the two models is better based on the values of R^2 and s shown on the printouts because the dependent variable is not the same for each model. (Recall the "warning" given at the end of Section 4.11.) An informal procedure for comparing the two models requires that we obtain predicted values for $\ln(y)$ using the prediction equation for Model 5, transform these predicted log values back to sale prices using the equation, $e^{\widehat{\ln(y)}}$, then recalculate the R_a^2 and s values for Model 5 using the formulas given in Chapter 4. We used the programming language of SAS to compute these values; they are compared to the values of R_a^2 and s for Model 4 in Table CS2.2.

Figure CS2.7 Portion of SAS regression output for Model 5

```
                      Dependent Variable: LNSALES

               Number of Observations Read      176
               Number of Observations Used      176


                       Analysis of Variance

                              Sum of        Mean
   Source            DF      Squares       Square    F Value    Pr > F

   Model             15     56.28950      3.75263    110.73    <.0001
   Error            160      5.42256      0.03389
   Corrected Total  175     61.71206


          Root MSE              0.18410    R-Square     0.9121
          Dependent Mean       6.11775    Adj R-Sq     0.9039
          Coeff Var            3.00920
```

Table CS2.2 Comparison of R_a^2 and s for Models 4 and 5

	R_a^2	s
Model 4:	.936 (from Figure CS2.6)	103.1 (from Figure CS2.6)
Model 5:	.904 (using transformation)	127.6 (using transformation)

You can see that Model 4 outperforms Model 5 with respect to both statistics. Model 4 has a slightly higher R_a^2 value and a lower model standard deviation. These results support our decision to build a model with sale price y rather than $\ln(y)$ as the dependent variable.

Interpreting the Prediction Equation

Substituting the estimates of the Model 4 parameters (Figure CS2.6) into the prediction equation, we have

$$\hat{y} = 22.79 + .11x_1 + 1.21x_2 + .0049x_1x_2 + 132.5x_3 + 72.3x_4 - 133.4x_5$$
$$- .94x_1x_3 + 1.08x_1x_4 + 1.60x_1x_5 - .25x_2x_3 - .45x_2x_4 + .17x_2x_5 + .0002x_1x_2x_3$$
$$- .004x_1x_2x_4 - .005x_1x_2x_5$$

We have noted that the model yields four response surfaces, one for each neighborhood. One way to interpret the prediction equation is to first find the equation of the response surface for each neighborhood. Substituting the appropriate values of the neighborhood dummy variables, x_3, x_4, and x_5, into the equation and combining like terms, we obtain the following:

Cheval: $(x_3 = 1, x_4 = x_5 = 0)$

$$\hat{y} = (22.79 + 132.5) + (.11 - .94)x_1 + (1.21 - .25)x_2 + (.0049 + .0002)x_1x_2$$
$$= 155.29 - .83x_1 + .96x_2 + .005x_1x_2$$

Davis Isles: $(x_3 = 0, x_4 = 1, x_5 = 0)$

$$\hat{y} = (22.79 + 72.3) + (.11 + 1.08)x_1 + (1.21 - .45)x_2 + (.0049 - .004)x_1x_2$$
$$= 95.09 + 1.19x_1 + .76x_2 + .001x_1x_2$$

Hyde Park: $(x_3 = x_4 = 0, x_5 = 1)$

$$\hat{y} = (22.79 - 133.4) + (.11 + 1.60)x_1 + (1.21 + .17)x_2 + (.0049 - .00495)x_1x_2$$
$$= -110.61 + 1.71x_1 + 1.38x_2 - .00005x_1x_2$$

Hunter's Green: $(x_3 = x_4 = x_5 = 0)$

$$\hat{y} = 22.79 + .11x_1 + 1.21x_2 + .0049x_1x_2$$

Note that each equation is in the form of an interaction model involving appraised land value x_1 and appraised improvements x_2. To interpret the β estimates of each interaction equation, we hold one independent variable fixed, say, x_1, and focus on the slope of the line relating y to x_2. For example, holding appraised land value constant at \$50,000 ($x_1 = 50$), the slope of the line relating y to x_2 for Hunter's Green (the base level neighborhood) is

$$\hat{\beta}_2 + \hat{\beta}_3 x_1 = 1.21 + .0049(50) = 1.455$$

Thus, for residential properties in Hunter's Green with appraised land value of \$50,000, the sale price will increase \$1,455 for every \$1,000 increase in appraised improvements.

Similar interpretations can be made for the slopes for other combinations of neighborhoods and appraised land value x_1. The estimated slopes for several of these combinations are computed and shown in Table CS2.3. Because of the interaction

Table CS2.3 Estimated dollar increase in sale price for $1,000 increase in appraised improvements

		Neighborhood			
		Cheval	Davis Isles	Hyde Park	Hunter's Green
	$50,000	1,210	810	1,377	1,455
APPRAISED LAND	$75,000	1,335	835	1,376	1,577
VALUE	$100,000	1,460	860	1,375	1,700
	$150,000	1,710	910	1,373	1,945

terms in the model, the increases in sale price for a $1,000 increase in appraised improvements, x_2, differ for each neighborhood and for different levels of land value, x_1.

Some trends are evident from Table CS2.3. For fixed appraised land value (x_1) the increase in sales price (y) for every $1,000 increase in appraised improvements (x_2) is smallest for Davis Isles and largest for Hunter's Green. For three of the four neighborhoods (Cheval, Davis Isles, and Hunter's Green), the slope increases as appraised land value increases. You can see this graphically in Figure CS2.8, which shows the slopes for the four neighborhoods when appraised land value is held fixed at $150,000.

Figure CS2.8 SAS graph of predicted sales for land value of $150,000

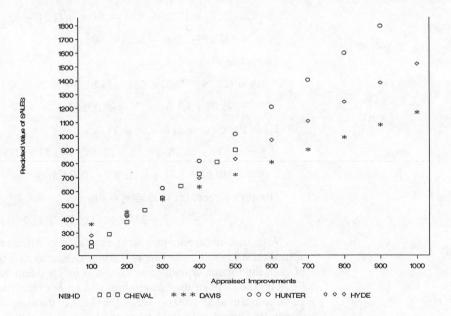

More information concerning the nature of the neighborhoods can be gleaned from the SAS printout shown in Figure CS2.9. This printout gives descriptive statistics (mean, standard deviation, minimum, and maximum) for sale price (y), land value (x_1), and improvements (x_2) for each of the neighborhoods. The mean sale prices confirm what realtors know to be true, that is, that neighborhoods Davis Isles and Hyde Park are two of the most expensive residential areas in the city. Most of the inhabitants are highly successful professionals or business entrepreneurs. In contrast, Cheval and Hunter's Green, although also relatively high-priced, are less

The MEANS Procedure

NBHD	N Obs	Variable	Mean	Std Dev	N	Minimum	Maximum
CHEVAL	44	SALES	455.3954545	223.6411182	44	219.9000000	1140.00
		LAND	95.3987727	40.4242457	44	45.3600000	207.0000000
		IMP	246.1400000	121.4848028	44	104.7660000	563.9150000
DAVISISLES	42	SALES	818.2357143	517.9731318	42	259.9000000	3200.00
		LAND	343.2939048	187.7635828	42	164.9000000	1004.59
		IMP	285.0496905	275.8944340	42	31.9290000	1129.65
HUNTERSGREEN	56	SALES	352.7928571	240.2715682	56	140.0000000	1550.00
		LAND	50.5408036	25.0642057	56	24.2500000	153.2760000
		IMP	214.4147143	128.1514669	56	91.8220000	906.4710000
HYDEPARK	34	SALES	665.3382353	451.5342525	34	152.0000000	2500.00
		LAND	238.4018529	76.4757030	34	118.8000000	533.2800000
		IMP	268.6203824	241.9964835	34	33.7930000	1221.20

Figure CS2.9 SAS descriptive statistics, by neighborhood

expensive residential areas inhabited primarily by either younger, upscale married couples or older, retired couples.

Examining Figure CS2.8 and Table CS2.3 again, notice that the less expensive neighborhoods tend to have the steeper slopes, and their estimated lines tend to be above the lines of the more expensive neighborhoods. Consequently, a property in Cheval is estimated to sell at a higher price than one in the more expensive Davis Isles when appraised land value is held fixed. This would suggest that the appraised values of properties in Davis Isles are too low (i.e., they are underappraised) compared with similar properties in Cheval. Perhaps a low appraised property value in Davis Isles corresponds to a smaller, less desirable lot; this may have a strong depressive effect on sale prices.

The two major points to be derived from an analysis of the sale price–appraised value lines are as follows:

1. The rate at which sale price increases with appraised value differs for different neighborhoods. This increase tends to be largest for the more expensive neighborhoods.

2. The lines for Cheval and Hunter's Green lie above those for Davis Isle and Hyde Park, indicating that properties in the higher-priced neighborhoods are being underappraised relative to sale price compared with properties in the lower-priced neighborhoods.

Predicting the Sale Price of a Property

How well do appraised land value x_1 and appraised improvements value x_2 predict residential property sale price? Recall that from Model 4 (Figure CS2.6), we obtained $R_a^2 = .936$, indicating that the model accounts for approximately 94% of the sample variability in the sale price values, y. This seems to indicate that the model provides a reasonably good fit to the data, but note that $s = 103.1$ (see Figure CS2.6). Our interpretation is that approximately 95% of the predicted sale price values will fall within $(2s)(\$1,000) = (2)(103.1)(\$1,000) = \$206,200$ of their actual values. This relatively large standard deviation may lead to large errors of prediction for some residential properties if the model is used in practice.

Output Statistics

Neighborhood	LAND	IMP	Predicted Value	Std Error Predict	95% CL Lower	95% CL Upper
CHEVAL	100	400	663.93	112.902	440.96	886.90
DAVISISLE	300	800	1276.76	110.396	1058.74	1494.79
HUNTERSGRN	75	300	505.89	106.098	296.36	715.42
HYDEPARK	250	750	1347.15	127.053	1096.24	1598.07

Figure CS2.10 SAS printout showing 95% prediction intervals for sale price

Figure CS2.10 is a portion of a SAS printout showing 95% prediction intervals for the sale prices of four residential properties, one selected from each neighborhood.

Note the large widths of the intervals. These wide prediction intervals cast doubt on whether the prediction equation could be of practical value in predicting property sale prices. We feel certain that a much more accurate predictor of sale price could be developed by relating y to the variables that describe the property (such as location, square footage, and number of bedrooms) and those that describe the market (mortgage interest rates, availability of money, and so forth).

Conclusions

The results of the regression analyses indicate that the relationships between property sale prices and appraised values are not consistent from one neighborhood to another. Furthermore, the widths of the prediction intervals are rather sizable, indicating that there is room for improvement in the methods used to determine appraised property values.

Follow-up Questions

1. Explain why the tests of model adequacy conducted in this case study give no assurance that Model 4 will be a successful predictor of sale price in the future.

2. After you have covered Section 5.11, return to this case study and use the data-splitting technique to assess the external validity of Model 4.

TAMSALES8
3. Recall that the data for this case study are described in Appendix E. The full data set, named TAM-

SALES8, contains sale price information for the four neighborhoods compared in this case study, as well as sale price information for four additional neighborhoods (Avila, Carrollwood, Tampa Palms, and Town & Country). Use the full data set to build a regression model for sale price of a residential property. Part of your analysis will involve a comparison of the eight neighborhoods.

PRINCIPLES OF MODEL BUILDING

Contents

Objectives

1. To demonstrate why the choice of the deterministic portion of a linear model is crucial to the acquisition of a good prediction equation.
2. To present models with only quantitative predictors.
3. To present models with only qualitative predictors.
4. To present models with both quantitative and qualitative predictors.
5. To present some basic procedures for building good linear models.

5.1 Introduction: Why Model Building Is Important

We have emphasized in both Chapters 3 and 4 that one of the first steps in the construction of a regression model is to hypothesize the form of the deterministic portion of the probabilistic model. This *model-building*, or model-construction, stage is the key to the success (or failure) of the regression analysis. If the hypothesized model does not reflect, at least approximately, the true nature of the relationship between the mean response $E(y)$ and the independent variables $x_1, x_2, \ldots, x_k$, the modeling effort will usually be unrewarded.

By **model building**, we mean writing a model that will provide a good fit to a set of data and that will give good estimates of the mean value of y and good predictions of future values of y for given values of the independent variables. To illustrate, several years ago, a nationally recognized educational research group issued a report concerning the variables related to academic achievement for a certain type of college student. The researchers selected a random sample of students and recorded a measure of academic achievement, y, at the end of the senior year together with data on an extensive list of independent variables, $x_1, x_2, \ldots, x_k$, that they thought were related to y. Among these independent variables were the student's IQ, scores on mathematics and verbal achievement examinations, rank in class, and so on. They fit the model

$$E(y) = \beta_0 + \beta_1 x_1 + \beta_2 x_2 + \cdots + \beta_k x_k$$

to the data, analyzed the results, and reached the conclusion that none of the independent variables was "significantly related" to y. The **goodness of fit** of the model, measured by the coefficient of determination R^2, was not particularly good, and t tests on individual parameters did not lead to rejection of the null hypotheses that these parameters equaled 0.

How could the researchers have reached the conclusion that there is no significant relationship, when it is evident, just as a matter of experience, that some of the independent variables studied are related to academic achievement? For example, achievement on a college mathematics placement test should be related to achievement in college mathematics. Certainly, many other variables will affect achievement—motivation, environmental conditions, and so forth—but generally speaking, there will be a positive correlation between entrance achievement test scores and college academic achievement. So, what went wrong with the educational researchers' study?

Although you can never discard the possibility of computing error as a reason for erroneous answers, most likely the difficulties in the results of the educational study were caused by the use of an improperly constructed model. For example, the model

$$E(y) = \beta_0 + \beta_1 x_1 + \beta_2 x_2 + \cdots + \beta_k x_k$$

assumes that the independent variables $x_1, x_2, \ldots, x_k$ affect mean achievement $E(y)$ independently of each other.* Thus, if you hold all the other independent variables constant and vary only x_1, $E(y)$ will increase by the amount β_1 for every unit increase in x_1. A 1-unit change in any of the other independent variables will increase $E(y)$ by the value of the corresponding β parameter for that variable.

Do the assumptions implied by the model agree with your knowledge about academic achievement? First, is it reasonable to assume that the effect of time spent on study is independent of native intellectual ability? We think not. No matter how much effort some students invest in a particular subject, their rate of achievement is low. For others, it may be high. Therefore, assuming that these two variables—effort and native intellectual ability—affect $E(y)$ independently of each other is likely to be an erroneous assumption. Second, suppose that x_5 is the amount of time a student devotes to study. Is it reasonable to expect that a 1-unit increase in x_5 will always produce the same change β_5 in $E(y)$? The changes in $E(y)$ for a 1-unit increase in x_5 might depend on the value of x_5 (e.g., the law of diminishing returns). Consequently, it is quite likely that the assumption of a constant rate of change in $E(y)$ for 1-unit increases in the independent variables will not be satisfied.

Clearly, the model

$$E(y) = \beta_0 + \beta_1 x_1 + \beta_2 x_2 + \cdots + \beta_k x_k$$

was a poor choice in view of the researchers' prior knowledge of some of the variables involved. Terms have to be added to the model to account for interrelationships among the independent variables and for curvature in the response function. Failure to include needed terms causes inflated values of SSE, nonsignificance in statistical tests, and, often, erroneous practical conclusions.

---

* Keep in mind that we are discussing the deterministic portion of the model and that the word *independent* is used in a mathematical rather than a probabilistic sense.

In this chapter, we discuss the most difficult part of a multiple regression analysis: the formulation of a good model for $E(y)$. Although many of the models presented in this chapter have already been introduced in optional Section 4.12, we assume the reader has little or no background in model building. This chapter serves as a basic reference guide to model building for teachers, students, and practitioners of multiple regression analysis.

5.2 The Two Types of Independent Variables: Quantitative and Qualitative

The independent variables that appear in a linear model can be one of two types. Recall from Chapter 1 that a *quantitative* variable is one that assumes numerical values corresponding to the points on a line. (Definition 1.4). An independent variable that is not quantitative, that is, one that is categorical in nature, is called *qualitative* (Definition 1.5).

The nicotine content of a cigarette, prime interest rate, number of defects in a product, and IQ of a student are all examples of quantitative independent variables. On the other hand, suppose three different styles of packaging, A, B, and C, are used by a manufacturer. This independent variable, style of packaging, is qualitative, since it is not measured on a numerical scale. Certainly, style of packaging is an independent variable that may affect sales of a product, and we would want to include it in a model describing the product's sales, y.

Definition 5.1 The different values of an independent variable used in regression are called its **levels**.

For a quantitative variable, the levels correspond to the numerical values it assumes. For example, if the number of defects in a product ranges from 0 to 3, the independent variable assumes four levels: 0, 1, 2, and 3.

The levels of a qualitative variable are not numerical. They can be defined only by describing them. For example, the independent variable style of packaging was observed at three levels: A, B, and C.

Example 5.1

In Chapter 4, we considered the problem of predicting executive salary as a function of several independent variables. Consider the following four independent variables that may affect executive salaries:

(a) Years of experience

(b) Gender of the employee

(c) Firm's net asset value

(d) Rank of the employee

For each of these independent variables, give its type and describe the nature of the levels you would expect to observe.

Solution

(a) The independent variable for the number of years of experience is quantitative, since its values are numerical. We would expect to observe levels ranging from 0 to 40 (approximately) years.

(b) The independent variable for gender is qualitative, since its levels can only be described by the nonnumerical labels "female" and "male."

(c) The independent variable for the firm's net asset value is quantitative, with a very large number of possible levels corresponding to the range of dollar values representing various firms' net asset values.

(d) Suppose the independent variable for the rank of the employee is observed at three levels: supervisor, assistant vice president, and vice president. Since we cannot assign a realistic measure of relative importance to each position, rank is a qualitative independent variable.　　　■

Quantitative and qualitative independent variables are treated differently in regression modeling. In the next section, we see how quantitative variables are entered into a regression model.

5.2 Exercises

5.1 Buy-side versus sell-side analysts' earnings forecasts. The *Financial Analysts Journal* (July/August 2008) published a study comparing the earnings forecasts of buy-side and sell-side analysts. A team of Harvard Business School professors used regression to model the relative optimism (y) of the analysts' 3-month horizon forecasts based on the following independent variables. Determine the type (quantitative or qualitative) of each variable.

(a) Whether the analyst worked for a buy-side firm or a sell-side firm.
(b) Number of days between forecast and fiscal year-end (i.e., forecast horizon).
(c) Number of quarters the analyst had worked with the firm.

5.2 Workplace bullying and intention to leave. Workplace bullying (e.g., work-related harassment, persistent criticism, withholding key information, spreading rumors, intimidation) has been shown to have a negative psychological effect on victims, often leading the victim to quit or resign. In *Human Resource Management Journal* (October 2008), researchers employed multiple regression to model bullying victims' intention to leave the firm as a function of perceived organizational support and level of workplace bullying. The dependent variable in the analysis, intention to leave (y), was measured on a quantitative scale. Identify the type (qualitative or quantitative) of the two key independent variables in the study, level of bullying (measured on a 50-point scale) and perceived organizational support (measured as "low," "neutral," or "high").

5.3 Expert testimony in homicide trials of battered women. The *Duke Journal of Gender Law and Policy* (Summer 2003) examined the impact of expert testimony on the outcome of homicide trials that involve battered woman syndrome. Multiple regression was employed to model the likelihood of changing a verdict from not guilty to guilty after deliberations, y, as a function of juror gender and whether or not expert testimony was given. Identify the independent variables in the model as quantitative or qualitative.

5.4 Chemical composition of rain water. The *Journal of Agricultural, Biological, and Environmental Statistics* (March 2005) presented a study of the chemical composition of rain water. The nitrate concentration, y (milligrams per liter), in a rain water sample was modeled as a function of two independent variables: water source (groundwater, subsurface flow, or overground flow) and silica concentration (milligrams per liter). Identify the type (quantitative or qualitative) for each independent variable.

5.5 Psychological response of firefighters. The *Journal of Human Stress* (Summer 1987) reported on a study of "psychological response of firefighters to chemical fire." The researchers used multiple regression to predict emotional distress as a function of the following independent variables. Identify each independent variable as quantitative or qualitative. For qualitative variables, suggest several levels that might be observed. For quantitative variables, give a range of values (levels) for which the variable might be observed.

(a) Number of preincident psychological symptoms
(b) Years of experience
(c) Cigarette smoking behavior
(d) Level of social support
(e) Marital status
(f) Age

(g) Ethnic status
(h) Exposure to a chemical fire
(i) Education level
(j) Distance lived from site of incident
(k) Gender

5.6 **Modeling a qualitative response.** Which of the assumptions about ε (Section 4.2) prohibit the use of a qualitative variable as a dependent variable? (We present a technique for modeling a qualitative dependent variable in Chapter 9.)

5.3 Models with a Single Quantitative Independent Variable

To write a prediction equation that provides a good model for a response (one that will eventually yield good predictions), we have to know how the response might vary as the levels of an independent variable change. Then we have to know how to write a mathematical equation to model it. To illustrate (with a simple example), suppose we want to model a student's score on a statistics exam, y, as a function of the single independent variable x, the amount of study time invested. It may be that exam score, y, increases in a straight line as the amount of study time, x, varies from 1 hour to 6 hours, as shown in Figure 5.1a. If this were the entire range of x-values for which you wanted to predict y, the model

$$E(y) = \beta_0 + \beta_1 x$$

would be appropriate.

Figure 5.1 Modeling exam score, y, as a function of study time, x

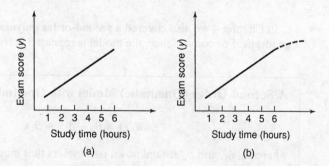

Now, suppose you want to expand the range of values of x to $x = 8$ or $x = 10$ hours of studying. Will the straight-line model

$$E(y) = \beta_0 + \beta_1 x$$

be satisfactory? Perhaps, but making this assumption could be risky. As the amount of studying, x, is increased, sooner or later the point of diminishing returns will be reached. That is, the increase in exam score for a unit increase in study time will decrease, as shown by the dashed line in Figure 5.1b. To produce this type of curvature, you must know the relationship between models and graphs, and how types of terms will change the shape of the curve.

A response that is a function of a single quantitative independent variable can often be modeled by the first few terms of a polynomial algebraic function. The equation relating the mean value of y to a polynomial of order p in one independent variable x is shown in the box.

A pth-Order Polynomial with One Independent Variable

$$E(y) = \beta_0 + \beta_1 x + \beta_2 x^2 + \beta_3 x^3 + \cdots + \beta_p x^p$$

where p is an integer and $\beta_0, \beta_1, \ldots, \beta_p$ are unknown parameters that must be estimated.

As we mentioned in Chapters 3 and 4, a **first-order polynomial** in x (i.e., $p = 1$),

$$E(y) = \beta_0 + \beta_1 x$$

graphs as a straight line. The β interpretations of this model are provided in the next box.

First-Order (Straight-Line) Model with One Independent Variable

$$E(y) = \beta_0 + \beta_1 x$$

Interpretation of model parameters

β_0: y-intercept; the value of $E(y)$ when $x = 0$
β_1: Slope of the line; the change in $E(y)$ for a 1-unit increase in x

---

In Chapter 4 we also covered a **second-order polynomial** model ($p = 2$), called a **quadratic**. For convenience, the model is repeated in the following box.

A Second-Order (Quadratic) Model with One Independent Variable

$$E(y) = \beta_0 + \beta_1 x + \beta_2 x^2$$

where β_0, β_1, and β_2 are unknown parameters that must be estimated.

Interpretation of model parameters

β_0: y-intercept; the value of $E(y)$ when $x = 0$
β_1: Shift parameter; changing the value of β_1 shifts the parabola to the right or left (increasing the value of β_1 causes the parabola to shift to the right)
β_2: Rate of curvature

Graphs of two quadratic models are shown in Figure 5.2. As we learned in Chapter 4, the quadratic model is the equation of a **parabola** that opens either upward, as in Figure 5.2a, or downward, as in Figure 5.2b. If the coefficient of x^2 is positive, it opens upward; if it is negative, it opens downward. The parabola may be shifted upward or downward, left or right. The least squares procedure uses only the portion of the parabola that is needed to model the data. For example, if you fit a parabola to the data points shown in Figure 5.3, the portion shown as a solid curve

Figure 5.2 Graphs for two second-order polynomial models

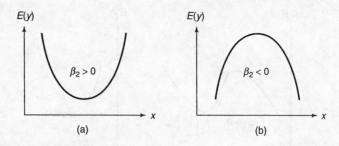

(a)   (b)

passes through the data points. The outline of the unused portion of the parabola is indicated by a dashed curve.

Figure 5.3 Example of the use of a quadratic model

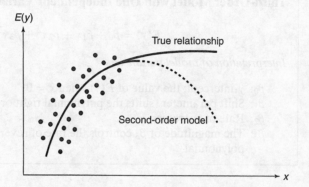

Figure 5.3 illustrates an important limitation on the use of prediction equations: The model is valid only over the range of x-values that were used to fit the model. For example, the response might rise, as shown in the figure, until it reaches a plateau. The second-order model might fit the data very well over the range of x-values shown in Figure 5.3, but would provide a very poor fit if data were collected in the region where the parabola turns downward.

How do you decide the order of the polynomial you should use to model a response if you have no prior information about the relationship between $E(y)$ and x? If you have data, construct a scatterplot of the data points, and see whether you can deduce the nature of a good approximating function. A pth-order polynomial, when graphed, will exhibit $(p - 1)$ peaks, troughs, or reversals in direction. Note that the graphs of the second-order model shown in Figure 5.2 each have $(p - 1) = 1$ peak (or trough). Likewise, a third-order model (shown in the box) will have $(p - 1) = 2$ peaks or troughs, as illustrated in Figure 5.4.

The graphs of most responses as a function of an independent variable x are, in general, curvilinear. Nevertheless, if the rate of curvature of the response curve is very small over the range of x that is of interest to you, a straight line might provide an excellent fit to the response data and serve as a very useful prediction equation. If the curvature is expected to be pronounced, you should try a second-order model. Third- or higher-order models would be used only where you expect more than one reversal in the direction of the curve. These situations are rare, except where the response is a function of time. Models for forecasting over time are presented in Chapter 10.

Figure 5.4 Graphs of two third-order polynomial models

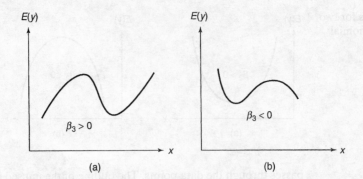

(a)   (b)

Third-Order Model with One Independent Variable

$$E(y) = \beta_0 + \beta_1 x + \beta_2 x^2 + \beta_3 x^3$$

Interpretation of model parameters

β_0: y-intercept; the value of $E(y)$ when $x = 0$

β_1: Shift parameter (shifts the polynomial right or left on the x-axis)

β_2: Rate of curvature

β_3: The magnitude of β_3 controls the rate of reversal of curvature for the polynomial

---

Example 5.2

To operate efficiently, power companies must be able to predict the peak power load at their various stations. Peak power load is the maximum amount of power that must be generated each day to meet demand. A power company wants to use daily high temperature, x, to model daily peak power load, y, during the summer months when demand is greatest. Although the company expects peak load to increase as the temperature increases, the *rate* of increase in $E(y)$ might not remain constant as x increases. For example, a 1-unit increase in high temperature from 100°F to 101°F might result in a larger increase in power demand than would a 1-unit increase from 80°F to 81°F. Therefore, the company postulates that the model for $E(y)$ will include a second-order (quadratic) term and, possibly, a third-order (cubic) term.

A random sample of 25 summer days is selected and both the peak load (measured in megawatts) and high temperature (in degrees) recorded for each day. The data are listed in Table 5.1.

(a) Construct a scatterplot for the data. What type of model is suggested by the plot?

(b) Fit the third-order model, $E(y) = \beta_0 + \beta_1 x + \beta_2 x^2 + \beta_3 x^3$, to the data. Is there evidence that the cubic term, $\beta_3 x^3$, contributes information for the prediction of peak power load? Test at $\alpha = .05$.

(c) Fit the second-order model, $E(y) = \beta_0 + \beta_1 x + \beta_2 x^2$, to the data. Test the hypothesis that the power load increases at an increasing rate with temperature. Use $\alpha = .05$.

(d) Give the prediction equation for the second-order model, part c. Are you satisfied with using this model to predict peak power loads?

POWERLOADS

Table 5.1 Power load data

Temperature °F	Peak Load megawatts	Temperature °F	Peak Load megawatts	Temperature °F	Peak Load megawatts
94	136.0	106	178.2	76	100.9
96	131.7	67	101.6	68	96.3
95	140.7	71	92.5	92	135.1
108	189.3	100	151.9	100	143.6
67	96.5	79	106.2	85	111.4
88	116.4	97	153.2	89	116.5
89	118.5	98	150.1	74	103.9
84	113.4	87	114.7	86	105.1
90	132.0				

Solution

(a) The scatterplot of the data, produced using MINITAB, is shown in Figure 5.5. The nonlinear, upward-curving trend indicates that a second-order model would likely fit the data well.

(b) The third-order model is fit to the data using MINITAB and the resulting printout is shown in Figure 5.6. The p-value for testing

$$H_0: \beta_3 = 0$$

$$H_a: \beta_3 \neq 0$$

highlighted on the printout is .911. Since this value exceeds $\alpha = .05$, there is insufficient evidence of a third-order relationship between peak load and high temperature. Consequently, we will drop the cubic term, $\beta_3 x^3$, from the model.

Figure 5.5 MINITAB scatterplot for power load data

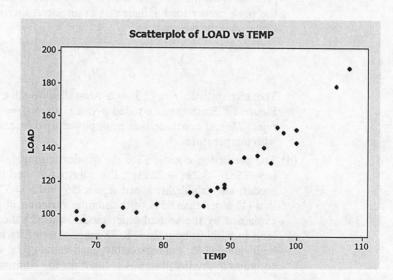

Figure 5.6 MINITAB output for third-order model of power load

```
The regression equation is
LOAD = 331 - 6.4 TEMP + 0.038 TEMP2 +0.000084 TEMP3

Predictor       Coef      SE Coef        T        P
Constant       331.3        477.1     0.69    0.495
TEMP           -6.39        16.79    -0.38    0.707
TEMP2         0.0378       0.1945     0.19    0.848
TEMP3      0.0000843    0.0007426     0.11    0.911

S = 5.501      R-Sq = 95.9%      R-Sq(adj) = 95.4%

Analysis of Variance

Source           DF         SS        MS        F        P
Regression        3    15012.2    5004.1   165.36    0.000
Residual Error   21      635.5      30.3
Total            24    15647.7
```

Figure 5.7 MINITAB output for second-order model of power load

```
The regression equation is
LOAD = 385 - 8.29 TEMP + 0.0598 TEMP2

Predictor       Coef      SE Coef        T        P
Constant      385.05        55.17     6.98    0.000
TEMP          -8.293        1.299    -6.38    0.000
TEMP2       0.059823     0.007549     7.93    0.000

S = 5.376      R-Sq = 95.9%      R-Sq(adj) = 95.6%

Analysis of Variance

Source           DF         SS        MS        F        P
Regression        2    15011.8    7505.9   259.69    0.000
Residual Error   22      635.9      28.9
Total            24    15647.7
```

(c) The second-order model is fit to the data using MINITAB and the resulting printout is shown in Figure 5.7. For this quadratic model, if β_2 is positive, then the peak power load y increases at an increasing rate with temperature x. Consequently, we test

$$H_0: \beta_2 = 0$$

$$H_a: \beta_2 > 0$$

The test statistic, $t = 7.93$, and two-tailed p-value are both highlighted in Figure 5.7. Since the one-tailed p-value, $p = 0/2 = 0$, is less than $\alpha = .05$, we reject H_0 and conclude that peak power load increases at an increasing rate with temperature.

(d) The prediction equation for the quadratic model, highlighted in Figure 5.7, is $\hat{y} = 385 - 8.29x + .0598x^2$. The adjusted-$R^2$ and standard deviation for the model, also highlighted, are $R_a^2 = .956$ and $s = 5.376$. These values imply that (1) more than 95% of the sample variation in peak power loads can be explained by the second-order model, and (2) the model can predict peak load to within about $2s = 10.75$ megawatts of its true value. Based on this high value of R_a^2 and reasonably small value of $2s$, we recommend using this equation to predict peak power loads for the company.

5.3   Exercises

5.7   Order of polynomials.   The accompanying graphs depict pth-order polynomials for one independent variable.

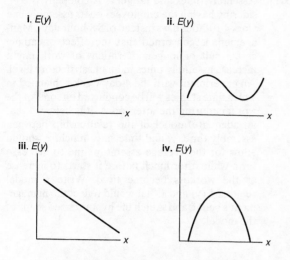

i. $E(y)$ ii. $E(y)$

iii. $E(y)$ iv. $E(y)$

(a) For each graph, identify the order of the polynomial.
(b) Using the parameters β_0, β_1, β_2, etc., write an appropriate model relating $E(y)$ to x for each graph.
(c) The signs ($+$ or $-$) of many of the parameters in the models of part b can be determined by examining the graphs. Give the signs of those parameters that can be determined.

5.8   Cooling method for gas turbines.   The *Journal of Engineering for Gas Turbines and Power* (January 2005) published a study of a high-pressure inlet fogging method for a gas turbine engine. The heat rate (kilojoules per kilowatt per hour) was measured for each in a sample of 67 gas turbines augmented with high-pressure inlet fogging. In addition, several other variables were measured, including cycle speed (revolutions per minute), inlet temperature (°C), exhaust gas temperature (°C), cycle pressure ratio, and air mass flow rate (kilograms per second). The data are saved in the GASTURBINE file. (The first and last five observations are listed in the table.) Consider using these variables as predictors of heat rate (y) in a regression model. Construct scatterplots relating heat rate to each of the independent variables. Based on the graphs, hypothesize a polynomial model relating y to each independent variable.

5.9   Study of tree frogs.   The optomotor responses of tree frogs were studied in the *Journal of Experimental Zoology* (September 1993). Microspectrophotometry was used to measure the threshold quantal flux (the light intensity at which the optomotor response was first observed) of tree frogs

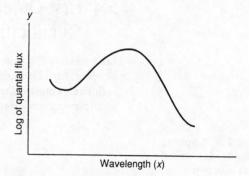

Log of quantal flux

Wavelength (x)

GASTURBINE (Data for first and last five gas turbines shown)

RPM	CPRATIO	INLET-TEMP	EXH-TEMP	AIRFLOW	HEATRATE
27245	9.2	1134	602	7	14622
14000	12.2	950	446	15	13196
17384	14.8	1149	537	20	11948
11085	11.8	1024	478	27	11289
14045	13.2	1149	553	29	11964
⋮					
18910	14.0	1066	532	8	12766
3600	35.0	1288	448	152	8714
3600	20.0	1160	456	84	9469
16000	10.6	1232	560	14	11948
14600	13.4	1077	536	20	12414

Source: Bhargava, R., and Meher-Homji, C. B. "Parametric analysis of existing gas turbines with inlet evaporative and overspray fogging," *Journal of Engineering for Gas Turbines and Power*, Vol. 127, No. 1, Jan. 2005.

tested at different spectral wavelengths. The data revealed the relationship between the log of quantal flux (y) and wavelength (x) shown in the graph on p. 271. Hypothesize a model for $E(y)$ that corresponds to the graph.

5.10 Tire wear and pressure. Underinflated or overinflated tires can increase tire wear and decrease gas mileage. A new tire was tested for wear at different pressures with the results shown in the table.

TIRES2

PRESSURE	MILEAGE
x, pounds per square inch	y, thousands
30	29
31	32
32	36
33	38
34	37
35	33
36	26

(a) Graph the data in a scatterplot.
(b) If you were given the information for $x = 30$, 31, 32, and 33 only, what kind of model would you suggest? For $x = 33$, 34, 35, and 36? For all the data?

5.11 Assembly times and fatigue. A company is considering having the employees on its assembly line work 4 10-hour days instead of 5 8-hour days. Management is concerned that the effect of fatigue as a result of longer afternoons of work might increase assembly times to an unsatisfactory level. An experiment with the 4-day week is planned in which time studies will be conducted on some of the workers during the afternoons. It is believed that an adequate model of the relationship between assembly time, y, and time since lunch, x, should allow for the average assembly time to decrease for a while after lunch before it starts to increase as the workers become tired. Write a model relating $E(y)$ and x that would reflect the management's belief, and sketch the hypothesized shape of the model.

5.4 First-Order Models with Two or More Quantitative Independent Variables

Like models for a single independent variable, models with two or more independent variables are classified as first-order, second-order, and so forth, but it is difficult (most often impossible) to graph the response because the plot is in a multidimensional space. For example, with one quantitative independent variable,

Figure 5.8 Response surface for first-order model with two quantitative independent variables

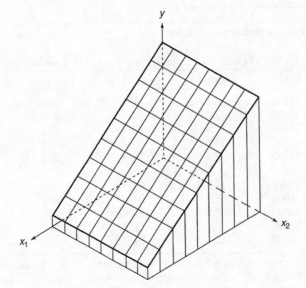

x, the response y traces a curve. But for two quantitative independent variables, x_1 and x_2, the plot of y traces a surface over the x_1, x_2-plane (see Figure 5.8). For three or more quantitative independent variables, the response traces a surface in a four- or higher-dimensional space. For these, we can construct two-dimensional graphs of y versus one independent variable for fixed levels of the other independent variables, or three-dimensional plots of y versus two independent variables for fixed levels of the remaining independent variables, but this is the best we can do in providing a graphical description of a response.

A **first-order model** in k quantitative variables is a first-order polynomial in k independent variables. For $k = 1$, the graph is a straight line. For $k = 2$, the response surface is a plane (usually tilted) over the x_1, x_2-plane.

If we use a first-order polynomial to model a response, we are assuming that there is no curvature in the response surface and that the variables affect the response independently of each other. For example, suppose the true relationship between the mean response and the independent variables x_1 and x_2 is given by the equation

$$E(y) = 1 + 2x_1 + x_2$$

First-Order Model in k Quantitative Independent Variables

$$E(y) = \beta_0 + \beta_1 x_1 + \beta_2 x_2 + \cdots + \beta_k x_k$$

where $\beta_0, \beta_1, \ldots, \beta_k$ are unknown parameters that must be estimated.

Interpretation of model parameters

β_0: y-intercept of $(k+1)$-dimensional surface; the value of $E(y)$ when $x_1 = x_2 = \cdots = x_k = 0$

β_1: Change in $E(y)$ for a 1-unit increase in x_1, when $x_2, x_3, \ldots, x_k$ are held fixed

β_2: Change in $E(y)$ for a 1-unit increase in x_2, when $x_1, x_3, \ldots, x_k$ are held fixed

$\vdots$

β_k: Change in $E(y)$ for a 1-unit increase in x_k, when $x_1, x_2, \ldots, x_{k-1}$ are held fixed

---

As in Section 4.3, we will graph this expression for selected values of x_2, say, $x_2 = 1, 2,$ and 3. The graphs, shown in Figure 5.9, are called **contour lines**. You can see from Figure 5.9 that regardless of the value of x_2, $E(y)$ graphs as a straight line with a slope of 2. Changing x_2 changes only the y-intercept (the constant in the equation). Consequently, assuming that a first-order model will adequately model a response is equivalent to assuming that a 1-unit change in one independent variable will have the same effect on the mean value of y regardless of the levels of the other independent variables. That is, *the contour lines will be parallel*. In Chapter 4, we stated that independent variables that have this property *do not interact*.

Except in cases where the ranges of levels for all independent variables are very small, the implication of no curvature in the response surface and the independence of variable effects on the response restrict the applicability of first-order models.

Figure 5.9 Contour lines of $E(y)$ for $x_2 = 1, 2, 3$ (first-order model)

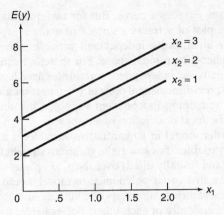

5.5 Second-Order Models with Two or More Quantitative Independent Variables

Second-order models with two or more independent variables permit curvature in the response surface. One important type of second-order term accounts for **interaction** between two variables.* Consider the two-variable model

$$E(y) = \beta_0 + \beta_1 x_1 + \beta_2 x_2 + \beta_3 x_1 x_2$$

This interaction model traces a ruled surface (twisted plane) in a three-dimensional space (see Figure 5.10). The second-order term $\beta_3 x_1 x_2$ is called the **interaction term**, and it permits the contour lines to be *nonparallel*.

Figure 5.10 Response surface for an interaction model (second-order)

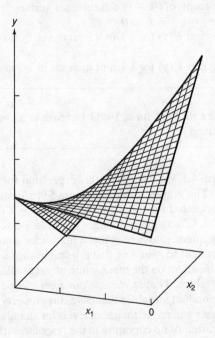

---

* The order of a term involving two or more *quantitative* independent variables is equal to the sum of their exponents. Thus, $\beta_3 x_1 x_2$ is a second-order term, as is $\beta_4 x_1^2$. A term of the form $\beta_i x_1 x_2 x_3$ is a third-order term.

In Section 4.10, we demonstrated the effect of interaction on the model:

$$E(y) = 1 + 2x_1 - x_2 + x_1x_2$$

The slope of the line relating x_1 to $E(y)$ is $\beta_1 + \beta_3x_2 = 2 + x_2$, while the y-intercept is $\beta_0 + \beta_2x_2 = 1 - x_2$.

Thus, when interaction is present in the model, both the y-intercept and the slope change as x_2 changes. Consequently, *the contour lines are not parallel. The presence of an interaction term implies that the effect of a 1-unit change in one independent variable will depend on the level of the other independent variable.* The contour lines for $x_2 = 1$, 2, and 3 are reproduced in Figure 5.11. You can see that when $x_2 = 1$, the line has slope $2 + 1 = 3$ and y-intercept $1 - 1 = 0$. But when $x_2 = 3$, the slope is $2 + 3 = 5$ and the y-intercept is $1 - 3 = -2$.

Figure 5.11 Contour lines of $E(y)$ for $x_2 = 1, 2, 3$ (first-order model plus interaction)

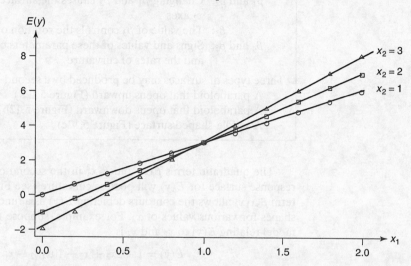

Interaction (Second-Order) Model with Two Independent Variables

$$E(y) = \beta_0 + \beta_1x_1 + \beta_2x_2 + \beta_3x_1x_2$$

Interpretation of Model Parameters

β_0: y-intercept; the value of $E(y)$ when $x_1 = x_2 = 0$

β_1 and β_2:   Changing β_1 and β_2 causes the surface to shift along the x_1 and x_2 axes

β_3:   Controls the rate of twist in the ruled surface (see Figure 5.10)

When one independent variable is held fixed, the model produces straight lines with the following slopes:

$\beta_1 + \beta_3x_2$:   Change in $E(y)$ for a 1-unit increase in x_1, when x_2 is held fixed

$\beta_2 + \beta_3x_1$:   Change in $E(y)$ for a 1-unit increase in x_2, when x_1 is held fixed

---

Definition 5.2 Two variables x_1 and x_2 are said to **interact** if the change in $E(y)$ for a 1-unit change in x_1 (when x_2 is held fixed) is dependent on the value of x_2.

We can introduce even more flexibility into a model by the addition of quadratic terms. The complete second-order model includes the constant β_0, all linear (first-order) terms, all two-variable interactions, and all quadratic terms. This complete second-order model for two quantitative independent variables is shown in the box.

Complete Second-Order Model with Two Independent Variables

$$E(y) = \beta_0 + \beta_1 x_1 + \beta_2 x_2 + \beta_3 x_1 x_2 + \beta_4 x_1^2 + \beta_5 x_2^2$$

Interpretation of Model Parameters

β_0: y-intercept; the value of $E(y)$ when $x_1 = x_2 = 0$

β_1 and β_2: Changing β_1 and β_2 causes the surface to shift along the x_1 and x_2 axes

β_3: The value of β_3 controls the rotation of the surface

β_4 and β_5: Signs and values of these parameters control the type of surface and the rates of curvature

Three types of surfaces may be produced by a second-order model.*
A paraboloid that opens upward (Figure 5.12a)
A paraboloid that opens downward (Figure 5.12b)
A saddle-shaped surface (Figure 5.12c)

The quadratic terms $\beta_4 x_1^2$ and $\beta_5 x_2^2$ in the second-order model imply that the response surface for $E(y)$ will possess curvature (see Figure 5.12). The interaction term $\beta_3 x_1 x_2$ allows the contours depicting $E(y)$ as a function of x_1 to have different shapes for various values of x_2. For example, suppose the complete second-order model relating $E(y)$ to x_1 and x_2 is

$$E(y) = 1 + 2x_1 + x_2 - 10x_1 x_2 + x_1^2 - 2x_2^2$$

Figure 5.12 Graphs of three second-order surfaces

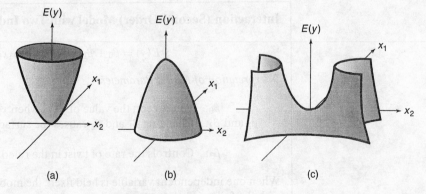

Then the contours of $E(y)$ for $x_2 = -1, 0$, and 1 are shown in Figure 5.13. When we substitute $x_2 = -1$ into the model, we get

$$E(y) = 1 + 2x_1 + x_2 - 10x_1 x_2 + x_1^2 - 2x_2^2$$
$$= 1 + 2x_1 - 1 - 10x_1(-1) + x_1^2 - 2(-1)^2$$
$$= -2 + 12x_1 + x_1^2$$

---

* The saddle-shaped surface (Figure 5.12c) is produced when $\beta_3^2 > 4\beta_4\beta_5$. For $\beta_3^2 < 4\beta_4\beta_5$, the paraboloid opens upward (Figure 5.12a) when $\beta_4 + \beta_5 > 0$ and opens downward (Figure 5.12b) when $\beta_4 + \beta_5 < 0$.

For $x_2 = 0$,

$$E(y) = 1 + 2x_1 + (0) - 10x_1(0) + x_1^2 - 2(0)^2$$
$$= 1 + 2x_1 + x_1^2$$

Similarly, for $x_2 = 1$,

$$E(y) = -8x_1 + x_1^2$$

Note how the shapes of the three contour curves in Figure 5.13 differ, indicating that the β parameter associated with the $x_1 x_2$ (interaction) term differs from 0.

The complete second-order model for three independent variables is shown in the next box.

Figure 5.13 Contours of $E(y)$ for $x_2 = -1, 0, 1$ (complete second-order model)

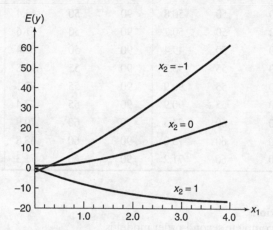

Complete Second-Order Model with Three Quantitative Independent Variables

$$E(y) = \beta_0 + \beta_1 x_1 + \beta_2 x_2 + \beta_3 x_3 + \beta_4 x_1 x_2 + \beta_5 x_1 x_3 + \beta_6 x_2 x_3$$
$$+ \beta_7 x_1^2 + \beta_8 x_2^2 + \beta_9 x_3^2$$

where $\beta_0, \beta_1, \beta_2, \ldots, \beta_9$ are unknown parameters that must be estimated.

This second-order model in three independent variables demonstrates how you would write a second-order model for any number of independent variables. Always include the constant β_0 and then all first-order terms corresponding to $x_1, x_2, \ldots$. Then add the interaction terms for all pairs of independent variables $x_1 x_2, x_1 x_3, x_2 x_3, \ldots$. Finally, include the second-order terms $x_1^2, x_2^2, \ldots$.

For any number, say, p, of quantitative independent variables, the response traces a surface in a $(p + 1)$-dimensional space, which is impossible to visualize. In spite of this handicap, the prediction equation can still tell us much about the phenomenon being studied.

Example 5.3

Many companies manufacture products that are at least partially produced using chemicals (e.g., steel, paint, gasoline). In many instances, the quality of the finished product is a function of the temperature and pressure at which the chemical reactions take place.

Suppose you wanted to model the quality, y, of a product as a function of the temperature, x_1, and the pressure, x_2, at which it is produced. Four inspectors independently assign a quality score between 0 and 100 to each product, and then the quality, y, is calculated by averaging the four scores. An experiment is conducted by varying temperature between 80° and 100°F and pressure between 50 and 60 pounds per square inch (psi). The resulting data ($n = 27$) are given in Table 5.2. Fit a complete second-order model to the data and sketch the response surface.

🔵 PRODQUAL

Table 5.2 Temperature, pressure, and quality of the finished product

$x_1,°F$	x_2, psi	y	$x_1,°F$	x_2, psi	y	$x_1,°F$	x_2, psi	y
80	50	50.8	90	50	63.4	100	50	46.6
80	50	50.7	90	50	61.6	100	50	49.1
80	50	49.4	90	50	63.4	100	50	46.4
80	55	93.7	90	55	93.8	100	55	69.8
80	55	90.9	90	55	92.1	100	55	72.5
80	55	90.9	90	55	97.4	100	55	73.2
80	60	74.5	90	60	70.9	100	60	38.7
80	60	73.0	90	60	68.8	100	60	42.5
80	60	71.2	90	60	71.3	100	60	41.4

Solution

The complete second-order model is

$$E(y) = \beta_0 + \beta_1 x_1 + \beta_2 x_2 + \beta_3 x_1 x_2 + \beta_4 x_1^2 + \beta_5 x_2^2$$

The data in Table 5.2 were used to fit this model. A portion of the SAS output is shown in Figure 5.14.

The least squares model is

$$\hat{y} = -5{,}127.90 + 31.10 x_1 + 139.75 x_2 - .146 x_1 x_2 - .133 x_1^2 - 1.14 x_2^2$$

Figure 5.14 SAS output for complete second-order model of quality

Dependent Variable: QUALITY

Analysis of Variance

Source	DF	Sum of Squares	Mean Square	F Value	Pr > F
Model	5	8402.26454	1680.45291	596.32	<.0001
Error	21	59.17843	2.81802		
Corrected Total	26	8461.44296			

Root MSE	1.67870	R-Square	0.9930
Dependent Mean	66.96296	Adj R-Sq	0.9913
Coeff Var	2.50690		

Parameter Estimates

Variable	DF	Parameter Estimate	Standard Error	t Value	Pr > |t|
Intercept	1	-5127.89907	110.29601	-46.49	<.0001
TEMP	1	31.09639	1.34441	23.13	<.0001
PRESSURE	1	139.74722	3.14005	44.50	<.0001
TEMPRESS	1	-0.14550	0.00969	-15.01	<.0001
TEMPSQ	1	-0.13339	0.00685	-19.46	<.0001
PRESSQ	1	-1.14422	0.02741	-41.74	<.0001

Figure 5.15 Graph of second-order least squares model for Example 5.3

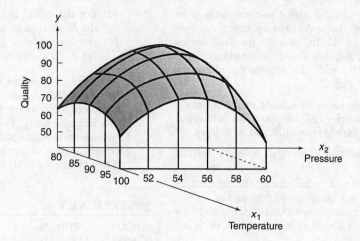

A three-dimensional graph of this prediction model is shown in Figure 5.15. The mean quality seems to be greatest for temperatures of about 85°–90°F and for pressures of about 55–57 pounds per square inch.[†] Further experimentation in these ranges might lead to a more precise determination of the optimal temperature–pressure combination.

A look at the adjusted coefficient of determination, $R_a^2 = .991$, the F-value for testing the entire model, $F = 596.32$, and the p-value for the test, $p = .0001$ (shaded in Figure 5.14), leaves little doubt that the complete second-order model is useful for explaining mean quality as a function of temperature and pressure. This, of course, will not always be the case. The additional complexity of second-order models is worthwhile only if a better model results. To determine whether the quadratic terms are important, we would test $H_0: \beta_4 = \beta_5 = 0$ using the F-test for comparing nested models outlined in Section 4.13.

5.5 Exercises

5.12 Signal-to-noise ratios of seismic waves. Chinese scientists have developed a method of boosting the signal-to-noise ratio of a seismic wave (*Chinese Journal of Geophysics*, Vol. 49, 2006). Suppose an exploration seismologist wants to develop a model to estimate the average signal-to-noise ratio of an earthquake's seismic wave, y, as a function of two independent variables:

$$x_1 = \text{Frequency (cycles per second)}$$

$$x_2 = \text{Amplitude of the wavelet}$$

(a) Identify the independent variables as quantitative or qualitative.

(b) Write the first-order model for $E(y)$.

(c) Write a model for $E(y)$ that contains all first-order and interaction terms. Sketch typical response curves showing $E(y)$, the mean signal-to-noise ratio, versus x_2, the amplitude of the wavelet, for different values of x_1 (assume that x_1 and x_2 interact).

(d) Write the complete second-order model for $E(y)$.

5.13 Signal-to-noise ratios of seismic waves (Cont'd). Refer to Exercise 5.12. Suppose the model from part c is fit, with the following result:

$$\hat{y} = 1 + .05x_1 + x_2 + .05x_1x_2$$

[†] Students with knowledge of calculus should note that we can determine the exact temperature and pressure that maximize quality in the least squares model by solving $\partial \hat{y}/\partial x_1 = 0$ and $\partial \hat{y}/\partial x_2 = 0$ for x_1 and x_2. These estimated optional values are $x_1 = 86.25°F$ and $x_2 = 55.58$ pounds per square inch. Remember, however, that these are only sample estimates of the coordinates for the optional value.

Graph the estimated signal-to-noise ratio $\hat{y}$ as a function of the wavelet amplitude, x_2, over the range $x_2 = 10$ to $x_2 = 50$ for frequencies of $x_1 = 1$, 5, and 10. Do these functions agree (approximately) with the graphs you drew for Exercise 5.12, part c?

5.14 **Signal-to-noise ratios of seismic waves (Cont'd).** Refer to Exercise 5.12. Suppose an additional independent variable is considered, as follows:

$x_3 = $ Time interval (seconds) between seismic waves

(a) Write the first-order model plus interaction for $E(y)$ as a function of x_1, x_2, and x_3.

(b) Write the complete second-order model plus interaction for $E(y)$ as a function of x_1, x_2, and x_3.

5.15 **Sustained-release drug study.** Researchers at the Upjohn Company utilized multiple regression analysis in the development of a sustained-release tablet. One of the objectives of the research was to develop a model relating the dissolution y of a tablet (i.e., the percentage of the tablet dissolved over a specified period of time) to the following independent variables:

$x_1 = $ Excipient level (i.e., amount of nondrug ingredient in the tablet)

$x_2 = $ Process variable (e.g., machine setting under which tablet is processed)

(a) Write the complete second-order model for $E(y)$.

(b) Write a model that hypothesizes straight-line relationships between $E(y)$, x_1, and x_2. Assume that x_1 and x_2 do not interact.

(c) Repeat part b, but add interaction to the model.

(d) For the model in part c, what is the slope of the $E(y)$, x_1 line for fixed x_2?

(e) For the model in part c, what is the slope of the $E(y)$, x_2 line for fixed x_1?

5.16 **Earnings of Mexican street vendors.** Refer to the *World Development* (February 1998) study of street vendors in the city of Puebla, Mexico, Exercise 4.6 (p. 184). Recall that the vendors' mean annual earnings, $E(y)$, was modeled as a first-order function of age (x_1) and hours worked (x_2). The data for the study are reproduced in the table.

STREETVEN

VENDOR NUMBER	ANNUAL EARNINGS, y	AGE, x_1	HOURS WORKED PER DAY, x_2
21	$2841	29	12
53	1876	21	8
60	2934	62	10
184	1552	18	10
263	3065	40	11
281	3670	50	11
354	2005	65	5
401	3215	44	8
515	1930	17	8
633	2010	70	6
677	3111	20	9
710	2882	29	9
800	1683	15	5
914	1817	14	7
997	4066	33	12

Source: Adapted from Smith, P. A., and Metzger, M. R. "The return to education: Street vendors in Mexico," *World Development*, Vol. 26, No. 2, Feb. 1998, pp. 289–296.

MINITAB Output for Exercise 5.16

```
The regression equation is
EARNINGS = 606 + 120 AGE - 140 HOURS + 2.66 AGEHOURS - 1.57 AGESQ + 8.1 HOURSQ

Predictor      Coef     SE Coef        T       P
Constant        606        2331     0.26   0.801
AGE          119.68       64.58     1.85   0.097
HOURS        -139.8       491.6    -0.28   0.783
AGEHOURS      2.662       3.420     0.78   0.456
AGESQ       -1.5710      0.6911    -2.27   0.049
HOURSQ         8.08       26.73     0.30   0.769

S = 482.8      R-Sq = 75.7%     R-Sq(adj) = 62.1%

Analysis of Variance

Source           DF        SS        MS       F       P
Regression        5   6520245   1304049    5.59   0.013
Residual Error    9   2098183    233131
Total            14   8618428
```

(a) Write a complete second-order model for mean annual earnings, $E(y)$, as a function of age (x_1) and hours worked (x_2).

(b) The model was fit to the data using MINITAB. Find the least squares prediction equation on the accompanying printout (bottom, p. 280).

(c) Is the model statistically useful for predicting annual earnings? Test using $\alpha = .05$.

(d) How would you test the hypothesis that the second-order terms in the model are not necessary for predicting annual earnings?

(e) Carry out the test, part d. Interpret the results.

🌐 QUASAR

5.17 Deep space survey of quasars. Refer to the *Astronomical Journal* (July 1995) study of quasars detected by a deep space survey, Exercise 4.11 (p. 186). Recall that several quantitative independent variables were used to model the quasar characteristic, rest frame equivalent width (y). The data for 25 quasars are saved in the QUASAR file.

(a) Write a complete second-order model for y as a function of redshift (x_1), lineflux (x_2), and AB_{1450} (x_4).

(b) Fit the model, part a, to the data using a statistical software package. Is the overall model statistically useful for predicting y?

(c) Conduct a test to determine if any of the curvilinear terms in the model, part a, are statistically useful predictors of y.

5.18 Goal congruence in top management teams. Do chief executive officers (CEOs) and their top managers always agree on the goals of the company? Goal importance congruence between CEOs and vice presidents (VPs) was studied in the *Academy of Management Journal* (February 2008). The researchers used regression to model a VP's attitude toward the goal of improving efficiency (y) as a function of the two quantitative independent variables, level of CEO leadership (x_1) and level

of congruence between the CEO and the VP (x_2). A complete second-order model in x_1 and x_2 was fit to data collected for $n = 517$ top management team members at U.S. credit unions.

(a) Write the complete second-order model for $E(y)$.

(b) The coefficient of determination for the model, part a, was reported as $R^2 = .14$. Interpret this value.

(c) The estimate of the β-value for the $(x_2)^2$ term in the model was found to be negative. Interpret this result, practically.

(d) A t-test on the β-value for the interaction term in the model, x_1x_2, resulted in a p-value of .02. Practically interpret this result, using $\alpha = .05$.

🌐 GASTURBINE

5.19 Cooling method for gas turbines. Refer to the *Journal of Engineering for Gas Turbines and Power* (January 2005) study of a high-pressure inlet fogging method for a gas turbine engine, Exercise 5.8 (p. 271). Consider a model for heat rate (kilojoules per kilowatt per hour) of a gas turbine as a function of cycle speed (revolutions per minute) and cycle pressure ratio. Recall that the data are saved in the GASTURBINE file.

(a) Write a complete second-order model for heat rate (y).

(b) Fit the model to the data and give the least squares prediction equation.

(c) Conduct a global F-test for overall model adequacy.

(d) Based on the prediction equation, graph the relationship between heat rate and cycle pressure ratio when cycle speed is held constant at 5,000 rpm.

(e) Repeat part d when cycle speed is held constant at 15,000 rpm.

(f) Compare the two graphs, parts d and e. What do you observe?

5.6 Coding Quantitative Independent Variables (Optional)

In fitting higher-order polynomial regression models (e.g., second- or third-order models), it is often a good practice to code the quantitative independent variables. For example, suppose one of the independent variables in a regression analysis is level of competence in performing a task, C, measured on a 20-point scale, and C is observed at three levels: 5, 10, and 15. We can code (or transform) the competence measurements using the formula

$$x = \frac{C - 10}{5}$$

Then the coded levels $x = -1, 0$, and 1 correspond to the original C levels 5, 10, and 15.

In a general sense, *coding* means transforming a set of independent variables (qualitative or quantitative) into a new set of independent variables. For example, if we observe two independent variables,

$$C = \text{Competence level}$$

$$S = \text{Satisfaction level}$$

then we can transform C and S into two new coded variables, x_1 and x_2, where x_1 and x_2 are related to C and S by two functional equations:

$$x_1 = f_1(C, S) \qquad x_2 = f_2(C, S)$$

The functions f_1 and f_2, which are frequently expressed as equations, establish a one-to-one correspondence between combinations of levels of C and S with combinations of the coded values of x_1 and x_2.

Since qualitative independent variables are not numerical, it is necessary to code their values to fit the regression model. (We demonstrate the coding scheme in Section 5.7.) However, you might ask why we would bother to code the quantitative independent variables. There are two related reasons for coding quantitative variables. At first glance, it would appear that a computer would be oblivious to the values assumed by the independent variables in a regression analysis, but this is not the case. To calculate the estimates of the model parameters using the method of least squares, the computer must invert a matrix of numbers, called the **coefficient** (or **information**) **matrix** (see Appendix B). Considerable rounding error may occur during the inversion process if the numbers in the coefficient matrix vary greatly in absolute value. This can produce sizable errors in the computed values of the least squares estimates, $\hat{\beta}_0, \hat{\beta}_1, \hat{\beta}_2, \ldots$. Coding makes it computationally easier for the computer to invert the matrix, thus leading to more accurate estimates.

A second reason for coding quantitative variables pertains to a problem we will discuss in detail in Chapter 7: the problem of independent variables (x's) being intercorrelated (called **multicollinearity**). When polynomial regression models (e.g., second-order models) are fit, the problem of multicollinearity is unavoidable, especially when higher-order terms are fit. For example, consider the quadratic model

$$E(y) = \beta_0 + \beta_1 x + \beta_2 x^2$$

If the range of the values of x is narrow, then the two variables, $x_1 = x$ and $x_2 = x^2$, will generally be highly correlated. As we point out in Chapter 7, the likelihood of rounding errors in the regression coefficients is increased in the presence of these highly correlated independent variables.

The following procedure is a good way to cope with the problem of rounding error:

1. Code the quantitative variable so that the new coded origin is in the center of the coded values. For example, by coding competence level, C, as

$$x = \frac{C - 10}{5}$$

we obtain coded values $-1, 0, 1$. This places the coded origin, 0, in the middle of the range of coded values (-1 to 1).

2. Code the quantitative variable so that the range of the coded values is approximately the same for all coded variables. You need not hold exactly to this requirement. The range of values for one independent variable could be double

or triple the range of another without causing any difficulty, but it is not desirable to have a sizable disparity in the ranges, say, a ratio of 100 to 1.

When the data are observational (the values assumed by the independent variables are uncontrolled), the coding procedure described in the box satisfies, reasonably well, these two requirements. The coded variable u is similar to the standardized normal z statistic of Section 1.6. Thus, the u-value is the deviation (the distance) between an x-value and the mean of the x-values, $\bar{x}$, expressed in units of s_x.* Since we know that most (approximately 95%) measurements in a set will lie within 2 standard deviations of their mean, it follows that most of the coded u-values will lie in the interval -2 to $+2$.

Coding Procedure for Observational Data

Let

$$x = \text{Uncoded quantitative independent variable}$$

$$u = \text{Coded quantitative independent variable}$$

Then if x takes values $x_1, x_2, \ldots, x_n$ for the n data points in the regression analysis, let

$$u_i = \frac{x_i - \bar{x}}{s_x}$$

where s_x is the standard deviation of the x-values, that is,

$$s_x = \sqrt{\frac{\sum_{i=1}^{n}(x_i - \bar{x})^2}{n - 1}}$$

If you apply this coding to each quantitative variable, the range of values for each will be approximately -2 to $+2$. The variation in the absolute values of the elements of the coefficient matrix will be moderate, and rounding errors generated in finding the inverse of the matrix will be reduced. Additionally, the correlation between x and x^2 will be reduced.†

Example 5.4

Carbon dioxide–baited traps are typically used by entomologists to monitor mosquito populations. An article in the *Journal of the American Mosquito Control Association* (March 1995) investigated whether temperature influences the number of mosquitoes caught in a trap. Six mosquito samples were collected on each of 9 consecutive days. For each day, two variables were measured: $x = $ average temperature (in degrees Centigrade) and $y = $ mosquito catch ratio (the number of mosquitoes caught in each sample divided by the largest sample caught). The data are reported in Table 5.3.

The researchers are interested in relating catch ratio y to average temperature x. Suppose we consider using a quadratic model, $E(y) = \beta_0 + \beta_1 x + \beta_2 x^2$.

---

* The divisor of the deviation, $x - \bar{x}$, need not equal s_x exactly. Any number approximately equal to s_x would suffice. Other candidate denominators are the range (R), $R/2$, and the interquartile range (IQR).
† Another by-product of coding is that the β coefficients of the model have slightly different interpretations. For example, in the model $E(y) = \beta_0 + \beta_1 u$, where $u = (x - 10)/5$, the change in y for every 1-unit increase in x is not β_1, but $\beta_1/5$. In general, for first-order models with coded independent quantitative variables, the slope associated with x_i is represented by β_i/s_{x_i}, where s_{x_i} is the divisor of the coded x_i.

MOSQUITO

Table 5.3 Data for Example 5.4

Date	Average Temperature, x	Catch Ratio, y
July 24	16.8	.66
25	15.0	.30
26	16.5	.46
27	17.7	.44
28	20.6	.67
29	22.6	.99
30	23.3	.75
31	18.2	.24
Aug. 1	18.6	.51

Source: Petric, D., et al. "Dependence of CO_2-baited suction trap captures on temperature variations," *Journal of the American Mosquito Control Association*, Vol. 11, No. 1, Mar. 1995, p. 8.

(a) Fit the model to the data in Table 5.3 and conduct t-tests on β_1 and β_2.

(b) Calculate and interpret the correlation between x and x^2.

(c) Give the equation relating the coded variable u to the temperature x using the coding system for observational data. Then calculate the coded values, u, for the $n = 9$ x-values and find the sum of the coded values. Also, find the correlation between u and u^2.

(d) Fit the model $E(y) = \beta_0 + \beta_1 u + \beta_2 u^2$ to the data. Compare the results to those in part a.

Solution

(a) We used MINITAB to fit the quadratic model to the data. The printout is shown in Figure 5.16. Note that the p-values (shaded) for testing H_0: $\beta_1 = 0$ and H_0: $\beta_2 = 0$ are .749 and .624, respectively. Consequently, a researcher may be tempted to conclude that temperature (x) is neither curvilinearly nor linearly related to catch ratio (y).

(b) The correlation coefficient for x and x^2, also obtained using MINITAB, is shown (highlighted) at the bottom of the printout in Figure 5.16. You can see that these two independent variables are highly correlated ($r = .998$). As we will address in Chapter 7, this high correlation case leads to extreme round-off errors in the parameter estimates as well as standard errors that are inflated. As a consequence, the t-values for testing the β parameters are smaller than expected, leading to nonsignificant results. Before concluding that temperature is unrelated to catch ratio, a prudent analyst will code the values of the quantitative variable in the model to reduce the level of correlation.

(c) The MINITAB printout, Figure 5.17, provides summary statistics for temperature, x. The values of $\bar{x}$ and s_x (highlighted) are

$$\bar{x} = 18.811 \quad \text{and} \quad s_x = 2.812.$$

Figure 5.16 MINITAB printout for the quadratic model, Example 5.4

Regression Analysis: RATIO versus TEMP, TEMPSQ

```
The regression equation is
RATIO = 1.09 - 0.119 TEMP + 0.00471 TEMPSQ


Predictor        Coef    SE Coef       T       P
Constant        1.091      3.380    0.32   0.758
TEMP          -0.1186     0.3537   -0.34   0.749
TEMPSQ       0.004705   0.009103    0.52   0.624


S = 0.170451   R-Sq = 60.4%   R-Sq(adj) = 47.2%


Analysis of Variance

Source           DF        SS        MS      F      P
Regression        2   0.26563   0.13282   4.57  0.062
Residual Error    6   0.17432   0.02905
Total             8   0.43996
```

Correlations: TEMP, TEMPSQ

```
Pearson correlation of TEMP and TEMPSQ = 0.998
P-Value = 0.000
```

Figure 5.17 MINITAB descriptive statistics for temperature, x

Descriptive Statistics: TEMP

```
Variable   N     Mean   StDev   Minimum   Median   Maximum
TEMP       9   18.811   2.812    15.000   18.200    23.300
```

Then the equation relating the coded variable u to x is

$$u = \frac{x - 18.8}{2.8}$$

For example, when temperature $x = 16.8$

$$u = \frac{x - 18.8}{2.8} = \frac{16.8 - 18.8}{2.8} = -.71$$

Similarly, when $x = 15.0$

$$u = \frac{x - 18.8}{2.8} = \frac{15.0 - 18.8}{2.8} = -1.36$$

Table 5.4 gives the coded values for all $n = 9$ observations. [*Note*: You can see that all the $n = 9$ values for u lie in the interval from -2 to $+2$.]

If you ignore rounding error, the sum of the $n = 9$ values for u will equal 0. This is because the sum of the deviations of a set of measurements about their mean is always equal to 0.

(d) We again used MINITAB to fit the quadratic model with coded temperature (u) to the data. The printout is shown in Figure 5.18. First, notice that the

Table 5.4 Coded values of x, Example 5.4

Temperature, x	Coded Values, u
16.8	−.71
15.0	−1.36
16.5	−.82
17.7	−.39
20.6	.64
22.6	1.36
23.3	1.61
18.2	−.21
18.6	−.07

Figure 5.18 MINITAB printout for the quadratic model with coded temperature

Correlations: U, USQ

```
Pearson correlation of U and USQ = 0.441
P-Value = 0.235
```

Regression Analysis: RATIO versus U, USQ

```
The regression equation is
RATIO = 0.525 + 0.164 U + 0.0372 USQ


Predictor     Coef   SE Coef      T      P
Constant   0.52469   0.08558   6.13  0.001
U          0.16423   0.06713   2.45  0.050
USQ        0.03721   0.07198   0.52  0.624


S = 0.170451   R-Sq = 60.4%   R-Sq(adj) = 47.2%


Analysis of Variance

Source          DF        SS        MS      F      P
Regression       2   0.26563   0.13282   4.57  0.062
Residual Error   6   0.17432   0.02905
Total            8   0.43996
```

correlation coefficient for u and u^2 (highlighted at the top of the printout) is .441. Thus, the potential of rounding errors on the parameter estimates should be greatly reduced. Now, examine the t-values and p-values (shaded) for testing H_0: $\beta_1 = 0$ and H_0: $\beta_2 = 0$. You can see that the results for testing the quadratic term, β_2, are identical as those shown in Figure 5.16. However, the test for the first-order term, β_1, is now statistically significant at $\alpha = .05$. Therefore, there is sufficient evidence of at least a linear relationship between catch ratio (y) and temperature (x). ∎

Other methods of coding quantitative variables have been developed to reduce rounding errors and multicollinearity. One of the more complex coding systems involves fitting **orthogonal polynomials**. An orthogonal system of coding guarantees that the coded independent variables will be uncorrelated. For a discussion of orthogonal polynomials, consult the references given at the end of this chapter.

5.6 Exercises

5.20 Tire wear and pressure. Suppose you want to use the coding system for observational data to fit a second-order model to the tire pressure–automobile mileage data in Exercise 5.10 (p. 272), which are repeated in the table.

TIRES2

PRESSURE	MILEAGE
x, pounds per square inch	y, thousands
30	29
31	32
32	36
33	38
34	37
35	33
36	26

(a) Give the equation relating the coded variable u to pressure, x, using the coding system for observational data.
(b) Calculate the coded values, u.
(c) Calculate the coefficient of correlation r between the variables x and x^2.
(d) Calculate the coefficient of correlation r between the variables u and u^2. Compare this value to the value computed in part c.
(e) Fit the model

$$E(y) = \beta_0 + \beta_1 u + \beta_2 u^2$$

using available statistical software. Interpret the results.

5.21 Comparing first- and second-year sales. As part of the first-year evaluation for new salespeople,

SALES

FIRST-YEAR SALES	SECOND-YEAR SALES
x, thousands of dollars	y, thousands of dollars
75.2	99.3
91.7	125.7
100.3	136.1
64.2	108.6
81.8	102.0
110.2	153.7
77.3	108.8
80.1	105.4

a large food-processing firm projects the second-year sales for each salesperson based on his or her sales for the first year. Data for $n = 8$ salespeople are shown in the table.

(a) Give the equation relating the coded variable u to first-year sales, x, using the coding system for observational data.
(b) Calculate the coded values, u.
(c) Calculate the coefficient of correlation r between the variables x and x^2.
(d) Calculate the coefficient of correlation r between the variables u and u^2. Compare this value to the value computed in part c.

5.22 Failure times of silicon wafer microchips. Refer to the National Semiconductor experiment with tin-lead solder bumps used to manufacture silicon wafer integrated circuit chips, Exercise 4.40

WAFER

TEMPERATURE (°C)	TIME TO FAILURE (HOURS)
165	200
162	200
164	1200
158	500
158	600
159	750
156	1200
157	1500
152	500
147	500
149	1100
149	1150
142	3500
142	3600
143	3650
133	4200
132	4800
132	5000
134	5200
134	5400
125	8300
123	9700

Source: Gee, S., & Nguyen, L. "Mean time to failure in wafer level-CSP packages with SnPb and SnAgCu solder bumps," International Wafer Level Packaging Conference, San Jose, CA, Nov. 3–4, 2005 (adapted from Figure 7).

(p. 207) Recall that the failure times of the microchips (in hours) was determined at different solder temperatures (degrees Centigrade). The data are reproduced in the table (p. 287). The researchers want to predict failure time (y) based on solder temperature (x) using the quadratic model, $E(y) = \beta_0 + \beta_1 x + \beta_2 x^2$. First, demonstrate the potential for extreme round-off errors in the parameter estimates for this model. Then, propose and fit an alternative model to the data, one that should greatly reduce the round-off error problem.

5.23 Study of tree frogs. Refer to the *Experimental Zoology* study of the optomotor responses of tree frogs, Exercise 5.9 (p. 271). Recall that the researchers modeled the light intensity at which optomotor response was first observed (y) as a function of spectral wavelength (x). In the solution to the exercise you hypothesized the third-order model, $E(y) = \beta_0 + \beta_1 x + \beta_2 x^2 + \beta_3 x^3$. Suppose data on $n = 10$ tree frogs were collected, with spectral wavelength (measured in microns) set at $x = 1$ for the first frog, $x = 2$ for the second, $x = 3$ for the third, ..., and $x = 10$ for the 10th frog.

(a) Demonstrate the potential for round-off errors in the parameter estimates by finding the correlation between x and x^2, x and x^3, and x^2 and x^3.

(b) Provide a formula for the coded independent variable u that will reduce level of multicollinearity in the independent variables.

(c) Refer to part b. Find the correlation between u and u^2, u and u^3, and u^2 and u^3. What do you observe? Is the level of correlation reduced for all three pairs?

5.7 Models with One Qualitative Independent Variable

Suppose we want to write a model for the mean performance, $E(y)$, of a diesel engine as a function of type of fuel. (For the purpose of explanation, we will ignore other independent variables that might affect the response.) Further suppose there are three fuel types available: a petroleum-based fuel (P), a coal-based fuel (C), and a blended fuel (B). The fuel type is a single qualitative variable with three levels corresponding to fuels P, C, and B. Note that with a qualitative independent variable, we cannot attach a quantitative meaning to a given level. All we can do is describe it.

To simplify our notation, let μ_P be the mean performance for fuel P, and let μ_C and μ_B be the corresponding mean performances for fuels C and B. Our objective is to write a single prediction equation that will give the mean value of y for the three fuel types. A coding scheme that yields useful β-interpretations is the following:

$$E(y) = \beta_0 + \beta_1 x_1 + \beta_2 x_2$$

where

$$x_1 = \begin{cases} 1 & \text{if fuel P is used} \\ 0 & \text{if not} \end{cases}$$

$$x_2 = \begin{cases} 1 & \text{if fuel C is used} \\ 0 & \text{if not} \end{cases}$$

The values of x_1 and x_2 for each of the three fuel types are shown in Table 5.5.

The variables x_1 and x_2 are not meaningful independent variables as in the case of the models containing quantitative independent variables. Instead, they are **dummy (indicator) variables** that make the model work. To see how they work, let $x_1 = 0$ and $x_2 = 0$. This condition will apply when we are seeking the mean response for fuel B (neither fuel P nor C is used; hence, it must be B). Then the mean value

Table 5.5 Mean response for the model with three diesel fuel types

Fuel Type	x_1	x_2	Mean Response, $E(y)$
Blended (B)	0	0	$\beta_0 = \mu_B$
Petroleum (P)	1	0	$\beta_0 + \beta_1 = \mu_P$
Coal (C)	0	1	$\beta_0 + \beta_2 = \mu_C$

of y when fuel B is used is

$$\mu_B = E(y) = \beta_0 + \beta_1(0) + \beta_2(0)$$
$$= \beta_0$$

This tells us that the mean performance level for fuel B is β_0. Or, it means that

$$\beta_0 = \mu_B.$$

Now suppose we want to represent the mean response, $E(y)$, when fuel P is used. Checking the dummy variable definitions, we see that we should let $x_1 = 1$ and $x_2 = 0$:

$$\mu_P = E(y) = \beta_0 + \beta_1(1) + \beta_2(0)$$
$$= \beta_0 + \beta_1$$

or, since $\beta_0 = \mu_B$,

$$\mu_P = \mu_B + \beta_1$$

Then it follows that the interpretation of β_1 is

$$\beta_1 = \mu_P - \mu_B$$

which is the difference in the mean performance levels for fuels P and B.

Finally, if we want the mean value of y when fuel C is used, we let $x_1 = 0$ and $x_2 = 1$:

$$\mu_C = E(y) = \beta_0 + \beta_1(0) + \beta_2(1)$$
$$= \beta_0 + \beta_2$$

or, since $\beta_0 = \mu_B$,

$$\mu_C = \mu_B + \beta_2$$

Then it follows that the interpretation of β_2 is

$$\beta_2 = \mu_C - \mu_B$$

Note that we were able to describe *three levels* of the qualitative variable with only *two dummy variables*, because the mean of the base level (fuel B, in this case) is accounted for by the intercept β_0.

Now, carefully examine the model for a single qualitative independent variable with three levels, because we will use exactly the same pattern for any number of levels. Arbitrarily select one level to be the base level, then setup dummy variables for the remaining levels. This setup always leads to the interpretation of the parameters given in the box.

> ### Procedure for Writing a Model with One Qualitative Independent Variable at k Levels (A, B, C, D, ...)
>
> $$E(y) = \beta_0 + \beta_1 x_1 + \beta_2 x_2 + \cdots + \beta_{k-1} x_{k-1}$$
>
> where
>
> $$x_i = \begin{cases} 1 & \text{if qualitative variable at level } i + 1 \\ 0 & \text{otherwise} \end{cases}$$
>
> The number of dummy variables for a single qualitative variable is always 1 less than the number of levels for the variable. Then, assuming the base level is A, the mean for each level is
>
> $$\mu_A = \beta_0$$
> $$\mu_B = \beta_0 + \beta_1$$
> $$\mu_C = \beta_0 + \beta_2$$
> $$\mu_D = \beta_0 + \beta_3$$
> $$\vdots$$
>
> β *Interpretations*:
>
> $$\beta_0 = \mu_A$$
> $$\beta_1 = \mu_B - \mu_A$$
> $$\beta_2 = \mu_C - \mu_A$$
> $$\beta_3 = \mu_D - \mu_A$$
> $$\vdots$$

Example 5.5

A large consulting firm markets a computerized system for monitoring road construction bids to various state departments of transportation. Since the high cost of maintaining the system is partially absorbed by the firm, the firm wants to compare the mean annual maintenance costs accrued by system users in three different states: Kansas, Kentucky, and Texas. A sample of 10 users is selected from each state installation and the maintenance cost accrued by each is recorded, as shown in Table 5.6.

(a) Propose a model for mean cost, $E(y)$, that will allow the firm to compare the three state users of the system.

(b) Do the data provide sufficient evidence (at $\alpha = .05$) to indicate that the mean annual maintenance costs accrued by system users differ for the three state installations?

(c) Find and interpret a 95% confidence interval for the difference between the mean cost in Texas and the mean cost in Kansas.

⊚ BIDMAINT

Table 5.6 Annual maintenance costs		
State Installation		
Kansas	Kentucky	Texas
$ 198	$ 563	$ 385
126	314	693
443	483	266
570	144	586
286	585	178
184	377	773
105	264	308
216	185	430
465	330	644
203	354	515
Totals $2,796	$3,599	$4,778

Solution

(a) The model relating $E(y)$ to the single qualitative variable, state installation, is

$$E(y) = \beta_0 + \beta_1 x_1 + \beta_2 x_2$$

where

$$x_1 = \begin{cases} 1 & \text{if Kentucky} \\ 0 & \text{if not} \end{cases} \qquad x_2 = \begin{cases} 1 & \text{if Texas} \\ 0 & \text{if not} \end{cases}$$

[Note: Base level = Kansas]
and

$$\beta_1 = \mu_2 - \mu_1$$
$$\beta_2 = \mu_3 - \mu_1$$

where μ_1, μ_2, and μ_3 are the mean responses for Kansas, Kentucky, and Texas, respectively.

(b) Testing the null hypothesis that the means for three states are equal (i.e., $\mu_1 = \mu_2 = \mu_3$) is equivalent to testing

$$H_0: \beta_1 = \beta_2 = 0$$

because if $\beta_1 = \mu_2 - \mu_1 = 0$ and $\beta_2 = \mu_3 - \mu_1 = 0$, then μ_1, μ_2, and μ_3 must be equal. The alternative hypothesis is

H_a: At least one of the parameters, β_1 or β_2, differs from 0.

There are two ways to conduct this test. We can fit the complete model shown previously and the reduced model (discarding the terms involving β_1 and β_2),

$$E(y) = \beta_0$$

Model Summary

Model	R	R Square	Adjusted R Square	Std. Error of the Estimate
1	.453[a]	.205	.146	168.948

a. Predictors: (Constant), X2, X1

ANOVA[b]

Model		Sum of Squares	df	Mean Square	F	Sig.
1	Regression	198772.5	2	99386.233	3.482	.045[a]
	Residual	770670.9	27	28543.367		
	Total	969443.4	29			

a. Predictors: (Constant), X2, X1

b. Dependent Variable: COST

Coefficients[a]

Model		Unstandardized Coefficients		Standardized Coefficients	t	Sig.	95% Confidence Interval for B	
		B	Std. Error	Beta			Lower Bound	Upper Bound
1	(Constant)	279.600	53.426		5.233	.000	169.979	389.221
	X1	80.300	75.556	.211	1.063	.297	-74.728	235.328
	X2	198.200	75.556	.520	2.623	.014	43.172	353.228

a. Dependent Variable: COST

Figure 5.19 SPSS printout for dummy variable model, Example 5.5

and conduct the nested model F-test described in Section 4.13 (we leave this as an exercise for you). Or, we can use the global F-test of the complete model (Section 4.6), which tests the null hypothesis that all parameters in the model, with the exception of β_0, equal 0. Either way you conduct the test, you will obtain the same computed value of F. The SPSS printout for fitting the complete model,

$$E(y) = \beta_0 + \beta_1 x_1 + \beta_2 x_2$$

is shown in Figure 5.19. The value of the F statistic (shaded) for testing the complete model is $F = 3.482$, the p-value for the test (also shaded) is $p = .045$. Since our choice of α, $\alpha = .05$, exceeds the p-value, we reject H_0 and conclude that at least one of the parameters, β_1 or β_2, differs from 0. Or equivalently, we conclude that the data provide sufficient evidence to indicate that the mean user maintenance cost does vary among the three state installations.

(c) Since $\beta_2 = \mu_3 - \mu_1 =$ the difference between the mean costs of Texas and Kansas, we want a 95% confidence interval for β_2. The interval, highlighted on Figure 5.19, is (43.172, 353.228). Consequently, we are 95% confident that the difference, $\mu_3 - \mu_1$, falls in our interval. This implies that the mean cost of users in Texas is anywhere from \$43.17 to \$353.23 higher than the mean cost of Kansas users.

5.8 Models with Two Qualitative Independent Variables

We demonstrate how to write a model with two qualitative independent variables and then, in Section 5.9, we explain how to use this technique to write models with any number of qualitative independent variables.

Let us return to the example used in Section 5.7, where we wrote a model for the mean performance, $E(y)$, of a diesel engine as a function of one qualitative independent variable, fuel type. Now suppose the performance is also a function of engine brand and we want to compare the top two brands. Therefore, this second qualitative independent variable, brand, will be observed at two levels. To simplify our notation, we will change the symbols for the three fuel types from B, D, C, to F_1, F_2, F_3, and we will let B_1 and B_2 represent the two brands. The six population means of performance measurements (measurements of y) are symbolically represented by the six cells in the two-way table shown in Table 5.7. Each μ subscript corresponds to one of the six fuel type–brand combinations.

Table 5.7 The six combinations of fuel type and diesel engine brand

		Brand	
		B_1	B_2
	F_1	μ_{11}	μ_{12}
FUEL TYPE	F_2	μ_{21}	μ_{22}
	F_3	μ_{31}	μ_{32}

First, we write a model in its simplest form—where the two qualitative variables affect the response independently of each other. To write the model for mean performance, $E(y)$, we start with a constant β_0 and then add *two* dummy variables for the three levels of fuel type in the manner explained in Section 5.7. These terms, which are called the **main effect terms** for fuel type, F, account for the effect of F on $E(y)$ when fuel type, F, and brand, B, affect $E(y)$ independently. Then,

$$E(y) = \beta_0 + \overbrace{\beta_1 x_1 + \beta_2 x_2}^{\substack{\text{Main effect} \\ \text{terms for } F}}$$

where

$$x_1 = \begin{cases} 1 & \text{if fuel type } F_2 \text{ was used} \\ 0 & \text{if not} \end{cases}$$

$$x_2 = \begin{cases} 1 & \text{if fuel type } F_3 \text{ was used} \\ 0 & \text{if not} \end{cases}$$

Now let level B_1 be the base level of the brand variable. Since there are two levels of this variable, we will need only one dummy variable to include the brand in the model:

$$E(y) = \beta_0 + \overbrace{\beta_1 x_1 + \beta_2 x_2}^{\substack{\text{Main effect} \\ \text{terms for } F}} + \overbrace{\beta_3 x_3}^{\substack{\text{Main effect} \\ \text{term for } B}}$$

where the dummy variables x_1 and x_2 are as defined previously and

$$x_3 = \begin{cases} 1 & \text{if engine brand } B_2 \text{ was used} \\ 0 & \text{if engine brand } B_1 \text{ was used} \end{cases}$$

If you check the model, you will see that by assigning specific values to x_1, x_2, and x_3, you create a model for the mean value of y corresponding to one of the cells of Table 5.7. We illustrate with two examples.

Example 5.6

Give the values of x_1, x_2, and x_3 and the model for the mean performance, $E(y)$, when using fuel type F_1 in engine brand B_1.

Solution

Checking the coding system, you will see that F_1 and B_1 occur when $x_1 = x_2 = x_3 = 0$. Then,

$$E(y) = \beta_0 + \beta_1 x_1 + \beta_2 x_2 + \beta_3 x_3$$
$$= \beta_0 + \beta_1(0) + \beta_2(0) + \beta_3(0)$$
$$= \beta_0$$

Therefore, the mean value of y at levels F_1 and B_1, which we represent as μ_{11}, is

$$\mu_{11} = \beta_0$$

Example 5.7

Give the values of x_1, x_2, and x_3 and the model for the mean performance, $E(y)$, when using fuel type F_3 in engine brand B_2.

Solution

Checking the coding system, you will see that for levels F_3 and B_2,

$$x_1 = 0 \qquad x_2 = 1 \qquad x_3 = 1$$

Then, the mean performance for fuel F_3 used in engine brand B_2, represented by the symbol μ_{32} (see Table 5.7), is

$$\mu_{32} = E(y) = \beta_0 + \beta_1 x_1 + \beta_2 x_2 + \beta_3 x_3$$
$$= \beta_0 + \beta_1(0) + \beta_2(1) + \beta_3(1)$$
$$= \beta_0 + \beta_2 + \beta_3$$

Note that in the model described previously, we assumed the qualitative independent variables for fuel type and engine brand affect the mean response, $E(y)$, independently of each other. This type of model is called a **main effects model** and is shown in the box (p. 295). Changing the level of one qualitative variable will have the same effect on $E(y)$ for any level of the second qualitative variable. In other words, the effect of one qualitative variable on $E(y)$ is independent (in a mathematical sense) of the level of the second qualitative variable.

When two independent variables affect the mean response independently of each other, you may obtain the pattern shown in Figure 5.20. Note that the difference in mean performance between any two fuel types (levels of F) is the same, *regardless* of the engine brand used. That is, the main effects model assumes that the relative effect of fuel type on performance is the same in both engine brands.

If F and B do not affect $E(y)$ independently of each other, then the response function might appear as shown in Figure 5.21. Note the difference between the mean response functions for Figures 5.20 and 5.21. When F and B affect the mean response in a dependent manner (Figure 5.21), the response functions differ for each brand. This means that you cannot study the effect of one variable on $E(y)$ without considering the level of the other. When this situation occurs, we say that the qualitative independent variables **interact**. The interaction model is shown in the box (p. 296). In this example, interaction might be expected if one fuel type tends to perform better in engine B_1, whereas another performs better in engine B_2.

Figure 5.20 Main effects model: Mean response as a function of F and B when F and B affect $E(y)$ independently

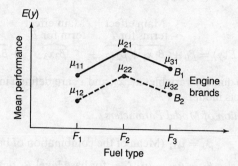

Main Effects Model with Two Qualitative Independent Variables, One at Three Levels (F_1, F_2, F_3) and the Other at Two Levels (B_1, B_2)

$$E(y) = \beta_0 + \overbrace{\beta_1 x_1 + \beta_2 x_2}^{\text{Main effect terms for } F} + \overbrace{\beta_3 x_3}^{\text{Main effect term for } B}$$

where

$$x_1 = \begin{cases} 1 & \text{if } F_2 \\ 0 & \text{if not} \end{cases} \qquad x_2 = \begin{cases} 1 & \text{if } F_3 \\ 0 & \text{if not} \end{cases} \qquad (F_1 \text{ is base level})$$

$$x_3 = \begin{cases} 1 & \text{if } B_2 \\ 0 & \text{if } B_1 \quad (\text{base level}) \end{cases}$$

Interpretation of Model Parameters

$\beta_0 = \mu_{11}$ (Mean of the combination of base levels)

$\beta_1 = \mu_{2j} - \mu_{1j}$, for any level $B_j (j = 1, 2)$

$\beta_2 = \mu_{3j} - \mu_{1j}$, for any level $B_j (j = 1, 2)$

$\beta_3 = \mu_{i2} - \mu_{i1}$, for any level $F_i (i = 1, 2, 3)$

Figure 5.21 Interaction model: Mean response as a function of F and B when F and B interact to affect $E(y)$

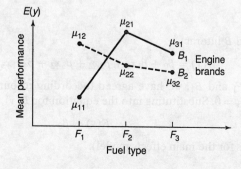

Interaction Model with Two Qualitative Independent Variables, One at Three Levels (F_1, F_2, F_3) and the Other at Two Levels (B_1, B_2)

$$E(y) = \beta_0 + \overbrace{\beta_1 x_1 + \beta_2 x_2}^{\substack{\text{Main effect} \\ \text{terms for } F}} + \overbrace{\beta_3 x_3}^{\substack{\text{Main effect} \\ \text{term for } B}} + \overbrace{\beta_4 x_1 x_3 + \beta_5 x_2 x_3}^{\substack{\text{Interaction} \\ \text{terms}}}$$

where the dummy variables x_1, x_2, and x_3 are defined in the same way as for the main effects model.

Interpretation of Model Parameters

$$\beta_0 = \mu_{11} \text{ (Mean of the combination of base levels)}$$

$$\beta_1 = \mu_{21} - \mu_{11} \text{ (i.e., for base level } B_1 \text{ only)}$$

$$\beta_2 = \mu_{31} - \mu_{11} \text{ (i.e., for base level } B_1 \text{ only)}$$

$$\beta_3 = \mu_{12} - \mu_{11} \text{ (i.e., for base level } F_1 \text{ only)}$$

$$\beta_4 = (\mu_{22} - \mu_{12}) - (\mu_{21} - \mu_{11})$$

$$\beta_5 = (\mu_{32} - \mu_{12}) - (\mu_{31} - \mu_{11})$$

When qualitative independent variables interact, the model for $E(y)$ must be constructed so that it is able (if necessary) to give a different mean value, $E(y)$, for every cell in Table 5.7. We do this by adding **interaction terms** to the main effects model. These terms will involve all possible two-way cross-products between each of the two dummy variables for F, x_1, and x_2, and the one dummy variable for B, x_3. The number of interaction terms (for two independent variables) will equal the number of main effect terms for the one variable times the number of main effect terms for the other.

When F and B interact, the model contains six parameters: the two main effect terms for F, one main effect term for B, $(2)(1) = 2$ interaction terms, and β_0. This will make it possible, by assigning the various combinations of values to the dummy variables x_1, x_2, and x_3, to give six different values for $E(y)$ that will correspond to the means of the six cells in Table 5.7.

Example 5.8

In Example 5.6, we gave the mean response when fuel F_1 was used in engine B_1, where we assumed that F and B affected $E(y)$ independently (no interaction). Now give the value of $E(y)$ for the model where F and B interact to affect $E(y)$.

Solution

When F and B interact,

$$E(y) = \beta_0 + \beta_1 x_1 + \beta_2 x_2 + \beta_3 x_3 + \beta_4 x_1 x_3 + \beta_5 x_2 x_3$$

For levels F_1 and B_1, we have agreed (according to our system of coding) to let $x_1 = x_2 = x_3 = 0$. Substituting into the equation for $E(y)$, we have

$$E(y) = \beta_0$$

(the same as for the main effects model).   ■

Example 5.9

In Example 5.7, we gave the mean response for fuel type F_3 and brand B_2, when F and B affected $E(y)$ independently. Now assume that F and B interact, and give the value for $E(y)$ when fuel F_3 is used in engine brand B_2.

Solution

When F and B interact,

$$E(y) = \beta_0 + \beta_1 x_1 + \beta_2 x_2 + \beta_3 x_3 + \beta_4 x_1 x_3 + \beta_5 x_2 x_3$$

To find $E(y)$ for F_3 and B_2, we set $x_1 = 0$, $x_2 = 1$, and $x_3 = 1$:

$$E(y) = \beta_0 + \beta_1(0) + \beta_2(1) + \beta_3(1) + \beta_4(0)(1) + \beta_5(1)(1)$$
$$= \beta_0 + \beta_2 + \beta_3 + \beta_5$$

This is the value of μ_{32} in Table 5.7. Note the difference in $E(y)$ for the model assuming independence between F and B versus this model, which assumes interaction between F and B. The difference is β_5.

Example 5.10

The performance, y (measured as mass burning rate per degree of crank angle), for the six combinations of fuel type and engine brand is shown in Table 5.8. The number of test runs per combination varies from one for levels (F_1, B_2) to three for levels (F_1, B_1). A total of 12 test runs are sampled.

(a) Assume the interaction between F and B is negligible. Fit the model for $E(y)$ with interaction terms omitted.

(b) Fit the complete model for $E(y)$ allowing for the fact that interactions might occur.

(c) Use the prediction equation for the model, part a to estimate the mean engine performance when fuel F_3 is used in brand B_2. Then calculate the sample mean for this cell in Table 5.8. Repeat for the model, part b. Explain the discrepancy between the sample mean for levels (F_3, B_2) and the estimate(s) obtained from one or both of the two prediction equations.

⊙ DIESEL

Table 5.8 Performance data for combinations of fuel type and diesel engine brand

		Brand	
		B_1	B_2
FUEL TYPE	F_1	65 73 68	36
	F_2	78 82	50 43
	F_3	48 46	61 62

Solution

(a) A portion of the SAS printout for main effects model

$$E(y) = \beta_0 + \underbrace{\beta_1 x_1 + \beta_2 x_2}_{\text{Main effect terms for } F} + \underbrace{\beta_3 x_3}_{\text{Main effect term for } B}$$

is given in Figure 5.22. The least squares prediction equation is (after rounding):

$$\hat{y} = 64.45 + 6.70 x_1 - 2.30 x_2 - 15.82 x_3$$

Dependent Variable: PERFORM

Analysis of Variance

Source	DF	Sum of Squares	Mean Square	F Value	Pr > F
Model	3	858.25758	286.08586	1.51	0.2838
Error	8	1512.40909	189.05114		
Corrected Total	11	2370.66667			

Root MSE	13.74959	R-Square	0.3620
Dependent Mean	59.33333	Adj R-Sq	0.1228
Coeff Var	23.17346		

Parameter Estimates

Variable	DF	Parameter Estimate	Standard Error	t Value	Pr > |t|
Intercept	1	64.45455	7.18049	8.98	<.0001
X1	1	6.70455	9.94093	0.67	0.5190
X2	1	-2.29545	9.94093	-0.23	0.8232
X3	1	-15.81818	8.29131	-1.91	0.0928

Output Statistics

Obs	FUELBRND	Dep Var PERFORM	Predicted Value	Std Error Mean Predict	95% CL Mean		Residual
1	F1B1	65.0000	64.4545	7.1805	47.8963	81.0128	0.5455
2	F1B1	73.0000	64.4545	7.1805	47.8963	81.0128	8.5455
3	F1B1	68.0000	64.4545	7.1805	47.8963	81.0128	3.5455
4	F1B2	36.0000	48.6364	9.2700	27.2598	70.0130	-12.6364
5	F2B1	78.0000	71.1591	8.0280	52.6464	89.6718	6.8409
6	F2B1	82.0000	71.1591	8.0280	52.6464	89.6718	10.8409
7	F2B2	50.0000	55.3409	8.0280	36.8282	73.8536	-5.3409
8	F2B2	43.0000	55.3409	8.0280	36.8282	73.8536	-12.3409
9	F3B1	48.0000	62.1591	8.0280	43.6464	80.6718	-14.1591
10	F3B1	46.0000	62.1591	8.0280	43.6464	80.6718	-16.1591
11	F3B2	61.0000	46.3409	8.0280	27.8282	64.8536	14.6591
12	F3B2	62.0000	46.3409	8.0280	27.8282	64.8536	15.6591

Sum of Residuals		0
Sum of Squared Residuals		1512.40909
Predicted Residual SS (PRESS)		3615.37520

Figure 5.22 SAS printout for main effects model, Example 5.10

(b) The SAS printout for the complete model is given in Figure 5.23. Recall that the complete model is

$$E(y) = \beta_0 + \beta_1 x_1 + \beta_2 x_2 + \beta_3 x_3 + \beta_4 x_1 x_3 + \beta_5 x_2 x_3$$

The least squares prediction equation is (after rounding):

$$\hat{y} = 68.67 + 11.33x_1 - 21.67x_2 - 32.67x_3 - .83x_1 x_3 + 47.17x_2 x_3$$

(c) To obtain the estimated mean response for cell (F_3, B_2), we let $x_1 = 0$, $x_2 = 1$, and $x_3 = 1$. Then, for the main effects model, we find

$$\hat{y} = 64.45 + 6.70(0) - 2.30(1) - 15.82(1) = 46.34$$

The 95% confidence interval for the true mean performance (shaded in Figure 5.22) is $(27.83, 64.85)$.

For the complete model, we find

$$\hat{y} = 68.67 + 11.33(0) - 21.67(1) - 32.67(1) - .83(0)(1) + 47.17(1)(1) = 61.50$$

The 95% confidence interval for true mean performance (shaded in Figure 5.23) is $(55.69, 67.31)$. The mean for the cell (F_3, B_2) in Table 5.8 is

$$\bar{y}_{32} = \frac{61 + 62}{2} = 61.5$$

```
                        Dependent Variable: PERFORM
                         Analysis of Variance

                                     Sum of            Mean
Source                    DF         Squares          Square    F Value    Pr > F

Model                     5       2303.00000        460.60000     40.84    0.0001
Error                     6         67.66667         11.27778
Corrected Total          11       2370.66667


            Root MSE              3.35824     R-Square     0.9715
            Dependent Mean       59.33333     Adj R-Sq     0.9477
            Coeff Var             5.65996

                         Parameter Estimates

                         Parameter       Standard
        Variable    DF    Estimate         Error     t Value    Pr > |t|

        Intercept    1    68.66667        1.93888      35.42     <.0001
        X1           1    11.33333        3.06564       3.70     0.0101
        X2           1   -21.66667        3.06564      -7.07     0.0004
        X3           1   -32.66667        3.87776      -8.42     0.0002
        X1X3         1    -0.83333        5.12980      -0.16     0.8763
        X2X3         1    47.16667        5.12980       9.19     <.0001

                         Output Statistics

                  Dep Var    Predicted     Std Error
Obs   FUELBRND     PERFORM      Value     Mean Predict       95% CL Mean        Residual

 1     F1B1       65.0000     68.6667       1.9389      63.9224    73.4109     -3.6667
 2     F1B1       73.0000     68.6667       1.9389      63.9224    73.4109      4.3333
 3     F1B1       68.0000     68.6667       1.9389      63.9224    73.4109     -0.6667
 4     F1B2       36.0000     36.0000       3.3582      27.7827    44.2173     -7.11E-15
 5     F2B1       78.0000     80.0000       2.3746      74.1895    85.8105     -2.0000
 6     F2B1       82.0000     80.0000       2.3746      74.1895    85.8105      2.0000
 7     F2B2       50.0000     46.5000       2.3746      40.6895    52.3105      3.5000
 8     F2B2       43.0000     46.5000       2.3746      40.6895    52.3105     -3.5000
 9     F3B1       48.0000     47.0000       2.3746      41.1895    52.8105      1.0000
10     F3B1       46.0000     47.0000       2.3746      41.1895    52.8105     -1.0000
11     F3B2       61.0000     61.5000       2.3746      55.6895    67.3105     -0.5000
12     F3B2       62.0000     61.5000       2.3746      55.6895    67.3105      0.5000


              Sum of Residuals                       0
              Sum of Squared Residuals        67.66667
              Predicted Residual SS (PRESS)  213.50000
```

Figure 5.23 SAS printout for interaction model, Example 5.10

which is precisely what is estimated by the complete (interaction) model. However, the main effects model yields a different estimate, 46.34. The reason for the discrepancy is that the main effects model assumes the two qualitative independent variables affect $E(y)$ independently of each other. That is, the change in $E(y)$ produced by a change in levels of one variable is the same regardless of the level of the other variable. In contrast, the complete model contains six parameters $(\beta_0, \beta_1, \ldots, \beta_5)$ to describe the six cell populations, so that each population cell mean will be estimated by its sample mean. Thus, the complete model estimate for any cell mean is equal to the observed (sample) mean for that cell. ∎

Example 5.10 demonstrates an important point. If we were to ignore the least squares analysis and calculate the six sample means in Table 5.8 directly, we would obtain estimates of $E(y)$ exactly the same as those obtained by a least squares analysis for the case where the interaction between F and B is assumed to exist. We would not obtain the same estimates if the model assumes that interaction does not exist.

Also, the estimates of means raise important questions. Do the data provide sufficient evidence to indicate that F and B interact? For our example, does the effect of fuel type on diesel engine performance depend on which engine brand is

Figure 5.24 MINITAB graph of sample means for engine performance

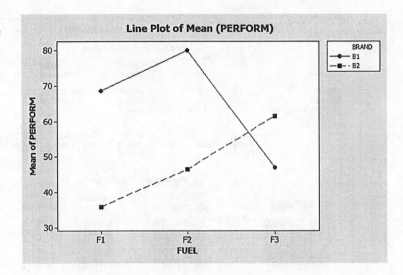

used? A MINITAB plot of all six sample means, shown in Figure 5.24, seems to indicate interaction, since fuel types F_1 and F_2 appear to operate more effectively in engine brand B_1, whereas the mean performance of F_3 is higher in brand B_2. Can these sample facts be reliably generalized to conclusions about the populations?

To answer this question, we will want to perform a test for interaction between the two qualitative independent variables, fuel type and engine brand. Since allowance for interaction between fuel type and brand in the complete model was provided by the addition of the terms $\beta_4 x_1 x_3$ and $\beta_5 x_2 x_3$, it follows that the null hypothesis that the independent variables fuel type and brand do not interact is equivalent to the hypothesis that the terms $\beta_4 x_1 x_3$ and $\beta_5 x_2 x_3$ are not needed in the model for $E(y)$—or equivalently, that $\beta_4 = \beta_5 = 0$. Conversely, the alternative hypothesis that fuel type and brand do interact is equivalent to stating that at least one of the two parameters, β_4 or β_5, differs from 0.

The appropriate procedure for testing a portion of the model parameters, a nested model F-test, was discussed in Section 4.13. The F-test is carried out as follows:

$$H_0: \beta_4 = \beta_5 = 0$$

H_a: At least one of β_4 and β_5 differs from 0

$$\text{Test statistic: } F = \frac{(\text{SSE}_R - \text{SSE}_C)/g}{\text{SSE}_C/[n - (k + 1)]}$$

where

$$\text{SSE}_R = \text{SSE for reduced model (main effects model)}$$
$$\text{SSE}_C = \text{SSE for complete model (interaction model)}$$
$$g = \text{Number of } \beta\text{'s tested}$$
$$= \text{Numerator df for the } F \text{ statistic}$$
$$n - (k + 1) = \text{df for error for complete model}$$
$$= \text{Denominator df for the } F \text{ statistic}$$

Figure 5.25 SAS printout for nested model F-test of interaction

```
Test INTERACT Results for Dependent Variable PERFORM

                            Mean
Source          DF         Square      F Value     Pr > F

Numerator        2       722.37121      64.05      <.0001
Denominator      6        11.27778
```

For this example, we have

$$SSE_R = 1{,}512.41 \text{ (highlighted on Figure 5.22)}$$
$$SSE_C = 67.67 \text{ (highlighted on Figure 5.23)}$$
$$g = 2 \quad \text{and} \quad n - (k+1) = 6$$

Then

$$F = \frac{(1{,}512.41 - 67.67)/2}{67.67/6}$$
$$= \frac{722.37}{11.28}$$
$$= 64.05$$

The test statistic, $F = 64.05$, is highlighted on the SAS printout of the analysis, Figure 5.25. The p-value of the test, also highlighted, is less than .0001. Thus, we are confident (at $\alpha = .05$) in concluding that the interaction terms contribute to the prediction of y, engine performance. Equivalently, there is sufficient evidence to conclude that factors F and B do interact.

5.8 Exercises

5.24 **Expert testimony in homicide trials of battered women.** Refer to the *Duke Journal of Gender Law and Policy* (Summer 2003) study of the impact of expert testimony on the outcome of homicide trials involving battered woman syndrome, Exercise 5.3 (p. 264). Recall that multiple regression was employed to model the likelihood of changing a verdict from not guilty to guilty after deliberations, y, as a function of juror gender (male or female) and expert testimony given (yes or no).

(a) Write a main effects model for $E(y)$ as a function of gender and expert testimony. Interpret the β coefficients in the model.

(b) Write an interaction model for $E(y)$ as a function of gender and expert testimony. Interpret the β coefficients in the model.

(c) Based on data collected on individual juror votes from past trials, the article reported that "when expert testimony was present, women jurors were more likely than men to change a verdict from not guilty to guilty after deliberations." Assume that when no expert testimony

was present, male jurors were more likely than women to change a verdict from not guilty to guilty after deliberations. Which model, part a or part b, hypothesizes the relationships reported in the article? Illustrate the model with a sketch.

5.25 **Psychological response of firefighters.** Refer to the *Journal of Human Stress* study of firefighters, Exercise 5.5 (p. 264). Consider using the qualitative variable, level of social support, as a predictor of emotional stress y. Suppose that four social support levels were studied: none, low, moderate, and high.

(a) Write a model for $E(y)$ as a function of social support at four levels.

(b) Interpret the β parameters in the model.

(c) Explain how to test for differences among the emotional stress means for the four social support levels.

5.26 **Milk production of shaded cows.** Because of the hot, humid weather conditions in Florida, the growth rates of beef cattle and the milk production of dairy cows typically decline during the summer. However, agricultural and environmental engineers have found that a well-designed shade structure can significantly increase the milk production of dairy cows. In one experiment, 30 cows were selected and divided into three groups of 10 cows each. Group 1 cows were provided with a man-made shade structure, group 2 cows with tree shade, and group 3 cows with no shade. Of interest was the mean milk production (in gallons) of the cows in each group.

(a) Identify the independent variables in the experiment.

(b) Write a model relating the mean milk production, $E(y)$, to the independent variables. Identify and code all dummy variables.

(c) Interpret the β parameters of the model.

5.27 **Quality of Bordeaux wine.** Refer to the *Economic Journal* (May 2008) study of the factors that yield a quality Bordeaux wine, Exercise 4.51 (p. 223). Recall that a quantitative measure of wine quality (y) was modeled as a function of several qualitative independent variables, including grape-picking method (manual or automated) and soil type (clay, gravel, or sand).

(a) Write an interaction model relating wine quality to the two qualitative independent variables. Let "automated" and "sand" represent the base levels for the two variables, respectively.

(b) Interpret the value of β_0 in the model.

(c) In terms of the β's, what is the mean quality of wine produced from grapes picked manually from clay soil?

(d) In terms of the β's, what is the difference between the mean quality of wine produced from grapes picked manually and grapes picked with an automated method when the soil type is sand?

5.28 **Insomnia and education.** Many workers suffer from stress and chronic insomnia. Is insomnia related to education status? Researchers at the Universities of Memphis, Alabama at Birmingham, and Tennessee investigated this question in the *Journal of Abnormal Psychology* (February 2005). Adults living in Tennessee were selected to participate in the study using a random-digit telephone dialing procedure. In addition to insomnia status (normal sleeper or chronic insomnia), the researchers classified each participant into one of four education categories (college graduate, some college, high school graduate, and high school dropout). The dependent variable (y) of interest to the researchers was a quantitative measure of

daytime functioning called the Fatigue Severity Scale (FSS), with values ranging from 0 to 5.

(a) Write a main effects model for $E(y)$ as a function of insomnia status and education level. Construct a graph similar to Figure 5.24 that represents the effects hypothesized by the model.

(b) Write an interaction model for $E(y)$ as a function of insomnia status and education level. Construct a graph similar to Figure 5.24 that represents the effects hypothesized by the model.

(c) The researchers discovered that the mean FSS for people with insomnia is greater than the mean FSS for normal sleepers, but that this difference is the same at all education levels. Based on this result, which of the two models best represents the data?

5.29 **Impact of flavor name on consumer choice.** Do consumers react favorably to products with ambiguous colors or names? Marketing researchers investigated this phenomenon in the *Journal of Consumer Research* (June 2005). As a "reward" for participating in an unrelated experiment, 100 consumers were told they could have some jelly beans available in several cups on a table. Some consumers were offered jelly beans with common descriptive flavor names (e.g., watermelon green), while the others were offered jelly beans with ambiguous flavor names (e.g., monster green). Also, some of the consumers took the jelly beans and immediately left (low cognitive load condition), while others were peppered with additional questions designed to distract them while they were taking their jelly beans (high cognitive load condition). The researchers modeled the number (y) of jelly beans taken by each consumer as a function of two qualitative variables: Flavor Name (common or ambiguous) and Cognitive Load (low or high). The sample mean, $\bar{y}$, for each of the four combinations of the qualitative variables is shown in the accompanying table.

	AMBIGUOUS	COMMON
Low Load	18.0	7.8
High Load	6.1	6.3

Source: Miller, E. G., & Kahn, B. E. "Shades of meaning: The effect of color and flavor names on consumer choice," *Journal of Consumer Research*, Vol. 32, June 2005 (Table 1). Reprinted with permission of the University of Chicago Press.

(a) Write an interaction model for $E(y)$ as a function of Flavor Name and Cognitive Load.

(b) Use the information in the table to estimate the β's in the interaction model.

(c) How would you determine whether the impact of flavor name on $E(y)$ depends on cognitive load? Explain.

5.30 Modeling faculty salary. The administration of a midwestern university commissioned a salary equity study to help establish benchmarks for faculty salaries. The administration utilized the following regression model for annual salary, y: $E(y) = \beta_0 + \beta_1 x$, where $x = 0$ if lecturer, 1 if assistant professor, 2 if associate professor, and 3 if full professor. The administration wanted to use the model to compare the mean salaries of professors in the different ranks. Explain the flaw in the model. Propose an alternative model that will achieve the administration's objective.

5.9 Models with Three or More Qualitative Independent Variables

We construct models with three or more qualitative independent variables in the same way that we construct models for two qualitative independent variables, except that we must add three-way interaction terms if we have three qualitative independent variables, three-way and four-way interaction terms for four independent variables, and so on. In this section, we explain what we mean by three-way and four-way interactions, and we demonstrate the procedure for writing the model for any number, say, k, of qualitative independent variables. The pattern used to write the model is shown in the box.

Recall that a two-way interaction term is formed by multiplying the dummy variable associated with one of the main effect terms of one (call it the first) independent variable by the dummy variable from a main effect term of another (the second) independent variable. Three-way interaction terms are formed in a similar way, by forming the product of three dummy variables, one from a main effect term from each of the three independent variables. Similarly, four-way interaction terms are formed by taking the product of four dummy variables, one from a main effect term from each of four independent variables. We illustrate with three examples.

Pattern of the Model Relating $E(y)$ to k Qualitative Independent Variables

$E(y) = \beta_0 + $ Main effect terms for all independent variables

$+$ All two-way interaction terms between pairs of independent variables

$+$ All three-way interaction terms between different groups of three independent variables

$+$

$\vdots$

$+$ All k-way interaction terms for the k independent variables

---

Example 5.11 Refer to Examples 5.6–5.10, where we modeled the performance, y, of a diesel engine as a function of fuel type (F_1, F_2, and F_3) and brand (B_1 and B_2). Now consider a third qualitative independent variable, injection system (S_1 and S_2).

Write a model for $E(y)$ that includes all main effect and interaction terms for the independent variables.

Solution

First write a model containing the main effect terms for the three variables:

$$E(y) = \beta_0 + \overbrace{\beta_1 x_1 + \beta_2 x_2}^{\substack{\text{Main effects} \\ \text{terms for Fuel}}} + \overbrace{\beta_3 x_3}^{\substack{\text{Main effect} \\ \text{terms for Brand}}} + \overbrace{\beta_4 x_4}^{\substack{\text{Main effect} \\ \text{terms for System}}}$$

where

$$x_1 = \begin{cases} 1 & \text{if level } F_2 \\ 0 & \text{if not} \end{cases} \qquad x_3 = \begin{cases} 1 & \text{if level } B_2 \\ 0 & \text{if not} \end{cases} \qquad x_4 = \begin{cases} 1 & \text{if level } S_2 \\ 0 & \text{if level } S_1 \text{(base)} \end{cases}$$

$$x_2 = \begin{cases} 1 & \text{if level } F_3 \\ 0 & \text{if not} \end{cases}$$

$$\text{(Base level} = F_1) \qquad \text{(Base level} = B_1)$$

The next step is to add two-way interaction terms. These will be of three types—those for the interaction between Fuel and Brand, between Fuel and System, and between Brand and System. Thus,

$$E(y) = \beta_0 + \overbrace{\beta_1 x_1 + \beta_2 x_2}^{\substack{\text{Main effects} \\ \text{Fuel}}} + \overbrace{\beta_3 x_3}^{\substack{\text{Main effect} \\ \text{Brand}}} + \overbrace{\beta_4 x_4}^{\substack{\text{Main effect} \\ \text{System}}}$$

$$+ \underbrace{\beta_5 x_1 x_3 + \beta_6 x_2 x_3}_{\substack{\text{Fuel} \times \text{Brand} \\ \text{interaction}}} + \underbrace{\beta_7 x_1 x_4 + \beta_8 x_2 x_4}_{\substack{\text{Fuel} \times \text{System} \\ \text{interaction}}} + \underbrace{\beta_9 x_3 x_4}_{\substack{\text{Brand} \times \text{System} \\ \text{interaction}}}$$

Finally, since there are three independent variables, we must include terms for the interaction of Fuel, Brand, and System. These terms are formed as the products of dummy variables, one from each of the Fuel, Brand, and System main effect terms. The complete model for $E(y)$ is

$$E(y) = \beta_0 + \overbrace{\beta_1 x_1 + \beta_2 x_2}^{\substack{\text{Main effects} \\ \text{Fuel}}} + \overbrace{\beta_3 x_3}^{\substack{\text{Main effect} \\ \text{Brand}}} + \overbrace{\beta_4 x_4}^{\substack{\text{Main effect} \\ \text{System}}}$$

$$+ \underbrace{\beta_5 x_1 x_3 + \beta_6 x_2 x_3}_{\text{Fuel} \times \text{Brand interaction}}$$

$$+ \underbrace{\beta_7 x_1 x_4 + \beta_8 x_2 x_4}_{\text{Fuel} \times \text{System interaction}}$$

$$+ \underbrace{\beta_9 x_3 x_4}_{\text{Brand} \times \text{System interaction}}$$

$$+ \underbrace{\beta_{10} x_1 x_3 x_4 + \beta_{11} x_2 x_3 x_4}_{\text{3-way interaction}}$$

Note that the complete model in Example 5.11 contains 12 parameters, one for each of the $3 \times 2 \times 2$ combinations of levels for Fuel, Brand, and System. There are 12 linearly independent linear combinations of these parameters, *one corresponding to each of the means of the $3 \times 2 \times 2$ combinations*. We illustrate with another example.

Example 5.12

Refer to Example 5.11 and give the expression for the mean value of performance y for engines using Fuel type F_2, of Brand B_1, and with injection System S_2.

Solution

Check the coding for the dummy variables (given in Example 5.11) and you will see that they assume the following values:

$$\text{For level } F_2: x_1 = 1, x_2 = 0$$

$$\text{For level } B_1: x_3 = 0,$$

$$\text{For level } S_2: x_4 = 1$$

Substituting these values into the expression for $E(y)$, we obtain

$$E(y) = \beta_0 + \beta_1(1) + \beta_2(0) + \beta_3(0) + \beta_4(1)$$
$$+ \beta_5(1)(0) + \beta_6(0)(0)$$
$$+ \beta_7(1)(1) + \beta_8(0)(1)$$
$$+ \beta_9(0)(1)$$
$$+ \beta_{10}(1)(0)(1) + \beta_{11}(0)(0)(1) = \beta_0 + \beta_1 + \beta_4 + \beta_7$$

Thus, the mean value of y observed at levels F_2, B_1, and S_2 is $\beta_0 + \beta_1 + \beta_4 + \beta_7$. You could find the mean values of y for the other 11 combinations of levels of Fuel, Brand, and System by substituting the appropriate values of the dummy variables into the expression for $E(y)$ in the same manner. Each of the 12 means is a unique linear combination of the 12 β parameters in the model. ∎

Example 5.13

Suppose you want to test the hypothesis that the three qualitative independent variables discussed in Example 5.11 do not interact, (i.e.), the hypothesis that the effect of any one of the variables on $E(y)$ is independent of the level settings of the other two variables. Formulate the appropriate test of hypothesis about the model parameters.

Solution

No interaction among the three qualitative independent variables implies that the main effects model,

$$E(y) = \beta_0 + \overbrace{\beta_1 x_1 + \beta_2 x_2}^{\substack{\text{Main effects}\\ \text{Fuel}}} + \overbrace{\beta_3 x_3}^{\substack{\text{Main effect}\\ \text{Brand}}} + \overbrace{\beta_4 x_4}^{\substack{\text{Main effect}\\ \text{System}}}$$

is appropriate for modeling $E(y)$ or, equivalently, that all interaction terms should be excluded from the model. This situation will occur if

$$\beta_5 = \beta_6 = \cdots = \beta_{11} = 0$$

Consequently, we will test the null hypothesis

$$H_0: \beta_5 = \beta_6 = \cdots = \beta_{11} = 0$$

against the alternative hypothesis that at least one of these β parameters differs from 0, or equivalently, that some interaction among the independent variables exists. This nested model F-test was described in Section 4.13. ∎

The examples in this section demonstrate a basic principle of model building. If you are modeling a response, y, and you believe that several qualitative independent variables affect the response, then you must know how to enter these variables into your model. You must understand the implication of the interaction (or lack of it) among a subset of independent variables and how to write the appropriate terms in the model to account for it. Failure to write a good model for your response will usually lead to inflated values of the SSE and s^2 (with a consequent loss of information), and it also can lead to biased estimates of $E(y)$ and biased predictions of y.

5.10 Models with Both Quantitative and Qualitative Independent Variables

Perhaps the most interesting data analysis problems are those that involve both quantitative and qualitative independent variables. For example, suppose mean performance of a diesel engine is a function of one qualitative independent variable, fuel type at levels F_1, F_2, and F_3, and one quantitative independent variable, engine speed in revolutions per minute (rpm). We will proceed to build a model in stages, showing graphically the interpretation that we would give to the model at each stage. This will help you see the contribution of various terms in the model.

At first we assume that the qualitative independent variable has no effect on the response (i.e., the mean contribution to the response is the same for all three fuel types), but the mean performance, $E(y)$, is related to engine speed. In this case, one response curve, which might appear as shown in Figure 5.26, would be sufficient to characterize $E(y)$ for all three fuel types. The following second-order model would likely provide a good approximation to $E(y)$:

$$E(y) = \beta_0 + \beta_1 x_1 + \beta_2 x_1^2$$

where x_1 is speed in rpm. This model has some distinct disadvantages. If differences in mean performance exist for the three fuel types, they cannot be detected (because the model does not contain any parameters representing differences among fuel types). Also, the differences would inflate the SSE associated with the fitted model and consequently would increase errors of estimation and prediction.

The next stage in developing a model for $E(y)$ is to assume that the qualitative independent variable, fuel type, does affect mean performance, but the effect on $E(y)$ is independent of speed. In other words, the assumption is that the two

Figure 5.26 Model for $E(y)$ as a function of engine speed

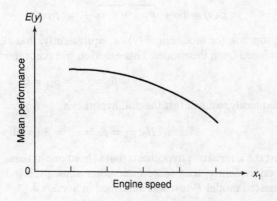

independent variables do not interact. This model is obtained by adding main effect terms for fuel type to the second-order model we used in the first stage. Therefore, using the methods of Sections 5.7 and 5.8, we choose F_1 as the base level and add two terms to the model corresponding to levels F_2 and F_3:

$$E(y) = \beta_0 + \beta_1 x_1 + \beta_2 x_1^2 + \beta_3 x_2 + \beta_4 x_3$$

where

$$x_1 = \text{Engine speed} \quad x_2 = \begin{cases} 1 & \text{if } F_2 \\ 0 & \text{if not} \end{cases} \quad x_3 = \begin{cases} 1 & \text{if } F_3 \\ 0 & \text{if not} \end{cases}$$

What effect do these terms have on the graph for the response curve(s)? Suppose we want to model $E(y)$ for level F_1. Then we let $x_2 = 0$ and $x_3 = 0$. Substituting into the model equation, we have

$$E(y) = \beta_0 + \beta_1 x_1 + \beta_2 x_1^2 + \beta_3(0) + \beta_4(0)$$
$$= \beta_0 + \beta_1 x_1 + \beta_2 x_1^2$$

which would graph as a second-order curve similar to the one shown in Figure 5.26.

Now suppose that we use one of the other two fuel types, for example, F_2. Then $x_2 = 1$, $x_3 = 0$, and

$$E(y) = \beta_0 + \beta_1 x_1 + \beta_2 x_1^2 + \beta_3(1) + \beta_4(0)$$
$$= (\beta_0 + \beta_3) + \beta_1 x_1 + \beta_2 x_1^2$$

This is the equation of exactly the same parabola that we obtained for fuel type F_1 except that the y-intercept has changed from β_0 to $(\beta_0 + \beta_3)$. Similarly, the response curve for F_3 is

$$E(y) = (\beta_0 + \beta_4) + \beta_1 x_1 + \beta_2 x_1^2$$

Therefore, the three response curves for levels F_1, F_2, and F_3 (shown in Figure 5.27) are identical except that they are shifted vertically upward or downward in relation to each other. The curves depict the situation when the two independent variables do not interact, that is, the effect of speed on mean performance is the same regardless of the fuel type used, and the effect of fuel type on mean performance is the same for all speeds (the relative distances between the curves is constant).

This noninteractive second-stage model has drawbacks similar to those of the simple first-stage model. It is highly unlikely that the response curves for the three fuel types would be identical except for differing y-intercepts. Because the model does not contain parameters that measure interaction between engine speed and

Figure 5.27 Model for $E(y)$ as a function of fuel type and engine speed (no interaction)

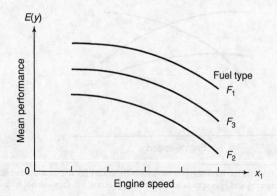

fuel type, we cannot test to see whether a relationship exists. Also, if interaction does exist, it will cause the SSE for the fitted model to be inflated and will consequently increase the errors of estimating model parameters $E(y)$.

This leads us to the final stage of the model-building process—adding interaction terms to allow the three response curves to differ in shape:

$$E(y) = \beta_0 + \overbrace{\beta_1 x_1 + \beta_2 x_1^2}^{\substack{\text{Main effect} \\ \text{terms for} \\ \text{engine speed}}} + \overbrace{\beta_3 x_2 + \beta_4 x_3}^{\substack{\text{Main effect} \\ \text{terms for} \\ \text{fuel type}}}$$

$$\overbrace{+ \beta_5 x_1 x_2 + \beta_6 x_1 x_3 + \beta_7 x_1^2 x_2 + \beta_8 x_1^2 x_3}^{\text{Interaction terms}}$$

where

$$x_1 = \text{Engine speed} \quad x_2 = \begin{cases} 1 & \text{if } F_2 \\ 0 & \text{if not} \end{cases} \quad x_3 = \begin{cases} 1 & \text{if } F_3 \\ 0 & \text{if not} \end{cases}$$

Notice that this model graphs as three different second-order curves.* If fuel type F_1 is used, we substitute $x_2 = x_3 = 0$ into the formula for $E(y)$, and all but the first three terms equal 0. The result is

$$E(y) = \beta_0 + \beta_1 x_1 + \beta_2 x_1^2$$

If F_2 is used, $x_2 = 1$, $x_3 = 0$, and

$$E(y) = \beta_0 + \beta_1 x_1 + \beta_2 x_1^2 + \beta_3(1) + \beta_4(0)$$
$$+ \beta_5 x_1(1) + \beta_6 x_1(0) + \beta_7 x_1^2(1) + \beta_8 x_1^2(0)$$
$$= (\beta_0 + \beta_3) + (\beta_1 + \beta_5)x_1 + (\beta_2 + \beta_7)x_1^2$$

The y-intercept, the coefficient of x_1, and the coefficient of x_1^2 differ from the corresponding coefficients in $E(y)$ at level F_1. Finally, when F_3 is used, $x_2 = 0$, $x_3 = 1$, and the result is

$$E(y) = (\beta_0 + \beta_4) + (\beta_1 + \beta_6)x_1 + (\beta_2 + \beta_8)x_1^2$$

A graph of the model for $E(y)$ might appear as shown in Figure 5.28. Compare this figure with Figure 5.26, where we assumed the response curves were identical for all three fuel types, and with Figure 5.27, where we assumed no interaction between

Figure 5.28 Graph of $E(y)$ as a function of fuel type and engine speed (interaction)

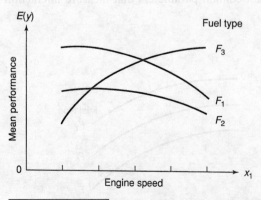

---

* Note that the model remains a second-order model for the quantitative independent variable x_1. The terms involving $x_1^2 x_2$ and $x_1^2 x_3$ appear to be third-order terms, but they are not because x_2 and x_3 are dummy variables.

the independent variables. Note in Figure 5.28 that the second-order curves may be completely different.

Now that you know how to write a model for two independent variables—one qualitative and one quantitative—we ask a question. Why do it? Why not write a separate second-order model for each level of fuel type where $E(y)$ is a function of engine speed only? *One reason we write the single model representing all three response curves is so that we can test to determine whether the curves are different.* For example, we might want to know whether the effect of fuel type depends on engine speed. Thus, one fuel type might be especially efficient at low engine speeds, but less so at high speeds. The reverse might be true for one of the other two fuel types. The hypothesis that the independent variables, fuel type and engine speed, affect the response independently of one another (a case of no interaction) is equivalent to testing the hypothesis that $\beta_5 = \beta_6 = \beta_7 = \beta_8 = 0$ [i.e., that the model in Figure 5.27 adequately characterizes $E(y)$] using the partial F-test discussed in Section 4.13. *A second reason for writing a single model is that we obtain a pooled estimate of σ^2, the variance of the random error component ε.* If the variance of ε is truly the same for each fuel type, the pooled estimate is superior to calculating three estimates by fitting a separate model for each fuel type.

In conclusion, suppose you want to write a model relating $E(y)$ to several quantitative and qualitative independent variables. Proceed in exactly the same manner as for two independent variables, one qualitative and one quantitative. First, write the model (using the methods of Sections 5.4 and 5.5) that you want to use to describe the quantitative independent variables. Then introduce the main effect and interaction terms for the qualitative independent variables. This gives a model that represents a set of identically shaped response surfaces, one corresponding to each combination of levels of the qualitative independent variables. If you could visualize surfaces in multidimensional space, their appearance would be analogous to the response curves in Figure 5.27. To complete the model, add terms for the interaction between the quantitative and qualitative variables. This is done by interacting *each* qualitative variable term with *every* quantitative variable term. We demonstrate with an example.

Example 5.14

A marine biologist wished to investigate the effects of three factors on the level of the contaminant DDT found in fish inhabiting a polluted lake. The factors were

1. Species of fish (two levels)
2. Location of capture (two levels)
3. Fish length (centimeters)

Write a model for the DDT level, y, found in contaminated fish.

Solution

The response y is affected by two qualitative factors (species and location), each at two levels, and one quantitative factor (length). Fish of each of the two species, S_1 and S_2, could be captured at each of the two locations, L_1 and L_2, giving $2 \times 2 = 4$ possible combinations—call them (S_1, L_1), (S_1, L_2), (S_2, L_1), (S_2, L_2). For each of these combinations, you would obtain a curve that graphs DDT level as a function of the quantitative factor x_1, fish length (see Figure 5.29). The stages in writing the model for the response y shown in Figure 5.29 are listed here.

STAGE 1 *Write a model relating y to the quantitative factor(s).* It is likely that an increase in the value of the single quantitative factor x_1, length, will yield an increase in DDT level. However, this increase is likely to be slower for larger fish,

Figure 5.29 A graphical portrayal of three factors—two qualitative and one quantitative—on DDT level

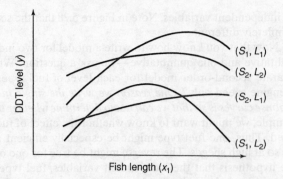

and will eventually level off once a fish reaches a certain length, thus producing the curvature shown in Figure 5.29. Consequently, we will model the mean DDT level, $E(y)$, with the second-order model.

$$E(y) = \beta_0 + \beta_1 x_1 + \beta_2 x_1^2$$

This is the model we would use if we were certain that the DDT curves were identical for all species–location combinations (S_i, L_j). The model would appear as shown in Figure 5.30a.

STAGE 2   *Add the terms, both main effect and interaction, for the qualitative factors.*

$$E(y) = \beta_0 + \overbrace{\beta_1 x_1 + \beta_2 x_1^2}^{\text{Terms for quantitative factor}}$$

$$+ \underbrace{\beta_3 x_2}_{\substack{\text{Main effect} \\ S}} + \underbrace{\beta_4 x_3}_{\substack{\text{Main effect} \\ L}} + \underbrace{\beta_5 x_2 x_3}_{\substack{SL \text{ interaction}}}$$

where

$$x_2 = \begin{cases} 1 & \text{if species } S_2 \\ 0 & \text{if species } S_1 \end{cases} \qquad x_3 = \begin{cases} 1 & \text{if location } L_2 \\ 0 & \text{if location } L_1 \end{cases}$$

This model implies that the DDT curves are identically shaped for each of the (S_i, L_j) combinations but that they possess different y-intercepts, as shown in Figure 5.30b.

STAGE 3   *Add terms to allow for interaction between the quantitative and qualitative factors.* This is done by interacting every pair of terms—one quantitative and

Figure 5.30 DDT curves for stages 1 and 2

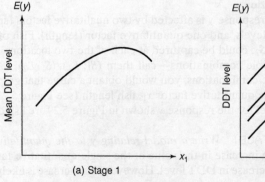

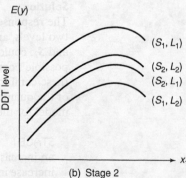

(a) Stage 1   (b) Stage 2

one qualitative. Thus, the complete model, which graphs as four different-shaped second-order curves (see Figure 5.29), is

$$E(y) = \overbrace{\beta_0 + \beta_1 x_1 + \beta_2 x_1^2}^{\text{First-stage terms}}$$

$$\overbrace{+ \beta_3 x_2 + \beta_4 x_3 + \beta_5 x_2 x_3}^{\text{Second-stage terms}}$$

$$\overbrace{+ \beta_6 x_1 x_2 + \beta_7 x_1 x_3 + \beta_8 x_1 x_2 x_3 + \beta_9 x_1^2 x_2 + \beta_{10} x_1^2 x_3 + \beta_{11} x_1^2 x_2 x_3}^{\text{Third-stage terms}}$$

Example 5.15

Use the model in Example 5.14 to find the equation relating $E(y)$ to x_1 for species S_1 and location L_2.

Solution

Checking the coding for the model, we see (noted at the second stage) that when DDT level y is measured on a fish of species S_1 captured at location L_2, we set $x_2 = 0$ and $x_3 = 1$. Substituting these values into the complete model, we obtain

$$E(y) = \beta_0 + \beta_1 x_1 + \beta_2 x_1^2$$

$$+ \beta_3 x_2 + \beta_4 x_3 + \beta_5 x_2 x_3$$

$$+ \beta_6 x_1 x_2 + \beta_7 x_1 x_3 + \beta_8 x_1 x_2 x_3 + \beta_9 x_1^2 x_2 + \beta_{10} x_1^2 x_3 + \beta_{11} x_1^2 x_2 x_3$$

$$= \beta_0 + \beta_1 x_1 + \beta_2 x_1^2$$

$$+ \beta_3(0) + \beta_4(1) + \beta_5(0)(1)$$

$$+ \beta_6 x_1(0) + \beta_7 x_1(1) + \beta_8 x_1(0)(1)$$

$$+ \beta_9 x_1^2(0) + \beta_{10} x_1^2(1) + \beta_{11} x_1^2(0)(1)$$

$$= (\beta_0 + \beta_4) + (\beta_1 + \beta_7)x_1 + (\beta_2 + \beta_{10})x_1^2$$

Note that this equation graphs as a portion of a parabola with y-intercept equal to $(\beta_0 + \beta_4)$. The coefficient of x_1 is $(\beta_1 + \beta_7)$, and the curvature coefficient (the coefficient of x_1^2) is $(\beta_2 + \beta_{10})$.

Example 5.16

Suppose you have two qualitative independent variables, A and B, and A is at two levels and B is at three levels. You also have two quantitative independent variables, C and D, each at three levels. Further suppose you plan to fit a second-order response surface as a function of the quantitative independent variables C and D, and that you want your model for $E(y)$ to allow for different shapes of the second-order surfaces for the six (2×3) combinations of levels corresponding to the qualitative independent variables A and B. Write a model for $E(y)$.

Solution

STAGE 1   *Write the second-order model corresponding to the two quantitative independent variables.* If we let

$$x_1 = \text{Level for independent variable } C$$

$$x_2 = \text{Level for independent variable } D$$

then

$$E(y) = \beta_0 + \beta_1 x_1 + \beta_2 x_2 + \beta_3 x_1 x_2 + \beta_4 x_1^2 + \beta_5 x_2^2$$

This is the model you would use if you believed that the six response surfaces, corresponding to the six combinations of levels of A and B, were identical.

STAGE 2 *Add the main effect and interaction terms for the qualitative independent variables.* These are

$$+ \underbrace{\beta_6 x_6}_{\substack{\text{Main effect} \\ \text{term for } A}} + \underbrace{\beta_7 x_7 + \beta_8 x_8}_{\substack{\text{Main effect} \\ \text{terms for } B}} + \underbrace{\beta_9 x_6 x_7 + \beta_{10} x_6 x_8}_{\substack{AB \text{ interaction} \\ \text{terms}}}$$

where $x_6 = \begin{cases} 1 & \text{if at level } A_2 \\ 0 & \text{if not} \end{cases}$ $x_7 = \begin{cases} 1 & \text{if at level } B_2 \\ 0 & \text{if not} \end{cases}$ $x_8 = \begin{cases} 1 & \text{if at level } B_3 \\ 0 & \text{if not} \end{cases}$

The addition of these terms to the model produces six identically shaped second-order surfaces, one corresponding to each of the six combinations of levels of A and B. They differ only in their y-intercepts.

STAGE 3 *Add terms that allow for interaction between the quantitative and qualitative independent variables.* This is done by interacting each of the five qualitative independent variable terms (both main effect and interaction) with each term (except β_0) of the quantitative first-stage model. Thus,

$$E(y) = \beta_0 + \beta_1 x_1 + \beta_2 x_2 + \beta_3 x_1 x_2 + \beta_4 x_1^2 + \beta_5 x_2^2 \qquad \left.\begin{matrix} \\ \end{matrix}\right\} \begin{matrix} \text{First-stage} \\ \text{model} \end{matrix}$$

$$+ \underbrace{\beta_6 x_6}_{\substack{\text{Main effect} \\ A}} + \underbrace{\beta_7 x_7 + \beta_8 x_8}_{\substack{\text{Main effect} \\ B}} + \underbrace{\beta_9 x_6 x_7 + \beta_{10} x_6 x_8}_{AB \text{ interaction}} \quad \left.\begin{matrix} \\ \\ \\ \end{matrix}\right\} \begin{matrix} \text{Portion} \\ \text{added to} \\ \text{form second-} \\ \text{stage model} \end{matrix}$$

$$+ \beta_{11} x_6 x_1 + \beta_{12} x_6 x_2 + \beta_{13} x_6 x_1 x_2 + \beta_{14} x_6 x_1^2 + \beta_{15} x_6 x_2^2 \quad \left.\begin{matrix} \\ \\ \end{matrix}\right\} \begin{matrix} \text{Interacting } x_6 \\ \text{with the} \\ \text{quantitative} \\ \text{terms} \end{matrix}$$

$$+ \beta_{16} x_7 x_1 + \beta_{17} x_7 x_2 + \beta_{18} x_7 x_1 x_2 + \beta_{19} x_7 x_1^2 + \beta_{20} x_7 x_2^2 \quad \left.\begin{matrix} \\ \\ \end{matrix}\right\} \begin{matrix} \text{Interacting } x_7 \\ \text{with the} \\ \text{quantitative} \\ \text{terms} \end{matrix}$$

$$+ \cdots \qquad\qquad\qquad\qquad\qquad\qquad\qquad\qquad\qquad\quad \cdots$$

$$\begin{aligned} &+ \beta_{31} x_6 x_8 x_1 + \beta_{32} x_6 x_8 x_2 + \beta_{33} x_6 x_8 x_1 x_2 \\ &+ \beta_{34} x_6 x_8 x_1^2 + \beta_{35} x_6 x_8 x_2^2 \end{aligned} \quad \left.\begin{matrix} \\ \\ \end{matrix}\right\} \begin{matrix} \text{Interacting} \\ x_6 x_8 \text{ with the} \\ \text{quantitative} \\ \text{terms} \end{matrix}$$

Note that the complete model contains 36 terms, one for β_0, five needed to complete the second-order model in the two quantitative variables, five for the two qualitative variables, and $5 \times 5 = 25$ terms for the interactions between the quantitative and qualitative variables.   ■

To see how the model gives different second-order surfaces—one for each combination of the levels of variables A and B—consider the next example.

Example 5.17

Refer to Example 5.16. Find the response surface that portrays $E(y)$ as a function of the two quantitative independent variables C and D for the (A_1, B_2) combination of levels of the qualitative independent variables.

Solution

Checking the coding, we see that when y is observed at the first level of A (level A_1) and the second level of B (level B_2), the dummy variables take the following values: $x_6 = 0$, $x_7 = 1$, $x_8 = 0$. Substituting these values into the formula for the complete model (and deleting the terms that equal 0), we obtain

$$E(y) = \beta_0 + \beta_1 x_1 + \beta_2 x_2 + \beta_3 x_1 x_2 + \beta_4 x_1^2 + \beta_5 x_2^2 + \beta_7 + \beta_{16} x_1 + \beta_{17} x_2$$
$$+ \beta_{18} x_1 x_2 + \beta_{19} x_1^2 + \beta_{20} x_2^2$$
$$= (\beta_0 + \beta_7) + (\beta_1 + \beta_{16}) x_1 + (\beta_2 + \beta_{17}) x_2 + (\beta_3 + \beta_{18}) x_1 x_2$$
$$+ (\beta_4 + \beta_{19}) x_1^2 + (\beta_5 + \beta_{20}) x_2^2$$

Note that this is the equation of a second-order model for $E(y)$. It graphs the response surface for $E(y)$ when the qualitative independent variables A and B are at levels A_1 and B_2. ∎

5.10 Exercises

5.31 Workplace bullying and intention to leave. Refer to the *Human Resource Management Journal* (October 2008) study of workplace bullying, Exercise 5.2 (p. 264). Recall that the researchers employed multiple regression to model a bullying victim's intention to leave the firm (y) as a function of level of bullying (measured on a 50-point scale) and perceived organizational support (measured as "low," "neutral," or "high").

(a) Write a complete second-order model for $E(y)$ as a function of the independent variables.
(b) In terms of the β's in the model, part a, give the mean value of intention to leave for a victim who reports a bullying level of 25 points and who perceives organizational support as low.
(c) How would you test whether the terms in the model that allow for a curvilinear relationship between intent to leave and level of bullying are statistically useful?
(d) Write a first-order model for $E(y)$ as a function of the independent variables that incorporates interaction.
(e) Refer to the model, part d. Demonstrate that the model graphs out as three nonparallel straight lines, one line for each level of perceived organizational support. As part of your answer, give the slope of each line in terms of the β's.

5.32 Chemical composition of rain water. Refer to the *Journal of Agricultural, Biological, and Envi-*

ronmental Statistics (March 2005) study of the chemical composition of rain water, Exercise 5.4 (p. 264). Recall that the researchers want to model the nitrate concentration, y (milligrams per liter), in a rain water sample as a function of two independent variables: water source (groundwater, subsurface flow, or overground flow) and silica concentration (milligrams per liter).

(a) Write a first-order model for $E(y)$ as a function of the independent variables. Assume that the rate of increase of nitrate concentration with silica concentration is the same for all three water sources. Sketch the relationships hypothesized by the model on a graph.
(b) Write a first-order model for $E(y)$ as a function of the independent variables, but now assume that the rate of increase of nitrate concentration with silica concentration differs for the three water sources. Sketch the relationships hypothesized by the model on a graph.

5.33 Quality of Bordeaux wine. Refer to the *Economic Journal* (May 2008) study of the factors that yield a quality Bordeaux wine, Exercise 5.27 (p. 302). In addition to grape-picking method (manual or automated) and soil type (clay, gravel, or sand), slope orientation (east, south, west, southeast, or southwest) was also used to model wine quality (y).

(a) Write a complete model for $E(y)$ as a function of the three qualitative independent variables.
(b) Interpret the value of β_0 in the model.

(c) In terms of the β's, what is the mean quality of wine produced from grapes picked manually from clay soil with an east slope orientation?

(d) In terms of the β's, what is the difference between the mean quality of wine produced from grapes picked manually and grapes picked with an automated method when the soil type is sand with a southwest slope orientation?

(e) For any slope orientation, assume that the difference between the mean quality of wine produced from grapes picked manually and grapes picked with an automated method is the same for clay, gravel, or sand soil. Under this assumption, which β's in the model will equal 0?

GASTURBINE

5.34 Cooling method for gas turbines. Refer to the *Journal of Engineering for Gas Turbines and Power* (January 2005) study of a high-pressure inlet fogging method for a gas turbine engine, Exercise 5.19 (p. 281). Recall that you analyzed a model for heat rate (kilojoules per kilowatt per hour) of a gas turbine as a function of cycle speed (revolutions per minute) and cycle pressure ratio. Now consider a qualitative predictor, engine type, at three levels (traditional, advanced, and aeroderivative).

(a) Write a complete second-order model for heat rate (y) as a function of cycle speed, cycle pressure ratio, and engine type.

(b) Demonstrate that the model graphs out as three second-order response surfaces, one for each level of engine type.

(c) Fit the model to the data in the GASTUR-BINE file and give the least squares prediction equation.

(d) Conduct a global F-test for overall model adequacy.

(e) Conduct a test to determine whether the second-order response surface is identical for each level of engine type.

5.35 Lead in fern moss. A study of the atmospheric pollution on the slopes of the Blue Ridge Mountains (Tennessee) was conducted. The file LEAD-MOSS contains the levels of lead found in 70 fern moss specimens (in micrograms of lead per gram of moss tissue) collected from the mountain slopes, as well as the elevation of the moss specimen (in feet) and the direction (1 if east, 0 if west) of the slope face. The first five and last five observations of the data set are listed in the table.

(a) Write the equation of a first-order model relating mean lead level, $E(y)$, to elevation (x_1) and slope face (x_2). Include interaction between elevation and slope face in the model.

LEADMOSS

SPECIMEN	LEAD LEVEL	ELEVATION	SLOPE FACE
1	3.475	2000	0
2	3.359	2000	0
3	3.877	2000	0
4	4.000	2500	0
5	3.618	2500	0
⋮	⋮	⋮	⋮
66	5.413	2500	1
67	7.181	2500	1
68	6.589	2500	1
69	6.182	2000	1
70	3.706	2000	1

Source: Schilling, J. "Bioindication of atmospheric heavy metal deposition in the Blue Ridge using the moss, *Thuidium delicatulum*," Master of Science thesis, Spring 2000.

(b) Graph the relationship between mean lead level and elevation for the different slope faces that is hypothesized by the model, part a.

(c) In terms of the β's of the model, part a, give the change in lead level for every one foot increase in elevation for moss specimens on the east slope.

(d) Fit the model, part a, to the data using an available statistical software package. Is the overall model statistically useful for predicting lead level? Test using $\alpha = .10$.

(e) Write the equation of the complete second-order model relating mean lead level, $E(y)$, to elevation (x_1) and slope face (x_2).

5.36 Improving SAT scores. Refer to the *Chance* (Winter 2001) study of students who paid a private tutor (or coach) to help them improve their Standardized Assessment Test (SAT) scores, Exercise 4.50 (p. 223). Recall that multiple regression was used to estimate the effect of coaching on SAT–Mathematics scores, where

$$y = \text{SAT–Math score}$$

$$x_1 = \text{score on PSAT}$$

$$x_2 = \{1 \text{ if student was coached, } 0 \text{ if not}\}$$

(a) Write a complete second-order model for $E(y)$ as a function of x_1 and x_2.

(b) Give the equation of the curve relating $E(y)$ to x_1 for noncoached students. Identify the y-intercept, shift parameter, and rate of curvature in the equation.

(c) Repeat part b for students who have been coached on the SAT.

(d) How would you test to determine if coaching has an effect on SAT–Math scores?

5.37 **Using glass to encapsulate waste.** Since glass is not subject to radiation damage, encapsulation of waste in glass is considered to be one of the most promising solutions to the problem of low-level nuclear waste in the environment. However, glass undergoes chemical changes when exposed to extreme environmental conditions, and certain of its constituents can leach into the surroundings. In addition, these chemical reactions may weaken the glass. These concerns led to a study undertaken jointly by the Department of Materials Science and Engineering at the University of Florida and the U.S. Department of Energy to assess the utility of glass as a waste encapsulant material.[†] Corrosive chemical solutions (called corrosion baths) were prepared and applied directly to glass samples containing one of three types of waste (TDS-3A, FE, and AL); the chemical reactions were observed over time. A few of the key variables measured were

y = Amount of silicon (in parts per million) found in solution at end of experiment. (This is both a measure of the degree of

breakdown in the glass and a proxy for the amount of radioactive species released) into the environment.

x_1 = Temperature (°C) of the corrosion bath

$$x_2 = \begin{cases} 1 & \text{if waste} \\ & \text{type TDS-3A} \\ 0 & \text{if not} \end{cases} \quad x_3 = \begin{cases} 1 & \text{if waste} \\ & \text{type FE} \\ 0 & \text{if not} \end{cases}$$

Waste type AL is the base level. Suppose we want to model amount y of silicon as a function of temperature (x_1) and type of waste (x_2, x_3).

(a) Write a model that proposes parallel straight-line relationships between amount of silicon and temperature, one line for each of the three waste types.

(b) Add terms for the interaction between temperature and waste type to the model of part a.

(c) Refer to the model of part b. For each waste type, give the slope of the line relating amount of silicon to temperature.

(d) Explain how you could test for the presence of temperature–waste type interaction.

5.11 External Model Validation (Optional)

Regression analysis is one of the most widely used statistical tools for estimation and prediction. All too frequently, however, a regression model deemed to be an adequate predictor of some response y performs poorly when applied in practice. For example, a model developed for forecasting new housing starts, although found to be statistically useful based on a test for overall model adequacy, may fail to take into account any extreme changes in future home mortgage rates generated by new government policy. This points out an important problem. *Models that fit the sample data well may not be successful predictors of y when applied to new data.* For this reason, it is important to assess the **validity** of the regression model in addition to its **adequacy** before using it in practice.

In Chapter 4, we presented several techniques for checking *model adequacy* (e.g., tests of overall model adequacy, partial F-tests, R_a^2, and s). In short, checking model adequacy involves determining whether the regression model adequately fits the *sample data*. **Model validation**, however, involves an assessment of how the fitted regression model will perform in practice—that is, how successful it will be when applied to new or future data. A number of different model validation techniques have been proposed, several of which are briefly discussed in this section. You will need to consult the references for more details on how to apply these techniques.

1. *Examining the predicted values*: Sometimes, the predicted values $\hat{y}$ of the fitted regression model can help to identify an invalid model. Nonsensical or unreasonable predicted values may indicate that the form of the model is incorrect or that the β coefficients are poorly estimated. For example, a model for a binary response y, where y is 0 or 1, may yield predicted probabilities that

---

[†] The background information for this exercise was provided by Dr. David Clark, Department of Materials Science and Engineering, University of Florida.

are negative or greater than 1. In this case, the user may want to consider a model that produces predicted values between 0 and 1 in practice (One such model, called the *logistic regression model*, is covered in Chapter 9.) On the other hand, if the predicted values of the fitted model all seem reasonable, the user should refrain from using the model in practice until further checks of model validity are carried out.

2. *Examining the estimated model parameters*: Typically, the user of a regression model has some knowledge of the relative size and sign (positive or negative) of the model parameters. This information should be used as a check on the estimated β coefficients. Coefficients with signs opposite to what is expected or with unusually small or large values or unstable coefficients (i.e., coefficients with large standard errors) forewarn that the final model may perform poorly when applied to new or different data.

3. *Collecting new data for prediction*: One of the most effective ways of validating a regression model is to use the model to predict y for a new sample. By directly comparing the predicted values to the observed values of the new data, we can determine the accuracy of the predictions and use this information to assess how well the model performs in practice.

Several measures of model validity have been proposed for this purpose. One simple technique is to calculate the percentage of variability in the new data explained by the model, denoted $R^2_{\text{prediction}}$, and compare it to the coefficient of determination R^2 for the least squares fit of the final model. Let $y_1, y_2, \ldots, y_n$ represent the n observations used to build and fit the final regression model and $y_{n+1}, y_{n+2}, \ldots, y_{n+m}$ represent the m observations in the new data set. Then

$$R^2_{\text{prediction}} = 1 - \left\{ \frac{\displaystyle\sum_{i=n+1}^{n+m} (y_i - \hat{y}_i)^2}{\displaystyle\sum_{i=n+1}^{n+m} (y_i - \bar{y})^2} \right\}$$

where $\hat{y}_i$ is the predicted value for the ith observation using the β estimates from the fitted model and $\bar{y}$ is the sample mean of the original data.* If $R^2_{\text{prediction}}$ compares favorably to R^2 from the least squares fit, we will have increased confidence in the usefulness of the model. However, if a significant drop in R^2 is observed, we should be cautious about using the model for prediction in practice.

A similar type of comparison can be made between the mean square error, MSE, for the least squares fit and the mean squared prediction error

$$\text{MSE}_{\text{prediction}} = \frac{\displaystyle\sum_{i=n+1}^{n+m} (y_i - \hat{y}_i)^2}{m - (k+1)}$$

where k is the number of β coefficients (excluding β_0) in the model. Whichever measure of model validity you decide to use, the number of observations in the new data set should be large enough to reliably assess the model's prediction performance. Montgomery, Peck, and Vining (2006), for example, recommend 15–20 new observations, *at minimum*.

---

* Alternatively, the sample mean of the new data may be used.

4. *Data-splitting (cross-validation)*: For those applications where it is impossible or impractical to collect new data, the original sample data can be split into two parts, with one part used to estimate the model parameters and the other part used to assess the fitted model's predictive ability. **Data-splitting** (or **cross-validation**, as it is sometimes known) can be accomplished in a variety of ways. A common technique is to randomly assign half the observations to the estimation data set and the other half to the prediction data set.[†] Measures of model validity, such as $R^2_{prediction}$ or $MSE_{prediction}$, can then be calculated. Of course, a sufficient number of observations must be available for data-splitting to be effective. For the estimation and prediction data sets of equal size, it has been recommended that the entire sample consist of *at least* $n = 2k + 25$ observations, where k is the number of β parameters in the model [see Snee (1977)].

5. *Jackknifing*: In situations where the sample data set is too small to apply data-splitting, a method called the **jackknife** can be applied. Let $y_{(i)}$ denote the predicted value for the ith observation obtained when the regression model is fit with the data point for y_i omitted (or deleted) from the sample. The jackknife method involves leaving each observation out of the data set, one at a time, and calculating the difference, $y_i - \hat{y}_{(i)}$, for all n observations in the data set. Measures of model validity, such as R^2 and MSE, are then calculated:

$$R^2_{jackknife} = 1 - \frac{\Sigma(y_i - \hat{y}_{(i)})^2}{\Sigma(y_i - \bar{y})^2}$$

$$MSE_{jackknife} = \frac{\Sigma(y_i - \hat{y}_{(i)})^2}{n - (k + 1)}$$

The numerator of both $R^2_{jackknife}$ and $MSE_{jackknife}$ is called the **prediction sum of squares**, or **PRESS**. In general, PRESS will be larger than the SSE of the fitted model. Consequently, $R^2_{jackknife}$ will be smaller than the R^2 of the fitted model and $MSE_{jackknife}$ will be larger than the MSE of the fitted model. These jackknife measures, then, give a more conservative (and more realistic) assessment of the ability of the model to predict future observations than the usual measures of model adequacy.

Example 5.18

In Chapter 4 (Example 4.10), we presented a model for executive salary (y) developed by Towers, Perrin, Forster & Crosby, an international management consulting firm. The multiplicative model fit was

$$E\{\ln(y)\} = \beta_0 + \beta_1 x_1 + \beta_2 x_2 + \beta_3 x_3 + \beta_4 x_4 + \beta_5 x_5 + \beta_6 x_1^2 + \beta_7 x_3 x_4$$

where $x_1 =$ years of experience, $x_2 =$ years of education, $x_3 = \{1$ if male, 0 if female$\}$, $x_4 =$ number of employees supervised, and $x_5 =$ corporate assets. Since the consulting firm intends to use the model in evaluating executive salaries at a variety of companies, it is important to validate the model externally. Apply one of the model validation techniques discussed in this section to the data for the $n = 100$ executives saved in the EXECSAL file.

 EXECSAL

Solution

With $n = 100$ observations, a data-splitting method could be employed to validate the model. For example, 80 observations (randomly selected from the sample) could

---

[†] Random splits are usually applied in cases where there is no logical basis for dividing the data. Consult the references for other, more formal data-splitting techniques.

be used to estimate the prediction equation, and the remaining 20 observations used to validate the results. Ideally, we would like to have more observations (say, 50) in the validation subset. However, this would dramatically reduce the sample size for the estimation subset and possibly lead to less reliable results.

As an alternative, we use the jackknifing (one-observation-out-at-a-time) approach to model validation. Most statistical software packages have routines that automatically perform the jackknifing and produce the PRESS statistic. Figure 5.31 is a MINITAB printout for the multiplicative model fit to the data in the EXECSAL file. The value of PRESS (highlighted) is .482665. We compare this jackknifed value of SSE to the total sum of squares (also highlighted on the printout) value of 6.68240 as follows:

$$R^2_{\text{jackknife}} = 1 - (\text{PRESS}/\text{SSTotal}) = 1 - (.482665/6.68240) = .92777$$

(Note that $R^2_{\text{jackknife}}$ is also highlighted in Figure 5.31.) Since $R^2_{\text{jackknife}}$ is only slightly smaller than the R^2 of the original model fit (.94), the consulting firm has increased confidence in using the model for evaluation of executive salaries.   ◼

Figure 5.31 MINITAB printout for the multiplicative model of executive salary

Regression Analysis: LNSAL versus EXP, EDUC, ...

```
The regression equation is
LNSAL = 9.86 + 0.0436 EXP + 0.0309 EDUC + 0.117 GENDER + 0.000326 NUMSUP
        + 0.00239 ASSETS - 0.000635 EXPSQ + 0.000302 GEN_SUP


Predictor        Coef      SE Coef        T       P
Constant      9.86182      0.09703   101.64   0.000
EXP          0.043643     0.003761    11.60   0.000
EDUC         0.030936     0.002950    10.49   0.000
GENDER        0.11661      0.03696     3.16   0.002
NUMSUP     0.00032594   0.00007850     4.15   0.000
ASSETS      0.0023911    0.0004439     5.39   0.000
EXPSQ      -0.0006348    0.0001383    -4.59   0.000
GEN_SUP    0.00030196   0.00009238     3.27   0.002


S = 0.0659583   R-Sq = 94.0%   R-Sq(adj) = 93.6%

PRESS = 0.482665   R-Sq(pred) = 92.78%


Analysis of Variance

Source          DF       SS       MS        F       P
Regression       7  6.28215  0.89745   206.29   0.000
Residual Error  92  0.40025  0.00435
Total           99  6.68240
```

The appropriate model validation technique(s) you employ will vary from application to application. Keep in mind that a favorable result is still no guarantee that the model will always perform successfully in practice. However, we have much greater confidence in a validated model than in one that simply fits the sample data well.

Quick Summary/Guides

KEY FORMULAS

Coding Quantitative x's

$u = (x - \bar{x})/s_x$, where $\bar{x}$ and s are the mean and standard deviation of x

Cross-validation

$$R^2_{\text{prediction}} = 1 - \sum_{i=n+1}^{n+m} (y_i - \hat{y}_i)^2 \bigg/ \sum_{i=n+1}^{n+m} (y_i - \bar{y})^2$$

$$\text{MSE}_{\text{prediction}} = \sum_{i=n+1}^{n+m} (y_i - \hat{y}_i)^2 \bigg/ [m - (k+1)]$$

$$R^2_{\text{jackknife}} = 1 - \sum_{i=1}^{n} (y_i - \hat{y}_{(i)})^2 \bigg/ \sum_{i=1}^{n} (y_i - \bar{y})^2$$

$$\text{MSE}_{\text{jackknife}} = \sum_{i=1}^{n} (y_i - \hat{y}_{(i)})^2 \bigg/ [n - (k+1)]$$

KEY IDEAS

Steps in Model Building

1. Identify the response (*dependent*) variable y
2. Classify each potential predictor *(independent)* variable as *quantitative* or *qualitative*
3. Define *dummy variables* to represent the qualitative independent variables
4. Consider *higher-order* terms (e.g., x^2, x^3) for quantitative variables
5. Possibly *code the quantitative variables* in higher-order polynomials
6. Consider *interaction* terms for both quantitative and qualitative independent variables
7. Compare *nested* models using *partial F-tests* to arrive at a final model
8. Consider validation of the final model using data-splitting or jackknifing

Procedure for Writing a Complete Second-order Model

1. Enter terms for all *quantitative* x's, including interactions and second-order terms

2. Enter terms for all *qualitative* x's, including main effects, two-way, three-way, . . . , and k-way interactions

3. Enter terms for all possible *quantitative by qualitative interactions,* that is interact all terms in Step 1 with all terms in Step 2

Models with One Quantitative x

First-order: $E(y) = \beta_0 + \beta_1 x$
Second-order: $E(y) = \beta_0 + \beta_1 x + \beta_2 x^2$
pth-order: $E(y) = \beta_0 + \beta_1 x + \beta_2 x^2 + \cdots + \beta_p x^p$

Models with Two Quantitative x's

First-order: $E(y) = \beta_0 + \beta_1 x_1 + \beta_2 x_2$
Second-order, interaction only: $E(y) = \beta_0 + \beta_1 x_1 + \beta_2 x_2 + \beta_3 x_1 x_2$
Complete second-order: $E(y) = \beta_0 + \beta_1 x_1 + \beta_2 x_2 + \beta_3 x_1 x_2 + \beta_4 x_1^2 + \beta_5 x_2^2$

Models with Three Quantitative x's

First-order: $E(y) = \beta_0 + \beta_1 x_1 + \beta_2 x_2 + \beta_3 x_3$
Second-order, interaction only: $E(y) = \beta_0 + \beta_1 x_1 + \beta_2 x_2 + \beta_3 x_3 + \beta_4 x_1 x_2 + \beta_5 x_1 x_3 + \beta_6 x_2 x_3$
Complete second-order: $E(y) = \beta_0 + \beta_1 x_1 + \beta_2 x_2 + \beta_3 x_3 + \beta_4 x_1 x_2 + \beta_5 x_1 x_3 + \beta_6 x_2 x_3 + \beta_7 x_1^2 + \beta_8 x_2^2 + \beta_9 x_3^2$

Model with One Qualitative x (k levels)

$E(y) = \beta_0 + \beta_1 x_1 + \beta_2 x_2 + \beta_3 x_3 + \cdots + \beta_{k-1} x_k,$

where $x_1 = \{1$ if level 1, 0 if not$\}$, $x_2 = \{1$ if level 2, 0 if not$\}$, . . . $x_{k-1} = \{1$ if level $k-1$, 0 if not$\}$

Models with Two Qualitative x's (one at two levels, one at three levels)

Main effects: $E(y) = \beta_0 + \beta_1 x_1 + \beta_2 x_2 + \beta_3 x_3,$

where x_1 represents the dummy variable for the qualitative variable at two levels;

x_2 and x_3 represent the dummy variables for the qualitative variable at three levels

Interaction: $E(y) = \beta_0 + \beta_1 x_1 + \beta_2 x_2 + \beta_3 x_3 + \beta_4 x_1 x_2 + \beta_5 x_1 x_3$

Models with One Quantitative x and One Qualitative x (at three levels)

First-order, no interaction: $E(y) = \beta_0 + \beta_1 x_1 + \beta_2 x_2 + \beta_3 x_3,$

where x_2 and x_3 represent the dummy variables for the qualitative variable at three levels

First-order, interaction: $E(y) = \beta_0 + \beta_1 x_1 + \beta_2 x_2 + \beta_3 x_3 + \beta_4 x_1 x_2 + \beta_5 x_1 x_3$

Second-order, no interaction: $E(y) = \beta_0 + \beta_1 x_1 + \beta_2 x_1^2 + \beta_3 x_2 + \beta_4 x_3$

Complete Second-order: $E(y) = \beta_0 + \beta_1 x_1 + \beta_2 x_1^2 + \beta_3 x_2 + \beta_4 x_3 + \beta_5 x_1 x_2 + \beta_6 x_1 x_3 + \beta_7 x_1^2 x_2 + \beta_8 x_1^2 x_3$

Models with Two Quantitative *x*'s and Two Qualitative *x*'s (both at two levels)

First-order, no interaction: $E(y) = \beta_0 + \beta_1 x_1 + \beta_2 x_2 + \beta_3 x_3 + \beta_4 x_4,$

where x_3 and x_4 represent the dummy variables for the qualitative variables

Complete first-order:

$$E(y) = \beta_0 + \beta_1 x_1 + \beta_2 x_2 + \beta_3 x_3 + \beta_4 x_4 + \beta_5 x_3 x_4$$
$$+ \beta_6 x_1 x_3 + \beta_7 x_1 x_4 + \beta_8 x_1 x_3 x_4$$
$$+ \beta_9 x_2 x_3 + \beta_{10} x_2 x_4 + \beta_{11} x_2 x_3 x_4$$

Second-order, no QN × QL interaction:

$$E(y) = \beta_0 + \beta_1 x_1 + \beta_2 x_2 + \beta_3 x_1 x_2 + \beta_4 x_1^2 + \beta_5 x_2^2$$
$$+ \beta_6 x_3 + \beta_7 x_4 + \beta_8 x_3 x_4$$

Complete Second-order:

$$E(y) = \beta_0 + \beta_1 x_1 + \beta_2 x_2 + \beta_3 x_1 x_2 + \beta_4 x_1^2$$
$$+ \beta_5 x_2^2 + \beta_6 x_3 + \beta_7 x_4 + \beta_8 x_3 x_4$$
$$+ \beta_9 x_1 x_3 + \beta_{10} x_1 x_4 + \beta_{11} x_1 x_3 x_4$$
$$+ \beta_{12} x_2 x_3 + \beta_{13} x_2 x_4 + \beta_{14} x_2 x_3 x_4$$
$$+ \beta_{15} x_1 x_2 x_3 + \beta_{16} x_1 x_2 x_4 + \beta_{17} x_1 x_2 x_3 x_4$$
$$+ \beta_{18} x_1^2 x_3 + \beta_{19} x_1^2 x_4 + \beta_{20} x_1^2 x_3 x_4$$
$$+ \beta_{21} x_2^2 x_3 + \beta_{22} x_2^2 x_4 + \beta_{23} x_2^2 x_3 x_4$$

Supplementary Exercises

[Exercises from the optional sections are identified by an asterisk (*).]

5.38. Winning marathon times. *Chance* (Winter 2000) published a study of men's and women's winning times in the Boston Marathon. The researchers built a model for predicting winning time (y) of the marathon as a function of year (x_1) in which the race is run and gender (x_2) of the winning runner. Classify each variable in the model as quantitative or qualitative.

5.39. Snow goose feeding trial. Writing in the *Journal of Applied Ecology* (Vol. 32, 1995), botanists used multiple regression to model the weight change (y) of a baby snow goose subjected to a feeding trial. Three independent variables measured are listed below. Classify each as quantitative or qualitative.

(a) Diet type (plant or duck chow)
(b) Digestion efficiency (percentage)
(c) Amount of acid-fiber detergent added to diet (percentage)

5.40. CEOs of global corporations. *Business Horizons* (January–February 1993) conducted a comprehensive study of 800 chief executive officers who run the country's largest global corporations. The purpose of the study was to build a profile of the CEOs based on characteristics of each CEO's social background. Several of the variables measured for each CEO are listed here. Classify each variable as quantitative or qualitative.

(a) State of birth   (b) Age
(c) Education level   (d) Tenure with firm
(e) Total compensation (f) Area of expertise

5.41. Data in psychiatric client files. Psychiatrists keep data in psychiatric client files that contain important information on each client's background. The data in these files could be used to predict the probability that therapy will be successful. Identify the independent variables listed here as qualitative or quantitative.

(a) Age   (b) Years in therapy
(c) Highest educational degree (d) Job classification
(e) Religious preference   (f) Marital status
(g) IQ   (h) Gender

5.42. Using lasers for solar lighting. Refer to the *Journal of Applied Physics* (September 1993) study of solar lighting with semiconductor lasers, Exercise 4.80 (p. 244). The data for the analysis are repeated in the table on the next page.

(a) Give the equation relating the coded variable u to waveguide, x, using the coding system for observational data.
(b) Calculate the coded values, u.
(c) Calculate the coefficient of correlation r between the variables x and x^2.
(d) Calculate the coefficient of correlation r between the variables u and u^2. Compare this value to the value computed in part c.
(e) Fit the model

$$E(y) = \beta_0 + \beta_1 u + \beta_2 u^2$$

using available statistical software. Interpret the results.

LASERS

THRESHOLD CURRENT	WAVEGUIDE Al MOLE FRACTION
y, A/cm^2	x
273	.15
175	.20
146	.25
166	.30
162	.35
165	.40
245	.50
314	.60

Source: Unnikrishnan, S., and Anderson, N. G. "Quantum-well lasers for direct solar photopumping," *Journal of Applied Physics*, Vol. 74, No. 6, Sept. 15, 1993, p. 4226 (data adapted from Figure 2).

5.43. **Starting salaries of graduates.** Each semester, the University of Florida's Career Resource Center collects information on the job status and starting salary of graduating seniors. Data recently collected over a 2-year period included over 900 seniors who had found employment at the time of graduation. This information was used to model starting salary y as a function of two qualitative independent variables: college at five levels (Business Administration, Engineering, Liberal Arts & Sciences, Journalism, and Nursing) and gender at two levels (male and female). A main effects model relating starting salary, y, to both college and gender is

$$E(y) = \beta_0 + \beta_1 x_1 + \beta_2 x_2 + \beta_3 x_3 + \beta_4 x_4 + \beta_5 x_5$$

where

$$x_1 = \begin{cases} 1 & \text{if Business Administration} \\ 0 & \text{if not} \end{cases}$$

$$x_2 = \begin{cases} 1 & \text{if Engineering} \\ 0 & \text{if not} \end{cases}$$

$$x_3 = \begin{cases} 1 & \text{if Liberal Arts \& Sciences} \\ 0 & \text{if not} \end{cases}$$

$$x_4 = \begin{cases} 1 & \text{if Journalism} \\ 0 & \text{if not} \end{cases}$$

$$x_5 = \begin{cases} 1 & \text{if female} \\ 0 & \text{if male} \end{cases}$$

(a) Write the equation relating mean starting salary, $E(y)$, to college, for male graduates only.

(b) Interpret β_1 in the model, part a.
(c) Interpret β_2 in the model, part a.
(d) Interpret β_3 in the model, part a.
(e) Interpret β_4 in the model, part a.
(f) Write the equation relating mean starting salary, $E(y)$, to college, for female graduates only.
(g) Interpret β_1 in the model, part f. Compare to your answer, part b.
(h) Interpret β_2 in the model, part f. Compare to your answer, part c.
(i) Interpret β_3 in the model, part f. Compare to your answer, part d.
(j) Interpret β_4 in the model, part f. Compare to your answer, part e.
(k) For a given college, interpret the value of β_5 in the model.
(l) A multiple regression analysis revealed the following statistics for the β_5 term in the model: $\hat{\beta}_5 = -1,142.17$, $s_{\hat{\beta}_5} = 419.58$, t (for H_0: $\beta_5 = 0$) = -2.72, p-value = .0066. Make a statement about whether gender has an effect on average starting salary.

5.44. **Starting salaries of graduates (cont'd).** Refer to Exercise 5.43.

(a) Write an interaction model relating starting salary, y, to both college and gender. Use the dummy variables assignments made in Exercise 5.43.
(b) Interpret β_1 in the model, part a.
(c) Interpret β_2 in the model, part a.
(d) Interpret β_3 in the model, part a.
(e) Interpret β_4 in the model, part a.
(f) Interpret β_5 in the model, part a.
(g) Explain how to test to determine whether the difference between the mean starting salaries of male and female graduates depends on college.

5.45. **Ages heights of elementary schoolchildren.** Refer to the *Archives of Disease in Childhood* (April 2000) study of whether height influences a child's progression through elementary school, Exercise 4.84 (p. 245). Recall that Australian schoolchildren were divided into equal thirds (tertiles) based on age (youngest third, middle third, and oldest third). The average heights of the three groups (where all height measurements were standardized using z-scores), by gender, are repeated in the table (next page).

(a) Write a main effects model for the mean standardized height, $E(y)$, as a function of age tertile and gender.
(b) Interpret the β's in the main effects model, part a.

	YOUNGEST TERTILE MEAN HEIGHT	MIDDLE TERTILE MEAN HEIGHT	OLDEST TERTILE MEAN HEIGHT
BOYS	0.33	0.33	0.16
GIRLS	0.27	0.18	0.21

Source: Wake, M., Coghlan, D., & Hesketh, K. "Does height influence progression through primary school grades?" *The Archives of Disease in Childhood*, Vol. 82, Apr. 2000 (Table 3), with permission from BMJ Publishing Group Ltd.

(c) Write a model for the mean standardized height, $E(y)$, that includes interaction between age tertile and gender.

(d) Use the information in the table to find estimates of the β's in the interaction model, part c.

(e) How would you test the hypothesis that the difference between the mean standardized heights of boys and girls is the same across all three age tertiles?

5.46. Winning marathon times (cont'd). Refer to the *Chance* (Winter 2000) study of winning Boston Marathon times, Exercise 5.38 (p. 320). The independent variables used to model winning time y are:

x_1 = year in which race is run (expressed as number of years since 1880)

x_2 = {1 if winning runner is male, 0 if female}

(a) Write a first-order, main effects model for $E(y)$ as a function of year and gender.

(b) Interpret the β parameters of the model, part a.

(c) Write a model for $E(y)$ as a function of year and gender that hypothesizes different winning time-year slopes for male and female runners. Sketch the model.

(d) Now consider the independent quantitative variable

x_3 = number of marathons run prior to the Boston Marathon during the year

Write a complete second-order model that relates $E(y)$ to x_1 and x_3.

(e) Add the main effect term for gender to the model of part d.

(f) Add terms to the model of part e to allow for interaction between quantitative and qualitative terms.

(g) Under what circumstances will the response curves of the model of part f possess the same shape but have different y-intercepts when x_1 is held constant?

(h) Under what circumstances will the response curves of the model of part f be parallel lines when x_1 is held constant?

(i) Under what circumstances will the response curves of the model of part f be identical when x_1 is held constant?

5.47. Modeling product sales. A company wants to model the total weekly sales, y, of its product as a function of the variables packaging and location. Two types of packaging, P_1 and P_2, are used in each of four locations, L_1, L_2, L_3, and L_4.

(a) Write a main effects model to relate $E(y)$ to packaging and location. What implicit assumption are we making about the interrelationships between sales, packaging, and location when we use this model?

(b) Now write a model for $E(y)$ that includes interaction between packaging and location. How many parameters are in this model (remember to include β_0)? Compare this number to the number of packaging–location combinations being modeled.

(c) Suppose the main effects and interaction models are fit for 40 observations on weekly sales. The values of SSE are

SSE for main effects model $= 422.36$

SSE for interaction model $= 346.65$

Determine whether the data indicate that the interaction between location and packaging is important in estimating mean weekly sales. Use $\alpha = .05$. What implications does your conclusion have for the company's marketing strategy?

5.48. Durability of car paint. An automobile manufacturer is experimenting with a new type of paint that is supposed to help the car maintain its new-car look. The durability of this paint depends on the length of time the car body is in the oven after it has been painted. In the initial experiment, three groups of 10 car bodies each were baked for three different lengths of time—12, 24, and 36 hours—at the standard temperature setting. Then, the paint finish of each of the 30 cars was analyzed to determine a durability rating, y.

(a) Write a quadratic model relating the mean durability, $E(y)$, to the length of baking.

(b) Could a cubic model be fit to the data? Explain.

(c) Suppose the research and development department develops three new types of paint to be tested. Thus, 90 cars are to be tested—30 for each type of paint. Write the complete second-order model for $E(y)$ as a function of the type of paint and bake time.

5.49. **"Sheepskin screening" in recruiting.** Economic research has long established evidence of a positive correlation between earnings and educational attainment (*Economic Inquiry*, January 1984). However, it is unclear whether higher wage rates for better educated workers reflect an individual's added value or merely the employer's use of higher education as a screening device in the recruiting process. One version of this "sheepskin screening" hypothesis supported by many economists is that wages will rise faster with extra years of education when the extra years culminate in a certificate (e.g., master's or Ph.D. degree, CPA certificate, or actuarial degree).

(a) Write a first-order, main effects model for mean wage rate $E(y)$ of an employer as a function of employee's years of education and whether or not the employee is certified.

(b) Write a first-order model for $E(y)$ that corresponds to the "sheepskin screening" hypothesis.

(c) Write the complete second-order model for $E(y)$ as a function of the two independent variables.

*5.50 **Coding quantitative data.** Use the coding system for observational data to fit a second-order model to the data on demand y and price p given in the following table. Show that the inherent multicollinearity problem with fitting a polynomial model is reduced when the coded values of p are used.

DEMAND

DEMAND y, pounds	1,120	999	932	884	807	760	701	688
PRICE p, dollars	3.00	3.10	3.20	3.30	3.40	3.50	3.60	3.70

5.51. **Diesel engine performance.** An experiment was conducted to evaluate the performances of a diesel engine run on synthetic (coal-derived) and petroleum-derived fuel oil (*Journal of Energy Resources Technology*, March 1990). The petroleum-derived fuel used was a number 2 diesel fuel (DF-2) obtained from Phillips Chemical Company. Two synthetic fuels were used: a blended fuel (50% coal-derived and 50% DF-2) and a blended fuel with advanced timing. The brake power (kilowatts) and fuel type were varied in test runs, and engine performance was measured. The following table gives the experimental results for the performance measure, mass burning rate per degree of crank angle.

SYNFUELS

BRAKE POWER, x_1	FUEL TYPE	MASS BURNING RATE, y
4	DF-2	13.2
4	Blended	17.5
4	Advanced Timing	17.5
6	DF-2	26.1
6	Blended	32.7
6	Advanced Timing	43.5
8	DF-2	25.9
8	Blended	46.3
8	Advanced Timing	45.6
10	DF-2	30.7
10	Blended	50.8
10	Advanced Timing	68.9
12	DF-2	32.3
12	Blended	57.1

Source: Litzinger, T. A., and Buzza, T. G. "Performance and emissions of a diesel engine using a coal-derived fuel," *Journal of Energy Resources Technology*, Vol. 112, Mar. 1990, p. 32, Table 3.

The researchers fit the interaction model

$$E(y) = \beta_0 + \beta_1 x_1 + \beta_2 x_2 + \beta_3 x_3 + \beta_4 x_1 x_2 + \beta_5 x_1 x_3$$

where

$$y = \text{Mass burning rate}$$
$$x_1 = \text{Brake power (kW)}$$
$$x_2 = \begin{cases} 1 & \text{if DF-2 fuel} \\ 0 & \text{if not} \end{cases}$$
$$x_3 = \begin{cases} 1 & \text{if blended fuel} \\ 0 & \text{if not} \end{cases}$$

The results are shown in the SAS printout (p. 324).

(a) Conduct a test to determine whether brake power and fuel type interact. Test using $\alpha = .01$.

(b) Refer to the model, part a. Give the estimates of the slope of the y vrs. x_1 line for each of the three fuel types.

5.52. **Potency of a drug.** Eli Lilly and Company has developed three methods (G, R_1, and R_2) for estimating the shelf life of its drug products based on potency. One way to compare the three methods is to build a regression model for the dependent variable, estimated shelf life y (as a percentage of true shelf life), with potency of the drug (x_1) as a quantitative predictor and method as a qualitative predictor.

(a) Write a first-order, main effects model for $E(y)$ as a function of potency (x_1) and method.

SAS Output for Exercise 5.51

Dependent Variable: BURNRATE

Analysis of Variance

Source	DF	Sum of Squares	Mean Square	F Value	Pr > F
Model	5	3253.97929	650.79586	25.65	<.0001
Error	8	203.01000	25.37625		
Corrected Total	13	3456.98929			

Root MSE	5.03748	R-Square	0.9413	
Dependent Mean	36.29286	Adj R-Sq	0.9046	
Coeff Var	13.88010			

Parameter Estimates

Variable	DF	Parameter Estimate	Standard Error	t Value	Pr > |t|
Intercept	1	-10.83000	8.27743	-1.31	0.2271
POWER	1	7.81500	1.12642	6.94	0.0001
X2	1	19.35000	10.68612	1.81	0.1078
X3	1	12.79000	10.68612	1.20	0.2656
POWERX2	1	-5.67500	1.37957	-4.11	0.0034
POWERX3	1	-2.95000	1.37957	-2.14	0.0649

Test INTERACT Results for Dependent Variable BURNRATE

Source	DF	Mean Square	F Value	Pr > F
Numerator	2	223.03750	8.79	0.0096
Denominator	8	25.37625		

(b) Interpret the β coefficients of the model, part a.

(c) Write a first-order model for $E(y)$ that will allow the slopes to differ for the three methods.

(d) Refer to part c. For each method, write the slope of the $y-x_1$ line in terms of the β's.

5.53. Modeling industry performance. The performance of an industry is often measured by the level of excess (or unutilized) capacity within the industry. Researchers examined the relationship between excess capacity y and several market variables in 273 U.S. manufacturing industries (*Quarterly Journal of Business and Economics*, Summer 1986). Two qualitative independent variables considered in the study were Market concentration (low, moderate, and high) and Industry type (producer or consumer).

(a) Write the main effects model for $E(y)$ as a function of the two qualitative variables.

(b) Interpret the β coefficients in the main effects model.

(c) Write the model for $E(y)$ that includes interaction between market concentration and industry type.

(d) Interpret the β coefficients in the interaction model.

(e) How would you test the hypothesis that the difference between the mean excess capacity levels of producer and consumer industry types is the same across all three market concentrations?

*5.54. **Analyzing the data in Example 5.3.** Use the coding system for observational data to fit a complete second-order model to the data of Example 5.3, which are repeated below.

(a) Give the coded values u_1 and u_2 for x_1 and x_2, respectively.

(b) Compare the coefficient of correlation between x_1 and x_1^2 with the coefficient of correlation between u_1 and u_1^2.

(c) Compare the coefficient of correlation between x_2 and x_2^2 with the coefficient of correlation between u_2 and u_2^2.

(d) Give the prediction equation.

PRODQUAL

x_1	x_2	y	x_1	x_2	y	x_1	x_2	y
80	50	50.8	90	50	63.4	100	50	46.6
80	50	50.7	90	50	61.6	100	50	49.1
80	50	49.4	90	50	63.4	100	50	46.4
80	55	93.7	90	55	93.8	100	55	69.8
80	55	90.9	90	55	92.1	100	55	72.5
80	55	90.9	90	55	97.4	100	55	73.2
80	60	74.5	90	60	70.9	100	60	38.7
80	60	73.0	90	60	68.8	100	60	42.5
80	60	71.2	90	60	71.3	100	60	41.4

5.55. Racial make-up of labor markets. Research conducted at Ohio State University focused on the factors that influence the allocation of black and white men in labor market positions (*American Sociological Review*, June 1986). Data collected for each of 837 labor market positions were used to build a regression model for y, defined as the natural logarithm of the ratio of the proportion of blacks employed in a labor market position to the corresponding proportion of whites employed (called the *black–white log odds ratio*). Positive values of y indicate that blacks have a greater likelihood of employment than whites. Several independent variables were considered, including the following:

x_1 = Market power (a quantitative measure of the size and visibility of firms in the labor market)

x_2 = Percentage of workers in the labor market who are union members

$$x_3 = \begin{cases} 1 & \text{if labor market position} \\ & \text{includes craft occupations} \\ 0 & \text{if not} \end{cases}$$

(a) Write the first-order main effects model for $E(y)$ as a function of x_1, x_2, and x_3.

(b) One theory hypothesized by the researchers is that the mean log odds ratio $E(y)$ is smaller for craft occupations than for noncraft occupations. (That is, the likelihood of black employment is less for craft occupations.) Explain how to test this hypothesis using the model in part a.

(c) Write the complete second-order model for $E(y)$ as a function of x_1, x_2, and x_3.

(d) Using the model in part c, explain how to test the hypothesis that level of market power x_1 has no effect on black–white log odds ratio y.

(e) Holding x_2 fixed, sketch the contour lines relating y to x_1 for the following model:

$$E(y) = \beta_0 + \beta_1 x_1 + \beta_2 x_2 + \beta_3 x_3 + \beta_4 x_1 x_3 + \beta_5 x_2 x_3$$

References

Draper, N., and Smith, H. *Applied Regression Analysis*, 3rd ed. New York: Wiley, 1998.

Geisser, S. "The predictive sample reuse method with applications," *Journal of the American Statistical Association*, Vol. 70, 1975.

Graybill, F. A. *Theory and Application of the Linear Model*. North Scituate, Mass.: Duxbury, 1976.

Kutner, M., Nachtsheim, C., Neter, J., and Li, W. *Applied Linear Statistical Models*, 5th ed. New York: McGraw-Hill/Irwin, 2005.

Mendenhall, W. *Introduction to Linear Models and the Design and Analysis of Experiments*. Belmont, Calif.: Wadsworth, 1968.

Montgomery, D., Peck, E., and Vining, G. *Introduction to Linear Regression Analysis*, 4th ed. New York: Wiley, 2006.

Snee, R., "Validation of regression models: Methods and examples," *Technometrics*, Vol. 19, 1977.

Chapter 6

VARIABLE SCREENING METHODS

Contents

Objectives

1. To introduce methods designed to select the most important independent variables for modeling the mean response, $E(y)$.

2. To learn when these methods are appropriate to apply.

6.1 Introduction: Why Use a Variable Screening Method?

Researches often will collect a data set with a large number of independent variables, each of which is a potential predictor of some dependent variable, y. The problem of deciding which x's in a large set of independent variables to include in a multiple regression model for $E(y)$ is common, for instance, when the dependent variable is profit of a firm, a college student's grade point average, or an economic variable reflecting the state of the economy (e.g., inflation rate).

Consider the problem of predicting the annual salary y of an executive. In Example 4.10 (p. 217) we examined a model with several predictors of y. Suppose we have collected data for 10 potential predictors of an executive's salary. Assume that the list includes seven quantitative x's and three qualitative x's (each of the qualitative x's at two levels). Now consider a complete second-order model for $E(y)$. From our discussion of model building in Chapter 5, we can write the model as follows:

$$E(y) = \beta_0 + \underbrace{\beta_1 x_1 + \beta_2 x_2 + \beta_3 x_3 + \beta_4 x_4 + \beta_5 x_5 + \beta_6 x_6 + \beta_7 x_7}$$

(first-order terms for quantitative variables)

$$+ \beta_8 x_1 x_2 + \beta_9 x_1 x_3 + \beta_{10} x_1 x_4 + \beta_{11} x_1 x_5 + \beta_{12} x_1 x_6 + \beta_{13} x_1 x_7$$

$$+ \underbrace{\beta_{14} x_2 x_3 + \beta_{15} x_2 x_4 + \beta_{16} x_2 x_5 + \beta_{17} x_2 x_6 + \beta_{18} x_2 x_7 + \cdots + \beta_{28} x_6 x_7}$$

(two-way interaction terms for quantitative variables)

$$+ \underbrace{\beta_{29} x_1^2 + \beta_{30} x_2^2 + \beta_{31} x_3^2 + \beta_{32} x_4^2 + \beta_{33} x_5^2 + \beta_{34} x_6^2 + \beta_{35} x_7^2}$$

(quadratic [second-order] terms for quantitative variables)

$$+ \beta_{36}x_8 + \beta_{37}x_9 + \beta_{38}x_{10} \quad \text{(dummy variables for qualitative variables)}$$

$$+ \underbrace{\beta_{39}x_8x_9 + \beta_{40}x_8x_{10} + \beta_{41}x_9x_{10} + \beta_{42}x_8x_9x_{10}}$$

(interaction terms for qualitative variables)

$$+ \beta_{43}x_1x_8 + \beta_{44}x_2x_8 + \beta_{45}x_3x_8 + \beta_{46}x_4x_8 + \beta_{47}x_5x_8 + \beta_{48}x_6x_8 + \beta_{49}x_7x_8$$

$$+ \beta_{50}x_1x_2x_8 + \beta_{51}x_1x_3x_8 + \beta_{52}x_1x_4x_8 + \cdots + \beta_{70}x_6x_7x_8$$

$$+ \underbrace{\beta_{71}x_1^2x_8 + \beta_{72}x_2^2x_8 + \cdots + \beta_{77}x_7^2x_8}$$

(interactions between quantitative terms and qualitative variable x_8)

$$+ \underbrace{\beta_{78}x_1x_9 + \beta_{79}x_2x_9 + \beta_{80}x_3x_9 + \cdots + \beta_{112}x_7^2x_9}$$

(interactions between quantitative terms and qualitative variable x_9)

$$+ \underbrace{\beta_{113}x_1x_{10} + \beta_{114}x_2x_{10} + \beta_{115}x_3x_{10} + \cdots + \beta_{147}x_7^2x_{10}}$$

(interactions between quantitative terms and qualitative variable x_{10})

$$+ \underbrace{\beta_{148}x_1x_8x_9 + \beta_{149}x_2x_8x_9 + \beta_{150}x_3x_8x_9 + \cdots + \beta_{182}x_7^2x_8x_9}$$

(interactions between quantitative terms and qualitative term x_8x_9)

$$+ \underbrace{\beta_{183}x_1x_8x_{10} + \beta_{184}x_2x_8x_{10} + \beta_{185}x_3x_8x_{10} + \cdots + \beta_{217}x_7^2x_8x_{10}}$$

(interactions between quantitative terms and qualitative term x_8x_{10})

$$+ \underbrace{\beta_{218}x_1x_9x_{10} + \beta_{219}x_2x_9x_{10} + \beta_{220}x_3x_9x_{10} + \cdots + \beta_{252}x_7^2x_9x_{10}}$$

(interactions between quantitative terms and qualitative term x_9x_{10})

$$+ \underbrace{\beta_{253}x_1x_8x_9x_{10} + \beta_{254}x_2x_8x_9x_{10} + \beta_{255}x_3x_8x_9x_{10} + \cdots + \beta_{287}x_7^2x_8x_9x_{10}}$$

(interactions between quantitative terms and qualitative term $x_8x_9x_{10}$)

To fit this model, we would need to collect data for, at minimum, 289 executives! Otherwise, we will have 0 degrees of freedom for estimating σ^2, the variance of the random error term. Even if we could obtain a data set this large, the task of interpreting the β parameters in the model is a daunting one. This model, with its numerous multivariable interactions and squared terms, is way too complex to be of use in practice.

In this chapter, we consider two systematic methods designed to reduce a large list of potential predictors to a more manageable one. These techniques, known as **variable screening procedures**, objectively determine which independent variables in the list are the most important predictors of y and which are the least important predictors. The most widely used method, *stepwise regression*, is discussed in Section 6.2, while another popular method, the *all-possible-regressions-selection* procedure, is the topic of Section 6.3. In Section 6.4, several caveats of these methods are identified.

6.2 Stepwise Regression

One of the most widely used variable screening methods is known as **stepwise regression**. To run a stepwise regression, the user first identifies the dependent variable (response) y, and the set of potentially important independent variables,

$x_1, x_2, \ldots, x_k$, where k is generally large. [*Note*: This set of variables could include both first-order and higher-order terms as well as interactions.] The data are entered into the computer software, and the stepwise procedure begins.

Step 1. The software program fits all possible one-variable models of the form

$$E(y) = \beta_0 + \beta_1 x_i$$

to the data, where x_i is the ith independent variable, $i = 1, 2, \ldots, k$. For each model, the test of the null hypothesis

$$H_0: \beta_1 = 0$$

against the alternative hypothesis

$$H_a: \beta_1 \neq 0$$

is conducted using the t-test (or the equivalent F-test) for a single β parameter. The independent variable that produces the largest (absolute) t-value is declared the best one-variable predictor of y.* Call this independent variable x_1.

Step 2. The stepwise program now begins to search through the remaining $(k - 1)$ independent variables for the best two-variable model of the form

$$E(y) = \beta_0 + \beta_1 x_1 + \beta_2 x_i$$

This is done by fitting all two-variable models containing x_1 (the variable selected in the first step) and each of the other $(k - 1)$ options for the second variable x_i. The t-values for the test $H_0: \beta_2 = 0$ are computed for each of the $(k - 1)$ models (corresponding to the remaining independent variables, $x_i, i = 2, 3, \ldots, k$), and the variable having the largest t is retained. Call this variable x_2.

Before proceeding to Step 3, the stepwise routine will go back and check the t-value of $\hat{\beta}_1$ after $\hat{\beta}_2 x_2$ has been added to the model. If the t-value has become nonsignificant at some specified α level (say $\alpha = .05$), the variable x_1 is removed and a search is made for the independent variable with a β parameter that will yield the most significant t-value in the presence of $\hat{\beta}_2 x_2$.

The reason the t-value for x_1 may change from step 1 to step 2 is that the meaning of the coefficient $\hat{\beta}_1$ changes. In step 2, we are approximating a complex response surface in two variables with a plane. The best-fitting plane may yield a different value for $\hat{\beta}_1$ than that obtained in step 1. Thus, both the value of $\hat{\beta}_1$ and its significance usually changes from step 1 to step 2. For this reason, stepwise procedures that recheck the t-values at each step are preferred.

Step 3. The stepwise regression procedure now checks for a third independent variable to include in the model with x_1 and x_2. That is, we seek the best model of the form

$$E(y) = \beta_0 + \beta_1 x_1 + \beta_2 x_2 + \beta_3 x_i$$

To do this, the computer fits all the $(k - 2)$ models using x_1, x_2, and each of the $(k - 2)$ remaining variables, x_i, as a possible x_3. The criterion is again to include the independent variable with the largest t-value. Call this best third variable x_3. The better programs now recheck the t-values corresponding to

---

*Note that the variable with the largest t-value is also the one with the largest (absolute) Pearson product moment correlation, r (Section 3.7), with y.

the x_1 and x_2 coefficients, replacing the variables that yield nonsignificant t-values. This procedure is continued until no further independent variables can be found that yield significant t-values (at the specified α level) in the presence of the variables already in the model.

The result of the stepwise procedure is a model containing only those terms with t-values that are significant at the specified α level. Thus, in most practical situations only several of the large number of independent variables remain. However, it is very important *not* to jump to the conclusion that all the independent variables important for predicting y have been identified or that the unimportant independent variables have been eliminated. Remember, the stepwise procedure is using only *sample estimates* of the true model coefficients (β's) to select the important variables. An extremely large number of single β parameter t-tests have been conducted, and the probability is very high that one or more errors have been made in including or excluding variables. That is, we have very probably included some unimportant independent variables in the model (Type I errors) and eliminated some important ones (Type II errors).

There is a second reason why we might not have arrived at a good model. When we choose the variables to be included in the stepwise regression, we may often omit high-order terms (to keep the number of variables manageable). Consequently, we may have initially omitted several important terms from the model. Thus, we should recognize stepwise regression for what it is: an objective *variable screening* procedure.

Successful model builders will now consider second-order terms (for quantitative variables) and other interactions among variables screened by the stepwise procedure. It would be best to develop this response surface model with a second set of data independent of that used for the screening, so the results of the stepwise procedure can be partially verified with new data. This is not always possible, however, because in many modeling situations only a small amount of data is available.

Do not be deceived by the impressive-looking t-values that result from the stepwise procedure—it has retained only the independent variables with the largest t-values. Also, be certain to consider second-order terms in systematically developing the prediction model. Finally, if you have used a first-order model for your stepwise procedure, remember that it may be greatly improved by the addition of higher-order terms.

Caution

Be wary of using the results of stepwise regression to make inferences about the relationship between $E(y)$ and the independent variables in the resulting first-order model. First, an extremely large number of t-tests have been conducted, leading to a high probability of making one or more Type I or Type II errors. Second, it is typical to enter only first-order and main effect terms as candidate variables in the stepwise model. Consequently, the final stepwise model will not include any higher-order or interaction terms. Stepwise regression should be used only when necessary, that is, when you want to determine which of a large number of potentially important independent variables should be used in the model-building process.

Example 6.1

Refer to Example 4.10 (p. 217) and the multiple regression model for executive salary. A preliminary step in the construction of this model is the determination of the most important independent variables. For one firm, 10 potential independent variables (seven quantitative and three qualitative) were measured in a sample of

⌬ EXECSAL2

Table 6.1 Independent variables in the executive salary example

Independent Variable	Description
x_1	Experience (years)—quantitative
x_2	Education (years)—quantitative
x_3	Gender (1 if male, 0 if female)—qualitative
x_4	Number of employees supervised—quantitative
x_5	Corporate assets (millions of dollars)—quantitative
x_6	Board member (1 if yes, 0 if no)—qualitative
x_7	Age (years)—quantitative
x_8	Company profits (past 12 months, millions of dollars)—quantitative
x_9	Has international responsibility (1 if yes, 0 if no)—qualitative
x_{10}	Company's total sales (past 12 months, millions of dollars)—quantitative

100 executives. The data, described in Table 6.1, are saved in the EXECSAL2 file. Since it would be very difficult to construct a complete second-order model with all of the 10 independent variables, use stepwise regression to decide which of the 10 variables should be included in the building of the final model for the natural log of executive salaries.

Solution

We will use stepwise regression with the main effects of the 10 independent variables to identify the most important variables. The dependent variable y is the natural logarithm of the executive salaries. The MINITAB stepwise regression printout is shown in Figure 6.1. MINITAB automatically enters the constant term (β_0) into the model in the first step. The remaining steps follow the procedure outlined earlier in this section.

In Step 1, MINITAB fits all possible one-variable models of the form,

$$E(y) = \beta_0 + \beta_1 x_i.$$

You can see from Figure 6.1 that the first variable selected is x_1, years of experience. Thus, x_1 has the largest (absolute) t-value associated with a test of $H_0: \beta_1 = 0$. This value, $t = 12.62$, is highlighted on the printout.

Next (step 2), MINITAB fits all possible two-variable models of the form,

$$E(y) = \beta_0 + \beta_1 x_1 + \beta_2 x_i.$$

(Note that the variable selected in the first step, x_1, is automatically included in the model.) The variable with the largest (absolute) t-value associated with a test of $H_0: \beta_2 = 0$ is the dummy variable for gender, x_3. This t-value, $t = 7.10$, is also highlighted on the printout.

In Step 3, all possible three-variable models of the form

$$E(y) = \beta_0 + \beta_1 x_1 + \beta_2 x_3 + \beta_3 x_i$$

are fit. (Note that x_1 and x_3 are included in the model.) MINITAB selects x_4, number of employees supervised, based on the value $t = 7.32$ (highlighted on the printout) associated with a test of $H_0: \beta_3 = 0$.

Figure 6.1 MINITAB stepwise regression results for executive salaries

Stepwise Regression: Y versus X1, X2, X3, X4, X5, X6, X7, X8, X9, X10

Alpha-to-Enter: 0.15 Alpha-to-Remove: 0.15

Response is Y on 10 predictors, with N = 100

Step	1	2	3	4	5
Constant	11.091	10.968	10.783	10.278	9.962
X1	0.0278	0.0273	0.0273	0.0273	0.0273
T-Value	12.62	15.13	18.80	24.68	26.50
P-Value	0.000	0.000	0.000	0.000	0.000
X3		0.197	0.233	0.232	0.225
T-Value		7.10	10.17	13.30	13.74
P-Value		0.000	0.000	0.000	0.000
X4			0.00048	0.00055	0.00052
T-Value			7.32	10.92	11.06
P-Value			0.000	0.000	0.000
X2				0.0300	0.0291
T-Value				8.38	8.72
P-Value				0.000	0.000
X5					0.00196
T-Value					3.95
P-Value					0.000
S	0.161	0.131	0.106	0.0807	0.0751
R-Sq	61.90	74.92	83.91	90.75	92.06
R-Sq(adj)	61.51	74.40	83.41	90.36	91.64
Mallows Cp	343.9	195.5	93.8	16.8	3.6
PRESS	2.66387	1.78796	1.17124	0.695637	0.610197
R-Sq(pred)	60.14	73.24	82.47	89.59	90.87

In Steps 4 and 5, the variables x_2 (years of education) and x_5 (corporate assets), respectively, are selected for inclusion into the model. The t-values for the tests of the appropriate β's are highlighted in Figure 6.1. MINITAB stopped after five steps because none of the other independent variables met the criterion for admission to the model. As a default, MINITAB (and most other statistical software packages) uses $\alpha = .15$ in the tests conducted. In other words, if the p-value associated with a test of a β-coefficient is greater than $\alpha = .15$, then the corresponding variable is not included in the model.

The results of the stepwise regression suggest that we should concentrate on the five variables, x_1, x_2, x_3, x_4, and x_5, in our final modeling effort. Models with curvilinear terms as well as interactions should be proposed and evaluated (as demonstrated in Chapter 5) to determine the best model for predicting executive salaries. ▬

There are several other stepwise regression techniques designed to select the most important independent variables. One of these, called **forward selection**, is nearly identical to the stepwise procedure previously outlined. The only difference is that the forward selection technique provides no option for rechecking the t-values corresponding to the x's that have entered the model in an earlier step. Thus, stepwise regression is preferred to forward selection in practice.

Another technique, called **backward elimination**, initially fits a model containing terms for all potential independent variables. That is, for k independent variables,

```
                        The REG Procedure
                          Model: MODEL1
                       Dependent Variable: Y

                   Backward Elimination: Step 5

                    Parameter    Standard
        Variable    Estimate      Error    Type II SS   F Value   Pr > F

        Intercept    9.96193     0.10106    54.83329    9717.56   <.0001
        X1           0.02728     0.00103     3.96275     702.28   <.0001
        X2           0.02909     0.00334     0.42894      76.02   <.0001
        X3           0.22469     0.01635     1.06565     188.85   <.0001
        X4           0.00052442  0.00004740  0.69078     122.42   <.0001
        X5           0.00196     0.00049718  0.08790      15.58   0.0002

              Bounds on condition number: 1.1016, 26.17
```

```
     All variables left in the model are significant at the 0.0500 level.


                    Summary of Backward Elimination

              Variable   Number    Partial    Model
     Step     Removed    Vars In   R-Square   R-Square   C(p)    F Value   Pr > F

       1       X10          9       0.0001     0.9228    9.1091    0.11    0.7420
       2       X7           8       0.0001     0.9227    7.1956    0.09    0.7683
       3       X8           7       0.0002     0.9225    5.4499    0.26    0.6117
       4       X6           6       0.0005     0.9220    4.0235    0.59    0.4444
       5       X9           5       0.0014     0.9206    3.6279    1.66    0.2011
```

Figure 6.2 SAS backward stepwise regression for executive salaries

the model $E(y) = \beta_0 + \beta_1 x_1 + \beta_2 x_2 + \cdots + \beta_k x_k$ is fit in step 1. The variable with the smallest t (or F) statistic for testing $H_0: \beta_i = 0$ is identified and dropped from the model if the t-value is less than some specified critical value. The model with the remaining $(k-1)$ independent variables is fit in step 2, and again, the variable associated with the smallest nonsignificant t-value is dropped. This process is repeated until no further nonsignificant independent variables can be found.

For example, applying the backward elimination method to the executive salary data of Example 6.1 yields the results shown in the SAS printout in Figure 6.2. At the bottom of the printout you can see that the variables x_{10}, x_7, x_8, x_6, and x_9 (in that order) were removed from the model, leaving x_1–x_5 as the selected independent variables. Thus, for this example, the backward elimination and stepwise methods yield identical results. This will not always be the case, however. In fact, the backward elimination method can be an advantage when at least one of the candidate independent variables is a qualitative variable at three or more levels (requiring at least two dummy variables), since the backward procedure tests the contribution of each dummy variable after the others have been entered into the model. The real disadvantage of using the backward elimination technique is that you need a sufficiently large number of data points to fit the initial model in Step 1.

6.3 All-Possible-Regressions Selection Procedure

In Section 6.2, we presented stepwise regression as an objective screening procedure for selecting the most important predictors of y. Other, more subjective, variable selection techniques have been developed in the literature for the purpose of identifying important independent variables. The most popular of these procedures are those that consider all possible regression models given the set of potentially important predictors. Such a procedure is commonly known as an **all-possible-regressions selection procedure**. The techniques differ with respect to the criteria for selecting the "best" subset of variables. In this section, we describe four criteria widely used in practice, then give an example illustrating the four techniques.

R^2 Criterion

Consider the set of potentially important variables, $x_1, x_2, x_3, \ldots, x_k$. We learned in Section 4.7 that the multiple coefficient of determination

$$R^2 = 1 - \frac{\text{SSE}}{\text{SS(Total)}}$$

will increase when independent variables are added to the model. Therefore, the model that includes all k independent variables

$$E(y) = \beta_0 + \beta_1 x_1 + \beta_2 x_2 + \cdots + \beta_k x_k$$

will yield the largest R^2. Yet, we have seen examples (Chapter 5) where adding terms to the model does not yield a significantly better prediction equation. The objective of the R^2 criterion is to find a subset model (i.e., a model containing a subset of the k independent variables) so that adding more variables to the model will yield only small increases in R^2. In practice, the best model found by the R^2 criterion will rarely be the model with the largest R^2. Generally, you are looking for a simple model that is as good as, or nearly as good as, the model with all k independent variables. But unlike that in stepwise regression, the decision about when to stop adding variables to the model is a subjective one.

Adjusted R^2 or MSE Criterion

One drawback to using the R^2 criterion, you will recall, is that the value of R^2 does not account for the number of β parameters in the model. If enough variables are added to the model so that the sample size n equals the total number of β's in the model, you will force R^2 to equal 1. Alternatively, we can use the adjusted R^2. It is easy to show that R_a^2 is related to MSE as follows:

$$R_a^2 = 1 - (n-1)\left[\frac{\text{MSE}}{\text{SS(Total)}}\right]$$

Note that R_a^2 increases only if MSE decreases [since SS(Total) remains constant for all models]. Thus, an equivalent procedure is to search for the model with the minimum, or near minimum, MSE.

C_p Criterion

A third option is based on a quantity called the **total mean square error (TMSE)** for the fitted regression model:

$$\text{TMSE} = E\left\{\sum_{i=1}^{n}[\hat{y}_i - E(y_i)]^2\right\} = \sum_{i=1}^{n}[E(\hat{y}_i) - E(y_i)]^2 + \sum_{i=1}^{n}\text{Var}(\hat{y}_i)$$

where $E(\hat{y}_i)$ is the mean response for the subset (fitted) regression model and $E(y_i)$ is the mean response for the true model. The objective is to compare the TMSE for the subset regression model with σ^2, the variance of the random error for the true model, using the ratio

$$\Gamma = \frac{\text{TMSE}}{\sigma^2}$$

Small values of Γ imply that the subset regression model has a small total mean square error relative to σ^2. Unfortunately, both TMSE and σ^2 are unknown, and we must rely on sample estimates of these quantities. It can be shown (proof omitted) that a good estimator of the ratio Γ is given by

$$C_p = \frac{\text{SSE}_p}{\text{MSE}_k} + 2(p+1) - n$$

where n is the sample size, p is the number of independent variables in the subset model, k is the total number of potential independent variables, SSE_p is the SSE for the subset model, and MSE_k is the MSE for the model containing all k independent variables. The statistical software packages discussed in this text have routines that calculate the C_p statistic. In fact, the C_p value is automatically printed at each step in the SAS and MINITAB stepwise regression printouts (see Figure 6.1).

The C_p criterion selects as the best model the subset model with (1) a small value of C_p (i.e., a small total mean square error) and (2) a value of C_p near $p+1$, a property that indicates that slight or no bias exists in the subset regression model.*

Thus, the C_p criterion focuses on minimizing total mean square error and the regression bias. If you are mainly concerned with minimizing total mean square error, you will want to choose the model with the smallest C_p value, as long as the bias is not large. On the other hand, you may prefer a model that yields a C_p value slightly larger than the minimum but that has slight (or no) bias.

PRESS Criterion

A fourth criterion used to select the best subset regression model is the PRESS statistic, introduced in Section 5.11. Recall that the PRESS (or, prediction sum of squares) statistic for a model is calculated as follows:

$$\text{PRESS} = \sum_{i=1}^{n} [y_i - \hat{y}_{(i)}]^2$$

where $\hat{y}_{(i)}$ denotes the predicted value for the ith observation obtained when the regression model is fit with the data point for the ith observation omitted (or deleted) from the sample.[†] Thus, the candidate model is fit to the sample data n times, each time omitting one of the data points and obtaining the predicted value of y for that data point. Since small differences $y_i - \hat{y}_{(i)}$ indicate that the model is predicting well, we desire a model with a small PRESS.

Computing the PRESS statistic may seem like a tiresome chore, since repeated regression runs (a total of n runs) must be made for each candidate model. However, most statistical software packages have options for computing PRESS automatically.[‡]

---

* A model is said to be *unbiased* if $E(\hat{y}) = E(y)$. We state (without proof) that for an unbiased regression model, $E(C_p) \approx p + 1$. In general, subset models will be biased since $k - p$ independent variables are omitted from the fitted model. However, when C_p is near $p + 1$, the bias is small and can essentially be ignored.

† The quantity $y_i - \hat{y}_{(i)}$ is called the "deleted" residual for the ith observation. We discuss deleted residuals in more detail in Chapter 8.

‡ PRESS can also be calculated using the results from a regression run on all n data points. The formula is

$$\text{PRESS} = \sum_{i=1}^{n} \left(\frac{y_i - \hat{y}_i}{1 - h_{ii}} \right)^2$$

where h_{ii} is a function of the independent variables in the model. In Chapter 8, we show how h_{ii} (called *leverage*) can be used to detect influential observations.

Plots aid in the selection of the best subset regression model using the all-possible-regressions procedure. The criterion measure, either R^2, MSE, C_p, or PRESS, is plotted on the vertical axis against p, the number of independent variables in the subset model, on the horizontal axis. We illustrate all three variable selection techniques in an example.

Example 6.2

Refer to Example 6.1 and the data on executive salaries. Recall that we want to identify the most important independent variables for predicting the natural log of salary from the list of 10 variables given in Table 6.1. Apply the all-possible-regressions selection procedure to find the most important independent variables.

Solution

We entered the executive salary data into MINITAB and used MINITAB's all-possible-regressions selection routine to obtain the printout shown in Figure 6.2. For $p = 10$ independent variables, there exists 1,023 possible subset first-order models. Although MINITAB fits all of these models, the output in Figure 6.3 shows only the results for the "best" model for each value of p. From the printout, you can see that the best one-variable model includes x_1 (years of experience); the best two-variable model includes x_1 and x_3 (gender); the best three-variable model includes x_1, x_3, and x_4 (number supervised); and so on.

These "best subset" models are summarized in Table 6.2. In addition to the variables included in each model, the table gives the values of R^2, adjusted-R^2, MSE, C_p, and PRESS. To determine which subset model to select, we plot these quantities against the number of variables, p. The MINITAB graphs for R^2, adjusted-R^2, C_p, and PRESS are shown in Figures 6.4a–d, respectively.

In Figure 6.4a, we see that the R^2 values tend to increase in very small amounts for models with more than $p = 5$ predictors. A similar pattern is shown in Figure 6.4b for R_a^2. Thus, both the R^2 and R_a^2 criteria suggest that the model containing the five predictors x_1, x_2, x_3, x_4, and x_5 is a good candidate for the best subset regression model.

Figure 6.4c shows the plotted C_p values and the line $C_p = p + 1$. Notice that the subset models with $p \geq 5$ independent variables all have relatively small C_p values and vary tightly about the line $C_p = p + 1$. This implies that these models

Figure 6.3 MINITAB all-possible-regressions selection results for executive salaries

---

Best Subsets Regression: Y versus X1, X2, ...

Response is Y

```
                                          X
                                X X X X X X X X X 1
Vars  R-Sq  R-Sq(adj)   C-p       S   1 2 3 4 5 6 7 8 9 0
   1  61.9    61.5     343.9  0.16119  X
   2  74.9    74.4     195.5  0.13145  X   X
   3  83.9    83.4      93.8  0.10583  X   X X
   4  90.7    90.4      16.8  0.080676 X X X X
   5  92.1    91.6       3.6  0.075118 X X X X X
   6  92.2    91.7       4.0  0.074857 X X X X X       X
   7  92.3    91.7       5.4  0.075022 X X X X X       X
   8  92.3    91.6       7.2  0.075326 X X X X X   X X
   9  92.3    91.5       9.1  0.075707 X X X X X X X X
  10  92.3    91.4      11.0  0.076084 X X X X X X X X X X
```

Table 6.2 Results for best subset models

Number of Predictors p	Variables in the Model	R^2	adj-R^2	MSE	C_p	PRESS
1	x_1	.619	.615	.0260	343.9	2.664
2	x_1, x_3	.749	.744	.0173	195.5	1.788
3	x_1, x_3, x_4	.839	.834	.0112	93.8	1.171
4	x_1, x_2, x_3, x_4	.907	.904	.0065	16.8	.696
5	x_1, x_2, x_3, x_4, x_5	.921	.916	.0056	3.6	.610
6	$x_1, x_2, x_3, x_4, x_5, x_9$	.922	.917	.0056	4.0	.610
7	$x_1, x_2, x_3, x_4, x_5, x_6, x_9$	.923	.917	.0056	5.4	.620
8	$x_1, x_2, x_3, x_4, x_5, x_6, x_8, x_9$	.923	.916	.0057	7.2	.629
9	$x_1, x_2, x_3, x_4, x_5, x_6, x_7, x_8, x_9$	.923	.915	.0057	9.1	.643
10	$x_1, x_2, x_3, x_4, x_5, x_6, x_7, x_8, x_9, x_{10}$	.923	.914	.0058	11.0	.654

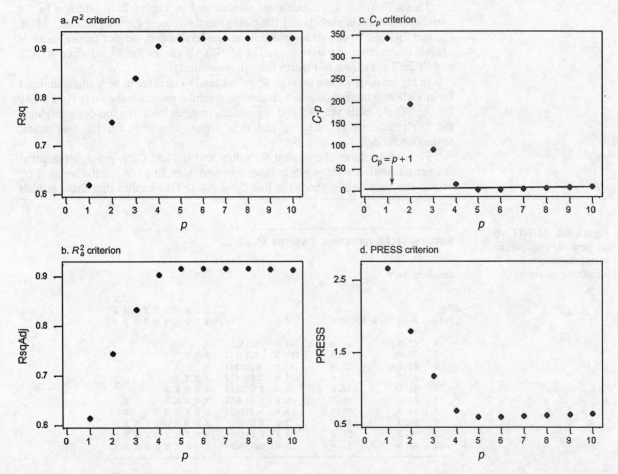

Figure 6.4 MINITAB plots of all-possible-regressions selection criteria for Example 6.2

have a small total mean square error and a negligible bias. The model corresponding to $p = 4$, although certainly outperforming the models $p \leq 3$, appears to fall short of the larger models according to the C_p criterion. From Figure 6.4d you can see that the PRESS is smallest for the five-variable model with $x_1, x_2, x_3, x_4,$ and x_5 (PRESS = .610).

According to all four criteria, the variables $x_1, x_2, x_3, x_4,$ and x_5 should be included in the group of the most important predictors. ▪

In summary, variable selection procedures based on the all-possible-regressions selection criterion will assist you in identifying the most important independent variables for predicting y. Keep in mind, however, that these techniques lack the objectivity of a stepwise regression procedure. Furthermore, you should be wary of concluding that the best model for predicting y has been found, since, in practice, interactions and higher-order terms are typically omitted from the list of potential important predictors.

6.4 Caveats

Both stepwise regression and the all-possible-regressions selection procedure are useful *variable screening methods*. Many regression analysts, however, tend to apply these procedures as *model-building methods*. Why? The stepwise (or best subset) model will often have a high value of R^2 and all the β coefficients in the model will be significantly different from 0 with small p-values (see Figure 6.1). And, with very little work (other than collecting the data and entering it into the computer), you can obtain the model using a statistical software package. Consequently, it is extremely tempting to use the stepwise model as the *final* model for predicting and making inferences about the dependent variable, y.

We conclude this chapter with several caveats and some advice on using stepwise regression and the all-possible-regressions selection procedure. Be wary of using the stepwise (or best subset) model as the final model for predicting y for several reasons. First, recall that either procedure tends to fit an extremely large number of models and perform an extremely large number of tests (objectively, in stepwise regression, and subjectively, in best subsets regression). Thus, the *probability of making at least one Type I error or at least one Type II error is often quite high*. That is, you are very likely to either include at least one unimportant independent variable or leave out at least one important independent variable in the final model!

Second, *analysts typically do not include higher-order terms or interactions in the list of potential predictors for stepwise regression*. Therefore, if no real model building is performed, the final model will be a first-order, main effects model. Most real-world relationships between variables are not linear, and these relationships often are moderated by another variable (i.e., interaction exists). In Chapter 8, we learn that higher-order terms are often revealed through residual plotting.

Third, even *if the analyst includes some higher-order terms and interactions as potential predictors, the stepwise and best subsets procedures will more than likely select a nonsensical model*. For example, consider the stepwise model

$$E(y) = \beta_0 + \beta_1 x_1 + \beta_2 x_2 x_5 + \beta_3 x_3^2.$$

The model includes an interaction for x_2 and x_5, but omits the main effects for these terms, and it includes a quadratic term for x_3 but omits the first-order (shift parameter) term. Also, this strategy requires the analyst to "guess" or have an intuitive feel for which interactions may be the most important. If all possible interactions and squared terms are included in the list of potential predictors, the problem discussed in Section 6.1 (lacking sufficient data to estimate all the model parameters) will arise, especially in the all-possible-regressions selection method.

Finally, make sure you *do not abandon your common sense or intuition when applying stepwise regression.* For example, suppose in the model for executive salary (Example 6.1) the qualitative variable gender was not selected by the stepwise method you employed. Also, suppose one of the issues of interest is to investigate a claim of gender bias at the firm (i.e., that the salaries for male executives tend to be higher than the salaries of female executives with the same level of education and experience). Even though gender was not selected, this does not necessarily imply that there is no gender bias. Remember that the stepwise method tests the gender term as a main effect only; there are no interactions or higher-order terms involving gender in the stepwise model. Consequently, if gender does interact with another variable (e.g., if the increase of salary with experience is higher for males than for females, indicating a gender by experience interaction), the stepwise method may not detect it. A good analyst, understanding the different ways that a gender bias may occur, would make sure to include the gender variable in any model-building effort. This will allow you to test for a gender bias by conducting a partial F-test for all the gender terms (including interactions) in the model.

All of these problems can be avoided if we use stepwise or all-possible-regressions as they were originally intended—as objective methods of screening independent variables from a long list of potential predictors of y. Once the "most important" variables have been selected—some selected by stepwise regression and maybe others simply by your knowledge of the data and substantive theories about the variables, common sense, or intuition—then *begin* the model-building phase of the analysis using the methods in Chapter 5.

Quick Summary/Guides

KEY FORMULAS

$R_a^2 = 1 - (n-1)[MSE/SS(Total)]$

$C_p = (SSE_p/MSE_k) + 2(p+1) - n$

$PRESS = \sum_{i=1}^{n}(y_i - y_{(i)})^2$

KEY IDEAS

Variable Screening Methods

1. *Stepwise regression* (forward, backward, or stepwise techniques)
2. *All-possible-regressions selection procedure*

All-Possible-Regressions Selection Criteria

1. R^2
2. R_a^2
3. MSE
4. C_p
5. PRESS

Potential Caveats in Using Variable Screening Methods to Determine the "Final" Model

1. High probabilities of Type I and Type II errors
2. No higher-order terms or interactions
3. Nonsensical terms in the model
4. Important independent variables omitted that interact with other x's

Supplementary Exercises

6.1. Selecting the best one-variable predictor. There are six independent variables, $x_1, x_2, x_3, x_4, x_5,$ and x_6, that might be useful in predicting a response y. A total of $n = 50$ observations are available, and it is decided to employ stepwise regression to help in selecting the independent variables that appear to be useful. The computer fits all possible one-variable models of the form

$$E(y) = \beta_0 + \beta_1 x_i$$

where x_i is the ith independent variable, $i = 1, 2, \ldots, 6$. The information in the table is provided from the computer printout.

INDEPENDENT VARIABLE	$\hat{\beta}_i$	$s_{\hat{\beta}_i}$
x_1	1.6	.42
x_2	−.9	.01
x_3	3.4	1.14
x_4	2.5	2.06
x_5	−4.4	.73
x_6	.3	.35

(a) Which independent variable is declared the best one-variable predictor of y? Explain.
(b) Would this variable be included in the model at this stage? Explain.
(c) Describe the next phase that a stepwise procedure would execute.

6.2. Accuracy of software effort estimates. Periodically, software engineers must provide estimates of their effort in developing new software. In the *Journal of Empirical Software Engineering* (Vol. 9, 2004), multiple regression was used to predict the accuracy of these effort estimates. The dependent variable, defined as the relative error in estimating effort,

$$y = (\text{Actual effort} - \text{Estimated effort})/$$
$$(\text{Actual effort})$$

was determined for each in a sample of $n = 49$ software development tasks. Eight independent variables were evaluated as potential predictors of relative error using stepwise regression. Each of these was formulated as a dummy variable, as shown below.

Company role of estimator: $x_1 = 1$ if developer, 0 if project leader
Task complexity: $x_2 = 1$ if low, 0 if medium/high
Contract type: $x_3 = 1$ if fixed price, 0 if hourly rate
Customer importance: $x_4 = 1$ if high, 0 if low/medium

Customer priority: $x_5 = 1$ if time-of-delivery, 0 if cost or quality
Level of knowledge: $x_6 = 1$ if high, 0 if low/medium
Participation: $x_7 = 1$ if estimator participates in work, 0 if not
Previous accuracy: $x_8 = 1$ if more than 20% accurate, 0 if less than 20% accurate

(a) In Step 1 of the stepwise regression, how many different one-variable models are fit to the data?
(b) In Step 1, the variable x_1 is selected as the "best" one-variable predictor. How is this determined?
(c) In Step 2 of the stepwise regression, how many different two-variable models (where x_1 is one of the variables) are fit to the data?
(d) The only two variables selected for entry into the stepwise regression model were x_1 and x_8. The stepwise regression yielded the following prediction equation:

$$\hat{y} = .12 - .28x_1 + .27x_8$$

Give a practical interpretation of the β estimates multiplied by x_1 and x_8.
(e) Why should a researcher be wary of using the model, part d, as the final model for predicting effort (y)?

6.3. Bus Rapid Transit study. Bus Rapid Transit (BRT) is a rapidly growing trend in the provision of public transportation in America. The Center for Urban Transportation Research (CUTR) at the University of South Florida conducted a survey of BRT customers in Miami (*Transportation Research Board* Annual Meeting, January 2003). Data on the following variables (all measured on a 5-point scale, where 1 = "very unsatisfied" and 5 = "very satisfied") were collected for a sample of over 500 bus riders: overall satisfaction with BRT (y), safety on bus (x_1), seat availability (x_2), dependability (x_3), travel time (x_4), cost (x_5), information/maps (x_6), convenience of routes (x_7), traffic signals (x_8), safety at bus stops (x_9), hours of service (x_{10}), and frequency of service (x_{11}). CUTR analysts used stepwise regression to model overall satisfaction (y).

(a) How many models are fit at Step 1 of the stepwise regression?
(b) How many models are fit at Step 2 of the stepwise regression?
(c) How many models are fit at Step 11 of the stepwise regression?

(d) The stepwise regression selected the following eight variables to include in the model (in order of selection): x_{11}, x_4, x_2, x_7, x_{10}, x_1, x_9, and x_3. Write the equation for $E(y)$ that results from stepwise.

(e) The model, part d, resulted in $R^2 = .677$. Interpret this value.

(f) Explain why the CUTR analysts should be cautious in concluding that the "best" model for $E(y)$ has been found.

6.4. Yield strength of steel alloy. Industrial engineers at the University of Florida used regression modeling as a tool to reduce the time and cost associated with developing new metallic alloys (*Modelling and Simulation in Materials Science and Engineering*, Vol. 13, 2005). To illustrate, the engineers built a regression model for the tensile yield strength (y) of a new steel alloy. The potential important predictors of yield strength are listed below.

x_1 = Carbon amount (% weight)
x_2 = Manganese amount (% weight)
x_3 = Chromium amount (% weight)
x_4 = Nickel amount (% weight)
x_5 = Molybdenum amount (% weight)
x_6 = Copper amount (% weight)
x_7 = Nitrogen amount (% weight)
x_8 = Vanadium amount (% weight)
x_9 = Plate thickness (millimeters)
x_{10} = Solution treating (milliliters)
x_{11} = Aging temperature (degrees, Celsius)

(a) The engineers used stepwise regression in order to search for a parsimonious set of predictor variables. Do you agree with this decision? Explain.

(b) The stepwise regression selected the following independent variables: x_1 = Carbon, x_2 = Manganese, x_3 = Chromium, x_5 = Molybdenum, x_6 = Copper, x_8 = Vanadium, x_9 = Plate thickness, x_{10} = Solution treating, and x_{11} = Ageing temperature. Based on this information, determine the total number of first-order models that were fit in the stepwise routine.

(c) Refer to part b. All these variables were statistically significant in the stepwise model, with $R^2 = .94$. Consequently, the engineers used the estimated stepwise model to predict yield strength. Do you agree with this decision? Explain.

6.5. Modeling marine life in the gulf. A marine biologist was hired by the EPA to determine whether the hot-water runoff from a particular power plant located near a large gulf is having an adverse effect on the marine life in the area. The biologist's goal is to acquire a prediction equation for the number of marine animals located at certain designated areas, or stations, in the gulf. Based on past experience, the EPA considered the following environmental factors as predictors for the number of animals at a particular station:

x_1 = Temperature of water (TEMP)
x_2 = Salinity of water (SAL)
x_3 = Dissolved oxygen content of water (DO)
x_4 = Turbidity index, a measure of the turbidity of the water (TI)
x_5 = Depth of the water at the station (ST_DEPTH)
x_6 = Total weight of sea grasses in sampled area (TGRSWT)

As a preliminary step in the construction of this model, the biologist used a stepwise regression procedure to identify the most important of these six variables. A total of 716 samples were taken at different stations in the gulf, producing the SAS printout shown at the top of the next page. (The response measured was y, the logarithm of the number of marine animals found in the sampled area.)

(a) According to the SAS printout, which of the six independent variables should be used in the model?

(b) Are we able to assume that the marine biologist has identified all the important independent variables for the prediction of y? Why?

(c) Using the variables identified in part a, write the first-order model with interaction that may be used to predict y.

(d) How would the marine biologist determine whether the model specified in part c is better than the first-order model?

(e) Note the small value of R^2. What action might the biologist take to improve the model?

6.6. Clerical staff work hours. In any production process in which one or more workers are engaged in a variety of tasks, the total time spent in production varies as a function of the size of the work pool and the level of output of the various activities. For example, in a large metropolitan department store, the number of hours worked (y) per day by the clerical staff may depend on the following variables:

x_1 = Number of pieces of mail processed (open, sort, etc.)
x_2 = Number of money orders and gift certificates sold
x_3 = Number of window payments (customer charge accounts) transacted
x_4 = Number of change order transactions processed
x_5 = Number of checks cashed

SAS Output for Exercise 6.5

The REG Procedure
Model: MODEL1
Dependent Variable: LOGNUM

Summary of Stepwise Selection

Step	Variable Entered	Variable Removed	Number Vars In	Partial R-Square	Model R-Square	C(p)	F Value	Pr > F
1	ST_DEPTH		1	0.1223	0.1223	51.57	99.47	<.0001
2	TGRSWT		2	0.0924	0.1821	1.52	79.38	<.0001
3	TI		3	0.0368	0.1870	3.51	54.59	<.0001
4	DO		4	0.0134	0.1889	1.03	41.40	<.0001
5		DO	3	0.0368	0.1870	3.51	54.59	<.0001

x_6 = Number of pieces of miscellaneous mail processed on an "as available" basis

x_7 = Number of bus tickets sold

The table of observations on p. 342 gives the output counts for these activities on each of 52 working days.

(a) Conduct a stepwise regression analysis of the data using an available statistical software package.

(b) Interpret the β estimates in the resulting stepwise model.

(c) What are the dangers associated with drawing inferences from the stepwise model?

6.7. Clerical staff work hours (cont'd). Refer to the data on units of production and time worked for a department store clerical staff in Exercise 6.6. For this exercise, consider only the independent variables x_1, x_2, x_3, and x_4 in an all-possible-regressions select procedure.

(a) How many models for $E(y)$ are possible, if the model includes (i) one variable, (ii) two variables, (iii) three variables, and (iv) four variables?

(b) For each case in part a, use a statistical software package to find the maximum R^2, minimum MSE, minimum C_p, and minimum PRESS.

(c) Plot each of the quantities R^2, MSE, C_p, and PRESS in part b against p, the number of predictors in the subset model.

(d) Based on the plots in part c, which variables would you select for predicting total hours worked, y?

6.8. Collusive bidding in road construction. Road construction contracts in the state of Florida are awarded on the basis of competitive, sealed bids; the contractor who submits the lowest bid price wins the contract. During the 1980s, the Office of the Florida Attorney General (FLAG) suspected

numerous contractors of practicing bid collusion (i.e., setting the winning bid price above the fair, or competitive, price in order to increase profit margin). By comparing the bid prices (and other important bid variables) of the fixed (or rigged) contracts to the competitively bid contracts, FLAG was able to establish invaluable benchmarks for detecting future bid-rigging. FLAG collected data for 279 road construction contracts. For each contract, the following variables shown in the table below were measured. (The data are saved in the file named FLAG2.)

(a) Consider building a model for the low-bid price (y). Apply stepwise regression to the data to find the independent variables most suitable for modeling y.

(b) Interpret the β estimates in the resulting stepwise regression model.

(c) What are the dangers associated with drawing inferences from the stepwise model?

FLAG2

1. Price of contract ($) bid by lowest bidder
2. Department of Transportation (DOT) engineer's estimate of fair contract price ($)
3. Ratio of low (winning) bid price to DOT engineer's estimate of fair price
4. Status of contract (1 if fixed, 0 if competitive)
5. District (1, 2, 3, 4, or 5) in which construction project is located
6. Number of bidders on contract
7. Estimated number of days to complete work
8. Length of road project (miles)
9. Percentage of costs allocated to liquid asphalt
10. Percentage of costs allocated to base material
11. Percentage of costs allocated to excavation
12. Percentage of costs allocated to mobilization
13. Percentage of costs allocated to structures
14. Percentage of costs allocated to traffic control
15. Subcontractor utilization (1 if yes, 0 if no)

CLERICAL

OBS.	DAY OF WEEK	y	x_1	x_2	x_3	x_4	x_5	x_6	x_7
1	M	128.5	7781	100	886	235	644	56	737
2	T	113.6	7004	110	962	388	589	57	1029
3	W	146.6	7267	61	1342	398	1081	59	830
4	Th	124.3	2129	102	1153	457	891	57	1468
5	F	100.4	4878	45	803	577	537	49	335
6	S	119.2	3999	144	1127	345	563	64	918
7	M	109.5	11777	123	627	326	402	60	335
8	T	128.5	5764	78	748	161	495	57	962
9	W	131.2	7392	172	876	219	823	62	665
10	Th	112.2	8100	126	685	287	555	86	577
11	F	95.4	4736	115	436	235	456	38	214
12	S	124.6	4337	110	899	127	573	73	484
13	M	103.7	3079	96	570	180	428	59	456
14	T	103.6	7273	51	826	118	463	53	907
15	W	133.2	4091	116	1060	206	961	67	951
16	Th	111.4	3390	70	957	284	745	77	1446
17	F	97.7	6319	58	559	220	539	41	440
18	S	132.1	7447	83	1050	174	553	63	1133
19	M	135.9	7100	80	568	124	428	55	456
20	T	131.3	8035	115	709	174	498	78	968
21	W	150.4	5579	83	568	223	683	79	660
22	Th	124.9	4338	78	900	115	556	84	555
23	F	97.0	6895	18	442	118	479	41	203
24	S	114.1	3629	133	644	155	505	57	781
25	M	88.3	5149	92	389	124	405	59	236
26	T	117.6	5241	110	612	222	477	55	616
27	W	128.2	2917	69	1057	378	970	80	1210
28	Th	138.8	4390	70	974	195	1027	81	1452
29	F	109.5	4957	24	783	358	893	51	616
30	S	118.9	7099	130	1419	374	609	62	957
31	M	122.2	7337	128	1137	238	461	51	968
32	T	142.8	8301	115	946	191	771	74	719
33	W	133.9	4889	86	750	214	513	69	489
34	Th	100.2	6308	81	461	132	430	49	341
35	F	116.8	6908	145	864	164	549	57	902
36	S	97.3	5345	116	604	127	360	48	126
37	M	98.0	6994	59	714	107	473	53	726
38	T	136.5	6781	78	917	171	805	74	1100
39	W	111.7	3142	106	809	335	702	70	1721
40	Th	98.6	5738	27	546	126	455	52	502
41	F	116.2	4931	174	891	129	481	71	737
42	S	108.9	6501	69	643	129	334	47	473
43	M	120.6	5678	94	828	107	384	52	1083
44	T	131.8	4619	100	777	164	834	67	841
45	W	112.4	1832	124	626	158	571	71	627
46	Th	92.5	5445	52	432	121	458	42	313
47	F	120.0	4123	84	432	153	544	42	654
48	S	112.2	5884	89	1061	100	391	31	280
49	M	113.0	5505	45	562	84	444	36	814
50	T	138.7	2882	94	601	139	799	44	907
51	W	122.1	2395	89	637	201	747	30	1666
52	Th	86.6	6847	14	810	230	547	40	614

Source: Adapted from *Work Measurement*, by G. L. Smith, Grid Publishing Co., Columbus, Ohio, 1978 (Table 3.1).

⊚ GASTURBINE (Data for first and last five gas turbines shown)

ENGINE	SHAFTS	RPM	CPRATIO	INLET-TEMP	EXH-TEMP	AIRFLOW	POWER	HEATRATE
Traditional	1	27245	9.2	1134	602	7	1630	14622
Traditional	1	14000	12.2	950	446	15	2726	13196
Traditional	1	17384	14.8	1149	537	20	5247	11948
Traditional	1	11085	11.8	1024	478	27	6726	11289
Traditional	1	14045	13.2	1149	553	29	7726	11964
...								
Aероderiv	2	18910	14.0	1066	532	8	1845	12766
Aероderiv	3	3600	35.0	1288	448	152	57930	8714
Aероderiv	3	3600	20.0	1160	456	84	25600	9469
Aероderiv	2	16000	10.6	1232	560	14	3815	11948
Aероderiv	1	14600	13.4	1077	536	20	4942	12414

Source: Bhargava, R., and Meher-Homji, C. B. "Parametric analysis of existing gas turbines with inlet evaporative and overspray fogging," *Journal of Engineering for Gas Turbines and Power*, Vol. 127, No. 1, Jan. 2005.

6.9. Collusive bidding in road construction (cont'd). Apply the all-possible-regressions selection method to the FLAG2 data in Exercise 6.8. Are the variables in the "best subset" model the same as those selected by stepwise regression?

6.10. Cooling method for gas turbines. Refer to the *Journal of Engineering for Gas Turbines and Power* (January 2005) study of a high-pressure inlet fogging method for a gas turbine engine, Exercise 5.8 (p. 271). A number of independent variables were used to predict the heat rate (kilojoules per kilowatt per hour) for each in a sample of 67 gas turbines augmented with high-pressure inlet fogging. The independent variables available are engine type (traditional, advanced, or aероderivative), number of shafts, cycle speed (revolutions per minute), cycle pressure ratio, inlet temperature (°C), exhaust gas temperature (°C), air mass flow rate (kilograms per second), and horsepower (Hp units). The data are saved in the GASTURBINE file. (The first and last five observations are listed in the table above.)

(a) Use stepwise regression (with stepwise selection) to find the "best" predictors of heat rate.

(b) Use stepwise regression (with backward elimination) to find the "best" predictors of heat rate.

(c) Use all-possible-regressions-selection to find the "best" predictors of heat rate.

(d) Compare the results, parts a–c. Which independent variables consistently are selected as the "best" predictors?

(e) Explain how you would use the results, parts a–c, to develop a model for heat rate.

6.11. Groundwater contamination in wells. In New Hampshire, about half the counties mandate the use of reformulated gasoline. This has lead to an increase in the contamination of groundwater with methyl *tert*-butyl ether (MTBE). *Environmental Science and Technology* (January 2005) reported on the factors related to MTBE contamination in public and private New Hampshire wells. Data were collected for a sample of 223 wells and saved

⊚ MTBE (Selected observations)

pH	DISSOXY	INDUSTRY	WELLCLASS	AQUIFER	DEPTH	DISTANCE	MTBE
7.87	0.58	0	Private	Bedrock	60.960	2386.29	0.20
8.63	0.84	0	Private	Bedrock	36.576	3667.69	0.20
7.11	8.37	0	Private	Bedrock	152.400	2324.15	0.20
8.08	0.71	4.21	Public	Bedrock	155.448	1379.81	0.20
7.28	3.33	1.67	Public	Bedrock	60.960	891.93	0.36
6.31	3.07	0.00	Public	Unconsol	13.411	1941.97	0.20
7.53	0.32	18.79	Public	Bedrock	121.920	65.93	15.61
7.45	0.17	21.36	Public	Bedrock	152.400	142.26	32.80

Source: Ayotte, J. D., Argue, D. M., and McGarry, F. J. "Methyl *tert*-butyl ether occurrence and related factors in public and private Wells in southeast New Hampshire," *Environmental Science and Technology*, Vol. 39, No. 1, Jan. 2005. Reprinted with permission from *Environmental Science and Technology*.

in the MTBE file. (Selected observations are listed in the table at the bottom of p. 343.) The list of potential predictors of MTBE level (micrograms per liter) include well class (public or private), aquifer (bedrock or unconsolidated), pH level (standard units), well depth (meters), amount of dissolved oxygen (milligrams per liter), distance from well to nearest fuel source (meters), and percentage of adjacent land allocated to industry. Apply a variable screening method to the data to find a small subset of the independent variables that are the best predictors of MTBE level.

Reference

Kutner, M., Nachtsheim, C., Neter, J., and Li, W. *Applied Linear Statistical Models*, 5th ed. New York: McGraw-Hill/Irwin, 2005.

DEREGULATION OF THE INTRASTATE TRUCKING INDUSTRY

The Problem

We illustrate the modeling techniques outlined in Chapters 5 and 6 with an actual study from engineering economics. Consider the problem of modeling the price charged for motor transport service (e.g., trucking) in Florida. In the early 1980s, several states removed regulatory constraints on the rate charged for intrastate trucking services. (Florida was the first state to embark on a deregulation policy on July 1, 1980.) Prior to this time, the state determined price schedules for motor transport service with review and approval by the Public Service Commission. Once approved, individual carriers were not allowed to deviate from these official rates. The objective of the regression analysis is twofold: (1) assess the impact of deregulation on the prices charged for motor transport service in the state of Florida, and (2) estimate a model of the supply price for predicting future prices.

The Data

The data employed for this purpose ($n = 134$ observations) were obtained from a population of over 27,000 individual shipments in Florida made by major intrastate carriers before and after deregulation. The shipments of interest were made by one particular carrier whose trucks originated from either the city of Jacksonville or Miami. The dependent variable of interest is y, the natural logarithm of the price (measured in 1980 dollars) charged per ton-mile. The independent variables available for predicting y are listed and described in Table CS3.1. These data are saved in the TRUCKING file.

Note the first three variables in Table CS3.1 are quantitative in nature, while the last four variables are all qualitative in nature. Of course, these qualitative independent variables will require the creation of the appropriate number of dummy variables: 1 dummy variable for city of origin, 1 for market size, 1 for deregulation, and 2 for product classification. In the case study, the intent is to build a model that does not differentiate among products. Hence, we will not include the dummy variables for product classification.

Variable Screening

One strategy to finding the best model for y is to use a "build-down" approach, that is, start with a complete second-order model and conduct tests to eliminate terms in the model that are not statistically useful. However, a complete second-order model

Table CS3.1 Independent variables for predicting trucking prices

Variable Name	Description
DISTANCE	Miles traveled (in hundreds)
WEIGHT	Weight of product shipped (in 1,000 pounds)
PCTLOAD	Percent of truck load capacity
ORIGIN	City of origin (JAX or MIA)
MARKET	Size of market destination (LARGE or SMALL)
DEREG	Deregulation in effect (YES or NO)
PRODUCT	Product classification (100, 150, or 200)—Value roughly corresponds to the value-to-weight ratios of the goods being shipped (more valuable goods are categorized in the higher classification)

with three quantitative predictors and three 0–1 dummy variables will involve 80 terms. (Check this as an exercise.) Since the sample size is $n = 134$, there are too few degrees of freedom available for fitting this model. Hence, we require a screening procedure to find a subset of the independent variables that best predict y.

We employed stepwise regression (Chapter 6) to obtain the "best" predictors of the natural logarithm of supply price. The SAS stepwise regression printout is shown in Figure CS3.1. This analysis leads us to select the following variables to begin the model-building process:

1. Distance shipped (hundreds of miles)
2. Weight of product shipped (thousands of pounds)
3. Deregulation in effect (yes or no)
4. Origin of shipment (Miami or Jacksonville)

Summary of Stepwise Selection

Step	Variable Entered	Variable Removed	Number Vars In	Partial R-Square	Model R-Square	C(p)	F Value	Pr > F
1	DISTANCE		1	0.2969	0.2969	417.090	55.74	<.0001
2	DEREG		2	0.3127	0.6096	175.795	104.91	<.0001
3	WEIGHT		3	0.1897	0.7993	30.1997	122.84	<.0001
4	ORIGIN		4	0.0362	0.8355	4.0122	28.40	<.0001

Figure CS3.1 Portion of SAS stepwise regression output

Distance shipped and weight of product are quantitative variables since they each assume numerical values (miles and pounds, respectively) corresponding to the points on a line. Deregulation and origin are qualitative, or categorical, variables that we must describe with dummy (or coded) variables. The variable assignments are given as follows:

$$x_1 = \text{Distance shipped}$$

$$x_2 = \text{Weight of product}$$

$$x_3 = \begin{cases} 1 & \text{if deregulation in effect} \\ 0 & \text{if not} \end{cases}$$

$$x_4 = \begin{cases} 1 & \text{if originate in Miami} \\ 0 & \text{if originate in Jacksonville} \end{cases}$$

Note that in defining the dummy variables, we have arbitrarily chosen "no" and "Jacksonville" to be the base levels of deregulation and origin, respectively.

Model Building

We begin the model-building process by specifying four models. These models, named Models 1–4, are shown in Table CS3.2. Notice that Model 1 is the complete second-order model. Recall from Section 5.10 that the complete second-order model contains quadratic (curvature) terms for quantitative variables and interactions among the quantitative and qualitative terms. For the trucking data, Model 1 traces a parabolic surface for mean natural log of price, $E(y)$, as a function of distance (x_1) and weight (x_2), and the response surfaces differ for the $2 \times 2 = 4$ combinations of the levels of deregulation (x_3) and origin (x_4). Generally, the complete second-order model is a good place to start the model-building process since most real-world relationships are curvilinear. (Keep in mind, however, that you must have a sufficient number of data points to find estimates of all the parameters in the model.) Model 1 is fit to the data for the 134 shipments in the TRUCKING file using SAS. The results are shown in Figure CS3.2. Note that the p-value for the global model F-test is less than .0001, indicating that the complete second-order model is statistically useful for predicting trucking price.

Model 2 contains all the terms of Model 1, except that the quadratic terms (i.e., terms involving x_1^2 and x_2^2) are dropped. This model also proposes four different response surfaces for the combinations of levels of deregulation and origin, but the surfaces are twisted planes (see Figure 5.10) rather than paraboloids. A direct

---

Table CS3.2 Hypothesized models for natural log of trucking price

Model 1: $\quad E(y) = \beta_0 + \beta_1 x_1 + \beta_2 x_2 + \beta_3 x_1 x_2 + \beta_4 x_1^2 + \beta_5 x_2^2 + \beta_6 x_3 + \beta_7 x_4 + \beta_8 x_3 x_4 + \beta_9 x_1 x_3$
$\qquad\qquad + \beta_{10} x_1 x_4 + \beta_{11} x_1 x_3 x_4 + \beta_{12} x_2 x_3 + \beta_{13} x_2 x_4 + \beta_{14} x_2 x_3 x_4 + \beta_{15} x_1 x_2 x_3$
$\qquad\qquad + \beta_{16} x_1 x_2 x_4 + \beta_{17} x_1 x_2 x_3 x_4 + \beta_{18} x_1^2 x_3 + \beta_{19} x_1^2 x_4 + \beta_{20} x_1^2 x_3 x_4 + \beta_{21} x_2^2 x_3$
$\qquad\qquad + \beta_{22} x_2^2 x_4 + \beta_{23} x_2^2 x_3 x_4$

Model 2: $\quad E(y) = \beta_0 + \beta_1 x_1 + \beta_2 x_2 + \beta_3 x_1 x_2 + \beta_6 x_3 + \beta_7 x_4 + \beta_8 x_3 x_4 + \beta_9 x_1 x_3 + \beta_{10} x_1 x_4 + \beta_{11} x_1 x_3 x_4$
$\qquad\qquad + \beta_{12} x_2 x_3 + \beta_{13} x_2 x_4 + \beta_{14} x_2 x_3 x_4 + \beta_{15} x_1 x_2 x_3 + \beta_{16} x_1 x_2 x_4 + \beta_{17} x_1 x_2 x_3 x_4$

Model 3: $\quad E(y) = \beta_0 + \beta_1 x_1 + \beta_2 x_2 + \beta_3 x_1 x_2 + \beta_4 x_1^2 + \beta_5 x_2^2 + \beta_6 x_3 + \beta_7 x_4 + \beta_8 x_3 x_4$

Model 4: $\quad E(y) = \beta_0 + \beta_1 x_1 + \beta_2 x_2 + \beta_3 x_1 x_2 + \beta_4 x_1^2 + \beta_5 x_2^2 + \beta_6 x_3 + \beta_7 x_4 + \beta_8 x_3 x_4$
$\qquad\qquad + \beta_9 x_1 x_3 + \beta_{10} x_1 x_4 + \beta_{11} x_1 x_3 x_4 + \beta_{12} x_2 x_3 + \beta_{13} x_2 x_4 + \beta_{14} x_2 x_3 x_4$
$\qquad\qquad + \beta_{15} x_1 x_2 x_3 + \beta_{16} x_1 x_2 x_4 + \beta_{17} x_1 x_2 x_3 x_4$

Model 5: $\quad E(y) = \beta_0 + \beta_1 x_1 + \beta_2 x_2 + \beta_3 x_1 x_2 + \beta_4 x_1^2 + \beta_5 x_2^2 + \beta_6 x_3 + \beta_9 x_1 x_3 + \beta_{12} x_2 x_3 + \beta_{15} x_1 x_2 x_3$

Model 6: $\quad E(y) = \beta_0 + \beta_1 x_1 + \beta_2 x_2 + \beta_3 x_1 x_2 + \beta_4 x_1^2 + \beta_5 x_2^2 + \beta_7 x_4 + \beta_{10} x_1 x_4 + \beta_{13} x_2 x_4 + \beta_{16} x_1 x_2 x_4$

Model 7: $\quad E(y) = \beta_0 + \beta_1 x_1 + \beta_2 x_2 + \beta_3 x_1 x_2 + \beta_4 x_1^2 + \beta_5 x_2^2 + \beta_6 x_3 + \beta_7 x_4 + \beta_9 x_1 x_3 + \beta_{10} x_1 x_4$
$\qquad\qquad + \beta_{12} x_2 x_3 + \beta_{13} x_2 x_4 + \beta_{15} x_1 x_2 x_3 + \beta_{16} x_1 x_2 x_4$

Figure CS3.2 SAS
regression printout for
Model 1

```
                    Dependent Variable: LNPRICE

                       Analysis of Variance

                              Sum of         Mean
Source              DF       Squares       Square    F Value    Pr > F

Model               23      83.90934      3.64823     65.59    <.0001
Error              110       6.11860      0.05562
Corrected Total    133      90.02794


        Root MSE              0.23585    R-Square     0.9320
        Dependent Mean      10.57621    Adj R-Sq     0.9178
        Coeff Var            2.22997


                       Parameter Estimates

                    Parameter     Standard
Variable      DF     Estimate       Error     t Value   Pr > |t|

Intercept      1     12.51593      0.95441      13.11    <.0001
X1             1     -0.89923      0.73410      -1.22     0.2232
X2             1      0.02421      0.02886       0.84     0.4034
X1X2           1     -0.02071      0.00673      -3.08     0.0026
X1SQ           1      0.15145      0.13455       1.13     0.2628
X2SQ           1     -0.00010196   0.00076963   -0.13     0.8948
X3             1     -1.12650      1.49104      -0.76     0.4516
X4             1      0.27615      0.96332       0.29     0.7749
X3X4           1      0.49697      1.50290       0.33     0.7415
X1X3           1      0.48205      1.15882       0.42     0.6782
X1X4           1      0.06958      0.73882       0.09     0.9251
X1X3X4         1     -0.54037      1.16440      -0.46     0.6435
X2X3           1     -0.09486      0.04477      -2.12     0.0363
X2X4           1     -0.05261      0.03528      -1.49     0.1387
X2X3X4         1      0.06826      0.05220       1.31     0.1937
X1X2X3         1      0.02207      0.01078       2.05     0.0429
X1X2X4         1      0.02355      0.00709       3.32     0.0012
X1X2X3X4       1     -0.02694      0.01127      -2.39     0.0185
X1SQX3         1     -0.11674      0.21918      -0.53     0.5954
X1SQX4         1     -0.07276      0.13510      -0.54     0.5913
X1SQX3X4       1      0.13424      0.21984       0.61     0.5427
X2SQX3         1      0.00043756   0.00119       0.37     0.7127
X2SQX4         1      0.00011095   0.00107       0.10     0.9174
X2SQX3X4       1     -0.00027597   0.00157      -0.18     0.8609
```

comparison of Models 1 and 2 will allow us to test for the importance of the curvature terms.

Model 3 contains all the terms of Model 1, except that the quantitative–qualitative interaction terms are omitted. This model proposes four curvilinear paraboloids corresponding to the four deregulation–origin combinations, that differ only with respect to the y-intercept. By directly comparing Models 1 and 3, we can test for the importance of all the quantitative–qualitative interaction terms.

Model 4 is identical to Model 1, except that it does not include any interactions between the quadratic terms and the two qualitative variables, deregulation (x_3) and origin (x_4). Although curvature is included in this model, the rates of curvature for both distance (x_1) and weight (x_2) are the same for all levels of deregulation and origin.

Figure CS3.3 shows the results of the nested model F-tests described in the above paragraphs. Each of these tests is summarized as follows:

Test for Significance of All Quadratic Terms (Model 1 vs. Model 2)

H_0: $\beta_4 = \beta_5 = \beta_{18} = \beta_{19} = \beta_{20} = \beta_{21} = \beta_{22} = \beta_{23} = 0$
H_a: At least one of the quadratic β's in Model 1 differs from 0

$F = 13.61$, p-value $< .0001$ (shaded at the top of Figure CS3.3)

Figure CS3.3 SAS nested model F-tests for terms in Model 1

```
Test QUADRATIC Results for Dependent Variable LNPRICE

                          Mean
Source         DF        Square      F Value    Pr > F

Numerator       8       0.75727       13.61      <.0001
Denominator   110       0.05562


Test QN_QL_INTERACT Results for
      Dependent Variable LNPRICE

                          Mean
Source         DF        Square      F Value    Pr > F

Numerator      15       0.25574        4.60      <.0001
Denominator   110       0.05562


Test QL_QUAD_INTERACT Results
  for Dependent Variable LNPRICE

                          Mean
Source         DF        Square      F Value    Pr > F

Numerator       6       0.01407        0.25      0.9572
Denominator   110       0.05562
```

Conclusion: There is sufficient evidence (at $\alpha = .01$) of curvature in the relationships between $E(y)$ and distance (x_1) and weight (x_2). Model 1 is a statistically better predictor of trucking price than Model 2.

Test for Significance of All Quantitative–Qualitative Interaction Terms (Model 1 vs. Model 3)

H_0: $\beta_9 = \beta_{10} = \beta_{11} = \beta_{12} = \beta_{13} = \beta_{14} = \beta_{15} = \beta_{16} = \beta_{17} = \beta_{18} = \beta_{19} = \beta_{20} = \beta_{21} = \beta_{22} = \beta_{23} = 0$

H_a: At least one of the QN $\times$ QL interaction β's in Model 1 differs from 0

$F = 4.60$, p-value $< .0001$ (shaded in the middle of Figure CS3.3)

Conclusion: There is sufficient evidence (at $\alpha = .01$) of interaction between the quantitative variables, distance (x_1) and weight (x_2), and the qualitative variables, deregulation (x_3) and origin (x_4). Model 1 is a statistically better predictor of trucking price than Model 3.

Test for Significance of Qualitative–Quadratic Interaction (Model 1 vs. Model 4)

H_0: $\beta_{18} = \beta_{19} = \beta_{20} = \beta_{21} = \beta_{22} = \beta_{23} = 0$

H_a: At least one of the qualitative–quadratic interaction β's in Model 1 differs from 0

$F = .25$, p-value $= .9572$ (shaded at the bottom of Figure CS3.3)

Conclusion: There is insufficient evidence (at $\alpha = .01$) of interaction between the quadratic terms for distance (x_1) and weight (x_2), and the qualitative variables, deregulation (x_3) and origin (x_4). Since these terms are not statistically useful, we will drop these terms from Model 1 and conclude that Model 4 is a statistically better predictor of trucking price.*

Based on the three nested-model F-tests, we found Model 4 to be the "best" of the first four models. The SAS printout for Model 4 is shown in Figure CS3.4. Looking at the results of the global F-test (p-value less than .0001), you can see that the overall model is statistically useful for predicting trucking price. Also, $R^2_{adj} = .9210$ implies that about 92% of the sample variation in the natural log of trucking price can be explained by the model. Although these model statistics are impressive, we may be able to find a simpler model that fits the data just as well.

Table CS3.2 gives three additional models. Model 5 is identical to Model 4, but all terms for the qualitative variable origin(x_4) have been dropped. A comparison of Model 4 to Model 5 will allow us to determine whether origin really has an impact on trucking price. Similarly, Model 6 is identical to Model 4, but now all terms for the qualitative variable deregulation (x_3) have been dropped. By comparing Model 4 to Model 6, we can determine whether deregulation has an impact on trucking price. Finally, we propose Model 7, which is obtained by dropping all the qualitative–qualitative interaction terms. A comparison of Model 4 to Model 7 will

Figure CS3.4 SAS regression printout for Model 4

```
                    Dependent Variable: LNPRICE

                       Analysis of Variance

                              Sum of          Mean
Source              DF       Squares        Square    F Value   Pr > F

Model               17      83.82495       4.93088     92.21   <.0001
Error              116       6.20299       0.05347
Corrected Total    133      90.02794


           Root MSE              0.23124    R-Square    0.9311
           Dependent Mean       10.57621    Adj R-Sq    0.9210
           Coeff Var             2.18646


                       Parameter Estimates

                    Parameter      Standard
Variable     DF      Estimate        Error     t Value   Pr > |t|

Intercept     1      12.08482       0.25871      46.71    <.0001
X1            1      -0.55296       0.09648      -5.73    <.0001
X2            1       0.01889       0.02120       0.89     0.3748
X1X2          1      -0.02041       0.00649      -3.15     0.0021
X1SQ          1       0.08738       0.00827      10.56    <.0001
X2SQ          1       0.00008202    0.00037354    0.22     0.8266
X3            1      -0.38504       0.40009      -0.96     0.3379
X4            1       0.76041       0.27135       2.80     0.0059
X3X4          1      -0.35311       0.42661      -0.83     0.4095
X1X3          1      -0.13163       0.14172      -0.93     0.3549
X1X4          1      -0.33374       0.08995      -3.71     0.0003
X1X3X4        1       0.18215       0.14746       1.24     0.2192
X2X3          1      -0.08259       0.02956      -2.79     0.0061
X2X4          1      -0.04830       0.02049      -2.36     0.0201
X2X3X4        1       0.05937       0.03217       1.85     0.0675
X1X2X3        1       0.02136       0.01043       2.05     0.0428
X1X2X4        1       0.02320       0.00685       3.39     0.0010
X1X2X3X4      1      -0.02601       0.01090      -2.39     0.0186
```

---

* There is always danger in dropping terms from the model. Essentially, we are accepting H_0: $\beta_{18} = \beta_{19} = \beta_{20} = \cdots = \beta_{23} = 0$ when $P(\text{Type II error}) = P(\text{Accepting } H_0 \text{ when } H_0 \text{ is false}) = \beta$ is unknown. In practice, however, many researchers are willing to risk making a Type II error rather than use a more complex model for $E(y)$ when simpler models that are nearly as good as predictors (and easier to apply and interpret) are available. Note that we used a relatively large amount of data ($n = 134$) in fitting our models and that R^2_{adj} for Model 4 is actually larger than R^2_{adj} for Model 1. If the quadratic interaction terms are, in fact, important (i.e., we have made a Type II error), there is little lost in terms of explained variability in using Model 4.

Figure CS3.5 SAS nested model F-tests for terms in Model 4

```
Test ORIGIN Results for Dependent Variable LNPRICE

                                  Mean
Source            DF             Square      F Value    Pr > F

Numerator          8            0.18987        3.55     0.0010
Denominator      116            0.05347
```

```
Test DEREG Results for Dependent Variable LNPRICE

                                  Mean
Source            DF             Square      F Value    Pr > F

Numerator          8            4.03417       75.44     <.0001
Denominator      116            0.05347
```

```
            Test ORG_DEREG_INTERACTION Results
                for Dependent Variable LNPRICE

                                  Mean
Source            DF             Square      F Value    Pr > F

Numerator          4            0.11367        2.13     0.0820
Denominator      116            0.05347
```

allow us to see whether deregulation and origin interact to effect the natural log of trucking price.

Figure CS3.5 shows the results of the nested model F-tests described above. A summary of each of these tests follows.

Test for Significance of All Origin Terms (Model 4 vs. Model 5)

$H_0\colon \beta_7 = \beta_8 = \beta_{10} = \beta_{11} = \beta_{13} = \beta_{14} = \beta_{16} = \beta_{17} = 0$

$H_a\colon$ At least one of the origin β's in Model 4 differs from 0

$$F = 3.55, \quad p\text{-value} = .001 \text{ (shaded at the top of Figure CS3.5)}$$

Conclusion: There is sufficient evidence (at $\alpha = .01$) to indicate that origin (x_4) has an impact on trucking price. Model 4 is a statistically better predictor of trucking price than Model 5.

Test for Significance of All Deregulation Terms (Model 4 vs. Model 6)

$H_0\colon \beta_6 = \beta_8 = \beta_9 = \beta_{11} = \beta_{12} = \beta_{14} = \beta_{15} = \beta_{17} = 0$

$H_a\colon$ At least one of the deregulation β's in Model 4 differs from 0

$$F = 75.44, \quad p\text{-value} < .0001 \text{ (shaded in the middle of Figure CS3.5)}$$

Conclusion: There is sufficient evidence (at $\alpha = .01$) to indicate that deregulation (x_3) has an impact on trucking price. Model 4 is a statistically better predictor of trucking price than Model 6.

Test for Significance of All Deregulation–Origin Interaction Terms (Model 4 vs. Model 7)

$H_0\colon \beta_8 = \beta_{11} = \beta_{14} = \beta_{17} = 0$

$H_a\colon$ At least one of the QL $\times$ QL interaction β's in Model 4 differs from 0

$$F = 2.13, \quad p\text{-value} = .0820 \text{ (shaded at the bottom of Figure CS3.5)}$$

Figure CS3.6 SAS regression printout for Model 7

Dependent Variable: LNPRICE

Analysis of Variance

Source	DF	Sum of Squares	Mean Square	F Value	Pr > F
Model	13	83.37026	6.41310	115.59	<.0001
Error	120	6.65767	0.05548		
Corrected Total	133	90.02794			

Root MSE	0.23554	R-Square	0.9260	
Dependent Mean	10.57621	Adj R-Sq	0.9180	
Coeff Var	2.22710			

Parameter Estimates

Variable	DF	Parameter Estimate	Standard Error	t Value	Pr > |t|
Intercept	1	12.19150	0.21583	56.49	<.0001
X1	1	-0.59800	0.08425	-7.10	<.0001
X2	1	-0.00598	0.01857	-0.32	0.7480
X1X2	1	-0.01078	0.00530	-2.03	0.0442
X1SQ	1	0.08575	0.00834	10.28	<.0001
X2SQ	1	0.00014207	0.00037728	0.38	0.7072
X3	1	-0.78192	0.12900	-6.06	<.0001
X4	1	0.67679	0.21035	3.22	0.0017
X1X3	1	0.03991	0.03999	1.00	0.3203
X1X4	1	-0.27464	0.07267	-3.78	0.0002
X2X3	1	-0.02094	0.01045	-2.00	0.0473
X2X4	1	-0.02619	0.01610	-1.63	0.1063
X1X2X3	1	-0.00332	0.00303	-1.10	0.2757
X1X2X4	1	0.01298	0.00544	2.39	0.0186

Conclusion: There is insufficient evidence (at $\alpha = .01$) to indicate that deregulation (x_3) and origin (x_4) interact. Thus, we will drop these interaction terms from Model 4 and conclude that Model 7 is a statistically better predictor of trucking price.

In summary, the nested model F-tests suggest that Model 7 is the best for modeling the natural log of trucking price. The SAS printout for Model 7 is shown in Figure CS3.6. The β-estimates used for making predictions of trucking price are highlighted on the printout.

A note of caution: Just as with t-tests on individual β parameters, you should avoid conducting too many partial F-tests. Regardless of the type of test (t test or F-test), the more tests that are performed, the higher the overall Type I error rate will be. In practice, you should limit the number of models that you propose for $E(y)$ so that the overall Type I error rate α for conducting partial F-tests remains reasonably small.[†]

Impact of Deregulation

In addition to estimating a model of the supply price for prediction purposes, a goal of the regression analysis was to assess the impact of deregulation on the trucking prices. To do this, we examine the β-estimates in Model 7, specifically the β's associated with the deregulation dummy variable, x_3. From Figure CS3.6, the prediction equation is:

$$\hat{y} = 12.192 - .598x_1 - .00598x_2 - .01078x_1x_2 + .086x_1^2 + .00014x_2^2$$
$$+ .677x_4 - .275x_1x_4 - .026x_2x_4 + .013x_1x_2x_4$$
$$- .782x_3 + .0399x_1x_3 - .021x_2x_3 - .0033x_1x_2x_3$$

[†] A technique suggested by Bonferroni is often applied to maintain control of the overall Type I error rate α. If c tests are to be performed, then conduct each individual test at significance level α/c. This will guarantee an overall Type I error rate less than or equal to α. For example, conducting each of $c = 5$ tests at the $.05/5 = .01$ level of significance guarantees an overall $\alpha \leq .05$.

Note that the terms in the equation were rearranged so that the β's associated with the deregulation variable are shown together at the end of the equation. Because of some interactions, simply examining the signs of the β-estimates can be confusing and lead to erroneous conclusions.

A good way to assess the impact of deregulation is to hold fixed all but one of the other independent variables in the model. For example, suppose we fix the weight of the shipment at 15,000 pounds and consider only shipments originating from Jacksonville (i.e., set $x_2 = 15$ and $x_4 = 0$). Substituting these values into the prediction equation and combining like terms yields:

$$\hat{y} = 12.192 - .598x_1 - .00598(15) - .01078x_1(15) + .086x_1^2 + .00014(15)^2$$
$$+ .677(0) - .275x_1(0) - .026(15)(0) + .013x_1(15)(0)$$
$$- .782x_3 + .0399x_1x_3 - .021(15)x_3 - .0033x_1(15)x_3$$
$$= 12.134 - .760x_1 + .086x_1^2 - 1.097x_3 - .0096x_1x_3$$

To see the impact of deregulation on the estimated curve relating log price (y) to distance shipped (x_1), compare the prediction equations for the two conditions, $x_3 = 0$ (regulated prices) and $x_3 = 1$ (deregulation):

$$\text{Regulated}(x_3 = 0) : \hat{y} = 12.134 - .760x_1 + .086x_1^2 - 1.097(0) - .0096x_1(0)$$
$$= 12.134 - .760x_1 + .086x_1^2$$
$$\text{Deregulation}(x_3 = 1) : \hat{y} = 12.134 - .760x_1 + .086x_1^2 - 1.097(1) - .0096x_1(1)$$
$$= 11.037 - .7696x_1 + .086x_1^2$$

Figure CS3.7 SAS graph of the prediction equation for log price

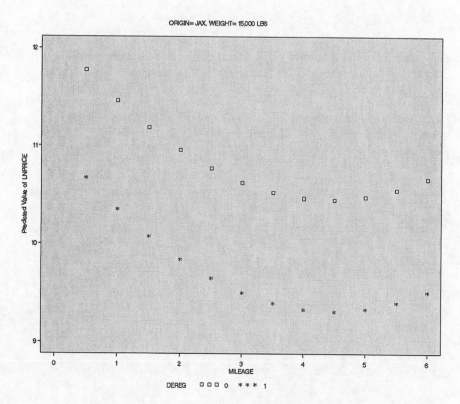

Notice that the y-intercept for the regulated prices (12.134) is larger than the y-intercept for the deregulated prices (11.037). Also, although the equations have the same rate of curvature, the estimated shift parameters differ.

These prediction equations are portrayed graphically in the SAS printout in Figure CS3.7. The graph clearly shows the impact of deregulation on the prices charged when the carrier leaves from Jacksonville with a cargo of 15,000 pounds. As expected from economic theory, the curve for the regulated prices lies above the curve for deregulated prices.

Follow-up Questions

1. In Figure CS3.7, give an expression (in terms of the estimated β's from Model 7) for the difference between the predicted regulated price and predicted deregulated price for any fixed value of mileage.

2. Demonstrate the impact of deregulation on price charged using the estimated β's from Model 7 in a fashion similar to the case study, but now hold origin fixed at Miami and weight fixed at 10,000 pounds.

3. The data file TRUCKING4 contains data on trucking prices for four Florida carriers (A, B, C, and D). These carriers are identified by the variable CARRIER. (Note: Carrier B is the carrier analyzed in the case study.) Using Model 7 as a base model, add terms that allow for different response curves for the four carriers. Conduct the appropriate test to determine if the curves differ.

Some Regression Pitfalls

Contents

Objectives

1. To identify several potential problems you may encounter when constructing a model for a response y.

2. To help you recognize when these problems exist so that you can avoid some of the pitfalls of multiple regression analysis.

7.1 Introduction

Multiple regression analysis is recognized by practitioners as a powerful tool for modeling a response y and is therefore widely used. But it is also one of the most abused statistical techniques. The ease with which a multiple regression analysis can be run with statistical computer software has opened the door to many data analysts who have but a limited knowledge of multiple regression and statistics. In practice, building a model for some response y is rarely a simple, straightforward process. There are a number of pitfalls that trap the unwary analyst. In this chapter, we discuss several problems that you should be aware of when constructing a multiple regression model.

7.2 Observational Data versus Designed Experiments

One problem encountered in using a regression analysis is caused by the type of data that the analyst is often forced to collect. Recall, from Section 2.4, that the data for regression can be either *observational* (where the values of the independent variables are uncontrolled) or *experimental* (where the x's are controlled via a designed experiment). Whether data are observational or experimental is important for the following reasons. First, as you will subsequently learn in Chapter 11, the quantity of information in an experiment is controlled not only by the *amount of data*, but also by the *values of the predictor variables* $x_1, x_2, \ldots, x_k$. Consequently, if you can design the experiment (sometimes this is physically impossible), you may be able to increase greatly the amount of information in the data at no additional cost.

Second, the use of observational data creates a problem involving **randomization**. When an experiment has been designed and we have decided on the various settings of the independent variables to be used, the experimental units are then randomly assigned in such a way that each combination of the independent variables has an equal chance of receiving experimental units with unusually high (or low) readings. (We illustrate this method of randomization in Chapter 12.) This procedure tends to average out any variation within the experimental units. The result is that if the difference between two sample means is statistically significant, then you can infer (with probability of Type I error equal to α) that the population means differ. But more important, you can infer that this difference was due to the settings of the predictor variables, which is what you did to make the two populations different. Thus, you can infer a cause-and-effect relationship.

If the data are observational, a statistically significant relationship between a response y and a predictor variable x does not imply a cause-and-effect relationship. It simply means that x contributes information for the prediction of y, and nothing more. This point is aptly illustrated in the following example.

Example 7.1

USA Today (April 16, 2002) published an article, "Cocaine Use During Pregnancy Linked to Development Problems." The article reports on the results of a study in which researchers gave IQ tests to two groups of infants born in Ohio—218 whose mothers admitted using cocaine during pregnancy and 197 who were not exposed to cocaine. The mothers (and their infants) who participated in the study were volunteers. About 80% of the mothers in each group were minorities, and both groups of women used a variety of legal and illegal substances, from alcohol to marijuana. The researchers found that "babies whose mothers use cocaine during pregnancy score lower on early intelligence tests... than those not exposed to the drug." The two variables measured in the study—IQ score at age 2, y, and cocaine use during pregnancy, x (where $x = 1$ if mother admits cocaine use during pregnancy and $x = 0$ if not)—were found to be negatively correlated. Although the quotation does not use the word *cause*, it certainly implies to the casual reader that a low IQ results from cocaine use during pregnancy.

(a) Are the data collected in the study observational or experimental?

(b) Identify any weaknesses in the study.

Solution

(a) The mothers (and their infants) represent the experimental units in the study. Since no attempt was made to control the value of x, cocaine use during pregnancy, the data are observational.

(b) The pitfalls provided by the study are apparent. First, the response y (IQ of infant at age 2) is related to only a single variable, x (cocaine use during pregnancy). Second, since the mothers who participated in the study were *not* randomly assigned to one of the two groups—an obviously impossible task—a real possibility exists that mothers with lower IQ and/or lower socioeconomic status tended to fall in the cocaine-use group. In other words, perhaps the study is simply showing that mothers from lower socioeconomic groups, who may not provide the daily care and nurturing that more fortunate children receive, are more likely to have babies with lower IQ scores. Also, many of the babies in the study were premature—a factor known to hinder a baby's development. Which variable—cocaine use during pregnancy, socioeconomic status, or premature status—is the cause of an infant's low IQ is impossible to determine based on the observational data collected in the study. This demonstrates the primary weakness of observational experiments. ▬

Example 7.2

Consider a study presented at an *American Heart Association Conference* (November 2005) to gauge whether animal-assisted therapy can improve the physiological responses of patients with heart failure. In the study, 76 heart patients were randomly assigned to one of three groups. Each patient in group T was visited by a human volunteer accompanied by a trained dog; each patient in group V was visited by a volunteer only; and the patients in group C were not visited at all. The anxiety level of each patient was measured (in points) both before and after the visits. The researchers discovered that the sample mean drop in anxiety level for patients in the three groups, T, V, and C, were 10.5, 3.9, and 1.4, respectively.

(a) Write a regression model for mean drop in anxiety level, $E(y)$, as a function of patient group.

(b) Use the results of the study to estimate the β's in the model.

(c) Are the data collected in the study observational or experimental?

(d) If the differences in the mean drops in anxiety levels are found to be statistically significant, can the researchers conclude that animal-assisted therapy is a promising treatment for improving the physiological responses of patients with heart failure?

Solution

(a) Since the independent variable in this study, patient group, is qualitative at three levels (T, V, and C), we create two dummy variables:

$$x_1 = \{1 \text{ if group T, 0 if not}\},$$

$$x_2 = \{1 \text{ if group V, 0 if not}\}, \text{ (Base level} = \text{group C)}$$

Then the model is: $E(y) = \beta_0 + \beta_1 x_1 + \beta_2 x_2$

(b) Let μ_j represent $E(y)$ for patient group j. For a model with dummy variables, we know (Section 5.7) that

$\beta_0 = \mu_C$ (the mean drop in anxiety for the base level, group C)

$\beta_1 = \mu_T - \mu_C$ (the difference in mean drops between group T and group C)

$\beta_2 = \mu_V - \mu_C$ (the difference in mean drops between group V and group C)

Since the sample mean drops in anxiety levels are $\bar{y}_T = 10.5$, $\bar{y}_V = 3.9$, and $\bar{y}_C = 1.4$, the estimated β's are:

$$\hat{\beta}_0 = \bar{y}_C = 1.4$$

$$\hat{\beta}_1 = \bar{y}_T - \bar{y}_C = 10.5 - 1.4 = 9.1$$

$$\hat{\beta}_2 = \bar{y}_V - \bar{y}_C = 3.9 - 1.4 = 2.5$$

(c) Clearly, the values of the independent variable (patient group) in the model were controlled by the researchers. Each experimental unit (heart patient) in the study was randomly assigned to one of the three groups, T, V, and C. Consequently, the data are experimental.

(d) Because the patients were randomly assigned to the groups, the previsit anxiety levels of patients in the animal-assisted therapy group (T) are likely to be about the same, on average, as the previsit anxiety levels of patients in the other two groups. In other words, the randomization is likely to remove any inherent differences in the previsit anxiety levels of patients in the three groups. Also, if any differences do exist after randomization, the researchers account for these differences by using "drop" in anxiety level after treatment as the dependent variable, y. Consequently, if the researchers do discover statistically significant differences in the mean drop in anxiety levels of the groups, the differences are likely to be due to the type of therapy each group received. In fact, the estimate of $\beta_1 = \mu_T - \mu_C$ was found to be positive and statistically different from 0. This led the study presenters to conclude that animal-assisted therapy is a promising treatment for improving the physiological responses of patients with heart failure. ■

The point of the previous examples is twofold. If you can control the values of the independent variables in an experiment, it pays to do so. If you cannot control them, you can still learn much from a regression analysis about the relationship between a response y and a set of predictors. In particular, a prediction equation that provides a good fit to your data will almost always be useful. However, **you must be careful about deducing cause-and-effect relationships between the response and the predictors in an observational experiment**.

Caution

> With observational data, a statistically significant relationship between a response y and a predictor variable x *does not necessarily* imply a cause-and-effect relationship.

Learning about the design of experiments is useful even if most of your applications of regression analysis involve observational data. Learning how to design an experiment and control the information in the data will improve your ability to assess the quality of observational data. We introduce experimental design in Chapter 11 and present methods for analyzing the data in a designed experiment in Chapter 12.

7.3 Parameter Estimability and Interpretation

Suppose we want to fit the first-order model

$$E(y) = \beta_0 + \beta_1 x$$

to relate a developmentally challenged child's creativity score y to the child's flexibility score x. Now, suppose we collect data for three such challenged children, and each child has a flexibility score of 5. The data are shown in Figure 7.1. You can see the problem: The parameters of the straight-line model cannot be estimated when all the data are concentrated at a single x-value. Recall that it takes two points (x-values) to fit a straight line. Thus, the parameters are not estimable when only one x-value is observed.

A similar problem would occur if we attempted to fit the second-order model

$$E(y) = \beta_0 + \beta_1 x + \beta_2 x^2$$

to a set of data for which only one *or two* different x-values were observed (see Figure 7.2). At least three different x-values must be observed before a second-order model can be fitted to a set of data (i.e., before all three parameters are estimable). In general, the number of levels of a quantitative independent variable x must be at least one more than the order of the polynomial in x that you want to fit. If two values of x are too close together, you may not be able to estimate a parameter because of

Figure 7.1 Creativity and flexibility data for three children

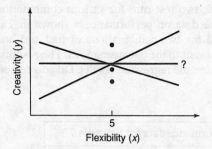

rounding error encountered in fitting the model. Remember, also, that the sample size n must be sufficiently large so that the degrees of freedom for estimating σ^2, df(Error) $= n - (k + 1)$, exceeds 0. In other words, n must exceed the number of β parameters in the model, $k + 1$. The requirements for fitting a pth-order polynomial regression model are shown in the box.

Figure 7.2 Only two different x-values observed—the second-order model is not estimable

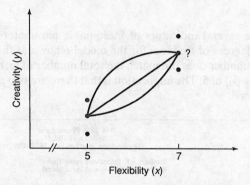

Requirements for Fitting a pth-Order Polynomial Regression Model

$$E(y) = \beta_0 + \beta_1 x + \beta_2 x^2 + \cdots + \beta_p x^p$$

1. The number of levels of x must be greater than or equal to $(p + 1)$.
2. The sample size n must be greater than $(p + 1)$ to allow sufficient degrees of freedom for estimating σ^2.

Most variables are not controlled by the researcher, but the independent variables are usually observed at a sufficient number of levels to permit estimation of the model parameters. However, when the statistical software you use is unable to fit a model, the problem is probably inestimable parameters. This pitfall is demonstrated in the next example.

Example 7.3

In Example 5.10 (p. 297), we investigated a model for diesel engine performance y (measured as mass burning rate per degree of crank angle), as a function of two qualitative predictors: fuel type at three levels (F_1, F_2, and F_3) and engine brand at two levels (B_1 and B_2). Consider the interaction model

$$E(y) = \beta_0 + \beta_1 x_1 + \beta_2 x_2 + \beta_3 x_3 + \beta_4 x_1 x_3 + \beta_5 x_2 x_3$$

where $x_1 = \{1$ if F_1, 0 if not$\}$, $x_2 = \{1$ if F_2, 0 if not$\}$, (Base level $= F_3$)

$\qquad x_3 = \{1$ if B_1, 0 if $B_2\}$

Now, suppose two test runs for various combinations of fuel type and engine brand yielded the data on performance y shown in Table 7.1. Note that there is no data collected for two combinations of fuel and brand (F_1B_2 and F_3B_1). The interaction model is fit to the data using SAS. The resulting SAS printout is shown in Figure 7.3. Examine the regression results. Do you detect a problem with parameter estimability?

🔘 DIESEL2

Table 7.1 Performance data for Example 7.3

		Brand	
		B_1	B_2
Fuel Type	F_1	73, 68	
	F_2	78, 82	50, 43
	F_3		61, 62

Solution

There are several indicators of inestimable parameters in Figure 7.3. First, note that the degrees of freedom for the model (shown at the top of the SAS printout) is 3. This number does not match the total number of terms in the interaction model (excluding β_0) of 5. The implication is that there are two β's in the model that cannot be estimated.

Figure 7.3 SAS output for interaction model, Example 7.3

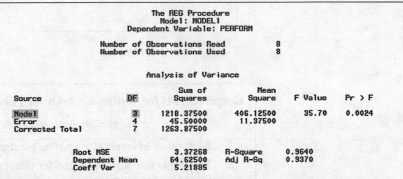

```
                              The REG Procedure
                                Model: MODEL1
                          Dependent Variable: PERFORM

                       Number of Observations Read          8
                       Number of Observations Used          8


                             Analysis of Variance

                                  Sum of        Mean
       Source            DF      Squares       Square   F Value   Pr > F

       Model              3   1218.37500    406.12500     35.70   0.0024
       Error              4     45.50000     11.37500
       Corrected Total    7   1263.87500


              Root MSE             3.37268     R-Square     0.9640
              Dependent Mean      64.62500     Adj R-Sq     0.9370
              Coeff Var            5.21885

NOTE: Model is not full rank. Least-squares solutions for the parameters are not unique. Some
      statistics will be misleading. A reported DF of 0 or B means that the estimate is biased.
NOTE: The following parameters have been set to 0, since the variables are a linear combination
      of other variables as shown.

                             X1X3 =   X1
                             X2X3 =  -X1 + X3

                             Parameter Estimates

                       Parameter     Standard
       Variable    DF    Estimate       Error   t Value   Pr > |t|
       Intercept    1    61.50000     2.38485     25.79    <.0001
       X1           B   -24.50000     4.76970     -5.14    0.0068
       X2           1   -15.00000     3.37268     -4.45    0.0113
       X3           B    33.50000     3.37268      9.93    0.0006
       X1X3         0           0                      .         .
       X2X3         0           0                      .         .
```

Second, the two notes in the middle of the SAS printout explicitly warn the analyst of a problem with parameter estimability. The second note identifies the two parameters that cannot be estimated. These inestimable β's are those associated with the two Fuel-by-Brand interaction terms, x_1x_3 and x_2x_3.

Third, the β estimates themselves (shown at the bottom of the SAS printout) are an indication of this regression pitfall. The two interaction β's are both set equal to 0 and there are no reported standard errors, t-values, and p-values for these β's. The prudent analyst is careful not to infer that there is no interaction between fuel type and engine brand. Rather, the most one can say is that there is not enough information in the sample data to estimate (and consequently, test for) interaction.

This problem occurs because of the missing data for the two combinations of fuel type and brand in Table 7.1. To see this, recall that interaction between fuel type and brand implies that the difference between the mean performance levels of the two brands will not be the same at each level of fuel type. Now examine the data in Table 7.1. Does the difference in the means for B_1 and B_2 depend on fuel type? For fuel type F_2, the estimated difference is

$$\bar{y}_{B_1} - \bar{y}_{B_2} = (78 + 82)/2 - (50 + 43)/2 = 80 - 46.5 = 33.5$$

However, due to the missing data, it is impossible to calculate a similar difference for fuel types F_1 and F_3. In fact, SAS reports the estimate of β_3 as 33.5. Why? From our discussion in Section 5.8 on interaction models with qualitative independent variables, we can show that β_3 represents the difference between the mean performance levels of brands B_1 and B_2 when fuel type is set at the base level (F_3)—not when fuel type is set at F_2. However, once SAS sets the interaction β's equal to 0 (due to lack of information), then the implicit assumption is that there is no interaction and the difference between brands is the same for all fuel types. ■

Given that the parameters of the model are estimable, it is important to interpret the parameter estimates correctly. A typical misconception is that $\hat{\beta}_i$ always measures the effect of x_i on $E(y)$, *independently* of the other x variables in the model. This may be true for some models, but it is not true in general. We see in Section 7.4 that when the independent variables are correlated, the values of the estimated β coefficients are often misleading. Even if the independent variables are uncorrelated, the presence of interaction changes the meaning of the parameters. For example, the underlying assumption of the first-order model

$$E(y) = \beta_0 + \beta_1 x_1 + \beta_2 x_2$$

is, in fact, that x_1 and x_2 affect the mean response $E(y)$ independently. Recall from Sections 4.3 and 5.4 that the slope parameter β_1 measures the rate of change of y for a 1-unit increase in x_1, for any given value of x_2. However, if the relationship between $E(y)$ and x_1 depends on x_2 (i.e., if x_1 and x_2 interact), then the interaction model

$$E(y) = \beta_0 + \beta_1 x_1 + \beta_2 x_2 + \beta_3 x_1 x_2$$

is more appropriate. For the interaction model, we showed that the effect of x_1 on $E(y)$ (i.e., the slope) is not measured by a single β parameter, but by $\beta_1 + \beta_3 x_2$.

Generally, the interpretation of an individual β parameter becomes increasingly difficult as the model becomes more complex. As we learned in Chapter 5, the individual β's of higher-order models usually have no practical interpretation.

Another misconception about the parameter estimates is that the magnitude of $\hat{\beta}_i$ determines the importance of x_i; that is, the larger (in absolute value) the $\hat{\beta}_i$, the more important the independent variable x_i is as a predictor of y. We learned in Chapter 4, however, that the standard error of the estimate $s_{\hat{\beta}_i}$ is critical in making inferences about the true parameter value. To reliably assess the importance of an individual term in the model, we conduct a test of H_0: $\beta_i = 0$ or construct a confidence interval for β_i using formulas that reflect the magnitude of $s_{\hat{\beta}_i}$.

In addition to the parameter estimates, $\hat{\beta}_i$, some statistical software packages report the **standardized regression coefficients**,

$$\hat{\beta}_i^* = \hat{\beta}_i \left(\frac{s_{x_i}}{s_y} \right)$$

where s_{x_i} and s_y are the standard deviations of the x_i and y-values, respectively, in the sample. Unlike $\hat{\beta}_i$, $\hat{\beta}_i^*$ is scaleless. These standardized regression coefficients make it more feasible to compare parameter estimates since the units are the same. However, the problems with interpreting standardized regression coefficients are much the same as those mentioned previously. Therefore, you should be wary of using a standardized regression coefficient as the sole determinant of an x variable's importance. The next example illustrates this point.

Example 7.4

Refer to the problem of modeling the auction price y of antique grandfather clocks, Examples 4.1–4.6. In Example 4.1, we fit the model

$$E(y) = \beta_0 + \beta_1 x_1 + \beta_2 x_2$$

where x_1 = age of the clock and x_2 = number of bidders. The SPSS printout for the regression analysis is shown in Figure 7.4. Locate the standardized β coefficients on the printout and interpret them.

Figure 7.4 SPSS regression printout for grandfather clock model

Model Summary

Model	R	R Square	Adjusted R Square	Std. Error of the Estimate
1	.945a	.892	.885	133.485

a. Predictors: (Constant), BIDDERS, AGE

ANOVAb

Model		Sum of Squares	df	Mean Square	F	Sig.
1	Regression	4283063	2	2141531.480	120.188	.000a
	Residual	516726.5	29	17818.157		
	Total	4799790	31			

a. Predictors: (Constant), BIDDERS, AGE

b. Dependent Variable: PRICE

Coefficientsa

Model		Unstandardized Coefficients		Standardized Coefficients		
		B	Std. Error	Beta	t	Sig.
1	(Constant)	-1338.951	173.809		-7.704	.000
	AGE	12.741	.905	.887	14.082	.000
	BIDDERS	85.953	8.729	.620	9.847	.000

a. Dependent Variable: PRICE

Solution

The standardized β coefficients are highlighted on the SPSS printout in the column labeled **Beta**. These values are

$$\hat{\beta}_1^* = .887 \quad \text{and} \quad \hat{\beta}_2^* = .620$$

Compare these values to the unstandardized β coefficients (in the **B** column):

$$\hat{\beta}_1 = 12.741 \quad \text{and} \quad \hat{\beta}_2 = 85.953$$

Based on the fact that $\hat{\beta}_2$ is nearly seven times larger than $\hat{\beta}_1$, we might be tempted to say that number of bidders (x_2) is a more important predictor of auction price than age of the clock (x_1). Once we standardize the β's (i.e., take the units of measurement and variation into account), we see that the opposite may, in fact, be true since $\hat{\beta}_1^*$ exceeds $\hat{\beta}_2^*$. Of course, from Example 4.6 we know that the two independent variables, x_1 and x_2, interact to affect y. Consequently, both age and number of bidders are important for predicting auction price and we should resist inferring that one of the variables is more important than the other. ■

7.4 Multicollinearity

Often, two or more of the independent variables used in the model for $E(y)$ will contribute redundant information. That is, the independent variables will be correlated with each other. For example, suppose we want to construct a model to predict the gasoline mileage rating, y, of a truck as a function of its load, x_1, and the horsepower, x_2, of its engine. In general, you would expect heavier loads to require greater horsepower and to result in lower mileage ratings. Thus, although both x_1 and x_2 contribute information for the prediction of mileage rating, some of the information is overlapping, because x_1 and x_2 are correlated. When the independent variables are correlated, we say that **multicollinearity** exists. In practice, it is not uncommon to observe correlations among the independent variables. However, a few problems arise when serious multicollinearity is present in the regression analysis.

Definition 7.1 Multicollinearity exists when two or more of the independent variables used in regression are moderately or highly correlated.

First, high correlations among the independent variables (i.e., **extreme** multicollinearity) increase the likelihood of rounding errors in the calculations of the β estimates, standard errors, and so forth.* Second, the regression results may be confusing and misleading.

To illustrate, if the gasoline mileage rating model

$$E(y) = \beta_0 + \beta_1 x_1 + \beta_2 x_2$$

were fitted to a set of data, we might find that the t-values for both $\hat{\beta}_1$ and $\hat{\beta}_2$ (the least squares estimates) are nonsignificant. However, the F-test for H_0: $\beta_1 = \beta_2 = 0$ would probably be highly significant. The tests may seem to be contradictory, but really they are not. The t-tests indicate that the contribution of one variable, say, $x_1 = $ load , is not significant after the effect of $x_2 = $ horsepower has been discounted (because x_2 is also in the model). The significant F-test, on the other hand, tells us that at least one of the two variables is making a contribution to the prediction of y (i.e., β_1, β_2, or both differ from 0). In fact, both are probably contributing, but the contribution of one overlaps with that of the other.

---

*The result is due to the fact that, in the presence of severe multicollinearity, the computer has difficulty inverting the information matrix ($X'X$). See Appendix B for a discussion of the ($X'X$) matrix and the mechanics of a regression analysis.

Multicollinearity can also have an effect on the signs of the parameter estimates. More specifically, a value of $\hat{\beta}_i$ may have the opposite sign from what is expected. For example, we expect the signs of both of the parameter estimates for the gasoline mileage rating model to be negative, yet the regression analysis for the model might yield the estimates $\hat{\beta}_1 = .2$ and $\hat{\beta}_2 = -.7$. The positive value of $\hat{\beta}_1$ seems to contradict our expectation that heavy loads will result in lower mileage ratings. We mentioned in the previous section, however, that it is dangerous to interpret a β coefficient when the independent variables are correlated. Because the variables contribute redundant information, the effect of load x_1 on mileage rating is measured only partially by $\hat{\beta}_1$. Also, we warned in Section 7.2 that we cannot establish a cause-and-effect relationship between y and the predictor variables based on observational data. By attempting to interpret the value $\hat{\beta}_1$, we are really trying to establish a cause-and-effect relationship between y and x_1 (by suggesting that a heavy load x_1 will *cause* a lower mileage rating y).

How can you avoid the problems of multicollinearity in regression analysis? One way is to conduct a designed experiment so that the levels of the x variables are uncorrelated (see Section 7.2). Unfortunately, time and cost constraints may prevent you from collecting data in this manner. For these and other reasons, much of the data collected in scientific studies are observational. Since observational data frequently consist of correlated independent variables, you will need to recognize when multicollinearity is present and, if necessary, make modifications in the analysis.

Several methods are available for detecting multicollinearity in regression. A simple technique is to calculate the coefficient of correlation r between each pair of independent variables in the model. If one or more of the r values is close to 1 or -1, the variables in question are highly correlated and a severe multicollinearity problem may exist.* Other indications of the presence of multicollinearity include those mentioned in the beginning of this section—namely, nonsignificant t-tests for the individual β parameters when the F-test for overall model adequacy is significant, and estimates with opposite signs from what is expected.

A more formal method for detecting multicollinearity involves the calculation of **variance inflation factors** for the individual β parameters. One reason why the t-tests on the individual β parameters are nonsignificant is that the standard errors of the estimates, $s_{\hat{\beta}_i}$, are inflated in the presence of multicollinearity. When the dependent and independent variables are appropriately transformed,† it can be shown that

$$s_{\hat{\beta}_i}^2 = s^2 \left(\frac{1}{1 - R_i^2} \right)$$

where s^2 is the estimate of σ^2, the variance of ε, and R_i^2 is the multiple coefficient of determination for the model that regresses the independent variable x_i on the remaining independent variables $x_1, x_2, \ldots, x_{i-1}, x_{i+1}, \ldots, x_k$. The quantity $1/(1 - R_i^2)$ is called the *variance inflation factor* for the parameter β_i, denoted $(\text{VIF})_i$. Note that $(\text{VIF})_i$ will be large when R_i^2 is large—that is, when the independent variable x_i is strongly related to the other independent variables.

Various authors maintain that, in practice, a severe multicollinearity problem exists if the largest of the variance inflation factors for the β's is greater than 10 or, equivalently, if the largest multiple coefficient of determination, R_i^2, is greater

---

* Remember that r measures only the pairwise correlation between x-values. Three variables, x_1, x_2, and x_3, may be highly correlated as a group, but may not exhibit large pairwise correlations. Thus, multicollinearity may be present even when all pairwise correlations are not significantly different from 0.

† The transformed variables are obtained as

$$y_i^* = (y_i - \bar{y})/s_y \quad x_{1i}^* = (x_{1i} - \bar{x}_1)/s_1 \quad x_{2i}^* = (x_{2i} - \bar{x}_2)/s_2$$

and so on, where $\bar{y}, \bar{x}_1, \bar{x}_2, \ldots$, and $s_y, s_1, s_2, \ldots$, are the sample means and standard deviations, respectively, of the original variables.

than .90.* Several of the statistical software packages discussed in this text have options for calculating variance inflation factors.[†]

The methods for detecting multicollinearity are summarized in the accompanying box. We illustrate the use of these statistics in Example 7.5.

Detecting Multicollinearity in the Regression Model

$$E(y) = \beta_0 + \beta_1 x_1 + \beta_2 x_2 + \cdots + \beta_k x_k$$

The following are indicators of multicollinearity:

1. Significant correlations between pairs of independent variables in the model
2. Nonsignificant t-tests for all (or nearly all) the individual β parameters when the F-test for overall model adequacy H_0: $\beta_1 = \beta_2 = \cdots = \beta_k = 0$ is significant
3. Opposite signs (from what is expected) in the estimated parameters
4. A variance inflation factor (VIF) for a β parameter greater than 10, where

$$(\text{VIF})_i = \frac{1}{1 - R_i^2}, \quad i = 1, 2, \ldots, k$$

and R_i^2 is the multiple coefficient of determination for the model

$$E(x_i) = \alpha_0 + \alpha_1 x_1 + \alpha_2 x_2 + \cdots + \alpha_{i-1} x_{i-1} + \alpha_{i+1} x_{i+1} + \cdots + \alpha_k x_k$$

Example 7.5

The Federal Trade Commission (FTC) annually ranks varieties of domestic cigarettes according to their tar, nicotine, and carbon monoxide contents. The U.S. surgeon general considers each of these three substances hazardous to a smoker's health. Past studies have shown that increases in the tar and nicotine contents of a cigarette are accompanied by an increase in the carbon monoxide emitted from the cigarette smoke. Table 7.2 lists tar, nicotine, and carbon monoxide contents (in milligrams) and weight (in grams) for a sample of 25 (filter) brands tested in a recent year. Suppose we want to model carbon monoxide content, y, as a function of tar content, x_1, nicotine content, x_2, and weight, x_3, using the model

$$E(y) = \beta_0 + \beta_1 x_1 + \beta_2 x_2 + \beta_3 x_3$$

The model is fit to the 25 data points in Table 7.2. A portion of the resulting SAS printout is shown in Figure 7.5. Examine the printout. Do you detect any signs of multicollinearity?

Solution

First, notice that a test of the global utility of the model

$$H_0: \beta_1 = \beta_2 = \beta_3 = 0$$

is highly significant. The F-value (shaded on the printout) is very large ($F = 78.98$), and the observed significance level of the test (also shaded) is small ($p < .0001$). Therefore, we can reject H_0 for any α greater than .0001 and conclude that at

---

* See, for example, Montgomery, Peck, and Vining (2006) or Kutner, Nachtsheim, Neter, and Li (2005).

[†] Some software packages calculate an equivalent statistic, called the **tolerance**. The tolerance for a β coefficient is the reciprocal of the variance inflation factor, that is,

$$(\text{TOL})_i = \frac{1}{(\text{VIF})_i} = 1 - R_i^2$$

For $R_i^2 > .90$ (the extreme multicollinearity case), $(\text{TOL})_i < .10$. These computer packages allow the user to set tolerance limits, so that any independent variable with a value of $(\text{TOL})_i$ below the tolerance limit will not be allowed to enter into the model.

⊛FTCCIGAR

Table 7.2 FTC cigarette data for Example 7.5

Brand	Tar x_1, milligrams	Nicotine x_2, milligrams	Weight x_3, grams	Carbon Monoxide y, milligrams
Alpine	14.1	.86	.9853	13.6
Benson & Hedges	16.0	1.06	1.0938	16.6
Bull Durham	29.8	2.03	1.1650	23.5
Camel Lights	8.0	.67	.9280	10.2
Carlton	4.1	.40	.9462	5.4
Chesterfield	15.0	1.04	.8885	15.0
Golden Lights	8.8	.76	1.0267	9.0
Kent	12.4	.95	.9225	12.3
Kool	16.6	1.12	.9372	16.3
L&M	14.9	1.02	.8858	15.4
Lark Lights	13.7	1.01	.9643	13.0
Marlboro	15.1	.90	.9316	14.4
Merit	7.8	.57	.9705	10.0
Multifilter	11.4	.78	1.1240	10.2
Newport Lights	9.0	.74	.8517	9.5
Now	1.0	.13	.7851	1.5
Old Gold	17.0	1.26	.9186	18.5
Pall Mall Light	12.8	1.08	1.0395	12.6
Raleigh	15.8	.96	.9573	17.5
Salem Ultra	4.5	.42	.9106	4.9
Tareyton	14.5	1.01	1.0070	15.9
True	7.3	.61	.9806	8.5
Viceroy Rich Lights	8.6	.69	.9693	10.6
Virginia Slims	15.2	1.02	.9496	13.9
Winston Lights	12.0	.82	1.1184	14.9

Source: Federal Trade Commission.

least one of the parameters, β_1, β_2, and β_3, is nonzero. The t-tests for two of the three individual β's, however, are nonsignificant. (The p-values for these tests are shaded on the printout.) Unless tar is the only one of the three variables useful for predicting carbon monoxide content, these results are the first indication of a potential multicollinearity problem.

A second clue to the presence of multicollinearity is the negative value for $\hat{\beta}_2$ and $\hat{\beta}_3$ (shaded on the printout),

$$\hat{\beta}_2 = -2.63 \quad \text{and} \quad \hat{\beta}_3 = -.13$$

From past studies, the FTC expects carbon monoxide content y to increase when either nicotine content x_2 or weight x_3 increases—that is, the FTC expects *positive* relationships between y and x_2, and y and x_3, not negative ones.

Figure 7.5 SAS regression printout for FTC model

```
                              Dependent Variable: CO

                    Number of Observations Read        25
                    Number of Observations Used        25

                           Analysis of Variance

                                    Sum of        Mean
        Source              DF     Squares       Square    F Value   Pr > F

        Model                3   495.25781    165.08594     78.98   <.0001
        Error               21    43.89259      2.09012
        Corrected Total     24   539.15040


                Root MSE              1.44573    R-Square     0.9186
                Dependent Mean      12.52800    Adj R-Sq     0.9070
                Coeff Var           11.53996

                            Parameter Estimates

                        Parameter    Standard                        Variance
        Variable    DF   Estimate       Error   t Value  Pr > |t|   Inflation

        Intercept    1    3.20219     3.46175      0.93    0.3655           0
        TAR          1    0.96257     0.24224      3.97    0.0007    21.63071
        NICOTINE     1   -2.63166     3.90056     -0.67    0.5072    21.89992
        WEIGHT       1   -0.13048     3.88534     -0.03    0.9735     1.33386
```

A more formal procedure for detecting multicollinearity is to examine the variance inflation factors. Figure 7.5 shows the variance inflation factors (shaded) for each of the three parameters under the column labeled **Variance Inflation**. Note that the variance inflation factors for both the tar and nicotine parameters are greater than 10. The variance inflation factor for the tar parameter, $(\text{VIF})_1 = 21.63$, implies that a model relating tar content x_1 to the remaining two independent variables, nicotine content x_2 and weight x_3, resulted in a coefficient of determination

$$R_1^2 = 1 - \frac{1}{(\text{VIF})_1}$$

$$= 1 - \frac{1}{21.63} = .954$$

All signs indicate that a serious multicollinearity problem exists. To confirm our suspicions, we used SAS to find the coefficient of correlation r for each of the three pairs of independent variables in the model. These values are highlighted on the SAS printout, Figure 7.6. You can see that tar content x_1 and nicotine content x_2 appear to be highly correlated ($r = .977$), whereas weight x_3 appears to be moderately correlated with both tar content ($r = .491$) and nicotine content ($r = .500$). In fact, all three sample correlations are significantly different from 0 based on the small p-values, also shown in Figure 7.6.

Figure 7.6 SAS correlation matrix for independent variables in FTC model

```
        Pearson Correlation Coefficients, N = 25
              Prob > |r| under H0: Rho=0

                      TAR      NICOTINE     WEIGHT

        TAR        1.00000     0.97661     0.49077
                                <.0001      0.0127

        NICOTINE   0.97661     1.00000     0.50018
                    <.0001                  0.0109

        WEIGHT     0.49077     0.50018     1.00000
                    0.0127      0.0109
```

Once you have detected that a multicollinearity problem exists, there are several alternative measures available for solving the problem. The appropriate measure to take depends on the severity of the multicollinearity and the ultimate goal of the regression analysis.

Some researchers, when confronted with highly correlated independent variables, choose to include only one of the correlated variables in the final model. One way of deciding which variable to include is to use *stepwise regression*, a topic discussed in Chapter 6. Generally, only one (or a small number) of a set of multicollinear independent variables will be included in the regression model by the stepwise regression procedure since this procedure tests the parameter associated with each variable in the presence of all the variables already in the model. For example, in fitting the gasoline mileage rating model introduced earlier, if at one step the variable representing truck load is included as a significant variable in the prediction of the mileage rating, the variable representing horsepower will probably never be added in a future step. Thus, if a set of independent variables is thought to be multicollinear, some screening by stepwise regression may be helpful.

If you are interested only in using the model for estimation and prediction, you may decide not to drop any of the independent variables from the model. In the presence of multicollinearity, we have seen that it is dangerous to interpret the individual β's for the purpose of establishing cause and effect. However, confidence intervals for $E(y)$ and prediction intervals for y generally remain unaffected *as long as the values of the independent variables used to predict y follow the same pattern of multicollinearity exhibited in the sample data*. That is, you must take strict care to ensure that the values of the x variables fall within the experimental region. (We discuss this problem in further detail in Section 7.5.) Alternatively, if your goal is to establish a cause-and-effect relationship between y and the independent variables, you will need to conduct a designed experiment to break up the pattern of multicollinearity.

Solutions to Some Problems Created by Multicollinearity

1. Drop one or more of the correlated independent variables from the final model. A screening procedure such as stepwise regression is helpful in determining which variables to drop.

2. If you decide to keep all the independent variables in the model:

   (a) Avoid making inferences about the individual β parameters (such as establishing a cause-and-effect relationship between y and the predictor variables).

   (b) Restrict inferences about $E(y)$ and future y-values to values of the independent variables that fall within the experimental region (see Section 7.5).

3. If your ultimate objective is to establish a cause-and-effect relationship between y and the predictor variables, use a designed experiment (see Chapters 11 and 12).

4. To reduce rounding errors in polynomial regression models, code the independent variables so that first-, second-, and higher-order terms for a particular x variable are not highly correlated (see Section 5.6).

5. To reduce rounding errors and stabilize the regression coefficients, use ridge regression to estimate the β parameters (see Section 9.7).

When fitting a polynomial regression model, for example, the second-order model

$$E(y) = \beta_0 + \beta_1 x + \beta_2 x^2$$

the independent variables $x_1 = x$ and $x_2 = x^2$ will often be correlated. If the correlation is high, the computer solution may result in extreme rounding errors. For this model, the solution is not to drop one of the independent variables but to transform the x variable in such a way that the correlation between the coded x and x^2 values is substantially reduced. Coding the independent quantitative variables as described in optional Section 5.6 is a useful technique for reducing the multicollinearity inherent with polynomial regression models.

Another, more complex, procedure for reducing the rounding errors caused by multicollinearity involves a modification of the least squares method, called **ridge regression**. In ridge regression, the estimates of the β coefficients are biased [i.e., $E(\beta_i) \neq \beta_i$] but have significantly smaller standard errors than the unbiased β estimates yielded by the least squares method. Thus, the β estimates for the ridge regression are more stable than the corresponding least squares estimates. Ridge regression is a topic discussed in optional Chapter 9. A summary of the solutions is given in the box on page 368.

7.5 Extrapolation: Predicting Outside the Experimental Region

Research economists develop highly technical models to relate the state of the economy to various economic indices and other independent variables. Many of these models are multiple regression models, where, for example, the dependent variable y might be next year's growth in gross domestic product (GDP) and the independent variables might include this year's rate of inflation, this year's Consumer Price Index, and so forth. In other words, the model might be constructed to predict next year's economy using this year's knowledge.

Unfortunately, these models are almost unanimously unsuccessful in predicting a recession. Why? One of the problems is that these regression models are used to predict y for values of the independent variables that are outside the region in which the model was developed. For example, the inflation rate in the late 1960s ranged from 6% to 8%. When the double-digit inflation of the early 1970s became a reality, some researchers attempted to use the same models to predict the growth in GDP 1 year hence. As you can see in Figure 7.7, the model may be very accurate for predicting y when x is in the range of experimentation, but the use of the model outside that range is a dangerous (although sometimes unavoidable) practice. A $100(1 - \alpha)\%$ prediction interval for GDP when the inflation rate is, say, 10%, will be less reliable than the stated confidence coefficient $(1 - \alpha)$. How much less is unknown.

Figure 7.7 Using a regression model outside the experimental region

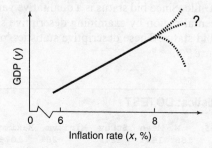

For a single independent variable x, the experimental region is simply the range of the values of x in the sample. Establishing the experimental region for a multiple

regression model that includes a number of independent variables may be more difficult. For example, consider a model for GDP (y) using inflation rate (x_1) and prime interest rate (x_2) as predictor variables. Suppose a sample of size $n = 5$ was observed, and the values of x_1 and x_2 corresponding to the five values for GDP were $(1, 10)$, $(1.25, 12)$, $(2.25, 10.25)$, $(2.5, 13)$, and $(3, 11.5)$. Notice that x_1 ranges from 1% to 3% and x_2 ranges from 10% to 13% in the sample data. You may think that the experimental region is defined by the ranges of the individual variables (i.e., $1 \le x_1 \le 3$ and $10 \le x_2 \le 13$). However, the levels of x_1 and x_2 *jointly* define the region. Figure 7.8 shows the experimental region for our hypothetical data. You can see that an observation with levels $x_1 = 3$ and $x_2 = 10$ clearly falls outside the experimental region, yet is within the ranges of the individual x-values. Using the model to predict GDP for this observation—called **hidden extrapolation**—may lead to unreliable results.

Figure 7.8 Experimental region for modeling GDP (y) as a function of inflation rate (x_1) and prime interest rate (x_2)

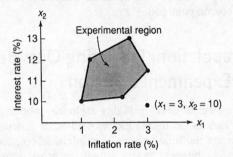

Example 7.6

In Section 4.14 (p. 235), we presented a study of bid collusion in the Florida highway construction industry. Recall that we modeled the cost of a road construction contract (y) as a function of the Department of Transportation (DOT) engineer's cost estimate $(x_1$, in thousands of dollars) and bid status (x_2), where $x_2 = 1$ if the contract was fixed, 0 if competitive. Based on data collected for $n = 235$ contracts (and saved in the *FLAG* file), the interaction model $E(y) = \beta_0 + \beta_1 x_1 + \beta_2 x_2 + \beta_3 x_1 x_2$ was found to be the best model for predicting contract cost. Find the experimental region for the model, then give values of the independent variables that fall outside this region.

Solution

For this regression analysis, the experimental region is defined as the values of the independent variables, DOT estimate (x_1) and bid status (x_2), that span the sample data in the *FLAG* file. Since bid status is a qualitative variable at two levels, we can find the experimental region by examining descriptive statistics for DOT estimate at each level of bid status. These descriptive statistics, produced using MINITAB, are shown in Figure 7.9.

Figure 7.9 MINITAB descriptive statistics for independent variables, Example 7.6

Descriptive Statistics: DOTEST

Variable	STATUS	N	Mean	StDev	Minimum	Maximum
DOTEST	0	185	1494	2249	28	10744
	1	50	803	1080	66	5448

Examining Figure 7.9, you can see that when the bids are fixed $(x_2 = 1)$, DOT estimate (x_1) ranges from a minimum of 66 thousand dollars to a maximum of 5,448 thousand dollars. In contrast, for competitive bids $(x_2 = 0)$, DOT estimate (x_1) ranges from 28 thousand dollars to 10,744 thousand dollars. These two ranges define the experimental region for the analysis. Consequently, the DOT should avoid making cost predictions for fixed contracts that have DOT estimates outside the interval ($66,000, $5,448,000) and for competitive contracts that have DOT estimates outside the interval ($28,000, $10,744,000).

7.6 Variable Transformations

The word *transform* means to change the form of some object or thing. Consequently, the phrase *variable transformation* means that we have done, or plan to do, something to change the form of the variable. For example, if one of the independent variables in a model is the price p of a commodity, we might choose to introduce this variable into the model as $x = 1/p$, $x = \sqrt{p}$, or $x = e^{-p}$. Thus, if we were to let $x = \sqrt{p}$, we would compute the square root of each price value, and these square roots would be the values of x that would be used in the regression analysis.

Transformations are performed on the y-values to make them more nearly satisfy the assumptions of Section 4.2 and, sometimes, to make the deterministic portion of the model a better approximation to the mean value of the transformed response. Transformations of the values of the independent variables are performed solely for the latter reason—that is, to achieve a model that provides a better approximation to $E(y)$. In this section, we discuss transformations on the dependent and independent variables to achieve a good approximation to $E(y)$. (Transformations on the y-values for the purpose of satisfying the assumptions are discussed in Chapter 8.)

Suppose you want to fit a model relating the demand y for a product to its price p. Also, suppose the product is a nonessential item, and you expect the mean demand to decrease as price p increases and then to decrease more slowly as p gets larger (see Figure 7.10). What function of p will provide a good approximation to $E(y)$?

To answer this question, you need to know the graphs of some elementary mathematical functions—there is a one-to-one relationship between mathematical functions and graphs. If we want to model a relationship similar to the one indicated in Figure 7.10, we need to be able to select a mathematical function that will possess a graph similar to the curve shown.

Figure 7.10 Hypothetical relation between demand y and price p

Portions of some curves corresponding to mathematical functions that decrease as p increases are shown in Figure 7.11. Of the seven models shown, the curves in

Figure 7.11c, 7.11d, 7.11f, and 7.11g will probably provide the best approximations to $E(y)$. These four graphs all show $E(y)$ decreasing and approaching (but never reaching) 0 as p increases. Figures 7.11c and 7.11d suggest that the independent variable, price, should be transformed using either $x = 1/p$ or $x = e^{-p}$. Then you might try fitting the model

$$E(y) = \beta_0 + \beta_1 x$$

using the transformed data. Or, as suggested by Figures 7.11f and 7.11g, you might try the transformation $x = \ln(p)$ and fit either of the models

$$E(y) = \beta_0 + \beta_1 x$$

or

$$E\{\ln(y)\} = \beta_0 + \beta_1 x$$

The functions shown in Figure 7.11 produce curves that either rise or fall depending on the sign of the parameter β_1 in parts a, c, d, e, f, and g, and on β_2 and the portion of the curve used in part b. When you choose a model for a regression analysis, you do not have to specify the sign of the parameter(s). The least squares procedure will choose as estimates of the parameters those that minimize the sum of squares of the residuals. Consequently, if you were to fit the model shown in Figure 7.11c to a set of y-values that increase in value as p increases, your least squares estimate of β_1 would be negative, and a graph of y would produce a curve similar to curve 2 in Figure 7.11c. If the y-values decrease as p increases, your estimate of β_1 will be positive and the curve will be similar to curve 1 in Figure 7.11c. All the curves in Figure 7.11 shift upward or downward depending on the value of β_0.

Figure 7.11 Graphs of some mathematical functions relating $E(y)$ to p

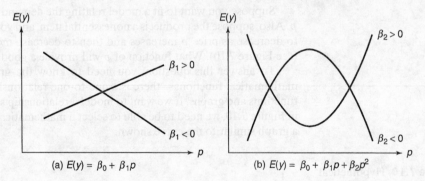

(a) $E(y) = \beta_0 + \beta_1 p$

(b) $E(y) = \beta_0 + \beta_1 p + \beta_2 p^2$

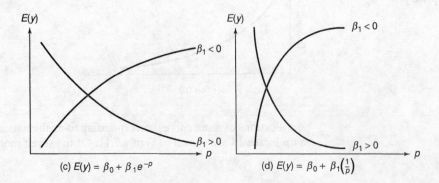

(c) $E(y) = \beta_0 + \beta_1 e^{-p}$

(d) $E(y) = \beta_0 + \beta_1\left(\frac{1}{p}\right)$

Figure 7.11 (*continued*)

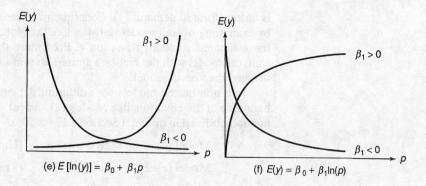

(e) $E[\ln(y)] = \beta_0 + \beta_1 p$

(f) $E(y) = \beta_0 + \beta_1 \ln(p)$

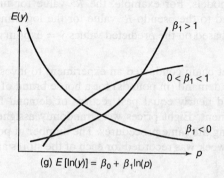

(g) $E[\ln(y)] = \beta_0 + \beta_1 \ln(p)$

Example 7.7

Refer to the models and graphs in Figure 7.11. Consider a situation where there is no *a priori* theory on the true relationship between demand (y) and price (p). Consequently, you will fit the models and compare them to determine the "best" model for $E(y)$.

(a) Identify the models that are nested. How would you compare these models?

(b) Identify the non-nested models. How would you compare these models?

Solution

(a) Nested models, by definition, have the same form for the dependent variable on the left-hand side of the equation. Also, for two nested models, the "complete" model has the same terms (independent variables) on the right-hand side of the equation as the "reduced" model, plus more. Thus, the only two nested models in Figure 7.11 are Models (a) and (b).

$$\text{Model (a): } E(y) = \beta_0 + \beta_1 p \qquad \text{(Reduced model)}$$

$$\text{Model (b): } E(y) = \beta_0 + \beta_1 p + \beta_2 p^2 \qquad \text{(Complete model)}$$

These two models can be compared by testing $H_0: \beta_2 = 0$ using a partial F-test (or a t-test).

(b) Any two of the remaining models shown in Figure 7.11 are non-nested. For example, Models (c) and (d) are non-nested models. Some other non-nested models are Models (a) and (c), Models (e) and (g), and Models (g) and (f). The procedure for comparing non-nested models will depend on whether or not the dependent variable on the left-hand side of the equation is the same. For example, for Models (a), (c), (d), and (f), the dependent variable is untransformed demand (y). Consequently, these models can be compared by examining overall model statistics like the global F-test, adjusted-R^2, and the estimated standard deviation s. Presuming the global F-test is significant, the model with the highest adjusted-R^2 and smallest value of s would be deemed the "best" model.

Two non-nested models with different dependent variables on the left-hand side of the equation, like Models (a) and (e), can be compared using the method outlined in optional Section 4.12 (p. 209).

$$\text{Model (a): } E(y) = \beta_0 + \beta_1 p \qquad \text{(Untransformed } y\text{)}$$

$$\text{Model (e): } E[\ln(y)] = \beta_0 + \beta_1 p \qquad \text{(Transformed } y\text{)}$$

The key is to calculate a statistic like R^2 or adjusted-R^2 that can be compared across models. For example, the R^2 value for untransformed Model (a) is compared to the pseudo-R^2 value for the log-transformed Model (e), where $R^2_{\ln(y)}$ is based on the predicted values $\hat{y} = \exp\{\widehat{\ln(y)}\}$.

Example 7.8

A supermarket chain conducted an experiment to investigate the effect of price p on the weekly demand (in pounds) for a house brand of coffee. Eight supermarket stores that had nearly equal past records of demand for the product were used in the experiment. Eight prices were randomly assigned to the stores and were advertised using the same procedures. The number of pounds of coffee sold during the following week was recorded for each of the stores and is shown in Table 7.3.

COFFEE

Table 7.3 Data for Example 7.8

Demand y, pounds	Price p, dollars
1,120	3.00
999	3.10
932	3.20
884	3.30
807	3.40
760	3.50
701	3.60
688	3.70

(a) Previous research by the supermarket chain indicates that weekly demand (y) decreases with price (p), but at a decreasing rate. This implies that model (d), Figure 7.11, is appropriate for predicting demand. Fit the model

$$E(y) = \beta_0 + \beta_1 x$$

to the data, letting $x = 1/p$.

(b) Do the data provide sufficient evidence to indicate that the model contributes information for the prediction of demand?

(c) Find a 95% confidence interval for the mean demand when the price is set at $3.20 per pound. Interpret this interval.

Solution

(a) The first step is to calculate $x = 1/p$ for each data point. These values are given in Table 7.4. The MINITAB printout* (Figure 7.12) gives

$$\hat{\beta}_0 = -1,180.5 \qquad \hat{\beta}_1 = 6,808.1$$

and

$$\hat{y} = -1,180.5 + 6,808.1x$$
$$= -1,180.5 + 6,808.1\left(\frac{1}{p}\right)$$

Table 7.4 Values of transformed price

y	$x = 1/p$
1,120	.3333
999	.3226
932	.3125
884	.3030
807	.2941
760	.2857
701	.2778
688	.2703

Figure 7.12 MINITAB regression printout for Example 7.8

```
The regression equation is
DEMAND = - 1180 + 6808 X

Predictor      Coef      SE Coef        T        P
Constant    -1180.5        107.7    -10.96    0.000
X            6808.1        358.4     19.00    0.000

S = 20.90      R-Sq = 98.4%     R-Sq(adj) = 98.1%

Analysis of Variance

Source          DF         SS         MS        F        P
Regression       1     157718     157718   360.94    0.000
Residual Error   6       2622        437
Total            7     160340

Predicted Values for New Observations

New Obs    Fit     SE Fit        95.0% CI            95.0% PI
1       947.05       8.66   ( 925.86, 968.24)   ( 891.67, 1002.43)

Values of Predictors for New Observations

New Obs        X
1         0.3125
```

---

* MINITAB uses full decimal accuracy for $x = 1/p$. Hence, the results shown in Figure 7.12 differ from results that would be calculated using the four-decimal values for $x = 1/p$ shown in the table.

(You can verify that the formulas of Section 3.3 give the same answers.) A graph of this prediction equation is shown in Figure 7.13.

(b) To determine whether x contributes information for the prediction of y, we test H_0: $\beta_1 = 0$ against the alternative hypothesis H_a: $\beta_1 \neq 0$. The test statistic, shaded in Figure 7.12, is $t = 19.0$. We wish to detect either $\beta_1 > 0$ or $\beta_1 < 0$, thus we will use a two-tailed test. Since the two-tailed p-value shown on the printout, .000, is less than $\alpha = .05$, we reject H_0: $\beta_1 = 0$ and conclude that $x = 1/p$ contributes information for the prediction of demand y.

(c) For price $p = 3.20$, $x = 1/p = .3125$. The bottom of the MINITAB printout gives a 95% confidence interval for the mean demand $E(y)$ when price is $p = \$3.20$ (i.e., $x = .3125$). The interval (shaded) is (925.86, 968.24). Thus, we are 95% confident that mean demand will fall between 926 and 968 pounds when the price is set at \$3.20. ∎

Figure 7.13 Graph of the demand–price curve for Example 7.8

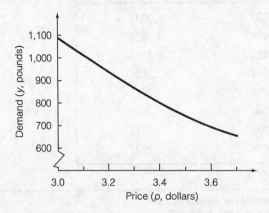

This discussion is intended to emphasize the importance of data transformation and to explain its role in model building. Keep in mind that the symbols $x_1, x_2, \ldots, x_k$ that appear in the linear models of this text can be transformations on the independent variables you have observed. These transformations, coupled with the model-building methods of Chapter 5, allow you to use a great variety of mathematical functions to model the mean $E(y)$ for data.

Quick Summary

Key Formulas

pth-order polynomial

levels of $x \geq (p + 1)$

Standardized beta for x_i

$\hat{\beta}_i^* = \hat{\beta}_i(s_x/s_y)$

Variance inflation factor for x_i

$\text{VIF}_i = 1/(1 - R_i^2)$, where R_i^2 is R^2 for the model
$E(x_i) = \beta_0 + \beta_1 x_1 + \beta_2 x_2 + \cdots + \beta_{i-1}x_{i-1} + \beta_{i+1}x_{i+1}$
$+ \cdots + \beta_k x_k$

Key Ideas

Establishing cause and effect

1. It is dangerous to deduce a cause-and-effect relationship with *observational data*

2. Only with a properly *designed experiment* can you establish a cause and effect

Parameter estimability

Insufficient data for levels of either a quantitative or qualitative independent variable can result in inestimable regression parameters.

Multicollinearity

1. Occurs when two or more independent variables are correlated.

2. Indicators of multicollinearity:

   (a) Highly correlated x's

   (b) Significant global F-test, but all t-tests on individual β's are nonsignificant

   (c) Signs of β's opposite from what is expected

   (d) VIF exceeding 10

3. Model modifications for solving multicollinearity:

   (a) Drop one or more of the highly correlated x's

   (b) Keep all x's in the model, but avoid making inferences on the β's

   (c) Code quantitative x's to reduce correlation between x and x^2

   (d) Use *ridge regression* to estimate the β's

Extrapolation

1. Occurs when you predict y for values of the independent variables that are outside the experimental region.

2. Be wary of *hidden extrapolation* (where values of the x's fall within the range of each individual x, but fall outside the experimental region defined jointly by the x's)

Variable transformations

Transforming y and/or the x's in a model can provide a better model fit.

Supplementary Exercises

7.1. Extrapolation. Why is it dangerous to predict y for values of independent variables that fall outside the experimental region?

7.2. Multicollinearity.

(a) Discuss the problems that result when multicollinearity is present in a regression analysis.

(b) How can you detect multicollinearity?

(c) What remedial measures are available when multicollinearity is detected?

7.3. Data transformations. Refer to Example 7.8. Can you think of any other transformations on price that might provide a good fit to the data? Try them and answer the questions in Example 7.8 again.

7.4. Women in top management. The *Journal of Organizational Culture, Communications and Conflict* (July 2007) published a study on women in upper-management positions at U.S. firms. Observational data ($n = 252$ months) were collected for several variables in an attempt to model the number of females in managerial positions (y). The independent variables included the number of females with a college degree (x_1), the number of female high school graduates with no college degree (x_2), the number of males in managerial positions (x_3), the number of males with a college degree (x_4), and the number of male high school graduates with no college degree (x_5).

(a) The correlation relating number of females in managerial positions and number of females

with a college degree was determined to be $r = .983$. Can the researchers conclude that an increase in the number of females with a college degree will cause the number of females in managerial positions to increase? Explain.

(b) The correlation relating number of males in managerial positions and number of males with a college degree was determined to be $r = .722$. What potential problem can occur in the regression analysis? Explain.

7.5. Urban/rural ratings of counties. Refer to the *Professional Geographer* (February 2000) study of urban and rural counties in the western United States, Exercise 4.16 (p. 190). Recall that six independent variables—total county population (x_1), population density (x_2), population concentration (x_3), population growth (x_4), proportion of county land in farms (x_5), and 5-year change in agricultural land base (x_6)—were used to model the urban/rural rating (y) of a county. Prior to running the multiple regression analysis, the researchers were concerned about possible multicollinearity in the data. The correlation matrix (shown on the next page) is a table of correlations between all pairs of the independent variables.

(a) Based on the correlation matrix, is there any evidence of extreme multicollinearity?

(b) Refer to the multiple regression results in the table given in Exercise 4.16 (p.190). Based on the reported tests, is there any evidence of extreme multicollinearity?

INDEPENDENT VARIABLE		x_1	x_2	x_3	x_4	x_5
x_1	Total population					
x_2	Population density	.20				
x_3	Population concentration	.45	.43			
x_4	Population growth	−.05	−.14	−.01		
x_5	Farm land	−.16	−.15	−.07	−.20	
x_6	Agricultural change	−.12	−.12	−.22	−.06	−.06

Source: Berry, K. A., et al. "Interpreting what is rural and urban for western U.S. counties," *Professional Geographer*, Vol. 52, No. 1, Feb. 2000 (Table 2).

🔘 PONDICE

7.6. Characteristics of sea ice melt ponds. Surface albedo is defined as the ratio of solar energy directed upward from a surface over energy incident upon the surface. Surface albedo is a critical climatological parameter of sea ice. The National Snow and Ice Data Center (NSIDC) collects data on the albedo, depth, and physical characteristics of ice melt ponds in the Canadian Arctic, including ice type (classified as first-year ice, multiyear ice, or landfast ice). Data for 504 ice melt ponds located in the Barrow Strait in the Canadian Arctic are saved in the PONDICE file. Environmental engineers want to model the broadband surface albedo level, y, of the ice as a function of pond depth, x_1 (meters), and ice type, represented by the dummy variables $x_2 = \{1$ if first-year ice, 0 if not$\}$ and $x_3 = \{1$ if multiyear ice, 0 if not$\}$. Ultimately, the engineers will use the model to predict the surface albedo level of an ice melt pond. Access the data in the PONDICE file and identify the experimental region for the engineers. What advice do you give them about the use of the prediction equation?

7.7. Personality and aggressive behavior. *Psychological Bulletin* (Vol. 132, 2006) reported on a study linking personality and aggressive behavior. Four of the variables measured in the study were aggressive behavior, irritability, trait anger, and narcissism. Pairwise correlations for these four variables are given below.

Aggressive behavior–Irritability: .77
Aggressive behavior–Trait anger: .48
Aggressive behavior–Narcissism: .50
Irritability–Trait anger: .57
Irritability–Narcissism: .16
Trait anger–Narcissism: .13

(a) Suppose aggressive behavior is the dependent variable in a regression model and the other variables are independent variables. Is there evidence of extreme multicollinearity? Explain.

(b) Suppose narcissism is the dependent variable in a regression model and the other variables are independent variables. Is there evidence of extreme multicollinearity? Explain.

7.8. Steam processing of peat. A bioengineer wants to model the amount (y) of carbohydrate solubilized during steam processing of peat as a function of temperature (x_1), exposure time (x_2), and pH value (x_3). Data collected for each of 15 peat samples were used to fit the model

$$E(y) = \beta_0 + \beta_1 x_1 + \beta_2 x_2 + \beta_3 x_3$$

A summary of the regression results follows:

$$\hat{y} = -3,000 + 3.2x_1 - .4x_2 - 1.1x_3 \qquad R^2 = .93$$

$$s_{\hat{\beta}_1} = 2.4 \qquad s_{\hat{\beta}_2} = .6 \qquad s_{\hat{\beta}_3} = .8$$

$$r_{12} = .92 \qquad r_{13} = .87 \qquad r_{23} = .81$$

Based on these results, the bioengineer concludes that none of the three independent variables, x_1, x_2, and x_3, is a useful predictor of carbohydrate amount, y. Do you agree with this statement? Explain.

7.9. Salaries of top university researchers. The provost of a top research university wants to know what salaries should be paid to the college's top researchers, based on years of experience. An independent consultant has proposed the quadratic model

$$E(y) = \beta_0 + \beta_1 x + \beta_2 x^2$$

where

y = Annual salary (thousands of dollars)

x = Years of experience

To fit the model, the consultant randomly sampled three researchers at other research universities and recorded the information given in the accompanying table. Give your opinion regarding the adequacy of the proposed model.

	y	x
Researcher 1	60	2
Researcher 2	45	1
Researcher 3	82	5

7.10. FDA investigation of a meat-processing plant. A particular meat-processing plant slaughters steers and cuts and wraps the beef for its customers. Suppose a complaint has been filed with the Food and

Drug Administration (FDA) against the processing plant. The complaint alleges that the consumer does not get all the beef from the steer he purchases. In particular, one consumer purchased a 300-pound steer but received only 150 pounds of cut and wrapped beef. To settle the complaint, the FDA collected data on the live weights and dressed weights of nine steers processed by a reputable meat-processing plant (not the firm in question). The results are listed in the table.

🔵 STEERS

LIVE WEIGHT x, pounds	DRESSED WEIGHT y, pounds
420	280
380	250
480	310
340	210
450	290
460	280
430	270
370	240
390	250

(a) Fit the model $E(y) = \beta_0 + \beta_1 x$ to the data.

(b) Construct a 95% prediction interval for the dressed weight y of a 300-pound steer.

(c) Would you recommend that the FDA use the interval obtained in part b to determine whether the dressed weight of 150 pounds is a reasonable amount to receive from a 300-pound steer? Explain.

7.11. FTC cigarette study. Refer to the FTC cigarette data of Example 7.5 (p. 365). The data are saved in the FTCCIGAR file.

(a) Fit the model $E(y) = \beta_0 + \beta_1 x_1$ to the data. Is there evidence that tar content x_1 is useful for predicting carbon monoxide content y?

(b) Fit the model $E(y) = \beta_0 + \beta_2 x_2$ to the data. Is there evidence that nicotine content x_2 is useful for predicting carbon monoxide content y?

(c) Fit the model $E(y) = \beta_0 + \beta_3 x_3$ to the data. Is there evidence that weight x_3 is useful for predicting carbon monoxide content y?

(d) Compare the signs of $\hat{\beta}_1$, $\hat{\beta}_2$, and $\hat{\beta}_3$ in the models of parts a, b, and c, respectively, to the signs of the $\hat{\beta}$'s in the multiple regression model fit in Example 7.5. Is the fact that the $\hat{\beta}$'s change dramatically when the independent variables are removed from the model an indication of a serious multicollinearity problem?

7.12. Demand for car motor fuel. An economist wants to model annual per capita demand, y, for

passenger car motor fuel in the United States as a function of the two quantitative independent variables, average real weekly earnings (x_1) and average price of regular gasoline (x_2). Data on these three variables for the years 1985–2008 are available in the *2009 Statistical Abstract of the United States*. Suppose the economist fits the model $E(y) = \beta_0 + \beta_1 x_1 + \beta_2 x_2$ to the data. Would you recommend that the economist use the least squares prediction equation to predict per capita consumption of motor fuel in 2011? Explain.

7.13. Accuracy of software effort estimates. Refer to the *Journal of Empirical Software Engineering* (Vol. 9, 2004) study of software engineers' effort in developing new software, Exercise 6.2 (p. 339). Recall that the researcher modeled the relative error in estimating effort (y) as a function of two qualitative independent variables: *company role of estimator* ($x_1 = 1$ if developer, 0 if project leader) and *previous accuracy* ($x_8 = 1$ if more than 20% accurate, 0 if less than 20% accurate). A stepwise regression yielded the following prediction equation:

$$\hat{y} = .12 - .28x_1 + .27x_8$$

(a) The researcher is concerned that the sign of $\hat{\beta}_1$ in the model is the opposite from what is expected. (The researcher expects a project leader to have a smaller relative error of estimation than a developer.) Give at least one reason why this phenomenon occurred.

(b) Now, consider the interaction model $E(y) = \beta_0 + \beta_1 x_1 + \beta_2 x_8 + \beta_3 x_1 x_8$. Suppose that there is no data collected for project leaders with less than 20% accuracy. Are all the β's in the interaction model estimable? Explain.

7.14. Yield strength of steel alloy. Refer to Exercise 6.4 (p. 340) and the *Modelling and Simulation in Materials Science and Engineering* (Vol. 13, 2005) study in which engineers built a regression model for the tensile yield strength (y) of a new steel alloy. The potential important predictors of yield strength are listed below. The engineers discovered that the independent variable Nickel (x_4) was highly correlated with each of the other 10 potential independent variables. Consequently, Nickel was dropped from the model. Do you agree with this decision? Explain.

x_1 = Carbon amount (% weight)
x_2 = Manganese amount (% weight)
x_3 = Chromium amount (% weight)
x_4 = Nickel amount (% weight)
x_5 = Molybdenum amount (% weight)
x_6 = Copper amount (% weight)

$x_7 =$ Nitrogen amount (% weight)

$x_8 =$ Vanadium amount (% weight)

$x_9 =$ Plate thickness (millimeters)

$x_{10} =$ Solution treating (milliliters)

$x_{11} =$ Ageing temperature (degrees, Celcius)

📀 FLAG2

7.15. Collusive bidding in road construction. Refer to the Florida Attorney General (FLAG) Office's investigation of bid-rigging in the road construction industry, Exercise 6.8 (p. 341). Recall that FLAG wants to model the price (y) of the contract bid by lowest bidder in hopes of preventing price-fixing in the future. Consider the independent variables selected by the stepwise regression run in Exercise 6.8. Do you detect any multicollinearity in these variables? If so, do you recommend that all of these variables be used to predict low-bid price, y?

7.16. Fitting a quadratic model. How many levels of x are required to fit the model $E(y) = \beta_0 + \beta_1 x + \beta_2 x^2$? How large a sample size is required to have sufficient degrees of freedom for estimating σ^2?

7.17. Fitting an interaction model. How many levels of x_1 and x_2 are required to fit the model $E(y) = \beta_0 + \beta_1 x_1 + \beta_2 x_2 + \beta_3 x_1 x_2$? How large a sample size is required to have sufficient degrees of freedom for estimating σ^2?

7.18. Fitting a complete second-order model. How many levels of x_1 and x_2 are required to fit the model $E(y) = \beta_0 + \beta_1 x_1 + \beta_2 x_2 + \beta_3 x_1 x_2 + \beta_4 x_1^2 + \beta_5 x_2^2$? How large a sample is required to have sufficient degrees of freedom for estimating σ^2?

📀 GASTURBINE

7.19. Cooling method for gas turbines. Refer to the *Journal of Engineering for Gas Turbines and Power* (January 2005) study of a high-pressure inlet fogging method for a gas turbine engine, Exercise 6.10 (p. 343). Recall that a number of independent variables were used to predict the heat rate (kilojoules per kilowatt per hour) for each in a sample of 67 gas turbines augmented with high-pressure inlet fogging. For this exercise, consider a first-order model for heat rate as a function of the quantitative independent variables' cycle speed (revolutions per minute), cycle pressure ratio, inlet temperature (°C), exhaust gas temperature (°C), air mass flow rate (kilograms per second), and horsepower (Hp units). Theoretically, the heat rate should increase as cycle speed increases. In contrast, theory states that the heat rate will decrease as any of the other independent variables increase. The model was fit to the data in the GASTURBINE file with the results shown in the accompanying MINITAB printout. Do you detect any signs of multicollinearity? If so, how should the model be modified?

MINITAB Output for Exercise 7.19

```
The regression equation is
HEATRATE = 14314 + 0.0806 RPM - 6.8 CPRATIO - 9.51 INLET-TEMP + 14.2 EXH-TEMP
          - 2.55 AIRFLOW + 0.00426 POWER


Predictor       Coef    SE Coef       T       P       VIF
Constant       14314       1112   12.87   0.000
RPM          0.08058    0.01611    5.00   0.000     4.015
CPRATIO        -6.78      30.38   -0.22   0.824     5.213
INLET-TEMP    -9.507      1.529   -6.22   0.000    13.852
EXH-TEMP      14.155      3.469    4.08   0.000     7.351
AIRFLOW       -2.553      1.746   -1.46   0.149    49.136
POWER        0.004257   0.004217   1.01   0.317    49.765


S = 458.757    R-Sq = 92.5%    R-Sq(adj) = 91.7%


Analysis of Variance

Source          DF          SS         MS        F       P
Regression       6   155269735   25878289   122.96   0.000
Residual Error  60    12627473     210458
Total           66   167897208
```

7.20. Log–log transformation. Consider the data shown in the table below.

🖫 EX7_20

x	54	42	28	38	25	70	48	41	20	52	65
y	6	16	33	18	41	3	10	14	45	9	5

(a) Plot the points on a scatterplot. What type of relationship appears to exist between x and y?

(b) For each observation calculate $\ln x$ and $\ln y$. Plot the log-transformed data points on a scatterplot. What type of relationship appears to exist between $\ln x$ and $\ln y$?

(c) The scatterplot from part b suggests that the transformed model

$$\ln y = \beta_0 + \beta_1 \ln x + \varepsilon$$

may be appropriate. Fit the transformed model to the data. Is the model adequate? Test using $\alpha = .05$.

(d) Use the transformed model to predict the value of y when $x = 30$. [*Hint*: Use the inverse transformation $y = e^{\ln y}$.]

7.21. Multicollinearity in real estate data. D. Hamilton illustrated the multicollinearity problem with an example using the data shown in the accompanying table. The values of x_1, x_2, and y in the table at right represent appraised land value, appraised improvements value, and sale price, respectively, of a randomly selected residential property. (All measurements are in thousands of dollars.)

(a) Calculate the coefficient of correlation between y and x_1. Is there evidence of a linear relationship between sale price and appraised land value?

(b) Calculate the coefficient of correlation between y and x_2. Is there evidence of a linear relationship between sale price and appraised improvements?

🖫 HAMILTON

x_1	x_2	y	x_1	x_2	y
22.3	96.6	123.7	30.4	77.1	128.6
25.7	89.4	126.6	32.6	51.1	108.4
38.7	44.0	120.0	33.9	50.5	112.0
31.0	66.4	119.3	23.5	85.1	115.6
33.9	49.1	110.6	27.6	65.9	108.3
28.3	85.2	130.3	39.0	49.0	126.3
30.2	80.4	131.3	31.6	69.6	124.6
21.4	90.5	114.4			

Source: Hamilton, D. "Sometimes $R^2 > r_{yx_1}^2 + r_{yx_2}^2$: Correlated variables are not always redundant," *American Statistician*, Vol. 41, No. 2, May 1987, pp. 129–132.

(c) Based on the results in parts a and b, do you think the model $E(y) = \beta_0 + \beta_1 x_1 + \beta_2 x_2$ will be useful for predicting sale price?

(d) Use a statistical computer software package to fit the model in part c, and conduct a test of model adequacy. In particular, note the value of R^2. Does the result agree with your answer to part c?

(e) Calculate the coefficient of correlation between x_1 and x_2. What does the result imply?

(f) Many researchers avoid the problems of multicollinearity by always omitting all but one of the "redundant" variables from the model. Would you recommend this strategy for this example? Explain. (Hamilton notes that, in this case, such a strategy "can amount to throwing out the baby with the bathwater.")

7.22. Socialization of doctoral students. *Teaching Sociology* (July 1995) developed a model for the professional socialization of graduate students working toward a Ph.D. in sociology. One of the dependent variables modeled was professional confidence, y, measured on a 5-point scale. The

Matrix of correlations for Exercise 7.22

INDEPENDENT VARIABLE	(1)	(2)	(3)	(4)	(5)	(6)	(7)	(8)	(9)	(10)
(1) Father's occupation	1.000	.363	.099	−.110	−.047	−.053	−.111	.178	.078	.049
(2) Mother's education	.363	1.000	.228	−.139	−.216	.084	−.118	.192	.125	.068
(3) Race	.099	.228	1.000	.036	−.515	.014	−.120	.112	.117	.337
(4) Sex	−.110	−.139	.036	1.000	.165	−.256	.173	−.106	−.117	.073
(5) Foreign status	−.047	−.216	−.515	.165	1.000	−.041	.159	−.130	−.165	−.171
(6) Undergraduate GPA	−.053	.084	.014	−.256	−.041	1.000	.032	.028	−.034	.092
(7) Year GRE taken	−.111	−.118	−.120	.173	.159	.032	1.000	−.086	−.602	.016
(8) Verbal GRE score	.178	.192	.112	−.106	−.130	.028	−.086	1.000	.132	.087
(9) Years in graduate program	.078	.125	.117	−.117	−.165	−.034	−.602	.132	1.000	−.071
(10) First-year graduate GPA	.049	.068	.337	.073	−.171	.092	.016	.087	−.071	1.000

Source: Keith, B., and Moore, H. A. "Training sociologists: An assessment of professional socialization and the emergence of career aspirations," *Teaching Sociology*, Vol. 23, No. 3, July 1995, p. 205 (Table 1).

model included over 20 independent variables and was fit to data collected for a sample of 309 sociology graduate students. One concern is whether multicollinearity exists in the data. A matrix of Pearson product moment correlations for 10 of the independent variables is shown on p. 381. [*Note*: Each entry in the table is the correlation coefficient r between the variable in the corresponding row and corresponding column.]

(a) Examine the correlation matrix and find the independent variables that are moderately or highly correlated.
(b) What modeling problems may occur if the variables, part a, are left in the model? Explain.

7.23. Fourth-order polynomial. To model the relationship between y, a dependent variable, and x, an independent variable, a researcher has taken one measurement on y at each of five different x-values. Drawing on his mathematical expertise, the researcher realizes that he can fit the fourth-order polynomial model

$$E(y) = \beta_0 + \beta_1 x + \beta_2 x^2 + \beta_3 x^3 + \beta_4 x^4$$

and it will pass exactly through all five points, yielding SSE $= 0$. The researcher, delighted with the "excellent" fit of the model, eagerly sets out to use it to make inferences. What problems will the researcher encounter in attempting to make inferences?

References

Draper, N., and Smith, H. *Applied Regression Analysis*, 2nd ed. New York: Wiley, 1981.

Kutner, M., Nachtsheim, C., Neter, J., and Li, W. *Applied Linear Statistical Models*. 5th ed. New York: McGraw-Hill/Irwin, 2005.

Montgomery, D., Peck, E., and Vining, G. *Introduction to Linear Regression Analysis*. 4th ed. New York: Wiley, 2006.

Mosteller, F., and Tukey, J. W. *Data Analysis and Regression: A Second Course in Statistics*. Reading, Mass.: Addison-Wesley, 1977.

RESIDUAL ANALYSIS

Chapter
8

Contents

Objectives

1. To show how residuals can be used to detect departures from the model assumptions.

2. To suggest some procedures for coping with these problems.

8.1 Introduction

We have repeatedly stated that the validity of many of the inferences associated with a regression analysis depends on the error term, ε, satisfying certain assumptions. Thus, when we test a hypothesis about a regression coefficient or a set of regression coefficients, or when we form a prediction interval for a future value of y, we must assume that (1) ε is normally distributed, (2) with a mean of 0, (3) the variance σ^2 is constant, and (4) all pairs of error terms are uncorrelated.* Fortunately, experience has shown that least squares regression analysis produces reliable statistical tests and confidence intervals as long as the departures from the assumptions are not too great. The objective of this chapter is to provide you with both graphical tools and statistical tests that will aid in identifying significant departures from the assumptions. In addition, these tools will help you evaluate the utility of the model and, in some cases, may suggest modifications to the model that will allow you to better describe the mean response.

First, in Section 8.2, we define a *regression residual* and give some properties of residuals that will aid in detecting problems with the assumptions. In Section 8.3, we show how to plot the residuals to reveal model inadequacies. In Section 8.4, we examine the use of these plots and a simple test to detect unequal variances at different levels of the independent variable(s). A graphical analysis of residuals for checking the normality assumption is the topic of Section 8.5. In Section 8.6, residual plots are used to detect outliers (i.e., observations that are unusually large or small relative to the others); procedures for measuring the influence these outliers may

---

* We assumed (Section 4.2) that the random errors associated with the linear model were independent. If two random variables are independent, it follows (proof omitted) that they will be uncorrelated. The reverse is generally untrue, except for normally distributed random variables. If two normally distributed random variables are uncorrelated, it can be shown that they are also independent.

have on the fitted regression model are also presented. Finally, we discuss the use of residuals to test for time series correlation of the error term in Section 8.7.

8.2 Regression Residuals

The error term in a multiple regression model is, in general, not observable. To see this, consider the model

$$y = \beta_0 + \beta_1 x_1 + \cdots + \beta_k x_k + \varepsilon$$

and solve for the error term:

$$\varepsilon = y - (\beta_0 + \beta_1 x_1 + \cdots + \beta_k x_k)$$

Although you will observe values of the dependent variable and the independent variables $x_1, x_2, \ldots, x_k$, you will not know the true values of the regression coefficients $\beta_0, \beta_1, \ldots, \beta_k$. Therefore, the exact value of ε cannot be calculated.

After we use the data to obtain least squares estimates $\hat{\beta}_0, \hat{\beta}_1, \ldots, \hat{\beta}_k$ of the regression coefficients, we can estimate the value of ε associated with each y-value using the corresponding **regression residual**, that is, the deviation between the observed and the predicted value of y:

$$\hat{\varepsilon}_i = y_i - \hat{y}_i$$

To accomplish this, we must substitute the values of $x_1, x_2, \ldots, x_k$ into the prediction equation for each data point to obtain $\hat{y}$, and then this value must be subtracted from the observed value of y.

> **Definition 8.1** The **regression residual** is the observed value of the dependent variable minus the predicted value, or
>
> $$\hat{\varepsilon} = y - \hat{y} = y - (\hat{\beta}_0 + \hat{\beta}_1 x_1 + \cdots + \hat{\beta}_k x_k)$$

The actual error ε and residual $\hat{\varepsilon}$ for a simple straight-line model are shown in Figure 8.1.

Figure 8.1 Actual random error ε and regression residual $\hat{\varepsilon}$

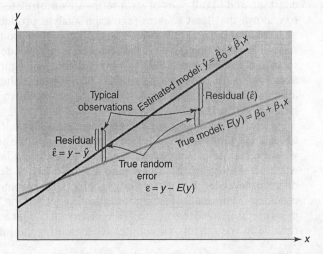

Because residuals estimate the true random error, they are used to check the regression assumptions. Such checks are generally referred to as **residual analyses**. Two useful properties of residuals are given in the next box (page 385).

Properties of Regression Residuals

1. *The mean of the residuals is equal to 0.* This property follows from the fact that the sum of the differences between the observed y-values and their least squares predicted $\hat{y}$ values is equal to 0.

$$\sum_{i=1}^{n} \hat{\varepsilon}_i = \sum_{i=1}^{n} (y_i - \hat{y}_i) = 0$$

2. *The standard deviation of the residuals is equal to the standard deviation of the fitted regression model, s.* This property follows from the fact that the sum of the squared residuals is equal to SSE, which when divided by the error degrees of freedom is equal to the variance of the fitted regression model, s^2. The square root of the variance is both the standard deviation of the residuals and the standard deviation of the regression model.

$$\sum_{i=1}^{n} \varepsilon_i^2 = \sum (y_i - \hat{y}_i)^2 = SSE$$

$$s = \sqrt{\frac{\sum \varepsilon_i^2}{n - (k+1)}} = \sqrt{\frac{SSE}{n - (k+1)}}$$

 OLYMPIC

Table 8.1 Data for 20 Olympic athletes

Athlete	Fat intake x, milligrams	Cholesterol y, milligrams/liter
1	1,290	1,182
2	1,350	1,172
3	1,470	1,264
4	1,600	1,493
5	1,710	1,571
6	1,840	1,711
7	1,980	1,804
8	2,230	1,840
9	2,400	1,956
10	2,930	1,954
11	1,200	1,055
12	1,375	1,241
13	1,410	1,254
14	1,560	1,377
15	1,600	1,465
16	1,950	1,741
17	2,050	1,810
18	2,250	1,893
19	2,600	1,972
20	2,760	1,935

Example 8.1

The data in Table 8.1 represent the level of cholesterol (in milligrams per liter) and average daily intake of saturated fat (in milligrams) for a sample of 20 Olympic athletes. Consider a regression model relating cholesterol level y to fat intake x. Calculate the regression residuals for

(a) the straight-line (first-order) model

(b) the quadratic (second-order) model

For both models, show that the sum of the residuals is 0.

Solution

(a) The SAS printout for the regression analysis of the first-order model,

$$y = \beta_0 + \beta_1 x + \varepsilon$$

is shown in Figure 8.2. The least squares model highlighted is

$$\hat{y} = 515.705 + .5692x$$

Figure 8.2 SAS printout for first-order model

The REG Procedure
Model: MODEL1
Dependent Variable: CHOLESTEROL

Number of Observations Read   20
Number of Observations Used   20

Analysis of Variance

Source	DF	Sum of Squares	Mean Square	F Value	Pr > F
Model	1	1617913	1617913	122.37	<.0001
Error	18	237980	13221		
Corrected Total	19	1855893			

Root MSE	114.98304	R-Square	0.8718	
Dependent Mean	1584.50000	Adj R-Sq	0.8646	
Coeff Var	7.25674			

Parameter Estimates

Variable	DF	Parameter Estimate	Standard Error	t Value	Pr > |t|
Intercept	1	515.70497	99.97886	5.16	<.0001
FAT	1	0.56919	0.05145	11.06	<.0001

Output Statistics

Obs	FAT	Dependent Variable	Predicted Value	Residual
1	1290	1182	1250	-67.9590
2	1350	1172	1284	-112.1104
3	1470	1264	1352	-88.4131
4	1600	1493	1426	66.5923
5	1710	1571	1489	81.9815
6	1840	1711	1563	147.9869
7	1980	1804	1643	161.3004
8	2230	1840	1785	55.0031
9	2400	1956	1882	74.2409
10	2930	1954	2183	-229.4293
11	1200	1055	1199	-143.7320
12	1375	1241	1298	-57.3401
13	1410	1254	1318	-64.2617
14	1560	1377	1404	-26.6401
15	1600	1465	1426	38.5923
16	1950	1741	1626	115.3761
17	2050	1810	1683	127.4572
18	2250	1893	1796	96.6193
19	2600	1972	1996	-23.5969
20	2760	1935	2087	-151.6672

Sum of Residuals	0
Sum of Squared Residuals	237980
Predicted Residual SS (PRESS)	327395

Thus, the residual for the first observation, $x = 1,290$, and $y = 1,182$, is obtained by first calculating the predicted value

$$\hat{y} = 515.705 + .5692(1,290) = 1249.96$$

and then subtracting from the observed value:

$$\hat{\varepsilon} = y - \hat{y} = 1,182 - 1,249.96$$
$$= -67.96$$

Similar calculations for the other nine observations produce the residuals highlighted in Figure 8.2. Note that the sum of the residuals (shown at the bottom of the SAS printout) is 0.

(b) The SAS printout for the second-order model

$$y = \beta_0 + \beta_1 x + \beta_2 x^2 + \varepsilon$$

Figure 8.3 SAS printout for quadratic (second-order) model

The REG Procedure
Model: MODEL1
Dependent Variable: CHOLESTEROL

Number of Observations Read   20
Number of Observations Used   20

Analysis of Variance

Source	DF	Sum of Squares	Mean Square	F Value	Pr > F
Model	2	1834730	917365	736.92	<.0001
Error	17	21163	1244.86963		
Corrected Total	19	1855893			

Root MSE	35.28271	R-Square	0.9886	
Dependent Mean	1584.50000	Adj R-Sq	0.9873	
Coeff Var	2.22674			

Parameter Estimates

Variable	DF	Parameter Estimate	Standard Error	t Value	Pr > |t|
Intercept	1	-1159.35021	130.57920	-8.88	<.0001
FAT	1	2.34394	0.13540	17.31	<.0001
FATSQ	1	-0.00043899	0.00003326	-13.20	<.0001

Output Statistics

Obs	FAT	Dependent Variable	Predicted Value	Residual
1	1290	1182	1134	48.1962
2	1350	1172	1205	-32.9035
3	1470	1264	1338	-73.6207
4	1600	1493	1467	25.8697
5	1710	1571	1565	5.8741
6	1840	1711	1667	43.7577
7	1980	1804	1761	43.3803
8	2230	1840	1885	-44.5632
9	2400	1956	1938	18.4995
10	2930	1954	1940	14.3268
11	1200	1055	1021	33.7723
12	1375	1241	1234	7.4044
13	1410	1254	1273	-18.8425
14	1560	1377	1429	-51.8616
15	1600	1465	1467	-2.1303
16	1950	1741	1742	-1.0590
17	2050	1810	1801	9.1448
18	2250	1893	1892	0.8919
19	2600	1972	1967	4.7060
20	2760	1935	1966	-30.8428

Sum of Residuals	0
Sum of Squared Residuals	21163
Predicted Residual SS (PRESS)	28812

is shown in Figure 8.3. The least squares model is

$$\hat{y} = -1,159.35 + 2.34394x - .000439x^2$$

For the first observation, $x = 1,290$ and $y = 1,182$, the predicted cholesterol level is

$$\hat{y} = -1,159.35 + 2.34394(1,290) - .000439(1,290)^2$$
$$= 1,133.8$$

and the regression residual is

$$\hat{\varepsilon} = y - \hat{y} = 1,182 - 1,133.8$$
$$= 48.2$$

All the regression residuals for the second-order model are highlighted in Figure 8.3.* Again, you can see (bottom of SAS printout) that the sum of the residuals is 0.   ◼

As we will see in the sections that follow, graphical displays of regression residuals are useful for detecting departures from the assumptions. For example, the regression residual can be plotted on the vertical axis against one of the independent variables on the horizontal axis, or against the predicted value $\hat{y}$ (which is a linear function of the independent variables). If the assumptions concerning the error term ε are satisfied, we expect to see residual plots that have no trends, no dramatic increases or decreases in variability, and only a few residuals (about 5%) more than 2 estimated standard deviations ($2s$) of ε above or below 0.

8.3 Detecting Lack of Fit

Consider the general linear model

$$y = \beta_0 + \beta_1 x_1 + \beta_2 x_2 + \cdots + \beta_k x_k + \varepsilon$$

Assume that this model is correctly specified (i.e., that the terms in the model accurately represent the true relationship of y with the independent variables). In Section 4.2, the first assumption we made about the random error term ε was that $E(\varepsilon) = 0$ for *any* given set of values of $x_1, x_2, \ldots, x_k$. Recall that this assumption implies that

$$E(y) = \beta_0 + \beta_1 x_1 + \beta_2 x_2 + \cdots + \beta_k x_k$$

Now, suppose an analyst hypothesizes a *misspecified model* with mean denoted by $E_m(y)$ so that $E(y) \neq E_m(y)$. The hypothesized equation for the misspecified model is $y = E_m(y) + \varepsilon$; thus, $\varepsilon = y - E_m(y)$. It is easy to see that, for the misspecified model, $E(\varepsilon) = E(y) - E_m(y) \neq 0$. That is, for misspecified models, the assumption of $E(\varepsilon) = 0$ will be violated.

Although a grossly misspecified model fit to a sample data set will most likely yield a poor fit, this will not always be true. A misspecified model may yield a significant global F-value; without further investigation, an unwary analyst may not detect the model's "lack of fit."

In this section we demonstrate how residual plots can detect lack of fit (i.e., whether the deterministic portion of the model, $E(y)$, is misspecified. As shown in the box, if the residual plot portrays a strong trend or pattern, it is likely that the

---

*The residuals shown in Figures 8.2 and 8.3 have been generated using statistical software. Therefore, the results reported here will differ slightly from hand-calculated residuals because of rounding error.

assumption $E(\varepsilon) = 0$ is violated. We will also learn that the nature of the trend or pattern will provide insight into how to modify the misspecified model.

Detecting Model Lack of Fit with Residuals

1. Plot the residuals, $\hat{\varepsilon}$, on the vertical axis against each of the independent variables, $x_1, x_2, \ldots, x_k$, on the horizontal axis.

2. Plot the residuals, $\hat{\varepsilon}$, on the vertical axis against the predicted value, $\hat{y}$, on the horizontal axis.

3. In each plot, look for trends, dramatic changes in variability, and/or more than 5% of residuals that lie outside 2s of 0. Any of these patterns indicates a problem with model fit.

Example 8.2

Refer to Example 8.1 and the cholesterol data saved in the OLYMPIC file. In part a, we used SAS to obtain residuals for the first-order model $E(y) = \beta_0 + \beta_1 x$, where y = cholesterol level and x = fat intake of Olympic athletes. Plot the residuals for this model against fat intake (placing x along the horizontal axis). What does the plot suggest about a potential lack of fit of the first-order model? How would you modify the model?

Solution

The residuals for the first-order model (shown on the SAS printout, Figure 8.2) are plotted versus fat intake, x, in the SAS graph in Figure 8.4. The distinctive aspect of this plot is the parabolic distribution of the residuals about their mean (i.e., all residuals tend to be positive for athletes with intermediate levels of fat intake and negative for the athletes with either relatively high or low levels of fat intake).

Figure 8.4 SAS plot of residuals for the first-order model

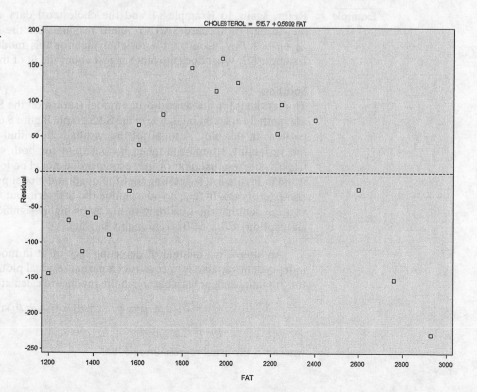

This obvious trend in the residuals suggests strongly that the assumption $E(\varepsilon) = 0$ is violated due to a misspecified model.

In addition, the parabolic trend suggests that the addition of a second-order (quadratic) term may improve the fit of the model. Why? Examine Figure 8.5, a MINITAB scatterplot of the cholesterol data with the least squares line superimposed on the data. First note the clear curvilinear trend of the data. This curvilinear trend leads to negative residuals (y-values that are below the predicted values or fitted line) for high and low levels of fat intake, x, and positive residuals (y-values that are above the fitted line) for intermediate levels of x. By fitting a second-order model to the data we can eliminate this residual trend.

Figure 8.5 MINITAB plot of cholesterol data with least squares line

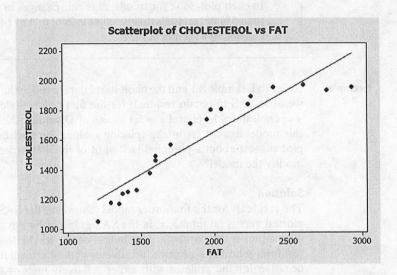

Example 8.3

Refer, again, to Example 8.1 and the cholesterol data saved in the OLYMPIC file. In part b, we used SAS to obtain residuals for the quadratic model $E(y) = \beta_0 + \beta_1 x + \beta_2 x^2$. Construct a residual plot for this model similar to the one in Example 8.2. What does the plot suggest about the fit of the quadratic model?

Solution

The residuals for the second-order model (shown on the SAS printout, Figure 8.3) are plotted versus fat intake, x, in the SAS graph, Figure 8.6. There are no distinctive patterns in this plot. Also, all of the residuals lie within $2s$ of the mean (0), and the variability around the mean is consistent for both small and large x-values. Clearly, the residual plot for the second-order model no longer shows the parabolic trend in Figure 8.4, suggesting that the quadratic model provides a better fit to the cholesterol data. In fact, if you examine the SAS printout for the model, Figure 8.3, you can see that the quadratic term (β_2) is highly significant. For this model, the assumption of $E(\varepsilon) = 0$ is reasonably satisfied.

An alternative method of detecting lack of fit in models with more than one independent variable is to construct a partial residual plot. The **partial residuals** for the jth independent variable, x_j, in the model are calculated as follows:

$$\hat{\varepsilon}^* = y - (\hat{\beta}_0 + \hat{\beta}_1 x_1 + \hat{\beta}_2 x_2 + \cdots + \hat{\beta}_{j-1} x_{j-1} + \hat{\beta}_{j+1} x_{j+1} + \cdots + \hat{\beta}_k x_k)$$

$$= \hat{\varepsilon} + \hat{\beta}_j x_j$$

where $\hat{\varepsilon}$ is the usual regression residual.

Figure 8.6 SAS plot of residuals for the quadratic model

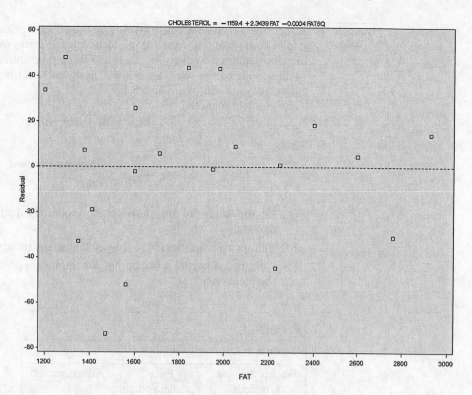

Partial residuals measure the influence of x_j on the dependent variable y *after the effects of the other independent variables $(x_1, x_2, \ldots, x_{j-1}, x_{j+1}, \ldots, x_k)$ have been removed or accounted for.* If the partial residuals $\hat{\varepsilon}^*$ are regressed against x_j in a straight-line model, the resulting least squares slope is equal to $\hat{\beta}_j$—the β estimate obtained from the full model. Therefore, when the partial residuals are plotted against x_j, the points are scattered around a line with slope equal to $\hat{\beta}_j$. Unusual deviations or patterns around this line indicate lack of fit for the variable x_j.

A plot of the partial residuals versus x_j often reveals more information about the relationship between y and x_j than the usual residual plot. In particular, a partial residual plot usually indicates more precisely how to modify the model,[†] as the next example illustrates.

Definition 8.2 The set of **partial regression residuals** for the jth independent variable x_j is calculated as follows:

$$\hat{\varepsilon}^* = y - (\hat{\beta}_0 + \hat{\beta}_1 x_1 + \hat{\beta}_2 x_2 + \cdots$$
$$+ \hat{\beta}_{j-1} x_{j-1} + \hat{\beta}_{j+1} x_{j+1} + \cdots + \hat{\beta}_k x_k)$$
$$= \hat{\varepsilon} + \hat{\beta}_j x_j$$

where $\hat{\varepsilon} = y - \hat{y}$ is the usual regression residual (see Definition 8.1).

[†] Partial residual plots display the correct functional form of the predictor variables across the relevant range of interest, except in cases where severe multicollinearity exists. See Mansfield and Conerly (1987) for an excellent discussion of the use of residual and partial residual plots.

Example 8.4   A supermarket chain wants to investigate the effect of price p on the weekly demand y for a house brand of coffee at its stores. Eleven prices were randomly assigned to the stores and were advertised using the same procedures. A few weeks later, the chain conducted the same experiment using no advertisements. The data for the entire study are shown in Table 8.2.

Consider the model

$$E(y) = \beta_0 + \beta_1 p + \beta_2 x_2$$

where

$$x_2 = \begin{cases} 1 & \text{if advertisement used} \\ 0 & \text{if not} \end{cases}$$

(a) Fit the model to the data. Is the model adequate for predicting weekly demand y?

(b) Plot the residuals versus p. Do you detect any trends?

(c) Construct a partial residual plot for the independent variable p. What does the plot reveal?

COFFEE2

Table 8.2 Data for Example 8.4

Weekly demand y, pounds	Price p, dollars/pound	Advertisement x_2
1190	3.0	1
1033	3.2	1
897	3.4	1
789	3.6	1
706	3.8	1
595	4.0	1
512	4.2	1
433	4.4	1
395	4.6	1
304	4.8	1
243	5.0	1
1124	3.0	0
974	3.2	0
830	3.4	0
702	3.6	0
619	3.8	0
529	4.0	0
451	4.2	0
359	4.4	0
296	4.6	0
247	4.8	0
194	5.0	0

(d) Fit the model $E(y) = \beta_0 + \beta_1 x_1 + \beta_2 x_2$, where $x_1 = 1/p$. Has the predictive ability of the model improved?

Solution

(a) The SPSS printout for the regression analysis is shown in Figure 8.7. The F-value for testing model adequacy (i.e., H_0: $\beta_1 = \beta_2 = 0$), is given on the printout (shaded) as $F = 373.71$ with a corresponding p-value (also shaded) of .000. Thus, there is sufficient evidence (at $\alpha = .01$) that the model contributes information for the prediction of weekly demand, y. Also, the coefficient of determination is $R^2 = .975$, meaning that the model explains approximately 97.5% of the sample variation in weekly demand.

Figure 8.7 SPSS regression printout for demand model

Model Summary[b]

Model	R	R Square	Adjusted R Square	Std. Error of the Estimate
1	.988[a]	.975	.973	49.876

a. Predictors: (Constant), X2, PRICE
b. Dependent Variable: DEMAND

ANOVA[b]

Model		Sum of Squares	df	Mean Square	F	Sig.
1	Regression	1859299	2	929649.475	373.710	.000[a]
	Residual	47264.868	19	2487.625		
	Total	1906564	21			

a. Predictors: (Constant), X2, PRICE
b. Dependent Variable: DEMAND

Coefficients[a]

Model		Unstandardized Coefficients		Standardized Coefficients	t	Sig.
		B	Std. Error	Beta		
1	(Constant)	2400.182	68.914		34.829	.000
	PRICE	-456.295	16.813	-.980	-27.139	.000
	X2	70.182	21.267	.119	3.300	.004

a. Dependent Variable: DEMAND

Recall from Example 7.8, however, that we fit a model with the transformed independent variable $x_1 = 1/p$. That is, we expect the relationship between weekly demand y and price p to be decreasing in a curvilinear fashion and approaching (but never reaching) 0 as p increases. (See Figure 7.11d.) If such a relationship exists, the model (with untransformed price), although statistically useful for predicting demand y, will be inadequate in a practical setting.

(b) The regression residuals for the model in part a are saved in SPSS and plotted against price p in Figure 8.8. Notice that the plot reveals a clear parabolic trend, implying a lack of fit. Thus, the residual plot supports our hypothesis that the weekly demand–price relationship is curvilinear, not linear. However, the appropriate transformation on price (i.e., $1/p$) is not evident from the plot. In fact, the nature of the curvature in Figure 8.8 may lead you to conclude that the addition of the quadratic term, $\beta_3 p^2$, to the model will solve the problem. In general, a residual plot will detect curvature if it exists, but may not reveal the appropriate transformation.

Figure 8.8 SPSS plot of residuals against price for demand model

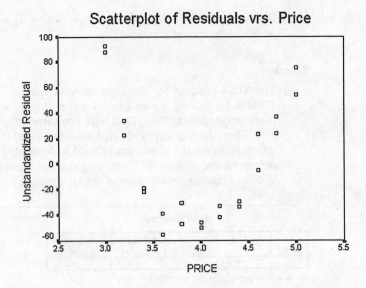

(c) Most statistical packages have options for automatic calculation of partial residuals. The partial residual plot for the independent variable p is shown in the SPSS printout, Figure 8.9. SPSS (as does SAS) finds the partial residuals for price, p, by finding (separately) the residuals of the dependent variable y when regressed against advertising x_2, and the residuals of price, p, when regressed against x_2. A plot of these residuals will look similar to a plot of the partial residuals of Definition 8.2 against price, p. You can see that the partial residual plot of Figure 8.9 also reveals a curvilinear trend, but, in addition, displays the correct functional form of weekly demand–price relationship. Notice that the curve is decreasing and approaching (but never reaching) 0 as p increases. This suggests that the appropriate transformation on price is either $1/p$ or e^{-p} (see Figures 7.11c and 7.11d).

Figure 8.9 SPSS partial residual plot for price

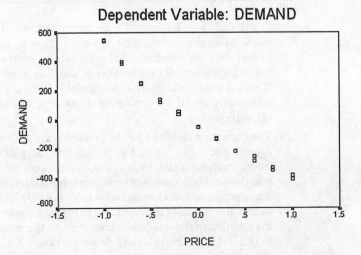

(d) Using the transformation $x_1 = 1/p$, we refit the model to the data, and the resulting SPSS printout is shown in Figure 8.10. The small p-value (.0001) for testing H_0: $\beta_1 = \beta_2 = 0$ indicates that the model is adequate for predicting y. Although the coefficient of determination increased only slightly (from $R^2 = .975$ to $R^2 = .999$), the model standard deviation decreased significantly (from $s = 49.876$ to $s = 11.097$). Thus, whereas the model with untransformed price can predict weekly demand for coffee to within $2s = 2(50) = 100$ pounds, the transformed model can predict demand to within $2(11) = 22$ pounds.   ∎

Figure 8.10 SPSS regression printout for demand model with transformed price

Model Summary

Model	R	R Square	Adjusted R Square	Std. Error of the Estimate
1	.999a	.999	.999	11.097

a. Predictors: (Constant), X2, X1

ANOVAb

Model		Sum of Squares	df	Mean Square	F	Sig.
1	Regression	1904224	2	952111.957	7731.141	.000a
	Residual	2339.904	19	123.153		
	Total	1906564	21			

a. Predictors: (Constant), X2, X1

b. Dependent Variable: DEMAND

Coefficientsa

Model		Unstandardized Coefficients		Standardized Coefficients	t	Sig.
		B	Std. Error	Beta		
1	(Constant)	-1217.343	14.898		-81.711	.000
	X1	6986.507	56.589	.992	123.460	.000
	X2	70.182	4.732	.119	14.831	.000

a. Dependent Variable: DEMAND

Residual (or partial residual) plots are useful for indicating potential model improvements, but they are no substitute for formal statistical tests of model terms to determine their importance. Thus, a true test of whether the second-order term contributes to the cholesterol model (Example 8.1) is the t-test of the null hypothesis H_0: $\beta_2 = 0$. The appropriate test statistic, shown in the printout of Figure 8.2, indicates that the second-order term does contribute information for the prediction of cholesterol level y. We have confidence in this statistical inference because we know the probability α of committing a Type I error (concluding a term is important when, in fact, it is not). In contrast, decisions based on residual plots are subjective, and their reliability cannot be measured. Therefore, we suggest that such plots be used only as indicators of *potential* problems. The final judgment on model adequacy should be based on appropriate statistical tests.[‡]

[‡] A more general procedure for determining whether the straight-line model adequately fits the data tests the null hypothesis H_0: $E(y) = \beta_0 + \beta_1 x$ against the alternative H_a: $E(y) \neq \beta_0 + \beta_1 x$. You can see that this test, called a test for *lack of fit*, does not restrict the alternative hypothesis to second-order models. Lack-of-fit tests are appropriate when the x-values are replicated (i.e., when the sample data include two or more observations for several different levels of x). When the data are observational, however, replication rarely occurs. (Note that none of the values of x is repeated in Table 8.1.) For details on how to conduct tests for lack of fit, consult the references given at the end of this chapter.

8.3 Exercises

8.1 Finding and plotting residuals. Consider the data on x and y shown in the table.

(a) Fit the model $E(y) = \beta_0 + \beta_1 x$ to the data.
(b) Calculate the residuals for the model.
(c) Plot the residuals versus x. Do you detect any trends? If so, what does the pattern suggest about the model?

⊙ EX8_1

x	-2	-2	-1	-1	0	0	1	1	2	2	3	3
y	1.1	1.3	2.0	2.1	2.7	2.8	3.4	3.6	4.0	3.9	3.8	3.6

8.2 Finding and plotting residuals. Consider the data on x and y shown in the table.

(a) Fit the model $E(y) = \beta_0 + \beta_1 x$ to the data.
(b) Calculate the residuals for the model.
(c) Plot the residuals versus x. Do you detect any trends? If so, what does the pattern suggest about the model?

⊙ EX8_2

x	2	4	7	10	12	15	18	20	21	25
y	5	10	12	22	25	27	39	50	47	65

MINITAB Output for Exercise 8.4

```
The regression equation is
VOLUME = 98.6 - 0.256 PRESSURE

Predictor      Coef     SE Coef        T       P
Constant     98.6149     0.4037    244.26   0.000
PRESSURE    -0.255594   0.008646    -29.56   0.000

S = 0.6484     R-Sq = 99.0%     R-Sq(adj) = 98.9%

Analysis of Variance

Source          DF       SS        MS        F        P
Regression       1     367.34    367.34    873.87   0.000
Residual Error   9       3.78      0.42
Total           10     371.12
```

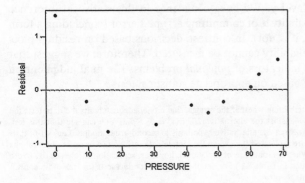

Residuals Versus PRESSURE
(response is VOLUME)

8.3 New tire wear tests. Refer to Example 3.2 (p. 120). Recall that a manufacturer of a new tire tested the tire for wear at different pressures with the results shown in the table.

(a) Fit the straight-line model $y = \beta_0 + \beta_1 x + \varepsilon$ to the data.
(b) Calculate the residuals for the model.
(c) Plot the residuals versus x. Do you detect any trends? If so, what does the pattern suggest about the model?
(d) Fit the quadratic model $y = \beta_0 + \beta_1 x + \beta_2 x^2 + \varepsilon$ to the data using an available statistical software package. Has the addition of the quadratic term improved model adequacy?

⊙ TIRES

PRESSURE x, pounds per sq. inch	MILEAGE y, thousands
30	29.5
30	30.2
31	32.1
31	34.5
32	36.3
32	35.0
33	38.2
33	37.6
34	37.7
34	36.1
35	33.6
35	34.2
36	26.8
36	27.4

8.4 Elasticity of moissanite. Moissanite is a popular abrasive material because of its extreme hardness. Another important property of moissanite is elasticity. The elastic properties of the material were investigated in the *Journal of Applied Physics* (September 1993). A diamond anvil cell was used to compress a mixture of moissanite, sodium chloride, and gold in a ratio of 33:99:1 by volume. The compressed volume, y, of the mixture (relative to the zero-pressure volume) was measured at each of 11 different pressures (GPa). The results are displayed in the table (p. 397). A MINITAB printout for the straight-line regression model $E(y) = \beta_0 + \beta_1 x$ and a MINITAB residual plot are displayed at left.

(a) Calculate the regression residuals.
(b) Plot the residuals against x. Do you detect a trend?
(c) Propose an alternative model based on the plot, part b.
(d) Fit and analyze the model you proposed in part c.

MOISSANITE

COMPRESSED VOLUME y, %	PRESSURE x, GPa
100	0
96	9.4
93.8	15.8
90.2	30.4
87.7	41.6
86.2	46.9
85.2	51.6
83.3	60.1
82.9	62.6
82.9	62.6
81.7	68.4

Source: Bassett, W. A., Weathers, M. S., and Wu, T. C. "Compressibility of SiC up to 68.4 GPa," *Journal of Applied Physics*, Vol. 74. No. 6, Sept. 15, 1993, p. 3825 (Table 1). Reprinted with permission from Journal of Applied Physics. Copyright © 1993, American Institute of Physics.

8.5 Failure times of silicon wafer microchips. Refer to the National Semiconductor study on using tin-lead solder bumps to manufacture silicon wafer integrated circuit chips, Exercise 4.40 (p. 207). The failure times of 22 microchips (in hours) were determined at different solder temperatures (degrees

WAFER

TEMPERATURE (°C)	TIME TO FAILURE (hours)
165	200
162	200
164	1200
158	500
158	600
159	750
156	1200
157	1500
152	500
147	500
149	1100
149	1150
142	3500
142	3600
143	3650
133	4200
132	4800
132	5000
134	5200
134	5400
125	8300
123	9700

Source: Gee, S., & Nguyen, L. "Mean time to failure in wafer level–CSP packages with SnPb and SnAgCu solder bumps," *International Wafer Level Packaging Conference*, San Jose, CA, Nov. 3–4, 2005 (adapted from Figure 7).

Centigrade), with the data repeated in the accompanying table. Recall that the researchers want to predict failure time (y) based on solder temperature (x). Fit the straight-line model, $E(y) = \beta_0 + \beta_1 x$, to the data and plot the residuals against solder temperature. Do you detect a lack of fit? In Exercise 4.40c you discovered upward curvature in the relationship between failure time and solder temperature. Does this result agree with the residual plot? Explain.

8.6 Demand for a rare gem. A certain type of rare gem serves as a status symbol for many of its owners. In theory, then, the demand for the gem would increase as the price increases, decreasing at low prices, leveling off at moderate prices, and increasing at high prices, because obtaining the gem at a high price confers high status on the owner. Although a quadratic model would seem to match the theory, the model proposed to explain the demand for the gem by its price is the first-order model

$$y = \beta_0 + \beta_1 x + \varepsilon$$

where y is the demand (in thousands) and x is the retail price per carat (dollars). This model was fit to the 12 data points given in the table, and the results of the analysis are shown in the SAS printout (p. 398).

(a) Use the least squares prediction equation to verify the values of the regression residuals shown on the printout.
(b) Plot the residuals against retail price per carat, x.
(c) Can you detect any trends in the residual plot? What does this imply?

GEM

x	100	700	450	150	500	800	70	50	300	350	750	700
y	130	150	60	120	50	200	150	160	50	40	180	130

8.7 Erecting boiler drums. Refer to the study of man-hours required to erect boiler drums, Exercise 4.74 (p. 241). Recall that the data on 35 boilers were used to fit the model

$$E(y) = \beta_0 + \beta_1 x_1 + \beta_2 x_2 + \beta_3 x_3 + \beta_4 x_4$$

where

y = Man-hours
x_1 = Boiler capacity (lb/hr)
x_2 = Design pressure (psi)
$x_3 = \begin{cases} 1 & \text{if industry erected} \\ 0 & \text{if not} \end{cases}$
$x_4 = \begin{cases} 1 & \text{if steam drum} \\ 0 & \text{if mud drum} \end{cases}$

The data are saved in the BOILERS file.

(a) Find the residuals for the model.

(b) Plot the residuals versus x_1. Do you detect any trends? If so, what does the pattern suggest about the model?

(c) Plot the residuals versus x_2. Do you detect any trends? If so, what does the pattern suggest about the model?

(d) Plot the partial residuals for x_1. Interpret the result.

(e) Plot the partial residuals for x_2. Interpret the result.

SAS Output for Exercise 8.6

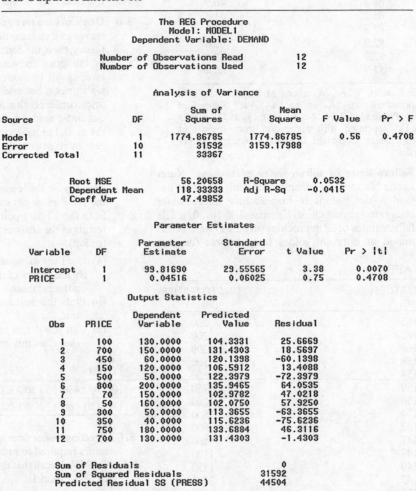

```
                           The REG Procedure
                             Model: MODEL1
                      Dependent Variable: DEMAND

              Number of Observations Read          12
              Number of Observations Used          12

                         Analysis of Variance

                                Sum of       Mean
     Source              DF    Squares      Square    F Value    Pr > F

     Model                1  1774.86785  1774.86785      0.56    0.4708
     Error               10    31592     3159.17988
     Corrected Total     11    33367


              Root MSE             56.20658    R-Square     0.0532
              Dependent Mean      118.33333    Adj R-Sq    -0.0415
              Coeff Var            47.49852

                         Parameter Estimates

                      Parameter     Standard
     Variable    DF    Estimate       Error    t Value    Pr > |t|

     Intercept    1    99.81690     29.55565      3.38      0.0070
     PRICE        1     0.04516      0.06025      0.75      0.4708

                         Output Statistics

                      Dependent    Predicted
          Obs  PRICE   Variable      Value     Residual

           1    100    130.0000    104.3331     25.6669
           2    700    150.0000    131.4303     18.5697
           3    450     60.0000    120.1398    -60.1398
           4    150    120.0000    106.5912     13.4088
           5    500     50.0000    122.3979    -72.3979
           6    800    200.0000    135.9465     64.0535
           7     70    150.0000    102.9782     47.0218
           8     50    160.0000    102.0750     57.9250
           9    300     50.0000    113.3655    -63.3655
          10    350     40.0000    115.6236    -75.6236
          11    750    180.0000    133.6884     46.3116
          12    700    130.0000    131.4303     -1.4303

          Sum of Residuals                            0
          Sum of Squared Residuals                31592
          Predicted Residual SS (PRESS)           44504
```

8.4 Detecting Unequal Variances

Recall that one of the assumptions necessary for the validity of regression inferences is that the error term ε have constant variance σ^2 for all levels of the independent variable(s). Variances that satisfy this property are called **homoscedastic**. Unequal variances for different settings of the independent variable(s) are said to be **heteroscedastic**. Various statistical tests for heteroscedasticity have been developed. However, plots of the residuals will frequently reveal the presence of heteroscedasticity. In this section we show how residual plots can be used to detect departures from the assumption of equal variances, and then give a simple test for heteroscedasticity. In addition, we suggest some modifications to the model that may remedy the situation.

When data fail to be homoscedastic, the reason is often that the variance of the response y is a function of its mean $E(y)$. Some examples follow:

1. If the response y is a count that has a Poisson distribution, the variance will be equal to the mean $E(y)$. Poisson data are usually counts per unit volume, area, time, etc. For example, the number of sick days per month for an employee would very likely be a Poisson random variable. If the variance of a response is proportional to $E(y)$, the regression residuals produce a pattern about $\hat{y}$, the least squares estimate of $E(y)$, like that shown in Figure 8.11.

Figure 8.11 A plot of residuals for poisson data

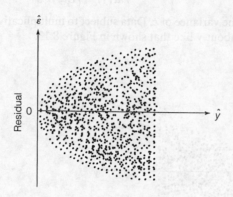

2. Many responses are proportions (or percentages) generated by **binomial experiments**. For instance, the proportion of a random sample of 100 convicted felons who are repeat offenders is an example of a binomial response. Binomial proportions have variances that are functions of both the true proportion (the mean) and the sample size. In fact, if the observed proportion $y_i = \hat{p}_i$ is generated by a binomial distribution with sample size n_i and true probability p_i, the variance of y_i is

$$\text{Var}(y_i) = \frac{p_i(1 - p_i)}{n_i} = \frac{E(y_i)[1 - E(y_i)]}{n_i}$$

Residuals for binomial data produce a pattern about $\hat{y}$ like that shown in Figure 8.12.

Figure 8.12 A plot of residuals for binomial data (proportions or percentages)

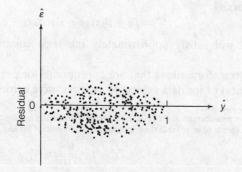

3. The random error component has been assumed to be **additive** in all the models we have constructed. An additive error is one for which the response is equal to the mean $E(y)$ *plus* random error,

$$y = E(y) + \varepsilon$$

Another useful type of model, especially for business and economic data, is the **multiplicative** model. In this model, the response is written as the *product* of its mean and the random error component, that is,

$$y = [E(y)]\varepsilon$$

The variance of this response will grow proportionally to the *square* of the mean, that is,

$$\text{Var}(y) = [E(y)]^2 \sigma^2$$

where σ^2 is the variance of ε. Data subject to multiplicative errors produce a pattern of residuals about $\hat{y}$ like that shown in Figure 8.13.

Figure 8.13 A plot of residuals for data subject to multiplicative errors

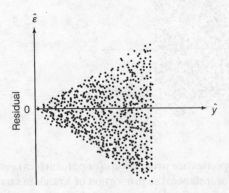

When the variance of y is a function of its mean, we can often satisfy the least squares assumption of homoscedasticity by transforming the response to some new response that has a constant variance. These are called **variance-stabilizing transformations**. For example, if the response y is a count that follows a Poisson distribution, the square root transform $\sqrt{y}$ can be shown to have approximately constant variance.* Consequently, if the response is a Poisson random variable, we would let

$$y^* = \sqrt{y}$$

and fit the model

$$y^* = \beta_0 + \beta_1 x_1 + \cdots + \beta_k x_k + \varepsilon$$

This model will satisfy approximately the least squares assumption of homoscedasticity.

Similar transformations that are appropriate for percentages and proportions (binomial data) or for data subject to multiplicative errors are shown in Table 8.3.

---

* The square root transformation for Poisson responses is derived by finding the integral of $1/\sqrt{E(y)}$. In general, it can be shown (proof omitted) that the appropriate transformation for any response y is

$$y^* = \int \frac{1}{\sqrt{V(y)}} dy$$

where $V(y)$ is an expression for the variance of y.

Table 8.3 Stabilizing transformations for heteroscedastic responses

Type of Response	Variance	Stabilizing Transformation
Poisson	$E(y)$	$\sqrt{y}$
Binomial proportion	$\dfrac{E(y)[1 - E(y)]}{n}$	$\sin^{-1}\sqrt{y}$
Multiplicative	$[E(y)]^2\sigma^2$	$\ln(y)$

The transformed responses will satisfy (at least approximately) the assumption of homoscedasticity.

Example 8.5

The data in Table 8.4 are the salaries, y, and years of experience, x, for a sample of 50 social workers. The second-order model $E(y) = \beta_0 + \beta_1 x + \beta_2 x^2$ was fit to the data using MINITAB. The printout for the regression analysis is shown in Figure 8.14 followed by a plot of the residuals versus $\hat{y}$ in Figure 8.15. Interpret the results.

⊙ SOCWORK

Table 8.4 Salary and experience data for 50 social workers

Years of Experience x	Salary y	Years of Experience x	Salary y	Years of Experience x	Salary y
7	$26,075	21	$43,628	28	$99,139
28	79,370	4	16,105	23	52,624
23	65,726	24	65,644	17	50,594
18	41,983	20	63,022	25	53,272
19	62,309	20	47,780	26	65,343
15	41,154	15	38,853	19	46,216
24	53,610	25	66,537	16	54,288
13	33,697	25	67,447	3	20,844
2	22,444	28	64,785	12	32,586
8	32,562	26	61,581	23	71,235
20	43,076	27	70,678	20	36,530
21	56,000	20	51,301	19	52,745
18	58,667	18	39,346	27	67,282
7	22,210	1	24,833	25	80,931
2	20,521	26	65,929	12	32,303
18	49,727	20	41,721	11	38,371
11	33,233	26	82,641		

Solution

The printout in Figure 8.14 suggests that the second-order model provides an adequate fit to the data. The R_a^2 value, .808, indicates that the model explains almost

81% of the total variation of the y-values about $\bar{y}$. The global F-value, $F = 103.99$, is highly significant ($p = .000$), indicating that the model contributes information for the prediction of y. However, an examination of the salary residuals plotted against the estimated mean salary, $\hat{y}$, as shown in Figure 8.15, reveals a potential problem. Note the "cone" shape of the residual variability; the size of the residuals increases as the estimated mean salary increases. This residual plot indicates that a multiplicative model may be appropriate. We explore this possibility further in the next example.

Figure 8.14 MINITAB regression printout for second-order model of salary

```
The regression equation is
SALARY = 20242 + 522 EXP + 53.0 EXPSQ

Predictor      Coef      SE Coef        T        P
Constant      20242         4423     4.58    0.000
EXP           522.3        616.7     0.85    0.401
EXPSQ         53.01        19.57     2.71    0.009

S = 8123      R-Sq = 81.6%      R-Sq(adj) = 80.8%

Analysis of Variance

Source           DF           SS           MS        F        P
Regression        2  13723582237   6861791118   103.99    0.000
Residual Error   47   3101279310     65984666
Total            49  16824861546
```

Figure 8.15 MINITAB residual plot for second-order model of salary

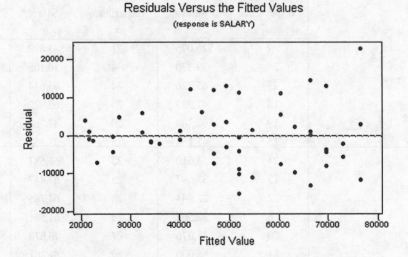

Residuals Versus the Fitted Values
(response is SALARY)

Refer, again, to the salary and experience data in Table 8.4. Use the natural log transformation on the dependent variable, and relate $\ln(y)$ to years of experience, x, using the second-order model

$$\ln(y) = \beta_0 + \beta_1 x + \beta_2 x^2 + \varepsilon$$

Evaluate the adequacy of the model.

Solution

The MINITAB printout in Figure 8.16 gives the regression analysis for the $n = 50$ measurements. The prediction equation used in computing the residuals is

$$\widehat{\ln(y)} = 9.84289 + .04969x + .0000094x^2$$

Example 8.6

Figure 8.16 MINITAB regression printout for second-order model of natural log of salary

```
The regression equation is
LOGY = 9.84 + 0.0497 EXP +0.000009 EXPSQ


Predictor        Coef       SE Coef         T        P
Constant      9.84289       0.08479    116.08    0.000
EXP           0.04969       0.01182      4.20    0.000
EXPSQ       0.0000094     0.0003753      0.03    0.980


S = 0.1557     R-Sq = 86.4%     R-Sq(adj) = 85.8%

Analysis of Variance

Source          DF        SS          MS         F        P
Regression       2    7.2122      3.6061    148.67    0.000
Residual Error  47    1.1400      0.0243
Total           49    8.3522
```

The residual plot in Figure 8.17 indicates that the logarithmic transformation has significantly reduced the heteroscedasticity.[†] Note that the cone shape is gone; there is no apparent tendency of the residual variance to increase as mean salary increases. We therefore are confident that inferences using the logarithmic model are more reliable than those using the untransformed model.

To evaluate model adequacy, we first note that $R_a^2 = .858$ and that about 86% of the variation in ln(salary) is accounted for by the model. The global F-value ($F = 148.67$) and its associated p-value ($p = .000$) indicate that the model significantly improves upon the sample mean as a predictor of ln(salary).

Although the estimate of β_2 is very small, we should check to determine whether the data provide sufficient evidence to indicate that the second-order term contributes information for the prediction of ln(salary). The test of

$$H_0: \beta_2 = 0$$

$$H_a: \beta_2 \neq 0$$

is conducted using the t statistic shown in Figure 8.16, $t = .03$. The p-value of the test, .98, is greater than $\alpha = .10$. Consequently, there is insufficient evidence to indicate that the second-order term contributes to the prediction of ln(salary).

Figure 8.17 MINITAB residual plot for second-order model of natural log of salary

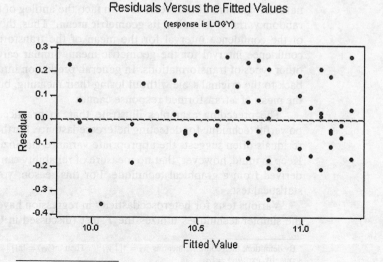

Residuals Versus the Fitted Values
(response is LOGY)

[†] A printout of the residuals is omitted.

There is no indication that the second-order model is an improvement over the straight-line model,

$$\ln(y) = \beta_0 + \beta_1 x + \varepsilon$$

for predicting ln(salary).

The MINITAB printout for the first-order model (Figure 8.18) shows that the prediction equation for the first-order model is

$$\widehat{\ln y} = 9.84133 + .049978x$$

The value $R_a^2 = .861$ is approximately the same as the value of R_a^2 obtained for the second-order model. The F statistic, computed from the mean squares in Figure 8.18, $F = 303.65$ (p-value $= .000$) indicates that the model contributes significantly to the prediction of $\ln(y)$. ▪

Figure 8.18 MINITAB regression printout for first-order model of natural log of salary

```
The regression equation is
LOGY = 9.84 + 0.0500 EXP

Predictor      Coef      SE Coef       T          P
Constant     9.84132     0.05636    174.63     0.000
EXP          0.049979    0.002868    17.43     0.000

S = 0.1541     R-Sq = 86.4%     R-Sq(adj) = 86.1%

Analysis of Variance

Source          DF        SS         MS         F         P
Regression       1      7.2121     7.2121     303.65    0.000
Residual Error  48      1.1401     0.0238
Total           49      8.3522
```

When the transformed model of Example 8.6 is used to predict the value of $\ln(y)$, the predicted value of y is the antilog, $\hat{y} = e^{\widehat{\ln y}}$. The endpoints of the prediction interval are similarly transformed back to the original scale, and the interval will retain its meaning. In repeated use, the intervals will contain the observed y-value $100(1 - \alpha)\%$ of the time.

Unfortunately, you cannot take antilogs to find the confidence interval for the mean value $E(y)$. The reason for this is that the mean value of $\ln(y)$ is not equal to the natural logarithm of the mean of y. In fact, the antilog of the logarithmic mean of a random variable y is called its **geometric mean**.* Thus, the antilogs of the endpoints of the confidence interval for the mean of the transformed response will give a confidence interval for the geometric mean. Similar care must be exercised with other types of transformations. In general, prediction intervals can be transformed back to the original scale without losing their meaning, but confidence intervals for the mean of a transformed response cannot.

The preceding examples illustrate that, in practice, residual plots can be a powerful technique for detecting heteroscedasticity. Furthermore, the pattern of the residuals often suggests the appropriate variance-stabilizing transformation to use. Keep in mind, however, that no measure of reliability can be attached to inferences derived from a graphical technique. For this reason, you may want to rely on a statistical test.

Various tests for heteroscedasticity in regression have been developed. One of the simpler techniques utilizes the F-test (discussed in Chapter 1) for comparing

---

* By definition, the geometric mean is $\bar{y}_G = \{\prod y_i\}^{1/n}$. Then $\ln(\bar{y}_G) = \ln\{\prod y_i\}^{1/n} = (1/n) \sum \ln(y_i) = \overline{\ln y}$. Consequently, $\exp\{\overline{\ln y}\} = \bar{y}_G$.

population variances. The procedure requires that you divide the sample data in half and fit the regression model to each half. If the regression model fit to one-half the observations yields a significantly smaller or larger MSE than the model fitted to the other half, there is evidence that the assumption of equal variances for all levels of the x variables in the model is being violated. (Recall that MSE, or mean square for error, estimates σ^2, the variance of the random error term.) Where you divide the data depends on where you suspect the differences in variances to be. We illustrate this procedure with an example.

Example 8.7

Refer to the data in Table 8.4 and the analysis of the model relating a social worker's salary (y) to years of experience (x). The residual plot for the quadratic model

$$E(y) = \beta_0 + \beta_1 x + \beta_2 x^2$$

indicates that the assumption of equal variances may be violated (see Figure 8.15). Conduct a statistical test of hypothesis to determine whether heteroscedasticity exists. Use $\alpha = .05$.

Solution

The residual plot shown in Figure 8.15 reveals that the residuals associated with larger values of predicted salary tend to be more variable than the residuals associated with smaller values of predicted salary. Therefore, we will divide the sample observations based on the values of $\hat{y}$, or, equivalently, the value of x (since, for the fitted model, $\hat{y}$ increases as x increases). An examination of the data in Table 8.4 reveals that approximately one-half of the 50 observed values of years of experience, x, fall below $x = 20$. Thus, we will divide the data into two subsamples as follows:

SUBSAMPLE 1	SUBSAMPLE 2
$x < 20$	$x \geq 20$
$n_1 = 24$	$n_2 = 26$

Figures 8.19a and 8.19b give the SAS printouts for the quadratic model fit to subsample 1 and subsample 2, respectively. The value of MSE is shaded in each printout.

Figure 8.19a SAS regression printout for second-order model of salary: Subsample 1 (years of experience < 20)

```
                        Dependent Variable: SALARY

                          Analysis of Variance

                                   Sum of          Mean
Source              DF            Squares         Square    F Value    Pr > F

Model                2         3231196786     1615598393      51.16    <.0001
Error               21          663116951       31576998
Corrected Total     23         3894313737

                Root MSE            5619.34139    R-Square     0.8297
                Dependent Mean            37153    Adj R-Sq     0.8135
                Coeff Var              15.12498


                          Parameter Estimates

                              Parameter        Standard
        Variable     DF        Estimate           Error    t Value    Pr > |t|

        Intercept     1           20372      3817.87780       5.34     <.0001
        EXP           1       263.17043       861.86817       0.31      0.7631
        EXPSQ         1        76.76912        40.01402       1.92      0.0687
```

Figure 8.19b SAS regression printout for second-order model of salary: Subsample 2 (years of experience > 20)

Dependent Variable: SALARY

Analysis of Variance

Source	DF	Sum of Squares	Mean Square	F Value	Pr > F
Model	2	2930781596	1465390798	15.47	<.0001
Error	23	2178353538	94711023		
Corrected Total	25	5109135134			

Root MSE	9731.95887	R-Square	0.5736	
Dependent Mean	62187	Adj R-Sq	0.5366	
Coeff Var	15.64951			

Parameter Estimates

Variable	DF	Parameter Estimate	Standard Error	t Value	Pr > |t|
Intercept	1	-19330	168798	-0.11	0.9098
EXP	1	3004.76433	14415	0.21	0.8367
EXPSQ	1	16.85876	304.22229	0.06	0.9563

The null and alternative hypotheses to be tested are

$$H_0: \frac{\sigma_1^2}{\sigma_2^2} = 1 \text{ (Assumption of equal variances satisfied)}$$

$$H_a: \frac{\sigma_1^2}{\sigma_2^2} \neq 1 \text{ (Assumption of equal variances violated)}$$

where

σ_1^2 = Variance of the random error term, ε, for subpopulation 1 (i.e., $x < 20$)

σ_2^2 = Variance of the random error term, ε, for subpopulation 2 (i.e., $x \geq 20$)

The test statistic for a two-tailed test is given by:

$$F = \frac{\text{Larger } s^2}{\text{Smaller } s^2} = \frac{\text{Larger MSE}}{\text{Smaller MSE}} \quad \text{(see Section 1.11)}$$

where the distribution of F is based on $v_1 = $ df(error) associated with the larger MSE and $v_2 = $ df(error) associated with the smaller MSE. Recall that for a quadratic model, df(error) $= n - 3$.

From the printouts shown in Figure 8.19a and 8.19b, we have

$$\text{MSE}_1 = 31,576,998 \quad \text{and} \quad \text{MSE}_2 = 94,711,023$$

Therefore, the test statistic is

$$F = \frac{\text{MSE}_2}{\text{MSE}_1} = \frac{94,711,023}{31,576,998} = 3.00$$

Since the MSE for subsample 2 is placed in the numerator of the test statistic, this F-value is based on $n_2 - 3 = 26 - 3 = 23$ numerator df and $n_1 - 3 = 24 - 3 = 21$ denominator df. For a two-tailed test at $\alpha = .05$, the critical value for $v_1 = 23$ and $v_2 = 21$ (found in Table 5 of Appendix C) is approximately $F_{.025} = 2.37$.

Since the observed value, $F = 3.00$, exceeds the critical value, there is sufficient evidence (at $\alpha = .05$) to indicate that the error variances differ.[‡] Thus, this test supports the conclusions reached by using the residual plots in the preceding examples. ∎

[‡] Most statistical tests require that the observations in the sample be independent. For this F-test, the observations are the residuals. Even if the standard least squares assumption of independent errors is satisfied, the regression residuals will be correlated. Fortunately, when n is large compared to the number of β parameters in the model, the correlation among the residuals is reduced and, in most cases, can be ignored.

The test for heteroscedasticity outlined in Example 8.7 is easy to apply when only a single independent variable appears in the model. For a multiple regression model that contains several different independent variables, the choice of the levels of the x variables for dividing the data is more difficult, if not impossible. If you require a statistical test for heteroscedasticity in a multiple regression model, you may need to resort to other, more complex, tests.[§] Consult the references at the end of this chapter for details on how to conduct these tests.

8.4 Exercises

8.8 Plotting residuals. Refer to Exercise 8.1 (p. 396). Plot the residuals for the first-order model versus $\hat{y}$. Do you detect any trends? If so, what does the pattern suggest about the model?

8.9 Plotting residuals. Refer to Exercise 8.2 (p. 396). Plot the residuals for the first-order model versus $\hat{y}$. Do you detect any trends? If so, what does the pattern suggest about the model?

◎ ASWELLS

8.10 Arsenic in groundwater. Refer to the *Environmental Science and Technology* (January 2005) study of the reliability of a commercial kit to test for arsenic in groundwater, Exercise 4.12 (p. 187). Recall that you fit a first-order model for arsenic level (y) as a function of latitude (x_1), longitude (x_2), and depth (x_3) to data saved in the ASWELLS file. Check the assumption of a constant error variance by plotting the model residuals against predicted arsenic level. Interpret the results.

◎ GASTURBINE

8.11 Cooling method for gas turbines. Refer to the *Journal of Engineering for Gas Turbines and Power* (January 2005) study of a high-pressure inlet fogging method for a gas turbine engine, Exercise 7.19 (p. 380). Now consider the interaction model for heat rate (y) of a gas turbine as a function of cycle speed (x_1) and cycle pressure ratio (x_2), $E(y) = \beta_0 + \beta_1 x_1 + \beta_2 x_2 + \beta_3 x_1 x_2$. Fit the model to the data saved in the GASTURBINE file, then plot the residuals against predicted heat rate. Is the assumption of a constant error variance reasonably satisfied? If not, suggest how to modify the model.

8.12 Fair market value of Hawaiian properties. Prior to 1980, private homeowners in Hawaii had to lease the land their homes were built on because the law (dating back to the islands' feudal period) required that land be owned only by the big estates. After 1980, however, a new law instituted

◎ HAWAII

PROPERTY	LEASED FEE VALUE y, thousands of dollars	SIZE x, thousands
1	70.7	13.5
2	52.7	9.6
3	87.6	17.6
4	43.2	7.9
5	103.8	11.5
6	45.1	8.2
7	86.8	15.2
8	73.3	12.0
9	144.3	13.8
10	61.3	10.0
11	148.0	14.5
12	85.0	10.2
13	171.2	18.7
14	97.5	13.2
15	158.1	16.3
16	74.2	12.3
17	47.0	7.7
18	54.7	9.9
19	68.0	11.2
20	75.2	12.4

condemnation proceedings so that citizens could buy their own land. To comply with the 1980 law, one large Hawaiian estate wanted to use regression analysis to estimate the fair market value of its land. Its first proposal was the quadratic model

$$E(y) = \beta_0 + \beta_1 x + \beta_2 x^2$$

where

y = Leased fee value (i.e., sale price of property)

x = Size of property in square feet

Data collected for 20 property sales in a particular neighborhood, given in the table above, were

---

[§] For example, consider fitting the absolute values of the residuals as a function of the independent variables in the model, that is, fit the regression model $E\{|\hat{\varepsilon}|\} = \beta_0 + \beta_1 x_1 + \beta_2 x_2 + \cdots + \beta_k x_k$. A nonsignificant global F implies that the assumption of homoscedasticity is satisfied. A significant F, however, indicates that changing the values of the x's will lead to a larger (or smaller) residual variance.

used to fit the model. The least squares prediction equation is

$$\hat{y} = -44.0947 + 11.5339x - .06378x^2$$

(a) Calculate the predicted values and corresponding residuals for the model.
(b) Plot the residuals versus $\hat{y}$. Do you detect any trends? If so, what does the pattern suggest about the model?
(c) Conduct a test for heteroscedasticity. [*Hint*: Divide the data into two subsamples, $x \le 12$ and $x > 12$, and fit the model to both subsamples.]
(d) Based on your results from parts b and c, how should the estate proceed?

8.13 Assembly line breakdowns. Breakdowns of machines that produce steel cans are very costly. The more breakdowns, the fewer cans produced, and the smaller the company's profits. To help anticipate profit loss, the owners of a can company would like to find a model that will predict the number of breakdowns on the assembly line. The model proposed by the company's statisticians is the following:

$$y = \beta_0 + \beta_1 x_1 + \beta_2 x_2 + \beta_3 x_3 + \beta_4 x_4 + \varepsilon$$

where y is the number of breakdowns per 8-hour shift,

$$x_1 = \begin{cases} 1 & \text{if afternoon shift} \\ 0 & \text{otherwise} \end{cases} \quad x_2 = \begin{cases} 1 & \text{if midnight shift} \\ 0 & \text{otherwise} \end{cases}$$

x_3 is the temperature of the plant (°F), and x_4 is the number of inexperienced personnel working on the assembly line. After the model is fit using the least squares procedure, the residuals are plotted against $\hat{y}$, as shown in the accompanying figure.

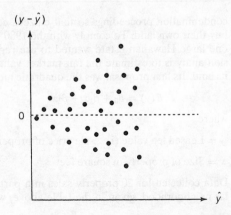

$(y - \hat{y})$

0

$\hat{y}$

(a) Do you detect a pattern in the residual plot? What does this suggest about the least squares assumptions?

(b) Given the nature of the response variable y and the pattern detected in part a, what model adjustments would you recommend?

8.14 Assembly line breakdowns (cont'd) Refer to Exercise 8.13. The regression analysis for the transformed model

$$y^* = \sqrt{y} = \beta_0 + \beta_1 x_1 + \beta_2 x_2 + \beta_3 x_3 + \beta_4 x_4 + \varepsilon$$

produces the prediction equation

$$\hat{y}^* = 1.3 + .008x_1 - .13x_2 + .0025x_3 + .26x_4$$

(a) Use the equation to predict the number of breakdowns during the midnight shift if the temperature of the plant at that time is 87°F and if there is only one inexperienced worker on the assembly line.
(b) A 95% prediction interval for y^* when $x_1 = 0$, $x_2 = 0$, $x_3 = 90°F$, and $x_4 = 2$ is (1.965, 2.125). For those same values of the independent variables, find a 95% prediction interval for y, the number of breakdowns per 8-hour shift.
(c) A 95% confidence interval for $E(y^*)$ when $x_1 = 0$, $x_2 = 0$, $x_3 = 90°F$, and $x_4 = 2$ is (1.987, 2.107). Using only the information given in this problem, is it possible to find a 95% confidence interval for $E(y)$? Explain.

8.15 Purchasing a service contract. The manager of a retail appliance store wants to model the proportion of appliance owners who decide to purchase a service contract for a specific major appliance. Since the manager believes that the proportion y decreases with age x of the appliance (in years), he will fit the first-order model

$$E(y) = \beta_0 + \beta_1 x$$

A sample of 50 purchasers of new appliances are contacted about the possibility of purchasing a service contract. Fifty owners of 1-year-old machines and 50 owners each of 2-, 3-, and 4-year-old machines are also contacted. One year later, another survey is conducted in a similar manner. The proportion y of owners deciding to purchase the service policy is shown in the table.

🔵 APPLIANCE

Age of Appliance x, years	0	0	1	1	2	2	3	3	4	4
Proportion Buying Service Contract, y	.94	.96	.7	.76	.6	.4	.24	.3	.12	.1

(a) Fit the first-order model to the data.
(b) Calculate the residuals and construct a residual plot versus $\hat{y}$.
(c) What does the plot from part b suggest about the variance of y?
(d) Explain how you could stabilize the variances.

(e) Refit the model using the appropriate variance-stabilizing transformation. Plot the residuals for the transformed model and compare to the plot obtained in part b. Does the assumption of homoscedasticity appear to be satisfied?

8.5 Checking the Normality Assumption

Recall from Section 4.2 that all the inferential procedures associated with a regression analysis are based on the assumptions that, for any setting of the independent variables, the random error ε is normally distributed with mean 0 and variance σ^2, and all pairs of errors are independent. Of these assumptions, the normality assumption is the least restrictive when we apply regression analysis in practice. That is, moderate departures from the assumption of normality have very little effect on Type I error rates associated with the statistical tests and on the confidence coefficients associated with the confidence intervals.

Although tests are available to check the normality assumption (see, e.g., Stephens, 1974), these tests tend to have low *power* when the assumption is violated.* Consequently, we discuss only graphical techniques in this section. The simplest way to determine whether the data violate the assumption of normality is to construct a frequency or relative frequency distribution for the residuals using the computer. If this distribution is not badly skewed, you can feel reasonably confident that the measures of reliability associated with your inferences are as stated in Chapter 4. This visual check is not foolproof because we are lumping the residuals together for all settings of the independent variables. It is conceivable (but not likely) that the distribution of residuals might be skewed to the left for some values of the independent variables and skewed to the right for others. Combining these residuals into a single relative frequency distribution could produce a distribution that is relatively symmetric. But, as noted above, we think that this situation is unlikely and that this graphical check is very useful.

To illustrate, consider the $n = 50$ residuals obtained from the model of ln(salary) in Example 8.6.[†] A MINITAB histogram and stem-and-leaf plot of the residuals are shown in Figures 8.20 and 8.21, respectively. Both graphs show that the distribution

Figure 8.20 MINITAB histogram of residuals from log model of salary

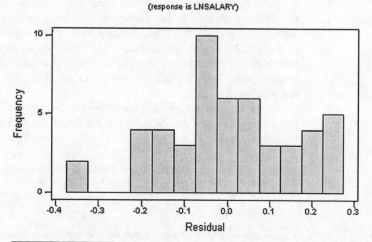

Histogram of the Residuals
(response is LNSALARY)

---
* The *power* of a test is defined as P(Reject $H_0|H_0$ false). Tests for normality test the null hypothesis H_0: errors are normally distributed. Thus, low power implies that the chance of detecting nonnormal errors when, in fact, they exist is low.

[†] A printout of the residuals is omitted.

Figure 8.21 MINITAB
stem-and-leaf plot of
residuals from log model
of salary

```
Stem-and-leaf of RESIDUAL   N  = 50
Leaf Unit = 0.010

     1    -3 5
     2    -3 3
     2    -2
     5    -2 000
    10    -1 86665
    12    -1 11
    18    -0 976655
    (8)   -0 44442221
    24     0 0122344
    17     0 5778
    13     1 044
    10     1 688
```

is mound-shaped and reasonably symmetric. Consequently, it is unlikely that the normality assumption would be violated using these data.

A third graphical technique for checking the assumption of normality is to construct a **normal probability plot**. In a normal probability plot, the residuals are graphed against the expected values of the residuals under the assumption of normality. (These expected values are sometimes called **normal scores**.) When the errors are, in fact, normally distributed, a residual value will approximately equal its expected value. Thus, a linear trend on the normal probability plot suggests that the normality assumption is nearly satisfied, whereas a nonlinear trend indicates that the assumption is most likely violated.

Most statistical software packages have procedures for constructing normal probability plots.*

Figure 8.22 shows the MINITAB normal probability plot for the residuals of Example 8.6. Notice that the points fall reasonably close to a straight line, indicating that the normality assumption is most likely satisfied.

Nonnormality of the distribution of the random error ε is often accompanied by heteroscedasticity. Both these situations can frequently be rectified by applying the variance-stabilizing transformations of Section 8.4. For example, if the relative frequency distribution (or stem-and-leaf display) of the residuals is highly skewed to the right (as it would be for Poisson data), the square-root transformation on y will stabilize (approximately) the variance and, at the same time, will reduce skewness in the distribution of the residuals. Thus, for any given setting of the independent variables, the square-root transformation will reduce the larger values of y to a greater extent than the smaller ones. This has the effect of reducing or eliminating the positive skewness.

For situations in which the errors are homoscedastic but nonnormal, normalizing transformations are available. This family of transformations on the dependent variable includes $\sqrt{y}$ and $\log(y)$ (Section 8.4), as well as such simple transformations

---

* To find normal scores for residuals without the aid of statistical software, first list the residuals in ascending order, where $\hat{\varepsilon}_i$ represents the ith ordered residual. Then, for each residual, calculate the corresponding tail area (of the standard normal distribution),

$$A = \frac{i - .375}{n + .25}$$

where n is the sample size. Finally, calculate the estimated value of $\hat{\varepsilon}_i$ under normality (i.e., the normal score) using the following formula:

$$E(\hat{\varepsilon}_i) \approx \sqrt{\text{MSE}}[Z(A)]$$

where MSE = mean square error for the fitted model and $Z(A)$ = value of the standard normal distribution (z-value) that cuts off an area of A in the lower tail of the distribution.

Figure 8.22 MINITAB normal probability plot of residuals from log model of salary

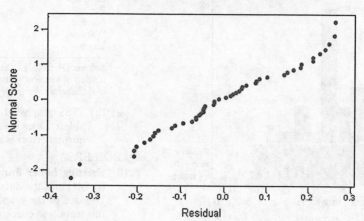

Normal Probability Plot of the Residuals

(response is LNSALARY)

as y^2, $1/\sqrt{y}$, and $1/y$. Box and Cox (1964) have developed a procedure for selecting the appropriate transformation to use. Consult the references to learn details of the Box–Cox approach.

Keep in mind that regression is *robust* with respect to nonnormal errors. That is, the inferences derived from the regression analysis tend to remain valid even when the assumption of normal errors is not exactly satisfied. However, if the errors come from a "heavy-tailed" distribution, the regression results may be sensitive to a small subset of the data and lead to invalid inferences. Consequently, you may want to search for a normalizing transformation only when the distribution of the regression residuals is highly skewed.

8.5 Exercises

8.16 Nonnormal residuals and robustness. What does it mean to say that least squares regression is robust in the presence of nonnormal residuals?

8.17 Populations of worldcities. In *Economic Development and Cultural Change*, L. De Cola conducted an extensive investigation of the geopolitical and socioeconomic processes that shape the urban size distributions of the world's nations. One of the goals of the study was to determine the factors that influence population size in each nation's largest city. Using data collected for a sample of 126 countries, De Cola fit the following log model:

$$E(y) = \beta_0 + \beta_1 x_1 + \beta_2 x_2 + \beta_3 x_3 + \beta_4 x_4$$
$$+ \beta_5 x_5 + \beta_6 x_6 + \beta_7 x_7 + \beta_8 x_8 + \beta_9 x_9 + \beta_{10} x_{10}$$

where

y = Log of population (in thousands) of largest city in country

x_1 = Log of area (in thousands of square kilometers) of country

x_2 = Log of radius (in hundreds of kilometers) of country

x_3 = Log of national population (in thousands)

x_4 = Percentage annual change in national population (1960–1970)

x_5 = Log of energy consumption per capita (in kilograms of coal equivalent)

x_6 = Percentage of nation's population in urban areas

x_7 = Log of population (in thousands) of second largest city in country

$x_8 = \begin{cases} 1 & \text{if seaport city} \\ 0 & \text{if not} \end{cases}$

$x_9 = \begin{cases} 1 & \text{if capital city} \\ 0 & \text{if not} \end{cases}$

$x_{10} = \begin{cases} 1 & \text{if city data are for metropolitan area} \\ 0 & \text{if not} \end{cases}$

[*Note:* All logarithms are to the base 10.]

SPSS Output for Exercise 8.17

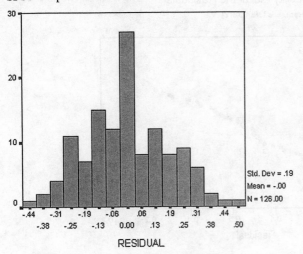

Normal P-P Plot of RESIDUAL

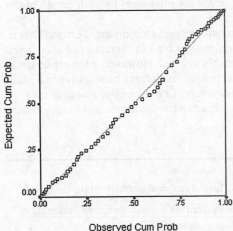

CITY	RESIDUAL	RANK
Bangkok	.510	126
Paris	.228	110
London	.033	78
Warsaw	−.132	32
Lagos	−.392	2

Source: De Cola, L. "Statistical determinants of the population of a nation's largest city," *Economic Development and Cultural Change*, Vol. 3. No. 1, Oct. 1984, pp. 71–98.

(b) SPSS graphs of all the residuals are shown at left. Does it appear that the assumption of normal errors is satisfied?

🖸 BOILERS

8.18 Erecting boiler drums. Refer to Exercise 4.74. (p. 241) and the data saved in the BOILERS file. Use one of the graphical techniques described in this section to check the normality assumption.

🖸 ASWELLS

8.19 Arsenic in groundwater. Refer to the *Environmental Science and Technology* (January 2005) study of the reliability of a commercial kit to test for arsenic in groundwater, Exercise 8.10 (p. 407). Use a residual graph to check the assumption of normal errors for the first-order model for arsenic level (y). Is the normality assumption reasonably satisfied? If not, suggest how to modify the model.

🖸 GASTURBINE

8.20 Cooling method for gas turbines. Refer to the *Journal of Engineering for Gas Turbines and Power* (January 2005) study of a high-pressure inlet fogging method for a gas turbine engine, Exercise 8.11 (p. 407). Use a residual graph to check the assumption of normal errors for the interaction model for heat rate (y). Is the normality assumption reasonably satisfied? If not, suggest how to modify the model.

🖸 HAWAII

8.21 Fair market value of Hawaiian properties. Refer to Exercise 8.12. (p. 407) and the data saved in the HAWAII file. Use one of the graphical techniques described in this section to check the normality assumption.

(a) The residuals for five cities selected from the total sample are given in the next table. Select one of these cities and calculate the estimated expected residual under the assumption of normality. (See the footnote, p. 410.). Assume MSE = .01.

8.6 Detecting Outliers and Identifying Influential Observations

We begin this section by defining a **standardized residual** as the value of the residual divided by the model standard deviation s. Since we assume that the residuals have a mean of 0 and a standard deviation estimated by s, you can see from Definition 8.3 that a standardized residual is simply the z-score for a residual (see Section 1.6).

Definition 8.3 The **standardized residual**, denoted z_i, for the ith observation is the residual for the observation divided by s, that is,
$$z_i = \hat{\varepsilon}_i/s = (y_i - \hat{y}_i)/s$$

Although we expect almost all the regression residuals to fall within three standard deviations of their mean of 0, sometimes one or several residuals fall outside this interval. Observations with residuals that are extremely large or small (say, more than 3 standard deviations from 0) are called **outliers**. Consequently, observations with standardized residuals that exceed 3 in absolute value are considered outliers.

> **Definition 8.4** An observation with a residual that is larger than 3s (in absolute value)—or, equivalently, a standardized residual that is larger than 3 (in absolute value)—is considered to be an **outlier**.

Note: As an alternative to standardized residuals, some software packages compute **studentized residuals**, so named because they follow an approximate Student's *t*-distribution.

> **Definition 8.5** The **studentized residual**, denoted z_i^*, for the ith observation is
> $$z_i^* = \frac{\hat{\varepsilon}_i}{s\sqrt{1-h_i}} = \frac{(y_i - \hat{y}_i)}{s\sqrt{1-h_i}}$$
> where h_i (called *leverage*) is defined in Definition 8.6 (pg. 418).

Outliers are usually attributable to one of several causes. The measurement associated with the outlier may be invalid. For example, the experimental procedure used to generate the measurement may have malfunctioned, the experimenter may have misrecorded the measurement, or the data may have been coded incorrectly for entry into the computer. Careful checks of the experimental and coding procedures should reveal this type of problem if it exists, so that we can eliminate erroneous observations from a data set, as the next example illustrates.

Example 8.8

Table 8.5 presents the sales, y, in thousands of dollars per week, for fast-food outlets in each of four cities. The objective is to model sales, y, as a function of traffic flow, adjusting for city-to-city variations that might be due to size or other market conditions. We expect a first-order (linear) relationship to exist between mean sales, $E(y)$, and traffic flow. Furthermore, we believe that the level of mean sales will differ from city to city, but that the change in mean sales per unit increase in traffic flow

FASTFOOD

Table 8.5 Data for fast-food sales

City	Traffic flow thousands of cars	Weekly sales y, thousands of dollars	City	Traffic flow thousands of cars	Weekly sales y, thousands of dollars
1	59.3	6.3	3	75.8	8.2
1	60.3	6.6	3	48.3	5.0
1	82.1	7.6	3	41.4	3.9
1	32.3	3.0	3	52.5	5.4
1	98.0	9.5	3	41.0	4.1
1	54.1	5.9	3	29.6	3.1
1	54.4	6.1	3	49.5	5.4
1	51.3	5.0	4	73.1	8.4
1	36.7	3.6	4	81.3	9.5
2	23.6	2.8	4	72.4	8.7
2	57.6	6.7	4	88.4	10.6
2	44.6	5.2	4	23.2	3.3

will remain the same for all cities (i.e., that the factors Traffic Flow and City do not interact). The model is therefore

$$E(y) = \beta_0 + \beta_1 x_1 + \beta_2 x_2 + \beta_3 x_3 + \beta_4 x_4$$

where $x_1 = \begin{cases} 1 & \text{if city 1} \\ 0 & \text{other} \end{cases}$ $x_2 = \begin{cases} 1 & \text{if city 2} \\ 0 & \text{other} \end{cases}$

$x_3 = \begin{cases} 1 & \text{if city 3} \\ 0 & \text{other} \end{cases}$ $x_4 = \text{Traffic flow}$

(a) Fit the model to the data and evaluate overall model adequacy.

(b) Plot the residuals from the model to check for any outliers.

(c) Based on the results, part b, make the necessary model modifications and reevaluate model fit.

Figure 8.23 MINITAB regression printout for model of fast-food sales

```
The regression equation is
SALES = - 16.5 + 1.11 X1 + 6.1 X2 + 14.5 X3 + 0.363 TRAFFIC

Predictor       Coef     SE Coef        T        P
Constant      -16.46       13.16    -1.25    0.226
X1             1.106        8.423     0.13    0.897
X2             6.14        11.68      0.53    0.605
X3            14.490        9.288     1.56    0.135
TRAFFIC        0.3629       0.1679    2.16    0.044

S = 14.86      R-Sq = 25.9%      R-Sq(adj) = 10.4%

Analysis of Variance

Source          DF        SS        MS        F        P
Regression       4    1469.8     367.4     1.66    0.200
Residual Error  19    4194.2     220.7
Total           23    5664.0

Obs    TRAFFIC    SALES       Fit    SE Fit   Residual   St Resid
  1      59.3      6.30      6.17      4.95       0.13       0.01
  2      60.3      6.60      6.53      4.96       0.07       0.01
  3      82.1      7.60     14.44      6.32      -6.84      -0.51
  4      32.3      3.00     -3.63      6.65       6.63       0.50
  5      98.0      9.50     20.21      8.25     -10.71      -0.87
  6      54.1      5.90      4.28      5.01       1.62       0.12
  7      54.4      6.10      4.39      5.01       1.71       0.12
  8      51.3      5.00      3.26      5.11       1.74       0.12
  9      36.7      3.60     -2.04      6.18       5.64       0.42
 10      23.6      2.80     -1.75      9.11       4.55       0.39
 11      57.6      6.70     10.59      8.97      -3.89      -0.33
 12      44.6      5.20      5.87      8.59      -0.67      -0.06
 13      75.8     82.00     25.54      7.27      56.46       4.36R
 14      48.3      5.00     15.56      5.62     -10.56      -0.77
 15      41.4      3.90     13.05      5.73      -9.15      -0.67
 16      52.5      5.40     17.08      5.66     -11.68      -0.85
 17      41.0      4.10     12.91      5.75      -8.81      -0.64
 18      29.6      3.10      8.77      6.43      -5.67      -0.42
 19      49.5      5.40     15.99      5.62     -10.59      -0.77
 20      73.1      8.40     10.07      6.71      -1.67      -0.13
 21      81.3      9.50     13.04      7.03      -3.54      -0.27
 22      72.4      8.70      9.81      6.69      -1.11      -0.08
 23      88.4     10.60     15.62      7.50      -5.02      -0.39
 24      23.2      3.30     -8.04     10.00      11.34       1.03

R denotes an observation with a large standardized residual
```

Solution

(a) The MINITAB printout for the regression analysis is shown in Figure 8.23. The regression analysis indicates that the first-order model in traffic flow is inadequate for explaining mean sales. The coefficient of determination, R^2 (highlighted), indicates that only 25.9% of the total sum of squares of deviations of the sales y about their mean $\bar{y}$ is accounted for by the model. The global F-value, 1.66 (also shaded), does not indicate that the model is useful for predicting sales. The observed significance level is only .200.

(b) Plots of the studentized residuals against traffic flow and city are shown in Figures 8.24 and 8.25, respectively. The dashed horizontal line locates the mean (0) for the residuals. As you can see, the plots of the residuals are very revealing. Both the plot of residuals against traffic flow in Figure 8.24 and the plot of residuals against city in Figure 8.25 indicate the presence of an outlier. One observation in city 3 (observation 13), with traffic flow of 75.8, has a studentized residual value of 4.36. (This observation is shaded on the MINITAB printout in Figure 8.23). A further check of the observation associated with this residual reveals that the sales value entered into the computer, 82.0, does not agree with the corresponding value of sales, 8.2, that

Figure 8.24 MINITAB plot of residuals versus traffic flow

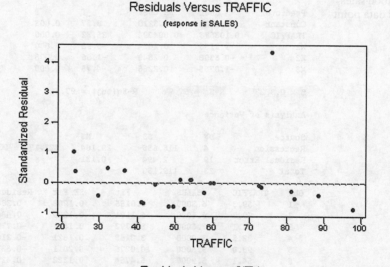

Figure 8.25 MINITAB plot of residuals versus city

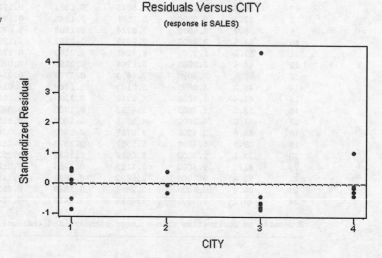

appears in Table 8.5. The decimal point was evidently dropped when the data were entered into MINITAB.

(c) If the correct y-value, 8.2, is substituted for the 82.0, we obtain the regression analysis shown in Figure 8.26. Plots of the studentized residuals against traffic flow and city are respectively shown in Figures 8.27 and 8.28. The corrected MINITAB printout indicates the dramatic effect that a single outlier can have on a regression analysis. The value of R^2 is now .979, and the F-value that tests the adequacy of the model, 222.17, verifies the strong predictive capability of the model. Further analysis reveals that significant differences exist in the mean sales among cities, and that the estimated mean weekly sales increase by \$104 for every 1,000-car increase in traffic flow ($\hat{\beta}_4 = .104$). The 95% confidence interval for β_4 is

$$\hat{\beta}_4 \pm t_{.025}s_{\hat{\beta}_4} = .104 \pm (2.093)(.0041) = .104 \pm .009$$

Thus, a 95% confidence interval for the mean increase in sales per 1,000-car increase in traffic flow is \$95 to \$113.

Figure 8.26 MINITAB regression printout for model of fast-food sales with corrected data point

```
The regression equation is
SALES = 1.08 + 0.104 TRAFFIC - 1.22 X1 - 0.531 X2 - 1.08 X3

Predictor        Coef      SE Coef          T        P
Constant       1.0834       0.3210       3.37    0.003
TRAFFIC      0.103673     0.004094      25.32    0.000
X1            -1.2158       0.2054      -5.92    0.000
X2            -0.5308       0.2848      -1.86    0.078
X3            -1.0765       0.2265      -4.75    0.000

S = 0.3623       R-Sq = 97.9%      R-Sq(adj) = 97.5%

Analysis of Variance

Source           DF           SS          MS        F        P
Regression        4      116.656      29.164   222.17    0.000
Residual Error   19        2.494       0.131
Total            23      119.150

Obs    TRAFFIC      SALES         Fit      SE Fit    Residual   St Resid
  1       59.3     6.3000      6.0155      0.1208     0.2845       0.83
  2       60.3     6.6000      6.1191      0.1209     0.4809       1.41
  3       82.1     7.6000      8.3792      0.1541    -0.7792      -2.38R
  4       32.3     3.0000      3.2163      0.1621    -0.2163      -0.67
  5       98.0     9.5000     10.0276      0.2011    -0.5276      -1.75
  6       54.1     5.9000      5.4764      0.1222     0.4236       1.24
  7       54.4     6.1000      5.5075      0.1221     0.5925       1.74
  8       51.3     5.0000      5.1861      0.1245    -0.1861      -0.55
  9       36.7     3.6000      3.6724      0.1507    -0.0724      -0.22
 10       23.6     2.8000      2.9993      0.2222    -0.1993      -0.70
 11       57.6     6.7000      6.5242      0.2188     0.1758       0.61
 12       44.6     5.2000      5.1765      0.2095     0.0235       0.08
 13       75.8     8.2000      7.8653      0.1773     0.3347       1.06
 14       48.3     5.0000      5.0143      0.1369    -0.0143      -0.04
 15       41.4     3.9000      4.2989      0.1398    -0.3989      -1.19
 16       52.5     5.4000      5.4497      0.1380    -0.0497      -0.15
 17       41.0     4.1000      4.2575      0.1402    -0.1575      -0.47
 18       29.6     3.1000      3.0756      0.1569     0.0244       0.07
 19       49.5     5.4000      5.1387      0.1370     0.2613       0.78
 20       73.1     8.4000      8.6619      0.1635    -0.2619      -0.81
 21       81.3     9.5000      9.5120      0.1714    -0.0120      -0.04
 22       72.4     8.7000      8.5893      0.1632     0.1107       0.34
 23       88.4    10.6000     10.2481      0.1829     0.3519       1.13
 24       23.2     3.0000      3.4886      0.2438    -0.1886      -0.70

R denotes an observation with a large standardized residual
```

Figure 8.27 MINITAB plot of residuals versus traffic flow for model with corrected data point

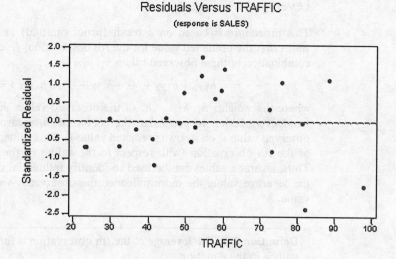

Figure 8.28 MINITAB plot of residuals versus city for model with corrected data point

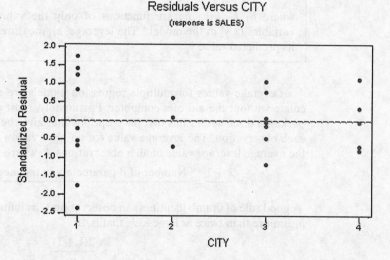

Outliers cannot always be explained by data entry or recording errors. Extremely large or small residuals may be attributable to skewness (nonnormality) of the probability distribution of the random error, chance, or unassignable causes. Although some analysts advocate elimination of outliers, regardless of whether cause can be assigned, others encourage the correction of only those outliers that can be traced to specific causes. The best philosophy is probably a compromise between these extremes. For example, before deciding the fate of an outlier, you may want to determine how much influence it has on the regression analysis. When an accurate outlier (i.e., an outlier that is not due to recording or measurement error) is found to have a dramatic effect on the regression analysis, it may be the model and not the outlier that is suspect. Omission of important independent variables or higher-order terms could be the reason why the model is not predicting well for the outlying observation. Several sophisticated numerical techniques are available for identifying outlying influential observations. We conclude this section with a brief discussion of some of these methods and an example.

Leverage

This procedure is based on a result (proof omitted) in regression analysis that states that the predicted value for the ith observation, $\hat{y}_i$, can be written as a linear combination of the n observed values $y_1, y_2, \ldots, y_n$:

$$\hat{y}_i = h_1 y_1 + h_2 y_2 + \cdots + h_i y_i + \cdots + h_n y_n, \ i = 1, 2, \ldots, n$$

where the weights $h_1, h_2, \ldots, h_n$ of the observed values are functions of the independent variables. In particular, the coefficient h_i measures the influence of the observed value y_i on its own predicted value $\hat{y}_i$. This value, h_i, is called the **leverage** of the ith observation (with respect to the values of the independent variables). Thus, leverage values can be used to identify influential observations—the larger the leverage value, the more influence the observed y-value has on its predicted value.

Definition 8.6 The **leverage** of the ith observation is the weight, h_i, associated with y_i in the equation

$$\hat{y}_i = h_1 y_1 + h_2 y_2 + h_3 y_3 + \cdots + h_i y_i + \cdots + h_n y_n$$

where $h_1, h_2, h_3 \ldots h_n$ are functions of only the values of the independent variables (x's) in the model.* The leverage, h_i, measures the influence of y_i on its predicted value $\hat{y}_i$.

Leverage values for multiple regression models are extremely difficult to calculate without the aid of a computer. Fortunately, most of the statistical software packages discussed in this text have options that give the leverage associated with each observation. The leverage value for an observation is usually compared with the average leverage value of all n observations, $\bar{h}$, where

$$\bar{h} = \frac{k+1}{n} = \frac{\text{Number of } \beta \text{ parameters in the model, including } \beta_0}{n}$$

A good rule of thumb identifies[†] an observation y_i as influential if its leverage value h_i is more than twice as large as $\bar{h}$, that is, if

$$h_i > \frac{2(k+1)}{n}$$

Rule of Thumb for Detecting Influence with Leverage
The observed value of y_i is influential if

$$h_i > \frac{2(k+1)}{n}$$

where h_i is the leverage for the ith observation and $k =$ the number of β's in the model (excluding β_0).

---

*In matrix notation, the leverage values are the diagonals of the H matrix (called the "hat" matrix), where $H = X(X'X)^{-1}X'$. See Appendix A for details on matrix multiplication and definition of the X matrix in regression.
† The proof of this result is beyond the scope of this text. Consult the references given at the end of this chapter. [See Kutner, Nachtsheim, Neter, and Li (2005).]

The Jackknife

Another technique for identifying influential observations requires that you delete the observations one at a time, each time refitting the regression model based on only the remaining $n - 1$ observations. This method is based on a statistical procedure called the **jackknife**,* which in gaining increasing acceptance among practitioners. The basic principle of the jackknife when applied to regression is to compare the regression results using all n observations to the results with the ith observation deleted to ascertain how much influence a particular observation has on the analysis. Using the jackknife, several alternative influence measures can be calculated.

The **deleted residual**, $d_i = y_i - \hat{y}_{(i)}$, measures the difference between the observed value y_i and the predicted value $\hat{y}_{(i)}$ based on the model with the ith observation deleted. [The notation (i) is generally used to indicate that the observed value y_i was deleted from the regression analysis.] An observation with an unusually large (in absolute value) deleted residual is considered to have large influence on the fitted model.

> **Definition 8.7** A **deleted residual**, denoted d_i, is the difference between the observed response y_i and the predicted value $\hat{y}_{(i)}$ obtained when the data for the ith observation is deleted from the analysis, that is,
>
> $$d_i = y_i - \hat{y}_{(i)}$$

A measure closely related to the deleted residual is the difference between the predicted value based on the model fit to all n observations and the predicted value obtained when y_i is deleted [i.e., $\hat{y}_i - \hat{y}_{(i)}$]. When the difference $\hat{y}_i - \hat{y}_{(i)}$ is large relative to the predicted value $\hat{y}_i$, the observation y_i is said to influence the regression fit.

A third way to identify an influential observation using the jackknife is to calculate, for each β parameter in the model, the difference between the parameter estimate based on all n observations and that based on only $n - 1$ observations (with the observation in question deleted). Consider, for example, the straightline model $E(y) = \beta_0 + \beta_1 x$. The differences $\hat{\beta}_0 - \hat{\beta}_0^{(i)}$ and $\hat{\beta}_1 - \hat{\beta}_1^{(i)}$ measure how influential the ith observation y_i is on the parameter estimates. [Using the (i) notation defined previously, $\hat{\beta}^{(i)}$ represents the estimate of the β coefficient when the ith observation is omitted from the analysis.] If the parameter estimates change drastically (i.e., if the absolute differences are large), y_i is deemed an influential observation.

Each of the statistical software packages discussed in this text has a jackknife routine that produces one or more of the measures we described.

Cook's Distance:

A measure of the overall influence an outlying observation has on the estimated β coefficients was proposed by R. D. Cook (1979). Cook's distance, D_i, is calculated for the ith observation as follows:

$$D_i = \frac{(y_i - \hat{y}_i)^2}{(k + 1)\text{MSE}} \left[\frac{h_i}{(1 - h_i)^2} \right]$$

---

* The procedure derives its name from the Boy Scout jackknife, which serves as a handy tool in a variety of situations when specialized techniques may not be applicable. [See Belsley, Kuh, and Welsch (2004).]

Note that D_i depends on both the residual $(y_i - \hat{y}_i)$ and the leverage h_i for the ith observation. Although not obvious from the formula, D_i is a summary measure of the distances between $\hat{\beta}_0$ and $\hat{\beta}_0^{(i)}$, $\hat{\beta}_1$ and $\hat{\beta}_1^{(i)}$, $\hat{\beta}_2$ and $\hat{\beta}_2^{(i)}$, and so on. A large value of D_i indicates that the observed y_i value has strong influence on the estimated β coefficients (since the residual, the leverage, or both will be large). Values of D_i can be compared to the values of the F distribution with $v_1 = k + 1$ and $v_2 = n - (k + 1)$ degrees of freedom. Usually, an observation with a value of D_i that falls at or above the 50th percentile of the F distribution is considered to be an influential observation. Like the other numerical measures of influence, options for calculating Cook's distance are available in most statistical software packages.

Example 8.9   We now return to the fast-food sales model, Example 8.8, in which we detected an outlier using residual plots. Recall that the outlier was due to an error in coding the weekly sales value for observation 13 (denoted y_{13}). The SAS regression analysis is run with options for producing influence diagnostics. (An **influence diagnostic** is a number that measures how much influence an observation has on the regression analysis.) The resulting SAS printout is shown in Figure 8.29. Locate and interpret the measures of influence for y_{13} on the printout.

Solution
The influence diagnostics are shown in the bottom portion of the SAS printout in Figure 8.29. Leverage values for each observation are given under the column heading **Hat Diag H**. The leverage value for y_{13} (shaded on the printout) is $h_{13} = .2394$, whereas the average leverage for all $n = 24$ observations is

$$\bar{h} = \frac{k + 1}{n}$$

$$= \frac{5}{24}$$

$$= .2083$$

Since the leverage value .2394 does not exceed $2\bar{h} = .4166$, we would not identify y_{13} as an influential observation. At first, this result may seem confusing since we already know the dramatic effect the incorrectly coded value of y_{13} had on the regression analysis. Remember, however, that the leverage values, $h_1, h_2, \ldots, h_{24}$, are functions of the independent variables only. Since we know the values of x_1, x_2, x_3, and x_4 were coded correctly, the relatively small leverage value of .2394 simply indicates that observation 13 is not an outlier with respect to the values of the independent variables.

A better overall measure of the influence of y_{13} on the fitted regression model is Cook's distance, D_{13}. Recall that Cook's distance is a function of both leverage and the magnitude of the residual. This value, $D_{13} = 1.196$ (shaded), is given in the column labeled **Cook's D** located on the right side of the printout. You can see that the value is extremely large relative to the other values of D_i in the printout. [In fact, $D_{13} = 1.196$ falls in the 65th percentile of the F distribution with $v_1 = k + 1 = 5$ and $v_2 = n - (k + 1) = 24 - 5 = 19$ degrees of freedom.] This implies that the observed value y_{13} has substantial influence on the estimates of the model parameters.

A statistic related to the deleted residual of the jackknife procedure is the **Studentized deleted residual** given under the column heading **Rstudent**.

Figure 8.29 SAS regression analysis with influence diagnostics for fast-food sales model

Dependent Variable: SALES

Analysis of Variance

Source	DF	Sum of Squares	Mean Square	F Value	Pr > F
Model	4	1469.76287	367.44072	1.66	0.1996
Error	19	4194.22671	220.74877		
Corrected Total	23	5663.98958			

Root MSE	14.85762	R-Square	0.2595	
Dependent Mean	9.07083	Adj R-Sq	0.1036	
Coeff Var	163.79550			

Parameter Estimates

Variable	DF	Parameter Estimate	Standard Error	t Value	Pr > |t|
Intercept	1	-16.45925	13.16400	-1.25	0.2264
X1	1	1.10609	8.42257	0.13	0.8969
X2	1	6.14277	11.67997	0.53	0.6050
X3	1	14.48962	9.28839	1.56	0.1353
TRAFFIC	1	0.36287	0.16791	2.16	0.0437

Output Statistics

Obs	Dep Var SALES	Predicted Value	Std Error Mean Predict	Residual	Std Error Residual	Student Residual	-2-1 0 1 2	Cook's D
1	6.3000	6.1652	4.9535	0.1348	14.008	0.00962	| |	0.000
2	6.6000	6.5281	4.9596	0.0719	14.005	0.00513	| |	0.000
3	7.6000	14.4387	6.3195	-6.8387	13.447	-0.509	| *|	0.011
4	3.0000	-3.6324	6.6491	6.6324	13.287	0.499	| |	0.012
5	9.5000	20.2084	8.2476	-10.7084	12.358	-0.866	| *|	0.067
6	5.9000	4.2783	5.0130	1.6217	13.986	0.116	| |	0.000
7	6.1000	4.3871	5.0054	1.7129	13.989	0.122	| |	0.000
8	5.0000	3.2622	5.1069	1.7378	13.952	0.125	| |	0.000
9	3.6000	-2.0357	6.1807	5.6357	13.511	0.417	| |	0.007
10	2.8000	-1.7527	9.1137	4.5527	11.734	0.388	| |	0.018
11	6.7000	10.5850	8.9723	-3.8850	11.843	-0.328	| |	0.012
12	5.2000	5.8677	8.5897	-0.6677	12.123	-0.0551	| |	0.000
13	82.0000	25.5362	7.2703	56.4638	12.957	4.358	| ******|	1.196
14	5.0000	15.5571	5.6157	-10.5571	13.755	-0.767	| *|	0.020
15	3.9000	13.0533	5.7339	-9.1533	13.707	-0.668	| *|	0.016
16	5.4000	17.0812	5.6598	-11.6812	13.737	-0.850	| *|	0.025
17	4.1000	12.9082	5.7479	-8.8082	13.701	-0.643	| *|	0.015
18	3.1000	8.7714	6.4338	-5.6714	13.392	-0.423	| |	0.008
19	5.4000	15.9926	5.6193	-10.5926	13.754	-0.770	| *|	0.020
20	8.4000	10.0668	6.7066	-1.6668	13.258	-0.126	| |	0.001
21	9.5000	13.0423	7.0271	-3.5423	13.091	-0.271	| |	0.004
22	8.7000	9.8128	6.6916	-1.1128	13.265	-0.0839	| |	0.000
23	10.6000	15.6187	7.5002	-5.0187	12.826	-0.391	| |	0.010
24	3.3000	-8.0406	9.9965	11.3406	10.992	1.032	| **|	0.176

Obs	RStudent	Hat Diag H	Cov Ratio	DFFITS	DFBETAS Intercept	X1	X2	X3	TRAFFIC
1	0.009366	0.1112	1.4742	0.0033	-0.0001	0.0020	0.0000	0.0000	0.0001
2	0.004998	0.1114	1.4747	0.0018	-0.0001	0.0011	0.0000	0.0000	0.0001
3	-0.4984	0.1809	1.4939	-0.2342	0.1256	-0.1339	-0.0539	-0.0510	-0.1455
4	0.4891	0.2003	1.5339	0.2447	0.1410	0.0780	-0.0604	-0.0572	-0.1633
5	-0.8606	0.3081	1.5482	-0.5743	0.3965	-0.2848	-0.1700	-0.1609	-0.4592
6	0.1129	0.1138	1.4735	0.0405	0.0054	0.0224	-0.0023	-0.0022	-0.0063
7	0.1192	0.1135	1.4724	0.0427	0.0053	0.0237	-0.0023	-0.0022	-0.0062
8	0.1213	0.1181	1.4799	0.0444	0.0094	0.0234	-0.0040	-0.0038	-0.0108
9	0.4079	0.1731	1.5134	0.1866	0.0964	0.0680	-0.0413	-0.0391	-0.1116
10	0.3791	0.3763	2.0190	0.2945	0.0859	-0.0178	0.1667	-0.0348	-0.0995
11	-0.3202	0.3647	2.0048	-0.2426	0.0614	-0.0127	-0.1967	-0.0249	-0.0711
12	-0.0536	0.3342	1.9667	-0.0380	0.0017	-0.0004	-0.0286	-0.0007	-0.0020
13	179.3101	0.2394	0.0000	100.6096	-55.1618	11.4109	23.6508	69.3701	63.8990
14	-0.7589	0.1429	1.3061	-0.3098	-0.0000	-0.0000	0.0000	-0.1873	0.0000
15	-0.6578	0.1489	1.3673	-0.2752	-0.0480	0.0099	0.0206	-0.1435	0.0556
16	-0.8439	0.1451	1.2625	-0.3477	0.0374	-0.0077	-0.0160	-0.2237	-0.0433
17	-0.6327	0.1497	1.3806	-0.2654	-0.0489	0.0101	0.0209	-0.1370	0.0566
18	-0.4141	0.1875	1.5381	-0.1990	-0.0838	0.0173	0.0359	-0.0710	0.0971
19	-0.7616	0.1430	1.3049	-0.3111	0.0096	-0.0020	-0.0041	-0.1919	-0.0112
20	-0.1224	0.2038	1.6389	-0.0619	-0.0237	0.0469	0.0318	0.0409	-0.0084
21	-0.2639	0.2237	1.6557	-0.1417	-0.0278	0.0974	0.0591	0.0797	-0.0461
22	-0.0817	0.2028	1.6408	-0.0412	-0.0164	0.0314	0.0215	0.0276	-0.0049
23	-0.3824	0.2548	1.6888	-0.2236	-0.0104	0.1378	0.0743	0.1054	-0.1037
24	1.0336	0.4527	1.7946	0.9400	0.9216	-0.6183	-0.6154	-0.6930	-0.7023

Sum of Residuals	0
Sum of Squared Residuals	4194.22671
Predicted Residual SS (PRESS)	7303.48386

> **Definition 8.8** The **Studentized deleted residual**, denoted d_i^*, is calculated by dividing the deleted residual d_i by its standard error s_{d_i}:
>
> $$d_i^* = \frac{d_i}{s_{d_i}}$$

Under the assumptions of Section 4.2, the Studentized deleted residual d_i^* has a sampling distribution that is approximated by a Student's t distribution with $(n-1)-(k+1)$ df. Note that the Studentized deleted residual for y_{13} (shaded on the printout) is $d_{13}^* = 179.3101$. This extremely large value is another indication that y_{13} is an influential observation.

The **Dffits** column gives the difference between the predicted value when all 24 observations are used and when the ith observation is deleted. The difference, $\hat{y}_i - \hat{y}_{(i)}$, is divided by its standard error so that the differences can be compared more easily. For observation 13, this scaled difference (shaded on the printout) is 100.6096, an extremely large value relative to the other differences in predicted values. Similarly, the changes in the parameter estimates when observation 13 is deleted are given in the **Dfbetas** columns (shaded) immediately to the right of **Dffits** on the printout. (Each difference is also divided by the appropriate standard error.) The large magnitude of these differences provides further evidence that y_{13} is very influential on the regression analysis. ▪

Several techniques designed to limit the influence an outlying observation has on the regression analysis are available. One method produces estimates of the β's that minimize the sum of the absolute deviations, $\sum_{i=1}^n |y_i - \hat{y}_i|$.* Because the deviations $(y_i - \hat{y}_i)$ are not squared, this method places less emphasis on outliers than the method of least squares. Regardless of whether you choose to eliminate an outlier or dampen its influence, careful study of residual plots and influence diagnostics is essential for outlier detection.

8.6 Exercises

8.22 Influence diagnostics. Give three different methods of identifying influential observations in regression. Will each of the methods always detect the same influential observations? Explain.

8.23 Detecting outliers. Refer to the data and model in Exercise 8.1 (p. 396). The MSE for the model is .1267. Plot the residuals versus $\hat{y}$. Identify any outliers on the plot.

8.24 Detecting outliers. Refer to the data and model in Exercise 8.2 (p. 396). The MSE for the model is 17.2557. Plot the residuals versus $\hat{y}$. Identify any outliers on the plot.

🌐 QUASAR

8.25 Deep space survey of quasars. Refer to the *Astronomical Journal* study of quasars detected by a deep space survey, Exercise 4.11 (p. 186).

The following model was fit to data collected on 90 quasars:

$$E(y) = \beta_0 + \beta_1 x_1 + \beta_2 x_2 + \beta_3 x_3 + \beta_4 x_4$$

where y = rest frame equivalent width
x_1 = redshift
x_2 = line flux
x_3 = line luminosity
x_4 = AB$_{1450}$ magnitude

A portion of the SPSS spreadsheet showing influence diagnostics is shown on p. 423. Do you detect any influential observations?

🌐 GFCLOCKS

8.26 Prices of antique clocks. Refer to the grandfather clock example, Example 4.1 (p. 183). The least squares model used to predict auction price, y,

---

* The method of absolute deviations requires linear programming techniques that are beyond the scope of this text. Consult the references given at the end of the chapter for details on how to apply this method.

SPSS Output for Exercise 8.25

	quasar	zdelres	cooksd	leverage	zdffits
1	1	-1.49140	.11570	.17631	-.78353
2	2	-.71142	.00685	.02192	-.18278
3	3	2.79116	.40406	.21782	1.64508
4	4	-.77375	.07423	.33791	-.60306
5	5	-.72387	.01380	.07392	-.25955
6	6	.46196	.01327	.18994	.25244
7	7	.58823	.00896	.07133	.20820
8	8	10.35824	2.40433	.37436	8.71278
9	9	.04929	.00011	.14268	.02330
10	10	-.06962	.00022	.13774	-.03237
11	11	-.03333	.00011	.27204	-.02245
12	12	.08399	.00098	.35741	.06821
13	13	.10748	.00068	.17873	.05687
14	14	-.19835	.00073	.04081	-.05881
15	15	-.75670	.00828	.02607	-.20126
16	16	-.50920	.01986	.22942	-.30922
17	17	-.41044	.00528	.09069	-.15914
18	18	-.26351	.00283	.12271	-.11616
19	19	-.62721	.01395	.10675	-.26011
20	20	-.16098	.00152	.17770	-.08492
21	21	.11697	.00031	.05825	.03861
22	22	.58346	.02083	.18833	.31738
23	23	-.52181	.01355	.15343	-.25554
24	24	-.66774	.03060	.21018	-.38571
25	25	.14011	.00038	.04355	.04231

from age of the clock, x_1, and number of bidders, x_2, was determined to be

$$\hat{y} = -1,339 + 12.74x_1 + 85.95x_2$$

(a) Use this equation to calculate the residuals of each of the prices given in Table 4.1 (p. 171).
(b) Calculate the mean and the variance of the residuals. The mean should equal 0, and the variance should be close to the value of MSE given in the SAS printout shown in Figure 4.3 (p. 172).
(c) Find the proportion of the residuals that fall outside 2 estimated standard deviations ($2s$) of 0 and outside $3s$.
(d) Rerun the analysis and request influence diagnostics. Interpret the measures of influence given on the printout.

8.27 Populations of world cities. Refer to the study of the population of the world's largest cities in Exercise 8.17 (p. 411). A multiple regression model for the natural logarithm of the population (in thousands) of each country's largest city was fit to data collected on 126 nations and resulted in $s = .19$. A SPSS stem-and-leaf plot of the standardized residuals is shown above (next column). Identify any outliers on the plot.

8.28 Modeling an employee's work-hours missed. A large manufacturing firm wants to determine

SPSS Output for Exercise 8.27

```
STDRESID Stem-and-Leaf Plot

 Frequency     Stem &  Leaf

    2.00       -2 .  04
    5.00       -1 .  55579
   14.00       -1 .  00122222233344
   19.00       -0 .  5555666666677778999
   23.00       -0 .  00011111111222334444444
   24.00        0 .  000000000011111111222233444
   14.00        0 .  55555556777789
   15.00        1 .  000001112222234
    7.00        1 .  5566679
    2.00        2 .  03
    1.00        2 .  6

Stem width:      1.00
Each leaf:       1 case(s)
```

whether a relationship exists between y, the number of work-hours an employee misses per year, and x, the employee's annual wages. A sample of 15 employees produced the data in the accompanying table.

(a) Fit the first-order model, $E(y) = \beta_0 + \beta_1 x$, to the data.
(b) Plot the regression residuals. What do you notice?
(c) After searching through its employees' files, the firm has found that employee #13 had been fired but that his name had not been removed from the active employee payroll. This explains the large accumulation of work-hours missed (543) by that employee. In view of this fact, what is your recommendation concerning this outlier?

MISSWORK

EMPLOYEE	WORK-HOURS MISSED y	ANNUAL WAGES x, thousands of dollars
1	49	12.8
2	36	14.5
3	127	8.3
4	91	10.2
5	72	10.0
6	34	11.5
7	155	8.8
8	11	17.2
9	191	7.8
10	6	15.8
11	63	10.8
12	79	9.7
13	543	12.1
14	57	21.2
15	82	10.9

(d) Measure how influential the observation for employee #13 is on the regression analysis.

(e) Refit the model to the data, excluding the outlier, and compare the results to those in part a.

⊙ ASWELLS

8.29 Arsenic in groundwater. Refer to the *Environmental Science and Technology* (January 2005) study of the reliability of a commercial kit to test for arsenic in groundwater, Exercise 8.10 (p. 407). Identify any outliers in the first-order model for arsenic level (y). Are any of these outliers influential data points? If so, what are your recommendations?

⊙ GASTURBINE

8.30 Cooling method for gas turbines. Refer to the *Journal of Engineering for Gas Turbines and Power* (January 2005) study of a high-pressure inlet fogging method for a gas turbine engine, Exercise 8.11 (p. 407). Identify any outliers in the interaction model for heat rate (y). Are any of these outliers influential data points? If so, what are your recommendations?

8.7 Detecting Residual Correlation: The Durbin–Watson Test

Many types of data are observed at regular time intervals. The Consumer Price Index (CPI) is computed and published monthly, the profits of most major corporations are published quarterly, and the *Fortune* 500 list of largest corporations is published annually. Data like these, which are observed over time, are called **time series**. We will often want to construct regression models where the data for the dependent and independent variables are time series.

Regression models of time series may pose a special problem. Because time series tend to follow economic trends and seasonal cycles, the value of a time series at time t is often indicative of its value at time $(t + 1)$. That is, the value of a time series at time t is **correlated** with its value at time $(t + 1)$. If such a series is used as the dependent variable in a regression analysis, the result is that the random errors are correlated. This leads to *standard errors of the β-estimates that are seriously underestimated* by the formulas given previously. Consequently, we cannot apply the standard least squares inference-making tools and have confidence in their validity. Modifications of the methods, which allow for correlated residuals in time series regression models, are presented in Chapter 10. In this section, we present a method of testing for the presence of residual correlation.

Consider the time series data in Table 8.6, which gives sales data for the 35-year history of a company. The SAS printout shown in Figure 8.30 gives the regression analysis for the first-order linear model

$$y = \beta_0 + \beta_1 t + \varepsilon$$

This model seems to fit the data very well, since $R^2 = .98$ and the F-value (1,615.72) that tests the adequacy of the model is significant. The hypothesis that the coefficient β_1 is positive is accepted at any α level less than .0001 ($t = 40.2$ with 33 df).

The residuals $\hat{\varepsilon} = y - (\hat{\beta}_0 + \hat{\beta}_1 t)$ are plotted in Figure 8.31. Note that there is a distinct tendency for the residuals to have long positive and negative runs. That is, if the residual for year t is positive, there is a tendency for the residual for year $(t + 1)$ to be positive. These cycles are indicative of possible positive correlation between residuals.

For most economic time series models, we want to test the null hypothesis

$$H_0\text{: No residual correlation}$$

against the alternative

$$H_a\text{: Positive residual correlation}$$

SALES35

Table 8.6 A firm's annual sales revenue (thousands of dollars)

Year t	Sales y	Year t	Sales y	Year t	Sales y
1	4.8	13	48.4	25	100.3
2	4.0	14	61.6	26	111.7
3	5.5	15	65.6	27	108.2
4	15.6	16	71.4	28	115.5
5	23.1	17	83.4	29	119.2
6	23.3	18	93.6	30	125.2
7	31.4	19	94.2	31	136.3
8	46.0	20	85.4	32	146.8
9	46.1	21	86.2	33	146.1
10	41.9	22	89.9	34	151.4
11	45.5	23	89.2	35	150.9
12	53.5	24	99.1		

Figure 8.30 SAS regression printout for model of annual sales

```
                    Dependent Variable: SALES
                       Analysis of Variance

                            Sum of          Mean
Source              DF      Squares        Square    F Value    Pr > F

Model                1        65875         65875    1615.72    <.0001
Error               33   1345.45355      40.77132
Corrected Total     34        67221


          Root MSE              6.38524    R-Square     0.9800
          Dependent Mean      77.72286    Adj R-Sq     0.9794
          Coeff Var            8.21540

                       Parameter Estimates

                     Parameter     Standard
Variable    DF        Estimate        Error    t Value    Pr > |t|

Intercept    1         0.40151      2.20571       0.18      0.8567
T            1         4.29563      0.10687      40.20      <.0001


          Durbin-Watson D                    0.821
          Number of Observations                35
          1st Order Autocorrelation          0.590
```

since the hypothesis of positive residual correlation is consistent with economic trends and seasonal cycles.

The **Durbin–Watson d statistic** is used to test for the presence of residual correlation. This statistic is given by the formula

$$d = \frac{\sum_{t=2}^{n}(\hat{\varepsilon}_t - \hat{\varepsilon}_{t-1})^2}{\sum_{t=1}^{n}\hat{\varepsilon}_t^2}$$

Figure 8.31 SAS plot of residuals for model of annual sales

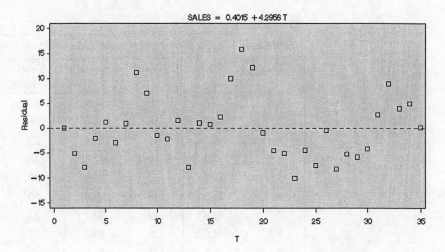

where n is the number of observations and $(\hat{\varepsilon}_t - \hat{\varepsilon}_{t-1})$ represents the difference between a pair of successive residuals. By expanding the numerator of d, we can also write

$$d = \frac{\sum\limits_{t=2}^{n} \hat{\varepsilon}_t^2}{\sum\limits_{t=1}^{n} \hat{\varepsilon}_t^2} + \frac{\sum\limits_{t=2}^{n} \hat{\varepsilon}_{t-1}^2}{\sum\limits_{t=1}^{n} \hat{\varepsilon}_t^2} - \frac{2\sum\limits_{t=2}^{n} \hat{\varepsilon}_t \hat{\varepsilon}_{t-1}}{\sum\limits_{t=1}^{n} \hat{\varepsilon}_t^2} \approx 2 - \frac{2\sum\limits_{t=2}^{n} \hat{\varepsilon}_t \hat{\varepsilon}_{t-1}}{\sum\limits_{t=1}^{n} \hat{\varepsilon}_t^2}$$

If the residuals are uncorrelated,

$$\sum_{t=2}^{n} \hat{\varepsilon}_t \hat{\varepsilon}_{t-1} \approx 0$$

indicating no relationship between $\hat{\varepsilon}_t$ and $\hat{\varepsilon}_{t-1}$, the value of d will be close to 2. If the residuals are highly positively correlated,

$$\sum_{t=2}^{n} \hat{\varepsilon}_t \hat{\varepsilon}_{t-1} \approx \sum_{t=2}^{n} \hat{\varepsilon}_t^2$$

(since $\hat{\varepsilon}_t \approx \hat{\varepsilon}_{t-1}$), and the value of d will be near 0:

$$d \approx 2 - \frac{2\sum\limits_{t=2}^{n} \hat{\varepsilon}_t \hat{\varepsilon}_{t-1}}{\sum\limits_{t=1}^{n} \hat{\varepsilon}_t^2} \approx 2 - \frac{2\sum\limits_{t=2}^{n} \hat{\varepsilon}_t^2}{\sum\limits_{t=1}^{n} \hat{\varepsilon}_t^2} \approx 2 - 2 = 0$$

If the residuals are very negatively correlated, then $\hat{\varepsilon}_t \approx -\hat{\varepsilon}_{t-1}$, so that

$$\sum_{t=2}^{n} \hat{\varepsilon}_t \hat{\varepsilon}_{t-1} \approx - \sum_{t=2}^{n} \hat{\varepsilon}_t^2$$

and d will be approximately equal to 4. Thus, d ranges from 0 to 4, with interpretations as summarized in the box.

> **Definition 8.9** The **Durbin–Watson d statistic** is calculated as follows:
>
> $$d = \frac{\sum\limits_{t=2}^{n}(\hat{\varepsilon}_t - \hat{\varepsilon}_{t-1})^2}{\sum\limits_{t=1}^{n}\hat{\varepsilon}_t^2}$$
>
> The d statistic has the following properties:
>
> 1. Range of d: $0 \leq d \leq 4$
> 2. If residuals are uncorrelated, $d \approx 2$.
> 3. If residuals are positively correlated, $d < 2$, and if the correlation is very strong, $d \approx 0$.
> 4. If residuals are negatively correlated, $d > 2$, and if the correlation is very strong, $d \approx 4$.

As an option, we requested SAS to produce the value of d for the annual sales model. The value, $d = .821$, is highlighted at the bottom of Figure 8.30. To determine if this value is close enough to 0 to conclude that positive residual correlation exists in the population, we need to find the rejection region for the test. Durbin and Watson (1951) have given tables for the lower-tail values of the d statistic, which we show in Table 8 ($\alpha = .05$) and Table 9 ($\alpha = .01$) in Appendix C.

Part of Table 8 is reproduced in Table 8.7. For the sales example, we have $k = 1$ independent variable and $n = 35$ observations. Using $\alpha = .05$ for the one-tailed test for positive residual correlation, the table values (shaded) are $d_L = 1.40$ and $d_U = 1.52$. The meaning of these values is illustrated in Figure 8.32. Because of the complexity of the sampling distribution of d, it is not possible to specify a single point that acts as a boundary between the rejection and nonrejection regions, as we did for the z, t, F, and other test statistics. Instead, an upper (d_U) and lower (d_L) bound are specified so that a d value less than d_L definitely *does* provide strong evidence of positive residual correlation at $\alpha = .05$ (recall that small d values indicate positive correlation), a d value greater than d_U does *not* provide evidence of positive correlation at $\alpha = .05$, but a value of d between d_L and d_U *might* be

Table 8.7 Reproduction of part of Table 8 in Appendix C ($\alpha = .05$)

n	$k=1$ d_L	$k=1$ d_U	$k=2$ d_L	$k=2$ d_U	$k=3$ d_L	$k=3$ d_U	$k=4$ d_L	$k=4$ d_U	$k=5$ d_L	$k=5$ d_U
31	1.36	1.50	1.30	1.57	1.23	1.65	1.16	1.74	1.09	1.83
32	1.37	1.50	1.31	1.57	1.24	1.65	1.18	1.73	1.11	1.82
33	1.38	1.51	1.32	1.58	1.26	1.65	1.19	1.73	1.13	1.81
34	1.39	1.51	1.33	1.58	1.27	1.65	1.21	1.73	1.15	1.81
35	1.40	1.52	1.34	1.58	1.28	1.65	1.22	1.73	1.16	1.80
36	1.41	1.52	1.35	1.59	1.29	1.65	1.24	1.73	1.18	1.80
37	1.42	1.53	1.36	1.59	1.31	1.66	1.25	1.72	1.19	1.80
38	1.43	1.54	1.37	1.59	1.32	1.66	1.26	1.72	1.21	1.79
39	1.43	1.54	1.38	1.60	1.33	1.66	1.27	1.72	1.22	1.79
40	1.44	1.54	1.39	1.60	1.34	1.66	1.29	1.72	1.23	1.79

Figure 8.32 Rejection region for the Durbin–Watson d-test: scale example

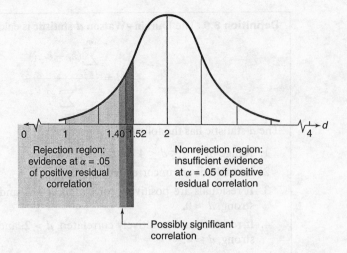

Rejection region: evidence at $\alpha = .05$ of positive residual correlation

Nonrejection region: insufficient evidence at $\alpha = .05$ of positive residual correlation

Possibly significant correlation

significant at the $\alpha = .05$ level. If $d_L < d < d_U$, more information is needed before we can reach any conclusion about the presence of residual correlation. A summary of the Durbin–Watson d-test is given in the box.

As indicated in the printout for the sales example (Figure 8.30), the computed value of d, .821, is less than the tabulated value of d_L, 1.40. Thus, we conclude that the residuals of the straight-line model for sales are positively correlated.

Durbin–Watson d-Test for Residual Correlation

LOWER-TAILED TEST	TWO-TAILED TEST	UPPER-TAILED TEST
H_0: No residual correlation	H_0: No residual correlation	H_0: No residual correlation
H_a: Positive residual correlation	H_a: Positive or negative residual correlation	H_a: Negative residual correlation

Test statistic:

$$d = \frac{\sum_{t=2}^{n}(\hat{\varepsilon}_t - \hat{\varepsilon}_{t-1})^2}{\sum_{t=1}^{n}\hat{\varepsilon}_t^2}$$

Rejection region:	*Rejection region*:	*Rejection region*:
$d < d_{L,\alpha}$	$d < d_{L,\alpha/2}$ or $(4-d) < d_{L,\alpha/2}$	$(4-d) < d_{L,\alpha}$
Nonrejection region:	*Nonrejection region*:	*Nonrejection region*:
$d > d_{U,\alpha}$	$d > d_{U,\alpha/2}$ or $(4-d) > d_{U,\alpha/2}$	$(4-d) > d_{U,\alpha}$
Inconclusive region:	*Inconclusive region*:	*Inconclusive region*:
$d_{L,\alpha} \leq (4-d) \leq d_{U,\alpha}$	Any other result	$d_{L,\alpha} < (4-d) < d_{U,\alpha}$

where $d_{L,\alpha}$ and $d_{U,\alpha}$ are the lower and upper tabulated values, respectively, corresponding to k independent variables and n observations.

Assumption: The residuals are normally distributed.

Tests for negative correlation and two-tailed tests can be conducted by making use of the symmetry of the sampling distribution of the d statistic about its mean, 2 (see Figure 8.32). That is, we compare $(4 - d)$ to d_L and d_U and conclude that the residuals are negatively correlated if $(4 - d) < d_L$, that there is insufficient evidence to conclude that the residuals are negatively correlated if $(4 - d) > d_U$, and that the test for negative residual correlation is *possibly* significant if $d_L < (4 - d) < d_U$.

Once strong evidence of residual correlation has been established, as in the case of the sales example, doubt is cast on the least squares results and any inferences drawn from them. In Chapter 10, we present a time series model that accounts for the correlation of the random errors. The residual correlation can be taken into account in a time series model, thereby improving both the fit of the model and the reliability of model inferences.

8.7 Exercises

8.31 Correlated regression errors. What are the consequences of running least squares regression when the errors are correlated?

8.32 Durbin–Watson d values. Find the values of d_L and d_U from Tables 8 and 9 in Appendix D for each of the following situations:

(a) $n = 30$, $k = 3$, $\alpha = .05$
(b) $n = 40$, $k = 1$, $\alpha = .01$
(c) $n = 35$, $k = 5$, $\alpha = .05$

8.33 Retail bank's deposit shares. Exploratory research published in the *Journal of Professional Services Marketing* (Vol. 5, 1990) examined the relationship between deposit share of a retail bank and several marketing variables. Quarterly deposit share data were collected for five consecutive years for each of nine retail banking institutions. The model analyzed took the following form:

$$y_t = \beta_0 + \beta_1 P_{t-1} + \beta_2 S_{t-1} + \beta_3 D_{t-1} + \varepsilon_t$$

where

y_t = Deposit share of bank in quarter
    $t, t = 1, 2, \ldots, 20$

P_{t-1} = Expenditures on promotion-
    related activities in quarter $t - 1$

S_{t-1} = Expenditures on service-
    related activities in quarter $t - 1$

D_{t-1} = Expenditures on distribution-
    related activities in quarter $t - 1$

A separate model was fit for each bank with the results shown in the table.

(a) Interpret the value of R^2 for each bank.
(b) Test the overall adequacy of the model for each bank.

(c) Conduct the Durbin–Watson d-test for each bank. Interpret the practical significance of the tests.

BANK	R^2	p-VALUE FOR GLOBAL F	DURBIN–WATSON d
1	.914	.000	1.3
2	.721	.004	3.4
3	.926	.000	2.7
4	.827	.000	1.9
5	.270	.155	.85
6	.616	.012	1.8
7	.962	.000	2.5
8	.495	.014	2.3
9	.500	.011	1.1

Note: The values of d shown are approximated based on other information provided in the article.

8.34 Forecasting car sales. Forecasts of automotive vehicle sales in the United States provide the basis for financial and strategic planning of large automotive corporations. The following forecasting model was developed for y, total monthly passenger car and light truck sales (in thousands):

$$E(y) = \beta_0 + \beta_1 x_1 + \beta_2 x_2 + \beta_3 x_3 + \beta_4 x_4 + \beta_5 x_5$$

where

x_1 = Average monthly retail price of
    regular gasoline

x_2 = Annual percentage change in GNP
    per quarter

x_3 = Monthly consumer confidence index

x_4 = Total number of vehicles scrapped
    (millions) per month

x_5 = Vehicle seasonality

The model was fitted to monthly data collected over a 12-year period (i.e., $n = 144$ months) with the following results:

$$\hat{y} = -676.42 - 1.93x_1 + 6.54x_2 + 2.02x_3$$

$$+ .08x_4 + 9.82x_5$$

$$R^2 = .856$$

Durbin–Watson $d = 1.01$

(a) Is there sufficient evidence to indicate that the model contributes information for the prediction of y? Test using $\alpha = .05$.

(b) Is there sufficient evidence to indicate that the regression errors are positively correlated? Test using $\alpha = .05$.

(c) Comment on the validity of the inference concerning model adequacy in light of the result of part b.

8.35 Buying power of the dollar. The consumer purchasing value of the dollar from 1970 to 2007 is illustrated by the data in the table below. The buying power of the dollar (compared with 1982) is listed for each year. The first-order model

$$y_t = \beta_0 + \beta_1 t + \varepsilon$$

was fit to the data using the method of least squares. The MINITAB printout and a plot of the regression residuals are shown at right.

BUYPOWER

YEAR	t	VALUE, y_t	YEAR	t	VALUE, y_t
1970	1	2.545	1989	20	0.880
1971	2	2.469	1990	21	0.839
1972	3	2.392	1991	22	0.822
1973	4	2.193	1992	23	0.812
1974	5	1.901	1993	24	0.802
1975	6	1.718	1994	25	0.797
1976	7	1.645	1995	26	0.782
1977	8	1.546	1996	27	0.762
1978	9	1.433	1997	28	0.759
1979	10	1.289	1998	29	0.765
1980	11	1.136	1999	30	0.752
1981	12	1.041	2000	31	0.725
1982	13	1.000	2001	32	0.711
1983	14	0.984	2002	33	0.720
1984	15	0.964	2003	34	0.698
1985	16	0.955	2004	35	0.673
1986	17	0.969	2005	36	0.642
1987	18	0.949	2006	37	0.623
1988	19	0.926	2007	38	0.600

Source: U.S. Bureau of the Census. *Statistical Abstract of the United States*, 2009.

(a) Examine the plot of the regression residuals against t. Is there a tendency for the residuals to have long positive and negative runs? To what do you attribute this phenomenon?

(b) Locate the Durbin–Watson d statistic on the printout and test the null hypothesis that the time series residuals are uncorrelated. Use $\alpha = .10$.

(c) What assumption(s) must be satisfied in order for the test of part b to be valid?

MINITAB output for Exercise 8.35

Regression Analysis: VALUE versus T

```
The regression equation is
VALUE = 1.94 - 0.0425 T


Predictor      Coef     SE Coef       T       P
Constant    1.94066     0.09302   20.86   0.000
T          -0.042545    0.004158  -10.23   0.000


S = 0.281076   R-Sq = 74.4%   R-Sq(adj) = 73.7%

Analysis of Variance

Source           DF       SS       MS       F       P
Regression        1   8.2713   8.2713  104.70   0.000
Residual Error   36   2.8441   0.0790
Total            37  11.1155


Durbin-Watson statistic = 0.0581054
```

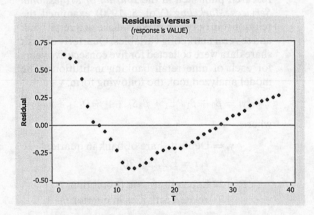

Residuals Versus T
(response is VALUE)

8.36 Life insurance policies in force. The next table represents all life insurance policies (in millions) in force on the lives of U.S. residents for the years 1980 through 2006.

(a) Fit the simple linear regression model, $E(Y_t) = \beta_0 + \beta_1 t$, to the data for the years 1980 to 2006 ($t = 1, 2, \ldots, 27$). Interpret the results.

(b) Find and plot the regression residuals against t. Does the plot suggest the presence of autocorrelation? Explain.

(c) Conduct the Durbin–Watson test (at $\alpha = .05$) to test formally for the presence of positively autocorrelated regression errors.

(d) Comment on the validity of the inference concerning model adequacy in light of the result of part b.

LIFEINS

YEAR	NO. OF POLICIES (in millions)	YEAR	NO. OF POLICIES (in millions)
1980	402	1994	366
1981	400	1995	370
1982	390	1996	355
1983	387	1997	351
1984	385	1998	358
1985	386	1999	367
1986	391	2000	369
1987	395	2001	377
1988	391	2002	375
1989	394	2003	379
1990	389	2004	373
1991	375	2005	373
1992	366	2006	375
1993	363		

Source: U.S. Bureau of the Census. *Statistical Abstract of the United States*, 2009.

8.37 **Marketing a new cold medicine.** A pharmaceutical company based in New Jersey recently introduced a new cold medicine called Coldex. (For proprietary reasons, the actual name of the product is withheld.) It is now sold in drugstores and supermarkets across the United States. Monthly sales for the first two years the product was on the market are reported in the next table. Consider the simple linear regression model, $E(y_t) = \beta_0 + \beta_1 t$, where y_t is the sales in month t.

(a) Fit the simple linear model to the data. Is the model statistically useful for predicting monthly sales?

(b) Construct a plot of the regression residuals against month, t. Does the plot suggest the presence of residual correlation? Explain.

(c) Use the Durbin–Watson test to formally test for correlated errors.

8.38 **Forecasting foreign exchange rates.** T. C. Chiang considered several time series forecasting models of future foreign exchange rates for U.S. currency (*Journal of Financial Research*, Summer 1986). One popular theory among financial analysts is that the forward (90-day) exchange rate is a useful

COLDEX

YEAR	MONTH	t	SALES, y_t
1	Jan.	1	3,394
	Feb.	2	4,010
	Mar.	3	924
	Apr.	4	205
	May	5	293
	Jun.	6	1,130
	Jul.	7	1,116
	Aug.	8	4,009
	Sep.	9	5,692
	Oct.	10	3,458
	Nov.	11	2,849
	Dec.	12	3,470
2	Jan.	13	4,568
	Feb.	14	3,710
	Mar.	15	1,675
	Apr.	16	999
	May	17	986
	Jun.	18	1,786
	Jul.	19	2,253
	Aug.	20	5,237
	Sep.	21	6,679
	Oct.	22	4,116
	Nov.	23	4,109
	Dec.	24	5,124

Source: Personal communication from Carol Cowley, Carla Marchesini, and Ginny Wilson, Rutgers University, Graduate School of Management.

predictor of the future spot exchange rate. Using monthly data on exchange rates for the British pound for $n = 81$ months, Chiang fitted the model

$$E(y_t) = \beta_0 + \beta_1 x_{t-1}$$

where

$y_t = \ln \text{(spot rate) in month } t$
$x_t = \ln \text{(forward rate) in month } t$

The method of least squares yielded the following results:

$$\hat{y}_t = -.009 + .986 x_{t-1} \quad (t = 41.9)$$

$$s = .0249 \quad R^2 = .957 \quad \text{Durbin–Watson } d = .962$$

(a) Is the model useful for predicting future spot exchange rates for the British pound? Test using $\alpha = .05$.

(b) Interpret the values of s and R^2.

(c) Is there evidence of positive autocorrelation among the residuals? Test using $\alpha = .05$.

(d) Based on the results of parts a–c, would you recommend using the least squares model to forecast spot exchange rates?

Quick Summary

KEY SYMBOLS & FORMULAS

Residual
$\hat{\varepsilon} = y - \hat{y}$

Partial residuals for x_j
$\hat{\varepsilon}^* = \hat{\varepsilon} + \hat{\beta}_j x_j$

Standardized residual
$z_i = \hat{\varepsilon}_i / s$

Studentized residual
$z_i^* = \dfrac{(y_i - \hat{y}_i)}{s\sqrt{1 - h_i}}$

Leverage for x_j
h_j, where
$\hat{y}_i = h_1 y_1 + h_2 y_2 + \cdots + h_j y_j + \cdots + h_n y_n$

Jackknifed predicted value
$\hat{y}_{(i)}$

Deleted residual
$d_i = y_i - \hat{y}_{(i)}$

Standard deviation of deleted residual s_{d_i}

Studentized deleted residual
$d_i^* = (y_i - \hat{y}_{(i)}) / s_{d_i}$

Cook's distance
$$D_i = \frac{(y_i - \hat{y}_i)^2}{(k+1)\,\text{MSE}} \left[\frac{h_i}{(1 - h_i)^2} \right]$$

Durbin–Watson statistic
$$d = \frac{\sum\limits_{t=2}^{h} (\hat{\varepsilon}_t - \hat{\varepsilon}_{t-1})^2}{\sum\limits_{t=1}^{h} \hat{\varepsilon}_t^2}$$

KEY IDEAS

Properties of residuals
1. mean of 0
2. standard deviation equal to standard deviation of regression model, s

---

Checking assumptions using residual plots

Assumption	Graph
1) $E(\varepsilon) = 0$	Plot $\hat{\varepsilon}$ vs. x_i (look for *trends*)
2) $\text{Var}(\varepsilon)$ is constant	Plot $\hat{\varepsilon}$ vs. $\hat{y}$ (look for *heteroscedastic patterns*)
3) Normal ε's	Histogram, stem/leaf plot, or normal probability plot for $\hat{\varepsilon}$ (look for extreme *skewness*)
4) Independent ε's	Plot $\hat{\varepsilon}$ vs. time (look for *oscillating trend*)

Variance-stabilizing transformation on y

Type of data	Transformation
Poisson	$y^* = \sqrt{y}$
Binomial	$y^* = \sin^{-1}\sqrt{y}$
Multiplicative	$y^* = \ln(y)$

Outlier: a standardized residual larger than 3 in absolute value (i.e., $|\hat{\varepsilon}/s| > 3$). Always investigate outliers before deleting them from the analysis.

Detecting influence

Leverage: jth observation is influential if leverage value $h_j > 2(k+1)/n$

Jackknife: jth observation is influential if studentized deleted residual, d_i^*, is large (e.g., exceeds 3)

Cook's distance: jth observation is influential if D_i is large (e.g., exceeds the 50th percentile of the F distribution for the model)

Testing for residual correlation: apply the *Durbin–Watson* test.

Supplementary Exercises

8.39. Interpreting residual plots. Identify the problem(s) in each of the five residual plots show on page 433.

8.40. Cost of modifying a naval fleet. A naval base is considering modifying or adding to its fleet of 48

standard aircraft. The final decision regarding the type and number of aircraft to be added depends on a comparison of cost versus effectiveness of the modified fleet. Consequently, the naval base would

Plots for Exercise 8.39

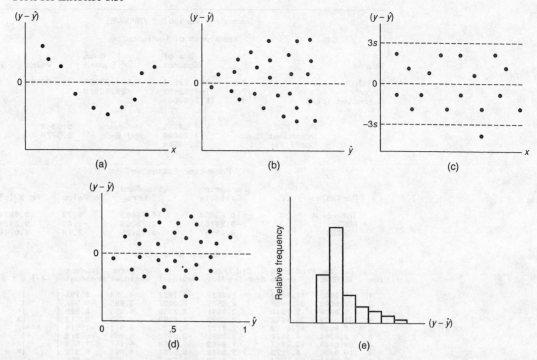

(a)

(b)

(c)

(d)

(e)

like to model the projected percentage increase y in fleet effectiveness by the end of the decade as a function of the cost x of modifying the fleet. A first proposal is the quadratic model

$$E(y) = \beta_0 + \beta_1 x + \beta_2 x^2$$

The data provided in the accompanying table were collected on 10 naval bases of a similar size that recently expanded their fleets. The data were used to fit the model, and the SAS printout of the multiple regression analysis is reproduced on p. 434.

(a) Construct a residual plot versus x. Do you detect any trends? Any outliers?

NAVALFLEET

PERCENTAGE IMPROVEMENT y	COST OF MODIFYING FLEET x, millions of dollars
18	125
32	160
9	80
37	162
6	110
3	90
30	140
10	85
25	150
2	50

(b) Interpret the influence diagnostics shown on the printout. Are there any observations that have large influence on the analysis?

8.41. Analysis of urban air samples. Chemical engineers at Tokyo Metropolitan University analyzed urban air specimens for the presence of low-molecular-weight dicarboxylic acids (*Environmental Science and Engineering*, October 1993). The dicarboxylic acid (as a percentage of total carbon) and oxidant

AIRTOKYO

DICARBOXYLIC ACID, %	OXIDANT, ppm	DICARBOXYLIC ACID, %	OXIDANT, ppm
.85	78	.50	32
1.45	80	.38	28
1.80	74	.30	25
1.80	78	.70	45
1.60	60	.80	40
1.20	62	.90	45
1.30	57	1.22	41
.20	49	1.00	34
.22	34	1.00	25
.40	36		

Source: Kawamura, K., and Ikushima, K. "Seasonal changes in the distribution of dicarboxylic acids in the urban atmosphere," *Environmental Science and Technology*, Vol. 27, No. 10, Oct. 1993, p. 2232 (data extracted from Figure 4).

SAS Regression Output for Exercise 8.40

Dependent Variable: IMPROVE

Analysis of Variance

Source	DF	Sum of Squares	Mean Square	F Value	Pr > F
Model	2	1368.77501	684.38750	33.08	0.0003
Error	7	144.82499	20.68928		
Corrected Total	9	1513.60000			

Root MSE	4.54855	R-Square	0.9043	
Dependent Mean	17.20000	Adj R-Sq	0.8770	
Coeff Var	26.44504			

Parameter Estimates

Variable	DF	Parameter Estimate	Standard Error	t Value	Pr > |t|
Intercept	1	10.65904	14.55009	0.73	0.4876
COST	1	-0.28161	0.28088	-1.00	0.3494
COSTSQ	1	0.00267	0.00125	2.13	0.0706

Output Statistics

Obs	Dep Var IMPROVE	Predicted Value	Std Error Mean Predict	Residual	Std Error Residual	Student Residual	-2-1 0 1 2	Cook's D
1	18.0000	17.2073	2.0627	0.7927	4.054	0.196	| | |	0.003
2	32.0000	34.0037	2.6525	-2.0037	3.695	-0.542	| *| |	0.051
3	9.0000	5.2310	2.0601	3.7690	4.055	0.929	| |* |	0.074
4	37.0000	35.1612	2.8397	1.8388	3.553	0.518	| |* |	0.057
5	6.0000	12.0128	2.2177	-6.0128	3.971	-1.514	| ***| |	0.238
6	3.0000	6.9572	2.0835	-3.9572	4.043	-0.979	| *| |	0.085
7	30.0000	23.6042	1.8468	6.3958	4.157	1.539	| |*** |	0.156
8	10.0000	6.0273	2.0492	3.9727	4.061	0.978	| |* |	0.081
9	25.0000	28.5367	2.0070	-3.5367	4.082	-0.866	| *| |	0.060
10	2.0000	3.2586	4.1920	-1.2586	1.765	-0.713	| *| |	0.955

Output Statistics

Obs	RStudent	Hat Diag H	Cov Ratio	DFFITS	Intercept	DFBETAS COST	COSTSQ
1	0.1815	0.2057	1.9665	0.0924	-0.0594	0.0657	-0.0639
2	-0.5129	0.3401	2.1156	-0.3682	-0.1192	0.1505	-0.1871
3	0.9190	0.2051	1.3457	0.4669	0.0286	0.0632	-0.1088
4	0.4886	0.3898	2.3148	0.3904	0.1473	-0.1817	0.2199
5	-1.7093	0.2377	0.6336	-0.9545	0.5920	-0.7014	0.7211
6	-0.9753	0.2098	1.2924	-0.5025	0.1474	-0.2357	0.2724
7	1.7511	0.1649	0.5511	0.7780	-0.3140	0.3143	-0.2585
8	0.9748	0.2030	1.2818	0.4919	-0.0653	0.1582	-0.2007
9	-0.8490	0.1947	1.4030	-0.4174	0.0129	0.0093	-0.0503
10	-0.6854	0.8494	8.4082	-1.6276	-1.4594	1.2838	-1.1539

concentrations (in ppm) for 19 air specimens collected from urban Tokyo are listed in the table (p. 433). Consider a straight-line model relating dicarboxylic acid percentage (y) to oxidant concentration (x). Assess the validity of the regression assumptions.

8.42. Revenue generated by a new pill. A leading pharmaceutical company that produces a new hypertension pill would like to model annual revenue generated by this product. Company researchers utilized data collected over the previous 15 years to fit the model

$$E(y_t) = \beta_0 + \beta_1 x_t + \beta_2 t$$

where

y_t = Revenue in year t (in millions of dollars)

x_t = Cost per pill in year t

t = Year $(1, 2, \ldots, 15)$

A company statistician suspects that the assumption of independent errors may be violated and that, in fact, the regression residuals are positively correlated. Test this claim using $\alpha = .05$ if the Durbin–Watson d Statistic is $d = .776$.

🔷 FACTORS

8.43. Caring for hospital patients. Refer to the study of coronary care patients, Exercise 3.68 (p. 150)

Recall that simple linear regression was used to model length of stay, y, to number of patient factors, x. Conduct a complete residual analysis for the model. (The data is saved in the FACTORS file.)

8.44. Forecasting absentee rate. The foreman of a printing shop that has been in business five years is scheduling his workload for next year, and he must estimate the number of employees available for work. He asks the company statistician to forecast next year's absentee rate. Since it is known that quarterly fluctuations exist, the following model is proposed:

$$y = \beta_0 + \beta_1 x_1 + \beta_2 x_2 + \beta_3 x_3 + \varepsilon$$

where

$$y = \text{Absentee rate} = \frac{\text{Total employees absent}}{\text{Total employees}}$$

$$x_1 = \begin{cases} 1 & \text{if quarter 1 (January–March)} \\ 0 & \text{if not} \end{cases}$$

$$x_2 = \begin{cases} 1 & \text{if quarter 2 (April–June)} \\ 0 & \text{if not} \end{cases}$$

$$x_3 = \begin{cases} 1 & \text{if quarter 3 (July–September)} \\ 0 & \text{if not} \end{cases}$$

(a) Fit the model to the data given in the next table.
(b) Consider the nature of the response variable, y. Do you think that there may be possible violations of the usual assumptions about ε? Explain.
(c) Suggest an alternative model that will approximately stabilize the variance of the error term ε.
(d) Fit the alternative model. Check R^2 to determine whether model adequacy has improved.

PRINTSHOP

YEAR	QUARTER 1	QUARTER 2	QUARTER 3	QUARTER 4
1	.06	.13	.28	.07
2	.12	.09	.19	.09
3	.08	.18	.41	.07
4	.05	.13	.23	.08
5	.06	.07	.30	.05

8.45. Breeding of thoroughbreds. The breeding ability of a thoroughbred horse is sometimes a more important consideration to prospective buyers than racing ability. Usually, the longer a horse lives, the greater its value for breeding purposes. Before marketing a group of horses, a breeder would like to be able to predict their life spans. The breeder believes that the gestation period of a thoroughbred horse may be an indicator of its life span. The information in the next table was supplied to the breeder by various stables in the area. (Note that the horse has the greatest variation of gestation period of any species due to seasonal and feed factors.) Consider the first-order model

$$y = \beta_0 + \beta_1 x + \varepsilon,$$

where y = life span (in years) and x = gestation period (in months).

HORSES

HORSE	GESTATION PERIOD x, days	LIFE SPAN y, years
1	403	30
2	279	22
3	307	7
4	416	31
5	265	21
6	356	27
7	298	25

(a) Fit the model to the data.
(b) Check model adequacy by interpreting the F and R^2 statistics.
(c) Construct a plot of the residuals versus x, gestation period.
(d) Check for residuals that lie outside the interval $0 \pm 2s$ or $0 \pm 3s$.
(e) The breeder has been informed that the short life span of horse #3 (7 years) was due to a very rare disease. Omit the data for horse #3 and refit the least squares line. Has the omission of this observation improved the model?

8.46. Analysis of television market share. The data in the table are the monthly market shares for a product over most of the past year. The least squares line relating market share to television advertising expenditure is found to be

$$\hat{y} = -1.56 + .687x$$

TVSHARE

MONTH	MARKET SHARE y,%	TELEVISION ADVERTISING EXPENDITURE x, thousands of dollars
January	15	23
February	17	27
March	17	25
May	13	21
June	12	20
July	14	24
September	16	26
October	14	23
December	15	25

(a) Calculate and plot the regression residuals in the manner outlined in this section.

(b) The response variable y, market share, is recorded as a percentage. What does this lead you to believe about the least squares assumption of homoscedasticity? Does the residual plot substantiate this belief?

(c) What variance-stabilizing transformation is suggested by the trend in the residual plot? Refit the first-order model using the transformed responses. Calculate and plot these new regression residuals. Is there evidence that the transformation has been successful in stabilizing the variance of the error term, ε?

WATEROIL

8.47. Recovering oil from a water/oil mix. Refer to the *Journal of Colloid and Interface Science* study of the voltage (y) required to separate water from oil, Exercise 4.14 (p. 189). A first-order model for y included seven independent variables. Conduct a complete residual analysis of the model using the data saved in the WATEROIL file.

8.48. Public perceptions of health risks. Refer to the *Journal of Experimental Psychology: Learning, Memory, and Cognition* (July 2005) study of the ability of people to judge risk of an infectious disease, Exercise 4.42 (p. 208). Recall that the researchers asked German college students to estimate the number of people infected with a certain disease in a typical year. The median estimates as well as the actual incidence rate for each in a sample of 24 infections are reproduced in the table. In Exercise 4.42 you fit the quadratic model, $E(y) = \beta_0 + \beta_1 x + \beta_2 x^2$, where y = actual incidence rate and x = estimated rate. Identify any outliers in this regression analysis. Are any of these outliers influential? If so, how should the researchers proceed?

FLAG

8.49. Bid-rigging in highway construction. Refer to the multiple regression example in Section 4.14 (p. 235). Recall that the cost (y) of constructing a

INFECTION

INFECTION	INCIDENCE RATE	ESTIMATE
Polio	0.25	300
Diphtheria	1	1000
Trachoma	1.75	691
Rabbit Fever	2	200
Cholera	3	17.5
Leprosy	5	0.8
Tetanus	9	1000
Hemorrhagic Fever	10	150
Trichinosis	22	326.5
Undulant Fever	23	146.5
Well's Disease	39	370
Gas Gangrene	98	400
Parrot Fever	119	225
Typhoid	152	200
Q Fever	179	200
Malaria	936	400
Syphilis	1514	1500
Dysentery	1627	1000
Gonorrhea	2926	6000
Meningitis	4019	5000
Tuberculosis	12619	1500
Hepatitis	14889	10000
Gastroenteritis	203864	37000
Botulism	15	37500

Source: Hertwig, R., Pachur, T., and Kurzenhauser, S. "Judgments of risk frequencies: Tests of possible cognitive mechanisms," *Journal of Experimental Psychology: Learning, Memory, and Cognition,* Vol. 31, No. 4, July 2005 (Table 1). Copyright © 2005 American Psychological Association, reprinted with permission.

road was modeled as a function of DOT engineers' estimated cost (x_1) and a dummy variable (x_2) for whether the winning bid was fixed or competitively bid. Access the data saved in the FLAG file to carry out a complete residual analysis for the model $E(y) = \beta_0 + \beta_1 x_1 + \beta_2 x_2 + \beta_3 x_1 x_2$. Do you recommend making any model modifications?

References

Barnett, V., and Lewis, T. *Outliers in Statistical Data.* 3rd ed. New York: Wiley, 1994.

Belsley, D. A., Kuh, E., and Welsch, R. E. *Regression Diagnostics: Identifying Influential Data and Sources of Collinearity.* New York: Wiley, 2004.

Box, G. E. P., and Cox, D. R. "An analysis of transformations." *Journal of the Royal Statistical Society, Series B,* 1964, Vol. 26, pp. 211–243.

Breusch, T. S., and Pagan, A. R. (1979), "A simple test for heteroscedasticity and random coefficient variation." *Econometrica,* Vol. 47, pp. 1287–1294.

Cook, R. D. "Influential observations in linear regression." *Journal of the American Statistical Association,* 1979, Vol. 74, pp. 169–174.

Cook, R. D., and Weisberg, S. *Residuals and Influence in Regression,* New York: Chapman and Hall, 1982.

Draper, N., and Smith, H. *Applied Regression Analysis*, 3rd ed. New York: Wiley, 1998.

Durbin, J., and Watson, G. S. "Testing for serial correlation in least squares regression, I." *Biometrika*, 1950, Vol. 37, pp. 409–428.

Durbin, J., and Watson, G. S. "Testing for serial correlation in least squares regression, II." *Biometrika*, 1951, Vol. 38, pp. 159–178.

Durbin, J., and Watson, G. S. "Testing for serial correlation in least squares regression, III." *Biometrika*, 1971, Vol. 58, pp. 1–19.

Granger, C. W. J., and Newbold, P. *Forecasting Economic Time Series*. New York: Academic Press, 1977.

Kutner, M., Nachtsheim, C., Neter, J., and Li, W. *Applied Linear Statistical Models*, 5th ed. New York: McGraw-Hill/Irwin, 2005.

Larsen, W. A., and McCleary, S. J. "The use of partial residual plots in regression analysis." *Technometrics*, Vol. 14, 1972, pp. 781–790.

Mansfield, E. R., and Conerly, M. D. "Diagnostic value of residual and partial residual plots." *American Statistician*, Vol. 41, No. 2, May 1987, pp. 107–116.

Mendenhall, W. *Introduction to Linear Models and the Design and Analysis of Experiments*. Belmont, Calif.: Wadsworth, 1968.

Montgomery, D. C., Peck, E. A., and Vining, G. G. *Introduction to Linear Regression Analysis* 4th ed. New York: Wiley, 2006.

Stephens, M. A. "EDF statistics for goodness of fit and some comparisons." *Journal of the American Statistical Association*, 1974, Vol. 69, pp. 730–737.

AN ANALYSIS OF RAIN LEVELS IN CALIFORNIA

The Problem

For this case study, we focus on an application of regression analysis in the science of geography. Writing in the journal *Geography* (July 1980), P. J. Taylor sought to describe the method of multiple regression to the research geographer "in a completely nontechnical manner." Taylor chose to investigate the variation in average annual precipitation in California—"a typical research problem that would be tackled using multiple regression analysis." In this case study, we use Taylor's data to build a model for average annual precipitation, y. Then we examine the residuals, the deviations between the predicted and the actual precipitation levels, to detect (as Taylor did) an important independent variable omitted from the regression model.

The Data

The state of California operates numerous meteorological stations. One of the many functions of each station is to monitor rainfall on a daily basis. This information is then used to produce an average annual precipitation level for each station.

Table CS4.1 lists average annual precipitation levels (in inches) for a sample of 30 meteorological stations scattered throughout the state. (These are the data analyzed by Taylor.) In addition to average annual precipitation (y), the table lists three independent variables that are believed (by California geographers) to have the most impact on the amount of rainfall at each station, as follows:

1. Altitude of the station (x_1, feet)
2. Latitude of the station (x_2, degrees)
3. Distance of the station from the Pacific coast (x_3, miles)

A Model for Average Annual Precipitation

As an initial attempt in explaining the average annual precipitation in California, Taylor considered the following first-order model:

Model I

$$E(y) = \beta_0 + \beta_1 x_1 + \beta_2 x_2 + \beta_3 x_3$$

CALIRAIN

Table CS4.1 Data for 30 meteorological stations in California

Station	Average Annual Precipitation y, inches	Altitude x_1, feet	Latitude x_2, degrees	Distance From Coast x_3, miles	Shadow x_4
1. Eureka	39.57	43	40.8	1	W
2. Red Bluff	23.27	341	40.2	97	L
3. Thermal	18.20	4,152	33.8	70	L
4. Fort Bragg	37.48	74	39.4	1	W
5. Soda Springs	49.26	6,752	39.3	150	W
6. San Francisco	21.82	52	37.8	5	W
7. Sacramento	18.07	25	38.5	80	L
8. San Jose	14.17	95	37.4	28	L
9. Giant Forest	42.63	6,360	36.6	145	W
10. Salinas	13.85	74	36.7	12	L
11. Fresno	9.44	331	36.7	114	L
12. Pt. Piedras	19.33	57	35.7	1	W
13. Paso Robles	15.67	740	35.7	31	L
14. Bakersfield	6.00	489	35.4	75	L
15. Bishop	5.73	4,108	37.3	198	L
16. Mineral	47.82	4,850	40.4	142	W
17. Santa Barbara	17.95	120	34.4	1	W
18. Susanville	18.20	4,152	40.3	198	L
19. Tule Lake	10.03	4,036	41.9	140	L
20. Needles	4.63	913	34.8	192	L
21. Burbank	14.74	699	34.2	47	W
22. Los Angeles	15.02	312	34.1	16	W
23. Long Beach	12.36	50	33.8	12	W
24. Los Banos	8.26	125	37.8	74	L
25. Blythe	4.05	268	33.6	155	L
26. San Diego	9.94	19	32.7	5	W
27. Daggett	4.25	2,105	34.1	85	L
28. Death Valley	1.66	−178	36.5	194	L
29. Crescent City	74.87	35	41.7	1	W
30. Colusa	15.95	60	39.2	91	L

Model 1 assumes that the relationship between average annual precipitation y and each independent variable is linear, and the effect of each x on y is independent of the other x's (i.e., no interaction).

Figure CS4.1 MINITAB
regression printout for
Model 1

Regression Analysis: Precip versus Altitude, Latitude, Distance

```
The regression equation is
Precip = - 102 + 0.00409 Altitude + 3.45 Latitude - 0.143 Distance


Predictor      Coef    SE Coef      T      P     VIF
Constant    -102.36      29.21   -3.50  0.002
Altitude   0.004091   0.001218    3.36  0.002   1.536
Latitude     3.4511     0.7949    4.34  0.000   1.058
Distance   -0.14286    0.03634   -3.93  0.001   1.493


S = 11.0980   R-Sq = 60.0%   R-Sq(adj) = 55.4%


Analysis of Variance

Source          DF      SS      MS      F      P
Regression       3  4809.4  1603.1  13.02  0.000
Residual Error  26  3202.3   123.2
Total           29  8011.7
```

The model is fit to the data of Table CS4.1, resulting in the MINITAB printout shown in Figure CS4.1. The key numbers on the printout are shaded and interpreted as follows:

Global $F = 13.02$ (p-value $= .000$): At any significant level $\alpha > .0001$, we reject the null hypothesis $H_0{:}\beta_1 = \beta_2 = \beta_3 = 0$. Thus, there is sufficient evidence to indicate that the model is "statistically" useful for predicting average annual precipitation, y.

$R_a^2 = .554$: After accounting for sample size and number of β parameters in the model, approximately 55% of the sample variation in average annual precipitation levels is explained by the first-order model with altitude (x_1), latitude (x_2), and distance from Pacific coast (x_3).

$s = 11.098$: Approximately 95% of the actual average annual precipitation levels at the stations will fall within $2s = 22.2$ inches of the values predicted by the first-order model.

$\hat{\beta}_1 = .00409$: Holding latitude (x_2) and distance from coast (x_3) constant, we estimate average annual precipitation (y) of a station to increase .0041 inch for every 1-foot increase in the station's altitude (x_1).

$\hat{\beta}_2 = 3.451$: Holding altitude (x_1) and distance from coast (x_3) constant, we estimate average annual precipitation (y) of a station to increase 3.45 inches for every 1-degree increase in the station's latitude (x_2).

$\hat{\beta}_3 = -.143$: Holding altitude (x_1) and latitude (x_2) constant, we estimate average annual precipitation (y) of a station to decrease .143 inch for every 1-mile increase in the station's distance from the Pacific coast (x_3).

Note also that (1) t-tests for the three independent variables in the model are all highly significant (p-value $< .01$) and (2) the VIF's for these variables are all small (indicating very little multicollinearity). Therefore, it appears that the first-order model is adequate for predicting a meteorological station's average annual precipitation.

Can we be certain, without further analysis, that additional independent variables or higher-order terms will not improve the prediction equation? The answer, of course, is no. In the next section, we use a residual analysis to help guide us to a better model.

A Residual Analysis of the Model

The residuals of Model 1 are analyzed using the graphs discussed in Chapter 8. The SPSS printout in Figure CS4.2 shows both a histogram and a normal probability plot for the standardized residuals. Both graphs appear to support the regression assumption of normally distributed errors.

Figure CS4.2 SPSS histogram and normal probability plot of Model 1 residuals

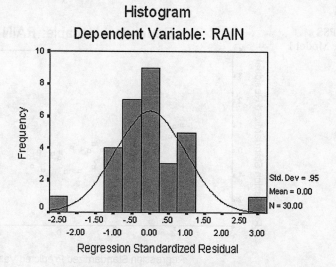

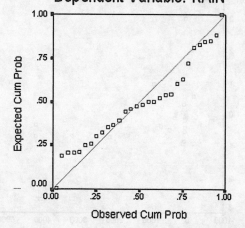

The SPSS printouts shown in Figure CS4.3 are plots of the residuals versus predicted rainfall, $\hat{y}$, and against each of the independent variables. Other than

one or two unusual observations (outliers), the plots exhibit no distinctive patterns or trends. Consequently, transformations on the independent variables for the purposes of improving the fit of the model or for stabilizing the error variance do not seem to be necessary.

On the surface, the residual plots seem to imply that no adjustments to the first-order model can be made to improve the prediction equation. Taylor, however, used his knowledge of geography and regression to examine Figure CS4.3 more closely. He found that the residuals shown in Figure CS4.3 actually exhibit a fairly consistent pattern. Taylor noticed that stations located on the west-facing slopes of the California mountains invariably had positive residuals whereas stations on the leeward side of the mountains had negative residuals.

Figure CS4.3 SPSS residual plots for Model 1

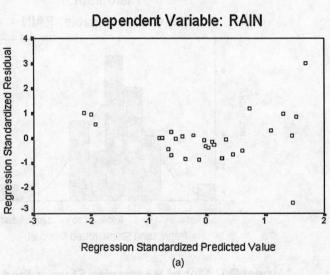

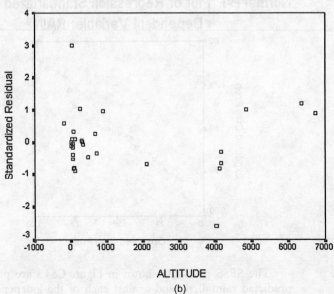

Figure CS4.3 *continued*

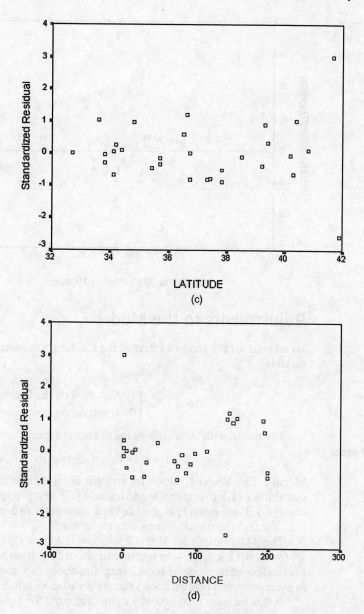

LATITUDE

(c)

DISTANCE

(d)

To see what Taylor observed more clearly, we plotted the residuals of Model 1 against $\hat{y}$ using either "W" or "L" as a plotting symbol. Stations numbered 1, 4, 5, 6, 9, 12, 16, 17, 21, 22, 23, 26, and 29 in Table CS4.1 were assigned a "W" since they all are located on west-facing slopes, whereas the remaining stations were assigned an "L" since they are leeward-facing. The revised residual plot, with a "W" or "L" assigned to each point, is shown in the SPSS printout, Figure CS4.4. You can see that with few exceptions, the "W" points have positive residuals (implying that the least squares model underpredicted the level of precipitation), whereas the "L" points have negative residuals (implying that the least squares model overpredicted the level of precipitation). In Taylor's words, the results shown in Figure CS4.4 "suggest a very clear shadow effect of the mountains, for which California is known." Thus, it appears we can improve the fit of the model by adding a variable that represents the shadow effect.

Figure CS4.4 SPSS plot of residuals for Model 1 with shadow effect

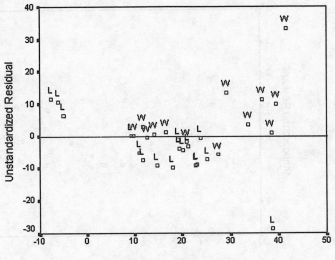

Adjustments to the Model

To account for the shadow effect of the California mountains, consider the dummy variable

$$\text{Shadow: } x_4 = \begin{cases} 1 & \text{if station on the leeward side} \\ 0 & \text{if station on the westward side} \end{cases}$$

The model with the shadow effect takes the form

Model 2

$$E(y) = \beta_0 + \beta_1 x_1 + \beta_2 x_2 + \beta_3 x_3 + \beta_4 x_4$$

Model 2, like Model 1, allows for straight-line relationships between precipitation and altitude (x_1), precipitation and latitude (x_2), and precipitation and distance from coast (x_3). The y-intercepts of these lines, however, will depend on the shadow effect (i.e., whether the station is leeward or westward).

The SAS printout for Model 2 is shown in Figure CS4.5. Note that the adjusted R^2 for Model 2 is .6963—an increase of about 15% from Model 1. This implies that the shadow-effect model (Model 2) explains about 15% more of the sample variation in average annual precipitation than the no-shadow-effect model (Model 1).

Is this increase a statistically significant one? To answer this question, we test the contribution for the shadow effect by testing

$$H_0: \beta_4 = 0 \quad \text{against} \quad H_a: \beta_4 \neq 0$$

The test statistic, shaded in Figure CS4.5, is $t = -3.63$ and the two-tailed p-value (also shaded) is $p = .0013$. Thus, there is sufficient evidence (at $\alpha = .01$) to conclude that $\beta_4 \neq 0$, that is, the shadow-effect term contributes to the prediction of average annual precipitation.

Can Model 2 be improved by adding interaction terms? Consider Model 3:

Model 3

$$E(y) = \beta_0 + \beta_1 x_1 + \beta_2 x_2 + \beta_3 x_3 + \beta_4 x_4 + \beta_5 x_1 x_4 + \beta_6 x_2 x_4 + \beta_7 x_3 x_4$$

Note that Model 3 includes interactions between the shadow effect (x_4) and each of the quantitative independent variables. This model allows the slopes of the lines relating y to x_1, y to x_2, and y to x_3 to depend on the shadow effect (x_4). The SAS printout for Model 3 is given in Figure CS4.6.

Figure CS4.5 SAS regression printout for Model 2

Dependent Variable: RAIN

Analysis of Variance

Source	DF	Sum of Squares	Mean Square	F Value	Pr > F
Model	4	5913.81738	1478.45434	17.62	<.0001
Error	25	2097.83621	83.91345		
Corrected Total	29	8011.65359			

Root MSE	9.16043	R-Square	0.7382	
Dependent Mean	19.80733	Adj R-Sq	0.6963	
Coeff Var	46.24766			

Parameter Estimates

Variable	DF	Parameter Estimate	Standard Error	t Value	Pr > |t|
Intercept	1	-97.89872	24.13791	-4.06	0.0004
ALTITUDE	1	0.00221	0.00113	1.95	0.0627
LATITUDE	1	3.45376	0.65609	5.26	<.0001
DISTANCE	1	-0.05365	0.03879	-1.38	0.1789
SHADOW	1	-15.85771	4.37100	-3.63	0.0013

Figure CS4.6 SAS printout for Model 3

Dependent Variable: RAIN

Analysis of Variance

Source	DF	Sum of Squares	Mean Square	F Value	Pr > F
Model	7	6921.64039	988.80577	19.96	<.0001
Error	22	1090.01319	49.54605		
Corrected Total	29	8011.65359			

Root MSE	7.03890	R-Square	0.8639	
Dependent Mean	19.80733	Adj R-Sq	0.8207	
Coeff Var	35.53682			

Parameter Estimates

Variable	DF	Parameter Estimate	Standard Error	t Value	Pr > |t|
Intercept	1	-160.70358	25.78066	-6.23	<.0001
ALTITUDE	1	0.00453	0.00418	1.08	0.2897
LATITUDE	1	5.14128	0.69811	7.36	<.0001
DISTANCE	1	-0.13008	0.17571	-0.74	0.4669
SHADOW	1	127.14457	37.57712	3.38	0.0027
SHAD_ALT	1	-0.00372	0.00433	-0.86	0.3992
SHAD_LAT	1	-3.78713	1.01911	-3.72	0.0012
SHAD_DIS	1	0.07079	0.17842	0.40	0.6954

Test SHAD_INTERACT Results for Dependent Variable RAIN

Source	DF	Mean Square	F Value	Pr > F
Numerator	3	335.94100	6.78	0.0021
Denominator	22	49.54605		

To determine whether these interaction terms are important, we test

$$H_0: \beta_5 = \beta_6 = \beta_7 = 0$$

$$H_a: \text{At least one of the } \beta\text{'s} \neq 0$$

The test is carried out by comparing Models 2 and 3 with the nested model partial F-test in Section 4.13. The F-test statistic, shaded at the bottom of Figure CS4.6, is $F = 6.78$ and the associated p-value (also shaded) is $p = .0021$.

Consequently, there is sufficient evidence (at $\alpha = .01$) to reject H_0 and conclude that at least one of the interaction β's is nonzero. This implies that Model 3, with the interaction terms, is a better predictor of average annual precipitation than Model 2.

The improvement of Model 3 over the other two models can be seen practically by examining R_a^2 and s on the printout, Figure CS4.6. For Model 3, $R_a^2 = .8207$, an increase of about 12% from Model 2 and 27% from Model 1. The standard deviation of Model 3 is $s = 7.04$, compared to $s = 9.16$ for Model 2 and $s = 11.1$ for Model 1. Thus, in practice, we expect Model 3 to predict average annual precipitation of a meteorological station to within about 14 inches of its true value. (This is compared to a bound on the error of prediction of about 22 inches for Model 1 and 18 inches for Model 2.) Clearly, a model that incorporates the shadow effect and its interactions with altitude, latitude, and distance from coast is a more useful predictor of average annual precipitation, y.

Conclusions

We have demonstrated how a residual analysis can help the analyst find important independent variables that were originally omitted from the regression model. This technique, however, requires substantial knowledge of the problem, data, and potentially important predictor variables. Without knowledge of the presence of a shadow effect in California, Taylor could not have enhanced the residual plot, Figure CS4.3, and consequently would not have seen its potential for improving the fit of the model.

Follow-up Questions

1. Conduct an outlier analysis of the residuals for Model 1. Identify any influential observations, and suggest how to handle these observations. (The data are saved in the CALIRAIN file.)

2. Determine whether interactions between the quantitative variables, altitude (x_1), latitude (x_2), and distance from coast (x_3), will improve the fit of the model.

Reference

Taylor, P. J. "A pedagogic application of multiple regression analysis." *Geography*, July 1980, Vol. 65, pp. 203–212.

AN INVESTIGATION OF FACTORS AFFECTING THE SALE PRICE OF CONDOMINIUM UNITS SOLD AT PUBLIC AUCTION

The Problem

This case involves a partial investigation of the factors that affect the sale price of oceanside condominium units. It represents an extension of an analysis of the same data by Herman Kelting (1979), who demonstrated that regression analysis could be a powerful tool for appraising the emerging condominium markets on the East coast of Florida.

The sales data were obtained for a new oceanside condominium complex consisting of two adjacent and connected eight-floor buildings. The complex contains 200 units of equal size (approximately 500 square feet each). The locations of the buildings relative to the ocean, the swimming pool, the parking lot, and so on, are shown in Figure CS5.1. There are several features of the complex that you should note. The units facing south, called *oceanview*, face the beach and ocean. In addition, units in building 1 have a good view of the pool. Units to the rear of the building, called *bayview*, face the parking lot and an area of land that ultimately borders a bay. The view from the upper floors of these units is primarily of wooded, sandy terrain. The bay is very distant and barely visible.

The only elevator in the complex is located at the east end of building 1, as are the office and the game room. People moving to or from the higher-floor units in building 2 would likely use the elevator and move through the passages to their units. Thus, units on the higher floors and at a greater distance from the elevator would be less convenient; they would require greater effort in moving baggage, groceries, and so on, and would be farther from the game room, the office, and the swimming pool. These units also possess an advantage: There would be the least amount of traffic through the hallways in the area, and hence they are the most private.

Lower-floor oceanside units are most suited to active people; they open onto the beach, ocean, and pool. They are within easy reach of the game room and they are easily reached from the parking area.

Checking Figure CS5.1, you will see that some of the units in the center of the complex, units numbered _11 and _14, have part of their view blocked. We would expect this to be a disadvantage. We will show you later that this expectation is true for the oceanview units and that these units sold at a lower price than adjacent oceanview units.

The condominium complex was completed during a recession; sales were slow and the developer was forced to sell most of the units at auction approximately

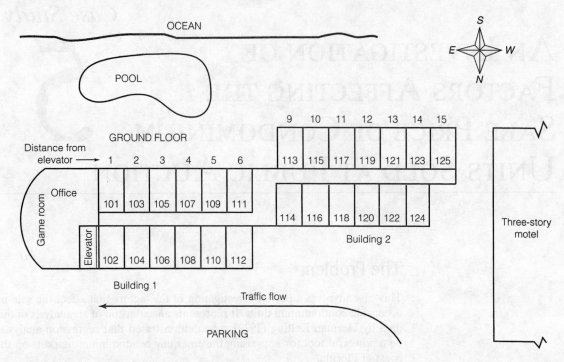

Figure CS5.1 Layout of condominium complex

18 months after opening. Many unsold units were furnished by the developer and rented prior to the auction.

This condominium complex was particularly suited to our study. The single elevator located at one end of the complex produces a remarkably high level of both inconvenience and privacy for the people occupying units on the top floors in building 2. Consequently, the data provide a good opportunity to investigate the relationship between sale price, height of the unit (floor number), distance of the unit from the elevator, and presence or absence of an ocean view. The presence or absence of furniture in each of the units also enables us to investigate the effect of the availability of furniture on sale price. Finally, the auction data are completely buyer-specified and hence consumer-oriented in contrast to most other real estate sales data, which are, to a high degree, seller- and broker-specified.

The Data

In addition to the sale price (measured in hundreds of dollars) the following data were recorded for each of the 106 units sold at the auction:

1. *Floor height* The floor location of the unit; this variable, x_1, could take values $1, 2, \ldots, 8$.

2. *Distance from elevator* This distance, measured along the length of the complex, was expressed in number of condominium units. An additional two units of distance was added to the units in building 2 to account for the walking distance in the connecting area between the two buildings. Thus, the distance of unit 105 from the elevator would be 3, and the distance between unit 113 and the elevator would be 9. This variable, x_2, could take values $1, 2, \ldots, 15$.

3. *View of ocean* The presence or absence of an ocean view was recorded for each unit and entered into the model with a dummy variable, x_3, where $x_3 = 1$ if the unit possessed an ocean view and $x_3 = 0$ if not. Note that units not possessing an ocean view would face the parking lot.

4. *End unit* We expected the partial reduction of view of end units on the ocean side (numbers ending in 11) to reduce their sale price. The ocean view of these end units is partially blocked by building 2. This qualitative variable was entered into the model with a dummy variable, x_4, where $x_4 = 1$ if the unit has a unit number ending in 11 and $x_4 = 0$ if not.

5. *Furniture* The presence or absence of furniture was recorded for each unit. This qualitative variable was entered into the model using a single dummy variable, x_5, where $x_5 = 1$ if the unit was furnished and $x_5 = 0$ if not.

CONDO
The raw data used in this analysis are saved in the CONDO file.

The Models

This case involves five independent variables, two quantitative (floor height x_1 and distance from elevator x_2) and three qualitative (view of ocean, end unit, and furniture). We postulated four models relating mean sale price to these five factors. The models, numbered 1–4, are developed in sequence, Model 1 being the simplest and Model 4, the most complex. Each of Models 2 and 3 contains all the terms of the preceding models along with new terms that we think will improve their predictive ability. Thus, Model 2 contains all the terms contained in Model 1 plus some new terms, and hence it should predict mean sale price as well as or better than Model 1. Similarly, Model 3 should predict as well as or better than Model 2. Model 4 does not contain all the terms contained in Model 3, but that is only because we have entered floor height into Model 4 as a qualitative independent variable. Consequently, Model 4 contains all the predictive ability of Model 3, and it could be an improvement over Model 3 if our theory is correct. The logic employed in this sequential model-building procedure is explained in the following discussion.

The simplest theory that we might postulate is that the five factors affect the price in an independent manner and that the effect of the two quantitative factors on sale price is linear. Thus, we envision a set of planes, each identical except for their y-intercepts. We would expect sale price planes for oceanview units to be higher than those with a bay view, those corresponding to end units (_11) would be lower than for non–end units, and those with furniture would be higher than those without.

Model 1

First-order, main effects

$$E(y) = \beta_0 + \beta_1 x_1 + \beta_2 x_2 + \beta_3 x_3 + \beta_4 x_4 + \beta_5 x_5$$

where

$x_1 = $ Floor height $(x_1 = 1, 2, \ldots, 8)$

$x_2 = $ Distance from elevator $(x_2 = 1, 2, \ldots, 15)$

$$x_3 = \begin{cases} 1 & \text{if an ocean view} \\ 0 & \text{if not} \end{cases} \quad x_4 = \begin{cases} 1 & \text{if an end unit} \\ 0 & \text{if not} \end{cases} \quad x_5 = \begin{cases} 1 & \text{if furnished} \\ 0 & \text{if not} \end{cases}$$

The second theory that we considered was that the effects on sale price of floor height and distance from elevator might not be linear. Consequently, we constructed Model 2, which is similar to Model 1 except that second-order terms are included for x_1 and x_2. This model envisions a single second-order response surface for $E(y)$ in x_1 and x_2 that possesses identically the same shape, regardless of the view, whether the unit is an end unit, and whether the unit is furnished. Expressed in other terminology, Model 2 assumes that there is no interaction between any of the qualitative factors (view of ocean, end unit, and furniture) and the quantitative factors (floor height and distance from elevator).

Model 2   Second-order, main effects

$$E(y) = \beta_0 + \overbrace{\beta_1 x_1 + \beta_2 x_2 + \beta_3 x_1 x_2 + \beta_4 x_1^2 + \beta_5 x_2^2}^{\text{Second-order model in floor and distance}}$$

$$+ \overbrace{\beta_6 x_3}^{\text{View of ocean}} + \overbrace{\beta_7 x_4}^{\text{End unit}} + \overbrace{\beta_8 x_5}^{\text{Furniture}}$$

Model 2 may possess a serious shortcoming. It assumes that the shape of the second-order response surface relating mean sale price $E(y)$ to x_1 and x_2 is identical for oceanview and bayview units. Since we think that there is a strong possibility that completely different preference patterns may govern the purchase of these two groups of units, we will construct a model that provides for two completely different second-order response surfaces—one for oceanview units and one for bayview units. Furthermore, we will assume that the effects of the two qualitative factors, end unit and furniture, are additive (i.e., their presence or absence will simply shift the mean sale price response surface up or down by a fixed amount). Thus, Model 3 is given as follows:

Model 3   Second-order, view interactions

$$E(y) = \beta_0 + \overbrace{\beta_1 x_1 + \beta_2 x_2 + \beta_3 x_1 x_2 + \beta_4 x_1^2 + \beta_5 x_2^2}^{\text{Second-order model in floor and distance}}$$

$$+ \overbrace{\beta_6 x_3}^{\text{View of ocean}} + \overbrace{\beta_7 x_4}^{\text{End unit}} + \overbrace{\beta_8 x_5}^{\text{Furniture}}$$

$$+ \overbrace{\beta_9 x_1 x_3 + \beta_{10} x_2 x_3 + \beta_{11} x_1 x_2 x_3 + \beta_{12} x_1^2 x_3 + \beta_{13} x_2^2 x_3}^{\substack{\text{Interaction of the second-order model} \\ \text{with view of ocean}}}$$

As a fourth possibility, we constructed a model similar to Model 3 but entered floor height as a qualitative factor at eight levels. This requires seven dummy variables:

$$x_6 = \begin{cases} 1 & \text{if first floor} \\ 0 & \text{if not} \end{cases}$$

$$x_7 = \begin{cases} 1 & \text{if second floor} \\ 0 & \text{if not} \end{cases}$$

$$\vdots$$

$$x_{12} = \begin{cases} 1 & \text{if seventh floor} \\ 0 & \text{if not} \end{cases}$$

Thus, Model 4 is:

Model 4   Floor height dummy variable model

$$E(y) = \beta_0 + \underbrace{\beta_1 x_2 + \beta_2 x_2^2}_{\text{Second-order in distance}}$$

$$+ \underbrace{\beta_3 x_3}_{\text{View}} + \underbrace{\beta_4 x_4}_{\text{End unit}} + \underbrace{\beta_5 x_5}_{\text{Furnished}}$$

$$+ \underbrace{\beta_6 x_6 + \beta_7 x_7 + \beta_8 x_8 + \beta_9 x_9 + \beta_{10} x_{10} + \beta_{11} x_{11} + \beta_{12} x_{12}}_{\text{Floor main effects}}$$

$$+ \underbrace{\beta_{13} x_2 x_6 + \beta_{14} x_2 x_7 + \cdots + \beta_{19} x_2 x_{12} + \beta_{20} x_2^2 x_6 + \beta_{21} x_2^2 x_7 + \cdots}_{}$$
$$\underbrace{+ \beta_{26} x_2^2 x_{12}}_{\text{Distance–floor interactions}}$$

$$+ \underbrace{\beta_{27} x_2 x_3 + \beta_{28} x_2^2 x_3}_{\text{Distance–view interactions}} + \underbrace{\beta_{29} x_3 x_6 + \beta_{30} x_3 x_7 + \cdots + \beta_{35} x_3 x_{12}}_{\text{Floor–view interactions}}$$

$$+ \beta_{36} x_2 x_3 x_6 + \beta_{37} x_2 x_3 x_7 + \cdots \beta_{42} x_2 x_3 x_{12}$$
$$+ \underbrace{\beta_{43} x_2^2 x_3 x_6 + \beta_{44} x_2^2 x_3 x_7 + \cdots \beta_{49} x_2^2 x_3 x_{12}}_{\text{Distance–view–floor interactions}}$$

The reasons for entering floor height as a qualitative factor are twofold:

1. Higher-floor units have better views but less accessibility to the outdoors. This latter characteristic could be a particularly undesirable feature for these units.
2. The views of some lower-floor bayside units were blocked by a nearby three-story motel.

If our supposition is correct and if these features would have a depressive effect on the sale price of these units, then the relationship between floor height and mean sale price would not be second-order (a smooth curvilinear relationship). Entering floor height as a qualitative factor would permit a better fit to this irregular relationship and improve the prediction equation. Thus, Model 4 is identical to Model 3 except that Model 4 contains seven main effect terms for floor height (in contrast to two for Model 3), and it also contains the corresponding interactions of these variables with the other variables included in Model 3.* We will subsequently show that there was no evidence to indicate that Model 4 contributes more information for the prediction of y than Model 3.

---

*Some of the terms in Model 4 were not estimable because sales were not consummated for some combinations of the independent variables. This is why the SAS printout in Figure CS5.5 shows only 41 df for the model.

The Regression Analyses

This section gives the regression analyses for the four models described previously. You will see that our approach is to build the model in a sequential manner. In each case, we use a partial F-test to see whether a particular model contributes more information for the prediction of sale price than its predecessor.

This procedure is more conservative than a step-down procedure. In a step-down approach you would assume Model 4 to be the appropriate model, then test and possibly delete terms. But deleting terms can be particularly risky because, in doing so, you are tacitly accepting the null hypothesis. Thus, you risk deleting important terms from the model and do so with an unknown probability of committing a Type II error.

Do not be unduly influenced by the individual t-tests associated with an analysis. As you will see, it is possible for a set of terms to contribute information for the prediction of y when none of their respective t-values are statistically significant. This is because the t-test focuses on the contribution of a single term, given that all the other terms are retained in the model. Therefore, if a set of terms contributes overlapping information, it is possible that none of the terms individually would be statistically significant, even when the set as a whole contributes information for the prediction of y.

The SAS regression analysis computer printouts for fitting Models 1, 2, 3, and 4 to the data are shown in Figures CS5.2, CS5.3, CS5.4, and CS5.5, respectively. A summary containing the key results (e.g., R_a^2 and s) for these models is provided in Table CS5.1.

Examining the SAS printout for the first-order model (Model 1) in Figure CS5.2, you can see that the value of the F statistic for testing the null hypothesis

$$H_0: \quad \beta_1 = \beta_2 = \cdots = \beta_5 = 0$$

is 47.98. This is statistically significant at a level of $\alpha = .0001$. Consequently, there is ample evidence to indicate that the overall model contributes information for the prediction of y. At least one of the five factors contributes information for the prediction of sale price.

Figure CS5.2 SAS regression printout for Model 1

Dependent Variable: PRICE100

Analysis of Variance

Source	DF	Sum of Squares	Mean Square	F Value	Pr > F
Model	5	23534	4706.82958	47.98	<.0001
Error	100	9810.07851	98.10079		
Corrected Total	105	33344			

Root MSE	9.90458	R-Square	0.7058	
Dependent Mean	191.81132	Adj R-Sq	0.6911	
Coeff Var	5.16371			

Parameter Estimates

Variable	DF	Parameter Estimate	Standard Error	t Value	Pr > |t|
Intercept	1	177.70349	4.16842	42.63	<.0001
FLOOR	1	-0.71514	0.53077	-1.35	0.1809
DIST	1	-0.87325	0.24495	-3.57	0.0006
VIEW	1	31.27285	2.23121	14.02	<.0001
END	1	-17.80782	3.98195	-4.47	<.0001
FURNISH	1	9.98376	2.05150	4.87	<.0001

Table CS5.1 A summary of the regressions for Models 1, 2, 3, and 4

Model	df(Model)	SSE	df(Error)	MSE	R_a^2	s
1	5	9810	100	98.1	.691	9.90
2	8	9033	97	93.1	.707	9.65
3	13	7844	92	85.3	.732	9.23
4	41	5207	64	81.4	.744	9.02

If you examine the t-tests for the individual parameters, you will see that they are all statistically significant except the test for β_1, the parameter associated with floor height x_1 (p-value $= .1809$). The failure of floor height x_1 to reveal itself as an important information contributor goes against our intuition, and it demonstrates the pitfalls that can attend an unwary attempt to interpret the results of t-tests in a regression analysis. Intuitively, we would expect floor height to be an important factor. You might argue that units on the higher floors possess a better view and hence should command a higher mean sale price. Or, you might argue that units on the lower floors have greater accessibility to the pool and ocean and, consequently, should be in greater demand. Why, then, is the t-test for floor height not statistically significant? The answer is that both of the preceding arguments are correct, one for the oceanside and one for the bayside. Thus, you will subsequently see that there is an interaction between floor height and view of ocean. Oceanview units on the lower floors sell at higher prices than oceanview units on the higher floors. In contrast, bayview units on the higher floors command higher prices than bayview units on the lower floors. These two contrasting effects tend to cancel (because we have not included interaction terms in the model) and thereby give the false impression that floor height is not an important variable for predicting mean sale price.

But, of course, we are looking ahead. Our next step is to determine whether Model 2 is better than Model 1.

Are floor height x_1 and distance from elevator x_2 related to sale price in a curvilinear manner, that is, should we be using a second-order response surface instead of a first-order surface to relate $E(y)$ to x_1 and x_2? To answer this question, we examine the drop in SSE from Model 1 to Model 2. The null hypothesis "Model 2 contributes no more information for the prediction of y than Model 1" is equivalent to testing

$$H_0: \quad \beta_3 = \beta_4 = \beta_5 = 0$$

where β_3, β_4, and β_5 appear in Model 2. The F statistic for the test, based on 3 and 97 df, is

$$F = \frac{(\text{SSE}_1 - \text{SSE}_2)/\#\beta\text{'s in } H_0}{\text{MSE}_2} = \frac{(9810 - 9033)/3}{93.1}$$

$$= 2.78$$

This value is shown (shaded) at the bottom of Figure CS5.3 as well as the corresponding p-value of the test. Since $\alpha = .05$ exceeds p-value $= .0452$, we reject H_0. There is evidence to indicate that Model 2 contributes more information for the prediction of y than Model 1. This tells us that there is evidence of curvature in the response surfaces relating mean sale price, $E(y)$, to floor height x_1 and distance from elevator x_2.

You will recall that the difference between Models 2 and 3 is that Model 3 allows for two different-shaped second-order surfaces, one for oceanview units and

Figure CS5.3 SAS
regression printout for
Model 2

```
                        Dependent Variable: PRICE100

                           Analysis of Variance

                                  Sum of         Mean
Source                DF        Squares        Square    F Value    Pr > F

Model                  8          24311    3038.84678      32.63    <.0001
Error                 97     9033.45215      93.12837
Corrected Total      105          33344


           Root MSE            9.65030    R-Square     0.7291
           Dependent Mean    191.81132    Adj R-Sq     0.7067
           Coeff Var           5.03114


                           Parameter Estimates

                     Parameter      Standard
Variable      DF     Estimate          Error    t Value    Pr > |t|

Intercept      1    194.59573        7.66067      25.40    <.0001
FLOOR          1     -6.83830        2.45493      -2.79     0.0064
DIST           1     -2.64122        1.22794      -2.15     0.0340
FLR_DIS        1      0.04384        0.13564       0.32     0.7472
FLOORSQ        1      0.58394        0.23786       2.45     0.0159
DISTSQ         1      0.11426        0.07714       1.48     0.1418
VIEW           1     30.42124        2.19627      13.85    <.0001
END            1    -16.80585        4.10416      -4.09    <.0001
FURNISH        1     11.27207        2.05390       5.49    <.0001


     Test SECOND_ORDER Results for Dependent Variable PRICE100

                                  Mean
Source                DF        Square    F Value    Pr > F

Numerator              3     258.87545       2.78    0.0452
Denominator           97      93.12837
```

another for bayview units; Model 2 employs a single surface to represent both types of units. Consequently, we wish to test the null hypothesis that "a single second-order surface adequately characterizes the relationship between $E(y)$, floor height x_1, and distance from elevator x_2 for both oceanview and bayview units" [i.e., Model 2 adequately models $E(y)$] against the alternative hypothesis that you need two different second-order surfaces [i.e, you need Model 3]. Thus,

$$H_0: \quad \beta_9 = \beta_{10} = \cdots = \beta_{13} = 0$$

where $\beta_9, \beta_{10}, \ldots, \beta_{13}$ are parameters in Model 3. The F statistic for this test, based on 5 and 92 df, is

$$
\begin{aligned}
F &= \frac{(\text{SSE}_2 - \text{SSE}_3)/\#\beta\text{'s in } H_0}{\text{MSE}_3} \\
&= \frac{(9033 - 7844)/5}{85.3} \\
&= 2.79
\end{aligned}
$$

This value and its associated p-value are shown (shaded) at the bottom of Figure CS5.4. Since $\alpha = .05$ exceeds p-value $= .0216$, we reject H_0 and conclude that there is evidence to indicate that we need two different second-order surfaces to relate $E(y)$ to x_1 and x_2, one each for oceanview and bayview units.

Finally, we question whether Model 4 will provide an improvement over Model 3, that is, will we gain information for predicting y by entering floor height into the model as a qualitative factor at eight levels? Although Models 3 and 4 are not "nested" models, we can compare them using a "partial" F-test. The F statistic to test the null hypothesis "Model 4 contributes no more information for predicting y than does Model 3" compares the drop in SSE from Model 3 to Model 4 with s_4^2.

Figure CS5.4 SAS regression printout for Model 3

Dependent Variable: PRICE100

Analysis of Variance

Source	DF	Sum of Squares	Mean Square	F Value	Pr > F
Model	13	25500	1961.56185	23.01	<.0001
Error	92	7843.92233	85.26003		
Corrected Total	105	33344			

Root MSE	9.23364	R-Square	0.7648	
Dependent Mean	191.81132	Adj R-Sq	0.7315	
Coeff Var	4.81392			

Parameter Estimates

Variable	DF	Parameter Estimate	Standard Error	t Value	Pr > |t|
Intercept	1	142.05564	27.95942	5.08	<.0001
FLOOR	1	8.79374	9.41542	0.93	0.3528
DIST	1	-4.00545	3.03226	-1.32	0.1898
FLR_DIS	1	0.32600	0.34146	0.95	0.3422
FLOORSQ	1	-0.58321	0.80662	-0.72	0.4715
DISTSQ	1	0.11069	0.16598	0.67	0.5065
VIEW	1	84.52959	28.85730	2.93	0.0043
END	1	-16.30753	4.00143	-4.08	<.0001
FURNISH	1	12.56003	2.04377	6.15	<.0001
FLR_VIEW	1	-14.11343	9.74820	-1.45	0.1511
DIS_VIEW	1	0.36326	3.34173	0.11	0.9137
F_D_VIEW	1	-0.29131	0.37258	-0.78	0.4363
FLRSQ_VU	1	0.95473	0.84502	1.13	0.2615
DISSQ_VU	1	0.06503	0.19068	0.34	0.7338

Test VIEW_INTERACT Results for
Dependent Variable PRICE100

Source	DF	Mean Square	F Value	Pr > F
Numerator	5	237.90596	2.79	0.0216
Denominator	92	85.26003		

This F statistic, based on 28 df (the difference in the numbers of parameters in Models 4 and 3) and 64 df, is

$$F = \frac{(SSE_3 - SSE_4)/\#\beta's \text{ in } H_0}{MSE_4}$$

$$= \frac{(7844 - 5207)/28}{81.4}$$

$$= 1.16$$

Since the p-value of the test is not shown in Figure CS5.5, we will conduct the test by finding the rejection region. Checking Table 4 in Appendix C, you will find that the value for $F_{.05}$, based on 28 and 64 df, is approximately 1.65. Since the computed

Figure CS5.5 Portion of SAS regression printout for Model 4

Dependent Variable: PRICE100

Analysis of Variance

Source	DF	Sum of Squares	Mean Square	F Value	Pr > F
Model	41	28137	686.27992	8.44	<.0001
Error	64	5206.74955	81.35546		
Corrected Total	105	33344			

Root MSE	9.01973	R-Square	0.8438	
Dependent Mean	191.81132	Adj R-Sq	0.7438	
Coeff Var	4.70240			

F is less than this value, there is not sufficient evidence to indicate that Model 4 is a significant improvement over Model 3.

Having checked the four theories discussed previously, we have evidence to indicate that Model 3 is the best of the four models. The R_a^2 for the model implies that about 73% of the sample variation in sale prices can be explained. We examine the prediction equation for Model 3 and see what it tells us about the relationship between the mean sale price $E(y)$ and the five factors used in our study; but first, it is important that we examine the residuals for Model 3 to determine whether the standard least squares assumptions about the random error term are satisfied.

An Analysis of the Residuals from Model 3

The four standard least squares assumptions about the random error term ε are (from Chapter 4) the following:

1. The mean is 0.
2. The variance (σ^2) is constant for all settings of the independent variables.
3. The errors follow a normal distribution.
4. The errors are independent.

If one or more of these assumptions are violated, any inferences derived from the Model 3 regression analysis are suspect. It is unlikely that assumption 1 (0 mean) is violated because the method of least squares guarantees that the mean of the residuals is 0. The same can be said for assumption 4 (independent errors) since the sale price data are not a time series. However, verifying assumptions 2 and 3 requires a thorough examination of the residuals from Model 3.

Recall that we can check for heteroscedastic errors (i.e., errors with unequal variances) by plotting the residuals against the predicted values. This residual plot is shown in Figure CS5.6. If the variances were not constant, we would expect to see a cone-shaped pattern (since the response is sale price) in Figure CS5.6, with the spread of the residuals increasing as $\hat{y}$ increases. Note, however, that except for one point that appears at the top of the graph, the residuals appear to be randomly scattered around 0. Therefore, assumption 2 (constant variance) appears to be satisfied.

Figure CS5.6 SAS plot of standardized residuals from Model 3

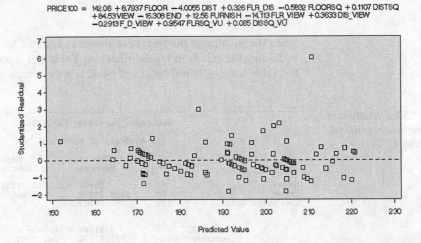

PRICE100 = 142.06 + 8.7937 FLOOR − 4.0065 DIST + 0.326 FLR_DIS − 0.5832 FLOORSQ + 0.1107 DISTSQ + 84.53 VIEW − 16.308 END + 12.56 FURNISH − 14.113 FLR_VIEW + 0.3633 DIS_VIEW − 0.2913 F_D_VIEW + 0.9547 FLRSQ_VU + 0.065 DISSQ_VU

To check the normality assumption (assumption 3), we have generated a SAS histogram of the residuals in Figure CS5.7. It is very evident that the distribution of residuals is not normal, but skewed to the right. At this point, we could opt to use a transformation on the response (similar to the variance-stabilizing transformations discussed in Section 8.4) to normalize the residuals. However, a nonnormal error distribution is often due to the presence of a single outlier. If this outlier is eliminated (or corrected), the normality assumption may then be satisfied.

Figure CS5.7 SAS histogram of residuals from Model 3

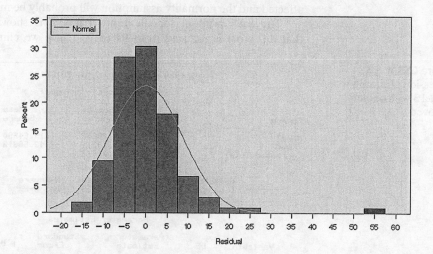

Figure CS5.6 is a plot of the *standardized* residuals. In Section 8.5, we defined outliers to have standardized residuals that exceed 3 (in absolute value) and suspect outliers as standardized residuals that fall between 2 and 3 (in absolute value). Therefore, we can use Figure CS5.6 to detect outliers. Note that there is one outlier and one suspect outlier, both with large *positive* standardized residuals (approximately 7 and 3, respectively). Should we automatically eliminate these two observations from the analysis and refit Model 3? Although many analysts adopt such an approach, we should carefully examine the observations before deciding to eliminate them. We may discover a correctable recording (or coding) error, or we may find that the outliers are very influential and are due to an inadequate model (in which case, it is the model that needs fixing, not the data).

An examination of the SAS printout of the Model 3 residuals (not shown) reveals that the two observations in question are identified by observation numbers 35 and 49 (where the observations are numbered from 1 to 106). The sale prices, floor heights, and so forth, for these two data points were found to be recorded and coded correctly. To determine how influential these outliers are on the analysis, influence diagnostics were generated using SAS. The results are summarized in Table CS5.2.

Table CS5.2 Influence diagnostics for two outliers in Model 3

Observation	Response y	Predicted Value $\hat{y}$	Residual $y - \hat{y}$	Leverage h	Cook's Distance D
35	265	210.77	54.23	.0605	.169
49	210	184.44	25.56	.1607	.125

Based on the "rules of thumb" given in Section 8.6, neither observation has strong influence on the analysis. Both leverage (h) values fall below $2(k + 1)/n = 2(14)/106 = .264$, indicating that the observations are not influential with respect to their x-values; and both Cook's D values fall below .96 [the 50th percentile of an F distribution with $\nu_1 = k + 1 = 14$ and $\nu_2 = n - (k + 1) = 106 - 14 = 92$ degrees of freedom], implying that they do not exhibit strong overall influence on the regression results (e.g., the β estimates). Consequently, if we remove these outliers from the data and refit Model 3, the least squares prediction equation will not be greatly affected and the normality assumption will probably be more nearly satisfied.

The SAS printout for the refitted model is shown in Figure CS5.8. Note that df(Error) is reduced from 92 to 90 (since we eliminated the two outlying

Figure CS5.8 SAS regression printout for Model 3 with outliers removed

Dependent Variable: PRICE100

Analysis of Variance

Source	DF	Sum of Squares	Mean Square	F Value	Pr > F
Model	13	23655	1819.64590	41.77	<.0001
Error	90	3921.13212	43.56813		
Corrected Total	103	27577			

Root MSE	6.60062	R-Square	0.8578
Dependent Mean	190.93269	Adj R-Sq	0.8373
Coeff Var	3.45704		

Parameter Estimates

Variable	DF	Parameter Estimate	Standard Error	t Value	Pr > |t|
Intercept	1	151.73426	20.09147	7.55	<.0001
FLOOR	1	3.76115	6.84884	0.55	0.5843
DIST	1	-1.93178	2.22898	-0.87	0.3884
FLR_DIS	1	0.25242	0.24477	1.03	0.3052
FLOORSQ	1	-0.15321	0.58763	-0.26	0.7949
DISTSQ	1	0.00420	0.12159	0.03	0.9725
VIEW	1	77.52403	20.73006	3.74	0.0003
END	1	-14.99258	2.86477	-5.23	<.0001
FURNISH	1	10.89332	1.47165	7.40	<.0001
FLR_VIEW	1	-10.91058	7.08792	-1.54	0.1272
DIS_VIEW	1	-1.79122	2.44305	-0.73	0.4653
F_D_VIEW	1	-0.24675	0.26701	-0.92	0.3579
FLRSQ_VU	1	0.72313	0.61542	1.18	0.2431
DISSQ_VU	1	0.20033	0.13878	1.44	0.1523

Figure CS5.9 SAS plot of standardized residuals from Model 3 with outliers removed

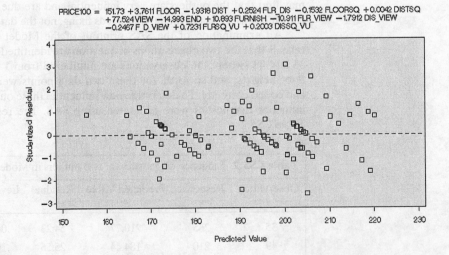

PRICE100 = 151.73 + 3.7611 FLOOR − 1.9318 DIST + 0.2524 FLR_DIS − 0.1532 FLOORSQ + 0.0042 DISTSQ + 77.524 VIEW − 14.993 END + 10.893 FURNISH − 10.911 FLR_VIEW − 1.7912 DIS_VIEW − 0.2467 F_D_VIEW + 0.7231 FLRSQ_VU + 0.2003 DISSQ_VU

observations), and the β estimates remain relatively unchanged. However, the model standard deviation is decreased from 9.23 to 6.60 and the R^2_{adj} is increased from .73 to .84, implying that the refitted model will yield more accurate predictions of sale price. A residual plot for the refitted model is shown in Figure CS5.9 and a histogram of the residuals in Figure CS5.10. The residual plot (Figure CS5.9) reveals no evidence of outliers, and the histogram of the residuals (Figure CS5.10) is now approximately normal.

Figure CS5.10 SAS histogram of residuals from Model 3 with outliers removed

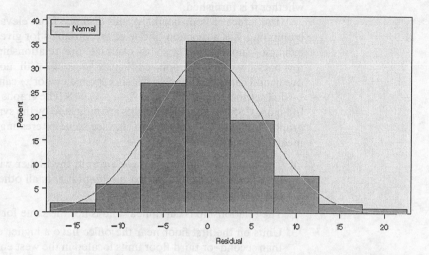

What the Model 3 Regression Analysis Tells Us

We have settled on Model 3 (with two observations deleted) as our choice to relate mean sale price $E(y)$ to five factors: the two quantitative factors, floor height x_1 and distance from elevator x_2; and the three qualitative factors, view of ocean, end unit, and furniture. This model postulates two different second-order surfaces relating mean sale price $E(y)$ to x_1 and x_2, one for oceanview units and one for bayview units. The effect of each of the two qualitative factors, end unit (numbered _11) and furniture, is to produce a change in mean sale price that is identical for all combinations of values of x_1 and x_2. In other words, assigning a value of 1 to one of the dummy variables increases (or decreases) the estimated mean sale price by a fixed amount. The net effect is to push the second-order surface upward or downward, with the direction depending on the level of the specific qualitative factor. The estimated increase (or decrease) in mean sale price because of a given qualitative factor is given by the estimated value of the β parameter associated with its dummy variable.

For example, the prediction equation (with rounded values given for the parameter estimates) obtained from Figure CS5.8 is

$$\hat{y} = 151.7 + 3.76x_1 - 1.93x_2$$
$$+.25x_1x_2 - .15x_1^2 + .004x_2^2$$
$$+77.52x_3 - 14.99x_4 + 10.89x_5$$
$$-10.91x_1x_3 - 1.79x_2x_3 - .25x_1x_2x_3$$
$$+.72x_1^2x_3 + .20x_2^2x_3$$

Since the dummy variables for end unit and furniture are, respectively, x_4 and x_5, the estimated changes in mean sale price for these qualitative factors are

$$\text{End unit } (x_4 = 1): \quad \hat{\beta}_7 \times (\$100) = -\$1,499$$

$$\text{Furnished } (x_5 = 1): \quad \hat{\beta}_8 \times (\$100) = +\$1,089$$

Thus, if you substitute $x_4 = 1$ into the prediction equation, the estimated mean decrease in sale price for an end unit is $1,499, regardless of the view, floor, and whether it is furnished.

The effect of floor height x_1 and distance from elevator x_2 can be determined by plotting $\hat{y}$ as a function of one of the variables for given values of the other. For example, suppose we wish to examine the relationship between $\hat{y}$, x_1, and x_2 for bayview ($x_3 = 0$), non–end units ($x_4 = 0$) with no furniture ($x_5 = 0$). The prediction curve relating $\hat{y}$ to distance from elevator x_2 can be graphed for each floor by first setting $x_1 = 1$, then $x_1 = 2, \ldots, x_1 = 8$. The graphs of these curves are shown in Figure CS5.11. The floor heights are indicated by the symbols at the bottom of the graph. In Figure CS5.11, we can also see some interesting patterns in the estimated mean sale prices:

1. The higher the floor of a bayview unit, the higher will be the mean sale price. Low floors look out onto the parking lot and, all other variables held constant, are least preferred.

2. The relationship is curvilinear and is not the same for each floor.

3. Units on the first floor near the office have a higher estimated mean sale price than second- or third-floor units located in the west end of the complex. Perhaps the reason for this is that these units are close to the pool and the game room, and these advantages outweigh the disadvantage of a poor view.

4. The mean sale price decreases as the distance from the elevator and center of activity increases for the lower floors, but the decrease is less as you move upward, floor to floor. Finally, note that the estimated mean sale price increases substantially for units on the highest floor that are farthest away from the elevator. These units are subjected to the least human traffic and are, therefore,

Figure CS5.11 SAS graph of predicted price versus distance (bayview units)

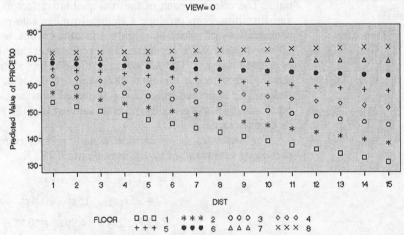

the most private. Consequently, a possible explanation for their high price is that buyers place a higher value on the privacy provided by the units than the negative value that they assign to their inconvenience. One additional explanation for the generally higher estimated sale price for units at the ends of the complex may be that they possess more windows.

A similar set of curves is shown in Figure CS5.12 for oceanview units ($x_3 = 1$). You will note some amazing differences between these curves and those for the bayview units in Figure CS5.11 (these differences explain why we needed two separate second-order surfaces to describe these two sets of units). The preference for floors is completely reversed on the ocean side of the complex: the lower the floor, the higher the estimated mean sale price. Apparently, people selecting the oceanview units are primarily concerned with accessibility to the ocean, pool, beach, and game room. Note that the estimated mean sale price is highest near the elevator. It drops and then rises as you reach the units farthest from the elevator. An explanation for this phenomenon is similar to the one that we used for the bayside units. Units near the elevator are more accessible and nearer to the recreational facilities. Those farthest from the elevators afford the greatest privacy. Units near the center of the complex offer reduced amounts of both accessibility *and* privacy. Notice that units adjacent to the elevator command a higher estimated mean sale price than those near the west end of the complex, suggesting that accessibility has a greater influence on price than privacy.

Figure CS5.12 SAS graph of predicted price versus distance (oceanview units)

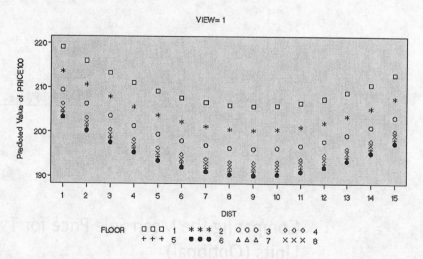

Rather than examine the graphs of $\hat{y}$ as a function of distance from elevator x_2, you may want to see how $\hat{y}$ behaves as a function of floor height x_1 for units located at various distances from the elevator. These estimated mean sale price curves are shown for bayview units in Figure CS5.13 and for oceanview units in Figure CS5.14. To avoid congestion in the graphs, we have shown only the curves for distances $x_2 = 1$, 5, 10, and 15. The symbols representing these distances are shown at the bottom of the graphs. We leave it to you and to the real estate experts to deduce the practical implications of these curves.

Figure CS5.13 SAS graph
of predicted price versus
floor height (bayview units)

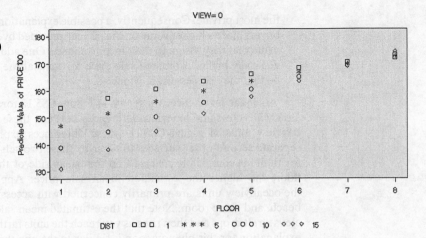

Figure CS5.14 SAS graph
of predicted price versus
floor height (ocean-
view units)

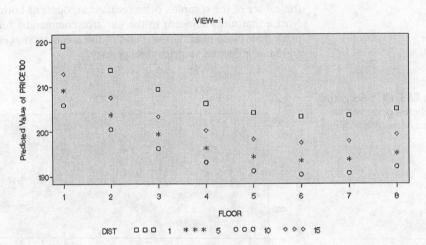

Comparing the Mean Sale Price for Two Types of Units (Optional)

[*Note*: This section requires an understanding of the mechanics of a multiple regression analysis presented in Appendix B.]

Comparing the mean sale price for two types of units might seem a useless endeavor, considering that all the units have been sold and that we will never be able to sell the units in the same economic environment that existed at the time the data were collected. Nevertheless, this information might be useful to a real estate appraiser or to a developer who is pricing units in a new and similar condominium complex. We will assume that a comparison is useful and show you how it can be accomplished.

Suppose you want to estimate the difference in mean sale price for units in two different locations and with or without furniture. For example, suppose you wish to estimate the difference in mean sale price between the first-floor oceanview and bayview units located at the east end of building 1 (units 101 and 102 in Figure CS5.1). Both these units afford a maximum of accessibility, but they possess different views. We assume that both are furnished. The estimate of the mean sale price $E(y)$ for the first-floor, bayview unit will be the value of $\hat{y}$ when $x_1 = 1$, $x_2 = 1$, $x_3 = 0$, $x_4 = 0$, $x_5 = 1$. Similarly, the estimated value of $E(y)$ for the first-floor, oceanview unit is obtained by substituting $x_1 = 1$, $x_2 = 1$, $x_3 = 1$, $x_4 = 0$, and $x_5 = 1$ into the prediction equation.

We will let $\hat{y}_o$ and $\hat{y}_b$ represent the estimated mean sale prices for the first-floor oceanview and bayview units, respectively. Then the estimator of the difference in mean sale prices for the two units is

$$\ell = \hat{y}_o - \hat{y}_b$$

We represent this estimator by the symbol ℓ, because it is a linear function of the parameter estimators $\hat{\beta}_0, \hat{\beta}_1, \ldots, \hat{\beta}_{13}$, that is,

$$\hat{y}_o = \hat{\beta}_0 + \hat{\beta}_1(1) + \hat{\beta}_2(1) + \hat{\beta}_3(1)(1) + \hat{\beta}_4(1)^2 + \hat{\beta}_5(1)^2$$
$$+ \hat{\beta}_6(1) + \hat{\beta}_7(0) + \hat{\beta}_8(1) + \hat{\beta}_9(1)(1) + \hat{\beta}_{10}(1)(1)$$
$$+ \hat{\beta}_{11}(1)(1)(1) + \hat{\beta}_{12}(1)^2(1) + \hat{\beta}_{13}(1)^2(1)$$

$$\hat{y}_b = \hat{\beta}_0 + \hat{\beta}_1(1) + \hat{\beta}_2(1) + \hat{\beta}_3(1)(1) + \hat{\beta}_4(1)^2 + \hat{\beta}_5(1)^2$$
$$+ \hat{\beta}_6(0) + \hat{\beta}_7(0) + \hat{\beta}_8(1) + \hat{\beta}_9(1)(0) + \hat{\beta}_{10}(1)(0)$$
$$+ \hat{\beta}_{11}(1)(1)(0) + \hat{\beta}_{12}(1)^2(0) + \hat{\beta}_{13}(1)^2(0)$$

then

$$\ell = \hat{y}_o - \hat{y}_b = \hat{\beta}_6 + \hat{\beta}_9 + \hat{\beta}_{10} + \hat{\beta}_{11} + \hat{\beta}_{12} + \hat{\beta}_{13}$$

A 95% confidence interval for the mean value of a linear function of the estimators $\hat{\beta}_0, \hat{\beta}_1, \ldots, \hat{\beta}_k$, given in Section B.7 of Appendix B, is

$$\ell \pm (t_{.025})s\sqrt{\mathbf{a}'(\mathbf{X}'\mathbf{X})^{-1}\mathbf{a}}$$

where in our case, $\ell = \hat{y}_o - \hat{y}_b$ is the estimate of the difference in mean values for the two units, $E(y_o) - E(y_b)$; s is the least squares estimate of the standard deviation from the regression analysis of Model 3 (Figure CS5.8); and $(\mathbf{X}'\mathbf{X})^{-1}$, the inverse matrix for the Model 3 regression analysis, is shown in Figure CS5.15. The $\mathbf{a}$ matrix is a column matrix containing elements $a_0, a_1, a_2, \ldots, a_{13}$, where $a_0, a_1, a_2, \ldots, a_{13}$ are the coefficients of $\hat{\beta}_0, \hat{\beta}_1, \ldots, \hat{\beta}_{13}$ in the linear function ℓ, that is,

$$\ell = a_0\hat{\beta}_0 + a_1\hat{\beta}_1 + \cdots + a_{13}\hat{\beta}_{13}$$

Since our linear function is

$$\ell = \hat{\beta}_6 + \hat{\beta}_9 + \hat{\beta}_{10} + \hat{\beta}_{11} + \hat{\beta}_{12} + \hat{\beta}_{13}$$

it follows that $a_6 = a_9 = a_{10} = a_{11} = a_{12} = a_{13} = 1$ and $a_0 = a_1 = a_2 = a_3 = a_4 = a_5 = a_7 = a_8 = 0$.

Substituting the values of $\hat{\beta}_6, \hat{\beta}_9, \hat{\beta}_{10}, \ldots, \hat{\beta}_{13}$ (given in Figure CS5.8) into ℓ, we have

$$\ell = \hat{y}_o - \hat{y}_b = \hat{\beta}_6 + \hat{\beta}_9 + \hat{\beta}_{10} + \hat{\beta}_{11} + \hat{\beta}_{12} + \hat{\beta}_{13}$$
$$= 77.52 - 10.91 - 1.79 - .25 + .72 + .20 = 65.49$$

or, $6,549.

The value of $t_{.025}$ needed for the confidence interval is approximately equal to 1.96 (because of the large number of degrees of freedom), and the value of s, given in Figure CS5.8, is $s = 6.60$. Finally, the matrix product $\mathbf{a}'(\mathbf{X}'\mathbf{X})^{-1}\mathbf{a}$ can be obtained by multiplying the $\mathbf{a}$ matrix (described in the preceding paragraph) and the $(\mathbf{X}'\mathbf{X})^{-1}$ matrix given in Figure CS5.15. It can be shown (proof omitted) that this matrix product is the sum of the elements of the $(\mathbf{X}'\mathbf{X})^{-1}$ matrix highlighted in Figure CS5.15. Substituting these values into the formula for the confidence interval, we obtain

$$\overbrace{\hat{y}_0 - \hat{y}_b}^{\ell} \pm t_{.025}s\sqrt{\mathbf{a}'(\mathbf{X}'\mathbf{X})^{-1}\mathbf{a}} = 65.49 \pm (1.96)(6.60)\sqrt{4.55}$$

$$= 65.49 \pm 27.58 = (37.91, 93.07)$$

Therefore, we estimate the difference in the mean sale prices of first-floor oceanview and bayview units (units 101 and 102) to lie within the interval \$3,791 to \$9,307.

You can use the technique described above to compare the mean sale prices for any pair of units.

Figure CS5.15 SAS printout of $(\mathbf{X}'\mathbf{X}^{-1})$ matrix for Model 3

Variable	Intercept	FLOOR	DIST	FLR_DIS	FLOORSQ
Intercept	9.2651939926	-2.992278509	-0.256509984	0.0170883115	0.2368571042
FLOOR	-2.992278509	1.0766279902	-0.005854674	0.0023211328	-0.090368025
DIST	-0.256509984	-0.005854674	0.1140368456	-0.008968213	0.004048291
FLR_DIS	0.0170883115	0.0023211328	-0.008968213	0.0013751281	-0.000755996
FLOORSQ	0.2368571042	-0.090368025	0.004048291	-0.000755996	0.0079255903
DISTSQ	0.0133588561	-0.001568449	-0.004451619	0.0000311222	0.0001270988
VIEW	-9.261052131	2.9935687455	0.2560234919	-0.017075652	-0.23705016
END	-0.003509088	-0.001093121	0.000412168	-0.000010725	0.0001635619
FURNISH	-0.036216143	-0.011281739	0.0042538497	-0.000110692	0.0016880681
FLR_VIEW	2.9939400548	-1.0761104	0.0056595128	-0.002316054	0.090290579
DIS_VIEW	0.2609271215	0.0072306621	-0.114555671	0.008981714	-0.004254178
F_D_VIEW	-0.016875095	-0.002254713	0.0089431694	-0.001374476	0.0007460576
FLRSQ_VU	-0.237301892	0.0902294686	-0.003996047	0.0007546363	-0.007904858
DISSQ_VU	-0.013735053	0.0014512596	0.0044958064	-0.000032272	-0.000109564

Variable	DISTSQ	VIEW	END	FURNISH	FLR_VIEW
Intercept	0.0133588561	-9.261052131	-0.003509088	-0.036216143	2.9939400548
FLOOR	-0.001568449	2.9935687455	-0.001093121	-0.011281739	-1.0761104
DIST	-0.004451619	0.2560234919	0.000412168	0.0042538497	0.0056595128
FLR_DIS	0.0000311222	-0.017075652	-0.000010725	-0.000110692	-0.002316054
FLOORSQ	0.0001270988	-0.23705016	0.0001635619	0.0016880681	0.090290579
DISTSQ	0.0003393423	-0.013335485	-0.000019801	-0.00020436	0.0015778251
VIEW	-0.013335485	9.8635213822	-0.003460162	0.0305310915	-3.162628076
END	-0.000019801	-0.003460162	0.188369971	0.0048165173	0.0123690474
FURNISH	-0.00020436	0.0305310915	0.0048165173	0.0497096845	0.0090011283
FLR_VIEW	0.0015778251	-3.162628076	0.0123690474	0.0090011283	1.1531049299
DIS_VIEW	0.0044765442	-0.325308972	-0.020075044	-0.010316741	-0.004258676
F_D_VIEW	-0.000029919	0.0221320773	-0.000318277	-0.000181965	0.0015446723
FLRSQ_VU	-0.000129609	0.2497183294	-0.000886821	-0.001077559	-0.097391471
DISSQ_VU	-0.000341465	0.0159493901	0.0015437548	0.0007207221	-0.001457361

Variable	DIS_VIEW	F_D_VIEW	FLRSQ_VU	DISSQ_VU
Intercept	0.2609271215	-0.016875095	-0.237301892	-0.013735053
FLOOR	0.0072306621	-0.002254713	0.0902294686	0.0014512596
DIST	-0.114555671	0.0089431694	-0.003996047	0.0044958064
FLR_DIS	0.008981714	-0.001374476	0.0007546363	-0.000032272
FLOORSQ	-0.004254178	0.0007460576	-0.007904858	-0.000109564
DISTSQ	0.0044765442	-0.000029919	-0.000129609	-0.000341465
VIEW	-0.325308972	0.0221320773	0.2497183294	0.0159493901
END	-0.020075044	-0.000318277	-0.000886821	0.0015437548
FURNISH	-0.010316741	-0.000181965	-0.001077559	0.0007207221
FLR_VIEW	-0.004258676	0.0015446723	-0.097391471	-0.001457361
DIS_VIEW	0.1369922565	-0.00972204	0.0044120185	-0.005677362
F_D_VIEW	-0.00972204	0.0016363944	-0.000836949	-0.000017251
FLRSQ_VU	0.0044120185	-0.000836949	0.0086931319	0.0001387393
DISSQ_VU	-0.005677362	-0.000017251	0.0001387393	0.0004420689

Conclusions

You may be able to propose a better model for mean sale price than Model 3, but we think that Model 3 provides a good fit to the data. Furthermore, it reveals some interesting information on the preferences of buyers of ocean-side condominium units.

Lower floors are preferred on the ocean side; the closer the units lie to the elevator and pool, the higher the estimated price. Some preference is given to the privacy of units located in the upper floor west-end.

Higher floors are preferred on the bayview side (the side facing away from the ocean), with maximum preference given to units near the elevator (convenient and close to activities) and, to a lesser degree, to the privacy afforded by the west-end units.

Follow-up Questions

1. Of the 209 units in the condominium complex, 106 were sold at auction and the remainder were sold (some more than once) at the developer's fixed price. This case study analyzed the data for the 106 units sold at auction. The data in the CONDO file includes all 209 units. Fit Models 1, 2, and 3 to the data for all 209 units.

2. Postulate some models that you think might be an improvement over Model 3. For example, consider a qualitative variable for sales method (auction or fixed price). Fit these models to the CONDO data set. Test to see whether they do, in fact, contribute more information for predicting sale price than Model 3.

Reference

Kelting, H. "Investigation of condominium sale prices in three market scenarios: Utility of stepwise, interactive, multiple regression analysis and implications for design and appraisal methodology." Unpublished paper, University of Florida, Gainesville, 1979.

SPECIAL TOPICS IN REGRESSION (OPTIONAL)

Contents

Objectives

1. To introduce some special regression techniques for problems that require more advanced methods of analysis

2. To learn when it is appropriate to apply these special methods

9.1 Introduction

The procedures presented in Chapters 3–8 provide the tools basic to a regression analysis. An understanding of these techniques will enable you to successfully apply regression analysis to a variety of problems encountered in practice. Some studies, however, may require more sophisticated techniques. In this chapter, we introduce several special topics in regression for the advanced student. A method that allows you to fit a different deterministic model over different ranges of an independent variable, called *piecewise linear regression*, is the topic of Section 9.2. In Section 9.3, we cover *inverse prediction*, a technique for predicting the value of an independent variable from a fixed value of the response. An alternative to solving the heteroscedacity problem, called *weighted least squares*, is the topic of Section 9.4. In Sections 9.5 and 9.6 we present some methods of modeling a qualitative dependent variable, including *logistic regression*. *Ridge regression*, a method for handling the multicollinearity problem, is presented in Section 9.7. Finally, in Sections 9.8 and 9.9, we discuss some *robust* and *nonparametric regression* methods.

9.2 Piecewise Linear Regression

Occasionally, the slope of the linear relationship between a dependent variable y and an independent variable x may differ for different intervals over the range of x. Consequently, the straight-line model $E(y) = \beta_0 + \beta_1 x_1$, which proposes one slope (represented by β_1) over the entire range of x, will be a poor fit to the data. For example, it is known that the compressive strength y of concrete depends on the

Figure 9.1 Relationship between compressive strength (y) and water/cement ratio (x)

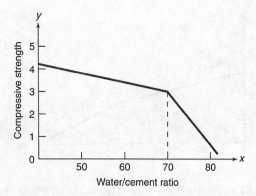

proportion x of water mixed with the cement. A certain type of concrete, when mixed in batches with varying water/cement ratios (measured as a percentage), may yield compressive strengths (measured in pounds per square inch) that follow the pattern shown in Figure 9.1. Note that the compressive strength decreases at a much faster rate for batches with water/cement ratios greater than 70%. That is, the slope of the relationship between compressive strength (y) and water/cement ratio (x) changes when $x = 70$.

A model that proposes different straight-line relationships for different intervals over the range of x is called a **piecewise linear regression model**. As its name suggests, the linear regression model is fit in pieces. For the concrete example, the piecewise model would consist of two pieces, $x \le 70$ and $x > 70$. The model can be expressed as follows:

$$y = \beta_0 + \beta_1 x_1 + \beta_2 (x_1 - 70) x_2 + \varepsilon$$

where

$$x_1 = \text{Water/cement ratio}(x)$$

$$x_2 = \begin{cases} 1 & \text{if } x_1 > 70 \\ 0 & \text{if } x_1 \le 70 \end{cases}$$

The value of the dummy variable x_2 controls the values of the slope and y-intercept for each piece. For example, when $x_1 \le 70$, then $x_2 = 0$ and the equation is given by

$$y = \beta_0 + \beta_1 x_1 + \beta_2 (x_1 - 70)(0) + \varepsilon$$

$$= \underbrace{\beta_0}_{y\text{-intercept}} + \underbrace{\beta_1}_{\text{Slope}} x_1 + \varepsilon$$

Conversely, if $x_1 > 70$, then $x_2 = 1$ and we have

$$y = \beta_0 + \beta_1 x_1 + \beta_2 (x_1 - 70)(1) + \varepsilon$$

$$= \beta_0 + \beta_1 x_1 + \beta_2 x_1 - 70\beta_2 + \varepsilon$$

or

$$y = \underbrace{(\beta_0 - 70\beta_2)}_{y\text{-intercept}} + \underbrace{(\beta_1 + \beta_2)}_{\text{Slope}} x_1 + \varepsilon$$

Thus, β_1 and $(\beta_1 + \beta_2)$ represent the slopes of the lines for the two intervals of x, $x \le 70$ and $x > 70$, respectively. Similarly, β_0 and $(\beta_0 - 70\beta_2)$ represent the respective y-intercepts. The slopes and y-intercepts of the two lines are illustrated

Figure 9.2 Slopes and
y-intercepts for piecewise
linear regression

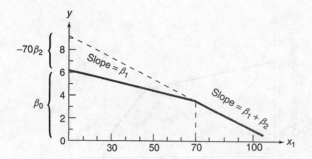

graphically in Figure 9.2. [*Note*: The value at which the slope changes, 70 in this example, is often referred to as a **knot value**. Typically, the knot values of a piecewise regression are established according to theory. In the absence of theory the knot values are unknown and must be estimated from the sample data. This is often accomplished by visually inspecting the scatterplot for the data and locating the points on the x-axis at which the slope appears to change.]

We can fit piecewise regression models by using the standard multiple regression procedures of most statistical software packages to make the appropriate transformations on the independent variables.

**Example
9.1**

Consider the data on compressive strength (y) and water/cement ratio (x) for 18 batches of concrete recorded in Table 9.1. (The water/cement ratio is computed by dividing the weight of water used in the mix by the weight of the cement.) Fit the following piecewise linear regression model to the data of Table 9.1. Interpret the results.

$$\text{Piecewise Model: } E(y) = \beta_0 + \beta_1 x_1 + \beta_2 x_2^*$$

🔘 CEMENT

Table 9.1 Data on compressive strength and water/cement ratios for 18 batches of cement

Batch	Compressive Strength y, pounds per square inch	Water/Cement Ratio x_1, percent	Batch	Compressive Strength y, pounds per square inch	Water/Cement Ratio x_1, percent
1	4.67	47	10	2.21	73
2	3.54	68	11	4.10	60
3	2.25	75	12	1.13	85
4	3.82	65	13	1.67	80
5	4.50	50	14	1.59	75
6	4.07	55	15	3.91	63
7	.76	82	16	3.15	70
8	3.01	72	17	4.37	50
9	4.29	52	18	3.75	57

Figure 9.3 SAS piecewise linear regression printout

```
                          Dependent Variable: STRENGTH

                              Analysis of Variance

                                    Sum of          Mean
        Source            DF        Squares         Square      F Value    Pr > F

        Model              2       24.71775       12.35888       114.44    <.0001
        Error             15        1.61990        0.10799
        Corrected Total   17       26.33765


                  Root MSE              0.32862      R-Square     0.9385
                  Dependent Mean        3.15500      Adj R-Sq     0.9303
                  Coeff Var            10.41595


                              Parameter Estimates

                                 Parameter      Standard
        Variable         DF       Estimate        Error      t Value    Pr > |t|

        Intercept         1        7.79198        0.67696      11.51     <.0001
        X1                1       -0.06633        0.01123      -5.90     <.0001
        X2STAR            1       -0.10119        0.02812      -3.60      0.0026
```

where

$$x_2^* = (x_1 - 70)x_2 \quad \text{and} \quad x_2 = \begin{cases} 1 & \text{if } x_1 > 70 \\ 0 & \text{if } x_1 \le 70 \end{cases}$$

Solution

The SAS printout for the piecewise linear regression is shown in Figure 9.3. The least squares prediction equation (shaded) is:

$$\hat{y} = 7.79198 - .06633x_1 - .10119x_2^*$$

The global F-value (114.44) and corresponding p-value (.0001) imply that the overall model is statistically useful for predicting compressive strength, y. Since $R_a^2 = .9303$, we know the model explains about 93% of the sample variation in compressive strength; and $s = .32862$ implies that the model can predict compressive strength to within about $2s = .66$ pound per square inch of its true value. Also note that t-tests on β_1 and β_2 indicate that both are significantly different from 0. Consequently, it appears that the piecewise model is both statistically and practically useful for predicting compressive strength.

From the estimated β's we can determine the rate of change of compressive strength with water/cement ratio, x_1, over the two ranges. When $x_1 \le 70$ (i.e., $x_2 = 0$), the slope of the line is $\hat{\beta}_1 = -.06633$. Thus, for ratios less than or equal to 70%, we estimate that mean compressive strength will decrease by .066 pound per square inch for every 1% increase in water/cement ratio. Alternatively, when $x_1 > 70$ (i.e., $x_2 = 1$), the slope of the line is $\hat{\beta}_1 + \hat{\beta}_2 = -.06633 + (-.10119) = -.16752$. Hence, for ratios greater than 70%, we estimate that mean compressive strength will decrease by .168 pound per square inch for every 1% increase in water/cement ratio. ∎

Piecewise regression is not limited to two pieces, nor is it limited to straight lines. One or more of the pieces may require a quadratic or higher-order fit. Also, piecewise regression models can be proposed to allow for discontinuities or jumps in the regression function. Such models require additional dummy variables to be introduced. Several different piecewise linear regression models relating y to an independent variable x are shown in the following box.

Piecewise Linear Regression Models Relating y to an Independent Variable x_1

TWO STRAIGHT LINES (CONTINUOUS):

$$E(y) = \beta_0 + \beta_1 x_1 + \beta_2(x_1 - k)x_2$$

where

$k = $ Knot value (i.e., the value of the independent variable x_1

at which the slope changes)

$$x_2 = \begin{cases} 1 & \text{if } x_1 > k \\ 0 & \text{if not} \end{cases}$$

	$x_1 \leq k$	$x_1 > k$
y-intercept	β_0	$\beta_0 - k\beta_2$
Slope	β_1	$\beta_1 + \beta_2$

THREE STRAIGHT LINES (CONTINUOUS):

$$E(y) = \beta_0 + \beta_1 x_1 + \beta_2(x_1 - k_1)x_2 + \beta_3(x_1 - k_2)x_3$$

where k_1 and k_2 are knot values of the independent variable x_1, $k_1 < k_2$, and

$$x_2 = \begin{cases} 1 & \text{if } x_1 > k_1 \\ 0 & \text{if not} \end{cases} \quad x_3 = \begin{cases} 1 & \text{if } x_1 > k_2 \\ 0 & \text{if not} \end{cases}$$

	$x_1 \leq k_1$	$k_1 < x_1 \leq k_2$	$x_1 > k_2$
y-intercept	β_0	$\beta_0 - k_1\beta_2$	$\beta_0 - k_1\beta_2 - k_2\beta_3$
Slope	β_1	$\beta_1 + \beta_2$	$\beta_1 + \beta_2 + \beta_3$

TWO STRAIGHT LINES (DISCONTINUOUS):

$$E(y) = \beta_0 + \beta_1 x_1 + \beta_2(x_1 - k)x_2 + \beta_3 x_2$$

where

$k = $ Knot value (i.e., the value of the independent variable x_1 at which the slope changes—also the point of discontinuity)

$$x_2 = \begin{cases} 1 & \text{if } x_1 > k \\ 0 & \text{if not} \end{cases}$$

	$x_1 \leq k$	$x_1 > k$
y-intercept	β_0	$\beta_0 - k\beta_2 + \beta_3$
Slope	β_1	$\beta_1 + \beta_2$

Tests of model adequacy, tests and confidence intervals on individual β parameters, confidence intervals for $E(y)$, and prediction intervals for y for piecewise regression models are conducted in the usual manner.

READSCORES

Example 9.2

The Academic Technology Services department at UCLA provides tutorials on the use of statistical software for students. Data on the ages and reading test scores for 130 children and young adults were used to demonstrate piecewise regression. Consider the straight-line model, $E(y) = \beta_0 + \beta_1 x_1$, where $y =$ reading score (points) and $x_1 =$ age (years).

(a) Fit the straight-line model to the data saved in the *READSCORES* file. Graph the least squares prediction equation and assess the adequacy of the model.

(b) Consider a quadratic model as an alternative model for predicting reading score. Assess the fit of this model.

(c) Now, consider using a piecewise linear regression model to predict reading score. Use the graph, part a, to estimate the knot value for a piecewise linear regression model.

(d) Fit the piecewise linear regression model to the data and assess model adequacy. Compare the results to the straight-line model, part a, and the quadratic model, part b.

Solution

(a) We used MINITAB to analyze the data saved in the *READSCORES* file. The data are plotted in Figure 9.4 and the MINITAB regression results are shown in Figure 9.5. The least squares prediction equation (highlighted in Figure 9.5 and shown in Figure 9.4) is

$$\hat{y} = -13.4 + 2.96x_1$$

The p-value for testing $H_0 : \beta_1 = 0$ is approximately 0, indicating that the model is statistically useful for predicting reading score (y). Also, the coefficient of determination, $r^2 = .689$, indicates that nearly 70% of the sample variation in reading scores can be explained by the linear relationship with age (x_1). In addition, the estimated model standard deviation, $s = 10$, implies that we can predict reading score to within about 20 points of its true value using the model.

Although the straight-line model fits the data adequately, the graph in Figure 9.4 shows a lot of variation of the points around the least squares line. There appears to be nonlinearity in the data, which cannot be accounted for

Figure 9.4 MINITAB graph of reading scores

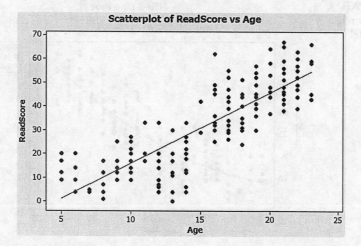

Figure 9.5 MINITAB analysis of straight-line model for reading score

---

Regression Analysis: ReadScore versus Age

The regression equation is
ReadScore = - 13.4 + 2.96 Age

```
Predictor       Coef   SE Coef       T       P
Constant      -13.447    2.844    -4.73   0.000
Age            2.9586    0.1757   16.84   0.000
```

S = 10.0555   R-Sq = 68.9%   R-Sq(adj) = 68.7%

Analysis of Variance

```
Source            DF      SS      MS       F       P
Regression         1   28684   28684   283.69   0.000
Residual Error   128   12942     101
Total            129   41627
```

---

by the straight-line model. Consequently, it is worth seeking a model with a better fit.

(b) The quadratic model, $E(y) = \beta_0 + \beta_1 x_1 + \beta_2 x_1^2$, was also fit to the data using MINITAB. A graph of the least squares curve and model results are shown in Figure 9.6. Note that r^2 has increased to .713, while s has decreased to 9.7; thus, the quadratic model does show a slightly improved fit to the data over the straight-line model. However, the variation of the data around the least squares curve displayed in Figure 9.6 is large enough that we will continue to search for a better model.

(c) Examining either Figure 9.4 or Figure 9.6 more closely, you'll notice that at about age 15, the reading scores of the children change dramatically. In fact, it appears that the rate of change of reading score with age is much different for children under 14 years or younger than for children 15 years or older. That is, the slope of the line appears to be different for the two groups of children. If so, a piecewise linear regression model may fit the data much better than the

Figure 9.6 MINITAB graph of reading scores, with quadratic model results

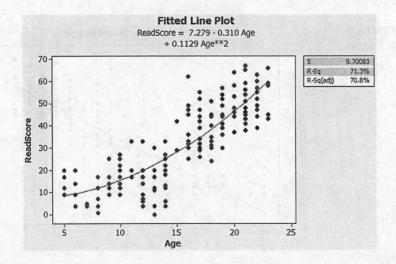

Fitted Line Plot
ReadScore = 7.279 - 0.310 Age
+ 0.1129 Age**2

S	9.70083
R-Sq	71.3%
R-Sq(adj)	70.8%

models of parts a and b. The estimated knot value for this piecewise regression model is at age equal to 14 years.

(d) Because of the dramatic jump in reading scores starting at age 15 years, the piecewise linear model will be discontinuous. According to the formulas in the box for two straight lines with a discontinuity at $x_1 = 14$, the model takes the form

$$E(y) = \beta_0 + \beta_1 x_1 + \beta_2(x_1 - 14)x_2 + \beta_3 x_2$$

where x_1 = Age and x_2 = {1 if $x_1 > 14$, 0 if not}. With this model, the two straight lines have the following equations:

$$\text{Age} \leq 14 \text{ (i.e., } x_2 = 0) : E(y) = \beta_0 + \beta_1 x_1 + \beta_2(x_1 - 14)(0) + \beta_3(0)$$

$$= \underbrace{\beta_0}_{y\text{-intercept}} + \underbrace{\beta_1}_{\text{slope}} x_1$$

$$\text{Age} > 14 \text{ (i.e., } x_2 = 1) : E(y) = \beta_0 + \beta_1 x_1 + \beta_2(x_1 - 14)(1) + \beta_3(1)$$

$$= \underbrace{(\beta_0 - 14\beta_2 + \beta_3)}_{y\text{-intercept}} + \underbrace{(\beta_1 + \beta_2)}_{\text{slope}} x_1$$

We used SAS to fit this piecewise model to the data; the SAS printout is shown in Figure 9.7. The global F-test (highlighted) indicates that the model is statistically

The GLM Procedure

Dependent Variable: SCORE

Source	DF	Sum of Squares	Mean Square	F Value	Pr > F
Model	3	31840.24688	10613.41563	136.64	<.0001
Error	126	9786.64542	77.67179		
Corrected Total	129	41626.89231			

R-Square	Coeff Var	Root MSE	SCORE Mean
0.764896	27.46191	8.813160	32.09231

Source	DF	Type I SS	Mean Square	F Value	Pr > F
AGE	1	28684.44113	28684.44113	369.30	<.0001
AGE14X2	1	1433.31177	1433.31177	18.45	<.0001
X2	1	1722.49398	1722.49398	22.18	<.0001

Source	DF	Type III SS	Mean Square	F Value	Pr > F
AGE	1	179.911904	179.911904	2.32	0.1305
AGE14X2	1	750.958276	750.958276	9.67	0.0023
X2	1	1722.493985	1722.493985	22.18	<.0001

Parameter	Estimate	Standard Error	t Value	Pr > |t|
intercept>14	-3.38756934	8.72307637	-0.39	0.6984
slope>14	2.53221448	0.45207342	5.60	<.0001

Parameter	Estimate	Standard Error	t Value	Pr > |t|
Intercept	8.31468778	4.45715945	1.87	0.0644
AGE	0.62843479	0.41291620	1.52	0.1305
AGE14X2	1.90377969	0.61226642	3.11	0.0023
X2	14.95065849	3.17477622	4.71	<.0001

Figure 9.7 SAS analysis of piecewise linear model for reading score

Figure 9.8 MINITAB graph of predicted reading scores using piecewise regression

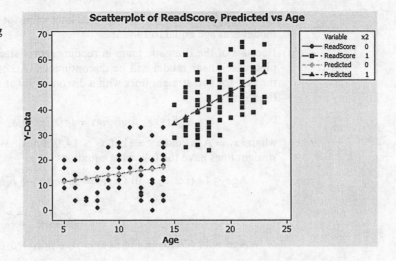

useful for predicting reading score. The t-test for the slope of the line for ages 14 years or under (i.e., the test for $H_0 : \beta_1 = 0$, highlighted at the bottom of the printout) is not statistically significant (p-value $= .1305$). However, the t-test for the slope for ages 15 years or older (i.e., the test for $H_0 : \beta_1 + \beta_2 = 0$, highlighted in the "Slope > 14" row) is statistically significant (p-value $< .0001$). The estimated slopes for the two lines are shown on the MINITAB graph, in Figure 9.8.

Note, also, that for the piecewise linear model $R^2 = .765$ and $s = 8.8$. Both these values are improvements over the corresponding values for the straight-line model (part a) and the quadratic model (part b). ∎

9.2 Exercises

9.1 Piecewise regression: Two lines. Consider a two-piece linear relationship between y and x with no discontinuity and a slope change at $x = 15$.

  (a) Specify the appropriate piecewise linear regression model for y.

  (b) In terms of the β coefficients, give the y-intercept and slope for observations with $x \leq 15$, for observations with $x > 15$.

  (c) Explain how you could determine whether the two slopes proposed by the model are, in fact, different.

9.2 Piecewise regression: Three lines. Consider a three-piece linear relationship between y and x with no discontinuity and slope changes at $x = 1.45$ and $x = 5.20$.

  (a) Specify the appropriate piecewise linear regression model for y.

  (b) In terms of the β coefficients, give the y-intercept and slope for each of the following intervals: $x \leq 1.45$, $1.45 < x \leq 5.20$, and $x > 5.20$.

  (c) Explain how you could determine whether at least two of the three slopes proposed by the model are, in fact, different.

9.3 Precewise regression with discontinuity. Consider a two-piece linear relationship between y and x with discontinuity and slope change at $x = 320$.

  (a) Specify the appropriate piecewise linear regression model for y.

  (b) In terms of the β coefficients, give the y-intercept and slope for observations with $x \leq 320$, for observations with $x > 320$.

  (c) Explain how you could determine whether the two straight lines proposed by the model are, in fact, different.

9.4 Living cells in plant tissue. In biology, researchers often conduct growth experiments to determine the number of cells that will grow in a petri dish after a certain period of time. The data in the next table represent the number of living cells in plant tissue after exposing the specimen to heat for a certain number of hours.

GROWTH

HOURS	CELLS
1	2
5	3
10	4
20	5
30	4
40	6
45	8
50	9
60	10
70	18
80	16
90	18
100	20
110	14
120	12
130	13
140	9

(a) Plot number of cells, y, against number of hours, x. What trends are apparent?

(b) Propose a piecewise linear model for number of cells. Give the value of x that represents the knot value.

(c) Fit the model, part b, to the data.

(d) Test the overall adequacy of the model using $\alpha = .05$.

(e) What is the estimated rate of growth when heating the tissue for less than 70 hours? More than 70 hours?

(f) Conduct a test (at $\alpha = .05$) to determine if the two estimates, part e, are statistically different.

9.5 Customer satisfaction study. In a paper presented at the 2009 IM Conference in China, a group of university finance professors examined the relationship between customer satisfaction (y) of a product and product performance (x), with performance measured on a 10-point scale. The researchers discovered that the linear relationship varied over different performance ranges. Consequently, they fit a piecewise linear regression model that hypothesized three continuous lines—one over the range $0 < x \leq 4$, one over the range $4 < x \leq 7$, and one over the range $7 < x \leq 10$.

(a) Identify the knot values for the piecewise model.

(b) Write the equation of the hypothesized model.

(c) In terms of the model parameters, give an expression for the slope of the line for each of the three performance ranges.

(d) For the attribute "offering appropriate service," the estimated slopes over the three ranges of performance level are 5.05, .59, and 1.45, respectively. Interpret these values.

BEDLOAD

9.6 Sediment transport in streams. Coarse sediments in a stream that are carried by intermittent contact with the streambed by rolling, sliding, and bouncing are called *bedload*. The rate at which the bedload is transported in a stream can vary dramatically, depending on flow conditions, stream slope, water depth, weight of the sediment, and so on. In a 2007 technical report for the U.S. Department of Agriculture, Forest Service, Rocky Mountain Research Station, piecewise regression was used to model bedload transport rate (kilograms per second) as a function of stream discharge rate (cubic meters

MINITAB Output for Exercise 9.6

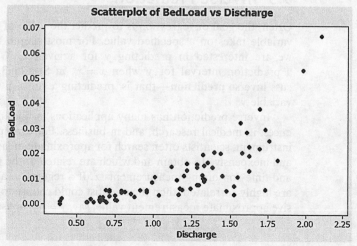

per second). The data for the study, collected from Hayden Creek in Colorado during snowmelt runoff, is saved in the BEDLOAD file.

(a) A graph of the data is shown in the MINITAB printout (p. 475). Analysts at the Rocky Mountain Research Station theorize that the rate of change of bedload transport with stream discharge rate increases when the discharge rate is 1.5 cubic meters per second or higher. Is there visual evidence of this from the graph?

(b) Hypothesize a piecewise linear regression model for bedload transport rate with a break point (knot) at a discharge rate of 1.5 cubic meters per second.

(c) Fit the piecewise model to the data in the BEDLOAD file and give an overall assessment of model adequacy.

(d) What is the estimated change in bedload transport rate for every 1 cubic meter per second increase in discharge rate for flows less than 1.5 cubic meters per second? For flows of 1.5 cubic meters per second or higher?

9.7 **Adhesion of plasma-sprayed coatings.** Materials engineers tested the adhesion properties of plasma-sprayed coatings (*Journal of Thermal Spray Technology*, December 1999). Twenty steel specimens were sand blasted then sprayed with a coat of plasma. Each steel piece was compressed with a heavy load until a crack in the coating was produced. The load required to produce the crack and the crack length was recorded for each. The experimental data (natural logarithms) are listed in the table. The goal is to predict crack length from the load.

(a) Plot the data. Do you think a piecewise linear model will fit the data well? If so, estimate the knot value.

(b) Give the equation of hypothesized piecewise linear regression model.

(c) Fit the model to the data and assess model adequacy.

💿 PLASMA

SPECIMEN	LN(CRACK LENGTH)	LN(LOAD)
1	3.35	2.25
2	3.45	2.25
3	3.60	2.75
4	3.75	2.90
5	3.80	2.90
6	3.85	2.90
7	3.90	3.20
8	4.10	3.55
9	4.50	3.70
10	4.55	3.70
11	4.70	3.75
12	4.80	3.75
13	4.75	3.75
14	4.90	3.80
15	4.85	3.80
16	4.85	3.80
17	5.05	3.85
18	5.10	3.85
19	5.25	3.90
20	5.15	3.90

Source: Godoy, C., and Batista, J. C. A. "Adhesion evaluation of plasma sprayed coatings using piecewise linear regression analysis," *Journal of Thermal Spray Technology*, Vol. 8, No. 4, December 1999 (adapted from Figure 9.4).

9.3 Inverse Prediction

Often, the goal of regression is to predict the value of one variable when another variable takes on a specified value. For most simple linear regression problems, we are interested in predicting y for a given x. We provided a formula for a prediction interval for y when $x = x_p$ in Section 3.9. In this section, we discuss **inverse prediction**—that is, predicting x for a given value of the dependent variable y.

Inverse prediction has many applications in the engineering and physical sciences, in medical research, and in business. For example, when calibrating a new instrument, scientists often search for approximate measurements y, which are easy and inexpensive to obtain and which are related to the precise, but more expensive and time-consuming measurements x. If a regression analysis reveals that x and y are highly correlated, then the scientist could choose to use the quick and inexpensive approximate measurement value, say, $y = y_p$, to estimate the unknown precise measurement x. (In this context, the problem of inverse prediction is sometimes referred to as a **linear calibration** problem.) Physicians often use inverse prediction to determine the required dosage of a drug. Suppose a regression analysis conducted on patients with high blood pressure showed that a linear relationship exists between decrease in blood pressure y and dosage x of a new drug. Then a physician treating a new patient may want to determine what dosage x to administer to reduce the patient's blood pressure by an amount $y = y_p$. To illustrate inverse prediction in a business setting, consider a firm that sells a particular product. Suppose the firm's monthly market share y is linearly related to its monthly television advertising expenditure x. For a particular month, the firm may want to know how much it must spend on advertising x to attain a specified market share $y = y_p$.

The classical approach to inverse prediction is first to fit the familiar straight-line model

$$y = \beta_0 + \beta_1 x + \varepsilon$$

to a sample of n data points and obtain the least squares prediction equation

$$\hat{y} = \hat{\beta}_0 + \hat{\beta}_1 x$$

Solving the least squares prediction equation for x, we have

$$x = \frac{\hat{y} - \hat{\beta}_0}{\hat{\beta}_1}$$

Now let y_p be an observed value of y in the future with unknown x. Then a point estimate of x is given by

$$\hat{x} = \frac{y_p - \hat{\beta}_0}{\hat{\beta}_1}$$

Although no exact expression for the standard error of $\hat{x}$ (denoted $s_{\hat{x}}$) is known, we can algebraically manipulate the formula for a prediction interval for y given x (see Section 3.9) to form a prediction interval for x given y. It can be shown (proof omitted) that an approximate $(1 - \alpha)$ 100% prediction interval for x when $y = y_p$ is

$$\hat{x} \pm t_{\alpha/2} s_{\hat{x}} \approx \hat{x} \pm t_{\alpha/2} \left(\frac{s}{\hat{\beta}_1} \right) \sqrt{1 + \frac{1}{n} + \frac{(\hat{x} - \bar{x})^2}{SS_{xx}}}$$

where the distribution of t is based on $(n - 2)$ degrees of freedom, $s = \sqrt{MSE}$, and

$$SS_{xx} = \sum x^2 - n(\bar{x})^2$$

This approximation is appropriate as long as the quantity

$$D = \left(\frac{t_{\alpha/2} s}{\hat{\beta}_1} \right)^2 \cdot \frac{1}{SS_{xx}}$$

is small. The procedure for constructing an approximate confidence interval for x in inverse prediction is summarized in the box on p. 478.

Example 9.3

A firm that sells copiers advertises regularly on television. One goal of the firm is to determine the amount it must spend on television advertising in a single month to gain a market share of 10%. For one year, the firm varied its monthly television advertising expenditures (x) and at the end of each month determined its market share (y). The data for the 12 months are recorded in Table 9.2.

Inverse Prediction: Approximate $(1 - \alpha)$ 100% Prediction Interval for x When $y = y_p$ in Simple Linear Regression

$$\hat{x} \pm t_{\alpha/2} \left(\frac{s}{\hat{\beta}_1} \right) \sqrt{1 + \frac{1}{n} + \frac{(\hat{x} - \bar{x})^2}{SS_{xx}}}$$

where

$$\hat{x} = \frac{y_p - \hat{\beta}_0}{\hat{\beta}_1}$$

$\hat{\beta}_0$ and $\hat{\beta}_1$ are the y-intercept and slope, respectively, of the least squares line

$$n = \text{Sample size}$$

$$\bar{x} = \frac{\sum x}{n}$$

$$SS_{xx} = \sum x^2 - n(\bar{x})^2$$

$$s = \sqrt{MSE}$$

and the distribution of t is based on $(n - 2)$ degrees of freedom.
Note: The approximation is appropriate when the quantity

$$D = \left(\frac{t_{\alpha/2}s}{\hat{\beta}_1} \right)^2 \cdot \frac{1}{SS_{xx}}$$

is small.*

🔘 COPIERS

Table 9.2 A firm's market share and television advertising expenditure for 12 months, Example 9.3

Month	Market Share y, percent	Television Advertising Expenditure x, \$ thousands
January	7.5	23
February	8.5	25
March	6.5	21
April	7.0	24
May	8.0	26
June	6.5	22
July	9.5	27
August	10.0	31
September	8.5	28
October	11.0	32
November	10.5	30
December	9.0	29

---

*Kutner, Nachtsheim, Neter, and Li (2005) and others suggest using the approximation when D is less than .1.

(a) Fit the straight-line model $y = \beta_0 + \beta_1 x + \varepsilon$ to the data.

(b) Is there evidence that television advertising expenditure x is linearly related to market share y? Test using $\alpha = .05$.

(c) Use inverse prediction to estimate the amount that must be spent on television advertising in a particular month for the firm to gain a market share of $y = 10\%$. Construct an approximate 95% prediction interval for monthly television advertising expenditure x.

Solution

(a) The MINITAB printout for the simple linear regression is shown in Figure 9.9. The least squares line (shaded) is:

$$\hat{y} = -1.975 + .397x$$

The least squares line is plotted along with 12 data points in Figure 9.10.

Figure 9.9 MINITAB analysis of straight-line model, Example 9.3

Regression Analysis: MKTSHARE versus TVADEXP

```
The regression equation is
MKTSHARE = - 1.97 + 0.397 TVADEXP


Predictor      Coef   SE Coef       T       P
Constant     -1.975     1.163   -1.70   0.120
TVADEXP     0.39685   0.04351    9.12   0.000


S = 0.520361    R-Sq = 89.3%    R-Sq(adj) = 88.2%


Analysis of Variance

Source          DF      SS      MS       F       P
Regression       1  22.521  22.521   83.17   0.000
Residual Error  10   2.708   0.271
Total           11  25.229
```

Figure 9.10 MINITAB scatterplot of data and least squares line, Example 9.3

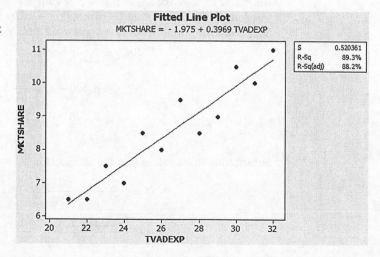

Fitted Line Plot
MKTSHARE = - 1.975 + 0.3969 TVADEXP

S	0.520361
R-Sq	89.3%
R-Sq(adj)	88.2%

(b) To determine whether television advertising expenditure x is linearly related to market share y, we test the hypothesis

$H_0: \beta_1 = 0$
$H_a: \beta_1 \neq 0$

The value of the test statistic, shaded in Figure 9.9, is $t = 9.12$, and the associated p-value of the test is $p = .000$ (also shaded). Thus, there is sufficient evidence (at $\alpha = .01$), to indicate that television advertising expenditure x and market share y are linearly related.

Caution

You should avoid using inverse prediction when there is insufficient evidence to reject the null hypothesis $H_0: \beta_1 = 0$. Inverse predictions made when x and y are *not* linearly related may lead to nonsensical results. Therefore, you should always conduct a test of model adequacy to be sure that x and y are linearly related before you carry out an inverse prediction.

(c) Since the model is found to be adequate, we can use the model to predict x from y. For this example, we want to estimate the television advertising expenditure x that yields a market share of $y_p = 10\%$.

Substituting $y_p = 10$, $\hat{\beta}_0 = -1.975$, and $\hat{\beta}_1 = .397$ into the formula for $\hat{x}$ given in the box (p. 478), we have

$$\hat{x} = \frac{y_p - \hat{\beta}_0}{\hat{\beta}_1}$$

$$= \frac{10 - (-1.975)}{.397}$$

$$= 30.16$$

Thus, we estimate that the firm must spend \$30,160 on television advertising in a particular month to gain a market share of 10%.

Before we construct an approximate 95% prediction interval for x, we check to determine whether the approximation is appropriate, that is, whether the quantity

$$D = \left(\frac{t_{\alpha/2}s}{\hat{\beta}_1}\right)^2 \cdot \frac{1}{SS_{xx}}$$

is small. For $\alpha = .05$, $t_{\alpha/2} = t_{.025} = 2.228$ for $n - 2 = 12 - 2 = 10$ degrees of freedom. From the printout (Figure 9.9), $s = .520$ and $\hat{\beta}_1 = .397$. The value of SS_{xx} is not shown on the printout and must be calculated as follows:

$$SS_{xx} = \sum x^2 - n(\bar{x})^2$$

$$= 8,570 - 12\left(\frac{318}{12}\right)^2$$

$$= 8,570 - 8,427$$

$$= 143$$

Substituting these values into the formula for D, we have

$$D = \left(\frac{t_{\alpha/2}s}{\hat{\beta}_1}\right)^2 \cdot \frac{1}{SS_{xx}}$$

$$= \left[\frac{(2.228)(.520)}{.397}\right]^2 \cdot \frac{1}{143} = .06$$

Since the value of D is small (i.e., less than .1), we may use the formula for the approximate 95% prediction interval given in the box (page 478):

$$\hat{x} \pm t_{\alpha/2} \left(\frac{s}{\hat{\beta}_1} \right) \sqrt{1 + \frac{1}{n} + \frac{(\hat{x} - \bar{x})^2}{SS_{xx}}}$$

$$30.16 \pm (2.228) \frac{(.520)}{.397} \sqrt{1 + \frac{1}{12} + \frac{\left(30.16 - \frac{318}{12} \right)^2}{143}}$$

$$30.16 \pm (2.92)(1.085)$$

$$30.16 \pm 3.17$$

or (26.99, 33.33). Therefore, using the 95% prediction interval, we estimate that the amount of monthly television advertising expenditure required to gain a market share of 10% falls between \$26,999 and \$33,330.

Caution

> When making inverse predictions, the analyst should be wary of the extrapolation problem. That is, avoid making predictions on x for values of y_p that are outside the range of the sample data.

Another approach to the inverse prediction problem is to regress x on y, that is, fit the model (called the **inverse estimator model**)

$$x = \beta_0 + \beta_1 y + \varepsilon$$

and then use the standard formula for a prediction interval given in Section 3.9. However, in theory this method requires that x be a random variable. In many applications, the value of x is set in advance (i.e., controlled) and therefore is *not* a random variable. (For example, the firm in Example 9.3 selected the amount x spent on advertising *prior* to each month.) Thus, the inverse model above may violate the standard least squares assumptions given in Chapter 4. Some researchers advocate the use of the inverse model despite this caution, whereas others have developed different estimators of x using a modification of the classical approach. Consult the references given at the end of this chapter for details on the various alternative methods of inverse prediction.

9.3 Exercises

9.8 **Wind turbine blade stress.** Refer to the *Wind Engineering* (January 2004) study of the use of timber in high-efficiency small wind turbine blades, Exercise 3.10 (p. 101). Recall that simple linear regression analysis was used to model $y =$ blade stress (in MPa) as a function of $x =$ natural logarithm of number of blade cycles. The model was fit to $n = 20$ data points for each of two types of timber, with the results shown next:

Radiata Pine: $\hat{y} = 97.37 - 2.50x$
Hoop Pine: $\hat{y} = 122.03 - 2.36x$

(a) For radiata pine, use inverse prediction to estimate the number of cycle blades required to yield a blade stress of 85 MPa. (Remember, x is measured as the natural logarithm of number of cycle blades.)

(b) For hoop pine, use inverse prediction to estimate the number of cycle blades required to yield a blade stress of 110 MPa.

9.9 **Feeding behavior of blackbream fish.** Refer to the *Brain and Behavior Evolution* (April 2000) study of the feeding behavior of blackbream fish, Exercise 3.12 (p. 102). Recall that the zoologists recorded the number of aggressive strikes of two blackbream fish feeding at the bottom of an aquarium in the 10-minute period following the addition of food. The table listing the weekly number of strikes and age of the fish (in days) is reproduced below. Use inverse prediction to estimate the age (x) of a blackbream fish, which was observed to have $y = 65$ aggressive strikes while feeding in another aquarium. Construct a 99% prediction interval around the estimate and interpret the result.

📀 BLACKBREAM

WEEK	NUMBER OF STRIKES (y)	AGE OF FISH (x, days)
1	85	120
2	63	136
3	34	150
4	39	155
5	58	162
6	35	169
7	57	178
8	12	184
9	15	190

Source: Shand J., et al. "Variability in the location of the retinal ganglion cell area centralis is correlated with ontogenetic changes in feeding behavior in the Blackbream, *Acanthopagrus* 'butcher'," *Brain, Behavior and Evolution*, Vol. 55, No. 4, April 2000 (Figure H).

9.10 **Sweetness of orange juice.** Refer to the study of the relationship between the "sweetness" rating of orange juice and the amount of water-soluble pectin in the juice, Exercise 3.13 (p. 102). The data for 24 production runs is reproduced in the table, followed by a MINITAB printout of the least squares regression results for the straight-line model relating sweetness index (y) to pectin amount (x). Use inverse prediction to estimate the amount of pectin required to yield a sweetness index of $y = 5.8$. Construct a 95% prediction interval around the estimate and interpret the result.

📀 OJUICE

RUN	SWEETNESS INDEX	PECTIN (ppm)
1	5.2	220
2	5.5	227
3	6.0	259
4	5.9	210
5	5.8	224
6	6.0	215
7	5.8	231
8	5.6	268
9	5.6	239
10	5.9	212
11	5.4	410
12	5.6	256
13	5.8	306
14	5.5	259
15	5.3	284
16	5.3	383
17	5.7	271
18	5.5	264
19	5.7	227
20	5.3	263
21	5.9	232
22	5.8	220
23	5.8	246
24	5.9	241

MINITAB Output for Exercise 9.10

```
The regression equation is
SWEET = 6.25 - 0.00231 PECTIN

Predictor      Coef      SE Coef       T       P
Constant      6.2521      0.2366     26.42   0.000
PECTIN      -0.0023106   0.0009049   -2.55   0.018

S = 0.2150      R-Sq = 22.9%     R-Sq(adj) = 19.4%

Analysis of Variance

Source          DF        SS         MS        F       P
Regression       1      0.30140    0.30140    6.52    0.018
Residual Error  22      1.01693    0.04622
Total           23      1.31833
```

9.11 Spreading rate of spilled liquid. Refer to the *Chemical Engineering Progress* (January 2005) study of the rate at which 50 gallons of methanol liquid will spread across a level outdoor surface, Exercise 3.16 (p. 103). Experimental data on the mass (in pounds) of the spill after a specified period of time for several times (in minutes) are reproduced in the accompanying table. In Exercise 3.16 you used simple linear regression to model spillage mass (y) as a function of time (x). The engineers would like to know the amount of time required for the mass to reach 5 pounds. Use inverse prediction to find a 95% confidence interval for the amount of time.

🌐 LIQUIDSPILL

TIME (MINUTES)	MASS (POUNDS)
0	6.64
1	6.34
2	6.04
4	5.47
6	4.94
8	4.44
10	3.98
12	3.55
14	3.15
16	2.79
18	2.45
20	2.14
22	1.86
24	1.60
26	1.37
28	1.17
30	0.98
35	0.60
40	0.34
45	0.17
50	0.06
55	0.02
60	0.00

Source: Barry, J. "Estimating rates of spreading and evaporation of volatile liquids," *Chemical Engineering Progress*, Vol. 101, No. 1, January 2005.

9.12 Residential fire damage. The data in Table 3.8 (p. 136) are reproduced here (next column). Use inverse prediction to estimate the distance from nearest fire station, x, for a residential fire that caused $y = \$18,200$ in damages. Construct a 90% prediction interval for x.

9.13 New drug study. A pharmaceutical company has developed a new drug designed to reduce a smoker's reliance on tobacco. Since certain

🌐 FIREDAM

DISTANCE FROM FIRE STATION	FIRE DAMAGE
x, miles	y, thousands of dollars
3.4	26.2
1.8	17.8
4.6	31.3
2.3	23.1
3.1	27.5
5.5	36.0
.7	14.1
3.0	22.3
2.6	19.6
4.3	31.3
2.1	24.0
1.1	17.3
6.1	43.2
4.8	36.4
3.8	26.1

dosages of the drug may reduce one's pulse rate to dangerously low levels, the product-testing division of the pharmaceutical company wants to model the relationship between decrease in pulse rate y (beats/minute) and dosage x (cubic centimeters). Different dosages of the drug were administered to eight randomly selected patients, and 30 minutes later the decrease in each patient's pulse rate was recorded, with the results given in the table below.

🌐 PULSEDRUG

PATIENT	DOSAGE x, cubic centimeters	DECREASE IN PULSE RATE y, beats/minute
1	2.0	12
2	4.0	20
3	1.5	6
4	1.0	3
5	3.0	16
6	3.5	20
7	2.5	13
8	3.0	18

(a) Fit the straight-line model $E(y) = \beta_0 + \beta_1 x$ to the data.

(b) Conduct a test for model adequacy. Use $\alpha = .05$.

(c) Use inverse prediction to estimate the appropriate dosage x to administer to reduce a patient's pulse rate $y = 10$ beats per minute. Construct an approximate 95% prediction interval for x.

9.4 Weighted Least Squares

In Section 8.3 we considered the problem of *heteroscedastic errors* (i.e., regression errors that have a nonconstant variance). We learned that the problem is typically solved by applying a *variance-stabilizing transformation* (e.g., $\sqrt{y}$ or the natural log of y) to the dependent variable. In situations where these types of transformations are not effective in stabilizing the error variance, alternative methods are required.

In this section, we consider a technique called **weighted least squares.** Weighted least squares has applications in the following areas:

1. Stabilizing the variance of ε to satisfy the standard regression assumption of homoscedasticity.
2. Limiting the influence of outlying observations in the regression analysis.
3. Giving greater weight to more recent observations in time series analysis (the topic of Chapter 10).

Although the applications are related, our discussion of weighted least squares in this section is directed toward the first application.

Consider the general linear model

$$y = \beta_0 + \beta_1 x_1 + \beta_2 x_2 + \cdots + \beta_k x_k + \varepsilon$$

To obtain the least squares estimates of the unknown β parameters, recall (from Section 4.3) that we minimize the quantity

$$\text{SSE} = \sum_{i=1}^{n}(y_i - \hat{y}_i)^2 = \sum_{i=1}^{n}[y_i - (\hat{\beta}_0 + \hat{\beta}_1 x_{1i} + \hat{\beta}_2 x_{2i} + \cdots + \hat{\beta}_k x_{ki})]^2$$

with respect to $\hat{\beta}_0, \hat{\beta}_1, \ldots, \hat{\beta}_k$. The least squares criterion weighs each observation equally in determining the estimates of the β's. With **weighted least squares** we want to weigh some observations more heavily than others. To do this we minimize

$$\text{WSSE} = \sum_{i=1}^{n} w_i(y_i - \hat{y}_i)^2$$

$$= \sum_{i=1}^{n} w_i[y_i - (\hat{\beta}_0 + \hat{\beta}_1 x_{1i} + \hat{\beta}_2 x_{2i} + \cdots + \hat{\beta}_k x_{ki})]^2$$

where w_i is the weight assigned to the ith observation. The resulting parameter estimates are called **weighted least squares estimates**. [Note that the ordinary least squares procedure assigns a weight of $w_i = 1$ to each observation.]

Definition 9.1 **Weighted least squares regression** is the procedure that obtains estimates of the β's by minimizing $\text{WSSE} = \sum_{i=1}^{n} w_i(y_i - \hat{y}_i)^2$, where w_i is the weight assigned to the ith observation. The β-estimates are called **weighted least squares estimates**.

Definition 9.2 The **weighted least squares residuals** are obtained by computing the quantity

$$\sqrt{w_i}(y_i - \hat{y}_i)$$

for each observation, where $\hat{y}_i$ is the predicted value of y obtained using the weight w_i in a weighted least squares regression.

The regression routines of most statistical software packages have options for conducting a weighted least squares analysis. However, the weights w_i must be specified. When using weighted least squares as a variance-stabilizing technique, the weight for the ith observation should be the reciprocal of the variance of that observation's error term, σ_i^2, that is,

$$w_i = \frac{1}{\sigma_i^2}$$

In this manner, *observations with larger error variances will receive less weight (and hence have less influence on the analysis) than observations with smaller error variances.*

In practice, the actual variances σ_i^2 will usually be unknown. Fortunately, in many applications, the error variance σ_i^2 is proportional to one or more of the levels of the independent variables. This fact will allow us to determine the appropriate weights to use. For example, in a simple linear regression problem, suppose we know that the error variance σ_i^2 increases proportionally with the value of the independent variable x_i, that is,

$$\sigma_i^2 = k x_i$$

where k is some unknown constant. Then the appropriate (albeit unknown) weight to use is

$$w_i = \frac{1}{k x_i}$$

Fortunately, it can be shown (proof omitted) that k can be ignored and the weights can be assigned as follows:

$$w_i = \frac{1}{x_i}$$

If the functional relationship between σ_i^2 and x_i is not known prior to conducting the analysis, the weights can be estimated based on the results of an ordinary (unweighted) least squares fit. For example, in simple linear regression, one approach is to divide the regression residuals into several groups of approximately equal size based on the value of the independent variable x and calculate the variance of the observed residuals in each group. An examination of the relationship between the residual variances and several different functions of x (such as x, x^2, and $\sqrt{x}$) may reveal the appropriate weights to use.

Example 9.4

A Department of Transportation (DOT) official is investigating the possibility of collusive bidding among the state's road construction contractors. One aspect of the investigation involves a comparison of the winning (lowest) bid price y on a job with the length x of new road construction, a measure of job size. The data listed in Table 9.3 were supplied by the DOT for a sample of 11 new road construction jobs with approximately the same number of bidders.

(a) Use the method of least squares to fit the straight-line model

$$E(y) = \beta_0 + \beta_1 x$$

(b) Calculate and plot the regression residuals against x. Do you detect any evidence of heteroscedasticity?

(c) Use the method described in the preceding paragraph to find the approximate weights necessary to stabilize the error variances with weighted least squares.

Determining the Weights in Weighted Least Squares for Simple Linear Regression

1. Divide the data into several groups according to the values of the independent variable, x. The groups should have approximately equal sample sizes.

   (a) If the data is replicated (i.e., there are multiple observations for each value of x) and balanced (i.e., the same number of observations for each value of x), then create one group for each value of x.

   (b) If the data is not replicated, group the data according to "nearest neighbors," that is, ranges of x (e.g., $0 \leq x < 5$, $5 \leq x < 10$, $10 \leq x < 15$, etc.).

2. Determine the sample mean ($\bar{x}$) and sample variance (s^2) of the residuals in each group.

3. For each group, compare the residual variance, s^2, to different functions of $\bar{x}$ [e.g., $f(\bar{x}) = \bar{x}$, $f(\bar{x}) = \bar{x}^2$, and $f(\bar{x}) = \sqrt{\bar{x}}$], by calculating the ratio $s^2/f(\bar{x})$.

4. Find the function of $\bar{x}$ for which the ratio is nearly constant across groups.

5. The appropriate weights for the groups are $1/f(\bar{x})$.

 DOT11

Table 9.3 Sample data for new road construction jobs, Example 9.4

Job	Length of Road x, miles	Winning Bid Price y, $ thousands	Job	Length of Road x, miles	Winning Bid Price y, $ thousands
1	2.0	10.1	7	7.0	71.1
2	2.4	11.4	8	11.5	132.7
3	3.1	24.2	9	10.9	108.0
4	3.5	26.5	10	12.2	126.2
5	6.4	66.8	11	12.6	140.7
6	6.1	53.8			

(d) Carry out the weighted least squares analysis using the weights determined in part c.

(e) Plot the weighted least squares residuals (see Definition 9.2) against x to determine whether the variances have stabilized.

Solution

(a) The simple linear regression analysis was conducted using MINITAB; the resulting printout is given in Figure 9.11. The least squares line (shaded on the printout) is

$$\hat{y} = -15.112 + 12.0687x$$

Note that the model is statistically useful (reject H_0: $\beta_1 = 0$) at $p = .000$.

Figure 9.11 MINITAB printout of straight-line model, Example 9.4

```
The regression equation is
BIDPRICE = - 15.1 + 12.1 LENGTH

Predictor        Coef      SE Coef          T        P
Constant      -15.112        3.342      -4.52    0.001
LENGTH        12.0687        0.4138      29.16    0.000

S = 5.374      R-Sq = 99.0%      R-Sq(adj) = 98.8%

Analysis of Variance

Source           DF          SS         MS        F        P
Regression        1       24558      24558   850.45    0.000
Residual Error    9         260         29
Total            10       24818

Obs    LENGTH    BIDPRICE        Fit    SE Fit    Residual    St Resid
  1       2.0       10.10       9.02      2.65        1.08        0.23
  2       2.4       11.40      13.85      2.52       -2.45       -0.52
  3       3.1       24.20      22.30      2.31        1.90        0.39
  4       3.5       26.50      27.13      2.19       -0.63       -0.13
  5       6.4       66.80      62.13      1.64        4.67        0.91
  6       6.1       53.80      58.51      1.67       -4.71       -0.92
  7       7.0       71.10      69.37      1.62        1.73        0.34
  8      11.5      132.70     123.68      2.45        9.02        1.89
  9      10.9      108.00     116.44      2.27       -8.44       -1.73
 10      12.2      126.20     132.13      2.67       -5.93       -1.27
 11      12.6      140.70     136.95      2.81        3.75        0.82
```

(b) The regression residuals are calculated and reported in the bottom portion of the MINITAB printout. A plot of the residuals against the predictor variable x is shown in Figure 9.12. The residual plot clearly shows that the residual variance increases as length of road x increases, strongly suggesting the presence of heteroscedasticity. A procedure such as weighted least squares is needed to stabilize the variances.

(c) To apply weighted least squares, we must first determine the weights. Since it is not clear what function of x the error variance is proportional to, we will apply the procedure described in the box to estimate the weights.

Figure 9.12 MINITAB plot of residuals against road length, x

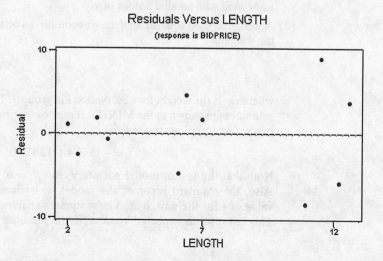

Residuals Versus LENGTH
(response is BIDPRICE)

First, we must divide the data into several groups according to the value of the independent variable x. Ideally, we want to form one group of data points for each different value of x. However, unless each value of x is replicated, not all of the group residual variances can be calculated. Therefore, we resort to grouping the data according to "nearest neighbors." One choice would be to use three groups, $2 \leq x \leq 4$, $6 \leq x \leq 7$, and $10 \leq x \leq 13$. These groups have approximately the same numbers of observations (namely, 4, 3, and 4 observations, respectively).

Next, we calculate the sample variance s_j^2 of the residuals included in each group. The three residual variances are given in Table 9.4. These variances are compared to three different functions of $\bar{x}$ ($\bar{x}, \bar{x}^2$, and $\sqrt{\bar{x}}$), as shown in Table 9.4, where $\bar{x}_j$ is the mean road length x for group j, $j = 1, 2, 3$.

Table 9.4 Comparison of residual variances to three functions of $\bar{x}$, Example 9.4

Group	Range of x	$\bar{x}_j$	s_j^2	$s_j^2/\bar{x}_j$	$s_j^2/\bar{x}_j^2$	$s_j^2/\sqrt{\bar{x}_j}$
1	$2 \leq x \leq 4$	2.75	3.722	1.353	.492	2.244
2	$6 \leq x \leq 7$	6.5	23.016	3.541	.545	9.028
3	$10 \leq x \leq 13$	11.8	67.031	5.681	.481	19.514

Note that the ratio $s_j^2/\bar{x}_j^2$ yields a value near .5 for each of the three groups. This result suggests that the residual variance of each group is proportional to $\bar{x}^2$, that is,

$$\sigma_j^2 = k\bar{x}_j^2, \quad j = 1, 2, 3$$

where k is approximately .5. Thus, a reasonable approximation to the weight for each group is

$$w_j = \frac{1}{\bar{x}_j^2}$$

With this weighting scheme, observations associated with large values of length of road x will have less influence on the regression residuals than observations associated with smaller values of x.

(d) A weighted least squares analysis was conducted on the data in Table 9.3 using the weights

$$w_{ij} = \frac{1}{\bar{x}_j^2}$$

where w_{ij} is the weight for observation i in group j. The weighted least squares estimates are shown in the MINITAB printout reproduced in Figure 9.13. The prediction equation (shaded) is

$$\hat{y} = -15.274 + 12.1204x$$

Note that the test of model adequacy, $H_0: \beta_1 = 0$, is significant at $p = .000$. Also, the standard error of the model, s, is significantly smaller than the value of s for the unweighted least squares analysis (.669 compared to 5.37). This last result is expected because, in the presence of heteroscedasticity, the

Figure 9.13 MINITAB printout of weighted least squares fit, Example 9.4

```
Weighted analysis using weights in WEIGHT

The regression equation is
BIDPRICE = - 15.3 + 12.1 LENGTH

Predictor       Coef      SE Coef        T         P
Constant      -15.274       1.601     -9.54     0.000
LENGTH        12.1204       0.3792     31.97    0.000

S = 0.6691      R-Sq = 99.1%     R-Sq(adj) = 99.0%

Analysis of Variance

Source          DF        SS         MS        F        P
Regression       1      457.48     457.48    1021.77   0.000
Residual Error   9        4.03       0.45
Total           10      461.51
```

unweighted least squares estimates are subject to greater sampling error than the weighted least squares estimates.

(e) A MINITAB plot of the weighted least squares residuals against x is shown in Figure 9.14. The lack of a discernible pattern in the residual plot suggests that the weighted least squares procedure has corrected the problem of unequal variances.

Figure 9.14 MINITAB plot of weighted residuals against road length, x

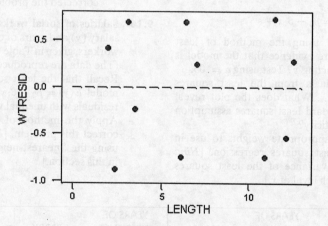

Before concluding this section, we mention that the "nearest neighbor" technique, illustrated in Example 9.4, will not always be successful in finding the optimal or near-optimal weights in weighted least squares. First, it may not be easy to identify the appropriate groupings of data points, especially if more than one independent variable is included in the regression. Second, the relationship between the residual variance and some preselected function of the independent variables may not reveal a consistent pattern over groups. In other words, unless the right function (or approximate function) of x is examined, the weights will be difficult to determine. More sophisticated techniques for choosing the weights in weighted least squares are available. Consult the references given at the end of this chapter for details on how to use these techniques.

9.4 Exercises

9.14 Determining the weights in WLS. Consider the straight-line model $y_i = \beta_0 + \beta_1 x_i + \varepsilon_i$. Give the appropriate weights w_i to use in a weighted least squares regression if the variance of the random error ε_i (i.e., σ_i^2) is proportional to

(a) x_i^2

(b) $\sqrt{x_i}$

(c) x_i

(d) $\frac{1}{n_i}$, where n_i is the number of observations at level x_i

(e) $\frac{1}{x_i}$

9.15 Quality control study. A machine that mass produces rubber gaskets can be set at one of three different speeds: 100, 150, or 200 gaskets per minute. As part of a quality control study, the machine was monitored several different times at each of the three speeds, and the number of defectives produced per hour was recorded. The data are provided in the table (next column). Since the number of defectives (y) is thought to be linearly related to speed (x), the following straight-line model is proposed:

$$y = \beta_0 + \beta_1 x + \varepsilon$$

(a) Fit the model using the method of least squares. Is there evidence that the model is useful for predicting y? Test using $\alpha = .05$.

(b) Plot the residuals from the least squares model against x. What does the plot reveal about the standard least squares assumption of homoscedasticity?

(c) Estimate the appropriate weights to use in a weighted least squares regression. [*Hint*: Calculate the variance of the least squares residuals at each level of x.]

GASKETS

MACHINE SPEED	NUMBER OF DEFECTIVES	MACHINE SPEED	NUMBER OF DEFECTIVES
x, gaskets per minute	y	x, gaskets per minute	y
100	15	150	35
100	23	150	24
100	11	200	26
100	14	200	48
100	18	200	27
150	19	200	38
150	29	200	39
150	20		

(d) Refit the model using weighted least squares. Compare the standard deviation of the weighted least squares slope to the standard deviation of the unweighted least squares slope.

(e) Plot the weighted residuals against x to determine whether using weighted least squares has corrected the problem of unequal variances.

9.16 Salaries of social workers. Refer to the data on salary (y) and years of experience (x) for 50 social workers, given in Table 8.4 of Example 8.5 (p. 401). (The data are reproduced below for convenience.) Recall that the least squares fit of the quadratic model $E(y) = \beta_0 + \beta_1 x + \beta_2 x^2$ yielded regression residuals with unequal variances (see Figure 8.15). Apply the method of weighted least squares to correct this problem. [*Hint*: Estimate the weights using the "nearest neighbor" technique outlined in this section.]

SOCWORK

YEARS OF EXPERIENCE x	SALARY y	YEARS OF EXPERIENCE x	SALARY y	YEARS OF EXPERIENCE x	SALARY y	YEARS OF EXPERIENCE x	SALARY y
7	$26,075	28	$64,785	7	$22,210	26	$65,343
28	79,370	26	61,581	2	20,521	19	46,216
23	65,726	27	70,678	18	49,727	16	54,288
18	41,983	20	51,301	11	33,233	3	20,844
19	62,309	18	39,346	21	43,628	12	32,586
15	41,154	1	24,833	4	16,105	23	71,235
24	53,610	26	65,929	24	65,644	20	36,530
13	33,697	20	41,721	20	63,022	19	52,745
2	22,444	26	82,641	20	47,780	27	67,282
8	32,562	28	99,139	15	38,853	25	80,931
20	43,076	23	52,624	25	66,537	12	32,303
21	56,000	17	50,594	25	67,447	11	38,371
18	58,667	25	53,272				

ASWELLS

9.17 Arsenic in groundwater. Refer to the *Environmental Science and Technology* (January 2005) study of the reliability of a commercial kit to test for arsenic in groundwater, Exercise 8.10 (p. 407). Consider the simple linear model, $E(y) = \beta_0 = \beta_1 x$, where y = arsenic level (ppm) and x = depth (feet).

(a) Fit the model to the data saved in the *ASWELLS* file. Construct a residual plot to check the assumption of a constant error variance. What do you observe?

(b) Classify the data into three groups according to depth (x) using the nearest neighbor approach. Then, use the technique outlined in this section to find the appropriate weights for a weighted least squares regression.

(c) Fit the model using weighted least squares and the weights from part b. Construct a plot of the weighted residuals to check the assumption of a constant error variance. What do you observe?

GASTURBINE

9.18 Cooling method for gas turbines. Refer to the *Journal of Engineering for Gas Turbines and Power* (January 2005) study of a high-pressure inlet fogging method for a gas turbine engine, Exercise 8.11 (p. 407). Consider the simple linear model, $E(y) = \beta_0 = \beta_1 x$, where y = heat rate and x = inlet temperature.

(a) Fit the model to the data saved in the *GASTURBINE* file. Construct a residual plot to check the assumption of a constant error variance. What do you observe?

(b) Classify the data into five groups according to inlet temperature (x) using the nearest neighbor approach. Then, use the technique outlined in this section to find the appropriate weights for a weighted least squares regression.

(c) Fit the model using weighted least squares and the weights from part b. Construct a plot of the weighted residuals to check the assumption of a constant error variance. What do you observe?

9.5 Modeling Qualitative Dependent Variables

For all models discussed in the previous sections of this text, the response (dependent) variable y is a *quantitative* variable. In this section, we consider models for which the response y is a **qualitative variable at two levels**, or, as it is sometimes called, a **binary variable**.

For example, a physician may want to relate the success or failure of a new surgical procedure to the characteristics (such as age and severity of the disease) of the patient. The value of the response of interest to the physician is either *yes*, the operation is a success, or *no*, the operation is a failure. Similarly, a state attorney general investigating collusive practices among bidders for road construction contracts may want to determine which contract-related variables (such as number of bidders, bid amount, and cost of materials) are useful indicators of whether a bid is fixed (i.e., whether the bid price is intentionally set higher than the fair market value). Here, the value of the response variable is either *fixed* bid or *competitive* bid.

Just as with qualitative independent variables, we use **dummy** (i.e., coded 0–1) **variables** to represent the qualitative response variable. For example, the response of interest to the entrepreneur is recorded as

$$y = \begin{cases} 1 & \text{if new business a success} \\ 0 & \text{if new business a failure} \end{cases}$$

where the assignment of 0 and 1 to the two levels is arbitrary. The linear statistical model takes the usual form

$$y = \beta_0 + \beta_1 x_1 + \beta_2 x_2 + \cdots + \beta_k x_k + \varepsilon$$

However, when the response is binary, the expected response

$$E(y) = \beta_0 + \beta_1 x_1 + \beta_2 x_2 + \cdots + \beta_k x_k$$

has a special meaning. It can be shown* that $E(y) = \pi$, where π is the probability that $y = 1$ for given values of $x_1, x_2, \ldots, x_k$. Thus, for the entrepreneur, the mean response $E(y)$ represents the probability that a new business with certain owner-related characteristics will be a success.

When the ordinary least squares approach is used to fit models with a binary response, several well-known problems are encountered. A discussion of these problems and their solutions follows.

Problem 1   *Nonnormal errors*: The standard least squares assumption of normal errors is violated since the response y and, hence, the random error ε can take on only two values. To see this, consider the simple model $y = \beta_0 + \beta_1 x + \varepsilon$. Then we can write

$$\varepsilon = y - (\beta_0 + \beta_1 x)$$

Thus, when $y = 1$, $\varepsilon = 1 - (\beta_0 + \beta_1 x)$ and when $y = 0$, $\varepsilon = -\beta_0 - \beta_1 x$.

When the sample size n is large, however, any inferences derived from the least squares prediction equation remain valid in most practical situations even though the errors are nonnormal.[†]

Problem 2   *Unequal variances*: It can be shown[‡] that the variance σ^2 of the random error is a function of π, the probability that the response y equals 1. Specifically,

$$\sigma^2 = V(\varepsilon) = \pi(1 - \pi)$$

Since, for the linear statistical model,

$$\pi = E(y) = \beta_0 + \beta_1 x_1 + \beta_2 x_2 + \cdots + \beta_k x_k$$

this implies that σ^2 is not constant and, in fact, depends on the values of the independent variables; hence, the standard least squares assumption of equal variances is also violated. One solution to this problem is to use weighted least squares (see Section 9.4), where the weights are inversely proportional to σ^2, that is,

$$w_i = \frac{1}{\sigma_i^2}$$

$$= \frac{1}{\pi_i(1 - \pi_i)}$$

Unfortunately, the true proportion

$$\pi_i = E(y_i)$$

$$= \beta_0 + \beta_1 x_{1i} + \beta_2 x_{2i} + \cdots + \beta_k x_{ki}$$

is unknown since $\beta_0, \beta_1, \ldots, \beta_k$ are unknown population parameters. However, a technique called **two-stage least squares** can be applied to circumvent this difficulty.

---

* The result is a straightforward application of the expectation theorem for a random variable. Let $\pi = P(y = 1)$ and $1 - \pi = P(y = 0)$, $0 \le \pi \le 1$. Then, by definition, $E(y) = \Sigma_y y_i \cdot p(y) = (1)P(y = 1) + (0)P(y = 0) = P(y = 1) = \pi$. Students familiar with discrete random variables will recognize y as the **Bernoulli random variable** (i.e., a binomial random variable with $n = 1$).

† This property is due to the asymptotic normality of the least squares estimates of the model parameters under very general conditions.

‡ Using the properties of expected values with the Bernoulli random variable, we obtain $V(y) = E(y^2) - [E(y)]^2 = \Sigma y^2 \cdot p(y) - (\pi)^2 = (1)^2 P(y = 1) + (0)^2 P(y = 0) - \pi^2 = \pi - \pi^2 = \pi(1 - \pi)$. Since in regression, $V(\varepsilon) = V(y)$, the result follows.

Two-stage least squares, as its name implies, involves conducting an analysis in two steps:

> **STAGE 1** Fit the regression model using the *ordinary least squares* procedure and obtain the predicted values $\hat{y}_i$, $i = 1, 2, \ldots, n$. Recall that $\hat{y}_i$ estimates π_i for the binary model.
>
> **STAGE 2** Refit the regression model using *weighted least squares*, where the estimated weights are calculated as follows:

$$w_i = \frac{1}{\hat{y}_i(1 - \hat{y}_i)}$$

Further iterations—revising the weights at each step—can be performed if desired. In most practical problems, however, the estimates of π_i obtained in stage 1 are adequate for use in weighted least squares.

Problem 3 *Restricting the predicted response to be between 0 and 1*: Since the predicted value $\hat{y}$ estimates $E(y) = \pi$, the probability that the response y equals 1, we would like $\hat{y}$ to have the property that $0 \le \hat{y} \le 1$. There is no guarantee, however, that the regression analysis will always yield predicted values in this range. Thus, the regression may lead to nonsensical results (i.e., negative estimated probabilities or predicted probabilities greater than 1). To avoid this problem, you may want to fit a model with a mean response function $E(y)$ that automatically falls between 0 and 1. (We consider one such model in the next section.)

In summary, the purpose of this section has been to identify some of the problems resulting from fitting a linear model with a binary response and to suggest ways in which to circumvent these problems. Another approach is to fit a model specially designed for a binary response, called a logistic regression model. Logistic regression models are the subject of Section 9.6.

9.5 Exercises

9.19 **Problems modeling a binary response.** Discuss the problems associated with fitting a multiple regression model where the response y is recorded as 0 or 1.

9.20 **Use of digital organizers.** A retailer of hand-held digital organizers conducted a study to relate ownership of the devices with annual income of heads of households. Data collected for a random sample of 20 households were used to fit the straight-line model $E(y) = \beta_0 + \beta_1 x$, where

$$y = \begin{cases} 1 & \text{if own a digital organizer} \\ 0 & \text{if not} \end{cases}$$

x = Annual income (in dollars)

The data are shown in the accompanying table. Fit the model using two-stage least squares. Is the model useful for predicting y? Test using $\alpha = .05$.

9.21 **Gender discrimination in hiring.** Suppose you are investigating allegations of gender discrimination in the hiring practices of a particular firm. An equal-rights group claims that females are

🖲 **PALMORG**

HOUSEHOLD	y	x	HOUSEHOLD	y	x
1	0	$36,300	11	1	$42,400
2	0	31,200	12	0	30,600
3	0	56,500	13	0	41,400
4	1	41,700	14	0	28,300
5	1	60,200	15	1	47,500
6	0	32,400	16	0	35,700
7	0	35,000	17	0	32,100
8	0	29,200	18	1	79,600
9	1	56,700	19	1	40,200
10	0	82,000	20	0	53,100

less likely to be hired than males with the same background, experience, and other qualifications. Data (shown in the next table) collected on 28 former applicants will be used to fit the model $E(y) = \beta_0 + \beta_1 x_1 + \beta_2 x_2 + \beta_3 x_3$, where

$$y = \begin{cases} 1 & \text{if hired} \\ 0 & \text{if not} \end{cases}$$

🔘 DISCRIM

HIRING STATUS	EDUCATION	EXPERIENCE	GENDER	HIRING STATUS	EDUCATION	EXPERIENCE	GENDER
y	x_1, years	x_2, years	x_3	y	x_1, years	x_2, years	x_3
0	6	2	0	1	4	5	1
0	4	0	1	0	6	4	0
1	6	6	1	0	8	0	1
1	6	3	1	1	6	1	1
0	4	1	0	0	4	7	0
1	8	3	0	0	4	1	1
0	4	2	1	0	4	5	0
0	4	4	0	0	6	0	1
0	6	1	0	1	8	5	1
1	8	10	0	0	4	9	0
0	4	2	1	0	8	1	0
0	8	5	0	0	6	1	1
0	4	2	0	1	4	10	1
0	6	7	0	1	6	12	0

x_1 = Years of higher education (4, 6, or 8)

x_2 = Years of experience

$x_3 = \begin{cases} 1 & \text{if male applicant} \\ 0 & \text{if female applicant} \end{cases}$

(a) Interpret each of the β's in the multiple regression model.

(b) Fit the multiple regression model using two-stage least squares. [*Hint*: Replace negative

predicted values from the first stage with a value of 0.01.]

(c) Conduct a test of model adequacy. Use $\alpha = .05$.

(d) Is there sufficient evidence to indicate that gender is an important predictor of hiring status? Test using $\alpha = .05$.

(e) Calculate a 95% confidence interval for the mean response $E(y)$ when $x_1 = 4$, $x_2 = 3$, and $x_3 = 0$. Interpret the interval.

9.6 Logistic Regression

Often, the relationship between a qualitative binary response y and a single predictor variable x is curvilinear. One particular curvilinear pattern frequently encountered in practice is the S-shaped curve shown in Figure 9.15. Points on the curve represent $\pi = P(y = 1)$ for each value of x. A model that accounts for this type of curvature is the **logistic** (or **logit**) **regression model**,

$$E(y) = \frac{\exp(\beta_0 + \beta_1 x)}{1 + \exp(\beta_0 + \beta_1 x)}$$

The logistic model was originally developed for use in **survival analysis**, where the response y is typically measured as 0 or 1, depending on whether the experimental unit (e.g., a patient) "survives." Note that the curve shown in Figure 9.15 has

Figure 9.15 Graph of $E(y)$ for the logistic model

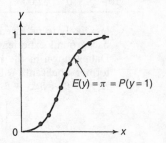

$E(y) = \pi = P(y = 1)$

asymptotes at 0 and 1—that is, the mean response $E(y)$ can never fall below 0 or above 1. Thus, the logistic model ensures that the estimated response $\hat{y}$ (i.e., the estimated probability that $y = 1$) lies between 0 and 1.

In general, the logistic model can be written as shown in the box.

Logistic Regression Model for a Binary Dependent Variable

$$E(y) = \frac{\exp(\beta_0 + \beta_1 x_1 + \beta_2 x_2 + \cdots + \beta_k x_k)}{1 + \exp(\beta_0 + \beta_1 x_1 + \beta_2 x_2 + \cdots + \beta_k x_k)}$$

where

$$y = \begin{cases} 1 & \text{if category A occurs} \\ 0 & \text{if category B occurs} \end{cases}$$

$$E(y) = P(\text{Category A occurs}) = \pi$$

$x_1, x_2, \ldots, x_k$ are quantitative or qualitative independent variables

---

Note that the general logistic model is not a linear function of the β parameters (see Section 4.1). Obtaining the parameter estimates of a **nonlinear regression model**, such as the logistic model, is a numerically tedious process and often requires sophisticated computer programs. In this section we briefly discuss two approaches to the problem, and give an example of a computer printout for the second.

1. *Least squares estimation using a transformation*: One method of fitting the model involves a transformation on the mean response $E(y)$. Recall (from Section 9.5) that for a binary response, $E(y) = \pi$, where π denotes the probability that $y = 1$. Then the logistic model

$$\pi = \frac{\exp(\beta_0 + \beta_1 x_1 + \cdots + \beta_k x_k)}{1 + \exp(\beta_0 + \beta_1 x_1 + \cdots + \beta_k x_k)}$$

implies (proof omitted) that

$$\ln\left(\frac{\pi}{1 - \pi}\right) = \beta_0 + \beta_1 x_1 + \cdots + \beta_k x_k$$

Set

$$\pi^* = \ln\left(\frac{\pi}{1 - \pi}\right)$$

The transformed logistic model

$$\pi^* = \beta_0 + \beta_1 x_1 + \cdots + \beta_k x_k$$

is now linear in the β's and the method of least squares can be applied.

Note: Since $\pi = P(y = 1)$, then $1 - \pi = P(y = 0)$. The ratio

$$\frac{\pi}{1 - \pi} = \frac{P(y = 1)}{P(y = 0)}$$

is known as the **odds** of the event, $y = 1$, occurring. (For example, if $\pi = .8$, then the odds of $y = 1$ occurring are $.8/.2 = 4$, or 4 to 1.) The transformed model, π^*, then, is a model for the natural logarithm of the odds of $y = 1$ occurring and is often called the **log-odds model**.

> **Definition 9.3** In logistic regression with a binary response y, we define the **odds of the event ($y = 1$) occurring** as follows:
>
> $$\text{Odds} = \frac{\pi}{1 - \pi} = \frac{P(y = 1)}{P(y = 0)}$$

Although the transformation succeeds in linearizing the response function, two other problems remain. First, since the true probability π is unknown, the values of the log-odds π^*, necessary for input into the regression, are also unknown. To carry out the least squares analysis, we must obtain estimates of π^* for each combination of the independent variables. A good choice is the estimator

$$\pi^* = \ln\left(\frac{\hat{\pi}}{1 - \hat{\pi}}\right)$$

where $\hat{\pi}$ is the sample proportion of 1's for the particular combination of x's. To obtain these estimates, however, *we must have replicated observations of the response y at each combination of the levels of the independent variables.* Thus, the least squares transformation approach is limited to replicated experiments, which occur infrequently in a practical business setting.

The second problem is associated with unequal variances. The transformed logistic model yields error variances that are inversely proportional to $\pi(1 - \pi)$. Since π, or $E(y)$, is a function of the independent variables, the regression errors are heteroscedastic. To stabilize the variances, weighted least squares should be used. This technique also requires that replicated observations be available for each combination of the x's and, in addition, that the number of observations at each combination be relatively large. If the experiment is replicated, with n_j (large) observations at each combination of the levels of the independent variables, then the appropriate weights to use are

$$w_j = n_j \hat{\pi}_j (1 - \hat{\pi}_j)$$

where

$$\hat{\pi}_j = \frac{\text{Number of 1's for combination } j \text{ of the } x\text{'s}}{n_j}$$

2. *Maximum likelihood estimation*: Estimates of the β parameters in the logistic model also can be obtained by applying a common statistical technique, called **maximum likelihood estimation**. Like the least squares estimators, the maximum likelihood estimators have certain desirable properties.[*] (In fact, when the errors of a linear regression model are normally distributed, the least squares estimates and maximum likelihood estimates are equivalent.) Many of the available statistical computer software packages use maximum likelihood estimation to fit logistic regression models. Therefore, one practical advantage of using the maximum likelihood method (rather than the transformation approach) to fit logistic regression models is that computer programs are readily available. Another advantage is that the data need not be replicated to apply maximum likelihood estimation.

The maximum likelihood estimates of the parameters of a logistic model have distributional properties that are different from the standard F and t distributions of least squares regression. Under certain conditions, the test statistics for testing

---

[*] For details on how to obtain maximum likelihood estimators and what their distributional properties are, consult the references given at the end of the chapter.

Figure 9.16 Several chi-square probability distributions

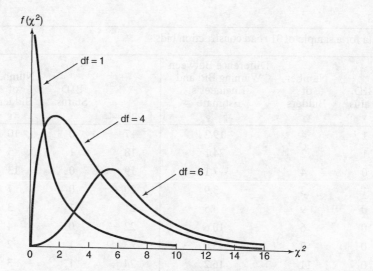

individual parameters and overall model adequacy have approximate **chi-square** (χ^2) **distributions**. The χ^2 distribution is similar to the F distribution in that it depends on degrees of freedom and is nonnegative, as shown in Figure 9.16. (Critical values of the χ^2 distribution for various values of α and degrees of freedom are given in Table 10, Appendix C.) We illustrate the application of maximum likelihood estimation for logistic regression with an example.

Example 9.5

Consider the problem of collusive (i.e., noncompetitive) bidding among road construction contractors. Recall (from Section 9.5) that contractors sometimes scheme to set bid prices higher than the fair market (or competitive) price. Suppose an investigator has obtained information on the bid status (fixed or competitive) for a sample of 31 contracts. In addition, two variables thought to be related to bid status are also recorded for each contract: number of bidders x_1 and the difference between the winning (lowest) bid and the estimated competitive bid (called the engineer's estimate) x_2, measured as a percentage of the estimate. The data appear in Table 9.5, with the response y recorded as follows:

$$y = \begin{cases} 1 & \text{if fixed bid} \\ 0 & \text{if competitive bid} \end{cases}$$

An appropriate model for $E(y) = \pi = P(y = 1)$ is the logistic model

$$\pi = \frac{\exp(\beta_0 + \beta_1 x_1 + \beta_2 x_2)}{1 + \exp(\beta_0 + \beta_1 x_1 + \beta_2 x_2)}$$

Alternatively, the model can be written $\pi^* = \beta_0 + \beta_1 x_1 + \beta_2 x_2$, where $\pi^* = \ln\left(\frac{\pi}{1-\pi}\right)$.

Solution

The model was fitted to the data using the logistic regression option of SAS. The resulting printout is shown in Figure 9.17. Interpret the results.

The maximum likelihood estimates of β_0, β_1, and β_2 (shaded in the printout) are $\hat{\beta}_0 = 1.4211$, $\hat{\beta}_1 = -.7553$, and $\hat{\beta}_2 = .1122$. Therefore, the prediction equation for the probability of a fixed bid [i.e., $\pi = P(y = 1)$] is

$$\hat{y} = \frac{\exp(1.4211 - .7553x_1 + .1122x_2)}{1 + \exp(1.4211 - .7553x_1 + .1122x_2)}$$

🔘 ROADBIDS

Table 9.5 Data for a sample of 31 road construction bids

Contract	BID Status y	Number of Bidders x_1	Difference Between Winning Bid and Engineer's Estimate x_2, %	Contract	BID Status y	Number of Bidders x_1	Difference Between Winning Bid and Engineer's Estimate x_2, %
1	1	4	19.2	17	0	10	6.6
2	1	2	24.1	18	1	5	−2.5
3	0	4	−7.1	19	0	13	24.2
4	1	3	3.9	20	0	7	2.3
5	0	9	4.5	21	1	3	36.9
6	0	6	10.6	22	0	4	11.7
7	0	2	−3.0	23	1	2	22.1
8	0	11	16.2	24	1	3	10.4
9	1	6	72.8	25	0	2	9.1
10	0	7	28.7	26	0	5	2.0
11	1	3	11.5	27	0	6	12.6
12	1	2	56.3	28	1	5	18.0
13	0	5	−.5	29	0	3	1.5
14	0	3	−1.3	30	1	4	27.3
15	0	3	12.9	31	0	10	−8.4
16	0	8	34.1				

Figure 9.17 SAS printout of logistic regression on bid status

Probability modeled is STATUS=1.

Model Convergence Status

Convergence criterion (GCONV=1E-8) satisfied.

Model Fit Statistics

Criterion	Intercept Only	Intercept and Covariates
AIC	43.381	28.843
SC	44.815	33.145
-2 Log L	41.381	22.843

Testing Global Null Hypothesis: BETA=0

Test	Chi-Square	DF	Pr > ChiSq
Likelihood Ratio	18.5377	2	<.0001
Score	13.4661	2	0.0012
Wald	6.4289	2	0.0402

Analysis of Maximum Likelihood Estimates

Parameter	DF	Estimate	Standard Error	Wald Chi-Square	Pr > ChiSq
Intercept	1	1.4211	1.2867	1.2198	0.2694
NUMBIDS	1	-0.7553	0.3388	4.9703	0.0258
DOTEST	1	0.1122	0.0514	4.7666	0.0290

Odds Ratio Estimates

Effect	Point Estimate	95% Wald Confidence Limits	
NUMBIDS	0.470	0.242	0.913
DOTEST	1.119	1.012	1.237

Association of Predicted Probabilities and Observed Responses

Percent Concordant	90.4	Somers' D	0.807
Percent Discordant	9.6	Gamma	0.807
Percent Tied	0.0	Tau-a	0.396
Pairs	228	c	0.904

Wald Confidence Interval for Parameters

Parameter	Estimate	95% Confidence Limits	
Intercept	1.4211	-1.1008	3.9431
NUMBIDS	-0.7553	-1.4193	-0.0913
DOTEST	0.1122	0.0115	0.2129

Wald Confidence Interval for Adjusted Odds Ratios

Effect	Unit	Estimate	95% Confidence Limits	
NUMBIDS	1.0000	0.470	0.242	0.913
DOTEST	1.0000	1.119	1.012	1.237

Classification Table

Prob Level	Correct Event	Correct Non-Event	Incorrect Event	Incorrect Non-Event	Percentages Correct	Sensi-tivity	Speci-ficity	False POS	False NEG
0.500	9	16	3	3	80.6	75.0	84.2	25.0	15.8

Predictions of STATUS

Obs	STATUS	NUMBIDS	DOTEST	Response Value	Probability	Lower 95% Confidence Limit	Upper 95% Confidence Limit
1	1	4	19.2	1	0.63510	0.32984	0.86023
2	1	2	24.1	1	0.93179	0.53645	0.99384
3	0	4	-7.1	1	0.08342	0.01043	0.44012
4	1	3	3.9	1	0.39958	0.15869	0.70133
5	0	9	4.5	1	0.00760	0.00016	0.26829
6	0	6	10.6	1	0.12771	0.02582	0.44709
7	0	2	-3.0	1	0.39506	0.10273	0.78837
8	0	11	16.2	1	0.00625	0.00007	0.36815
9	1	6	72.8	1	0.99368	0.35196	0.99998
10	0	7	28.7	1	0.34392	0.06138	0.80777
11	1	3	11.5	1	0.60957	0.31578	0.84081
12	1	2	56.3	1	0.99803	0.69691	0.99999
13	0	5	-0.5	1	0.08230	0.01254	0.38784
14	0	3	-1.3	1	0.27078	0.07453	0.63132
15	1	3	12.9	1	0.64625	0.34076	0.86589
16	0	8	34.1	1	0.31103	0.03168	0.86168
17	0	10	6.6	1	0.00453	0.00006	0.26606
18	0	5	-2.5	1	0.06686	0.00852	0.37405
19	0	13	24.2	1	0.00339	0.00001	0.45718
20	0	7	2.3	1	0.02639	0.00166	0.30643
21	1	3	36.9	1	0.96427	0.54748	0.99834
22	0	4	-11.7	1	0.05152	0.00412	0.41605
23	1	2	22.1	1	0.91607	0.51883	0.99103
24	1	3	10.4	1	0.57983	0.29466	0.82010
25	0	2	9.1	1	0.71739	0.33903	0.92627
26	0	5	2.0	1	0.10612	0.02005	0.40786
27	0	6	12.6	1	0.15486	0.03485	0.48182
28	1	5	18.0	1	0.41683	0.17873	0.70127
29	0	3	1.5	1	0.33705	0.11481	0.66587
30	1	4	27.3	1	0.81199	0.40058	0.96541
31	0	10	-8.4	1	0.00085	0.00000	0.15847

Figure 9.17 *Continued*

In general, the coefficient $\hat{\beta}_i$ in the logistic model estimates the change in the log-odds when x_i is increased by 1 unit, holding all other x's in the model fixed. The antilog of the coefficient, $e^{\hat{\beta}_i}$, then estimates the odds-ratio

$$\frac{\pi_{x+1}/(1 - \pi_{x+1})}{\pi_x/(1 - \pi_x)}$$

where π_x is the value of $P(y = 1)$ for a fixed value x.[†] Typically, analysts compute $(e^{\hat{\beta}_i}) - 1$, which is an estimate of the percentage increase (or decrease) in the odds $\pi = P(y = 1)/P(y = 0)$ for every 1-unit increase in x_i, holding the other x's fixed.

This leads to the following interpretations of the β estimates:

$\hat{\beta}_1 = -.7553$; $e^{\hat{\beta}_1} = .47$; $e^{\hat{\beta}_1} - 1 = -.53$: For each additional bidder (x_1), we estimate the odds of a fixed contract to *decrease* by 53%, holding **DOTEST** (x_2) fixed.

$\hat{\beta}_2 = .1122$; $e^{\hat{\beta}_2} = 1.12$; $e^{\hat{\beta}_2} - 1 = .12$: For every 1% increase in **DOTEST** (x_2), we estimate the odds of a fixed contract to *increase* by 12%, holding **NUMBIDS** (x_1) fixed.

Interpretations of β Parameters in the Logistic Model

$$\pi^* = \beta_0 + \beta_1 x_1 + \beta_2 x_2 + \cdots + \beta_k x_k$$

where

$$\pi^* = \ln\left(\frac{\pi}{1 - \pi}\right)$$

$$\pi = P(y = 1)$$

$\beta_i = $ Change in log-odds π^* for every 1-unit increase in x_i, holding all other x's fixed

$e^{\beta_i} - 1 = $ Percentage change in odds $\pi/(1 - \pi)$ for every 1-unit increase in x_i, holding all other x's fixed

---

The standard errors of the β estimates are given under the column **Standard Error**, and the (squared) ratios of the β estimates to their respective standard errors are given under the column **Wald Chi-Square**. As in regression with a linear model, this ratio provides a test statistic for testing the contribution of each variable to the model (i.e., for testing H_0: $\beta_i = 0$).[‡] The observed significance levels of the tests (i.e., the p-values) are given under the column **Pr > Chi-Square**. Note that both independent variables, **NUMBIDS** (x_1) and **DOTEST** (x_2), have p-values less than .03 (implying that we would reject H_0: $\beta_1 = 0$ and H_0: $\beta_2 = 0$ for $\alpha = .03$).

[†] To see this, consider the model $\pi^* = \beta_0 + \beta_1 x$, where $x = 1$ or $x = 0$. When $x = 1$, we have $\pi_1^* = \beta_0 + \beta_1$; when $x = 0$, $\pi_0^* = \beta_0$. Now replace π_i^* with $\ln[\pi_i/(1 + \pi_i)]$, and take the antilog of each side of the equation. Then we have $\pi_1/(1 - \pi_1) = e^{\beta_0} e^{\beta_1}$ and $\pi_0/(1 - \pi_0) = e^{\beta_0}$. Consequently, the odds-ratio is

$$\frac{\pi_1/(1 - \pi_1)}{\pi_0/(1 - \pi_0)} = e^{\beta_1}$$

[‡] In the logistic regression model, the ratio $(\hat{\beta}_i/s_{\hat{\beta}_i})^2$ has an approximate χ^2 distribution with 1 degree of freedom. Consult the references for more details about the χ^2 distribution and its use in logistic regression.

The test statistic for testing the overall adequacy of the logistic model (i.e., for testing H_0: $\beta_1 = \beta_2 = 0$), is given in the upper portion of the printout (shaded in the **Likelihood Ratio** row) as $\chi^2 = 18.5377$, with observed significance level (shaded) $p < .0001$.[§] Based on the p-value of the test, we can reject H_0 and conclude that at least one of the β coefficients is nonzero. Thus, the model is adequate for predicting bid status y.

Finally, the bottom portion of the printout gives predicted values and lower and upper 95% prediction limits for each observation used in the analysis. The 95% prediction interval for π for a contract with $x_1 = 3$ bidders and winning bid amount $x_2 = 11.5\%$ above the engineer's estimate is shaded on the printout. We estimate π, the probability of this particular contract being fixed, to fall between .31578 and .84081. Note that all the predicted values and limits lie between 0 and 1, a property of the logistic model. ∎

Example 9.6

Refer to Example 9.5. The bid-rigging investigator wants to improve the model for bid status (y) by considering higher-order terms for number of bidders (x_1) and percent difference between winning and DOT estimated price (x_2). Again, denote $\pi = P(y = 1)$ and $\pi^* = ln[\pi/(1 - \pi)]$. Consider the complete second-order logistic regression model

$$\pi^* = \beta_0 + \beta_1 x_1 + \beta_2 x_2 + \beta_3 x_1 x_2 + \beta_4 x_1^2 + \beta_5 x_2^2$$

Fit the model to the data in Table 9.4, then determine whether any of the higher-order terms significantly improve the model.

Solution

One way to determine whether any of the higher-order terms significantly improve the model is to test the null hypothesis,

$$H_0: \beta_3 = \beta_4 = \beta_5 = 0$$

This test is obtained using methodology similar to the technique for comparing nested models discussed in Section 4.13. Here, we have:

Complete model: $\pi^* = \beta_0 + \beta_1 x_1 + \beta_2 x_2 + \beta_3 x_1 x_2 + \beta_4 x_1^2 + \beta_5 x_2^2$

Reduced model: $\pi^* = \beta_0 + \beta_1 x_1 + \beta_2 x_2$

Rather than an F-statistic, the test statistic for testing a subset of betas in logistic regression is based on the chi-square distribution. In fact, the test statistic is the difference between the global χ^2 values for testing the overall adequacy of the two models, that is, $\chi^2 = \chi^2_{(Complete)} - \chi^2_{(Reduced)}$. The degrees of freedom for this chi-square value is based on the difference between the degrees of freedom for the two models, that is, $df = df_{(Complete)} - df_{(Reduced)}$.

We fit the complete model to the data using SAS. A portion of the resulting SAS printout is shown in Figure 9.18. The chi-square value for the complete model, highlighted on the printout, is $\chi^2_{(Complete)} = 27.56$, with 5 df. The chi-square value for the reduced model (obtained from Example 9.5) is $\chi^2_{(Reduced)} = 18.54$, with 2 df. Consequently, the test statistic for testing H_0: $\beta_3 = \beta_4 = \beta_5 = 0$ is

$$\text{Test statistic: } \chi^2 = \chi^2_{(Complete)} - \chi^2_{(Reduced)} = 27.56 - 18.54 = 9.02$$

[§] The test statistic has an approximate χ^2 distribution with $k = 2$ degrees of freedom, where k is the number of β parameters in the model (excluding β_0).

Figure 9.18 SAS analysis of complete second-order logistic model, Example 9.6

```
            Testing Global Null Hypothesis: BETA=0

Test                   Chi-Square       DF      Pr > ChiSq

Likelihood Ratio         27.5606        5         <.0001
Score                    14.7723        5         0.0114
Wald                      5.7009        5         0.3364

            Analysis of Maximum Likelihood Estimates

                                   Standard       Wald
Parameter   DF    Estimate          Error     Chi-Square   Pr > ChiSq

Intercept    1    -20.7253        13.6471       2.3063       0.1288
NUMBIDS      1      9.5926         6.7562       2.0159       0.1557
DOTEST       1      0.5403         0.3452       2.4491       0.1176
BID_EST      1     -0.0867         0.0808       1.1527       0.2830
NUMBIDSQ     1     -1.1214         0.8144       1.8964       0.1685
DOTESTSQ     1      0.00156        0.00456      0.1166       0.7327
```

For, say $\alpha = .01$, the critical value (obtained from Table 10, Appendix B) is: $\chi^2_{(2df)} = 11.3449$; thus, the rejection region for the test is:

$$\text{Rejection region: } \chi^2 > 11.3449$$

Since the test statistic value does not fall into the rejection region, our conclusion is as follows:

Conclusion: Insufficient evidence to reject H_0, that is, there is insufficient evidence to conclude that the higher-order terms for number of bidders (x_1) and percent difference between winning and DOT estimated price (x_2) are statistically useful predictors in the logistic regression model.

Based on the results of this nested-model test, the investigator will use the first-order (reduced) model, $\pi^* = \beta_0 + \beta_1 x_1 + \beta_2 x_2$, for predicting the probability of a fixed contract.   ◼

In summary, we have presented two approaches to fitting logistic regression models. If the data are replicated, you may want to apply the transformation approach. The maximum likelihood estimation approach can be applied to any data set, but you need access to a statistical software package (such as SAS or SPSS) with logistic regression procedures.

This section should be viewed only as an overview of logistic regression. Many of the details of fitting logistic regression models using either technique have been omitted. Before conducting a logistic regression analysis, we strongly recommend that you consult the references given at the end of this chapter.

9.6 Exercises

9.22 **Use of digital organizers.** Refer to Exercise 9.20 (p. 493) and the problem of modeling $y = \{1$ if own a digital organizer , 0 if not$\}$ as a function of annual income, x.

(a) Define π for this problem.

(b) Define "odds" for this problem.

(c) The data for the random sample of 20 households were used to fit the logit model

$$E(y) = \frac{\exp(\beta_0 + \beta_1 x)}{1 + \exp(\beta_0 + \beta_1 x)}$$

An SPSS printout of the logistic regression is presented on the next page. Interpret the results.

9.23 **Study of orocline development.** In *Tectonics* (October, 2004), geologists published their research on the formation of oroclines (curved mountain belts) in the central Appalachian mountains. A comparison was made of two nappes (sheets of rock that have moved over a large horizontal distance), one in Pennsylvania and the other in Maryland. Rock samples at the mountain rim of both locations were collected and the foliation intersection axes (FIA) preserved within large mineral grains was measured (in degrees) for each. Consider a logistic regression model for $y = \{1$ if Maryland nappe, 0 if Pennsylvania nappe$\}$ using only $x = $ FIA.

SPSS Output for Exercise 9.22

Omnibus Tests of Model Coefficients

		Chi-square	df	Sig.
Step 1	Step	2.929	1	.087
	Block	2.929	1	.087
	Model	2.929	1	.087

Model Summary

Step	-2 Log likelihood	Cox & Snell R Square	Nagelkerke R Square
1	22.969[a]	.136	.188

a. Estimation terminated at iteration number 4 because parameter estimates changed by less than .001.

Classification Table[a]

Observed			Predicted		
			OWN		
			0	1	Percentage Correct
Step 1	OWN	0	12	1	92.3
		1	5	2	28.6
	Overall Percentage				70.0

a. The cut value is .500

Variables in the Equation

		B	S.E.	Wald	df	Sig.	Exp(B)
Step 1[a]	INCOME	.000	.000	2.453	1	.117	1.000
	Constant	-3.112	1.675	3.454	1	.063	.044

a. Variable(s) entered on step 1: INCOME.

(a) Define π for this model.

(b) Write the equation of the model.

(c) Give a practical interpretation of the value of β_1 in the model.

(d) In terms of the β's, give the predicted probability that a rock sample is from the Maryland nappe if the FIA for the rock sample is 80°.

9.24 **Flight response of geese.** Offshore oil drilling near an Alaskan estuary has led to increased air traffic—mostly large helicopters—in the area. The U.S. Fish and Wildlife Service commissioned a study to investigate the impact these helicopters have on the flocks of Pacific brant geese, which inhabit the estuary in Fall before migrating (*Statistical Case Studies: A Collaboration between Academe and Industry*, 1998). Two large helicopters were flown repeatedly over the estuary at different altitudes and lateral distances from the flock. The flight responses of the geese (recorded

PACGEESE (*First 10 observations shown*)

OVERFLIGHT	ALTITUDE	LATERAL DISTANCE	FLIGHT RESPONSE
1	0.91	4.99	HIGH
2	0.91	8.21	HIGH
3	0.91	3.38	HIGH
4	9.14	21.08	LOW
5	1.52	6.60	HIGH
6	0.91	3.38	HIGH
7	3.05	0.16	HIGH
8	6.10	3.38	HIGH
9	3.05	6.60	HIGH
10	12.19	6.60	HIGH

Source: Erickson, W., Nick, T., and Ward, D. "Investigating flight response of Pacific brant to helicopters at Izembek Lagoon, Alaska by using logistic regression," *Statistical Case Studies: A Collaboration between Academe and Industry*, ASA-SIAM Series on Statistics and Applied Probability, 1998. Copyright © 1998 Society for Industrial and Applied Mathematics. Reprinted with permission. All rights reserved.

MINITAB Output for Exercise 9.24

Binary Logistic Regression: RESPONSE versus ALTITUDE, LATERAL

Link Function: Logit

Response Information

```
Variable   Value   Count
RESPONSE   1        285    (Event)
           0        179
           Total    464
```

Logistic Regression Table

```
                                                  Odds      95% CI
Predictor      Coef      SE Coef       Z      P   Ratio   Lower   Upper
Constant     2.39541   0.305507     7.84  0.000
ALTITUDE     0.196522  0.0674483    2.91  0.004   1.22    1.07    1.39
LATERAL     -0.238832  0.0224781  -10.63  0.000   0.79    0.75    0.82
```

Log-Likelihood = -179.815
Test that all slopes are zero: G = 259.181, DF = 2, P-Value = 0.000

---

as "low" or "high"), altitude (x_1 = hundreds of meters), and lateral distance (x_2 = hundreds of meters) for each of 464 helicopter overflights were recorded and are saved in the PACGEESE file. (The data for the first 10 overflights are shown in the table, p. 503.) MINITAB was used to fit the logistic regression model $\pi^* = \beta_0 + \beta_1 x_1 + \beta_2 x_2$, where y = {1 if high response, 0 if low response}, $\pi = P(y = 1)$, and $\pi^* = ln[\pi/(1 - \pi)]$. The resulting printout is shown above.

(a) Is the overall logit model statistically useful for predicting geese flight response? Test using $\alpha = .01$.

(b) Conduct a test to determine if flight response of the geese depends on altitude of the helicopter. Test using $\alpha = .01$.

(c) Conduct a test to determine if flight response of the geese depends on lateral distance of helicopter from the flock. Test using $\alpha = .01$.

(d) Predict the probability of a high flight response from the geese for a helicopter flying over the estuary at an altitude of $x_1 = 6$ hundred meters and at a lateral distance of $x_2 = 3$ hundred meters.

DISCRIM

9.25 Gender discrimination in hiring. Refer to Exercise 9.21 (p. 493). Use the data in the DISCRIM file to fit the logit model

$$E(y) = \frac{\exp(\beta_0 + \beta_1 x_1 + \beta_2 x_2 + \beta_3 x_3)}{1 + \exp(\beta_0 + \beta_1 x_1 + \beta_2 x_2 + \beta_3 x_3)}$$

where

$$y = \begin{cases} 1 & \text{if hired} \\ 0 & \text{if not} \end{cases}$$

x_1 = Years of higher education (4, 6, or 8)

x_2 = Years of experience

$$x_3 = \begin{cases} 1 & \text{if male applicant} \\ 0 & \text{if female applicant} \end{cases}$$

(a) Conduct a test of model adequacy. Use $\alpha = .05$.

(b) Is there sufficient evidence to indicate that gender is an important predictor of hiring status? Test using $\alpha = .05$.

(c) Find a 95% confidence interval for the mean response $E(y)$ when $x_1 = 4$, $x_2 = 0$, and $x_3 = 1$. Interpret the interval.

MTBE

9.26 Groundwater contamination in wells. Many New Hampshire counties mandate the use of reformulated gasoline, leading to an increase in groundwater contamination. Refer to the *Environmental Science and Technology* (January 2005) study of the factors related to methyl *tert*-butyl ether (MTBE) contamination in public and private New Hampshire wells, Exercise 6.11 (p. 343). Data were collected for a sample of 223 wells and are saved in the MTBE file. Recall that the list of potential predictors of MTBE level include well class (public or private), aquifer (bedrock or unconsolidated), pH level (standard units), well depth (meters), amount of dissolved oxygen (milligrams per liter), distance from well to nearest fuel source (meters), and percentage of adjacent land allocated to industry. For this exercise, consider the dependent variable $y = \{1$ if a "detectible level" of MTBE is found, 0 if the level of MTBE found is "below limit"$\}$. Using the independent variables indentified in Exercise 6.11, fit a logistic regression model for the probability of a "detectible level" of MTBE. Interpret the results of the logistic regression. Do you recommend using the model? Explain.

🖴 PONDICE

9.27 Characteristics of ice melt ponds. Refer to the National Snow and Ice Data Center (NSIDC) collection of data on ice melt ponds in the Canadian Arctic, Exercise 1.16 (p. 11). The data are saved in the PONDICE file. One variable of interest to environmental engineers studying the melt ponds is the type of ice observed for each pond. Recall that ice type is classified as first-year ice, multiyear ice, or landfast ice. In particular, for non-first-year ice, the engineers are searching for ice pond characteristics that help predict whether the ice type is multiyear or landfast ice. Three pond characteristics hypothesized to impact ice type are depth (x_1), broadband surface albedo (x_2), and visible surface albedo (x_3). The engineers want to use these three characteristics as predictors of ice type in a logistic regression analysis, where $y = \{1$ if landfast ice, 0 if multiyear ice$\}$.

(a) Define π for this logistic regression analysis.

(b) Write a first-order, main effects logit model for π as a function of x_1, x_2, and x_3.

(c) Fit the model, part b, to the data saved in the PONDICE file and give the prediction equation. (Note: Be sure to delete the observations in the data file that correspond to first-year ice.)

(d) Conduct a test of overall model adequacy.

(e) Write a logit model for π as a function of x_1, x_2, and x_3 that includes all possible pairwise interactions between the independent variables.

(f) Fit the model, part e, to the data saved in the PONDICE file and give the prediction equation.

(g) Compare the models, parts b and e, by conducting the appropriate nested model test. What do you conclude?

9.28 A new dental bonding agent. When bonding teeth, orthodontists must maintain a dry field. A new bonding adhesive (called "Smartbond") has been developed to eliminate the necessity of a dry field. However, there is concern that the new bonding adhesive may not stick to the tooth as well as the current standard, a composite adhesive (*Trends in Biomaterials and Artificial Organs*, January 2003). Tests were conducted on a sample of 10 extracted teeth bonded with the new adhesive and a sample of 10 extracted teeth bonded with the composite adhesive. The Adhesive Remnant Index (ARI), which measures the residual adhesive of a bonded tooth on a scale of 1 to 5, was determined for each of the 20 bonded teeth after 1 hour of drying. (Note: An ARI score of 1 implies all adhesive remains on the tooth, while a score of 5 means none of the adhesive remains on the tooth.) The data are listed in the accompanying table. Fit a logistic regression model for the probability of the new (Smartbond) adhesive based on the ARI value of the bonded tooth. Interpret the results.

🖴 BONDING

TOOTH	ADHESIVE	ARISCORE
1	Smartbond	1
2	Smartbond	1
3	Smartbond	2
4	Smartbond	2
5	Smartbond	2
6	Smartbond	2
7	Smartbond	2
8	Smartbond	2
9	Smartbond	2
10	Smartbond	2
11	Composite	1
12	Composite	2
13	Composite	2
14	Composite	2
15	Composite	2
16	Composite	2
17	Composite	3
18	Composite	3
19	Composite	3
20	Composite	4

Source: Sunny, J., and Vallathan, A. "A comparative *in vitro* study with new generation ethyl cyanoacrylate (Smartbond) and a composite bonding agent," *Trends in Biomaterials and Artificial Organs*, Vol. 16, No. 2, January 2003 (Table 6).

9.7 Ridge Regression

When the sample data for regression exhibit multicollinearity, the least squares estimates of the β coefficients may be subject to extreme roundoff error as well as inflated standard errors (see Section 7.5). Because their magnitudes and signs may change considerably from sample to sample, the least squares estimates are said to be *unstable*. A technique developed for stabilizing the regression coefficients in the presence of multicollinearity is **ridge regression**.

Ridge regression is a modification of the method of least squares to allow *biased* estimators of the regression coefficients. At first glance, the idea of biased estimation may not seem very appealing. But consider the sampling distributions of two different estimators of a regression coefficient β, one unbiased and the other biased, shown in Figure 9.19.

Figure 9.19a shows an unbiased estimator of β with a fairly large variance. In contrast, the estimator shown in Figure 9.19b has a slight bias but is much less variable. In this case, we would prefer the biased estimator over the unbiased estimator since it will lead to more precise estimates of the true β (i.e., narrower confidence intervals for β). One way to measure the "goodness" of an estimator of β is to calculate the **mean square error** of $\hat{\beta}$, denoted by MSE($\hat{\beta}$), where MSE($\hat{\beta}$) is defined as

$$MSE(\hat{\beta}) = E[(\hat{\beta} - \beta)^2]$$
$$= V(\hat{\beta}) + [E(\hat{\beta}) - \beta]^2$$

Figure 9.19 Sampling distributions of two estimators of a regression coefficient β

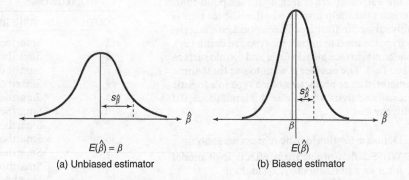

(a) Unbiased estimator   (b) Biased estimator

The difference $E(\hat{\beta}) - \beta$ is called the **bias** of $\hat{\beta}$. Therefore, MSE($\hat{\beta}$) is just the sum of the variance of $\hat{\beta}$ and the squared bias:

$$MSE(\hat{\beta}) = V(\hat{\beta}) + (\text{Bias in}\,\hat{\beta})^2$$

Let $\hat{\beta}_{LS}$ denote the least squares estimate of β. Then, since $E(\hat{\beta}_{LS}) = \beta$, the bias is 0 and

$$MSE(\hat{\beta}_{LS}) = V(\hat{\beta}_{LS})$$

We have previously stated that the variance of the least squares regression coefficient, and hence MSE($\hat{\beta}_{LS}$), will be quite large in the presence of multicollinearity. The idea behind ridge regression is to introduce a small amount of bias in the ridge estimator of β, denoted by $\hat{\beta}_R$, so that its mean square error is considerably smaller than the corresponding mean square error for least squares, that is,

$$MSE(\hat{\beta}_R) < MSE(\hat{\beta}_{LS})$$

In this manner, ridge regression will lead to narrower confidence intervals for the β coefficients, and hence, more stable estimates.

Although the mechanics of a ridge regression are beyond the scope of this text, we point out that some of the more sophisticated software packages (including SAS) are now capable of conducting this type of analysis. To obtain the ridge regression coefficients, the user must specify the value of a biasing constant c, where $c \geq 0$.* Researchers have shown that as the value of c increases, the bias in the ridge estimates increases while the variance decreases. The idea is to choose c so that the total mean square error for the ridge estimators is smaller than the total mean square error for the least squares estimates. Although such a c exists, the optimal value, unfortunately, is unknown.

Various methods for choosing the value of c have been proposed. One commonly used graphical technique employs a **ridge trace**. Values of the estimated ridge regression coefficients are calculated for different values of c ranging from 0 to 1 and are plotted. The plots for each of the independent variables in the model are overlaid to form the ridge trace. An example of ridge trace for a model with three independent variables is shown in Figure 9.20. Initially, the estimated coefficients may fluctuate dramatically as c is increased from 0 (especially if severe multicollinearity is present). Eventually, however, the ridge estimates will stabilize. After careful examination of the ridge trace, the analyst chooses the smallest value of c for which it appears that all the ridge estimates are stable. The choice of c, therefore, is subjective. (*Note*: Selecting c as small as possible minimizes the bias of the least squares estimators.)

Figure 9.20 Ridge trace of β coefficients of a model with three independent variables

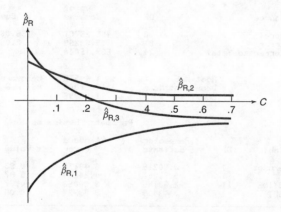

In addition to the ridge trace, some researchers examine the variance inflation factors (VIFs) at each value of the biasing constant, c. Like the estimated coefficients, the VIFs will begin to stabilize as the value of c increases. One choice is to select the constant c for which the VIFs are all small (say, about 1).

Once the value of c has been determined (using the ridge trace, VIFs, or some other analytical technique), the corresponding ridge estimates may be used in place of the least squares estimates. If the optimal (or near-optimal) value of c has been selected, the new estimates will have reduced variances (which lead to narrower confidence intervals for the β's). Also, some of the other problems associated with multicollinearity (e.g., incorrect signs on the β's) should have been corrected.

---

* In matrix notation, the ridge estimator $\hat{\beta}_R$ is calculated as follows:

$$\hat{\beta}_R = (X'X) + cI)^{-1}X'Y$$

When $c = 0$, the least squares estimator

$$\hat{\beta}_{LS} = (X'X)^{-1}X'Y$$

is obtained. See Appendix A for details on the matrix mechanics of a regression analysis.

FTCCIGAR

Example
9.7

Refer to Example 7.5 (p. 365) and the Federal Trade Commission (FTC) study of hazardous substances produced from cigarette smoke. The data in the **FTCCIGAR** file contains the tar content (x_1), nicotine content (x_2), weight (x_3), and carbon monoxide content (y) for each in a sample of 25 cigarette brands. Recall that we fit the model $E(y) = \beta_0 + \beta_1 x_1 + \beta_2 x_2 + \beta_3 x_3$ and discovered several signs of multicollinearity. The SAS printout, reproduced in Figure 9.21, shows several high VIF values as well as negative values for two of the beta estimates. (Remember, the FTC knows that all three independent variables have positive relationships with CO content.) Apply ridge regression to the data in the **FTCCIGAR** file as a remedy for the multicollinearity problem. Give an estimate of the optimal biasing constant c and use this value to find the ridge estimates of the betas in the model.

Figure 9.21 SAS
regression printout for
model of carbon monoxide
content, Example 9.7

```
                          The REG Procedure
                             Model: MODEL1
                         Dependent Variable: CO

                    Number of Observations Read        25
                    Number of Observations Used        25

                          Analysis of Variance

                                  Sum of        Mean
        Source           DF      Squares      Square    F Value   Pr > F

        Model             3    495.25781   165.08594     78.98   <.0001
        Error            21     43.89259     2.09012
        Corrected Total  24    539.15040


                Root MSE              1.44573    R-Square    0.9186
                Dependent Mean      12.52800    Adj R-Sq    0.9070
                Coeff Var           11.53996


                          Parameter Estimates

                      Parameter    Standard                          Variance
        Variable  DF   Estimate       Error    t Value  Pr > |t|    Inflation

        Intercept  1    3.20219     3.46175       0.93    0.3655           0
        TAR        1    0.96257     0.24224       3.97    0.0007    21.63071
        NICOTINE   1   -2.63166     3.90056      -0.67    0.5072    21.89992
        WEIGHT     1   -0.13048     3.88534      -0.03    0.9735     1.33386
```

Solution
We used the RIDGE option of the SAS regression procedure to produce a ridge trace for the model. The graph is shown in Figure 9.22. (Note that SAS uses the symbol k as the biasing constant rather than c.) You can see that when the constant is about $c = .1$, the beta estimates begin to stabilize. In addition, for this constant value the beta estimates for nicotine (x_2) and weight (x_3) are now positive and, hence, more meaningful.

The VIFs associated with each of the independent variables for different biasing constants are listed in the SAS printout, Figure 9.23a. Note that for the ridge constant of $c = .1$ (shaded), the VIFs are all about equal to 1. In addition, Figure 9.23b shows that for this biasing constant, the standard deviation of the model (RMSE) is $s = 1.585$ (highlighted), only slightly higher than the value of s for the original least squares model. Figure 9.23b also gives the ridge estimates of the betas for the model (highlighted). This results in the ridge regression prediction equation:

$$\hat{y} = 1.905 + .483x_1 + 4.29x_2 + .988x_3$$

Figure 9.22 SAS ridge trace for model of carbon monoxide content

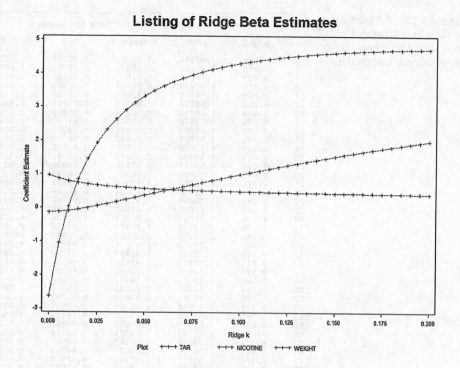

Figure 9.23a SAS listing of VIFs for ridge regression of carbon monoxide content

Listing of Ridge VIFs

Obs	_RIDGE_	TAR	NICOTINE	WEIGHT
2	0.000	21.6307	21.8999	1.33386
4	0.005	14.7765	14.9551	1.31300
6	0.010	10.7667	10.8923	1.29308
8	0.015	8.2197	8.3118	1.27385
10	0.020	6.5014	6.5709	1.25517
12	0.025	5.2873	5.3410	1.23700
14	0.030	4.3976	4.4397	1.21927
16	0.035	3.7260	3.7595	1.20197
18	0.040	3.2065	3.2333	1.18507
20	0.045	2.7962	2.8177	1.16856
22	0.050	2.4664	2.4838	1.15241
24	0.055	2.1972	2.2112	1.13661
26	0.060	1.9745	1.9859	1.12116
28	0.065	1.7882	1.7973	1.10603
30	0.070	1.6307	1.6378	1.09123
32	0.075	1.4962	1.5017	1.07673
34	0.080	1.3804	1.3846	1.06254
36	0.085	1.2799	1.2830	1.04863
38	0.090	1.1922	1.1943	1.03502
40	0.095	1.1151	1.1163	1.02167
42	0.100	1.0469	1.0474	1.00860
44	0.105	0.9863	0.9862	0.99578
46	0.110	0.9322	0.9315	0.98322
48	0.115	0.8836	0.8824	0.97091
50	0.120	0.8398	0.8382	0.95883
52	0.125	0.8002	0.7982	0.94699
54	0.130	0.7642	0.7619	0.93538
56	0.135	0.7313	0.7288	0.92399
58	0.140	0.7013	0.6985	0.91281
60	0.145	0.6738	0.6708	0.90185
62	0.150	0.6485	0.6453	0.89109
64	0.155	0.6251	0.6217	0.88053
66	0.160	0.6035	0.6000	0.87016
68	0.165	0.5834	0.5798	0.85999
70	0.170	0.5648	0.5610	0.85000
72	0.175	0.5475	0.5436	0.84019
74	0.180	0.5312	0.5273	0.83056
76	0.185	0.5161	0.5120	0.82110
78	0.190	0.5019	0.4978	0.81181
80	0.195	0.4885	0.4844	0.80268
82	0.200	0.4760	0.4718	0.79371

Figure 9.23b SAS listing of ridge beta estimates and root MSE for carbon monoxide content model

Listing of Ridge Beta Estimates

Obs	_RIDGE_	_RMSE_	Intercept	TAR	NICOTINE	WEIGHT
3	0.000	1.44573	3.20219	0.96257	-2.63166	-0.13048
5	0.005	1.45161	3.04372	0.86232	-1.06039	-0.12421
7	0.010	1.46274	2.91877	0.79175	0.03324	-0.09473
9	0.015	1.47470	2.81471	0.73925	0.83620	-0.05175
11	0.020	1.48610	2.72473	0.69857	1.44920	-0.00051
13	0.025	1.49656	2.64483	0.66604	1.93124	0.05596
15	0.030	1.50606	2.57248	0.63938	2.31922	0.11578
17	0.035	1.51466	2.50602	0.61707	2.63736	0.17770
19	0.040	1.52248	2.44431	0.59810	2.90226	0.24091
21	0.045	1.52965	2.38653	0.58173	3.12563	0.30482
23	0.050	1.53625	2.33207	0.56743	3.31602	0.36903
25	0.055	1.54238	2.28049	0.55481	3.47975	0.43323
27	0.060	1.54811	2.23142	0.54356	3.62167	0.49720
29	0.065	1.55350	2.18460	0.53346	3.74551	0.56079
31	0.070	1.55861	2.13980	0.52432	3.85418	0.62386
33	0.075	1.56347	2.09684	0.51600	3.95004	0.68633
35	0.080	1.56814	2.05557	0.50837	4.03496	0.74813
37	0.085	1.57263	2.01586	0.50136	4.11048	0.80920
39	0.090	1.57697	1.97761	0.49486	4.17786	0.86950
41	0.095	1.58119	1.94073	0.48883	4.23815	0.92902
43	0.100	1.58530	1.90513	0.48320	4.29223	0.98771
45	0.105	1.58933	1.87075	0.47793	4.34083	1.04558
47	0.110	1.59327	1.83751	0.47298	4.38460	1.10262
49	0.115	1.59716	1.80538	0.46832	4.42406	1.15881
51	0.120	1.60098	1.77430	0.46391	4.45969	1.21416
53	0.125	1.60477	1.74421	0.45974	4.49188	1.26866
55	0.130	1.60852	1.71509	0.45577	4.52097	1.32233
57	0.135	1.61223	1.68690	0.45199	4.54728	1.37517
59	0.140	1.61592	1.65959	0.44839	4.57108	1.42718
61	0.145	1.61959	1.63315	0.44495	4.59258	1.47837
63	0.150	1.62324	1.60753	0.44165	4.61201	1.52876
65	0.155	1.62689	1.58271	0.43848	4.62954	1.57834
67	0.160	1.63052	1.55867	0.43544	4.64534	1.62713
69	0.165	1.63414	1.53537	0.43252	4.65956	1.67515
71	0.170	1.63777	1.51281	0.42970	4.67233	1.72239
73	0.175	1.64139	1.49095	0.42697	4.68375	1.76887
75	0.180	1.64501	1.46978	0.42434	4.69395	1.81461
77	0.185	1.64863	1.44927	0.42180	4.70301	1.85961
79	0.190	1.65225	1.42941	0.41933	4.71103	1.90389
81	0.195	1.65588	1.41018	0.41694	4.71807	1.94745
83	0.200	1.65952	1.39156	0.41462	4.72421	1.99031

Caution

You should not assume that ridge regression is a panacea for multicollinearity or poor data. Although there are probably ridge regression estimates that are better than the least squares estimates when multicollinearity is present, the choice of the biasing constant c is crucial. Unfortunately, much of the controversy in ridge regression centers on how to find the optimal value of c. In addition, the exact distributional properties of the ridge estimators are unknown when c is estimated from the data. For these reasons, some statisticians recommend that ridge regression be used only as an exploratory data analysis tool for identifying unstable regression coefficients, and not for estimating parameters and testing hypotheses in a linear regression model.

9.8 Robust Regression

Consider the problem of fitting the linear regression model

$$y = \beta_0 + \beta_1 x_1 + \beta_2 x_2 + \cdots + \beta_k x_k + \varepsilon$$

by the method of least squares when the errors ε are nonnormal. In practice, moderate departures from the assumption of normality tend to have minimal effect on the validity of the least squares results (see Section 8.4). However, when the distribution of ε is **heavy-tailed** (longer-tailed) compared to the normal distribution, the method of least squares may not be appropriate. For example, the heavy-tailed error distribution shown in Figure 9.24 will most likely produce outliers with strong

Figure 9.24 Probability distribution of ε: normal versus heavy-tailed

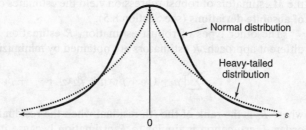

Normal distribution

Heavy-tailed distribution

influence on the regression analysis. Furthermore, since they tend to "pull" the least squares fit too much in their direction, these outliers will have smaller than expected residuals and, consequently, are more difficult to detect.

Robust regression procedures are available for errors that follow a nonnormal distribution. In the context of regression, the term *robust* describes a technique that yields estimates for the β's that are nearly as good as the least squares estimates when the assumption of normality is satisfied, and significantly better for a heavy-tailed distribution. Robust regression is designed to dampen the effect of outlying observations that otherwise would exhibit strong influence on the analysis. This has the effect of leaving the residuals of influential observations large so that they may be more easily identified.

A number of different robust regression procedures exist. They fall into one of three general classes: **M estimators, R estimators**, and **L estimators**. Of the three, robust techniques that produce M estimates of the β coefficients receive the most attention in the literature.

The M estimates of the β coefficients are obtained by minimizing the quantity

$$\sum_{i=1}^{n} f(\hat{\varepsilon}_i)$$

where

$$\hat{\varepsilon}_i = y_i - (\hat{\beta}_0 + \hat{\beta}_1 x_{1i} + \hat{\beta}_2 x_{2i} + \cdots + \hat{\beta}_k x_{ki})$$

are the unobservable residuals and $f(\hat{\varepsilon}_i)$ is some function of the residuals. Note that since we are minimizing

$$\sum_{i=1}^{n} f(\hat{\varepsilon}_i) = \sum_{i=1}^{n} \hat{\varepsilon}_i^2$$

$$= \sum_{i=1}^{n} [y_i - (\hat{\beta}_0 + \hat{\beta}_1 x_{1i} + \hat{\beta}_2 x_{2i} + \cdots + \hat{\beta}_k x_{ki})]^2$$

$$= \text{SSE}$$

the function $f(\hat{\varepsilon}_i) = \hat{\varepsilon}_i^2$ yields the ordinary least squares estimates and, therefore, is appropriate when the errors are normal. For errors with heavier-tailed distributions, the analyst chooses some other function $f(\hat{\varepsilon}_i)$ that places less weight on the errors in the tails of the distribution. For example, the function $f(\hat{\varepsilon}_i) = |\hat{\varepsilon}_i|$ is appropriate when the errors follow the heavy-tailed distribution pictured in Figure 9.24. Since we are minimizing

$$\sum_{i=1}^{n} f(\hat{\varepsilon}_i) = \sum_{i=1}^{n} |\hat{\varepsilon}_i|$$

$$= \sum_{i=1}^{n} |y_i - (\hat{\beta}_0 + \hat{\beta}_1 x_{1i} + \hat{\beta}_2 x_{2i} + \cdots + \hat{\beta}_k x_{ki})|$$

the M estimators of robust regression yield the estimates obtained from the **method of absolute deviations** (see Section 8.5).

The other types of robust estimation, R estimation and L estimation, take a different approach. R estimators are obtained by minimizing the quantity

$$\sum_{i=1}^{n}[y_i - (\hat{\beta}_0 + \hat{\beta}_1 x_{1i} + \hat{\beta}_2 x_{2i} + \cdots + \hat{\beta}_k x_{ki})]R_i$$

where R_i is the rank of the ith residual when the residuals are placed in ascending order. L estimation is similar to R estimation because it involves ordering of the data, but it uses measures of location (such as the sample median) to estimate the regression coefficients.

The numerical techniques for obtaining robust estimates (M, R, or L estimates) are quite complex and require sophisticated computer programs. Several statistical software packages, including SAS and R, now incorporate robust regression routines.

💿 FASTFOOD

Example 9.8

Refer to Example 8.8 (p. 413) and the data on weekly sales of fast-food outlets in four different cities. We fit the model $E(y) = \beta_0 + \beta_1 x_1 + \beta_2 x_2 + \beta_3 x_3 + \beta_4 x_4$, where

y = weekly sales (thousands of dollars)

$x_1 = \{1$ if city 1, 0 if not$\}$, $x_2 = \{1$ if city 2, 0 if not$\}$, $x_3 = \{1$ if city 3, 0 if not$\}$

x_4 = traffic flow (thousands of cars)

Recall that the data, saved in the FASTFOOD file, contained an influential observation due to a coding error. For the 13th observation in the data set, weekly sales was recorded as 82 rather than the correct value of 8.2. In this example, we assume that the weekly sales for this observation is, in fact, 82 thousand dollars. That is, suppose there is no coding error.

The SAS printout for the model is reproduced in Figure 9.25. Note that the overall model is not statistically useful for predicting weekly sales (global F p-value

Figure 9.25 SAS least squares regression printout for fast-food sales model

```
                              The REG Procedure
                                Model: MODEL1
                            Dependent Variable: SALES

                      Number of Observations Read          24
                      Number of Observations Used          24


                             Analysis of Variance

                                    Sum of         Mean
        Source            DF        Squares       Square     F Value    Pr > F

        Model              4     1469.76287    367.44072       1.66     0.1996
        Error             19     4194.22671    220.74877
        Corrected Total   23     5663.98958


               Root MSE             14.85762     R-Square     0.2595
               Dependent Mean        9.07083     Adj R-Sq     0.1036
               Coeff Var           163.79550


                             Parameter Estimates

                            Parameter      Standard
        Variable    DF       Estimate         Error    t Value    Pr > |t|

        Intercept    1      -16.45925      13.16400      -1.25      0.2264
        X1           1        1.10609       8.42257       0.13      0.8969
        X2           1        6.14277      11.67997       0.53      0.6050
        X3           1       14.48962       9.28839       1.56      0.1353
        TRAFFIC      1        0.36287       0.16791       2.16      0.0437
```

is .1996) and the value of R^2 is only about .26. We know that influence diagnostics (see Example 8.8) identify the 13th observation as an influential observation. However, since the data for this observation is recorded correctly, and if upon further investigation we determine that the observation is typical of weekly sales in the city, then it would not be prudent to delete the outlier from the analysis. Rather, we will apply a robust regression method that dampens the influence of this outlier.

Use robust regression with M-estimation to fit the model to the data in the altered FASTFOOD file. Evaluate the fit of the model.

Solution

We used the ROBUSTREG procedure in SAS to fit the model. (For this procedure, the default method of estimation is M-estimation.) The resulting SAS printout is shown in Figure 9.26. First, note that tests of significance for the beta estimates (highlighted on the printout) now show that all four independent variables are statistically useful for predicting weekly sales at $\alpha = .10$. Second, the value of R^2 (also highlighted) is now .799, a significant increase over the corresponding value from least squares regression. Finally, although the printout indicates that the 13th observation remains an outlier (highlighted in the "Diagnostics" portion of the printout), it is clear that the robust estimation method was successful in dampening the influence of this data point enough for us to recognize the utility of the model in predicting fast-food sales. ▬

Much of the current research in the area of robust regression is focused on the distributional properties of the robust estimators of the β coefficients. At present, there is little information available on robust confidence intervals, prediction intervals, and hypothesis testing procedures. For this reason, some researchers recommend that robust regression be used in conjunction with and as a check on the method of least squares. If the results of the two procedures are substantially the same, use the least squares fit since confidence intervals and tests on the regression coefficients can be made. On the other hand, if the two analyses yield quite different results, use the robust fit to identify any influential observations. A careful examination of these data points may reveal the problem with the least squares fit.

9.9 Nonparametric Regression Models

Recall, in Section 4.1, that the general multiple regression model assumes a "linear" form:

$$E(y) = \beta_0 + \beta_1 x_1 + \beta_2 x_2 + \beta_3 x_3 + \cdots + \beta_k x_k$$

That is, the model is proposed as a linear function of the unknown β's and the method of least squares is used to estimate these β's. In situations where the assumption of linearity may be violated (e.g., with a binary response y), **nonparametric regression models** are available.

In nonparametric regression, the analyst does not necessarily propose a specific functional relationship between y and x_i. Rather, the linear term $\beta_i x_i$ is replaced by a smooth function of x_i that is estimated by visually exploring the data. The general form of a nonparametric regression model is:

$$E(y) = s_0 + s_1(x_1) + s_2(x_2) + s_3(x_3) + \cdots + s_k(x_k),$$

where $s_1(x_1)$ is a smooth function relating y to x_1, $s_2(x_2)$ is a smooth function relating y to x_2, and so on. The smooth function $s_i(x_i)$, or **smoother** as it is commonly known,

Figure 9.26 SAS robust regression printout for fast-food sales model

```
                                    The ROBUSTREG Procedure
                                      Model Information

           Data Set                                        WORK.IN
           Dependent Variable                                SALES
           Number of Independent Variables                       4
           Number of Observations                              24
           Method                                     M Estimation


                    Number of Observations Read            24
                    Number of Observations Used            24


                                    Parameter Information

                        Parameter           Effect

                        Intercept           Intercept
                        X1                  X1
                        X2                  X2
                        X3                  X3
                        TRAFFIC             TRAFFIC


                                    Summary Statistics

                                                                        Standard
           Variable       Q1        Median        Q3       Mean        Deviation       MAD

           X1              0           0        1.0000      0.3750        0.4945          0
           X2              0           0           0        0.1250        0.3378          0
           X3              0           0        1.0000      0.2917        0.4643          0
           TRAFFIC     41.2000     53.3000    72.7500     55.4500       20.4089      21.4236
           SALES        4.0000      5.6500     8.0000      9.0708       15.6927       2.7428


                                    Parameter Estimates

                                      Standard     95% Confidence        Chi-
           Parameter  DF  Estimate     Error          Limits           Square   Pr > ChiSq

           Intercept   1    1.0464     0.3414     0.3772     1.7157       9.39      0.0022
           X1          1   -1.1755     0.2185    -1.6037    -0.7473      28.95     <.0001
           X2          1   -0.5141     0.3030    -1.1079     0.0796       2.88      0.0897
           X3          1   -1.1158     0.2409    -1.5880    -0.6436      21.45     <.0001
           TRAFFIC     1    0.1042     0.0044     0.0956     0.1127     572.14     <.0001
           Scale       1    0.3341


                                        Diagnostics

                                        Standardized
                                           Robust
                            Obs          Residual         Outlier

                             13          222.0105            *


                                     Diagnostics Summary

                        Observation
                        Type             Proportion        Cutoff

                        Outlier            0.0417          3.0000


                                      Goodness-of-Fit

                        Statistic            Value

                        R-Square            0.7989
                        AICR               33.1876
                        BICR               41.8817
                        Deviance            2.9012
```

summarizes the trend in the relationship between y and x_i. Figure 9.27 shows a scatterplot for a data set collected on number of exacerbations, y, and age, x, of patients with multiple sclerosis (MS). The smooth function, $s(x)$, that represents a possible trend in the data is shown on the graph.

Figure 9.27 Scatterplot of data for MS patients

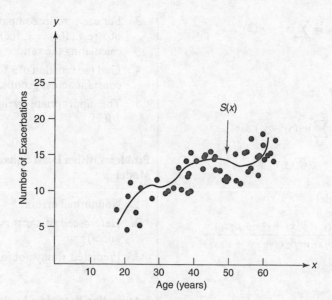

A variety of nonparametric methods have been proposed for estimating smoothers. These include **cubic smoothing splines** and **thin-plate smoothing splines**, where the term *spline* is used to describe a smoothed curve obtained through a complex mathematical optimization technique. Details on these nonparametric methods are beyond the scope of this text. You should consult the references at the end of the chapter if you wish to learn more about them.

As you might expect, as with robust regression, sophisticated computer software is required to fit a nonparametric regression model. Nonparametric regression procedures are available in SAS. (See the SAS Tutorial that accompanies this text.) Although a nonparametric model can be used as an alternative to the general linear model, the two models really serve two different analytic purposes. The linear statistical model in Chapters 3–8 emphasizes estimation and inferences for the model parameters. In contrast, nonparametric regression models are more suitable for exploring and visualizing relationships among the variables of interest.

Quick Summary

Key Formulas

Piecewise Linear Regression Models

2 Lines (Continuous)

$$E(y) = \beta_0 + \beta_1 x_1 + \beta_2(x_1 - k)x_2,$$

where $x_2 = \{1 \quad \text{if } x_1 > k, 0 \quad \text{if not}\}$

3 Lines (Continuous)

$$E(y) = \beta_0 = +\beta_1 x_1 + \beta_2(x_1 - k_1)x_2 + \beta_3(x_1 - k_2)x_3$$

where $x_2 = \{1 \text{ if } x_1 > k_1, 0 \quad \text{if not}\}$,
$x_3 = \{1 \text{ if } x_1 > k_2, 0 \text{ if not}\}$

2 Lines (Discontinuous)

$$E(y) = \beta_0 + \beta_1 x_1 + \beta_2(x_1 - k)x_2 + \beta_3 x_2,$$

where $x_2 = \{1 \quad \text{if } x_1 > k, 0 \quad \text{if not}\}$

Inverse Prediction–Prediction Interval for x when $y = y_p$

$$\hat{x} \pm (t_{\alpha/2}) \left(\frac{s}{\hat{\beta}_1}\right) \sqrt{1 + \frac{1}{n} + \frac{(\hat{x} - \bar{x})^2}{SS_{xx}}},$$

where $\hat{x} = \dfrac{y_p - \hat{\beta}_0}{\hat{\beta}_1}$, $SS_{xx} = \sum x^2 - n(\bar{x})^2$,

$s = \sqrt{MSE}$

Weighted Least Squares

$$WSSE = \sum_{i=1}^{n} w_i (y_i - \hat{y}_i)^2$$

$$Residual = \sqrt{w_i}(y_i - \hat{y}_i)$$

Logistic Regression Model

$$E(y) = \frac{\exp(\beta_0 + \beta_1 x_1 + \beta_2 x_2 + \cdots + \beta_k x_k)}{1 + \exp(\beta_0 + \beta_1 x_1 + \beta_2 x_2 + \cdots + \beta_k x_k)}$$

$$= P(y = 1) = \pi$$

$$Odds = \frac{\pi}{1 - \pi}$$

$$\pi^* = \ln\left(\frac{\pi}{1 - \pi}\right) = \beta_0 + \beta_1 x_1 + \beta_2 x_2 + \cdots + \beta_k x_k$$

Test Statistic for Comparing Nested Models

$$\chi^2 = \chi^2_{complete} - \chi^2_{reduced}, \ df = g$$

Inverse Prediction

1. Approximation is valid when
   $D = \left[\dfrac{t_{\alpha/2} \cdot s}{\hat{\beta}_1}\right]^2 \cdot \dfrac{1}{SS_{xx}}$ is small
2. Should be avoided when the linear model is not statistically useful for predicting y
3. Should be avoided when the value of y_p falls outside the range of the sample y-values

Weighted Least Squares (WLS)

1. Stabilizes the variance of random error ε
2. Limits the influence of outlying observations
3. Gives greater weight to more recent observations in time series data

Determining the Weights in WLS

1. Divide the residuals into several groups with approximately equal sample sizes.
2. Determine $\bar{x}$ and s^2 of the residuals in each group.

3. For each group, compare s^2 to different functions of $\bar{x}$ [e.g., $f(\bar{x}) = \bar{x}$, $f(\bar{x}) = \bar{x}^2$, and $f(\bar{x}) = \sqrt{\bar{x}}$], by calculating the ratio $s^2/f(\bar{x})$.
4. Find the function of $\bar{x}$ for which the ratio is nearly constant across groups.
5. The appropriate weights for the groups are $1/f(\bar{x})$.

Problems with a Least Squares Binary Regression Model

1. Nonnormal errors
2. Heteroscedastic errors (i.e., unequal error variances)
3. Predicted y may not fall between 0 and 1

Interpreting Betas in a Logistic Regression Model

β_i = change in log-odds for every 1-unit increase in x_i, holding all other x's fixed

$e^{\beta_i} - 1$ = percentage change in odds for every 1-unit increase in x_i, holding all other x's fixed

Ridge Regression

1. Stabilizes beta estimates in presence of multicollinearity
2. Leads to narrower confidence intervals (smaller variances) for the beta estimates
3. Introduces bias in the estimates of the betas

Estimating the Biasing Constant c in Ridge Regression

1. Calculate beta estimates for different values of c and formulate a graph of the results (i.e., a *ridge trace*)
2. Choose the smallest value of c that stabilizes the estimates and/or variance inflation factors

Robust Regression

1. Used when the error distribution is nonnormal (heavy-tailed)
2. Used when several influential observations (outliers) are present in the data
3. Little information available on confidence intervals and hypothesis tests

Methods of Estimation with Robust Regression

1. *M-estimation*: minimize a sum of a function of errors, where the function places less weight on the errors in the tail of the error probability distribution (e.g., *method of absolute deviations*)

2. *R-estimation*: uses ranks of residuals in the minimization process

3. *L-estimation*: uses a median rather than the mean as a measure of location in the minimization process

Nonparametric Regression

1. Employs a smooth function of x_i (i.e., a *smoother*) to relate y to x_i

2. Smoothers are estimated using *splines* (i.e., smooth curves obtained through complex mathematical optimization)

3. Most suitable for exploring relationships among the variables of interest

References

Agresti, A. *Categorical Data Analysis*. New York: Wiley, 1990.

Andrews, D. F. "A robust method for multiple linear regression." *Technometrics*, Vol. 16, 1974, pp. 523–531.

Chatterjee, S., and Mächler, M. "Robust regression: A weighted least squares approach." *Communications in Statistics, Theory, and Methods*, Vol. 26, 1995, pp. 1381–1394.

Cox, D. R. *The Analysis of Binary Data*. London: Methuen, 1970.

Conniffe, D., and Stone, J. "A critical review of ridge regression." *Statistician*, Vol. 22, 1974, pp. 181–187.

Draper, N. R., and Van Nostrand, R. C. "Ridge regression and James–Stein estimation: Review and comments." *Technometrics*. Vol. 21, 1979, p. 451.

Geisser, S. "The predictive sample reuse method with applications." *Journal of the American Statistical Association*, Vol. 70, 1975, pp. 320–328.

Gibbons, D. I., and McDonald, G. C. "A rational interpretation of the ridge trace." *Technometrics*, Vol. 26, 1984, pp. 339–346.

Graybill, F. A. *Theory and Application of the Linear Model*. North Scituate, Mass.: Duxbury Press, 1976.

Halperin, M., Blackwelder, W. C., and Verter, J. I. "Estimation of the multivariate logistic risk function: A comparison of the discriminant function and maximum likelihood approaches." *Journal of Chronic Diseases*, Vol. 24, 1971, pp. 125–158.

Härdle, W. *Applied Nonparametric Regression*. New York: Cambridge University Press, 1992.

Hastie, T., and Tibshirani, R. *Generalized Additive Models*. New York: Chapman and Hall, 1990.

Hauck, W. W., and Donner, A. "Wald's test as applied to hypotheses in logit analysis." *Journal of the American Statistical Association*, Vol. 72, 1977, pp. 851–853.

Hill, R. W., and Holland, P. W. "Two robust alternatives to least squares regression." *Journal of the American Statistical Association*, Vol. 72, 1977, pp. 828–833.

Hoerl, A. E., and Kennard, R. W. "Ridge regression: Biased estimation for nonorthogonal problems." *Technometrics*, Vol. 12, 1970, pp. 55–67.

Hoerl, A. E., Kennard, R. W., and Baldwin, K. F. "Ridge regression: Some simulations." *Communications in Statistics*, Vol. A5, 1976, pp. 77–88.

Hogg, R. V. "Statistical robustness: One view of its use in applications today." *American Statistician*, Vol. 33, 1979, pp. 108–115.

Hosmer, D. W., and Lemeshow, S. *Applied Logistic Regression*. (2nd ed.) New York: Wiley, 2000.

Huber, P. J. *Robust Statistics*. New York: Wiley, 1981.

Kutner, M., Nachtsheim, C., Neter, J., and Li, W., *Applied Linear Statistical Models*, 5th ed. New York: McGraw-Hill, 2005.

McDonald, G. C. "Ridge Regression." *WIREs Computational Statistics*, Vol. 1, 2009, pp. 93–100.

Montgomery, D., Peck, E. and Vining, G. *Introduction to Linear Regression Analysis*, 4th ed. New York: Wiley, 2006.

Mosteller, F., and Tukey, J. W. *Data Analysis and Regression: A Second Course in Statistics*. Reading, Mass.: Addison-Wesley, 1977.

Obenchain, R. L. "Classical F-tests and confidence intervals for ridge regression." *Technometrics*, Vol. 19, 1977, pp. 429–439.

Rousseeuw, P. J., and Leroy, A. M. *Robust Regression and Outlier Detection*. New York: Wiley, 1987.

Ruppert, D. "Computing S Estimators for Regression and Multivariate Location/Dispersion." *Journal of Computational and Graphical Statistics*, Vol. 1, 1992, pp. 253–270.

Snee, R. D. "Validation of regression models: Methods and examples." *Technometrics*, Vol. 19, 1977, pp. 415–428.

Stone, C. J. "Additive regression and other nonparametric models." *Annals of Statistics*, Vol. 13, 1985, p. 689–705.

Tsiatis, A. A. "A note on the goodness-of-fit test for the logistic regression model." *Biometrika*, Vol. 67, 1980, pp. 250–251.

Wahba, G. *Spline models for observational data*, Philadelphia: Society for Industrial and Applied Mathematics, 1990.

Walker, S. H., and Duncan, D. B. "Estimation of the probability of an event as a function of several independent variables." *Biometrika*, Vol. 54, 1967, pp. 167–179.

Yohai, V. J., and Zamar, R. H. "Optimal locally robust M estimate of regression." *Journal of Statistical Planning and Inference*, Vol. 64, 1997, pp. 309–323.

Yohai, V. J., Stahel, W. A., and Zamar, R. H. "A Procedure for Robust Estimation and Inference in Linear Regression." *Directions in Robust Statistics and Dragnostics, Part II*, New York: Springer-Verlag, 1991.

Zaman, A., Rousseeuw, P. J., and Orhan, M. "Econometric applications of high-breakdown robust regression techniques." *Econometrics Letters*, Vol. 71, 2001, pp. 1–8.

Chapter 10

INTRODUCTION TO TIME SERIES MODELING AND FORECASTING

Contents

Objectives

1. To introduce time series data.
2. To present descriptive methods for forecasting time series.
3. To present inferential models for forecasting time series.
4. To introduce time series models with autocorrelated errors.

10.1 What Is a Time Series?

In many business and economic studies, the response variable y is measured sequentially in time. For example, we might record the number y of new housing starts for each month in a particular region. This collection of data is called a **time series**. Other examples of time series are data collected on the quarterly number of highway deaths in the United States, the annual sales for a corporation, and the recorded month-end values of the prime interest rate.

> **Definition 10.1** A **time series** is a collection of data obtained by observing a response variable at periodic points in time.

> **Definition 10.2** If repeated observations on a variable produce a time series, the variable is called a **time series variable**. We use y_t to denote the value of the variable at time t.

519

If you were to develop a model relating the number of new housing starts to the prime interest rate over time, the model would be called a **time series model**, because both the dependent variable, new housing starts, and the independent variable, prime interest rate, are measured sequentially over time. Furthermore, time itself would probably play an important role in such a model, because the economic trends and seasonal cycles associated with different points in time would almost certainly affect both time series.

The construction of time series models is an important aspect of business and economic analyses, because many of the variables of most interest to business and economic researchers are time series. This chapter is an introduction to the very complex and voluminous body of material concerned with time series modeling and forecasting future values of a time series.

10.2 Time Series Components

Researchers often approach the problem of describing the nature of a time series y_t by identifying four kinds of change, or variation, in the time series values. These four components are commonly known as (1) secular trend, (2) cyclical effect, (3) seasonal variation, and (4) residual effect. The components of a time series are most easily identified and explained pictorially.

Figure 10.1a shows a **secular trend** in the time series values. The secular component describes the tendency of the value of the variable to increase or decrease over a long period of time. Thus, this type of change or variation is also known as the **long-term trend**. In Figure 10.1a, the long-term trend is of an increasing nature. However, this does not imply that the time series has always moved upward from month to month and from year to year. You can see that although the series fluctuates, the trend has been an increasing one over that period of time.

The **cyclical effect** in a time series, as shown in Figure 10.1b, generally describes the fluctuation about the secular trend that is attributable to business and economic conditions at the time. These fluctuations are sometimes called **business cycles**. During a period of general economic expansion, the business cycle lies above the secular trend, whereas during a recession, when business activity is likely to slump, the cycle lies below the secular trend. You can see that the cyclical variation does not follow any definite trend, but moves rather unpredictably.

The **seasonal variation** in a time series describes the fluctuations that recur during specific portions of each year (e.g., monthly or seasonally). In Figure 10.1c, you can

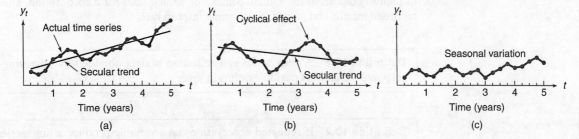

Figure 10.1 The components of a time series

see that the pattern of change in the time series within a year tends to be repeated from year to year, producing a wavelike or oscillating curve.

The final component, the **residual effect**, is what remains after the secular, cyclical, and seasonal components have been removed. This component is not systematic and may be attributed to unpredictable influences such as wars and political unrest, hurricanes and droughts, and the randomness of human actions. Thus, the residual effect represents the random error component of a time series.

Definition 10.3 The **secular trend** (T_t) of a time series is the tendency of the series to increase or decrease over a long period of time. It is also known as the **long-term trend**.

Definition 10.4 The **cyclical fluctuation** (C_t) of a time series is the wavelike or oscillating pattern about the secular trend that is attributable to business and economic conditions at the time. It is also known as a **business cycle**.

Definition 10.5 The **seasonal variation** (S_t) of a time series describes the fluctuations that recur during specific portions of the year (e.g., monthly or seasonally).

Definition 10.6 The **residual effect** (R_t) of a time series is what remains after the secular, cyclical, and seasonal components have been removed.

In many practical applications of time series, the objective is to *forecast* (predict) some *future value or values* of the series. To obtain forecasts, some type of model that can be projected into the future must be used to describe the time series. One of the most widely used models is the ***additive model***.*

$$y_t = T_t + C_t + S_t + R_t$$

where $T_t, C_t, S_t,$ and R_t represent the secular trend, cyclical effect, seasonal variation, and residual effect, respectively, of the time series variable y_t. Various methods exist for estimating the components of the model and forecasting the time series. These range from simple **descriptive techniques**, which rely on smoothing the pattern of the time series, to complex **inferential models**, which combine regression analysis with specialized time series models. Several descriptive forecasting techniques are presented in Section 10.3, and forecasting using the general linear regression model of Chapter 4 is discussed in Section 10.4. The remainder of the chapter is devoted to the more complex and more powerful time series models.

---

* Another useful model is the **multiplicative model** $y_t = T_t C_t S_t R_t$. Recall (Section 4.12) that this model can be written in the form of an additive model by taking natural logarithms:

$$\ln y_t = \ln T_t + \ln C_t + \ln S_t + \ln R_t$$

10.3 Forecasting Using Smoothing Techniques (Optional)

Various descriptive methods are available for identifying and characterizing a time series. Generally, these methods attempt to remove the rapid fluctuations in a time series so that the secular trend can be seen. For this reason, they are sometimes called **smoothing techniques**. Once the secular trend is identified, forecasts for future values of the time series are easily obtained. In this section, we present three of the more popular smoothing techniques.

Moving Average Method

A widely used smoothing technique is the **moving average method**. A moving average, M_t, at time t is formed by averaging the time series values over adjacent time periods. Moving averages aid in identifying the secular trend of a time series because the averaging modifies the effect of short-term (cyclical or seasonal) variation. That is, a plot of the moving averages yields a "smooth" time series curve that clearly depicts the long-term trend.

For example, consider the 2006–2009 quarterly power loads for a utility company located in a southern part of the United States, given in Table 10.1. A MINITAB graph of the quarterly time series, Figure 10.2, shows the pronounced seasonal variation (i.e., the fluctuation that recurs from year to year). The quarterly power

QTRPOWER

Table 10.1 Quarterly power loads, 2006–2009			
Year	Quarter	Time t	Power Load y_t, megawatts
2006	I	1	103.5
	II	2	94.7
	III	3	118.6
	IV	4	109.3
2007	I	5	126.1
	II	6	116.0
	III	7	141.2
	IV	8	131.6
2008	I	9	144.5
	II	10	137.1
	III	11	159.0
	IV	12	149.5
2009	I	13	166.1
	II	14	152.5
	III	15	178.2
	IV	16	169.0

loads are highest in the summer months (quarter III) with another smaller peak in the winter months (quarter I), and lowest during the spring and fall (quarters II and IV). To clearly identify the long-term trend of the series, we need to average, or "smooth out," these seasonal fluctuations. We apply the moving average method for this purpose.

Figure 10.2 MINITAB plot of quarterly power loads

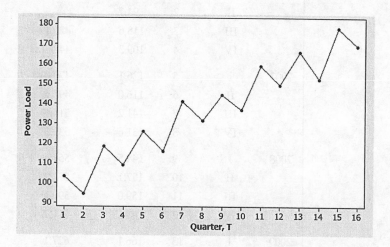

The first step in calculating a moving average for quarterly data is to sum the observed time values y_t—in this example, quarterly power loads—for the four quarters during the initial year 2006. Summing the values from Table 10.1, we have

$$y_1 + y_2 + y_3 + y_4 = 103.5 + 94.7 + 118.6 + 109.3$$

$$= 426.1$$

This sum is called a **4-point moving total**, which we denote by the symbol L_t. It is customary to use a subscript t to represent the time period at the midpoint of the four quarters in the total. Since for this sum, the midpoint is between $t = 2$ and $t = 3$, we use the conventional procedure of "dropping it down one line" to $t = 3$. Thus, our first 4-point moving total is $L_3 = 426.1$.

We find the next moving total by eliminating the first quantity in the sum, $y_1 = 103.5$, and adding the next value in the time series sequence, $y_5 = 126.1$. This enables us to keep four quarters in the total of adjacent time periods. Thus, we have

$$L_4 = y_2 + y_3 + y_4 + y_5 = 94.7 + 118.6 + 109.3 + 126.1 = 448.7$$

Continuing this process of "moving" the 4-point total over the time series until we have included the last value, we find

$$L_5 = y_3 + y_4 + y_5 + y_6 \quad = 118.6 + 109.3 + 126.1 + 116.0 = 470.0$$
$$L_6 = y_4 + y_5 + y_6 + y_7 \quad = 109.3 + 126.1 + 116.0 + 141.2 = 492.6$$
$$\vdots \qquad\qquad \vdots \qquad\qquad\qquad\qquad \vdots \quad \vdots$$
$$L_{15} = y_{13} + y_{14} + y_{15} + y_{16} = 166.1 + 152.5 + 178.2 + 169.0 = 665.8$$

The complete set of 4-point moving totals is given in the appropriate column in Table 10.2. Notice that three data points will be "lost" in forming the moving totals.

After the 4-point moving totals are calculated, the second step is to determine the **4-point moving average**, denoted by M_t, by dividing each of the moving totals

Table 10.2 4-Point moving average for the quarterly power load data

Year	Quarter	Time t	Power Load y_t	4-Point Moving Total L_t	4-Point Moving Average M_t	Ratio y_t/M_t
2006	I	1	103.5	—	—	—
	II	2	94.7	—	—	—
	III	3	118.6	426.1	106.5	1.113
	IV	4	109.3	448.7	112.2	.974
2007	I	5	126.1	470.0	117.5	1.073
	II	6	116.0	492.6	123.2	.942
	III	7	141.2	514.9	128.7	1.097
	IV	8	131.6	533.3	133.3	.987
2008	I	9	144.5	554.4	138.6	1.043
	II	10	137.1	572.2	143.1	.958
	III	11	159.0	590.1	147.5	1.078
	IV	12	149.5	611.7	152.9	.978
2009	I	13	166.1	627.1	156.8	1.059
	II	14	152.5	646.3	161.6	.944
	III	15	178.2	665.8	166.5	1.071
	IV	16	169.0	—	—	—

by 4. For example, the first three values of the 4-point moving average for the quarterly power load data are

$$M_3 = \frac{y_1 + y_2 + y_3 + y_4}{4} = \frac{L_3}{4} = \frac{426.1}{4} = 106.5$$

$$M_4 = \frac{y_2 + y_3 + y_4 + y_5}{4} = \frac{L_4}{4} = \frac{448.7}{4} = 112.2$$

$$M_5 = \frac{y_3 + y_4 + y_5 + y_6}{4} = \frac{L_5}{4} = \frac{470.0}{4} = 117.5$$

All of the 4-point moving averages are given in the appropriate column in Table 10.2.

Both the original power load time series and the 4-point moving average are graphed (using MINITAB) in Figure 10.3. Notice that the moving average has smoothed the time series, that is, the averaging has modified the effects of the short-term or seasonal variation. The plot of the 4-point moving average clearly depicts the secular (long-term) trend component of the time series.

In addition to identifying a long-term trend, moving averages provide us with a measure of the seasonal effects in a time series. The ratio between the observed power load y_t and the 4-point moving average M_t for each quarter measures the seasonal effect (primarily attributable to temperature differences) for that quarter. The ratios y_t/M_t are shown in the last column in Table 10.2. Note that the ratio is always greater than 1 in quarters I and III, and always less than 1 in quarters II and IV. The average of the ratios for a particular quarter, multiplied by 100, can be

Figure 10.3 MINITAB plot of quarterly power loads and 4-point moving average

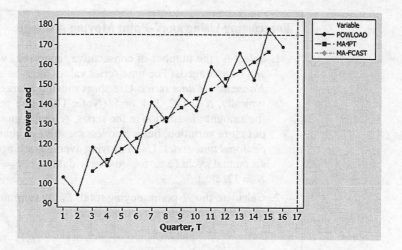

used to form a **seasonal index** for that quarter. For example, the seasonal index for quarter I is

$$100 \left(\frac{1.073 + 1.043 + 1.059}{3} \right) = 105.8$$

implying that the time series value in quarter I is, on the average, 105.8% of the moving average value for that time period.

To forecast a future value of the time series, simply extend the moving average M_t on the graph to the future time period.* For example, a graphical extension of the moving average for the quarterly power loads to quarter I of 2010 ($t = 17$) yields a moving average of approximately $M_{17} = 175$ (see Figure 10.3). Thus, if there were no seasonal variation in the time series, we would expect the power load for quarter I of 2010 to be approximately 175 megawatts. To adjust the forecast for seasonal variation, multiply the future moving average value $M_{17} = 175$ by the seasonal index for quarter I, then divide by 100:

$$F_{17} = M_{17} \left(\frac{\text{Seasonal index for quarter I}}{100} \right)$$

$$= 175 \left(\frac{105.8}{100} \right)$$

$$\approx 185$$

where F_{17} is the forecast of y_{17}. Therefore, the moving average forecast for the power load in quarter I of 2010 is approximately 185 megawatts.

Moving averages are not restricted to 4 points. For example, you may wish to calculate a 7-point moving average for daily data, a 12-point moving average for monthly data, or a 5-point moving average for yearly data. Although the choice of the number of points is arbitrary, you should search for the number N that yields a smooth series, but is not so large that many points at the end of the series are "lost." The method of forecasting with a general N-point moving average is outlined in the box.

---

* Some statistical software packages (e.g., MINITAB) will use the last moving average in the series as the value of the forecast for any future time period.

Forecasting Using an N-Point Moving Average

1. Select N, the number of consecutive time series values $y_1, y_2, \ldots, y_N$ that will be averaged. (The time series values must be equally spaced.)
   Nonseasonal time series: Use short moving averages to smooth the series, typically, $N = 1, 2, 3, 4,$ or 5. (Note: The value you select will depend on the amount of variation in the series. A longer moving average will smooth out more variation, but is also less sensitive to changes in the series.)
   Seasonal time series: Use a moving average of length equal to the length of an annual cycle (e.g., for quarterly data use $N = 4$, for monthly data use $N = 12$, etc.).

2. Calculate the N-point moving total, L_t, by summing the time series values over N adjacent time periods, where

$$L_t = \begin{cases} y_{t-(N-1)/2} + \cdots + y_t + \cdots + y_{t+(N-1)/2} & \text{if } N \text{ is odd} \\ y_{t-N/2} + \cdots + y_t + \cdots + y_{t+N/2-1} & \text{if } N \text{ is even} \end{cases}$$

3. Compute the N-point moving average, M_t, by dividing the corresponding moving total by N:

$$M_t = \frac{L_t}{N}$$

4. Graph the moving average M_t on the vertical axis with time t on the horizontal axis. (This plot should reveal a smooth curve that identifies the long-term trend of the time series.*) Extend the graph to a future time period to obtain the forecasted value of M_t.

5. For a future time period t, the forecast of y_t is

$$F_t = \begin{cases} M_t & \text{if little or no seasonal variation exists in the time series} \\ M_t \left(\dfrac{\text{Seasonal index}}{100} \right) & \text{otherwise} \end{cases}$$

where the seasonal index for a particular quarter (or month) is the average of past values of the ratios

$$\frac{y_t}{M_t}(100)$$

for that quarter (or month).

Exponential Smoothing

One problem with using a moving average to forecast future values of a time series is that values at the ends of the series are lost, thereby requiring that we subjectively extend the graph of the moving average into the future. No exact calculation of a forecast is available since the moving average at a future time period t requires that we know one or more future values of the series. **Exponential smoothing** is a technique that leads to forecasts that can be explicitly calculated. Like the moving average method, exponential smoothing deemphasizes (or smooths) most of the residual effects. However, exponential smoothing averages only past and current values of the time series.

---

* When the number N of points is small, the plot may not yield a very smooth curve. However, the moving average will be smoother (or less variable) than the plot of the original time series values.

To obtain an exponentially smoothed time series, we first need to choose a weight w, between 0 and 1, called the **exponential smoothing constant**. The exponentially smoothed series, denoted E_t, is then calculated as follows:

$$E_1 = y_1$$

$$E_2 = wy_2 + (1 - w)E_1$$

$$E_3 = wy_3 + (1 - w)E_2$$

$$\vdots \quad \vdots$$

$$E_t = wy_t + (1 - w)E_{t-1}$$

You can see that the exponentially smoothed value at time t is simply a weighted average of the current time series value, y_t, and the exponentially smoothed value at the previous time period, E_{t-1}. Smaller values of w give less weight to the current value, y_t, whereas larger values give more weight to y_t.

For example, suppose we want to smooth the quarterly power loads given in Table 10.1 using an exponential smoothing constant of $w = .7$. Then we have

$$E_1 = y_1 = 103.5$$

$$E_2 = .7y_2 + (1 - .7)E_1$$

$$= .7(94.7) + .3(103.5) = 97.34$$

$$E_3 = .7y_3 + (1 - .7)E_2$$

$$= .7(118.6) + .3(97.34) = 112.22$$

$$\vdots$$

The exponentially smoothed values (using $w = .7$) for all the quarterly power loads, obtained using MINITAB, are highlighted on the MINITAB printout, Figure 10.4. Both the actual and the smoothed time series values are graphed on the MINITAB printout, Figure 10.5.

Exponential smoothing forecasts are obtained by using the most recent exponentially smoothed value, E_t. In other words, if n is the last time period in which y_t is observed, then the forecast for a future time period t is given by

$$F_t = E_n$$

As you can see, the right-hand side of the forecast equation does not depend on t; hence, F_t is used to forecast *all* future values of y_t. The MINITAB printout, Figure 10.4, shows that the smoothed value for quarter 4 of 2009 ($t = 16$) is $E_{16} = 169.688$. Therefore, this value represents the forecast for the power load in quarter I of 2010 ($t = 17$), that is,

$$F_{17} = E_{16} = 169.688$$

This forecast is shown at the bottom of Figure 10.4 and graphically in Figure 10.5. The forecasts for quarter II of 2010 ($t = 18$), quarter III of 2010 ($t = 19$), and all other future time periods will be the same:

$$F_{18} = 169.688$$

$$F_{19} = 169.688$$

$$F_{20} = 169.688$$

$$\vdots$$

Figure 10.4 MINITAB printout of exponentially smoothed quarterly power loads

Single Exponential Smoothing for POWLOAD

Data   POWLOAD
Length 16

Smoothing Constant
Alpha 0.7

Accuracy Measures
MAPE   7.604
MAD   10.634
MSD   177.873

Time	POWLOAD	Smooth	Predict	Error
1	103.5	103.500	103.500	0.0000
2	94.7	97.340	103.500	-8.8000
3	118.6	112.222	97.340	21.2600
4	109.3	110.177	112.222	-2.9220
5	126.1	121.323	110.177	15.9234
6	116.0	117.597	121.323	-5.3230
7	141.2	134.119	117.597	23.6031
8	131.6	132.356	134.119	-2.5191
9	144.5	140.857	132.356	12.1443
10	137.1	138.227	140.857	-3.7567
11	159.0	152.768	138.227	20.7730
12	149.5	150.480	152.768	-3.2681
13	166.1	161.414	150.480	15.6196
14	152.5	155.174	161.414	-8.9141
15	178.2	171.292	155.174	23.0258
16	169.0	169.688	171.292	-2.2923

Forecasts

Period	Forecast	Lower	Upper
17	169.688	143.635	195.741

Figure 10.5 MINITAB plot of exponentially smoothed quarterly power loads

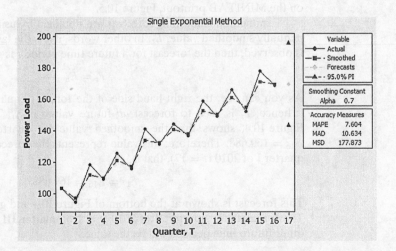

This points out one disadvantage of the exponential smoothing forecasting technique. Since the exponentially smoothed forecast is constant for all future values, any changes in trend and/or seasonality are not taken into account. Therefore, exponentially smoothed forecasts are appropriate only when the trend and seasonal components of the time series are relatively insignificant.

Forecasting Using Exponential Smoothing

1. The data consist of n equally spaced time series values,

$$y_1, y_2, \ldots, y_n.$$

2. Select a smoothing constant, w, between 0 and 1. (Smaller values of w give less weight to the current value of the series and yield a smoother series. Larger values of w give more weight to the current value of the series and yield a more variable series.)

3. Calculate the exponentially smoothed series, E_t, as follows:*

$$E_1 = y_1$$
$$E_2 = wy_2 + (1 - w)E_1$$
$$E_3 = wy_3 + (1 - w)E_2$$
$$\vdots$$
$$E_n = wy_n + (1 - w)E_{n-1}$$

4. The forecast for any future time period t is:

$$F_t = E_n, \quad t = n + 1, n + 2, \ldots$$

Holt–Winters Forecasting Model

One drawback to the exponential smoothing forecasting method is that the secular trend and seasonal components of a time series are not taken into account. The **Holt–Winters forecasting model** is an extension of the exponential smoothing method that explicitly recognizes the trend and seasonal variation in a time series.

Consider a time series with a trend component, but little or no seasonal variation. Then the Holt–Winters model for y_t is

$$E_t = wy_t + (1 - w)(E_{t-1} + T_{t-1})$$
$$T_t = v(E_t - E_{t-1}) + (1 - v)T_{t-1}$$

where E_t is the exponentially smoothed series, T_t is the trend component, and w and v are smoothing constants between 0 and 1. Note that the trend component T_t is a weighted average of the most recent change in the smoothed value (measured by the difference $E_t - E_{t-1}$) and the trend estimate of the previous time period (T_{t-1}). When seasonal variation is present in the time series, the Holt–Winters model takes the form

$$E_t = w(y_t/S_{t-p}) + (1 - w)(E_{t-1} + T_{t-1})$$
$$T_t = v(E_t - E_{t-1}) + (1 - v)T_{t-1}$$
$$S_t = u(y_t/E_t) + (1 - u)S_{t-P}$$

where S_t is the seasonal component, u is a constant between 0 and 1, and P is the number of time periods in a cycle (usually a year). The seasonal component S_t is a weighted average of the ratio y_t/E_t (i.e., the ratio of the actual time series value

---

* Optionally, the first "smoothed" value, E_1, can be computed as the average value of y_t over the entire series.

to the smoothed value) and the seasonal component for the previous cycle. For example, for the quarterly power loads, $P = 4$ (four quarters in a year) and the seasonal component for, say, quarter III of 2007 ($t = 7$) is a weighted average of the ratio y_7/E_7 and the seasonal component for quarter III of 2006 ($t = 3$). That is,

$$S_7 = u(y_7/E_7) + (1 - u)S_3$$

Forecasting Using the Holt–Winters Model

Trend Component Only

1. The data consist of n equally spaced time series values, $y_1, y_2, \ldots, y_n$.
2. Select smoothing constants w and v, where $0 \le w \le 1$ and $0 \le v \le 1$.
3. Calculate the exponentially smoothed component, E_t, and the trend component, T_t, for $t = 2, 3, \ldots, n$ as follows:

$$E_t = \begin{cases} y_2, & t = 2 \\ wy_t + (1-w)(E_{t-1} + T_{t-1}), \\ & t > 2 \end{cases}$$

$$T_t = \begin{cases} y_2 - y_1, & t = 2 \\ v(E_t - E_{t-1}) + (1-v)T_{t-1}, \\ & t > 2 \end{cases}$$

[*Note:* E_1 and T_1 are not defined.]

4. The forecast for a future time period t is given by

$$F_t = \begin{cases} E_n + T_n, & t = n+1 \\ E_n + 2T_n, & t = n+2 \\ \vdots \\ E_n + kT_n, & t = n+k \end{cases}$$

Trend and Seasonal Components

1. The data consist of n equally spaced time series values, $y_1, y_2, \ldots, y_n$.
2. Select smoothing constants w, v, and u, where $0 \le w \le 1$, $0 \le v \le 1$, and $0 \le u \le 1$.
3. Determine P, the number of time periods in a cycle. Usually, $P = 4$ for quarterly data and $P = 12$ for monthly data.
4. Calculate the exponentially smoothed component, E_t, the trend component, T_t, and the seasonal component, S_t, for $t = 2, 3, \ldots, n$ as follows:

$$E_t = \begin{cases} y_2, & t = 2 \\ wy_t + (1-w)(E_{t-1} + T_{t-1}), \\ & t = 3, 4, \ldots, P+2 \\ w(y_t/S_{t-P}) + (1-w) \\ \times(E_{t-1} + T_{t-1}), & t > P+2 \end{cases}$$

$$T_t = \begin{cases} y_2 - y_1, & t = 2 \\ v(E_t - E_{t-1}) + (1-v)T_{t-1}, \\ & t > 2 \end{cases}$$

$$S_t = \begin{cases} y_t/E_t, & t = 2, 3, \ldots, P+2 \\ u(y_t/E_t) + (1-u)S_{t-P}, \\ & t > P+2q \end{cases}$$

[*Note:* E_1, T_1, and S_1 are not defined.]

5. The forecast for a future time period t is given by

$$F_t = \begin{cases} (E_n + T_n)S_{n+1-P}, & t = n+1 \\ (E_n + 2T_n)S_{n+2-P}, & t = n+2 \\ \vdots \\ (E_n + kT_n)S_{n+k-P}, & t = n+k \end{cases}$$

Forecasts for future time periods, $t = n+1, n+2, \ldots$, using the Holt–Winters models are obtained by summing the most recent exponentially smoothed component with an estimate of the expected increase (or decrease) attributable to trend.

For seasonal models, the forecast is multiplied by the most recent estimate of the seasonal component (similar to the moving average method).

The Holt–Winters forecasting methodology is summarized in the box on the previous page. As Example 10.1 illustrates, these calculations can become quite tedious. Most time series analysts will utilize a statistical software package to apply the Holt–Winters forecasting method. Although there are slight variations in their initial computing formulas, SAS, MINITAB, and SPSS all have options for producing Holt–Winters forecasts.

Example 10.1

Refer to the 2006–2009 quarterly power loads listed in Table 10.1. Use the Holt–Winters forecasting model with both trend and seasonal components to forecast the utility company's quarterly power loads in 2010. Use the smoothing constants $w = .7$, $v = .5$, and $u = .5$.

Solution

First note that $P = 4$ for the quarterly time series. Following the formulas for E_t, T_t, and S_t given in the box, we calculate

$$E_2 = y_2 = 94.7$$

$$T_2 = y_2 - y_1 = 94.7 - 103.5 = -8.8$$

$$S_2 = y_2/E_2 = 94.7/94.7 = 1$$

$$E_3 = .7y_3 + (1 - .7)(E_2 + T_2)$$
$$= .7(118.6) + .3(94.7 - 8.8) = 108.8$$

$$T_3 = .5(E_3 - E_2) + (1 - .5)T_2$$
$$= .5(108.8 - 94.7) + .5(-8.8) = 2.6$$

$$S_3 = y_3/E_3 = 118.6/108.8 = 1.090$$

$$E_4 = .7y_4 + (1 - .7)(E_3 + T_3)$$
$$= .7(109.3) + .3(108.8 + 2.6) = 109.9$$

$$T_4 = .5(E_4 - E_3) + (1 - .5)T_3$$
$$= .5(109.9 - 108.8) + .5(2.6) = 1.9$$

$$S_4 = y_4/E_4 = 109.3/109.9 = .994$$

$$\vdots$$

The forecast for quarter I of 2010 (i.e., y_{17}) is given by

$$F_{17} = (E_{16} + T_{16})S_{17-4}$$
$$= (E_{16} + T_{16})S_{13} = (168.7 + 4.7)(1.044)$$
$$= 181.0$$

(Remember that beginning with $t = P + 3 = 7$, the formulas for E_t and S_t, shown in the box, are slightly different.) All the values of E_t, T_t, and S_t are given in Table 10.3. Similarly, the forecasts for y_{18}, y_{19}, and y_{20} (quarters II, III, and IV, respectively) are

$$F_{18} = (E_{16} + 2T_{16})S_{18-4}$$
$$= (E_{16} + 2T_{16})S_{14} = [168.7 + 2(4.7)](.959)$$
$$= 170.8$$

$$F_{19} = (E_{16} + 3T_{16})S_{19-4}$$
$$= (E_{16} + 3T_{16})S_{15} = [168.7 + 3(4.7)](1.095)$$
$$= 200.2$$

$$F_{20} = (E_{16} + 4T_{16})S_{20-4}$$
$$= (E_{16} + 4T_{16})S_{16} = [168.7 + 4(4.7)](.999)$$
$$= 187.3$$

Table 10.3 Holt–Winters components for quarterly power load data

Year	Quarter	Time t	Power Load y_t	E_t $(w = .7)$	T_t $(v = .5)$	S_t $(u = .5)$
2006	I	1	103.5	—	—	—
	II	2	94.7	94.7	−8.8	1.000
	III	3	118.6	108.8	2.6	1.090
	IV	4	109.3	109.9	1.9	.994
2007	I	5	126.1	121.8	6.9	1.035
	II	6	116.0	119.8	2.5	.968
	III	7	141.2	127.4	5.1	1.100
	IV	8	131.6	132.3	5.0	.995
2008	I	9	144.5	138.9	5.8	1.038
	II	10	137.1	142.6	4.8	.965
	III	11	159.0	145.4	3.8	1.097
	IV	12	149.5	149.9	4.2	.996
2009	I	13	166.1	158.2	6.3	1.044
	II	14	152.5	160.0	4.1	.959
	III	15	178.2	162.9	3.5	1.095
	IV	16	169.0	168.7	4.7	.999

With any of these forecasting methods, **forecast errors** can be computed once the future values of the time series have been observed. Forecast error is defined as the difference between the actual future value and predicted value at time t, $(y_t - F_t)$. Aggregating the forecast errors into a summary statistic is useful for assessing the overall accuracy of the forecasting method. Formulas for three popular measures of forecast accuracy, the **mean absolute percentage error (MAPE), mean absolute deviation (MAD)**, and **root mean squared error (RMSE)**, are given in the box (p. 533). Both MAPE and MAD are summary measures for the "center" of the distribution of forecast errors, while RMSE is a measure of the "variation" in the distribution.

Example 10.2   Refer to the quarterly power load data, Table 10.1. The exponential smoothing and Holt–Winters forecasts for the four quarters of 2010 are listed in Table 10.4, as are the actual quarterly power loads (not previously given) for the year. Compute MAD and RMSE for each of the two forecasting methods. Which method yields more accurate forecasts?

Measures of Overall Forecast Accuracy for m Forecasts

$$\text{Mean absolute percentage error: } \text{MAPE} = \frac{\sum_{t=1}^{m} \left| \frac{(y_t - F_t)}{y_t} \right|}{m} \times 100$$

$$\text{Mean absolute deviation: } \text{MAD} = \frac{\sum_{t=1}^{m} |y_t - F_t|}{m}$$

$$\text{Root mean square error: } \text{RMSE} = \sqrt{\frac{\sum_{t=1}^{m} (y_t - F_t)^2}{m}}$$

Solution

The first step is to calculate the forecast errors, $y_t - F_t$, for each method. For example, for the exponential smoothing forecast of quarter I ($t = 17$), $y_{17} = 181.5$ and $F_{17} = 169.7$. Thus, the forecast error is $y_{17} - F_{17} = 181.5 - 169.7 = 11.8$. The forecast errors for the remaining exponential smoothing forecasts and the Holt–Winters forecasts are also shown in Table 10.4.

Table 10.4 Forecasts and actual quarterly power loads for 2010

Quarter	Time t	Actual Power Load y_t	Exponential Smoothing		Holt–Winters	
			Forecast F_t	Error $(y_t - F_t)$	Forecast F_t	Error $(y_t - F_t)$
I	17	181.5	169.7	11.8	181.0	.5
II	18	175.2	169.7	5.5	170.8	4.4
III	19	195.0	169.7	25.3	200.2	−5.2
IV	20	189.3	169.7	19.6	187.3	2.0

The MAD and RMSE calculations for each method are as follows:

Exponential smoothing:

$$\text{MAPE} = \left\{ \frac{\left|\frac{11.8}{181.5}\right| + \left|\frac{5.5}{175.2}\right| + \left|\frac{25.3}{195.0}\right| + \left|\frac{19.6}{189.3}\right|}{4} \right\} \times 100 = 8.24\%$$

$$\text{MAD} = \frac{|11.8| + |5.5| + |25.3| + |19.6|}{4} = 15.55$$

$$\text{RMSE} = \sqrt{\frac{(11.8)^2 + (5.5)^2 + (25.3)^2 + (19.6)^2}{4}} = 17.27$$

Holt–Winters:

$$\text{MAPE} = \left\{ \frac{\left|\frac{.5}{181.5}\right| + \left|\frac{4.4}{175.2}\right| + \left|\frac{-5.2}{195.0}\right| + \left|\frac{2.0}{189.3}\right|}{4} \right\} \times 100 = 1.63\%$$

$$\text{MAD} = \frac{|.5| + |4.4| + |-5.2| + |2.0|}{4} = 3.03$$

$$\text{RMSE} = \sqrt{\frac{(.5)^2 + (4.4)^2 + (-5.2)^2 + (2.0)^2}{4}} = 3.56$$

The Holt–Winters values of MAPE, MAD, and RMSE are each about one-fifth of the corresponding exponential smoothed values. Overall, the Holt–Winters method clearly leads to more accurate forecasts than exponential smoothing. This, of course, is expected since the Holt–Winters method accounts for both long-term and seasonal variation in the power loads, whereas exponential smoothing does not.   ■

[*Note*: Most statistical software packages will automatically compute the values of MAPE, MAD, and RMSE (also called the *mean squared deviation,* or *MSD*) for all n observations in the data set. For example, see the highlighted portion at the top of the MINITAB printout, Figure 10.4.]

We conclude this section with a comment. A major disadvantage of forecasting with smoothing techniques (the moving average method, exponential smoothing, or the Holt–Winters models) is that no measure of the forecast error (or reliability) is known *prior* to observing the future value. Although forecast errors can be calculated *after* the future values of the time series have been observed (as in Example 10.2), we prefer to have some measure of the accuracy of the forecast *before* the actual values are observed. One option is to compute forecasts and forecast errors for all n observations in the data set and use these "past" forecast errors to estimate the standard deviation of all forecast errors (i.e., the *standard error of the forecast*). A rough estimate of this standard error is the value of RMSE, and an approximate 95% prediction interval for any future forecast is

$$F_t \pm 2(\text{RMSE})$$

(An interval like this is shown at the bottom of the MINITAB printout, Figure 10.4.) However, because the theoretical distributional properties of the forecast errors with smoothing methods are unknown, many analysts regard smoothing methods as descriptive procedures rather than as inferential ones.

In the preceding chapters, we learned that predictions with inferential regression models are accompanied by well-known measures of reliability. The standard errors of the predicted values allow us to construct 95% prediction intervals. We discuss inferential time series forecasting models in the remaining sections of this chapter.

10.3 Exercises

10.1 **Quarterly single-family housing starts.** The quarterly numbers of single-family housing starts (in thousands of dwellings) in the United States from 2004 through 2008 are recorded in the next table (p. 535).

(a) Plot the quarterly time series. Can you detect a long-term trend? Can you detect any seasonal variation?

(b) Calculate the 4-point moving average for the quarterly housing starts.

(c) Graph the 4-point moving average on the same set of axes you used for the graph in part a. Is the long-term trend more evident? What effects has the moving average method removed or smoothed?

(d) Calculate the seasonal index for the number of housing starts in quarter I.

(e) Calculate the seasonal index for the number of housing starts in quarter II.

(f) Use the moving average method to forecast the number of housing starts in quarters I and II of 2009.

QTRHOUSE

YEAR	QUARTER	HOUSING STARTS
2006	1	382
	2	433
	3	372
	4	278
2007	1	260
	2	333
	3	265
	4	188
2008	1	162
	2	194
	3	163
	4	103

Source: U.S. Bureau of the Census. *Statistical Abstract of the United States*, 2009.

10.2 **Quarterly housing starts (cont'd).** Refer to the quarterly housing starts data in Exercise 10.1.

(a) Calculate the exponentially smoothed series for housing starts using a smoothing constant of $w = .2$.

(b) Use the exponentially smoothed series from part a to forecast the number of housing starts in the first two quarters of 2009.

(c) Use the Holt–Winters forecasting model with both trend and seasonal components to forecast the number of housing starts in the first two quarters of 2009. Use smoothing constants $w = .2$, $v = .5$, and $u = .7$.

10.3 **Quarterly housing starts (cont'd).** Refer to Exercises 10.1 and 10.2. The actual numbers of housing starts (in thousands) for quarters I and II of 2009 are 78 and 124, respectively.

(a) Compare the accuracy of the moving average, exponential smoothing, and Holt–Winters forecasts using MAD.

(b) Repeat part a using RMSE.

(c) Comment on which forecasting method is more accurate.

10.4 **OPEC crude oil imports.** The data in the next table are the amounts of crude oil (millions of barrels) imported into the United States from the Organization of Petroleum Exporting Countries (OPEC) for the years 1990–2007.

(a) Plot the yearly time series. Can you detect a long-term trend?

(b) Calculate and plot a 3-point moving average for annual OPEC oil imports.

(c) Calculate and plot the exponentially smoothed series for annual OPEC oil imports using a smoothing constant of $w = .3$.

(d) Forecast OPEC oil imports in 2008 using the moving average method.

(e) Forecast OPEC oil imports in 2008 using exponential smoothing with $w = .3$.

(f) Forecast OPEC oil imports in 2008 using the Holt–Winters forecasting model with trend. Use smoothing constants $w = .3$ and $v = .8$.

(g) Actual OPEC crude oil imports in 2008 totaled 2,179 million barrels. Calculate the errors of the forecast, parts d–f. Which method yields the most accurate short-term forecast?

OPECOIL

YEAR	t	IMPORTS, Y_t
1990	1	1,283
1991	2	1,233
1992	3	1,247
1993	4	1,339
1994	5	1,307
1995	6	1,219
1996	7	1,258
1997	8	1,378
1998	9	1,522
1999	10	1,543
2000	11	1,659
2001	12	1,770
2002	13	1,490
2003	14	1,671
2004	15	1,948
2005	16	1,738
2006	17	1,745
2007	18	1,969

Source: U.S. Bureau of the Census. *Statistical Abstract of the United States*, 2009.

10.5 **Consumer Price Index.** The Consumer Price Index (CPI) measures the increase (or decrease) in the prices of goods and services relative to a base year. The CPI for the years 1990–2008 (using 1984 as a base period) is shown in the table on p. 536.

(a) Graph the time series. Do you detect a long-term trend?

(b) Calculate and plot a 5-point moving average for the CPI. Use the moving average to forecast the CPI in 2011.

(c) Calculate and plot the exponentially smoothed series for the CPI using a smoothing constant of $w = .4$. Use the exponentially smoothed values to forecast the CPI in 2011.

(d) Use the Holt–Winters forecasting model with trend to forecast the CPI in 2011. Use smoothing constants $w = .4$ and $v = .5$.

⊙ CPI

YEAR	CPI
1990	125.8
1991	129.1
1992	132.8
1993	136.8
1994	147.8
1995	152.4
1996	156.9
1997	160.5
1998	163.0
1999	166.6
2000	171.5
2001	177.1
2002	179.9
2003	184.0
2004	188.9
2005	195.3
2006	201.6
2007	207.3
2008	215.3

Source: Survey of Current Business, U.S. Department of Commerce, Bureau of Economic Analysis.

10.6 **S&P 500 Index.** Standard & Poor's 500 Composite Stock Index (S&P 500) is a stock market index. Like the Dow Jones Industrial Average, it is an indicator of stock market activity. The table below contains end-of-quarter values of the S&P 500 for the years 2001–2008.

⊙ SP500

YEAR	QUARTER	S&P 500	YEAR	QUARTER	S&P 500
2001	1	1,160.3	2005	1	1,180.6
	2	1,224.4		2	1,191.3
	3	1,040.9		3	1,228.8
	4	1,148.1		4	1,248.3
2002	1	1,147.4	2006	1	1,294.9
	2	989.8		2	1,270.2
	3	815.3		3	1,335.8
	4	879.8		4	1,418.3
2003	1	848.2	2007	1	1,420.9
	2	974.5		2	1,503.3
	3	996.0		3	1,526.7
	4	1,111.9		4	1,468.4
2004	1	1,126.2	2008	1	1,322.7
	2	1,140.8		2	1,280.0
	3	1,114.6		3	1,164.7
	4	1,211.9		4	903.3

Source: Standard & Poor's Statistical Service: Current Statistics, 2009; www.economagic.com

(a) Calculate a 4-point moving average for the quarterly S&P 500.
(b) Plot the quarterly index and the 4-point moving average on the same graph. Can you identify the long-term trend of the time series? Can you identify any seasonal variations about the secular trend?
(c) Use the moving average method to forecast the S&P 500 for the 1st quarter of 2009.
(d) Calculate and plot the exponentially smoothed series for the quarterly S&P 500 using a smoothing constant of $w = .3$.
(e) Use the exponential smoothing technique with $w = .3$ to forecast the S&P 500 for the 1st quarter of 2009.
(f) Use the Holt–Winters forecasting model with trend and seasonal components to forecast the S&P 500 for the 1st quarter of 2009. Use smoothing constants $w = .3$, $v = .8$, and $u = .5$.

10.7 **Yearly price of gold.** The price of gold is used by some financial analysts as a barometer of investors' expectations of inflation, with the price of gold tending to increase as concerns about inflation increase. The table below shows the average annual price of gold (in dollars per ounce) from 1990 through 2008.

⊙ GOLDYR

YEAR	PRICE	YEAR	PRICE
1990	384	2000	279
1991	362	2001	271
1992	344	2002	310
1993	360	2003	363
1994	384	2004	410
1995	384	2005	445
1996	388	2006	603
1997	331	2007	695
1998	294	2008	872
1999	279		

Source: World Gold Council, www.kitco.com.

(a) Calculate a 3-point moving average for the gold price time series. Plot the gold prices and the 3-point moving average on the same graph. Can you detect the long-term trend and any cyclical patterns in the time series?
(b) Use the moving averages to forecast the price of gold in 2006, 2007, and 2008.
(c) Calculate and plot the exponentially smoothed gold price series using a smoothing constant of $w = .8$.
(d) Use the exponentially smoothed series to forecast the price of gold in 2006, 2007, and 2008.
(e) Use the Holt–Winters forecasting model with trend to forecast the price of gold, 2006–2008. Use smoothing constants $w = .8$ and $v = .4$.
(f) Use the actual gold prices in 2006–2008 to assess the accuracy of the three forecasting methods, parts b, d, and e.

10.4 Forecasting: The Regression Approach

Many firms use past sales to forecast future sales. Suppose a wholesale distributor of sporting goods is interested in forecasting its sales revenue for each of the next 5 years. Since an inaccurate forecast may have dire consequences to the distributor, some measure of the forecast's reliability is required. To make such forecasts and assess their reliability, an **inferential time series forecasting model** must be constructed. The familiar general linear regression model in Chapter 4 represents one type of inferential model since it allows us to calculate prediction intervals for the forecasts.

To illustrate the technique of forecasting with regression, we'll reconsider the data on annual sales (in thousands of dollars) for a firm (say, the sporting goods distributor) in each of its 35 years of operation. The data, first presented in Table 8.6, is reproduced in Table 10.5. A SAS plot of the data (Figure 10.6) reveals a linearly

🔘 SALES 35

Table 10.5 A firm's yearly sales revenue (thousands of dollars)

t	y_t	t	y_t	t	y_t
1	4.8	13	48.4	25	100.3
2	4.0	14	61.6	26	111.7
3	5.5	15	65.6	27	108.2
4	15.6	16	71.4	28	115.5
5	23.1	17	83.4	29	119.2
6	23.3	18	93.6	30	125.2
7	31.4	19	94.2	31	136.3
8	46.0	20	85.4	32	146.8
9	46.1	21	86.2	33	146.1
10	41.9	22	89.9	34	151.4
11	45.5	23	89.2	35	150.9
12	53.5	24	99.1		

Figure 10.6 SAS scatterplot of sales data

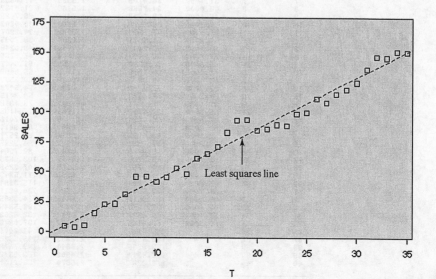

increasing trend, so the first-order (straight-line) model

$$E(y_t) = \beta_0 + \beta_1 t$$

seems plausible for describing the secular trend. The SAS printout for the model is shown in Figure 10.7. The model apparently provides an excellent fit to the data, with $R^2 = .98$, $F = 1{,}615.724$ (p-value $< .0001$), and $s = 6.38524$. The least squares prediction equation, whose coefficients are shaded in Figure 10.7, is

$$\hat{y}_t = \hat{\beta}_0 + \hat{\beta}_1 t = .401513 + 4.295630t$$

We can obtain sales forecasts and corresponding 95% prediction intervals for years 36–40 by using the formulas given in Section 3.9 or using statistical

Figure 10.7 SAS printout for straight-line model of yearly sales revenue

Dependent Variable: SALES

Analysis of Variance

Source	DF	Sum of Squares	Mean Square	F Value	Pr > F
Model	1	65875	65875	1615.72	<.0001
Error	33	1345.45355	40.77132		
Corrected Total	34	67221			

Root MSE	6.38524	R-Square	0.9800	
Dependent Mean	77.72286	Adj R-Sq	0.9794	
Coeff Var	8.21540			

Parameter Estimates

Variable	DF	Parameter Estimate	Standard Error	t Value	Pr > |t|
Intercept	1	0.40151	2.20571	0.18	0.8567
T	1	4.29563	0.10687	40.20	<.0001

Dependent Variable: SALES

Output Statistics

Obs	T	Dep Var SALES	Predicted Value	Std Error Mean Predict	95% CL Predict		Residual
1	1	4.8000	4.6971	2.1132	-8.9867	18.3809	0.1029
2	2	4.0000	8.9928	2.0220	-4.6339	22.6195	-4.9928
3	3	5.5000	13.2884	1.9325	-0.2844	26.8612	-7.7884
4	4	15.6000	17.5840	1.8448	4.0618	31.1062	-1.9840
5	5	23.1000	21.8797	1.7593	8.4047	35.3546	1.2203
6	6	23.3000	26.1753	1.6761	12.7443	39.6063	-2.8753
7	7	31.4000	30.4709	1.5959	17.0805	43.8614	0.9291
8	8	46.0000	34.7666	1.5189	21.4132	48.1199	11.2334
9	9	46.1000	39.0622	1.4457	25.7425	52.3819	7.0378
10	10	41.9000	43.3578	1.3769	30.0683	56.6473	-1.4578
11	11	45.5000	47.6534	1.3132	34.3907	60.9162	-2.1534
12	12	53.5000	51.9491	1.2554	38.7095	65.1887	1.5509
13	13	48.4000	56.2447	1.2043	43.0248	69.4646	-7.8447
14	14	61.6000	60.5403	1.1609	47.3365	73.7442	1.0597
15	15	65.6000	64.8360	1.1259	51.6447	78.0273	0.7640
16	16	71.4000	69.1316	1.1003	55.9493	82.3139	2.2684
17	17	83.4000	73.4272	1.0846	60.2503	86.6042	9.9728
18	18	93.6000	77.7229	1.0793	64.5477	90.8980	15.8771
19	19	94.2000	82.0185	1.0846	68.8415	95.1954	12.1815
20	20	85.4000	86.3141	1.1003	73.1318	99.4964	-0.9141
21	21	86.2000	90.6097	1.1259	77.4185	103.8010	-4.4097
22	22	89.9000	94.9054	1.1609	81.7016	108.1092	-5.0054
23	23	89.2000	99.2010	1.2043	85.9811	112.4209	-10.0010
24	24	99.1000	103.4966	1.2554	90.2571	116.7362	-4.3966
25	25	100.3000	107.7923	1.3132	94.5295	121.0550	-7.4923
26	26	111.7000	112.0879	1.3769	98.7984	125.3774	-0.3879
27	27	108.2000	116.3835	1.4457	103.0639	129.7032	-8.1835
28	28	115.5000	120.6792	1.5189	107.3258	134.0325	-5.1792
29	29	119.2000	124.9748	1.5959	111.5843	138.3653	-5.7748
30	30	125.2000	129.2704	1.6761	115.8394	142.7014	-4.0704
31	31	136.3000	133.5661	1.7593	120.0911	147.0410	2.7339
32	32	146.8000	137.8617	1.8448	124.3395	151.3839	8.9383
33	33	146.1000	142.1573	1.9325	128.5845	155.7301	3.9427
34	34	151.4000	146.4529	2.0220	132.8263	160.0796	4.9471
35	35	150.9000	150.7486	2.1132	137.0648	164.4324	0.1514
36	36	.	155.0442	2.2057	141.3001	168.7883	.
37	37	.	159.3398	2.2995	145.5322	173.1474	.
38	38	.	163.6355	2.3944	149.7613	177.5097	.
39	39	.	167.9311	2.4903	153.9872	181.8750	.
40	40	.	172.2267	2.5870	158.2101	186.2433	.

software. These values are given in the bottom portion of the SAS printout shown in Figure 10.7. For example, for $t = 36$, we have $\hat{y}_{36} = 155.0442$ with the 95% prediction interval (141.3001, 168.7883). That is, we predict that sales revenue in year $t = 36$ will fall between \$141,300 and \$168,788 with 95% confidence.

Note that the prediction intervals for $t = 36, 37, \ldots, 40$ widen as we attempt to forecast farther into the future. Intuitively, we know that the farther into the future we forecast, the less certain we are of the accuracy of the forecast since some unexpected change in business and economic conditions may make the model inappropriate. Since we have less confidence in the forecast for, say, $t = 40$ than for $t = 36$, it follows that the prediction interval for $t = 40$ must be wider to attain a 95% level of confidence. For this reason, time series forecasting (regardless of the forecasting method) is generally confined to the short term.

Multiple regression models can also be used to forecast future values of a time series with seasonal variation. We illustrate with an example.

Example 10.3

Refer to the 2006–2009 quarterly power loads listed in Table 10.1.

(a) Propose a model for quarterly power load, y_t, that will account for both the secular trend and seasonal variation present in the series.

(b) Fit the model to the data, and use the least squares prediction equation to forecast the utility company's quarterly power loads in 2010. Construct 95% prediction intervals for the forecasts.

Solution

(a) A common way to describe seasonal differences in a time series is with dummy variables.* For quarterly data, a model that includes both trend and seasonal components is

$$E(y_t) = \beta_0 + \underbrace{\beta_1 t}_{\substack{\text{Secular} \\ \text{trend}}} + \underbrace{\beta_2 Q_1 + \beta_3 Q_2 + \beta_4 Q_3}_{\text{Seasonal component}}$$

where

t = Time period, ranging from $t = 1$ for quarter I of 2006 to $t = 16$ for quarter IV of 2009

y_t = Power load (megawatts) in time t

$$Q_1 = \begin{cases} 1 & \text{if quarter I} \\ 0 & \text{if not} \end{cases} \qquad Q_2 = \begin{cases} 1 & \text{if quarter II} \\ 0 & \text{if not} \end{cases}$$

$$Q_3 = \begin{cases} 1 & \text{if quarter III} \\ 0 & \text{if not} \end{cases} \qquad \text{Base level} = \text{quarter IV}$$

The β coefficients associated with the seasonal dummy variables determine the mean increase (or decrease) in power load for each quarter, relative to the base level quarter, quarter IV.

(b) The model is fit to the data from Table 10.1 using the SAS multiple regression routine. The resulting SAS printout is shown in Figure 10.8. Note that the

---

* Another way to account for seasonal variation is with trigonometric (sine and cosine) terms. We discuss seasonal models with trigonometric terms in Section 10.7.

model appears to fit the data quite well: $R^2 = .9972$, indicating that the model accounts for 99.7% of the sample variation in power loads over the 4-year period; $F = 968.96$ strongly supports the hypothesis that the model has predictive utility (p-value $< .0001$); and the standard deviation, **Root MSE** = 1.53242, implies that the model predictions will usually be accurate to within approximately $\pm 2(1.53)$, or about ± 3.06 megawatts.

Forecasts and corresponding 95% prediction intervals for the 2010 power loads are reported in the bottom portion of the printout in Figure 10.8. For example, the forecast for power load in quarter I of 2010 is 184.7 megawatts with the 95% prediction interval (180.5, 188.9). Therefore, using a 95% prediction interval, we expect the power load in quarter I of 2010 to fall

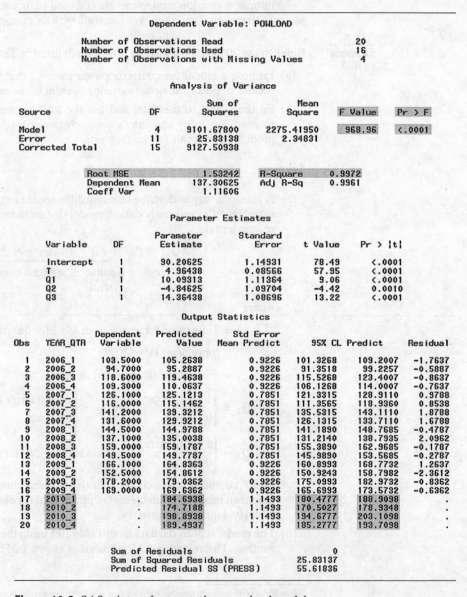

Dependent Variable: POWLOAD

Number of Observations Read	20
Number of Observations Used	16
Number of Observations with Missing Values	4

Analysis of Variance

Source	DF	Sum of Squares	Mean Square	F Value	Pr > F
Model	4	9101.67800	2275.41950	968.96	<.0001
Error	11	25.83138	2.34831		
Corrected Total	15	9127.50938			

Root MSE	1.53242	R-Square	0.9972
Dependent Mean	137.30625	Adj R-Sq	0.9961
Coeff Var	1.11606		

Parameter Estimates

Variable	DF	Parameter Estimate	Standard Error	t Value	Pr > |t|
Intercept	1	90.20625	1.14931	78.49	<.0001
T	1	4.96438	0.08566	57.95	<.0001
Q1	1	10.09313	1.11364	9.06	<.0001
Q2	1	-4.84625	1.09704	-4.42	0.0010
Q3	1	14.36438	1.08696	13.22	<.0001

Output Statistics

Obs	YEAR_QTR	Dependent Variable	Predicted Value	Std Error Mean Predict	95% CL Predict		Residual
1	2006_1	103.5000	105.2638	0.9226	101.3268	109.2007	-1.7637
2	2006_2	94.7000	95.2887	0.9226	91.3518	99.2257	-0.5887
3	2006_3	118.6000	119.4638	0.9226	115.5268	123.4007	-0.8637
4	2006_4	109.3000	110.0637	0.9226	106.1268	114.0007	-0.7637
5	2007_1	126.1000	125.1213	0.7851	121.3315	128.9110	0.9788
6	2007_2	116.0000	115.1462	0.7851	111.3565	118.9360	0.8538
7	2007_3	141.2000	139.3212	0.7851	135.5315	143.1110	1.8788
8	2007_4	131.6000	129.9212	0.7851	126.1315	133.7110	1.6788
9	2008_1	144.5000	144.9788	0.7851	141.1890	148.7685	-0.4787
10	2008_2	137.1000	135.0038	0.7851	131.2140	138.7935	2.0962
11	2008_3	159.0000	159.1787	0.7851	155.3890	162.9685	-0.1787
12	2008_4	149.5000	149.7787	0.7851	145.9890	153.5685	-0.2787
13	2009_1	166.1000	164.8363	0.9226	160.8993	168.7732	1.2637
14	2009_2	152.5000	154.8612	0.9226	150.9243	158.7982	-2.3612
15	2009_3	178.2000	179.0362	0.9226	175.0993	182.9732	-0.8362
16	2009_4	169.0000	169.6362	0.9226	165.6993	173.5732	-0.6362
17	2010_1	.	184.6938	1.1493	180.4777	188.9098	.
18	2010_2	.	174.7188	1.1493	170.5027	178.9348	.
19	2010_3	.	198.8938	1.1493	194.6777	203.1098	.
20	2010_4	.	189.4937	1.1493	185.2777	193.7098	.

Sum of Residuals	0
Sum of Squared Residuals	25.83137
Predicted Residual SS (PRESS)	55.61836

Figure 10.8 SAS printout for quarterly power load model

between 180.5 and 188.9 megawatts. Recall from Table 10.4 in Example 10.2 that the actual 2010 quarterly power loads are 181.5, 175.2, 195.0, and 189.3, respectively. Note that each of these falls within its respective 95% prediction interval shown in Figure 10.8. ■

Many descriptive forecasting techniques have proved their merit by providing good forecasts for particular applications. Nevertheless, the advantage of forecasting using the regression approach is clear: Regression analysis provides us with a measure of reliability for each forecast through prediction intervals. However, there are two problems associated with forecasting time series using a multiple regression model.

Problem 1   We are using the least squares prediction equation to forecast values outside the region of observation of the independent variable, t. For example, in Example 10.3, we are forecasting for values of t between 17 and 20 (the four quarters of 2010), even though the observed power loads are for t-values between 1 and 16. As noted in Chapter 7, it is risky to use a least squares regression model for prediction outside the range of the observed data because some unusual change—economic, political, etc.—may make the model inappropriate for predicting future events. Because forecasting always involves predictions about future values of a time series, this problem obviously cannot be avoided. However, it is important that the forecaster recognize the dangers of this type of prediction.

Problem 2   Recall the standard assumptions made about the random error component of a multiple regression model (Section 4.2). We assume that the errors have mean 0, constant variance, normal probability distributions, and are *independent*. The latter assumption is often violated in time series that exhibit short-term trends. As an illustration, refer to the plot of the sales revenue data shown in Figure 10.6. Notice that the observed sales tend to deviate about the least squares line in positive and negative runs. That is, if the difference between the observed sales and predicted sales in year t is positive (or negative), the difference in year $t + 1$ tends to be positive (or negative). Since the variation in the yearly sales is systematic, the implication is that the errors are correlated. In fact, the **Durbin–Watson test** for correlated errors (see Section 8.6.) supports this inference. Violation of this standard regression assumption could lead to unreliable forecasts.

Time series models have been developed specifically for the purpose of making forecasts when the errors are known to be correlated. These models include an **autoregressive term** for the correlated errors that result from cyclical, seasonal, or other short-term effects. Time series autoregressive models are the subject of Sections 10.5–10.11.

10.4 Exercises

10.8 Mortgage interest rates. The level at which commercial lending institutions set mortgage interest rates has a significant effect on the volume of buying, selling, and construction of residential and commercial real estate. The data in the next table (p. 542) are the annual average mortgage interest rates for conventional, fixed-rate, 30-year loans for the period 1985–2007.

(a) Fit the simple regression model

$$E(Y_t) = \beta_0 + \beta_1 t$$

where t is the number of years since 1985 (i.e., $t = 0, 1, \ldots, 22$).

(b) Forecast the average mortgage interest rate in 2010. Find a 95% prediction interval for this forecast.

⊕ INTRATE30

YEAR	INTEREST RATE(%)	YEAR	INTEREST RATE (%)
1985	11.85	1997	7.57
1986	11.33	1998	6.92
1987	10.46	1999	7.46
1988	10.86	2000	8.08
1989	12.07	2001	7.01
1990	9.97	2002	6.56
1991	11.14	2003	5.89
1992	8.27	2004	5.86
1993	7.17	2005	5.93
1994	8.28	2006	6.47
1995	7.86	2007	6.40
1996	7.76		

Source: U.S. Bureau of the Census. *Statistical Abstract of the United States*, 2009.

10.9 Price of natural gas. The annual prices of natural gas from 1990 to 2007 are listed in the table. A simple linear regression model, $E(Y_t) = \beta_0 + \beta_1 t$, where t is the number of years since 1990, is proposed to forecast the annual price of natural gas.

⊕ NATGAS

YEAR	PRICE	YEAR	PRICE	YEAR	PRICE
1990	5.80	1997	6.94	2004	10.75
1991	5.82	1998	6.82	2005	12.70
1992	5.89	1999	6.69	2006	13.75
1993	6.16	2000	7.76	2007	13.01
1994	6.41	2001	9.63		
1995	6.06	2002	7.89		
1996	6.34	2003	9.63		

Source: U.S. Bureau of the Census. *Statistical Abstract of the United States*, 2009.

(a) Give the least squares estimates of the β's and interpret their values.

(b) Evaluate the model's fit.

(c) Find and interpret 95% prediction intervals for the years 2008 and 2009.

(d) Describe the problems associated with using a simple linear regression model to predict time series data.

10.10 Life insurance policies in force. The next table represents all life insurance policies (in millions) in force on the lives of U.S. residents for the years 1980 through 2006.

(a) Use the method of least squares to fit a simple regression model to the data.

(b) Forecast the number of life insurance policies in force for 2007 and 2008.

(c) Construct 95% prediction intervals for the forecasts of part b.

(d) Check the accuracy of your forecasts by looking up the actual number of life insurance policies in force for 2007 and 2008 in the *Statistical Abstract of the United States.*

⊕ LIFEINS

YEAR	NO. OF POLICIES (IN MILLIONS)	YEAR	NO. OF POLICIES (IN MILLIONS)
1980	402	1994	366
1981	400	1995	370
1982	390	1996	355
1983	387	1997	351
1984	385	1998	358
1985	386	1999	367
1986	391	2000	369
1987	395	2001	377
1988	391	2002	375
1989	394	2003	379
1990	389	2004	373
1991	375	2005	373
1992	366	2006	375
1993	363		

Source: U.S. Bureau of the Census. *Statistical Abstract of the United States*, 2009; www.census.gov

10.11 Graphing calculator sales. The table below presents the quarterly sales index for one brand of graphing calculator at a campus bookstore. The quarters are based on an academic year, so the first quarter represents fall; the second, winter; the third, spring; and the fourth, summer.

Define the time variable as $t = 1$ for the first quarter of 2005, $t = 2$ for the second quarter of 2005, etc. Consider the following seasonal dummy variables:

$$Q_1 = \begin{cases} 1 & \text{if Quarter 1} \\ 0 & \text{otherwise} \end{cases}$$

$$Q_2 = \begin{cases} 1 & \text{if Quarter 2} \\ 0 & \text{otherwise} \end{cases}$$

$$Q_3 = \begin{cases} 1 & \text{if Quarter 3} \\ 0 & \text{otherwise} \end{cases}$$

⊕ GRAPHICAL

YEAR	FIRST QUARTER	SECOND QUARTER	THIRD QUARTER	FOURTH QUARTER
2005	438	398	252	160
2006	464	429	376	216
2007	523	496	425	318
2008	593	576	456	398
2009	636	640	526	498

(a) Write a regression model for $E(Y_t)$ as a function of t, Q_1, Q_2, and Q_3.

(b) Find and interpret the least squares estimates and evaluate the usefulness of the model.

(c) Which of the assumptions about the random error component is in doubt when a regression model is fit to time series data?

(d) Find the forecasts and the 95% prediction intervals for the 2010 quarterly sales. Interpret the result.

 SP500

10.12 S&P 500 Index. Refer to the quarterly S&P 500 values given in Exercise 10.6 (p. 536).

(a) Hypothesize a time series model to account for trend and seasonal variation.

(b) Fit the model in part a to the data.

(c) Use the least squares model from part b to forecast the S&P 500 for all four quarters of 2009. Obtain 95% prediction intervals for the forecasts.

10.13 Local area home sales. A realtor working in a large city wants to identify the secular trend in the weekly number of single-family houses sold by her firm. For the past 15 weeks she has collected data on her firm's home sales, as shown in the table (top, next column).

(a) Plot the time series. Is there visual evidence of a quadratic trend?

(b) The realtor hypothesizes the model $E(y_t) = \beta_0 + \beta_1 t + \beta_2 t^2$ for the secular trend of the weekly time series. Fit the model to the data, using the method of least squares.

(c) Plot the least squares model on the graph of part a. How well does the quadratic model describe the secular trend?

🔷 HOMESALES

WEEK t	HOMES SOLD y_t	WEEK t	HOMES SOLD y_t	WEEK t	HOMES SOLD y_t
1	59	6	137	11	88
2	73	7	106	12	75
3	70	8	122	13	62
4	82	9	93	14	44
5	115	10	86	15	45

(d) Use the model to forecast home sales in week 16 with a 95% prediction interval.

10.14 Hotel room occupancy rates. A traditional indicator of the economic health of the accommodations (hotel–motel) industry is the trend in room occupancy. Average monthly occupancies for 2 recent years are given in the table below for hotels and motels in the cities of Atlanta, Georgia, and Phoenix, Arizona. Let y_t = occupancy rate for Phoenix in month t.

(a) Propose a model for $E(y_t)$ that accounts for possible seasonal variation in the monthly series. [*Hint*: Consider a model with dummy variables for the 12 months, January, February, etc.]

(b) Fit the model of part a to the data.

(c) Test the hypothesis that the monthly dummy variables are useful predictors of occupancy rate. [*Hint*: Conduct a partial F-test.]

(d) Use the fitted least squares model from part b to forecast the Phoenix occupancy rate in January of year 3 with a 95% prediction interval.

(e) Repeat parts a–d for the Atlanta monthly occupancy rates.

10.15 Retirement security income. The Employee Retirement Income Security Act (ERISA) was originally established to enhance retirement

🔷 ROOMOCC

YEAR ONE MONTH	ROOMS OCCUPIED (%) ATLANTA	PHOENIX	YEAR TWO MONTH	ROOMS OCCUPIED (%) ATLANTA	PHOENIX
January	59	67	January	64	72
February	63	85	February	69	91
March	68	83	March	73	87
April	70	69	April	67	75
May	63	63	May	68	70
June	59	52	June	71	61
July	68	49	July	67	46
August	64	49	August	71	44
September	62	56	September	65	63
October	73	69	October	72	73
November	62	63	November	63	71
December	47	48	December	47	51

Source: Trends in the Hotel Industry.

security income. J. Ledolter (University of Iowa) and M. L. Power (Iowa State University) investigated the effects of ERISA on the growth in the number of private retirement plans (*Journal of Risk and Insurance*, December 1983). Using quarterly data ($n = 107$ quarters), Ledolter and Power fitted quarterly time series models for the number of pension qualifications and the number of profit-sharing plan qualifications. One of several models investigated was the quadratic model $E(y_t) = \beta_0 + \beta_1 t + \beta_2 t^2$, where y_t is the logarithm of the dependent variable (number of pension or number of profit-sharing qualifications) in quarter t. The results (modified for the purpose of this exercise) are summarized here:

Pension plan qualifications:

$$\hat{y}_t = 6.19 + .039t - .00024t^2$$

$$t \text{ (for } H_0: \beta_2 = 0) = -1.39$$

Profit-sharing plan qualifications:

$$\hat{y}_t = 6.22 + .035t - .00021t^2$$

$$t \text{ (for } H_0: \beta_2 = 0) = -1.61$$

(a) Is there evidence that the quarterly number of pension plan qualifications increases at a decreasing rate over time? Test using $\alpha = .05$. [*Hint:* Test $H_0: \beta_2 = 0$ against $H_a: \beta_2 < 0$.]

(b) Forecast the number of pension plan qualifications for quarter 108. [*Hint:* Since y_t is the logarithm of the number of pension plan qualifications, to obtain the forecast you must take the antilogarithm of $\hat{y}_{108}$, i.e., $e^{\hat{y}_{108}}$.]

(c) Is there evidence that the quarterly number of profit-sharing plan qualifications increases at a decreasing rate over time? Test using $\alpha = .05$. [*Hint:* Test $H_0: \beta_2 = 0$ against $H_a: \beta_2 < 0$.]

(d) Forecast the number of profit-sharing plan qualifications for quarter 108. [*Hint:* Since y_t is the logarithm of the number of profit-sharing plan qualifications, to obtain the forecast you must take the antilogarithm of $\hat{y}_{108}$, i.e., $e^{\hat{y}_{108}}$.]

10.5 Autocorrelation and Autoregressive Error Models

In Chapter 8, we presented the Durbin–Watson test for detecting correlated residuals in a regression analysis. Correlated residuals are quite common when the response is a *time series* variable. Correlation of residuals for a regression model with a time series response is called **autocorrelation**, because the correlation is between residuals from the *same* time series model at different points in time.

A special case of autocorrelation that has many applications to business and economic phenomena is the case in which neighboring residuals one time period apart (say, at times t and $t + 1$) are correlated. This type of correlation is called **first-order autocorrelation**. In general, correlation between time series residuals m time periods apart is mth-order autocorrelation.

> **Definition 10.7 Autocorrelation** is the correlation between time series residuals at different points in time. The special case in which neighboring residuals one time period apart (at times t and $t + 1$) are correlated is called **first-order autocorrelation**. In general, **mth-order autocorrelation** occurs when residuals at times t and $(t + m)$ are correlated.

To see how autocorrelated residuals affect the regression model, we will assume a model similar to the linear statistical model in Chapter 4,

$$y_t = E(y_t) + R_t$$

where $E(y_t)$ is the regression model

$$E(y_t) = \beta_0 + \beta_1 x_1 + \cdots + \beta_k x_k$$

and R_t represents the random residual. We assume that the residual R_t has mean 0 and constant variance σ^2, but that it is autocorrelated. The effect of autocorrelation on the general linear model depends on the pattern of the autocorrelation. One of the most common patterns is that the autocorrelation between residuals at consecutive time points is positive. Thus, when the residual at time t, R_t, indicates that the observed value y_t is more than the mean value $E(y_t)$, then the residual at time $(t+1)$ will have a tendency (probability greater than .5) to be positive. This would occur, for example, if you were to model a monthly economic index (e.g., the Consumer Price Index) with a straight-line model. In times of recession, the observed values of the index will tend to be less than the predictions of a straight line for most or all of the months during the period. Similarly, in extremely inflationary periods, the residuals are likely to be positive because the observed value of the index will lie above the straight-line model. In either case, the fact that residuals at consecutive time points tend to have the same sign implies that they are **positively correlated**.

A second property commonly observed for autocorrelated residuals is that the size of the autocorrelation between values of the residual R at two different points in time diminishes rapidly as the distance between the time points increases. Thus, the autocorrelation between R_t and R_{t+m} becomes smaller (i.e., weaker) as the distance m between the time points becomes larger.

A residual model that possesses this property—positive autocorrelation diminishing rapidly as distance between time points increases—is the **first-order autoregressive error model**:

$$R_t = \phi R_{t-1} + \varepsilon_t, \quad -1 < \phi < 1$$

where ε_t, a residual called **white noise**, is uncorrelated with any and all other residual components. Thus, the value of the residual R_t is equal to a constant multiple, ϕ (Greek letter "phi"), of the previous residual, R_{t-1}, plus random error. In general, the constant ϕ is between -1 and $+1$, and the numerical value of ϕ determines the sign (positive or negative) and strength of the autocorrelation. In fact, it can be shown (proof omitted) that the autocorrelation (abbreviated AC) between two residuals that are m time units apart, R_t, and R_{t+m}, is

$$AC(R_t, R_{t+m}) = \phi^m$$

Since the absolute value of ϕ will be less than 1, the autocorrelation between R_t and R_{t+m}, ϕ^m, will decrease as m increases. This means that neighboring values of R_t (i.e., $m=1$) will have the highest correlation, and the correlation diminishes rapidly as the distance m between time points is increased. This points to an interesting property of the autoregressive time series model. The autocorrelation function depends only on the distance m between R values, and not on the time t. Time series models that possess this property are said to be **stationary**.

Definition 10.8   A **stationary time series model** for regression residuals is one that has mean 0, constant variance, and autocorrelations that depend only on the distance between time points.

The autocorrelation function of first-order autoregressive models is shown for several values of ϕ in Figure 10.9. Note that positive values of ϕ yield positive autocorrelation for all residuals, whereas negative values of ϕ imply negative correlation for neighboring residuals, positive correlation between residuals two time points apart, negative correlation for residuals three time points apart, and so forth. The appropriate pattern will, of course, depend on the particular application, but the occurrence of a positive autocorrelation pattern is more common.

Although the first-order autoregressive error model provides a good representation for many autocorrelation patterns, more complex patterns can be described by

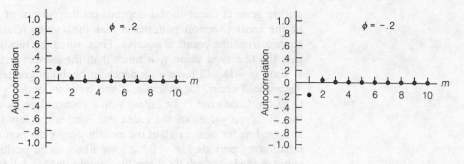

(a)   Weak autocorrelation

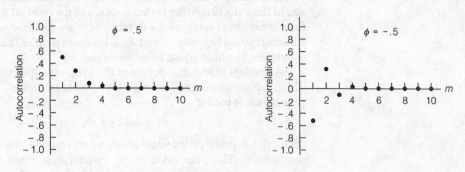

(b)   Moderate autocorrelation

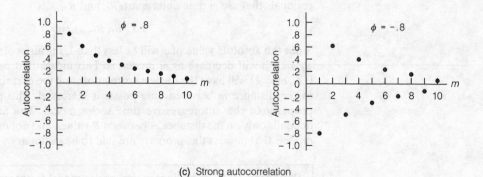

(c)   Strong autocorrelation

Figure 10.9   Autocorrelation functions for several first-order autoregressive error models: $R_t = \phi_1 R_{t-1} + \varepsilon_t$

higher-order autoregressive models. The general form of a pth-order autoregressive error model is

$$R_t = \phi_1 R_{t-1} + \phi_2 R_{t-2} + \cdots + \phi_p R_{t-p} + \varepsilon_t$$

The inclusion of p parameters, $\phi_1, \phi_2, \ldots, \phi_p$, permits more flexibility in the pattern of autocorrelations exhibited by a residual time series. When an autoregressive model is used to describe residual autocorrelations, the observed autocorrelations are used to estimate these parameters. Methods for estimating these parameters will be presented in Section 10.8.

10.5 Exercises

10.16 Finding autocorrelations. Suppose that the random component of a time series model follows the first-order autoregressive model $R_t = \phi R_{t-1} + \varepsilon_t$, where ε_t is a white-noise process. Consider four versions of this model: $\phi = .9$, $\phi = -.9$, $\phi = .2$, and $\phi = -.2$.

(a) Calculate the first 10 autocorrelations, $\text{AC}(R_t, R_{t+m})$, $m = 1, 2, 3, \ldots, 10$, for each of the four models.

(b) Plot the autocorrelations against the distance in time separating the R values (m) for each case.

(c) Examine the rate at which the correlation diminishes in each plot. What does this imply?

10.17 Non–1st-order autogressive model. When using time series to analyze quarterly data (data in which seasonal effects are present), it is highly possible that the random component of the model R_t also exhibits the same seasonal variation as the dependent variable. In these cases, the following non–first-order autoregressive model is sometimes postulated for the correlated error term, R_t:

$$R_t = \phi R_{t-4} + \varepsilon_t$$

where $|\phi| < 1$ and ε_t is a white-noise process. The autocorrelation function for this model is given by

$$\text{AC}(R_t, R_{t+m}) = \begin{cases} \phi^{m/4} & \text{if } m = 4, 8, 12, 16, 20, \ldots \\ 0 & \text{if otherwise} \end{cases}$$

(a) Calculate the first 20 autocorrelations ($m = 1, 2, \ldots, 20$) for the model with constant coefficient $\phi = .5$.

(b) Plot the autocorrelations against m, the distance in time separating the R values. Compare the rate at which the correlation diminishes with the first-order model $R_t = .5R_{t-1} + \varepsilon_t$.

10.18 Identifying the autoregressive model. Consider the autocorrelation pattern shown in the figure. Write a first-order autoregressive model that exhibits this pattern.

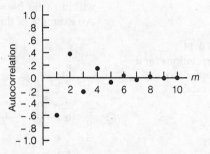

10.19 Writing the equation of an autoregressive model. Write the general form for a fourth-order autoregressive model.

10.6 Other Models for Autocorrelated Errors (Optional)

There are many models for autocorrelated residuals in addition to the autoregressive model, but the autoregressive model provides a good approximation for the autocorrelation pattern in many applications. Recall that the autocorrelations for autoregressive models diminish rapidly as the time distance m between the residuals increases. Occasionally, residual autocorrelations appear to change abruptly from nonzero for small values of m to 0 for larger values of m. For example, neighboring residuals ($m = 1$) may be correlated, whereas residuals that are farther apart ($m > 1$) are uncorrelated. This pattern can be described by the **first-order moving average model**

$$R_t = \varepsilon_t + \theta \varepsilon_{t-1}$$

Note that the residual R_t is a linear combination of the current and previous *uncorrelated* (white-noise) residuals. It can be shown that the autocorrelations for this model are

$$\text{AC}(R_t, R_{t+m}) = \begin{cases} \frac{\theta}{1+\theta^2} & \text{if } m = 1 \\ 0 & \text{if } m > 1 \end{cases}$$

This pattern is shown in Figure 10.10.

Figure 10.10
Autocorrelations for the
first-order moving average
model: $R_t = \varepsilon_t + \theta\varepsilon_{t-1}$

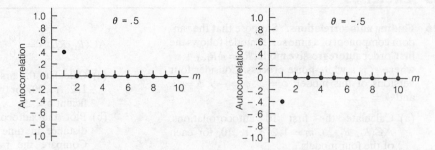

More generally, a qth-order moving average model is given by

$$R_t = \varepsilon_t + \theta_1\varepsilon_{t-1} + \theta_2\varepsilon_{t-2} + \cdots + \theta_q\varepsilon_{t-q}$$

Residuals within q time points are correlated, whereas those farther than q time points apart are uncorrelated. For example, a regression model for the quarterly earnings per share for a company may have residuals that are autocorrelated when within 1 year ($m = 4$ quarters) of one another, but uncorrelated when farther apart. An example of this pattern is shown in Figure 10.11.

Figure 10.11
Autocorrelations for a
fourth-order moving
average model

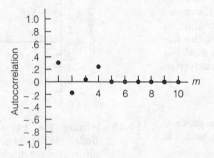

Some autocorrelation patterns require even more complex residual models. A more general model is a combination of the **autoregressive–moving average (ARMA) models,**

$$R_t = \phi_1 R_{t-1} + \cdots + \phi_p R_{t-p} + \varepsilon_t + \theta_1\varepsilon_{t-1} + \cdots + \theta_q\varepsilon_{t-q}$$

Like the autoregressive model, the ARMA model has autocorrelations that diminish as the distance m between residuals increases. However, the patterns that can be described by ARMA models are more general than those of either autoregressive or moving average models.

In Section 10.8, we present a method for estimating the parameters of an autoregressive residual model. The method for fitting time series models when the residual is either moving average or ARMA is more complicated, however. Consult the references at the end of the chapter for details of these methods.

10.7 Constructing Time Series Models

Recall that the general form of the times series model is

$$y_t = E(y_t) + R_t$$

We are assuming that the expected value of y_t is

$$E(y_t) = \beta_0 + \beta_1 x_1 + \beta_2 x_2 + \cdots + \beta_k x_k$$

where $x_1, x_2, \ldots, x_k$ are independent variables, which themselves may be time series, and the residual component, R_t, accounts for the pattern of autocorrelation in the residuals. Thus, a time series model consists of a pair of models: one model for the deterministic component $E(y_t)$ and one model for the autocorrelated residuals R_t.

Choosing the Deterministic Component

The deterministic portion of the model is chosen in exactly the same manner as the regression models of the preceding chapters except that some of the independent variables might be time series variables or might be trigonometric functions of time (such as $\sin t$ or $\cos t$). It is helpful to think of the deterministic component as consisting of the trend (T_t), cyclical (C_t), and seasonal (S_t) effects described in Section 10.2.

For example, we may want to model the number of new housing starts, y_t, as a function of the prime interest rate, x_t. Then, one model for the mean of y_t is

$$E(y_t) = \beta_0 + \beta_1 x_t$$

for which the mean number of new housing starts is a multiple β_1 of the prime interest rate, plus a constant β_0. Another possibility is a second-order relationship,

$$E(y_t) = \beta_0 + \beta_1 x_t + \beta_2 x_t^2$$

which permits the *rate* of increase in the mean number of housing starts to increase or decrease with the prime interest rate.

Yet another possibility is to model the mean number of new housing starts as a function of both the prime interest rate and the year, t. Thus, the model

$$E(y_t) = \beta_0 + \beta_1 x_t + \beta_2 t + \beta_3 x_t t$$

implies that the mean number of housing starts increases linearly in x_t, the prime interest rate, but the rate of increase depends on the year t. If we wanted to adjust for seasonal (cyclical) effects due to t, we might introduce time into the model using trigonometric functions of t. This topic is explained subsequently in greater detail.

Another important type of model for $E(y_t)$ is the **lagged independent variable model**. *Lagging* means that we are pairing observations on a dependent variable and independent variable at two different points in time, with the time corresponding to the independent variable lagging behind the time for the dependent variable. Suppose, for example, we believe that the monthly mean number of new housing starts is a function of the *previous* month's prime interest rate. Thus, we model y_t as a linear function of the lagged independent variable, prime interest rate, x_{t-1},

$$E(y_t) = \beta_0 + \beta_1 x_{t-1}$$

or, alternatively, as the second-order function,

$$E(y_t) = \beta_0 + \beta_1 x_{t-1} + \beta_2 x_{t-1}^2$$

For this example, the independent variable, prime interest rate x_t, is lagged 1 month behind the response y_t.

Many time series have distinct seasonal patterns. Retail sales are usually highest around Christmas, spring, and fall, with relative lulls in the winter and summer periods. Energy usage is highest in summer and winter, and lowest in spring and fall. Teenage unemployment rises in the summer months when schools are not in session, and falls near Christmas when many businesses hire part-time help.

When a time series' seasonality is exhibited in a relatively consistent pattern from year to year, we can model the pattern using trigonometric terms in the model for $E(y_t)$. For example, the model of a monthly series with mean $E(y_t)$ might be

$$E(y_t) = \beta_0 + \beta_1 \left(\cos \frac{2\pi}{12} t \right) + \beta_2 \left(\sin \frac{2\pi}{12} t \right)$$

This model would appear as shown in Figure 10.12. Note that the model is **cyclic**, with a **period** of 12 months. That is, the mean $E(y_t)$ completes a cycle every 12 months and then repeats the same cycle over the next 12 months. Thus, the **expected peaks and valleys** of the series remain the same from year to year. The coefficients β_1 and β_2 determine the **amplitude** and **phase shift** of the model. The amplitude is the magnitude of the seasonal effect, whereas the phase shift locates the peaks and valleys in time. For example, if we assume month 1 is January, the mean of the time series depicted in Figure 10.12 has a peak each April and a valley each October.

Figure 10.12 A seasonal time series model

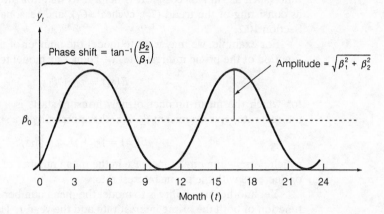

If the data are monthly or quarterly, we can treat the season as a qualitative independent variable (see Example 10.3), and write the model

$$E(y_t) = \beta_0 + \beta_1 S_1 + \beta_2 S_2 + \beta_3 S_3$$

where

$$S_1 = \begin{cases} 1 & \text{if season is spring (II)} \\ 0 & \text{otherwise} \end{cases} \qquad S_2 = \begin{cases} 1 & \text{if season is summer (III)} \\ 0 & \text{otherwise} \end{cases}$$

$$S_3 = \begin{cases} 1 & \text{if season is fall (IV)} \\ 0 & \text{otherwise} \end{cases}$$

Thus, S_1, S_2, and S_3 are dummy variables that describe the four levels of season, letting winter (I) be the base level. The β coefficients determine the mean value of y_t for each season, as shown in Figure 10.13. Note that for the dummy variable model and the trigonometric model, we assume the seasonal effects are approximately the same from year to year. If they tend to increase or decrease with time, an

Figure 10.13 Seasonal model for quarterly data using dummy variables

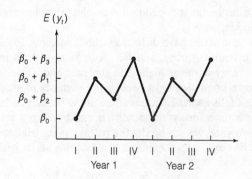

interaction of the seasonal effect with time may be necessary. (An example is given in Section 10.10.)

The appropriate form of the deterministic time series model will depend on both theory and data. Economic theory often provides several plausible models relating the mean response to one or more independent variables. The data can then be used to determine which, if any, of the models is best supported. The process is often an iterative one, beginning with preliminary models based on theoretical notions, using data to refine and modify these notions, collecting additional data to test the modified theories, and so forth.

Choosing the Residual Component

The appropriate form of the residual component, R_t, will depend on the pattern of autocorrelation in the residuals (see Sections 10.5 and 10.6). The autoregressive model in Section 10.5 is very useful for this aspect of time series modeling. The general form of an autoregressive model of order p is

$$R_t = \phi_1 R_{t-1} + \phi_2 R_{t-2} + \cdots + \phi_p R_{t-p} + \varepsilon_t$$

where ε_t is white noise (uncorrelated error). Recall that the name *autoregressive* comes from the fact that R_t is regressed on its own past values. As the order p is increased, more complex autocorrelation functions can be modeled. There are several other types of models that can be used for the random component, but the autoregressive model is very flexible and receives more application in business forecasting than the other models.

The simplest autoregressive error model is the **first-order autoregressive model**

$$R_t = \phi R_{t-1} + \varepsilon_t$$

Recall that the autocorrelation between residuals at two different points in time diminishes as the distance between the time points increases. Since many business and economic time series exhibit this property, the first-order autoregressive model is a popular choice for the residual component.

To summarize, we describe a general approach for constructing a time series:

1. Construct a regression model for the trend, seasonal, and cyclical components of $E(y_t)$. This model may be a polynomial in t for the trend (usually a straight-line or quadratic model) with trigonometric terms or dummy variables for the seasonal (cyclical) effects. The model may also include other time series variables as independent variables. For example, last year's rate of inflation may be used as a predictor of this year's gross domestic product (GDP).

2. Next, construct a model for the random component (residual effect) of the model. A model that is widely used in practice is the first-order autoregressive error model

$$R_t = \phi R_{t-1} + \varepsilon_t$$

When the pattern of autocorrelation is more complex, use the general pth-order autoregressive model

$$R_t = \phi_1 R_{t-1} + \phi_2 R_{t-2} + \cdots + \phi_p R_{t-p} + \varepsilon_t$$

3. Combine the two components so that the model can be used for forecasting:

$$y_t = E(y_t) + R_t$$

Prediction intervals are calculated to measure the reliability of the forecasts. In the following two sections, we demonstrate how time series models are fitted to data and used for forecasting. In Section 10.10, we present an example in which we fit a seasonal time series model to a set of data.

10.7   Exercises

10.20   Modeling stock price. Suppose you are interested in buying stock in the Pepsi Company (PepsiCo). Your broker has advised you that your best strategy is to sell the stock at the first substantial jump in price. Hence, you are interested in a short-term investment. Before buying, you would like to model the closing price of PepsiCo, y_t, over time (in days), t.

(a) Write a first-order model for the deterministic portion of the model, $E(y_t)$.

(b) If a plot of the daily closing prices for the past month reveals a quadratic trend, write a plausible model for $E(y_t)$.

(c) Since the closing price of PepsiCo on day $(t + 1)$ is very highly correlated with the closing price on day t, your broker suggests that the random error components of the model are not white noise. Given this information, postulate a model for the error term, R_t.

10.21   Modeling the GDP. An economist wishes to model the gross domestic product (GDP) over time (in years) and also as a function of certain personal consumption expenditures. Let t = time in years and let

y_t = GDP at time t
x_{1t} = Durable goods at time t
x_{2t} = Nondurable goods at time t
x_{3t} = Services at time t

(a) The economist believes that y_t is linearly related to the independent variables x_{1t}, x_{2t}, x_{3t}, and t. Write the first-order model for $E(y_t)$.

(b) Rewrite the model if interaction between the independent variables and time is present.

(c) Postulate a model for the random error component, R_t. Explain why this model is appropriate.

10.22   Overbooking airline flights. Airlines sometimes overbook flights because of "no-show" passengers (i.e., passengers who have purchased a ticket but fail to board the flight). An airline supervisor wishes to be able to predict, for a flight from Miami to New York, the monthly accumulation of no-show passengers during the upcoming year, using data from the past 3 years. Let y_t = Number of no-shows during month t.

(a) Using dummy variables, propose a model for $E(y_t)$ that will take into account the seasonal

(fall, winter, spring, summer) variation that may be present in the data.

(b) Postulate a model for the error term R_t.

(c) Write the full time series model for y_t (include random error terms).

(d) Suppose the airline supervisor believes that the seasonal variation in the data is not constant from year to year, in other words, that there exists interaction between time and season. Rewrite the full model with the interaction terms added.

10.23   Market price of hogs. A farmer is interested in modeling the daily price of hogs at a livestock market. The farmer knows that the price varies over time (days) and also is reasonably confident that a seasonal effect is present.

(a) Write a seasonal time series model with trigonometric terms for $E(y_t)$, where y_t = Selling price (in dollars) of hogs on day t.

(b) Interpret the β parameters.

(c) Include in the model an interaction between time and the trigonometric components. What does the presence of interaction signify?

(d) Is it reasonable to assume that the random error component of the model, R_t, is white noise? Explain. Postulate a more appropriate model for R_t.

10.24   The relationship between experience and productivity. Numerous studies have been conducted to examine the relationship between seniority and productivity in business. A problem encountered in such studies is that individual output is often difficult to measure. G. A. Krohn developed a technique for estimating the experience–productivity relationship when such a measure is available (*Journal of Business and Economic Statistics*, October 1983). Krohn modeled the batting average of a major league baseball player in year t (y_t) as a function of the player's age in year t (x_t) and an autoregressive error term (R_t).

(a) Write a model for $E(y_t)$ that hypothesizes, as did Krohn, a curvilinear relationship with x_t.

(b) Write a first-order autoregressive model for R_t.

(c) Use the models from parts a and b to write the full time series autoregressive model for y_t.

10.8 Fitting Time Series Models with Autoregressive Errors

We have proposed a general form for a time series model:

$$y_t = E(y_t) + R_t$$

where

$$E(y_t) = \beta_0 + \beta_1 x_1 + \cdots + \beta_k x_k$$

and, using an autoregressive model for R_t,

$$R_t = \phi R_{t-1} + \phi_2 R_{t-2} + \cdots + \phi_p R_{t-p} + \varepsilon_t$$

We now want to develop estimators for the parameters $\beta_0, \beta_1, \ldots, \beta_k$ of the regression model, and for the parameters $\phi_1, \phi_2, \ldots, \phi_p$ of the autoregressive model. The ultimate objective is to use the model to obtain forecasts (predictions) of future values of y_t, as well as to make inferences about the structure of the model itself.

We will introduce the techniques of fitting a time series model with a simple example. Refer to the data in Table 10.5, the annual sales for a firm in each of its 35 years of operation. Recall that the objective is to forecast future sales in years 36–40. In Section 10.4, we used a simple straight-line model for the mean sales

$$E(y_t) = \beta_0 + \beta_1 t$$

to make the forecasts.

The SAS printout showing the least squares estimates of β_0 and β_1 is reproduced in Figure 10.14. Although the model is useful for predicting annual sales (p-value for H_0: $\beta_1 = 0$ is less than .0001), the Durbin–Watson statistic is $d = .821$, which is less than the tabulated value, $d_L = 1.40$ (Table 8 in Appendix C), for $\alpha = .05, n = 35$, and $k = 1$ independent variable. Thus, there is evidence that the residuals are positively correlated. The MINITAB plot of the least squares residuals over time, in Figure 10.15, shows the pattern of positive autocorrelation. The residuals tend to cluster in positive and negative runs; if the residual at time t is positive, the residual at time $(t + 1)$ tends to be positive.

Figure 10.14 SAS printout for model of annual sales revenue

```
                          Dependent Variable: SALES

                            Analysis of Variance

                                   Sum of          Mean
Source                DF          Squares        Square    F Value   Pr > F

Model                  1            65875         65875    1615.72   <.0001
Error                 33       1345.45355      40.77132
Corrected Total       34            67221


              Root MSE            6.38524     R-Square     0.9800
              Dependent Mean     77.72286     Adj R-Sq     0.9794
              Coeff Var           8.21540



                            Parameter Estimates

                          Parameter      Standard
          Variable   DF     Estimate        Error    t Value   Pr > |t|

          Intercept   1      0.40151      2.20571       0.18     0.8567
          T           1      4.29563      0.10687      40.20     <.0001


              Durbin-Watson D                          0.821
              Number of Observations                      35
              1st Order Autocorrelation                0.590
```

Figure 10.15 MINITAB
residual plot annual
sales model

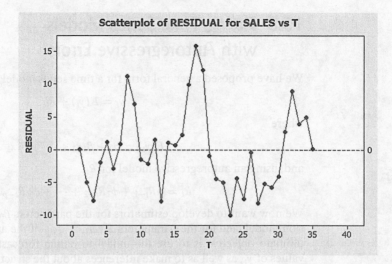

What are the consequences of fitting the least squares model when autocorrelated residuals are present? Although *the least squares estimators of β_0 and β_1 remain unbiased* even if the residuals are autocorrelated, that is, $E(\hat{\beta}_0) = \beta_0$ and $E(\hat{\beta}_1) = \beta_1$, the *standard errors given by least squares theory are usually smaller than the true standard errors* when the residuals are positively autocorrelated. Consequently, t-values computed by the methods in Chapter 4 (which apply when the errors are uncorrelated) will usually be inflated and will lead to a higher Type I error rate (α) than the value of α selected for a test. Thus, the application of standard least squares techniques to time series often produces misleading statistical test results that result in overoptimistic evaluations of a model's predictive ability. There is a second reason for seeking methods that specifically take into account the autocorrelated residuals. If we can successfully model the residual autocorrelation, we should achieve a smaller MSE and correspondingly narrower prediction intervals than those given by the least squares model.

To account for the autocorrelated residual, we postulate a first-order autoregressive model,

$$R_t = \phi R_{t-1} + \varepsilon_t$$

Thus, we use the pair of models

$$y_t = \beta_0 + \beta_1 t + R_t$$
$$R_t = \phi R_{t-1} + \varepsilon_t$$

to describe the yearly sales of the firm. To estimate the parameters of the time series model (β_0, β_1, and ϕ), a modification of the least squares method is required. To do this, we use a *transformation* that is much like the variance-stabilizing transformations discussed in Chapter 8.

First, we multiply the model

$$y_t = \beta_0 + \beta_1 t + R_t \tag{10.1}$$

by ϕ at time $(t - 1)$ to obtain

$$\phi y_{t-1} = \phi \beta_0 + \phi \beta_1 (t - 1) + \phi R_{t-1} \tag{10.2}$$

Taking the difference between equations (10.1) and (10.2), we have

$$y_t - \phi y_{t-1} = \beta_0(1 - \phi) + \beta_1[t - \phi(t - 1)] + (R_t - \phi R_{t-1})$$

or, since $R_t = \phi R_{t-1} + \varepsilon_t$, then

$$y_t^* = \beta_0^* + \beta_1 t^* + \varepsilon_t$$

where $y_t^* = y_t - \phi y_{t-1}$, $t^* = t - \phi(t - 1)$, and $\beta_0^* = \beta_0(1 - \phi)$. Thus, we can use the transformed dependent variable y_t^* and transformed independent variable t^* to obtain least squares estimates of β_0^* and β_1. The residual ε_t is uncorrelated, so that the assumptions necessary for the least squares estimators are all satisfied. The estimator of the original intercept, β_0, can be calculated by

$$\hat{\beta}_0 = \frac{\hat{\beta}_0^*}{1 - \phi}$$

This transformed model appears to solve the problem of first-order autoregressive residuals. However, making the transformation requires knowing the value of the parameter ϕ. Also, we lose the initial observation, since the values of y_t^* and t^* can be calculated only for $t \geq 2$. The methods for estimating ϕ and adjustments for the values at $t = 1$ will not be detailed here. [See Anderson (1971) or Fuller (1996).] Instead, we will present output from the SAS computer package, which both performs the transformation and estimates the model parameters, β_0, β_1, and ϕ.

The SAS printout of the straight-line, autoregressive time series model fit to the sales data is shown in Figure 10.16. The estimates of β_0 and β_1 in the deterministic component of the model (highlighted at the bottom of the printout) are $\hat{\beta}_0 = .4058$ and $\hat{\beta}_1 = 4.2959$. The estimate of the first-order autoregressive parameter ϕ (highlighted in the middle of the printout) is $-.589624$. However, the SAS time series model is defined so that ϕ takes the *opposite* sign from the value specified in our model. Consequently, you must multiply the estimate shown on the SAS printout by (-1) to obtain the estimate of ϕ for our model: $\hat{\phi} = (-1)(-.589264) = .589264$. Therefore, the fitted models are:

$$\hat{y}_t = .4058 + 4.2959t + \hat{R}_t, \quad \hat{R}_t = .589624\hat{R}_{t-1}$$

Note, also, that there are two R^2 values shown at the bottom of the SAS printout, Figure 10.16. The quantity labeled as **Regress R-Square** is not the value of R^2 based on the original time series variable, y_t. Instead, it is based on the values of the transformed variable, y_t^*. When we refer to R^2 in this chapter, we will always mean the value of R^2 based on the original time series variable. This value, which usually will be larger than the R^2 for the transformed time series variable, is given on the printout as **Total R-Square.** Thus, the time series autoregressive model yields

$$MSE = 27.42767$$

and

$$R^2 = .9869.$$

A comparison of the least squares (Figure 10.7) and autoregressive (Figure 10.16) computer printouts is given in Table 10.6. Note that the autoregressive model reduces MSE and increases R^2. The values of the estimators β_0 and β_1 change very little, but the estimated standard errors are considerably increased, thereby decreasing the t-value for testing H_0: $\beta_1 = 0$. The implication that the linear relationship between sales y_t and year t is of significant predictive value is the same using either method.

Figure 10.16 SAS printout for annual sales model with autoregressive errors

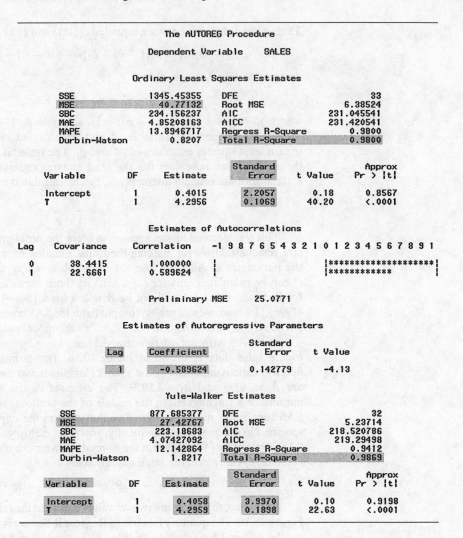

The AUTOREG Procedure

Dependent Variable   SALES

Ordinary Least Squares Estimates

SSE	1345.45355	DFE	33
MSE	40.77132	Root MSE	6.38524
SBC	234.156237	AIC	231.045541
MAE	4.85208163	AICC	231.420541
MAPE	13.8946717	Regress R-Square	0.9800
Durbin-Watson	0.8207	Total R-Square	0.9800

Variable	DF	Estimate	Standard Error	t Value	Approx Pr > |t|
Intercept	1	0.4015	2.2057	0.18	0.8567
T	1	4.2956	0.1069	40.20	<.0001

Estimates of Autocorrelations

Lag	Covariance	Correlation	-1 9 8 7 6 5 4 3 2 1 0 1 2 3 4 5 6 7 8 9 1
0	38.4415	1.000000	|********************|
1	22.6661	0.589624	|***********|

Preliminary MSE   25.0771

Estimates of Autoregressive Parameters

Lag	Coefficient	Standard Error	t Value
1	-0.589624	0.142779	-4.13

Yule-Walker Estimates

SSE	877.685377	DFE	32
MSE	27.42767	Root MSE	5.23714
SBC	223.18683	AIC	218.520786
MAE	4.07427092	AICC	219.29498
MAPE	12.142864	Regress R-Square	0.9412
Durbin-Watson	1.8217	Total R-Square	0.9869

Variable	DF	Estimate	Standard Error	t Value	Approx Pr > |t|
Intercept	1	0.4058	3.9970	0.10	0.9198
T	1	4.2959	0.1898	22.63	<.0001

Table 10.6 Comparison of least squares and time series results

	Least Squares	Autoregressive
R^2	.980	.987
MSE	40.77	27.43
$\hat{\beta}_0$	.4015	.4058
$\hat{\beta}_1$	4.2956	4.2959
Standard error ($\hat{\beta}_0$)	2.2057	3.9970
Standard error ($\hat{\beta}_1$)	.1069	.1898
t statistic for $H_0: \beta_1 = 0$	40.20	22.63
	$(p < .0001)$	$(p < .0001)$
$\hat{\phi}$	—	.5896
t statistic for $H_0: \phi = 0$	—	4.13

However, you can see that the underestimation of standard errors by using least squares in the presence of residual autocorrelation could result in the inclusion of unimportant independent variables in the model, since the t-values will usually be inflated.

Is there evidence of positive autocorreled residuals? An approximate t-test* of the hypothesis $H_0: \phi = 0$ yields a $t = 4.13$. With 32 df, this value is significant at less than $\alpha = .01$. Thus, the result of the Durbin–Watson d test is confirmed: There is adequate evidence of positive residual autocorrelation.[†] Furthermore, the first-order autoregressive model appears to describe this residual correlation well.

The steps for fitting a time series model to a set of data are summarized in the box. Once the model is estimated, the model can be used to forecast future values of the time series y_t.

Steps for Fitting Time Series Models

1. Use the least squares approach to obtain initial estimates of the β parameters. Do *not* use the t- or F-tests to assess the importance of the parameters, since the estimates of their standard errors may be biased (often underestimated).

2. Analyze the residuals to determine whether they are autocorrelated. The Durbin–Watson test is one technique for making this determination.

3. If there is evidence of autocorrelation, construct a model for the residuals. The autoregressive model is one useful model. Consult the references at the end of the chapter for more types of residual models and for methods of identifying the most suitable model.

4. Reestimate the β parameters, taking the residual model into account. This involves a simple transformation if an autoregressive model is used; several statistical software packages have computer routines to accomplish this.

10.8 Exercises

10.25 Quarterly GDP values. The gross domestic product (GDP) is a measure of total U.S. output and is, therefore, an important indicator of the U.S. economy. The quarterly GDP values (in billions of dollars) from 2004 to 2008 are given in the next table (p. 558). Let y_t be the GDP in quarter t, $t = 1, 2, 3, \ldots, 20$.

(a) Hypothesize a time series model for quarterly GDP that includes a straight-line long-term trend and autocorrelated residuals.

(b) The SAS printout for the time series model $y_t = \beta_0 + \beta_1 t + \phi R_{t-1} + \varepsilon_t$ is shown at the bottom of p. 558. Write the least squares prediction equation.

(c) Interpret the estimates of the model parameters, β_0, β_1, and ϕ.

(d) Interpret the values of R^2 and S.

---

* An explanation of this t-test has been omitted. Consult the references at the end of the chapter for details of this test.

[†] This result is to be expected since it can be shown (proof omitted) that $\hat{\phi} \approx 1 - d/2$, where d is the value of the Durbin–Watson statistic.

💿 GDP

YEAR	QUARTER	GDP
2004	1	11,406
	2	11,610
	3	11,779
	4	11,949
2005	1	12,155
	2	12,298
	3	12,538
	4	12,695
2006	1	12,960
	2	13,134
	3	13,250
	4	13,370
2007	1	13,511
	2	13,738
	3	13,951
	4	14,031
2008	1	14,151
	2	14,295
	3	14,413
	4	14,200

Source: U.S. Department of Commerce, Bureau of Economic Analysis, 2009; www.bea.gov.

💿 INTRATE30

10.26 Mortgage interest rates. Refer to Exercise 10.8 (p. 541) and the data on mortgage interest rates.

(a) Hypothesize a time series model for annual average mortgage interest rate, y_t, that takes into account the residual autocorrelation.

(b) Fit the autoregressive time series model, part a. Interpret the estimates of the model parameters.

10.27 Monthly gold prices. The fluctuation of gold prices is a reflection of the strength or weakness of the U.S. dollar. The next table (p. 559) shows monthly gold prices from January 2001 to December 2008. Suppose we want to model the monthly price, y_t, as a function of t, where t represents month (i.e., $t = 1, 2, 3, \ldots, 96$).

(a) Construct a scatterplot of the data. Do you observe a long-term trend?

(b) Propose a time series model that includes a long-term quadratic trend and autocorrelated residuals.

SAS Output for Exercise 10.25

The AUTOREG Procedure

Dependent Variable   GDP

Ordinary Least Squares Estimates

SSE	297252.367	DFE	18
MSE	16514	Root MSE	128.50689
SBC	254.881096	AIC	252.889631
MAE	85.8141203	AICC	253.595514
MAPE	0.65182465	Regress R-Square	0.9834
Durbin-Watson	0.6170	Total R-Square	0.9834

Variable	DF	Estimate	Standard Error	t Value	Approx Pr > |t|
Intercept	1	11363	59.6955	190.35	<.0001
T	1	162.7266	4.9833	32.65	<.0001

Estimates of Autocorrelations

Lag	Covariance	Correlation	-1 9 8 7 6 5 4 3 2 1 0 1 2 3 4 5 6 7 8 9 1
0	14862.6	1.000000	|********************|
1	5563.1	0.374304	|*******|

Preliminary MSE   12780.3

Estimates of Autoregressive Parameters

Lag	Coefficient	Standard Error	t Value
1	-0.374304	0.224905	-1.66

Yule-Walker Estimates

SSE	225352.96	DFE	17
MSE	13256	Root MSE	115.13495
SBC	252.489499	AIC	249.502303
MAE	69.2335112	AICC	251.002303
MAPE	0.52383884	Regress R-Square	0.9724
Durbin-Watson	0.7900	Total R-Square	0.9874

Variable	DF	Estimate	Standard Error	t Value	Approx Pr > |t|
Intercept	1	11374	79.4734	143.12	<.0001
T	1	160.2342	6.5421	24.49	<.0001

GOLDMON

MONTH	2001	2002	2003	2004	2005	2006	2007	2008
Jan	265.5	281.7	356.9	414.0	424.2	549.9	631.2	889.6
Feb	261.9	295.5	359.0	405.3	423.4	555.0	664.7	922.3
Mar	263.0	294.0	340.6	406.7	434.2	557.1	654.9	968.4
Apr	260.5	302.7	328.2	403.0	428.9	610.6	679.4	909.7
May	272.4	314.5	355.7	383.4	421.9	676.5	666.9	888.7
Jun	270.2	321.2	356.5	392.0	430.7	596.2	655.5	889.5
Jul	267.5	313.3	351.0	398.1	424.5	633.8	665.3	939.8
Aug	272.4	310.3	359.8	400.5	437.9	632.6	665.4	839.0
Sep	283.4	319.2	378.9	405.3	456.0	598.2	712.7	829.9
Oct	283.1	316.6	378.9	420.5	469.9	585.8	754.6	806.6
Nov	276.2	319.2	389.9	439.4	476.7	627.8	806.3	760.9
Dec	275.9	333.4	407.6	441.7	509.8	629.8	803.2	816.1

Source: Standard & Poor's Statistics, 2009; www.kitco.com Current Statistics, 2009.

(c) Fit the time series model, part b. Identify and interpret (i) the estimates of the model parameters, (ii) the value of R^2, and (iii) the test for a quadratic long-term trend.

10.28 Retirement security income. Refer to Exercise 10.15 (p. 543) and the study on the long-term effects of the Employment Retirement Income Security Act (ERISA). Ledolter and Power also fitted quarterly time series models for the number of pension plan terminations and the number of profit-sharing plan terminations. To account for residual correlation,

they fit straight-line autoregressive models of the form, $y_t = \beta_0 + \beta_1 t + \phi R_{t-1} + \varepsilon_t$. The results were as follows:

Pension plan: $\hat{y}_t = 3.54 + .039t + .40\hat{R}_{t-1}$

Profit-sharing plan: $\hat{y}_t = 3.45 + .038t + .22\hat{R}_{t-1}$

(a) Interpret the estimates of the model parameters for pension plan terminations.
(b) Interpret the estimates of the model parameters for profit-sharing plan terminations.

10.9 Forecasting with Time Series Autoregressive Models

Often, the ultimate objective of fitting a time series model is to forecast future values of the series. We will demonstrate the techniques for the simple model

$$y_t = \beta_0 + \beta_1 x_t + R_t$$

with the first-order autoregressive residual

$$R_t = \phi R_{t-1} + \varepsilon_t$$

Suppose we use the data $(y_1, x_1), (y_2, x_2), \ldots, (y_n, x_n)$ to obtain estimates of β_0, β_1, and ϕ, using the method presented in Section 10.8. We now want to forecast the value of y_{n+1}. From the model,

$$y_{n+1} = \beta_0 + \beta_1 x_{n+1} + R_{n+1}$$

where

$$R_{n+1} = \phi R_n + \varepsilon_{n+1}$$

Combining these, we obtain

$$y_{n+1} = \beta_0 + \beta_1 x_{n+1} + \phi R_n + \varepsilon_{n+1}$$

From this equation, we obtain the forecast of y_{n+1}, denoted F_{n+1}, by estimating each of the unknown quantities and setting ε_{n+1} to its expected value of 0:*

$$F_{n+1} = \hat{\beta}_0 + \hat{\beta}_1 x_{n+1} + \hat{\phi} \hat{R}_n$$

where $\hat{\beta}_0$, $\hat{\beta}_1$, and $\hat{\phi}$ are the estimates based on the time series model–fitting approach presented in Section 10.8. The estimate $\hat{R}_n$ of the residual R_n is obtained by noting that

$$R_n = y_n - (\beta_0 + \beta_1 x_n)$$

so that

$$\hat{R}_n = y_n - (\hat{\beta}_0 + \hat{\beta}_1 x_n)$$

The two-step-ahead forecast of y_{n+2} is similarly obtained. The true value of y_{n+2} is

$$y_{n+2} = \beta_0 + \beta_1 x_{n+2} + R_{n+2}$$
$$= \beta_0 + \beta_1 x_{n+2} + \phi R_{n+1} + \varepsilon_{n+2}$$

and the forecast at $t = n + 2$ is

$$F_{n+2} = \hat{\beta}_0 + \hat{\beta}_1 x_{n+2} + \hat{\phi} \hat{R}_{n+1}$$

The residual R_{n+1} (and all future residuals) can now be obtained from the recursive relation

$$R_{n+1} = \phi R_n + \varepsilon_{n+1}$$

so that

$$\hat{R}_{n+1} = \hat{\phi} \hat{R}_n$$

Thus, the forecasting of future y-values is an iterative process, with each new forecast making use of the previous residual to obtain the estimated residual for the future time period. The general forecasting procedure using time series models with first-order autoregressive residuals is outlined in the next box.

Example 10.4

Suppose we want to forecast the sales of the company for the data in Table 10.5. In Section 10.8, we fit the regression–autoregression pair of models

$$y_t = \beta_0 + \beta_1 t + R_t \quad R_t = \phi R_{t-1} + \varepsilon_t$$

Using 35 years of sales data, we obtained the estimated models

$$\hat{y}_t = .4058 + 4.2959t + \hat{R}_t \quad \hat{R}_t = .5896 \hat{R}_{t-1}$$

Combining these, we have

$$\hat{y}_t = .4058 + 4.2959t + .5896 \hat{R}_{t-1}$$

(a) Use the fitted model to forecast sales in years $t = 36, 37$, and 38.

(b) Find approximate 95% prediction intervals for the forecasts.

---

*Note that the forecast requires the value of x_{n+1}. When x_t is itself a time series, the future value x_{n+1} will generally be unknown and must also be estimated. Often, $x_t = t$ (as in Example 10.4). In this case, the future time period (e.g., $t = n + 1$) is known and no estimate is required.

Forecasting Using Time Series Models with First-Order Autoregressive Residuals

$$y_t = \beta_0 + \beta_1 x_{1t} + \beta_2 x_{2t} + \cdots + \beta_k x_{kt} + R_t$$

$$R_t = \phi R_{t-1} + \varepsilon_t$$

Step 1. Use a statistical software package to obtain the estimated model

$$\hat{y}_t = \hat{\beta}_0 + \hat{\beta}_1 x_{1t} + \hat{\beta}_2 x_{2t} + \cdots + \hat{\beta}_k x_{kt} + \hat{R}_t, t = 1, 2, \ldots, n$$

$$\hat{R}_t = \hat{\phi} \hat{R}_{t-1}$$

Step 2. Compute the estimated residual for the last time period in the data (i.e., $t = n$) as follows:

$$\hat{R}_n = y_n - \hat{y}_n$$

$$= y_n - (\hat{\beta}_0 + \hat{\beta}_1 x_{1n} + \hat{\beta}_2 x_{2n} + \cdots + \hat{\beta}_k x_{kn})$$

Step 3. The forecast of the value y_{n+1} (i.e., the one-step-ahead forecast) is

$$F_{n+1} = \hat{\beta}_0 + \hat{\beta}_1 x_{1n+1} + \hat{\beta}_2 x_{2,n+1} + \cdots + \hat{\beta}_k x_{k,n+1} + \hat{\phi} \hat{R}_n$$

where $\hat{R}_n$ is obtained from step 2.

Step 4. The forecast of the value y_{n+2} (i.e., the two-step-ahead forecast) is

$$F_{n+2} = \hat{\beta}_0 + \hat{\beta}_1 x_{1,n+2} + \hat{\beta}_2 x_{2,n+2} + \cdots + \hat{\beta}_k x_{k,n+2} + (\hat{\phi})^2 \hat{R}_n$$

where $\hat{R}_{n+1}$ is obtained from step 3.
In general, the m-step-ahead forecast is

$$F_{n+m} = \hat{\beta}_0 + \hat{\beta}_1 x_{1,n+m} + \hat{\beta}_2 x_{2,n+m} + \cdots + \hat{\beta}_k x_{k,n+m} + (\hat{\phi})^m \hat{R}_n$$

Solution

(a) The forecast for the 36th year requires an estimate of the last residual R_{35},

$$\hat{R}_{35} = y_{35} - [\hat{\beta}_0 + \hat{\beta}_1(35)]$$

$$= 150.9 - [.4058 + 4.2959(35)]$$

$$= .1377$$

Then the one-step-ahead forecast (i.e., the sales forecast for year 36) is

$$F_{36} = \hat{\beta}_0 + \hat{\beta}_1(36) + \hat{\phi}\hat{R}_{35}$$

$$= .4058 + 4.2959(36) + (.5896)(.1377)$$

$$= 155.14$$

Using the formula in the box, the two-step-ahead forecast (i.e., the sales forecast for year 37) is

$$F_{37} = \hat{\beta}_0 + \hat{\beta}_1(37) + (\hat{\phi})^2 \hat{R}_{35}$$

$$= .4058 + 4.2959(37) + (.5896)^2(.1377)$$

$$= 159.40$$

Similarly, the three-step-ahead forecast (i.e., the sales forecast for year 38) is

$$F_{38} = \hat{\beta}_0 + \hat{\beta}_1(38) + (\hat{\phi})^3 \hat{R}_{35}$$

$$= .4058 + 4.2959(38) + (.5896)^3(.1377)$$

$$= 163.68$$

Some statistical software packages (e.g., SAS) have options for computing forecasts using the autoregressive model. The three forecasted values, F_{36}, F_{37}, and F_{38}, are shown (shaded) at the bottom of the SAS printout, Figure 10.17, in the FORECAST column.

Figure 10.17 SAS printout of forecasts of annual sales revenue using straight-line model with autoregressive errors

T	SALES	FORECAST	LCL95	UCL95
1	4.8	4.702	-10.644	12.515
2	4.0	9.056	-3.980	16.489
3	5.5	10.347	-2.509	20.469
4	15.6	12.994	0.308	24.456
5	23.1	20.712	8.186	28.451
6	23.3	26.897	14.521	32.456
7	31.4	28.778	16.542	36.472
8	46.0	35.317	23.210	40.500
9	46.1	45.689	33.699	44.542
10	41.9	47.511	35.627	48.601
11	45.5	46.797	35.009	52.678
12	53.5	50.683	38.977	56.776
13	48.4	57.163	45.527	60.898
14	61.6	55.919	44.341	65.047
15	65.6	65.465	53.933	69.225
16	71.4	69.586	58.086	73.435
17	83.4	74.769	63.289	77.678
18	93.6	83.607	72.134	81.956
19	94.2	91.384	79.904	86.270
20	85.4	93.501	82.001	90.619
21	86.2	90.075	78.543	95.001
22	89.9	92.310	80.733	99.414
23	89.2	96.254	84.619	103.858
24	99.1	97.605	85.899	108.327
25	100.3	105.205	93.416	112.821
26	111.7	107.675	95.792	117.335
27	108.2	116.160	104.170	121.869
28	115.5	115.859	103.752	126.418
29	119.2	121.927	109.690	130.982
30	125.2	125.871	113.495	135.558
31	136.3	131.172	118.645	140.145
32	146.8	139.480	126.793	144.742
33	146.1	147.434	134.577	149.347
34	151.4	148.784	135.748	153.959
35	150.9	153.672	140.449	158.577
36	.	155.140	141.720	163.201
37	.	159.403	144.397	167.830
38	.	163.679	148.034	172.463

We can proceed in this manner to generate sales forecasts as far into the future as desired. However, the potential for error increases as the distance into the future increases. Forecast errors are traceable to three primary causes, as follows:

1. The form of the model may change at some future time. This is an especially difficult source of error to quantify, since we will not usually know when or if the model changes, or the extent of the change. The possibility of a change in the model structure is the primary reason we have consistently urged you to avoid predictions outside the observed range of the independent variables. However, time series forecasting leaves us little choice—by definition, the forecast will be a prediction at a future time.

2. A second source of forecast error is the uncorrelated residual, ε_t, with variance σ^2. For a first-order autoregressive residual, the forecast variance of the one-step-ahead prediction is σ^2, whereas that for the two-step-ahead prediction is $\sigma^2(1 + \phi^2)$, and, in general, for m steps ahead, the forecast variance[†] is $\sigma^2(1 + \phi^2 + \phi^4 + \cdots + \phi^{2(m-1)})$. Thus, the forecast variance increases as the

---

[†] See Fuller (1996).

distance is increased. These variances allow us to form approximate 95% prediction intervals for the forecasts (see the next box).

3. A third source of variability is that attributable to the error of estimating the model parameters. This is generally of less consequence than the others, and is usually ignored in forming prediction intervals.

(a) To obtain a prediction interval, we first estimate σ^2 by the MSE, the mean square for error from the time series regression analysis. For the sales data, we form an *approximate* 95% prediction interval for the sales in year 36:

$$F_{36} \pm 2\sqrt{\text{MSE}}$$

$$155.1 \pm 2\sqrt{27.42767}$$

$$155.1 \pm 10.5$$

or (144.6, 165.6). Thus, we forecast that the sales in year 36 will be between $145,000 and $165,000.

The approximate 95% prediction interval for year 37 is

$$F_{37} \pm 2\sqrt{\text{MSE}(1 + \phi^2)}$$

$$159.4 \pm 2\sqrt{27.42767[1 + (.5896)^2]}$$

$$159.4 \pm 12.2$$

or (147.2, 171.6). Note that this interval is wider than that for the one-step-ahead forecast. The intervals will continue to widen as we attempt to forecast farther ahead.

The formulas for computing *exact* 95% prediction intervals using the time series autoregressive model are complex and beyond the scope of this text. However, we can use statistical software to obtain them. The exact 95% prediction intervals for years 36–38 are shown at the bottom of the SAS printout, Figure 10.17, in the **LCL95** and **UCL95** columns. Note that the exact prediction intervals are wider than the approximate intervals. We again stress that the accuracy of these forecasts and intervals depends on the assumption that the model structure does not change during the forecasting period. If, for example, the company merges with another company during year 37, the structure of the sales model will almost surely change, and therefore, prediction intervals past year 37 are probably useless. ▄

Approximate 95% Forecasting Limits Using Time Series Models with First-Order Autoregressive Residuals

One-Step-Ahead Forecast:

$$\hat{y}_{n+1} \pm 2\sqrt{\text{MSE}}$$

Two-Step-Ahead Forecast:

$$\hat{y}_{n+2} \pm 2\sqrt{\text{MSE}(1 + \hat{\phi}^2)}$$

Three-Step-Ahead Forecast:

$$\hat{y}_{n+3} \pm 2\sqrt{\text{MSE}(1 + \hat{\phi}^2 + \hat{\phi}^4)}$$

$$\vdots$$

m-Step-Ahead Forecast:

$$\hat{y}_{n+m} \pm 2\sqrt{\text{MSE}(1 + \hat{\phi}^2 + \hat{\phi}^4 + \cdots + \hat{\phi}^{2(m-1)})}$$

[*Note*: MSE estimates σ^2, the variance of the uncorrelated residual ε_t.]

It is important to note that the forecasting procedure makes explicit use of the residual autocorrelation. The result is a better forecast than would be obtained using the standard least squares procedure in Chapter 4 (which ignores residual correlation). Generally, this is reflected by narrower prediction intervals for the time series forecasts than for the least squares prediction.[‡] The end result, then, of using a time series model when autocorrelation is present is that you obtain more reliable estimates of the β coefficients, smaller residual variance, and more accurate prediction intervals for future values of the time series.

10.9 Exercises

10.29 Calculating forecasts. The quarterly time series model $y_t = \beta_0 + \beta_1 t + \beta_2 t^2 + \phi R_{t-1} + \varepsilon_t$ was fit to data collected for $n = 48$ quarters, with the following results:

$$\hat{y}_t = 220 + 17t - .3t^2 + .82\hat{R}_{t-1}$$

$$y_{48} = 350 \quad \text{MSE} = 10.5$$

(a) Calculate forecasts for y_t for $t = 49$, $t = 50$, and $t = 51$.
(b) Construct approximate 95% prediction intervals for the forecasts obtained in part a.

10.30 Calculating forecasts. The annual time series model $y_t = \beta_0 + \beta_1 t + \phi R_{t-1} + \varepsilon_t$ was fit to data collected for $n = 30$ years with the following results:

$$\hat{y}_t = 10 + 2.5t + .64\hat{R}_{t-1}$$

$$y_{30} = 82 \quad \text{MSE} = 4.3$$

(a) Calculate forecasts for y_t for $t = 31$, $t = 32$, and $t = 33$.
(b) Construct approximate 95% prediction intervals for the forecasts obtained in part a.

10.31 Quarterly GDP values. Use the fitted time series model in Exercise 10.25 (p. 557) to forecast GDP for the four quarters of 2009 and calculate approximate 95% forecast limits. Go to your library (or search the Internet) to find the actual GDP values for 2009. Do the forecast intervals contain the actual 2009 GDP values?

10.32 Mortgage interest rates. Use the fitted time series model in Exercise 10.26 (p. 558) to forecast average mortgage interest rate for 2008. Place approximate 95% confidence bounds on the forecast.

10.33 Monthly gold prices. Use the fitted time series model in Exercise 10.27 (p. 558) to forecast gold price in January and February of 2009. Calculate approximate 95% prediction intervals for the

forecasts. Do these intervals contain the actual gold prices? (Go to your library or search the Internet to find the actual gold prices in January and February of 2009.) If not, give a plausible explanation.

10.34 Yearly price of gold. Refer to the gold price series, 1990–2008, Exercise 10.7 (p. 536). The data is saved in the GOLDYR file.

(a) Hypothesize a deterministic model for $E(y_t)$ based on a plot of the time series.
(b) Do you expect the random error term of the model, part a, to be uncorrected? Explain.
(c) Hypothesize a model for the correlated error term, R_t.
(d) Combine the two models, parts a and c, to form a time series forecasting model.
(e) Fit the time series model, part d, to the data.
(f) Use the fitted time series model from part e to forecast the gold price in 2009. Place an approximate 95% prediction interval around the forecast.

10.35 Retirement security income. Refer to Exercise 10.28 (p. 559). The values of MSE for the quarterly time series models of retirement plan terminations are as follows:

Pension plan termination: MSE $= .0440$

Profit-sharing plan termination: MSE $= .0402$

(a) Forecast the number of pension plan terminations for quarter 108. Assume that $y_{107} = 7.5$. [*Hint:* Remember that the forecasted number of pension plan terminations is $e^{\hat{y}_{108}}$.]
(b) Place approximate 95% confidence bounds on the forecast obtained in part a. [*Hint:* First, calculate upper and lower confidence limits for y_{108}, then take antilogarithms.]
(c) Repeat parts a and b for the number of profit-sharing plan terminations in quarter 108. Assume that $y_{107} = 7.6$.

[‡] When n is large, approximate 95% prediction intervals obtained from the standard least squares procedure reduce to $\hat{y}_t \pm 2\sqrt{\text{MSE}}$ for *all* future values of the time series. These intervals may actually be narrower than the more accurate prediction intervals produced from the time series analysis.

10.10 Seasonal Time Series Models: An Example

We have used a simple regression model to illustrate the methods of model estimation and forecasting when the residuals are autocorrelated. In this section, we present a more realistic example that requires a **seasonal model** for $E(y_t)$, as well as an autoregressive model for the residual.

Critical water shortages have dire consequences for both business and private sectors of communities. Forecasting water usage for months in advance is essential to avoid such shortages. Suppose a community has monthly water usage records over the past 15 years. A plot of the last 6 years of the time series, y_t, is shown in Figure 10.18. Note that both an increasing trend and a seasonal pattern appear prominent in the data. The water usage seems to peak during the summer months and decline during the winter months. Thus, we might propose the following model:

$$E(y_t) = \beta_0 + \beta_1 t + \beta_2 \left(\cos \frac{2\pi}{12} t \right) + \beta_3 \left(\sin \frac{2\pi}{12} t \right)$$

Since the amplitude of the seasonal effect (i.e., the magnitude of the peaks and valleys) appears to increase with time, we include in the model an interaction between time and trigonometric components, to obtain

$$E(y_t) = \beta_0 + \beta_1 t + \beta_2 \left(\cos \frac{2\pi}{12} t \right) + \beta_3 \left(\sin \frac{2\pi}{12} t \right)$$
$$+ \beta_4 t \left(\cos \frac{2\pi}{12} t \right) + \beta_5 t \left(\sin \frac{2\pi}{12} t \right)$$

Figure 10.18 Water usage time series

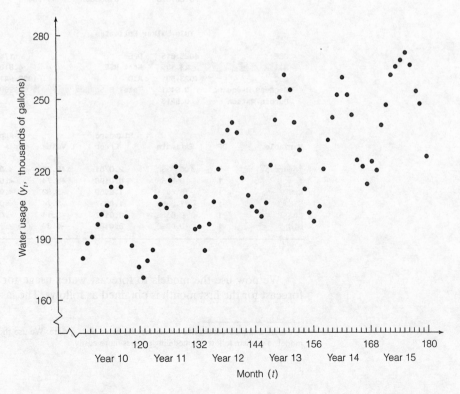

The model for the random component R_t must allow for short-term cyclic effects. For example, in an especially hot summer, if the water usage, y_t, exceeds the expected usage, $E(y_t)$, for July, we would expect the same thing to happen in August. Thus, we propose a first-order autoregressive model[†] for the random component:

$$R_t = \phi R_{t-1} + \varepsilon_t$$

We now fit the models to the time series y_t, where y_t is expressed in thousands of gallons. The SAS printout is shown in Figure 10.19. The estimated models are given by

$$\hat{y}_t = 100.0832 + .8263t - 10.8011 \left(\cos \frac{2\pi}{12}t \right) - 7.0858 \left(\sin \frac{2\pi}{12}t \right)$$

$$- .0556t \left(\cos \frac{2\pi}{12}t \right) - .0296t \left(\sin \frac{2\pi}{12}t \right) + \hat{R}_t$$

$$\hat{R}_t = .6617 \hat{R}_{t-1}$$

with MSE = 23.135. The R^2 value of .99 indicates that the models provide a good fit to the data.

Figure 10.19 SAS printout for time series model of water usage

The AUTOREG Procedure

Dependent Variable   USAGE

Estimates of Autoregressive Parameters

Lag	Coefficient	Standard Error	t Value
1	-0.661679	0.055885	-11.84

Yule-Walker Estimates

SSE	4025.513	DFE	174
MSE	23.135	Root MSE	4.810
SBC	1023.591	AIC	1002.941
Regress R-Square	0.9431	Total R-Square	0.9900
Durbin-Watson	0.5216		

Variable	DF	Estimate	Standard Error	t Value	Approx Pr > |t|
Intercept	1	100.0832	2.0761	48.21	<.0001
T	1	0.8263	0.0198	41.74	<.0001
COS	1	-10.8011	1.8559	-5.82	<.0001
SIN	1	-7.0858	1.8957	-3.74	.0003
COS_T	1	-0.0556	0.0177	-3.14	.0020
SIN_T	1	-0.0296	0.0182	-1.63	.1049

We now use the models to forecast water usage for the next 12 months. The forecast for the first month is obtained as follows. The last residual value (obtained

---

[†] A more complex time series model may be more appropriate. We use the simple first-order autoregressive model so you can follow the modeling process more easily.

from a portion of the printout not shown) is $\hat{R}_{180} = -1.3247$. Then the formula for the one-step-ahead forecast is

$$F_{181} = \hat{\beta}_0 + \hat{\beta}_1(181) + \hat{\beta}_2\left(\cos\frac{2\pi}{12}181\right) + \hat{\beta}_3\left(\sin\frac{2\pi}{12}181\right)$$

$$+ \hat{\beta}_4(181)\left(\cos\frac{2\pi}{12}181\right) + \hat{\beta}_5(181)\left(\sin\frac{2\pi}{12}181\right) + \hat{\phi}\,\hat{R}_{180}$$

Substituting the values of $\hat{\beta}_0, \hat{\beta}_1, \ldots, \hat{\beta}_5$, and $\hat{\phi}$ shown in Figure 10.20, we obtain $F_{181} = 238.0$. Approximate 95% prediction bounds on this forecast are given by $\pm 2\sqrt{\text{MSE}} = \pm 2\sqrt{23.135} = \pm 9.6$.[‡] That is, we expect our forecast for 1 month ahead to be within 9,600 gallons of the actual water usage. This forecasting process is then repeated for the next 11 months. The forecasts and their bounds are shown in Figure 10.20. Also shown are the actual values of water usage during year 16. Note that the forecast prediction intervals widen as we attempt to forecast farther into the future. This property of the prediction intervals makes long-term forecasts very unreliable.

The variety and complexity of time series modeling techniques are overwhelming. We have barely scratched the surface here. However, if we have convinced you that time series modeling is a useful and powerful tool for business forecasting, we have accomplished our purpose. The successful construction of time series models requires much experience, and entire texts are devoted to the subject (see the references at the end of the chapter).

We conclude with a warning: Many oversimplified forecasting methods have been proposed. They usually consist of graphical extensions of a trend or seasonal pattern to future time periods. Although such pictorial techniques are easy to understand and therefore are intuitively appealing, they should be avoided. There

Figure 10.20 Forecasts of water usage

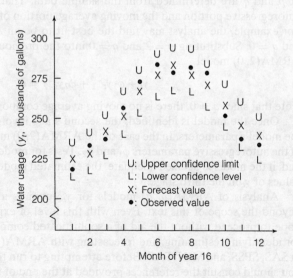

[‡] We are ignoring the errors in the parameter estimates in calculating the forecast reliability. These errors should be small for a series of this length.

is no measure of reliability for these forecasts, and thus the risk associated with making decisions based on them is very high.

10.11 Forecasting Using Lagged Values of the Dependent Variable (Optional)

In Section 10.7, we discussed a variety of choices for the deterministic component, $E(y_t)$, of the time series models. All these models were functions of independent variables, such as t, x_t, x_{t-1}, and seasonal dummy variables. Often, the forecast of y_t can be improved by adding *lagged values of the dependent variable* to the model. For example, since the price y_t of a stock on day t is highly correlated with the price on the previous day (i.e., on day $t - 1$), a useful model for $E(y_t)$ is

$$E(y_t) = \beta_0 + \beta_1 y_{t-1}$$

Models with lagged values of y_t tend to violate the standard regression assumptions outlined in Section 4.2; thus, they must be fitted using specialized methods.

Box and Jenkins (1977) developed a method of analyzing time series models based on past values of y_t and past values of the random error ε_t. The general model, denoted **ARMA(p, q)**, takes the form

$$y_t + \phi_1 y_{t-1} + \phi_2 y_{t-2} + \cdots + \phi_p y_{t-p} = \varepsilon_t + \theta_1 \varepsilon_{t-1} + \theta_2 \varepsilon_{t-2} + \cdots + \theta_q \varepsilon_{t-q}$$

Note that the left side of the equation is a **pth-order autoregressive model** for y_t (see Section 10.5), whereas the right side of the equation is a **qth-order moving average model** for the random error ε_t (see Section 10.6).

The analysis of an ARMA(p, q) model is divided into three stages: (1) identification, (2) estimation, and (3) forecasting. In the identification stage, the values of p and q are determined from the sample data. That is, the order of both the autoregressive portion and the moving average portion of the model are identified.[*] For example, the analyst may find the best fit to be an ARMA model with $p = 2$ and $q = 0$. Substituting $p = 2$ and $q = 0$ into the previous equation, we obtain the ARMA(2, 0) model

$$y_t + \phi_1 y_{t-1} + \phi_2 y_{t-2} = \varepsilon_t$$

Note that since $q = 0$, there is no moving average component to the model.

Once the model is identified, the second stage involves obtaining estimates of the model's parameters. In the case of the ARMA(2, 0) model, we require estimates of the autoregressive parameters ϕ_1 and ϕ_2. Tests for model adequacy are conducted, and, if the model is deemed adequate, the estimated model is used to forecast future values of y_t in the third stage.

Analysis of ARMA(p, q) models for y_t requires a level of expertise that is beyond the scope of this text. Even with this level of expertise, the analyst cannot hope to proceed without the aid of a sophisticated computer program. Procedures for identifying, estimating, and forecasting with ARMA(p, q) models are available in SAS, SPSS, and MINITAB. Before attempting to run these procedures, however, you should consult the references provided at the end of this chapter.

---

[*] This step involves a careful examination of a plot of the sample autocorrelations. Certain patterns in the plot (such as those shown in Figures 10.9–10.12) allow the analyst to identify p and q.

Quick Summary

KEY FORMULAS

Time series model

$$y_t = T_t + C_t + S_t + R_t$$

Exponential smoothing

$$E_t = wY_t + (1 - w)E_{t-1}$$

Forecast: $F_{n+k} = E_n$

Holt–Winter's method

$$E_t = wY_t + (1 - w)(E_{t-1} + T_{t-1}),$$
$$T_t = v(E_t - E_{t-1}) + (1 - v)T_{t-1}$$

Forecast: $F_{n+k} = E_n + T_k$

Moving average

$M_t = L_t/N$, where

$$L_1 = \begin{cases} y_{t-(N-1)/2} + \cdots + y_t + \cdots + y_{t+(N-1)/2} & \text{if } N \text{ is odd} \\ y_{t-N/2} + \cdots + y_t + \cdots + y_{t+N/2-1} & \text{if } N \text{ is even} \end{cases}$$

$$\text{Forecast: } F_{n+k} = \begin{cases} M_n & \text{if no seasonal} \\ & \text{variation} \\ M_n(\text{Seasonal} & \text{if seasonal variation} \\ \text{index}/100) \end{cases}$$

Mean absolute deviation

$$\text{MAD} = \sum_{t=1}^{t=m} |y_t - F_t|/m$$

Mean absolute percentage error

$$\text{MAPE} = \frac{\sum_{t=1}^{t=m} \left| \dfrac{y_t - F_t}{y_t} \right|}{m} \times 100$$

Root mean squared error

$$\text{RMSE} = \sqrt{\sum_{t=1}^{t=m} \frac{(y_t - F_t)^2}{m}}$$

AR(p) error model

$$R_t = \phi_1 R_{t-1} + \phi_2 R_{t-2} + \cdots + \phi_p R_{t-p} + \varepsilon_t$$

MA(q) error model

$$R_t = \varepsilon_t + \theta_1 \varepsilon_{t-1} + \theta_2 \varepsilon_{t-2} + \cdots + \theta_q \varepsilon_{t-q}$$

95% forecast limits using AR(1) error model

$$\widehat{y_{n+m}} \pm 2\sqrt{\text{MSE}(1 + \phi^2 + \phi^4 \cdots + \phi^{2(m-1)})}$$

KEY SYMBOLS

y_t	time series value at time t
T_t	secular (long-term) trend at time t
C_t	cyclical fluctuation (business cycle) at time t
S_t	seasonal variation at time t
R_t	residual effect at time t
M_t	moving average value at time t
N	number of points (time series values) summed in moving average
E_t	exponentially smoothed value at time t
MAD	mean absolute deviation of forecast errors
MAPE	mean absolute percentage forecast error
RMSE	root mean squared forecast error
AC	autocorrelation between two time series residuals

KEY IDEAS

Time series components

1. *Secular* (long-term) *trend*
2. *Cyclical effect* (business cycle)
3. *Seasonal variation*
4. *Residual effect*

Time series forecasting methods

1. Descriptive: *Moving average method* (accounts for secular and seasonal trends)
   Exponential smoothing (apply if little or no secular or seasonal trends)
   Holt–Winter's method (accounts for short-term and seasonal trends)
2. Inferential: *Least squares regression* (allows for 95% prediction intervals)
   Time series models with autocorrelated error component

Measures of forecast accuracy

1. MAD
2. MAPE
3. RMSE

Problems with least squares regression forecasting

1. Prediction outside the experimental region for time

2. Errors are autocorrelated

Autocorrelation

Correlation between time series errors at different points in time.

Supplementary Exercises

10.36. Mortgage interest rates. Refer to Exercise 10.8 (p. 541). The data on the annual average mortgage interest rates for conventional, fixed-rate, 30-year loans for the period 1985–2007 are reproduced below. Forecast the 2010 average mortgage interest rate using each of the methods listed here. Compare the results to the forecast obtained in Exercise 10.8b.

INTRATE30

YEAR	INTEREST RATE (%)	YEAR	INTEREST RATE (%)
1985	11.85	1997	7.57
1986	11.33	1998	6.92
1987	10.46	1999	7.46
1988	10.86	2000	8.08
1989	12.07	2001	7.01
1990	9.97	2002	6.56
1991	11.14	2003	5.89
1992	8.27	2004	5.86
1993	7.17	2005	5.93
1994	8.28	2006	6.47
1995	7.86	2007	6.40
1996	7.76		

Source: U.S. Bureau of the Census. *Statistical Abstract of the United States,* 2009.

(a) A 3-point moving average
(b) The exponential smoothing technique $(w = .2)$
(c) The Holt–Winters model with trend $(w = .2$ and $v = .5)$
(d) A straight-line, first-order autoregressive model (Obtain an approximate 95% prediction interval.)

10.37. Prices of high-technology stocks. Consider the 2008 monthly closing prices (i.e., closing prices on the last day of each month) given in the next table for three high-technology company stocks listed on the New York Stock Exchange—IBM, Intel, and Microsoft.

(a) Calculate and plot a 3-point moving average for the IBM stock price time series. Can you detect the secular trend? Does there appear to be a seasonal pattern?
(b) Use the moving average from part a to forecast the IBM stock price in January 2009.

HITECH

YEAR	MONTH	TIME	IBM	INTEL	MICROSOFT
2008	Jan	1	107.11	21.10	32.60
	Feb	2	113.86	19.97	27.20
	Mar	3	115.14	21.18	28.38
	Apr	4	120.70	22.26	28.52
	May	5	129.43	23.18	28.32
	Jun	6	118.53	21.48	27.51
	Jul	7	127.98	22.19	25.72
	Aug	8	121.73	22.87	27.29
	Sep	9	116.96	18.73	26.69
	Oct	10	92.97	16.03	22.33
	Nov	11	81.60	13.80	20.22
	Dec	12	84.16	14.66	19.44

Source: Standard & Poor's *NYSE Daily Stock Price Record,* 2008; http://moneycentral.msn.com.

(c) For the IBM stock prices, calculate and plot the exponentially smoothed series using $w = .6$.
(d) Obtain the forecast for IBM stock price in January 2009 using the exponential smoothing technique.
(e) Obtain the forecast for IBM stock price in January 2009 using the Holt–Winters model with trend and seasonal components and smoothing constants $w = .6$, $v = .7$, and $u = .5$.
(f) Propose a time series model for monthly IBM stock price that accounts for secular trend, seasonal variation (if any), and residual autocorrelation.
(g) Fit the time series model specified in part f using an available software package.
(h) Use the time series model to forecast IBM stock price in January 2009. Obtain an approximate 95% prediction interval for the forecast.
(i) Repeat parts a–h for Intel stock prices.
(j) Repeat parts a–h for Microsoft stock prices.

ROOMOCC

10.38. Hotel room occupancy rates. Refer to the data on average monthly occupancies of hotels and motels in the cities of Atlanta and Phoenix, Exercise 10.14 (p. 543). In part a of Exercise 10.14, you hypothesized a model for mean occupancy (y_r) that accounted for seasonal variation in the series.

(a) Modify the model of part a in Exercise 10.14 to account for first-order residual correlation.

(b) Fit the model in part a to the data for each city. Interpret the results.

(c) Would you recommend using the model to forecast monthly occupancy rates in year 3? Explain.

10.39. U.S. beer production. The accompanying table shows U.S. beer production for the years 1980–2007. Suppose you are interested in forecasting U.S. beer production in 2010.

(a) Construct a time series plot for the data. Do you detect a long-term trend?

(b) Hypothesize a model for y_t that incorporates the trend.

(c) Fit the model to the data using the method of least squares.

(d) Plot the least squares model from part a and extend the curve to forecast y_{31}, the U.S. beer production (in millions of barrels) in 2010. How reliable do you think this forecast is?

⊙ USBEER

YEAR	t	BEER	YEAR	t	BEER	YEAR	t	BEER
1980	1	188	1990	11	204	2000	21	199
1981	2	194	1991	12	203	2001	22	199
1982	3	194	1992	13	202	2002	23	200
1983	4	195	1993	14	203	2003	24	195
1984	5	193	1994	15	202	2004	25	198
1985	6	193	1995	16	199	2005	26	197
1986	7	195	1996	17	201	2006	27	198
1987	8	195	1997	18	199	2007	28	199
1988	9	198	1998	19	198			
1989	10	200	1999	20	198			

Source: U.S. Beer Institute; *2008 Brewer's Almanac.*

(e) Calculate and plot the residuals for the model obtained in part a. Is there visual evidence of residual autocorrelation?

(f) How could you test to determine whether residual autocorrelation exists? If you have access to a computer package, carry out the test. Use $\alpha = .05$.

(g) Hypothesize a time series model that will account for the residual autocorrelation. Fit the model to the data and interpret the results.

(h) Compute a 95% prediction interval for y_{31}, the U.S. beer production in 2010. Why is this forecast preferred to that in part b?

10.40. Testing for autocorrelation. Suppose you were to fit the time series model

$$E(y_t) = \beta_0 + \beta_1 t + \beta_2 t^2$$

to quarterly time series data collected over a 10-year period ($n = 40$ quarters).

(a) Set up the test of hypothesis for positively autocorrelated residuals. Specify H_0, H_a, the test statistic, and the rejection region. Use $\alpha = .05$.

(b) Suppose the Durbin–Watson d statistic is calculated to be 1.14. What is the appropriate conclusion?

10.41. Modeling a firm's monthly income. Suppose a CPA firm wants to model its monthly income, y_t. The firm is growing at an increasing rate, so that the mean income will be modeled as a second-order function of t. In addition, the mean monthly income increases significantly each year from January through April because of processing tax returns.

(a) Write a model for $E(y_t)$ to reflect both the second-order function of time, t, and the January–April jump in mean income.

(b) Suppose the size of the January–April jump grows each year. How could this information be included in the model? Assume that 5 years of monthly data are available.

⊙ OPECOIL

10.42. OPEC crude oil imports. Refer to the data on annual OPEC oil imports, Exercise 10.4 (p. 535).

(a) Plot the time series.

(b) Hypothesize a straight-line autoregressive time series model for annual amount of imported crude oil, y_t.

(c) Fit the proposed model to the data. Interpret the results.

(d) From the output, write the modified least squares prediction equation for y_t.

(e) Forecast the amount of foreign crude oil imported into the United States from OPEC in 2010. Place approximate 95% prediction bounds on the forecast value.

10.43. London stock rates. An analysis of seasonality in returns of stock traded on the London Stock Exchange was published in the *Journal of Business* (Vol. 60, 1987). One of the objectives was to determine whether the introduction of a capital gains tax in 1965 affected rates of return. The following model was fitted to data collected over the years 1956–1980:

$$y_t = \beta_0 + \beta_1 D_1 + \varepsilon_t$$

where y_t is the difference between the April rates of return of the two stocks on the exchange with the largest and smallest returns in year t, and D_t is a dummy variable that takes on the value 1 in the posttax period (1966–1980) and the value 0 in the pretax period (1956–1965).

(a) Interpret the value of β_1.

(b) Interpret the value of β_0.

(c) The least squares prediction equation was found to be $\hat{y}_t = -.55 + 3.08D_t$. Use the equation to estimate the mean difference in April rates of returns of the two stocks during the pretax period.

(d) Repeat part c for the posttax period.

References

Abraham, B., and Ledholter, J. *Statistical Methods for Forecasting*. New York: Wiley, 1983 (paperback, 2005).

Anderson, T. W. *The Statistical Analysis of Time Series*. New York: Wiley, 1971 (paperback, 1994).

Ansley, C. F., Kohn, R., and Shively, T. S. "Computing p-values for the generalized Durbin–Watson and other invariant test statistics." *Journal of Econometrics*, Vol. 54, 1992.

Baillie, R. T., and Bollerslev, T. "Prediction in dynamic models with time-dependent conditional variances." *Journal of Econometrics*, Vol. 52, 1992.

Box, G. E. P., Jenkins, G. M., and Reinsel, G. C. *Time Series Analysis: Forecasting and Control*, 4th ed. New York: Wiley, 2008.

Chipman, J. S. "Efficiency of least squares estimation of linear trend when residuals are autocorrelated." *Econometrica*, Vol. 47, 1979.

Cochrane, D., and Orcutt, G. H. "Application of least squares regression to relationships containing autocorrelated error terms." *Journal of the American Statistical Association*, Vol. 44, 1949, 32–61.

Durbin, J., and Watson, G. S. "Testing for serial correlation in least squares regression, I." *Biometrika*, Vol. 37, 1950, 409–428.

Durbin, J., and Watson, G. S. "Testing for serial correlation in least squares regression, II." *Biometrika*, Vol. 38, 1951, 159–178.

Durbin, J., and Watson, G. S. "Testing for serial correlation in least squares regression, III." *Biometrika*, Vol. 58, 1971, 1–19.

Engle, R. F., Lilien, D. M., and Robins, R. P., "Estimating time varying risk in the term structure: The ARCH-M model." *Econometrica*, Vol. 55, 1987.

Evans, M. *Practical Business Forecasting*. New York: Wiley-Blackwell, 2002.

Fuller, W. A. *Introduction to Statistical Time Series*, 2nd ed. New York: Wiley, 1996.

Gallant, A. R., and Goebel, J. J. "Nonlinear regression with autoregressive errors." *Journal of the American Statistical Association*, Vol. 71, 1976.

Godfrey, L. G. "Testing against general autoregressive and moving average error models when the regressors include lagged dependent variables." *Econometrica*, Vol. 46, 1978, 1293–1301.

Granger, C. W. J., and Newbold, P. *Forecasting Economic Time Series*, 2nd ed. New York: Academic Press, 1986.

Greene, W. H. *Econometric Analysis*, 6th ed. Upper Saddle River, N.J.: Prentice Hall, 2008.

Hamilton, J. D. *Time Series Analysis*. Princeton, NJ.: Princeton University Press, 1994.

Harvey, A. *Time Series Models*, 2nd ed. Cambridgemass.: MIT Press, 1993.

Johnston, J. *Econometric Methods*, 2nd ed. New York: McGraw-Hill, Inc., 1972.

Jones, R. H. "Maximum likelihood fitting of ARMA models to time series with missing observations." *Technometrics*, Vol. 22, 1980.

Judge, G. G., Griffiths, W. E., Hill, R. C., and Lee, T. C. *The Theory and Practice of Econometrics*, 2nd ed. New York: Wiley, 1985.

Maddala, G. S. *Introduction to Econometrics*, 3rd ed. New York: Wiley, 2001.

Makridakis, S., et al. *The Forecasting Accuracy of Major Time Series Methods*. New York: Wiley, 1984.

McLeod, A. I., and Li, W. K. "Diagnostic checking ARMA time series models using squared-residual autocorrelations." *Journal of Time Series Analysis*, Vol. 4, 1983.

Nelson, C. R. *Applied Time Series Analysis for Managerial Forecasting*. San Francisco: Holden-Day, 1990.

Nelson, D. B. "Stationarity and persistence in the GARCH(1,1) model." *Econometric Theory*, Vol. 6, 1990.

Nelson, D. B., and Cao, C. Q. "Inequality constraints in the univariate GARCH model." *Journal of Business & Economic Statistics*, Vol. 10, 1992.

Park, R. E., and Mitchell, B. M. "Estimating the autocorrelated error model with trended data." *Journal of Econometrics*, Vol. 13, 1980.

Shively, T. S. "Fast evaluation of the distribution of the Durbin–Watson and other invariant test statistics in time series regression. *Journal of the American Statistical Association*, Vol. 85, 1990.

Theil, H. *Principles of Econometrics*. New York: Wiley, 1971.

White, K. J. "The Durbin–Watson test for autocorrelation in nonlinear models." *Review of Economics and Statistics*, Vol. 74, 1992.

Willis, R. E. *A Guide to Forecasting for Planners*. Englewood Cliffs, NJ.: Prentice Hall, 1987.

MODELING DAILY PEAK ELECTRICITY DEMANDS

The Problem

To operate effectively, power companies must be able to predict daily peak demand for electricity. *Demand* (or *load*) is defined as the rate (measured in megawatts) at which electric energy is delivered to customers. Since demand is normally recorded on an hourly basis, daily peak demand refers to the maximum hourly demand in a 24-hour period. Power companies are continually developing and refining statistical models of daily peak demand.

Models of daily peak demand serve a twofold purpose. First, the models provide short-term *forecasts* that will assist in the economic planning and dispatching of electric energy. Second, models that relate peak demand to one or more weather variables provide estimates of historical peak demands under a set of alternative weather conditions. That is, since changing weather conditions represent the primary source of variation in peak demand, the model can be used to answer the often-asked question, "What would the peak daily demand have been had normal weather prevailed?" This second application, commonly referred to as *weather normalization*, is mainly an exercise in *backcasting* (i.e., adjusting historical data) rather than forecasting (Jacob, 1985).

Since peak demand is recorded over time (days), the dependent variable is a time series and one approach to modeling daily peak demand is to use a time series model. This case study presents key results of a study designed to compare several alternative methods of modeling daily peak demand for the Florida Power Corporation (FPC). For this case study, we focus on two time series models and a multiple regression model proposed in the original FPC study. Then we demonstrate how to forecast daily peak demand using one of the time series models. (We leave the problem of backcasting as an exercise.)

The Data

The data for the study consist of daily observations on peak demand recorded by the FPC for one year beginning November 1 and ending October 31 of the next calendar year, and several factors that are known to influence demand.* It is typically assumed that demand consists of two components, (1) a non–weather-sensitive "base" demand that is not influenced by temperature changes and (2) a weather-sensitive demand that is highly responsive to changes in temperature.

---

*For reasons of confidentiality, the data cannot be made available for this text.

The principal factor that affects the usage of non–weather-sensitive appliances (such as refrigerators, generators, light, and computers) is the *day of the week*. Typically, Saturdays have lower peak demands than weekdays due to decreased commercial and industrial activity, whereas Sundays and holidays exhibit even lower peak demand levels as commercial and industrial activity declines even further.

The single most important factor affecting the usage of weather-sensitive appliances (such as heaters and air conditioners) is *temperature*. During the winter months, as temperatures drop below comfortable levels, customers begin to operate their electric heating units, thereby increasing the level of demand placed on the system. Similarly, during the summer months, as temperatures climb above comfortable levels, the use of air conditioning drives demand upward. Since the FPC serves 32 counties along west-central and northern Florida, it was necessary to obtain temperature conditions from multiple weather stations. This was accomplished by identifying three primary weather stations within the FPC service area and recording the temperature value at the hour of peak demand each day at each station. A weighted average of these three daily temperatures was used to represent coincident temperature (i.e., temperature at the hour of peak demand) for the entire FPC service area, where the weights were proportional to the percentage of total electricity sales attributable to the weather zones surrounding each of the three weather stations.

To summarize, the independent variable (y_t) and the independent variables recorded for each of the 365 days of the year were as follows:

Dependent Variable:

y_t = Peak demand (in megawatts) observed on day t

Independent Variables:

Day of the week: Weekday, Saturday, or Sunday/holiday

Temperature: Coincident temperature (in degrees), i.e., the temperature recorded at the hour of the peak demand on day t, calculated as a weighted average of three daily temperatures

The Models

In any modeling procedure, it is often helpful to graph the data in a scatterplot. Figure CS6.1 shows a graph of the daily peak demand (y_t) from November 1 through October 31. The effects of seasonal weather on peak demand are readily apparent from the figure. One way to account for this seasonal variation is to include dummy variables for months or trigonometric terms in the model (refer to Section 10.7). However, since temperature is such a strong indicator of the weather, the FPC chose a simpler model with temperature as the sole seasonal weather variable.

Figure CS6.2 is a scatterplot of daily peak demands versus coincident temperature. Note the nonlinear relationship that exists between the two variables. During the cool winter months, peak demand is inversely related to temperature; lower temperatures cause increased usage of heating equipment, which in turn causes higher peak demands. In contrast, the summer months reveal a positive relationship between peak demand and temperature; higher temperatures yield higher peak demands because of greater usage of air conditioners. You might think that a second-order (quadratic) model would be a good choice to account for the

Figure CS6.1 Daily peak
megawatt demands,
November–October

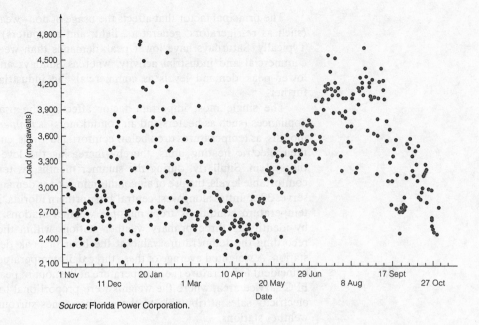

Source: Florida Power Corporation.

U-shaped distribution of peak demands shown in Figure CS6.2. The FPC, however, rejected such a model for two reasons:

1. A quadratic model yields a symmetrical shape (i.e., a parabola) and would, therefore, not allow independent estimates of the winter and summer peak demand–temperature relationship.

Figure CS6.2 Daily peak
demand versus
temperature,
November–October

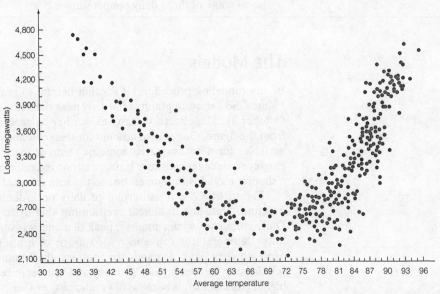

Source: Florida Power Corporation.

Figure CS6.3 Theoretical relationship between daily peak demand and temperature

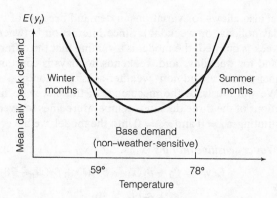

2. In theory, there exists a mild temperature range where peak demand is assumed to consist solely of the non–weather-sensitive base demand component. For this range, a temperature change will not spur any additional heating or cooling and, consequently, has no impact on demand. The lack of linearity in the bottom portion of the U-shaped parabola fitted by the quadratic model would yield overestimates of peak demand at the extremes of the mild temperature range and underestimates for temperatures in the middle of this range (see Figure CS6.3).

The solution was to model daily peak demand with a piecewise linear regression model (see Section 9.2). This approach has the advantage of allowing the peak demand–temperature relationship to vary between some prespecified temperature ranges, as well as providing a mechanism for joining the separate pieces.

Using the piecewise linear specification as the basic model structure, the following model of daily peak demand was proposed:

Model 1

$$y_t = \beta_0 + \underbrace{\beta_1(x_{1t} - 59)x_{2t} + \beta_2(x_{1t} - 78)x_{3t}}_{\text{Temperature}} + \underbrace{\beta_3 x_{4t} + \beta_4 x_{5t}}_{\text{Day of the week}} + \varepsilon_t$$

where

$x_{1t} = $ Coincident temperature on day t

$$x_{2t} = \begin{cases} 1 & \text{if } x_{1t} < 59 \\ 0 & \text{if not} \end{cases} \qquad x_{3t} = \begin{cases} 1 & \text{if } x_{1t} > 78 \\ 0 & \text{if not} \end{cases}$$

$$x_{4t} = \begin{cases} 1 & \text{if Saturday} \\ 0 & \text{if not} \end{cases} \qquad x_{5t} = \begin{cases} 1 & \text{if Sunday/holiday} \\ 0 & \text{if not} \end{cases} \qquad \text{(Base level = Weekday)}$$

$\varepsilon_t = $ Uncorrelated error term

Model 1 proposes three different straight-line relationships between peak demand (y_t) and coincident temperature x_{1t}, one for each of the three temperature ranges corresponding to winter months (less than 59°), non–weather-sensitive months (between 59° and 78°), and summer months (greater than 78°).* The

---

* The temperature values, 59° and 78°, identify where the winter- and summer-sensitive portions of demand join the base demand component. These "knot values" were determined from visual inspection of the graph in Figure CS6.2.

model also allows for variations in demand because of day of the week (Saturday, Sunday/holiday, or weekday). Since interaction between temperature and day of the week is omitted, the model is assuming that the differences between mean peak demand for weekdays and weekends/holidays is constant for the winter-sensitive, summer-sensitive, and non–weather-sensitive months.

We will illustrate the mechanics of the piecewise linear terms by finding the equations of the three demand–temperature lines for weekdays (i.e., $x_{4t} = x_{5t} = 0$). Substituting $x_{4t} = 0$ and $x_{5t} = 0$ into the model, we have

Winter-sensitive months ($x_{1t} < 59°$, $x_{2t} = 1$, $x_{3t} = 0$):

$$E(y_t) = \beta_0 + \beta_1(x_{1t} - 59)(1) + \beta_2(x_{1t} - 78)(0) + \beta_3(0) + \beta_4(0)$$

$$= \beta_0 + \beta_1(x_{1t} - 59)$$

$$= (\beta_0 - 59\beta_1) + \beta_1 x_{1t}$$

Summer-sensitive months ($x_{1t} > 78°$, $x_{2t} = 0$, $x_{3t} = 1$):

$$E(y_t) = \beta_0 + \beta_1(x_{1t} - 59)(0) + \beta_2(x_{1t} - 78)(1) + \beta_3(0) + \beta_4(0)$$

$$= \beta_0 + \beta_2(x_{1t} - 78)$$

$$= (\beta_0 - 78\beta_2) + \beta_2 x_{1t}$$

Non–weather-sensitive months ($59° \leq x_{1t} \leq 78°$, $x_{2t} = x_{3t} = 0$):

$$E(y_t) = \beta_0 + \beta_1(x_{1t} - 59)(0) + \beta_2(x_{1t} - 78)(0) + \beta_3(0) + \beta_4(0)$$

$$= \beta_0$$

Note that the slope of the demand–temperature line for winter-sensitive months (when $x_{1t} < 59$) is β_1 (which we expect to be negative), whereas the slope for summer-sensitive months (when $x_{1t} > 78$) is β_2 (which we expect to be positive). The intercept term β_0 represents the mean daily peak demand observed in the non–weather-sensitive period (when $59 \leq x_{1t} \leq 78$). Notice also that the peak demand during non–weather-sensitive days does not depend on temperature (x_{1t}).

Model 1 is a multiple regression model that relies on the standard regression assumptions of independent errors (ε_t uncorrelated). This may be a serious short-coming in view of the fact that the data are in the form of a time series. To account for possible autocorrelated residuals, two time series models were proposed:

Model 2

$$y_t = \beta_0 + \beta_1(x_{1t} - 59)x_{2t} + \beta_2(x_{1t} - 78)x_{3t} + \beta_3 x_{4t} + \beta_4 x_{5t} + R_t$$

$$R_t = \phi_1 R_{t-1} + \varepsilon_t$$

Model 2 proposes a regression–autoregression pair of models for daily peak demand (y_t). The deterministic component, $E(y_t)$, is identical to the deterministic component of Model 1; however, a first-order autoregressive model is chosen for the random error component. Recall (from Section 10.5) that a first-order autoregressive model is appropriate when the correlation between residuals diminishes as the distance between time periods (in this case, days) increases.

Model 3

$$y_t = \beta_0 + \beta_1(x_{1t} - 59)x_{2t} + \beta_2(x_{1t} - 78)x_{3t} + \beta_3 x_{4t} + \beta_4 x_{5t} + R_t$$

$$R_t = \phi_1 R_{t-1} + \phi_2 R_{t-2} + \phi_5 R_{t-5} + \phi_7 R_{t-7} + \varepsilon_t$$

Model 3 extends the first-order autoregressive error model of Model 2 to a seventh-order autoregressive model with lags at 1, 2, 5, and 7. In theory, the peak demand on day t will be highly correlated with the peak demand on day $t + 1$. However, there also may be significant correlation between demand 2 days, 5 days, and/or 1 week (7 days) apart. This more general error model is proposed to account for any residual correlation that may occur as a result of the week-to-week variation in peak demand, in addition to the day-to-day variation.

The Regression and Autoregression Analyses

The multiple regression computer printout for Model 1 is shown in Figure CS6.4, and a plot of the least squares fit is shown in Figure CS6.5. The model appears to provide a good fit to the data, with $R^2 = .8307$ and $F = 441.73$ (significant at $p = .0001$). The value $s = 245.585$ implies that we can expect to predict daily peak demand accurate to within $2s \approx 491$ megawatts of its true value. However, we must be careful not to conclude at this point that the model is useful for predicting peak demand. Recall that in the presence of autocorrelated residuals, the standard errors of the regression coefficients are underestimated, thereby inflating the corresponding t statistics for testing H_0: $\beta_i = 0$. At worst, this could lead to the false conclusion that a β parameter is significantly different from 0; at best, the results, although significant, give an overoptimistic view of the predictive ability of the model.

Figure CS6.4 SAS least squares regression printout for daily peak demand, Model 1

The REG Procedure
Dependent Variable: LOAD

Analysis of Variance

Source	DF	Sum of Squares	Mean Square	F Value	Pr > F
Model	4	106565982	26641496	441.73	<.0001
Error	360	21712247	60311.8		
Corrected Total	364	128278229			

Root MSE	245.585	R-Square	0.8307
Dependent Mean	3191.863	Adj R-Sq	0.8289
Coeff Var	7.694		

Parameter Estimates

Variable	DF	Parameter Estimate	Standard Error	t Value	Pr > |t|
Intercept	1	2670.171	21.2518	125.64	<.0001
AVTW	1	-82.040	2.9419	-27.89	<.0001
AVTS	1	114.443	3.0505	37.52	<.0001
SAT	1	-164.932	37.9902	-4.34	0.0001
SUN	1	-285.114	35.3283	-8.07	<.0001

Durbin-Watson D	0.705
Number of Observations	365
1st Order Autocorrelation	0.648

Figure CS6.5 Daily peak demand versus temperature: Actual versus fitted piecewise linear model

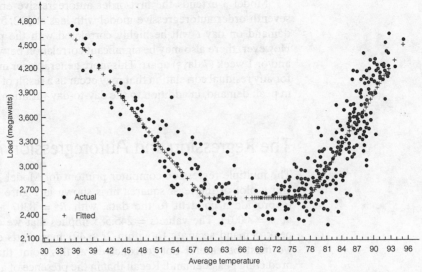

Source: Florida Power Corporation.

To determine whether the residuals of the multiple regression model are positively autocorrelated, we conduct the Durbin–Watson test:

H_0: Uncorrelated residuals

H_a: Positive residual correlation

Recall that the Durbin–Watson test is designed specifically for detecting first-order autocorrelation in the residuals, R_t. Thus, we can write the null and alternative hypotheses as

$$H_0: \phi_1 = 0$$
$$H_a: \phi_1 > 0$$

where $R_t = \phi_1 R_{t-1} + \varepsilon_t$, and ε_t = uncorrelated error (white noise).

The test statistic, shaded at the bottom of the printout in Figure CS6.4, is $d = .705$. Recall that small values of d lead us to reject $H_0: \phi_1 = 0$ in favor of the alternative $H_a: \phi_1 > 0$. For $\alpha = .01$, $n = 365$, and $k = 4$ (the number of β parameters in the model, excluding β_0), the lower bound on the critical value (obtained from Table 9 in Appendix C) is approximately $d_L = 1.46$. Since the value of the test statistic, $d = .705$, falls well below the lower bound, there is strong evidence at $\alpha = .01$ of positive (first-order) autocorrelated residuals. Thus, we need to incorporate terms for residual autocorrelation into the model.

The time series printouts for Models 2 and 3 are shown in Figures CS6.6 and CS6.7, respectively. A summary of the results for all three models is given in Table CS6.1.

Table CS6.1 Summary of Results for Models 1, 2, and 3			
	Model 1	Model 2	Model 3
R^2	.8307	.9225	.9351
MSE	60,312	27,687	23,398
s	245.585	166.4	153.0

Figure CS6.6 SAS printout for autoregressive time series model of daily peak demand, Model 2

```
                          The AUTOREG Procedure
                        Dependent Variable: LOAD

                      Estimates of Autocorrelations

Lag  Covariance Correlation  -1 9 8 7 6 5 4 3 2 1 0 1 2 3 4 5 6 7 8 9 1
  0    59485.6   1.00000 |                    |********************|
  1    38519.4   0.64754 |                    |**************      |


                   Preliminary MSE =   34542.75

              Estimates of the Autoregressive Parameters

                                      Standard
               Lag    Coefficient      Error     t Value
                 1     -0.647554      0.039887    -16.23

                       Yule-Walker Estimates

          SSE               9939789    DFE                 358
          MSE               27687.4    Root MSE         166.39
          Regress R-Square   0.7626    Total R-Square   0.9225


                                    Standard              Approx
        Variable   DF    Estimate     Error    t Value   Pr > |t|

        Intercept   1    2812.967    29.8791     94.15    <.0001
        AVTW        1     -65.337     2.6639    -24.53     <.0001
        AVTS        1      83.455     3.8531     21.66     <.0001
        SAT         1    -130.828    22.4136     -5.84     <.0001
        SUN         1    -275.551    21.3737    -12.89     <.0001
```

The addition of the first-order autoregressive error term in Model 2 yielded a drastic improvement to the fit of the model. The value of R^2 increased from .83 for Model 1 to .92, and the standard deviation s decreased from 245.6 to 166.4. These results support the conclusion reached by the Durbin–Watson test—namely, that the first-order autoregressive lag parameter ϕ_1 is significantly different from 0.

Does the more general autoregressive error model (Model 3) provide a better approximation to the pattern of correlation in the residuals than the first-order autoregressive model (Model 2)? To test this hypothesis, we would need to test $H_0: \phi_2 = \phi_5 = \phi_7 = 0$. Although we omit discussion of tests on autoregressive parameters in this text,* we can arrive at a decision from a pragmatic point of view by again comparing the values of R^2 and s for the two models. The more complex autoregressive model proposed by Model 3 yields a slight increase in R^2 (.935 compared to .923 for Model 2) and a slight decrease in the value of s (153.0 compared to 166.4 for Model 2). The additional lag parameters, although they may be statistically significant, may not be practically significant. The practical analyst may decide that the first-order autoregressive process proposed by Model 2 is the more desirable option since it is easier to use to forecast peak daily demand (and therefore more explainable to managers) while yielding approximate prediction errors (as measured by $2s$) that are only slightly larger than those for Model 3.

For the purposes of illustration, we use Model 2 to forecast daily peak demand in the following section.

---

* For details of tests on autoregressive parameters, see Fuller (1996).

Figure CS6.7 SAS printout for seventh-order autoregressive time series model of daily peak demand, Model 3

```
                    The AUTOREG Procedure
                  Dependent Variable: LOAD

                Estimates of Autocorrelations

Lag  Covariance Correlation  -1 9 8 7 6 5 4 3 2 1 0 1 2 3 4 5 6 7 8 9 1
 0    59485.6    1.00000  |                    |********************|
 1    38519.4    0.64754  |                    |*************        |
 2    35741.0    0.60083  |                    |***********          |
 3    32863.2    0.55254  |                    |***********          |
 4    29917.9    0.50294  |                    |**********           |
 5    31340.9    0.52686  |                    |**********           |
 6    30061.9    0.50535  |                    |**********           |
 7    31503.0    0.52967  |                    |**********           |

             Preliminary MSE =   28841.4

        Estimates of the Autoregressive Paramenters

                              Standard
          Lag   Coefficient     Error      t Value
           1    -0.367936     0.049902      -7.37
           2    -0.207028     0.051722      -4.00
           3     0.000000     0.000000
           4     0.000000     0.000000
           5    -0.135264     0.049072      -2.76
           6     0.000000     0.000000
           7    -0.153385     0.048430      -3.17

                  Yule-Walker Estimates

      SSE            8329842     DFE              356
      MSE            23398.4     Root MSE      152.97
      Regress R-Square  0.8112   Total R-Square  0.9351

                         Standard              Approx
Variable   DF   Estimate    Error   t Value   Pr > |t|

Intercept   1   2809.950   58.2346    48.25     <.0001
AVTW        1    -71.282    2.2621    -31.51     <.0001
AVTS        1     79.120    4.1806     18.93     <.0001
SAT         1   -150.524   23.4728     -6.41     <.0001
SUN         1   -262.273   21.6832    -12.10     <.0001
```

Forecasting Daily Peak Electricity Demand

Suppose the FPC decided to use Model 2 to forecast daily peak demand for the first seven days of November of the following year. The estimated model,* obtained from Figure CS6.6, is given by

$$\hat{y}_t = 2{,}812.967 - 65.337(x_{1t} - 59)x_{2t} + 83.455(x_{1t} - 78)x_{3t}$$

$$-130.828x_{4t} - 275.551x_{5t} + \hat{R}_t$$

$$\hat{R}_t = .6475\hat{R}_{t-1}$$

The forecast for November 1 of the next year ($t = 366$), requires an estimate of the residual R_{365}, where $\hat{R}_{365} = y_{365} - \hat{y}_{365}$. The last day of the November–October time period ($t = 365$) was October 31, a Monday. On this day the peak demand was recorded as $y_{365} = 2{,}752$ megawatts and the coincident temperature as $x_{1,365} = 77°$. Substituting the appropriate values of the dummy variables into the equation for $\hat{y}_t$

---

* Remember that the estimate of ϕ_1 is obtained by multiplying the value reported on the SAS printout by (-1).

(i.e., $x_{2t} = 0$, $x_{3t} = 0$, $x_{4t} = 0$, and $x_{5t} = 0$), we have

$$\hat{R}_{365} = y_{365} - \hat{y}_{365}$$

$$= 2{,}752 - [2{,}812.967 - 65.337(77 - 59)(0) + 83.455(77 - 78)(0)$$

$$-130.828(0) - 275.551(0)]$$

$$= 2{,}752 - 2{,}812.967 = -60.967$$

Then the formula for calculating the forecast for Tuesday, November 1, is

$$\hat{y}_{366} = 2{,}812.967 - 65.337(x_{1,366} - 59)x_{2,366} + 83.455(x_{1,366} - 78)x_{3,366}$$

$$-130.828x_{4,366} - 275.551x_{5,366} + \hat{R}_{366}$$

where

$$\hat{R}_{366} = \hat{\phi}_1 \hat{R}_{365} = (.6475)(-60.967) = -39.476$$

Note that the forecast requires an estimate of coincident temperature on that day, $\hat{x}_{1,366}$. If the FPC wants to forecast demand under normal weather conditions, then this estimate can be obtained from historical data for that day. Or, the FPC may choose to rely on a meteorologist's weather forecast for that day. For this example, assume that $\hat{x}_{1,366} = 76°$ (the actual temperature recorded by the FPC). Then $x_{2,366} = x_{3,366} = 0$ (since $59 \le \hat{x}_{1,366} \le 78$) and $x_{4,366} = x_{5,366} = 0$ (since the target day is a Tuesday). Substituting these values and the value of $\hat{R}_{366}$ into the equation, we have

$$\hat{y}_{366} = 2{,}812.967 - 65.337(76 - 59)(0) + 83.455(76 - 78)(0)$$

$$-130.828(0) - 275.551(0) - 39.476$$

$$= 2{,}773.49$$

Similarly, a forecast for Wednesday, November 2 (i.e., $t = 367$), can be obtained:

$$\hat{y}_{367} = 2{,}812.967 - 65.337(x_{1,367} - 59)x_{2,367} + 83.455(x_{1,367} - 78)x_{3,367}$$

$$-130.828x_{3,367} - 275.551x_{4,367} + \hat{R}_{367}$$

where $\hat{R}_{367} = \hat{\phi}_1 \hat{R}_{366} = (.6475)(-39.476) = -25.561$ and $x_{3,367} = x_{4,367} = 0$. For an estimated coincident temperature of $\hat{x}_{1,367} = 77°$ (again, this is the actual temperature recorded on that day), we have $x_{2,367} = 0$ and $x_{3,367} = 0$. Substituting these values into the prediction equation, we obtain

$$\hat{y}_{367} = 2{,}812.967 - 65.337(77 - 59)(0) + 83.455(77 - 78)(0)$$

$$-130.828(0) - 275.551(0) - 25.561$$

$$= 2{,}812.967 - 25.561$$

$$= 2{,}787.41$$

Approximate 95% prediction intervals for the two forecasts are calculated as follows:

Tuesday, Nov. 1:

$$\hat{y}_{366} \pm 1.96\sqrt{MSE}$$

$$= 2{,}773.49 \pm 1.96\sqrt{27{,}687.44}$$

$$= 2{,}773.49 \pm 326.14 \text{ or } (2{,}447.35, \ 3{,}099.63)$$

Wednesday, Nov. 2:

$$\hat{y}_{367} \pm 1.96\sqrt{\text{MSE}(1 + \hat{\phi}_1^2)}$$

$$= 2{,}787.41 \pm 1.96\sqrt{(27{,}687.44)[1 + (.6475)^2]}$$

$$= 2{,}787.41 \pm 388.53 \text{ or } (2{,}398.88, 3{,}175.94)$$

The forecasts, approximate 95% prediction intervals, and actual daily peak demands (recorded by the FPC) for the first seven days of November of the next year are given in Table CS6.2. Note that actual peak demand y_t falls within the corresponding prediction interval for all seven days. Thus, the model appears to be useful for making short-term forecasts of daily peak demand. Of course, if the prediction intervals were extremely wide, this result would be of no practical value. For example, the forecast error $y_t - \hat{y}_t$, measured as a percentage of the actual value y_t, may be large even when y_t falls within the prediction interval. Various techniques, such as the percent forecast error, are available for evaluating the accuracy of forecasts. Consult the references given at the end of Chapter 10 for details on these techniques.

Table CS6.2 Forecasts and actual peak demands for the first seven days of November, following year

Date	Day t	Forecast $\hat{y}_t$	Approximate 95% Prediction Interval	Actual Demand y_t	Actual Temperature x_{1t}
Tues., Nov. 1	366	2,773.49	(2,447.35, 3,099.63)	2,799	76
Wed., Nov. 2	367	2,787.41	(2,398.88, 3,175.94)	2,784	77
Thurs., Nov. 3	368	2,796.42	(2,384.53, 3,208.31)	2,845	77
Fri., Nov. 4	369	2,802.25	(2,380.92, 3,223.58)	2,701	76
Sat., Nov. 5	370	2,675.20	(2,249.97, 3,100.43)	2,512	72
Sun., Nov. 6	371	2,532.92	(2,106.07, 2,959.77)	2,419	71
Mon., Nov. 7	372	2,810.06	(2,382.59, 3,237.53)	2,749	68

Conclusions

This case study presents a time series approach to modeling and forecasting daily peak demand observed at Florida Power Corporation. A graphical analysis of the data provided the means of identifying and formulating a piecewise linear regression model relating peak demand to temperature and day of the week. The multiple regression model, although providing a good fit to the data, exhibited strong signs of positive residual autocorrelation.

Two autoregressive time series models were proposed to account for the autocorrelated errors. Both models were shown to provide a drastic improvement in model adequacy. Either could be used to provide reliable short-term forecasts of daily peak demand or for weather normalization (i.e., estimating the peak demand if normal weather conditions had prevailed).

Follow-up Questions

1. All three models discussed in this case study make the underlying assumption that the peak demand–temperature relationship is independent of day of the week. Write a model that includes interaction between temperature and day of the week. Show the effect the interaction has on the straight-line relationships between peak demand and temperature. Explain how you could test the significance of the interaction terms.

2. Consider the problem of using Model 2 for weather normalization. Suppose the temperature on Saturday, March 5 (i.e., $t = 125$), was $x_{1,125} = 25°$, unusually cold for that day. Normally, temperatures range from 40° to 50° on March 5 in the FPC service area. Substitute $x_{1,125} = 45°$ into the prediction equation to obtain an estimate of the peak demand expected if normal weather conditions had prevailed on March 5. Calculate an approximate 95% prediction interval for the estimate. [*Hint:* Use $\hat{y}_{125} \pm 1.96\sqrt{\text{MSE}}$.]

References

Fuller, W. A. *Introduction to Statistical Time Series*, 2nd ed. New York: Wiley, 1996.

Jacob, M. F. "A time series approach to modeling daily peak electricity demands." Paper presented at the SAS Users Group International Annual Conference, Reno, Nevada, 1985.

Chapter 11

PRINCIPLES OF EXPERIMENTAL DESIGN

Contents

Objectives

1. To present an overview of experiments designed to compare two or more population means

2. To explain the statistical principles of experimental design

11.1 Introduction

In Chapter 7, we learned that a regression analysis of observational data has some limitations. In particular, establishing a cause-and-effect relationship between an independent variable x and the response y is difficult since the values of other relevant independent variables—both those in the model and those omitted from the model—are not controlled. Recall that experimental data are data collected with the values of the x's set in advance of observing y (i.e., the values of the x's are controlled). With experimental data, we usually select the x's so that we can compare the mean responses, $E(y)$, for several different combinations of the x values.

The procedure for selecting sample data with the x's set in advance is called the **design of the experiment**. The statistical procedure for comparing the population means is called an **analysis of variance**. The objective of this chapter is to introduce some key aspects of experimental design. The analysis of the data from such experiments using an analysis of variance is the topic of Chapter 12.

11.2 Experimental Design Terminology

The study of experimental design originated with R. A. Fisher in the early 1900s in England. During these early years, it was associated solely with agricultural experimentation. The need for experimental design in agriculture was very clear: It takes a full year to obtain a single observation on the yield of a new variety of most crops. Consequently, the need to save time and money led to a study of ways to obtain more information using smaller samples. Similar motivations led to its subsequent acceptance and wide use in all fields of scientific experimentation. Despite this fact, the terminology associated with experimental design clearly indicates its early association with the biological sciences.

We will call the process of collecting sample data an **experiment** and the (*dependent*) variable to be measured, the **response** y. The planning of the sampling procedure is called the **design** of the experiment. The object upon which the response measurement y is taken is called an **experimental unit**.

Definition 11.1   The process of collecting sample data is called an **experiment**.

Definition 11.2   The plan for collecting the sample is called the **design of the experiment**.

Definition 11.3   The variable measured in the experiment is called the **response variable**.

Definition 11.4   The object upon which the response y is measured is called an **experimental unit**.

Independent variables that may be related to a response variable y are called **factors**. The value—that is, the intensity setting—assumed by a factor in an experiment is called a **level**. The combinations of levels of the factors for which the response will be observed are called **treatments**.

Definition 11.5   The independent variables, quantitative or qualitative, that are related to a response variable y are called **factors**.

Definition 11.6   The intensity setting of a factor (i.e., the value assumed by a factor in an experiment) is called a **level**.

Definition 11.7   A **treatment** is a particular combination of levels of the factors involved in an experiment.

**Example
11.1**

A Designed Experiment

A marketing study is conducted to investigate the effects of brand and shelf location on weekly coffee sales. Coffee sales are recorded for each of two brands (brand A and brand B) at each of three shelf locations (bottom, middle, and top). The

$2 \times 3 = 6$ combinations of brand and shelf location were varied each week for a period of 18 weeks. Figure 11.1 is a layout of the design. For this experiment, identify

(a) the experimental unit

(b) the response, y

(c) the factors

(d) the factor levels

(e) the treatments

Figure 11.1 Layout for designed experiment of Example 11.1

	SHELF LOCATION		
	Bottom	Middle	Top
A	Week 1 9 14	Week 2 7 16	Week 4 12 17
BRAND			
B	Week 5 10 13	Week 3 8 18	Week 6 11 15

Solution

(a) Since the data will be collected each week for a period of 18 weeks, the experimental unit is 1 week.

(b) The variable of interest (i.e., the response) is $y =$ weekly coffee sales. Note that weekly coffee sales is a quantitative variable.

(c) Since we are interested in investigating the effect of brand and shelf location on sales, brand and shelf location are the factors. Note that both factors are qualitative variables, although, in general, they may be quantitative or qualitative.

(d) For this experiment, brand is measured at two levels (A and B) and shelf location at three levels (bottom, middle, and top).

(e) Since coffee sales are recorded for each of the six brand–shelf location combinations (brand A, bottom), (brand A, middle), (brand A, top), (brand B, bottom), (brand B, middle), and (brand B, top), then the experiment involves six treatments (see Figure 11.1). The term *treatments* is used to describe the factor level combinations to be included in an experiment because many experiments involve "treating" or doing something to alter the nature of the experimental unit. Thus, we might view the six brand–shelf location combinations as treatments on the experimental units in the marketing study involving coffee sales.   ▬

Now that you understand some of the terminology, it is helpful to think of the design of an experiment in four steps.

Step 1. Select the factors to be included in the experiment, and identify the parameters that are the object of the study. Usually, the target parameters are the population means associated with the factor level combinations (i.e., treatments).

Step 2. Choose the treatments (the factor level combinations to be included in the experiment).

Step 3. Determine the number of observations (sample size) to be made for each treatment. [This will usually depend on the standard error(s) that you desire.]

Step 4. Plan how the treatments will be assigned to the experimental units. That is, decide on which design to use.

By following these steps, you can control the quantity of information in an experiment. We explain how this is done in Section 11.3.

11.3 Controlling the Information in an Experiment

The problem of acquiring good experimental data is analogous to the problem faced by a communications engineer. The receipt of any signal, verbal or otherwise, depends on the volume of the signal and the amount of background noise. The greater the volume of the signal, the greater will be the amount of information transmitted to the receiver. Conversely, the amount of information transmitted is reduced when the background noise is great. These intuitive thoughts about the factors that affect the information in an experiment are supported by the following fact: The standard errors of most estimators of the target parameters are proportional to σ (a measure of data variation or noise) and inversely proportional to the sample size (a measure of the volume of the signal). To illustrate, take the simple case where we wish to estimate a population mean μ by the sample mean $\bar{y}$. The standard error of the sampling distribution of $\bar{y}$ is

$$\sigma_{\bar{y}} = \frac{\sigma}{\sqrt{n}} \quad \text{(see Section 1.7)}$$

For a fixed sample size n, the smaller the value of σ, which measures the **variability (noise)** in the population of measurements, the smaller will be the standard error $\sigma_{\bar{y}}$. Similarly, by increasing the sample size n (**volume of the signal**) in a given experiment, you decrease $\sigma_{\bar{y}}$.

The first three steps in the design of an experiment—selecting the factors and treatments to be included in an experiment and specifying the sample sizes—determine the volume of the signal. You must select the treatments so that the observed values of y provide information on the parameters of interest. Then the larger the treatment sample sizes, the greater will be the quantity of information in the experiment. We present an example of a volume-increasing experiment in Section 11.5.

Is it possible to observe y and obtain no information on a parameter of interest? The answer is yes. To illustrate, suppose that you attempt to fit a first-order model

$$E(y) = \beta_0 + \beta_1 x$$

to a set of $n = 10$ data points, all of which were observed for a single value of x, say, $x = 5$. The data points might appear as shown in Figure 11.2. Clearly, there is no possibility of fitting a line to these data points. The only way to obtain information on β_0 and β_1 is to observe y for *different* values of x. Consequently, the $n = 10$ data points in this example contain absolutely no information on the parameters β_0 and β_1.

Step 4 in the design of an experiment provides an opportunity to reduce the noise (or experimental error) in an experiment. As we illustrate in Section 11.4, known sources of data variation can be reduced or eliminated by **blocking**—that

Figure 11.2 Data set with $n = 10$ responses, all at $x = 5$

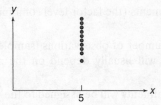

is, observing all treatments within relatively homogeneous **blocks** of experimental material. When the treatments are compared within each block, any background noise produced by the block is canceled, or eliminated, allowing us to obtain better estimates of treatment differences.

Summary of Steps in Experimental Design

Volume-increasing:
1. Select the factors.
2. Choose the treatments (factor level combinations).
3. Determine the sample size for each treatment.

Noise-reducing:
4. Assign the treatments to the experimental units.

In summary, it is useful to think of experimental designs as being either "noise reducers" or "volume increasers." We will learn, however, that most designs are multifunctional. That is, they tend to both reduce the noise and increase the volume of the signal at the same time. Nevertheless, we will find that specific designs lean heavily toward one or the other objective.

11.4 Noise-Reducing Designs

Noise reduction in an experimental design (i.e., the removal of extraneous experimental variation) can be accomplished by an appropriate assignment of treatments to the experimental units. The idea is to compare treatments within blocks of relatively homogeneous experimental units. The most common design of this type is called a **randomized block design**.

To illustrate, suppose we want to compare the mean performance times of female long-distance runners using three different training liquids (e.g., fructose drinks, glucose drinks, and water) 1 hour prior to running a race. Thus, we want to compare the three means μ_A, μ_B, and μ_C, where μ_i is the mean performance time for liquid i. One way to design the experiment is to select 15 female runners (where the runners are the experimental units) and randomly assign one of the three liquids (treatments) to each runner. A diagram of this design, called a **completely randomized design** (since the treatments are randomly assigned to the experimental units), is shown in Table 11.1.

Definition 11.8 A **completely randomized design** to compare p treatments is one in which the treatments are randomly assigned to the experimental units.

Table 11.1 Completely randomized design with $p = 3$ treatments

Runner	Treatment (Liquid) Assigned
1	B
2	A
3	B
4	C
5	C
6	A
7	B
8	C
9	A
10	A
11	C
12	A
13	B
14	C
15	B

This design has the obvious disadvantage that the performance times would vary greatly depending on the fitness level of the athlete, the athlete's age, and so on. A better design—one that contains more information on the mean performance times—would be to use only five runners and require each athlete to run three long-distance races, drinking a different liquid before each race. This *randomized block* procedure acknowledges the fact that performance time in a long-distance race varies substantially from runner to runner. By comparing the three performance times for each runner, we eliminate runner-to-runner variation from the comparison.

The randomized block design that we have just described is diagrammed in Figure 11.3. The figure shows that there are five runners. Each runner can be viewed as a **block** of three experimental units—one corresponding to the use of each of

Figure 11.3 Diagram for a randomized block design containing $b = 5$ blocks and $p = 3$ treatments

Blocks (Runners)   Treatments (Liquids)

Block			
1	B	A	C
2	A	C	B
3	B	C	A
4	A	B	C
5	A	C	B

the training liquids, A, B, and C. The blocks are said to be **randomized** because the treatments (liquids) are randomly assigned to the experimental units within a block. For our example, the liquids drunk prior to a race would be assigned in random order to avoid bias introduced by other unknown and unmeasured variables that may affect a runner's performance time.

In general, a randomized block design to compare p treatments will contain b relatively homogeneous blocks, with each block containing p experimental units. Each treatment appears once in every block, with the p treatments randomly assigned to the experimental units within each block.

> **Definition 11.9** A **randomized block design** to compare p treatments involves b blocks, each containing p relatively homogeneous experimental units. The p treatments are randomly assigned to the experimental units within each block, with one experimental unit assigned per treatment.

Example 11.2

Suppose you want to compare the abilities of four real estate appraisers, A, B, C, and D. One way to make the comparison would be to randomly allocate a number of pieces of real estate—say, 40—10 to each of the four appraisers. Each appraiser would appraise the property, and you would record y, the difference between the appraised and selling prices expressed as a percentage of the selling price. Thus, y measures the appraisal error expressed as a percentage of selling price, and the treatment allocation to experimental units that we have described is a completely randomized design.

(a) Discuss the problems with using a completely randomized design for this experiment.

(b) Explain how you could employ a randomized block design.

Solution

(a) The problem with using a completely randomized design for the appraisal experiment is that the comparison of mean percentage errors will be influenced by the nature of the properties. Some properties will be easier to appraise than others, and the variation in percentage errors that can be attributed to this fact will make it more difficult to compare the treatment means.

(b) To eliminate the effect of property-to-property variability in comparing appraiser means, you could select only 10 properties and require each appraiser to appraise the value of each of the 10 properties. Although in this case there is probably no need for randomization, it might be desirable to randomly assign the order (in time) of the appraisals. This randomized block design, consisting of $p = 4$ treatments and $b = 10$ blocks, would appear as shown in Figure 11.4.   ■

Each experimental design can be represented by a general linear model relating the response y to the factors (treatments, blocks, etc.) in the experiment. When the factors are qualitative in nature (as is often the case), the model includes dummy variables. For example, consider the completely randomized design portrayed in Table 11.1. Since the experiment involves three treatments (liquids), we require two dummy variables. The model for this completely randomized design would appear as follows:

$$y = \beta_0 + \beta_1 x_1 + \beta_2 x_2 + \varepsilon$$

Figure 11.4 Diagram for a randomized block design: Example 11.2

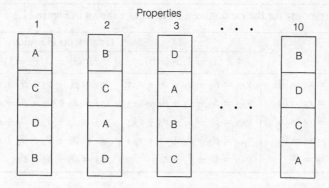

where

$$x_1 = \begin{cases} 1 & \text{if liquid A} \\ 0 & \text{if not} \end{cases} \qquad x_2 = \begin{cases} 1 & \text{if liquid B} \\ 0 & \text{if not} \end{cases}$$

We have arbitrarily selected liquid C as the base level. From our discussion of dummy-variable models in Chapter 5, we know that the mean responses for the three liquids are

$$\mu_A = \beta_0 + \beta_1$$
$$\mu_B = \beta_0 + \beta_2$$
$$\mu_C = \beta_0$$

Recall that $\beta_1 = \mu_A - \mu_C$ and $\beta_2 = \mu_B - \mu_C$. Thus, to estimate the differences between the treatment means, we require estimates of β_1 and β_2.

Similarly, we can write the model for the randomized block design in Figure 11.3 as follows:

$$y = \beta_0 + \underbrace{\beta_1 x_1 + \beta_2 x_2}_{\text{Treatment effects}} + \underbrace{\beta_3 x_3 + \beta_4 x_4 + \beta_5 x_5 + \beta_6 x_6}_{\text{Block effects}} + \varepsilon$$

where

$$x_1 = \begin{cases} 1 & \text{if liquid A} \\ 0 & \text{if not} \end{cases} \qquad x_2 = \begin{cases} 1 & \text{if liquid B} \\ 0 & \text{if not} \end{cases} \qquad x_3 = \begin{cases} 1 & \text{if runner 1} \\ 0 & \text{if not} \end{cases}$$

$$x_4 = \begin{cases} 1 & \text{if runner 2} \\ 0 & \text{if not} \end{cases} \qquad x_5 = \begin{cases} 1 & \text{if runner 3} \\ 0 & \text{if not} \end{cases} \qquad x_6 = \begin{cases} 1 & \text{if runner 4} \\ 0 & \text{if not} \end{cases}$$

In addition to the treatment terms, the model includes four dummy variables representing the five blocks (runners). Note that we have arbitrarily selected runner 5 as the base level. Using this model, we can write each response y in the experiment of Figure 11.3 as a function of β's, as shown in Table 11.2.

For example, to obtain the model for the response y for treatment A in block 1 (denoted y_{A1}), we substitute $x_1 = 1$, $x_2 = 0$, $x_3 = 1$, $x_4 = 0$, $x_5 = 0$, and $x_6 = 0$ into the equation. The resulting model is

$$y_{A1} = \beta_0 + \beta_1 + \beta_3 + \varepsilon_{A1}$$

Now we will use Table 11.2 to illustrate how a randomized block design reduces experimental noise. Since each treatment appears in each of the five blocks, there are

Table 11.2 The response for the randomized block design shown in Figure 11.3

Blocks (Runners)	Treatments (Liquids)		
	$A\ (x_1 = 1, x_2 = 0)$	$B\ (x_1 = 0, x_2 = 1)$	$C\ (x_1 = 0, x_2 = 0)$
$1\ (x_3 = 1, x_4 = x_5 = x_6 = 0)$	$y_{A1} = \beta_0 + \beta_1 + \beta_3 + \varepsilon_{A1}$	$y_{B1} = \beta_0 + \beta_2 + \beta_3 + \varepsilon_{B1}$	$y_{C1} = \beta_0 + \beta_3 + \varepsilon_{C1}$
$2\ (x_4 = 1, x_3 = x_5 = x_6 = 0)$	$y_{A2} = \beta_0 + \beta_1 + \beta_4 + \varepsilon_{A2}$	$y_{B2} = \beta_0 + \beta_2 + \beta_4 + \varepsilon_{B2}$	$y_{C2} = \beta_0 + \beta_4 + \varepsilon_{C2}$
$3\ (x_5 = 1, x_3 = x_4 = x_6 = 0)$	$y_{A3} = \beta_0 + \beta_1 + \beta_5 + \varepsilon_{A3}$	$y_{B3} = \beta_0 + \beta_2 + \beta_5 + \varepsilon_{B3}$	$y_{C3} = \beta_0 + \beta_5 + \varepsilon_{C3}$
$4\ (x_6 = 1, x_3 = x_4 = x_5 = 0)$	$y_{A4} = \beta_0 + \beta_1 + \beta_6 + \varepsilon_{A4}$	$y_{B4} = \beta_0 + \beta_2 + \beta_6 + \varepsilon_{B4}$	$y_{C4} = \beta_0 + \beta_6 + \varepsilon_{C4}$
$5\ (x_3 = x_4 = x_5 = x_6 = 0)$	$y_{A5} = \beta_0 + \beta_1 + \varepsilon_{A5}$	$y_{B5} = \beta_0 + \beta_2 + \varepsilon_{B5}$	$y_{C5} = \beta_0 + \varepsilon_{C5}$

five measured responses per treatment. Averaging the five responses for treatment A shown in Table 11.2, we obtain

$$\bar{y}_A = \frac{y_{A1} + y_{A2} + y_{A3} + y_{A4} + y_{A5}}{5}$$

$$= [(\beta_0 + \beta_1 + \beta_3 + \varepsilon_{A1}) + (\beta_0 + \beta_1 + \beta_4 + \varepsilon_{A2}) + (\beta_0 + \beta_1 + \beta_5 + \varepsilon_{A3})$$
$$+ (\beta_0 + \beta_1 + \beta_6 + \varepsilon_{A4}) + (\beta_0 + \beta_1 + \varepsilon_{A5})]/5$$

$$= \frac{5\beta_0 + 5\beta_1 + (\beta_3 + \beta_4 + \beta_5 + \beta_6) + (\varepsilon_{A1} + \varepsilon_{A2} + \varepsilon_{A3} + \varepsilon_{A4} + \varepsilon_{A5})}{5}$$

$$= \beta_0 + \beta_1 + \frac{(\beta_3 + \beta_4 + \beta_5 + \beta_6)}{5} + \bar{\varepsilon}_A$$

Similarly, the mean responses for treatments B and C are obtained:

$$\bar{y}_B = \frac{y_{B1} + y_{B2} + y_{B3} + y_{B4} + y_{B5}}{5}$$

$$= \beta_0 + \beta_2 + \frac{(\beta_3 + \beta_4 + \beta_5 + \beta_6)}{5} + \bar{\varepsilon}_B$$

$$\bar{y}_C = \frac{y_{C1} + y_{C2} + y_{C3} + y_{C4} + y_{C5}}{5}$$

$$= \beta_0 + \frac{(\beta_3 + \beta_4 + \beta_5 + \beta_6)}{5} + \bar{\varepsilon}_C$$

Since the objective is to compare treatment means, we are interested in the differences $\bar{y}_A - \bar{y}_B$, $\bar{y}_A - \bar{y}_C$, and $\bar{y}_B - \bar{y}_C$, which are calculated as follows:

$$\bar{y}_A - \bar{y}_B = [\beta_0 + \beta_1 + (\beta_3 + \beta_4 + \beta_5 + \beta_6)/5 + \bar{\varepsilon}_A]$$
$$- [\beta_0 + \beta_2 + (\beta_3 + \beta_4 + \beta_5 + \beta_6)/5 + \bar{\varepsilon}_B]$$
$$= (\beta_1 - \beta_2) + (\bar{\varepsilon}_A - \bar{\varepsilon}_B)$$

$$\bar{y}_A - \bar{y}_C = [\beta_0 + \beta_1 + (\beta_3 + \beta_4 + \beta_5 + \beta_6)/5 + \bar{\varepsilon}_A]$$
$$- [\beta_0 + (\beta_3 + \beta_4 + \beta_5 + \beta_6)/5 + \bar{\varepsilon}_C]$$
$$= \beta_1 + (\bar{\varepsilon}_A - \bar{\varepsilon}_C)$$

$$\bar{y}_B - \bar{y}_C = [\beta_0 + \beta_2 + (\beta_3 + \beta_4 + \beta_5 + \beta_6)/5 + \bar{\varepsilon}_B]$$
$$- [\beta_0 + (\beta_3 + \beta_4 + \beta_5 + \beta_6)/5 + \bar{\varepsilon}_C]$$
$$= \beta_2 + (\bar{\varepsilon}_B - \bar{\varepsilon}_C)$$

For each pairwise comparison, the block β's (β_3, β_4, β_5, and β_6) cancel out, leaving only the treatment β's (β_1 and β_2). That is, the experimental noise resulting from differences between blocks is eliminated when treatment means are compared. The quantities $\bar{\varepsilon}_A - \bar{\varepsilon}_B$, $\bar{\varepsilon}_A - \bar{\varepsilon}_C$, and $\bar{\varepsilon}_B - \bar{\varepsilon}_C$ are the errors of estimation and represent the noise that tends to obscure the true differences between the treatment means.

What would occur if we employed the completely randomized design in Table 11.1 rather than the randomized block design? Since each runner is assigned to drink a single liquid, each treatment does not appear in each block. Consequently, when we compare the treatment means, the runner-to-runner variation (i.e., the block effects) will not cancel. For example, the difference between $\bar{y}_A$ and $\bar{y}_C$ would be

$$\bar{y}_A - \bar{y}_C = \beta_1 + \underbrace{(\text{Block }\beta\text{'s that do not cancel}) + (\bar{\varepsilon}_A - \bar{\varepsilon}_C)}_{\text{Error of estimation}}$$

Thus, for the completely randomized design, the error of estimation will be increased by an amount involving the block effects (β_3, β_4, β_5, and β_6) that do not cancel. These effects, which inflate the error of estimation, cancel out for the randomized block design, thereby reducing the noise in the experiment.

Example 11.3

Refer to Example 11.2 and the randomized block design employed to compare the mean percentage error rates for the four appraisers. The design is illustrated in Figure 11.4.

(a) Write the model for the randomized block design.

(b) Interpret the β parameters of the model, part a.

(c) How can we use the model, part a, to test for differences among the mean percentage error rates of the four appraisers?

Solution

(a) The experiment involves a qualitative factor (Appraisers) at four levels, which represent the treatments. The blocks for the experiment are the 10 properties. Therefore, the model is

$$E(y) = \beta_0 + \underbrace{\beta_1 x_1 + \beta_2 x_2 + \beta_3 x_3}_{\text{Treatments (Appraisers)}} + \underbrace{\beta_4 x_4 + \beta_5 x_5 + \cdots + \beta_{12} x_{12}}_{\text{Blocks (Properties)}}$$

where

$$x_1 = \begin{cases} 1 & \text{if appraiser A} \\ 0 & \text{if not} \end{cases} \quad x_2 = \begin{cases} 1 & \text{if appraiser B} \\ 0 & \text{if not} \end{cases} \quad x_3 = \begin{cases} 1 & \text{if appraiser C} \\ 0 & \text{if not} \end{cases}$$

$$x_4 = \begin{cases} 1 & \text{if property 1} \\ 0 & \text{if not} \end{cases} \quad x_5 = \begin{cases} 1 & \text{if property 2} \\ 0 & \text{if not,} \end{cases} \cdots \quad x_{12} = \begin{cases} 1 & \text{if property 9} \\ 0 & \text{if not} \end{cases}$$

(b) Note that we have arbitrarily selected appraiser D and property 10 as the base levels. Following our discussion in Section 5.8, the interpretations of the β's are

$\beta_1 = \mu_A - \mu_D$ for a given property
$\beta_2 = \mu_B - \mu_D$ for a given property
$\beta_3 = \mu_C - \mu_D$ for a given property

$\beta_4 = \mu_1 - \mu_{10}$ for a given appraiser

$\beta_5 = \mu_2 - \mu_{10}$ for a given appraiser

$\vdots$

$\beta_{12} = \mu_9 - \mu_{10}$ for a given appraiser

(c) One way to determine whether the means for the four appraisers differ is to test the null hypothesis

$$H_0 : \mu_A = \mu_B = \mu_C = \mu_D$$

From our β interpretations in part b, this hypothesis is equivalent to testing

$$H_0 : \beta_1 = \beta_2 = \beta_3 = 0$$

To test this hypothesis, we drop the treatment β's (β_1, β_2, and β_3) from the complete model and fit the reduced model

$$E(y) = \beta_0 + \beta_4 x_4 + \beta_5 x_5 + \cdots + \beta_{12} x_{12}$$

Then we conduct the nested model partial F-test (see Section 4.13), where

$$F = \frac{(\text{SSE}_{\text{Reduced}} - \text{SSE}_{\text{Complete}})/3}{\text{MSE}_{\text{Complete}}}$$

The randomized block design represents one of the simplest types of noise-reducing designs. Other, more complex designs that employ the principle of blocking remove trends or variation in two or more directions. The **Latin square design** is useful when you want to eliminate two sources of variation (i.e., when you want to block in two directions). **Latin cube designs** allow you to block in three directions. A further variation in blocking occurs when the block contains fewer experimental units than the number of treatments. By properly assigning the treatments to a specified number of blocks, you can still obtain an estimate of the difference between a pair of treatments free of block effects. These are known as **incomplete block designs**. Consult the references for details on how to set up these more complex block designs.

11.4 Exercises

11.1 Quantity of information in an experiment.

  (a) What two factors affect the quantity of information in an experiment?

  (b) How do block designs increase the quantity of information in an experiment?

11.2 Accounting and Machiavellianism. A study of Machiavellian traits in accountants was published in *Behavioral Research in Accounting* (January 2008). Recall (from Exercise 1.6, p. 4) that *Machiavellian* describes negative character traits such as manipulation, cunning, duplicity, deception, and bad faith. A Machiavellian ("Mach") rating score was determined for each in a sample of accounting alumni of a large southwestern university. The accountants were then classified as having high, moderate, or low Mach rating scores. For one portion of the study, the researcher investigated the impact of both Mach score classification and gender on the average income of

an accountant. For this experiment, identify each of the following:

  (a) experimental unit
  (b) response variable
  (c) factors
  (d) levels of each factor
  (e) treatments

11.3 Taste preferences of cockatiels. *Applied Animal Behaviour Science* (October 2000) published a study of the taste preferences of caged cockatiels. A sample of birds bred at the University of California, Davis, were randomly divided into three experimental groups. Group 1 was fed purified water in bottles on both sides of the cage. Group 2 was fed water on one side and a liquid sucrose (sweet) mixture on the opposite side of the cage. Group 3 was fed water on one side and a liquid sodium chloride (sour) mixture on the opposite side of the cage. One variable of interest to the researchers was total consumption of liquid by each cockatiel.

(a) What is the experimental unit for this study?
(b) Is the study a designed experiment? What type of design is employed?
(c) What are the factors in the study?
(d) Give the levels of each factor.
(e) How many treatments are in the study? Identify them.
(f) What is the response variable?
(g) Write the regression model for the designed experiment.

11.4 Peer mentor training at a firm. Peer mentoring occurs when a more experienced employee provides one-on-one support and knowledge sharing with a less experienced employee. The *Journal of Managerial Issues* (Spring 2008) published a study of the impact of peer mentor training at a large software company. Participants were 222 employees who volunteered to attend a one-day peer mentor training session. One variable of interest was the employee's level of competence in peer mentoring (measured on a 7-point scale). The competence level of each trainee was measured at three different times in the study: one week before training, two days after training, and two months after training. One goal of the experiment was to compare the mean competence levels of the three time periods.

(a) Identify the response variable.
(b) Identify the factors (and factor levels) in the experiment.
(c) How many treatments are included in the experiment?
(d) What type of experimental design is employed?
(e) Identify the blocks in the experiment.

11.5 Treatment means in a block design. Refer to the randomized block design of Examples 11.2 and 11.3.

(a) Write the model for each observation of percentage appraisal error y for appraiser B. Sum the observations to obtain the average for appraiser B.
(b) Repeat part a for appraiser D.
(c) Show that $(\bar{y}_B - \bar{y}_D) = \beta_2 + (\bar{\varepsilon}_B - \bar{\varepsilon}_D)$. Note that the β's for blocks cancel when computing this difference.

11.6 Drift ratios of buildings. A commonly used index to estimate the reliability of a building subjected to lateral loads is the drift ratio. Sophisticated computer programs such as STAAD-III have been developed to estimate the drift ratio based on variables such as beam stiffness, column stiffness, story height, moment of inertia, etc. Civil engineers at SUNY, Buffalo, and the University of Central Florida performed an experiment to compare drift ratio estimates using STAAD-III with the estimates produced by a new, simpler micro-computer program called DRIFT (*Microcomputers in Civil Engineering*, 1993). Data for a 21-story building were used as input to the programs. Two runs were made with STAAD-III: Run 1 considered axial deformation of the building columns, and run 2 neglected this information. The goal of the analysis was to compare the mean drift ratios (where drift is measured as lateral displacement) estimated by the three computer runs.

(a) Identify the treatments in the experiment.
(b) Because lateral displacement will vary greatly across building levels (floors), a randomized block design will be used to reduce the level-to-level variation in drift. Explain, diagrammatically, the setup of the design if all 21 levels are to be included in the study.
(c) Write the linear model for the randomized block design.

11.5 Volume-Increasing Designs

In this section, we focus on how the proper choice of the treatments associated with *two or more factors* can increase the "volume" of information extracted from the experiment. The volume-increasing designs we discuss are commonly known as **factorial designs** because they involve careful selection of the combinations of **factor levels** (i.e., treatments) in the experiment.

Consider a utility company that charges its customers a lower rate for using electricity during off-peak (less demanded) hours. The company is experimenting with several time-of-day pricing schedules. Two factors (i.e., independent variables) that the company can manipulate to form the schedule are price ratio, x_1, measured as the ratio of peak to off-peak prices, and peak period length, x_2, measured in hours.

Suppose the utility company wants to investigate pricing ratio at two levels, 200% and 400%, and peak period length at two levels, 6 and 9 hours. The company will measure customer satisfaction, y, for several different schedules (i.e., combinations of x_1 and x_2) with the goal of comparing the mean satisfaction levels of the schedules. How should the company select the treatments for the experiment?

One method of selecting the combined levels of price ratio and peak period length to be assigned to the experimental units (customers) would be to use the "one-at-a-time" approach. According to this procedure, one independent variable is varied while the remaining independent variables are held constant. This process is repeated only once for each of the independent variables in the experiment. This plan would *appear* to be extremely logical and consistent with the concept of blocking introduced in Section 11.4—that is, making comparisons within relatively homogeneous conditions—but this is not the case, as we demonstrate.

The one-at-a-time approach applied to price ratio (x_1) and peak period length (x_2) is illustrated in Figure 11.5. When length is held constant at $x_2 = 6$ hours, we observe the response y at a ratio of $x_1 = 200\%$ and $x_1 = 400\%$, thus yielding one pair of y values to estimate the average change in customer satisfaction as a result of changing the pricing ratio (x_1). Also, when pricing ratio is held constant at $x_1 = 200\%$, we observe the response y at a peak period length of $x_2 = 9$ hours. This observation, along with the one at (200%, 6 hours), allows us to estimate the average change in customer satisfaction due to a change in peak period length (x_2). The three treatments just described, (200%, 6 hours), (400%, 6 hours), and (200%, 9 hours), are indicated as points in Figure 11.5. The figure shows two measurements (points) for each treatment. This is necessary to obtain an estimate of the standard deviation of the differences of interest.

Figure 11.5 One-at-a-time approach to selecting treatments

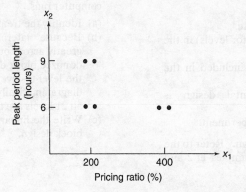

A second method of selecting the factor level combinations would be to choose the same three treatments as implied by the one-at-a-time approach and then to choose the fourth treatment at (400%, 9 hours) as shown in Figure 11.6. In other words, we have varied both variables, x_1 and x_2, at the same time.

Which of the two designs yields more information about the treatment differences? Surprisingly, the design of Figure 11.6, with only four observations, yields more accurate information than the one-at-a-time approach with its six observations. First, note that both designs yield two estimates of the difference between the mean response y at $x_1 = 200\%$ and $x_1 = 400\%$ when peak period length (x_2) is held constant, and both yield two estimates of the difference between the mean response y at $x_2 = 6$ hours and $x_2 = 9$ hours when pricing ratio (x_1) is held constant. But what if the

Figure 11.6 Selecting all possible treatments

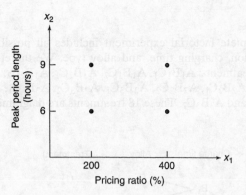

difference between the mean response y at $x_1 = 200\%$ and at $x_1 = 400\%$ depends on which level of x_2 is held fixed? That is, what if pricing ratio (x_1) and peak period length (x_2) *interact*? Then, we require estimates of the mean difference ($\mu_{200} - \mu_{400}$) when $x_2 = 6$ and the mean difference ($\mu_{200} - \mu_{400}$) when $x_2 = 9$. Estimates of both these differences are obtainable from the second design, Figure 11.6. However, since no estimate of the mean response for $x_1 = 400$ and $x_2 = 9$ is available from the one-at-a-time method, the interaction will go undetected for this design!

The importance of interaction between independent variables was emphasized in Section 4.10 and Chapter 5. If interaction is present, we cannot study the effect of one variable (or factor) on the response y independent of the other variable. Consequently, we require experimental designs that provide information on factor interaction.

Designs that accomplish this objective are called **factorial experiments**. A **complete factorial experiment** is one that includes all possible combinations of the levels of the factors as treatments. For the experiment on time-of-day pricing, we have two levels of pricing ratio (200% and 400%) and two levels of peak period length (6 and 9 hours). Hence, a complete factorial experiment will include $(2 \times 2) = 4$ treatments, as shown in Figure 11.6, and is called a **2 × 2 factorial design**.

> **Definition 11.10** A **factorial design** is a method for selecting the treatments (i.e., the factor level combinations) to be included in an experiment. A complete factorial experiment is one in which the treatments consist of all factor level combinations.

If we were to include a third factor, say, season, at four levels, then a complete factorial experiment would include all $2 \times 2 \times 4 = 16$ combinations of pricing ratio, peak period length, and season. The resulting collection of data would be called a **2 × 2 × 4 factorial design**.

Example 11.4

Suppose you plan to conduct an experiment to compare the yield strengths of nickel alloy tensile specimens charged in a sulfuric acid solution. In particular, you want to investigate the effect on mean strength of three factors: nickel composition at three levels (A_1, A_2, and A_3), charging time at three levels (B_1, B_2, and B_3), and alloy type at two levels (C_1 and C_2). Consider a complete factorial experiment. Identify the treatments for this $3 \times 3 \times 2$ factorial design.

Solution

The complete factorial experiment includes all possible combinations of nickel composition, charging time, and alloy type. We therefore would include the following treatments: $A_1B_1C_1$, $A_1B_1C_2$, $A_1B_2C_1$, $A_1B_2C_2$, $A_1B_3C_1$, $A_1B_3C_2$, $A_2B_1C_1$, $A_2B_1C_2$, $A_2B_2C_1$, $A_2B_2C_2$, $A_2B_3C_1$, $A_2B_3C_2$, $A_3B_1C_1$, $A_3B_1C_2$, $A_3B_2C_1$, $A_3B_2C_2$, $A_3B_3C_1$, and $A_3B_3C_2$. These 18 treatments are diagrammed in Figure 11.7.   ■

Figure 11.7 The 18 treatments for the $3 \times 3 \times 2$ factorial in Example 11.4

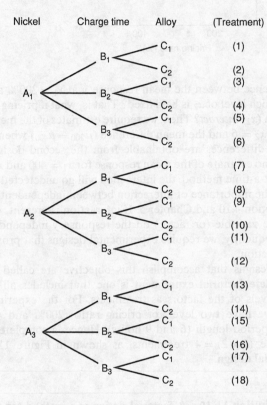

The linear statistical model for a factorial design includes terms for each of the factors in the experiment—called **main effects**—and terms for factor interactions. For example, the model for the 2×2 factorial for the time-of-day pricing experiment includes a first-order term for the quantitative factor, pricing ratio (x_1); a first-order term for the quantitative factor, peak period length (x_2); and an interaction term:

$$y = \beta_0 + \underbrace{\beta_1 x_1 + \beta_2 x_2}_{\text{Main effects}} + \underbrace{\beta_3 x_1 x_2}_{\text{Interaction}} + \varepsilon$$

In general, the regression model for a complete factorial design for k factors contains terms for the following:

The main effects for each of the k factors

Two-way interaction terms for all pairs of factors

Three-way interaction terms for all combinations of three factors

⋮

k-way interaction terms of all combinations of k factors

If the factors are qualitative, then we set up dummy variables and proceed as in the next example.

Example 11.5

Write the model for the $3 \times 3 \times 2$ factorial experiment in Example 11.4.

Solution

Since the factors are qualitative, we set up dummy variables as follows:

$$x_1 = \begin{cases} 1 & \text{if nickel A}_1 \\ 0 & \text{if not} \end{cases} \qquad x_2 = \begin{cases} 1 & \text{if nickel A}_2 \\ 0 & \text{if not} \end{cases}$$

$$x_3 = \begin{cases} 1 & \text{if charge B}_1 \\ 0 & \text{if not} \end{cases} \qquad x_4 = \begin{cases} 1 & \text{if charge B}_2 \\ 0 & \text{if not} \end{cases}$$

$$x_5 = \begin{cases} 1 & \text{if alloy C}_1 \\ 0 & \text{if alloy C}_2 \end{cases}$$

Then the appropriate model is

$$y = \beta_0 + \underbrace{\beta_1 x_1 + \beta_2 x_2}_{\text{Nickel main effects}} + \underbrace{\beta_3 x_3 + \beta_4 x_4}_{\text{Charge main effects}} + \underbrace{\beta_5 x_5}_{\text{Alloy main effect}}$$

$$+ \underbrace{\beta_6 x_1 x_3 + \beta_7 x_1 x_4 + \beta_8 x_2 x_3 + \beta_9 x_2 x_4}_{\text{Nickel} \times \text{Charge}} + \underbrace{\beta_{10} x_1 x_5 + \beta_{11} x_2 x_5}_{\text{Nickel} \times \text{Alloy}}$$

$$+ \underbrace{\beta_{12} x_3 x_5 + \beta_{13} x_4 x_5}_{\text{Charge} \times \text{Alloy}}$$

$$+ \underbrace{\beta_{14} x_1 x_3 x_5 + \beta_{15} x_1 x_4 x_5 + \beta_{16} x_2 x_3 x_5 + \beta_{17} x_2 x_4 x_5}_{\text{Nickel} \times \text{Charge} \times \text{Alloy}}$$

Note that the number of parameters in the model for the $3 \times 3 \times 2$ factorial design in Example 11.5 is 18, which is equal to the number of treatments contained in the experiment. This is always the case for a complete factorial experiment. Consequently, if we fit the complete model to a single replication of the factorial treatments (i.e., one y observation measured per treatment), we will have no degrees of freedom available for estimating the error variance, σ^2. One way to solve this problem is to add additional data points to the sample. Researchers usually accomplish this by **replicating** the complete set of factorial treatments. That is, we collect two or more observed y values for each treatment in the experiment. This provides sufficient degrees of freedom for estimating σ^2.

One potential disadvantage of a complete factorial experiment is that it may require a large number of treatments. For example, an experiment involving 10 factors each at two levels would require $2^{10} = 1,024$ treatments! This might occur in an exploratory study where we are attempting to determine which of a large set of factors affect the response y. Several volume-increasing designs are available that employ only a fraction of the total number of treatments in a complete factorial experiment. For this reason, they are called **fractional factorial experiments**. Fractional factorials permit the estimation of the β parameters of lower-order terms (e.g., main effects and two-way interactions); however, β estimates of certain higher-order terms (e.g., three-way and four-way interactions) will be the same as some lower-order terms, thus confounding the results of the experiment. Consequently, a great deal of expertise is required to run and interpret fractional factorial experiments. Consult the references for details on fractional factorials and other more complex, volume-increasing designs.

11.5 Exercises

11.7 Quantity of information in an experiment. In what sense does a factorial experiment increase the quantity of information in an experiment?

11.8 Baker's versus brewer's yeast. The *Electronic Journal of Biotechnology* (December 15, 2003) published an article on a comparison of two yeast extracts, baker's yeast and brewer's yeast. Brewer's yeast is a surplus by-product obtained from a brewery, hence it is less expensive than primary-grown baker's yeast. Samples of both yeast extracts were prepared at four different temperatures (45, 48, 51, and 54°C), and the autolysis yield (recorded as a percentage) was measured for each of the yeast–temperature combinations. The goal of the analysis is to investigate the impact of yeast extract and temperature on mean autolysis yield.

(a) Identify the factors (and factor levels) in the experiment.
(b) Identify the response variable.
(c) How many treatments are included in the experiment?
(d) What type of experimental design is employed?

11.9 Exam performance study. In *Teaching of Psychology* (August 1998), a study investigated whether final exam performance is affected by whether students take a practice test. Students in an introductory psychology class at Pennsylvania State University were initially divided into three groups based on their class standing: low, medium, or high. Within each group, the students were randomly assigned either to attend a review session or to take a practice test before the final exam. Six groups were formed: low, review; low, practice exam; medium, review; medium, practice exam; high, review; and high, practice exam: One goal of the study was to compare the mean final exam scores of the six groups of students.

(a) What is the experimental unit for this study?
(b) Is the study a designed experiment? What type of design is employed?
(c) What are the factors in the study?
(d) Give the levels of each factor.
(e) How many treatments are in the study? Identify them.
(f) What is the response variable?

11.10 Testing a new pain-reliever tablet. Paracetamol is the active ingredient in drugs designed to relieve mild to moderate pain and fever. The properties of paracetamol tablets derived from khaya gum were studied in the *Tropical Journal of Pharmaceutical Research* (June 2003). Three factors believed to impact the properties of paracetamol tablets are (1) the nature of the binding agent, (2) the concentration of the binding agent, and (3) the relative density of the tablet. In the experiment, binding agent was set at two levels (khaya gum and PVP), binding concentration at two levels (.5% and 4.0%), and relative density at two levels (low and high). One of the dependent variables investigated in the study was tablet dissolution time, that is, the amount of time (in minutes) for 50% of the tablet to dissolve. The goal of the study was to determine the effect of binding agent, binding concentration, and relative density on mean dissolution time.

(a) Identify the dependent (response) variable in the study.
(b) What are the factors investigated in the study? Give the levels of each.
(c) How many treatments are possible in the study? List them.

11.11 Reaction time in a visual search. Many cognitively demanding jobs (e.g., air traffic controller, radar/sonar operator) require efficient processing of visual information. Researchers at Georgia Tech investigated the variables that affect the reaction time of subjects performing a visual search task (*Human Factors*, June 1993). College students were trained on computers using one of two methods: continuously consistent or adjusted consistent. Each student was then assigned to one of six different practice sessions. Finally, the consistency of the search task was manipulated at four degrees: 100%, 67%, 50%, or 33%. The goal of the researcher was to compare the mean reaction times of students assigned to each of the (training method) × (practice session) × (task consistency) = $2 \times 6 \times 4 = 48$ experimental conditions.

(a) List the factors involved in the experiment.
(b) For each factor, state whether it is quantitative or qualitative.
(c) How many treatments are involved in this experiment? List them.

11.12 Two-factor factorial design. Consider a factorial design with two factors, A and B, each at three levels. Suppose we select the following treatment (factor level) combinations to be included in the experiment: A_1B_1, A_2B_1, A_3B_1, A_1B_2, and A_1B_3.

(a) Is this a complete factorial experiment? Explain.

(b) Explain why it is impossible to investigate AB interaction in this experiment.

11.13 Models for a factorial design. Write the complete factorial model for:

(a) A 2×3 factorial experiment where both factors are qualitative.

(b) A $2 \times 3 \times 3$ factorial experiment where the factor at two levels is quantitative and the other two factors are qualitative.

11.14 Selecting levels of a factor. Suppose you wish to investigate the effect of three factors on a response y. Explain why a factorial selection of treatments is better than varying each factor, one at a time, while holding the remaining two factors constant.

11.15 Drawback to a block design. Why is the randomized block design a poor design to use to investigate the effect of two qualitative factors on a response y?

11.6 Selecting the Sample Size

We demonstrated how to select the sample size for estimating a single population mean or comparing two population means in Sections 1.8 and 1.10. We now show you how this problem can be solved for designed experiments.

As mentioned in Section 11.3, a measure of the quantity of information in an experiment that is pertinent to a particular population parameter is the standard error of the estimator of the parameter. A more practical measure is the half-width of the parameter's confidence interval, which will, of course, be a function of the standard error. For example, the half-width of a confidence interval for a population mean (given in Section 1.8) is

$$(t_{\alpha/2})s_{\bar{y}} = t_{\alpha/2}\left(\frac{s}{\sqrt{n}}\right)$$

Similarly, the half-width of a confidence interval for the slope β_1 of a straight-line model relating y to x (given in Section 3.6) is

$$(t_{\alpha/2})s_{\hat{\beta}_1} = t_{\alpha/2}\left(\frac{s}{\sqrt{SS_{xx}}}\right) = t_{\alpha/2}\left(\sqrt{\frac{SSE}{n-2}}\right)\left(\frac{1}{\sqrt{SS_{xx}}}\right)$$

In both cases, the half-width is a function of the total number of data points in the experiment; each interval half-width gets smaller as the total number of data points n increases. The same is true for a confidence interval for a parameter β_i of a general linear model, for a confidence interval for $E(y)$, and for a prediction interval for y. Since each designed experiment can be represented by a linear model, this result can be used to select, approximately, the number of replications (i.e., the number of observations measured for each treatment) in the experiment.

For example, consider a designed experiment consisting of three treatments, A, B, and C. Suppose we want to estimate $(\mu_B - \mu_C)$, the difference between the treatment means for B and C. From our knowledge of linear models for designed experiments, we know this difference will be represented by one of the β parameters in the model, say, β_2. The confidence interval for β_2 for a single replication of the experiment is

$$\hat{\beta}_2 \pm (t_{\alpha/2})s_{\hat{\beta}_2}$$

If we repeat exactly the same experiment r times (we call this **r replications**), it can be shown (proof omitted) that the confidence interval for β_2 will be

$$\hat{\beta}_2 \pm B \quad \text{where } B = t_{\alpha/2}\left(\frac{s_{\hat{\beta}_2}}{\sqrt{r}}\right)$$

To find r, we first set the half-width of the interval to the largest value, B, we are willing to tolerate. Then we approximate $(t_{\alpha/2})$ and $s_{\hat{\beta}_2}$ and solve for the number of replications r.

Example 11.6

Consider a 2×2 factorial experiment to investigate the effect of two factors on the light output y of flashbulbs used in cameras. The two factors (and their levels) are: $x_1 =$ Amount of foil contained in the bulb (100 and 200 milligrams) and $x_2 =$ Speed of sealing machine (1.2 and 1.3 revolutions per minute). The complete model for the 2×2 factorial experiment is

$$E(y) = \beta_0 + \beta_1 x_1 + \beta_2 x_2 + \beta_3 x_1 x_2$$

How many replicates of the 2×2 factorial are required to estimate β_3, the interaction β, to within .3 of its true value using a 95% confidence interval?

Solution

To solve for the number of replicates, r, we want to solve the equation

$$t_{\alpha/2}\left(\frac{s_{\hat{\beta}_3}}{\sqrt{r}}\right) = B$$

You can see that we need to have an estimate of $s_{\hat{\beta}_3}$, the standard error of $\hat{\beta}_3$, for a single replication. Suppose it is known from a previous experiment conducted by the manufacturer of the flashbulbs that $s_{\hat{\beta}_3} \approx .2$. For a 95% confidence interval, $\alpha = .05$ and $\alpha/2 = .025$. Since we want the half-width of the interval to be $B = .3$, we have

$$t_{.025}\left(\frac{.2}{\sqrt{r}}\right) = .3$$

The degrees of freedom for $t_{.025}$ will depend on the sample size $n = (2 \times 2)r = 4r$; consequently, we must approximate its value. In fact, since the model includes four parameters, the degrees of freedom for t will be $\text{df(Error)} = n - 4 = 4r - 4 = 4(r - 1)$. At minimum, we require two replicates; hence, we will have at least $4(2 - 1) = 4$ df. In Table 2 in Appendix D, we find $t_{.025} = 2.776$ for df $= 4$. We will use this conservative estimate of t in our calculations.

Substituting $t = 2.776$ into the equation, we have

$$\frac{2.776(.2)}{\sqrt{r}} = .3$$

$$\sqrt{r} = \frac{(2.776)(.2)}{.3} = 1.85$$

$$r = 3.42$$

Since we can run either three or four replications (but not 3.42), we should choose four replications to be reasonably certain that we will be able to estimate the interaction parameter, β_3, to within .3 of its true value. The 2×2 factorial with four replicates would be laid out as shown in Table 11.3.

Table 11.3 2×2 factorial, with four replicates

		Amount of Foil, x_1	
		100	200
MACHINE SPEED,	1.2	4 observations on y	4 observations on y
x_2	1.3	4 observations on y	4 observations on y

11.6 Exercises

11.16 Replication in an experiment. Why is replication important in a complete factorial experiment?

11.17 Estimating the number of replicates. Consider a 2×2 factorial. How many replications are required to estimate the interaction β to within two units with a 90% confidence interval? Assume that the standard error of the estimate of the interaction β (based on a single replication) is approximately 3.

11.18 Finding the number of blocks. For a randomized block design with b blocks, the estimated standard error of the estimated difference between any two treatment means is $s\sqrt{2/b}$. Use this formula to determine the number of blocks required to estimate $(\mu_A - \mu_B)$, the difference between two treatment means, to within 10 units using a 95% confidence interval. Assume $s \approx 15$.

11.7 The Importance of Randomization

All the basic designs presented in this chapter involve **randomization** of some sort. In a completely randomized design and a basic factorial experiment, the treatments are randomly assigned to the experimental units. In a randomized block design, the blocks are randomly selected and the treatments within each block are assigned in random order. Why randomize? The answer is related to the assumptions we make about the random error ε in the linear model. Recall (Section 4.2) our assumption that ε follows a normal distribution with mean 0 and constant variance σ^2 for fixed settings of the independent variables (i.e., for each of the treatments). Furthermore, we assume that the random errors associated with repeated observations are independent of each other in a probabilistic sense.

Experimenters rarely know all of the important variables in a process, nor do they know the true functional form of the model. Hence, the functional form chosen to fit the true relation is only an approximation, and the variables included in the experiment form only a subset of the total. The random error, ε, is thus a composite error caused by the failure to include all of the important factors as well as the error in approximating the function.

Although many unmeasured and important independent variables affecting the response y do not vary in a completely random manner during the conduct of a designed experiment, we hope their behavior is such that their cumulative effect varies in a random manner and satisfies the assumptions upon which our inferential procedures are based. *The randomization in a designed experiment has the effect of randomly assigning these error effects to the treatments and assists in satisfying the assumptions on ε.*

Quick Summary

KEY FORMULAS

Finding the Number of Replicates, r, for Estimating a Difference in Means Represented by β_j

$$r = \left[\frac{(t_{\alpha/2})(s_{\hat{\beta}_j})}{B} \right]^2$$

where B = half-width of the $100(1 - \alpha)\%$ confidence interval

$t_{\alpha/2}$ is based on df(Error) for model
$s_{\hat{\beta}_j}$ = standard error of $\hat{\beta}_j$

KEY IDEAS

Experimental Design: A plan for collecting the data

Steps in Experimental Design

1. Select the *factors*

2. Choose the factor level combinations (*treatments*)

3. Determine the number of *replicates* for each treatment

4. Assign the *treatments* to the *experimental units*

Two Methods for Assigning Treatments

1. *Completely randomized design*

2. *Randomized block design*

Amount of Information in Experimental Design is controlled by manipulating:

1. *Volume* of the signal

2. *Noise* (random variation) in the data

Volume-increasing Design: *Complete factorial experiment* (select all combinations of factor levels)

Noise-reducing Design: *Randomized block design* (assign treatments to homogeneous blocks of experimental units)

Supplementary Exercises

11.19. Quantity of information. How do you measure the quantity of information in a sample that is pertinent to a particular population parameter?

11.20. Volume of the signal. What steps in the design of an experiment affect the volume of the signal pertinent to a particular population parameter?

11.21. Reducing noise. In what step in the design of an experiment can you possibly reduce the variation produced by extraneous and uncontrolled variables?

11.22. Comparing designs. Explain the difference between a completely randomized design and a randomized block design. When is a randomized block design more advantageous?

11.23. Experimental treatments. Consider a two-factor factorial experiment where one factor is set at two levels and the other factor is set at four levels. How many treatments are included in the experiment? List them.

11.24. Complete factorial design. Write the complete factorial model for a $2 \times 2 \times 4$ factorial experiment where both factors at two levels are quantitative and the third factor at four levels is qualitative. If you conduct one replication of this experiment, how many degrees of freedom will be available for estimating σ^2?

11.25. Complete factorial design (continued). Refer to Exercise 11.24. Write the model for y assuming that you wish to enter main-effect terms for the factors, but no terms for factor interactions. How many degrees of freedom will be available for estimating σ^2?

11.26. Beer brand market shares. Retail store audits are periodic audits of a sample of retail sales to monitor inventory and purchases of a particular product. Such audits are often used by marketing researchers to estimate market share. A study was conducted to compare market shares of beer brands estimated by two different auditing methods.

(a) Identify the treatments in the experiment.

(b) Because of brand-to-brand variation in estimated market share, a randomized block design will be used. Explain how the treatments might be assigned to the experimental units if 10 beer brands are to be included in the study.

(c) Write the linear model for the randomized block design.

11.27. Time to perform a fire fighting task. Researchers investigated the effect of gender (male or female) and weight (light or heavy) on the length of time required by firefighters to perform a particular firefighting task (*Human Factors*, 1982). Eight firefighters were selected in each of the four gender–weight categories. Each firefighter was required to perform a certain task. The time (in minutes) needed to perform the task was recorded for each.

(a) List the factors involved in the experiment.

(b) For each factor, state whether it is quantitative or qualitative.

(c) How many treatments are involved in this experiment? List them.

References

Box G. E. P., Hunter, W. G., and Hunter, J. S. *Statistics for Experimenters: Design, Innovation, and Discovery*, 2nd ed. New York: Wiley, 2005.

Cochran, W. G., and Cox, G. M. *Experimental Designs*, 2nd ed. New York: Wiley, 1992 (paperback).

Davies, O. L. *The Design and Analysis of Industrial Experiments*, 2nd ed. New York: Hafner, 1967.

Kirk, R. E. *Experimental Design*: *Procedures for Behavioral Sciences*, 3rd ed. Pacific Grove, Calif.: Brooks/Cole, 1995.

Kutner, M., Nachtsheim, C., Neter, J., and Li, W. *Applied Linear Statistical Models*, 5th ed. Homewood, Ill.: Richard D. Irwin, 2004.

Mendenhall, W. *Introduction to Linear Models and the Design and Analysis of Experiments*. Belmont, Calif.: Wadsworth, 1968.

Winer, B. J., Brown, D. R., and Michels, K. M. *Statistical Principles in Experimental Design*, 3rd ed. New York: McGraw-Hill, 1991.

THE ANALYSIS OF VARIANCE FOR DESIGNED EXPERIMENTS

Chapter

12

Contents

Objectives

1. To present a method for analyzing data collected from designed experiments for comparing two or more population means.
2. To define the relationship of the analysis of variance to regression analysis and to identify their common features.

3. To apply ANOVA tests in completely randomized, randomized block, and factorial designs.

12.1 Introduction

Once the data for a designed experiment have been collected, we will want to use the sample information to make inferences about the population means associated with the various treatments. The method used to compare the treatment means is traditionally known as **analysis of variance**, or **ANOVA**. The analysis of variance procedure provides a set of formulas that enable us to compute test statistics and confidence intervals required to make these inferences.

The formulas—one set for each experimental design—were developed in the early 1900s, well before the invention of computers. The formulas are easy to use, although the calculations can become quite tedious. However, you will recall from Chapter 11 that a linear model is associated with each experimental design. Consequently, the same inferences derived from the ANOVA calculation formulas can be obtained by properly analyzing the model using a regression analysis and the computer.

In this chapter, the main focus is on the regression approach to analyzing data from a designed experiment. Several common experimental designs—some of which were presented in Chapter 11—are analyzed. We also provide the ANOVA calculation formulas for each design and show their relationship to regression. First, we provide the logic behind an analysis of variance and these formulas in Section 12.2.

12.2 The Logic Behind an Analysis of Variance

The concept behind an analysis of variance can be explained using the following simple example.

Consider an experiment with a single factor at two levels (i.e., two treatments). Suppose we want to decide whether the two treatment means differ based on the means of two independent random samples, each containing $n_1 = n_2 = 5$ measurements, and that the y-values appear as in Figure 12.1. Note that the five circles on the left are plots of the y-values for sample 1 and the five solid dots on the right are plots of the y-values for sample 2. Also, observe the horizontal lines that pass through the means for the two samples $\bar{y}_1$ and $\bar{y}_2$. Do you think the plots provide sufficient evidence to indicate a difference between the corresponding population means?

Figure 12.1 Plots of data for two samples

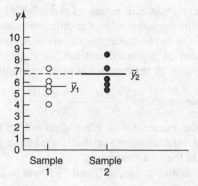

If you are uncertain whether the population means differ for the data in Figure 12.1, examine the situation for two different samples in Figure 12.2a. We think that you will agree that for these data, it appears that the population means differ. Examine a third case in Figure 12.2b. For these data, it appears that there is little or no difference between the population means.

What elements of Figures 12.1 and 12.2 did we intuitively use to decide whether the data indicate a difference between the population means? The answer to the question is that we visually compared the distance (the variation) *between* the sample means to the variation *within* the y-values for each of the two samples. Since the difference between the sample means in Figure 12.2a is large relative to the

Figure 12.2 Plots of data for two cases

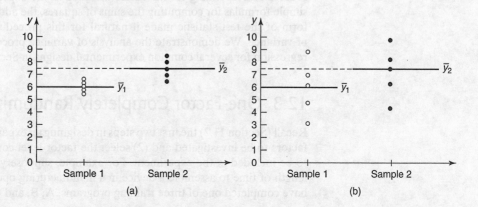

within-sample variation, we inferred that the population means differ. Conversely, in Figure 12.2b, the variation between the sample means is small relative to the within-sample variation, and therefore there is little evidence to imply that the means are significantly different.

The variation within samples is measured by the pooled s^2 that we computed for the independent random samples t-test in Section 1.9, namely,

Within-sample variation: $\quad s^2 = \dfrac{\sum\limits_{i=1}^{n_1}(y_{i1} - \bar{y}_1)^2 + \sum\limits_{i=1}^{n_2}(y_{i2} - \bar{y}_2)^2}{n_1 + n_2 - 2}$

$$= \dfrac{\text{SSE}}{n_1 + n_2 - 2}$$

where y_{i1} is the ith observation in sample 1 and y_{i2} is the ith observation in sample 2. The quantity in the numerator of s^2 is often denoted **SSE**, the **sum of squared errors**. As with regression analysis, SSE measures unexplained variability. But in this case, it measures variability *unexplained* by the differences between the sample means.

A measure of the between-sample variation is given by the weighted sum of squares of deviations of the individual sample means about the mean for all 10 observations, $\bar{y}$, divided by the number of samples minus 1, that is,

Between-sample variation: $\quad \dfrac{n_1(\bar{y}_1 - \bar{y})^2 + n_2(\bar{y}_2 - \bar{y})^2}{2 - 1} = \dfrac{\text{SST}}{1}$

The quantity in the numerator is often denoted **SST**, the **sum of squares for treatments**, since it measures the variability *explained* by the differences between the sample means of the two treatments.

For this experimental design, SSE and SST sum to a known total, namely,

$$\text{SS(Total)} = \sum(y_i - \bar{y})^2$$

[*Note*: SS(Total) is equivalent to SS_{yy} in regression.] Also, the ratio

$$F = \dfrac{\text{Between-sample variation}}{\text{Within-sample variation}}$$

$$= \dfrac{\text{SST}/1}{\text{SSE}/(n_1 + n_2 - 2)}$$

has an F distribution with $\nu_1 = 1$ and $\nu_2 = n_1 + n_2 - 2$ degrees of freedom (df) and therefore can be used to test the null hypothesis of no difference between the treatment means. The additivity property of the sums of squares led early researchers to view this analysis as a **partitioning** of $\text{SS(Total)} = \Sigma(y_i - \bar{y})^2$ into sources corresponding to the factors included in the experiment and to SSE. The simple formulas for computing the sums of squares, the additivity property, and the form of the test statistic made it natural for this procedure to be called **analysis of variance**. We demonstrate the analysis of variance procedure and its relation to regression for several common experimental designs in Sections 12.3–12.6.

12.3 One-Factor Completely Randomized Designs

Recall (Section 11.2) the first two steps in designing an experiment: (1) decide on the factors to be investigated and (2) select the factor level combinations (treatments) to be included in the experiment. For example, suppose you wish to compare the length of time to assemble a device in a manufacturing operation for workers who have completed one of three training programs, A, B, and C. Then this experiment involves a single factor, training program, at three levels, A, B, and C. Since training program is the only factor, these levels (A, B, and C) represent the treatments. Now we must decide the sample size for each treatment (step 3) and figure out how to assign the treatments to the experimental units, namely, the specific workers (step 4).

As we learned in Chapter 11, the most common assignment of treatments to experimental units is called a **completely randomized design**. To illustrate, suppose we wish to obtain equal amounts of information on the mean assembly times for the three training procedures (i.e., we decide to assign equal numbers of workers to each of the three training programs). Also, suppose we use the procedure in Section 1.8 (Example 1.13) to select the sample size and determine the number of workers in each of the three samples to be $n_1 = n_2 = n_3 = 10$. Then a completely randomized design is one in which the $n_1 + n_2 + n_3 = 30$ workers are *randomly assigned*, 10 to each of the three treatments. *A random assignment is one in which any one assignment is as probable as any other.* This eliminates the possibility of bias that might occur if the workers were assigned in some systematic manner. For example, a systematic assignment might accidentally assign most of the manually dexterous workers to training program A, thus underestimating the true mean assembly time corresponding to A.

Example 12.1 illustrates how a **random number generator** can be used to assign the 30 workers to the three treatments.

Example 12.1

Consider a completely randomized design involving three treatment groups (say, training programs A, B, and C) and $n = 30$ experimental units (say, workers in a manufacturing operation). Use a random number generator to assign the treatments to the experimental units for this design.

Solution

The first step is to number the experimental units (workers) from 1 to 30. Then, the first 10 numbers randomly selected from the 30 will be assigned to training program A, the second 10 randomly selected numbers will be assigned to training program B, and the remaining 10 numbers will be assigned to training program C. Either a random number table (e.g., Table 7, Appendix D) or computer software can be used to make the random assignments. Figure 12.3 is a MINITAB worksheet showing the random assignments made with the MINITAB "Random Data" function. The column named "Random" shows the randomly selected order of the workers. You can see that MINITAB randomly assigned workers numbered 17, 21, 9, 11, 18, 3, 24, 4, 29, and 5 to training program A, workers numbered 27, 16, 30, 15, 20, 19, 13, 22, 25, and 6 to training program B, and workers numbered 8, 26, 23, 1, 12, 14, 28, 7, 10, and 2 to training program C. ◼

Example 12.2

Suppose a beverage bottler wished to compare the effect of three different advertising displays on the sales of a beverage in supermarkets. Identify the experimental units you would use for the experiment, and explain how you would employ a completely randomized design to collect the sales data.

Solution

Presumably, the bottler has a list of supermarkets that market the beverage in a number of different cities. If we decide to measure the sales increase (or decrease) as the monthly dollar increase in sales (over the previous month) for a given supermarket, then the experimental unit is a 1-month unit of time in a specific supermarket. Thus, we would randomly select a 1-month period of time for each of $n_1 + n_2 + n_3$ supermarkets and assign n_1 supermarkets to receive display D_1, n_2 to receive D_2, and n_3 to receive display D_3. ◼

Figure 12.3 MINITAB random assignments of workers to training programs

↓	C1	C2	C3	C4	C5
	Worker	Random	ProgramA	ProgramB	ProgramC
1	1	17	17	27	8
2	2	21	21	16	26
3	3	9	9	30	23
4	4	11	11	15	1
5	5	18	18	20	12
6	6	3	3	19	14
7	7	24	24	13	28
8	8	4	4	22	7
9	9	29	29	25	10
10	10	5	5	6	2
11	11	27			
12	12	16			
13	13	30			
14	14	15			
15	15	20			
16	16	19			
17	17	13			
18	18	22			
19	19	25			
20	20	6			
21	21	8			
22	22	26			
23	23	23			
24	24	1			
25	25	12			
26	26	14			
27	27	28			
28	28	7			
29	29	10			
30	30	2			

In some experimental situations, we are unable to assign the treatment to the experimental units randomly because of the nature of the experimental units themselves. For example, suppose we want to compare the mean annual salaries of professors in three College of Liberal Arts departments: chemistry, mathematics, and sociology. Then the treatments—chemistry, mathematics, and sociology—cannot be "assigned" to the professors (experimental units). A professor is a member of the chemistry, mathematics, or sociology (or some other) department and cannot be arbitrarily assigned one of the treatments. Rather, we view the treatments (departments) as populations from which we will select independent random samples of experimental units (professors). A completely randomized design involves a comparison of the means for a number, say, p, of treatments, based on independent random samples of $n_1, n_2, \ldots, n_p$ observations, drawn from populations associated with treatments $1, 2, \ldots, p$, respectively. We repeat our definition of a completely randomized design (given in Section 11.4) with this modification. The general layout for a completely randomized design is shown in Figure 12.4.

Definition 12.1 A **completely randomized design** to compare p treatment means is one in which the treatments are randomly assigned to the experimental units, or in which independent random samples are drawn from each of the p populations.

After collecting the data from a completely randomized design, we want to make inferences about p population means where μ_i is the mean of the population of measurements associated with treatment i, for $i = 1, 2, \ldots, p$. The null hypothesis

Figure 12.4 Layout for a completely randomized design

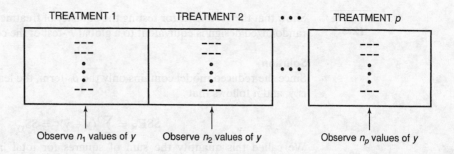

Observe n_1 values of y Observe n_2 values of y Observe n_p values of y

to be tested is that the p treatment means are equal (i.e., $H_0: \mu_1 = \mu_2 = \cdots = \mu_p$) and the alternative hypothesis we wish to detect is that at least two of the treatment means differ. The appropriate linear model for the response y is

$$E(y) = \beta_0 + \beta_1 x_1 + \beta_2 x_2 + \cdots + \beta_{p-1} x_{p-1}$$

where

$$x_1 = \begin{cases} 1 & \text{if treatment 2} \\ 0 & \text{if not} \end{cases} \quad x_2 = \begin{cases} 1 & \text{if treatment 3} \\ 0 & \text{if not} \end{cases} \cdots x_{p-1} = \begin{cases} 1 & \text{if treatment } p \\ 0 & \text{if not} \end{cases}$$

and (arbitrarily) treatment 1 is the base level. Recall that this 0–1 system of coding implies that

$$
\begin{aligned}
\beta_0 &= \mu_1 \\
\beta_1 &= \mu_2 - \mu_1 \\
\beta_2 &= \mu_3 - \mu_1 \\
&\;\;\vdots \\
\beta_{p-1} &= \mu_p - \mu_1
\end{aligned}
$$

The null hypothesis that the p population means are equal is equivalent to the null hypothesis that all the treatment differences equal 0, that is,

$$H_0: \beta_1 = \beta_2 = \cdots = \beta_{p-1} = 0$$

To test this hypothesis using regression, we use the technique in Section 4.13; that is, we compare the sum of squares for error, SSE_R, for the nested *reduced* model

$$E(y) = \beta_0$$

to the sum of squares for error, SSE_C, for the *complete* model

$$E(y) = \beta_0 + \beta_1 x_1 + \beta_2 x_2 + \cdots + \beta_{p-1} x_{p-1}$$

using the F statistic

$$
\begin{aligned}
F &= \frac{(SSE_R - SSE_C)/(\text{Number of } \beta \text{ parameters in } H_0)}{SSE_C/[n - (\text{Number of } \beta \text{ parameters in the complete model})]} \\[2mm]
&= \frac{(SSE_R - SSE_C)/(p-1)}{SSE_C/(n-p)} \\[2mm]
&= \frac{(SSE_R - SSE_C)/(p-1)}{MSE_C}
\end{aligned}
$$

where F is based on $\nu_1 = (p-1)$ and $\nu_2 = (n-p)$ df. If F exceeds the upper critical value, F_α, we reject H_0 and conclude that at least one of the treatment differences, $\beta_1, \beta_2, \ldots, \beta_{p-1}$, differs from zero (i.e., we conclude that at least two treatment means differ).

Example 12.3

Show that the F statistic for testing the equality of treatment means in a completely randomized design is equivalent to a global F-test of the complete model.

Solution

Since the reduced model contains only the β_0 term, the least squares estimate of β_0 is $\bar{y}$, and it follows that

$$\text{SSE}_R = \sum(y - \bar{y})^2 = \text{SS}_{yy}$$

We called this quantity the sum of squares for total in Chapter 4. The difference $(\text{SSE}_R - \text{SSE}_C)$ is simply $(\text{SS}_{yy} - \text{SSE})$ for the complete model. Since in regression $(\text{SS}_{yy} - \text{SSE}) = \text{SS (Model)}$, and the complete model has $(p - 1)$ terms (excluding β_0),

$$F = \frac{(\text{SSE}_R - \text{SSE}_C)/(p-1)}{\text{MSE}_C} = \frac{\text{SS (Model)}/(p-1)}{\text{MSE}} = \frac{\text{MS (Model)}}{\text{MSE}}$$

Thus, it follows that the test statistic for testing the null hypothesis,

$$H_0: \mu_1 = \mu_2 = \cdots = \mu_p$$

in a completely randomized design is the same as the F statistic for testing the global utility of the complete model for this design.   ■

The regression approach to analyzing data from a completely randomized design is summarized in the next box. Note that the test requires several assumptions about the distributions of the response y for the p treatments and that these *assumptions are necessary regardless of the sizes of the samples*. (We have more to say about these assumptions in Section 12.9.)

Model and F-Test for a Completely Randomized Design with p Treatments

Complete model: $E(y) = \beta_0 + \beta_1 x_1 + \beta_2 x_2 + \cdots + \beta_{p-1} x_{p-1}$

where $\quad x_1 = \begin{cases} 1 & \text{if treatment 2} \\ 0 & \text{if not} \end{cases} \qquad x_2 = \begin{cases} 1 & \text{if treatment 3} \\ 0 & \text{if not} \end{cases}, \ldots,$

$\qquad\qquad x_{p-1} = \begin{cases} 1 & \text{if treatment } p \\ 0 & \text{if not} \end{cases}$

$H_0: \beta_1 = \beta_2 = \cdots = \beta_{p-1} = 0$ (i.e., $H_0: \mu_1 = \mu_2 = \cdots = \mu_p$)

H_a: At least one of the β parameters listed in H_0 differs from 0
(i.e., H_a: At least two means differ)

Test statistic: $\quad F = \dfrac{\text{MS(Model)}}{\text{MSE}}$

Rejection region: $\quad F > F_\alpha$, where the distribution of F is based on $v_1 = p - 1$ and $v_2 = (n - p)$ degrees of freedom.

Assumptions: 1. All p population probability distributions corresponding to the p treatments are normal.
2. The population variances of the p treatments are equal.

Example 12.4

Sociologists often conduct experiments to investigate the relationship between socioeconomic status and college performance. Socioeconomic status is generally partitioned into three groups: lower class, middle class, and upper class. Consider the problem of comparing the mean grade point average of those college freshmen associated with the lower class, those associated with the middle class, and those associated with the upper class. The grade point averages (GPAs) for random samples of seven college freshmen associated with each of the three socioeconomic classes were selected from a university's files at the end of the academic year. The data are recorded in Table 12.1. Do the data provide sufficient evidence to indicate a difference among the mean freshmen GPAs for the three socioeconomic classes? Test using $\alpha = .05$.

GPA3

Table 12.1 Grade point averages for three socioeconomic groups

	Lower Class	Middle Class	Upper Class
	2.87	3.23	2.25
	2.16	3.45	3.13
	3.14	2.78	2.44
	2.51	3.77	2.54
	1.80	2.97	3.27
	3.01	3.53	2.81
	2.16	3.01	1.36
Sample means:	$\bar{y}_1 = 2.521$	$\bar{y}_2 = 3.249$	$\bar{y}_3 = 2.543$

Solution

This experiment involves a single factor, socioeconomic class, at three levels. Thus, we have a completely randomized design with $p = 3$ treatments. Let μ_L, μ_M, and μ_U represent the mean GPAs for students in the lower, middle, and upper socioeconomic classes, respectively. Then we want to test

H_0: $\mu_L = \mu_M = \mu_U$

against

H_a: At least two of the three treatment means differ.

The appropriate linear model for $p = 3$ treatments is

Complete model: $E(y) = \beta_0 + \beta_1 x_1 + \beta_2 x_2$

where

$$x_1 = \begin{cases} 1 & \text{if middle socioeconomic class} \\ 0 & \text{if not} \end{cases}$$

$$x_2 = \begin{cases} 1 & \text{if upper socioeconomic class} \\ 0 & \text{if not} \end{cases}$$

Thus, we want to test H_0: $\beta_1 = \beta_2 = 0$.

The SAS regression analysis for the complete model is shown in Figure 12.5. The F statistic for testing the overall adequacy of the model (shaded on the printout) is $F = 4.58$, where the distribution of F is based on $v_1 = (p - 1) = 3 - 1 = 2$ and

Figure 12.5 SAS regression printout for the completely randomized design in Example 12.4

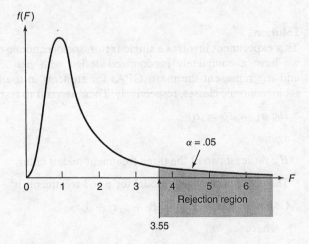

Dependent Variable: GPA

Analysis of Variance

Source	DF	Sum of Squares	Mean Square	F Value	Pr > F
Model	2	2.39687	1.19843	4.58	0.0247
Error	18	4.71111	0.26173		
Corrected Total	20	7.10798			

Root MSE	0.51159	R-Square	0.3372	
Dependent Mean	2.77095	Adj R-Sq	0.2636	
Coeff Var	18.46275			

Parameter Estimates

Variable	DF	Parameter Estimate	Standard Error	t Value	Pr > |t|
Intercept	1	2.52143	0.19336	13.04	<.0001
X1	1	0.72714	0.27346	2.66	0.0160
X2	1	0.02143	0.27346	0.08	0.9384

$v_2 = (n - p) = 21 - 3 = 18$ df. For $\alpha = .05$, the critical value (obtained from Table 4 in Appendix C) is $F_{.05} = 3.55$ (see Figure 12.6).

Since the computed value of F, 4.58, exceeds the critical value, $F_{.05} = 3.55$, we reject H_0 and conclude (at the $\alpha = .05$ level of significance) that the mean GPA for college freshmen differs in at least two of the three socioeconomic classes. We can arrive at the same conclusion by noting that $\alpha = .05$ is greater than the p-value (.0247) shaded on the printout. ■

Figure 12.6 Rejection region for Example 12.4; numerator df = 2, denominator df = 18, $\alpha = .05$

$f(F)$

$\alpha = .05$

3.55

Rejection region

F

The analysis of the data in Example 12.4 can also be accomplished using ANOVA computing formulas. In Section 12.2, we learned that an analysis of variance partitions SS(Total) $= \Sigma(y - \bar{y})^2$ into two components, SSE and SST (see Figure 12.7).

Recall that the quantity SST denotes the sum of squares for treatments and measures the variation explained by the differences between the treatment means. The sum of squares for error, SSE, is a measure of the unexplained variability, obtained by calculating a pooled measure of the variability *within* the p samples.

Figure 12.7 Partitioning of SS(Total) for a completely randomized design

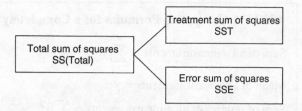

If the treatment means truly differ, then SSE should be substantially smaller than SST. We compare the two sources of variability by forming an F statistic:

$$F = \frac{\text{SST}/(p-1)}{\text{SSE}/(n-p)} = \frac{\text{MST}}{\text{MSE}}$$

where n is the total number of measurements. The numerator of the F statistic, $\text{MST} = \text{SST}/(p-1)$, denotes **mean square for treatments** and is based on $(p-1)$ degrees of freedom—one for each of the p treatments minus one for the estimation of the overall mean. The denominator of the F statistic, $\text{MSE} = \text{SSE}/(n-p)$, denotes **mean square for error** and is based on $(n-p)$ degrees of freedom—one for each of the n measurements minus one for each of the p treatment means being estimated. We have already demonstrated that this F statistic is identical to the global F-value for the regression model specified previously.

For completeness, we provide the computing formulas for an analysis of variance in the box, p. 618.

Example 12.5

Refer to Example 12.4. Analyze the data in Table 12.1 using the ANOVA "sums of squares" approach. Use $\alpha = .05$.

Solution

Rather than performing the tedious calculations by hand (we leave this for the student as an exercise), we use a statistical software package with an ANOVA routine. All three of the software packages discussed in this text (SAS, MINITAB, and SPSS) have procedures that automatically compute the ANOVA sums of squares and the ANOVA F statistic.

The MINITAB ANOVA printout is shown in Figure 12.8. The value of the test statistic (shaded on the printout) is $F = 4.58$. Note that this is identical to the F-value obtained using the regression approach in Example 12.4. The p-value of the test (also shaded) is $p = .025$. (Likewise, this quantity is identical to that in Example 12.4.) Since $\alpha = .05$ exceeds this p-value, we have sufficient evidence to conclude that the treatments differ.

Figure 12.8 MINITAB ANOVA printout for the completely randomized design, Example 12.5

```
One-way ANOVA: GPA versus GROUP

Analysis of Variance for GPA
Source    DF       SS       MS       F       P
GROUP      2    2.397    1.198    4.58   0.025
Error     18    4.711    0.262
Total     20    7.108

                              Individual 95% CIs For Mean
                              Based on Pooled StDev
Level     N     Mean    StDev  --------+---------+---------+---------+--------
1         7   2.5214   0.5041  (--------*---------)
2         7   3.2486   0.3526                     (-------*-------)
3         7   2.5429   0.6377  (-------*-------)
                              --------+---------+---------+---------+--------
Pooled StDev =  0.5116                   2.50      3.00      3.50
```

ANOVA Computing Formulas for a Completely Randomized Design

Sum of all n measurements $= \displaystyle\sum_{i=1}^{n} y_i$

Mean of all n measurements $= \bar{y}$

Sum of squares of all n measurements $= \displaystyle\sum_{i=1}^{n} y_i^2$

$\text{CM} = $ Correction for mean

$$= \frac{(\text{Total of all observations})^2}{\text{Total number of observations}} = \frac{\left(\displaystyle\sum_{i=1}^{n} y_i\right)^2}{n}$$

$\text{SS(Total)} = $ Total sum of squares

$$= (\text{Sum of squares of all observations}) - \text{CM}$$

$$= \sum_{i=1}^{n} y_i^2 - \text{CM}$$

$\text{SST} = $ Sum of squares for treatments

$$= \left(\begin{array}{c} \text{Sum of squares of treatment totals with} \\ \text{each square divided by the number of} \\ \text{observations for that treatment} \end{array}\right) - \text{CM}$$

$$= \frac{T_1^2}{n_1} + \frac{T_2^2}{n_2} + \cdots + \frac{T_p^2}{n_p} - \text{CM}$$

$\text{SSE} = $ Sum of squares for error

$$= \text{SS(Total)} - \text{SST}$$

$\text{MST} = $ Mean square for treatments

$$= \frac{\text{SST}}{p-1}$$

$\text{MSE} = $ Mean square for error

$$= \frac{\text{SSE}}{n-p}$$

$$F = \frac{\text{MST}}{\text{MSE}}$$

The results of an analysis of variance are often summarized in tabular form. The general form of an ANOVA table for a completely randomized design is shown in the next box. The column head **SOURCE** refers to the source of variation, and for each source, **df** refers to the degrees of freedom, **SS** to the sum of squares, **MS** to the mean square, and F to the F statistic comparing the treatment mean square to the error mean square. Table 12.2 is the ANOVA summary table corresponding to the analysis of variance data for Example 12.5, obtained from the MINITAB printout.

ANOVA Summary Table for a Completely Randomized Design

SOURCE	df	SS	MS	F
Treatments	$p-1$	SST	$\text{MST} = \frac{\text{SST}}{p-1}$	$F = \frac{\text{MST}}{\text{MSE}}$
Error	$n-p$	SSE	$\text{MSE} = \frac{\text{SSE}}{n-p}$	
Total	$n-1$	SS(Total)		

Table 12.2 ANOVA summary table for Example 12.5

Source	df	SS	MS	F
Socioeconomic class	2	2.40	1.198	4.58
Error	18	4.71	.262	
Total	20	7.11		

Because the completely randomized design involves the selection of independent random samples, we can find a confidence interval for a single treatment mean using the method in Section 1.8 or for the difference between two treatment means using the methods in Section 1.10. The estimate of σ^2 will be based on the pooled sum of squares within all p samples, that is,

$$\text{MSE} = s^2 = \frac{\text{SSE}}{n-p}$$

This is the same quantity that is used as the denominator for the analysis of variance F-test. The formulas for the confidence intervals in Sections 1.8 and 1.10 are reproduced in the box.

Confidence Intervals for Means: Completely Randomized Design

Single treatment mean (say, treatment i): $\bar{y}_i \pm t_{\alpha/2}\left(\dfrac{s}{\sqrt{n_i}}\right)$

Difference between two treatment means (say, treatments i and j):

$$(\bar{y}_i - \bar{y}_j) \pm t_{\alpha/2}s\sqrt{\frac{1}{n_i} + \frac{1}{n_j}}$$

where $\bar{y}_i$ is the sample mean response for population (treatment) i, $s = \sqrt{\text{MSE}}$, and $t_{\alpha/2}$ is the tabulated value of t (Table 2 in Appendix D) that locates $\alpha/2$ in the upper tail of the t distribution with $(n-p)$ df (the degrees of freedom associated with error in the ANOVA).

Example 12.6

Refer to Example 12.4. Find a 95% confidence interval for μ_L, the mean GPA of freshmen from the lower socioeconomic class.

Solution

From Table 12.2, MSE = .262. Then

$$s = \sqrt{\text{MSE}} = \sqrt{.262} = .512$$

The sample mean GPA for freshmen students from the lower class is

$$\bar{y}_L = \frac{17.65}{7} = 2.521$$

where 17.65 is the total of GPAs for the lower socioeconomic class obtained from Table 12.1. The tabulated value of $t_{.025}$ for 18 df (the same as for MSE) is 2.101. Therefore, a 95% confidence interval for μ_L, the mean GPA of college freshmen in the lower class, is

$$\bar{y}_L \pm (t_{\alpha/2}) \frac{s}{\sqrt{n}} = 2.521 \pm (2.101) \frac{.512}{\sqrt{7}}$$

$$= 2.521 \pm .407$$

or (2.114, 2.928). This interval is shown graphically (shaded) on the MINITAB printout, Figure 12.8 (p. 617).

Note that this confidence interval is relatively wide—probably too wide to be of any practical value (considering that GPA is measured on a 4-point scale). The interval is this wide because of the large amount of variation within each socioeconomic class. For example, the GPA for freshmen in the lower class varies from 1.8 to 3.14. The more variable the data, the larger the value of s in the confidence interval and the wider the confidence interval. Consequently, if you want to obtain a more accurate estimate of treatment means with a narrower confidence interval, you will have to select larger samples of freshmen from within each socioeconomic class. ▬

Although we can use the formula given in the box to compare two treatment means in ANOVA, unless the two treatments are selected a priori (i.e., prior to conducting the ANOVA), we must apply one of the methods for comparing means presented in Sections 12.7 and 12.8 to obtain valid results.

12.3 Exercises

12.1 Taste preferences of cockatiels. Refer to the completely randomized design in Exercise 11.3 (p. 584). Recall that the researchers want to compare the mean liquid consumptions of cockatiels in three feeding groups. Use the random number table (Table 7 in Appendix D) to randomly assign the cockatiels to the three groups. Assume there are 15 cockatiels in the study.

12.2 ANOVA-summary table. A partially completed ANOVA table for a completely randomized design is shown here.

SOURCE	df	SS	MS	F
Treatments	4	24.7	—	—
Error	—	—	—	
Total	34	62.4		

(a) Complete the ANOVA table.
(b) How many treatments are involved in the experiment?
(c) Do the data provide sufficient evidence to indicate a difference among the treatment means? Test using $\alpha = .10$.

12.3 Two-group completely randomized design. The data for a completely randomized design with two treatments are shown in the table on p. 621.

(a) Give the linear model appropriate for analyzing the data using regression.
(b) Fit the model, part a, to the data and conduct the analysis. [*Hint*: You do not need a computer to fit the model. Use the formulas provided in Chapter 3.]

⦿ EX12_3

TREATMENT 1	TREATMENT 2
10	12
7	8
8	13
11	10
10	10
9	11
9	

12.4 Two-group completely randomized design (cont'd). Refer to Exercise 12.3.

(a) Calculate MST for the data using the ANOVA formulas. What type of variability is measured by this quantity?

(b) Calculate MSE for the data using the ANOVA formulas. What type of variability is measured by this quantity?

(c) How many degrees of freedom are associated with MST?

(d) How many degrees of freedom are associated with MSE?

(e) Compute the test statistic appropriate for testing $H_0: \mu_1 = \mu_2$ against the alternative that the two treatment means differ, using a significance level of $\alpha = .05$. (Compare the value to the test statistic obtained using regression in part b, Exercise 12.3.)

(f) Summarize the results from parts a–e in an ANOVA table.

(g) Specify the rejection region, using a significance level of $\alpha = .05$.

(h) State the proper conclusion.

12.5 Two-group completely randomized design (cont'd). Exercises 12.3 and 12.4 involve a test of the null hypothesis $H_0: \mu_1 = \mu_2$ based on independent random sampling (recall the definition of a completely randomized design). This test was conducted in Section 1.10 using a Student's t statistic.

(a) Use the Student's t-test to test the hypothesis $H_0: \mu_1 = \mu_2$ against the alternative hypothesis $H_a: \mu_1 \neq \mu_2$. Test using $\alpha = .05$.

(b) It can be shown (proof omitted) that an F statistic with $v_1 = 1$ numerator degree of freedom and v_2 denominator degrees of freedom is equal to t^2, where t is a Student's t statistic based on v_2 degrees of freedom. Square the value of t calculated in part a, and show that it is equal to the value of F calculated in Exercises 12.3b and 12.4e.

(c) Is the analysis of variance F-test for comparing two population means a one- or a two-tailed test of $H_0: \mu_1 = \mu_2$? [*Hint*: Although the t-test can be used to test either for

$H_a: \mu_1 < \mu_2$, or for $H_a: \mu_1 < \mu_2$, the alternative hypothesis for the F-test is H_a: The two means are different.]

12.6 Making high-stakes insurance decisions. The *Journal of Economic Psychology* (September 2008) published the results of a high-stakes experiment where subjects (university students) were asked how much they would pay for insuring a valuable painting. The painting was threatened by both fire and theft, hence, the need for insurance. Of interest was the amount the subject was willing to pay (WTP) for insurance (thousands of dollars). For one part of the experiment, a total of 252 subjects were randomly assigned to one of three groups. Group 1 subjects ($n_1 = 84$) were informed of the hazards (both fire and theft) but were not told the exact probabilities of the hazards occurring. These subjects provided a separate WTP value for fire and theft. Group 2 subjects ($n_2 = 84$) were also informed of the hazards (fire/theft) and were not told the exact probabilities of the hazards occurring. However, these subjects provided a single WTP value covering both fire and theft. Group 3 subjects ($n_3 = 84$) were told of the hazards in sequential order (fire first, then theft). After being given the exact probability of fire occurring, the subjects provided a WTP value for fire. Then they were given the exact probability of theft occurring and were asked to provide a WTP value for theft. The researchers investigated whether the mean total WTP value differed for the three groups.

(a) Explain why the experimental design employed is a completely randomized design.

(b) Identify the dependent (response) variable and treatments for the design.

(c) Give the null and alternative hypotheses of interest to the researchers.

(d) Use a random number generator to randomly assign each of the 252 subjects to one of the three groups. Be sure to assign 84 subjects to each group.

12.7 Robots trained to behave like ants. Robotics researchers investigated whether robots could be trained to behave like ants in an ant colony (*Nature*, August 2000). Robots were trained and randomly assigned to "colonies" (i.e., groups) consisting of 3, 6, 9, or 12 robots. The robots were assigned the task of foraging for "food" and to recruit another robot when they identified a resource-rich area. One goal of the experiment was to compare the mean energy expended (per robot) of the four different colony sizes.

(a) What type of experimental design was employed?

(b) Identify the treatments and the dependent variable.

(c) Set up the null and alternative hypotheses of the test.

(d) The following ANOVA results were reported: $F = 7.70$, numerator df $= 3$, denominator df $= 56$, p-value $< .001$. Conduct the test at a significance level of $\alpha = .05$ and interpret the result.

TVADRECALL

12.8 Study of recall of TV commercials. Television advertisers seek to promote their products on TV programs that attract the most viewers. Do TV shows with violence and sex impair memory for commercials? To answer this question, Iowa State professors B. Bushman and A. Bonacci conducted a designed experiment in which 324 adults were randomly assigned to one of three viewer groups, 108 in each group (*Journal of Applied Psychology,* June 2002). One group watched a TV program (e.g., "Tour of Duty") with a violent content code (V) rating; the second group viewed a show (e.g., "Strip Mall") with a sex content code (S) rating; and the last group watched a neutral TV program (e.g., "Candid Camera") with neither a V nor an S rating. Nine commercials were imbedded into each TV show. After viewing the program, each participant was scored on their recall of the brand names in the commercial messages, with scores ranging from 0 (no brands recalled) to 9 (all brands recalled). The data (simulated from information provided in the article) are saved in the *TVADRECALL* file. The researchers compared the mean recall scores of the three viewing groups with an analysis of variance for a completely randomized design.

(a) Identify the experimental units in the study.

(b) Identify the dependent (response) variable in the study.

(c) Identify the factor and treatments in the study.

(d) The sample mean recall scores for the three groups were $\bar{y}_V = 2.08$, $\bar{y}_S = 1.71$, and $\bar{y}_{Neutral} = 3.17$. Explain why one should not draw an inference about differences in the population mean recall scores based on only these summary statistics.

(e) An ANOVA on the data in the *TVADRECALL* file yielded the results shown in the MINITAB printout below. Locate the test statistic and p-value on the printout.

(f) Interpret the results, part e, using $\alpha = .01$. What can the researchers conclude about the three groups of TV ad viewers?

MINITAB Output for Exercise 12.8

One-way ANOVA: VIOLENT, SEX, NEUTRAL

```
Source    DF        SS       MS       F       P
Factor     2    123.27    61.63    20.45    0.000
Error    321    967.35     3.01
Total    323   1090.62

S = 1.736    R-Sq = 11.30%    R-Sq(adj) = 10.75%
```

12.9 Heights of grade school repeaters. The *Archives of Disease in Childhood* (April 2000) published a study of whether height influences a child's progression through elementary school. Within each grade, Australian schoolchildren were divided into equal thirds (tertiles) based on age (youngest third, middle third, and oldest third). The researchers compared the average heights of the three groups using an analysis of variance. (All height measurements were standardized using z-scores.) A summary of the results for all grades combined, by gender, is shown in the table at the bottom of the page.

(a) What is the null hypothesis for the ANOVA of the boys' data?

(b) Write the linear model appropriate for analyzing the data.

(c) Interpret the results of the test, part a. Use $\alpha = .05$.

(d) Repeat parts a–c for the girls' data.

(e) Summarize the results of the hypothesis tests in the words of the problem.

12.10 Contingent valuation of homes in contaminated areas. Contingent valuation (CV) is a method of estimating property values that uses survey responses from potential homeowners. CV surveys were employed to determine the impact of

	SAMPLE SIZE	YOUNGEST TERTILE MEAN HEIGHT	MIDDLE TERTILE MEAN HEIGHT	OLDEST TERTILE MEAN HEIGHT	F-VALUE	p-VALUE
Boys	1439	0.33	0.33	0.16	4.57	0.01
Girls	1409	0.27	0.18	0.21	0.85	0.43

Source: Reproduced from *Archives of Disease in Childhood,* "Does height influence progression through primary school grades?" Melissa Wake, David Coghlan, and Kylie Hesketh, Vol. 82, Issue 4, April 2000 (Table 3), with permission from BMJ Publishing Group Ltd.

contamination on property values in the *Journal of Real Estate Research* (Vol. 27, 2005). Homeowners were randomly selected from each of seven states—Kentucky, Pennsylvania, Ohio, Alabama, Illinois, South Carolina, and Texas. Each homeowner was asked to estimate the property value of a home located in an area contaminated by petroleum leaking from underground storage tanks (LUST). The dependent variable of interest was the LUST discount percentage (i.e., the difference between the current home value and estimated LUST value, as a percentage). The researchers were interested in comparing the mean LUST discount percentages across the seven states.

(a) Give the null and alternative hypotheses of interest to the researchers.

(b) An ANOVA summary table is shown below. Use the information provided to conduct the hypothesis test, part a. Use $\alpha = .10$.

SOURCE	df	SS	MS	*F*-VALUE	*p*-VALUE
States	6	.1324	.0221	1.60	.174
Error	59	.8145	.0138		
Total	65	.9469			

12.11 Hazardous organic solvents. The *Journal of Hazardous Materials* (July 1995) published the results of a study of the chemical properties of three different types of hazardous organic solvents used to clean metal parts: aromatics, chloroalkanes, and esters. One variable studied was sorption rate, measured as mole percentage. Independent samples of solvents from each type were tested and their sorption rates were recorded, as shown in the next table. A MINITAB analysis of variance of the data is provided below.

(a) Construct an ANOVA table from the MINITAB printout.

MINITAB Output for Exercise 12.11

```
One-way ANOVA: SORPRATE versus SOLVENT

Analysis of Variance for SORPRATE
Source     DF        SS        MS        F        P
SOLVENT     2    3.3054    1.6527    24.51    0.000
Error      29    1.9553    0.0674
Total      31    5.2608

                          Individual 95% CIs For Mean
                          Based on Pooled StDev
Level       N      Mean     StDev  ----+---------+---------+---------+--
1           9    0.9422    0.1683                        (----*-----)
2           8    1.0063    0.4010                          (------*-----)
3          15    0.3300    0.2076  (----*----)
                                   ----+---------+---------+---------+--
Pooled StDev =    0.2597           0.30      0.60      0.90      1.20
```

(b) Is there evidence of differences among the mean sorption rates of the three organic solvent types? Test using $\alpha = .10$.

SORPRATE

AROMATICS		CHLOROALKANES		ESTERS		
1.06	.95	1.58	1.12	.29	.43	.06
.79	.65	1.45	.91	.06	.51	.09
.82	1.15	.57	.83	.44	.10	.17
.89	1.12	1.16	.43	.61	.34	.60
1.05				.55	.53	.17

Source: Reprinted from *Journal of Hazardous Materials*, Vol. 42, No. 2, J. D. Ortego et al., "A review of polymeric geosynthetics used in hazardous waste facilities." p. 142 (Table 9), July 1995, Elsevier Science-NL, Sara Burgerhartstraat 25, 1055 KV Amsterdam, The Netherlands.

12.12 Restoring self-control when intoxicated. Does coffee or some other form of stimulation really allow a person suffering from alcohol intoxication to "sober-up"? Psychologists from the University of Waterloo investigated these theories in *Experimental and Clinical Psychopharmacology* (February 2005). A sample of 44 healthy male college students participated in the experiment. Each student was asked to memorize a list of 40 words (20 words on a green list and 20 words on a red list). The students were then randomly assigned to one of four different treatment groups (11 students in each group). Students in three of the groups were each given two alcoholic beverages to drink prior to performing a word completion task. Group A received only the alcoholic drinks; group AC had caffeine powder dissolved in their drinks; group AR received a monetary award for correct responses on the word completion task; group P (the placebo group) were told that they would receive alcohol but instead received two drinks containing a carbonated beverage (with a few drops of alcohol on the surface to provide an alcoholic scent). After consuming their drinks and resting for 25 minutes, the students performed the word completion task. Their scores (simulated, based on summary information from the article) are reported in the table on p. 624. (Note: A task score represents the difference between the proportion of correct responses on the green list of words and the proportion of incorrect responses on the red list of words.)

(a) What type of experimental design is employed in this study?

(b) Analyze the data for the researchers, using $\alpha = .05$. Are there differences among the mean task scores for the four groups?

(c) What assumptions must be met to ensure the validity of the inference, part b?

🔘 DRINKERS

AR	AC	A	P
.51	.50	.16	.58
.58	.30	.10	.12
.52	.47	.20	.62
.47	.36	.29	.43
.61	.39	-.14	.26
.00	.22	.18	.50
.32	.20	-.35	.44
.53	.21	.31	.20
.50	.15	.16	.42
.46	.10	.04	.43
.34	.02	-.25	.40

Source: Grattan-Miscio, K. E., and Vogel-Sprott, M. "Alcohol, intentional control, and inappropriate behavior: Regulation by caffeine or an incentive," *Experimental and Clinical Psychopharmacology,* Vol. 13, No. 1, February 2005 (Table 1). Copyright © 2005 American Psychological Association, reprinted with permission.

12.13 **The "name game."** Psychologists at Lancaster University (United Kingdom) evaluated three methods of name retrieval in a controlled setting (*Journal of Experimental Psychology—Applied,* June 2000). A sample of 139 students was randomly divided into three groups, and each group of students used a different method to learn the names of the other students in the group. Group 1 used the "simple name game," where the first

🔘 NAMEGAME

SIMPLE NAME GAME

24	43	38	65	35	15	44	44	18	27	0	38	50	31
7	46	33	31	0	29	0	0	52	0	29	42	39	26
51	0	42	20	37	51	0	30	43	30	99	39	35	19
24	34	3	60	0	29	40	40						

ELABORATE NAME GAME

39	71	9	86	26	45	0	38	5	53	29	0	62	0
1	35	10	6	33	48	9	26	83	33	12	5	0	0
25	36	39	1	37	2	13	26	7	35	3	8	55	50

PAIRWISE INTRODUCTIONS

5	21	22	3	32	29	32	0	4	41	0	27	5	9
66	54	1	15	0	26	1	30	2	13	0	2	17	14
5	29	0	45	35	7	11	4	9	23	4	0	8	2
18	0	5	21	14									

Source: Morris, P. E., and Fritz, C. O. "The name game: Using retrieval practice to improve the learning of names," *Journal of Experimental Psychology—Applied,* Vol. 6, No. 2, June 2000 (data simulated from Figure 1). Copyright © 2000, American Psychological Association, reprinted with permission.

student states his/her full name, the second student announces his/her name and the name of the first student, the third student says his/her name and the names of the first two students, etc. Group 2 used the "elaborate name game," a modification of the simple name game where the students not only state their names but also their favorite activity (e.g., sports). Group 3 used "pairwise introductions," where students are divided into pairs and each student must introduce the other member of the pair. One year later, all subjects were sent pictures of the students in their group and asked to state the full name of each. The researchers measured the percentage of names recalled for each student respondent. The data (simulated based on summary statistics provided in the research article) are shown in the accompanying table. Conduct an analysis of variance to determine whether the mean percentages of names recalled differ for the three name retrieval methods. Use $\alpha = .05$.

12.14 **Is honey a cough remedy?** Pediatric researchers at Pennsylvania State University carried out a designed study to test whether a teaspoon of honey before bed calms a child's cough and published their results in *Archives of Pediatrics and Adolescent Medicine* (December 2007). (This experiment was first described in Exercise 1.21, p. 17.) A sample of 105 children who were ill with an upper respiratory tract infection and their parents participated in the study. On the first night, the parents rated their children's cough symptoms on a scale from 0 (no problems at all) to 6 (extremely severe) in five different areas. The total symptoms score (ranging from 0 to 30 points) was the variable of interest for the 105 patients. On the second night, the parents were instructed to give their sick child a dosage of liquid "medicine" prior to bedtime. Unknown to the parents, some were given a dosage of dextromethorphan (DM)—an over-the-counter cough medicine—while others were given a similar dose of honey. Also, a third group of parents (the control group) gave their sick children no dosage at all. Again, the parents rated their children's cough symptoms, and the improvement in total cough symptoms score was determined for each child. The data (improvement scores) for the study are shown in the table (page 625). The goal of the researchers was to compare the mean improvement scores for the three treatment groups.

(a) Identify the type of experimental design employed. What are the treatments?

(b) Conduct an analysis of variance on the data and interpret the results.

HONEYCOUGH

Honey Dosage:	12	11	15	11	10	13	10	4	15	16	9	14	10	6	10	8	11	12	12	8		
	12	9	11	15	10	15	9	13	8	12	10	8	9	5	12							
DM Dosage:	4	6	9	4	7	7	7	9	12	10	11	6	3	4	9	12	7	6	8	12	12	4	12
	13	7	10	13	9	4	4	10	15	9												
No Dosage (Control):	5	8	6	1	0	8	12	8	7	7	1	6	7	7	12	7	9	7	9	5	11	9	5
	6	8	8	6	7	10	9	4	8	7	3	1	4	3								

Source: Paul, I. M., et al. "Effect of honey, dextromethorphan, and no treatment on nocturnal cough and sleep quality for coughing children and their parents," *Archives of Pediatrics and Adolescent Medicine*, Vol. 161, No. 12, Dec. 2007 (data simulated).

12.15 **Animal-assisted therapy for heart patients.** Refer to the *American Heart Association Conference* (November 2005) study to gauge whether animal-assisted therapy can improve the physiological responses of heart failure patients, Exercise 1.30 (p. 24). Recall that 76 heart patients were randomly assigned to one of three groups. Each patient in group T was visited by a human volunteer accompanied by a trained dog; each patient in group V was visited by a volunteer only; and the patients in group C were not visited at all. The anxiety level of each patient was measured (in points) both before and after the visits. The accompanying table gives summary statistics for the drop in anxiety level for patients in the three groups. The mean drops in anxiety levels of the three groups of patients were compared using an analysis of variance. Although the ANOVA table was not provided in the article, sufficient information is provided to reconstruct it.

	SAMPLE SIZE	MEAN DROP	STD. DEV.	GROUP TOTAL
Group T: Volunteer + trained dog	26	10.5	7.6	273.0
Group V: Volunteer only	25	3.9	7.5	97.5
Group C: Control group (no visit)	25	1.4	7.5	35.0

Source: Cole, K., et al. "Animal assisted therapy decreases hemodynamics, plasma epinephrine and state anxiety in hospitalized heart failure patients," *American Heart Association Conference,* Dallas, Texas, Nov. 2005.

(a) Use the sum of the group totals and the total sample size to compute CM.

(b) Use the individual group totals and sample sizes to compute SST.

(c) Compute SSE using the pooled sum of squares formula:

$$\text{SSE} = \sum(y_{i1} - \bar{y}_1)^2 + \sum(y_{i2} - \bar{y}_2)^2$$
$$+ \sum(y_{i3} - \bar{y}_3)^2$$
$$= (n_1 - 1)s_1^2 + (n_2 - 1)s_2^2 + (n_3 - 1)s_3^2$$

(d) Find SS(Total).

(e) Construct an ANOVA table for the data.

(f) Do the data provide sufficient evidence to indicate differences in the mean drops in anxiety levels by patients in the three groups? Test using $\alpha = .05$.

(g) Give the linear model appropriate for analyzing the data using regression.

(h) Use the information in the table to find the least squares prediction equation.

12.16 **Estimating the age of glacial drifts.** The *American Journal of Science* (January 2005) conducted a study of the chemical makeup of buried tills (glacial drifts) in Wisconsin. The ratio of the elements aluminum (Al) and beryllium (Be) in sediment is related to the duration of burial. The Al/Be ratios for a sample of 26 buried till specimens were determined. The till specimens were obtained from five different boreholes (labeled UMRB-1, UMRB-2, UMRB-3, SWRA, and SD). The data are shown in the table. Conduct an analysis of variance of the data. Is there sufficient evidence to indicate differences among the mean Al/Be ratios for the five boreholes? Test using $\alpha = .10$.

TILLRATIO

UMRB-1:	3.75	4.05	3.81	3.23	3.13	3.30	3.21
UMRB-2:	3.32	4.09	3.90	5.06	3.85	3.88	
UMBR-3:	4.06	4.56	3.60	3.27	4.09	3.38	3.37
SWRA:	2.73	2.95	2.25				
SD:	2.73	2.55	3.06				

Source: Adapted from *American Journal of Science*, Vol. 305, No. 1, Jan. 2005, p. 16 (Table 2).

ACCHW

12.17 **Homework assistance for accounting students.** The *Journal of Accounting Education* (Vol. 25, 2007) published a study comparing different methods of assisting accounting students with their homework. A total of 75 junior-level accounting majors who were enrolled in Intermediate Financial Accounting participated in the experiment. The students took a pretest on a topic not covered in class, then each was given a homework problem to solve on the same topic. A completely randomized design was employed, with students randomly assigned to receive one of three different levels of assistance on the homework: (1) the completed solution, (2) check figures at various steps of the solution, and (3) no help at all. After finishing the homework, the students were all given a posttest on the subject. The response variable of interest to the researchers was the knowledge gain (or test score improvement), measured as the difference between the posttest and pretest scores. The data (simulated from descriptive statistics published in the article) are saved in the ACCHW file.

(a) Give the null and alternative hypothesis tested in an analysis of variance of the data.

(b) Summarize the results of the analysis in an ANOVA table.

(c) Interpret the results practically.

12.4 Randomized Block Designs

Randomized block design is a commonly used noise-reducing design. Recall (Definition 10.9) that a randomized block design employs groups of homogeneous experimental units (matched as closely as possible) to compare the means of the populations associated with p treatments. The general layout of a randomized block design is shown in Figure 12.9. Note that there are b blocks of relatively homogeneous experimental units. Since each treatment must be represented in each block, the blocks each contain p experimental units. Although Figure 12.9 shows the p treatments in order within the blocks, in practice they would be assigned to the experimental units in random order (hence the name **randomized block design**).

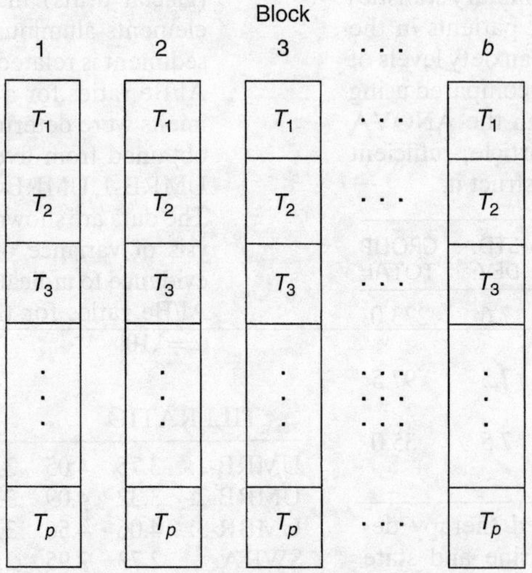

Figure 12.9 General form of a randomized block design (treatment is denoted by T_p)

The complete model for a randomized block design contains $(p-1)$ dummy variables for treatments and $(b-1)$ dummy variables for blocks. Therefore, the total number of terms in the model, excluding β_0, is $(p-1)+(b-1) = p+b-2$, as shown here.

Complete model:

$$E(y) = \beta_0 + \underbrace{\beta_1 x_1 + \beta_2 x_2 + \cdots + \beta_{p-1} x_{p-1}}_{\text{Treatment effects}} + \underbrace{\beta_p x_p + \cdots + \beta_{p+b-2} x_{p+b-2}}_{\text{Block effects}}$$

where

$$x_1 = \begin{cases} 1 & \text{if treatment 2} \\ 0 & \text{if not} \end{cases},$$

$$x_2 = \begin{cases} 1 & \text{if treatment 3} \\ 0 & \text{if not} \end{cases}, \ldots, x_{p-1} = \begin{cases} 1 & \text{if treatment } p \\ 0 & \text{if not} \end{cases}$$

$$x_p = \begin{cases} 1 & \text{if block 2} \\ 0 & \text{if not} \end{cases}, x_{p+1} = \begin{cases} 1 & \text{if block 3} \\ 0 & \text{if not} \end{cases}, \ldots, x_{p+b-2} = \begin{cases} 1 & \text{if block } b \\ 0 & \text{if not} \end{cases}$$

Note that the model does *not* include terms for treatment–block interaction. The reasons are twofold. First, the addition of these terms would leave 0 degrees of freedom for estimating σ^2. Second, the failure of the mean difference between a pair of treatments to remain the same from block to block is, by definition, experimental error. In other words, in a randomized block design, treatment–block interaction and experimental error are synonymous.

The primary objective of the analysis is to compare the p treatment means, $\mu_1, \mu_2, \ldots, \mu_p$. That is, we want to test the null hypothesis

$$H_0: \ \mu_1 = \mu_2 = \mu_3 = \cdots = \mu_p$$

Recall (Section 11.3) that this is equivalent to testing whether all the treatment parameters in the complete model are equal to 0, that is,

$$H_0: \ \beta_1 = \beta_2 = \cdots = \beta_{p-1} = 0$$

To perform this test using regression, we drop the treatment terms and fit the reduced model:

Reduced model for testing treatments

$$E(y) = \beta_0 + \underbrace{\beta_p x_p + \beta_{p+1} x_{p+1} + \cdots + \beta_{p+b-2} x_{p+b-2}}_{\text{Block effects}}$$

Then we compare the SSEs for the two models, SSE_R and SSE_C, using the "partial" F statistic:

$$F = \frac{(\text{SSE}_R - \text{SSE}_C)/\text{Number of } \beta\text{'s tested}}{\text{MSE}_C}$$

$$= \frac{(\text{SSE}_R - \text{SSE}_C)/(p - 1)}{\text{MSE}_C}$$

A significant F-value implies that the treatment means differ.

Occasionally, experimenters want to determine whether blocking was effective in removing the extraneous source of variation (i.e., whether there is evidence of a difference among block means). In fact, if there are no differences among block means, the experimenter will lose information by blocking because blocking reduces the number of degrees of freedom associated with the estimated variance of the model, s^2. If blocking is *not* effective in reducing the variability, then the block parameters in the complete model will all equal 0 (i.e., there will be no differences among block means). Thus, we want to test

$$H_0: \ \beta_p = \beta_{p+1} = \cdots = \beta_{p+b-2} = 0$$

Another reduced model, with the block β's dropped, is fitted:

Reduced model for testing blocks

$$E(y) = \beta_0 + \underbrace{\beta_1 x_1 + \beta_2 x_2 + \cdots + \beta_{p-1} x_{p-1}}_{\text{Treatment effects}}$$

The SSE for this second reduced model is compared to the SSE for the complete model in the usual fashion. A significant F-test implies that blocking is effective in removing (or reducing) the targeted extraneous source of variation.

These two tests are summarized in the following box.

Models and ANOVA F-Tests for a Randomized Block Design with p Treatments and b Blocks

Complete model:

$$E(y) = \beta_0 + \overbrace{\beta_1 x_1 + \cdots + \beta_{p-1} x_{p-1}}^{(p-1)\text{treatment terms}} + \overbrace{\beta_p x_p + \cdots + \beta_{p+b-2} x_{p+b-2}}^{(b-1)\text{block terms}}$$

where

$$x_1 = \begin{cases} 1 & \text{if treatment 2} \\ 0 & \text{if not} \end{cases} \quad \cdots \quad x_{p-1} = \begin{cases} 1 & \text{if treatment } p \\ 0 & \text{if not} \end{cases}$$

$$x_p = \begin{cases} 1 & \text{if block 2} \\ 0 & \text{if not} \end{cases} \quad \cdots \quad x_{p+b-2} = \begin{cases} 1 & \text{if block } b \\ 0 & \text{if not} \end{cases}$$

TEST FOR COMPARING TREATMENT MEANS

$H_0: \beta_1 = \beta_2 = \cdots = \beta_{p-1} = 0$

(i.e., H_0: The p treatment means are equal.)

H_a: At least one of the β parameters listed in H_0 differs from 0

(i.e., H_a: At least two treatment means differ.)

Reduced model: $E(y) = \beta_0 + \beta_p x_p + \cdots + \beta_{p+b-2} x_{p+b-2}$

Test statistic:
$$F = \frac{(\text{SSE}_R - \text{SSE}_C)/(p-1)}{\text{SSE}_C/(n-p-b+1)}$$

$$= \frac{(\text{SSE}_R - \text{SSE}_C)/(p-1)}{\text{MSE}_C}$$

where

$$\text{SSE}_R = \text{SSE for reduced model}$$
$$\text{SSE}_C = \text{SSE for complete model}$$
$$\text{MSE}_C = \text{MSE for complete model}$$

Rejection region: $F > F_\alpha$ where F is based on $\nu_1 = (p-1)$ and
$$\nu_2 = (n-p-b+1)\text{degrees of freedom}$$

TEST FOR COMPARING BLOCK MEANS

$H_0: \beta_p = \beta_{p+1} = \cdots = \beta_{p+b-2} = 0$

(i.e., H_0: The b block means are equal.)

H_a: At least one of the β parameters listed in H_0 differs from 0.

(i.e., H_a: At least two block means differ.)

Reduced model: $E(y) = \beta_0 + \beta_1 x_1 + \beta_2 x_2 + \cdots + \beta_{p-1} x_{p-1}$

Test statistic:
$$F = \frac{(\text{SSE}_R - \text{SSE}_C)/(b-1)}{\text{SSE}_C/(n-p-b+1)}$$

$$= \frac{(\text{SSE}_R - \text{SSE}_C)/(b-1)}{\text{MSE}_C}$$

where

$$\text{SSE}_R = \text{SSE for reduced model}$$
$$\text{SSE}_C = \text{SSE for complete model}$$
$$\text{MSE}_C = \text{MSE for complete model}$$

Rejection region: $F > F_\alpha$ where F is based on $\nu_1 = (b - 1)$ and $\nu_2 = (n - p - b + 1)$ degrees of freedom

Assumptions:

1. The probability distribution of the difference between any pair of treatment observations within a block is approximately normal.

2. The variance of the difference is constant and the same for all pairs of observations.

Example 12.7

Prior to submitting a bid for a construction job, cost engineers prepare a detailed analysis of the estimated labor and materials costs required to complete the job. This estimate will depend on the engineer who performs the analysis. An overly large estimate will reduce the chance of acceptance of a company's bid price, whereas an estimate that is too low will reduce the profit or even cause the company to lose money on the job. A company that employs three job cost engineers wanted to compare the mean level of the engineers' estimates. Due to the known variability in costs that exists from job to job, each engineer estimated the cost of the same four randomly selected jobs. The data (in hundreds of thousands of dollars) for this randomized block design are shown in Table 12.3.

💿 COSTENG

Table 12.3 Data for the randomized block design in Example 12.7

		Job				
		1	2	3	4	Treatment Means
Engineer	1	4.6	6.2	5.0	6.6	5.60
	2	4.9	6.3	5.4	6.8	5.85
	3	4.4	5.9	5.4	6.3	5.50
Block Means		4.63	6.13	5.27	6.57	

(a) Explain why a randomized block design was employed and give the appropriate linear model for this design.

(b) Perform an analysis of variance on the data, and test to determine whether there is sufficient evidence to indicate differences among treatment means. Test using $\alpha = .05$.

(c) Test to determine whether blocking on jobs was successful in reducing the job-to-job variation in the estimates. Use $\alpha = .05$.

Solution

(a) The data for this experiment were collected according to a randomized block design because estimates of the same job were expected to be more nearly alike than estimates between jobs due to the known variability in job costs. Thus, the experiment involves three treatments (engineers) and four blocks (jobs).

The complete model for this design is

$$E(y) = \beta_0 + \underbrace{\beta_1 x_1 + \beta_2 x_2}_{\text{Treatments (engineers)}} + \underbrace{\beta_3 x_3 + \beta_4 x_4 + \beta_5 x_5}_{\text{Blocks (jobs)}}$$

where

$$y = \text{Cost estimate}$$

$$x_1 = \begin{cases} 1 & \text{if engineer 2} \\ 0 & \text{if not} \end{cases} \qquad x_2 = \begin{cases} 1 & \text{if engineer 3} \\ 0 & \text{if not} \end{cases}$$

Base level = Engineer 1

$$x_3 = \begin{cases} 1 & \text{if block 2} \\ 0 & \text{if not} \end{cases} \quad x_4 = \begin{cases} 1 & \text{if block 3} \\ 0 & \text{if not} \end{cases} \quad x_5 = \begin{cases} 1 & \text{if block 4} \\ 0 & \text{if not} \end{cases}$$

Base level = Block 1

(b) We conducted an analysis of variance of the data using SAS. The SAS printout for the complete model is shown in Figure 12.10. Note that $SSE_C = .18667$ and $MSE_C = .03111$ (shaded on the printout).

Figure 12.10 SAS regression printout for randomized block design complete model, Example 12.7

Dependent Variable: COST

Analysis of Variance

Source	DF	Sum of Squares	Mean Square	F Value	Pr > F
Model	5	7.02333	1.40467	45.15	0.0001
Error	6	0.18667	0.03111		
Corrected Total	11	7.21000			

Root MSE	0.17638	R-Square	0.9741	
Dependent Mean	5.65000	Adj R-Sq	0.9525	
Coeff Var	3.12183			

Parameter Estimates

Variable	DF	Parameter Estimate	Standard Error	t Value	Pr > |t|
Intercept	1	4.58333	0.12472	36.75	<.0001
X1	1	0.25000	0.12472	2.00	0.0919
X2	1	-0.10000	0.12472	-0.80	0.4533
X3	1	1.50000	0.14402	10.42	<.0001
X4	1	0.63333	0.14402	4.40	0.0046
X5	1	1.93333	0.14402	13.42	<.0001

Test ENGINEER Results for Dependent Variable COST

Source	DF	Mean Square	F Value	Pr > F
Numerator	2	0.13000	4.18	0.0730
Denominator	6	0.03111		

Test JOBS Results for Dependent Variable COST

Source	DF	Mean Square	F Value	Pr > F
Numerator	3	2.25444	72.46	<.0001
Denominator	6	0.03111		

To test for differences among the treatment means, we will test

$$H_0: \mu_1 = \mu_2 = \mu_3$$

where μ_i = mean cost estimate of engineer i. This is equivalent to testing

$$H_0: \beta_1 = \beta_2 = 0$$

in the complete model. We fit the reduced model

$$E(y) = \beta_0 + \underbrace{\beta_3 x_3 + \beta_4 x_4 + \beta_5 x_5}_{\text{Blocks (jobs)}}$$

The SAS printout for this reduced model is shown in Figure 12.11. Note that $SSE_R = .44667$ (shaded on the printout). The remaining elements of the test follow.

Figure 12.11 SAS regression printout for randomized block design reduced model for testing treatments

```
                         Dependent Variable: COST

                            Analysis of Variance

                                  Sum of            Mean
Source                DF         Squares          Square    F Value    Pr > F

Model                  3         6.76333         2.25444      40.38    <.0001
Error                  8         0.44667         0.05583
Corrected Total       11         7.21000


              Root MSE              0.23629     R-Square     0.9380
              Dependent Mean        5.65000     Adj R-Sq     0.9148
              Coeff Var             4.18214


                          Parameter Estimates

                        Parameter        Standard
Variable      DF         Estimate           Error    t Value    Pr > |t|

Intercept      1          4.63333         0.13642      33.96     <.0001
X3             1          1.50000         0.19293       7.77     <.0001
X4             1          0.63333         0.19293       3.28      0.0111
X5             1          1.93333         0.19293      10.02     <.0001
```

Test statistic:

$$F = \frac{(SSE_R - SSE_C)/(p-1)}{MSE_C} = \frac{(.44667 - .18667)/2}{.03111} = 4.18$$

Rejection region: $F > 5.14$, where $F_{.05} = 5.14$ (from Table 4, Appendix C) is based on $\nu_1 = (p-1) = 2$ df and $\nu_2 = (n - p - b + 1) = 6$ df.

Conclusion: Since $F = 4.18$ is less than the critical value, 5.14, there is insufficient evidence, at the $\alpha = .05$ level of significance, to indicate differences among the mean estimates for the three cost engineers.

As an option, SAS will conduct this nested model F-test. The test statistic, $F = 4.18$, is highlighted on the middle of the SAS complete model printout, Figure 12.10. The p-value of the test (also highlighted) is $p = .0730$. Since this value exceeds $\alpha = .05$, our conclusion is confirmed—there is insufficient evidence to reject H_0.

(c) To test for the effectiveness of blocking on jobs, we test

$$H_0: \beta_3 = \beta_4 = \beta_5 = 0$$

in the complete model specified in part a. The reduced model is

$$E(y) = \beta_0 + \underbrace{\beta_1 x_1 + \beta_2 x_2}_{\text{Treatments (engineers)}}$$

Figure 12.12 SAS regression printout for randomized block design reduced model for testing blocks

```
                              Dependent Variable: COST

                               Analysis of Variance

                                       Sum of              Mean
Source                   DF            Squares            Square    F Value    Pr > F

Model                     2            0.26000           0.13000       0.17    0.8477
Error                     9            6.95000           0.77222
Corrected Total          11            7.21000


              Root MSE                 0.87876    R-Square      0.0361
              Dependent Mean           5.65000    Adj R-Sq     -0.1781
              Coeff Var               15.55331


                            Parameter Estimates

                        Parameter        Standard
Variable         DF      Estimate           Error     t Value    Pr > |t|

Intercept         1       5.60000         0.43938       12.75      <.0001
X1                1       0.25000         0.62138        0.40      0.6968
X2                1      -0.10000         0.62138       -0.16      0.8757
```

The SAS printout for this second reduced model is shown in Figure 12.12. Note that $SSE_R = 6.95$ (shaded on the printout). The elements of the test follow.

Test statistic:

$$F = \frac{(SSE_R - SSE_C)/(b-1)}{MSE_C} = \frac{(6.95 - .18667)/3}{.03111} = 72.46$$

Rejection region: $F > 4.76$, where $F_{.05} = 4.76$ (from Table 4, Appendix D) is based on $\nu_1 = (b-1) = 3$ df and $\nu_2 = (n - p - b + 1) = 6$ df.

Conclusion: Since $F = 72.46$ exceeds the critical value 4.76, there is sufficient evidence (at $\alpha = .05$) to indicate differences among the block (job) means. It appears that blocking on jobs was effective in reducing the job-to-job variation in cost estimates.

We also requested SAS to perform this nested model F-test for blocks. The results, $F = 72.46$ and p-value $< .0001$, are shaded at the bottom of the SAS complete model printout, Figure 12.10. The small p-value confirms our conclusion; there is sufficient evidence at ($\alpha = .05$) to reject H_0.

Caution

The result of the test for the equality of block means must be interpreted with care, especially when the calculated value of the F-test statistic does not fall in the rejection region. This does not necessarily imply that the block means are the same (i.e., that blocking is unimportant). Reaching this conclusion would be equivalent to accepting the null hypothesis, a practice we have carefully avoided because of the unknown probability of committing a Type II error (i.e., of accepting H_0 when H_a is true). In other words, even when a test for block differences is inconclusive, we may still want to use the randomized block design in similar future experiments. If the experimenter believes that the experimental units are more homogeneous within blocks than among blocks, he or she should use the randomized block design regardless of whether the test comparing the block means shows them to be different.

Figure 12.13 Partitioning of the total sum of squares for the randomized block design

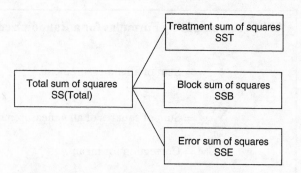

The traditional analysis of variance approach to analyzing the data collected from a randomized block design is similar to the completely randomized design. The partitioning of SS(Total) for the randomized block design is most easily seen by examining Figure 12.13. Note that SS(Total) is now partitioned into *three* parts:

$$SS(Total) = SSB + SST + SSE$$

The formulas for calculating SST and SSB follow the same pattern as the formula for calculating SST for the completely randomized design.

From these quantities, we obtain mean square for treatments, MST, mean square for blocks, MSB, and mean square for error, MSE, as shown in the next box (p. 634). The test statistics are

$$F = \frac{MST}{MSE} \text{ for testing treatments}$$

$$F = \frac{MSB}{MSE} \text{ for testing blocks}$$

These *F*-values are equivalent to the "partial" *F* statistics of the regression approach.

Example 12.8

Refer to Example 12.7. Perform an analysis of variance of the data in Table 12.3 using the ANOVA sums of squares approach.

Solution

Rather than perform the calculations by hand (again, we leave this as an exercise for the student), we utilize a statistical software package. The SPSS printout of the ANOVA is displayed in Figure 12.14. The *F*-value for testing treatments, $F = 4.179$, and the *F*-value for testing blocks, $F = 72.464$, are both shaded on the printout. Note that these values are identical to the *F*-values computed using the regression approach, Example 12.7. The *p*-values of the tests (also shaded) lead to the same conclusions reached in Example 12.7. For example, the *p*-value for the test of treatment differences, $p = .073$, exceeds $\alpha = .05$; thus, there is insufficient evidence of differences among the treatment means. ■

As with a completely randomized design, the sources of variation and their respective degrees of freedom, sums of squares, and mean squares for a randomized block design are shown in an ANOVA summary table. The general format of such

ANOVA Computing Formulas for a Randomized Block Design

$$\sum_{i=1}^{n} y_i = \text{Sum of all } n \text{ measurements}$$

$$\sum_{i=1}^{n} y_i^2 = \text{Sum of squares of all } n \text{ measurements}$$

$$\text{CM} = \text{Correction for mean}$$

$$= \frac{(\text{Total of all measurements})^2}{\text{Total number of measurements}} = \frac{\left(\sum_{i=1}^{n} y_i\right)^2}{n}$$

$$\text{SS(Total)} = \text{Total sum of squares}$$

$$= (\text{Sum of squares of all measurements}) - \text{CM}$$

$$= \sum_{i=1}^{n} y_i^2 - \text{CM}$$

$$\text{SST} = \text{Sum of squares for treatments}$$

$$= \left(\begin{array}{c}\text{Sum of squares of treatment totals with} \\ \text{each square divided by } b, \text{ the number of} \\ \text{measurements for that treatment}\end{array}\right) - \text{CM}$$

$$= \frac{T_1^2}{b} + \frac{T_2^2}{b} + \cdots + \frac{T_p^2}{b} - \text{CM}$$

$$\text{SSB} = \text{Sum of squares for blocks}$$

$$= \left(\begin{array}{c}\text{Sum of squares for block totals with} \\ \text{each square divided by } p, \text{ the number} \\ \text{of measurements in that block}\end{array}\right) - \text{CM}$$

$$= \frac{B_1^2}{p} + \frac{B_2^2}{p} + \cdots + \frac{B_b^2}{p} - \text{CM}$$

$$\text{SSE} = \text{Sum of squares for error} = \text{SS(Total)} - \text{SST} - \text{SSB}$$

$$\text{MST} = \text{Mean square for treatments} = \frac{\text{SST}}{p - 1}$$

$$\text{MSB} = \text{Mean square for blocks} = \frac{\text{SSB}}{b - 1}$$

$$\text{MSE} = \text{Mean square for error} = \frac{\text{SSE}}{n - p - b + 1}$$

$$F = \frac{\text{MST}}{\text{MSE}} \text{ for testing treatments}$$

$$F = \frac{\text{MSB}}{\text{MSE}} \text{ for testing blocks}$$

Figure 12.14 SPSS ANOVA printout for randomized block design

Tests of Between-Subjects Effects

Dependent Variable: COST

Source	Type III Sum of Squares	df	Mean Square	F	Sig.
Corrected Model	7.023[a]	5	1.405	45.150	.000
Intercept	383.070	1	383.070	12312.964	.000
ENGINEER	.260	2	.130	4.179	.073
JOB	6.763	3	2.254	72.464	.000
Error	.187	6	3.111E-02		
Total	390.280	12			
Corrected Total	7.210	11			

a. R Squared = .974 (Adjusted R Squared = .953)

a table for a randomized block design is shown in the next box; the ANOVA table for the data in Table 12.3 is shown in Table 12.4. (These quantities are shaded on the SPSS printout, Figure 12.14.) Note that the degrees of freedom for the three sources of variation, treatments, blocks, and error sum to the degrees of freedom for SS(Total). Similarly, the sums of squares for the three sources will always sum to SS(Total).

General Format of ANOVA Table for a Randomized Block Design

Source	df	SS	MS	F
Treatments	$p - 1$	SST	$MST = \dfrac{SST}{p-1}$	$F = \dfrac{MST}{MSE}$
Blocks	$b - 1$	SSB	$MSB = \dfrac{SSB}{b-1}$	$F = \dfrac{MSB}{MSE}$
Error	$n - p - b + 1$	SSE	$MSE = \dfrac{SSE}{n-p-b+1}$	
Total	$n - 1$	SS(Total)		

Table 12.4 ANOVA summary table for Example 12.8

Source	df	SS	MS	F
Treatments (Engineers)	2	.260	.130	4.18
Blocks (Jobs)	3	6.763	2.254	72.46
Error	6	.187	.031	
Total	11	7.210		

Confidence intervals for the difference between a pair of treatment means or block means for a randomized block design are shown in the following box.

> **Confidence Intervals for the Difference $(\mu_i - \mu_j)$ Between a Pair of Treatment Means or Block Means**
>
> $$\text{Treatment means: } (\overline{T}_i - \overline{T}_j) \pm t_{\alpha/2} s \sqrt{\frac{2}{b}}$$
>
> $$\text{Block means: } (\overline{B}_i - \overline{B}_j) \pm t_{\alpha/2} s \sqrt{\frac{2}{p}}$$
>
> where
>
> $$b = \text{Number of blocks}$$
> $$p = \text{Number of treatments}$$
> $$s = \sqrt{\text{MSE}}$$
> $$\overline{T}_i = \text{Sample mean for treatment } i$$
> $$\overline{B}_i = \text{Sample mean for block } i$$
>
> and $t_{\alpha/2}$ is based on $(n - p - b + 1)$ degrees of freedom

Example 12.9

Refer to Example 12.7. Find a 90% confidence interval for the difference between the mean level of estimates for engineers 1 and 2.

Solution

From Example 12.7, we know that $b = 4$, $\overline{T}_1 = 5.60$, $\overline{T}_2 = 5.85$, and $s^2 = \text{MSE}_C = .03111$. The degrees of freedom associated with s^2 (and, therefore, with $t_{\alpha/2}$) is 6. Therefore, $s = \sqrt{s^2} = \sqrt{.03111} = .176$ and $t_{\alpha/2} = t_{.05} = 1.943$. Substituting these values into the formula for the confidence interval for $(\mu_1 - \mu_2)$, we obtain

$$(\overline{T}_1 - \overline{T}_2) \pm (t_{\alpha/2}) s \sqrt{\frac{2}{b}}$$

$$(5.60 - 5.85) \pm (1.943)(.176)\sqrt{\frac{2}{4}}$$

$$-.25 \pm .24$$

or $-.49$ to $-.01$. Since each unit represents \$100,000, we estimate the difference between the mean level of job estimates for estimators 1 and 2 to be enclosed by the interval, $-\$49,000$ to $-\$1,000$. [*Note*: At first glance, this result may appear to contradict the result of the F-test for comparing treatment means. However, the observed significance level of the F-test (.07) implies that significant differences exist between the means at $\alpha = .10$, which is consistent with the fact that 0 is not within the 90% confidence interval.]

There is one very important point to note when you block the treatments in an experiment. Recall from Section 11.3 that the block effects cancel. This fact enables us to calculate confidence intervals for the difference between treatment means using the formulas given in the box. However, if a sample treatment mean is used to estimate *a single treatment mean*, the block effects do not cancel. *Therefore, the only way that you can obtain an unbiased estimate of a single treatment mean (and corresponding confidence interval) in a blocked design is to randomly select the blocks from a large collection (population) of blocks and to treat the block effect as a second random component, in addition to random error.* Designs that contain two or more random components are called *nested designs* and are beyond the scope of this text. For more information on this topic, see the references at the end of this chapter.

12.4 Exercises

12.18 ANOVA Summary table. The analysis of variance for a randomized block design produced the ANOVA table entries shown here.

SOURCE	df	SS	MS	F
Treatments	3	27.1	—	—
Blocks	5	—	14.90	—
Error	—	33.4	—	
Total	—	—		

The sample means for the four treatments are as follows:

$$\bar{y}_A = 9.7 \quad \bar{y}_B = 12.1 \quad \bar{y}_C = 6.2 \quad \bar{y}_D = 9.3$$

(a) Complete the ANOVA table.
(b) Do the data provide sufficient evidence to indicate a difference among the treatment means? Test using $\alpha = .01$.
(c) Do the data provide sufficient evidence to indicate that blocking was a useful design strategy to employ for this experiment? Explain.
(d) Find a 95% confidence interval for $(\mu_A - \mu_B)$.
(e) Find a 95% confidence interval for $(\mu_B - \mu_D)$.

12.19 Making high-stakes insurance decisions. Refer to the *Journal of Economic Psychology* (September 2008) study on high-stakes insurance decisions, Exercise 12.6 (p. 621). A second experiment involved only the Group 2 subjects. In part A of the experiment, these 84 subjects were informed of the hazards (both fire and theft) of owning a valuable painting, but were not told the exact probabilities of the hazards occurring. The subjects then provided an amount they were willing to pay (WTP) for insuring the painting. In part B of the experiment, these same subjects were informed of the exact probabilities of the hazards (fire and theft) of owning a valuable sculpture. The subjects then provided a WTP amount for insuring the sculpture. The researchers were interested in comparing the mean WTP amounts for the painting and the sculpture.

(a) Explain why the experimental design employed is a randomized block design.
(b) Identify the dependent (response) variable, treatments, and blocks for the design.
(c) Give the null and alternative hypotheses of interest to the researchers.

12.20 Peer mentor training at a firm. Peer mentoring occurs when a more experienced employee provides one-on-one support and knowledge sharing with a less experienced employee. The *Journal of Managerial Issues* (Spring 2008) published a study of the impact of peer mentor training at a large software company. Participants were 222 employees who volunteered to attend a one-day peer mentor training session. One variable of interest was the employee's level of competence in peer mentoring (measured on a 7-point scale). The competence level of each trainee was measured at three different times in the study: one week before training, two days after training, and two months after training. One goal of the experiment was to compare the mean competence levels of the three time periods.

(a) Explain why this data should be analyzed using a randomized block design. As part of your answer, identify the blocks and the treatments.
(b) A partial ANOVA table for the experiment is shown below. Explain why there is enough information in the table to make conclusions.

SOURCE	df	SS	MS	F-VALUE	p-VALUE
Time Period	2	—	—	—	0.001
Blocks	221	—	—	—	0.001
Error	442	—	—		
Total	665	—			

(c) State the null hypothesis of interest to the researcher.
(d) Make the appropriate conclusion.

12.21 Plants and stress reduction. Plant therapists believe that plants can reduce the stress levels of humans. A Kansas State University study was conducted to investigate this phenomenon. Two weeks before final exams, 10 undergraduate students took part in an experiment to determine what effect the presence of a live plant, a photo of a plant, or absence of a plant has on the student's ability to relax while isolated in a dimly lit room. Each student participated in three sessions—one with a live plant, one with a plant photo, and one with no plant (control).* During each session, finger temperature was measured at 1-minute intervals for 20 minutes. Since increasing finger temperature indicates an increased level of relaxation, the maximum temperature (in degrees) was used as the response variable. The data for

---

* The experiment is simplified for this exercise. The actual experiment involved 30 students who participated in 12 sessions.

PLANTS

STUDENT	LIVE PLANT	PLANT PHOTO	NO PLANT (CONTROL)
1	91.4	93.5	96.6
2	94.9	96.6	90.5
3	97.0	95.8	95.4
4	93.7	96.2	96.7
5	96.0	96.6	93.5
6	96.7	95.5	94.8
7	95.2	94.6	95.7
8	96.0	97.2	96.2
9	95.6	94.8	96.0
10	95.6	92.6	96.6

Source: Elizabeth Schreiber, Department of Statistics, Kansas State University, Manhattan, Kansas.

SPSS Output for Exercise 12.21

Tests of Between-Subjects Effects

Dependent Variable: FINGTEMP

Source	Type III Sum of Squares	df	Mean Square	F	Sig.
Corrected Model	18.537[a]	11	1.685	.523	.863
Intercept	272176.875	1	272176.875	84413.380	.000
PLANT	.122	2	6.100E-02	.019	.981
STUDENT	18.415	9	2.046	.635	.754
Error	58.038	18	3.224		
Total	272253.450	30			
Corrected Total	76.575	29			

a. R Squared = .242 (Adjusted R Squared = -.221)

the experiment, provided in the table at left, were analyzed using the ANOVA procedure of SPSS. Use the accompanying SPSS printout to make the proper inference.

12.22 Studies on treating Alzheimer's disease. The quality of the research methodology used in journal articles that investigate the effectiveness of Alzheimer's disease (AD) treatments was examined in *eCAM* (November 2006). For each in a sample of 13 research papers, the quality of the methodology on each of nine dimensions was measured using the Wong Scale, with scores ranging from 1 (low quality) to 3 (high quality). The data in the table below gives the individual dimension scores for the sample of 13 papers. (Note: The researchers labeled the nine dimensions as What-A, What-B, What-C, Who-A, Who-B, Who-C, How-A, How-B, and How-C.)

(a) One goal of the study is to compare the mean Wong scores of the nine research methodology dimensions. Set up the null and alternative hypotheses for this test.

(b) The researchers used a completely randomized design ANOVA to analyze the data. Explain why a randomized block ANOVA is a more appropriate method.

(c) The SAS output for a randomized block ANOVA of the data (with Dimensions as treatments and Papers as blocks) is shown below. Interpret the p-values of the tests shown.

TREATAD2

PAPER	WHAT-A	WHAT-B	WHAT-C	WHO-A	WHO-B	WHO-C	HOW-A	HOW-B	HOW-C
1	3	3	2	2	2	3	2	2	3
2	3	3	2	1	3	2	3	2	2
3	2	3	3	1	2	2	1	2	3
4	2	3	3	2	1	3	2	1	2
5	2	2	2	2	2	3	1	3	2
6	2	3	2	1	1	1	2	2	1
7	2	3	2	2	1	3	2	2	2
8	2	2	2	1	2	3	2	2	3
9	1	3	2	2	2	2	1	1	1
10	2	3	2	1	2	2	2	3	3
11	2	2	1	1	1	2	3	2	3
12	2	2	3	2	1	3	1	2	3
13	3	3	2	2	1	3	2	2	3

Source: Chiappelli, F. et al. "Evidence-based research in complementary and alternative medicine III: Treatment of patients with Alzheimer's disease," *eCAM,* Vol. 3, No. 4, Nov. 2006 (Table 1).

SAS Output for Exercise 12.22

Source	DF	Type III SS	Mean Square	F Value	Pr > F
DIMENSION	8	15.86324786	1.98290598	5.40	<.0001
PAPER	12	6.44444444	0.53703704	1.46	0.1520

12.23 **Light to dark transition of genes.** Refer to the *Journal of Bacteriology* (July 2002) study of the sensitivity of bacteria genes to light, Exercise 1.73 (p. 63). Recall that scientists isolated 103 genes of the bacterium responsible for photosynthesis and respiration. Each gene was grown to mid-exponential phase in a growth incubator in "full light," then exposed to three alternative light/dark conditions: "full dark" (lights extinguished for 24 hours), "transient light" (lights turned back on for 90 minutes), and "transient dark" (lights turned back off for an additional 90 minutes). At the end of each light/dark condition, the standardized growth measurement was determined for each of the 103 genes. The complete data set is saved in the GENEDARK file. (Data for the first 10 genes are shown in the accompanying table.) Assume that the goal of the experiment is to compare the mean standardized growth measurements for the three light/dark conditions.

GENEDARK (*First 10 observations shown*)

GENE ID	FULL-DARK	TR-LIGHT	TR-DARK
SLR2067	−0.00562	1.40989	−1.28569
SLR1986	−0.68372	1.83097	−0.68723
SSR3383	−0.25468	−0.79794	−0.39719
SLL0928	−0.18712	−1.20901	−1.18618
SLR0335	−0.20620	1.71404	−0.73029
SLR1459	−0.53477	2.14156	−0.33174
SLL1326	−0.06291	1.03623	0.30392
SLR1329	−0.85178	−0.21490	0.44545
SLL1327	0.63588	1.42608	−0.13664
SLL1325	−0.69866	1.93104	−0.24820

Source: Gill, R. T., et al. "Genome-wide dynamic transcriptional profiling of the light to dark transition in *Synechocystis Sp.* PCC6803," *Journal of Bacteriology*, Vol. 184, No. 13, July 2002.

(a) Explain why the data should be analyzed as a randomized block design.

(b) Specify the null and alternative hypothesis for comparing the light/dark condition means.

(c) Using a statistical software package, conduct the test, part b. Interpret the results at $\alpha = .05$.

12.24 **Absentee rates at a jeans plant.** A plant that manufactures denim jeans in the United Kingdom recently introduced a computerized automated handling system. The new system delivers garments to the assembly line operators by means of an overhead conveyor. While the automated system minimizes operator handling time, it inhibits operators from working ahead and taking breaks from their machine. A study in *New Technology, Work, and Employment* (July 2001) investigated the impact of the new handling system on worker absentee rates at the jeans plant. One theory is that the mean absentee rate will vary by day of the week, as operators decide to indulge in one-day absences to relieve work pressure. Nine weeks were randomly selected and the absentee rate (percentage of workers absent) determined for each day (Monday through Friday) of the work week. The data are listed in the table. Conduct a complete analysis of the data to determine whether the mean absentee rate differs across the five days of the work week.

JEANS

WEEK	MONDAY	TUESDAY	WEDNESDAY	THURSDAY	FRIDAY
1	5.3	0.6	1.9	1.3	1.6
2	12.9	9.4	2.6	0.4	0.5
3	0.8	0.8	5.7	0.4	1.4
4	2.6	0.0	4.5	10.2	4.5
5	23.5	9.6	11.3	13.6	14.1
6	9.1	4.5	7.5	2.1	9.3
7	11.1	4.2	4.1	4.2	4.1
8	9.5	7.1	4.5	9.1	12.9
9	4.8	5.2	10.0	6.9	9.0

Source: Boggis, J. J. "The eradication of leisure," *New Technology, Work, and Employment*, Volume 16, Number 2, July 2001 (Table 3).

12.25 **Massage therapy for boxers.** Eight amateur boxers participated in an experiment to investigate the effect of massage on boxing performance (*British Journal of Sports Medicine*, April 2000). The punching power of each boxer (measured in Newtons) was recorded in the round following each of four different interventions: (M1) in round 1 following a pre-bout sports massage, (R1) in round 1 following a pre-bout period of rest, (M5) in round 5 following a sports massage between rounds, and (R5) in round 5 following a period of rest between rounds. Based on information provided in the article, the data in the table (p. 640) were obtained. The main goal of the experiment is to compare the punching power means of the four interventions.

(a) Give the complete model appropriate for this design.

(b) Give the reduced model appropriate for testing for differences in the punching power means of the four interventions.

(c) Give the reduced model appropriate for determining whether blocking by boxers was effective in removing an unwanted source of variability.

🔘 BOXING

BOXER	INTERVENTION			
	M1	R1	M5	R5
1	1243	1244	1291	1262
2	1147	1053	1169	1177
3	1247	1375	1309	1321
4	1274	1235	1290	1285
5	1177	1139	1233	1238
6	1336	1313	1366	1362
7	1238	1279	1275	1261
8	1261	1152	1289	1266

Source: Reproduced from *British Journal of Sports Medicine,* "Effects of massage on physiological restoration, perceived recovery, and repeated sports performance," Brian Hemmings, Marcus Smith, Jan Graydon, and Rosemary Dyson, Vol. 34, Issue 2, April 2000 (Table 3), with permission from BMJ Publishing Group Ltd.

12.26 Massage therapy for boxers (cont'd). Refer to Exercise 12.25. The models of parts a, b, and c were fit to data in the table using MINITAB. The MINITAB printouts are displayed here.

(a) Construct an ANOVA summary table.

(b) Is there evidence of differences in the punching power means of the four interventions? Use $\alpha = .05$.

(c) Is there evidence of a difference among the punching power means of the boxers? That is, is there evidence that blocking by boxers was effective in removing an unwanted source of variability? Use $\alpha = .05$.

MINITAB Printout for Complete Model in Exercise 12.25a

```
The regression equation is
POWER = 1260 + 18.0 B1 - 106 B2 + 71.0 B3 + 29.0 B4 - 45.3 B5 + 102 B6
      + 21.3 B7 - 31.1 I1 + 6.3 I2 - 47.7 I3

Predictor      Coef     SE Coef        T        P
Constant    1260.16       20.84    60.48    0.000
B1            18.00       25.13     0.72    0.482
B2          -105.50       25.13    -4.20    0.000
B3            71.00       25.13     2.83    0.010
B4            29.00       25.13     1.15    0.261
B5           -45.25       25.13    -1.80    0.086
B6           102.25       25.13     4.07    0.001
B7            21.25       25.13     0.85    0.407
I1           -31.12       17.77    -1.75    0.094
I2             6.25       17.77     0.35    0.729
I3           -47.75       17.77    -2.69    0.014

S = 35.54      R-Sq = 83.4%     R-Sq(adj) = 75.4%

Analysis of Variance

Source          DF        SS        MS        F        P
Regression      10    132798     13280    10.51    0.000
Residual Error  21     26525      1263
Total           31    159323
```

MINITAB Printout for Complete Model in Exercise 12.25b

```
The regression equation is
POWER = 1242 + 18.0 B1 - 106 B2 + 71.0 B3 + 29.0 B4 - 45.3 B5 + 102 B6
      + 21.3 B7

Predictor      Coef     SE Coef        T        P
Constant    1242.00       20.99    59.18    0.000
B1            18.00       29.68     0.61    0.550
B2          -105.50       29.68    -3.55    0.002
B3            71.00       29.68     2.39    0.025
B4            29.00       29.68     0.98    0.338
B5           -45.25       29.68    -1.52    0.140
B6           102.25       29.68     3.45    0.002
B7            21.25       29.68     0.72    0.481

S = 41.97      R-Sq = 73.5%     R-Sq(adj) = 65.7%

Analysis of Variance

Source          DF        SS        MS        F        P
Regression       7    117044     16721     9.49    0.000
Residual Error  24     42279      1762
Total           31    159323
```

MINITAB Printout for Complete Model in Exercise 12.25c

```
The regression equation is
POWER = 1271 - 31.1 I1 + 6.3 I2 - 47.7 I3

Predictor      Coef     SE Coef        T        P
Constant    1271.50       25.32    50.22    0.000
I1           -31.12       35.80    -0.87    0.392
I2             6.25       35.80     0.17    0.863
I3           -47.75       35.80    -1.33    0.193

S = 71.61      R-Sq = 9.9%     R-Sq(adj) = 0.2%

Analysis of Variance

Source          DF        SS        MS        F        P
Regression       3     15754      5251     1.02    0.397
Residual Error  28    143569      5127
Total           31    159323
```

12.27 Anticorrosive behavior of steel coated with epoxy. Organic coatings that use epoxy resins are widely used for protecting steel and metal against weathering and corrosion. Researchers at National Technical University (Athens, Greece) examined the steel anticorrosive behavior of different epoxy coatings formulated with zinc pigments in an attempt to find the epoxy coating with the best corrosion inhibition (*Pigment and Resin Technology*, Vol. 32, 2003). The experimental units were flat, rectangular panels cut from steel sheets. Each panel was coated with one of four different coating systems, S1, S2, S3, and S4. Three panels were prepared for each coating system. (These panels are labeled S1-A, S1-B, S1-C, S2-A, S2-B, ..., S4-C.) Each coated panel was immersed in de-ionized and de-aerated water and then tested for corrosion. Since exposure time is likely to have a strong influence on anticorrosive behavior, the researchers attempted to remove this extraneous source of variation through the experimental design. Exposure times were fixed at 24 hours, 60 days, and 120 days. For each of the coating systems, one panel was exposed to water for 24 hours, one exposed to water for 60 days, and one exposed to water for 120 days in random order. Following exposure, the corrosion rate

(nanoamperes per square centimeter) was determined for each panel. The lower the corrosion rate, the greater the anticorrosion performance of the coating system. The data are shown in the accompanying table. Are there differences among the epoxy treatment means? [Hint: Analyze the data as a randomized block design with three blocks (Times) and four treatments (Systems).]

EPOXY

EXPOSURE TIME	SYSTEM S1 (PANEL)	SYSTEM S2 (PANEL)	SYSTEM S3 (PANEL)	SYSTEM S4 (PANEL)
24 hours	6.7 (A)	7.5 (C)	8.2 (C)	6.1 (B)
60 days	8.7 (C)	9.1 (A)	10.5 (B)	8.3 (A)
120 days	11.8 (B)	12.6 (B)	14.5 (A)	11.8 (C)

Source: Kouloumbi, N. et al. "Anticorrosion performance of epoxy coatings on steel surface exposed to de-ionized water," *Pigment and Resin Technology*, Vol. 32, No. 2, 2003 (Table II).

12.5 Two-Factor Factorial Experiments

In Section 11.4, we learned that factorial experiments are volume-increasing designs conducted to investigate the effect of two or more independent variables (factors) on the mean value of the response y. In this section, we focus on the analysis of two-factor factorial experiments.

Suppose, for example, we want to relate the mean number of defects on a finished item—say, a new desk top—to two factors, type of nozzle for the varnish spray gun and length of spraying time. Suppose further that we want to investigate the mean number of defects per desk for three types (three levels) of nozzles (N_1, N_2, and N_3) and for two lengths (two levels) of spraying time (S_1 and S_2). If we choose the treatments for the experiment to include all combinations of the three levels of nozzle type with the two levels of spraying time (i.e., we observe the number of defects for the factor level combinations N_1S_1, N_1S_2, N_2S_1, N_2S_2, N_3S_1, N_3S_2), our design is called a **complete 3 × 2 factorial experiment**. Note that the design will contain $3 \times 2 = 6$ treatments.

Factorial experiments, you will recall, are useful methods for selecting treatments because they permit us to make inferences about factor interactions. The complete model for the 3×2 factorial experiment contains $(3 - 1) = 2$ main effect terms for nozzles, $(2 - 1) = 1$ main effect term for spray time, and $(3 - 1)(2 - 1) = 2$ nozzle–spray time interaction terms:

$$E(y) = \beta_0 + \underbrace{\beta_1 x_1 + \beta_2 x_2}_{\substack{\text{Main effects} \\ \text{Nozzle}}} + \underbrace{\beta_3 x_3}_{\substack{\text{Main effect} \\ \text{Spray time}}} + \underbrace{\beta_4 x_1 x_3 + \beta_5 x_2 x_3}_{\substack{\text{Interaction} \\ \text{Nozzle} \times \text{Spray time}}}$$

The main effects for the independent variables (factors) in the model are typically represented by dummy variables, regardless of type (quantitative or qualitative). However, if the factors are quantitative, the main effects can be represented by terms such as x, x^2, x^3, and so on. (We illustrate this approach later in this section.) In our 3×2 factorial experiment, we define the dummy variables for nozzle type and spraying time as follows:

$$x_1 = \begin{cases} 1 & \text{if nozzle } N_1 \\ 0 & \text{if not} \end{cases} \qquad x_2 = \begin{cases} 1 & \text{if nozzle } N_2 \\ 0 & \text{if not} \end{cases} \qquad \text{Base level} = N_3$$

$$x_3 = \begin{cases} 1 & \text{if spraying time } S_1 \\ 0 & \text{if spraying time } S_2 \end{cases}$$

Note that the model for the 3×2 factorial contains a total of $3 \times 2 = 6$ β parameters. If we observe only a single value of the response y for each of the $3 \times 2 = 6$ treatments, then $n = 6$ and df(Error) for the complete model is $(n - 6) = 0$. Consequently, for a factorial experiment, *the number r of observations per factor level combination (i.e., the number of replications of the factorial experiment) must always be 2 or more.* Otherwise, no degrees of freedom are available for estimating σ^2.

To test for factor interaction, we drop the interaction terms and fit the reduced model:

$$E(y) = \beta_0 + \underbrace{\beta_1 x_1 + \beta_2 x_2}_{\substack{\text{Main effects} \\ \text{Nozzle}}} + \underbrace{\beta_3 x_3}_{\substack{\text{Main effect} \\ \text{Spray time}}}$$

The null hypothesis of no interaction, H_0: $\beta_4 = \beta_5 = 0$, is tested by comparing the SSEs for the two models in a partial F statistic. This test for interaction is summarized, in general, in the box on pages 643–4.

Tests for factor main effects are conducted in a similar manner. The main effect terms of interest are dropped from the complete model and the reduced model is fitted. The SSEs for the two models are compared in the usual fashion.

Before we work through a numerical example of an analysis of variance for a factorial experiment, we must understand the practical significance of the tests for factor interaction and factor main effects. We illustrate these concepts in Example 12.10.

Example 12.10

A company that stamps gaskets out of sheets of rubber, plastic, and other materials wants to compare the mean number of gaskets produced per hour for two different types of stamping machines. Practically, the manufacturer wants to determine whether one machine is more productive than the other. Even more important is determining whether one machine is more productive in making rubber gaskets while the other is more productive in making plastic gaskets. To answer these questions, the manufacturer decides to conduct a 2×3 factorial experiment using three types of gasket material, B_1, B_2, and B_3, with each of the two types of stamping machines, A_1 and A_2. Each machine is operated for three 1-hour time periods for each of the gasket materials, with the 18 1-hour time periods assigned to the six machine–material combinations in random order. (The purpose of the randomization is to eliminate the possibility that uncontrolled environmental factors might bias the results.) Suppose we have calculated and plotted the six treatment means. Two hypothetical plots of the six means are shown in Figures 12.15a and 12.15b. The three means for stamping machine A_1 are connected by solid line segments and the corresponding three means for machine A_2 by dashed line segments. What do these plots imply about the productivity of the two stamping machines?

Models and ANOVA F-Test for Interaction in a Two-Factor Factorial Experiment with Factor A at a Levels and Factor B at b Levels

Complete model:

$$E(y) = \beta_0 + \overbrace{\beta_1 x_1 + \cdots + \beta_{a-1} x_{a-1}}^{\text{Main effect } A \text{ terms}} + \overbrace{\beta_a x_a + \cdots + \beta_{a+b-2} x_{a+b-2}}^{\text{Main effect } B \text{ terms}}$$

$$+ \overbrace{\beta_{a+b-1} x_1 x_a + \beta_{a+b} x_1 x_{a+1} + \cdots + \beta_{ab-1} x_{a-1} x_{a+b-2}}^{AB \text{ interaction terms}}$$

where*

$$x_1 = \begin{cases} 1 & \text{if level 2 of factor } A \\ 0 & \text{if not} \end{cases} \quad \cdots$$

$$x_{a-1} = \begin{cases} 1 & \text{if level } a \text{ of factor } A \\ 0 & \text{if not} \end{cases}$$

$$x_a = \begin{cases} 1 & \text{if level 2 of factor } B \\ 0 & \text{if not} \end{cases} \quad \cdots$$

$$x_{a+b-2} = \begin{cases} 1 & \text{if level } b \text{ of factor } B \\ 0 & \text{if not} \end{cases}$$

H_0: $\beta_{a+b-1} = \beta_{a+b} = \cdots = \beta_{ab-1} = 0$

    (i.e., H_0: No interaction between factors A and B.)

H_a: At least one of the β parameters listed in H_0 differs from 0.

    (i.e., H_a: Factors A and B interact.)

Reduced model:

$$E(y) = \beta_0 + \overbrace{\beta_1 x_1 + \cdots + \beta_{a-1} x_{a-1}}^{\text{Main effect A terms}} + \overbrace{\beta_a x_a + \cdots + \beta_{a+b-2} x_{a+b-2}}^{\text{Main effect B terms}}$$

Test statistic:
$$F = \frac{(\text{SSE}_R - \text{SSE}_C)/[(a-1)(b-1)]}{\text{SSE}_C/[ab(r-1)]}$$

$$= \frac{(\text{SSE}_R - \text{SSE}_C)/[(a-1)(b-1)]}{\text{MSE}_C}$$

where

    $\text{SSE}_R = \text{SSE for reduced model}$

    $\text{SSE}_C = \text{SSE for complete model}$

    $\text{MSE}_C = \text{MSE for complete model}$

    $r = $ Number of replications (i.e., number of y measurements per cell of the $a \times b$ factorial)

Rejection region: $F > F_\alpha$, where F is based on $\nu_1 = (a-1)(b-1)$ and $\nu_2 = ab(r-1)$ df

Assumptions: 1. The population probability distribution of the observations for any factor level combination is approximately normal.
2. The variance of the probability distribution is constant and the same for all factor level combinations.

*Note: The independent variables, $x_1, x_2, \ldots, x_{a+b-2}$, are defined for an experiment in which both factors represent *qualitative* variables. When a factor is *quantitative*, you may choose to represent the main effects with quantitative terms such as x, x^2, x^3, and so forth.

Solution

Figure 12.15a suggests that machine A_1 produces a larger number of gaskets per hour, regardless of the gasket material, and is therefore superior to machine A_2. On the average, machine A_1 stamps more cork (B_1) gaskets per hour than rubber or plastic, but the *difference* in the mean numbers of gaskets produced by the two machines remains approximately the same, regardless of the gasket material. Thus, the difference in the mean number of gaskets produced by the two machines is *independent* of the gasket material used in the stamping process.

In contrast to Figure 12.15a, Figure 12.15b shows the productivity for machine A_1 to be greater than that for machine A_2, when the gasket material is cork (B_1) or plastic (B_3). But the means are reversed for rubber (B_2) gasket material. For this material, machine A_2 produces, on the average, more gaskets per hour than machine A_1. Thus, Figure 12.15b illustrates a situation where the mean value of the response variable *depends* on the combination of the factor levels. When this situation occurs, we say that the factors *interact*. Thus, one of the most important objectives of a factorial experiment is to detect factor interaction if it exists. ▬

Figure 12.15 Hypothetical plot of the means for the six machine–material combinations

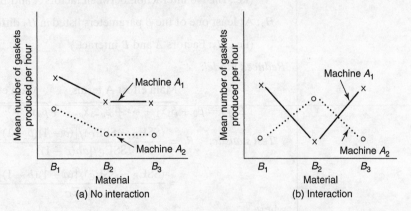

(a) No interaction (b) Interaction

Definition 12.2 In a factorial experiment, when the difference in the mean levels of factor A depends on the different levels of factor B, we say that the factors A and B **interact**. If the difference is independent of the levels of B, then there is **no interaction** between A and B.

Tests for main effects are relevant only when no interaction exists between factors. Generally, the test for interaction is performed first. *If there is evidence of factor interaction, then we will not perform the tests on the main effects.* (See Figure 12.16.) Rather, we want to focus attention on the individual cell (treatment) means, perhaps locating one that is the largest or the smallest.

Figure 12.16 Testing guidelines for a two-factor factorial experiment

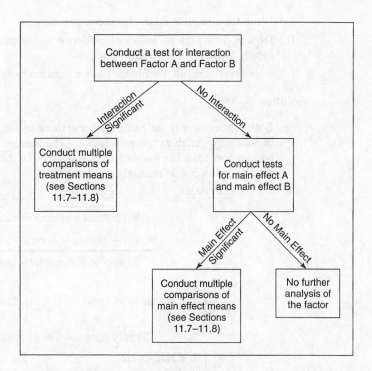

Example 12.11

A manufacturer, whose daily supply of raw materials is variable and limited, can use the material to produce two different products in various proportions. The profit per unit of raw material obtained by producing each of the two products depends on the length of a product's manufacturing run and, hence, on the amount of raw material assigned to it. Other factors, such as worker productivity and machine breakdown, affect the profit per unit as well, but their net effect on profit is random and uncontrollable. The manufacturer has conducted an experiment to investigate the effect of the level of supply of raw materials, S, and the ratio of its assignment, R, to the two product manufacturing lines on the profit y per unit of raw material. The ultimate goal would be to be able to choose the best ratio R to match each day's supply of raw materials, S. The levels of supply of the raw material chosen for the experiment were 15, 18, and 21 tons; the levels of the ratio of allocation to the two product lines were 1:2, 1:1, and 2:1. The response was the profit (in dollars) per unit of raw material supply obtained from a single day's production. Three replications of a complete 3×3 factorial experiment were conducted in a random sequence (i.e., a completely randomized design). The data for the 27 days are shown in Table 12.5.

⊛ RAWMATERIAL

Table 12.5 Data for Example 12.11

		\multicolumn{3}{c}{Raw Material Supply (S), tons}		
		15	18	21
Ratio of Raw Material Allocation (R)	1:2	23, 20, 21	22, 19, 20	19, 18, 21
	1:1	22, 20, 19	24, 25, 22	20, 19, 22
	2:1	18, 18, 16	21, 23, 20	20, 22, 24

(a) Write the complete model for the experiment.

(b) Do the data present sufficient evidence to indicate an interaction between supply S and ratio R? Use $\alpha = .05$.

(c) Based on the result, part b, should we perform tests for main effects?

Solution

(a) Both factors, supply and ratio, are set at three levels. Consequently, we require two dummy variables for each factor. (The number of main effect terms will be one less than the number of levels for a factor.) The complete factorial model for this 3×3 factorial experiment is

$$E(y) = \beta_0 + \underbrace{\beta_1 x_1 + \beta_2 x_2}_{\text{Supply main effects}} + \underbrace{\beta_3 x_3 + \beta_4 x_4}_{\text{Ratio main effects}}$$

$$+ \underbrace{\beta_5 x_1 x_3 + \beta_6 x_1 x_4 + \beta_7 x_2 x_3 + \beta_8 x_2 x_4}_{\text{Supply} \times \text{Ratio interaction}}$$

where

$$x_1 = \begin{cases} 1 & \text{if supply is 15 tons} \\ 0 & \text{if not} \end{cases} \qquad x_2 = \begin{cases} 1 & \text{if supply is 18 tons} \\ 0 & \text{if not} \end{cases}$$

(Supply base level = 21 tons)

$$x_3 = \begin{cases} 1 & \text{if ratio is 1:2} \\ 0 & \text{if not} \end{cases} \qquad x_4 = \begin{cases} 1 & \text{if ratio is 1:1} \\ 0 & \text{if not} \end{cases}$$

(Ratio base level = 2:1)

Note that the interaction terms for the model are constructed by taking the products of the various main effect terms, one from each factor. For example, we included terms involving the products of x_1 with x_3 and x_4. The remaining interaction terms were formed by multiplying x_2 by x_3 and by x_4.

(b) To test the null hypothesis that supply and ratio do not interact, we must test the null hypothesis that the interaction terms are not needed in the linear model of part a:

$H_0: \beta_5 = \beta_6 = \beta_7 = \beta_8 = 0$

This requires that we fit the reduced model

$$E(y) = \beta_0 + \beta_1 x_1 + \beta_2 x_2 + \beta_3 x_3 + \beta_4 x_4$$

and perform the partial F-test outlined in Section 4.13. The test statistic is

$$F = \frac{(\text{SSE}_R - \text{SSE}_C)/4}{\text{MSE}_C}$$

where

$$\text{SSE}_R = \text{SSE for reduced model}$$
$$\text{SSE}_C = \text{SSE for complete model}$$
$$\text{MSE}_C = \text{MSE for complete model}$$

The complete model of part a and the reduced model here were fit to the data in Table 12.5 using SAS. The SAS printouts are displayed in Figures 12.17a and 12.17b. The pertinent quantities, shaded on the printouts, are:

$\text{SSE}_C = 43.33333$ (see Figure 12.17a)
$\text{MSE}_C = 2.40741$ (see Figure 12.17a)
$\text{SSE}_R = 89.55556$ (see Figure 12.17b)

Figure 12.17a SAS regression printout for complete factorial model

Dependent Variable: PROFIT

Number of Observations Read   27
Number of Observations Used   27

Analysis of Variance

Source	DF	Sum of Squares	Mean Square	F Value	Pr > F
Model	8	74.66667	9.33333	3.88	0.0081
Error	18	43.33333	2.40741		
Corrected Total	26	118.00000			

Root MSE	1.55158	R-Square	0.6328
Dependent Mean	20.66667	Adj R-Sq	0.4696
Coeff Var	7.50766		

Parameter Estimates

Variable	DF	Parameter Estimate	Standard Error	t Value	Pr > |t|
Intercept	1	22.00000	0.89581	24.56	<.0001
X1	1	-2.66667	1.26686	-2.10	0.0496
X2	1	-1.66667	1.26686	-1.32	0.2048
X3	1	-4.66667	1.26686	-3.68	0.0017
X4	1	-0.66667	1.26686	-0.53	0.6051
X1X3	1	6.66667	1.79161	3.72	0.0016
X1X4	1	1.66667	1.79161	0.93	0.3645
X2X3	1	4.66667	1.79161	2.60	0.0179
X2X4	1	4.00000	1.79161	2.23	0.0385

Test INTERACTION Results for Dependent Variable PROFIT

Source	DF	Mean Square	F Value	Pr > F
Numerator	4	11.55556	4.80	0.0082
Denominator	18	2.40741		

Figure 12.17b SAS regression printout for reduced (main effects) factorial model

Dependent Variable: PROFIT

Number of Observations Read   27
Number of Observations Used   27

Analysis of Variance

Source	DF	Sum of Squares	Mean Square	F Value	Pr > F
Model	4	28.44444	7.11111	1.75	0.1757
Error	22	89.55556	4.07071		
Corrected Total	26	118.00000			

Root MSE	2.01760	R-Square	0.2411
Dependent Mean	20.66667	Adj R-Sq	0.1031
Coeff Var	9.76258		

Parameter Estimates

Variable	DF	Parameter Estimate	Standard Error	t Value	Pr > |t|
Intercept	1	20.11111	0.86824	23.16	<.0001
X1	1	0.11111	0.95111	0.12	0.9081
X2	1	1.22222	0.95111	1.29	0.2121
X3	1	-0.88889	0.95111	-0.93	0.3601
X4	1	1.22222	0.95111	1.29	0.2121

Substituting these values into the formula for the test statistic, we obtain

$$F = \frac{(\text{SSE}_R - \text{SSE}_C)/4}{\text{MSE}_C} = \frac{(89.55556 - 43.33333)/4}{2.40741} = 4.80$$

This "partial" F-value is shaded at the bottom of the SAS printout, Figure 12.17a, as is the p-value of the test, .0082. Since $\alpha = .05$ exceeds the p-value, we reject H_0 and conclude that supply and ratio interact.

(c) The presence of interaction tells you that the mean profit depends on the particular combination of levels of supply S and ratio R. Consequently, there is little point in checking to see whether the means differ for the three levels of supply or whether they differ for the three levels of ratio (i.e., we will not perform the tests for main effects). For example, the supply level that gave the highest mean profit (over all levels of R) might not be the same supply—ratio level combination that produces the largest mean profit per unit of raw material.

The traditional analysis of variance approach to analyzing a complete two-factor factorial with factor A at a levels and factor B at b levels utilizes the fact that the total sum of squares, SS(Total), can be partitioned into four parts, SS(A), SS(B), SS(AB), and SSE (see Figure 12.18). The first two sums of squares, SS(A) and SS(B), are called **main effect sums of squares** to distinguish them from the **interaction sum of squares**, SS(AB).

Figure 12.18 Partitioning of the total sum of squares for a complete two-factor factorial experiment

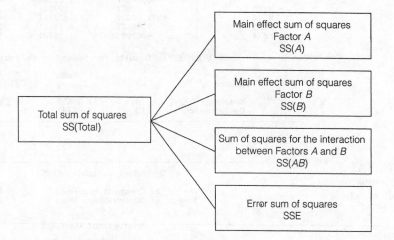

Since the sums of squares and the degrees of freedom for the analysis of variance are additive, the ANOVA table appears as shown in the following box.

ANOVA Table for an $a \times b$ Factorial Design with r Observations per Cell

SOURCE	df	SS	MS	F
Main effects A	$(a-1)$	SS(A)	MS(A) = SS(A)/$(a-1)$	MS(A) / MSE
Main effects B	$(b-1)$	SS(B)	MS(B) = SS(B)/$(b-1)$	MS(B) / MSE
AB interaction	$(a-1)(b-1)$	SS(AB)	MS(AB) = SS(AB)/$[(a-1)(b-1)]$	MS(AB) / MSE
Error	$ab(r-1)$	SSE	MSE = SSE/$[ab(r-1)]$	
Total	$abr-1$	SS(Total)		

(Note: $n = abr$).

Note that the F statistics for testing factor main effects and factor interaction are obtained by dividing the appropriate mean square by MSE. The numerator df for the test of interest will equal the df of the source of variation being tested; the denominator df will always equal df(Error). These F-tests are equivalent to the F-tests obtained by fitting complete and reduced models in regression.*

For completeness, the formulas for calculating the ANOVA sums of squares for a complete two-factor factorial experiment are given in the next box.

ANOVA Computing Formulas for a Two-Factor Factorial Experiment

$$CM = \text{Correction for the mean}$$

$$= \frac{(\text{Total of all } n \text{ measurements})^2}{n}$$

$$= \frac{\left(\sum_{i=1}^{n} y_i\right)^2}{n}$$

$$SS(\text{Total}) = \text{Total sum of squares}$$

$$= \text{Sum of squares of all } n \text{ measurements} - CM$$

$$= \sum_{i=1}^{n} y_i^2 - CM$$

$$SS(A) = \text{Sum of squares for main effects, independent variable 1}$$

$$= \left(\begin{array}{c} \text{Sum of squares of the totals } A_1, A_2, \ldots, A_a \\ \text{divided by the number of measurements} \\ \text{in a single total, namely, } br \end{array}\right) - CM$$

$$= \frac{\sum_{i=1}^{a} A_i^2}{br} - CM$$

$$SS(B) = \text{Sum of squares for main effects, independent variable 2}$$

$$= \left(\begin{array}{c} \text{Sum of squares of the totals } B_1, B_2, \ldots, B_b \\ \text{divided by the number of measurements} \\ \text{in a single total, namely, } ar \end{array}\right) - CM$$

$$= \frac{\sum_{i=1}^{b} B_i^2}{ar} - CM$$

---

* The ANOVA F-tests for main effects shown in the ANOVA summary table are equivalent to those of the regression approach only when the reduced model includes interaction terms. Since we usually test for main effects only after determining that interaction is nonsignificant, some statisticians favor dropping the interaction terms from both the complete and reduced models prior to conducting the main effect tests. For example, to test for main effect A, the complete model includes terms for main effects A and B, whereas the reduced model includes terms for main effect B only. To obtain the equivalent result using the ANOVA approach, the sums of squares for AB interaction and error are "pooled" and a new MSE is computed, where

$$MSE = \frac{SS(AB) + SSE}{n - a - b + 1}$$

$SS(AB) =$ Sum of squares for AB interaction

$$= \left(\begin{array}{c} \text{Sum of squares of the cell totals} \\ AB_{11}, AB_{12}, \ldots, AB_{ab} \text{ divided by} \\ \text{the number of measurements} \\ \text{in a single total, namely, } r \end{array} \right) - SS(A) - SS(B) - CM$$

$$= \frac{\sum_{i=1}^{b} \sum_{i=1}^{a} AB_{ij}^2}{r} - SS(A) - SS(B) - CM$$

where

$a =$ Number of levels of independent variable 1

$b =$ Number of levels of independent variable 2

$r =$ Number of measurements for each pair of levels of independent variables 1 and 2

$n =$ Total number of measurements

$\quad = a \times b \times r$

$A_i =$ Total of all measurements of independent variable 1 at level i $(i = 1, 2, \ldots, a)$

$B_j =$ Total of all measurements of independent variable 2 at level j $(j = 1, 2, \ldots, b)$

$AB_{ij} =$ Total of all measurements at the ith level of independent variable 1 and at the jth level of independent variable 2 $(i = 1, 2, \ldots, a;$ $j = 1, 2, \ldots, b)$

Example 12.12

Refer to Example 12.11.

(a) Construct an ANOVA summary table for the analysis.

(b) Conduct the test for supply × ratio interaction using the traditional analysis of variance sums of squares approach.

(c) Illustrate the nature of the interaction by plotting the sample profit means as in Figure 12.15. Interpret the results.

Solution

(a) Although the formulas given in the previous box are straightforward, they can become quite tedious to use. Therefore, we use a statistical software package to conduct the ANOVA. A SAS printout of the ANOVA is displayed in Figure 12.19a. The value of SS(Total), given in the SAS printout under **Sum of Squares** in the **Corrected Total** row, is SS(Total) = 118. The sums of squares, mean squares, and F-values for the factors S, R, and $S \times R$

interaction are given under the **Anova SS**, **Mean Square**, and **F-Value** columns, respectively, in the bottom portion of the printout. These values are shown in Table 12.6.

Table 12.6 ANOVA table for Example 12.12

Source	df	SS	MS	F
Supply	2	20.22	10.11	4.20
Ratio	2	8.22	4.11	1.71
Supply × Ratio interaction	4	46.22	11.56	4.80
Error	18	43.33	2.41	
Total	26	118.00		

Figure 12.19a SAS ANOVA printout for complete factorial design

```
                              The ANOVA Procedure
Dependent Variable: PROFIT

                                        Sum of
    Source                 DF          Squares    Mean Square   F Value   Pr > F

    Model                   8       74.6666667      9.3333333      3.88   0.0081
    Error                  18       43.3333333      2.4074074
    Corrected Total        26      118.0000000

                 R-Square     Coeff Var      Root MSE     PROFIT Mean

                 0.632768      7.507656      1.551582       20.66667

    Source                 DF        Anova SS    Mean Square   F Value   Pr > F

    SUPPLY                  2      20.22222222   10.11111111      4.20   0.0318
    RATIO                   2       8.22222222    4.11111111      1.71   0.2094
    SUPPLY*RATIO            4      46.22222222   11.55555556      4.80   0.0082
```

(b) To test the hypothesis that supply and ratio interact, we use the test statistic

$$F = \frac{\text{MS}(SR)}{\text{MSE}} = \frac{11.56}{2.41} = 4.80 \text{ (shaded on the SAS printout, Figure 12.19a)}$$

The p-value of the test (also shaded on the SAS printout) is .0082. (Both of these values are identical to the values obtained using regression in Example 12.11.) Since the p-value is less than the selected value of $\alpha = .05$, we conclude that supply and ratio interact.

(c) A MINITAB plot of the sample profit means, shown in Figure 12.19b, illustrates the nature of the supply × ratio interaction. From the graph, you can see that the difference between the profit means for any two levels of ratio (e.g., for $R = .5$ and $R = 2$) is not the same for the different levels of supply. For example, at $S = 15$, the mean is largest for $R = .5$ and smallest for $R = 2$; however, at $S = 21$, the mean is largest for $R = 2$ and smallest for $R = .5$. Consequently, the ratio of raw material allocation that yields the greatest profit will depend on the supply available.

Figure 12.19b MINITAB plot of profit means illustrating interaction

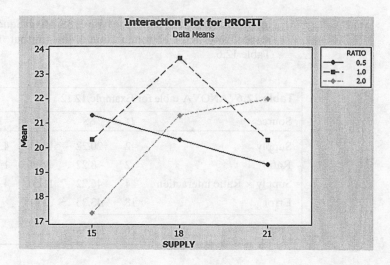

Confidence intervals for a single treatment mean and for the difference between two treatment means in a factorial experiment are provided in the following boxes.

$100(1 - \alpha)\%$ Confidence Interval for the Mean of a Single Treatment: Factorial Experiment

$$\bar{y}_{ij} \pm (t_{\alpha/2})\left(\frac{s}{\sqrt{r}}\right)$$

where

$\bar{y}_{ij}$ = Sample mean for the treatment identified by level i of the first factor

and level j of the second factor

r = Number of measurements per treatment

$s = \sqrt{MSE}$

and $t_{\alpha/2}$ is based on $ab(r - 1)$ df.

$100(1 - \alpha)\%$ Confidence Interval for the Difference Between a Pair of Treatment Means: Factorial Experiment

$$(\bar{y}_1 - \bar{y}_2) \pm (t_{\alpha/2})s\sqrt{\frac{2}{r}}$$

where

$\bar{y}_1$ = Sample mean of the r measurements for the first treatment

$\bar{y}_2$ = Sample mean of the r measurements for the second treatment

$s = \sqrt{MSE}$ and $t_{\alpha/2}$ is based on $ab(r - 1)$ df.

Example 12.13

Refer to Examples 12.11 and 12.12.

(a) Find a 95% confidence interval to estimate the mean profit per unit of raw materials when $S = 18$ tons and the ratio of production is $R = 1{:}1$.

(b) Find a 95% confidence interval to estimate the difference in mean profit per unit of raw materials when $(S = 18, R = 1{:}2)$ and $(S = 18, R = 1{:}1)$.

Solution

(a) A 95% confidence interval for the mean $E(y)$ when supply $S = 18$ and $R = 1{:}1$ is

$$\bar{y} \pm (t_{.025}) \left(\frac{s}{\sqrt{r}} \right)$$

where $\bar{y}$ is the mean of the $r = 3$ values of y for $S = 18$, $R = 1{:}1$ (obtained from Table 12.6), and $t_{.025} = 2.101$ is based on 18 df. Substituting, we obtain

$$\frac{71}{3} \pm (2.101) \left(\frac{1.55}{\sqrt{3}} \right)$$

$$23.67 \pm 1.88$$

Therefore, our interval estimate for the mean profit per unit of raw material where $S = 18$ and $R = 1{:}1$ is $21.79 to $25.55.

(b) A 95% confidence interval for the difference in mean profit per unit of raw material for two different combinations of levels of S and R is

$$(\bar{y}_1 - \bar{y}_2) \pm (t_{.025}) s \sqrt{\frac{2}{r}}$$

where $\bar{y}_1$ and $\bar{y}_2$ represent the means of the $r = 3$ replications for the factor level combinations $(S = 18, R = 1{:}2)$ and $(S = 18, R = 1{:}1)$, respectively. From Table 12.6, the sums of the three measurements for these two treatments are 61 and 71. Substituting, we obtain

$$\left(\frac{61}{3} - \frac{71}{3} \right) \pm (2.101)(1.55) \sqrt{\frac{2}{3}}$$

$$-3.33 \pm 2.66$$

Therefore, the interval estimate for the difference in mean profit per unit of raw material for the two factor level combinations is $(-$5.99, -$0.67)$. The negative values indicate that we estimate the mean for $(S = 18, R = 1{:}2)$ to be less than the mean for $(S = 18, R = 1{:}1)$ by between $0.67 and $5.99. ◼

Throughout this chapter, we have presented two methods for analyzing data from a designed experiment: the regression approach and the traditional ANOVA approach. In a factorial experiment, the two methods yield identical results when both factors are qualitative; however, regression will provide more information when at least one of the factors is quantitative and if we represent the main effects with quantitative terms like x, x^2, and so on. For example, the analysis of variance in Example 12.12 enables us to estimate the mean profit per unit of supply for *only* the nine combinations of supply–ratio levels. It will not permit us to estimate the mean response for some other combination of levels of the independent variables not included among the nine used in the factorial experiment. Alternatively, the prediction equation obtained from a regression analysis with quantitative terms enables us to estimate the mean profit per unit of supply when ($S = 17$, $R = 1:1$). We could not obtain this estimate from the analysis of variance in Example 12.12.

The prediction equation found by regression analysis also contributes other information not provided by traditional analysis of variance. For example, we might wish to estimate the rate of change in the mean profit, $E(y)$, for unit changes in S, R, or both for specific values of S and R. Or, we might want to determine whether the third- and fourth-order terms in the complete model really contribute additional information for the prediction of profit, y.

We illustrate some of these applications in the next several examples.

Example 12.14

Refer to the 3×3 factorial design and the data in Example 12.11. Since both factors, supply and ratio, are quantitative in nature, we can represent the main effects of the complete factorial model using quantitative terms such as x, x^2, x^3, and so on, rather than dummy variables. Like with dummy variables, the number of quantitative main effects will be one less than the number of levels for a quantitative factor. The logic follows from our discussion about estimating model parameters in Section 7.3. At two levels, the quantitative main effect is x; at three levels, the quantitative main effects are x and x^2.

(a) Specify the complete model for the factorial design using quantitative main effects for supply and ratio.

(b) Fit the model to the data in Table 12.5 and show that the F-test for interaction produced on the printout is equivalent to the corresponding test produced using dummy variables for main effects.

Solution

(a) Now both supply (15, 18, and 21 tons) and ratio (1:2, 1:1, and 2:1) are set at three levels; consequently, each factor will have two quantitative main effects. If we let x_1 represent the actual level of supply of raw material (in tons) and let x_2 represent the actual fraction for ratio of allocation (e.g., 1/2, 1, and 2), then the main effects are x_1 and x_1^2 for supply and x_2 and x_2^2 for ratio. Consequently, the complete factorial model for mean profit, $E(y)$, is

$$E(y) = \beta_0 + \underbrace{\beta_1 x_1 + \beta_2 x_1^2}_{\substack{\text{Supply} \\ \text{main effects}}} + \underbrace{\beta_3 x_2 + \beta_4 x_2^2}_{\substack{\text{Ratio} \\ \text{main effects}}}$$
$$+ \underbrace{\beta_5 x_1 x_2 + \beta_6 x_1 x_2^2 + \beta_7 x_1^2 x_2 + \beta_8 x_1^2 x_2^2}_{\text{Supply} \times \text{Ratio interaction}}$$

Note that the number of terms (main effects and interactions) in the model is equivalent to the dummy variable model in Example 12.1.

(b) The SAS printout for the complete model, part a, is shown in Figure 12.20. First, note that SSE = 43.33333 and MSE = 2.40741 (highlighted) are equivalent to the corresponding values shown on the printout for the dummy variable model, Figure 12.17a. Second, the partial F-value ($F = 4.80$) for testing the null hypothesis of no interaction (H_0: $\beta_5 = \beta_6 = \beta_7 = \beta_8 = 0$), highlighted in the middle of the printout, is equivalent to the corresponding test shown in Figure 12.17a. Thus, whether you conduct the test for factor interaction using regression with dummy variables, regression with quantitative main effects, or with the traditional ANOVA approach, the results will be identical. ∎

Figure 12.20 SAS regression printout for complete factorial model with quantitative main effects

Dependent Variable: PROFIT

Analysis of Variance

Source	DF	Sum of Squares	Mean Square	F Value	Pr > F
Model	8	74.66667	9.33333	3.88	0.0081
Error	18	43.33333	2.40741		
Corrected Total	26	118.00000			

Root MSE	1.55158	R-Square	0.6328	
Dependent Mean	20.66667	Adj R-Sq	0.4696	
Coeff Var	7.50766			

Parameter Estimates

Variable	DF	Parameter Estimate	Standard Error	t Value	Pr > |t|
Intercept	1	245.33333	130.49665	1.88	0.0764
SUPPLY	1	-25.07407	14.71842	-1.70	0.1057
SUPPSQ	1	0.67901	0.40837	1.66	0.1137
RATIO	1	-534.33333	252.45535	-2.12	0.0485
RATSQ	1	192.66667	97.17011	1.98	0.0629
RAT_SUPP	1	60.55556	28.47387	2.13	0.0475
S_RATSQ	1	-22.14815	10.95960	-2.02	0.0584
R_SUPPSQ	1	-1.66667	0.79003	-2.11	0.0492
RSQ_SSQ	1	0.61728	0.30408	2.03	0.0574

Test INTERACT Results for Dependent Variable PROFIT

Source	DF	Mean Square	F Value	Pr > F
Numerator	4	11.55556	4.80	0.0082
Denominator	18	2.40741		

Test HIGHORDR Results for Dependent Variable PROFIT

Source	DF	Mean Square	F Value	Pr > F
Numerator	3	3.71958	1.55	0.2373
Denominator	18	2.40741		

Example 12.15

Refer to Example 12.14. Do the data provide sufficient information to indicate that third- and fourth-order terms in the complete factorial model contribute information for the prediction of y? Use $\alpha = .05$.

Solution

If the response to the question is yes, then at least one of the parameters, β_6, β_7, or β_8, of the complete factorial model differs from 0 (i.e., they are needed in the model). Consequently, the null hypothesis is

H_0: $\beta_6 = \beta_7 = \beta_8 = 0$

and the alternative hypothesis is

H_a: At least one of the three β's is nonzero.

To test this hypothesis, we compute the drop in SSE between the appropriate reduced and complete model.

For this application the complete model is the complete factorial model in Example 12.14:

Complete model: $E(y) = \beta_0 + \beta_1 x_1 + \beta_2 x_1^2 + \beta_3 x_2 + \beta_4 x_2^2 + \beta_5 x_1 x_2$
$$+ \beta_6 x_1 x_2^2 + \beta_7 x_1^2 x_2 + \beta_8 x_1^2 x_2^2$$

The reduced model is this complete model minus the third- and fourth-order terms, that is, the reduced model is the second-order model shown here:

Reduced model: $E(y) = \beta_0 + \beta_1 x_1 + \beta_2 x_1^2 + \beta_3 x_2 + \beta_4 x_2^2 + \beta_5 x_1 x_2$

Recall (from Figure 12.20) that the SSE and MSE for the complete model are $SSE_C = 43.3333$ and $MSE_C = 2.4074$. A SAS printout of the regression analysis of the reduced model is shown in Figure 12.21. The SSE for the reduced model (shaded) is $SSE_R = 54.49206$.

Consequently, the test statistic required to conduct the nested model F-test is

Test statistic:

$$F = \frac{(SSE_R - SSE_C)/(\text{Number of } \beta\text{'s tested})}{MSE_C} = \frac{(54.49206 - 43.3333)/3}{2.4074}$$

$$= 1.55$$

This "partial" F-value can also be obtained using SAS options and is given at the bottom of the SAS complete model printout, Figure 12.20, as well as the p-value of the test, .2373.

Conclusion: Since $\alpha = .05$ is less than p-value $= .2373$, we cannot reject the null hypothesis that $\beta_6 = \beta_7 = \beta_8 = 0$. That is, there is insufficient evidence (at $\alpha = .05$) to indicate that the third- and fourth-order terms associated with β_6, β_7, and β_8 contribute information for the prediction of y. Since the complete factorial model contributes no more information about y than the reduced (second-order) model, we recommend using the reduced model in practice. ▬

Figure 12.21 SAS regression printout for reduced (second-order) factorial model

Dependent Variable: PROFIT

Analysis of Variance

Source	DF	Sum of Squares	Mean Square	F Value	Pr > F
Model	5	63.50794	12.70159	4.89	0.0040
Error	21	54.49206	2.59486		
Corrected Total	26	118.00000			

Root MSE	1.61086	R-Square	0.5382
Dependent Mean	20.66667	Adj R-Sq	0.4283
Coeff Var	7.79447		

Parameter Estimates

Variable	DF	Parameter Estimate	Standard Error	t Value	Pr > |t|
Intercept	1	-27.81481	23.80152	-1.17	0.2557
SUPPLY	1	5.94444	2.64418	2.25	0.0354
RATIO	1	-7.76190	5.04523	-1.54	0.1389
RAT_SUPP	1	0.74603	0.20295	3.68	0.0014
SUPPSQ	1	-0.18519	0.07307	-2.53	0.0193
RATSQ	1	-2.29630	1.33939	-1.71	0.1012

Example 12.16

Use the second-order model in Example 12.15 and find a 95% confidence interval for the mean profit per unit supply of raw material when $S = 17$ and $R = 1$.

Solution

The portion of the SAS printout for the second-order model with 95% confidence intervals for $E(y)$ is shown in Figure 12.22.

Figure 12.22 SAS printout of confidence intervals for reduced (second-order) factorial model

Obs	SUPPLY	RATIO	Dep Var PROFIT	Predicted Value	Std Error Mean Predict	95% CL Mean		Residual
1	15	0.5	23.0000	20.8254	0.8033	19.1549	22.4959	2.1746
2	15	0.5	20.0000	20.8254	0.8033	19.1549	22.4959	-0.8254
3	15	0.5	21.0000	20.8254	0.8033	19.1549	22.4959	0.1746
4	18	0.5	22.0000	21.4444	0.6932	20.0029	22.8860	0.5556
5	18	0.5	19.0000	21.4444	0.6932	20.0029	22.8860	-2.4444
6	18	0.5	20.0000	21.4444	0.6932	20.0029	22.8860	-1.4444
7	21	0.5	19.0000	18.7302	0.8033	17.0596	20.4007	0.2698
8	21	0.5	18.0000	18.7302	0.8033	17.0596	20.4007	-0.7302
9	21	0.5	21.0000	18.7302	0.8033	17.0596	20.4007	2.2698
10	15	1	22.0000	20.8175	0.7006	19.3605	22.2744	1.1825
11	15	1	20.0000	20.8175	0.7006	19.3605	22.2744	-0.8175
12	15	1	19.0000	20.8175	0.7006	19.3605	22.2744	-1.8175
13	18	1	24.0000	22.5556	0.6932	21.1140	23.9971	1.4444
14	18	1	25.0000	22.5556	0.6932	21.1140	23.9971	2.4444
15	18	1	22.0000	22.5556	0.6932	21.1140	23.9971	-0.5556
16	21	1	20.0000	20.9603	0.7006	19.5034	22.4173	-0.9603
17	21	1	19.0000	20.9603	0.7006	19.5034	22.4173	-1.9603
18	21	1	22.0000	20.9603	0.7006	19.5034	22.4173	1.0397
19	15	2	18.0000	17.3571	0.8590	15.5707	19.1436	0.6429
20	15	2	18.0000	17.3571	0.8590	15.5707	19.1436	0.6429
21	15	2	16.0000	17.3571	0.8590	15.5707	19.1436	-1.3571
22	18	2	21.0000	21.3333	0.6932	19.8917	22.7749	-0.3333
23	18	2	23.0000	21.3333	0.6932	19.8917	22.7749	1.6667
24	18	2	20.0000	21.3333	0.6932	19.8917	22.7749	-1.3333
25	21	2	20.0000	21.9762	0.8590	20.1897	23.7627	-1.9762
26	21	2	22.0000	21.9762	0.8590	20.1897	23.7627	0.0238
27	21	2	24.0000	21.9762	0.8590	20.1897	23.7627	2.0238
28	17	1	.	22.3466	0.6625	20.9687	23.7244	.

The confidence interval for $E(y)$ when $S = 17$ and $R = 1$ is given in the last row of the printout. You can see that the interval is (20.9687, 23.7244). Thus, we estimate (with confidence coefficient equal to .95) that the mean profit per unit of supply will lie between $20.97 and $23.72 when $S = 17$ tons and $R = 1$. Beyond this immediate result, this example illustrates the power and versatility of a regression analysis. In particular, there is no way to obtain this estimate from the traditional analysis of variance in Example 12.12. However, a computerized regression package can be easily programmed to include the confidence interval automatically. ▪

12.5 Exercises

12.28 ANOVA summary table. The analysis of variance for a 3×2 factorial experiment, with four observations per treatment, produced the ANOVA summary table entries shown here.

SOURCE	df	SS	MS	F
A	—	100	—	—
B	1	—	—	—
AB	2	—	2.5	—
Error	—	—	2.0	
Total	—	700		

(a) Complete the ANOVA summary table.
(b) Test for interaction between factor A and factor B. Use $\alpha = .05$.
(c) Test for differences in main effect means for factor A. Use $\alpha = .05$.
(d) Test for differences in main effect means for factor B. Use $\alpha = .05$.

12.29 Removing bacteria from water. A coagulation–microfiltration process for removing bacteria from water was investigated in *Environmental Science and Engineering* (September 1, 2000). Chemical engineers at Seoul National University performed a designed experiment to estimate the effect of both the level of the coagulant and acidity (pH) level on the coagulation efficiency of the process. Six levels of coagulant (5, 10, 20, 50, 100, and 200 milligrams per liter) and six pH levels (4.0, 5.0, 6.0, 7.0, 8.0, and 9.0) were employed. Water specimens collected from the Han River in Seoul, Korea, were placed in jars, and each jar randomly assigned to receive one of the $6 \times 6 = 36$ combinations of coagulant level and pH level.

(a) What type of experimental design was applied in this study?
(b) Give the factors, factor levels, and treatments for the study.

12.30 **Mussel settlement patterns on algae.** Mussel larvae are in great abundance in the drift material that washes up on Ninety Mile Beach in New Zealand. These larvae tend to settle on algae. Environmentalists at the University of Auckland investigated the impact of algae type on the abundance of mussel larvae in drift material (*Malacologia*, February 8, 2002). Drift material from three different wash-up events on Ninety Mile Beach were collected; for each wash-up, the algae was separated into four strata: coarse-branching, medium-branching, fine-branching, and hydroid algae. Two samples were randomly selected for each of the $3 \times 4 = 12$ event/strata combinations, and the mussel density (percent per square centimeter) was measured for each. The data was analyzed as a complete 3×4 factorial design. The ANOVA summary table is shown below.

SOURCE	df	F	p-VALUE
Event	2	.35	>.05
Strata	3	217.33	<.05
Interaction	6	1.91	>.05
Error	12		
Total	23		

(a) Identify the factors (and levels) in this experiment.

(b) How many treatments are included in the experiment?

(c) How many replications are included in the experiment?

(d) What is the total sample size for the experiment?

(e) What is the response variable measured?

(f) Which ANOVA F-test should be conducted first? Conduct this test (at $\alpha = .05$) and interpret the results.

(g) If appropriate, conduct the F-tests (at $\alpha = .05$) for the main effects. Interpret the results.

12.31 **Learning from picture book reading.** *Developmental Psychology* (November 2006) published an article that examined toddlers' ability to learn from reading picture books. The experiment involved 36 children at each of three different ages: 18, 24, and 30 months. The children were randomly assigned into one of three different reading book conditions: book with color photographs (Photos), book with colored pencil drawings (Drawings), and book with no photographs or drawings (Control). Thus, a 3×3 factorial experiment was employed (with age at three levels and reading book condition at three levels). After a book reading session, the children were scored on their ability to reenact the target actions in the book. Scores ranged from 0 (low) to

3 (high). An ANOVA of the reenactment scores is summarized in the table.

SOURCE	df	F	p-VALUE
Age	–	11.93	<.001
Book	–	23.64	<.001
Age × Book	–	2.99	<.05
Error	–		
Total	107		

(a) Fill in the missing degrees of freedom (df) values in the table.

(b) How many treatments are investigated in this experiment? List them.

(c) Conduct a test for AGE × BOOK interaction at $\alpha = .05$. Interpret the result practically.

(d) Based on the test, part c, do you need to conduct tests for AGE and BOOK main effects?

12.32 **Virtual reality–based rehabilitation systems.** In *Robotica* (Vol. 22, 2004), researchers described a study of the effectiveness of display devices for three virtual reality (VR)–based hand rehabilitation systems. Display device A is a projector, device B is a desktop computer monitor, and device C is a head-mounted display. Twelve nondisabled right-handed male subjects were randomly assigned to the three VR display devices, four subjects in each group. Additionally, within each group two subjects were randomly assigned to use an auxiliary lateral image and two subjects were not. Consequently, a 3×2 factorial design was employed, with VR display device at three levels (A, B, or C) and auxiliary lateral image at two levels (yes or no). Each subject carried out a "pick-and-place" procedure using the assigned VR system, and the collision frequency (number of collisions between moved objects) was measured.

(a) Give the sources of variation and associated degrees of freedom in an ANOVA summary table for this design.

(b) How many treatments are investigated in this experiment?

(c) The factorial ANOVA resulted in the following p-values: Display main effect (.045), Auxiliary lateral image main effect (.003), and Interaction (.411). Interpret, practically, these results. Use $\alpha = .05$ for each test you conduct.

12.33 **Insomnia and education.** Many workers suffer from stress and chronic insomnia. Is insomnia related to education status? Researchers at the Universities of Memphis, Alabama at Birmingham, and Tennessee investigated this question in the *Journal of Abnormal Psychology* (February 2005). Adults living in Tennessee were selected to participate in the study using a random-digit telephone dialing procedure. In addition to insomnia status (normal sleeper or chronic insomnia), the researchers classified each participant into one of four education categories (college graduate, some college, high school graduate, and high school dropout). One dependent variable of interest to the researchers was a quantitative measure of daytime functioning called the Fatigue Severity Scale (FSS). The data was analyzed as a 2×4 factorial experiment, with Insomnia status and Education level as the two factors.

(a) Determine the number of treatments for this study. List them.

(b) The researchers reported that "the Insomnia $\times$ Education interaction was not statistically significant." Practically interpret this result. (Illustrate with a graph.)

(c) The researchers discovered that the sample mean FSS for people with insomnia was greater than the sample mean FSS for normal sleepers and that this difference was statistically significant. Practically interpret this result.

(d) The researchers reported that the main effect of Education was statistically significant. Practically interpret this result.

12.34 **Impact of paper color on exam scores.** A study published in *Teaching Psychology* (May 1998) examined how external clues influence student performance. Introductory psychology students were randomly assigned to one of four different midterm examinations. Form 1 was printed on blue paper and contained difficult questions, while form 2 was also printed on blue paper but contained simple questions. Form 3 was printed on red paper, with difficult questions; form 4 was printed on red paper with simple questions. The researchers were interested in the impact that Color (red or blue) and Question (simple or difficult) had on mean exam score.

(a) What experimental design was employed in this study? Identify the factors and treatments.

(b) Give the complete model appropriate for analyzing the data for this experiment.

(c) The researchers conducted an ANOVA and found a significant interaction between Color and Question (p-value $< .03$). Interpret this result.

(d) The sample mean scores (percentage correct) for the four exam forms are listed in the next table. Plot the four means on a graph to illustrate the Color $\times$ Question interaction.

FORM	COLOR	QUESTION	MEAN SCORE
1	Blue	Difficult	53.3
2	Blue	Simple	80.0
3	Red	Difficult	39.3
4	Red	Simple	73.6

12.35 **Strengthening tin–lead solder joints.** The chemical element antimony is sometimes added to tin–lead solder to replace the more expensive tin and to reduce the cost of soldering. A factorial experiment was conducted to determine how antimony affects the strength of the tin–lead solder joint (*Journal of Materials Science*, May 1986). Tin–lead solder specimens were prepared using one of four possible cooling methods (water-quenched, WQ; oil-quenched, OQ; air-blown, AB; and furnace-cooled, FC) and with one of four possible amounts of antimony (0%, 3%, 5%, and 10%) added to the composition. Three solder joints were randomly assigned to each of the $4 \times 4 = 16$ treatments and the shear strength of each measured. The experimental results, shown in the accompanying table, were subjected to an ANOVA using MINITAB. The printout is shown on p. 660.

🔵 TINLEAD

AMOUNT OF ANTIMONY % WEIGHT	COOLING METHOD	SHEAR STRENGTH, MPa		
0	WQ	17.6	19.5	18.3
0	OQ	20.0	24.3	21.9
0	AB	18.3	19.8	22.9
0	FC	19.4	19.8	20.3
3	WQ	18.6	19.5	19.0
3	OQ	20.0	20.9	20.4
3	AB	21.7	22.9	22.1
3	FC	19.0	20.9	19.9
5	WQ	22.3	19.5	20.5
5	OQ	20.9	22.9	20.6
5	AB	22.9	19.7	21.6
5	FC	19.6	16.4	20.5
10	WQ	15.2	17.1	16.6
10	OQ	16.4	19.0	18.1
10	AB	15.8	17.3	17.1
10	FC	16.4	17.6	17.6

Source: With kind permission from Springer Science Business Media: Tomlinson, W. J., and Cooper, G. A. "Fracture mechanism of brass/Sn-Pb-Sb solder joints and the effect of production variables on the joint strength." *Journal of Materials Science*, Vol. 21, No. 5, May 1986, p. 1731 (Table II). Copyright © 1986 Chapman and Hall.

(a) Using dummy variables, write the appropriate linear model for analyzing the data.

(b) Construct an ANOVA summary table for the experiment.

(c) Conduct a test to determine whether the two factors, amount of antimony and cooling method, interact. Use $\alpha = .01$.

(d) Interpret the result obtained in part c.

(e) If appropriate, conduct the tests for main effects. Use $\alpha = .01$.

(f) Rewrite the linear model using quantitative terms for the factor, Antimony.

(g) Fit the model, part f, to the data and give the prediction equation.

(h) Conduct a test to determine whether the second-and third-order terms for Antimony can be dropped from the model, part f.

MINITAB output for Exercise 12.35

Two-way ANOVA: STRENGTH versus ANTIMONY, METHOD

```
Analysis of Variance for STRENGTH
Source       DF      SS      MS      F      P
ANTIMONY      3   104.19   34.73   20.12  0.000
METHOD        3    28.63    9.54    5.53  0.004
Interaction   9    25.13    2.79    1.62  0.152
Error        32    55.25    1.73
Total        47   213.20
```

12.36 Commercial eggs produced from different housing systems. In the production of commercial eggs in Europe, four different types of housing systems for the chickens are used: cage, barn, free range, and organic. The characteristics of eggs produced from the four housing systems were investigated in *Food Chemistry* (Vol. 106, 2008). Twenty-four commercial grade A eggs were randomly selected—six from each of the four types of housing systems. Of the six eggs selected from each housing system, three were Medium weight class eggs and three were Large weight class eggs. The data on whipping capacity (percent overrun) for the 24 sampled eggs are shown in the next table. The researchers want to investigate the effect of both housing system and weight class on the mean whipping capacity of the eggs. In particular, they want to know whether the difference between the mean whipping capacity of medium and large eggs depends on the housing system.

(a) Identify the factors and treatments for this experiment.

(b) Use statistical software to conduct an ANOVA on the data. Report the results in an ANOVA table.

(c) Is there evidence of interaction between housing system and weight class? Test using $\alpha = .05$. What does this imply, practically?

(d) Interpret the main effect test for housing system (using $\alpha = .05$). What does this imply, practically?

(e) Interpret the main effect test for weight class (using $\alpha = .05$). What does this imply, practically?

EGGS2

HOUSING	WTCLASS	OVERRUN (%)
CAGE	M	495, 462, 488
	L	502, 472, 474
FREE	M	513, 510, 510
	L	520, 531, 521
BARN	M	515, 516, 514
	L	526, 501, 508
ORGANIC	M	532, 511, 527
	L	530, 544, 531

12.37 The thrill of a close game. Do women enjoy the thrill of a close basketball game as much as men? To answer this question, male and female undergraduate students were recruited to participate in an experiment (*Journal of Sport and Social Issues*, February 1997). The students watched one of eight live televised games of a recent NCAA basketball tournament. (None of the games involved a home team to which the students could be considered emotionally committed.) The "suspense" of each game was classified into one of four categories according to the closeness of scores at the game's conclusions: minimal (15 point or greater differential), moderate (10–14 point differential), substantial (5–9 point differential), and extreme (1–4 point differential). After the game, each student rated his or her enjoyment on an 11-point scale ranging from 0 (not at all) to 10 (extremely). The enjoyment rating data were analyzed as a 4×2 factorial design, with suspense (four levels) and gender (two levels) as the two factors. The $4 \times 2 = 8$ treatment means are shown in the accompanying table.

	GENDER	
SUSPENSE	MALE	FEMALE
Minimal	1.77	2.73
Moderate	5.38	4.34
Substantial	7.16	7.52
Extreme	7.59	4.92

Source: Gan, Su-lin, et al. "The thrill of a close game: Who enjoys it and who doesn't" *Journal of Sport and Social Issues,* Vol 21, No. 1, Feb. 1997, pp. 59–60.

(a) Plot the treatment means in a graph similar to Figure 12.15. Does the pattern of means suggest interaction between suspense and gender? Explain.

(b) The ANOVA F-test for interaction yielded the following results: numerator df $= 3$, denominator df $= 68$, $F = 4.42$, p-value $= .007$. What can you infer from these results?

(c) Based on the test, part b, is the difference between the mean enjoyment levels of males and females the same, regardless of the suspense level of the game?

12.38 Violent lyrics and aggressiveness. In the *Journal of Personality and Social Psychology* (May 2003), psychologists investigated the potentially harmful effects of violent music lyrics. The researchers theorized that listening to a song with violent lyrics will lead to more violent thoughts and actions. A total of 60 undergraduate college students participated in one experiment designed by the researchers. Half of the students were volunteers and half were required to participate as part of their introductory psychology class. Each student listened to a song by the group "Tool"—half the students were randomly assigned a song with violent lyrics and half assigned a song with nonviolent lyrics. Consequently, the experiment used a 2×2 factorial design with the factors Song (violent, nonviolent) and Pool (volunteer, psychology class). After listening to the song, each student was given a list of word pairs and asked to rate the similarity of each word in the pair on a 7-point scale. One word in each pair was aggressive in meaning (e.g., *choke*) and the other was ambiguous (e.g., *night*). An aggressive cognition score was assigned based on the average word-pair scores. (The higher the score, the more the subject associated an ambiguous word with a violent word.) The data (simulated) are shown below. Conduct a complete analysis of variance on the data.

LYRICS

	VOLUNTEER					PSYCHOLOGY CLASS				
VIOLENT SONG	4.1	3.5	3.4	4.1	3.7	3.4	3.9	4.2	3.2	4.3
	2.8	3.4	4.0	2.5	3.0	3.3	3.1	3.2	3.8	3.1
	3.4	3.5	3.2	3.1	3.6	3.8	4.1	3.3	3.8	4.5
NON-VIOLENT SONG	2.4	2.4	2.5	2.6	3.6	2.5	2.9	2.9	3.0	2.6
	4.0	3.3	3.7	2.8	2.9	2.4	3.5	3.3	3.7	3.3
	3.2	2.5	2.9	3.0	2.4	2.8	2.5	2.8	2.0	3.1

12.39 Exam performance study. *Teaching of Psychology* (August 1998) published a study on whether a practice test helps students prepare for a final exam. Students in an introductory psychology class were grouped according to their class standing and whether they attended a review session or took a practice test prior to the final exam. The experimental design was a 3×2 factorial design, with Class Standing at three levels (low, medium, or high) and Exam Preparation at two levels (practice exam or review session). There were 22 students in each of the $3 \times 2 = 6$ treatment groups. After completing the final exam, each student rated her or his exam preparation on an 11-point scale ranging from 0 (not helpful at all) to 10 (extremely helpful). The data for this experiment (simulated from summary statistics provided in the article) are saved in the PRACEXAM file. The first five and last five observations in the data set are listed below. Conduct a complete analysis of variance of the helpfulness ratings data, including (if warranted) multiple comparisons of means. Do your findings support the research conclusion that "students at all levels of academic ability benefit from a ... practice exam"?

PRACEXAM (*First and last 5 observations*)

EXAM PREPARATION	CLASS STANDING	HELPFULNESS RATING
PRACTICE	LOW	6
PRACTICE	LOW	7
PRACTICE	LOW	7
PRACTICE	LOW	5
PRACTICE	LOW	3
⋮	⋮	⋮
REVIEW	HI	5
REVIEW	HI	2
REVIEW	HI	5
REVIEW	HI	4
REVIEW	HI	3

Source: Balch, W. R. "Practice versus review exams and final exam performance," *Teaching of Psychology*, Vol. 25, No. 3, Aug. 1998 (adapted from Table 1).

12.40 Rate of combustion of graphite. As part of a study on the rate of combustion of artificial graphite in humid air flow, researchers conducted an experiment to investigate oxygen diffusivity through a water vapor mixture. A 3×9 factorial experiment was conducted with mole fraction of water (H_2O) at three levels and temperature of the nitrogen–water mixture at nine levels. The data are shown on p. 662.

(a) Explain why the traditional analysis of variance (using the ANOVA formulas) is inappropriate for the analysis of these data.

(b) Plot the data to determine if a first- or second-order model for mean oxygen diffusivity, $E(y)$, is more appropriate.

WATERVAPOR

TEMPERATURE	MOLE FRACTION OF H_2O		
°K	.0022	.017	.08
1,000	1.68	1.69	1.72
1,100	1.98	1.99	2.02
1,200	2.30	2.31	2.35
1,300	2.64	2.65	2.70
1,400	3.00	3.01	3.06
1,500	3.38	3.39	3.45
1,600	3.78	3.79	3.85
1,700	4.19	4.21	4.27
1,800	4.63	4.64	4.71

Source: Reprinted from *Combustion and Flame,* Vol. 50, Kiyoshi Matsui, Hiroshi Tsuji, and Atsushi Makino, "The effects of vapor concentration on the rate of combustion of an artificial graphite in humid air flow," pp. 12, Copyright © 1983, with permission from Elsevier.

(c) Write an interaction model relating mean oxygen diffusivity, $E(y)$, to temperature x_1 and mole fraction x_2.

(d) Suppose that temperature and mole fraction of H_2O do not interact. What does this imply about the relationship between $E(y)$ and x_1 and x_2?

(e) Do the data provide sufficient information to indicate that temperature and mole fraction of H_2O interact? Use the accompanying

MINITAB Output for Exercise 12.40

```
The regression equation is
OXYDIFF = - 2.10 + 0.00368 TEMP - 0.24 MOLE + 0.00073 TEMPMOLE

Predictor       Coef      SE Coef        T        P
Constant     -2.09528     0.09035    -23.19    0.000
TEMP        0.00368411   0.00006347    58.05    0.000
MOLE           -0.238       1.913     -0.12    0.902
TEMPMOLE     0.000733     0.001344      0.55    0.591

S = 0.06081    R-Sq = 99.7%    R-Sq(adj) = 99.6%

Analysis of Variance

Source            DF       SS         MS         F        P
Regression         3    24.7733    8.2578    2233.31    0.000
Residual Error    23     0.0850    0.0037
Total             26    24.8583


Predicted Values for New Observations

New Obs    Fit      SE Fit        95.0% CI              95.0% PI
1       2.7062     0.0139    ( 2.6774,  2.7350)   ( 2.5772,  2.8352)

Values of Predictors for New Observations

New Obs    TEMP     MOLE    TEMPMOLE
1          1300    0.0170      22.1
```

MINITAB printout to conduct the test at $\alpha = .05$.

(f) Give the least squares prediction equation for $E(y)$.

(g) Substitute into the prediction equation to predict the mean diffusivity when the temperature of the process is $1,300°K$ and the mole fraction of water is .017.

(h) Locate the 95% confidence interval for mean diffusivity when the temperature of the process is $1,300°K$ and the mole fraction of water is .017 shown on the MINITAB printout. Interpret the result.

12.41 Impact of flavor name on consumer choice. Do consumers react favorably to products with ambiguous colors or names? This was the research question of interest in an article published in the *Journal of Consumer Research* (June 2005). As a "reward" for participating in an unrelated experiment, 100 consumers were told they could have some jelly beans available in several cups on a table. Half the consumers were assigned to take jelly beans with common descriptive flavor names (e.g., watermelon green), while the other half were assigned to take jelly beans with ambiguous flavor names (e.g., monster green). Within each group, half of the consumers took the jelly beans and left (low cognitive load condition), while the other half were distracted with additional questions designed to distract them while they were taking their jelly beans (high cognitive load condition). Consequently, a 2×2 factorial experiment was employed—with Flavor Name (common or ambiguous) and Cognitive Load (low or high) as the two factors—with 25 consumers assigned to each of four treatments. The dependent variable of interest was the number of jelly beans taken by each consumer. The means and standard deviations of the four treatments are shown in the accompanying table.

	AMBIGUOUS		COMMON	
	MEAN	STD. DEV.	MEAN	STD. DEV.
Low Load	18.0	15.0	7.8	9.5
High Load	6.1	9.5	6.3	10.0

Source: Miller, E. G. and Kahn, B. E. "Shades of meaning: The effect of color and flavor names on consumer choice," *Journal of Consumer Research,* Vol. 32, June 2005 (Table 1). Reprinted with permission of the University of Chicago Press.

(a) Give the equation of the complete factorial model for this 2×2 factorial experiment.

(b) Use the summary statistics provided in the table to derive the estimates of the model parameters. (*Hint*: First, write each parameter as a function of the treatment means.)

(c) Calculate the total of the $n = 25$ measurements for each of the four categories, then compute the correction for mean, CM.

(d) Calculate the sums of squares for Load, Name, and Load × Name interaction.

(e) Calculate the sample variance for each treatment. Then calculate the sum of squares of deviations within each sample for the four treatments.

(f) Calculate SSE. (*Hint*: SSE is the pooled sum of squares for the deviations, part e.)

(g) Now that you know SS(Load), SS(Name), SS(Load × Name), and SSE, find SS(Total).

(h) Summarize the calculations in an ANOVA table and interpret the results. Use $\alpha = .05$ for any inferential techniques you employ. Illustrate your conclusions graphically.

(i) State any assumptions required for the inferences to be valid.

12.6 More Complex Factorial Designs (Optional)

In this optional section, we present some useful factorial designs that are more complex than the basic two-factor factorial in Section 12.5. These designs fall under the general category of a ***k*-way classification of data**. A *k*-way classification of data arises when we run all combinations of the levels of *k* independent variables. These independent variables can be factors or blocks.

For example, consider a replicated $2 \times 3 \times 3$ factorial experiment in which the $2 \times 3 \times 3 = 18$ treatments are assigned to the experimental units according to a completely randomized design. Since every combination of the three factors (a total of 18) is examined, the design is often called a three-way classification of data. Similarly, a *k*-way classification of data would result if we randomly assign the treatments of a $(k - 1)$-factor factorial experiment to the experimental units of a randomized block design. For example, if we assigned the $2 \times 3 = 6$ treatments of a complete 2×3 factorial experiment to blocks containing six experimental units each, the data would be arranged in a three-way classification (i.e., according to the two factors and the blocks).

The formulas required for calculating the sums of squares for main effects and interactions for an analysis of variance for a *k*-way classification of data are complicated and are, therefore, not given here. If you are interested in the computational formulas, see the references. As with the designs in the previous three sections, we provide the appropriate linear model for these more complex designs and use either regression or the standard ANOVA output of a statistical software package to analyze the data.

Example 12.17

Consider a $2 \times 3 \times 3$ factorial experiment with qualitative factors and $r = 3$ experimental units randomly assigned to each treatment.

(a) Write the appropriate linear model for the design.

(b) Indicate the sources of variation and their associated degrees of freedom in a partial ANOVA table.

Solution

(a) Denote the three qualitative factors as A, B, and C, with A at two levels, and B and C at three levels. Then the linear model for the experiment will contain one parameter corresponding to main effects for A, two each for B and C, $(1)(2) = 2$ each for the AB and AC interactions, $(2)(2) = 4$ for the BC interaction, and $(1)(2)(2) = 4$ for the three-way ABC interaction. Three-way interaction terms measure the failure of two-way interaction effects to remain the same from one level to another level of the third factor.

$$E(y) = \beta_0 + \underbrace{\beta_1 x_1}_{\substack{\text{Main effect} \\ A}} + \underbrace{\beta_2 x_2 + \beta_3 x_3}_{\substack{\text{Main effects} \\ B}} + \underbrace{\beta_4 x_4 + \beta_5 x_5}_{\substack{\text{Main effects} \\ C}}$$

$$+ \underbrace{\beta_6 x_1 x_2 + \beta_7 x_1 x_3}_{A \times B \text{ interaction}} + \underbrace{\beta_8 x_1 x_4 + \beta_9 x_1 x_5}_{A \times C \text{ interaction}}$$

$$+ \underbrace{\beta_{10} x_2 x_4 + \beta_{11} x_2 x_5 + \beta_{12} x_3 x_4 + \beta_{13} x_3 x_5}_{B \times C \text{ interaction}}$$

$$+ \underbrace{\beta_{14} x_1 x_2 x_4 + \beta_{15} x_1 x_3 x_4 + \beta_{16} x_1 x_2 x_5 + \beta_{17} x_1 x_3 x_5}_{A \times B \times C \text{ interaction}}$$

where

$$x_1 = \begin{cases} 1 & \text{if level 1 of A} \\ 0 & \text{if level 2 of A} \end{cases} \qquad x_2 = \begin{cases} 1 & \text{if level 1 of B} \\ 0 & \text{if not} \end{cases}$$

$$x_3 = \begin{cases} 1 & \text{if level 2 of B} \\ 0 & \text{if not} \end{cases} \qquad x_4 = \begin{cases} 1 & \text{if level 1 of C} \\ 0 & \text{if not} \end{cases}$$

$$x_5 = \begin{cases} 1 & \text{if level 2 of C} \\ 0 & \text{if not} \end{cases}$$

(b) The sources of variation and the respective degrees of freedom corresponding to these sets of parameters are shown in Table 12.7.

Table 12.7 Table of sources and degrees of freedom for Example 12.17

Source	df
Main effect A	1
Main effect B	2
Main effect C	2
AB interaction	2
AC interaction	2
BC interaction	4
ABC interaction	4
Error	36
Total	53

The degrees of freedom for SS(Total) will always equal $(n - 1)$—that is, n minus 1 degree of freedom for β_0. Since the degrees of freedom for all sources must sum to the degrees of freedom for SS(Total), it follows that the degrees of freedom for error will equal the degrees of freedom for SS(Total), minus the sum of the degrees of freedom for main effects and interactions, that is, $(n - 1) - 17$. Our experiment will contain three observations for each of the $2 \times 3 \times 3 = 18$ treatments; therefore, $n = (18)(3) = 54$, and the degrees of freedom for error will equal $53 - 17 = 36$.

If data for this experiment were analyzed using statistical software, the printout would show the analysis of variance table that we have constructed and would include the associated mean squares, values of the F-test statistics, and their observed significance levels. Each F statistic would represent the ratio of the source mean square to MSE $= s^2$. ∎

Example 12.18

A transistor manufacturer conducted an experiment to investigate the effects of three factors on productivity (measured in thousands of dollars of items produced) per 40-hour week. The factors were as follows:

(a) Length of work week (two levels): five consecutive 8-hour days or four consecutive 10-hour days

(b) Shift (two levels): day or evening shift

(c) Number of coffee breaks (three levels): 0, 1, or 2

The experiment was conducted over a 24-week period with the $2 \times 2 \times 3 = 12$ treatments assigned randomly to the 24 weeks. The data for this completely randomized design are shown in Table 12.8. Perform an analysis of variance for the data.

TRANSISTOR1

Table 12.8 Data for Example 12.18

		Day Shift			Night Shift		
		Coffee Breaks			Coffee Breaks		
		0	1	2	0	1	2
Length	4 days	94	105	96	90	102	103
of		97	106	91	89	97	98
Work	5 days	96	100	82	81	90	94
Week		92	103	88	84	92	96

Solution

The data were subjected to an analysis of variance. The SAS printout is shown in Figure 12.23. Pertinent sections of the SAS printout are boxed and numbered, as follows:

1. The value of SS(Total), shown in the **Corrected Total** row of box 1, is 1,091.833333. The number of degrees of freedom associated with this quantity is $(n - 1) = (24 - 1) = 23$. Box 1 gives the partitioning (the analysis of variance) of this quantity into two sources of variation. The first source, **Model**, corresponds to the 11 parameters (all except β_0) in the model. The second source is **Error**. The degrees of freedom, sums of squares, and mean squares for these quantities are shown in their respective columns. For example, MSE $= 6.833333$. The F statistic for testing

$$H_0: \beta_1 = \beta_2 = \ldots = \beta_{11} = 0$$

is based on $\nu_1 = 11$ and $\nu_2 = 12$ degrees of freedom and is shown on the printout as $F = 13.43$. The observed significance level, shown under **Pr > F**, is less than .0001. This small observed significance level presents ample evidence to indicate that at least one of the three independent variables—shifts, number of days in a work week, or number of coffee breaks per day—contributes information for the prediction of mean productivity.

The ANOVA Procedure

Dependent Variable: PRODUCT

①

Source	DF	Sum of Squares	Mean Square	F Value	Pr > F
Model	11	1009.833333	91.803030	13.43	<.0001
Error	12	82.000000	6.833333		
Corrected Total	23	1091.833333			

④

R-Square	Coeff Var	Root MSE ③	PRODUCT Mean
0.924897	2.768647	2.614065	94.41667

②

Source	DF	Anova SS	Mean Square	F Value	Pr > F
SHIFT	1	48.1666667	48.1666667	7.05	0.0210
DAYS	1	204.1666667	204.1666667	29.88	0.0001
SHIFT*DAYS	1	8.1666667	8.1666667	1.20	0.2958
BREAKS	2	334.0833333	167.0416667	24.45	<.0001
SHIFT*BREAKS	2	385.5833333	192.7916667	28.21	<.0001
BREAKS*DAYS	2	8.0833333	4.0416667	0.59	0.5689
SHIFT*BREAKS*DAYS	2	21.5833333	10.7916667	1.58	0.2461

Figure 12.23 SAS ANOVA printout for $2 \times 2 \times 3$ factorial

2. To determine which sets of parameters are actually contributing information for the prediction of y, we examine the breakdown (box 2) of SS(Model) into components corresponding to the sets of parameters for main effects **SHIFT, DAYS**, and **BREAKS**, and parameters for two-way interactions, **SHIFT*DAYS**, **SHIFT*BREAKS**, and **DAYS*BREAKS**. The last **Model** source of variation corresponds to the set of all three-way **SHIFT*DAYS*BREAKS** parameters. Note that the degrees of freedom for these sources sum to 11, the number of degrees of freedom for **Model**. Similarly, the sum of the component sums of squares is equal to SS(Model). Box 2 does not give the mean squares associated with the sources, but it does give the F-values associated with testing hypotheses concerning the set of parameters associated with each source. Box 2 also gives the observed significance levels of these tests. You can see that there is ample evidence to indicate the presence of a **SHIFT*BREAKS** interaction. The F-tests associated with all three main effect parameter sets are also statistically significant at the $\alpha = .05$ level of significance. The practical implication of these results is that there is evidence to indicate that all three independent variables—shift, number of work days per week, and number of coffee breaks per day—contribute information for the prediction of productivity. The presence of a **SHIFT*BREAKS** interaction means that the effect of the number of breaks on productivity is not the same from shift to shift. Thus, the specific number of coffee breaks that might achieve maximum productivity on one shift might be different from the number of breaks that would achieve maximum productivity on the other shift.

3. Box 3 gives the value of $s = \sqrt{\text{MSE}} = 2.614065$. This value would be used to construct a confidence interval to compare the difference between two of the 12 treatment means. The confidence interval for the difference between a pair of means, $(\mu_i - \mu_j)$, would be

$$(\bar{y}_i - \bar{y}_j) \pm t_{\alpha/2} s \sqrt{\frac{2}{r}}$$

where r is the number of replications of the factorial experiment within a completely randomized design. There were $r = 2$ observations for each of the 12 treatments (factor level combinations) in this example.

4. Box 4 gives the value of R^2, a measure of how well the model fits the experimental data. It is of value primarily when the number of degrees of freedom for error is large—say, at least 5 or 6. The larger the number of degrees of freedom for error, the greater will be its practical importance. The value of R^2 for this analysis, .924897, indicates that the model provides a fairly good fit to the data. It also suggests that the model could be improved by adding new predictor variables or, possibly, by including higher-order terms in the variables originally included in the model.

Example 12.19

In a manufacturing process, a plastic rod is produced by heating a granular plastic to a molten state and then extruding it under pressure through a nozzle. An experiment was conducted to investigate the effect of two factors, extrusion temperature (°F) and pressure (pounds per square inch), on the rate of extrusion (inches per second) of the molded rod. A complete 2×2 factorial experiment (i.e., with each factor at two levels) was conducted. Three batches of granular plastic were used for the experiment, with each batch (viewed as a block) divided into four equal parts. The four portions of granular plastic for a given batch were randomly assigned to the four treatments; this was repeated for each of the three batches, resulting in a 2×2 factorial experiment laid out in three blocks. The data are shown in Table 12.9. Perform an analysis of variance for these data.

🔵 RODMOLD

Table 12.9 Data for Example 12.19

		Batch (Block)					
		1		2		3	
		Pressure		Pressure		Pressure	
		40	60	40	60	40	60
Temperature	200°	1.35	1.74	1.31	1.67	1.40	1.86
	300°	2.48	3.63	2.29	3.30	2.14	3.27

Solution

This experiment consists of a three-way classification of the data corresponding to batches (blocks), pressure, and temperature. The analysis of variance for this 2×2 factorial experiment (four treatments) laid out in a randomized block design (three blocks) yields the sources and degrees of freedom shown in Table 12.10.

The linear model for the experiment is

$$E(y) = \beta_0 + \overbrace{\beta_1 x_1}^{\substack{\text{Main} \\ \text{effect} \\ \text{P}}} + \overbrace{\beta_2 x_2}^{\substack{\text{Main} \\ \text{effect} \\ \text{T}}} + \overbrace{\beta_3 x_1 x_2}^{\substack{\text{PT} \\ \text{inter-} \\ \text{action}}} + \overbrace{\beta_4 x_3 + \beta_5 x_4}^{\substack{\text{Block} \\ \text{terms}}}$$

Table 12.10 Table of sources and degrees of freedom for Example 12.19

Source	df
Pressure (P)	1
Temperature (T)	1
Blocks	2
Pressure × Temperature interaction	1
Error	6
Total	11

where

$$x_1 = \text{Pressure} \qquad x_2 = \text{Temperature}$$

$$x_3 = \begin{cases} 1 & \text{if block 2} \\ 0 & \text{otherwise} \end{cases} \qquad x_4 = \begin{cases} 1 & \text{if block 3} \\ 0 & \text{otherwise} \end{cases}$$

The SPSS printout for the analysis of variance is shown in Figure 12.24. The F-test for the overall model (shaded at the top of the printout) is highly significant (p-value = .000). Thus, there is ample evidence to indicate differences among the block means or the treatment means or both. Proceeding to the breakdown of the model sources, you can see that the values of the F statistics for pressure, temperature, and the temperature × pressure interaction are all highly significant (i.e., their observed significance levels are very small). Therefore, all of the terms ($\beta_1 x_1$, $\beta_2 x_2$, and $\beta_3 x_1 x_2$) contribute information for the prediction of y.

The treatments in the experiment were assigned according to a randomized block design. Thus, we expected the extrusion of the plastic to vary from batch to batch. However, the F-test for testing differences among block (batch) means is not statistically significant (p-value = .2654); there is insufficient evidence to indicate a difference in the mean extrusion of the plastic from batch to batch. Blocking does not appear to have increased the amount of information in the experiment. ▬

Figure 12.24 SPSS ANOVA printout for Example 12.19

Tests of Between-Subjects Effects

Dependent Variable: RATE

Source	Type III Sum of Squares	df	Mean Square	F	Sig.
Corrected Model	7.149[a]	5	1.430	83.226	.000
Intercept	58.256	1	58.256	3390.818	.000
PRESSURE	1.687	1	1.687	98.222	.000
TEMP	5.044	1	5.044	293.590	.000
PRESSURE * TEMP	.361	1	.361	20.985	.004
BATCH	5.732E-02	2	2.866E-02	1.668	.265
Error	.103	6	1.718E-02		
Total	65.509	12			
Corrected Total	7.252	11			

a. R Squared = .986 (Adjusted R Squared = .974)

Many other complex designs, such as fractional factorials, Latin square designs, split-plot designs, and incomplete blocks designs, fall under the general k-way classification of data. Consult the references for the layout of these designs and the linear models appropriate for analyzing them.

12.6 Exercises

12.42 ANOVA summary table. An experiment was conducted to investigate the effects of three factors—paper stock, bleaching compound, and coating type—on the whiteness of fine bond paper. Three paper stocks (factor A), four types of bleaches (factor B), and two types of coatings (factor C) were used for the experiment. Six paper specimens were prepared for each of the $3 \times 4 \times 2$ stock–bleach–coating combinations and a measure of whiteness was recorded.

(a) Construct an analysis of variance table showing the sources of variation and the respective degrees of freedom.

(b) Suppose MSE = .14, MS(AB) = .39, and the mean square for all interactions combined is .73. Do the data provide sufficient evidence to indicate any interactions among the three factors? Test using $\alpha = .05$.

(c) Do the data present sufficient evidence to indicate an AB interaction? Test using $\alpha = .05$. From a practical point of view, what is the significance of an AB interaction?

(d) Suppose SS(A) = 2.35, SS(B) = 2.71, and SS(C) = .72. Find SS(Total). Then find R^2 and interpret its value.

12.43 Testing a new pain-reliever tablet. Paracetamol is the active ingredient in drugs designed to relieve mild to moderate pain and fever. The properties of paracetamol tablets derived from khaya gum were studied in the *Tropical Journal of Pharmaceutical Research* (June 2003). Three factors believed to impact the properties of paracetamol tablets are (1) the nature of the binding agent, (2) the concentration of the binding agent, and (3) the relative density of the tablet. In the experiment, binding agent was set at two levels (khaya gum and PVP), binding concentration at two levels (.5% and 4.0%), and relative density at two levels (low and high). One of the dependent variables investigated in the study was tablet dissolution time, that is, the amount of time (in minutes) for 50% of the tablet to dissolve. The goal of the study was to determine the effect of binding agent, binding concentration, and relative density on mean dissolution time.

(a) Identify the dependent (response) variable in the study.

(b) What are the factors investigated in the study? Give the levels of each.

(c) What type of experimental design was employed?

(d) How many treatments are possible in the study? List them.

(e) Write the linear model appropriate for analyzing the data.

(f) The sample mean dissolution times for the treatments associated with the factors binding agent and relative density when the other factor (binding concentration) is held fixed at .5% are: $\bar{y}_{Gum/Low} = 4.70$, $\bar{y}_{Gum/High} = 7.95$, $\bar{y}_{PVP/Low} = 3.00$, and $\bar{y}_{PVP/High} = 4.10$. Do the results suggest there is an interaction between binding agent and relative density? Explain.

12.44 Computer-mediated communication study. Computer-mediated communication (CMC) is a form of interaction that heavily involves technology (e.g., instant-messaging, e-mail). The *Journal of Computer-Mediated Communication* (April 2004) published a study to compare relational intimacy in people interacting via CMC to people meeting face-to-face (FTF). Participants were 48 undergraduate students, of which half were randomly assigned to the CMC group (communicating with the "chat" mode of instant-messaging software) and half assigned to the FTF group (meeting in a conference room). Subjects within each group were randomly assigned to either a high equivocality (HE) or low equivocality (LE) task that required communication with their group members. In addition, the researchers counterbalanced gender, so that each group–task combination had an equal number of females and males; these subjects were then divided into male–male pairs, female–female pairs, and male–female pairs. Consequently, there were two pairs of subjects assigned to each of the 2 (groups) $\times$ 2 (tasks) $\times$ 3 (gender pairs) = 12 treatments. A layout of the design is shown below. The variable of interest, relational intimacy score, was measured (on a 7-point scale) for each subject pair.

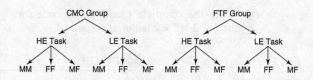

(a) Write the complete model appropriate for the $2 \times 2 \times 3$ factorial design.

(b) Give the sources of variation and associated degrees of freedom for an ANOVA table for this design.

(c) The researchers found no significant three-way interaction. Interpret this result practically.

(d) The researchers found a significant two-way interaction between Group and Task. Interpret this result practically.

(e) The researchers found no significant main-effect or interactions for Gender pair. Interpret this result practically.

12.45 Flotation of sulfured copper materials. The *Brazilian Journal of Chemical Engineering* (Vol. 22, 2005) published a study to compare two foaming agents in the flotation of sulfured copper materials process. The two agents were surface-active-bio-oil (SABO) and pine-oil (PO). A $2 \times 2 \times 2 \times 2$ factorial design was used to investigate the effect of four factors on the percentage of copper in the flotation concentrate. The four factors are foaming agent (SABO or PO), agent-to-mineral mass ratio (low or high), collector-to-mineral mass ratio (low or high), and liquid-to-solid ratio (low or high). Percentage copper measurements (y) were obtained for each of the $2 \times 2 \times 2 \times 2 = 16$ treatments. The data are listed in the table.

⊙ FOAM

AGENT-TO-MINERAL MASS RATIO	COLLECTOR-TO-MINERAL MASS RATIO	LIQUID-TO-SOLID RATIO	% COPPER SABO	PO
L	L	L	6.11	6.96
H	L	L	6.17	7.31
L	H	L	6.60	7.37
H	H	L	7.15	7.52
L	L	H	6.24	7.17
H	L	H	6.98	7.48
L	H	H	7.19	7.57
H	H	H	7.59	7.78

Source: Brossard, L. E. et al. "The surface-active bio oil solution in sulfured copper mineral benefit," *Brazilian Journal of Chemical Engineering,* Vol. 22, No. 1, 2005 (Table 3).

(a) Write the complete model appropriate for the $2 \times 2 \times 2 \times 2$ factorial design.

(b) Note that there is no replication in the experiment (that is, there is only one observation for each of the 16 treatments). How will this impact the analysis of the model, part a?

(c) Write a model for $E(y)$ that includes only main effects and two-way interaction terms.

(d) Fit the model, part c, to the data. Give the least squares prediction equation.

(e) Conduct tests (at $\alpha = .05$) for the interaction terms. Interpret the results.

(f) Is it advisable to conduct any main effect tests? If so, perform the analysis. If not, explain why.

12.46 Yield strength of nickel alloys. In increasingly severe oil-well environments, oil producers are interested in high-strength nickel alloys that are corrosion-resistant. Since nickel alloys are especially susceptible to hydrogen embrittlement, an experiment was conducted to compare the yield strengths of nickel alloy tensile specimens cathodically charged in a 4% sulfuric acid solution saturated with carbon disulfide, a hydrogen recombination poison. Two alloys were combined: inconel alloy (75% nickel composition) and incoloy (30% nickel composition). The alloys were tested under two material conditions (cold rolled and cold drawn), each at three different charging times (0, 25, and 50 days). Thus, a $2 \times 2 \times 3$ factorial experiment was conducted, with alloy type at two levels, material condition at two levels, and charging time at three levels. Two hydrogen-charged tensile specimens were prepared for each of the $2 \times 2 \times 3 = 12$ factor level combinations. Their yield strengths (kilograms per square inch) are recorded in the table. The SAS analysis of variance printout for the data is shown on p. 671.

(a) Is there evidence of any interactions among the three factors? Test using $\alpha = .05$. [*Note*: This means that you must test all the interaction parameters. The drop in SSE appropriate for the test would be the sum of all interaction sums of squares.]

(b) Now examine the F-tests shown on the printout for the individual interactions. Which, if

⊙ NICKEL

		ALLOY TYPE						
		INCONEL				INCOLOY		
		COLD ROLLED		COLD DRAWN		COLD ROLLED		COLD DRAWN
CHARGING TIME	0 days	53.4	52.6	47.1	49.3	50.6	49.9	30.9	31.4
	25 days	55.2	55.7	50.8	51.4	51.6	53.2	31.7	33.3
	50 days	51.0	50.5	45.2	44.0	50.5	50.2	29.7	28.1

SAS Output for Exercise 12.46

The ANOVA Procedure

Dependent Variable: YIELD

Source	DF	Sum of Squares	Mean Square	F Value	Pr > F
Model	11	1931.734583	175.612235	258.73	<.0001
Error	12	8.145000	0.678750		
Corrected Total	23	1939.879583			

R-Square	Coeff Var	Root MSE	YIELD Mean
0.995801	1.801942	0.823863	45.72083

Source	DF	Anova SS	Mean Square	F Value	Pr > F
ALLOY	1	552.0004167	552.0004167	813.26	<.0001
MATCOND	1	956.3437500	956.3437500	1408.98	<.0001
ALLOY*MATCOND	1	339.7537500	339.7537500	500.56	<.0001
TIME	2	71.0408333	35.5204167	52.33	<.0001
ALLOY*TIME	2	7.9858333	3.9929167	5.88	0.0166
MATCOND*TIME	2	4.1725000	2.0862500	3.07	0.0836
ALLOY*MATCOND*TIME	2	0.4375000	0.2187500	0.32	0.7306

any, of the interactions are statistically significant at the .05 level of significance?

12.47 Yield strength of nickel alloys (cont'd). Refer to Exercise 12.46. Since charging time is a quantitative factor, we could plot the strength y versus charging time x_1 for each of the four combinations of alloy type and material condition. This suggests that a prediction equation relating mean strength $E(y)$ to charging time x_1 may be useful. Consider the model

$$E(y) = \beta_0 + \beta_1 x_1 + \beta_2 x_1^2 + \beta_3 x_2 + \beta_4 x_3 + \beta_5 x_2 x_3$$
$$+ \beta_6 x_1 x_2 + \beta_7 x_1 x_3 + \beta_8 x_1 x_2 x_3$$
$$+ \beta_9 x_1^2 x_2 + \beta_{10} x_1^2 x_3 + \beta_{11} x_1^2 x_2 x_3$$

where

$x_1 = $ Charging time

$$x_2 = \begin{cases} 1 & \text{if inconel alloy} \\ 0 & \text{if incoloy alloy} \end{cases} \quad x_3 = \begin{cases} 1 & \text{if cold rolled} \\ 0 & \text{if cold drawn} \end{cases}$$

(a) Using the model terms, give the relationship between mean strength $E(y)$ and charging time x_1 for cold-drawn incoloy alloy.
(b) Using the model terms, give the relationship between mean strength $E(y)$ and charging time x_1 for cold-drawn inconel alloy.

(c) Using the model terms, give the relationship between mean strength $E(y)$ and charging time x_1 for cold-rolled inconel alloy.
(d) Fit the model to the data and find the least-squares prediction equation.
(e) Refer to part d. Find the prediction equations for each of the four combinations of alloy type and material condition.
(f) Refer to part d. Plot the data points for each of the four combinations of alloy type and material condition. Graph the respective prediction equations.

12.48 Yield strength of nickel alloys (cont'd). Refer to Exercises 12.46–12.47. If the relationship between mean strength $E(y)$ and charging time x_1 is the same for all four combinations of alloy type and material condition, the appropriate model for $E(y)$ is

$$E(y) = \beta_0 + \beta_1 x_1 + \beta_2 x_1^2$$

Fit the model to the data. Use the regression results, together with the information from Exercise 12.47, to decide whether the data provide sufficient evidence to indicate differences among the second-order models relating $E(y)$ to x_1 for the four categories of alloy type and material condition. Test using $\alpha = .05$.

12.7 Follow-Up Analysis: Tukey's Multiple Comparisons of Means

Many practical experiments are conducted to determine the largest (or the smallest) mean in a set. For example, suppose a chemist has developed five chemical solutions for removing a corrosive substance from metal. The chemist would want to determine which solution will remove the greatest amount of corrosive substance in a single application. Similarly, a production engineer might want to determine which among six machines or which among three foremen achieves the highest mean productivity per hour. A stockbroker might want to choose one stock, from among four, that yields the highest mean return, and so on.

Once differences among, say, five treatment means have been detected in an ANOVA, choosing the treatment with the largest mean might appear to be a simple matter. We could, for example, obtain the sample means $\bar{y}_1, \bar{y}_2, \ldots, \bar{y}_5$, and compare them by constructing a $(1 - \alpha)100\%$ confidence interval for the difference between each pair of treatment means. However, there is a problem associated with this procedure: **A confidence interval for $\mu_i - \mu_j$, with its corresponding value of α, is valid only when the two treatments (i and j) to be compared are selected prior to experimentation**. After you have looked at the data, you cannot use a confidence interval to compare the treatments for the largest and smallest sample means because they will always be farther apart, on the average, than any pair of treatments selected at random. Furthermore, **if you construct a series of confidence intervals, each with a chance α of indicating a difference between a pair of means if no difference exists, then the risk of making at least one Type I error in the series of inferences will be larger than the value of α specified for a single interval**.

There are a number of procedures for comparing and ranking a group of treatment means as part of a **follow-up** (or **post-hoc**) **analysis** to the ANOVA. The one that we present in this section, known as **Tukey's method for multiple comparisons**, utilizes the Studentized range

$$q = \frac{\bar{y}_{\max} - \bar{y}_{\min}}{s/\sqrt{n}}$$

(where $\bar{y}_{\max}$ and $\bar{y}_{\min}$ are the largest and smallest sample means, respectively) to determine whether the difference in any pair of sample means implies a difference in the corresponding treatment means. The logic behind this **multiple comparisons procedure** is that if we determine a critical value for the difference between the largest and smallest sample means, $|\bar{y}_{\max} - \bar{y}_{\min}|$, one that implies a difference in their respective treatment means, then any other pair of sample means that differ by as much as or more than this critical value would also imply a difference in the corresponding treatment means. Tukey's (1949) procedure selects this critical distance, ω, so that the probability of making one or more Type I errors (concluding that a difference exists between a pair of treatment means if, in fact, they are identical) is α. Therefore, the risk of making a Type I error applies to the whole procedure, that is, to the comparisons of all pairs of means in the experiment, rather than to a single comparison. Consequently, the value of α selected by the researchers is called an **experimentwise error rate** (in contrast to a **comparisonwise error rate**).

Tukey's procedure relies on the assumption that the p sample means are based on independent random samples, *each containing an equal number n_t of observations*. (When the number of observations per treatment are equal, researchers often refer to this as a **balanced design**.) Then if $s = \sqrt{MSE}$ is the computed standard deviation for the analysis, the distance ω is

$$\omega = q_\alpha(p, v) \frac{s}{\sqrt{n_t}}$$

The tabulated statistic $q_\alpha(p, v)$ is the critical value of the Studentized range, the value that locates α in the upper tail of the q distribution. This critical value depends on α, the number of treatment means involved in the comparison, and v, the number of degrees of freedom associated with MSE, as shown in the box. Values of $q_\alpha(p, v)$ for $\alpha = .05$ and $\alpha = .01$ are given in Tables 11 and 12, respectively, in Appendix D.

Tukey's Multiple Comparisons Procedure: Equal Sample Sizes

1. Select the desired experimentwise error rate, α.
2. Calculate

$$\omega = q_\alpha(p, v)\frac{s}{\sqrt{n_t}}$$

where

   p = Number of sample means (i.e., number of treatments)
   $s = \sqrt{MSE}$
   v = Number of degrees of freedom associated with MSE
   n_t = Number of observations in each of the p samples (i.e., number of observations per treatment)
   $q_\alpha(p, v)$ = Critical value of the Studentized range (Tables 11 and 12 in Appendix D)

3. Calculate and rank the p sample means.
4. For each treatment pair, calculate the difference between the treatment means and compare the difference to ω.
5. Place a bar over those pairs of treatment means that differ by less than ω. A pair of treatments not connected by an overbar (i.e., differing by more than ω) implies a difference in the corresponding population means.

Note: The confidence level associated with all inferences drawn from the analysis is $(1 - \alpha)$.

Example 12.20

Refer to the ANOVA for the completely randomized design, Examples 12.4 and 12.5. Recall that we rejected the null hypothesis of no differences among the mean GPAs for the three socioeconomic groups of college freshmen. Use Tukey's method to compare the three treatment means.

Solution

Step 1. For this follow-up analysis, we will select an experimentwise error rate of $\alpha = .05$.

Step 2. From previous examples, we have ($p = 3$) treatments, $v = 18$ df for error, $s = \sqrt{MSE} = .512$, and $n_t = 7$ observations per treatment. The critical value of the Studentized range (obtained from Table 11, Appendix D) is $q_{.05}(3, 18) = 3.61$. Substituting these values into the formula for ω, we obtain

$$\omega = q_{.05}(3, 18)\left(\frac{s}{\sqrt{n_t}}\right) = 3.61\left(\frac{.512}{\sqrt{7}}\right) = .698$$

Step 3. The sample means for the three socioeconomic groups (obtained from Table 12.1) are, in order of magnitude,

$$\bar{y}_L = 2.521 \quad \bar{y}_U = 2.543 \quad \bar{y}_M = 3.249$$

Step 4. The differences between treatment means are

$$\bar{y}_M - \bar{y}_L = 3.249 - 2.521 = .728$$

$$\bar{y}_M - \bar{y}_U = 3.249 - 2.534 = .715$$

$$\bar{y}_U - \bar{y}_L = 2.534 - 2.521 = .013$$

Step 5. Based on the critical difference $\omega = .70$, the three treatment means are ranked as follows:

Sample means:	2.521	2.543	3.249
Treatments:	Lower	Upper	Middle

From this information, we infer that the mean freshman GPA for the middle class is significantly larger than the means for the other two classes, since $\bar{y}_M$ exceeds both $\bar{y}_L$ and $\bar{y}_U$ by more than the critical value. However, the lower and upper classes are connected by a horizontal line since $|\bar{y}_L - \bar{y}_U|$ is less than ω. This indicates that the means for these treatments are not significantly different.

In summary, the Tukey analysis reveals that the mean GPA for the middle class of students is significantly larger than the mean GPAs of either the upper or lower classes, but that the means of the upper and lower classes are not significantly different. These inferences are made with an overall confidence level of $(1-\alpha) = .95$.

As Example 12.20 illustrates, Tukey's multiple comparisons of means procedure involves quite a few calculations. Most analysts utilize statistical software packages

Figure 12.25a SAS printout of Tukey's multiple comparisons of means, Example 12.20

```
                          The ANOVA Procedure

              Tukey's Studentized Range (HSD) Test for GPA

NOTE: This test controls the Type I experimentwise error rate, but it generally has a higher
                        Type II error rate than REGWQ.


                Alpha                                    0.05
                Error Degrees of Freedom                   18
                Error Mean Square                    0.261729
                Critical Value of Studentized Range   3.60930
                Minimum Significant Difference         0.6979


        Means with the same letter are not significantly different.


            Tukey Grouping          Mean       N     CLASS

                       A           3.2486       7     MIDDLE

                       B           2.5429       7     UPPER
                       B
                       B           2.5214       7     LOWER


                          The ANOVA Procedure

              Tukey's Studentized Range (HSD) Test for GPA

   NOTE: This test controls the Type I experimentwise error rate.


                Alpha                                    0.05
                Error Degrees of Freedom                   18
                Error Mean Square                    0.261729
                Critical Value of Studentized Range   3.60930
                Minimum Significant Difference         0.6979


     Comparisons significant at the 0.05 level are indicated by ***.


                              Difference      Simultaneous
                  CLASS        Between       95% Confidence
                Comparison      Means            Limits

             MIDDLE - UPPER     0.7057      0.0078   1.4036   ***
             MIDDLE - LOWER     0.7271      0.0292   1.4251   ***
             UPPER  - MIDDLE   -0.7057     -1.4036  -0.0078   ***
             UPPER  - LOWER     0.0214     -0.6765   0.7193
             LOWER  - MIDDLE   -0.7271     -1.4251  -0.0292   ***
             LOWER  - UPPER    -0.0214     -0.7193   0.6765
```

Figure 12.25b MINITAB printout of Tukey's multiple comparisons of means, Example 12.20

```
Tukey's pairwise comparisons

    Family error rate = 0.0500
    Individual error rate = 0.0200

Critical value = 3.61

Intervals for (column level mean) - (row level mean)

                    1               2

    2         -1.4252
              -0.0291

    3         -0.7195          0.0077
               0.6766          1.4038
```

Figure 12.25c SPSS printout of Tukey's multiple comparisons of means, Example 12.20

Post Hoc Tests

Multiple Comparisons

Dependent Variable: GPA
Tukey HSD

(I) GROUP	(J) GROUP	Mean Difference (I-J)	Std. Error	Sig.	95% Confidence Interval Lower Bound	95% Confidence Interval Upper Bound
1	2	-.7271*	.27346	.040	-1.4251	-.0292
	3	-.0214	.27346	.997	-.7193	.6765
2	1	.7271*	.27346	.040	.0292	1.4251
	3	.7057*	.27346	.047	.0078	1.4036
3	1	.0214	.27346	.997	-.6765	.7193
	2	-.7057*	.27346	.047	-1.4036	-.0078

*. The mean difference is significant at the .05 level.

Homogeneous Subsets

GPA

Tukey HSD[a]

GROUP	N	Subset for alpha = .05 — 1	Subset for alpha = .05 — 2
1	7	2.5214	
3	7	2.5429	
2	7		3.2486
Sig.		.997	1.000

Means for groups in homogeneous subsets are displayed.
a. Uses Harmonic Mean Sample Size = 7.000.

to conduct Tukey's method. The SAS, MINITAB, and SPSS printouts of the Tukey analysis for Example 12.20 are shown in Figures 12.25a, 12.25b, and 12.25c, respectively. Optionally, SAS presents the results in one of two forms. In the top printout, Figure 12.25a, SAS lists the treatment means vertically in descending order. Treatment means connected by the same letter (A, B, C, etc.) in the left column are *not* significantly different. You can see from Figure 12.25a that the middle class has a different letter (A) than the upper and lower classes (assigned the letter B). In the bottom printout of Figure 12.25a, SAS lists the Tukey confidence intervals for $(\mu_i - \mu_j)$, for all possible treatment pairs, i and j. Intervals that include 0 imply that the two treatments compared are not significantly different. The only interval at the bottom of Figure 12.25a that includes 0 is the one involving the upper and lower classes; hence, the GPA means for these two treatments are not significantly different. All the confidence intervals involving the middle class indicate that the middle class mean GPA is larger than either the upper or lower class mean.

Both MINITAB and SPSS present the Tukey comparisons in the form of confidence intervals for pairs of treatment means. Figures 12.25b and 12.25c (top) show the lower and upper endpoints of a confidence interval for $(\mu_1 - \mu_2), (\mu_1 - \mu_3)$, and $(\mu_2 - \mu_3)$, where "1" represents the lower class, "2" represents the middle class, and "3" represents the upper class. SPSS, like SAS, also produces a list of the treatment means arranged in subsets. The bottom of Figure 12.25c shows the means for treatments 1 and 3 (lower and upper classes) in the same subset, implying that these two means are not significantly different. The mean for treatment 2 (middle class) is in a different subset; hence, its treatment mean is significantly different than the others.

Example 12.21

Refer to Example 12.18. In a simpler experiment, the transistor manufacturer investigated the effects of just two factors on productivity (measured in thousands of dollars of items produced) per 40-hour week. The factors were:

Length of work week (two levels): five consecutive 8-hour days or four consecutive 10-hour days

Number of coffee breaks (three levels): 0, 1, or 2

The experiment was conducted over a 12-week period with the $2 \times 3 = 6$ treatments assigned in a random manner to the 12 weeks. The data for this two-factor factorial experiment are shown in Table 12.11.

(a) Perform an analysis of variance for the data.

(b) Compare the six population means using Tukey's multiple comparisons procedure. Use $\alpha = .05$.

🌐 TRANSISTOR2

Table 12.11 Data for Example 12.21

		Coffee Breaks		
		0	1	2
Length of Work Week	4 days	101	104	95
		102	107	92
	5 days	95	109	83
		93	110	87

Solution

(a) The SAS printout of the ANOVA for the 2×3 factorial is shown in Figure 12.26. Note that the test for interaction between the two factors, length (L) and breaks (B), is significant at $\alpha = .01$. (The p-value, .0051, is shaded on the printout.) Since interaction implies that the level of length (L) that yields the highest mean productivity may differ across different levels of breaks (B), we ignore the tests for main effects and focus our investigation on the individual treatment means.

The GLM Procedure

Dependent Variable: PRODUCT

Source	DF	Sum of Squares	Mean Square	F Value	Pr > F
Model	5	811.6666667	162.3333333	48.70	<.0001
Error	6	20.0000000	3.3333333		
Corrected Total	11	831.6666667			

R-Square	Coeff Var	Root MSE	PRODUCT Mean
0.975952	1.859839	1.825742	98.16667

Source	DF	Type III SS	Mean Square	F Value	Pr > F
DAYS	1	48.0000000	48.0000000	14.40	0.0090
BREAKS	2	667.1666667	333.5833333	100.07	<.0001
DAYS*BREAKS	2	96.5000000	48.2500000	14.47	0.0051

Least Squares Means
Adjustment for Multiple Comparisons: Tukey

DAYS	BREAKS	PRODUCT LSMEAN	LSMEAN Number
4	0	101.500000	1
4	1	105.500000	2
4	2	93.500000	3
5	0	94.000000	4
5	1	109.500000	5
5	2	85.000000	6

Least Squares Means for effect DAYS*BREAKS
Pr > |t| for H0: LSMean(i)=LSMean(j)

Dependent Variable: PRODUCT

i/j	1	2	3	4	5	6
1		0.3573	0.0329	0.0437	0.0329	0.0008
2	0.3573		0.0046	0.0057	0.3573	0.0002
3	0.0329	0.0046		0.9997	0.0010	0.0250
4	0.0437	0.0057	0.9997		0.0012	0.0192
5	0.0329	0.3573	0.0010	0.0012		<.0001
6	0.0008	0.0002	0.0250	0.0192	<.0001	

Figure 12.26 SAS ANOVA printout for Example 12.21

(b) The sample means for the six factor level combinations are highlighted in the middle of the SAS printout, Figure 12.26. Since the sample means represent measures of productivity in the manufacture of transistors, we want to find the length of work week and number of coffee breaks that yield the highest mean productivity.

In the presence of interaction, SAS displays the results of the Tukey multiple comparisons by listing the p-values for comparing all possible treatment mean pairs. These p-values are shown at the bottom of Figure 12.26. First, we demonstrate how to conduct the multiple comparisons using the formulas in the box. Then we explain (in notes) how to use p-values reported in the SAS to rank the means.

The first step in the ranking procedure is to calculate ω for $p = 6$ (we are ranking six treatment means), $n_t = 2$ (two observations per treatment), $\alpha = .05$, and $s = \sqrt{\text{MSE}} = \sqrt{3.33} = 1.83$ (where MSE is shaded in Figure 12.26). Since MSE is based on $\nu = 6$ degrees of freedom, we have

$$q_{.05}(6, 6) = 5.63$$

and

$$\omega = q_{.05}(6, 6) \left(\frac{s}{\sqrt{n_t}} \right)$$

$$= (5.63) \left(\frac{1.83}{\sqrt{2}} \right)$$

$$= 7.27$$

Therefore, population means corresponding to pairs of sample means that differ by more than $\omega = 7.27$ will be judged to be different. The six sample means are ranked as follows:

Sample means	85.0	93.5	94.0	101.5	105.5	109.5
Treatments (Length, Breaks)	(5, 2)	(4, 2)	(5, 0)	(4, 0)	(4, 1)	(5, 1)
Number on SAS printout:	6	3	4	1	2	5

Using $\omega = 7.27$ as a yardstick to determine differences between pairs of treatments, we have placed connecting bars over those means that *do not* significantly differ. The following conclusions can be drawn:

1. There is evidence of a difference between the population mean of the treatment corresponding to a 5-day work week with two coffee breaks (with the smallest sample mean of 85.0) and every other treatment mean. Therefore, we can conclude that the 5-day, two-break work week yields the lowest mean productivity among all length–break combinations.
   [*Note:* This inference can also be derived from the p-values shown under the mean 6 column at the bottom of the SAS printout, Figure 12.26. Each p-value (obtained using Tukey's adjustment) is used to compare the (5,2) treatment mean with each of the other treatment means. Since all the p-values are less than our selected experimentwise error rate of $\alpha = .05$, the (5,2) treatment mean is significantly different than each of the other means.]

2. The population mean of the treatment corresponding to a 5-day, one-break work week (with the largest sample mean of 109.5) is significantly larger than the treatments corresponding to the four smallest sample means. However, there is no evidence of a difference between the 5-day, one-break treatment mean and the 4-day, one-break treatment mean (with a sample mean of 105.5).
   [*Note:* This inference is supported by the Tukey-adjusted p-values shown under the mean 5 column—the column for the (5,1) treatment—in Figure 12.26. The only p-value that is *not* smaller than .05 is the one comparing mean 5 to mean 2, where mean 2 represents the (4,1) treatment.]

3. There is no evidence of a difference between the 4-day, one-break treatment mean (with a sample mean of 105.5) and the 4-day, zero-break treatment mean (with a sample mean of 101.5). Both of these treatments, though, have significantly larger means than the treatments corresponding to the three smallest sample means.
   [*Note:* This inference is supported by the Tukey-adjusted p-values shown under the mean 2 column—the column for the (4,1) treatment—in Figure 12.26. The p-value comparing mean 2 to mean 1, where mean 1 represents the (4,0) treatment, exceeds $\alpha = .05$.]

4. There is no evidence of a difference between the treatments corresponding to the sample means 93.5 and 94.0, i.e., between the (4,2) and (5,0) treatment means.

[*Note*: This inference can also be obtained by observing that the Tukey-adjusted p-value shown in Figure 12.26 under the mean 4 column—the column for the (5,0) treatment—and in the mean 3 row—the row for the (4,2) treatment—is greater than $\alpha = .05$.]

In summary, the treatment means appear to fall into four groups, as follows:

	TREATMENTS (LENGTH, BREAKS)
Group 1 (lowest mean productivity)	(5, 2)
Group 2	(4, 2) and (5, 0)
Group 3	(4, 0) and (4, 1)
Group 4 (highest mean productivity)	(4, 1) and (5, 1)

Notice that it is unclear where we should place the treatment corresponding to a 4-day, one-break work week because of the overlapping bars above its sample mean, 105.5. That is, although there is sufficient evidence to indicate that treatments (4, 0) and (5, 1) differ, neither has been shown to differ significantly from treatment (4, 1). Tukey's method guarantees that the probability of making one or more Type I errors in these pairwise comparisons is only $\alpha = .05$. ◼

Remember that Tukey's multiple comparisons procedure requires the sample sizes associated with the treatments to be equal. This, of course, will be satisfied for the randomized block designs and factorial experiments described in Sections 12.4 and 12.5, respectively. The sample sizes, however, may not be equal in a completely randomized design (Section 12.3). In this case a modification of Tukey's method (sometimes called the Tukey–Kramer method) is necessary, as described in the box (p. 680). The technique requires that the critical difference ω_{ij} be calculated for each pair of treatments (i, j) in the experiment and pairwise comparisons made based on the appropriate value of ω_{ij}. However, when Tukey's method is used with unequal sample sizes, the value of α selected a priori by the researcher only approximates the true experimentwise error rate. In fact, when applied to unequal sample sizes, the procedure has been found to be more conservative (i.e., less likely to detect differences between pairs of treatment means when they exist) than in the case of equal sample sizes. For this reason, researchers sometimes look to alternative methods of multiple comparisons when the sample sizes are unequal. Two of these methods are presented in optional Section 12.8.

In general, multiple comparisons of treatment means should be performed only as a follow-up analysis to the ANOVA, that is, only after we have conducted the appropriate analysis of variance F-test(s) and determined that sufficient evidence exists of differences among the treatment means. Be wary of conducting multiple comparisons when the ANOVA F-test indicates no evidence of a difference among a small number of treatment means—this may lead to confusing and contradictory results.*

Warning

In practice, it is advisable to avoid conducting multiple comparisons of a small number of treatment means when the corresponding ANOVA F-test is nonsignificant; otherwise, confusing and contradictory results may occur.

---

* When a large number of treatments are to be compared, a borderline, nonsignificant F-value (e.g., .05 < p-value < .10) may mask differences between some of the means. In this situation, it is better to ignore the F-test and proceed directly to a multiple comparisons procedure.

Tukey's Approximate Multiple Comparisons Procedure for Unequal Sample Sizes

1. Calculate for each treatment pair (i, j)

$$\omega_{ij} = q_\alpha(p, v)\frac{s}{\sqrt{2}}\sqrt{\frac{1}{n_i} + \frac{1}{n_j}}$$

where

p = Number of sample means

$s = \sqrt{MSE}$

v = Number of degrees of freedom associated with MSE

n_i = Number of observations in sample for treatment i

n_j = Number of observations in sample for treatment j

$q_\alpha(p, v)$ = Critical value of the Studentized range

(Tables 11 and 12 of Appendix D)

2. Rank the p sample means and place a bar over any treatment pair (i, j) that differs by less than ω_{ij}. Any pair of sample means not connected by an overbar (i.e., differing by more than ω) implies a difference in the corresponding population means.

Note: This procedure is approximate, that is, the value of α selected by the researcher approximates the true probability of making at least one Type I error.

12.7 Exercises

12.49 **Robots trained to behave like ants.** Refer to the *Nature* (August 2000) study of robots trained to behave like ants, Exercise 12.7 (p. 621). Multiple comparisons of mean energy expended for the four colony sizes were conducted using an experimentwise error rate of .05. The results are summarized below.

Sample mean:	.97	.95	.93	.80
Group size:	3	6	9	12

(a) How many pairwise comparisons are conducted in this analysis?

(b) Interpret the results shown in the table.

12.50 **Peer mentor training at a firm.** Refer to the *Journal of Managerial Issues* (Spring 2008) study of the impact of peer mentor training at a large software company, Exercise 12.20 (p. 637). A randomized block design (with trainees as blocks) was set up to compare the mean competence levels of trainees measured at three different times:

one week before training, two days after training, and two months after training. A multiple comparisons of means for the three time periods (using Tukey's method and an experimentwise error rate of .10) is summarized below. Fully interpret the results.

Sample mean:	3.65	4.14	4.17
Time period:	*Before*	*2 months after*	*2 days after*

12.51 **Mussel settlement patterns on algae.** Refer to the *Malacologia* (February 8, 2002) study of the impact of algae type on the abundance of mussel larvae in drift material, Exercise 12.30 (p. 658). Recall that algae was categorized into four strata—coarse-branching, medium-branching, fine-branching, and hydroid algae—and the average mussel density (percent per square centimeter) was determined for each. Tukey multiple comparisons of the four algae strata means (at $\alpha = .05$) are summarized on p. 681. Which means are significantly different?

Multiple comparisons for Exercise 12.51

Mean abundance ($\%/cm^2$):	9	10	27	55
Algae stratum:	Coarse	Medium	Fine	Hydroid

12.52 Learning from picture book reading. Refer to the *Developmental Psychology* (November, 2006) study of toddlers' ability to learn from reading picture books, Exercise 12.31 (p. 658). Recall that a 3×3 factorial experiment was employed, with age at three levels and reading book condition at three levels. At each age level, the researchers performed Tukey multiple comparisons of the reading book condition mean scores at $\alpha = .05$. The results are summarized in the table below. What can you conclude from this analysis? Support your answer with a plot of the means.

	.40	.75	1.20
AGE = 18 months:	Control	Drawings	Photos

	.60	1.61	1.63
AGE = 24 months:	Control	Drawings	Photos

	.50	2.20	2.21
AGE = 30 months:	Control	Drawings	Photos

12.53 End-user computing study. The *Journal of Computer Information Systems* (Spring 1993) published the results of a study of end-user computing. Data on the ratings of 18 specific end-user computing (EUC) policies were obtained

	EUC POLICY	MEAN RATING
1.	Organizational value	2.439
2.	Training	2.683
3.	Goals	2.854
4.	Justify applications	3.098
5.	Relation with MIS	3.293
6.	Hardware movement	3.366
7.	Accountability	3.390
8.	Justify data	3.561
9.	Ownership of files	3.756
10.	In-house software	3.854
11.	Copyright infringement	3.878
12.	Compatibility	4.000
13.	Document files	4.000
14.	Role of networking	4.049
15.	Data confidentiality	4.073
16.	Data security	4.219
17.	Hardware standards	4.293
18.	Software purchases	4.317

Source: Mitchell, R. B., and Neal, R. "Status of planning and control systems in the end-user computing environment," *Journal of Computer Information Systems*, Vol. 33, No. 3, Spring 1993, p. 29 (Table 4).

for each of 82 managers. (Managers rated policies on a 5-point scale, where 1 = no value and 5 = necessity.) The goal was to compare the mean ratings of the 18 EUC policies; thus, a randomized block design with 18 treatments (policies) and 82 blocks (managers) was used. Since the ANOVA F-test for treatments was significant at $\alpha = .01$, a follow-up analysis was conducted. The mean ratings for the 18 EUC policies are reported in the table. Using an overall significance level of $\alpha = .05$, the Tukey critical difference for comparing the 18 means was determined to be $\omega = .32$.

(a) Determine the pairs of EUC policy means that are significantly different.

(b) According to the researchers, the group of policies receiving the highest rated values have mean ratings of 4.0 and above. Do you agree with this assessment?

12.54 Insomnia and education. Refer to the *Journal of Abnormal Psychology* (February 2005) study relating daytime functioning to insomnia and education status, Exercise 12.33 (p. 658). In a 2×4 factorial experiment, with insomnia status at two levels (normal sleeper or chronic insomnia) and education at four levels (college graduate, some college, high school graduate, and high school dropout), only the main effect for education was statistically significant. Recall that the dependent variable was measured on the Fatigue Severity Scale (FSS). In a follow-up analysis, the sample mean FSS values for the four education levels were compared using Tukey's method ($\alpha = .05$), with the results shown below. What do you conclude?

Mean:	3.3	3.6	3.7	4.2
Education:	College graduate	Some college	HS graduate	HS dropout

🖥 TINLEAD

12.55 Strengthening tin-lead solder joints. Refer to Exercise 12.35 (p. 659). Use Tukey's multiple comparisons procedure to compare the mean shear strengths for the four antimony amounts. Identify the means that appear to differ. Use $\alpha = .01$.

🖥 EGGS2

12.56 Commercial eggs produced from different housing systems. Refer to the *Food Chemistry* (Vol. 106, 2008) study of four different types of egg housing systems, Exercise 12.36 (p. 659). Recall that you discovered that the mean whipping capacity (percent overflow) differed for cage, barn, free range, and organic egg housing systems. A multiple comparisons of means was conducted using Tukey's method with an experimentwise

SPSS Output for Exercise 12.56

Multiple Comparisons

OVERRUN

Tukey HSD

(I) HOUSING	(J) HOUSING	Mean Difference (I-J)	Std. Error	Sig.	95% Confidence Interval	
					Lower Bound	Upper Bound
BARN	CAGE	31.17*	6.355	.001	12.95	49.38
	FREE	-4.17	6.355	.912	-22.38	14.04
	ORGANIC	-15.83	6.355	.100	-34.04	2.38
CAGE	BARN	-31.17*	6.355	.001	-49.38	-12.95
	FREE	-35.33*	6.355	.000	-53.54	-17.12
	ORGANIC	-47.00*	6.355	.000	-65.21	-28.79
FREE	BARN	4.17	6.355	.912	-14.04	22.38
	CAGE	35.33*	6.355	.000	17.12	53.54
	ORGANIC	-11.67	6.355	.295	-29.88	6.54
ORGANIC	BARN	15.83	6.355	.100	-2.38	34.04
	CAGE	47.00*	6.355	.000	28.79	55.21
	FREE	11.67	6.355	.295	-5.54	29.88

Based on observed means.

The error term is Mean Square(Error) = 121.542.

*. The mean difference is significant at the 0.05 level.

error rate of .05. The results are displayed in the SPSS printout above.

(a) Locate the confidence interval for ($\mu_{CAGE} - \mu_{BARN}$) on the printout and interpret the result.

(b) Locate the confidence interval for ($\mu_{CAGE} - \mu_{FREE}$) on the printout and interpret the result.

(c) Locate the confidence interval for ($\mu_{CAGE} - \mu_{ORGANIC}$) on the printout and interpret the result.

(d) Locate the confidence interval for ($\mu_{BARN} - \mu_{FREE}$) on the printout and interpret the result.

(e) Locate the confidence interval for ($\mu_{BARN} - \mu_{ORGANIC}$) on the printout and interpret the result.

(f) Locate the confidence interval for ($\mu_{FREE} - \mu_{ORGANIC}$) on the printout and interpret the result.

(g) Based on the results, parts a–f, provide a ranking of the housing system means. Include the experimentwise error rate as a statement of reliability.

TREATAD2

12.57 Studies on treating Alzheimer's disease. Refer to the *eCAM* (November 2006) study of the quality of the research methodology used in journal articles that investigate the effectiveness of Alzheimer's disease (AD) treatments,

Exercise 12.22 (p. 638). Using 13 research papers as blocks, a randomized block design was employed to compare the mean quality scores of the nine research methodology dimensions, What-A, What-B, What-C, Who-A, Who-B, Who-C, How-A, How-B, and How-C.

(a) The SAS printout on p. 683 reports the results of a Tukey multiple comparisons of the nine Dimension means. Which pairs of means are significantly different?

(b) Refer to part a. The experimentwise error rate used in the analysis is .05. Interpret this value.

DRINKERS

12.58 Restoring self-control when intoxicated. Refer to the *Experimental and Clinical Psychopharmacology* (February 2005) study of restoring self-control while intoxicated, Exercise 12.12 (p. 623). The researchers theorized that if caffeine can really restore self-control, then students in Group AC (alcohol plus caffeine group) will perform the same as students in Group P (placebo group) on the word completion task. Similarly, if an incentive can restore self-control, then students in Group AR (alcohol plus reward group) will perform the same as students in Group P. Finally, the researchers theorized that students in Group A (alcohol only group) will perform

SAS Output for Exercise 12.57

```
            Tukey's Studentized Range (HSD) Test for WONG

NOTE: This test controls the Type I experimentwise error rate, but it generally has a higher Type
                              II error rate than REGWQ.


        Alpha                                    0.05
        Error Degrees of Freedom                   96
        Error Mean Square                    0.367165
        Critical Value of Studentized Range   4.48830
        Minimum Significant Difference         0.7543


     Means with the same letter are not significantly different.

        Tukey Grouping           Mean     N   DIMENSION

                    A          2.6923     13   WHAT-B
                    A
             B      A          2.5385     13   WHO-C
             B      A
             B      A          2.3846     13   HOW-C
             B      A
             B      A    C     2.1538     13   WHAT-C
             B      A    C
             B      A    C     2.1538     13   WHAT-A
             B      A    C
             B      A    C     2.0000     13   HOW-B
             B           C
             B           C     1.9231     13   HOW-A
             B           C
                         C     1.6154     13   WHO-B
                         C
                         C     1.5385     13   WHO-A
```

worse on the word completion task than students in any of the other three groups. Access the data in the DRINKERS file and conduct Tukey's multiple comparisons of the means, using an experimentwise error rate of .05. Are the researchers' theories supported?

12.8 Other Multiple Comparisons Methods (Optional)

In this optional section, we present two alternatives to Tukey's method of multiple comparisons of treatment means. The choice of methods will depend on the type of experimental design used and the particular error rate that the researcher wants to control.

Scheffé Method

Recall that Tukey's method of multiple comparisons is designed to control the experimentwise error rate (i.e., the probability of making at least one Type I error in the comparison of *all pairs* of treatment means in the experiment). Therefore, Tukey's method should be applied when you are interested in pairwise comparisons only.

Scheffé (1953) developed a more general procedure for comparing all possible linear combinations of the treatment means, called **contrasts**.

> **Definition 12.3** A **contrast** L is a linear combination of the p treatment means in a designed experiment, that is,
>
> $$L = \sum_{i=1}^{p} c_i \mu_i$$
>
> where the constants $c_1, c_2, \ldots, c_p$ sum to 0, that is, $\sum_{i=1}^{p} c_i = 0$.

For example, in an experiment with four treatments (A, B, C, D), you might want to compare the following contrasts, where μ_i represents the population mean for treatment i:

$$L_1 = \frac{\mu_A + \mu_B}{2} - \frac{\mu_C + \mu_D}{2}$$

$$L_2 = \mu_A - \mu_D$$

$$L_3 = \frac{\mu_B + \mu_C + \mu_D}{3} - \mu_A$$

The contrast L_2 involves a comparison of a pair of treatment means, whereas L_1 and L_3 are more complex comparisons of the treatments. Thus, pairwise comparisons are special cases of general contrasts.

As in Tukey's method, the value of α selected by the researcher using Scheffé's method applies to the procedure as a whole, that is, to the comparisons of all possible contrasts (not just those considered by the researcher). Unlike Tukey's method, however, the probability of at least one Type I error, α, is exact regardless of whether the sample sizes are equal. For this reason, some researchers prefer Scheffé's method to Tukey's method in the case of unequal samples, even if only pairwise comparisons of treatment means are made. The Scheffé method for general contrasts is outlined in the box.

Scheffé's Multiple Comparisons Procedure for General Contrasts

1. For each contrast $L = \sum_{i=1}^{p} c_i \mu_i$, calculate

$$\hat{L} = \sum_{i=1}^{p} c_i \bar{y}_i \quad \text{and} \quad S = \sqrt{(p-1)(F_\alpha)(\text{MSE}) \sum_{i=1}^{p} \left(\frac{c_i^2}{n_i}\right)}$$

where

p = Number of sample (treatment) means

MSE = Mean squared error

n_i = Number of observations in sample for treatment i

$\bar{y}_i$ = Sample mean for treatment i

F_α = Critical value of F distribution with $p - 1$ numerator df and v denominator df (Tables 3, 4, 5, and 6 in Appendix D)

v = Number of degrees of freedom associated with MSE

2. Calculate the confidence interval $\hat{L} \pm S$ for each contrast. The confidence coefficient, $1 - \alpha$, applies to the procedure as a whole (i.e., to the entire set of confidence intervals for all possible contrasts).

In the special case of all pairwise comparisons in an experiment with four treatments, the relevant contrasts reduce to $L_1 = \mu_A - \mu_B$, $L_2 = \mu_A - \mu_C$, $L_3 = \mu_A - \mu_D$, and so forth. Notice that for each of these contrasts $\sum c_i^2/n_i$ reduces to $(1/n_i + 1/n_j)$, where n_i and n_j are the sizes of treatments i and j, respectively.

[For example, for contrast L_1, $c_1 = 1$, $c_2 = -1$, $c_3 = c_4 = 0$, and $\sum c_i^2/n_i = (1/n_1 + 1/n_2)$.] Consequently, the formula for S in the general contrast method can be simplified and pairwise comparisons made using the technique in Section 12.7. The Scheffé method for pairwise comparisons of treatment means is shown in the next box.

Scheffé's Multiple Comparisons Procedure for Pairwise Comparisons of Treatment Means

1. Calculate Scheffé's critical difference for each pair of treatments (i, j):

$$S_{ij} = \sqrt{(p-1)(F_\alpha)(\text{MSE})\left(\frac{1}{n_i} + \frac{1}{n_j}\right)}$$

where

$$
\begin{aligned}
p &= \text{Number of sample (treatment) means} \\
\text{MSE} &= \text{Mean squared error} \\
n_i &= \text{Number of observations in sample} \\
&\quad \text{for treatment } i \\
n_j &= \text{Number of observations in sample} \\
&\quad \text{for treatment } j \\
F_\alpha &= \text{Critical value of } F \text{ distribution with } p-1 \\
&\quad \text{numerator df and } \nu \text{ denominator df} \\
&\quad \text{(Tables 3, 4, 5, and 6 in Appendix D)} \\
\nu &= \text{Number of degrees of freedom} \\
&\quad \text{associated with MSE}
\end{aligned}
$$

2. Rank the p sample means and place a bar over any treatment pair (i, j) that differs by less than S_{ij}. Any pair of sample means not connected by an overbar implies a difference in the corresponding population means.

Example 12.22

Refer to the completely randomized design for comparing the mean GPAs of three socioeconomic classes of college freshmen, Examples 12.4 and 12.20. In Example 12.20, we used Tukey's method to rank the three treatment means. Conduct the multiple comparisons using Scheffé's method at $\alpha = .05$. Use the fact that MSE $= .262$, $p = 3$ treatments, $\nu = \text{df(Error)} = 18$, and the three treatment means are $\bar{y}_{\text{Lower}} = 2.521$, $\bar{y}_{\text{Middle}} = 3.249$, and $\bar{y}_{\text{Upper}} = 2.543$.

Solution

Recall that in this completely randomized design, there are seven observations per treatment; thus, $n_i = n_j = 7$ for all treatment pairs (i, j).

Since the values of p, F, and MSE are fixed, the critical difference S_{ij} will be the same for all treatment pairs (i, j). Substituting in the appropriate values into the formula given in the box, the critical difference is:

$$S_{ij} = \sqrt{(p-1)(F_{.05})\text{MSE}\left(\frac{1}{n_i} + \frac{1}{n_i}\right)}$$

$$= \sqrt{(2)(3.55)(.262)\left(\frac{1}{7} + \frac{1}{7}\right)} = .729$$

Treatment means differing by more than $S = .729$ will imply a significant difference between the corresponding population means. Now, the difference between the largest sample mean and the smallest sample mean, $(3.249 - 2.521) = .728$, is less than $S = .729$. Consequently, we obtain the following rankings:

Sample means:	2.521	2.543	3.249
Treatments:	Lower	Upper	Middle

Thus, Scheffé's method fails to detect any significant differences among the GPA means of the three socioeconomic classes at an experimentwise error rate of $\alpha = .05$. [*Note*: This inference is confirmed by the results shown on the SAS printout of the Scheffé analysis, Figure 12.27. All three classes have the same "Scheffé grouping" letter (A).]   ▬

Figure 12.27 SAS printout of Scheffé's multiple comparisons of GPA means, Example 12.22

```
                    The ANOVA Procedure

                  Scheffe's Test for GPA

NOTE: This test controls the Type I experimentwise error rate.

        Alpha                            0.05
        Error Degrees of Freedom           18
        Error Mean Square            0.261729
        Critical Value of F          3.55456
        Minimum Significant Difference 0.7291


Means with the same letter are not significantly different.

  Scheffe Grouping        Mean     N    CLASS

                A       3.2486     7    MIDDLE
                A
                A       2.5429     7    UPPER
                A
                A       2.5214     7    LOWER
```

Note that in Example 12.22, the Scheffé method produced a critical difference of $S = .729$—a value larger than Tukey's critical difference of $\omega = .698$ (Example 12.20). This implies that Tukey's method produces narrower confidence intervals than Scheffé's method for differences in pairs of treatment means. Therefore, if only pairwise comparisons of treatments are to be made, Tukey's is the preferred method as long as the sample sizes are equal. The Scheffé method, on the other hand, yields narrower confidence intervals (i.e., smaller critical differences) for situations in which the goal of the researchers is to make comparisons of general contrasts.

Bonferroni Approach

As noted previously, Tukey's multiple comparisons procedure is approximate in the case of unequal sample sizes. That is, the value of α selected a priori only approximates the true probability of making at least one Type I error. The Bonferroni approach is an exact method that is applicable in either the equal or the unequal sample size case (see Miller, 1981). Furthermore, Bonferroni's procedure covers all possible comparisons of treatments, including pairwise comparisons, general contrasts, or combinations of pairwise comparisons and more complex contrasts.

The Bonferroni approach is based on the following result (proof omitted): If g comparisons are to be made, each with confidence coefficient $1 - \alpha/g$, then the overall probability of making one or more Type I errors (i.e., the experimentwise error rate) is at most α. That is, the set of intervals constructed using the Bonferroni method yields an overall confidence level of at least $1 - \alpha$. For example, if you want to construct $g = 2$ confidence intervals with an experimentwise error rate of at most $\alpha = .05$, then each individual interval must be constructed using a confidence level of $1 - .05/2 = .975$.

The Bonferroni approach for general contrasts is shown in the next box. When applied only to pairwise comparisons of treatments, the Bonferroni approach can be carried out as shown in the box on p. 688.

Bonferroni Multiple Comparisons Procedure for General Contrasts

1. For each contrast $L = \sum_{i=1}^{p} c_i \mu_i$, calculate

$$\hat{L} = \sum_{i=1}^{p} c_i \bar{y}_i$$

and

$$B = t_{\alpha/(2g)} s \sqrt{\sum_{i=1}^{p} \left(\frac{c_i^2}{n_i} \right)}$$

where

$p =$ Number of sample (treatment) means
$g =$ Number of contrasts
$s = \sqrt{\text{MSE}}$
$v =$ Number of degrees of freedom associated with MSE
$n_i =$ Number of observations in sample for treatment i
$\bar{y}_i =$ Sample mean for treatment i
$t_{\alpha/(2g)} =$ Critical value of t distribution with v df and tail area $\alpha/(2g)$

2. Calculate the confidence interval $\hat{L} \pm B$ for each contrast. The confidence coefficient for the procedure as a whole (i.e., for the entire set of confidence intervals) is *at least* $(1 - \alpha)$.

Example 12.23

Refer to Example 12.22. Use Bonferroni's method to perform pairwise comparisons of the three treatment means. Use $\alpha = .05$.

Solution
From Example 12.22, we have $p = 3$, $s = \sqrt{.262} = .512$, $v = 18$, and $n_i = n_j = 7$ for all treatment pairs (i, j). For $p = 3$ means, the number of pairwise comparisons to be made is

$$g = \frac{p(p - 1)}{2} = \frac{3(2)}{2} = 3$$

Bonferroni Multiple Comparisons Procedure for Pairwise Comparisons of Treatment Means

1. Calculate for each treatment pair (i, j)

$$B_{ij} = t_{\alpha/(2g)}s\sqrt{\frac{1}{n_i} + \frac{1}{n_j}}$$

where

p = Number of sample (treatment) means in the experiment

g = Number of pairwise comparisons

[*Note*: If all pairwise comparisons are to be made, then $g = p(p-1)/2$.]

$s = \sqrt{\text{MSE}}$

ν = Number of degrees of freedom associated with MSE

n_i = Number of observations in sample for treatment i

n_j = Number of observations in sample for treatment j

$t_{\alpha/(2g)}$ = Critical value of t distribution with ν df and tail area

$\alpha/(2g)$(Table 2 in Appendix D)

2. Rank the sample means and place a bar over any treatment pair (i, j) whose sample means differ by less than B_{ij}. Any pair of means not connected by an overbar implies a difference in the corresponding population means. *Note*: The level of confidence associated with all inferences drawn from the analysis is at least $(1 - \alpha)$.

Thus, we need to find the critical value, $t_{\alpha/(2g)} = t_{.05/[2(3)]} = t_{.0083}$, for the t distribution based on $\nu = 18$ df. This value, although not shown in Table 2 in Appendix D, is approximately 2.64.* Substituting $t_{.0083} \approx 2.64$ into the equation for Bonferroni's critical difference B_{ij}, we have

$$B_{ij} \approx (t_{.0083})s\sqrt{\frac{1}{n_i} + \frac{1}{n_j}} = (2.64)(.512)\sqrt{\frac{1}{7} + \frac{1}{7}} = .722$$

for any treatment pair (i, j).

Using the value $B_{ij} = .722$ to detect significant differences between treatment means, we obtain the following results:

Sample means:	2.521	2.543	3.249
Treatments:	Lower	Upper	Middle

---

* We obtained the value using the SAS probability generating function for a Student's t distribution.

You can see that the middle class mean GPA is significantly larger than the mean for the lower class. However, there is no significant difference between the pair of means for the lower and upper classes and no significant difference between the pair of means for the upper and middle classes. In other words, it is unclear whether the upper class mean GPA should be grouped with the treatment with the largest mean (middle class) or the treatment with the smallest mean (lower class). [*Note:* These inferences are supported by the results shown in the SAS printout of the Bonferroni analysis, Figure 12.28. The middle and upper class treatments have the same "Bonferroni grouping" letter (A), while the upper and lower class treatments have the same letter (B).] All inferences derived from this analysis can be made at an overall confidence level of at least $(1 - \alpha) = .95$. ∎

The ANOVA Procedure

Bonferroni (Dunn) t Tests for GPA

NOTE: This test controls the Type I experimentwise error rate, but it generally has a higher Type II error rate than REGWQ.

Alpha	0.05
Error Degrees of Freedom	18
Error Mean Square	0.261729
Critical Value of t	2.63914
Minimum Significant Difference	0.7217

Means with the same letter are not significantly different.

Bon Grouping		Mean	N	CLASS
	A	3.2486	7	MIDDLE
	A			
B	A	2.5429	7	UPPER
B				
B		2.5214	7	LOWER

Figure 12.28 SAS printout of Bonferroni's multiple comparisons of GPA means, Example 12.23

When applied to pairwise comparisons of treatments, the Bonferroni method, like the Scheffé procedure, produces wider confidence intervals (reflected by the magnitude of the critical difference) than Tukey's method. (In Example 12.23, Bonferroni's critical difference is $B \approx .722$ compared to Tukey's $\omega = .698$.) Therefore, if only pairwise comparisons are of interest, Tukey's procedure is again superior. However, if the sample sizes are unequal or more complex contrasts are to be compared, the Bonferroni technique may be preferred. Unlike the Tukey and Scheffé methods, however, Bonferroni's procedure requires that you know in advance how many contrasts are to be compared. Also, the value needed to calculate the critical difference B, $t_{\alpha/(2g)}$, may not be available in the t tables provided in most texts, and you will have to estimate it.

In this section, we have presented two alternatives to Tukey's multiple comparisons procedure. The technique you select will depend on several factors, including the sample sizes and the type of comparisons to be made. Keep in mind, however, that many other methods of making multiple comparisons are available, and one or more of these techniques may be more appropriate to use in your particular application. Consult the references given at the end of this chapter for details on other techniques.

12.8 Exercises

12.59 **Heights of grade school repeaters.** Refer to The *Archives of Disease in Childhood* (April 2000) comparison of the average heights of the three groups of Australian schoolchildren based on age, Exercise 12.9 (p. 622). The three height means for boys were ranked using the Bonferroni method at $\alpha = .05$. The results are summarized below. (Recall that all height measurements were standardized using z-scores.)

Sample means:	0.16	0.33	0.33
Age group:	Oldest	Youngest	Middle

(a) Is there a significant difference between the standardized height means for the oldest and youngest boys?

(b) Is there a significant difference between the standardized height means for the oldest and middle-aged boys?

(c) Is there a significant difference between the standardized height means for the youngest and middle-aged boys?

(d) What is the experimentwise error rate for the inferences made in parts a–c? Interpret this value.

(e) The researchers did not perform a Bonferroni analysis of the height means for the three groups of girls. Explain why not.

12.60 **Hazardous organic solvents.** Refer to the *Journal of Hazardous Materials* study of mean sorption rates for three types of organic solvents, Exercise 12.11 (p. 623). SAS was used to produce Bonferroni confidence intervals for all pairs of treatment means. The SAS printout is shown below.

(a) Find and interpret the experimentwise error rate shown on the printout.

(b) Locate and interpret the confidence interval for the difference between the mean sorption rates of aromatics and esters.

(c) Use the confidence intervals to determine which pairs of treatment means are significantly different.

12.61 **Dental anxiety study.** Does recalling a traumatic dental experience increase your level of anxiety at the dentist's office? In a study published in *Psychological Reports* (August 1997), researchers at Wittenberg University randomly assigned 74 undergraduate psychology students to one of three experimental conditions. Subjects in the "Slide" condition viewed 10 slides of scenes from a dental office. Subjects in the "Questionnaire" condition completed a full dental history questionnaire; one of the questions asked them to describe their worst dental experience. Subjects in the "Control" condition received no actual treatment. All students then completed the Dental Fear Scale, with scores ranging from 27 (no fear) to 135 (extreme fear). The sample dental fear means for the Slide, Questionnaire, and Control groups were reported as 43.1, 53.8, and 41.8, respectively.

(a) A completely randomized design ANOVA was carried out on the data, with the following results: $F = 4.43$, p-value $< .05$. Interpret these results.

(b) According to the article, a Bonferroni ranking of the three dental fear means (at $\alpha = .05$) "indicated a significant difference between the mean scores on the Dental Fear Scale for the Control and Questionnaire groups,

SAS Output for Exercise 12.60

```
              Bonferroni (Dunn) t Tests for SORPRATE

NOTE:This test controls the Type I experimentwise error rate, but it generally has a higher Type
              II error rate than Tukey's for all pairwise comparisons.


                    Alpha                          0.05
                    Error Degrees of Freedom         29
                    Error Mean Square          0.067426
                    Critical Value of t         2.54091


      Comparisons significant at the 0.05 level are indicated by ***.


                                   Difference
                      SOLVENT       Between          Simultaneous 95%
                      Comparison     Means        Confidence Limits

                    CHLOR - AROMA    0.06403    -0.25657   0.38462
                    CHLOR - ESTER    0.67625     0.38740   0.96510    ***
                    AROMA - CHLOR   -0.06403    -0.38462   0.25657
                    AROMA - ESTER    0.61222     0.33403   0.89041    ***
                    ESTER - CHLOR   -0.67625    -0.96510  -0.38740    ***
                    ESTER - AROMA   -0.61222    -0.89041  -0.33403    ***
```

but not for the means between the Control and Slide groups." Summarize these results in a chart that connects means that are not significantly different.

🔘 HONEYCOUGH

12.62 Is honey a cough remedy? Refer to the *Archives of Pediatrics and Adolescent Medicine* (December 2007) study of treatments for children's cough symptoms, Exercise 12.14 (p. 624). Do you agree with the statement (extracted from the article), "honey may be a preferable treatment for the cough and sleep difficulty associated with childhood upper respiratory tract infection"? Perform a multiple comparisons of means to answer the question. Justify your choice of a multiple comparisons method.

12.63 Guilt in decision making. The effect of guilt emotion on how a decision-maker focuses on the problem was investigated in the January 2007 issue of the *Journal of Behavioral Decision Making*. A sample of 77 volunteer students participated in one portion of the experiment, where each was randomly assigned to one of three emotional states (guilt, anger, or neutral) through a reading/writing task. (Note: Twenty-six students were assigned to the "guilt" state, 26 to the "anger" state, and 25 to the "neutral" state.) Immediately after the task, the students were presented with a decision problem where the stated option has predominantly negative features (e.g., spending money on repairing a very old car). Prior to making the decision, the researchers asked each subject to list possible, more attractive, alternatives. The researchers then compared the mean number of alternatives listed across the three emotional states with an analysis of variance for a completely randomized design. A partial ANOVA summary table is shown below.

SOURCE	df	F-VALUE	p-VALUE
Emotional State	2	22.68	0.001
Error	74		
Total	76		

(a) What conclusion can you draw from the ANOVA results?

(b) Explain why it is inappropriate to apply Tukey's multiple comparisons procedure to compare the treatment means.

(c) The Bonferroni multiple comparisons of means procedure was applied to the data using an experimentwise error rate of .05. Explain what the value, .05, represents.

(d) The multiple comparisons yielded the results shown in the next column. What conclusion can you draw?

Sample mean:	1.90	2.17	4.75
Emotional state:	*Angry*	*Neutral*	*Guilt*

12.64 Chemical properties of whole wheat breads. Whole wheat breads contain a high amount of phytic acid, which tends to lower the absorption of nutrient minerals. The *Journal of Agricultural and Food Chemistry* (January 2005) published the results of a study to determine if sourdough can increase the solubility of whole wheat bread. Four types of bread were prepared from whole meal flour: (1) yeast added, (2) sourdough added, (3) no yeast or sourdough added (control), and (4) lactic acid added. Data were collected on the soluble magnesium level (percent of total magnesium) during fermentation for dough samples of each bread type, and analyzed using a one-way ANOVA. The four mean soluble magnesium levels were compared in pairwise fashion using Bonferroni's method.

(a) How many pairwise comparisons are made in the Bonferroni analysis?

(b) The experimentwise error rate for the analysis was .05. Interpret this value.

(c) Based on the experimentwise error rate, part b, what comparisonwise error rate is used with each comparison made in the Bonferroni analysis?

(d) The results of the Bonferroni analysis are summarized below. Which treatment(s) yielded the significantly highest mean soluble magnesium level? The lowest?

Mean:	7%	12.5%	22%	27.5%
Bread type:	Control	Yeast	Lactate	Sourdough

12.65 Estimating the age of glacial drifts. Refer to the *American Journal of Science* (January 2005) study of the chemical makeup of buried tills (glacial drifts) in Wisconsin, Exercise 12.16 (p. 625). The data are repeated in the table. Use a multiple comparisons procedure to compare the mean Al/Be ratios for the five boreholes (labeled UMRB-1, UMRB-2, UMRB-3, SWRA, and SD),

🔘 TILLRATIO

UMRB-1:	3.75	4.05	3.81	3.23	3.13	3.30	3.21
UMRB-2:	3.32	4.09	3.90	5.06	3.85	3.88	
UMBR-3:	4.06	4.56	3.60	3.27	4.09	3.38	3.37
SWRA:	2.73	2.95	2.25				
SD:	2.73	2.55	3.06				

Source: Adapted from *American Journal of Science*, Vol. 305, No. 1, Jan. 2005, p. 16 (Table 2).

with an experiment wise error rate of .10. Identify the means that appear to differ.

12.66 **Training program evaluation.** A field experiment was conducted at a not-for-profit research and development organization to examine the expectations, attitudes, and decisions of employees with regard to training programs (*Academy of Management Journal*, September 1987). In particular, the study was aimed at determining how managers' evaluations of a training program were affected by the prior information they received and by the degree of choice they had in entering the program. These two factors, prior information and degree of choice, were each varied at two levels. The prior information managers received about the training program was either a realistic preview of the program and its benefits or a traditional announcement that tended to exaggerate the workshop's benefits. Degree of choice was either low (mandatory attendance) or high (little pressure from supervisors to attend). Twenty-one managers were randomly assigned to each

of the $2 \times 2 = 4$ experimental conditions; thus, a 2×2 factorial design was employed. At the end of the training program, each manager was asked to rate his or her satisfaction with the workshop on a 7-point scale (1 = no satisfaction, 7 = extremely satisfied). The ratings were subjected to an analysis of variance, with the results shown in the partial ANOVA summary table.

(a) Complete the ANOVA summary table.
(b) Conduct the appropriate ANOVA F-tests (use $\alpha = .05$). Interpret the results.
(c) The sample mean satisfaction ratings of managers for the four combinations of prior information and degree of choice are shown in the table below. Use Tukey's method to rank the four means. Use $\alpha = .06$.
(d) Use the Scheffé method to perform all pairwise comparisons of the four treatment means. Use $\alpha = .05$.
(e) Use the Bonferroni approach to perform all pairwise comparisons of the four treatment means. Use $\alpha = .05$.
(f) Compare the results, parts c–e.

SOURCE	df	SS	MS	F
Prior information (P)	1	—	1.55	—
Degree of choice (D)	1	—	22.26	—
PD interaction	1	—	.61	—
Error	80	—	1.43	
Total	83	—		

Source: Hicks, W. D., and Klimoski, R. J. "Entry into training programs and its effects on training outcomes: A field experiment," *Academy of Management Journal*, Vol. 30, No. 3, Sept. 1987, p. 548.

		PRIOR INFORMATION	
		REALISTIC PREVIEW	TRADITIONAL ANNOUNCEMENT
Degree of Choice	High	6.20	6.06
	Low	5.33	4.82

Source: Hicks, W. D., and Klimoski, R. J. "Entry into training programs and its effects on training outcomes: A field experiment," *Academy of Management Journal*, Vol. 30, No. 3, Sept. 1987, p. 548.

12.9 Checking ANOVA Assumptions

For each of the experiments and designs discussed in this chapter, we listed in the relevant boxes the assumptions underlying the analysis in the terminology of ANOVA. For example, in the box on page 614, the assumptions for a completely randomized design are that (1) the p probability distributions of the response y corresponding to the p treatments are normal and (2) the population variances of the p treatments are equal. Similarly, for randomized block designs and factorial designs, the data for the treatments must come from normal probability distributions with equal variances.

These assumptions are equivalent to those required for a regression analysis (see Section 4.2). The reason, of course, is that the probabilistic model for the response y that underlies each design is the familiar general linear regression model in Chapter 4. A brief overview of the techniques available for checking the ANOVA assumptions follows.

Detecting Nonnormal Populations

1. For each treatment, construct a histogram, stem-and-leaf display, or normal probability plot for the response, y. Look for highly skewed distributions. [*Note*: For relatively large samples (e.g., 20 or more observations per treatment), ANOVA, like regression, is **robust** with respect to the normality assumption. That is, slight departures from normality will have little impact on the validity of the inferences derived from the analysis. [If the sample size for each treatment is small, then these graphs will probably be of limited use.]

2. Formal statistical tests of normality (such as the **Shapiro–Wilk test** or **Kolmogorov–Smirnov test**) are also available. The null hypothesis is that the probability distribution of the response y is normal. These tests, however, are sensitive to slight departures from normality. Since in most scientific applications the normality assumption will not be satisfied exactly, these tests will likely result in a rejection of the null hypothesis and, consequently, are of limited use in practice. Consult the references for more information on these formal tests.

3. If the distribution of the response departs greatly from normality, a **normalizing transformation** may be necessary. For example, for highly skewed distributions, transformations on the response y such as $\log(y)$ or $\sqrt{y}$ tend to "normalize" the data since these functions "pull" the observations in the tail of the distribution back toward the mean.

Detecting Unequal Variances

1. For each treatment, construct a box plot or **frequency** (dot) plot for y and look for differences in spread (variability). If the variability of the response in each plot is about the same, then the assumption of equal variances is likely to be satisfied. [*Note*: *ANOVA is robust with respect to unequal variances for **balanced designs** (i.e., designs with equal sample sizes for each treatment)*.]

2. When the sample sizes are small for each treatment, only a few points are graphed on the frequency plots, making it difficult to detect differences in variation. In this situation, you may want to use one of several formal statistical tests of homogeneity of variances that are available. For p treatments, the null hypothesis is $H_0: \sigma_1^2 = \sigma_2^2 = \ldots = \sigma_p^2$, where σ_i^2 is the population variance of the response y corresponding to the ith treatment. If all p populations are approximately normal, **Bartlett's test for homogeneity of variances** can be applied. Bartlett's test works well when the data come from normal (or near normal) distributions. *The results, however, can be misleading for nonnormal data.* In situations where the response is clearly not normally distributed, **Levene's test** is more appropriate. The elements of these tests are shown in the accompanying boxes. Note that Bartlett's test statistic depends on whether the sample sizes are equal or unequal.

3. When unequal variances are detected, use one of the **variance-stabilizing transformations** of the response y discussed in Section 8.3.

Bartlett's Test of Homogeneity of Variance

H_0: $\sigma_1^2 = \sigma_2^2 = \ldots = \sigma_p^2$

H_a: At least two variances differ.

Test statistic (equal sample sizes):

$$B = \frac{(n-1)\left[p \ln \bar{s}^2 - \sum \ln s_i^2\right]}{1 + \dfrac{p+1}{3p(n-1)}}$$

where

$n = n_1 = n_2 = \cdots = n_p$

$s_i^2 =$ Sample variance for sample i

$\bar{s}^2 =$ Average of the p sample variances $= \left(\sum s_i^2\right)\Big/ p$

$\ln x =$ Natural logarithm (i.e., log to the base e) of the quantity x

Test statistic (unequal sample sizes):

$$B = \frac{\left[\sum(n_i - 1)\right] \ln \bar{s}^2 - \sum(n_i - 1) \ln s_i^2}{1 + \dfrac{1}{3(p-1)}\left\{\sum \dfrac{1}{(n_i - 1)} - \dfrac{1}{\sum(n_i - 1)}\right\}}$$

where

$n_i =$ Sample size for sample i

$s_i^2 =$ Sample variance for sample i

$\bar{s}^2 =$ Weighted average of the p sample variances $= \dfrac{\sum(n_i - 1)s_i^2}{\sum(n_i - 1)}$

$\ln x =$ Natural logarithm (i.e., log to the base e) of the quantity x

Rejection region: $B > \chi_\alpha^2$, where χ_α^2 locates an area α in the upper tail of a χ^2 distribution with $(p - 1)$ degrees of freedom

Assumptions: 1. Independent random samples are selected from the p populations.

2. All p populations are normally distributed.

Example 12.24   Refer to the ANOVA for the completely randomized design, Example 12.4 (p. 615). Recall that we found differences among the mean GPAs for the three socioeconomic classes of college freshmen. Check to see if the ANOVA assumptions are satisfied for this analysis.

Levene's Test of Homogeneity of Variance

$H_0: \sigma_1^2 = \sigma_2^2 = \ldots = \sigma_p^2$

H_a: At least two variances differ

Test statistic: $F = MST/MSE$

where MST and MSE are obtained from an ANOVA with p treatments conducted on the transformed response variable $y_i^* = |y_i - Med_p|$, and Med_p is the median of the response y-values for treatment p.

Rejection region: $F > F_\alpha$, where F_α locates an area α in the upper tail of an F distribution with $v_1 = (p - 1)$ df and $v_2 = (n - p)$ df

Assumptions: 1. Independent random samples are selected from the p treatment populations.
2. The response variable y is a continuous random variable.

Solution

First, we'll check the assumption of normality. For this design, there are only seven observations per treatment (class); consequently, constructing graphs (e.g., histograms or stem-and-leaf plots) for each treatment will not be very informative. Alternatively, we can combine the data for the three treatments and form a histogram for all 21 observations in the data set. A MINITAB histogram for the response variable, GPA, is shown in Figure 12.29. Clearly, the data fall into an approximately mound-shaped distribution. This result is supported by the MINITAB normal probability plot for GPA displayed in Figure 12.30. The results of a test for normality of the data is also shown (highlighted) in Figure 12.30. Since the p-value of the test exceeds .10, there is insufficient evidence (at $\alpha = .05$) to conclude that the data are nonnormal. Consequently, it appears that the GPAs come from a normal distribution.

Figure 12.29 MINITAB histogram for all the GPAs in the completely randomized design

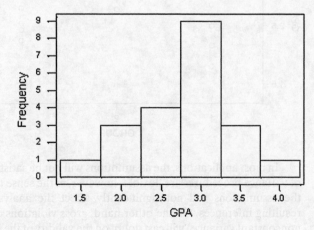

Figure 12.30 MINITAB normal probability plot and normality test for all the GPAs in the completely randomized design

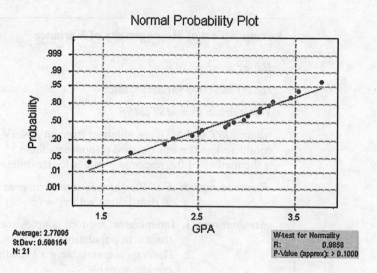

Next, we check the assumption of equal variances. MINITAB dot plots for GPA are displayed in Figure 12.31. Note that the variability of the response in each plot is about the same; thus, the assumption of equal variances appears to be satisfied. To formally test the hypothesis, H_0: $\sigma_1^2 = \sigma_2^2 = \sigma_3^2$, we conduct both Bartlett's and Levene's test for homogeneity of variances. Rather than use the computing formulas shown in the boxes, we resort to a statistical software package. The MINITAB printout of the test results are shown in Figure 12.32. The p-values for both tests are shaded on the printout. Since both p-values exceed at $\alpha = .05$, there is insufficient evidence to reject the null hypothesis of equal variances. Therefore, it appears that the assumption of equal variance is satisfied also.   ▬

Figure 12.31 MINITAB dot plots for the GPAs in the completely randomized design, by treatment group

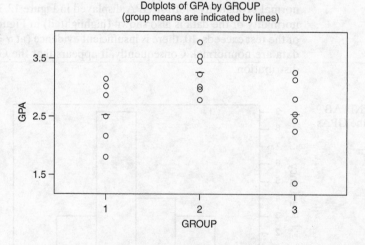

In most applications, the assumptions will not be satisfied exactly. These analysis of variance procedures are flexible, however, in the sense that slight departures from the assumptions will not significantly affect the analysis or the validity of the resulting inferences. On the other hand, gross violations of the assumptions (e.g., a nonconstant variance) will cast doubt on the validity of the inferences. Therefore, you should make it standard practice to verify that the assumptions are (approximately) satisfied.

Figure 12.32 MINITAB printout of tests for homogeneity of GPA variances for the completely randomized design

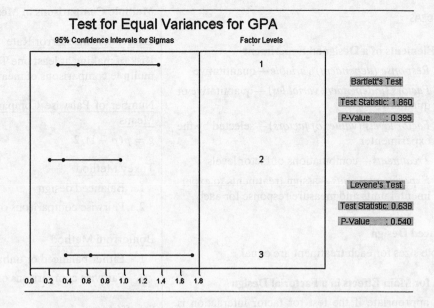

12.9 Exercises

12.67 Hazardous organic solvents. Check the assumptions for the completely randomized design ANOVA in Exercise 12.11 (p. 623).

12.68 The "name game." Check the assumptions for the completely randomized design ANOVA in Exercise 12.13 (p. 624).

12.69 Is honey a cough remedy? Check the assumptions for the completely randomized design ANOVA in Exercise 12.14 (p. 624).

12.70 Violent lyrics and aggressiveness. Check the assumptions for the factorial design ANOVA in Exercise 12.38 (p. 661).

12.71 Exam performance study. Check the assumptions for the factorial design ANOVA in Exercise 12.39 (p. 661).

Quick Summary

KEY SYMBOLS/NOTATION

ANOVA	Analysis of variance
SST	Sum of Squares for Treatments
MST	Mean Square for Treatments
SSB	Sum of Squares for Blocks
MSB	Mean Square for Blocks
SSE	Sum of Squares for Error
MSE	Mean Square for Error
$a \times b$ factorial	Factorial design with one factor at a levels and the other factor at b levels

SS(A)	Sum of Squares for main effect factor A
MS(A)	Mean Square for main effect factor A
SS(B)	Sum of Squares for main effect factor B
MS(B)	Mean Square for main effect factor B
SS(AB)	Sum of Squares for factor $A \times B$ interaction
MS(AB)	Mean Square for factor $A \times B$ interaction

KEY IDEAS

Key Elements of a Designed Experiment

1. *Response (dependent) variable*—quantitative
2. *Factors (independent variables)*—quantitative or qualitative
3. *Factor levels (values of factors)*—selected by the experimenter
4. *Treatments*—combinations of factor levels
5. *Experimental units*—assign treatments to experimental units and measure response for each

Balanced Design

Sample sizes for each treatment are equal.

Tests for Main Effects in a Factorial Design

Only appropriate if the test for factor interaction is nonsignificant.

Robust Method

Slight to moderate departures from normality do not have impact on validity of the ANOVA results.

Conditions Required for Valid *F*-test in a Completely Randomized Design

1. All p treatment populations are approximately normal.
2. $\sigma_1^2 = \sigma_2^2 = \cdots = \sigma_p^2$

Conditions Required for Valid *F*-tests in a Randomized Block Design

1. All treatment-block populations are approximately normal.
2. All treatment-block populations have the same variance.

Conditions Required for Valid *F*-tests in a Complete Factorial Design

1. All treatment populations are approximately normal.
2. All treatment populations have the same variance.

Multiple Comparisons of Means Methods

Experimentwise Error Rate

Risk of making at least one Type 1 error when making multiple comparisons of means in ANOVA

Number of Pairwise Comparisons with p Treatment Means

$g = p(p-1)/2$

Tukey Method

1. Balanced design
2. Pairwise comparisons of means

Bonferroni Method

1. Either balanced or unbalanced design
2. Pairwise comparisons of means

Scheffé Method

1. Either balanced or unbalanced design
2. General contrasts of means

Linear Model for a Completely Randomized Design with p Treatments

$$E(y) = \beta_0 + \beta_1 x_1 + \beta_2 x_2 + \cdots + \beta_{p-1} x_{p-1} \text{ where}$$

$$x_1 = \{1 \text{ if Treatment 1, 0 if not}\},$$

$$x_2 = \{1 \text{ if Treatment 2, 0 if not}\}, \ldots,$$

$$x_{p-1} = \{1 \text{ if Treatment } p-1, 0 \text{ if not}\}$$

Linear Model for a Randomized Block Design with p Treatments and b Blocks

$$E(y) = \beta_0 + \underbrace{\beta_1 x_1 + \beta_2 x_2 + \cdots + \beta_{p-1} x_{p-1}}_{\text{Treatment terms}}$$

$$+ \underbrace{\beta_p x_p + \beta_{p+1} x_{p+1} + \cdots + \beta_{p+b-2} x_{p+b-2}}_{\text{Block terms}}$$

where

$$x_1 = \{1 \text{ if Treatment 1, 0 if not}\},$$

$$x_2 = \{1 \text{ if Treatment 2, 0 if not}\}, \ldots,$$

$$x_{p-1} = \{1 \text{ if Treatment } p-1, 0 \text{ if not}\}$$
$$x_p = \{1 \text{ if Block 1, 0 if not}\},$$
$$x_{p+1} = \{1 \text{ if Block 2, 0 if not}\},\ldots,$$
$$x_{p+b-2} = \{1 \text{ if Block } b-1, 0 \text{ if not}\}$$

Linear Model for a Complete Factorial Block Design with Factor A at *a* levels and Factor B at *b* levels

$$E(y) = \beta_0 + \underbrace{\beta_1 x_1 + \beta_2 x_2 + \ldots + \beta_{a-1} x_{a-1}}_{\text{A main effect terms}}$$
$$+ \underbrace{\beta_a x_a + \beta_{a+1} x_{a+1} + \cdots + \beta_{a+b-2} x_{a+b-2}}_{\text{B main effect terms}}$$

$$+ \underbrace{\beta_{a+b-1} x_1 x_a + \beta_{a+b} x_1 x_{a+1} + \cdots + \beta_{ab-1} x_{a-1} x_{b-1}}_{\text{A} \times \text{B interaction terms}}$$

where

$$x_1 = \{1 \text{ if A level 1, 0 if not}\},$$
$$x_2 = \{1 \text{ if A level 2, 0 if not}\},\ldots,$$
$$x_{a-1} = \{1 \text{ if A level } a-1, 0 \text{ if not}\}$$
$$x_a = \{1 \text{ if B level 1, 0 if not}\},$$
$$x_{a+1} = \{1 \text{ if B level 2, 0 if not}\},\ldots,$$
$$x_{a+b-2} = \{1 \text{ if B level } b-1, 0 \text{ if not}\}$$

GUIDE TO SELECTING THE EXPERIMENTAL DESIGN

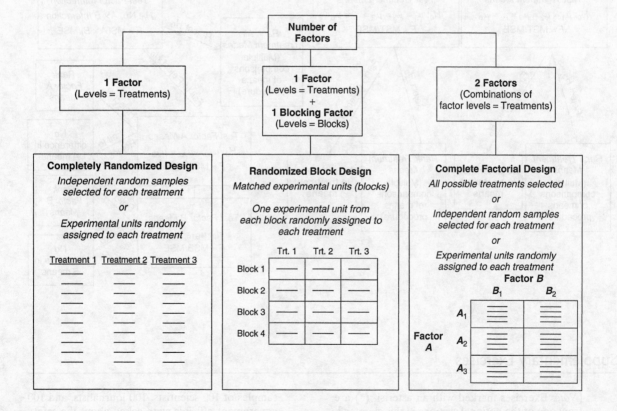

GUIDE TO CONDUCTING ANOVA F-TESTS

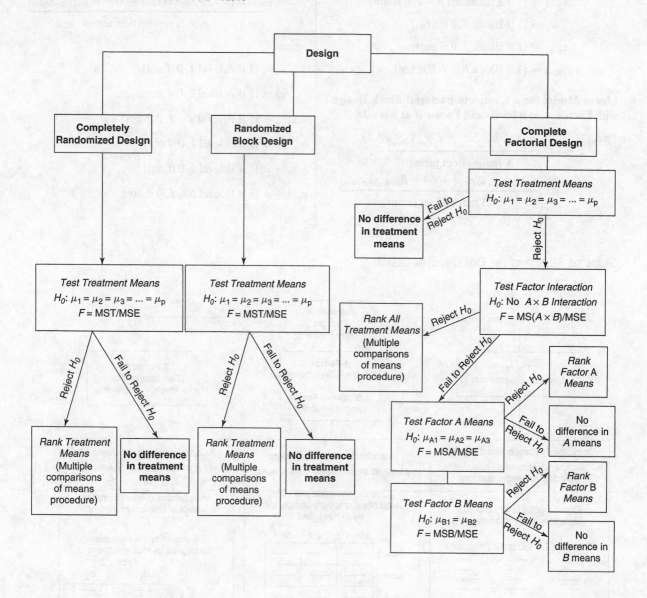

Supplementary Exercises

[*Note:* Exercises marked with an asterisk (*) are from one of the optional sections in this chapter.]

12.72. Nuclear power plant safety. An article in the *American Journal of Political Science* (January 1998) examined the attitudes of three groups of professionals that influence U.S. policy. Random samples of 100 scientists, 100 journalists, and 100 government officials were asked about the safety of nuclear power plants. Responses were made on a seven-point scale, where 1 = very unsafe and 7 = very safe. The mean safety scores for the groups are scientists, 4.1; journalists, 3.7; government officials, 4.2.

(a) Identify the response variable for this study.
(b) How many treatments are included in this study? Describe them.
(c) Specify the null and alternative hypotheses that should be used to investigate whether there are differences in the attitudes of scientists, journalists, and government officials regarding the safety of nuclear power plants.
(d) The MSE for the sample data is 2.355. At least how large must MST be in order to reject the null hypothesis of the test of part a using $\alpha = .05$?
(e) If the MST = 11.280, what is the approximate p-value of the test of part a?
(f) Determine the number of pairwise comparisons of treatment means that can be made in this study.
(g) Using an experimentwise error rate of $\alpha = .05$, Tukey's minimum significant difference for comparing means is .23. Use this information to conduct a multiple comparisons of the safety score means. Fully interpret the results.

12.73. Objectivity in auditing. Objectivity is an essential characteristic of auditing. A study was conducted to investigate whether prior involvement in audit program design impairs an external auditor's objectivity in making decisions about that program (*Accounting and Finance*, November 1993). A sample of 45 auditors was randomly divided into three equal-size groups, designated A/R, A/P, and Control. The A/R group designed an audit program for accounts receivable and evaluated an audit program for accounts payable designed by someone else. The A/P group did the reverse. Finally, the control group merely evaluated the audit programs for both accounts. All 45 auditors were then requested to allocate an additional 15 hours to investigate suspected irregularities in either one or both audit programs. The objective of the experiment was to compare the mean number of hours allocated to accounts receivable for the three groups.

(a) What type of design is used in this study?
(b) Identify the treatments for this design.
(c) A partial ANOVA table is shown below. Complete the table.
(d) Based on the results of the analysis, what inference can the researchers make?

SOURCE	df	SS	MS	F	p-VALUE
Groups	—	71.51	—	—	.01
Error	—	—	7.65		
Total	—	392.98			

(e) The means of the three groups of auditors were ranked using a Tukey multiple comparisons procedure at $\alpha = .05$, as shown here:

Mean number of hours allocated	6.7	7.6	9.7
Group	A/R	C	A/P

At the beginning of the study, the researchers theorized that the A/R group would allocate the least audit effort to receivables and that the A/P group would allocate the most. Formally stated, the researchers believed that $\mu_{AR} < \mu_C < \mu_{AP}$. Do the results support this theory? Explain.

12.74. Hair color and pain. Studies conducted at the University of Melbourne (Australia) indicate that there may be a difference between the pain thresholds of blondes and brunettes. Men and women of various ages were divided into four categories according to hair color: light blond, dark blond, light brunette, and dark brunette. The purpose of the experiment was to determine whether hair color is related to the amount of pain evoked by common types of mishaps and assorted types of trauma. Each person in the experiment was given a pain threshold score based on his or her performance in a pain sensitivity test (the higher the score, the higher the person's pain tolerance). SAS was used to conduct the analysis of variance of the data listed in the table. The SAS printout is also provided (top, p. 702).

🔘 HAIRPAIN

LIGHT BLOND	DARK BLOND	LIGHT BRUNETTE	DARK BRUNETTE
62	63	42	32
60	57	50	39
71	52	41	51
55	41	37	30
48	43		35

(a) Based on the given information, what type of experimental design appears to have been employed?
(b) Using the SAS printout, conduct a test to determine whether the mean pain thresholds differ among people possessing the four types of hair color. Use $\alpha = .05$.
(c) What is the observed significance level for the test in part b? Interpret it.
(d) What assumptions must be met in order to ensure the validity of the inferences you made in part b?

SAS Output for Exercise 12.74

The ANOVA Procedure

Dependent Variable: PAIN

Source	DF	Sum of Squares	Mean Square	F Value	Pr > F
Model	3	1360.726316	453.575439	6.79	0.0041
Error	15	1001.800000	66.786667		
Corrected Total	18	2362.526316			

R-Square	Coeff Var	Root MSE	PAIN Mean
0.575962	17.08184	8.172311	47.84211

Source	DF	Anova SS	Mean Square	F Value	Pr > F
COLOR	3	1360.726316	453.575439	6.79	0.0041

12.75. Facial expression study. What do people infer from facial expressions of emotion? This was the research question of interest in an article published in the *Journal of Nonverbal Behavior* (Fall 1996). A sample of 36 introductory psychology students was randomly divided into six groups. Each group was assigned to view one of six slides showing a person making a facial expression.[*] The six expressions were (1) angry, (2) disgusted, (3) fearful, (4) happy, (5) sad, and (6) neutral faces. After viewing the slides, the students rated the degree of dominance they inferred from the facial expression (on a scale ranging from -15 to $+15$). The data (simulated from summary information provided in the article) are listed in the table.

⊚ FACES

ANGRY	DISGUSTED	FEARFUL	HAPPY	SAD	NEUTRAL
2.10	.40	.82	1.71	.74	1.69
.64	.73	−2.93	−.04	−1.26	−.60
.47	−.07	−.74	1.04	−2.27	−.55
.37	−.25	.79	1.44	−.39	.27
1.62	.89	−.77	1.37	−2.65	−.57
−.08	1.93	−1.60	.59	−.44	−2.16

(a) Conduct an analysis of variance to determine whether the mean dominance ratings differ among the six facial expressions. Use $\alpha = .10$.

(b) Use Tukey's method to rank the dominance rating means of the six facial expressions. (Use an experimentwise error rate of $\alpha = .10$.)

⊚ OILSPILL

12.76. Refer to the *Marine Technology* (January 1995) study of major ocean oil spills by tanker vessels, Exercise 1.88 (p. 73). The spillage amounts (thousands of metric tons) and cause of accident for 48 tankers are saved in the OILSPILL file. (*Note*: Delete the two tankers with oil spills of unknown causes.)

(a) Conduct an analysis of variance (at $\alpha = .01$) to compare the mean spillage amounts for the four accident types: (1) collision, (2) grounding, (3) fire/explosion, and (4) hull failure. Interpret your results.

(b) Use Scheffé's method to rank the spillage amount means of the four accident types. (Use an experimentwise error rate of $\alpha = .01$.)

12.77. Effectiveness of hunting decoys. Using decoys is a common method of hunting waterfowl. A study in the *Journal of Wildlife Management* (July 1995) compared the effectiveness of three different decoy types—taxidermy-mounted decoys, plastic shell decoys, and full-bodied plastic decoys—in attracting Canada geese to sunken pit blinds. In order to account for an extraneous source of variation, three pit blinds were used as blocks in the experiment. Thus, a randomized block design with three treatments (decoy types) and three blocks (pit blinds) was employed. The response variable was the percentage of a goose flock to approach within 46 meters of the pit blind on a given day. The data are given in the table on p. 703.[†] A MINITAB printout of the analysis follows.

---

[*] In the actual experiment, each group viewed all six facial expression slides and the design employed was a Latin Square (beyond the scope of this text).

[†] The actual design employed in the study was more complex than the randomized block design shown here. In the actual study, each number in the table represents the mean daily percentage of goose flocks attracted to the blind, averaged over 13–17 days.

Data and MINITAB Output for Exercise 12.77

🔵 DECOY

BLIND	SHELL	FULL-BODIED	TAXIDERMY-MOUNTED
1	7.3	13.6	17.8
2	12.6	10.4	17.0
3	16.4	23.4	13.6

Source: Harrey, W. F., Hindman, L. J. and Rhodes, W. E. "Vulnerability of Canada geese to taxidermy-mounted decoys," *Journal of Wildlife Management,* Vol. 59, No. 3, July 1995, p. 475 (Table 1). Copyright ©The Wildlife Society.

Two-way ANOVA: PERCENT versus DECOY, BLIND

```
Analysis of Variance for PERCENT
Source    DF      SS      MS      F       P
DECOY      2    30.1    15.0    0.61    0.589
BLIND      2    44.1    22.1    0.89    0.479
Error      4    99.3    24.8
Total      8   173.6
```

(a) Find and interpret the F statistic for comparing the response means of the three decoy types.

(b) Perform the analysis, part a, by fitting and comparing the appropriate linear models. Verify that the results agree.

12.78. Dwarf shrubs and fire. Rugel's pawpaw (yellow squirrel banana) is an endangered species of a dwarf shrub. Biologists from Stetson University conducted an experiment to determine the effects of fire on the shrub's growth (*Florida Scientist,* Spring 1997). Twelve experimental plots of land were selected in a pasture where the shrub is abundant. Within each plot, three pawpaws were randomly selected and treated as follows: one shrub was subjected to fire, another to clipping, and the third was left unmanipulated (a control). After five months, the number of flowers produced by each of the 36 shrubs was determined. The objective of the study was to compare the mean number of flowers produced by pawpaws for the three treatments (fire, clipping, and control).

(a) Identify the type of experimental design employed, including the treatments, response variable, and experimental units.

(b) Illustrate the layout of the design using a graphic similar to Figure 12.9 (p. 626).

(c) Give the linear model appropriate for analyzing the data.

(d) The ANOVA of the data resulted in a test statistic of $F = 5.42$ for treatments with p-value $= .009$. Interpret this result.

(e) The three treatment means were compared using Tukey's method at $\alpha = .05$. Interpret the results shown below.

Mean number of flowers:	1.17	10.58	17.08
Treatment:	Control	Clipping	Burning

12.79. Humor in magazine ads. *Industrial Marketing Management* (1993) published the results of a study of humor in trade magazine advertisements. A sample of 665 ads were categorized according to nationality (U.S., British, or German) and industry (29 categories, ranging from accounting to travel). Then a panel of judges determined the degree of humor in each ad using a 5-point scale (where 1 = not at all humorous and 5 = very humorous). The data were analyzed using a 3×29 factorial ANOVA, where the factors were nationality (at three levels) and industry (at 29 levels). The results of the ANOVA are reported in the accompanying table.*

SOURCE	df	SS	MS	F	p-VALUE
Nationality (N)	2	1.44	.72	2.40	.087
Industry (I)	28	48.00	1.71	5.72	.000
$N \times I$	49	20.28	.41	1.38	.046

Source: Reprinted from *Industrial Marketing Management,* Vol. 22, Issue 1, Lynette S. McCullough and Ronald K. Taylor, "Humor in American, British, and German Ads," pp. 21, Copyright © 1993, with permission from Elsevier.

(a) Using $\alpha = .05$, interpret the ANOVA results. Comment on the order in which the ANOVA F-tests should be conducted and whether any of the tests should be ignored.

(b) According to the researchers, "British ads were more likely to be humorous than German or U.S. ads in the Graphics industry. German ads were least humorous in the Grocery and Mining industries, but funnier than U.S. ads in the Medical industry and funnier than British ads in the Packaging industry." Do these inferences agree or conflict with the conclusions reached in part a? Explain.

12.80. Estimating building drift ratio. A commonly used index to estimate the reliability of a building subjected to lateral loads is the drift ratio. Sophisticated software such as STAAD-III have been developed to estimate the drift ratio based on variables such as beam stiffness, column stiffness,

---

* As a result of missing data, the number of degrees of freedom for Nationality × Industry interaction is less than $2 \times 28 = 56$.

story height, moment of inertia, and so on. Civil engineers performed an experiment to compare drift ratio estimates using STAAD-III with the estimates produced by a new, simpler software program called DRIFT (*Microcomputers in Civil Engineering*, 1993). Data for a 21-story building were used as input to the programs. Two runs were made with STAAD-III: Run 1 considered axial deformation of the building columns, and run 2 neglected this information. The goal of the analysis is to compare the mean drift ratios (where drift is measured as lateral displacement) estimated by the three software runs (the two STAAD-III runs and DRIFT). The lateral displacements (in inches) estimated by the three software programs are recorded in the table for each of five building levels (1, 5, 10, 15, and 21). A MINITAB printout of the analysis of variance for the data is also shown in the next column.

⊚ STAAD

LEVEL	STAAD-III(1)	STAAD-III(2)	DRIFT
1	.17	.16	.16
5	1.35	1.26	1.27
10	3.04	2.76	2.77
15	4.54	3.98	3.99
21	5.94	4.99	5.00

Source: Valles, R. E., et al. "Simplified drift evaluation of wall-frame structures." *Microcomputers in Civil Engineering*, Vol. 8, 1993, p. 242 (Table 2).

(a) Identify the treatments in the experiment.
(b) Because lateral displacement will vary greatly across building levels (floors), a randomized block design will be used to reduce the level-to-level variation in drift. Explain, diagrammatically, the setup of the design if all 21 levels are to be included in the study.
(c) Using the information in the printout, compare the mean drift ratios estimated by the three programs.

MINITAB Output for Exercise 12.80

Two-way ANOVA: DRIFT versus PROGRAM, LEVEL

```
Analysis of Variance for DRIFT
Source     DF       SS        MS        F       P
PROGRAM     2   0.4664    0.2332     4.79   0.043
LEVEL       4  52.1812   13.0453   267.74   0.000
Error       8   0.3898    0.0487
Total      14  53.0374
```

12.81. **Mosquito insecticide study.** A species of Caribbean mosquito is known to be resistant against certain insecticides. The effectiveness of five different types of insecticides—temephos, malathion, fenitrothion, fenthion, and chlorpyrifos—in controlling this mosquito species was investigated in the *Journal of the American Mosquito Control Association* (March 1995). Mosquito larvae were collected from each of seven Caribbean locations. In a laboratory, the larvae from each location were divided into five batches and each batch was exposed to one of the five insecticides. The dosage of insecticide required to kill 50% of the larvae was recorded and divided by the known dosage for a susceptible mosquito strain. The resulting value is called the resistance ratio. (The higher the ratio, the more resistant the mosquito species is to the insecticide relative to the susceptible mosquito strain.) The resistance ratios for the study are listed in the table below. The researchers want to compare the mean resistance ratios of the five insecticides.

(a) Explain why the experimental design is a randomized block design. Identify the treatments and the blocks.
(b) Conduct a complete analysis of the data. Are any of the insecticides more effective than any of the others?

⊚ CARIBMOSQ

LOCATION	INSECTICIDE				
	TEMEPHOS	MALATHION	FENITROTHION	FENTHION	CHLORPYRIFOS
Anguilla	4.6	1.2	1.5	1.8	1.5
Antigua	9.2	2.9	2.0	7.0	2.0
Dominica	7.8	1.4	2.4	4.2	4.1
Guyana	1.7	1.9	2.2	1.5	1.8
Jamaica	3.4	3.7	2.0	1.5	7.1
St. Lucia	6.7	2.7	2.7	4.8	8.7
Suriname	1.4	1.9	2.0	2.1	1.7

Source: Rawlins, S. C., and Oh Hing Wan, J. "Resistance in some Caribbean population of *Aedes aegypti* to several insecticides," *Journal of the American Mosquito Control Association*, Vol. 11, No. 1, Mar. 1995 (Table 1).

12.82. Infants learning an artificial language. *Science* (January 1, 1999) reported on the ability of 7-month-old infants to learn an unfamiliar language. In one experiment, 16 infants were trained in an artificial language. Then, each infant was presented with two three-word sentences that consisted entirely of new words (e.g., "wo fe wo"). One sentence was consistent (i.e., constructed from the same grammar as in the training session) and one sentence was inconsistent (i.e., constructed from grammar in which the infant was not trained). The variable measured in each trial was the time (in seconds) the infant spent listening to the speaker, with the goal to compare the mean listening times of consistent and inconsistent sentences.

(a) The data were analyzed as a randomized block design with the 16 infants representing the blocks and the two sentence types (consistent and inconsistent) representing the treatments. Do you agree with this data analysis method? Explain.

(b) Refer to part a. The test statistic for testing treatments was $F = 25.7$ with an associated observed significance level of $p < .001$. Interpret this result.

(c) Explain why the data could also be analyzed as a paired difference experiment, with a test statistic of $t = 5.07$.

(d) The mean listening times and standard deviations for the two treatments are given here. Use this information to calculate the F statistic for comparing the treatment means in an ANOVA for a completely randomized design. Explain why this test statistic provides weaker evidence of a difference between treatment means than the test in part b.

	CONSISTENT SENTENCES	INCONSISTENT SENTENCES
Mean	6.3	9.0
Standard dev.	2.6	2.16

(e) Explain why there is no need to control the experimentwise error rate when ranking the treatment means for this experiment.

12.83. Alcohol and marital interactions. An experiment was conducted to examine the effects of alcohol on the marital interactions of husbands and wives (*Journal of Abnormal Psychology*, November 1998). A total of 135 couples participated in the experiment. The husband in each couple was classified as aggressive (60 husbands) or nonaggressive (75 husbands), based on an interview and his response to a questionnaire. Before the marital interactions of the couples were observed, each husband was randomly assigned to three groups:

receive no alcohol, receive several alcoholic mixed drinks, or receive placebos (nonalcoholic drinks disguised as mixed drinks). Consequently, a 2×3 factorial design was employed, with husband's aggressiveness at two levels (aggressive or nonaggressive) and husband's alcohol condition at three levels (no alcohol, alcohol, and placebo). The response variable observed during the marital interaction was severity of conflict (measured on a 100-point scale).

(a) A partial ANOVA table is shown below. Fill in the missing degrees of freedom.

(b) Interpret the p-value of the F-test for Aggressiveness.

(c) Interpret the p-value of the F-test for Alcohol Condition.

(d) The F-test for interaction was omitted from the article. Discuss the dangers of making inferences based on the tests, parts a and b, without knowing the result of the interaction test.

SOURCE	df	F	p-VALUE
Aggressiveness (A)	—	16.43	< .001
Alcohol Condition (C)	—	6.00	< .01
A × C	—	—	—
Error	129		
Total	—		

12.84. Time-of-day pricing for electricity. *Time-of-day pricing* is a plan by which customers are charged a lower rate for using electricity during off-peak (less demanded) hours. One experiment (reported in the *Journal of Consumer Research*, June 1982) was conducted to measure customer satisfaction with several time-of-day pricing schemes. The experiment consisted of two factors, price ratio (the ratio of peak to off-peak prices) and peak period length, each at three levels. The $3 \times 3 = 9$ combinations of price ratio and peak period length represent the nine time-of-day pricing schemes. For each pricing scheme, customers were randomly selected and asked to rate satisfaction with the plan using an index from 10 to 38, with 38 indicating extreme satisfaction. Suppose four customers were sampled for each pricing scheme. The table on p. 706 gives the satisfaction scores for these customers. [*Note*: The data are based on mean scores provided in the *Journal of Consumer Research* article.]

(a) Use a statistical software package to conduct an analysis of variance of the data. Report the results in an ANOVA table.

(b) Compute the nine customer satisfaction index means.

(c) Plot the nine means from part b on a graph similar to Figure 12.15 (p. 644). Does it appear

TIMEOFDAY

		PRICING RATIO					
		2:1		4:1		8:1	
PEAK PERIOD LENGTH	6 hours	25	28	31	29	24	28
		26	27	26	27	25	26
	9 hours	26	27	25	24	33	28
		29	30	30	26	25	27
	12 hours	22	20	33	27	30	31
		25	21	25	27	26	27

that the two factors, price ratio and peak period length, interact? Explain.

(d) Do the data provide sufficient evidence of interaction between price ratio and peak period length? Test using $\alpha = .05$.

(e) Do the data provide sufficient evidence that mean customer satisfaction differs for the three peak period lengths? Test using $\alpha = .05$.

(f) When is the test of part e appropriate?

(g) Find a 90% confidence interval for the mean customer satisfaction rating of a pricing scheme with a peak period length of 9 hours and pricing ratio of 2:1.

(h) Find a 95% confidence interval for the difference between the mean customer satisfaction ratings of pricing schemes 9 hours, 8:1 ratio and 6 hours, 8:1 ratio. Interpret the interval.

(i) Use Tukey's multiple comparisons procedure to compare the mean satisfaction scores for the three peak period lengths under each of the three pricing ratios. Identify the means that appear to differ under each pricing ratio. Use $\alpha = .01$.

12.85. **Are you lucky?** Parapsychologists define "lucky" people as individuals who report that seemingly chance events consistently tend to work out in their favor. A team of British psychologists designed a study to examine the effects of luckiness and competition on performance in a guessing task (*The Journal of Parapsychology*, March 1997). Each in a sample of 56 college students was classified as lucky, unlucky, or uncertain based on their responses to a Luckiness Questionnaire. In addition, the participants were randomly assigned to either a competitive or noncompetitive condition. All students were then asked to guess the outcomes of 50 flips of a coin. The response variable measured was percentage of coin-flips correctly guessed.

(a) An ANOVA for a 2×3 factorial design was conducted on the data. Identify the factors and their levels for this design.

(b) The results of the ANOVA are summarized in the next table. Fully interpret the results.

SOURCE	df	F	p-VALUE
Luckiness (L)	2	1.39	.26
Competition (C)	1	2.84	.10
L × C	2	0.72	.72
Error	50		
Total	55		

12.86. **Impact of vitamin B supplement.** In the *Journal of Nutrition* (July 1995), University of Georgia researchers examined the impact of a vitamin B supplement (nicotinamide) on the kidney. The experimental "subjects" were 28 Zucker rats—a species that tends to develop kidney problems. Half of the rats were classified as obese and half as lean. Within each group, half were randomly assigned to receive a vitamin B–supplemented diet and half were not. Thus, a 2×2 factorial experiment was conducted with seven rats assigned to each of the four combinations of size (lean or obese) and diet (supplemental or not). One of the response variables measured was weight (in grams) of the kidney at the end of a 20-week feeding period. The data (simulated from summary information provided in the journal article) are shown in the table.

VITAMINB

		DIET			
		REGULAR		VITAMIN B SUPPLEMENT	
RAT SIZE	LEAN	1.62	1.47	1.51	1.63
		1.80	1.37	1.65	1.35
		1.71	1.71	1.45	1.66
		1.81		1.44	
	OBESE	2.35	2.84	2.93	2.63
		2.97	2.05	2.72	2.61
		2.54	2.82	2.99	2.64
		2.93		2.19	

(a) Conduct an analysis of variance on the data. Summarize the results in an ANOVA table.

(b) Conduct the appropriate ANOVA F-tests at $\alpha = .01$. Interpret the results.

12.87. **Steam explosion of peat.** The steam explosion of peat yields fermentable carbohydrates that have a number of potentially important industrial uses. A study of the steam explosion process was initiated to determine the optimum conditions for the release of fermentable carbohydrate (*Biotechnology and Bioengineering*, February 1986). Triplicate samples of peat were treated for .5, 1.0, 2.0, 3.0, and 5.0 minutes at 170°, 200°, and 215°C, in the steam explosion process. Thus, the experiment consists of two factors—temperature at three levels and treatment time at five levels. The accompanying table gives the percentage of carbohydrate solubilized for each of the $3 \times 5 = 15$ peat samples.

🔘 PEAT

TEMPERATURE °C	TIME MINUTES	CARBOHYDRATE SOLUBILIZED %
170	.5	1.3
170	1.0	1.8
170	2.0	3.2
170	3.0	4.9
170	5.0	11.7
200	.5	9.2
200	1.0	17.3
200	2.0	18.1
200	3.0	18.1
200	5.0	18.8
215	.5	12.4
215	1.0	20.4
215	2.0	17.3
215	3.0	16.0
215	5.0	15.3

Source: Forsberg, C. W. et al. "The release of fermentable carbohydrate from peat by steam explosion and its use in the microbial production of solvents." *Biotechnology and Bioengineering,* Vol. 28, No.2, Feb. 1986 p. 179 (Table 1).

(a) What type of experimental design was used?
(b) Explain why the traditional analysis of variance formulas are inappropriate for the analysis of these data.
(c) Write a second-order model relating mean amount of carbohydrate solubilized, $E(y)$, to temperature (x_1) and time (x_2).
(d) Explain how you could test the hypothesis that the two factors, temperature (x_1) and time (x_2), interact.
(e) Fit the model and perform the test for interaction.

12.88. Effects of acid rain on soil pH. *Acid rain* is formed by the combination of water vapor in clouds with nitrous oxide and sulfur dioxide, which are among the emissions products of coal and oil combustion. To determine the effects of acid rain on soil pH in a natural ecosystem, engineers at the University of Florida's Institute of Food and Agricultural Sciences irrigated experimental plots near Gainesville, Florida, with rainwater at two pH levels, 3.7 and 4.5. The acidity of the soil was then measured at three different depths, 0–15, 15–30, and 30–46 centimeters. Tests were conducted during three different time periods. The resulting soil pH values are shown in the table. Treat the experiment as a 2×3 factorial laid out in three blocks, where the factors are acid rain at two pH levels

🔘 ACIDRAIN

		APRIL 3 ACID RAIN pH		JUNE 16 ACID RAIN pH		JUNE 30 ACID RAIN pH	
		3.7	4.5	3.7	4.5	3.7	4.5
	0–15	5.33	5.33	5.47	5.47	5.20	5.13
SOIL DEPTH, cm	15–30	5.27	5.03	5.50	5.53	5.33	5.20
	30–46	5.37	5.40	5.80	5.60	5.33	5.17

Source: "Acid rain linked to growth of coal-fired power," *Florida Agricultural Research,* 83, Vol. 2, No. 1, Winter 1983.

SAS Output for Exercise 12.88

The ANOVA Procedure

Dependent Variable: SOILPH

Source	DF	Sum of Squares	Mean Square	F Value	Pr > F
Model	7	0.48685556	0.06955079	6.99	0.0034
Error	10	0.09952222	0.00995222		
Corrected Total	17	0.58637778			

R-Square	Coeff Var	Root MSE	SOILPH Mean
0.830276	1.861595	0.099761	5.358889

Source	DF	Anova SS	Mean Square	F Value	Pr > F
DEPTH	2	0.06714444	0.03357222	3.37	0.0759
RAINPH	1	0.03042222	0.03042222	3.06	0.1110
DEPTH*RAINPH	2	0.00781111	0.00390556	0.39	0.6854
DATE	2	0.38147778	0.19073889	19.17	0.0004

and soil depth at three levels, and the blocks are the three time periods. The SAS printout for the analysis of variance is provided on p. 707.

(a) Write the equation of the linear model appropriate for analyzing the data.

(b) Is there evidence of an interaction between pH level of acid rain and soil depth? Test using $\alpha = .05$.

(c) Conduct a test to determine whether blocking over time was effective in removing an extraneous source of variation. Use $\alpha = .05$.

12.89. Cerebral specialization in Down syndrome adults. Most people are right-handed due to the propensity of the left hemisphere of the brain to control sequential movement. Similarly, the fact that some tasks are performed better with the left hand is likely due to the superiority of the right hemisphere of the brain to process the necessary information. Does such cerebral specialization in spatial processing occur in adults with Down syndrome? A 2×2 factorial experiment was conducted to answer this question (*American Journal on Mental Retardation*, May 1995). A sample of adults with Down syndrome was compared to a control group of normal individuals of a similar age. Thus, one factor was Group at two levels (Down syndrome and control) and the second factor was the Handedness (left or right) of the subject. All the subjects performed a task that typically yields a left-hand advantage. The response variable was "laterality index," measured on a range from -100 to 100 points. (A large positive index indicates a right-hand advantage, while a large negative index indicates a left-hand advantage.)

(a) Identify the treatments in this experiment.

(b) Construct a graph that would support a finding of no interaction between the two factors.

(c) Construct a graph that would support a finding of interaction between the two factors.

(d) The F-test for factor interaction yielded an observed significance level of $p < .05$. Interpret this result.

(e) Multiple comparisons of all pairs of treatment means yielded the rankings shown below. Interpret the results.

(f) The experimentwise error rate for the analysis, part e, was $\alpha = .05$. Interpret this value.

Mean laterality index:	-30	-4	$-.5$	$+.5$
Group/Handed:	Down/ Left	Control/ Right	Control/ Left	Down/ Right

12.90. Job satisfaction of truck drivers. Turnover among truck drivers is a major problem for both carriers and shippers. Since knowledge of driver-related

job attitudes is valuable for predicting and controlling future turnover, a study of the work-related attitudes of truck drivers was conducted (*Transportation Journal*, Fall 1993). The two factors considered in the study were career stage and time spent on road. Career stage was set at three levels: early (less than 2 years), mid-career (between 3 and 10 years), and late (more than 10 years). Road time was dichotomized as short (gone for one weekend or less) and long (gone for longer than one weekend). Data were collected on job satisfaction for drivers sampled in each of the $3 \times 2 = 6$ combinations of career stage and road time. [Job satisfaction was measured on a 5-point scale, where 1 = really dislike and 5 = really like.]

(a) Identify the response variable for this experiment.

(b) Identify the factors for this experiment.

(c) Identify the treatments for this experiment.

(d) The ANOVA table for the analysis is shown here. Fully interpret the results.

SOURCE	F-VALUE	p-VALUE
Career stage (*CS*)	26.67	$p \le .001$
Road time (*RT*)	.19	$p > .05$
CS $\times$ *RT*	1.59	$p < .05$

Source: McElroy, J. C., et al. "Career stage, time spent on the road, and truckload driver attitudes," *Transportation Journal*, Vol. 33, No. 1, Fall 1993, p. 10 (Table 2).

(e) The researchers theorized that the impact of road time on job satisfaction may be different depending on the career stage of the driver. Do the results support this theory?

(f) The researchers also theorized that career stage affects the job satisfaction of truck drivers. Do the results support this theory?

*(g) Since career stage was found to be the only significant factor in the 3×2 factorial ANOVA, the mean job satisfaction levels of the three career stages (early, middle, and late) were compared using the Bonferroni method. Find the adjusted α level to use in the analysis, if the researchers desire an overall significance level of $\alpha = .09$.

*(h) The sample mean job satisfaction levels for the three career stages are given here. Assuming equal sample sizes for each stage and a Bonferroni critical difference of $B = .06$, rank the means.

Mean job satisfaction	3.47	3.38	3.36
Career stage	Early	Middle	Late

12.91. Performance of different organization types. In business, the prevailing theory is that companies can be categorized into one of four types based on their strategic profile: *reactors*, which are dominated by industry forces; *defenders*, which specialize in lowering costs for established products while maintaining quality; *prospectors*, which develop new/improved products; and *analyzers*, which operate in two product areas—one stable and the other dynamic. The *American Business Review* (January 1990) reported on a study that proposes a fifth organization type, *balancers*, which operate in three product spheres—one stable and two dynamic. Each firm in a sample of 78 glassware firms was categorized as one of these five types, and the level of performance (process research and development ratio) of each was measured.

(a) A completely randomized design ANOVA of the data resulted in a significant (at $\alpha = .05$) F-value for treatments (organization types). Interpret this result.

(b) Multiple comparisons of the five mean performance levels (using Tukey's procedure at $\alpha = .05$) are summarized in the following table. Interpret the results.

Mean	.138	.235	.820	.826	.911
Type	Reactor	Prospector	Defender	Analyzer	Balancer

Source: Wright, P., et al. "Business performance and conduct of organization types: A study of select special-purpose and laboratory glassware firms," *American Business Review*, Jan. 1990, p. 95 (Table 4).

12.92. Removing water from paper. The percentage of water removed from paper as it passes through a dryer depends on the temperature of the dryer and the speed of the paper passing through it. A laboratory experiment was conducted to investigate the relationship between dryer temperature T at three levels (100°, 120°, and 140°F) and exposure time E (which is related to speed) also at three levels (10, 20, and 30 seconds). Four paper specimens were prepared for each of the $3 \times 3 = 9$ conditions. The data (percentage of water removed) are shown in the accompanying table. Carry out a complete analysis of the data using an available statistical software package.

⊚ DRYPAPER

		TEMPERATURE (T)					
		100		120		140	
	10	24	26	33	33	45	49
		21	25	36	32	44	45
EXPOSURE	20	39	34	51	50	67	64
TIME (E)		37	40	47	52	68	65
	30	58	55	75	71	89	87
		56	53	70	73	86	83

12.93. Auditor risk study. *Accounting Review* (January 1991) reported on a study of the effect of two factors, confirmation of accounts receivable and verification of sales transactions, on account misstatement risk by auditors. Both factors were held at the same two levels: completed or not completed. Thus, the experimental design is a 2×2 factorial design.

(a) Identify the factors, factor levels, and treatments for this experiment.

(b) Explain what factor interaction means for this experiment.

(c) A graph of the hypothetical mean misstatement risks for each of the $2 \times 2 = 4$ treatments is shown here. In this hypothetical case, does it appear that interaction exists?

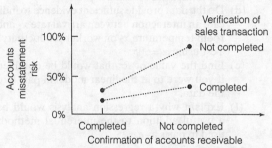

Source: Brown, C. E. and Solomon, I. "Configural information processing in auditing: The role of domain-specific knowledge," *Accounting Review*, Vol. 66, No. 1, Jan 1991, p. 105 (Figure 1).

12.94. Factors that impact productivity of an assembly line. A production manager who supervises an assembly operation wants to investigate the effect of the incoming rate (parts per minute), x_1, of components and room temperature, x_2, on the productivity (number of items produced per minute), y. The component parts approach the worker on a belt and return to the worker if not selected on the first trip past the assembly point. It is thought that an increase in the arrival rate of components has a positive effect on the assembly rate, up to a point, after which increases may annoy the assembler and reduce productivity. Similarly, it is suspected that lowering the room temperature is beneficial to a point, after which reductions may reduce productivity. The experimenter used the same assembly position for each worker. Thirty-two workers were used for the experiment, two each assigned to the 16 factor level combinations of a 4×4 factorial experiment. The data, in parts per minute averaged over a 5-minute period, are shown in the table at the top of p. 710.

(a) Perform an analysis of variance for the data. Display the computed quantities in an ANOVA table.

⊙ ASSEMBLY1

		RATE OF INCOMING COMPONENTS (x_1), PARTS PER MINUTE						
		40		50		60		70
ROOM	65	24.0, 23.8		25.6, 25.4		29.2, 29.4		28.4, 27.6
TEMPERATURE	70	25.0, 26.0		28.8, 28.8		31.6, 32.0		30.2, 30.0
(x_2),° F	75	25.6, 25.0		27.6, 28.0		29.8, 28.6		28.0, 27.0
	80	24.0, 24.6		27.6, 26.2		27.6, 28.6		26.0, 24.4

(b) Write the linear model implied by the analysis of variance. [*Hint:* For a quantitative variable recorded at four levels, main effects include terms for x, x^2, and x^3.]

(c) Do the data provide sufficient evidence to indicate differences among the mean responses for the 16 treatments of the 4×4 factorial experiment? Test using $\alpha = .05$.

(d) Do the data provide sufficient evidence to indicate an interaction between arrival rate x_1 and room temperature x_2 on worker productivity? Test using $\alpha = .05$.

(e) Find the value of R^2 that would be obtained if you were to fit the linear model in part b to the data.

(f) Explain why a regression analysis would be a useful addition to the inferential methods used in parts a–e.

12.95. **Factors that impact productivity of an assembly line (cont'd).** A second-order model would be a reasonable choice to model the data in Exercise 12.94. To simplify the analysis, we will code the arrival rate and temperature values as follows:

$$x_1 = \frac{\text{Arrival rate} - 55}{5} \qquad x_2 = \frac{\text{Temperature} - 72.5}{2.5}$$

A printout of the SAS regression analysis is shown below.

(a) Write the second-order model for the response. Note the difference between this model and the ANOVA model in Exercise 12.94b.

(b) Give the prediction relating the response y to the coded independent variables x_1 and x_2.

(c) Why does the SSE given in the computer printout differ from the SSE obtained in Exercise 12.94?

(d) Find the value of R^2 appropriate for your second-order model and interpret its value.

(e) Do the data provide sufficient evidence to indicate that the complete factorial model provides more information for predicting y than a second-order model?

*12.96 **Factors that impact productivity of an assembly line (cont'd).** Consider a second experiment designed to investigate the effect of arrival rate of product components, (x_1) and temperature of the room (x_2) on the length of time (y) required by individual workers to perform a product assembly operation. Each factor will be held at two levels: arrival rate at .5 and 1.0 component per second,

SAS Output for Exercise 12.95

Dependent Variable: PPM5

Analysis of Variance

Source	DF	Sum of Squares	Mean Square	F Value	Pr > F
Model	5	130.80680	26.16136	27.73	<.0001
Error	26	24.53320	0.94358		
Corrected Total	31	155.34000			

Root MSE	0.97138	R-Square	0.8421	
Dependent Mean	27.32500	Adj R-Sq	0.8117	
Coeff Var	3.55492			

Parameter Estimates

Variable	DF	Parameter Estimate	Standard Error	t Value	Pr > |t|
Intercept	1	29.85625	0.34876	85.61	<.0001
X1	1	0.56000	0.07679	7.29	<.0001
X2	1	-0.16250	0.07679	-2.12	0.0441
X1X2	1	-0.11350	0.03434	-3.30	0.0028
X1SQ	1	-0.27500	0.04293	-6.41	<.0001
X2SQ	1	-0.23125	0.04293	-5.39	<.0001

⊙ ASSEMBLY2

				WORKER									
				1	2	3	4	5	6	7	8	9	10
	70°F	ARRIVAL	.5	1.7	1.3	1.7	2.0	2.0	2.3	2.0	2.8	1.5	1.6
ROOM		RATE	1.0	.8	.8	1.5	1.2	1.2	1.7	1.1	1.5	.5	1.0
TEMPERATURE		(component	.5	1.3	1.5	2.3	1.6	2.2	2.1	1.8	2.4	1.3	1.8
	80°F	per second)	1.0	1.8	1.5	2.3	2.0	2.7	2.2	2.3	2.6	1.3	1.8

and temperature at 70° and 80°F. Thus, a 2×2 factorial experiment will be employed. To block out worker-to-worker variability, each of 10 randomly selected workers will be required to assemble the product under all four experimental conditions. Therefore, the four treatments (working conditions) will be assigned to the experimental units (workers) using a randomized block design, where the workers represent the blocks.

(a) Write the appropriate complete model for the randomized block design, with arrival rate and temperature as quantitative factors.

(b) The assembly time data for the 2×2 factorial experiment with a randomized block design are given in the table above. Use regression to determine if a difference exists among the four treatment means.

(c) Does the effect of a change in arrival rate on assembly time depend on temperature (i.e., do arrival rate and temperature interact)?

(d) Estimate the mean loss (or gain) in assembly time as arrival rate is increased from .5 to 1.0 component per second and temperature is held at 70°F. What inference can you make based on this estimate?

12.97. Availability of duck food in summer. Ducks inhabiting the Great Salt Lake marshes feed on a variety of animals, including water boatmen, brine shrimp, beetles, and snails. The changes in the availability of these animal species for ducks during the summer was investigated (*Wetlands*, March 1995). The goal was to compare the mean amount (measured as biomass) of a particular duck food species across four different summer time periods: (1) July 9–23, (2) July 24–Aug. 8, (3) Aug. 9–23, and (4) Aug. 24–31. Ten stations in the marshes were randomly selected, and the biomass density in a water specimen collected from each was measured. Biomass measurements (milligrams per square meter) were collected during each of the four summer time periods at each station, with stations treated as a blocking factor. Thus, the data were analyzed as a randomized block design.

(a) Fill in the missing degrees of freedom in the randomized block ANOVA table shown in the next column.

(b) The F-value (and corresponding p-value) shown in the ANOVA table, part a, were

computed from an analysis of biomass of water boatmen nymphs (a common duck food). Interpret these results.

SOURCE	df	F	p-VALUE
Time Period	—	11.25	.0001
Station	—	—	—
Error	—		
Total	39		

(c) A multiple comparisons of time period means was conducted using an experimentwise error rate of .05. The results are summarized below. Identify the time period(s) with the largest and smallest mean biomass.

Mean biomass
  (mg/m^2):   19   54.5   90   148
Time period: 8/24–8/31   8/9–8/23   7/24–8/8   7/9–7/23

*12.98 Light output of flashbulbs. A $2 \times 2 \times 2 \times 2 = 2^4$ factorial experiment was conducted to investigate the effect of four factors on the light output, y, of flashbulbs. Two observations were taken for each of the factorial treatments. The factors are amount of foil contained in a bulb (100 and 120 milligrams); speed of sealing machine (1.2 and 1.3 revolutions per minute); shift (day or night); and machine operator (A or B). The data for the two replications of the 2^4 factorial experiment are shown in the table on p. 712. To simplify computations, define

$$x_1 = \frac{\text{Amount of foil} - 110}{10}$$

$$x_2 = \frac{\text{Speed of machine} - 1.25}{.05}$$

so that x_1 and x_2 will take values -1 and $+1$. Also, define

$$x_3 = \begin{cases} -1 & \text{if night shift} \\ 1 & \text{if day shift} \end{cases} \quad x_4 = \begin{cases} -1 & \text{if machine operator B} \\ 1 & \text{if machine operator A} \end{cases}$$

(a) Write the complete factorial model for y as a function of x_1, x_2, x_3, and x_4.

(b) How many degrees of freedom will be available for estimating σ^2?

⊚ FLASHBULB

		AMOUNT OF FOIL			
		100 milligrams		120 milligrams	
		SPEED OF MACHINE			
		1.2 rpm	1.3 rpm	1.2 rpm	1.3 rpm
DAY	Operator B	6; 5	5; 4	16; 14	13; 14
SHIFT	Operator A	7; 5	6; 5	16; 17	16; 15
NIGHT	Operator B	8; 6	7; 5	15; 14	17; 14
SHIFT	Operator A	5; 4	4; 3	15; 13	13; 14

(c) Do the data provide sufficient evidence (at $\alpha = .05$) to indicate that any of the factors contribute information for the prediction of y?

(d) Identify the factors that appear to affect the amount of light y in the flashbulbs.

12.99. Detecting early failure of transformer parts. A trade-off study regarding the inspection and test of transformer parts was conducted by the quality department of a major defense contractor. The investigation was structured to examine the effects of varying inspection levels and incoming test times to detect early part failure or fatigue. The levels of inspection selected were full military inspection (A), reduced military specification level (B), and commercial grade (C). Operational burn-in test times chosen for this study were at 1-hour increments from 1 hour to 9 hours. The response was failures per thousand pieces obtained from samples taken from lot sizes inspected to a specified level and burned-in over a prescribed time length. Three replications were randomly sequenced under each condition, making this a complete 3×9 factorial experiment (a total of 81 observations). The data are shown in the table below. Analyze the data and interpret the results.

⊚ BURNIN

	INSPECTION LEVELS								
BURN-IN, hours	Full Military Specification, A			Reduced Military Specification, B			Commercial, C		
1	7.60	7.50	7.67	7.70	7.10	7.20	6.16	6.13	6.21
2	6.54	7.46	6.84	5.85	6.15	6.15	6.21	5.50	5.64
3	6.53	5.85	6.38	5.30	5.60	5.80	5.41	5.45	5.35
4	5.66	5.98	5.37	5.38	5.27	5.29	5.68	5.47	5.84
5	5.00	5.27	5.39	4.85	4.99	4.98	5.65	6.00	6.15
6	4.20	3.60	4.20	4.50	4.56	4.50	6.70	6.72	6.54
7	3.66	3.92	4.22	3.97	3.90	3.84	7.90	7.47	7.70
8	3.76	3.68	3.80	4.37	3.86	4.46	8.40	8.60	7.90
9	3.46	3.55	3.45	5.25	5.63	5.25	8.82	9.76	9.52

Source: Danny La Nuez, former graduate student, College of Business Administration, University of South Florida.

References

Box, G. E. P., Hunter, W. G., and Hunter, J. S. *Statistics for Experimenters: Design, Innovation, and Discovery*, 2nd ed. New York: Wiley, 2005.

Cochran, W. G., and Cox, G. M. *Experimental Designs*, 2nd ed. New York: Wiley, 1992 (paper back).

Hicks, C. R., and Turner, K. V. *Fundamental Concepts in the Design of Experiments*, 5th ed. New York: Oxford University Press, 1999.

Hochberg, Y., and Tamhane, A. C. *Multiple Comparison Procedures*. New York: Wiley, 1987

Hsu, J. C. *Multiple Comparisons, Theory and Methods.* New York: Chapman & Hall, 1996.

Johnson, R., and Wichern, D. *Applied Multivariate Statistical Methods*, 6th ed. Upper Saddle River, N. J.: Prentice Hall, 2007.

Kirk, R. E. *Experimental Design: Procedures for Behavioral Sciences* 3rd ed. Pacific Grove, Calif.: Brooks/Cole, 1995.

Kramer, C. Y. "Extension of multiple range tests to group means with unequal number of replications." *Biometrics*, Vol. 12, 1956, pp. 307–310.

Kutner, M., Nachtsheim, C., Neter, J., and Li, W. *Applied Linear Statistical Models*, 5th ed. Homewood, Ill.: Richard D. Irwin, 2004.

Levene, H. *Contributions to Probability and Statistics*. Stanford, Calif.: Stanford University Press, 1960, pp. 278–292.

Mason, R. L., Gunst, R. F., and Hess, J. L. *Statistical Design and Analysis of Experiments*, New York: John Wiley & Sons, 1989.

Mendenhall, W. *Introduction to Linear Models and the Design and Analysis of Experiments*. Belmont, Calif.: Wadsworth, 1968.

Miller, R. G. *Simultaneous Statistical Inference*, 2nd ed. New York: Springer-Verlag, 1981.

Montgomery, D. C. *Design and Analysis of Experiments*, 6th ed. New York: John Wiley & Sons, 2004.

Scheffe, H. "A method for judging all contrasts in the analysis of variance." *Biometrika*, Vol. 40, 1953, pp. 87–104.

Scheffe, H. *The Analysis of Variance*. New York: Wiley, 1959.

Searle, S. R., Casella, G., and McCulloch, C. E. *Variance Components*. New York: Wiley, 2006 (paperback).

Tukey, J. W. "Comparing individual means in the analysis of variance." *Biometrics*, Vol. 5, 1949, pp. 99–114.

Uusipaikka, E. "Exact simultaneous confidence intervals for multiple comparisons among three or four mean values." *Journal of the American Statistical Association*, Vol. 80, 1985, pp. 196–201.

Winer, B. J., Brown, D. R., and Michels, K. M. *Statistical Principals in Experimental Design*, 3rd ed. New York: McGraw-Hill, 1991.

RELUCTANCE TO TRANSMIT BAD NEWS: THE MUM EFFECT

The Problem

Over 30 years ago, psychologists S. Rosen and A. Tesser found that people were reluctant to transmit bad news to peers in a nonprofessional setting. Rosen and Tesser termed this phenomenon the "MUM effect."* Since that time, numerous studies have investigated the impact of the MUM effect in a professional setting (e.g., on doctor–patient relationships, organizational functioning, and group psychotherapy). The consensus: The reluctance to transmit bad news continues to be a major professional concern.

Why do people keep mum when given an opportunity to transmit bad news to others? Two theories have emerged from this research. The first maintains that the MUM effect is an *aversion to private discomfort*. To avoid discomforts such as empathy with the victim's distress or guilt feelings for their own good fortune, would-be communicators of bad news keep mum. The second theory is that the MUM effect is a *public display*. People experience little or no discomfort when transmitting bad news, but keep mum to avoid an unfavorable impression or to pay homage to a social norm.

 MUM The subject of this case study is an article by C. F. Bond and E. L. Anderson (*Journal of Experimental Social Psychology*, Vol. 23, 1987). Bond and Anderson conducted a controlled experiment to determine which of the two explanations for the MUM effect is more plausible. "If the MUM effect is an aversion to private discomfort," they state, "subjects should show the effect whether or not they are visible [to the victim]. If the effect is a public display, it should be stronger if the subject is visible than if the subject cannot be seen."

The Design

Forty undergraduates (25 males and 15 females) at Duke University participated in the experiment to fulfill an introductory psychology course requirement. Each subject was asked to administer an IQ test to another student and then provide the test taker with his or her percentile score. Unknown to the subject, the test taker was a confederate student working with the researchers.

The experiment manipulated two factors, *subject visibility* and *confederate success*, each at two levels. Subject visibility was manipulated by giving written

---

*Rosen, S., and Tesser, A. "On reluctance to communicate undesirable information: The MUM effect." *Journal of Communication*, Vol. 22, 1970, pp. 124–141.

MUM

Table CS7.1 2 × 2 Factorial design

		CONFEDERATE SUCCESS	
		Success	Failure
SUBJECT VISIBILITY	Visible (VS)	Subject 1 2 ⋮ 10	Subject 21 22 ⋮ 30 (VF)
	Not Visible (NS)	Subject 11 12 ⋮ 20	Subject 31 32 ⋮ 40 (NF)

instructions to each subject. Some subjects were told that they were *visible* to the test taker through a glass plate and the others were told that they were *not visible* through a one-way mirror. Confederate success was manipulated by supplying the subject with one of two bogus answer keys. With one answer key, the confederate would always seem to succeed at the test, placing him or her in the top 20% of all Duke undergraduates; when the other answer key was used, the confederate would always seem to fail, ranking in the bottom 20%.

Ten subjects were randomly assigned to each of the 2 × 2 = 4 experimental conditions; thus, a 2 × 2 factorial design with 10 replications was employed. The design is diagrammed in Table CS7.1. For convenience, we use the letters NS, NF, VS, and VF to represent the four treatments.

One of several behavioral variables that were measured during the experiment was *latency to feedback*, defined as time (in seconds) between the end of the test and delivery of feedback (i.e., the percentile score) from the subject to the test taker. This case focuses on an analysis of variance of the dependent variable, latency to feedback. The longer it takes the subject to deliver the score, presumably the greater the MUM effect. The experimental data are saved in the MUM file.* With an analysis of this data, the researchers hope to determine whether either one of the two factors, subject visibility or confederate success, has an impact on the MUM effect, and, if so, whether the factors are independent.

Analysis of Variance Models and Results

Since both factors, subject visibility and confederate success, are qualitative, the complete model for this 2 × 2 factorial experiment is written as follows.

$$\textit{Complete model}: \quad E(y) = \beta_0 + \underbrace{\beta_1 x_1}_{\substack{\text{Visibility} \\ \text{main} \\ \text{effect}}} + \underbrace{\beta_2 x_2}_{\substack{\text{Success} \\ \text{main} \\ \text{effect}}} + \underbrace{\beta_3 x_1 x_2}_{\substack{\text{Visibility} \times \text{Success} \\ \text{interaction}}}$$

---

* The data are simulated based on summary results reported in the journal article.

Figure CS7.1 SAS ANOVA printout for the 2×2 factorial

```
                              The GLM Procedure
Dependent Variable: FEEDBACK
                                    Sum of
    Source                DF       Squares      Mean Square    F Value    Pr > F
    Model                  3     37371.27500    12457.09167     37.90     <.0001
    Error                 36     11833.70000      328.71389
    Corrected Total       39     49204.97500

            R-Square      Coeff Var      Root MSE      FEEDBACK Mean
            0.759502      19.03961       18.13047        95.22500

    Source                DF     Type III SS    Mean Square    F Value    Pr > F
    SUBJECT                1      8381.02500     8381.02500      25.50     <.0001
    CONFED                 1      8151.02500     8151.02500      24.80     <.0001
    SUBJECT*CONFED         1     20839.22500    20839.22500      63.40     <.0001
```

where $y = $ Latency to feedback

$$x_1 = \begin{cases} 1 & \text{if subject visible} \\ 0 & \text{if not} \end{cases} \qquad x_2 = \begin{cases} 1 & \text{if confederate success} \\ 0 & \text{if confederate failure} \end{cases}$$

To test for factor interaction, we can compare the complete model to the reduced model

Reduced model: $E(y) = \beta_0 + \beta_1 x_1 + \beta_2 x_2$

using the partial F-test, or, equivalently, we can conduct a t-test on the interaction parameter, β_3. Either way, the null hypothesis to be tested is

H_0: $\beta_3 = 0$

Alternatively, we can use the ANOVA routine of a statistical software package to conduct the test.

The SAS ANOVA printout is shown in Figure CS7.1. From the highlighted portions of the printout, we form the ANOVA table displayed in Table CS7.2. The F statistic for testing the visibility–success interaction reported in the table is $F = 63.40$ with a p-value less than .0001. Therefore, we reject H_0 at $\alpha = .05$ and conclude that the two factors, subject visibility and confederate success, interact.

Practically, this result implies that the effect of confederate success on mean latency to feedback, $E(y)$, depends on whether the subject is visible. Similarly, the effect of subject visibility on $E(y)$ depends on the success or failure of the confederate student. In other words, we cannot examine the effect of one factor on latency to feedback without knowing the level of the second factor. Consequently,

Table CS7.2 ANOVA table for the 2 × 2 factorial experiment

Source	df	SS	MS	F	p-value
Subject visibility	1	8,381.025	8,381.025	25.50	<.0001
Confederate success	1	8,151.025	8,151.025	24.80	<.0001
Visibility × Success	1	20,839.225	20,839.225	63.40	<.0001
Error	36	11,833.700	328.714		
Total	39	49,204.975			

we ignore the F-test for factor main effects and focus on the nature of the differences among the means of the $2 \times 2 = 4$ experimental conditions.

Follow-up Analysis

The sample latency to feedback means (in seconds) for each of the four experimental conditions are highlighted on the SAS printout, Figure CS7.2. These four means are listed in Table CS7.3 and plotted using MINITAB in Figure CS7.3.

Figure CS7.2 SAS printout of Tukey multiple comparisons of feedback means

```
                    The GLM Procedure
                    Least Squares Means
          Adjustment for Multiple Comparisons: Tukey

                               FEEDBACK      LSMEAN
       SUBJECT   CONFED          LSMEAN      Number

       NotVis    Failure      72.200000          1
       NotVis    Success      89.300000          2
       Visibl    Failure     146.800000          3
       Visibl    Success      72.600000          4


       Least Squares Means for effect SUBJECT*CONFED
            Pr > |t| for H0: LSMean(i)=LSMean(j)

               Dependent Variable: FEEDBACK

  i/j           1            2            3            4

   1                      0.1696       <.0001       1.0000
   2         0.1696                    <.0001       0.1858
   3         <.0001       <.0001                    <.0001
   4         1.0000       0.1858       <.0001
```

Table CS7.3 Sample means for the four experimental conditions

		Confederate Success	
		Success	Failure
SUBJECT	Visible	72.6	146.8
	Not visible	89.3	72.2

Figure CS7.3 MINITAB plot of sample means for the 2×2 factorial

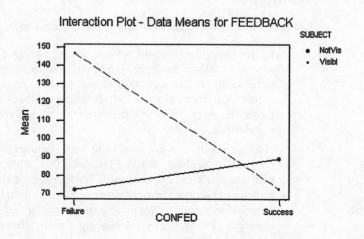

Interaction Plot – Data Means for FEEDBACK

We will conduct a follow-up analysis of the ANOVA by ranking the four treatment means. Since a balanced design is employed ($n = 10$ subjects per treatment), Tukey's method of multiple comparisons of means will be used.

Rather than use the formula for calculating Tukey's critical difference, ω (given in Section 12.7), we use the Tukey option for factorial experiments available in SAS. The results are shown at the bottom of the SAS printout, Figure CS7.2. Tukey-adjusted p-values that are less than $\alpha = .05$ are highlighted on the printout. These identify the significantly different pairs of treatment means: (3,2), (3,1) and (4,3), where 1, 2, 3, and 4 correspond to treatments NF, NS, VF, and VS, respectively. The implication is that mean 3 (the VF treatment mean) is significantly different from each of the other treatment means; however, none of the other treatment means are significantly different. The rankings are summarized in the traditional fashion in Table CS7.4.

Table CS7.4 Summary of Tukey's rankings of the treatment means

Sample mean:	72.2	72.6	89.3	146.8
Treatment:	NF	VS	NS	VF
(Mean #)	(1)	(4)	(2)	(3)

Another way to view the results in Table CS7.4 is to notice that there is no significant difference between the two nonvisible subject means for confederate success and failure (i.e., treatments NF and NS). However, there is a significant difference between the two visible subject means for confederate success and failure (i.e., treatments VF and VS). Thus, only when the subject is visible can we conclude that confederate success has an effect on the mean latency to feedback. Furthermore, since $\bar{y}_{VF} = 146.8$ is over twice as large as $\bar{y}_{VS} = 72.6$, the researchers conclude that "subjects appear reluctant to transmit bad news—but only when they are visible to the news recipient."

Conclusions

In their discussion of these results, the researchers conclude:

> In this experiment, subjects were required to give a test taker either success or failure feedback. While doing so, they presumed themselves to be visible to the test taker or visible to no one. Subjects who were visible took twice as long to deliver failure feedback as success feedback; those who were not visible delivered failure and success feedback with equal speed.
>
> These results are not consistent with the discomfort explanation as originally conceived. We had imagined that subjects might empathize with another's failure, that mere observation of the failure would be sufficient to arouse vicarious distress. We found no behavioral evidence of such discomfort. . . . We also imagined that subjects would be reluctant to induce discomfort by announcing a poor intelligence performance,

and that they would defer the announcement while checking the IQ test score. We found evidence of this deferral—but only when the subject could be seen. In private, subjects seemed blithe to others' misfortune—as quick to relay bad as good news. As the latency results suggest, there is no inherent discomfort in the transmission of bad news.

Follow-up Questions

1. Use Table CS7.2 to determine SSEs for the complete and reduced ANOVA models. Then use these values to obtain the F statistic for testing interaction.

2. The journal article made no mention of an analysis of the ANOVA residuals. Discuss the potential problems of an ANOVA when no residual analysis is conducted. Carry out this analysis of the data in the MUM file.

3. A second dependent variable measured in the study was *gaze*, defined as the proportion of time the subject was looking toward the confederate test taker on the other side of the glass plate. Gaze was measured at four points in time using videotape segments: early in the test, late in the test, during the wait for feedback, and after the feedback. Construct a complete model for analyzing gaze as a function of subject visibility, confederate success, and videotape segment. Identify the important tests to conduct.

Reference

Bond, C. F., and Anderson, E. L. "The reluctance to transmit bad news: Private discomfort or public display?" *Journal of Experimental Social Psychology*, Vol. 23, 1987, pp. 176–187.

Appendix

Derivation of the Least Squares Estimates of β_0 and β_1 in Simple Linear Regression

Consider the simple linear regression model, $E(y) = \beta_0 + \beta_1 x$. The least squares estimates, $\hat{\beta}_0$ and $\hat{\beta}_1$, are, by definition (see Chapter 3) the values that minimize

$$SSE = \sum (y_i - \hat{y}_i)^2 = \sum [y_i - (\hat{\beta}_0 + \hat{\beta}_1 x_i)]^2$$

The estimates are obtained by setting the two partial derivatives, $\partial SSE / \partial \hat{\beta}_0$ and $\partial SSE / \partial \hat{\beta}_1$, equal to 0 and solving the resulting simultaneous system of equations.

Step 1. Find the partial derivatives,

$$\frac{\partial SSE}{\partial \hat{\beta}_0} = \sum_{i=1}^{n} 2[y_i - (\hat{\beta}_0 + \hat{\beta}_1 x_i)](-1) = -2\sum y_i + 2\sum \hat{\beta}_0 + 2\sum \hat{\beta}_1 x_i$$

$$\frac{\partial SSE}{\partial \hat{\beta}_1} = \sum_{i=1}^{n} 2[y_i - (\hat{\beta}_0 + \hat{\beta}_1 x_i)](-x_i) = -2\sum x_i y_i + 2\sum \hat{\beta}_0 x_i + 2\sum \hat{\beta}_1 x_i^2$$

Step 2. Set the partial derivatives equal to 0 and simplify. After dividing each equation by -2, we obtain:

$$\sum_{i=1}^{n} y_i - \sum_{i=1}^{n} \hat{\beta}_0 - \hat{\beta}_1 \sum_{i=1}^{n} x_i = \sum_{i=1}^{n} y_i - n\hat{\beta}_0 - \hat{\beta}_1 \sum_{i=1}^{n} x_i = 0$$

$$\sum_{i=1}^{n} x_i y_i - \hat{\beta}_0 \sum_{i=1}^{n} x_i - \hat{\beta}_1 \sum_{i=1}^{n} x_i^2 = 0$$

or

$$n\hat{\beta}_0 + \hat{\beta}_1 \sum_{i=1}^{n} x_i = \sum_{i=1}^{n} y_i$$

$$\hat{\beta}_0 \sum_{i=1}^{n} x_i + \hat{\beta}_1 \sum_{i=1}^{n} x_i^2 = \sum_{i=1}^{n} x_i y_i$$

Step 3. Solve this pair of simultaneous linear equations for $\hat{\beta}_0$ and $\hat{\beta}_1$:

$$n\hat{\beta}_0 = \sum y_i - \hat{\beta}_1 \sum x_i \implies \hat{\beta}_0 = \frac{\sum y_i}{n} - \hat{\beta}_1 \frac{\sum x_i}{n}$$

$$\implies \boxed{\hat{\beta}_0 = \bar{y} - \hat{\beta}_1 \bar{x}}$$

$$\hat{\beta}_1 \sum x_i^2 = \sum x_i y_i - \hat{\beta}_0 \sum x_i = \sum x_i y_i - (\bar{y} - \hat{\beta}_1 \bar{x}) \sum x_i$$

$$= \sum x_i y_i - \bar{y} \sum x_i + \hat{\beta}_1 \bar{x} \sum x_i$$

$$= \sum x_i y_i - \frac{(\sum x_i)(\sum y_i)}{n} + \hat{\beta}_1 \frac{(\sum x_i)^2}{n}$$

$$\implies \hat{\beta}_1 \left\{ \sum x_i^2 - \frac{(\sum x_i)^2}{n} \right\} = \sum x_i y_i - \frac{(\sum x_i)(\sum y_i)}{n}$$

$$\implies \hat{\beta}_1 = \frac{\sum x_i y_i - \frac{(\sum x_i)(\sum y_i)}{n}}{\sum x_i^2 - \frac{(\sum x_i)^2}{n}} \implies \boxed{\hat{\beta}_1 = \frac{SS_{xy}}{SS_{xx}}}$$

Note: For regression through the origin, the estimate of the slope in the model $y = \beta_1 x + \varepsilon$ is derived as follows:

$$SSE = \sum (y_i - \hat{y}_i)^2 = \sum (y_i - \hat{\beta}_1 x_i)^2$$

$$= \sum (y_i^2 - 2\hat{\beta}_1 x_i y_i + \hat{\beta}_1^2 x_i^2)$$

$$= \sum y_i^2 - 2\hat{\beta}_1 \sum x_i y_i + \hat{\beta}_1^2 \sum x_i^2$$

$$\frac{\partial SSE}{\partial \beta_1} = -2 \sum x_i y_i + 2\hat{\beta}_1 \sum x_i^2 = 0$$

$$\implies \hat{\beta}_1 \sum x_i^2 = \sum x_i y_i$$

$$\implies \boxed{\hat{\beta}_1 = \frac{\sum x_i y_i}{\sum x_i^2}}$$

THE MECHANICS OF A MULTIPLE REGRESSION ANALYSIS

Contents

B.1 Introduction

The rationale behind a multiple regression analysis and the types of inferences it permits you to make are the subjects of Chapter 4. We noted that the method of least squares most often leads to a very difficult computational problem—namely, the solution of a set of $(k + 1)$ simultaneous linear equations in the unknown values of the estimates $\hat{\beta}_0, \hat{\beta}_1, \ldots, \hat{\beta}_k$—and that the formulas for the estimated standard errors $s_{\beta_0}, s_{\beta_1}, \ldots, s_{\beta_k}$ are too complicated to express as ordinary algebraic formulas. We circumvented both these problems easily. We relied on the least squares estimates, confidence intervals, tests, and so on, provided by a standard regression analysis software package. Thus, Chapter 4 provides a basic working knowledge of the types of inferences you might wish to make from a multiple regression analysis and explains how to interpret the results. If we can do this, why would we wish to know the actual process performed by the computer?

There are several answers to this question:

1. Some multiple regression statistical software packages do not print all the information you may want. As one illustration, we noted in Chapter 4 that very often the objective of a regression analysis is to develop a prediction equation that can be used to estimate the mean value of y (say, mean profit or mean yield) for given values of the predictor variables $x_1, x_2, \ldots, x_k$. Some software packages do not give the confidence interval for $E(y)$ or a prediction interval for y. Thus, you might need to know how to find the necessary quantities from the analysis and perform the computations yourself.

2. A multiple regression software package may possess the capability of computing some specific quantity that you desire, but you may find the instructions on how to "call" for this special calculation difficult to understand. It may be easier to identify the components required for your computation and do it yourself.

3. For some designed experiments, finding the least squares equations and solving them is a trivial operation. Understanding the process by which the least squares equations are generated and understanding how they are solved will help you understand how experimental design affects the results of a regression analysis. Thus, a knowledge of the computations involved in performing a regression analysis will help you to better understand the contents of Chapters 10 and 11.

To summarize, "knowing how it is done" is not essential for performing an ordinary regression analysis or interpreting its results. But "knowing how" helps, and it is essential for an understanding of many of the finer points associated with a multiple regression analysis. This appendix explains "how it is done" without getting into the unpleasant task of performing the computations for solving the least squares equations. This mechanical and tedious procedure can be left to a computer (the solutions are verifiable). We illustrate the procedure in Appendix C.

B.2 Matrices and Matrix Multiplication

Although it is very difficult to give the formulas for the multiple regression least squares estimators and for their estimated standard errors in ordinary algebra, it is easy to do so using **matrix algebra**. Thus, by arranging the data in particular rectangular patterns called **matrices** and by performing various operations with them, we can obtain the least squares estimates and their estimated standard errors. In this section and Sections B.3 and B.4, we define what we mean by a matrix and explain various operations that can be performed with matrices. We explain how to use this information to conduct a regression analysis in Section B.5.

Three matrices, $\mathbf{A}$, $\mathbf{B}$, and $\mathbf{C}$, are shown here. Note that each matrix is a rectangular arrangement of numbers with one number in every row–column position.

$$\mathbf{A} = \begin{bmatrix} 2 & 3 \\ 0 & 1 \\ -1 & 6 \end{bmatrix} \quad \mathbf{B} = \begin{bmatrix} 3 & 0 & 1 \\ -1 & 0 & 1 \\ 4 & 2 & 0 \end{bmatrix} \quad \mathbf{C} = \begin{bmatrix} 1 \\ 2 \\ 1 \end{bmatrix}$$

> **Definition B.1** A **matrix** is a rectangular array of numbers.*

The numbers that appear in a matrix are called **elements** of the matrix. If a matrix contains r rows and c columns, there will be an element in each of the row–column positions of the matrix, and the matrix will have $r \times c$ elements. For example, the matrix $\mathbf{A}$ shown previously contains $r = 3$ rows, $c = 2$ columns, and $rc = (3)(2) = 6$ elements, one in each of the six row–column positions.

> **Definition B.2** A number in a particular row–column position is called an **element** of the matrix.

Notice that the matrices $\mathbf{A}$, $\mathbf{B}$, and $\mathbf{C}$ contain different numbers of rows and columns. The numbers of rows and columns give the **dimensions** of a matrix.

---

* For our purpose, we assume that the numbers are real.

When we give a formula in matrix notation, the elements of a matrix will be represented by symbols. For example, if we have a matrix

$$A = \begin{bmatrix} a_{11} & a_{12} & a_{13} \\ a_{21} & a_{22} & a_{23} \end{bmatrix}$$

the symbol a_{ij} will denote the element in the ith row and jth column of the matrix. The first subscript always identifies the row and the second identifies the column in which the element is located. For example, the element a_{12} is in the first row and second column of the matrix A. The rows are always numbered from top to bottom, and the columns are always numbered from left to right.

Definition B.3 A matrix containing r rows and c columns is said to be an $r \times c$ **matrix** where r and c are the **dimensions** of the matrix.

Definition B.4 If $r = c$, a matrix is said to be a **square matrix**.

Matrices are usually identified by capital letters, such as A, B, C, corresponding to the letters of the alphabet employed in ordinary algebra. The difference is that in ordinary algebra, a letter is used to denote a single real number, whereas in matrix algebra, *a letter denotes a rectangular array of numbers*. The operations of matrix algebra are very similar to those of ordinary algebra—you can add matrices, subtract them, multiply them, and so on. But since we are concerned only with the applications of matrix algebra to the solution of the least squares equations, we define only the operations and types of matrices that are pertinent to that subject.

The most important operation for us is matrix multiplication, which requires **row–column multiplication**. To illustrate this process, suppose we wish to find the product AB, where

$$A = \begin{bmatrix} 2 & 1 \\ 4 & -1 \end{bmatrix} \qquad B = \begin{bmatrix} 2 & 0 & 3 \\ -1 & 4 & 0 \end{bmatrix}$$

We will always multiply the rows of A (the matrix on the left) by the columns of B (the matrix on the right). The product formed by the first row of A times the first column of B is obtained by multiplying the elements in corresponding positions and summing these products. Thus, the first row, first column product, shown diagrammatically here, is

$$(2)(2) + (1)(-1) = 4 - 1 = 3$$

$$AB = \begin{bmatrix} 2 & 1 \\ 4 & -1 \end{bmatrix} \begin{bmatrix} 2 & 0 & 3 \\ -1 & 4 & 0 \end{bmatrix} = \begin{bmatrix} 3 & & \\ & & \end{bmatrix}$$

Similarly, the first row, second column product is

$$(2)(0) + (1)(4) = 0 + 4 = 4$$

So far we have

$$AB = \begin{bmatrix} 3 & 4 & \\ & & \end{bmatrix}$$

To find the complete matrix product $\mathbf{AB}$, all we need to do is find each element in the $\mathbf{AB}$ matrix. Thus, we will define an element in the ith row, jth column of $\mathbf{AB}$ as the product of the ith row of $\mathbf{A}$ and the jth column of $\mathbf{B}$. We complete the process in Example B.1.

Example B.1

Find the product $\mathbf{AB}$, where

$$\mathbf{A} = \begin{bmatrix} 2 & 1 \\ 4 & -1 \end{bmatrix} \qquad \mathbf{B} = \begin{bmatrix} 2 & 0 & 3 \\ -1 & 4 & 0 \end{bmatrix}$$

Solution

If we represent the product $\mathbf{AB}$ as

$$\mathbf{C} = \begin{bmatrix} c_{11} & c_{12} & c_{13} \\ c_{21} & c_{22} & c_{23} \end{bmatrix}$$

we have already found $c_{11} = 3$ and $c_{12} = 4$. Similarly, the element c_{21}, the element in the second row, first column of $\mathbf{AB}$, is the product of the second row of $\mathbf{A}$ and the first column of $\mathbf{B}$:

$$(4)(2) + (-1)(-1) = 8 + 1 = 9$$

Proceeding in a similar manner to find the remaining elements of $\mathbf{AB}$, we have

$$\mathbf{AB} = \begin{bmatrix} 2 & 1 \\ 4 & -1 \end{bmatrix} \begin{bmatrix} 2 & 0 & 3 \\ -1 & 4 & 0 \end{bmatrix} = \begin{bmatrix} 3 & 4 & 6 \\ 9 & -4 & 12 \end{bmatrix}$$

Now, try to find the product $\mathbf{BA}$, using matrices $\mathbf{A}$ and $\mathbf{B}$ from Example B.1. You will observe two very important differences between multiplication in matrix algebra and multiplication in ordinary algebra:

1. You cannot find the product $\mathbf{BA}$ because you cannot perform row–column multiplication. You can see that the dimensions do not match by placing the matrices side by side.

$$\underset{2 \times 3 \quad\quad 2 \times 2}{\mathbf{BA}} \qquad \text{does not exist}$$

The number of elements (3) in a row of $\mathbf{B}$ (the matrix on the left) does not match the number of elements (2) in a column of $\mathbf{A}$ (the matrix on the right). Therefore, you cannot perform row–column multiplication, and the matrix product $\mathbf{BA}$ does not exist. The point is, not all matrices can be multiplied. You can find products for matrices $\mathbf{A}$ and $\mathbf{B}$ only when $\mathbf{A}$ is $r \times d$ and $\mathbf{B}$ is $d \times c$. That is:

Requirement for Matrix Multiplication

$$\underset{r \times d \quad\quad d \times c}{\mathbf{AB}}$$

The two inner dimension numbers must be equal. The dimensions of the product will always be given by the outer dimension numbers:

Dimensions of AB Are $r \times c$

$$\underbrace{\mathbf{AB}}_{r \times d \quad d \times c}$$

2. The second difference between ordinary and matrix multiplication is that in ordinary algebra, $ab = ba$. In matrix algebra, **AB** usually does not equal **BA**. In fact, as noted in item 1, **BA** may not even exist.

> **Definition B.5** The product **AB** of an $r \times d$ matrix **A** and a $d \times c$ matrix **B** is an $r \times c$ matrix **C**, where the element $c_{ij} (i = 1, 2, \ldots, r; j = 1, 2, \ldots, c)$ of **C** is the product of the ith row of **A** and the jth column of **B**.

Example B.2

Given the matrices below, find **IA** and **IB**.

$$\mathbf{A} = \begin{bmatrix} 2 \\ 1 \\ 3 \end{bmatrix} \qquad \mathbf{B} = \begin{bmatrix} 3 & 0 \\ 1 & 2 \\ 4 & -1 \end{bmatrix} \qquad \mathbf{I} = \begin{bmatrix} 1 & 0 & 0 \\ 0 & 1 & 0 \\ 0 & 0 & 1 \end{bmatrix}$$

Solution

Notice that the product

$$\underbrace{\mathbf{IA}}_{3 \times 3 \quad 3 \times 1}$$

exists and that it will be of dimensions 3×1. Performing the row–column multiplication yields

$$\mathbf{IA} = \begin{bmatrix} 1 & 0 & 0 \\ 0 & 1 & 0 \\ 0 & 0 & 1 \end{bmatrix} \begin{bmatrix} 2 \\ 1 \\ 3 \end{bmatrix} = \begin{bmatrix} 2 \\ 1 \\ 3 \end{bmatrix}$$

Similarly,

$$\underbrace{\mathbf{IB}}_{3 \times 3 \quad 3 \times 2}$$

exists and is of dimensions 3×2. Performing the row—column multiplications, we obtain

$$\mathbf{IB} = \begin{bmatrix} 1 & 0 & 0 \\ 0 & 1 & 0 \\ 0 & 0 & 1 \end{bmatrix} \begin{bmatrix} 3 & 0 \\ 1 & 2 \\ 4 & -1 \end{bmatrix} = \begin{bmatrix} 3 & 0 \\ 1 & 2 \\ 4 & -1 \end{bmatrix}$$

Notice that the **I** matrix possesses a special property. We have **IA** = **A** and **IB** = **B**. We comment further on this property in Section B.3.

B.2 Exercises

B.1 Given the matrices **A**, **B**, and **C**:

$$A = \begin{bmatrix} 3 & 0 \\ -1 & 4 \end{bmatrix} \quad B = \begin{bmatrix} 2 & 1 \\ 0 & -1 \end{bmatrix}$$

$$C = \begin{bmatrix} 1 & 0 & 3 \\ -2 & 1 & 2 \end{bmatrix}$$

(a) Find **AB**.   (b) Find **AC**.   (c) Find **BA**.

B.2 Given the matrices **A**, **B**, and **C**:

$$A = \begin{bmatrix} 3 & 1 & 3 \\ 2 & 0 & 4 \\ -4 & 1 & 2 \end{bmatrix} \quad B = \begin{bmatrix} 1 & 0 & 2 \end{bmatrix} \quad C = \begin{bmatrix} 3 \\ 0 \\ 2 \end{bmatrix}$$

(a) Find **AC**.   (b) Find **BC**.
(c) Is it possible to find **AB**? Explain.

B.3 Suppose **A** is a 3 × 2 matrix and **B** is a 2 × 4 matrix.

(a) What are the dimensions of **AB**?
(b) Is it possible to find the product **BA**? Explain.

B.4 Suppose matrices **B** and **C** are of dimensions 1 × 3 and 3 × 1, respectively.

(a) What are the dimensions of the product **BC**?
(b) What are the dimensions of **CB**?
(c) If **B** and **C** are the matrices shown in Exercise A.2, find **CB**.

B.5 Given the matrices **A**, **B**, and **C**:

$$A = \begin{bmatrix} 1 & 0 & 0 \\ 0 & 3 & 0 \\ 0 & 0 & 2 \end{bmatrix}$$

$$B = \begin{bmatrix} 2 & 3 \\ -3 & 0 \\ 4 & -1 \end{bmatrix} \quad C = \begin{bmatrix} 3 & 0 & 2 \end{bmatrix}$$

(a) Find **AB**.   (b) Find **CA**.   (c) Find **CB**.

B.6 Given the matrices:

$$A = \begin{bmatrix} 3 & 0 & -1 & 2 \end{bmatrix} \quad B = \begin{bmatrix} 2 \\ -1 \\ 0 \\ 3 \end{bmatrix}$$

(a) Find **AB**.   (b) Find **BA**.

B.3 Identity Matrices and Matrix Inversion

In ordinary algebra, the number 1 is the identity element for the multiplication operation. That is, 1 is the element such that any other number, say, c, multiplied by the identity element is equal to c. Thus, $4(1) = 4, (-5)(1) = -5$, and so forth.

The corresponding identity element for multiplication in matrix algebra, identified by the symbol **I**, is a matrix such that

$$AI = IA = A \quad \text{for any matrix } A$$

The difference between identity elements in ordinary algebra and matrix algebra is that in ordinary algebra there is only one identity element, the number 1. In matrix algebra, the identity matrix must possess the correct dimensions for the product **IA** to exist. Consequently, there is an infinitely large number of identity matrices—all square and possessing the same pattern. The 1 × 1, 2 × 2, and 3 × 3 identity matrices are

$$\underset{1 \times 1}{I} = [1] \qquad \underset{2 \times 2}{I} = \begin{bmatrix} 1 & 0 \\ 0 & 1 \end{bmatrix} \qquad \underset{3 \times 3}{I} = \begin{bmatrix} 1 & 0 & 0 \\ 0 & 1 & 0 \\ 0 & 0 & 1 \end{bmatrix}$$

In Example B.2, we demonstrated the fact that this matrix satisfies the property

$$IA = A$$

Definition B.6 If $\mathbf{A}$ is any matrix, then a matrix $\mathbf{I}$ is defined to be an **identity matrix** if $\mathbf{AI} = \mathbf{IA} = \mathbf{A}$. The matrices that satisfy this definition possess the pattern

$$\mathbf{I} = \begin{bmatrix} 1 & 0 & 0 & \cdots & 0 \\ 0 & 1 & 0 & \cdots & 0 \\ 0 & 0 & 1 & \cdots & 0 \\ \cdot & \cdot & \cdot & \cdots & \cdot \\ \cdot & \cdot & \cdot & \cdots & \cdot \\ \cdot & \cdot & \cdot & \cdots & \cdot \\ 0 & 0 & 0 & \cdots & 1 \end{bmatrix}$$

Example B.3

If $\mathbf{A}$ is the matrix shown here, find $\mathbf{IA}$ and $\mathbf{AI}$.

$$\mathbf{A} = \begin{bmatrix} 3 & 4 & -1 \\ 1 & 0 & 2 \end{bmatrix}$$

Solution

$$\underset{\underset{2 \times 2}{\nearrow} \quad \underset{2 \times 3}{\nwarrow}}{\overset{\mathbf{IA}}{}} = \begin{bmatrix} 1 & 0 \\ 0 & 1 \end{bmatrix} \begin{bmatrix} 3 & 4 & -1 \\ 1 & 0 & 2 \end{bmatrix} = \begin{bmatrix} 3 & 4 & -1 \\ 1 & 0 & 2 \end{bmatrix} = \mathbf{A}$$

$$\underset{\underset{2 \times 3}{\nearrow} \quad \underset{3 \times 3}{\nwarrow}}{\overset{\mathbf{AI}}{}} = \begin{bmatrix} 3 & 4 & -1 \\ 1 & 0 & 2 \end{bmatrix} \begin{bmatrix} 1 & 0 & 0 \\ 0 & 1 & 0 \\ 0 & 0 & 1 \end{bmatrix} = \begin{bmatrix} 3 & 4 & -1 \\ 1 & 0 & 2 \end{bmatrix} = \mathbf{A}$$

Notice that the identity matrices used to find the products $\mathbf{IA}$ and $\mathbf{AI}$ were of different dimensions. This was necessary for the products to exist. ∎

The identity element assumes importance when we consider the process of division and its role in the solution of equations. In ordinary algebra, division is essentially multiplication using the reciprocals of elements. For example, the equation

$$2x = 6$$

can be solved by dividing both sides of the equation by 2, *or* it can be solved by *multiplying* both sides of the equation by $\frac{1}{2}$, which is the reciprocal of 2. Thus,

$$\left(\frac{1}{2} \right) 2x = \frac{1}{2}(6)$$

$$x = 3$$

What is the reciprocal of an element? It is the element such that the reciprocal times the element is equal to the identity element. Thus, the reciprocal of 3 is $\frac{1}{3}$ because

$$3 \left(\frac{1}{3} \right) = 1$$

The identity matrix plays the same role in matrix algebra. Thus, the reciprocal of a matrix $\mathbf{A}$, called the **inverse of A** and denoted by the symbol $\mathbf{A}^{-1}$, is a matrix such that $\mathbf{AA}^{-1} = \mathbf{A}^{-1}\mathbf{A} = \mathbf{I}$.

Inverses are defined only for square matrices, but not all square matrices possess inverses. Those that do have inverses play an important role in solving the least squares equations and in other aspects of a regression analysis. We will show you one important application of the inverse matrix in Section B.4. The procedure for finding the inverse of a matrix is demonstrated in Appendix C.

Definition B.7 The square matrix $\mathbf{A}^{-1}$ is said to be the **inverse** of the square matrix $\mathbf{A}$ if

$$\mathbf{A}^{-1}\mathbf{A} = \mathbf{A}\mathbf{A}^{-1} = \mathbf{I}$$

The procedure for finding an inverse matrix is computationally quite tedious and is performed most often using a computer. There are several exceptions. For example, finding the inverse of one type of matrix, called a **diagonal matrix**, is easy. A diagonal matrix is one that has nonzero elements down the **main diagonal** (running from top left of the matrix to bottom right) and 0 elements elsewhere. Thus, the identity matrix is a diagonal matrix (with 1's along the main diagonal), as are the following matrices:

$$\mathbf{A} = \begin{bmatrix} 3 & 0 & 0 \\ 0 & 1 & 0 \\ 0 & 0 & 2 \end{bmatrix} \qquad \mathbf{B} = \begin{bmatrix} 5 & 0 & 0 & 0 \\ 0 & 2 & 0 & 0 \\ 0 & 0 & 1 & 0 \\ 0 & 0 & 0 & 5 \end{bmatrix}$$

Definition B.8 A **diagonal matrix** is one that contains nonzero elements on the main diagonal and 0 elements elsewhere.

You can verify that the inverse of

$$\mathbf{A} = \begin{bmatrix} 3 & 0 & 0 \\ 0 & 1 & 0 \\ 0 & 0 & 2 \end{bmatrix} \quad \text{is} \quad \mathbf{A}^{-1} = \begin{bmatrix} \frac{1}{3} & 0 & 0 \\ 0 & 1 & 0 \\ 0 & 0 & \frac{1}{2} \end{bmatrix}$$

That is, $\mathbf{A}\mathbf{A}^{-1} = \mathbf{I}$. In general, the inverse of a diagonal matrix is given by the following theorem, which is stated without proof:

Theorem B.1   The **inverse of a diagonal matrix**

$$\mathbf{D} = \begin{bmatrix} d_{11} & 0 & 0 & \cdots & 0 \\ 0 & d_{22} & 0 & \cdots & 0 \\ 0 & 0 & d_{33} & \cdots & 0 \\ \cdot & \cdot & \cdot & \cdots & \cdot \\ \cdot & \cdot & \cdot & \cdots & \cdot \\ \cdot & \cdot & \cdot & \cdots & \cdot \\ 0 & 0 & 0 & \cdots & d_{nn} \end{bmatrix} \quad \text{is} \quad \mathbf{D}^{-1} = \begin{bmatrix} 1/d_{11} & 0 & 0 & \cdots & 0 \\ 0 & 1/d_{22} & 0 & \cdots & 0 \\ 0 & 0 & 1/d_{33} & \cdots & 0 \\ \cdot & \cdot & \cdot & \cdots & \cdot \\ \cdot & \cdot & \cdot & \cdots & \cdot \\ \cdot & \cdot & \cdot & \cdots & \cdot \\ 0 & 0 & 0 & \cdots & 1/d_{nn} \end{bmatrix}$$

A second type of matrix that is easy to invert is a 2×2 matrix. The following theorem shows how to find the inverse of this type of matrix.

Theorem B.2   The **inverse of a 2 × 2 matrix**

$$A = \begin{bmatrix} a & b \\ c & d \end{bmatrix} \text{ is } A^{-1} = \begin{bmatrix} \frac{d}{ad-bc} & \frac{-b}{ad-bc} \\ \frac{-c}{ad-bc} & \frac{a}{ad-bc} \end{bmatrix}$$

You can verify that the inverse of

$$A = \begin{bmatrix} 1 & -2 \\ -2 & 6 \end{bmatrix} \text{ is } A^{-1} = \begin{bmatrix} 3 & 1 \\ 1 & \frac{1}{2} \end{bmatrix}$$

We demonstrate another technique for finding A^{-1} in Appendix C.

B.3 Exercises

B.7 Let $A = \begin{bmatrix} 3 & 0 & 2 \\ -1 & 1 & 4 \end{bmatrix}$.

(a) Give the identity matrix that will be used to obtain the product IA.

(b) Show that $IA = A$.

(c) Give the identity matrix that will be used to find the product AI.

(d) Show that $AI = A$.

B.8 Given the following matrices A and B, show that $AB = I$, that $BA = I$, and consequently, verify that $B = A^{-1}$.

$$A = \begin{bmatrix} 1 & 0 & 0 \\ 0 & 2 & 0 \\ 0 & 0 & 3 \end{bmatrix} \quad B = \begin{bmatrix} 1 & 0 & 0 \\ 0 & \frac{1}{2} & 0 \\ 0 & 0 & \frac{1}{3} \end{bmatrix}$$

B.9 If

$$A = \begin{bmatrix} 12 & 0 & 0 & 8 \\ 0 & 12 & 0 & 0 \\ 0 & 0 & 8 & 0 \\ 8 & 0 & 0 & 8 \end{bmatrix}$$

verify that

$$A^{-1} = \begin{bmatrix} \frac{1}{4} & 0 & 0 & -\frac{1}{4} \\ 0 & \frac{1}{12} & 0 & 0 \\ 0 & 0 & \frac{1}{8} & 0 \\ -\frac{1}{4} & 0 & 0 & \frac{3}{8} \end{bmatrix}$$

B.10 If

$$A = \begin{bmatrix} 3 & 0 & 0 \\ 0 & 5 & 0 \\ 0 & 0 & 7 \end{bmatrix}$$

show that

$$A^{-1} = \begin{bmatrix} \frac{1}{3} & 0 & 0 \\ 0 & \frac{1}{5} & 0 \\ 0 & 0 & \frac{1}{7} \end{bmatrix}$$

B.11 Verify Theorem B.1.

B.12 Verify Theorem B.2.

B.13 Find the inverse of

$$A = \begin{bmatrix} 2 & -1 \\ 2 & 3 \end{bmatrix}$$

B.4 Solving Systems of Simultaneous Linear Equations

Consider the following set of simultaneous linear equations in two unknowns:

$$2v_1 + v_2 = 7$$

$$v_1 - v_2 = 2$$

Note that the solution for these equations is $v_1 = 3$, $v_2 = 1$.

Now define the matrices

$$A = \begin{bmatrix} 2 & 1 \\ 1 & -1 \end{bmatrix} \quad V = \begin{bmatrix} v_1 \\ v_2 \end{bmatrix} \quad G = \begin{bmatrix} 7 \\ 2 \end{bmatrix}$$

Thus, $\mathbf{A}$ is the matrix of coefficients of v_1 and v_2, $\mathbf{V}$ is a column matrix containing the unknowns (written in order, top to bottom), and $\mathbf{G}$ is a column matrix containing the numbers on the right-hand side of the equal signs.

Now, the given system of simultaneous equations can be rewritten as a **matrix equation:**

$$\mathbf{AV} = \mathbf{G}$$

By a matrix equation, we mean that the product matrix, $\mathbf{AV}$, is equal to the matrix $\mathbf{G}$. *Equality of matrices means that corresponding elements are equal.* You can see that this is true for the expression $\mathbf{AV} = \mathbf{G}$, since

$$\underset{2 \times 2 \qquad 2 \times 1}{\overset{\mathbf{AV}}{\begin{bmatrix} 2 & 1 \\ 1 & -1 \end{bmatrix} \begin{bmatrix} v_1 \\ v_2 \end{bmatrix}}} = \begin{bmatrix} (2v_1 + v_2) \\ (v_1 - v_2) \end{bmatrix} = \underset{2 \times 1}{\mathbf{G}}$$

The matrix procedure for expressing a system of two simultaneous linear equations in two unknowns can be extended to express a set of k simultaneous equations in k unknowns. If the equations are written in the orderly pattern

$$a_{11}v_1 + a_{12}v_2 + \cdots + a_{1k}v_k = g_1$$
$$a_{21}v_1 + a_{22}v_2 + \cdots + a_{2k}v_k = g_2$$
$$\vdots \qquad \vdots \qquad \qquad \vdots \qquad \vdots$$
$$a_{k1}v_1 + a_{k2}v_2 + \cdots + a_{kk}v_k = g_k$$

then the set of simultaneous linear equations can be expressed as the matrix equation $\mathbf{AV} = \mathbf{G}$, where

$$\mathbf{A} = \begin{bmatrix} a_{11} & a_{12} & \cdots & a_{1k} \\ a_{21} & & \cdots & a_{2k} \\ \vdots & & & \vdots \\ a_{k1} & & \cdots & a_{kk} \end{bmatrix} \qquad \mathbf{V} = \begin{bmatrix} v_1 \\ v_2 \\ \vdots \\ v_k \end{bmatrix} \qquad \mathbf{G} = \begin{bmatrix} g_1 \\ g_2 \\ \vdots \\ g_k \end{bmatrix}$$

Now let us solve this system of simultaneous equations. (If they are uniquely solvable, it can be shown that $\mathbf{A}^{-1}$ exists.) Multiplying both sides of the matrix equation by $\mathbf{A}^{-1}$, we have

$$(\mathbf{A}^{-1})\mathbf{AV} = (\mathbf{A}^{-1})\mathbf{G}$$

But since $\mathbf{A}^{-1}\mathbf{A} = \mathbf{I}$, we have

$$(\mathbf{I})\mathbf{V} = \mathbf{A}^{-1}\mathbf{G}$$
$$\mathbf{V} = \mathbf{A}^{-1}\mathbf{G}$$

In other words, if we know $\mathbf{A}^{-1}$, we can find the solution to the set of simultaneous linear equations by obtaining the product $\mathbf{A}^{-1}\mathbf{G}$.

Matrix Solution to a Set of Simultaneous Linear Equations, $\mathbf{AV} = \mathbf{G}$

Solution: $\mathbf{V} = \mathbf{A}^{-1}\mathbf{G}$

Example B.4

Apply the result from the box to find the solution to the set of simultaneous linear equations

$$2v_1 + v_2 = 7$$
$$v_1 - v_2 = 2$$

Solution

The first step is to obtain the inverse of the coefficient matrix,

$$\mathbf{A} = \begin{bmatrix} 2 & 1 \\ 1 & -1 \end{bmatrix}$$

namely,

$$\mathbf{A}^{-1} = \begin{bmatrix} \frac{1}{3} & \frac{1}{3} \\ \frac{1}{3} & -\frac{2}{3} \end{bmatrix}$$

(This matrix can be found using a packaged computer program for matrix inversion or, for this simple case, you could use the procedure explained in Appendix C.) As a check, note that

$$\mathbf{A}^{-1}\mathbf{A} = \begin{bmatrix} \frac{1}{3} & \frac{1}{3} \\ \frac{1}{3} & -\frac{2}{3} \end{bmatrix} \begin{bmatrix} 2 & 1 \\ 1 & -1 \end{bmatrix} = \begin{bmatrix} 1 & 0 \\ 0 & 1 \end{bmatrix} = \mathbf{I}$$

The second step is to obtain the product $\mathbf{A}^{-1}\mathbf{G}$. Thus,

$$\mathbf{V} = \mathbf{A}^{-1}\mathbf{G} = \begin{bmatrix} \frac{1}{3} & \frac{1}{3} \\ \frac{1}{3} & -\frac{2}{3} \end{bmatrix} \begin{bmatrix} 7 \\ 2 \end{bmatrix} = \begin{bmatrix} 3 \\ 1 \end{bmatrix}$$

Since

$$\mathbf{V} = \begin{bmatrix} v_1 \\ v_2 \end{bmatrix} = \begin{bmatrix} 3 \\ 1 \end{bmatrix}$$

it follows that $v_1 = 3$ and $v_2 = 1$. You can see that these values of v_1 and v_2 satisfy the simultaneous linear equations and are the values that we specified as a solution at the beginning of this section. ∎

B.4 Exercises

B.14 Suppose the simultaneous linear equations

$$3v_1 + v_2 = 5$$
$$v_1 - v_2 = 3$$

are expressed as a matrix equation,

$$\mathbf{AV} = \mathbf{G}$$

(a) Find the matrices $\mathbf{A}$, $\mathbf{V}$, and $\mathbf{G}$.
(b) Verify that

$$\mathbf{A}^{-1} = \begin{bmatrix} \frac{1}{4} & \frac{1}{4} \\ \frac{1}{4} & -\frac{3}{4} \end{bmatrix}$$

[*Note*: A procedure for finding $\mathbf{A}^{-1}$ is given in Appendix C.]

(c) Solve the equations by finding $\mathbf{V} = \mathbf{A}^{-1}\mathbf{G}$.

B.15 For the simultaneous linear equations

$$10v_1 + 20v_3 - 60 = 0$$
$$20v_2 - 60 = 0$$
$$20v_1 + 68v_3 - 176 = 0$$

(a) Find the matrices $\mathbf{A}$, $\mathbf{V}$, and $\mathbf{G}$.
(b) Verify that

$$\mathbf{A}^{-1} = \begin{bmatrix} \frac{17}{70} & 0 & -\frac{1}{14} \\ 0 & \frac{1}{20} & 0 \\ -\frac{1}{14} & 0 & \frac{1}{28} \end{bmatrix}$$

(c) Solve the equations by finding $\mathbf{V} = \mathbf{A}^{-1}\mathbf{G}$.

B.5 The Least Squares Equations and Their Solutions

To apply matrix algebra to a regression analysis, we must place the data in matrices in a particular pattern. We will suppose that the linear model is

$$y = \beta_0 + \beta_1 x_1 + \beta_2 x_2 + \cdots + \beta_k x_k + \varepsilon$$

where (from Chapter 4) $x_1, x_2, \ldots, x_k$ could actually represent the squares, cubes, cross products, or other functions of predictor variables, and ε is a random error. We will assume that we have collected data for n observations (i.e., n values of y and corresponding values of $x_1, x_2, \ldots, x_k$) and that these are denoted as shown in the table:

Observation	y Value	x_1	x_2	$\cdots$	x_k
1	y_1	x_{11}	x_{21}		x_{k1}
2	y_2	x_{12}	x_{22}		x_{k2}
$\vdots$	$\vdots$	$\vdots$	$\vdots$	$\vdots$	$\vdots$
n	y_n	x_{1n}	x_{2n}		x_{kn}

Then the two data matrices $\mathbf{Y}$ and $\mathbf{X}$ are as shown in the next box.

The Data Matrices Y and X and the $\hat{\beta}$ Matrix

$$\mathbf{Y} = \begin{bmatrix} y_1 \\ y_2 \\ y_3 \\ \vdots \\ y_n \end{bmatrix} \quad \mathbf{X} = \begin{bmatrix} 1 & x_{11} & x_{21} & \cdots & x_{k1} \\ 1 & x_{12} & x_{22} & \cdots & x_{k2} \\ 1 & x_{13} & x_{23} & \cdots & x_{k3} \\ \vdots & \vdots & \vdots & & \vdots \\ 1 & x_{1n} & x_{2n} & \cdots & x_{kn} \end{bmatrix} \quad \hat{\boldsymbol{\beta}} = \begin{bmatrix} \hat{\beta}_0 \\ \hat{\beta}_1 \\ \hat{\beta}_2 \\ \vdots \\ \hat{\beta}_k \end{bmatrix}$$

Notice that the first column in the $\mathbf{X}$ matrix is a column of 1's. Thus, we are inserting a value of x, namely, x_0, as the coefficient of β_0, where x_0 is a variable always equal to 1. Therefore, there is one column in the $\mathbf{X}$ matrix for each β parameter. Also, remember that a particular data point is identified by specific rows of the $\mathbf{Y}$ and $\mathbf{X}$ matrices. For example, the y value y_3 for data point 3 is in the third row of the $\mathbf{Y}$ matrix, and the corresponding values of $x_1, x_2, \ldots, x_k$ appear in the third row of the $\mathbf{X}$ matrix.

The $\hat{\boldsymbol{\beta}}$ matrix shown in the box contains the least squares estimates (which we are attempting to obtain) of the coefficients $\beta_0, \beta_1, \ldots, \beta_k$ of the linear model

$$y = \beta_0 + \beta_1 x_1 + \beta_2 x_2 + \cdots + \beta_k x_k + \varepsilon$$

To write the least squares equation, we need to define what we mean by the **transpose of a matrix**. If

$$\mathbf{Y} = \begin{bmatrix} 5 \\ 1 \\ 0 \\ 4 \\ 2 \end{bmatrix} \quad \mathbf{X} = \begin{bmatrix} 1 & 0 \\ 1 & 1 \\ 1 & 4 \\ 1 & 2 \\ 1 & 6 \end{bmatrix}$$

then the transpose matrices of the $\mathbf{Y}$ and $\mathbf{X}$ matrices, denoted as $\mathbf{Y}'$ and $\mathbf{X}'$, respectively, are

$$\mathbf{Y}' = \begin{bmatrix} 5 & 1 & 0 & 4 & 2 \end{bmatrix} \quad \mathbf{X}' = \begin{bmatrix} 1 & 1 & 1 & 1 & 1 \\ 0 & 1 & 4 & 2 & 6 \end{bmatrix}$$

Definition B.9 The **transpose of a matrix** $\mathbf{A}$, denoted as $\mathbf{A}'$, is obtained by interchanging corresponding rows and columns of the $\mathbf{A}$ matrix. That is, the ith row of the $\mathbf{A}$ matrix becomes the ith column of the $\mathbf{A}'$ matrix.

Using the $\mathbf{Y}$ and $\mathbf{X}$ data matrices, their transposes, and the $\hat{\boldsymbol{\beta}}$ matrix, we can write the least squares equations (proof omitted) as:

Least Squares Matrix Equation

$$(\mathbf{X}'\mathbf{X})\hat{\boldsymbol{\beta}} = \mathbf{X}'\mathbf{Y}$$

Thus, $(\mathbf{X}'\mathbf{X})$ is the coefficient matrix of the least squares estimates $\hat{\beta}_0, \hat{\beta}_1, \ldots, \hat{\beta}_k$, and $\mathbf{X}'\mathbf{Y}$ gives the matrix of constants that appear on the right-hand side of the equality signs. In the notation of Section B.4,

$$\mathbf{A} = \mathbf{X}'\mathbf{X} \quad \mathbf{V} = \hat{\boldsymbol{\beta}} \quad \mathbf{G} = \mathbf{X}'\mathbf{Y}$$

The solution, which follows from Section B.4, is

Least Squares Matrix Solution

$$\hat{\boldsymbol{\beta}} = (\mathbf{X}'\mathbf{X})^{-1}\mathbf{X}'\mathbf{Y}$$

Thus, to solve the least squares matrix equation, the computer calculates $(\mathbf{X}'\mathbf{X})$, $(\mathbf{X}'\mathbf{X})^{-1}$, $\mathbf{X}'\mathbf{Y}$, and, finally, the product $(\mathbf{X}'\mathbf{X})^{-1}\mathbf{X}'\mathbf{Y}$. We will illustrate this process using the data for the advertising example from Section 3.3.

Example B.5   Find the least squares line for the data given in Table B.1.

Table B.1

Month	Advertising Expenditure x, hundreds of dollars	Sales Revenue y, thousands of dollars
1	1	1
2	2	1
3	3	2
4	4	2
5	5	4

Solution
The model is

$$y = \beta_0 + \beta_1 x_1 + \varepsilon$$

and the $\mathbf{Y}$, $\mathbf{X}$, and $\hat{\boldsymbol{\beta}}$ matrices are

$$\mathbf{Y} = \begin{bmatrix} 1 \\ 1 \\ 2 \\ 2 \\ 4 \end{bmatrix} \quad \mathbf{X} = \begin{bmatrix} \overset{x_0}{1} & \overset{x_1}{1} \\ 1 & 2 \\ 1 & 3 \\ 1 & 4 \\ 1 & 5 \end{bmatrix} \quad \hat{\boldsymbol{\beta}} = \begin{bmatrix} \hat{\beta}_0 \\ \hat{\beta}_1 \end{bmatrix}$$

Then,

$$\mathbf{X'X} = \begin{bmatrix} 1 & 1 & 1 & 1 & 1 \\ 1 & 2 & 3 & 4 & 5 \end{bmatrix} \begin{bmatrix} 1 & 1 \\ 1 & 2 \\ 1 & 3 \\ 1 & 4 \\ 1 & 5 \end{bmatrix} = \begin{bmatrix} 5 & 15 \\ 15 & 55 \end{bmatrix}$$

$$\mathbf{X'Y} = \begin{bmatrix} 1 & 1 & 1 & 1 & 1 \\ 1 & 2 & 3 & 4 & 5 \end{bmatrix} \begin{bmatrix} 1 \\ 1 \\ 2 \\ 2 \\ 4 \end{bmatrix} = \begin{bmatrix} 10 \\ 37 \end{bmatrix}$$

The last matrix that we need is $(\mathbf{X'X})^{-1}$. This matrix, which can be found by using Theorem B.2 (or by using the method in Appendix C), is

$$(\mathbf{X'X})^{-1} = \begin{bmatrix} 1.1 & -.3 \\ -.3 & .1 \end{bmatrix}$$

Then the solution to the least squares equation is

$$\hat{\beta} = (\mathbf{X'X})^{-1}\mathbf{X'Y} = \begin{bmatrix} 1.1 & -.3 \\ -.3 & .1 \end{bmatrix} \begin{bmatrix} 10 \\ 37 \end{bmatrix} = \begin{bmatrix} -.1 \\ .7 \end{bmatrix}$$

Thus, $\hat{\beta}_0 = -.1$, $\hat{\beta}_1 = .7$, and the prediction equation is

$$\hat{y} = -.1 + .7x$$

You can verify that this is the same answer as obtained in Section 3.3.

Example B.6

Table B.2 contains data on monthly electrical power usage and home size for a sample of $n = 10$ homes. Find the least squares solution for fitting the monthly power usage y to size of home x for the model

$$y = \beta_0 + \beta_1 x + \beta_2 x^2 + \varepsilon$$

Table B.2 Data for Power Usage Study

Size of Home x, square feet	Monthly Usage y, kilowatt-hours	Size of Home x, square feet	Monthly Usage y, kilowatt-hours
1,290	1,182	1,840	1,711
1,350	1,172	1,980	1,804
1,470	1,264	2,230	1,840
1,600	1,493	2,400	1,956
1,710	1,571	2,930	1,954

Solution

The $\mathbf{Y}$, $\mathbf{X}$, and $\hat{\beta}$ matrices are as follows:

$$\mathbf{Y} = \begin{bmatrix} 1,182 \\ 1,172 \\ 1,264 \\ 1,493 \\ 1,571 \\ 1,711 \\ 1,804 \\ 1,840 \\ 1,956 \\ 1,954 \end{bmatrix}$$

$$\mathbf{X} = \begin{bmatrix} x_0 & x & x^2 \\ 1 & 1,290 & 1,664,100 \\ 1 & 1,350 & 1,822,500 \\ 1 & 1,470 & 2,160,900 \\ 1 & 1,600 & 2,560,000 \\ 1 & 1,710 & 2,924,100 \\ 1 & 1,840 & 3,385,600 \\ 1 & 1,980 & 3,920,400 \\ 1 & 2,230 & 4,972,900 \\ 1 & 2,400 & 5,760,000 \\ 1 & 2,930 & 8,584,900 \end{bmatrix}$$

Then

$$\mathbf{X'X} = \begin{bmatrix} 10 & 18,800 & 37,755,400 \\ 18,800 & 37,755,400 & 8,093.9 \times 10^7 \\ 37,755,400 & 8,093.9 \times 10^7 & 1.843 \times 10^{14} \end{bmatrix}$$

$$\mathbf{X'Y} = \begin{bmatrix} 15,947 \\ 31,283,250 \\ 6.53069 \times 10^{10} \end{bmatrix}$$

and (obtained using a statistical software package):

$$(\mathbf{X'X})^{-1} = \begin{bmatrix} 26.9156 & -.027027 & 6.3554 \times 10^{-6} \\ -.027027 & 2.75914 \times 10^{-5} & -6.5804 \times 10^{-9} \\ 6.3554 \times 10^{-6} & -6.5804 \times 10^{-9} & 1.5934 \times 10^{-12} \end{bmatrix}$$

Finally, performing the multiplication, we obtain

$$\hat{\beta} = (\mathbf{X'X})^{-1}\mathbf{X'Y}$$

$$= \begin{bmatrix} 26.9156 & -.027027 & 6.3554 \times 10^{-6} \\ -.027027 & 2.75914 \times 10^{-5} & -6.5804 \times 10^{-9} \\ 6.3554 \times 10^{-6} & -6.5804 \times 10^{-9} & 1.5934 \times 10^{-12} \end{bmatrix} \begin{bmatrix} 15,947 \\ 31,283,250 \\ 6.53069 \times 10^{10} \end{bmatrix}$$

$$= \begin{bmatrix} -1,216.14389 \\ 2.39893 \\ -.00045 \end{bmatrix}$$

Thus,

$$\hat{\beta}_0 = -1,216.14389$$

$$\hat{\beta}_1 = 2.39893$$

$$\hat{\beta}_2 = -.00045$$

and the prediction equation is

$$\hat{y} = -1,216.14389 + 2.39893x - .00045x^2$$

The MINITAB printout for the regression analysis is shown in Figure B.1. Note that the β estimates we obtained agree with the shaded values. ∎

Figure B.1 MINITAB regression printout for power usage model

```
The regression equation is
Y = - 1216 + 2.40 X -0.000450 XSQ

Predictor        Coef      SE Coef          T        P
Constant      -1216.1        242.8      -5.01    0.002
X              2.3989       0.2458       9.76    0.000
XSQ        -0.00045004   0.00005908     -7.62    0.000

S = 46.80      R-Sq = 98.2%     R-Sq(adj) = 97.7%

Analysis of Variance

Source          DF          SS          MS         F        P
Regression       2      831070      415535    189.71    0.000
Residual Error   7       15333        2190
Total            9      846402
```

B.5 Exercises

B.16 Use the method of least squares to fit a straight line to the five data points:

x	-2	-1	0	1	2
y	4	3	3	1	-1

(a) Construct **Y** and **X** matrices for the data.
(b) Find **X'X** and **X'Y**.
(c) Find the least squares estimates $\hat{\beta} = (\mathbf{X'X})^{-1}\mathbf{X'Y}$. [*Note:* See Theorem B.1 for information on finding $(\mathbf{X'X})^{-1}$.]
(d) Give the prediction equation.

B.17 Use the method of least squares to fit the model $E(y) = \beta_0 + \beta_1 x$ to the six data points:

x	1	2	3	4	5	6
y	1	2	2	3	5	6

(a) Construct **Y** and **X** matrices for the data.
(b) Find **X'X** and **X'Y**.
(c) Verify that

$$(\mathbf{X'X})^{-1} = \begin{bmatrix} \frac{13}{15} & -\frac{7}{35} \\ -\frac{7}{35} & \frac{2}{35} \end{bmatrix}$$

(d) Find the $\hat{\beta}$ matrix.

(e) Give the prediction equation.

B.18 An experiment was conducted in which two y observations were collected for each of five values of x:

x	-2		-1		0		1		2	
y	1.1	1.3	2.0	2.1	2.7	2.8	3.4	3.6	4.1	4.0

Use the method of least squares to fit the second-order model, $E(y) = \beta_0 + \beta_1 x + \beta_2 x^2$, to the 10 data points.

(a) Give the dimensions of the **Y** and **X** matrices.
(b) Verify that

$$(\mathbf{X'X})^{-1} = \begin{bmatrix} \frac{17}{70} & 0 & -\frac{1}{14} \\ 0 & \frac{1}{20} & 0 \\ -\frac{1}{14} & 0 & \frac{1}{28} \end{bmatrix}$$

(c) Both **X'X** and $(\mathbf{X'X})^{-1}$ are symmetric matrices. What is a symmetric matrix?
(d) Find the $\hat{\beta}$ matrix and the least squares prediction equation.
(e) Plot the data points and graph the prediction equation.

B.6 Calculating SSE and s^2

You will recall that the variances of the estimators of all the β parameters and of $\hat{y}$ depend on the value of σ^2, the variance of the random error ε that appears in the linear model. Since σ^2 will rarely be known in advance, we must use the sample data to estimate its value.

> **Matrix Formulas for SSE and s^2**
>
> $$SSE = \mathbf{Y'Y} - \boldsymbol{\beta'}\mathbf{X'Y}$$
>
> $$s^2 = \frac{SSE}{n - \text{Number of } \beta \text{ parameters in model}}$$

We demonstrate the use of these formulas with the advertising–sales data in Example B.5.

Example B.7

Find the SSE for the advertising–sales data in Example B.5.

Solution

From Example B.5,

$$\hat{\boldsymbol{\beta}} = \begin{bmatrix} -.1 \\ .7 \end{bmatrix} \quad \text{and} \quad \mathbf{X'Y} = \begin{bmatrix} 10 \\ 37 \end{bmatrix}$$

Then,

$$\mathbf{Y'Y} = \begin{bmatrix} 1 & 1 & 2 & 2 & 4 \end{bmatrix} \begin{bmatrix} 1 \\ 1 \\ 2 \\ 2 \\ 4 \end{bmatrix} = 26$$

and

$$\hat{\boldsymbol{\beta}}'\mathbf{X'Y} = \begin{bmatrix} -.1 & .7 \end{bmatrix} \begin{bmatrix} 10 \\ 37 \end{bmatrix} = 24.9$$

So

$$SSE = \mathbf{Y'}Y - \hat{\boldsymbol{\beta}}'\mathbf{X'Y} = 26 - 24.9 = 1.1$$

(Note that this is the same answer as that obtained in Section 3.3.) Finally,

$$s^2 = \frac{SSE}{n - \text{Number of } \beta \text{ parameters in model}} = \frac{1.1}{5 - 2} = .367$$

This estimate is needed to construct a confidence interval for β_1, to test a hypothesis concerning its value, or to construct a confidence interval for the mean sales for a given advertising expenditure.

B.7 Standard Errors of Estimators, Test Statistics, and Confidence Intervals for $\beta_0, \beta_1, \ldots, \beta_k$

This appendix is important because all the relevant information pertaining to the standard errors of the sampling distributions of $\hat{\beta}_0, \hat{\beta}_1, \ldots, \hat{\beta}_k$ (and hence of $\hat{\mathbf{Y}}$) is contained in $(\mathbf{X'X})^{-1}$. Thus, if we denote the $(\mathbf{X'X})^{-1}$ matrix as

$$(\mathbf{X'X})^{-1} = \begin{bmatrix} c_{00} & c_{01} & \cdots & c_{0k} \\ c_{10} & c_{11} & \cdots & c_{1k} \\ c_{20} & c_{21} & \cdots & c_{2k} \\ \vdots & \vdots & \vdots & \vdots \\ c_{k0} & c_{k1} & \cdots & c_{kk} \end{bmatrix}$$

then it can be shown (proof omitted) that the standard errors of the sampling distributions of $\hat{\beta}_0, \hat{\beta}_1, \ldots, \hat{\beta}_k$ are

$$\sigma_{\hat{\beta}_0} = \sigma\sqrt{c_{00}}$$
$$\sigma_{\hat{\beta}_1} = \sigma\sqrt{c_{11}}$$
$$\sigma_{\hat{\beta}_2} = \sigma\sqrt{c_{22}}$$
$$\vdots$$
$$\sigma_{\hat{\beta}_k} = \sigma\sqrt{c_{kk}}$$

where σ is the standard deviation of the random error ε. In other words, the diagonal elements of $(\mathbf{X'X})^{-1}$ give the values of $c_{00}, c_{11}, \ldots, c_{kk}$ that are required for finding the standard errors of the estimators $\hat{\beta}_0, \hat{\beta}_1, \ldots, \hat{\beta}_k$. The estimated values of the standard errors are obtained by replacing σ by s in the formulas for the standard errors. Thus, the estimated standard error of $\hat{\beta}_1$ is $s_{\hat{\beta}_1} = s\sqrt{c_{11}}$.

The confidence interval for a single β parameter, β_i, is given in the next box.

Confidence Interval for β_i

$$\hat{\beta}_i \pm t_{\alpha/2} \text{ (Estimated standard error of } \hat{\beta}_i)$$

or

$$\hat{\beta}_i \pm (t_{\alpha/2})s\sqrt{c_{ii}}$$

where $t_{\alpha/2}$ is based on the number of degrees of freedom associated with s.

Similarly, the test statistic for testing the null hypothesis $H_0 : \beta_i = 0$ is as shown in the following box.

Test Statistic for H_0: $\beta_i = 0$

$$t = \frac{\hat{\beta}_i}{s\sqrt{c_{ii}}}$$

Example B.8

Refer to Example B.5 and find the estimated standard error for the sampling distribution of $\hat{\beta}_1$, the estimator of the slope of the line β_1. Then give a 95% confidence interval for β_1.

Solution

The $(\mathbf{X'X})^{-1}$ matrix for the least squares solution in Example B.5 was

$$(\mathbf{X'X})^{-1} = \begin{bmatrix} 1.1 & -.3 \\ -.3 & .1 \end{bmatrix}$$

Therefore, $c_{00} = 1.1$, $c_{11} = .1$, and the estimated standard error for $\hat{\beta}_1$ is

$$s_{\hat{\beta}_1} = s\sqrt{c_{11}} = \sqrt{.367}(\sqrt{.1}) = .192$$

The value for s, $\sqrt{.367}$, was obtained from Example B.7.

A 95% confidence interval for β_1 is

$$\hat{\beta}_1 \pm (t_{\alpha/2})s\sqrt{c_{11}}$$
$$.7 \pm (3.182)(.192) = (.09, 1.31)$$

The t value, $t_{.025}$, is based on $(n-2) = 3$ df. Observe that this is the same confidence interval as the one obtained in Section 3.6. ▨

Refer to Example B.6 and the least squares solution for fitting power usage y to the size of a home x using the model

$$y = \beta_0 + \beta_1 x + \beta_2 x^2 + \varepsilon$$

The MINITAB printout for the analysis is reproduced in Figure B.2.

(a) Compute the estimated standard error for $\hat{\beta}_1$, and compare this result with the value shaded in Figure B.2.

(b) Compute the value of the test statistic for testing $H_0: \beta_2 = 0$. Compare this with the value shaded in Figure B.2.

Figure B.2 MINITAB regression printout for power usage model

```
The regression equation is
Y = - 1216 + 2.40 X -0.000450 XSQ

Predictor        Coef      SE Coef          T        P
Constant       -1216.1        242.8      -5.01    0.002
X               2.3989       0.2458       9.76    0.000
XSQ        -0.00045004   0.00005908      -7.62    0.000

S = 46.80      R-Sq = 98.2%      R-Sq(adj) = 97.7%

Analysis of Variance

Source          DF        SS         MS        F        P
Regression       2     831070     415535   189.71    0.000
Residual Error   7      15333       2190
Total            9     846402
```

Solution

The fitted model is

$$\hat{y} = -1,216.14389 + 2.39893x - .00045x^2$$

The $(\mathbf{X}'\mathbf{X})^{-1}$ matrix, obtained in Example B.6, is

$$(\mathbf{X}'\mathbf{X})^{-1} = \begin{bmatrix} 26.9156 & -.027027 & 6.3554 \times 10^{-6} \\ -.027027 & 2.75914 \times 10^{-5} & -6.5804 \times 10^{-9} \\ 6.3554 \times 10^{-6} & -6.5804 \times 10^{-9} & 1.5934 \times 10^{-12} \end{bmatrix}$$

From $(\mathbf{X}'\mathbf{X})^{-1}$, we know that

$c_{00} = 26.9156$

$c_{11} = 2.75914 \times 10^{-5}$

$c_{22} = 1.5934 \times 10^{-12}$

and from the printout, $s = 46.80$.

(a) The estimated standard error of $\hat{\beta}_1$ is

$$s_{\hat{\beta}_1} = s\sqrt{c_{11}}$$

$$= (46.80)\sqrt{2.75914 \times 10^{-1}} = .2458$$

Notice that this agrees with the value of $s_{\hat{\beta}_1}$ shaded in the MINITAB printout (Figure B.2).

(b) The value of the test statistic for testing H_0: $\beta_2 = 0$ is

$$t = \frac{\hat{\beta}_2}{s\sqrt{c_{22}}} = \frac{-.00045}{(46.80)\sqrt{1.5934 \times 10^{-12}}} = -7.62$$

Notice that this value of the t statistic agrees with the value -7.62 shaded in the printout (Figure B.2). ◼

B.7 Exercises

B.19 Do the data given in Exercise B.16 provide sufficient evidence to indicate that x contributes information for the prediction of y? Test H_0: $\beta_1 = 0$ against H_a: $\beta_1 \neq 0$ using $\alpha = .05$.

B.20 Find a 90% confidence interval for the slope of the line in Exercise B.19.

B.21 The term in the second-order model $E(y) = \beta_0 + \beta_1 x + \beta_2 x^2$ that controls the curvature in its graph is $\beta_2 x^2$. If $\beta_2 = 0$, $E(y)$ graphs as a straight line. Do the data given in Exercise B.18 provide sufficient evidence to indicate curvature in the model for $E(y)$? Test H_0: $\beta_2 = 0$ against H_a: $\beta_2 \neq 0$ using $\alpha = .10$.

B.8 A Confidence Interval for a Linear Function of the β Parameters; a Confidence Interval for $E(y)$

Suppose we were to postulate that the mean value of the productivity, y, of a company is related to the size of the company, x, and that the relationship could be modeled by the expression

$$E(y) = \beta_0 + \beta_1 x + \beta_2 x^2$$

A graph of $E(y)$ might appear as shown in Figure B.3.

We might have several reasons for collecting data on the productivity and size of a set of n companies and for finding the least squares prediction equation,

$$\hat{y} = \hat{\beta}_0 + \hat{\beta}_1 x + \hat{\beta}_2 x^2$$

For example, we might wish to estimate the mean productivity for a company of a given size (say, $x = 2$). That is, we might wish to estimate

$$E(y) = \beta_0 + \beta_1 x + \beta_2 x^2$$
$$= \beta_0 + 2\beta_1 + 4\beta_2 \quad \text{where} \quad x = 2$$

Or we might wish to estimate the marginal increase in productivity, the slope of a tangent to the curve, when $x = 2$ (see Figure B.4). The marginal productivity for y

Figure B.3 Graph of mean productivity $E(y)$

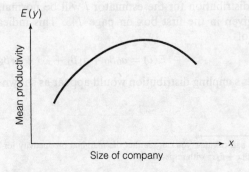

Mean productivity

Size of company

Figure B.4 Marginal productivity

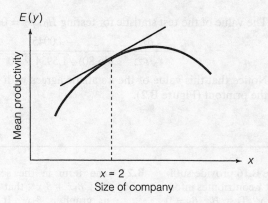

when $x = 2$ is the rate of change of $E(y)$ with respect to x, evaluated at $x = 2$.* The marginal productivity for a value of x, denoted by the symbol $dE(y)/dx$, can be shown (proof omitted) to be

$$\frac{dE(y)}{dx} = \beta_1 + 2\beta_2 x$$

Therefore, the marginal productivity at $x = 2$ is

$$\frac{dE(y)}{dx} = \beta_1 + 2\beta_2(2) = \beta_1 + 4\beta_2$$

For $x = 2$, both $E(y)$ and the marginal productivity are *linear* functions of the unknown parameters $\beta_0, \beta_1, \beta_2$ in the model. The problem we pose in this section is that of finding confidence intervals for linear functions of β parameters or testing hypotheses concerning their values. The information necessary to solve this problem is rarely given in a standard multiple regression analysis computer printout, but we can find these confidence intervals or values of the appropriate test statistics from knowledge of $(\mathbf{X}'\mathbf{X})^{-1}$.

For the model

$$y = \beta_0 + \beta_1 x_1 + \cdots + \beta_k x_k + \varepsilon$$

we can make an inference about a linear function of the β parameters, say,

$$a_0 \beta_0 + a_1 \beta_1 + \cdots + a_k \beta_k$$

where $a_0, a_1, \ldots, a_k$ are known constants. We will use the corresponding linear function of least squares estimates,

$$\ell = a_0 \hat{\beta}_0 + a_1 \hat{\beta}_1 + \cdots + a_k \hat{\beta}_k$$

as our best estimate of $a_0 \beta_0 + a_1 \beta_1 + \cdots + a_k \beta_k$.

Then, for the assumptions on the random error ε (stated in Section 4.2), the sampling distribution for the estimator l will be normal, with mean and standard error as given in the first box on page 743. This indicates that l is an unbiased estimator of

$$E(\ell) = a_0 \beta_0 + a_1 \beta_1 + \cdots + a_k \beta_k$$

and that its sampling distribution would appear as shown in Figure B.5.

---

* If you have had calculus, you can see that the marginal productivity for y given x is the first derivative of $E(y) = \beta_0 + \beta_1 x + \beta_2 x^2$ with respect to x.

Figure B.5 Sampling distribution for ℓ

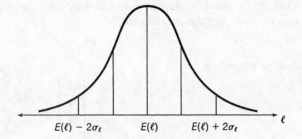

$$E(\ell) - 2\sigma_\ell \qquad E(\ell) \qquad E(\ell) + 2\sigma_\ell \qquad \ell$$

Mean and Standard Error of ℓ

$$E(\ell) = a_0\beta_0 + a_1\beta_1 + \cdots + a_k\beta_k$$

$$\sigma_l = \sqrt{\sigma^2 \mathbf{a}'(\mathbf{X}'\mathbf{X})^{-1}\mathbf{a}}$$

where σ^2 is the variance of ε, $(\mathbf{X}'\mathbf{X})^{-1}$ is the inverse matrix obtained in fitting the least squares model to the set of data, and

$$\mathbf{a} = \begin{bmatrix} a_0 \\ a_1 \\ a_2 \\ \vdots \\ a_k \end{bmatrix}$$

It can be demonstrated that a $100(1 - \alpha)\%$ confidence interval for $E(\ell)$ is as shown in the next box.

A $100(1 - \alpha)\%$ Confidence Interval for $E(\ell)$

$$\ell \pm t_{\alpha/2}\sqrt{s^2 \mathbf{a}'(\mathbf{X}'\mathbf{X})^{-1}\mathbf{a}}$$

where

$$E(\ell) = a_0\beta_0 + a_1\beta_1 + \cdots + a_k\beta_k$$

$$\ell = a_0\hat{\beta}_0 + a_1\hat{\beta}_1 + \cdots + a_k\hat{\beta}_k \qquad \mathbf{a} = \begin{bmatrix} a_0 \\ a_1 \\ a_2 \\ \vdots \\ a_k \end{bmatrix}$$

s^2 and $(\mathbf{X}'\mathbf{X})^{-1}$ are obtained from the least squares procedure, and $t_{\alpha/2}$ is based on the number of degrees of freedom associated with s^2.

The linear function of the β parameters that is most often the focus of our attention is

$$E(y) = \beta_0 + \beta_1 x_1 + \cdots + \beta_k x_k$$

That is, we want to find a confidence interval for $E(y)$ for specific values of $x_1, x_2, \ldots, x_k$. For this special case,

$$\ell = \hat{y}$$

and the $\mathbf{a}$ matrix is

$$\mathbf{a} = \begin{bmatrix} 1 \\ x_1 \\ x_2 \\ \vdots \\ x_k \end{bmatrix}$$

where the symbols $x_1, x_2, \ldots, x_k$ in the $\mathbf{a}$ matrix indicate the specific numerical values assumed by these variables. Thus, the procedure for forming a confidence interval for $E(y)$ is as shown in the box.

A $100(1 - \alpha)\%$ Confidence Interval for $E(y)$

$$\ell \pm t_{\alpha/2}\sqrt{s^2\mathbf{a}'(\mathbf{X}'\mathbf{X})^{-1}\mathbf{a}}$$

where

$$E(y) = \beta_0 + \beta_1 x_1 + \beta_2 x_2 + \cdots + \beta_k x_k$$

$$\ell = \hat{y} = \hat{\beta}_0 + \hat{\beta}_1 x_1 + \cdots + \hat{\beta}_k x_k \quad \mathbf{a} = \begin{bmatrix} 1 \\ x_1 \\ x_2 \\ \vdots \\ x_k \end{bmatrix}$$

s^2 and $(\mathbf{X}'\mathbf{X})^{-1}$ are obtained from the least squares analysis, and $t_{\alpha/2}$ is based on the number of degrees of freedom associated with s^2, namely, $n - (k + 1)$.

Example B.10

Refer to the data in Example B.5 for sales revenue y and advertising expenditure x. Find a 95% confidence interval for the mean sales revenue $E(y)$ when advertising expenditure is $x = 4$.

Solution

The confidence interval for $E(y)$ for a given value of x is

$$\hat{y} \pm t_{\alpha/2}\sqrt{s^2\mathbf{a}'(\mathbf{X}'\mathbf{X})^{-1}\mathbf{a}}$$

Consequently, we need to find and substitute the values of $\mathbf{a}'(\mathbf{X}'\mathbf{X})^{-1}\mathbf{a}$, $t_{\alpha/2}$, and $\hat{y}$ into this formula. Since we wish to estimate

$$E(y) = \beta_0 + \beta_1 x$$
$$= \beta_0 + \beta_1(4) \quad \text{when} \quad x = 4$$
$$= \beta_0 + 4\beta_1$$

it follows that the coefficients of β_0 and β_1 are $a_0 = 1$ and $a_1 = 4$, and thus,

$$\mathbf{a} = \begin{bmatrix} 1 \\ 4 \end{bmatrix}$$

From Examples B.5 and B.7, $\hat{y} = -.1 + .7x$,

$$(\mathbf{X'X})^{-1} = \begin{bmatrix} 1.1 & -.3 \\ -.3 & .1 \end{bmatrix}$$

and $s^2 = .367$. Then,

$$\mathbf{a'(X'X)^{-1}a} = \begin{bmatrix} 1 & 4 \end{bmatrix} \begin{bmatrix} 1.1 & -.3 \\ -.3 & .1 \end{bmatrix} \begin{bmatrix} 1 \\ 4 \end{bmatrix}$$

We first calculate

$$\mathbf{a'(X'X)^{-1}} = \begin{bmatrix} 1 & 4 \end{bmatrix} \begin{bmatrix} 1.1 & -.3 \\ -.3 & .1 \end{bmatrix} = \begin{bmatrix} -.1 & .1 \end{bmatrix}$$

Then,

$$\mathbf{a'(X'X)^{-1}a} = \begin{bmatrix} -.1 & .1 \end{bmatrix} \begin{bmatrix} 1 \\ 4 \end{bmatrix} = .3$$

The t value, $t_{.025}$, based on 3 df is 3.182. So, a 95% confidence interval for the mean sales revenue with an advertising expenditure of 4 is

$$\hat{y} \pm t_{\alpha/2} \sqrt{s^2 \mathbf{a'(X'X)^{-1}a}}$$

Since $\hat{y} = -.1 + .7x = -.1 + (.7)(4) = 2.7$, the 95% confidence interval for $E(y)$ when $x = 4$ is

$$2.7 \pm (3.182)\sqrt{(.367)(.3)}$$

$$2.7 \pm 1.1$$

Notice that this is exactly the same result as obtained in Example 3.4. ▬

Example B.11

An economist recorded a measure of productivity y and the size x for each of 100 companies producing cement. A regression model,

$$y = \beta_0 + \beta_1 x + \beta_2 x^2 + \varepsilon$$

fit to the $n = 100$ data points produced the following results:

$$\hat{y} = 2.6 + .7x - .2x^2$$

where x is coded to take values in the interval $-2 < x < 2$,[†] and

$$(\mathbf{X'X})^{-1} = \begin{bmatrix} .0025 & .0005 & -.0070 \\ .0005 & .0055 & 0 \\ -.0070 & 0 & .0050 \end{bmatrix} \quad s = .14$$

Find a 95% confidence interval for the marginal increase in productivity given that the coded size of a plant is $x = 1.5$.

Solution

The mean value of y for a given value of x is

$$E(y) = \beta_0 + \beta_1 x + \beta_2 x^2$$

Therefore, the marginal increase in y for $x = 1.5$ is

$$\frac{dE(y)}{dx} = \beta_1 + 2\beta_2 x$$

$$= \beta_1 + 2(1.5)\beta_2$$

[†] We give a formula for *coding* observational data in Section 5.6.

Figure B.6 A graph of $\hat{y} = 2.6 + .7x - .2x^2$

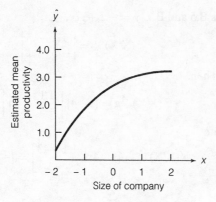

Or,

$$E(l) = \beta_1 + 3\beta_2 \quad \text{when} \quad x = 1.5$$

Note from the prediction equation, $\hat{y} = 2.6 + .7x - .2x^2$, that $\hat{\beta}_1 = .7$ and $\hat{\beta}_2 = -.2$. Therefore,

$$l = \hat{\beta}_1 + 3\hat{\beta}_2 = .7 + 3(-.2) = .1$$

and

$$\mathbf{a} = \begin{bmatrix} a_0 \\ a_1 \\ a_2 \end{bmatrix} = \begin{bmatrix} 0 \\ 1 \\ 3 \end{bmatrix}$$

We next calculate

$$\mathbf{a}'(\mathbf{X}'\mathbf{X})^{-1}\mathbf{a} = \begin{bmatrix} 0 & 1 & 3 \end{bmatrix} \begin{bmatrix} .0025 & .0005 & -.0070 \\ .0005 & .0055 & 0 \\ -.0070 & 0 & .0050 \end{bmatrix} \begin{bmatrix} 0 \\ 1 \\ 3 \end{bmatrix} = .0505$$

Then, since s is based on $n - (k + 1) = 100 - 3 = 97$ df, $t_{.025} \approx 1.96$, and a 95% confidence interval for the marginal increase in productivity when $x = 1.5$ is

$$\ell \pm t_{.025}\sqrt{(s^2)\mathbf{a}'(\mathbf{X}'\mathbf{X})^{-1}\mathbf{a}}$$

or

$$.1 \pm (1.96)\sqrt{(.14)^2(.0505)}$$

$$.1 \pm .062$$

Thus, the marginal increase in productivity, the slope of the tangent to the curve

$$E(y) = \beta_0 + \beta_1 x + \beta_2 x^2$$

is estimated to lie in the interval $.1 \pm .062$ at $x = 1.5$. A graph of $\hat{y} = 2.6 + .7x - .2x^2$ is shown in Figure B.6.

B.9 A Prediction Interval for Some Value of y to Be Observed in the Future

We have indicated in Sections 3.9 and 4.12 that two of the most important applications of the least squares predictor $\hat{y}$ are estimating the mean value of y (the topic of the preceding section) and predicting a new value of y, yet unobserved, for specific values of $x_1, x_2, \ldots, x_k$. The difference between these two inferential problems (when each would be pertinent) was explained in Chapters 3 and 4, but we give another example to make certain that the distinction is clear.

Suppose you are the manager of a manufacturing plant and that y, the daily profit, is a function of various process variables $x_1, x_2, \ldots, x_k$. Suppose you want to know how much money you would make *in the long run* if the x's are set at specific values. For this case, you would be interested in finding a confidence interval for the mean profit per day, $E(y)$. In contrast, suppose you planned to operate the plant for just one more day! Then you would be interested in predicting the value of y, the profit associated with tomorrow's production.

We have indicated that the error of prediction is always larger than the error of estimating $E(y)$. You can see this by comparing the formula for the prediction interval (shown in the next box) with the formula for the confidence interval for $E(y)$ that was given in Section B.8.

A $100(1 - \alpha)\%$ Prediction Interval for y

$$\hat{y} \pm t_{\alpha/2}\sqrt{s^2 + s^2\mathbf{a}'(\mathbf{X}'\mathbf{X})^{-1}\mathbf{a}} = \hat{y} \pm t_{\alpha/2}\sqrt{s^2[1 + \mathbf{a}'(\mathbf{X}'\mathbf{X})^{-1}\mathbf{a}]}$$

where

$$\hat{y} = \hat{\beta}_0 + \hat{\beta}_1 x_1 + \cdots + \hat{\beta}_k x_k$$

s^2 and $(\mathbf{X}'\mathbf{X})^{-1}$ are obtained from the least squares analysis,

$$\mathbf{a} = \begin{bmatrix} 1 \\ x_1 \\ x_2 \\ \vdots \\ x_k \end{bmatrix}$$

contains the numerical values of $x_1, x_2, \ldots, x_k$, and $t_{\alpha/2}$ is based on the number of degrees of freedom associated with s^2, namely, $n - (k + 1)$.

Example B.12

Refer to the sales–advertising expenditure example (Example B.10). Find a 95% prediction interval for the sales revenue next month, if it is known that next month's advertising expenditure will be $x = 4$.

Solution

The 95% prediction interval for sales revenue y is

$$\hat{y} \pm t_{\alpha/2}\sqrt{s^2[1 + \mathbf{a}'(\mathbf{X}'\mathbf{X})^{-1}\mathbf{a}]}$$

From Example B.10, when $x = 4$, $\hat{y} = -.1 + .7x = -.1 + (.7)(4) = 2.7$, $s^2 = .367$, $t_{.025} = 3.182$, and $\mathbf{a}'(\mathbf{X}'\mathbf{X})^{-1}\mathbf{a} = .3$. Then the 95% prediction interval for y is

$$2.7 \pm (3.182)\sqrt{(.367)(1 + .3)}$$

$$2.7 \pm 2.2$$

You will find that this is the same solution as obtained in Example 3.5.   ◼

B.9 Exercises

B.22 Refer to Exercise B.16. Find a 90% confidence interval for $E(y)$ when $x = 1$. Interpret the interval.

B.23 Refer to Exercise B.16. Suppose you plan to observe y for $x = 1$. Find a 90% prediction interval for that value of y. Interpret the interval.

B.24 Refer to Exercise B.17. Find a 90% confidence interval for $E(y)$ when $x = 2$. Interpret the interval.

B.25 Refer to Exercise B.17. Find a 90% prediction interval for a value of y to be observed in the future when $x = 2$. Interpret the interval.

B.26 Refer to Exercise B.18. Find a 90% confidence interval for the mean value of y when $x = 1$. Interpret the interval.

B.27 Refer to Exercise B.18. Find a 90% prediction interval for a value of y to be observed in the future when $x = 1$.

B.28 The productivity (items produced per hour) per worker on a manufacturing assembly line is expected to increase as piecework pay rate (in dollars) increases; it is expected to stabilize after a certain pay rate has been reached. The productivity of five different workers was recorded for each of five piecework pay rates, \$.80, \$.90, \$1.00, \$1.10, and \$1.20, thus giving $n = 25$ data points. A multiple regression analysis using a second-order model,

$$E(y) = \beta_0 + \beta_1 x + \beta_2 x^2$$

gave

$$\hat{y} = 2.08 + 8.42x - 1.65x^2$$

$$\text{SSE} = 26.62, \text{SS}_{yy} = 784.11, \text{ and}$$

$$(\mathbf{X'X})^{-1} = \begin{bmatrix} .020 & -.010 & .015 \\ -.010 & .040 & -.006 \\ .015 & -.006 & .028 \end{bmatrix}$$

(a) Find s^2.

(b) Find a 95% confidence interval for the mean productivity when the pay rate is \$1.10. Interpret this interval.

(c) Find a 95% prediction interval for the production of an individual worker who is paid at a rate of \$1.10 per piece. Interpret the interval.

(d) Find R^2 and interpret the value.

Summary

Except for the tedious process of inverting a matrix (discussed in Appendix C), we have covered the major steps performed by a computer in fitting a linear statistical model to a set of data using the method of least squares. We have also explained how to find the confidence intervals, prediction intervals, and values of test statistics that would be pertinent in a regression analysis.

In addition to providing a better understanding of a multiple regression analysis, the most important contributions of this appendix are contained in Sections B.8 and B.9. If you want to make a specific inference concerning the mean value of y or any linear function of the β parameters and if you are unable to obtain the results from the computer package you are using, you will find the contents of Sections B.8 and B.9 very useful. Since you will almost always be able to find a computer program package to find $(\mathbf{X'X})^{-1}$, you will be able to calculate the desired confidence interval(s) and so forth on your own.

Supplementary Exercises

B.29. Use the method of least squares to fit a straight line to the six data points:

x	−5	−3	−1	1	3	5
y	1.1	1.9	3.0	3.8	5.1	6.0

(a) Construct **Y** and **X** matrices for the data.

(b) Find **X'X** and **X'Y**.

(c) Find the least squares estimates,

$$\hat{\beta} = (\mathbf{X'X})^{-1}\mathbf{X'Y}$$

[*Note*: See Theorem A.1 for information on finding $(\mathbf{X'X})^{-1}$.]

(d) Give the prediction equation.

(e) Find SSE and s^2.

(f) Does the model contribute information for the prediction of y? Test $H_0: \beta_1 = 0$. Use $\alpha = .05$.

(g) Find r^2 and interpret its value.

(h) Find a 90% confidence interval for $E(y)$ when $x = .5$. Interpret the interval.

B.30. An experiment was conducted to investigate the effect of extrusion pressure P and temperature T on the strength y of a new type of plastic. Two plastic specimens were prepared for each of five combinations of pressure and temperature. The specimens were then tested in random order, and the breaking strength for each specimen was recorded. The independent variables were coded to simplify computations, that is,

$$x_1 = \frac{P - 200}{10} \qquad x_2 = \frac{T - 400}{25}$$

The $n = 10$ data points are listed in the table.

y	x_1	x_2
5.2; 5.0	-2	2
.3; $-.1$	-1	-1
$-1.2; -1.1$	0	-2
2.2; 2.0	1	-1
6.2; 6.1	2	2

(a) Give the **Y** and **X** matrices needed to fit the model $y = \beta_0 + \beta_1 x_1 + \beta_2 x_2 + \varepsilon$.

(b) Find the least squares prediction equation.

(c) Find SSE and s^2.

(d) Does the model contribute information for the prediction of y? Test using $\alpha = .05$.

(e) Find R^2 and interpret its value.

(f) Test the null hypothesis that $\beta_1 = 0$. Use $\alpha = .05$. What is the practical implication of the test?

(g) Find a 90% confidence interval for the mean strength of the plastic for $x_1 = -2$ and $x_2 = 2$.

(h) Suppose a single specimen of the plastic is to be installed in the engine mount of a Douglas DC-10 aircraft. Find a 90% prediction interval for the strength of this specimen if $x_1 = -2$ and $x_2 = 2$.

B.31. Suppose we obtained two replications of the experiment described in Exercise B.17; that is, two values of y were observed for each of the six values of x. The data are shown at the bottom of the page.

(a) Suppose (as in Exercise B.17) you wish to fit the model $E(y) = \beta_0 + \beta_1 x$. Construct **Y** and **X** matrices for the data. [*Hint:* Remember, the **Y** matrix must be of dimension 12×1.]

(b) Find **X'X** and **X'Y**.

(c) Compare the **X'X** matrix for two replications of the experiment with the **X'X** matrix obtained for a single replication (part b of Exercise B.17). What is the relationship between the elements in the two matrices?

(d) Observe the $(\mathbf{X'X})^{-1}$ matrix for a single replication (see part c of Exercise B.17). Verify that the $(\mathbf{X'X})^{-1}$ matrix for two replications contains elements that are equal to $\frac{1}{2}$ of the values of the corresponding elements in the $(\mathbf{X'X})^{-1}$ matrix for a single replication of the experiment. [*Hint:* Show that the product of the $(\mathbf{X'X})^{-1}$ matrix (for two replications) and the **X'X** matrix from part c equals the identity matrix **I**.]

(e) Find the prediction equation.

(f) Find SSE and s^2.

(g) Do the data provide sufficient information to indicate that x contributes information for the prediction of y? Test using $\alpha = .05$.

(h) Find r^2 and interpret its value.

B.32. Refer to Exercise B.31.

(a) Find a 90% confidence interval for $E(y)$ when $x = 4.5$. Interpret the interval.

(b) Suppose we wish to predict the value of y if, in the future, $x = 4.5$. Find a 90% prediction interval for y and interpret the interval.

B.33. Refer to Exercise B.31. Suppose you replicated the experiment described in Exercise B.17 three times; that is, you collected three observations on y for each value of x. Then $n = 18$.

(a) What would be the dimensions of the **Y** matrix?

(b) Write the **X** matrix for three replications. Compare with the **X** matrices for one and for two replications. Note the pattern.

(c) Examine the **X'X** matrices obtained for one and two replications of the experiment (obtained in Exercises B.17 and B.31, respectively). Deduce the values of the elements of the **X'X** matrix for three replications.

(d) Look at your answer to Exercise B.31, part d. Deduce the values of the elements in the $(\mathbf{X'X})^{-1}$ matrix for three replications.

(e) Suppose you wanted to find a 90% confidence interval for $E(y)$ when $x = 4.5$ based on three replications of the experiment. Find the value of $\mathbf{a'(X'X)^{-1}a}$ that appears in the confidence interval and compare with the value of $\mathbf{a'(X'X)^{-1}a}$ that would be obtained for a single replication of the experiment.

x	1		2		3		4		5		6	
y	1.1	.5	1.8	2.0	2.0	2.9	3.8	3.4	4.1	5.0	5.0	5.8

(f) Approximately how much of a reduction in the width of the confidence interval is obtained by using three versus two replications?

[*Note*: The values of *s* computed from the two sets of data will almost certainly be different.]

References

Draper, N. and Smith, H. *Applied Regression Analysis.* 3rd ed. New York: Wiley, 1998.

Graybill, F. A. *Theory and Application of the Linear Model.* North Scituate, Mass.: Duxbury, 1976.

Kutner, M. H., Nachtsheim, C. J., Neter, J., and Li, W. *Applied Linear Statistical Models*, 5th ed. New York: McGraw-Hill, 2005.

Mendenhall, W. *Introduction to Linear Models and the Design and Analysis of Experiments.* Belmont, Calif.: Wadsworth, 1968.

A PROCEDURE FOR INVERTING A MATRIX

There are several different methods for inverting matrices. All are tedious and time-consuming. Consequently, in practice, you will invert almost all matrices using a computer. The purpose of this section is to present one method for inverting small (2×2 or 3×3) matrices manually, thus giving you an appreciation of the enormous computing problem involved in inverting large matrices (and, consequently, in fitting linear models containing many terms to a set of data). In particular, you will be able to understand why rounding errors creep into the inversion process and, consequently, why two different computer programs might invert the same matrix and produce inverse matrices with slightly different corresponding elements.

The procedure we will demonstrate to invert a matrix **A** requires us to perform a series of operations on the rows of the **A** matrix. For example, suppose

$$\mathbf{A} = \begin{bmatrix} 1 & -2 \\ -2 & 6 \end{bmatrix}$$

We will identify two different ways to operate on a row of a matrix:*

1. We can multiply every element in one particular row by a constant, *c*. For example, we could operate on the first row of the **A** matrix by multiplying every element in the row by a constant, say, 2. Then the resulting row would be [2 − 4].

2. We can operate on a row by multiplying another row of the matrix by a constant and then adding (or subtracting) the elements of that row to elements in corresponding positions in the row operated upon. For example, we could operate on the first row of the **A** matrix by multiplying the second row by a constant, say, 2:

$$2[-2 \ 6] = [-4 \ 12]$$

Then we add this row to row 1:

$$[(1 - 4)(-2 + 12)] = [-3 \ 10]$$

Note one important point. We operated on the *first* row of the **A** matrix. Although we used the second row of the matrix to perform the operation, *the second row would remain unchanged*. Therefore, the row operation on the **A** matrix that we have just described would produce the new matrix,

$$\begin{bmatrix} -3 & 10 \\ -2 & 6 \end{bmatrix}$$

---

*We omit a third row operation, because it would add little and could be confusing.

Matrix inversion using row operations is based on an elementary result from matrix algebra. It can be shown (proof omitted) that performing a series of row operations on a matrix **A** is equivalent to multiplying **A** by a matrix **B** (i.e., row operations produce a new matrix, **BA**). This result is used as follows: Place the **A** matrix and an identity matrix **I** of the same dimensions side by side. Then perform the same series of row operations on both **A** and **I** until the **A** matrix has been changed into the identity matrix **I**. This means that you have multiplied both **A** and **I** by some matrix **B** such that:

$$\mathbf{A} = \begin{bmatrix} & & \\ & & \\ & & \end{bmatrix} \qquad \mathbf{I} = \begin{bmatrix} 1 & 0 & 0 & \cdots & 0 \\ 0 & 1 & 0 & \cdots & 0 \\ 0 & 0 & 1 & \cdots & 0 \\ \vdots & \vdots & \vdots & & \vdots \\ 0 & 0 & 0 & \cdots & 1 \end{bmatrix}$$

$$\downarrow \qquad \leftarrow \text{Row operations change } \mathbf{A} \text{ to } \mathbf{I} \rightarrow \qquad \downarrow$$

$$\mathbf{I} = \begin{bmatrix} & & \\ & & \\ & & \end{bmatrix} \qquad \mathbf{B} = \begin{bmatrix} & & \\ & & \\ & & \end{bmatrix}$$

$$\mathbf{BA} = \mathbf{I} \text{ and } \mathbf{BI} = \mathbf{B}$$

Since **BA** = **I**, it follows that $\mathbf{B} = \mathbf{A}^{-1}$. Therefore, as the **A** matrix is transformed by row operations into the identity matrix **I**, the identity matrix **I** is transformed into $\mathbf{A}^{-1}$, that is,

$$\mathbf{BI} = \mathbf{B} = \mathbf{A}^{-1}$$

We show you how this procedure works with two examples.

Example C.1

Find the inverse of the matrix

$$\mathbf{A} = \begin{bmatrix} 1 & -2 \\ -2 & 6 \end{bmatrix}$$

Solution

Place the **A** matrix and a 2×2 identity matrix side by side and then perform the following series of row operations (we indicate by an arrow the row operated upon in each operation):

$$\mathbf{A} = \begin{bmatrix} 1 & -2 \\ -2 & 6 \end{bmatrix} \qquad \mathbf{I} = \begin{bmatrix} 1 & 0 \\ 0 & 1 \end{bmatrix}$$

OPERATION 1:   Multiply the first row by 2 and add it to the second row:

$$\rightarrow \begin{bmatrix} 1 & -2 \\ 0 & 2 \end{bmatrix} \qquad \begin{bmatrix} 1 & 0 \\ 2 & 1 \end{bmatrix}$$

OPERATION 2:   Multiply the second row by $\frac{1}{2}$:

$$\rightarrow \begin{bmatrix} 1 & -2 \\ 0 & 1 \end{bmatrix} \qquad \begin{bmatrix} 1 & 0 \\ 1 & \frac{1}{2} \end{bmatrix}$$

OPERATION 3:   Multiply the second row by 2 and add it to the first row:

$$\rightarrow \begin{bmatrix} 1 & 0 \\ 0 & 1 \end{bmatrix} \quad \begin{bmatrix} 3 & 1 \\ 1 & \frac{1}{2} \end{bmatrix}$$

Thus,

$$\mathbf{A}^{-1} = \begin{bmatrix} 3 & 1 \\ 1 & \frac{1}{2} \end{bmatrix}$$

(Note that our solution matches the one obtained using Theorem B.2.)

The final step in finding an inverse is to check your solution by finding the product $\mathbf{A}^{-1}\,\mathbf{A}$ to see whether it equals the identity matrix $\mathbf{I}$. To check:

$$\mathbf{A}^{-1}\mathbf{A} = \begin{bmatrix} 3 & 1 \\ 1 & \frac{1}{2} \end{bmatrix} \begin{bmatrix} 1 & -2 \\ -2 & 6 \end{bmatrix}$$

$$= \begin{bmatrix} 1 & 0 \\ 0 & 1 \end{bmatrix}$$

Since this product is equal to the identity matrix, it follows that our solution for $\mathbf{A}^{-1}$ is correct.   ▬

Example C.2

Find the inverse of the matrix

$$\mathbf{A} = \begin{bmatrix} 2 & 0 & 3 \\ 0 & 4 & 1 \\ 3 & 1 & 2 \end{bmatrix}$$

Solution

Place an identity matrix alongside the $\mathbf{A}$ matrix and perform the row operations:

OPERATION 1:   Multiply row 1 by $\frac{1}{2}$:

$$\rightarrow \begin{bmatrix} 1 & 0 & \frac{3}{2} \\ 0 & 4 & 1 \\ 3 & 1 & 2 \end{bmatrix} \quad \begin{bmatrix} \frac{1}{2} & 0 & 0 \\ 0 & 1 & 0 \\ 0 & 0 & 1 \end{bmatrix}$$

OPERATION 2:   Multiply row 1 by 3 and subtract from row 3:

$$\begin{bmatrix} 1 & 0 & \frac{3}{2} \\ 0 & 4 & 1 \\ \rightarrow 0 & 1 & -\frac{5}{2} \end{bmatrix} \quad \begin{bmatrix} \frac{1}{2} & 0 & 0 \\ 0 & 1 & 0 \\ -\frac{3}{2} & 0 & 1 \end{bmatrix}$$

OPERATION 3:   Multiply row 2 by $\frac{1}{4}$:

$$\begin{bmatrix} 1 & 0 & \frac{3}{2} \\ \rightarrow 0 & 1 & \frac{1}{4} \\ 0 & 1 & -\frac{5}{2} \end{bmatrix} \quad \begin{bmatrix} \frac{1}{2} & 0 & 0 \\ 0 & \frac{1}{4} & 0 \\ -\frac{3}{2} & 0 & 1 \end{bmatrix}$$

OPERATION 4:   Subtract row 2 from row 3:

$$\begin{bmatrix} 1 & 0 & \frac{3}{2} \\ 0 & 1 & \frac{1}{4} \\ \rightarrow 0 & 0 & -\frac{11}{4} \end{bmatrix} \quad \begin{bmatrix} \frac{1}{2} & 0 & 0 \\ 0 & \frac{1}{4} & 0 \\ -\frac{3}{2} & -\frac{1}{4} & 1 \end{bmatrix}$$

OPERATION 5: Multiply row 3 by $-\frac{4}{11}$:

$$\rightarrow \begin{bmatrix} 1 & 0 & \frac{3}{2} \\ 0 & 1 & \frac{1}{4} \\ 0 & 0 & 1 \end{bmatrix} \quad \begin{bmatrix} \frac{1}{2} & 0 & 0 \\ 0 & \frac{1}{4} & 0 \\ \frac{12}{22} & \frac{1}{11} & -\frac{4}{11} \end{bmatrix}$$

OPERATION 6: Operate on row 2 by subtracting $\frac{1}{4}$ of row 3:

$$\rightarrow \begin{bmatrix} 1 & 0 & \frac{3}{2} \\ 0 & 1 & 0 \\ 0 & 0 & 1 \end{bmatrix} \quad \begin{bmatrix} \frac{1}{2} & 0 & 0 \\ -\frac{3}{22} & \frac{5}{22} & \frac{1}{11} \\ \frac{12}{22} & \frac{1}{11} & -\frac{4}{11} \end{bmatrix}$$

OPERATION 7: Operate on row 1 by subtracting $\frac{3}{2}$ of row 3:

$$\rightarrow \begin{bmatrix} 1 & 0 & 0 \\ 0 & 1 & 0 \\ 0 & 0 & 1 \end{bmatrix} \quad \begin{bmatrix} -\frac{7}{22} & -\frac{3}{22} & \frac{6}{11} \\ -\frac{3}{22} & \frac{5}{22} & \frac{1}{11} \\ \frac{6}{11} & \frac{1}{11} & -\frac{4}{11} \end{bmatrix} = \mathbf{A}^{-1}$$

To check the solution, we find the product:

$$\mathbf{A}^{-1}\mathbf{A} = \begin{bmatrix} -\frac{7}{22} & -\frac{3}{22} & \frac{6}{11} \\ -\frac{3}{22} & \frac{5}{22} & \frac{1}{11} \\ \frac{6}{11} & \frac{1}{11} & -\frac{4}{11} \end{bmatrix} \begin{bmatrix} 2 & 0 & 3 \\ 0 & 4 & 1 \\ 3 & 1 & 2 \end{bmatrix}$$

$$= \begin{bmatrix} 1 & 0 & 0 \\ 0 & 1 & 0 \\ 0 & 0 & 1 \end{bmatrix}$$

Since the product $\mathbf{A}^{-1}\mathbf{A}$ is equal to the identity matrix, it follows that our solution for $\mathbf{A}^{-1}$ is correct. ∎

Examples C.1 and C.2 indicate the strategy employed when performing row operations on the $\mathbf{A}$ matrix to change it into an identity matrix. Multiply the first row by a constant to change the element in the top left row into a 1. Then perform operations to change all elements in the first column into 0's. Then operate on the second row and change the second diagonal element into a 1. Then operate to change all elements in the second column beneath row 2 into 0's. Then operate on the diagonal element in row 3, and so on. When all elements on the main diagonal are 1's and all below the main diagonal are 0's, perform row operations to change the last column to 0; then the next-to-last, and so on, until you get back to the first column. The procedure for changing the off-diagonal elements to 0's is indicated diagrammatically as shown in Figure C.1.

The preceding instructions on how to invert a matrix using row operations suggest that the inversion of a large matrix would involve many multiplications, subtractions, and additions and, consequently, could produce large rounding errors in the calculations unless you carry a large number of significant figures in the calculations. This explains why two different multiple regression analysis computer programs may produce different estimates of the same β parameters, and it emphasizes the importance of carrying a large number of significant figures in all computations when inverting a matrix.

You can find other methods for inverting matrices in any linear algebra textbook. All work—exactly—in theory. It is only in the actual process of performing the calculations that the rounding errors occur.

Figure C.1 Diagram of matrix inversion steps

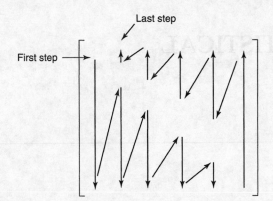

C.0   Exercise

C.1   Invert the following matrices and check your answers to make certain that $\mathbf{A}^{-1}\mathbf{A} = \mathbf{A}\mathbf{A}^{-1} = \mathbf{I}$:

(a) $\mathbf{A} = \begin{bmatrix} 3 & 2 \\ 4 & 5 \end{bmatrix}$
(b) $\mathbf{A} = \begin{bmatrix} 3 & 0 & -2 \\ 1 & 4 & 2 \\ 5 & 1 & 1 \end{bmatrix}$

(c) $\mathbf{A} = \begin{bmatrix} 1 & 0 & 1 \\ 0 & 2 & 1 \\ 1 & 1 & 3 \end{bmatrix}$
(d) $\mathbf{A} = \begin{bmatrix} 4 & 0 & 10 \\ 0 & 10 & 0 \\ 10 & 0 & 5 \end{bmatrix}$

[*Note*: No answers are given to these exercises. You will know whether your answer is correct if $\mathbf{A}^{-1}\mathbf{A} = \mathbf{I}$.]

USEFUL STATISTICAL TABLES

Contents

Table D.1   Normal curve areas

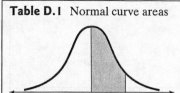

z	.00	.01	.02	.03	.04	.05	.06	.07	.08	.09
.0	.0000	.0040	.0080	.0120	.0160	.0199	.0239	.0279	.0319	.0359
.1	.0398	.0438	.0478	.0517	.0557	.0596	.0636	.0675	.0714	.0753
.2	.0793	.0832	.0871	.0910	.0948	.0987	.1026	.1064	.1103	.1141
.3	.1179	.1217	.1255	.1293	.1331	.1368	.1406	.1443	.1480	.1517
.4	.1554	.1591	.1628	.1664	.1700	.1736	.1772	.1808	.1844	.1879
.5	.1915	.1950	.1985	.2019	.2054	.2088	.2123	.2157	.2190	.2224
.6	.2257	.2291	.2324	.2357	.2389	.2422	.2454	.2486	.2517	.2549
.7	.2580	.2611	.2642	.2673	.2704	.2734	.2764	.2794	.2823	.2852
.8	.2881	.2910	.2939	.2967	.2995	.3023	.3051	.3078	.3106	.3133
.9	.3159	.3186	.3212	.3238	.3264	.3289	.3315	.3340	.3365	.3389
1.0	.3413	.3438	.3461	.3485	.3508	.3531	.3554	.3577	.3599	.3621
1.1	.3643	.3665	.3686	.3708	.3729	.3749	.3770	.3790	.3810	.3830
1.2	.3849	.3869	.3888	.3907	.3925	.3944	.3962	.3980	.3997	.4015
1.3	.4032	.4049	.4066	.4082	.4099	.4115	.4131	.4147	.4162	.4177
1.4	.4192	.4207	.4222	.4236	.4251	.4265	.4279	.4292	.4306	.4319
1.5	.4332	.4345	.4357	.4370	.4382	.4394	.4406	.4418	.4429	.4441
1.6	.4452	.4463	.4474	.4484	.4495	.4505	.4515	.4525	.4535	.4545
1.7	.4554	.4564	.4573	.4582	.4591	.4599	.4608	.4616	.4625	.4633
1.8	.4641	.4649	.4656	.4664	.4671	.4678	.4686	.4693	.4699	.4706
1.9	.4713	.4719	.4726	.4732	.4738	.4744	.4750	.4756	.4761	.4767
2.0	.4772	.4778	.4783	.4788	.4793	.4798	.4803	.4808	.4812	.4817
2.1	.4821	.4826	.4830	.4834	.4838	.4842	.4846	.4850	.4854	.4857
2.2	.4861	.4864	.4868	.4871	.4875	.4878	.4881	.4884	.4887	.4890
2.3	.4893	.4896	.4898	.4901	.4904	.4906	.4909	.4911	.4913	.4916
2.4	.4918	.4920	.4922	.4925	.4927	.4929	.4931	.4932	.4934	.4936
2.5	.4938	.4940	.4941	.4943	.4945	.4946	.4948	.4949	.4951	.4952
2.6	.4953	.4955	.4956	.4957	.4959	.4960	.4961	.4962	.4963	.4964
2.7	.4965	.4966	.4967	.4968	.4969	.4970	.4971	.4972	.4973	.4974
2.8	.4974	.4975	.4976	.4977	.4977	.4978	.4979	.4979	.4980	.4981
2.9	.4981	.4982	.4982	.4983	.4984	.4984	.4985	.4985	.4986	.4986
3.0	.4987	.4987	.4987	.4988	.4988	.4989	.4989	.4989	.4990	.4990

Source: Abridged from Table 1 of A. Hald, *Statistical Tables and Formulas* (New York: John Wiley & Sons, Inc.), 1952. Reproduced by permission of the publisher.

Table D.2 Critical values for Student's *t*

ν	$t_{.100}$	$t_{.050}$	$t_{.025}$	$t_{.010}$	$t_{.005}$	$t_{.001}$	$t_{.0005}$
1	3.078	6.314	12.706	31.821	63.657	318.31	636.62
2	1.886	2.920	4.303	6.965	9.925	22.326	31.598
3	1.638	2.353	3.182	4.541	5.841	10.213	12.924
4	1.533	2.132	2.776	3.747	4.604	7.173	8.610
5	1.476	2.015	2.571	3.365	4.032	5.893	6.869
6	1.440	1.943	2.447	3.143	3.707	5.208	5.959
7	1.415	1.895	2.365	2.998	3.499	4.785	5.408
8	1.397	1.860	2.306	2.896	3.355	4.501	5.041
9	1.383	1.833	2.262	2.821	3.250	4.297	4.781
10	1.372	1.812	2.228	2.764	3.169	4.144	4.587
11	1.363	1.796	2.201	2.718	3.106	4.025	4.437
12	1.356	1.782	2.179	2.681	3.055	3.930	4.318
13	1.350	1.771	2.160	2.650	3.012	3.852	4.221
14	1.345	1.761	2.145	2.624	2.977	3.787	4.140
15	1.341	1.753	2.131	2.602	2.947	3.733	4.073
16	1.337	1.746	2.120	2.583	2.921	3.686	4.015
17	1.333	1.740	2.110	2.567	2.898	3.646	3.965
18	1.330	1.734	2.101	2.552	2.878	3.610	3.922
19	1.328	1.729	2.093	2.539	2.861	3.579	3.883
20	1.325	1.725	2.086	2.528	2.845	3.552	3.850
21	1.323	1.721	2.080	2.518	2.831	3.527	3.819
22	1.321	1.717	2.074	2.508	2.819	3.505	3.792
23	1.319	1.714	2.069	2.500	2.807	3.485	3.767
24	1.318	1.711	2.064	2.492	2.797	3.467	3.745
25	1.316	1.708	2.060	2.485	2.787	3.450	3.725
26	1.315	1.706	2.056	2.479	2.779	3.435	3.707
27	1.314	1.703	2.052	2.473	2.771	3.421	3.690
28	1.313	1.701	2.048	2.467	2.763	3.408	3.674
29	1.311	1.699	2.045	2.462	2.756	3.396	3.659
30	1.310	1.697	2.042	2.457	2.750	3.385	3.646
40	1.303	1.684	2.021	2.423	2.704	3.307	3.551
60	1.296	1.671	2.000	2.390	2.660	3.232	3.460
120	1.289	1.658	1.980	2.358	2.617	3.160	3.373
∞	1.282	1.645	1.960	2.326	2.576	3.090	3.291

Source: This table is reproduced with the kind permission of the Trustees of *Biometrika* from E. S. Pearson and H. O. Hartley (eds.), *The Biometrika Tables for Statisticians*, Vol. 1, 3rd ed., *Biometrika*, 1996.

Table D.4 Critical values for the F Statistic: $F_{.05}$

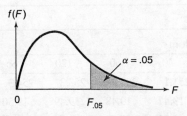

v_1	Numerator Degrees of Freedom								
v_2	1	2	3	4	5	6	7	8	9
1	161.4	199.5	215.7	224.6	230.2	234.0	236.8	238.9	240.5
2	18.51	19.00	19.16	19.25	19.30	19.33	19.35	19.37	19.38
3	10.13	9.55	9.28	9.12	9.01	8.94	8.89	8.85	8.81
4	7.71	6.94	6.59	6.39	6.26	6.16	6.09	6.04	6.00
5	6.61	5.79	5.41	5.19	5.05	4.95	4.88	4.82	4.77
6	5.99	5.14	4.76	4.53	4.39	4.28	4.21	4.15	4.10
7	5.59	4.74	4.35	4.12	3.97	3.87	3.79	3.73	3.68
8	5.32	4.46	4.07	3.84	3.69	3.58	3.50	3.44	3.39
9	5.12	4.26	3.86	3.63	3.48	3.37	3.29	3.23	3.18
10	4.96	4.10	3.71	3.48	3.33	3.22	3.14	3.07	3.02
11	4.84	3.98	3.59	3.36	3.20	3.09	3.01	2.95	2.90
12	4.75	3.89	3.49	3.26	3.11	3.00	2.91	2.85	2.80
13	4.67	3.81	3.41	3.18	3.03	2.92	2.83	2.77	2.71
14	4.60	3.74	3.34	3.11	2.96	2.85	2.76	2.70	2.65
15	4.54	3.68	3.29	3.06	2.90	2.79	2.71	2.64	2.59
16	4.49	3.63	3.24	3.01	2.85	2.74	2.66	2.59	2.54
17	4.45	3.59	3.20	2.96	2.81	2.70	2.61	2.55	2.49
18	4.41	3.55	3.16	2.93	2.77	2.66	2.58	2.51	2.46
19	4.38	3.52	3.13	2.90	2.74	2.63	2.54	2.48	2.42
20	4.35	3.49	3.10	2.87	2.71	2.60	2.51	2.45	2.39
21	4.32	3.47	3.07	2.84	2.68	2.57	2.49	2.42	2.37
22	4.30	3.44	3.05	2.82	2.66	2.55	2.46	2.40	2.34
23	4.28	3.42	3.03	2.80	2.64	2.53	2.44	2.37	2.32
24	4.26	3.40	3.01	2.78	2.62	2.51	2.42	2.36	2.30
25	4.24	3.39	2.99	2.76	2.60	2.49	2.40	2.34	2.28
26	4.23	3.37	2.98	2.74	2.59	2.47	2.39	2.32	2.27
27	4.21	3.35	2.96	2.73	2.57	2.46	2.37	2.31	2.25
28	4.20	3.34	2.95	2.71	2.56	2.45	2.36	2.29	2.24
29	4.18	3.33	2.93	2.70	2.55	2.43	2.35	2.28	2.22
30	4.17	3.32	2.92	2.69	2.53	2.42	2.33	2.27	2.21
40	4.08	3.23	2.84	2.61	2.45	2.34	2.25	2.18	2.12
60	4.00	3.15	2.76	2.53	2.37	2.25	2.17	2.10	2.04
120	3.92	3.07	2.68	2.45	2.29	2.17	2.09	2.02	1.96
∞	3.84	3.00	2.60	2.37	2.21	2.10	2.01	1.94	1.88

(continued overleaf)

Table D.4 (*Continued*)

ν_2 \ ν_1	Numerator Degrees of Freedom									
	10	12	15	20	24	30	40	60	120	∞
1	241.9	243.9	245.9	248.0	249.1	250.1	251.1	252.2	253.3	254.3
2	19.40	19.41	19.43	19.45	19.45	19.46	19.47	19.48	19.49	19.50
3	8.79	8.74	8.70	8.66	8.64	8.62	8.59	8.57	8.55	8.53
4	5.96	5.91	5.86	5.80	5.77	5.75	5.72	5.69	5.66	5.63
5	4.74	4.68	4.62	4.56	4.53	4.50	4.46	4.43	4.40	4.36
6	4.06	4.00	3.94	3.87	3.84	3.81	3.77	3.74	3.70	3.67
7	3.64	3.57	3.51	3.44	3.41	3.38	3.34	3.30	3.27	3.23
8	3.35	3.28	3.22	3.15	3.12	3.08	3.04	3.01	2.97	2.93
9	3.14	3.07	3.01	2.94	2.90	2.86	2.83	2.79	2.75	2.71
10	2.98	2.91	2.85	2.77	2.74	2.70	2.66	2.62	2.58	2.54
11	2.85	2.79	2.72	2.65	2.61	2.57	2.53	2.49	2.45	2.40
12	2.75	2.69	2.62	2.54	2.51	2.47	2.43	2.38	2.34	2.30
13	2.67	2.60	2.53	2.46	2.42	2.38	2.34	2.30	2.25	2.21
14	2.60	2.53	2.46	2.39	2.35	2.31	2.27	2.22	2.18	2.13
15	2.54	2.48	2.40	2.33	2.29	2.25	2.20	2.16	2.11	2.07
16	2.49	2.42	2.35	2.28	2.24	2.19	2.15	2.11	2.06	2.01
17	2.45	2.38	2.31	2.23	2.19	2.15	2.10	2.06	2.01	1.96
18	2.41	2.34	2.27	2.19	2.15	2.11	2.06	2.02	1.97	1.92
19	2.38	2.31	2.23	2.16	2.11	2.07	2.03	1.98	1.93	1.88
20	2.35	2.28	2.20	2.12	2.08	2.04	1.99	1.95	1.90	1.84
21	2.32	2.25	2.18	2.10	2.05	2.01	1.96	1.92	1.87	1.81
22	2.30	2.23	2.15	2.07	2.03	1.98	1.94	1.89	1.84	1.78
23	2.27	2.20	2.13	2.05	2.01	1.96	1.91	1.86	1.81	1.76
24	2.25	2.18	2.11	2.03	1.98	1.94	1.89	1.84	1.79	1.73
25	2.24	2.16	2.09	2.01	1.96	1.92	1.87	1.82	1.77	1.71
26	2.22	2.15	2.07	1.99	1.95	1.90	1.85	1.80	1.75	1.69
27	2.20	2.13	2.06	1.97	1.93	1.88	1.84	1.79	1.73	1.67
28	2.19	2.12	2.04	1.96	1.91	1.87	1.82	1.77	1.71	1.65
29	2.18	2.10	2.03	1.94	1.90	1.85	1.81	1.75	1.70	1.64
30	2.16	2.09	2.01	1.93	1.89	1.84	1.79	1.74	1.68	1.62
40	2.08	2.00	1.92	1.84	1.79	1.74	1.69	1.64	1.58	1.51
60	1.99	1.92	1.84	1.75	1.70	1.65	1.59	1.53	1.47	1.39
120	1.91	1.83	1.75	1.66	1.61	1.55	1.50	1.43	1.35	1.25
∞	1.83	1.75	1.67	1.57	1.52	1.46	1.39	1.32	1.22	1.00

DENOMINATOR DEGREES OF FREEDOM

Source: From M. Merrington and C. M. Thompson, "Tables of percentage points of the inverted beta (*F*)-distribution," *Biometrika*, 1943, 33, 73–88. Reproduced by permission of the *Biometrika* Trustees.

Table D.5 Critical values for the F Statistic: $F_{.025}$

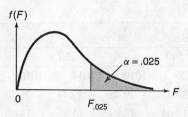

v_2 \ v_1	Numerator Degrees of Freedom								
	1	2	3	4	5	6	7	8	9
1	647.8	799.5	864.2	899.6	921.8	937.1	948.2	956.7	963.3
2	38.51	39.00	39.17	39.25	39.30	39.33	39.36	39.37	39.39
3	17.44	16.04	15.44	15.10	14.88	14.73	14.62	14.54	14.47
4	12.22	10.65	9.98	9.60	9.36	9.20	9.07	8.98	8.90
5	10.01	8.43	7.76	7.39	7.15	6.98	6.85	6.76	6.68
6	8.81	7.26	6.60	6.23	5.99	5.82	5.70	5.60	5.52
7	8.07	6.54	5.89	5.52	5.29	5.12	4.99	4.90	4.82
8	7.57	6.06	5.42	5.05	4.82	4.65	4.53	4.43	4.36
9	7.21	5.71	5.08	4.72	4.48	4.32	4.20	4.10	4.03
10	6.94	5.46	4.83	4.47	4.24	4.07	3.95	3.85	3.78
11	6.72	5.26	4.63	4.28	4.04	3.88	3.76	3.66	3.59
12	6.55	5.10	4.47	4.12	3.89	3.73	3.61	3.51	3.44
13	6.41	4.97	4.35	4.00	3.77	3.60	3.48	3.39	3.31
14	6.30	4.86	4.24	3.89	3.66	3.50	3.38	3.29	3.21
15	6.20	4.77	4.15	3.80	3.58	3.41	3.29	3.20	3.12
16	6.12	4.69	4.08	3.73	3.50	3.34	3.22	3.12	3.05
17	6.04	4.62	4.01	3.66	3.44	3.28	3.16	3.06	2.98
18	5.98	4.56	3.95	3.61	3.38	3.22	3.10	3.01	2.93
19	5.92	4.51	3.90	3.56	3.33	3.17	3.05	2.96	2.88
20	5.87	4.46	3.86	3.51	3.29	3.13	3.01	2.91	2.84
21	5.83	4.42	3.82	3.48	3.25	3.09	2.97	2.87	2.80
22	5.79	4.38	3.78	3.44	3.22	3.05	2.93	2.84	2.76
23	5.75	4.35	3.75	3.41	3.18	3.02	2.90	2.81	2.73
24	5.72	4.32	3.72	3.38	3.15	2.99	2.87	2.78	2.70
25	5.69	4.29	3.69	3.35	3.13	2.97	2.85	2.75	2.68
26	5.66	4.27	3.67	3.33	3.10	2.94	2.82	2.73	2.65
27	5.63	4.24	3.65	3.31	3.08	2.92	2.80	2.71	2.63
28	5.61	4.22	3.63	3.29	3.06	2.90	2.78	2.69	2.61
29	5.59	4.20	3.61	3.27	3.04	2.88	2.76	2.67	2.59
30	5.57	4.18	3.59	3.25	3.03	2.87	2.75	2.65	2.57
40	5.42	4.05	3.46	3.13	2.90	2.74	2.62	2.53	2.45
60	5.29	3.93	3.34	3.01	2.79	2.63	2.51	2.41	2.33
120	5.15	3.80	3.23	2.89	2.67	2.52	2.39	2.30	2.22
∞	5.02	3.69	3.12	2.79	2.57	2.41	2.29	2.19	2.11

(*continued overleaf*)

Table D.5 (*Continued*)

ν_2 \ ν_1	Numerator Degrees of Freedom									
	10	12	15	20	24	30	40	60	120	∞
1	968.6	976.7	984.9	993.1	997.2	1001	1006	1010	1014	1018
2	39.40	39.41	39.43	39.45	39.46	39.46	39.47	39.48	39.49	39.50
3	14.42	14.34	14.25	14.17	14.12	14.08	14.04	13.99	13.95	13.90
4	8.84	8.75	8.66	8.56	8.51	8.46	8.41	8.36	8.31	8.26
5	6.62	6.52	6.43	6.33	6.28	6.23	6.18	6.12	6.07	6.02
6	5.46	5.37	5.27	5.17	5.12	5.07	5.01	4.96	4.90	4.85
7	4.76	4.67	4.57	4.47	4.42	4.36	4.31	4.25	4.20	4.14
8	4.30	4.20	4.10	4.00	3.95	3.89	3.84	3.78	3.73	3.67
9	3.96	3.87	3.77	3.67	3.61	3.56	3.51	3.45	3.39	3.33
10	3.72	3.62	3.52	3.42	3.37	3.31	3.26	3.20	3.14	3.08
11	3.53	3.43	3.33	3.23	3.17	3.12	3.06	3.00	2.94	2.88
12	3.37	3.28	3.18	3.07	3.02	2.96	2.91	2.85	2.79	2.72
13	3.25	3.15	3.05	2.95	2.89	2.84	2.78	2.72	2.66	2.60
14	3.15	3.05	2.95	2.84	2.79	2.73	2.67	2.61	2.55	2.49
15	3.06	2.96	2.86	2.76	2.70	2.64	2.59	2.52	2.46	2.40
16	2.99	2.89	2.79	2.68	2.63	2.57	2.51	2.45	2.38	2.32
17	2.92	2.82	2.72	2.62	2.56	2.50	2.44	2.38	2.32	2.25
18	2.87	2.77	2.67	2.56	2.50	2.44	2.38	2.32	2.26	2.19
19	2.82	2.72	2.62	2.51	2.45	2.39	2.33	2.27	2.20	2.13
20	2.77	2.68	2.57	2.46	2.41	2.35	2.29	2.22	2.16	2.09
21	2.73	2.64	2.53	2.42	2.37	2.31	2.25	2.18	2.11	2.04
22	2.70	2.60	2.50	2.39	2.33	2.27	2.21	2.14	2.08	2.00
23	2.67	2.57	2.47	2.36	2.30	2.24	2.18	2.11	2.04	1.97
24	2.64	2.54	2.44	2.33	2.27	2.21	2.15	2.08	2.01	1.94
25	2.61	2.51	2.41	2.30	2.24	2.18	2.12	2.05	1.98	1.91
26	2.59	2.49	2.39	2.28	2.22	2.16	2.09	2.03	1.95	1.88
27	2.57	2.47	2.36	2.25	2.19	2.13	2.07	2.00	1.93	1.85
28	2.55	2.45	2.34	2.23	2.17	2.11	2.05	1.98	1.91	1.83
29	2.53	2.43	2.32	2.21	2.15	2.09	2.03	1.96	1.89	1.81
30	2.51	2.41	2.31	2.20	2.14	2.07	2.01	1.94	1.87	1.79
40	2.39	2.29	2.18	2.07	2.01	1.94	1.88	1.80	1.72	1.64
60	2.27	2.17	2.06	1.94	1.88	1.82	1.74	1.67	1.58	1.48
120	2.16	2.05	1.94	1.82	1.76	1.69	1.61	1.53	1.43	1.31
∞	2.05	1.94	1.83	1.71	1.64	1.57	1.48	1.39	1.27	1.00

Denominator Degrees of Freedom (row label ν_2)

Table D.6 Critical values for the F Statistic: $F_{.01}$

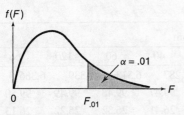

ν_1	Numerator Degrees of Freedom								
ν_2	1	2	3	4	5	6	7	8	9
1	4,052	4,999.5	5,403	5,625	5,764	5,859	5,928	5,982	6,022
2	98.50	99.00	99.17	99.25	99.30	99.33	99.36	99.37	99.39
3	34.12	30.82	29.46	28.71	28.24	27.91	27.67	27.49	27.35
4	21.20	18.00	16.69	15.98	15.52	15.21	14.98	14.80	14.66
5	16.26	13.27	12.06	11.39	10.97	10.67	10.46	10.29	10.16
6	13.75	10.92	9.78	9.15	8.75	8.47	8.26	8.10	7.98
7	12.25	9.55	8.45	7.85	7.46	7.19	6.99	6.84	6.72
8	11.26	8.65	7.59	7.01	6.63	6.37	6.18	6.03	5.91
9	10.56	8.02	6.99	6.42	6.06	5.80	5.61	5.47	5.35
10	10.04	7.56	6.55	5.99	5.64	5.39	5.20	5.06	4.94
11	9.65	7.21	6.22	5.67	5.32	5.07	4.89	4.74	4.63
12	9.33	6.93	5.95	5.41	5.06	4.82	4.64	4.50	4.39
13	9.07	6.70	5.74	5.21	4.86	4.62	4.44	4.30	4.19
14	8.86	6.51	5.56	5.04	4.69	4.46	4.28	4.14	4.03
15	8.68	6.36	5.42	4.89	4.56	4.32	4.14	4.00	3.89
16	8.53	6.23	5.29	4.77	4.44	4.20	4.03	3.89	3.78
17	8.40	6.11	5.18	4.67	4.34	4.10	3.93	3.79	3.68
18	8.29	6.01	5.09	4.58	4.25	4.01	3.84	3.71	3.60
19	8.18	5.93	5.01	4.50	4.17	3.94	3.77	3.63	3.52
20	8.10	5.85	4.94	4.43	4.10	3.87	3.70	3.56	3.46
21	8.02	5.78	4.87	4.37	4.04	3.81	3.64	3.51	3.40
22	7.95	5.72	4.82	4.31	3.99	3.76	3.59	3.45	3.35
23	7.88	5.66	4.76	4.26	3.94	3.71	3.54	3.41	3.30
24	7.82	5.61	4.72	4.22	3.90	3.67	3.50	3.36	3.26
25	7.77	5.57	4.68	4.18	3.85	3.63	3.46	3.32	3.22
26	7.72	5.53	4.64	4.14	3.82	3.59	3.42	3.29	3.18
27	7.68	5.49	4.60	4.11	3.78	3.56	3.39	3.26	3.15
28	7.64	5.45	4.57	4.07	3.75	3.53	3.36	3.23	3.12
29	7.60	5.42	4.54	4.04	3.73	3.50	3.33	3.20	3.09
30	7.56	5.39	4.51	4.02	3.70	3.47	3.30	3.17	3.07
40	7.31	5.18	4.31	3.83	3.51	3.29	3.12	2.99	2.89
60	7.08	4.98	4.13	3.65	3.34	3.12	2.95	2.82	2.72
120	6.85	4.79	3.95	3.48	3.17	2.96	2.79	2.66	2.56
∞	6.63	4.61	3.78	3.32	3.02	2.80	2.64	2.51	2.41

DENOMINATOR DEGREES OF FREEDOM

(*continued overleaf*)

Table D.6 (*Continued*)

ν_2 \ ν_1	Numerator Degrees of Freedom									
	10	12	15	20	24	30	40	60	120	∞
1	6,056	6,106	6,157	6,209	6,235	6,261	6,287	6,313	6,339	6,366
2	99.40	99.42	99.43	99.45	99.46	99.47	99.47	99.48	99.49	99.50
3	27.23	27.05	26.87	26.69	26.60	26.50	26.41	26.32	26.22	26.13
4	14.55	14.37	14.20	14.02	13.93	13.84	13.75	13.65	13.56	13.46
5	10.05	9.89	9.72	9.55	9.47	9.38	9.29	9.20	9.11	9.02
6	7.87	7.72	7.56	7.40	7.31	7.23	7.14	7.06	6.97	6.88
7	6.62	6.47	6.31	6.16	6.07	5.99	5.91	5.82	5.74	5.65
8	5.81	5.67	5.52	5.36	5.28	5.20	5.12	5.03	4.95	4.86
9	5.26	5.11	4.96	4.81	4.73	4.65	4.57	4.48	4.40	4.31
10	4.85	4.71	4.56	4.41	4.33	4.25	4.17	4.08	4.00	3.91
11	4.54	4.40	4.25	4.10	4.02	3.94	3.86	3.78	3.69	3.60
12	4.30	4.16	4.01	3.86	3.78	3.70	3.62	3.54	3.45	3.36
13	4.10	3.96	3.82	3.66	3.59	3.51	3.43	3.34	3.25	3.17
14	3.94	3.80	3.66	3.51	3.43	3.35	3.27	3.18	3.09	3.00
15	3.80	3.67	3.52	3.37	3.29	3.21	3.13	3.05	2.96	2.87
16	3.69	3.55	3.41	3.26	3.18	3.10	3.02	2.93	2.84	2.75
17	3.59	3.46	3.31	3.16	3.08	3.00	2.92	2.83	2.75	2.65
18	3.51	3.37	3.23	3.08	3.00	2.92	2.84	2.75	2.66	2.57
19	3.43	3.30	3.15	3.00	2.92	2.84	2.76	2.67	2.58	2.49
20	3.37	3.23	3.09	2.94	2.86	2.78	2.69	2.61	2.52	2.42
21	3.31	3.17	3.03	2.88	2.80	2.72	2.64	2.55	2.46	2.36
22	3.26	3.12	2.98	2.83	2.75	2.67	2.58	2.50	2.40	2.31
23	3.21	3.07	2.93	2.78	2.70	2.62	2.54	2.45	2.35	2.26
24	3.17	3.03	2.89	2.74	2.66	2.58	2.49	2.40	2.31	2.21
25	3.13	2.99	2.85	2.70	2.62	2.54	2.45	2.36	2.27	2.17
26	3.09	2.96	2.81	2.66	2.58	2.50	2.42	2.33	2.23	2.13
27	3.06	2.93	2.78	2.63	2.55	2.47	2.38	2.29	2.20	2.10
28	3.03	2.90	2.75	2.60	2.52	2.44	2.35	2.26	2.17	2.06
29	3.00	2.87	2.73	2.57	2.49	2.41	2.33	2.23	2.14	2.03
30	2.98	2.84	2.70	2.55	2.47	2.39	2.30	2.21	2.11	2.01
40	2.80	2.66	2.52	2.37	2.29	2.20	2.11	2.02	1.92	1.80
60	2.63	2.50	2.35	2.20	2.12	2.03	1.94	1.84	1.73	1.60
120	2.47	2.34	2.19	2.03	1.95	1.86	1.76	1.66	1.53	1.38
∞	2.32	2.18	2.04	1.88	1.79	1.70	1.59	1.47	1.32	1.00

Left column label: DENOMINATOR DEGREES OF FREEDOM

Source: From M. Merrington and C. M. Thompson, "Tables of percentage points of the inverted beta (*F*)-distribution," *Biometrika*, 1943, 33, 73–88. Reproduced by permission of the *Biometrika* Trustees.

Table D.7 Random numbers

Row \ Column	1	2	3	4	5	6	7	8	9	10	11	12	13	14
1	10480	15011	01536	02011	81647	91646	69179	14194	62590	36207	20969	99570	91291	90700
2	22368	46573	25595	85393	30995	89198	27982	53402	93965	34095	52666	19174	39615	99505
3	24130	48360	22527	97265	76393	64809	15179	24830	49340	32081	30680	19655	63348	58629
4	42167	93093	06243	61680	07856	16376	39440	53537	71341	57004	00849	74917	97758	16379
5	37570	39975	81837	16656	06121	91782	60468	81305	49684	60672	14110	06927	01263	54613
6	77921	06907	11008	42751	27756	53498	18602	70659	90655	15053	21916	81825	44394	42880
7	99562	72905	56420	69994	98872	31016	71194	18738	44013	48840	63213	21069	10634	12952
8	96301	91977	05463	07972	18876	20922	94595	56869	69014	60045	18425	84903	42508	32307
9	89579	14342	63661	10281	17453	18103	57740	84378	25331	12566	58678	44947	05585	56941
10	85475	36857	53342	53988	53060	59533	38867	62300	08158	17983	16439	11458	18593	64952
11	28918	69578	88231	33276	70997	79936	56865	05859	90106	31595	01547	85590	91610	78188
12	63553	40961	48235	03427	49626	69445	18663	72695	52180	20847	12234	90511	33703	90322
13	09429	93969	52636	92737	88974	33488	36320	17617	30015	08272	84115	27156	30613	74952
14	10365	61129	87529	85689	48237	52267	67689	93394	01511	26358	85104	20285	29975	89868
15	07119	97336	71048	08178	77233	13916	47564	81056	97735	85977	29372	74461	28551	90707
16	51085	12765	51821	51259	77452	16308	60756	92144	49442	53900	70960	63990	75601	40719
17	02368	21382	52404	60268	89368	19885	55322	44819	01188	65255	64835	44919	05944	55157
18	01011	54092	33362	94904	31273	04146	18594	29852	71585	85030	51132	01915	92747	64951
19	52162	53916	46369	58586	23216	14513	83149	98736	23495	64350	94738	17752	35156	35749
20	07056	97628	33787	09998	42698	06691	76988	13602	51851	46104	88916	19509	25625	58104
21	48663	91245	85828	14346	09172	30168	90229	04734	59193	22178	30421	61666	99904	32812
22	54164	58492	22421	74103	47070	25306	76468	26384	58151	06646	21524	15227	96909	44592
23	32639	32363	05597	24200	13363	38005	94342	28728	35806	06912	17012	64161	18296	22851
24	29334	27001	87637	87308	58731	00256	45834	15398	46557	41135	10367	07684	36188	18510
25	02488	33062	28834	07351	19731	92420	60952	61280	50001	67658	32586	86679	50720	94953
26	81525	72295	04839	96423	24878	82651	66566	14778	76797	14780	13300	87074	79666	95725
27	29676	20591	68086	26432	46901	20849	89768	81536	86645	12659	92259	57102	80428	25280
28	00742	57392	39064	66432	84673	40027	32832	61362	98947	96067	64760	64584	96096	98253
29	05366	04213	25669	26422	44407	44048	37937	63904	45766	66134	75470	66520	34693	90449
30	91921	26418	64117	94305	26766	25940	39972	22209	71500	64568	91402	42416	07844	69618
31	00582	04711	87917	77341	42206	35126	74087	99547	81817	42607	43808	76655	62028	76630
32	00725	69884	62797	56170	86324	88072	76222	36086	84637	93161	76038	65855	77919	88006
33	69011	65795	95876	55293	18988	27354	26575	08625	40801	59920	29841	80150	12777	48501
34	25976	57948	29888	88604	67917	48708	18912	82271	65424	69774	33611	54262	85963	03547
35	09763	83473	73577	12908	30883	18317	28290	35797	05998	41688	34952	37888	38917	88050
36	91576	42595	27958	30134	04024	86385	29880	99730	55536	84855	29080	09250	79656	73211
37	17955	56349	90999	49127	20044	59931	06115	20542	18059	02008	73708	83517	36103	42791
38	46503	18584	18845	49618	02304	51038	20655	58727	28168	15475	56942	53389	20562	87338
39	92157	89634	94824	78171	84610	82834	09922	25417	44137	48413	25555	21246	35509	20468
40	14577	62765	35605	81263	39667	47358	56873	56307	61607	49518	89656	20103	77490	18062
41	98427	07523	33362	64270	01638	92477	66969	98420	04880	45585	46565	04102	46880	45709

(*continued overleaf*)

Table D.7 (*Continued*)

Row \ Column	1	2	3	4	5	6	7	8	9	10	11	12	13	14
42	34914	63976	88720	82765	34476	17032	87589	40836	32427	70002	70663	88863	77775	69348
43	70060	28277	39475	46473	23219	53416	94970	25832	69975	94884	19661	72828	00102	66794
44	53976	54914	06990	67245	68350	82948	11398	42878	80287	88267	47363	46634	06541	97809
45	76072	29515	40980	07391	58745	25774	22987	80059	39911	96189	41151	14222	60697	59583
46	90725	52210	83974	29992	65831	38857	50490	83765	55657	14361	31720	57375	56228	41546
47	64364	67412	33339	31926	14883	24413	59744	92351	97473	89286	35931	04110	23726	51900
48	08962	00358	31662	25388	61642	34072	81249	35648	56891	69352	48373	45578	78547	81788
49	95012	68379	93526	70765	10592	04542	76463	54328	02349	17247	28865	14777	62730	92277
50	15664	10493	20492	38391	91132	21999	59516	81652	27195	48223	46751	22923	32261	85653
51	16408	81899	04153	53381	79401	21438	83035	92350	36693	31238	59649	91754	72772	02338
52	18629	81953	05520	91962	04739	13092	97662	24822	94730	06496	35090	04822	86774	98289
53	73115	35101	47498	87637	99016	71060	88824	71013	18735	20286	23153	72924	35165	43040
54	57491	16703	23167	49323	45021	33132	12544	41035	80780	45393	44812	12515	98931	91202
55	30405	83946	23792	14422	15059	45799	22716	19792	09983	74353	68668	30429	70735	25499
56	16631	35006	85900	98275	32388	52390	16815	69298	82732	38480	73817	32523	41961	44437
57	96773	20206	42559	78985	05300	22164	24369	54224	35083	19687	11052	91491	60383	19746
58	38935	64202	14349	82674	66523	44133	00697	35552	35970	19124	63318	29686	03387	59846
59	31624	76384	17403	53363	44167	64486	64758	75366	76554	31601	12614	33072	60332	92325
60	78919	19474	23632	27889	47914	02584	37680	20801	72152	39339	34806	08930	85001	87820
61	03931	33309	57047	74211	63445	17361	62825	39908	05607	91284	68833	25570	38818	46920
62	74426	33278	43972	10119	89917	15665	52872	73823	73144	88662	88970	74492	51805	99378
63	09066	00903	20795	95452	92648	45454	09552	88815	16553	51125	79375	97596	16296	66092
64	42238	12426	87025	14267	20979	04508	64535	31355	86064	29472	47689	05974	52468	16834
65	16153	08002	26504	41744	81959	65642	74240	56302	00033	67107	77510	70625	28725	34191
66	21457	40742	29820	96783	29400	21840	15035	34537	33310	06116	95240	15957	16572	06004
67	21581	57802	02050	89728	17937	37621	47075	42080	97403	48626	68995	43805	33386	21597
68	55612	78095	83197	33732	05810	24813	86902	60397	16489	03264	88525	42786	05269	92532
69	44657	66999	99324	51281	84463	60563	79312	93454	68876	25471	93911	25650	12682	73572
70	91340	84979	46949	81973	37949	61023	43997	15263	80644	43942	89203	71795	99533	50501
71	91227	21199	31935	27022	84067	05462	35216	14486	29891	68607	41867	14951	91696	85065
72	50001	38140	66321	19924	72163	09538	12151	06878	91903	18749	34405	56087	82790	70925
73	65390	05224	72958	28609	81406	39147	25549	48542	42627	45233	57202	94617	23772	07896
74	27504	96131	83944	41575	10573	08619	64482	73923	36152	05184	94142	25299	84387	34925
75	37169	94851	39117	89632	00959	16487	65536	49071	39782	17095	02330	74301	00275	48280
76	11508	70225	51111	38351	19444	66499	71945	05422	13442	78675	84081	66938	93654	59894
77	37449	30362	06694	54690	04052	53115	62757	95348	78662	11163	81651	50245	34971	52924
78	46515	70331	85922	38329	57015	15765	97161	17869	45349	61796	66345	81073	49106	79860
79	30986	81223	42416	58353	21532	30502	32305	86482	05174	07901	54339	58861	74818	46942
80	63798	64995	46583	09785	44160	78128	83991	42865	92520	83531	80377	35909	81250	54238
81	82486	84846	99254	67632	43218	50076	21361	64816	51202	88124	41870	52689	51275	83556
82	21885	32906	92431	09060	64297	51674	64126	62570	26123	05155	59194	52799	28225	85762

(*continued overleaf*)

Table D.7 (*Continued*)

Row \ Column	1	2	3	4	5	6	7	8	9	10	11	12	13	14
83	60336	98782	07408	53458	13564	59089	26445	29789	85205	41001	12535	12133	14645	23541
84	43937	46891	24010	25560	86355	33941	25786	54990	71899	15475	95434	98227	21824	19585
85	97656	63175	89303	16275	07100	92063	21942	18611	47348	20203	18534	03862	78095	50136
86	03299	01221	05418	38982	55758	92237	26759	86367	21216	98442	08303	56613	91511	75928
87	79626	06486	03574	17668	07785	76020	79924	25651	83325	88428	85076	72811	22717	50585
88	85636	68335	47539	03129	65651	11977	02510	26113	99447	68645	34327	15152	55230	93448
89	18039	14367	61337	06177	12143	46609	32989	74014	64708	00533	35398	58408	13261	47908
90	08362	15656	60627	36478	65648	16764	53412	09013	07832	41574	17639	82163	60859	75567
91	79556	29068	04142	16268	15387	12856	66227	38358	22478	73373	88732	09443	82558	05250
92	92608	82674	27072	32534	17075	27698	98204	63863	11951	34648	88022	56148	34925	57031
93	23982	25835	40055	67006	12293	02753	14827	23235	35071	99704	37543	11601	35503	85171
94	09915	96306	05908	97901	28395	14186	00821	80703	70426	75647	76310	88717	37890	40129
95	59037	33300	26695	62247	69927	76123	50842	43834	86654	70959	79725	93872	28117	19233
96	42488	78077	69882	61657	34136	79180	97526	43092	04098	73571	80799	76536	71255	64239
97	46764	86273	63003	93017	31204	36692	40202	35275	57306	55543	53203	18098	47625	88684
98	03237	45430	55417	63282	90816	17349	88298	90183	36600	78406	06216	95787	42579	90730
99	86591	81482	52667	61582	14972	90053	89534	76036	49199	43716	97548	04379	46370	28672
100	38534	01715	94964	87288	65680	43772	39560	12918	86537	62738	19636	51132	25739	56947

Source: Abridged from W. H. Beyer (ed.). *CRC Standard Mathematical Tables*, 24th edition. Cleveland: The Chemical Rubber Company, 1976.

Table D.8 Critical values for the Durbin–Watson d Statistic ($\alpha = .05$)

n	k = 1		k = 2		k = 3		k = 4		k = 5	
	d_L	d_U	d_L	d_U	d_L	d_U	d_L	d_U	d_L	d_U
15	1.08	1.36	.95	1.54	.82	1.75	.69	1.97	.56	2.21
16	1.10	1.37	.98	1.54	.86	1.73	.74	1.93	.62	2.15
17	1.13	1.38	1.02	1.54	.90	1.71	.78	1.90	.67	2.10
18	1.16	1.39	1.05	1.53	.93	1.69	.82	1.87	.71	2.06
19	1.18	1.40	1.08	1.53	.97	1.68	.86	1.85	.75	2.02
20	1.20	1.41	1.10	1.54	1.00	1.68	.90	1.83	.79	1.99
21	1.22	1.42	1.13	1.54	1.03	1.67	.93	1.81	.83	1.96
22	1.24	1.43	1.15	1.54	1.05	1.66	.96	1.80	.86	1.94
23	1.26	1.44	1.17	1.54	1.08	1.66	.99	1.79	.90	1.92
24	1.27	1.45	1.19	1.55	1.10	1.66	1.01	1.78	.93	1.90
25	1.29	1.45	1.21	1.55	1.12	1.66	1.04	1.77	.95	1.89
26	1.30	1.46	1.22	1.55	1.14	1.65	1.06	1.76	.98	1.88
27	1.32	1.47	1.24	1.56	1.16	1.65	1.08	1.76	1.01	1.86
28	1.33	1.48	1.26	1.56	1.18	1.65	1.10	1.75	1.03	1.85
29	1.34	1.48	1.27	1.56	1.20	1.65	1.12	1.74	1.05	1.84
30	1.35	1.49	1.28	1.57	1.21	1.65	1.14	1.74	1.07	1.83
31	1.36	1.50	1.30	1.57	1.23	1.65	1.16	1.74	1.09	1.83
32	1.37	1.50	1.31	1.57	1.24	1.65	1.18	1.73	1.11	1.82
33	1.38	1.51	1.32	1.58	1.26	1.65	1.19	1.73	1.13	1.81
34	1.39	1.51	1.33	1.58	1.27	1.65	1.21	1.73	1.15	1.81
35	1.40	1.52	1.34	1.58	1.28	1.65	1.22	1.73	1.16	1.80
36	1.41	1.52	1.35	1.59	1.29	1.65	1.24	1.73	1.18	1.80
37	1.42	1.53	1.36	1.59	1.31	1.66	1.25	1.72	1.19	1.80
38	1.43	1.54	1.37	1.59	1.32	1.66	1.26	1.72	1.21	1.79
39	1.43	1.54	1.38	1.60	1.33	1.66	1.27	1.72	1.22	1.79
40	1.44	1.54	1.39	1.60	1.34	1.66	1.29	1.72	1.23	1.79
45	1.48	1.57	1.43	1.62	1.38	1.67	1.34	1.72	1.29	1.78
50	1.50	1.59	1.46	1.63	1.42	1.67	1.38	1.72	1.34	1.77
55	1.53	1.60	1.49	1.64	1.45	1.68	1.41	1.72	1.38	1.77
60	1.55	1.62	1.51	1.65	1.48	1.69	1.44	1.73	1.41	1.77
65	1.57	1.63	1.54	1.66	1.50	1.70	1.47	1.73	1.44	1.77
70	1.58	1.64	1.55	1.67	1.52	1.70	1.49	1.74	1.46	1.77
75	1.60	1.65	1.57	1.68	1.54	1.71	1.51	1.74	1.49	1.77
80	1.61	1.66	1.59	1.69	1.56	1.72	1.53	1.74	1.51	1.77
85	1.62	1.67	1.60	1.70	1.57	1.72	1.55	1.75	1.52	1.77
90	1.63	1.68	1.61	1.70	1.59	1.73	1.57	1.75	1.54	1.78
95	1.64	1.69	1.62	1.71	1.60	1.73	1.58	1.75	1.56	1.78
100	1.65	1.69	1.63	1.72	1.61	1.74	1.59	1.76	1.57	1.78

Source: From J. Durbin and G. S. Watson, "Testing for serial correlation in least squares regression, II," *Biometrika,* 1951, 30, 159–178. Reproduced by permission of the *Biometrika* Trustees.

Table D.9 Critical values for the Durbin–Watson d Statistic ($\alpha = .01$)

n	$k = 1$ d_L	$k = 1$ d_U	$k = 2$ d_L	$k = 2$ d_U	$k = 3$ d_L	$k = 3$ d_U	$k = 4$ d_L	$k = 4$ d_U	$k = 5$ d_L	$k = 5$ d_U
15	.81	1.07	.70	1.25	.59	1.46	.49	1.70	.39	1.96
16	.84	1.09	.74	1.25	.63	1.44	.53	1.66	.44	1.90
17	.87	1.10	.77	1.25	.67	1.43	.57	1.63	.48	1.85
18	.90	1.12	.80	1.26	.71	1.42	.61	1.60	.52	1.80
19	.93	1.13	.83	1.26	.74	1.41	.65	1.58	.56	1.77
20	.95	1.15	.86	1.27	.77	1.41	.68	1.57	.60	1.74
21	.97	1.16	.89	1.27	.80	1.41	.72	1.55	.63	1.71
22	1.00	1.17	.91	1.28	.83	1.40	.75	1.54	.66	1.69
23	1.02	1.19	.94	1.29	.86	1.40	.77	1.53	.70	1.67
24	1.04	1.20	.96	1.30	.88	1.41	.80	1.53	.72	1.66
25	1.05	1.21	.98	1.30	.90	1.41	.83	1.52	.75	1.65
26	1.07	1.22	1.00	1.31	.93	1.41	.85	1.52	.78	1.64
27	1.09	1.23	1.02	1.32	.95	1.41	.88	1.51	.81	1.63
28	1.10	1.24	1.04	1.32	.97	1.41	.90	1.51	.83	1.62
29	1.12	1.25	1.05	1.33	.99	1.42	.92	1.51	.85	1.61
30	1.13	1.26	1.07	1.34	1.01	1.42	.94	1.51	.88	1.61
31	1.15	1.27	1.08	1.34	1.02	1.42	.96	1.51	.90	1.60
32	1.16	1.28	1.10	1.35	1.04	1.43	.98	1.51	.92	1.60
33	1.17	1.29	1.11	1.36	1.05	1.43	1.00	1.51	.94	1.59
34	1.18	1.30	1.13	1.36	1.07	1.43	1.01	1.51	.95	1.59
35	1.19	1.31	1.14	1.37	1.08	1.44	1.03	1.51	.97	1.59
36	1.21	1.32	1.15	1.38	1.10	1.44	1.04	1.51	.99	1.59
37	1.22	1.32	1.16	1.38	1.11	1.45	1.06	1.51	1.00	1.59
38	1.23	1.33	1.18	1.39	1.12	1.45	1.07	1.52	1.02	1.58
39	1.24	1.34	1.19	1.39	1.14	1.45	1.09	1.52	1.03	1.58
40	1.25	1.34	1.20	1.40	1.15	1.46	1.10	1.52	1.05	1.58
45	1.29	1.38	1.24	1.42	1.20	1.48	1.16	1.53	1.11	1.58
50	1.32	1.40	1.28	1.45	1.24	1.49	1.20	1.54	1.16	1.59
55	1.36	1.43	1.32	1.47	1.28	1.51	1.25	1.55	1.21	1.59
60	1.38	1.45	1.35	1.48	1.32	1.52	1.28	1.56	1.25	1.60
65	1.41	1.47	1.38	1.50	1.35	1.53	1.31	1.57	1.28	1.61
70	1.43	1.49	1.40	1.52	1.37	1.55	1.34	1.58	1.31	1.61
75	1.45	1.50	1.42	1.53	1.39	1.56	1.37	1.59	1.34	1.62
80	1.47	1.52	1.44	1.54	1.42	1.57	1.39	1.60	1.36	1.62
85	1.48	1.53	1.46	1.55	1.43	1.58	1.41	1.60	1.39	1.63
90	1.50	1.54	1.47	1.56	1.45	1.59	1.43	1.61	1.41	1.64
95	1.51	1.55	1.49	1.57	1.47	1.60	1.45	1.62	1.42	1.64
100	1.52	1.56	1.50	1.58	1.48	1.60	1.46	1.63	1.44	1.65

Source: From J. Durbin and G. S. Watson, "Testing for serial correlation in least squares regression, II," *Biometrika*, 1951, 30, 159–178. Reproduced by permission of the *Biometrika* Trustees.

Table D.10 Critical values for the χ^2 Statistic

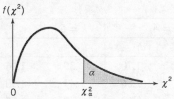

Degrees of Freedom	$\chi^2_{.995}$	$\chi^2_{.990}$	$\chi^2_{.975}$	$\chi^2_{.950}$	$\chi^2_{.900}$
1	.0000393	.0001571	.0009821	.0039321	.0157908
2	.0100251	.0201007	.0506356	.102587	.210720
3	.0717212	.114832	.215795	.351846	.584375
4	.206990	.297110	.484419	.710721	1.063623
5	.411740	.554300	.831211	1.145476	1.61031
6	.675727	.872085	1.237347	1.63539	2.20413
7	.989265	1.239043	1.68987	2.16735	2.83311
8	1.344419	1.646482	2.17973	2.73264	3.48954
9	1.734926	2.087912	2.70039	3.32511	4.16816
10	2.15585	2.55821	3.24697	3.94030	4.86518
11	2.60321	3.05347	3.81575	4.57481	5.57779
12	3.07382	3.57056	4.40379	5.22603	6.30380
13	3.56503	4.10691	5.00874	5.89186	7.04150
14	4.07468	4.66043	5.62872	6.57063	7.78953
15	4.60094	5.22935	6.26214	7.26094	8.54675
16	5.14224	5.81221	6.90766	7.96164	9.31223
17	5.69724	6.40776	7.56418	8.67176	10.0852
18	76.26481	7.01491	8.23075	9.39046	10.8649
19	6.84398	7.63273	8.90655	10.1170	11.6509
20	7.43386	8.26040	9.59083	10.8508	12.4426
21	8.03366	8.89720	10.28293	11.5913	13.2396
22	8.64272	9.54249	10.9823	12.3380	14.0415
23	9.26042	10.19567	11.6885	13.0905	14.8479
24	9.88623	10.8564	12.4011	13.8484	15.6587
25	10.5197	11.5240	13.1197	14.6114	16.4734
26	11.1603	12.1981	13.8439	15.3791	17.2919
27	11.8076	12.8786	14.5733	16.1513	18.1138
28	12.4613	13.5648	15.3079	16.9279	18.9392
29	13.1211	14.2565	16.0471	17.7083	19.7677
30	13.7867	14.9535	16.7908	18.4926	20.5992
40	20.7065	22.1643	24.4331	26.5093	29.0505
50	27.9907	29.7067	32.3574	34.7642	37.6886
60	35.5346	37.4848	40.4817	43.1879	46.4589
70	43.2752	45.4418	48.7576	51.7393	55.3290
80	51.1720	53.5400	57.1532	60.3915	64.2778
90	59.1963	61.7541	65.6466	69.1260	73.2912
100	67.3276	70.0648	74.2219	77.9295	82.3581
150	109.142	112.668	117.985	122.692	128.275
200	152.241	156.432	162.728	168.279	174.835
300	240.663	245.972	253.912	260.878	269.068
400	330.903	337.155	346.482	354.641	364.207
500	422.303	429.388	439.936	449.147	459.926

(*continued overleaf*)

Table D.10 (*Continued*)

Degrees of Freedom	$\chi^2_{.100}$	$\chi^2_{.050}$	$\chi^2_{.025}$	$\chi^2_{.010}$	$\chi^2_{.500}$
1	2.70554	3.84146	5.02389	6.63490	7.87944
2	4.60517	5.99147	7.37776	9.21034	10.5966
3	6.25139	7.81473	9.34840	11.3449	12.8381
4	7.77944	9.48773	11.1433	13.2767	14.8602
5	9.23635	11.0705	12.8325	15.0863	16.7496
6	10.6446	12.5916	14.4494	16.8119	18.5476
7	12.0170	14.0671	16.0128	18.4753	20.2777
8	13.3616	15.5073	17.5346	20.0902	21.9550
9	14.6837	16.9190	19.0228	21.6660	23.5893
10	15.9871	18.3070	20.4831	23.2093	25.1882
11	17.2750	19.6751	21.9200	24.7250	26.7569
12	18.5494	21.0261	23.3367	26.2170	28.2995
13	19.8119	22.3621	24.7356	27.6883	29.8194
14	21.0642	23.6848	26.1190	29.1413	31.3193
15	22.3072	24.9958	27.4884	30.5779	32.8013
16	23.5418	26.2962	28.8454	31.9999	34.2672
17	24.7690	27.5871	30.1910	33.4087	35.7185
18	25.9894	28.8693	31.5264	34.8053	37.1564
19	27.2036	30.1435	32.8523	36.1908	38.5822
20	28.4120	31.4104	34.1696	37.5662	39.9968
21	29.6151	32.6705	35.4789	38.9321	41.4010
22	30.8133	33.9244	36.7807	40.2894	42.7956
23	32.0069	35.1725	38.0757	41.6384	44.1813
24	33.1963	36.4151	39.3641	42.9798	45.5585
25	34.3816	37.6525	40.6465	44.3141	46.9278
26	36.5631	38.8852	41.9232	45.6417	48.2899
27	36.7412	40.1133	43.1944	46.9630	49.6449
28	37.9159	41.3372	44.4607	48.2782	50.9933
29	39.0875	42.5569	45.7222	49.5879	52.3356
30	40.2560	43.7729	46.9792	50.8922	53.6720
40	51.8050	55.7585	59.3417	63.6907	66.7659
50	63.1671	67.5048	71.4202	76.1539	79.4900
60	74.3970	79.0819	83.2976	88.3794	91.9517
70	85.5271	90.5312	95.0231	100.425	104.215
80	96.5782	101.879	106.629	112.329	116.321
90	107.565	113.145	118.136	124.116	128.299
100	118.498	124.342	129.561	135.807	140.169
150	172.581	179.581	185.800	193.208	198.360
200	226.021	233.994	241.058	249.445	255.264
300	331.789	341.395	349.874	359.906	366.844
400	436.649	447.632	457.305	468.724	476.606
500	540.930	553.127	563.852	576.493	585.207

Table D.11 Percentage Points of the Studentized Range $q(p, \nu)$, upper 5%

ν \ p	2	3	4	5	6	7	8	9	10	11
1	17.97	26.98	32.82	37.08	40.41	43.12	45.40	47.36	49.07	50.59
2	6.08	8.33	9.80	10.88	11.74	12.44	13.03	13.54	13.99	14.39
3	4.50	5.91	6.82	7.50	8.04	8.48	8.85	9.18	9.46	9.72
4	3.93	5.04	5.76	6.29	6.71	7.05	7.35	7.60	7.83	8.03
5	3.64	4.60	5.22	5.67	6.03	6.33	6.58	6.80	6.99	7.17
6	3.46	4.34	4.90	5.30	5.63	5.90	6.12	6.32	6.49	6.65
7	3.34	4.16	4.68	5.06	5.36	5.61	5.82	6.00	6.16	6.30
8	3.26	4.04	4.53	4.89	5.17	5.40	5.60	5.77	5.92	6.05
9	3.20	3.95	4.41	4.76	5.02	5.24	5.43	5.59	5.74	5.87
10	3.15	3.88	4.33	4.65	4.91	5.12	5.30	5.46	5.60	5.72
11	3.11	3.82	4.26	4.57	4.82	5.03	5.20	5.35	5.49	5.61
12	3.08	3.77	4.20	4.51	4.75	4.95	5.12	5.27	5.39	5.51
13	3.06	3.73	4.15	4.45	4.69	4.88	5.05	5.19	5.32	5.43
14	3.03	3.70	4.11	4.41	4.64	4.83	4.99	5.13	5.25	5.36
15	3.01	3.67	4.08	4.37	4.60	4.78	4.94	5.08	5.20	5.31
16	3.00	3.65	4.05	4.33	4.56	4.74	4.90	5.03	5.15	5.26
17	2.98	3.63	4.02	4.30	4.52	4.70	4.86	4.99	5.11	5.21
18	2.97	3.61	4.00	4.28	4.49	4.67	4.82	4.96	5.07	5.17
19	2.96	3.59	3.98	4.25	4.47	4.65	4.79	4.92	5.04	5.14
20	2.95	3.58	3.96	4.23	4.45	4.62	4.77	4.90	5.01	5.11
24	2.92	3.53	3.90	4.17	4.37	4.54	4.68	4.81	4.92	5.01
30	2.89	3.49	3.85	4.10	4.30	4.46	4.60	4.72	4.82	4.92
40	2.86	3.44	3.79	4.04	4.23	4.39	4.52	4.63	4.73	4.82
60	2.83	3.40	3.74	3.98	4.16	4.31	4.44	4.55	4.65	4.73
120	2.80	3.36	3.68	3.92	4.10	4.24	4.36	4.47	4.56	4.64
∞	2.77	3.31	3.63	3.86	4.03	4.17	4.29	4.39	4.47	4.55

(*continued overleaf*)

Table D.11 (*Continued*)

ν \ p	12	13	14	15	16	17	18	19	20
1	51.96	53.20	54.33	55.36	56.32	57.22	58.04	58.83	59.56
2	14.75	15.08	15.38	15.65	15.91	16.14	16.37	16.57	16.77
3	9.95	10.15	10.35	10.52	10.69	10.84	10.98	11.11	11.24
4	8.21	8.37	8.52	8.66	8.79	8.91	9.03	9.13	9.23
5	7.32	7.47	7.60	7.72	7.83	7.93	8.03	8.12	8.21
6	6.79	6.92	7.03	7.14	7.24	7.34	7.43	7.51	7.59
7	6.43	6.55	6.66	6.76	6.85	6.94	7.02	7.10	7.17
8	6.18	6.29	6.39	6.48	6.57	6.65	6.73	6.80	6.87
9	5.98	6.09	6.19	6.28	6.36	6.44	6.51	6.58	6.64
10	5.83	5.93	6.03	6.11	6.19	6.27	6.34	6.40	6.47
11	5.71	5.81	5.90	5.98	6.06	6.13	6.20	6.27	6.33
12	5.61	5.71	5.80	5.88	5.95	6.02	6.09	6.15	6.21
13	5.53	5.63	5.71	5.79	5.86	5.93	5.99	6.05	6.11
14	5.46	5.55	5.64	5.71	5.79	5.85	5.91	5.97	6.03
15	5.40	5.49	5.57	5.65	5.72	5.78	5.85	5.90	5.96
16	5.35	5.44	5.52	5.59	5.66	5.73	5.79	5.84	5.90
17	5.31	5.39	5.47	5.54	5.61	5.67	5.73	5.79	5.84
18	5.27	5.35	5.43	5.50	5.57	5.63	5.69	5.74	5.79
19	5.23	5.31	5.39	5.46	5.53	5.59	5.65	5.70	5.75
20	5.20	5.28	5.36	5.43	5.49	5.55	5.61	5.66	5.71
24	5.10	5.18	5.25	5.32	5.38	5.44	5.49	5.55	5.59
30	5.00	5.08	5.15	5.21	5.27	5.33	5.38	5.43	5.47
40	4.90	4.98	5.04	5.11	5.16	5.22	5.27	5.31	5.36
60	4.81	4.88	4.94	5.00	5.06	5.11	5.15	5.20	5.24
120	4.71	4.78	4.84	4.90	4.95	5.00	5.04	5.09	5.13
∞	4.62	4.68	4.74	4.80	4.85	4.89	4.93	4.97	5.01

Source: Biometrika Tables for Statisticians, Vol. 1, 3rd ed., edited by E. S. Pearson and H. O. Hartley (Cambridge University Press, 1966). Reproduced by permission of Professor E. S. Pearson and the *Biometrika* Trustees.

Table D.12 Percentage Points of the Studentized Range $q(p, \nu)$, Upper 1%

ν \ p	2	3	4	5	6	7	8	9	10	11
1	90.03	135.0	164.3	185.6	202.2	215.8	227.2	237.0	245.6	253.2
2	14.04	19.02	22.29	24.72	26.63	28.20	29.53	30.68	31.69	32.59
3	8.26	10.62	12.17	13.33	14.24	15.00	15.64	16.20	16.69	17.13
4	6.51	8.12	9.17	9.96	10.58	11.10	11.55	11.93	12.27	12.57
5	5.70	6.98	7.80	8.42	8.91	9.32	9.67	9.97	10.24	10.48
6	5.24	6.33	7.03	7.56	7.97	8.32	8.61	8.87	9.10	9.30
7	4.95	5.92	6.54	7.01	7.37	7.68	7.94	8.17	8.37	8.55
8	4.75	5.64	6.20	6.62	6.96	7.24	7.47	7.68	7.86	8.03
9	4.60	5.43	5.96	6.35	6.66	6.91	7.13	7.33	7.49	7.65
10	4.48	5.27	5.77	6.14	6.43	6.67	6.87	7.05	7.21	7.36
11	4.39	5.15	5.62	5.97	6.25	6.48	6.67	6.84	6.99	7.13
12	4.32	5.05	5.50	5.84	6.10	6.32	6.51	6.67	6.81	6.94
13	4.26	4.96	5.40	5.73	5.98	6.19	6.37	6.53	6.67	6.79
14	4.21	4.89	5.32	5.63	5.88	6.08	6.26	6.41	6.54	6.66
15	4.17	4.84	5.25	5.56	5.80	5.99	6.16	6.31	6.44	6.55
16	4.13	4.79	5.19	5.49	5.72	5.92	6.08	6.22	6.35	6.46
17	4.10	4.74	5.14	5.43	5.66	5.85	6.01	6.15	6.27	6.38
18	4.07	4.70	5.09	5.38	5.60	5.79	5.94	6.08	6.20	6.31
19	4.05	4.67	5.05	5.33	5.55	5.73	5.89	6.02	6.14	6.25
20	4.02	4.64	5.02	5.29	5.51	5.69	5.84	5.97	6.09	6.19
24	3.96	4.55	4.91	5.17	5.37	5.54	5.69	5.81	5.92	6.02
30	3.89	4.45	4.80	5.05	5.24	5.40	5.54	5.65	5.76	5.85
40	3.82	4.37	4.70	4.93	5.11	5.26	5.39	5.50	5.60	5.69
60	3.76	4.28	4.59	4.82	4.99	5.13	5.25	5.36	5.45	5.53
120	3.70	4.20	4.50	4.71	4.87	5.01	5.12	5.21	5.30	5.37
∞	3.64	4.12	4.40	4.60	4.76	4.88	4.99	5.08	5.16	5.23

(continued overleaf)

Table D.12 (*Continued*)

ν \ p	12	13	14	15	16	17	18	19	20
1	260.0	266.2	271.8	277.0	281.8	286.3	290.0	294.3	298.0
2	33.40	34.13	34.81	35.43	36.00	36.53	37.03	37.50	37.95
3	17.53	17.89	18.22	18.52	18.81	19.07	19.32	19.55	19.77
4	12.84	13.09	13.32	13.53	13.73	13.91	14.08	14.24	14.40
5	10.70	10.89	11.08	11.24	11.40	11.55	11.68	11.81	11.93
6	9.48	9.65	9.81	9.95	10.08	10.21	10.32	10.43	10.54
7	8.71	8.86	9.00	9.12	9.24	9.35	9.46	9.55	9.65
8	8.18	8.31	8.44	8.55	8.66	8.76	8.85	8.94	9.03
9	7.78	7.91	8.03	8.13	8.23	8.33	8.41	8.49	8.57
10	7.49	7.60	7.71	7.81	7.91	7.99	8.08	8.15	8.23
11	7.25	7.36	7.46	7.56	7.65	7.73	7.81	7.88	7.95
12	7.06	7.17	7.26	7.36	7.44	7.52	7.59	7.66	7.73
13	6.90	7.01	7.10	7.19	7.27	7.35	7.42	7.48	7.55
14	6.77	6.87	6.96	7.05	7.13	7.20	7.27	7.33	7.39
15	6.66	6.76	6.84	6.93	7.00	7.07	7.14	7.20	7.26
16	6.56	6.66	6.74	6.82	6.90	6.97	7.03	7.09	7.15
17	6.48	6.57	6.66	6.73	6.81	6.87	6.94	7.00	7.05
18	6.41	6.50	6.58	6.65	6.72	6.79	6.85	6.91	6.97
19	6.34	6.43	6.51	6.58	6.65	6.72	6.78	6.84	6.89
20	6.28	6.37	6.45	6.52	6.59	6.65	6.71	6.77	6.82
24	6.11	6.19	6.26	6.33	6.39	6.45	6.51	6.56	6.61
30	5.93	6.01	6.08	6.14	6.20	6.26	6.31	6.36	6.41
40	5.76	5.83	5.90	5.96	6.02	6.07	6.12	6.16	6.21
60	5.60	5.67	5.73	5.78	5.84	5.89	5.93	5.97	6.01
120	5.44	5.50	5.56	5.61	5.66	5.71	5.75	5.79	5.83
∞	5.29	5.35	5.40	5.45	5.49	5.54	5.57	5.61	5.65

Source: Biometrika Tables for Statisticians, Vol. 1, 3rd ed., edited by E. S. Pearson and H. O. Hartley (Cambridge University Press, 1966). Reproduced by permission of Professor E. S. Pearson and the *Biometrika* Trustees.

FILE LAYOUTS FOR CASE STUDY DATA SETS

Case Study 1: Legal Advertising—Does It Pay?		
💿 LEGALADV ($n = 48$ observations)		
Variable	Type	Description
MONTH	Numeric	Month
TOTADVEXP	Numeric	Advertising expenditure (thousands of dollars)
NEWPI	Numeric	Number of new personal injury cases
NEWWC	Numeric	Number of new worker's compensation cases
ADVEXP6	Numeric	Cumulative advertising expenditure over previous 6 months (thousands of dollars)

Case Study 2: Modeling the Sales Prices of Properties in Four Neighborhoods		
💿 TAMSALES4 ($n = 176$ observations)		
Variable	Type	Description
SALES	Numeric	Sales price (thousands of dollars)
LAND	Numeric	Land value (thousands of dollars)
IMP	Numeric	Value of improvements (thousands of dollars)
NBHD	Character	Neighborhood (HYDEPARK, DAVISISLES, CHEVAL, HUNTERSGREEN)
💿 TAMSALES8 ($n = 350$ observations)		
Variable	Type	Description
SALES	Numeric	Sales price (thousands of dollars)
LAND	Numeric	Land value (thousands of dollars)
IMP	Numeric	Value of improvements (thousands of dollars)
NBHD	Character	Neighborhood (HYDEPARK, DAVISISLES, CHEVAL, HUNTERSGREEN, AVILA, CRLLWOODVILL, TAMPAPALMS, TOWN&CNTRY)

Case Study 3: Deregulation of the Intrastate Trucking Industry

TRUCKING ($n = 134$ observations)

Variable	Type	Description
PRICPTM	Numeric	Price charged per ton-mile (dollars)
DISTANCE	Numeric	Distance traveled (hundreds of miles)
WEIGHT	Numeric	Weight of product shipped (thousands of pounds)
PCTLOAD	Numeric	Weight as a percentage of truck load capacity
ORIGIN	Character	City of origin (JAX or MIA)
MARKET	Character	Size of market destination (LARGE or SMALL)
DEREG	Character	Deregulation in effect (YES or NO)
PRODUCT	Numeric	Product classification (100, 150, or 200)
LNPRICE	Numeric	Natural logarithm of price charged

TRUCKING4 ($n = 448$ observations)

Variable	Type	Description
PRICPTM	Numeric	Price charged per ton-mile (dollars)
DISTANCE	Numeric	Distance traveled (hundreds of miles)
WEIGHT	Numeric	Weight of product shipped (thousands of pounds)
PCTLOAD	Numeric	Weight as a percentage of truck load capacity
ORIGIN	Character	City of origin (JAX or MIA)
MARKET	Character	Size of market destination (LARGE or SMALL)
DEREG	Character	Deregulation in effect (YES or NO)
PRODUCT	Numeric	Product classification (100, 150, or 200)
CARRIER	Character	Florida trucking carrier (A, B, C, or D)
LNPRICE	Numeric	Natural logarithm of price charged

Case Study 4: An Analysis of Rain Levels in California

CALIRAIN ($n = 30$ observations)

Variable	Type	Description
STATION	Numeric	Station number
NAME	Character	Name of meteorological station
PRECIP	Numeric	Average annual precipitation (inches)
ALTITUDE	Numeric	Altitude (feet)
LATITUDE	Numeric	Latitude (degrees)
DISTANCE	Numeric	Distance to Pacific Ocean (miles)
SHADOW	Character	Slope face (W = westward or L = leeward)

Case Study 5: An Investigation of Factors Affecting the Sales Price of Condominium Units Sold at Public Auction

💿 CONDO ($n = 209$ observations)

Variable	Type	Description
PRICE	Numeric	Price paid (hundreds of dollars)
FLOOR	Numeric	Floor height $(1, 2, 3, \ldots, 8)$
DISTELEV	Numeric	Distance from elevator $(1, 2, 3, \ldots, 15)$
VIEW	Numeric	View $(1 = \text{Ocean}, 0 = \text{Bay})$
ENDUNIT	Numeric	End unit $(1 = \text{end}, 0 = \text{not})$
FURNISH	Numeric	Furnished $(1 = \text{furnish}, 0 = \text{not})$
METHOD	Character	Sales method $(A = \text{auction}, F = \text{fixed price})$

Case Study 7: Reluctance to Transmit Bad News: The MUM Effect

💿 MUM ($n = 40$ observations)

Variable	Type	Description
FEEDBACK	Numeric	Latency to feedback (seconds)
SUBJECT	Character	Subject visibility (NOTVIS or VISIBLE)
CONFED	Character	Confederate success (FAILURE or SUCCESS)

Answers to Selected Exercises

Chapter 1

1.1 (a) Quantitative **(b)** Qualitative **(c)** Quantitative **(d)** Qualitative **(e)** Quantitative **(f)** Quantitative

1.3 (a) earthquakes **(b)** type of ground motion–qualitative; magnitude and acceleration–quantitative

1.5 (a) Qualitative **(b)** Qualitative **(c)** Quantitative **(d)** Quantitative **(e)** Quantitative **(f)** Quantitative **(g)** Quantitative **(h)** Qualitative

1.7 (a) Population–all decision makers; sample–155 volunteer students; variables–emotional state and whether or not to repair a very old car **(b)** subjects in the guilty-state group are less likely to repair an old car

1.9 (a) Amateur boxers **(b)** massage or rest group–qualitative; heart rate and blood lactate level–both quantitative **(c)** no difference in mean heart rates of two groups of boxers **(d)** no

1.11 (a) Population–all adults in Tennessee; sample–575 study participants **(b)** years of education–quantitative; insomnia–qualitative **(c)** less educated adults are more likely to have chronic insomnia

1.13 (a) Black–.203; White–.637; Sumatran–.017; Javan–.003; Indian–.140 **(c)** .84; .16

1.15 (a) pie chart **(b)** type of firearms owned **(c)** rifle (33 %), shotgun (21%), and revolver (20 %)

1.17 (d) public wells (40 %); private wells (21 %)

1.19 (b) yes

1.21 yes

1.23 (b) .98

1.25 (a) 2.12; average magnitude for the aftershocks is 2.12 **(b)** 6.7; difference between the largest and smallest magnitude is 6.7 **(c)** .66; about 95 % of the magnitudes fall in the interval mean ± 2(std. dev.) = (.8, 3.44) **(d)** μ = mean; σ = standard deviation

1.27 (a) $\bar{y}$ = 94.91, s = 4.83 **(b)** (85.25, 104.57) **(c)** .976; yes

1.29 (a) 1.47; average daily ammonia concentration is 1.47 ppm **(b)** .0640; about 95 % of the daily ammonia concentration levels fall within $\bar{y} \pm 2s$ = (1.34, 1.60) ppm **(c)** morning

1.31 (a) (−111, 149) **(b)** (−91, 105) **(c)** SAT-Math

1.33 (a) .9544 **(b)** .1587 **(c)** .1587 **(d)** .8185 **(e)** .1498 **(f)** .9974

1.35 (a) .9406 **(b)** .9406 **(c)** .1140

1.37 (a) .2690 **(b)** .2611 **(c)** (13.6, 62.2)

1.39 .8664

1.41 standard deviation for means of $n = 12$ measurements is smaller

1.43 (a) .10; .0141 **(b)** Central Limit Theorem **(c)** .017

1.45 (a) 1.13 ± .67 **(b)** yes

1.47 (a) 99.6 **(b)** (97.4, 101.8) **(c)** 95 % confident that the true mean Mach rating score is between 97.4 and 101.8 **(d)** yes

1.49 (17.1, 20.9); 99 % confident that the true mean quality of the methodology is between 17.1 and 20.9 Wong Scale points

1.51 (a) null hypothesis **(b)** alternative hypothesis **(c)** rejecting H_0 when H_0 is true **(d)** accepting H_0 when H_0 is false **(e)** probability of Type I error **(f)** probability of Type II error **(g)** observed significance level

1.53 (a) .025 **(b)** .05 **(c)** .005 **(d)** .0985 **(e)** .10 **(f)** .01

1.55 (a) $H_0 : \mu = 15, H_a : \mu < 15$ **(b)** concluding that the average level of mercury uptake is less than 15 ppm, when the average level is equal to 15 ppm **(c)** concluding that the average level of mercury uptake is equal to 15 ppm, when the average level is less than 15 ppm

1.57 (a) $z = -3.72$, reject H_0

1.59 (b) $p = .0214$, reject H_0 at $\alpha = .05$ **(c)** $p \approx 0$, reject H_0 at $\alpha = .05$

1.61 $t = 7.83$, reject H_0

1.63 Approx. normal (for large n) with mean $\mu_{\bar{y}_1 - \bar{y}_2} = (\mu_1 - \mu_2)$ and standard deviation

$$\sigma_{\bar{y}_1 - \bar{y}_2} = \sqrt{\frac{\sigma_1^2}{n_1} + \frac{\sigma_2^2}{n_2}}$$

1.65 $t = 1.08$, fail to reject H_0; insufficient evidence to conclude that the mean ratings for the two groups differ

1.67 $t = 2.68$, reject H_0; sufficient evidence to conclude that the mean performance level in the rudeness condition is less than the mean in the control group

1.69 (a) $t = -7.24$, reject H_0 **(b)** $t = -.50$, fail to reject H_0 **(c)** $t = -1.25$, fail to reject H_0

1.71 (a) each participant acted as a speaker and an audience member **(b)** $\mu_d = \mu_{speaker} - \mu_{audience}$ **(c)** No; need sample statistics for differences **(d)** reject $H_0 : \mu_d = 0$

1.73 (a) $t = -2.97$, p-value $= .016$, fail to reject H_0 **(b)** no **(c)** $t = .57$, p-value $= .58$, fail to reject H_0; no **(d)** $t = 3.23$, p-value $= .01$, fail to reject H_0; no

1.75 $F = 1.30$, fail to reject $H_0 : \sigma_{DM}^2/\sigma_H^2 = 1$

1.77 $F = 1.16$, fail to reject $H_0 : \sigma_N^2/\sigma_B^2 = 1$

1.79 $F = 5.87$, reject $H_0 : \sigma_1^2/\sigma_3^2 = 1$

1.81 (a) 5, 21.5, 4.637 **(b)** 16.75, 36.25, 6.021 **(c)** 4.857, 29.81, 5.460 **(d)** 4, 0, 0

1.83 (a) $z = -4$ **(b)** $z = .5$ **(c)** $z = 0$ **(d)** $z = 6$

1.85 (a) all men and women **(b)** 300 people from Gainesville, FL **(c)** inferential **(d)** qualitative

1.87 Relative frequencies: Burnished (.159), Monochrome (.550), Slipped (.066), Paint-curvilinear (.017), Paint-geometric (.197), Paint-natural (.005), Cycladic (.005), Conical (.002)

1.89 (c) Eclipse

1.91 (a) .3085 **(b)** .1587 **(c)** .1359 **(d)** .6915 **(e)** 0 **(f)** .9938

1.93 (a) .26 **(b)** .086 **(c)** $\bar{y} = 7.43, s = .82; (5.79, 9.06); 95\%$ (Empirical Rule) **(d)** $\bar{y} = 1.22, s = 5.11; (-9.00, 11.44); 95\%$ (Empirical Rule)

1.95 (a) .3156 **(b)** .1894

1.97 (a) no, $p \approx 0$ **(b)** μ and/or σ differ from stated values

1.99 (.61, 2.13); 99 % confident that mean weight of dry seeds in the crops of all spinifex pigeons is between 0.61 and 2.13 grams.

1.101 (a) 293; 119.78 **(c)** .0158

1.103 $t = 1.21$, p-value $= .257$, fail to reject $H_0 : \mu = 2{,}550$

1.105 (a) $z = -2.64$, reject $H_0 : \mu_{\text{Never}} - \mu_{\text{Repeat}} = 0$

1.107 (a) no; $z = 1.19$, fail to reject $H_0 : \mu_{\text{Males}} - \mu_{\text{Females}} = 0$ **(b)** $.5 \pm .690$ **(d)** $p = .2340$ **(d)** $F = 1.196$, reject $H_0 : \sigma^2_{\text{Males}}/\sigma^2_{\text{Females}} = 1$

1.109 (a) yes; $z = -6.02$, p-value $= 0$, reject $H_0 : \mu_S - \mu_{\text{TRI}} = 0$ **(b)** no; $F = 1.41$, fail to reject $H_0 : \sigma^2_{\text{TRI}}/\sigma^2_S = 1$

Chapter 3

3.3 (a) $\beta_0 = 2; \beta_1 = 2$ **(b)** $\beta_0 = 4; \beta_1 = 1$ **(c)** $\beta_0 = -2; \beta_1 = 4$ **(d)** $\beta_0 = -4; \beta_1 = -1$

3.5 (a) $\beta_1 = 2; \beta_0 = 3$ **(b)** $\beta_1 = 1; \beta_0 = 1$ **(c)** $\beta_1 = 3; \beta_0 = -2$ **(d)** $\beta_1 = 5; \beta_0 = 0$ **(e)** $\beta_1 = -2; \beta_0 = 4$

3.7 (a) $\hat{\beta}_0 = 2; \hat{\beta}_1 = -1.2$

3.9 (a) yes **(b)** positive **(c)** line is based on sample data

3.11 (a) no **(b)** no **(c)** yes **(d)** negative **(e)** yes; winners tend to punish less than non-winners

3.13 (a) $\hat{y} = 6.25 - .0023x$ **(c)** 5.56

3.15 $\hat{y} = .5704 + .0264x$; since $x = 0$ is nonsensical, no practical interpretation of $\hat{\beta}_0 = .5704$; for each one-position increase in order, estimated recall proportion increases by $\hat{\beta}_1 = .0264$

3.17 (a) .0313 **(b)** .1769

3.19 .2

3.21 (a) $y = \beta_0 + \beta_1 x + \varepsilon$ **(b)** $\hat{y} = 119.9 + .3456x$ **(d)** 635.2 **(e)** $\hat{y} \pm 1270.4$

3.23 (a) yes, $t = 6.71$ **(b)** yes, $t = -5.20$

3.25 $(-.0039, -.0008)$

3.27 (a) $t = 6.29$, reject H_0 **(b)** $.88 \pm .24$ **(c)** no evidence to say slope differs from 1

3.29 yes, $t = 2.86$

3.31 yes; $t = 3.02$, p-value $= .009/2 = .0045$, reject H_0: $\beta_1 = 0$ in favor of $H_a: \beta_1 > 0$

3.33 (a) $\hat{\beta}_0 = .5151, \hat{\beta}_1 = .000021$ **(b)** yes, p-value $= .008/2 = .004$ **(c)** Denver's elevation is very high **(d)** $\hat{\beta}_0 = .5154, \hat{\beta}_1 = .000020$ **(b)** no, p-value $= .332/2 = .166$

3.35 (a) .9583; .9183 **(b)** $-.9487$; .90

3.37 positive

3.39 (a) $y = \beta_0 + \beta_1 x + \varepsilon$ **(b)** moderate positive linear relationship between RMP and SET ratings **(c)** positive **(d)** reject H_0 at $\alpha = .05$ **(e)** .4624

3.41 (b) .1998; .0032; .3832; .0864; .9006 **(c)** reject H_0; fail to reject H_0; reject H_0; fail to reject H_0; reject H_0

3.43 (a) fail to reject H_0: $\rho = 0$ at $\alpha = .05$ **(c)** .25 **(d)** fail to reject H_0: $\rho = 0$ at $\alpha = .05$ **(f)** .0144

3.45 $r = .5702; r^2 = .3251$

3.47 (a) Negative **(b)** $r = -.0895$

3.49 (a) 4.06; .23 **(b)** $10.6 \pm .233$ **(c)** $8.9 \pm .319$ **(d)** $12.3 \pm .319$ **(e)** wider **(f)** 12.3 ± 1.046

3.51 (a) prediction interval for y with $x = 10$ **(b)** confidence interval for $E(y)$ with $x = 10$

3.53 for $x = 220, 95\%$ confidence interval for $E(y)$ is $(5.65, 5.84)$

3.55 (a) $(3.17, 3.85)$ **(b)** $(2.00, 5.02)$

3.57 $(0, 1535.3)$

3.59 $\hat{y} = 9470.5 + .192x; t = 12.69$, p-value $= 0$, reject H_0: $\beta_1 = 0$; $r^2 = .712$; $2s = 1724$; 95 % PI for y when $x = 12, 000$: $(10{,}033, 13{,}508)$

3.61 $\hat{y} = 62.5 - .35x; t = -12.56$, p-value $= 0$, reject H_0: $\beta_1 = 0$; $r^2 = .07$; $2s = 24.6$; 95 % PI for y when $x = 50$: $(21.1, 69.3)$

3.63 (a) $\hat{y} = -9.2667x$ **(b)** 12.8667; 3.2167; 1.7935 **(c)** yes; $t = -28.30$ **(d)** $-9.267 \pm .909$ **(e)** $-9.267 \pm .909$ **(f)** -9.267 ± 5.061

3.65 (a) $\hat{y} = 5.364x$ **(b)** yes; $t = 25.28$ **(c)** 18.77 ± 6.30

3.67 (a) $\hat{y} = 51.18x$ **(b)** yes; $t = 154.56$ **(c)** $\hat{y} = 1{,}855.45 + 47.07x$; yes, $t = 93.36$ **(d)** $y = \beta_0 + \beta_1 x + \varepsilon$

3.69 (a) $y = \beta_0 + \beta_1 x + \varepsilon$; negative **(b)** no

3.71 $\hat{y} = -.00105 + .00321x; t = 7.43$, reject H_0: $\beta_1 = 0$; $r^2 = .81$; $2s = .014$; 95 % PI for y when $x = 8$: $(.0085, .0408)$

3.73 $100(r^2)\%$ of sample variation in ESLR score can be explained by x (SG, SR, or ER score) in linear model

3.75 $\hat{y} = -3.05 + .108x; t = 4.00$, reject $H_0: \beta_1 = 0; r^2 = .57$; $2s = 2.18$

3.77 (b) $\hat{y} = 78.52 - .24x$ **(d)** $t = -2.31$, fail to reject H_0 **(e)** observation #5 is an outlier **(f)** $\hat{y} = 139.76 - .45x; t = -15.35$, reject H_0

3.79 (a) $y = \beta_0 + \beta_1 x + \varepsilon$ **(b)** $\beta_1 > 0$ **(d)** no; $t = -3.12$, fail to reject H_0

3.81 (a) yes; $t = 3.05$, reject $H_0: \rho = 0$ **(b)** mother, number of friends; $t = -2.43$ **(d)** only testing for linear relationships

3.83 (a) $\hat{y} = 14{,}012 - 1{,}783x$; no (at $\alpha = .05$), $t = -.98$ **(b)** $\hat{y} = 19{,}680 - 3{,}887x$; yes (at $\alpha = .05$), $t = -2.29$ **(c)** Predicting outside range of x $(1.52 - 4.11)$

Chapter 4

4.1 df $= n - $ (number of independent variables $+ 1$)

4.3 (a) $E(y) = \beta_0 + \beta_1 x_1 + \beta_2 x_2 + \beta_3 x_3$ **(b)** 8% of the sample variation in frequency of marijuana is explained by the model   **(c)** reject H_0: $\beta_1 = \beta_2 = \beta_3 = 0$ **(d)** reject H_0: $\beta_1 = 0$ **(e)** fail to reject H_0: $\beta_2 = 0$ **(f)** fail to reject H_0: $\beta_3 = 0$

4.5 (a) $\hat{y} = 3.70 + .34x_1 + .49x_2 + .72x_3 + 1.14x_4 + 1.51x_5 + .26x_6 - .14x_7 - .10x_8 - .10x_9$ **(b)** $t = -1.00$, fail to reject H_0 **(c)** $1.51 \pm .098$

4.7 (a) for every 1-unit increase in proportion of block with low density (x_1), population density will increase by 2; for every 1-unit increase in proportion of block with high density (x_2), population density will increase by 5   **(b)** 68.6% of the sample variation in population density is explained by the model   **(c)** H_0: $\beta_1 = \beta_2 = 0$ **(d)** $F = 133.27$ **(e)** reject H_0

4.9 (a) $\hat{y} = 1.81231 + .10875x_1 + .00017x_2$ **(b)** for every 1-mile increase in road length (x_1), number of crashes increase by .109; for every one-vehicle increase in AADT (x_2), number of crashes increase by .00017   **(c)** $.109 \pm .082$ **(d)** $.00017 \pm .00008$ **(e)** $\hat{y} = 1.20785 + .06343x_1 + .00056x_2$; for every 1-mile increase in road length (x_1), number of crashes increase by .063; for every one-vehicle increase in AADT (x_2), number of crashes increase by .00056; $.063 \pm .046$; $.00056 \pm .00031$

4.11 (a) $E(y) = \beta_0 + \beta_1x_1 + \beta_2x_2 + \beta_3x_3 + \beta_4x_4$ **(b)** $\hat{y} = 21{,}087.95 + 108.45x_1 + 557.91x_2 - 340.17x_3 + 85.68x_4$ **(d)** $t = 1.22$, p-value $= .238$, fail to reject H_0 **(e)** $R^2 = .912$, $R_a^2 = .894$; R_a^2 **(f)** $F = 51.72$, reject H_0

4.13 (a) $E(y) = \beta_0 + \beta_1x_1 + \beta_2x_2 + \beta_3x_3 + \beta_4x_4 + \beta_5x_5$ **(b)** $\hat{y} = 13{,}614 + .09x_1 - 9.20x_2 + 14.40x_3 + .35x_4 - .85x_5$ **(d)** 458.8; $\approx 95\%$ of sample heat rates fall within 917.6 kJ/kw-hr of model predicted values   **(e)** $.917$; 91.7% of the sample variation in heat rate is explained by the model   **(f)** yes, $F = 147.3$, p-value ≈ 0

4.15 (a) $F = 1.056$; do not reject H_0 **(b)** $.05$; 5% of the sample variation in IQ is explained by the model

4.17 (a) 36.2% of the sample variation in active caring score is explained by the model   **(b)** $F = 5.11$, reject H_0

4.21 (a) $(2.68, 5.82)$ **(b)** $(-3.04, 11.54)$

4.23 95% PI for y when $x_1 = 23.755$, $x_2 = 90.662$, $x_3 = 25.0$: $(24.03, 440.64)$

4.25 $(-1.233, 1.038)$

4.27 (a) $E(y) = \beta_0 + \beta_1x_1 + \beta_2x_2 + \beta_3x_1x_2$ **(b)** linear relationship between number of defects and turntable speed depends on blade position   **(c)** $\beta_3 < 0$

4.29 (a) $E(y) = \beta_0 + \beta_1x_1 + \beta_2x_2 + \beta_3x_1x_2$ **(b)** linear relationship between negative feelings score and number ahead in line depends on number behind in line   **(c)** fail to reject H_0: $\beta_3 = 0$ **(d)** $\beta_1 > 0$; $\beta_2 < 0$

4.31 (a) $E(y) = \beta_0 + \beta_1x_1 + \beta_2x_2 + \beta_3x_3 + \beta_4x_1x_3 + \beta_5x_2x_3$ **(b)** $\hat{y} = 10{,}845 - 1280.0x_1 + 217.4x_2 - 1549.2x_3 - 11.0x_1x_3 + 19.98x_2x_3$ **(c)** $t = -.93$, p-value $= .355$, fail to reject H_0 **(d)** $t = 1.78$, p-value $= .076$, fail to reject H_0 **(e)** no interaction

4.33 (a) slope depends on x_2 and x_5 **(b)** $F = 5.505$, $R^2 = .6792$, $s = .505$; yes

4.35 (a) $E(y) = \beta_0 + \beta_1x_1 + \beta_2x_1^2$ **(b)** β_2 **(c)** negative

4.37 (a) yes   **(b)** $t = 2.69$, p-value $= .031$, reject H_0

4.39 (a) $E(y) = \beta_0 + \beta_1x + \beta_2x^2$ **(b)** positive   **(c)** no; $E(y) = \beta_0 + \beta_1x_1$

4.41 (a) curvilinear trend   **(b)** $\hat{y} = 1.01 - 1.17x + .29x^2$ **(c)** yes; $t = 7.36$, p-value ≈ 0, reject H_0

4.43 (a) $E(y) = \beta_0 + \beta_1x_1 + \beta_2x_2$ **(b)** $E(y) = \beta_0 + \beta_1x_1 + \beta_2x_2 + \beta_3x_3 + \beta_4x_4$

4.45 (a) $E(y) = \beta_0 + \beta_1x_1$, where $x = \{1$ if A, 0 if B$\}$ **(b)** $E(y) = \beta_0 + \beta_1x_1 + \beta_2x_2 + \beta_3x_3$, where $x_1 = \{1$ if A, 0 if not$\}$, $x_2 = \{1$ if B, 0 if not$\}$, $x_3 = \{1$ if C, 0 if not$\}$; $\beta_0 = \mu_D$, $\beta_1 = \mu_A - \mu_D$, $\beta_2 = \mu_B - \mu_D$, $\beta_3 = \mu_C - \mu_D$

4.47 (c) Parallel lines

4.49 (b) second-order   **(c)** different shapes   **(d)** yes   **(e)** shift curves along the x_1-axis

4.51 (a) method: $x_1 = \{1$ if manual, 0 if automated$\}$; soil: $x_2 = \{1$ if clay, 0 if not$\}$, $x_3 = \{1$ if gravel, 0 if not$\}$; slope: $x_4 = \{1$ if east, 0 if not$\}$, $x_5 = \{1$ if south, 0 if not$\}$, $x_6 = \{1$ if west, 0 if not$\}$, $x_7 = \{1$ if southeast, 0 if not$\}$ **(b)** $E(y) = \beta_0 + \beta_1x_1$; $\beta_0 = \mu_{Auto}$, $\beta_1 = \mu_{Manual} - \mu_{Auto}$ **(c)** $E(y) = \beta_0 + \beta_1x_2 + \beta_2x_3$; $\beta_0 = \mu_{Sand}$, $\beta_1 = \mu_{Clay} - \mu_{Sand}$, $\beta_2 = \mu_{Gravel} - \mu_{Sand}$ **(d)** $E(y) = \beta_0 + \beta_1x_4 + \beta_2x_5 + \beta_3x_6 + \beta_4x_7$; $\beta_0 = \mu_{SW}$, $\beta_1 = \mu_E - \mu_{SW}$, $\beta_2 = \mu_S - \mu_{SW}$, $\beta_3 = \mu_W - \mu_{SW}$, $\beta_4 = \mu_{SE} - \mu_{SW}$

4.53 (a) $E(y) = \beta_0 + \beta_1x_1 + \beta_2x_2$, where $x_1 = \{1$ if Full solution, 0 if not$\}$, $x_2 = \{1$ if Check figures, 0 if not$\}$ **(b)** β_1 **(c)** $\hat{y} = 2.433 - .483x_1 + .287x_2$ **(c)** $F = .45$, p-value $= .637$, fail to reject H_0

4.55 (a) reject H_0: $\beta_1 = \beta_2 = \dots \beta_{12} = 0$; sufficient evidence to indicate the model is adequate for predicting log of card price   **(b)** fail to reject H_0: $\beta_1 = 0$ **(c)** reject H_0: $\beta_3 = 0$ **(d)** $E[\ln(y)] = \beta_0 + \beta_1x_4 + \beta_2x_5 + \beta_3x_6 + \dots + \beta_9x_{12} + \beta_{10}x_4x_5 + \beta_{11}x_4x_6 + \dots + \beta_{17}x_4x_{12}$

4.57 (a) Model 1: $t = 2.58$, reject H_0; Model 2: reject H_0: $\beta_1 = 0$ ($t = 3.32$), reject H_0: $\beta_2 = 0$ ($t = 6.47$), reject H_0: $\beta_3 = 0$ ($t = -4.77$), do not reject H_0: $\beta_4 = 0$ ($t = .24$); Model 3: reject H_0: $\beta_1 = 0$ ($t = 3.21$), reject H_0: $\beta_2 = 0$ ($t = 5.24$), reject H_0: $\beta_3 = 0$ ($t = -4.00$), do not reject H_0: $\beta_4 = 0$ ($t = 2.28$), do not reject H_0: $\beta_5 = 0$ ($t = .014$)   **(c)** Model 2

4.59 (a) $\hat{y} = 80.22 + 156.5x_1 - 42.3x_1^2 + 272.84x_2 + 760.1x_1x_2 + 47.0x_1^2x_2$ **(b)** yes, $F = 417.05$, p-value ≈ 0 **(c)** no; fail to reject H_0: $\beta_2 = 0$ and H_0: $\beta_5 = 0$

4.61 Nested models: a and b, a and d, a and e, b and c, b and d, b and e, c and e, d and e

4.63 (a) 10.1% (55.5%) of the sample variation in aggression score is explained by Model 1 (Model 2)   **(b)** H_0: $\beta_5 = \beta_6 = \beta_7 = \beta_8 = 0$ **(c)** yes   **(d)** reject H_0 **(e)** $E(y) = \beta_0 + \beta_1x_1 + \beta_2x_2 + \beta_3x_3 + \beta_4x_4 + \beta_5x_5 + \beta_6x_6 + \beta_7x_7 + \beta_8x_8 + \beta_9x_5x_6 + \beta_{10}x_5x_7 + \beta_{11}x_5x_8 + \beta_{12}x_6x_7 + \beta_{13}x_6x_8 + \beta_{14}x_7x_8$ **(f)** fail to reject H_0: $\beta_9 = \beta_{10} = \dots = \beta_{14} = 0$

4.65 (a) $E(y) = \beta_0 + \beta_1x_1 + \beta_2x_2 + \dots + \beta_{11}x_{11}$ **(b)** $E(y) = \beta_0 + \beta_1x_1 + \beta_2x_2 + \dots + \beta_{11}x_{11} + \beta_{12}x_1x_9 + \beta_{13}x_1x_{10} + \beta_{14}x_1x_{11} + \beta_{15}x_2x_9 + \beta_{16}x_2x_{10} + \beta_{17}x_2x_{11} + \dots + \beta_{35}x_8x_{11}$ **(c)** Test H_0: $\beta_{12} = \beta_{13} = \beta_{14} = \dots = \beta_{35} = 0$

4.67 (a) $E(y) = \beta_0 + \beta_1 x_1 + \beta_2 x_2 + \ldots + \beta_{10} x_{10}$ **(b)** $H_0: \beta_3 = \beta_4 = \ldots = \beta_{10} = 0$ **(c)** reject H_0 **(e)** 14 ± 5.88 **(f)** yes **(g)** $E(y) = \beta_0 + \beta_1 x_1 + \beta_2 x_2 + \ldots + \beta_{10} x_{10} + \beta_{11} x_2 x_1 + \beta_{12} x_2 x_3 + \ldots + \beta_{19} x_2 x_{10}$ **(h)** H_0: $\beta_{11} = \beta_{12} = \ldots = \beta_{19} = 0$; partial (nested model) F-test

4.69 (a) multiple t-tests result in an increased Type I error rate **(b)** $H_0: \beta_2 = \beta_5 = 0$ **(c)** fail to reject H_0

4.71 (a) yes; $t = 5.96$, reject H_0 **(b)** $t = .01$, do not reject H_0 **(c)** $t = 1.91$, reject H_0

4.73 (a) estimate of β_1 **(b)** prediction equation less reliable for values of x's outside the range of the sample data

4.75 (a) $E(y) = \beta_0 + \beta_1 x_1 + \beta_2 x_2 + \beta_3 x_3 + \beta_4 x_4 + \beta_5 x_5$ **(b)** $F = 34.47$, reject H_0 **(c)** $E(y) = \beta_0 + \beta_1 x_1 + \beta_2 x_2 + \ldots + \beta_7 x_7$ **(d)** 60.3% of the sample variation in GSI is explained by the model **(e)** reject H_0 for both

4.77 (a) $E(y) = \beta_0 + \beta_1 x_1 + \beta_2 x_2 + \beta_3 x_1 x_2$ **(b)** p-value = .02, reject H_0 **(c)** impact of intensity on test score depends on treatment and is measured by $(\beta_2 + \beta_3 x_1)$, not by β_2 alone

4.79 (a) Negative **(b)** $F = 1.60$, fail to reject H_0 **(c)** $F = 1.61$, fail to reject H_0

4.81 yes; $t = 4.20$, p-value = .004

4.83 (a) $E(y) = \beta_0 + \beta_1 x_1 + \beta_2 x_2$ where $x_1 = \{1$ if Communist, 0 if not$\}$, $x_2 = \{1$ if Democratic, 0 if not$\}$ **(b)** $\beta_0 = \mu_{\text{Dictator}}$, $\beta_1 = \mu_{\text{Communist}} - \mu_{\text{Dictator}}$, $\beta_2 = \mu_{\text{Democratic}} - \mu_{\text{Dictator}}$

4.85 no; income (x_1) not significant, air carrier dummies $(x_3 - x_{30})$ are significant

4.87 (a) quantitative **(b)** quantitative **(c)** qualitative **(d)** $E(y) = \beta_0 + \beta_1 x_1 + \beta_2 x_2 + \beta_3 x_3 + \beta_4 x_4$, where $x_1 = \{1$ if Benzene, 0 if not$\}$, $x_2 = \{1$ if Toluene, 0 if not$\}$, $x_3 = \{1$ if Chloroform, 0 if not$\}$, $x_4 = \{1$ if Methanol, 0 if not$\}$ **(e)** $\beta_0 = \mu_A$, $\beta_1 = \mu_B - \mu_A$, $\beta_2 = \mu_T - \mu_A$, $\beta_3 = \mu_C - \mu_A$, $\beta_4 = \mu_M - \mu_A$ **(f)** F-test with H_0: $\beta_1 = \beta_2 = \beta_3 = \beta_4 = 0$

4.89 (a) $H_0: \beta_5 = \beta_6 = \beta_7 = \beta_8 = 0$; males: $F > 2.37$; females: $F > 2.45$ **(c)** reject H_0 for both

Chapter 5

5.1 (a) Qualitative **(b)** Quantitative **(c)** Quantitative

5.3 Gender: qualitative; Testimony: qualitative

5.5 (a) Quantitative **(b)** Quantitative **(c)** Qualitative **(d)** Qualitative **(e)** Qualitative **(f)** Quantitative **(g)** Qualitative **(h)** Qualitative **(i)** Qualitative **(j)** Quantitative **(k)** Qualitative

5.7 (a) (i) 1; (ii) 3; (iii) 1; (iv) 2 **(b)** (i) $E(y) = \beta_0 + \beta_1 x$; (ii) $E(y) = \beta_0 + \beta_1 x + \beta_2 x^2 + \beta_3 x^3$; (iii) $E(y) = \beta_0 + \beta_1 x$; (iv) $E(y) = \beta_0 + \beta_1 x + \beta_2 x^2$ **(c)** (i) $\beta_1 > 0$; (ii) $\beta_3 > 0$; (iii) $\beta_1 < 0$; (iv) $\beta_2 < 0$

5.9 $E(y) = \beta_0 + \beta_1 x + \beta_2 x^2 + \beta_3 x^3$

5.11 $E(y) = \beta_0 + \beta_1 x + \beta_2 x^2$

5.15 (a) $E(y) = \beta_0 + \beta_1 x_1 + \beta_2 x_2 + \beta_3 x_1 x_2 + \beta_4 x_1^2 + \beta_5 x_2^2$ **(b)** $E(y) = \beta_0 + \beta_1 x_1 + \beta_2 x_2$ **(c)** $E(y) = \beta_0 + \beta_1 x_1 + \beta_2 x_2 + \beta_3 x_1 x_2$ **(d)** $\beta_1 + \beta_3 x_2$ **(e)** $\beta_2 + \beta_3 x_1$

5.17 (a) $E(y) = \beta_0 + \beta_1 x_1 + \beta_2 x_1^2 + \beta_3 x_2 + \beta_4 x_2^2 + \beta_5 x_4 + \beta_6 x_4^2 + \beta_7 x_1 x_2 + \beta_8 x_1 x_4 + \beta_9 x_2 x_4$ **(b)** $\hat{y} = 10,283.2 + 276.8 x_1 + .990 x_1^2 + 3325.2 x_2 + 266.6 x_2^2 + 1301.3 x_4 + 40.22 x_4^2 + 41.98 x_1 x_2 + 15.98 x_1 x_4 + 207.4 x_2 x_4$; yes, $F = 613.278$, reject H_0 **(c)** $F = 108.43$, reject H_0

5.19 (a) $E(y) = \beta_0 + \beta_1 x_1 + \beta_2 x_2 + \beta_3 x_1 x_2 + \beta_4 x_1^2 + \beta_5 x_2^2$ **(b)** $\hat{y} = 15,583 + .078 x_1 - 523 x_2 + .0044 x_1 x_2 - .0000002 x_1^2 + 8.84 x_2^2$ **(c)** $F = 93.55$, reject H_0 **(f)** graphs have similar shape

5.21 (a) $u = (x - 85.1)/14.81$ **(b)** $-.668, .446, 1.026, -1.411, -.223, 1.695, -.527, -.338$ **(c)** .9967 **(d)** .376 **(e)** $\hat{y} = 110.953 + 14.377 u + 7.425 u^2$

5.23 (a) .975, .928, .987 **(b)** $u = (x - 5.5)/3.03$ **(c)** 0, .923, 0; yes

5.25 (a) $E(y) = \beta_0 + \beta_1 x_1 + \beta_2 x_2 + \beta_3 x_3$, where $x_1 = \{1$ if low, 0 if not$\}$, $x_2 = \{1$ if moderate, 0 if not$\}$, $x_3 = \{1$ if high, 0 if not$\}$ **(b)** $\beta_0 = \mu_{\text{None}}$, $\beta_1 = \mu_{\text{Low}} - \mu_{\text{None}}$, $\beta_2 = \mu_{\text{Mod}} - \mu_{\text{None}}$, $\beta_3 = \mu_{\text{High}} - \mu_{\text{None}}$ **(c)** F-test of H_0: $\beta_1 = \beta_2 = \beta_3 = 0$

5.27 (a) $E(y) = \beta_0 + \beta_1 x_1 + \beta_2 x_2 + \beta_3 x_3 + \beta_4 x_1 x_2 + \beta_5 x_1 x_3$, where $x_1 = \{1$ if manual, 0 if automated$\}$, $x_2 = \{1$ if clay, 0 if not$\}$, $x_3 = \{1$ if gravel, 0 if not$\}$ **(b)** $\mu_{\text{Automated/Sand}}$ **(c)** $\beta_0 + \beta_1 + \beta_2 + \beta_4$ **(d)** β_1

5.29 (a) $E(y) = \beta_0 + \beta_1 x_1 + \beta_2 x_2 + \beta_3 x_1 x_2$, where $x_1 = \{1$ if common, 0 if ambiguous$\}$, $x_2 = \{1$ if low, 0 if high$\}$ **(b)** $\hat{\beta}_0 = 6.1$, $\hat{\beta}_1 = .2$, $\hat{\beta}_2 = 11.9$, $\hat{\beta}_3 = -10.4$ **(c)** t-test for $H_0: \beta_3 = 0$

5.31 (a) $E(y) = \beta_0 + \beta_1 x_1 + \beta_2 x_1^2 + \beta_3 x_2 + \beta_4 x_3 + \beta_5 x_1 x_2 + \beta_6 x_1 x_3 + \beta_7 x_1^2 x_2 + \beta_8 x_1^2 x_3$, where $x_1 =$ level of bullying, $x_2 = \{1$ if low, 0 if not$\}$, $x_3 = \{1$ if neutral, 0 if not$\}$ **(b)** $\beta_0 + 25 \beta_1 + 625 \beta_2 + \beta_3 + 25 \beta_5 + 625 \beta_7$ **(c)** nested F-test of H_0: $\beta_2 = \beta_7 = \beta_8 = 0$ **(d)** $E(y) = \beta_0 + \beta_1 x_1 + \beta_2 x_2 + \beta_3 x_3 + \beta_4 x_1 x_2 + \beta_5 x_1 x_3$ **(e)** low: $\beta_1 + \beta_4$; neutral: $\beta_1 + \beta_5$; high: β_1

5.33 (a) $E(y) = \beta_0 + \beta_1 x_1 + \beta_2 x_2 + \beta_3 x_3 + \beta_4 x_4 + \beta_5 x_5 + \beta_6 x_6 + \beta_7 x_7 + \beta_8 x_1 x_2 + \beta_9 x_1 x_3 + \beta_{10} x_1 x_4 + \beta_{11} x_1 x_5 + \beta_{12} x_1 x_6 + \beta_{13} x_1 x_7 + \beta_{14} x_2 x_4 + \beta_{15} x_2 x_5 + \beta_{16} x_2 x_6 + \beta_{17} x_2 x_7 + \beta_{18} x_3 x_4 + \beta_{19} x_3 x_5 + \beta_{20} x_3 x_6 + \beta_{21} x_3 x_7 + \beta_{22} x_1 x_2 x_4 + \beta_{23} x_1 x_2 x_5 + \beta_{24} x_1 x_2 x_6 + \beta_{25} x_1 x_2 x_7 + \beta_{26} x_1 x_3 x_4 + \beta_{27} x_1 x_3 x_5 + \beta_{28} x_1 x_3 x_6 + \beta_{29} x_1 x_3 x_7$, where $x_1 = \{1$ if manual, 0 if automated$\}$, $x_2 = \{1$ if clay, 0 if not$\}$, $x_3 = \{1$ if gravel, 0 if not$\}$, $x_4 = \{1$ if East, 0 if not$\}$, $x_5 = \{1$ if South, 0 if not$\}$, $x_6 = \{1$ if West, 0 if not$\}$, $x_7 = \{1$ if Southeast, 0 if not$\}$ **(b)** $\mu_{\text{Automated/Sand/SW}}$ **(c)** $\beta_0 + \beta_1 + \beta_2 + \beta_4 + \beta_8 + \beta_{10} + \beta_{14} + \beta_{22}$ **(d)** β_1 **(e)** $\beta_8 = \beta_9 = \beta_{22} = \beta_{23} = \beta_{24} = \beta_{25} = \beta_{26} = \beta_{27} = \beta_{28} = \beta_{29} = 0$

5.35 (a) $E(y) = \beta_0 + \beta_1 x_1 + \beta_2 x_2 + \beta_3 x_1 x_2$ **(c)** $\beta_1 + \beta_3$ **(d)** no; $F = .26$, p-value = .857, do not reject H_0 **(e)** $E(y) = \beta_0 + \beta_1 x_1 + \beta_2 x_1^2 + \beta_3 x_2 + \beta_4 x_1 x_2 + \beta_5 x_1^2 x_2$

5.37 (a) $E(y) = \beta_0 + \beta_1 x_1 + \beta_2 x_2 + \beta_3 x_3$ **(b)** $E(y) = \beta_0 + \beta_1 x_1 + \beta_2 x_2 + \beta_3 x_3 + \beta_4 x_1 x_2 + \beta_5 x_1 x_3$ **(c)** AL: β_1; TDS: $\beta_1 + \beta_4$; FE: $\beta_1 + \beta_5$ **(d)** nested F-test of $H_0: \beta_4 = \beta_5 = 0$

5.39 (a) Qualitative **(b)** Quantitative **(c)** Quantitative

5.41 (a) Quantitative **(b)** Quantitative **(c)** Qualitative **(d)** Qualitative **(e)** Qualitative **(f)** Qualitative **(g)** Quantitative **(h)** Qualitative

5.43 (a) $E(y) = \beta_0 + \beta_1 x_1 + \beta_2 x_2 + \beta_3 x_3 + \beta_4 x_4$ (b) $\mu_{BA} - \mu_N$ (c) $\mu_E - \mu_N$ (d) $\mu_{LAS} - \mu_N$ (e) $\mu_J - \mu_N$ (f) $E(y) = (\beta_0 + \beta_5) + \beta_1 x_1 + \beta_2 x_2 + \beta_3 x_3 + \beta_4 x_4$ (g) $\mu_{BA} - \mu_N$ (h) $\mu_E - \mu_N$ (i) $\mu_{LAS} - \mu_N$ (j) $\mu_J - \mu_N$ (k) $\mu_F - \mu_M$ l. Reject H_0: $\beta_5 = 0$; gender has an effect

5.45 (a) $E(y) = \beta_0 + \beta_1 x_1 + \beta_2 x_2 + \beta_3 x_3$, where $x_1 = \{1$ if boy, 0 if girl$\}$, $x_2 = \{1$ if youngest third, 0 if not$\}$, $x_3 = \{1$ if middle third, 0 if not$\}$ (b) $\beta_0 = \mu_{Girls/Oldest}$, $\beta_1 = \mu_{Boys} - \mu_{Girls}$, $\beta_2 = \mu_{Youngest} - \mu_{Oldest}$, $\beta_3 = \mu_{Middle} - \mu_{Oldest}$ (c) $E(y) = \beta_0 + \beta_1 x_1 + \beta_2 x_2 + \beta_3 x_3 + \beta_4 x_1 x_2 + \beta_5 x_1 x_3$ (d) .21, $-.05$, .06, $-.03$, .11, .20 (e) nested F-test of H_0: $\beta_4 = \beta_5 = 0$

5.47 (a) $E(y) = \beta_0 + \beta_1 x_1 + \beta_2 x_2 + \beta_3 x_3 + \beta_4 x_4$, where $x_1 = \{1$ if P_1, 0 if $P_2\}$, $x_2 = \{1$ if L_1, 0 if not$\}$, $x_3 = \{1$ if L_2, 0 if not$\}$, $x_4 = \{1$ if L_3, 0 if not$\}$ (b) 8; $E(y) = \beta_0 + \beta_1 x_1 + \beta_2 x_2 + \beta_3 x_3 + \beta_4 x_4 + \beta_5 x_1 x_2 + \beta_6 x_1 x_3 + \beta_7 x_1 x_4$ (c) $F = 2.33$, do not reject H_0

5.49 (a) $E(y) = \beta_0 + \beta_1 x_1 + \beta_2 x_2$, where $x_1 =$ years of education, $x_2 = \{1$ if Certificate, 0 if not$\}$ (b) $E(y) = \beta_0 + \beta_1 x_1 + \beta_2 x_2 + \beta_3 x_1 x_2$ (c) $E(y) = \beta_0 + \beta_1 x_1 + \beta_2 x_1^2 + \beta_3 x_2 + \beta_4 x_1 x_2 + \beta_5 x_1^2 x_2$

5.51 (a) $F = 8.79$, p-value $= .0096$; reject H_0 (b) DF-2: 2.14; blended: 4.865; adv. timing: 7.815

5.53 (a) $E(y) = \beta_0 + \beta_1 x_1 + \beta_2 x_2 + \beta_3 x_3$, where $x_1 = \{1$ if Low, 0 if not$\}$, $x_2 = \{1$ if Moderate, 0 if not$\}$, $x_3 = \{1$ if producer, 0 if consumer$\}$ (b) $\beta_0 = \mu_{High/Cons}$, $\beta_1 = \mu_{Low} - \mu_{High}$, $\beta_2 = \mu_{Mod} - \mu_{High}$, $\beta_3 = \mu_{Prod} - \mu_{Cons}$ (c) $E(y) = \beta_0 + \beta_1 x_1 + \beta_2 x_2 + \beta_3 x_3 + \beta_4 x_1 x_3 + \beta_5 x_2 x_3$ (d) $\beta_0 = \mu_{High/Cons}$, $\beta_1 = (\mu_{Low} - \mu_{High})$ for consumers, $\beta_2 = (\mu_{Mod} - \mu_{High})$ for consumers, $\beta_3 = (\mu_{Prod} - \mu_{Cons})$ for high, $\beta_4 = (\mu_{Low} - \mu_{High})_{Prod} - (\mu_{Low} - \mu_{High})_{Cons}$, $\beta_5 = (\mu_{Mod} - \mu_{High})_{Prod} - (\mu_{Mod} - \mu_{High})_{Cons}$, (e) nested F-test of H_0: $\beta_4 = \beta_5 = 0$

5.55 (a) $E(y) = \beta_0 + \beta_1 x_1 + \beta_2 x_2 + \beta_3 x_3$ (b) t-test of H_0: $\beta_3 = 0$ vs. H_a: $\beta_3 < 0$ (c) $E(y) = \beta_0 + \beta_1 x_1 + \beta_2 x_2 + \beta_3 x_1 x_2 + \beta_4 x_1^2 + \beta_5 x_2^2 + \beta_6 x_3 + \beta_7 x_1 x_3 + \beta_8 x_2 x_3 + \beta_9 x_1 x_2 x_3 + \beta_{10} x_1^2 x_3 + \beta_{11} x_2^2 x_3$ (d) nested F-test of H_0: $\beta_1 = \beta_3 = \beta_4 = \beta_7 = \beta_9 = \beta_{10} = 0$

Chapter 6

6.1 (a) x_2 (b) yes (c) fit all possible 2-variable models, $E(y) = \beta_0 + \beta_1 x_2 + \beta_2 x_j$

6.3 (a) 11 (b) 10 (c) 1 (d) $E(y) = \beta_0 + \beta_1 x_{11} + \beta_2 x_4 + \beta_3 x_2 + \beta_4 x_7 + \beta_5 x_{10} + \beta_6 x_1 + \beta_7 x_9 + \beta_8 x_3$ (e) 67.7% of sample variation in satisfaction is explained by the model (f) no interactions or higher-order terms tested

6.5 (a) x_5, x_6, and x_4 (b) no (c) $E(y) = \beta_0 + \beta_1 x_4 + \beta_2 x_5 + \beta_3 x_6 + \beta_4 x_4 x_5 + \beta_5 x_4 x_6 + \beta_6 x_5 x_6$ (d) nested F-test of H_0: $\beta_4 = \beta_5 = \beta_6 = 0$ (e) consider interaction and higher-order terms

6.7 (a) (i) 4; (ii) 6; (iii) 4; (iv) 1 (b) (i) .213, 193.8, 2.5, 10,507; (ii) .247, 189.1, 2.3, 10,494; (iii) .266, 188.2, 3.1, 10,489; (iv) .268, 191.7, 5.0, 10,710 (d) x_2, x_3, x_4

6.9 yes; DOT estimate and low-bid-estimate ratio

6.11 Stepwise: well depth and percent of land allocated to industry

Chapter 7

7.1 model less reliable

7.3 (a) $x = ln(p)$ (b) yes, $t = -15.89$ (c) (924.5, 975.5)

7.5 (a) no (b) no

7.7 (a) no (b) yes

7.9 Unable to test model adequacy since df(Error) $= n - 3 = 0$

7.11 (a) $\hat{y} = 2.74 + .80 x_1$; yes, $t = 15.92$ (b) $\hat{y} = 1.66 + 12.40 x_2$; yes, $t = 11.76$ (c) $\hat{y} = -11.80 + 25.07 x_3$; yes, $t = 2.51$ (d) yes

7.13 (a) multicollinearity (b) no, β_3 not estimable

7.15 no multicollinearity, use DOTEST and LBERATIO

7.17 Two levels each; $n > 4$

7.19 yes, high VIFs for Inlet-temp, Airflow and Power; include only one of these 3 variables

7.21 (a) .0025; no (b) .434; no (c) no (d) $\hat{y} = -45.154 + 3.097 x_1 + 1.032 x_2$, $F = 39{,}222.34$, reject H_0: $\beta_1 = \beta_2 = 0$; $R^2 = .9998$ (e) $-.8998$; high correlation (f) no

7.23 df(Error) $= 0$, s^2 undefined, no test of model adequacy

Chapter 8

8.1 (a) $\hat{y} = 2.588 + .541 x$ (b) $-.406, -.206, -.047, .053, .112, .212, .271, .471, .330, .230, -.411, -.611$ (c) Yes; needs curvature

8.3 (a) $\hat{y} = 40.35 - .207 x$ (b) $-4.64, -3.94, -1.83, .57, 2.58, 1.28, 4.69, 4.09, 4.39, 2.79, .50, 1.10, -6.09, -5.49$ (c) Yes; needs curvature (d) $\hat{y} = -1051 + 66.19 x - 1.006 x^2$; yes, $t = -11.80$

8.5 $\hat{y} = 30{,}856 - 191.57 x$; yes, quadratic trend; yes

8.7 (a) $-389, -178, 496, \ldots, 651$ (b) No trends (c) No trends (d) No trends (e) No trends

8.9 Yes; assumption of constant variance appears satisfied; add curvature

8.11 Yes

8.13 (a) Yes; assumption of equal variances violated (b) Use transformation $y^* = \sqrt{y}$

8.15 (a) $\hat{y} = .94 - .214 x$ (b) $0, .02, -.026, .034, .088, -.112, -.058, .002, .036, .016$ (c) Unequal variances (d) Use transformation $y^* = \sin^{-1}\sqrt{y}$ (e) $\hat{y}^* = 1.307 - .2496 x$; yes

8.17 (a) Lagos: $-.223$ (b) yes

8.19 No; remove outliers or normalizing transformation

8.21 Residuals are approximately normal

8.23 No outliers

8.25 Observations #8 and #3 are influential

8.27 No outliers

8.29 Several, including wells 4567 and 7893; both influential; possibly delete

8.31 Inflated t-statistics for testing model parameters

8.33 (b) Model adequate at $\alpha = .05$ for all banks except bank 5 (c) Reject H_0 (two-tailed at $\alpha = .05$) for banks 2, 5; fail to reject H_0 for banks 4, 6, 8; test inconclusive for banks 1, 3, 7, 9

8.35 (a) Yes; residual correlation   **(b)** $d = .058$, reject H_0 **(c)** Normal errors

8.37 (a) $\hat{y} = 1668.44 + 105.83t$; yes, $t = 2.11$, reject H_0 **(b)** Yes   **(c)** $d = .845$, reject H_0

8.39 (a) Misspecified model; quadratic term missing   **(b)** Unequal variances   **(c)** Outlier   **(d)** Unequal variances   **(e)** Nonnormal errors

8.41 Assumptions reasonably satisfied

8.43 Assumptions reasonably satisfied

8.45 (a) $\hat{y} = -3.94 + .082x$ **(b)** $R^2 = .372$; $F = 2.96$, p-value $= .146$, model not useful   **(d)** Obs. for horse #3 is outside $2s$ interval   **(e)** Yes; $R^2 = .970$; $F = 130.71$, p-value $= 0$

8.47 Possible violation of constant variance assumption

8.49 Violation of constant variance assumption; variance-stabilizing transformation $\ln(y)$

Chapter 9

9.1 (a) $E(y) = \beta_0 + \beta_1 x_1 + \beta_2 (x_1 - 15)x_2$, where $x_1 = x$ and $x_2 = \{1 \text{ if } x_1 > 15, \ 0 \text{ if not}\}$ **(b)** $x \le 15$: y-intercept $= \beta_0$, slope $= \beta_1$; $x > 15$: y-intercept $= \beta_0 - 15\beta_2$, slope $= \beta_1 + \beta_2$ **(c)** t-test for $H_0 : \beta_2 = 0$

9.3 (a) $E(y) = \beta_0 + \beta_1 x_1 + \beta_2 (x_1 - 320)x_2 + \beta_3 x_2$, where $x_1 = x$ and $x_2 = \{1 \text{ if } x_1 > 320, 0 \text{ if not}\}$ **(b)** $x \le 320$: y-intercept $= \beta_0$, slope $= \beta_1$; $x > 320$: y-intercept $= \beta_0 - 320\beta_2 + \beta_3$, slope $= \beta_1 + \beta_2$ **(c)** nested F-test for $H_0 : \beta_2 = \beta_3 = 0$

9.5 (a) 4 and 7   **(b)** $E(y) = \beta_0 + \beta_1 x_1 + \beta_2 (x_1 - 4)x_2 + \beta_3 (x_1 - 7)x_3$, where $x_1 = x$, $x_2 = \{1 \text{ if } x_1 > 4, 0 \text{ if not}\}$, $x_3 = \{1 \text{ if } x_1 > 7, 0 \ \text{ if not}\}$ **(c)** $x \le 4 : \beta_1$; $4 < x \le 7 : (\beta_1 + \beta_2)$; $x > 7 : (\beta_1 + \beta_2 + \beta_3)$ **(d)** for every 1-point increase in performance over range $x \le 4$, satisfaction increases 5.05 units

9.7 (a) yes; 3.55   **(b)** $E(y) = \beta_0 + \beta_1 x_1 + \beta_2 (x_1 - 3.55)x_2$, where $x_1 = $ load and $x_2 = \{1 \text{ if load} > 3.55, 0 \text{ if not}\}$ **(c)** $\hat{y} = 2.22 + .529x_1 + 2.63(x_1 - 3.55)x_2$; $R^2 = .994$, $F = 1371$, p-value $= 0$, reject H_0

9.9 135.1 ± 74.1 **9.11** 1.93 ± 16.45

9.13 (a) $\hat{y} = -2.03 + 6.06x$ **(b)** $t = 10.35$, reject H_0 **(c)** $1.985 \pm .687$

9.15 (a) $\hat{y} = -3.367 + .194x$; yes, $t = 4.52$ (p-value $= .0006$)   **(b)** Residuals: $-1.03, 6.97, -5.03, -2.03, 1.97, -6.73, 3.27, -5.73, 9.27, -1.73, -9.43, 12.57, -8.43, 2.57, 3.57$; as speed increases, variation tends to increase   **(c)** $w_i = 1/x_i^2$; $x = 100$: 20.7; $x = 150$: 44.3; $x = 200$: 84.3   **(d)** $\hat{y} = -3.057 + .192x$; $s = .044$

9.17 (a) $\hat{y} = 140.6 - .67x$; assumption violated   **(b)** $w = 1/(\bar{x})^2$

9.19 violation of constant error variance assumption; violation of assumption of normal errors; predicted y is not bounded between 0 and 1

9.21 (a) $\beta_1 = $ change in P(hired) for every 1-year increase in higher education; $\beta_2 = $ change in P(hired) for every 1-year

increase in experience; $\beta_3 = $ P(hired)$_{Males}$ - P(hired)$_{Females}$ **(b)** $\hat{y} = -.5279 + .0750x_1 + .0747x_2 + .3912x_3$ **(c)** $F = 21.79$, reject H_0 **(d)** yes; $t = 4.01$ **(e)** $(-.097, .089)$

9.23 (a) P(Maryland nappe)   **(b)** $\pi^* = \beta_0 + \beta_1 x$, where $\pi^* = ln\{\pi/(1 - \pi)\}$ **(c)** change in log-odds of a Maryland nappe for every 1-degree increase in FIA   **(d)** exp $(\beta_0 + 80\beta_1)/\{1 - \exp(\beta_0 + 80\beta_1)\}$

9.25 (a) $\chi^2 = 20.43$, reject H_0 **(b)** Yes; $\chi^2 = 4.63$ **(c)** $(.00048, .40027)$

9.27 (a) P(landfast ice)   **(b)** $\pi^* = \beta_0 + \beta_1 x_1 + \beta_2 x_2 + \beta_3 x_3$, where $\pi^* = ln\{\pi/(1 - \pi)\}$ **(c)** $\hat{\pi}^* = .30 + 4.13x_1 + 47.12x_2 - 31.14x_3$ **(d)** $\chi^2 = 70.45$, reject H_0 **(e)** $\pi^* = \beta_0 + \beta_1 x_1 + \beta_2 x_2 + \beta_3 x_3 + \beta_4 x_1 x_2 + \beta_5 x_1 x_3 + \beta_6 x_2 x_3$ **(f)** $\hat{\pi}^* = 6.10 - 3.00x_1 + 10.56x_2 - 39.69x_3 + 50.49x_1 x_2 - 6.14 x_1 x_3 + 56.24x_2 x_3$ **(g)** $\chi^2 = 32.19$, reject H_0; interaction model is better

Chapter 10

10.1 (a) yes; yes   **(b)** 366.3, 335.8, 310.8, 284.0, 261.5, 237.0, 202.3, 176.8, 155.5   **(c)** yes   **(d)** 81.9   **(e)** 113.5   **(f)** Quarter 1: 94.2; Quarter 2: 107.8

10.3 (a) Moving average: 16.2; Exponential smoothing: 107; Holt-Winters: 35.9   **(b)** Moving average: 16.2; Exponential smoothing: 109.4; Holt-Winters: 43.9   **(c)** Moving average

10.5 (a) Yes   **(b)** forecast $= 231.7$ **(c)** forecast $= 205.4$ **(d)** forecast $= 232.8$

10.7 (a) yes   **(b)** 2006: 495; 2007: 540 ; 2008: 585   **(d)** 2006: 435.6 ; 2007: 435.6; 2008: 435.6   **(e)** 2006: 480.85; 2007: 518.1; 2008: 555.35   **(f)** moving average most accurate

10.9 (a) $\hat{\beta}_0 = 4.36$: estimated price is \$4.36 in 1990; $\hat{\beta}_1 = .455$: for each additional year, price increases by \$.455   **(b)** $t = 8.43$, p-value $= 0$, reject H_0 **(c)** 2008: (9.742, 15.357); 2009: (10.151, 15.858)   **(d)** extrapolation; no account for cyclical trends

10.11 (a) $E(y_t) = \beta_0 + \beta_1 t + \beta_2 Q_1 + \beta_3 Q_2 + \beta_4 Q_3$ **(b)** $\hat{y}_t = 119.85 + 16.51t + 262.34Q_1 + 222.83Q_2 + 105.51Q_3$; $F = 117.82$, p-value $= 0$, reject H_0 **(c)** independent error   **(d)** Q_1: 728.95, (662.8, 795.1); Q_2: 705.95, (639.8 772.1); Q_3: 605.15, (539.0, 671.3); Q_4: 516.115, (450.0, 582.3)

10.13 (a) yes   **(b)** $\hat{y}_t = 39.49 + 19.13t - 1.315t^2$ **(d)** $(-31.25, 48.97)$

10.15 (a) no, $t = -1.39$ **(b)** 2003.48   **(c)** no, $t = -1.61$ **(d)** 1901.81

10.17 (a) 0, 0, 0, .5, 0, 0, 0, .25, 0, 0, 0, .125, 0, 0, 0, .0625, 0, 0, 0, .03125   **(b)** .5, .25, .125, .0625, .03125, .0156, ...

10.19 $R_t = \phi_1 R_{t-1} + \phi_2 R_{t-2} + \phi_3 R_{t-3} + \phi_4 R_{t-4} + \varepsilon_t$

10.21 (a) $E(y_t) = \beta_0 + \beta_1 x_{1t} + \beta_2 x_{2t} + \beta_3 x_{3t} + \beta_4 t$ **(b)** $E(y_t) = \beta_0 + \beta_1 x_{1t} + \beta_2 x_{2t} + \beta_3 x_{3t} + \beta_4 t + \beta_5 x_{1t} t + \beta_6 x_{2t} t + \beta_7 x_{3t} t$ **(c)** $R_t = \phi R_{t-1} + \varepsilon_t$

10.23 (a) $E(y_t) = \beta_0 + \beta_1 [\cos(2\pi/365)t] + \beta_2 [\sin(2\pi/365)t]$ **(c)** $E(y_t) = \beta_0 + \beta_1 [\cos(2\pi/365)t] + \beta_2 [\sin(2\pi/365)t] + \beta_3 t + \beta_4 t [\cos(2\pi/365)t] + \beta_5 t [\sin(2\pi/365)t]$ **(d)** no; $R_t = \phi R_{t-1} + \varepsilon_t$

10.25 (a) $y_t = \beta_0 + \beta_1 t + \phi R_{t-1} + \varepsilon_t$ **(b)** $\hat{y}_t = 11,374 + 160.23t + .3743\hat{R}_{t-1}$ **(d)** $R^2 = .9874$, $s = 115.13$

10.27 (a) yes, upward trend **(b)** $y_t = \beta_0 + \beta_1 t + \beta_2 t^2 + \phi R_{t-1} + \varepsilon_t$ **(c)** $\hat{y}_t = 263.14 + 1.145t + .056t^2 + .792\hat{R}_{t-1}$; $R^2 = .747$; t $= 3.66$, p-value $= .0004$, reject H_0

10.29 (a) $F_{49} = 336.91$; $F_{50} = 323.41$; $F_{51} = 309.46$ **(b)** $t = 49$: 336.91 ± 6.48; $t = 50$: 323.41 ± 8.34; $t = 51$: 309.46 ± 9.36

10.35 (a) 2136.2 **(b)** (1404.3, 3249.7) **(c)** 1944; (1301.8, 2902.9)

10.37 (a) yes; possibly **(b)** 64 **(d)** 86.6 **(e)** 73.5 **(f)** $y_t = \beta_0 + \beta_1 t + \beta_2 S_1 + \beta_3 S_2 + \beta_4 S_3 + \phi R_{t-1} + \varepsilon_t$, where $S_1 = \{1$ if Jan/Feb/Mar, 0 if not$\}$, $S_2 = \{1$ if Apr/May/Jun, 0 if not$\}$, $S_3 = \{1$ if Jul/Aug/Sep, 0 if not$\}$ **(g)** $\hat{y}_t = 101.69 - 1.47t + 13.32S_1 + 29.94S_2 + 32.45S_3 - .543\hat{R}_{t-1}$ **(h)** 96.1 ± 11

10.39 (a) yes, curvilinear trend **(b)** $y_t = \beta_0 + \beta_1 t + \beta_2 t^2 + \varepsilon_t$ **(c)** $\hat{y}_t = 189.03 + 1.38t - .041t^2$ **(e)** yes **(f)** Durbin–Watson test; $d = .96$, reject H_0 **(g)** $y_t = \beta_0 + \beta_1 t + \beta_2 t^2 + \phi R_{t-1} + \varepsilon_t$; **(h)** $\hat{y}_t = 188.7 + 1.39t - .04t^2 + .456\hat{R}_{t-1}$

10.41 (a) $E(y_t) = \beta_0 + \beta_1 t + \beta_2 t^2 + \beta_3 x$, where $x = \{1$ if Jan–Apr, 0 if not$\}$ **(b)** add interaction terms

10.43 (a) $\mu_{\text{Post}} - \mu_{\text{Pre}}$ **(b)** μ_{Pre} **(c)** $-.55$ **(d)** 2.53

Chapter 11

11.1 (a) Noise (variability) and volume (n) **(b)** remove noise from an extraneous source of variation

11.3 (a) cockatiel **(b)** yes; completely randomized design **(c)** experimental group **(d)** 1,2,3 **(e)** 3 **(f)** total consumption **(g)** $E(y) = \beta_0 + \beta_1 x_1 + \beta_2 x_2$, where $x_1 = \{1$ if group 1, 0 if not$\}$, $x_2 = \{1$ if group 2, 0 if not$\}$

11.5 (a) $y_{B1} = \beta_0 + \beta_2 + \beta_4 + \varepsilon_{B1}$; $y_{B2} = \beta_0 + \beta_2 + \beta_5 + \varepsilon_{B2}$; $\dots$; $y_{B,10} = \beta_0 + \beta_2 + \varepsilon_{B,10}$; $\bar{y}_B = \beta_0 + \beta_2 + (\beta_4 + \beta_5 + \cdots + \beta_{12})/10 + \bar{\varepsilon}_B$ **(b)** $y_{D1} = \beta_0 + \beta_4 + \varepsilon_{D1}$; $y_{D2} = \beta_0 + \beta_5 + \varepsilon_{D2}$; $\dots$; $y_{D,10} = \beta_0 + \varepsilon_{D,10}$; $\bar{y}_D = \beta_0 + (\beta_4 + \beta_5 + \cdots + \beta_{12})/10 + \bar{\varepsilon}_D$

11.7 ability to investigate factor interaction

11.9 (a) students **(b)** yes; factorial design **(c)** class standing and type of preparation **(d)** class standing: low, medium, high; type of preparation: review session and practice exam **(e)** (low, review), (medium, review), (high, review), (low, practice), (medium, practice), (high, practice) **(f)** final exam score

11.11 (a) training method, practice session, task consistency **(b)** task consistency is QL, others are QN **(c)** 48; (CC/1/100), (AC/1/100), (CC/2/100), (AC/2/100), $\dots$, (CC/6/33), (AC/6/33)

11.13 (a) $E(y) = \beta_0 + \beta_1 x_1 + \beta_2 x_2 + \beta_3 x_3 + \beta_4 x_1 x_2 + \beta_5 x_1 x_3$, where x_1 is dummy variable for QL factor A; x_2, x_3 are dummy variables for QL factor B **(b)** $E(y) = \beta_0 + \beta_1 x_1 + \beta_2 x_2 + \beta_3 x_3 + \beta_4 x_4 + \beta_5 x_5 + \beta_6 x_1 x_2 + \beta_7 x_1 x_3 + \beta_8 x_1 x_4 + \beta_9 x_1 x_5 + \beta_{10} x_2 x_4 + \beta_{11} x_2 x_5 + \beta_{12} x_3 x_4 + \beta_{13} x_3 x_5 + \beta_{14} x_1 x_2 x_4 + \beta_{15} x_1 x_2 x_5 + \beta_{16} x_1 x_3 x_4 + \beta_{17} x_1 x_3 x_5$, where $x_1 = $ QN factor A; x_2, x_3 are dummy variables for QL factor B; x_4, x_5 are dummy variables for QL factor C

11.15 Cannot investigate factor interaction

11.17 11

11.19 sample size (n) and standard deviation of estimator

11.21 Step 4

11.23 8 treatments: A_1B_1, A_1B_2, A_1B_3, A_1B_4, A_2B_1, A_2B_2, A_2B_3, A_2B_4

11.25 $E(y) = \beta_0 + \beta_1 x_1 + \beta_2 x_2 + \beta_3 x_3 + \beta_4 x_4 + \beta_5 x_5$; 10

11.27 (a) Sex and weight **(b)** Both qualitative **(c)** 4 ; (ML), (MH), (FL), and (FH)

Chapter 12

12.3 (a) $E(y) = \beta_0 + \beta_1 x$, where $x = \{1$ if treatment 1, 0 if treatment 2$\}$ **(b)** $\hat{y} = 10.667 - 1.524x$; $t = -1.775$, fail to reject H_0

12.5 (a) $t = -1.78$; do not reject H_0 **(c)** Two-tailed

12.7 (a) completely randomized design **(b)** colonies 3, 6, 9 and 12; energy expended **(c)** H_0: $\mu_1 = \mu_2 = \mu_3 = \mu_4$ **(d)** Reject H_0

12.9 (a) H_0: $\mu_1 = \mu_2 = \mu_3$ **(b)** $E(y) = \beta_0 + \beta_1 x_1 + \beta_2 x_2$ **(c)** Reject H_0 **(d)** Fail to reject H_0

12.11 (a)

Source	df	SS	MS	F
Solvent	2	3.3054	1.6527	24.51
Error	29	1.9553	0.0674	
Total	31	5.2607		

(b) $F = 24.51$, reject H_0

12.13 $F = 7.69$, reject H_0

12.15 $F = 9.97$, reject H_0

12.17 (a) H_0: $\mu_1 = \mu_2 = \mu_3$ **(b)**

Source	df	SS	MS	F
Level	2	6.643	3.322	0.45
Error	72	527.357	7.324	
Total	74	534.000		

(c) fail to reject H_0

12.19 (a) same subjects used in parts A and B **(b)** response = WTP; treatments = A and B; blocks: subjects **(c)** treatments: H_0: $\mu_A = \mu_B$

12.21 No evidence of a difference among the three plant session means; $F = .019$

12.23 (a) same genes examined across all three conditions **(b)** H_0: $\mu_{\text{Full-Dark}} = \mu_{\text{TR-Light}} = \mu_{\text{TR-Dark}}$ **(c)** $F = 5.33$, reject H_0

12.25 (a) $E(y) = \beta_0 + \beta_1 x_1 + \beta_2 x_2 + \beta_3 x_3 + \beta_4 x_4 + \beta_5 x_5 + \cdots + \beta_{10} x_{10}$, where x_1, x_2 and x_3 are dummy variables for interventions (treatments), $x_4, x_5, \dots, x_{10}$ are dummy variables for boxers (blocks) **(b)** $E(y) = \beta_0 + \beta_4 x_4 + \beta_5 x_5 + \cdots + \beta_{10} x_{10}$, where $x_4, x_5, \dots, x_{10}$ are dummy variables for boxers (blocks) **(c)** $E(y) = \beta_0 + \beta_1 x_1 + \beta_2 x_2 + \beta_3 x_3$, where x_1, x_2 and x_3 are dummy variables for interventions (treatments)

12.27 Yes, $F = 34.12$

12.29 (a) factorial design (b) level of coagulant (5, 10, 20, 50, 100, and 200); acidity level (4, 5, 6, 7, 8, and 9); 36 combinations

12.31 (a) df(AGE) = 2, df(BOOK) = 2, df(AGE × BOOK) = 4, df(ERROR)= 99 (b) $3 \times 3 = 9$ (c) reject H_0; sufficient evidence of interaction (d) no

12.33 (a) $2 \times 4 = 8$ (b) difference between mean FSS values for normal sleepers and insomniacs is independent of education level (c) $\mu_{\text{Insomia}} > \mu_{\text{Normal}}$ (d) Mean FSS values at different education levels are significantly different

12.35 (a) $E(y) = \beta_0 + \beta_1 x_1 + \beta_2 x_2 + \beta_3 x_3 + \beta_4 x_4 + \beta_5 x_1 x_3 + \beta_6 x_1 x_4 + \beta_7 x_2 x_3 + \beta_8 x_2 x_4$, where x_1 and x_2 are dummy variables for Antimony, x_3 and x_4 are dummy variables for Method (b)

Source	df	SS	MS	F
Amount	3	104.19	34.73	20.12
Method	3	28.63	9.54	5.53
Amount×Method	9	25.13	2.79	1.62
Error	32	55.25	1.73	
Total	47	213.20		

(c) Do not reject H_0; $F = 1.62$ (d) Difference in mean shear strengths for any two levels of antimony amount does not depend on cooling method (e) Amount: reject H_0, $F = 20.12$; Method: reject H_0, $F = 5.53$

12.37 (a) yes (b) reject H_0; evidence of interaction (c) no

12.39 Interaction: fail to reject H_0, $F = 1.77$; Preparation: reject H_0, $F = 14.40$; Standing: fail to reject H_0, $F = 2.17$

12.41 (a) $E(y) = \beta_0 + \beta_1 x_1 + \beta_2 x_2 + \beta_3 x_1 x_2$, where $x_1 = $ {1 if Low load, 0 if High load}, $x_2 = $ {1 if Ambiguous, 0 if Common} (b) $\hat{\beta}_0 = \bar{y}_{\text{High/Common}} = 6.3, \hat{\beta}_1 = \bar{y}_{\text{Low/Common}} - \bar{y}_{\text{High/Common}} = 1.5, \hat{\beta}_2 = \bar{y}_{\text{Ambig/High}} - \bar{y}_{\text{Common/High}} = -.2, \hat{\beta}_3 = (\bar{y}_{\text{Low/Common}} - \bar{y}_{\text{High/Common}}) - (\bar{y}_{\text{Low/Ambig}} - \bar{y}_{\text{High/Ambig}}) = -10.2$ (c) 9,120.25 (d) SS(Load) = 1,222.25; SS(Name)= 625; SS(Load × Name) = 676 (e) 5,400; 2,166; 2,166; 2,400 (f) 12,132 (g) 14,555.25 (h)

Source	df	SS	MS	F
Load	1	1122.25	1122.25	8.88
Name	1	625.00	625.00	4.95
Load×Name	1	676.00	675.00	5.35
Error	96	12,132.00	126.375	
Total	99	14,555.25		

12.43 (a) dissolution time (b) Binding agent (khaya gum, PVP), Concentration (.5%, 4%), and Relative density (low, high) (c) factorial design (d) 8 (e) $E(y) = \beta_0 + \beta_1 x_1 + \beta_2 x_2 + \beta_3 x_3 + \beta_4 x_1 x_2 + \beta_5 x_1 x_3 + \beta_6 x_2 x_3 + \beta_7 x_1 x_2 x_3$, where $x_1 = $ {1 if khaya gum, 0 if PVP}, $x_2 = $ {1 if .5%, 0 if 4 %}, and $x_3 = $ {1 if low density, 0 if high density} (f) yes

12.45 (a) $E(y) = \beta_0 + \beta_1 x_1 + \beta_2 x_2 + \beta_3 x_3 + \beta_4 x_4 + \beta_5 x_1 x_2 + \beta_6 x_1 x_3 + \beta_7 x_1 x_4 + \beta_8 x_2 x_3 + \beta_9 x_2 x_4 + \beta_{10} x_3 x_4 + \beta_{11} x_1 x_2 x_3 + \beta_{12} x_1 x_2 x_4 + \beta_{13} x_2 x_3 x_4 + \beta_{14} x_1 x_2 x_3 x_4$, where $x_1 = $ {1 if high

level of Agent-to-mineral, 0 if not}, $x_2 = $ {1 if high level of collector-to-mineral, 0 if not}, $x_3 = $ {1 if high level of liquid-to-solid, 0 if not}, $x_4 = $ {1 if foaming agent SABO, 0 if PO} (b) df(Error) = 0 (c) $E(y) = \beta_0 + \beta_1 x_1 + \beta_2 x_2 + \beta_3 x_3 + \beta_4 x_4 + \beta_5 x_1 x_2 + \beta_6 x_1 x_3 + \beta_7 x_1 x_4 + \beta_8 x_2 x_3 + \beta_9 x_2 x_4 + \beta_{10} x_3 x_4$ (d) $\hat{y} = 7.03 + .205 x_1 + .327 x_2 + .12 x_3 - 1.09 x_4 - .038 x_1 x_2 + .137 x_1 x_3 + .183 x_1 x_4 + .042 x_2 x_3 + .428 x_2 x_4 + .282 x_3 x_4$ (e) only significant interaction is Collector × Foaming agent (f) perform main effect tests for Agent-to-mineral mass ratio (not significant) and Liquid-to-solid ratio (not significant)

12.47 (a) $E(y) = \beta_0 + \beta_1 x_1 + \beta_2 x_1^2$ (b) $E(y) = (\beta_0 + \beta_3) + (\beta_1 + \beta_6) x_1 + (\beta_2 + \beta_9) x_1^2$ (c) $E(y) = (\beta_0 + \beta_3 + \beta_4 + \beta_5) + (\beta_1 + \beta_6 + \beta_7 + \beta_8) x_1 + (\beta_2 + \beta_9 + \beta_{10} + \beta_{11}) x_1^2$ (d) $\hat{y} = 31.15 + .153 x_1 - .00396 x_1^2 + 17.05 x_2 + 1.91 x_3 - 14.3 x_2 x_3 + .151 x_1 x_2 + .017 x_1 x_3 - .08 x_1 x_2 x_3 - .00356 x_1^2 x_2 + .0006 x_1^2 x_3 + .0012 x_1^2 x_2 x_3$ (e) Rolled/inconel: $\hat{y} = 53 + .241 x_1 - .00572 x_1^2$; Rolled/incoloy: $\hat{y} = 50.25 + .17 x_1 + .00336 x_1^2$; Drawn/inconel: $\hat{y} = 48.2 + .304 x_1 - .00752 x_1^2$; Drawn/incoloy: $\hat{y} = 31.15 + .153 x_1 - .00396 x_1^2$

12.49 (a) 6 (b) $\mu_{12} < (\mu_3, \mu_6, \mu_9)$

12.51 $\mu_{\text{Hydroid}} > \mu_{\text{Fine}} > (\mu_{\text{Medium}}, \mu_{\text{Coarse}})$

12.53 (a) Policy 1 mean differs from each of policies 3–18; 2 differs from 4–18; 3 differs from 5–18; 4 differs from 8–18; 5, 6, and 7 differ from 9–18; 8 differs from 12–18; 9, 10, and 11 differ from 16–18 (b) Yes

12.55 $\omega = 1.82$; $(\mu_5, \mu_3, \mu_0) > \mu_{10}$

12.57 (a) $\mu_{\text{What-B}} > (\mu_{\text{How-A}}, \mu_{\text{Who-B}}, \mu_{\text{Who-A}})$; $(\mu_{\text{Who-C}}, \mu_{\text{How-C}}) > (\mu_{\text{Who-B}}, \mu_{\text{Who-A}})$ (b) probability of making at least one Type I error in the multiple comparisons

12.59 (a) yes (b) yes (c) no (d) .05 (e) no significant differences in means found for girls

12.61 (a) Reject H_0 (b) $\mu_Q > (\mu_S, \mu_C)$

12.63 (a) mean number of alternatives differ for the 3 emotional states (b) design not balanced (c) probability of making at least one Type I error (d) $\mu_{\text{Guilt}} > (\mu_{\text{Neutral}}, \mu_{\text{Angry}})$

12.65 $(\mu_{\text{MRB-2}}, \mu_{\text{MRB-3}}) > (\mu_{\text{SD}}, \mu_{\text{SWRA}})$; $\mu_{\text{MRB-1}} > \mu_{\text{SWRA}}$

12.67 variance assumption violated

12.69 approximately satisfied

12.71 approximately satisfied

12.73 (a) completely randomized (b) A/R, A/P, Control (c) df(Groups) = 2, df(Error) = 42, SSE = 321.47, MST = 35.755, F = 4.67 (d) reject H_0: $\mu_{\text{A/R}} = \mu_{\text{A/P}} = \mu_{\text{Control}}$ (e) partially; $(\mu_{\text{A/R}}, \mu_{\text{Control}}) < \mu_{\text{A/P}}$

12.75 (a) $F = 3.96$, reject H_0; mean ratings differ (b) $\mu_H > (\mu_F, \mu_S)$; $\mu_A > \mu_S$

12.77 (a) $F = 0.61$, fail to reject H_0 (b) $F = 0.61$, fail to reject H_0

12.79 (a) Evidence of $N \times I$ interaction; ignore tests for main effects (b) Agree; interaction implies differences among N means depend on level of I

12.81 (a) Five insecticides; seven locations (b) Fail to reject H_0 at $\alpha = .10$, $F = 2.11$; no

12.83 **(a)** $df(A) = 1$, $df(C) = 2$, $df(A \times C) = 2$, $df(Total) = 134$ **(b)** Reject H_0 **(c)** Reject H_0 **(d)** main effect tests assume no interaction

12.85 **(a)** luckiness (lucky, unlucky, and uncertain); competition (competitive and noncompetitive) **(b)** Interaction: $F = .72$, do not reject H_0; Luckiness: $F = 1.39$, do not reject H_0; Competition: $F = 2.84$, do not reject H_0

12.87 **(a)** Factorial **(b)** No replications **(c)** $E(y) = \beta_0 + \beta_1 x_1 + \beta_2 x_2 + \beta_3 x_1 x_2 + \beta_4 x_1^2 + \beta_5 x_2^2$ **(d)** Test H_0: $\beta_3 = 0$ **(e)** $\hat{y} = -384.75 + 3.73 x_1 + 12.72 x_2 - .05 x_1 x_2 - .009 x_1^2 - .322 x_2^2$; $t = -2.05$, reject H_0 (p-value $= .07$)

12.89 **(a)** (Down, left), (Down, right), (Control, left), (Control, right) **(d)** reject H_0 **(e)** $\mu_{DL} < (\mu_{DC}, \mu_{NL}, \mu_{NC})$

12.91 **(a)** Reject H_0: $\mu_R = \mu_P = \mu_D = \mu_A = \mu_B$ **(b)** (μ_R, μ_P) $<$ (μ_D, μ_A, μ_B); $\mu_D < \mu_B$

12.93 **(a)** Factors (levels): accounts receivable (completed, not completed); verification (completed, not completed); treatments: CC, CN, NC, NN **(b)** difference between means of completed and not completed accounts receivable depends on verification level **(c)** Yes

12.95 **(a)** $E(y) = \beta_0 + \beta_1 x_1 + \beta_2 x_2 + \beta_3 x_1 x_2 + \beta_4 x_1^2 + \beta_5 x_2^2$ **(b)** $\hat{y} = 29.86 + .56 x_1 - .1625 x_2 - .1135 x_1 x_2 - .275 x_1^2 - .23125 x_2^2$ **(c)** two models are different **(d)** $R^2 = .842$ **(e)** Yes; $F = 5.67$

12.97 **(a)** $df(Time) = 3$, $df(Station) = 9$, $df(Error) = 27$ **(b)** Reject H_0; $F = 11.25$ **(c)** Largest: 7/9–7/23 or 7/24–8/8; smallest: 8/24–8/31

12.99 Evidence of interaction ($p = .0001$)

INDEX